A PUBLICATION OF THE
NATIONAL RESEARCH COUNCIL OF CANADA
MONOGRAPH PUBLISHING PROGRAM

Towards Sustainable Management of the Boreal Forest

Edited by

Philip J. Burton
Symbios Research & Restoration
P.O. Box 3398, 3868 13th Ave.
Smithers, British Columbia V0J 2N0, Canada

Christian Messier
Groupe de Recherche en Écologie Forestière interuniversitaire (GREFi)
Départment des sciences biologiques
Université de Québec à Montréal
C.P. 8888, Succursale Centre-Ville
Montréal, Québec H3C 3P8, Canada

Daniel W. Smith
Environmental Engineering and Science Program
Department of Civil and Environmental Engineering
304 Environmental Engineering Building, University of Alberta
Edmonton, Alberta T6G 2G7, Canada

Wiktor L. Adamowicz
Sustainable Forest Management Network and
Department of Rural Economy
515 General Services Building, University of Alberta
Edmonton, Alberta T6G 2H1, Canada

NRC·CNRC
NRC Research Press
Ottawa 2003

NRC No. 44463.
ISBN 0-660-18762-0

National Library of Canada cataloguing in publication data

Towards sustainable management of the boreal forest

Issued by the National Research Council of Canada.
NRC No. 44463.
ISBN 0-660-18762-0

Includes bibliographical references.
Includes an index.

1. Taigas – Management.	2. Taigas – Management – Canada.	3. Sustainable forestry.
4. Sustainable forestry – Canada.	5. Forest management.	
I. Burton, P. J .	II. National Research Council Canada.	

SD387.T68 2003   634.9'2   C2003-980256-6

NRC Monograph Publishing Program

Editor: P.B. Cavers (University of Western Ontario)

Editorial Board: H. Alper, OC, FRSC (University of Ottawa); G.L. Baskerville, FRSC (University of British Columbia); W.G.E. Caldwell, OC, FRSC (University of Western Ontario); S. Gubins (*Annual Reviews*); B.K. Hall, FRSC (Dalhousie University); P. Jefferson (Agriculture and Agri-Food Canada); W.H. Lewis (Washington University); A.W. May, OC (Memorial University of Newfoundland); G.G.E. Scudder, OC, FRSC (University of British Columbia); B.P. Dancik, Editor-in-Chief, NRC Research Press (University of Alberta)

Inquiries: Monograph Publishing Program, NRC Research Press, National Research Council of Canada, Ottawa, Ontario K1A 0R6, Canada. Web site: www.monographs.nrc-cnrc.gc.ca

Correct citation for this publication: Burton, P.J., Messier, C., Smith, D.W., Adamowicz, W.L. (*Editors*). 2003. *Towards Sustainable Management of the Boreal Forest*. NRC Research Press, Ottawa, Ontario, Canada. 1039 p.

Contents

Acknowledgements

The preparation of this book has been underwritten by the Sustainable Forest Management Network (SFMN), a Network of Centres of Excellence (NCE) receiving partial support from Canada's National Science and Engineering Research Council (NSERC) and the Social Sciences and Humanities Research Council (SSHRC). This book, and much of the research on which it reports, would not have been possible without this long-term support, and that of the SFMN's many governmental, industrial, First Nation, conservation, and institutional partners. A complete listing of the SFM Network Partners can be found on the SFMN web site (http://sfm-1.biology. ualberta.ca/english/network/epartners.htm).

The editorial team would like to thank an Advisory Committee of researchers and partners that helped guide the early design and scope of this work. In addition to the editorial team, this committee included Yves Bergeron, Gillian Binsted, Stan Boutin, Luigi Morgantini, Marc Stevenson, and Shawn Wasel. The editors would also like to thank the SFMN and its staff for institutional support, particularly the assistance provided by Marvin Abugov, Alison Boddy, Bruce MacLock, and Shirley Vandermey. Additional institutional support for the editorial team was provided by the University of Alberta, the Université du Québec à Montréal, Symbios Research & Restoration (Smithers, B.C.), Resources for the Future (Washington, D.C.), and Harvard University (Petersham, Massachusetts). Editorial assistance and indexing services were provided by Margaret Bakelaar, Gillian Binsted, Carla Burton, Anne Macadam, Julie Poulin, and their associates.

The quality of this publication has benefited immeasurably from an external review of individual chapters and the entire book. We would hereby like to thank the following international community of fellow researchers and managers for their constructive comments and suggestions: Jim Baker, Jim Beck, Tom Beierle, Phil Comeau, Nancy Diaz, Andrew Fall, John Ferguson, Lee Frelich, Dirk Jaeger, Eric Kasischke, Antero Luonsi, Dave MacLean, Ted Mao, Pat Marchak, Paul McFarlane, Michael McGonigle, Luigi Morgantini, Passi Puttonen, Jonathan Russell, Stella Swanson, Ron Trosper, Casey van Kooten, Gordon Weetman, John Zasada, and at least two reviewers who remain anonymous. Responsibility for incorporating or ignoring the recommendations generously provided by these reviewers remains that of the authors and editors.

Finally, the book editors and authors would like to acknowledge the support and guidance provided by the Monograph Program of the National Research Council (NRC) Research Press. In particular, we thank Gerry Neville, Carol McKinley, Diane Candler, and Nancy Daly for the guidance, patience, and extra hours they have devoted to this project.

This book offers many case studies, conclusions, and recommendations regarding progress towards sustainable forest management. Nevertheless, it must be emphasized that all opinions expressed are those of the authors, and do not represent the position of the SFMN or its partners, nor that of the National Research Council or any other government agency. Particular corporations, agencies, or products may be described in the text, but such references should be interpreted as examples only, and do not constitute a comprehensive list or exclusive endorsement.

Maps on pages 5, 6, 75, and 91 are taken from the Atlas of Canada website http://atlas.gc.ca © 2003, Her Majesty the Queen in Right of Canada, with permission of Natural Resources Canada.

Chapter 1

The current state of boreal forestry and the drive for change

P.J. Burton, C. Messier, G.F. Weetman, E.E. Prepas, W.L. Adamowicz, and R. Tittler

"The great northern forest may be the largest ecosystem on Earth. But size is no guarantee of survival."

Great Northern Forest, 1994, a film directed by Joseph Viszmeg, produced by Albert Karvonen and Jerry Krepakevich. Karvonen Films Ltd. and National Film Board of Canada

Boreal forests in Canada and around the world

Canada is covered by approximately one-tenth of the world's forests and about 40% of the world's boreal forests. Approximately 58% of Canada's total land area, 55% of its surface freshwater, 13% of its population, and 15% of its endangered and threatened species fall within the boreal region (CCEA 2002). It is estimated that boreal forests contribute approximately 60% of Canada's economic activity in forest products, and were therefore responsible for approximately 11.6 million m^3 of harvested roundwood

P.J. Burton.[1] Symbios Research & Restoration, P.O. Box 3398, Smithers, British Columbia V0J 2N0, Canada.
C. Messier. Dép. des sciences biologiques, Université du Québec à Montréal, Montréal, Quebec, Canada.
G.F. Weetman. Department of Forest Sciences, University of British Columbia, Vancouver, British Columbia, Canada.
E.E. Prepas. Faculty of Forestry and the Forest Environment, Lakehead University, Thunder Bay, Ontario, Canada.
W.L. Adamowicz. Sustainable Forest Management Network and Dept. of Rural Economy, University of Alberta, Edmonton, Alberta, Canada.
R. Tittler. Ottawa–Carleton Institute of Biology, Carleton University, Ottawa, Ontario, Canada.
[1]Author for correspondence. Telephone: 250-847-1261. e-mail: symbios@bulkley.net

Correct citation: Burton, P.J., Messier, C., Weetman, G.F., Prepas, E.E., Adamowicz, W.L., and Tittler, R. 2003. The current state of boreal forestry and the drive for change. Chapter 1. *In* Towards Sustainable Management of the Boreal Forest. *Edited by* P.J. Burton, C. Messier, D.W. Smith, and W.L. Adamowicz. NRC Research Press, Ottawa, Ontario, Canada. pp. 1–40.

production, \$26.6 billion[2] in exports, \$1.38 billion in government revenues from sale of timber, and 211 200 direct jobs nationwide in 1999 (CCFM 2002). With an employment multiplier of 2.4, this means that more than half a million Canadians are employed directly or indirectly in the processing of wood fibre from Canada's boreal forests, while many more make their living through the use of wood and paper products. Hundreds of communities and at least 3.6 million people reside within the boreal ecoregions of Canada (CCEA 2002), and millions more live in the boreal regions of Fennoscandia and Russia. So boreal forests and boreal forest products are important in Canada and in other northern nations, from ecological, social, and economic perspectives. Canada has a larger proportion of its boreal forests in a wild, never-been-cut state than any other country in the world. But what is to be the future of this forest ecosystem and the forest resources, the forest products industry, and forest communities that depend on it?

The circumboreal forest is the most extensive terrestrial biome in the world, encompassing some 14 million km[2] or 32% of the Earth's forest cover (Table 1.1). Dominated by open or closed forests of predominantly coniferous tree species, the boreal zone is found in the cold northern climates of Europe, Asia, and North America. With slow-growing, uniform wood having excellent properties for pulping (papermaking), dimensional lumber and plywood or other panelling, the boreal forest consists of one of the world's largest reserves of unexploited wood fibre. The biome provides important habitat for large numbers of wild birds and mammals because it consists of vast unpopulated landscapes interlaced with large rivers, lakes, and wetlands. It also supports a high diversity of mushrooms and other little-studied organism groups. There are more than 23 000 identified species of biota residing in the North American boreal biome (Zasada et al. 1997). The harsh climate has largely resulted in the region being less agricultural and less settled than many other parts of the world. Though home to Indigenous peoples and hardy souls who pride themselves for living "off the land" and on the frontier (Henry 2002), these northernmost forests also represent the epitome of wilderness for the world's city-dwellers. Commercial uses have historically been concentrated around mining, hydroelectricity generation, and the harvest of furs and timber. During the last two decades of the twentieth century, expansion of commercial forestry into broad areas of northern Canada and Russia has prompted concerns about mankind's ability to develop the resources of the boreal region in a sustainable manner. How can this wilderness be developed for broad social and economic benefits while protecting its environmental values and future options? Addressing this issue has become the focus of international scientific and public attention (e.g., Zasada et al. 1997; Schindler 1998; SSCAF 1999; Henry 2002; Schneider 2002; TRN 2002; CPAWS 2002–2003; CBI 2003) and has been the primary focus of Canada's Sustainable Forest Management Network (SFMN 2003).

The definition of "boreal forest" differs among various jurisdictions and with the purpose of any particular mapping and tabulation exercise. In general, we could call the boreal zone those arctic, subarctic, or northern mid-latitude regions that are dominated

---

[2]All \$ values reported in this book refer to Canadian dollars (CAD), which were trading at approximately 1.00 CAD = 0.74 USD (United States dollars) and 1.00 CAD = 0.64 EUR (Euros) in late June 2003.

Table 1.1. Distribution and status of the world's boreal forests. Statistics are quite reliable for Canadian provinces and territories, but were not organized in the same manner for other jurisdictions for which they are not readily available and should be considered less reliable.

Country	Province/state/ administrative region	Total area (km²)	Boreal ecozones*					
			Ecozone area (km²)	Forested/ wooded (km²)	"Timber productive" (km²)	Logged at least once (km²)	Converted to non-forest (km²)	"Protected" areas** (km²)
Canada	Yukon	482443[a]	474256[d]	279750[d]	51480[j]	36927[m]	13[d]	53885[n]
	NWT & Nunavut	3439296[a]	1056043[d]	675891[d]	137570[j]	70969[m]	4322[d]	179527[***]
	British Columbia	944735[a]	298748[d]	213425[d]	114620[j]	40551[m]	3855[d]	49913[o]
	Alberta	661848[a]	458486[d]	387722[d]	231380[j]	148885[m]	5118[d]	66022[p]
	Saskatchewan	651036[a]	411267[d]	313808[d]	126030[j]	80962[m]	3739[d]	11045[q]
	Manitoba	647797[a]	578536[d]	414598[d]	148390[j]	66336[m]	13379[d]	27458[m]
	Ontario	1076395[a]	905219[d]	814710[d]	296580[j]	227304[m]	15872[d]	82500[***]
	Quebec	1520560[a]	1234606[d]	927787[d]	429410[j]	298747[m]	5531[d]	12317[r]
	Newfoundland & Labrador	405212[a]	387095[d]	274904[d]	112490[b,k]	31614[m]	1046[d]	8543[s]
United States	Alaska	1593444[b]	656600[e]	522070[g]	91000[b,k]	30071[m]	—	62652[g]
Russian Federation	European Russia	3831800[c]	3458800[c]	2006040[c,h]	1498240[c]	1792397[c+]	—	10682[c]
	Western Siberia	2904900[c]	2341450[c]	1736380[c,h]	1078640[c]	1321385[c+]	—	20750[c]
	Eastern Siberia	7226000[c]	5136660[c]	3972660[c,h]	3352360[c]	2258530[c+]	—	71707[c]
	Russian Far East	3112700[c]	1547980[c]	1411130[c,h]	1059340[c]	1041414[c+]	—	18486[c]
Norway	—	324220[b]	87100[j]	82472[j]	72000[b,l]	78349[m]	—	5574[t]
Sweden	—	449964[b]	272600[j]	195365[j]	155365[b,l]	185597[m]	—	9814[t]
Finland	—	338145[b]	218800[j]	214963[j]	193467[b,l]	204215[m]	—	34419[t]
Iceland	—	103001[b]	1340[j]	310[j]	0[j]	307[m]	—	200[t]
Total:		29713496	19525586	14443985	9148362	8194559		725493

Notes:

*In Canada, boreal ecozones are considered to include the boreal, taiga, and Hudson plains, the boreal and taiga shield, and the boreal and taiga cordillera.

**Protected areas defined as IUCN categories I–V (often estimated from national/provincial totals).

***Values are the average of several estimates from disparate sources.

+Based on the midpoint of a range of estimates for "intact forest" in Russia for which minimum size criteria were also utilized in the source data.

[a]From *The National Atlas of Canada, 5th Edition*, "Facts About Canada", available at http://atlas.gc.ca/site/english/facts/surfareas.html [viewed 18 June 2003].

[b]From National Geographic Society, 1999, *Atlas of the World, 7th Edition*, available at http://plasma.nationalgeographic.com/mapmachine/ [viewed 18 June 2003].

[c]From Yaroshenko et al. (2001) and Aksenov et al. (2002); ecozone, forested, and productive values of regions differ among sources, and were selected so that area of ecozone > forested > productive land bases.

[d]From Statistics Canada (1992).   [e]From Bailey (1995).   [f]From TBFRA (2000).   [g]From Duffy et al. (1999).

[h]Area of Russian taiga forests (pre-tundra and northern taiga, middle taiga, and southern taiga) from Isaev and Shvidenko (2002).   [i]From FAO (2003).

[j]Timber productive areas in Canada from Canadian Council of Forest Ministers and National Forestry Database Program, available at http://nfdp.ccfm.org/cp95/data_ e/com19e.htm (though corresponds to forest regions, not ecozones).   [k]From Wheeler (n.d.).   [l]From Runesson (n.d.).

[m]Boreal ecoregions from maps and supporting documentation of WorldWildlifeFund, 2001, *Ecoregions*, available at http://www.worldwildlife.org/wildworld/profiles/.

[n]From Yukon *State of the Environment Report*, 1999, available at http://www.environmentyukon.gov.yk.ca/soe/content/chap3.pdf.   [o]From BCMWLAP (2001).   [p]From Schneider (2002).   [q]From Kulshreshtha et al. (2000).   [r]From Environment Québec (2003).   [s]From Ryan (n.d.).

[t]From United Nations, *Iceland Country Profile*, available at http://www.un.org/esa/earthsummit/icela-cp.htm#chap11.

by a cold climate and able to support only a few coniferous and broadleaf tree genera (Box 1.1). It is useful to distinguish geographic areas dominated by a boreal climate (broad regions), those covered by boreal forest and woodlands (i.e., excluding lakes, wetlands, and other untreed areas), and those supporting commercially valuable stands of boreal trees (or "operable" timber, a definition that varies according to local economic and technological conditions). Table 1.1 provides a breakdown of these categories for various political jurisdictions around the world. The Russian Federation clearly contains the greatest boreal resources, followed by Canada. Traits typical of boreal ecosystems, boreal forestry, and boreal communities (Box 1.1) can be found in Alaska, much of Canada, Fennoscandia, and in both European and Asiatic Russia. This book strives to be relevant to the entire circumpolar boreal biome, but it showcases research and recent thinking on the sustainable management of the boreal forests undergoing commercial development in Canada.

Our discussion of boreal ecology, boreal forestry, and boreal communities broadly refers to lands within the circumpolar boreal zone (Fig. 1.1), as presented by Hare and Ritchie (1972) and by Pruitt (1978). In Canada, the extent of the boreal forest region is defined by seven ecozones (Fig. 1.2), as classified by the Ecological Stratification Working Group (ESWG 1996), which builds on earlier classifications of forest regions (Rowe 1972) and ecoclimatic regions (EWG 1989). The exact boundaries of this zone are not important to the application of concepts explored in this book, for which the

---

Box 1.1. Most boreal forests are characterized by: *ie. not generally mountainous*

- gently rolling terrain, either on glacial till, or with shallow-soiled and infertile uplands alternating with wetlands and poorly drained organic soils (extensive organic terrain with scattered conifers, often called muskeg);
- a cold, continental climate with severe winter temperatures, a short growing season, and cold soils; permanently frozen soil (permafrost) becomes progressively more abundant as one goes north;
- forests dominated by relatively few tree species:
  - conifers usually of the pine (*Pinus*), larch (*Larix*), spruce (*Picea*), and fir (*Abies*) genera;
  - broadleaf species usually of the birch (*Betula*), poplar (*Populus*), willow (*Salix*), alder (*Alnus*), and rowan (*Sorbus*) genera;
- a cycle of natural disturbance and succession dominated by wildfire or insect outbreaks;
- slow tree growth that results in strong, tight-ringed wood with excellent properties for building construction, and uniform fibres greatly valued for paper making;
- slow decomposition rates, resulting in accumulations of organic matter in bogs and on the forest floor, and consequently strong nitrogen limitations to plant productivity;
- animal populations that may be strongly migratory on an annual basis or cyclic over longer periods;
- hordes of biting and sucking insects throughout much of a brief summer!
- Indigenous peoples tied to the land through millenia of hunting, gathering, trading and traditional ecological knowledge; and
- low human populations, few cities, marginal agricultural production, and long distances from major centres of commerce.

Fig. 1.1. Global distribution of the boreal forest (from NRCan 2002).

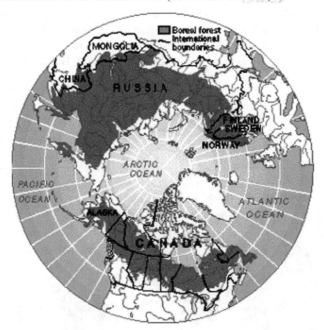

boreal designation can safely be extended to the sub-boreal plateaus of British Colum-bia, the northern conifer forests of Minnesota, Wisconsin, Michigan, and Maine, and even the sparse birch (*Betula*) woodlands of Iceland and the largely deforested High-lands of Scotland. Portions of northern Mongolia and northern China are also consid-ered to have boreal forests (Fig. 1.1; Fengyou and Jingwen 1995), though it has been argued that these might more properly be considered subalpine forests (Wang 1961; A.A. Maslov, Institute of Forest Science, Russian Academy of Sciences, Uspenskoe, Moscow Region, Russia, personal communication, August 2002). The ecology of these and other subalpine and montane coniferous forests (e.g., the more southerly coniferous forests of the Western Cordillera and Appalachian Mountains of North America, those of the southern Ural Mountains in Russia and the mountain forests of central Europe), though sharing biogeographic affinities with the boreal forests, is controlled more by elevation rather than by regional climate. In many ways, the coniferous montane and subalpine forests behave much like boreal forests, so results presented here will often be applicable to them.

We do not present new information on issues related to the ecology, management, or socio-economics of the open woodlands that constitute the forest–tundra transition zone and much of the boreal climatic region, a zone classified as "taiga" in Canada (Fig. 1.2). The distribution of trees is sporadic, timber volumes are low, regeneration is diffi-cult and growth is slow in the taiga ecozones, so those forests are not generally consid-ered to be commercially viable as sources of timber. Nevertheless, there is even development pressure in these sparse northern forests, though the long-term sustain-ability of timber harvesting at the very limit of the productive forest has been questioned by many scientists and environmentalists. Industrial forestry, and the work of Canada's

Fig. 1.2. The terrestrial ecozones of Canada. The "boreal zone" is generally considered to include the Taiga Cordillera (11), Boreal Cordillera (12), Taiga Plains (4), Boreal Plains (9), Taiga Shield (5), Boreal Shield (6), and Hudson Plains (15). Those are the zones for which statistics have been compiled in Table 1.1. Most of the commercial forestry activity and research discussed in this book is restricted to the Boreal Shield, Boreal Plains, and (to a lesser degree) the Boreal Cordillera ecozones.

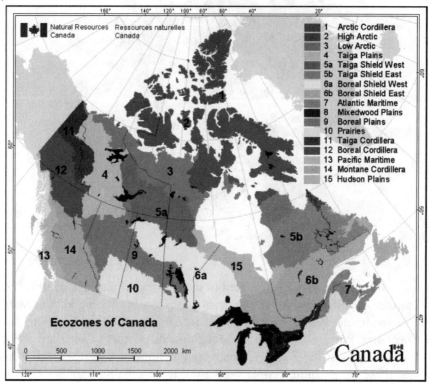

Map courtesy of Natural Resources Canada

Sustainable Forest Management Network on which this book reports, however, focuses on the southern boreal forest (Fig. 1.3) that is rapidly being developed and managed for fibre production. The principles developed in the following chapters should be applicable to any landscape currently dominated by unmanaged forests and natural ecological processes, and situated largely on public or other centrally coordinated land.

A brief history of boreal forestry

Pre-industrial uses

Human use of the world's boreal forests has been comparatively modest until very recently. Unlike the temperate and Mediterranean climates, boreal regions have never been favoured for agriculture and the growth of large urban populations. Most of the world's boreal forests are relatively recent occupants of their current land base, having established on land that had been glaciated as recently as 8000–12 000 years ago (Pielou 1991), though it has been suggested that much of eastern Siberia may have escaped

Fig. 1.3. Most of the commercial forestry activities and forest research in Canada have concentrated on the southern boreal forest, much of which is a biologically rich mosaic of coniferous and broadleaf trees, peatlands, and other wetlands. Shown here is a portion of mixedwood forest in the Mid-Boreal Lowlands ecoregion of the Boreal Plains ecozone, near Meadow Lake in northwestern Saskatchewan.

Photo courtesy of Millar Western Forest Products Ltd.

glaciation for the last 70 000 years (Velichko 1995). Aboriginal use of the forest followed immediately, with forest products used for subsistence levels of shelter, fuel, and food by small and scattered populations of people (Chap. 3). This is not to say that Aboriginal impacts on boreal forests were negligible, however, as hunting pressures may have been severe (Kay 1994) and prescribed fires may have been widely used for vegetation management (Lewis and Ferguson 1988). Most plant species found in the boreal flora of western Canada were used to some degree (Marles et al. 2000) and may have become locally rare (if depleted) or locally abundant (if nurtured) as a result. Similar patterns of subsistence hunting, supplemented by the herding of semi-domesticated reindeer (*Rangifer tarandus tarandus*) and gardening, characterized early use of the Fennoscandian and Russian forests.

The northern forests of both Eurasia and North America went through a phase in which the harvest and trade of furs constituted the first ventures into the cash economy, and was responsible for the establishment of trade routes and trading centres (Newman 1985). Population pressures and the use of iron tools resulted in greater colonization of the southern reaches of the boreal forests in the last millenium, with saws, axes, and plows facilitating the felling of trees, and the clearing and cultivation of land. Gardens expanded into fields of northern cereal grains (oats, barley, and rye) and pastures and hay meadows were created to support domesticated livestock. Nonetheless, such subsistence farming had little more impact on the boreal landscape than did subsistence hunting and gathering, except in specific regions (such as central Sweden) where populations and development were so pervasive as to dominate the landscape. It is worth noting, however, that even where settler populations were low, uncontrolled fire escapes (usually associated with land clearing) resulted in forest land being burned much more often or more severely than had historically occurred under unmanaged or Indigenous management conditions (Pyne 1982). Accurately characterizing natural, Aboriginal, and modern fire regimes remains a challenging aspect of determining historical templates for guiding ecosystem management in many parts of the world (Chap. 8).

Early timber harvesting

A now-classic description of the early roots of forestry and the early stages of forest use and management around the world was written by Bernard Fernow (1913), and a recent global history has been provided by Williams (2003). Epp (2000) provides a good summary of the history of Ontario forest utilization, while Hannelius and Kuusela (1995) provide a comparable history of forest use in Finland. Reed & Associates (1978), Apsey et al. (2000), and Reed (2001) provide more comprehensive reviews of the history of forest management and the timber trade in Canada; the following summary draws from these sources. The utilization of boreal forests for commercial supplies of timber did not begin in earnest until the 17th and 18th centuries around the Baltic Sea, and in the early 19th century in Canada. Repeated cycles of European wars and the demand by competing navies (for timber and pine tar) prompted massive forest exploitation on both sides of the Atlantic Ocean. First shipped as square timbers from along the major rivers, logs were soon processed into lumber instead, though in many cases development remained sparse in the boreal zones as logs were processed in mills located further south in larger population centres. Logs were transported to mills in rafts or booms on creeks, rivers, and lakes. Sawmills were first powered by flowing water, and then by steam in the latter half of the 19th century. Depending on the cycles of domestic economic growth and foreign demand, these timber supplies were utilized both for local construction and for export. Timber required for fortifications, firewood, mine supports, railway ties and railway trestles also constituted important draws upon local forest resources.

As in other parts of the world, these first periods of boreal utilization would largely be characterized as "unregulated exploitation" (Box 1.2; see also Kimmins 1991), though various processes of authorization had to be respected under both Aboriginal and European political systems. Inevitably, the need for forest protection (from overcutting and [or] from fire) and renewal prompted the regulation of timber harvesting and land clearing, or the beginnings of an "administrative stage of forestry" (Box 1.2; Kimmins 1991). In some parts of the world, special tracts of forest land were protected for religious or hunting purposes, but this doesn't seem to have been the case in the boreal forest. Rather, systems of licensing and documenting forest cutting were put in place and prohibitions may have been placed on burning without any real limits on what could be cut or when (Reed & Associates 1978). Deforested lands were often abandoned or sold off for farm land, though much of the land was so unsuitable for agriculture that it gradually reverted to forest anyway. The need to retain some land in a forested state rose to the level of public and political consciousness in the 1870s and 1880s in Canada, and resulted in the designation of the first forest reserves (Epp 2000). A key feature of forest management in Canada has been the decision to retain public ("Crown") ownership

---

Box 1.2. A history of forest use in most parts of the world.

1. *Subsistence* — local uses only, sustainable because withdrawals are at such low levels that they are easily replaced by natural growth and regeneration;

2. *Exploitation* — may be local or for export, but harvesting and land-use conversion exceeds ability of the system to renew itself;

3. *Regulated* — restrictions on forest harvesting and land-use conversion are devised and administered by the authorities; though still concerned primarily with logging rates, use may or may not be sustainable, depending on how effective the government is at constraining the rate or extent of unregulated exploitation, and how effective forest ecosystems are at self-renewal;

4. *Sustained yield timber production* — the scientific production of timber, based on the regulated management (including forest regeneration and the monitoring of stocks and their growth) of even-aged cohorts of trees; a sustainable flow of fibre volume is assured, but forests may not be sustaining many other values;

5. *Utilitarian multiple use* — other useful forest resources are added to the sustained yield basket of forest goods and services, as epitomized in the 1960 *Multiple Use Act* in the United States, which mandated the U.S. Forest Service to manage for "wood, water, wildlife, range, and recreation"; good potential for sustainability in principle, but rarely attained because of industry-driven development and attempts to maintain all values on each parcel of land;

6. *Site-specific (ecosystem based)* — recognition that all forests and all sites within a forest cannot be treated the same, with better identification and classification of ecosystems, silvics, and animal behaviour; utilitarian management practices are modified and diversified to achieve greater effectiveness;

7. *Biocentric* — utilitarian values are recognized, but so are a number of other values related to biodiversity protection and aesthetic or spiritual importance that have little commercial basis; rather, all life is seen as having value, and is worth protecting at some level; with no professional forest managers "speaking for" all the life forms and non-utilitarian values, this stage is marked by the growing demand for public participation in the forest management process; and

8. *Landscape design* — driven by biodiversity concerns rather than just the timber harvest scheduling problems of the past, the strategic deployment of forest management effort in space and time is undertaken; coinciding with a greater emphasis on "coarse filter" management of habitats rather than single species, this phase is much facilitated by geographic information system (GIS) and computer simulation technologies; it is realized that most forests have to be managed over large spatial scales and long time frames in order to produce desired goods and services in perpetuity.

All phases were not necessarily experienced in all jurisdictions, and not necessarily in this order. All phases persist to the modern day, and some (e.g., subsistence, regulated, and utilitarian multiple use) may be occurring simultaneously in different parts of the same jurisdiction.

---

over most forest lands. The principle of retaining Crown ownership of forest lands was a strong conclusion emanating from the First Annual Meeting of the Commission of Conservation in Ottawa (chaired by Clifford Sifton) in 1910. Thus even much of the early logging was done on leases, and the harvesting and management of Crown forest lands is conducted under various forms of private/corporate tenure on this public land to this day.

It is at this stage in the evolution of forestry, from about 1890 to 1930, that large tracts were first made available to pulp and paper companies in North America and in Scandinavia. This large-scale utilization of boreal spruce (*Picea* spp.), pine (*Pinus* spp.), and fir (*Abies* spp.) grew from the demand for newsprint in the United States, England, and Germany, coupled with the development of industrial papermaking from conifer fibres using chemical pulping processes. Timber licences were granted according to the size of the pulp mill and its feedstock requirements over its expected lifespan, with little regard for (or expectation of) forest regeneration or sustainability (Reed & Associates 1978). In many cases, transportation improvements and the regenerative capacity of northern forests allowed pulp mills to operate for longer than expected; some of these mills (or their replacements) and their associated woodland operations remain profitable to the present day. On the other hand, vast areas of logged and burned land generated concern for more stringent fire control and reforestation initiatives; forest reserves were justified as a means of slowing the rate of exploitation, not protecting environmental values. We see the legacy of these extensive, contractual timber harvesting rights associated with primary processing facilities (pulp mills or saw mills) persisting as the normal mode of forest industry involvement in the boreal forests of Canada and Russia. In other parts of the world, such as Norway, Sweden, and Finland, forest lands are managed by a wider array of individual, municipal, and corporate landowners, so forest woodlands management is not so tightly linked to a particular processing facility.

Scientific forestry

The next phase of forestry has typically involved the adoption of scientific principles of sustained-yield fibre production (Box 1.2). These principles were borrowed from Germany (e.g., Schlich 1896–1911; Fernow 1902), where a commitment to even-aged conifer plantations had already succeeded in producing timber over one or more rotations. With a focused emphasis on timber production, even-aged management, and maintaining a normalized portfolio of age classes in each forest estate, sustained-yield management still constitutes an important foundation of sustainable forestry around the world today. Though implemented as public policy much earlier in many European countries, Canadian provinces did not legislate the application of sustained-yield forestry to public lands until the 1940s and 1950s (Reed & Associates 1978). The perceived need to better balance the age-class structure of wild forests (i.e., to equalize the area/volume in each age class, and to eliminate the older, less productive age classes) remains embedded in much of the legislation and culture of forest management of all Canadian provinces.

Associated with the growth in affluence, mobility, and recreational time of the middle class in the 1950s and 1960s came growing recognition of other useful forest products and services. This broader view was still largely utilitarian or materialistic in its emphasis (Box 1.2). The traditional view of the forest manager was that sound timber management provided a necessary and sufficient condition for the protection and production of water, wildlife, and opportunities for livestock grazing and recreation as well.

While the rangeland and watershed utility of boreal forests never matched the importance of these values in the western United States (except in the case of reindeer herding in Fennoscandia), expectations for "multiple use" are enshrined in legislation guiding the management of most public forest lands in the northern hemisphere. In the boreal region, support for multiple use is generally interpreted as a commitment to maintaining habitat suitable for game animals and for recreational opportunities. Other industries such as cattle ranching, mining, transportation, hydroelectric generation, and petroleum extraction were (and continue to be) considered legitimate components of most managed forests (Pratt and Urquhart 1994). These additional non-forest land uses are generally tolerated by the forest managers because they have to be, while most other non-timber forest resources remained largely uninventoried and ignored, and certainly not actively managed or promoted.

Refinements to the sustained-yield and multiple-use phases of forestry emphasized recognition of the need to protect soil resources, avoid damage to streams, and included the development of site-specific guidelines to enhance silvicultural success of timber management (e.g., MacKinnon et al. 1992; Lundmark 1995). This "ecosystem-based" approach has also resulted in a broader interpretation of forest values and their long-term sustainability, but should not be confused with more recent biocentric approaches sometimes called "ecosystem management" (Boyce and Haney 1997; Kohm and Franklin 1997) or "ecological forest management" (Schneider 2002). Nevertheless, site-specific silvicultural planning has helped diversify the prescription of forest management practices. Learning the art and science of how to "read the site", its history, limitations, and potential has become one of the most challenging components in the education of foresters and the training of forest technicians.

The modern era

Kimmins (1991, 1995) terms the final or ultimate stage in the development of forestry as the "social stage of forestry", in which environmentally sound forestry satisfies diverse social needs. In practice, this post-modern phase can be broken down into two separate but related components: the recognition of more biocentric or non-utilitarian values in forests, and the recognition that sustaining all forest values requires planning and design over large areas and long time frames. So we name these stages the "biocentric" and the "landscape design" phases of forest management (Box 1.2). Encompassing the preceding stages of ecologically based sustained-yield timber production and utilitarian multiple-use forestry, the biocentric perspective also recognizes and promotes a much greater set of biological and social values in forest ecosystems. Emerging in the 1970s in most countries along with the demand for anti-pollution laws, more parks and ecological reserves, and more democratic participation in the land management process, the demand for the protection of biodiversity, old-growth forest, and wilderness is still very strong. Likewise, as computer tools became more powerful and more available, the ability to portray and simulate the impacts of forest management on multiple resources across the landscape over time has allowed managers to undertake a much more sophisticated kind of forest planning across entire regions and over time scales of centuries (see Chap. 14).

Sustainability of all forest values has emerged as a dominant paradigm for forest managers in the 1990s (see Chap. 2). Though implied in all earlier regulatory and sustained-yield perspectives, sustainability now explicitly becomes the gauge against which we evaluate all impacts and actions. The "maintenance of ecological integrity" may be considered a prerequisite to the utilization of resources (Grumbine 1994), but practitioners still differ widely in their interpretation of concepts such as "ecosystem health", "ecological integrity", and "environmental degradation". In many ways, use of the sustainability yardstick can be interpreted as the extension of the sustained-yield concept to various non-timber attributes of the forest landscape. If we are to follow strict requirements for sustainability of all forest values, then hunted and non-hunted wildlife, stream water quantity and quality, wild berries, mushrooms, and numerous other values would be harvested or otherwise disrupted only if their persistence is assured, and to the extent they can be renewed.

The history of forestry, like that of most other industries, is typified by increasing mechanization, automation, and consolidation over the last several decades of the twentieth century. Axes and whipsaws (Fig. 1.4a) were replaced by gasoline-powered chain saws in the 1950s (Swift 1983). Log yarding and short-hauling by oxen and horses (Fig. 1.4b) was largely phased out in the 1950s and 1960s. Sometime over these same decades, logging in northern forests changed from being primarily a seasonal (winter) activity for farmers and farm hands to being a year-round activity with a dedicated workforce (Bernsohn 1981). Most logging in all boreal regions today is highly mechanized (see Chap. 15): the cutting of trees is done with tracked fellerbunchers; large rubber-tired skidders or log forwarders transport the logs to landings or roadsides where loaders transfer them to trucks. Log drives on rivers, log booms on lakes, and logging railroads have largely been replaced by these ever larger and more powerful diesel

Fig. 1.4. Standard practices of timber harvesting in the pre-industrial and early industrial eras were very labour intensive, patchy, and left much residual structure but with little planning for forest regeneration or forest sustainability. (*a*) Hand falling a large white spruce using a two-handed saw; (*b*) logging sleigh loaded for a trip out of the bush to a lake or river log dump. Note the residual aspen trees in both photos. Due to reduced labour requirements, many northern communities may no longer be sustainable today if they are dependent primarily on the forest products industry.

(*a*)   (*b*)

Archival photos courtesy of Alberta Forest Products Association

trucks that transport logs hundreds of kilometres to mills, and also carry lumber or pulp thousands of kilometres to their markets in population centres. Better transportation also resulted in most logging camps being closed in favour of workers commuting to the bush from the mill towns, with a concomitant improvement in family life, health care, and educational opportunities (Marchak 1983).

Through most of the 1970s and 1980s, small dispersed sawmills (often family owned and run) scattered throughout the boreal hinterland were generally phased out in favour of large integrated mills, further contributing to a depopulation of many forest-based communities. These centralized sawmills included increasingly mechanized and computerized facilities and drying kilns for optimizing the value of solid wood products. Wasteful practices associated with the wide kerf of thick circular saws and the genera-tion of rounded slabs of waste wood were phased out as the old equipment was replaced by thinner bandsaws and chippers in the 1970s and 1980s. In regions which were able to support both sawmills and pulp mills, great efficiencies in the use of the timber sup-ply were achieved when sawmills started to square logs with chippers prior to sawing, with the rounded portions of boles being directed to the pulp industry as chips rather than being burned. Though timber licenses continued to be granted throughout the north as new automated mills were being established, small remote communities were often bypassed, and even the larger communities no longer benefited from the generation of large numbers of jobs in the forest industry. These changes, induced by centralization, mechanization, and automation, have been part of broader global trends. Some logging and processing jobs have been replaced by increased levels of professional and techni-cal involvement in forest planning and silviculture. But the lack of alternative employ-ment opportunities has been especially hard on unskilled workers and Native communities.

Before we explore, in a separate section, more recent developments leading to the "post-modern era", it is worth outlining the standard manner in which forestry was prac-ticed throughout the multiple-use era of the 1960s through the 1980s. These approaches remain largely in effect to this day, if only because of the forestry legislation, rules, and tenures in place from the previous era. This standard approach to forestry is taught in all forestry schools, has been common to large managed forests in temperate zones, and was modified only slightly for application in boreal climates. Similar procedures have been applied on both public and private land, though the scale of operations and the details of management and planning clearly varied (as they do today) with the size of the land holding.

An overview of standard forest management: the old "status quo"

The hierarchy of planning

There is generally a standard system of hierarchical forest planning that prevails in all boreal jurisdictions around the world (Box 1.3). Strategic forest-level planning is fol-lowed by tactical, spatially explicit development planning, and then by stand-level site-specific operational plans and stand management prescriptions (Tittler et al. 2001). Historically, many of the problems in forest management have developed because

Box 1.3. The commitment to sustainability and the hierarchy of forest planning found in major boreal jurisdictions around the world.

Although they vary among countries and provinces, standard forest planning processes are hierarchical, involving multiple scales and steps. Table 1.2 summarizes Tittler et al. (2001) in terms of (1) the commitment to the principles of sustainable forest management, and (2) the planning hierarchy applying to public lands in different boreal jurisdictions.

There are a number of commonalities among the forest planning processes in effect in most boreal jurisdictions:

1. Sustainable forest management (SFM) is now a goal in all jurisdictions, though how this is to be achieved is rarely spelled out. Most criteria for SFM (see Chap. 2 for discussion) are mentioned, though the maintenance of global carbon cycles is not usually addressed; forest regeneration is required under all planning systems.

2. The planning hierarchy is more complex on public lands than on private lands, and more clearly addresses multiple forest values and solicits public input.

3. Commercial forestry development often takes place in the context of some kind of regional land-use zoning, with various degrees of public input.

4. Some kind of forest-wide plan is used to set the rate of timber harvesting ("the rate of cut") and to lay out the general goals of forest management, usually without the benefit of a spatially explicit inventory of forest values.

5. As forest-level plans are gradually implemented to meet the stated goals of management, emphasis focuses on a few geographic areas in which roading, harvesting, and silviculture will be concentrated in the near term. These medium-scale (landscape or tactical) plans are always spatially explicit, and typically constitute the stage at which landscape ecology, analysis, and design must balance timber and conservation values in a managed forest.

6. Stand-level or annual operating plans may be simple logging plans or integrated logging and silvicultural prescriptions; they specify the on-the-ground implementation of forest management activities. These plans usually include a degree of detailed ecological site description and technical specifications for engineering and silvicultural actions.

From Tittler et al. (2001)

forest-level planning often proceeded without consensus on vision for a desired future forest, and the other levels of planning and development often proceeded with little direction from higher level plans or consideration of their true role in achieving management goals (Baskerville 1992; Bunnell 1992; Weetman 1994). These deficiencies need not suggest that the planning hierarchy (which is widespread and logical) is a poor one; rather, there tends to be poor understanding of its purpose and poor communication between its levels.

Forest planning in Canada has generally proceeded according to the following pattern (Tittler et al. 2001):

(1) the provincial government (which has ownership and control over the natural resources) passed forestry acts and regulations and developed forest management policies;

(2) based on these policies and those of other ministries and stakeholders (e.g., agriculture, fish and wildlife, environmental protection), the government then generated regional land-use plans;

Table 1.2. A comparison of sustainable forest management (SFM) criteria and the hierarchy of forest planning practiced in different boreal jurisdictions (summarized from Tittler et al. 2001).

Attribute	British Columbia	Alberta	Saskatchewan	Ontario	Quebec	Sweden	Finland	Russia
Commitment to SFM criteria:								
Biodiversity	Yes	Yes	No	Yes	Yes	Yes	Yes	Yes
Forest health	Yes	Yes	Yes	Yes	Yes	?	Yes	Yes
Soil and water	Yes	Yes	No	Yes	Yes	Yes	Yes	Yes
Productive capacity (wood and non-wood)	Yes	Yes	Yes	Yes	Yes	Yes	Yes	Yes
Global carbon cycles	No	No	No	?	Yes	No	No	Yes
Socio-economic functions	Yes	Yes	Yes	Yes	Yes	?	Yes	Yes
Planning hierarchy:								
Province-wide (policy & development strategy)	Not	Not	Yes	Not	Not	Not	Req. (at a regional scale)	Not
Regional (land use)	Yes	?	Yes	Vol.	Yes	Not	Not	Not
Forest estate or management area (forest plan)	Not	Yes	Yes (same as landscape)	Yes (same as landscape)	Yes	Vol.	Req.	Req. (same as landscape)
Landscape (tactical plan)	Yes	Yes	Yes (same as forest)	Yes (same as forest)	Yes	Vol.	Req.	Req. (same as forest)
Stand/cutblock (annual operating plan)	Yes	Yes	Yes	Yes	Yes	Vol.	Req.	Req.

Note: Not, not required and not generally undertaken; ?, not explicitly required, but may be addressed in other requirements; Req., required by law, always undertaken; Vol., voluntary, often undertaken.

(3) the provincial government carried out some level of forest inventory;

(4) timber rights were granted to industry in the form of licences, forest management agreements, or any one of a variety of other area-based or timber volume-based tenures; and

(5) as the industry then developed and implemented a range of strategic (ca. 20 year), tactical (ca. 5 year), and operational (annual) plans, government agencies continued to monitor and inspect industrial operations.

Each level of planning had to be consistent with the level above it. There may or may not have been extensive public input in the setting of policy, land-use boundaries, resource emphasis designations, and long-term development plans. Setting the annual allowable cut (AAC) for each forest management area or sustained-yield unit is a right generally retained by the province as custodian of the public resource and arbiter of competing forest values; AACs were typically re-determined every 5 or 10 years, and their levels are strong indicators of the degree of commitment to sustainability. Though partially based on a highly technical and quantitative process of timber supply analysis (Utzig and Macdonald 2000; CCI 2001), the setting of AACs typically has been a subjective and imprecise exercise, often with unknown or unstated assumptions and error margins. Provincial policies have varied considerably in their approach to setting AACs, however. For example, New Brunswick has implemented a policy, put in place in 1982, that the AAC, which is determined by each licensee using an approved methodology, must equal the non-declining long-run sustained yield (LRSY) based on an 80-year timber supply projection (Erdle 1998). In contrast, British Columbia governments have purposely adopted a policy of overcutting the LRSY, fully cognisant of future falldowns in timber supply, in order to pay for infrastructure (roads) and to normalize the forest age-class structure (Utzig and Macdonald 2000).

Strategic and tactical planning

In more detail, planning in each forest licence or forest management unit proceeded through the strategic, tactical, and operational levels as follows. Prior to this, the overall extent of a timber resource and how much of it was to be managed under a single licence or manager had to be determined. A "sustained yield unit" was defined (mapped) as a moderately sized ranger district, or the hinterland required to supply a specific pulp mill or sawmill. Forest inventory was based on aerial photographs backed by ground-based timber cruises to estimate the volume of commercial timber in all identified forest cover types. Excluding the area of parks, reserves, and other unsuitable or unavailable land (e.g., swamps and scrub), and based on the expected regeneration delay and the rotation length for desired crop trees (derived from retrospective sampling of fully stocked wild stands), an approximate AAC would be derived for that forest management unit.[3] Accessing all commercial quantities of timber, trying to "normalize" the forest structure into uniform acreages of each age class (typically 10 or 20 years), harvesting all stands over a given age, and utilization or conversion to commercial species were common objectives in such planning exercises. Fire prevention, fire fighting, and

---

[3]Excellent summaries of the AAC determination process are provided by Utzig and Macdonald (2000) and Cortex Consultants (CCI 2001).

spraying spruce budworm outbreaks with insecticide were considered important forest protection functions to maintain the timber supply and to regulate the age-class profile of the forest. It was standard practice to assume that all other forest values would be accommodated in a balanced forest age-class distribution, and that recreational needs could be met through improving motorized access. These policies and goals collectively constituted the framework of a "forest plan", essentially a strategic management plan, as described in Chap. 11.

Secondly, moving to the tactical level of a landscape or operating division, forest harvesting (and the road building required to support it) then generally proceeded in 5-year or 10-year steps without planning the long-term configuration of the landscape. Cognisant of the need to harvest older stands before the canopy broke up and stocking levels fell, the general rule was to find the oldest but still commercially viable stands as close to processing facilities as possible for annual harvest. But foresters of this era also prided themselves on promoting more productive second-growth stands, so earlier practices of selective logging were phased out in favour of clear cutting in order to enhance the soil warmth and light conditions required for optimal conifer seedling growth (Stiell 1976; Lavender et al. 1990; Keenan and Kimmins 1993). Staggered cutblock locations were employed in order to retain local wildlife habitat and visual values during a first pass of harvesters through a forest. Large contiguous clearcuts were considered not only unsightly, but unsuitable for the provision of adequate proximity to escape cover and thermal cover required by ungulates (Clark and Gilbert 1982). Cutblock layout tended to follow fairly standardized dimensions, often with straight edges and square boundaries in order to facilitate surveying, legal descriptions, and the calculation of stand areas; the "checkerboard" landscape is a typical product of this era (Franklin and Forman 1987). These considerations still figure in the tactical development of landscapes for sustainable forest management (Chap. 12).

Stand-level planning and practices

Standard stand-level management practices reflected the dominant silvicultural practices and paradigms of the era. Small cutblocks or strip cuts were employed with the hopes that natural regeneration from seeds blown in from the intact forest would be sufficient to regenerate the harvested area. This approach was effective in some cases, but was unreliable in others. Following public outcries regarding the inadequacy of reforestation (e.g., Swift 1983), forest regeneration was subsequently promoted using mechanical scarification or site preparation methods, broadcast burning of logging slash and the forest floor, and the planting of nursery-grown conifer seedlings produced from wild-collected seed. Additional silvicultural intervention was generally considered optional, though most jurisdictions required some assurance that commercial tree seedlings were of a sufficient density ("full stocking") and stature (i.e., deemed "free-growing") to produce a new stand of timber in the expected time frame. Additional brushing and spacing of stands might be conducted if government funds (generally acknowledged to be rural employment promotion programs) were made available. But efforts to add value through pruning, to undertake spacing to accelerate minimal piece-size attainment, or to "capture mortality" by thinning from below were rarely undertaken in the boreal forests of Canada (Weetman 1995). In contrast, thinning is practiced on approximately one-third of the managed stands (most of which are boreal) in Russia (Shutov 1995). Foresters in Fennoscandia undertook much more frequent stand interventions, with much land drained to improve conifer tree growth and most stands commercially thinned prior to the final harvest (Bernadotte and Gustafsson 1987; Hannelius and Kuusela 1995).

Stand manipulation for the enhancement or restoration of non-timber values was generally not practiced in the modern era. It was largely assumed that other forest users (e.g., trappers, recreationists) could have their needs met elsewhere on the land base or so long as individual campsites and cabins were protected. Forest development was generally considered to promote recreational opportunities by expanding access within the forest. Institutional (governmental) support for the forest industry typically included very generous, fixed, long-term stumpage fees paid for the harvesting of public timber, the subsidization of main forest roads, and the provision of other infrastructure to promote industrial development of the frontier. Long-term forest tenures with guarantees of favourable stumpage rates were considered collateral for the attraction of investment capital needed to attract private industry and its associated role in regional economic development (Pratt and Urquhart 1994, p. 152). Community involvement in forest management was largely through the provision of labour and the consumption of milltown benefits (Chap. 5), though notices of road building and logging activities might be posted on an annual basis. Monitoring systems were the responsibility of government agencies, largely to ensure compliance with rules and regulations. The entire system of forestry was characterized by industry–government cooperation in development of the timber resource, truncation of forest age classes, a homogenization of stand structures and compositions, minimal (or extensive rather than intensive) silviculture which was often ineffective in achieving forest regeneration (Swift 1983), a "top-down", "command-and-control" management style, and limited consideration of non-timber and social values.

Limitations

Lest we view pre-1990s foresters as backwards or unenlightened, it is important to recognize that forest managers of the time, like today, were trying to practice resource conservation in an atmosphere of business imperatives and political constraints, but without much of the public support and many of the tools available today. Forest inventories not only ignored animals and other non-timber forest products, but were often unreliable with respect to the trees too. Without ready access to computers and the appropriate training and culture of everyday computer use, there was no practical way to link basic growth and yield information to the forest inventory, coupled with harvesting and regeneration data, to effectively use dynamic inventory control as a formal management system. It wasn't until the 1960s and 1970s that forest ecologists in Canada devised systems of ecological classification applicable to the vast landscapes being managed, and it wasn't until the 1980s and 1990s that these frameworks were fully developed and linked to successional dynamics and silvicultural interpretations. Incomplete assignment of forest tenures and allowable cuts on Crown lands perpetuated the impression that forests were inexhaustible, with more wood always available in the next valley, so issues of sustainability and avoiding waste were often downplayed in corporate board rooms. Incomplete species utilization (e.g., little use of trembling aspen, *Populus tremuloides*), inefficient sawmills, and pulp mills completely dependent on pulpwood leases (rather than using chips from sawmills), unregulated use of logging roads (by hunters and poachers), and highly toxic contaminants in pulp mill effluents all represented aspects of wasteful and destructive resource use. A lack of mechanization in the logging sector, a dependence on river drives, and inadequate infrastructure foreclosed many management and development options. Likewise, poor amenities and concern with basic survival in many remote communities usually meant there was little public demand for participation in forest planning and little concern for environmental degradation. Foresters and managers recognized the desirability of multiple-use forest management, but had no way to really plan for it using the technologies available. Furthermore, non-timber forest values were always lower on the priority list than "getting the wood out" so long as public demand for protection of those values was low. Many of these technical and social factors have changed over the last 10–20 years.

A time of accelerated change

Social pressures

A number of socio-political developments coincided in the 1990s to mark the beginning of a period of revolutionary change in the management of the world's boreal forests. Whereas the history of forest exploitation and management as described above could be said to have progressed in an evolutionary manner and a reasonable rate, this all changed as the 20th century was coming to a close. Some of these trends were well established since the 1970s, while others were unforeseen. For example, the annual rate of timber harvest in all of Canada was relatively stable at approximately 6000 km^2/year from 1950 to 1970, but this increased to more than 10 000 km^2/year in the late 1980s, though government-approved AACs remained relatively constant at approximately 228 million m^3/year at a national level (SERP 1999). Such an increase in the level of forest development begs for a corresponding doubling of the effort to better understand and manage our northern forests, yet fiscal restraint was resulting in a loss of research and monitoring capabilities at both federal and provincial levels. In North America, corporate buyouts, consolidation, and downsizing in support of increased returns to shareholders also threatened the continuity of forest management culture and expertise acquired within forest products companies. In Fennoscandia, public dissatisfaction with the success of timber management at the expense of biodiversity conservation was leading to a reconsideration of standard forestry practices (Kouki 1994). Meanwhile, the collapse of the Soviet Union and the resulting economic and regulatory chaos resulted in accelerated logging in parts of Russia, but reduced investment in forest management activities (Shutov 1995).

Some key events, trends, and indicators of relevance to boreal forest management in the early 1990s are summarized in Boxes 1.4 and 1.5. These events and trends can be divided into two categories: those primarily of a political or public nature (Box 1.4); and

Box 1.4. Environmental and socio-political developments that raised public concerns about forest management in Canada in the 1990s.

- Blockades against the clearcut logging of coastal forests around Clayoquot Sound, on the west coast of Vancouver Island, took place from 1988 to 1993, with 856 people arrested in 1993, and calls for the boycott of forest products marketed by companies operating in the area (Drengson and Taylor 1997; FOCS 1998).
- Government strategies for the protection of representative old-growth forests and for the completion of provincial networks of parks were initiated in British Columbia ("Protected Areas Strategy", 1992), Ontario ("Lands For Life", 1997), New Brunswick ("Protected Areas Strategy", 1999), and nationwide (Gauthier 1992) in the early 1990s.
- Protests against logging of recreationally valuable wilderness lands around Lake Temagami (Ontario), against the wishes of the Temagami First Nation, raged from 1989 to 1997, featuring road blockades, hundreds of arrests, and arson (Ottertooth 2000).
- Polychlorinated dibenzodioxins (dioxins) and polychlorinated dibenzofurans (furans) were detected in the effluent of kraft pulp mills as a consequence of bleaching with chlorine compounds. Dioxins (especially 2,3,7,8-TCCD) and furans are highly toxic (carcinogenic and teratogenic), persistent, and they accumulate in the food chain. First publicized in 1987, governments responded in the early 1990s with regulations for marked reductions in allowable concentrations of these compounds in pulp mill effluents (Anonymous 1989; Cretney 1996).
- There was widespread environmental concern and public involvement in hearings, campaigns, and protests centred on the construction of the Alberta-Pacific Forest Industries pulp mill (then touted as the largest in the world) on the Athabasca River north of Edmonton, Alberta. Proposed in 1988, approved in 1990, the mill opened in 1993 (Pratt and Urquhart 1994).
- From 1988 to 1998, environmental and aboriginal support groups launched letter writing and boycott campaigns protesting the disenfranchisement of the Lubicon Lake Cree by petroleum and pulp companies in northern Alberta (Pratt and Urquhart 1994; WHA n.d.).
- Following on the heals of the Bruntland Commission's report on sustainable development (WCED 1987), the United Nations Council on the Environment and Development (UNCED) meeting in Rio de Janeiro in 1992 (popularly called "The Earth Summit") yielded four documents with major implications for the ways in which forestry is practised (IGC 1996):
  - Framework Convention on Climate Change, including the recognition that fossil fuel consumption has accentuated the atmospheric greenhouse effect and that these emissions must be reduced and offset;
  - Convention on Biological Diversity, requiring commitments to save endangered species and protect significant natural areas;
  - Agenda 21, presenting a set of 27 principles for defining equitable and sustainable relationships between man and his environment; and
  - Guiding Principles on Forests, including the management, conservation, and sustainable development of all types of forests and commitments to combat deforestation.
- Major publicity campaigns against clearcut logging were launched by the Sierra Club, Greenpeace, and other environmental non-governmental organizations (Fritz 1989; Devall 1993; McCrory et al. 1994).
- The Forest Stewardship Council was founded in 1993 through the joint initiative of numerous international ENGOs, to serve as a positive alternative to forest products boycotts: i.e., to identify and certify those products manufactured sustainably, rather than identifying and boycotting companies targeted by particular environmental or social justice campaigns (FSC 2001).

Box 1.4 (concluded).

- Growing concern over environmental damage associated with logging in the Pacific Northwest of the United States, leveraged by endangered species protection for the northern spotted owl (*Strix occidentalis caurina*), resulted in most logging in western U.S. National Forests being severely curtailed in 1993.

- Signatory nations agreed to respond to the Rio conventions with national strategies and action plans, such as the Canadian Biodiversity Strategy, released by the Canadian Biodiversity Working Group in 1994 (CBWG 1994).

- A second attempt at imposing a countervailing duty on Canadian exports to the United States was successful in 1986, aimed primarily at low Canadian stumpage fees. Replaced by export taxes and increases in stumpage, another countervail action was spawned in 1991, which led to the 1996 Softwood Lumber Agreement, with yet higher stumpage rates and a quota system to assuage American protectionist claims that forest companies harvesting on public land in Canada were being subsidized (Reed 2001).

- The Supreme Court of Canada, in its landmark Delgamuukw decision, defined Aboriginal title as a right to exclusively use and occupy land and set out the tests that an Aboriginal group must meet in order to prove its title. Since Aboriginal title is inalienable except to the Crown, all governments (federal and provincial) have a fiduciary duty to consult an Aboriginal group on government-authorized activities on the group's territories, even where treaties have been signed (Delgamuukw v. British Columbia 1997; Pape & Salter 1998; Persky 1998). The resulting consultation guidelines in B.C., however, have largely failed the tests of effectiveness, efficiency, and equitability (Lindsay and Smith 2001).

- The World Resources Institute indicated that 63% of Canada's original frontier forest (primarily boreal forests) has either been lost or is currently threatened. It notes that (WRI 1997; BFN n.d.):
  - Canada's forests are primarily managed for timber, though surveys of public values put a priority on non-timber uses;
  - almost 50% of the boreal forest is under tenure to a few large-scale forest companies;
  - 30% of the boreal forest is within 1 km of a road or access route; and
  - current harvesting rates appear unsustainable over the long term.

- The Canadian Senate Subcommittee on the Boreal Forest concluded that, "The world's boreal forest, a resource of which Canada is the major trustee, is under siege" (SSCAF 1999; BFN n.d.).

- Several television shows and magazine and newspaper articles (e.g., Lanken 1996) have highlighted the rate of industrial development (associated with forest harvesting, oil and gas exploration, oil sands mining) occurring in the boreal forests of Canada and Siberia, raising the profile of northern forests as the last great wilderness in the minds of many people in the industrialized nations.

- The film *L'Erreur Boreale*, directed by the Quebec songwriter Richard Desjardins and produced by the National Film Board of Canada, was shown throughout Quebec in 1999, articulating public concern and mistrust regarding a history of unsustainable logging in Quebec and industry-government complicity in their exclusive management of public lands.

- Several ENGOs (Canadian Parks and Wilderness Society, Kendall Foundation, Pew Charitable Trust, Taiga Rescue Network, Western Canada Wilderness Committee) initiated boreal campaigns in the late 1990s, lobbying for the protection of large tracts of boreal forest as wilderness parks.

- Ola Ullsten, former Prime Minister of Sweden and co-chair of the World Commission on Forests and Sustainable Development, stated in 1999 that, "New ways must be found to slow and ultimately reverse forest decline, and Canada has a special responsibility because it still has 20–25% of its primary forest" (BFN n.d.).

those developing primarily in the scientific research and forest management community (Box 1.5). Both sets of developments raised the profile of forest practices in the public eye, and contributed a sense of defensiveness (on one hand) and urgency (on the other) on the part of forest managers. Yet the environmental and political concerns and the scientific and professional developments were never that much out of step. Many management solutions proposed to address public concerns, however, have been largely unsupported by scientific research or cases of effective application on an operational level

In Quebec, a movie directed by a popular folk singer, Richard Desjardins, generated outrage in the public and havoc in government and industry (Box 1.4). In this movie, the worst images associated with the old (though still practiced) way of doing forestry were cleverly presented for all to see. The message was clear and strong: "the forest was being savagely exploited by a consortium of industry and government representatives". That message had a lot of impact because most of the public thought that Quebec's forest was so large that we could never run out of wood. For the first time, many citizens now wondered: Were we cutting too much? Were we cutting in a way that was damaging to the land and to its inhabitants? If we are in some way accountable for the well-being of our "brothers and sisters" of the natural world, are these standard forestry practices (such the clearcutting of large chunks of the boreal forest) the right thing to do?

A 1991 nationwide public opinion poll of forestry issues found that 79% of Canadians believed that clearcutting is a poor forest management practice, 81% thought chemicals used in forest management pose a hazard to human health and the environment, and 86% held the opinion that most old-growth forests in Canada should be protected (ERGL 1992). More recently, another national survey confirmed that 75% of Canadians agreed or strongly agreed with the statement that clearcutting has negative environmental effects, and 95% agreed or strongly agreed that forests should be managed to ensure healthy populations of all wild species of trees, other plants, and animals (Robinson and Hawley 1997). If resulting from an election or a referendum, such numbers would be considered an unambiguous mandate for change in any democratic society. Does this indicate that the public simply needs "educating", or is it the politicians, corporate officers, and foresters who are out of step with society? Had we, the professional researchers and managers of forests, collectively been blinded by our old philosophies of nature? Are forest managers behaving in a manner accountable to the owners of Crown land, to the citizens of our democracy? These questions are now being debated daily, but a new philosophy of man's place in nature is progressively replacing the old one.

In many cases, the technical issues have been easier to overcome than the social ones. Concerted federal-provincial cooperation and infusions of research money would typically develop and field-test some timely scientific–technical solutions over the course of approximately 5–15 years. These included challenges in regenerating northern forests dominated by cold soils and short growing seasons (Grossnickle 2000; Wagner and Columbo 2001), population declines in sensitive species such as woodland caribou (Armleder and Stevenson 1995; Dyer et al. 2001), poor integration of stand-level actions with forest-level goals (Baskerville 1992), uncoordinated and often contradictory rules and guidelines from government, and other unforeseen cumulative effects (Gosselink et al. 1990; Duinker 1994). Vast frontiers were being opened (e.g., in Alberta, Table 1.3, and in Siberia) at the same time as we came to realize "at a gut level" that even the wild northern forests have their limits. Expectation for non-timber values to be respected and for land management decision making to be shared coincided with a growing disillusionment with the fibre volume emphasis of classical forestry (Hammond 1991; Maser 1994; Kohm and Franklin 1997). Well-respected scientists continue to point out that current management policies and practices are generating unacceptable impacts on northern ecosystems and that we really don't understand many aspects of the boreal biome (Zasada et al. 1997; Schindler 1998; Krebs et al. 2001; Henry 2002; Schneider et al. 2003). Collectively, these developments have forced forest managers and the public in general to face the finiteness of our northern forests (and their resources and biodiversity), and to place a new premium on research and the pursuit of sustainability.

Time for a new paradigm

More fundamental than these developments, perhaps, has been the need for the profession of forestry to change as the view of nature in Canadian society has changed. The prevailing perception of our place as humans in the natural world has matured dramatically since the beginning of regulated forestry in the western world some 200 years ago. Up to the 18th century, humans tended to place themselves above the natural world; we considered ourselves not only superior in terms of intelligence, but not actually part of nature per se. Human languages are full of the distinction between "natural" or "wild" vs. "artificial" or "man-made" environments. Though Carolus von Linnaeus classified mankind as a species of animal, it was the work of Charles Darwin that firmly asserted that the laws of evolution and natural selection affected *Homo sapiens* as they did other species: our connection with the natural world was actually much stronger than we thought. This idea was hard to accept, but with it came a new thinking in terms of our dependence on the natural world. Animals were no longer perceived simply as "machines" (as suggested by René Descartes[4]) or material to be used by us for our own selfish purposes, but as "brothers" in the continual evolution of life on earth. New ways of looking at nature also emerged, with philosopher–naturalists such as Henry Thoreau, John Muir, and Aldo Leopold suggesting that we were, in fact, part of nature and that we needed to develop a new ethic for interacting with nature. Indeed, it has been suggested that no environmental problem can be solved without addressing both the ecological and social processes at play, and environmental management requires an understanding of "complex ecosocial systems" (Waltner-Toews et al. 2003).

The various environmental groups that have recently challenged the old dogmas of forestry have clearly emerged from this post-modern perception of our relationship with nature. In contrast, the classic forestry dogma is a logical extension of the mechanistic views of the universe espoused by philosophers like Descartes and Emmanuel Kant. The

---

[4] *"Descartes, a founding father of modern science, introduced the concept that mind and matter were distinct and that the natural world, far from being an oasis, was a machine to be dissected, discovered, and dominated by humans who were separate from insensate nature"* (Pincett 1993).

Box 1.5. Scientific, professional, and regulatory developments in forest management that prompted review of standard forest management approaches in Canada in the 1990s.

- Experience with second-pass harvesting and spatially explicit simulation models confirmed earlier warnings that rules for stand-level management do not result in desirable attributes at the landscape level (Franklin and Forman 1987; Nelson and Finn 1991; Jordan and Baskent 1992; Bunnell 1992; Baskerville 1994; Weetman 1994).

- Widespread population declines in neotropical migrant birds were found to be associated with forest fragmentation (Askins et al. 1990; Wilcove and Robinson 1990; Robinson et al. 1995).

- It was concluded that conservation of biodiversity at the landscape level of organization, consisting of gradients and mosaics of many community types, can only be achieved in large areas of wilderness (Noss 1990).

- Ecologists and forestry researchers suggested that timber harvesting should more closely emulate the patterns, composition, and residual structures of natural disturbances as a "coarse filter" approach to conserving the forest biodiversity which we cannot possibly inventory or adequately research (Hansen et al. 1991; Booth et al. 1993; Hunter 1993).

- In response to widespread public concern about forest management, the federal government in 1992, through the Canadian Forest Service, sponsored the designation of 10 "model forests" in a program designed to advance the concept of sustainable forest management, showcase integrated resource management, improve "best forestry practices", promote the use and demonstration of advanced technology for forest management, and communicate the results to the public (MFP 1998–2002).

- At about the same time as logging was severely constrained in the U.S. National Forests (Box 1.4), concepts of "new forestry", "ecosystem management", and "ecoforestry" emerged in the U.S.A. and grew in popularity as more environmentally sound alternatives to industrial even-aged timber management (Franklin 1989; Gillis 1990; Hammond 1991; Kessler et al. 1992; FEMAT 1993; Grumbine 1994).

- Concepts of "ecosystem health" and "ecological integrity" gained credence in the development of new paradigms for environmental management, replacing simpler management models based on productivity, population dynamics, or species diversity (Karr 1991; Costanza et al. 1992; Woodley et al. 1993; Angermeier and Karr 1994). Discussion and testing to aid in the selection of suitable indicators in a variety of ecosystems followed.

- The British Columbia government responded to growing demands for forestry reform and the threat of international boycotts by enacting The Forest Practices Code Act (1995). It adopted a highly prescriptive and regulatory approach to protecting all forest values, though was never fully implemented. Other provinces followed more open-ended industry-led initiatives for forestry reform, though the need for greater protection for biodiversity, non-timber forest values, and clean air and water was widely acknowledged.

- The Clayoquot Sound Scientific Panel recommended the adoption of "new forestry" and nature-emulating management principles as part of the compromise to facilitate both new protected areas and ongoing industrial forestry in a high-profile area of civil disobedience (blockades and mass arrests) on the British Columbia coast (CSSP 1995).

- Provincial governments undertook widespread public consultation on land-use issues, and forest companies undertook third-party audits ("forest certification") in the late 1990s as means of avoiding conflict, blockades, and boycotts.

Box 1.5. (concluded).

- Environment Canada identified large-scale resource development activities as the largest threat to biodiversity in the North American boreal forest ecosystem. These activities include large-scale commercial logging, mining, and hydroelectric development. Only 2–4% of Canada's boreal zone is strictly protected from such threats (SERP 1999; BFN n.d.).
- Canada's federal and provincial government ministers responsible for forest management reached a historical accord on the goal of sustainable forests: "Our goal is to maintain and enhance the long-term health of our forest ecosystems, for the benefit of all living things both nationally and globally, while providing environmental, economic, social and cultural opportunities for the benefit of present and future generations." A 5-year strategy and action plan (see details in Chap. 2) were released in support of this vision (CCFM 1998).

main objective of classical forestry can be viewed as the efficient, assembly-line-like production of timber, transforming the natural world to fill our material needs. Ideas such as the normalization of the forest age-class distribution, sustained yield and perpetual cutting of timber, management of solely even-aged stands, and the cutting cycle were developed to make the natural world conform to the industrial model of optimized productivity, risk reduction, and detailed management. It is no wonder that clashes between the old and new philosophies erupted with strong emotions and even violence. This was best exemplified by logging protests that took place in the coastal forests of Oregon, Washington, and British Columbia, where the clearcutting of large areas of ancient forests was totally unacceptable to many people (see Box 1.4). The clash was particularly antagonistic because most foresters believed that they were doing good work, improving on nature by transforming irregular, low-productivity, and over-mature forests into vigorous, more productive, well-managed forests. The old view of nature was clashing with the new, and nobody knew where these opposing world views were heading or how they could be reconciled.

The concept of sustainable development (WCED 1987) then provided a useful way forward. Though rejected by many environmentalists as being self-contradictory and merely an extension of utilitarian concepts for the conservation and wise use of resources, this idea offered the important test of "sustainability" as the yardstick against which resource management decisions were to be gauged (see Chap. 2). Armed with this test, could it be possible to use natural resources such as the wood growing in natural forests, without having to transform the resource base so much that its ecological integrity and ongoing sustainability (persistence) was threatened? Even countries such as Sweden, where a century of rigorously applied classical forestry was successful in transforming its forest lands into a showcase of boreal timber productivity, have been affected by these changes in our view of nature. Standing snags and fallen logs left to decay on the forest floor, perceived as wasted fibre and bad forestry just 10 years ago, are now promoted (De Jong et al. 1999).

Is it possible to use our northern forests in a sustainable way? Well, the contributors to this book by and large believe it is, but that progress in this regard has been spotty. Sustainable management of the boreal forest will require a drastic change in our view

Table 1.3. The recent surge in industrial forestry in northern Alberta.

Company	FMA* signed[a] (year)	FMA* area[b] (km²)	Annual allowable cut[b] (m³/year)	Average timber volume[c] (m³/ha)	Estimated[d] area of harvest (km²/year)	Primary mill capacity (woodstock) (m³/year)
Weldwood	1954	9964	2075040	117.9	120[e]	2857000[e]
Canadian Forest Prod.	1964	6504	1209000	120.9	40	1228000[f]
Weyerhauser (Grande Prairie)	1969	13541	3089760	125.6	98	—
West Fraser (Blue Ridge Lumber)	1975	6666	1025790	110.3	37	935000[g]
Weyerhauser (Slave Lake)	1986	7192	891540	76.2	47	—
Alberta Newsprint	1989	3736	641110	105.1	24	—
Daishowa–Marubeni	1989, 1996	24321	1618020	60.6	107	1800000[h,i]
Slave Lake Pulp	1990, 1996	6293[j]	965700	111.0	36[j]	680000[l]
Alberta-Pacific	1991	57331	3418620	50.2	140[k]	2500000[h]
Tolko (High Level Forest Products)	1996	35624	1562560	60.8	103	—
Millar Western	1997[l]	2882	538200	103.5	21	1400000[m]
Tolko (High Prairie)	1997[l]	2734	381330	66.9	23	—
Vanderwell Contractors	1997[l]	560	32550	46.5	2.8	—
Weyerhauser (Edson)	1997[l]	5118	568890	90.3	25	—

* FMA = Forest Management Agreement, an area-based tenure to be managed for sustained fibre yield, with planning and land stewardship responsibilities; 1 km² = 100 ha.

Sources:

[a] Wein (1999); http://www.rr.ualberta.ca/courses/encs204/content/3_ch5.htm.

[b] Personal communication from Darren Tapp, Alberta Sustainable Resource Development, 17 June 2002.

[c] Commercial volumes from Alberta Sustainable Resource Development, www3.gov.ab.ca/srd/land/lad/docs/Volumetabl.pdf

[d] Unless reported directly by licensees[j,k], the approximate area harvested each year is estimated from the AAC (in m³/year) divided by average timber volume (in m³/ha), divided by 100 to convert from ha to km², and multiplied by 0.4 (40%), the provincial average proportion which is commercially harvestable; in fact, not all of the AAC is utilized, and FMAs vary considerably in the proportion harvestable.

[e] http://www.weldwood.com/hinfr01/internet/hinnet.nsf

[f] Grand Prairie and Hines Creek facilities combined; personal communication from Lee Cooner, Canadian Forest Products, 2 July 2003.

[g] Robertson 2000; http://www.ualberta.ca/~gis1/wshop/paper.pdf

[h] Pratt and Urquhart (1994).

[i] http://www.dmi.ca/quickfacts.htm and http://www.dmi.ca/inAB.htm

[j] Personal communication from Gordon Sanders, Slave Lake Pulp, 27 June 2003.

[k] http://www.alpac.ca/Forest_Management/ForestManagementArea.htm

[l] Personal communication from Rick Keller, Alberta Environmental Protection.

[m] http://www.millarwestern.com/environment/envupdate_forest_mgmt.pdf

of nature, a new respect for non-commercial considerations, and the creative application of new techniques and approaches. Are we up to this new challenge? How can we stimulate the exploration, testing, and adoption of these solutions, if indeed they are solutions? This is a global quest, requiring an international sharing of ideas, data, and experience in all kinds of managed forests (Higman et al. 1999; Hunter 1999; Lindenmayer and Franklin 2003). Even if new ways of managing forests sustainably can be devised and tested, industrial and governmental "buy-in" and appropriate institutional structures will be needed in order to make them operational.

There is tremendous variation in the emphasis currently placed on non-timber values and community involvement by different provinces, forest districts, forest product companies, operating divisions, and individual foresters. Many foresters are in an awkward position, wanting to undertake the sustainable management of all forest values, but being obliged to look after the more immediate interests of their employers, the forest products companies or government agencies. So long as the companies are primarily in the fibre business, and so long as public servants are subject to political interference, forest management professionals will always be in a potential conflict of interest, and will need considerable support from their professional associations and peers in order to maintain their integrity in the fight for sustainable forest management.

Advancing sustainable forestry in boreal Canada

The Sustainable Forest Management Network

Several nationwide initiatives were undertaken in Canada in the 1990s to advance the science and management of forests. Foremost among these probably has been the creation of the Model Forest Program in 1992 (MFP 1998–2002), agreement by the Canadian Council of Forest Ministers on criteria and indicators of sustainable forest management in 1995 (CCFM 1998; Chap. 2), and the creation of the Sustainable Forest Management Network in 1995. It is the achievements of Canada's Sustainable Forest Management Network (SFMN) that are highlighted throughout this book. The Network was born from efforts to grapple with the emerging issues described in the previous section. In particular, it owes its existence to the willingness of individuals at the University of Alberta, Alberta-Pacific Forest Industries Inc., the National Research Council, the Province of Alberta, and Aboriginal governments to learn about ecosystem dynamics in the boreal landscape prior to first-pass timber harvesting. The SFMN is based on the premise that students, trained in an environment where all concerned parties were "at the table", would be the best ambassadors of sustainable forestry for future generations. Furthermore, the Network was created around scientists and other researchers who were not only from the traditional forestry schools, but also included scholars from biology, geography, economics, and other social sciences departments, denoting recognition that the SFM requires movement beyond the existing orthodoxies and would benefit from inputs from a wide range of disciplines. The SFMN has concentrated on the important scientific underpinnings for managing forests, while many of the Model Forests have wrestled with the implementation of SFM in management control systems (see their web page at http://www.modelforest.net for more background and links to individual Model Forests).

The seeds of the SFMN were sown at a University of Alberta field station (Meanook Biological Research Station), some 130 km north of Edmonton, located less than 100 km from the Alberta-Pacific pulp mill. In August 1991, an environmental manager from Alberta-Pacific was invited to present a seminar that opened the doors to the opportunity for large-scale integrated research on sustainable forest management (SFM) focused around university-level education in a multi-sector environment. Over the ensuing 3 years, the partnership between academic, private, and public sectors was built; the multi-sectoral initiative garnered its share of detractors and many unsuccessful applications were discussed and submitted for consideration to a wide range of potential funding sources. Then in the summer of 1994, the opportunity to participate in the second round of Networks of Centres of Excellence (NCE) emerged. These networks were to be funded by one or more of the Natural Sciences and Engineering Research Council of Canada (NSERC), Social Sciences and Humanities Research Council of Canada (SSHRC), and the National Research Council of Canada (NRC). Networks of Centres of Excellence (see http://www.nce.gc.ca/index.htm) were envisioned as vehicles for facilitating the cooperation of researchers and managers in academic institutions and private industry for working on problems of importance to the well-being of the nation.

Ultimately the Alberta-based initiative on SFM was expanded to include the Province of Alberta and a linkage with the more established forest industry in eastern Canada, initially through established networks in the province of Quebec. The goal was to build a genuine network of forest users, forest practitioners, and forest regulators that would help forest researchers educate the next generation of practitioners in Canada. The vision was to link landscape and watershed management with planning and practices as fibre was brought from the forest to the mill, and the process wastes returned to the surrounding environment. The SFMN proposal, subsequently expanded to include more than 80 investigators across the country, was successfully awarded funding in July 1995.

Promoting research as a practical means of solving complex problems poses a challenge in the world of Canadian government and business, though it can be demonstrated that research is a fundamental driver of productivity growth and economic development. Within Canada, forest management research has been funded by various government agencies as well as by the forest industry itself. However, funding levels have been well below world averages. Globally, private sector funding of research averages about 0.5% of revenues or sales; the average in Canada is about half of this at 0.25% (Binkley and Forgacs 1997). Total forest-related research and development (R&D) in Canada, including public research agencies, is somewhat less than 0.4% of sales (Binkley and Forgacs 1997). In the Fennoscandian countries, by way of contrast, average industrial R&D expenditures in the forest sector are approximately 1% of sales. The research expenditures of just two Swedish firms (SCA and STORA) in 1994 exceeded the combined expenditures by the entire Canadian industry, including the research institutes PAPRICAN, FERIC, and FORINTEK (Binkley and Forgacs 1997). It is gradually being appreciated that if Canada wishes to maintain a predominantly public forest and simultaneously create the new knowledge and technology for sustainable forest management and industrial growth, then the industry and the public should invest in R&D, especially that focused on issues that arise from Canada's unique blend of social and industrial institutions. The SFMN has helped create an invigorated research environment and culture of SFM by bringing industry, universities, governments, First Nations, and rural communities into a lasting partnership focused on mutual priorities.

The convergence of industrial, academic, political, environmental, and Aboriginal interests in pursuing a rational, fact-based, and equitable path for this next round of frontier development provided the window of opportunity that led to the birth of the SFMN. The Sustainable Forest Management Network (see http://sfm-1.biology.ualberta.ca/english/home/index.htm), one of Canada's 22 Networks of Centres of Excellence, provides interdisciplinary research on the management of the forests of Canada. It focuses on integrated and directed research in developing new planning and management tools for industry, as well as policy insights and improved institutions for government, to assist in ensuring that Canada's forests are managed sustainably. Headquartered at the University of Alberta in Edmonton, the Network currently operates with a $7 million annual budget that is judiciously allocated to research that meets the SFM Network mission. The network has a number of funding partners beyond the NCE program, including provincial governments, forest products companies, and First Nations. Thirty Canadian universities are represented in the Network. Over 100 researchers and 200 graduate students are involved in SFM Network research, and more than 200 graduate students have already completed their studies within the SFM Network (SFMN 2003).

Research emphasis

Natural disturbance was one of the cornerstones on which much of the land- and water-based research was built in the original SFMN application. Much research in the early stages of the Network focused on the extent to which natural disturbances could contribute to an understanding of management options and the development of practices that would conserve biodiversity and promote sustainability of northern forest environments. The SFMN initiative was based on the idea that baseline conditions must be understood for key biological processes, including the impact of natural (e.g. fire, insects, blowdown) disturbances, before we could intelligently manipulate components of the system (e.g., through timber harvesting) without unforeseen impacts on the environment. An understanding of natural disturbance regimes, natural regeneration, and natural stand development have long been considered essential underpinnings of scientific silviculture (Toumey and Korstian 1947; Oliver and Larson 1990), but these subjects had received comparatively little study in Canada's boreal forests.

There has also been growing interest in, and demand for, more naturalistic approaches to forestry, including the desire to explore the degree to which forest harvesting operations could better mimic natural disturbance processes (Fig. 1.5). Various versions of the natural disturbance emulation model had been proposed by silviculturists many decades ago (Troup 1928; Toumey and Korstian 1947), and more recently by forest ecologists concerned about wildlife and other non-timber forest values (e.g., Hansen et al. 1991; Hunter 1993; Attiwill 1994; and Bergeron and Harvey 1997). The Senate Committee on Agriculture and Forestry has recommended adoption of a natural

Fig. 1.5. Much of the early research conducted by the SFMN emphasized the investigation of natural disturbances such as wildfires and insect outbreaks: their biophysical processes, resulting patterns, their effects on biota and forest values, recovery over time, and how forest management might perpetuate, manipulate, or emulate them. Shown here is vegetation recovery soon after a wildfire near Whitecourt, Alberta.

Photo courtesy of Millar Western Forest Products Ltd.

landscape-based approach to managing the boreal forest (SSCAF 1999). But the emulation of natural disturbances as an effective paradigm for SFM had not been systematically tested, so this constituted a dominant theme of SFMN research. This early research emphasis is reflected in the contents of this book, especially Part 3 on forest ecology and management. A follow-up research focus has emerged to explore methods of intensive forestry that can compensate for reduced fibre harvesting where the "lighter touch" of nature-emulating forestry is implemented. Thus much of the discussion in subsequent chapters, and much of the ongoing research of the SFMN, explores the concept of "triad" zoning, with intensive silviculture on a small subset of the landbase designed to offset a substantial network of protected areas and extensive management for multiple forest values elsewhere on the landbase (Gladstone and Ledig 1990; Burton 1995; Hunter and Calhoun 1996; Binkley 1997; Erdle 1999).

Research funded through the SFMN was originally organized around four themes, each with its own set of disciplinary expertise:
- Socio-economic sustainability (economists and other social scientists);
- Ecological basis of sustainability (ecologists);
- Planning and practices (foresters and planners); and
- Minimal impact technologies (engineers).

This structure was eventually reorganized around three more interdisciplinary themes (see below), each designed to leave a "legacy" of knowledge, training, and reform. This shift came about as academic researchers worked on real-world issues with industrial, Aboriginal, and government-based partners, and better facilitated interdisciplinary and

---

Box 1.6. The major elements of sustainable forest management, as recommended by the Alberta Forest Management Science Council.

1. The conservation of ecological integrity of the forest is a necessary condition for the sound and sustainable management of the forest.
2. Defining a vision of a desired future forest is a necessary step in implementing a more sustainable forest management program.
3. Social and economic values are integral to the selection and attainment of the desired future forest.
4. The temporal and spatial scales used to manage the forest must be consistent with the scales of disturbances and processes inherent to the forest and social scales relevant to forest resource use.
5. Adaptive management monitors progress towards the desired future forest, continually improves the knowledge base, and adjusts actions to correct for deviations.

---

inter-sectoral team-building. For example, four researchers who were involved in the SFMN were invited to participate in developing guidelines for sustainability in the Alberta forestry sector (Box 1.6), and the interplay of research and policy was enhanced and better coordinated as a result. It is recognized that sustainability will only occur when users (the implementers of management policies and practices) interact with scientists and policymakers (often the agents of change).

This book

The three legacies around which the latter half of the first phase of the SFMN was organized, and on which the central three sections of this book are structured, were the following:
- Strategies for sustainable forest management (primarily socio-economic);
- Understanding disturbance (primarily ecological); and
- Impact minimization (primarily engineering).

The Network continues to evolve, with Phase II of its program now addressing all forests in Canada, and more intimately coordinated with other forestry research institutions in the country. The latest phase of SFMN research is currently structured to specifically emphasize (1) criteria and indicators of sustainability, and (2) the development and testing of a specific set of alternative institutions and strategies for sustainability (SFMN 2003).

The requirements for sustainable forestry are sometimes presented as the intersecting interests of environmental, social, and economic values (Zonneveld 1990; Salwasser et al. 1993), the "three pillars of sustainability". In charting the solution to this puzzle, however, the organization of this book follows the lead of the SFMN in identifying three somewhat different programs of research and action. Social and economic values are combined in the analysis of institutions, values, communities, and Aboriginal perspectives (termed "Legacy 2: Strategies for Sustainable Forest Management" by the SFMN). The environmental pillar is focused further on understanding the ecology of northern forests in order to better manage them (under the auspices of "Legacy 1: Understanding

Disturbance" in the SFMN research program; SFMN 2001). Finally, we additionally recognize the need for technological tools to both minimize detrimental impacts on the environment and as an instrument of improved efficiency and competitiveness in industry (referred to as "Legacy 3: Minimizing Impacts" in the SFMN). So this book starts out with an exploration of the goals of sustainable forest management, particularly as they apply to boreal forestry, and the premises for the main thrusts (or research legacies) of the SFMN and the book. This is followed by five chapters on the social and economic dimensions of sustainability. Many of the more technical results from ecological and silvicultural research are encompassed in seven chapters in the forest ecology and management section. Some environmental impacts of the forest products industry and strategies for their prevention and amelioration are grouped in six chapters on minimizing the impacts of forest use. The final section of the book provides a framework of adaptive management, case studies of implementation, and an overview of the challenges, options, and progress in achieving sustainability in Canadian forest management.

Detailed descriptions of research projects, workshops, and annual reports can be found at the SFMN website, http://sfm-1.biology.ualberta.ca. While this book showcases results of the SFMN research program and numerous Canadian case studies, it also pulls together some key concepts and advances in SFM from around the world, especially those applicable in boreal regions. Though not a blueprint for sustainable forest management, this book attempts to report on the state of the art, some new and as yet unproven ideas, and the most up-to-date basic and applied research on the components of SFM. The following chapters are laced with numerous examples, case studies, and scenarios that demonstrate how various elements of SFM are being effectively implemented already. While much has been learned and much is being applied, we still have a long way to go; this is a progress report on an ongoing (indeed, never-ending!) undertaking.

Acknowledgements

We thank Marvin Abugov, Marc-Olivier D'Astous, and Darren Tapp for helping us assemble the data presented in the tables. We are grateful for the assistance and constructive suggestions of Carla Burton, Peter Duinker, and three anonymous reviewers.

References

Aksenov, D., Dobrynin, D., Dubinin, M., Egorov, A., Isaev, A., Karpachevskiy, M., Laestadius, L., Potapov, P., Purekhovskiy, A., Turubanpva, S., and Yaroshenko, A. 2002. Atlas of Russia's intact forest landscapes. Russian NGOs Forest Club, Moscow, Russia. Available at http://www.forest.ru/eng/publications/intact/ [viewed 25 June 2003].

Angermeier, P.L., and Karr, J.R. 1994. Biological integrity versus biological diversity as policy directives. BioScience, **44**: 690–697.

Anonymous. 1989. Dioxin. Alkaline Paper Advocate. Vol. 2. No. 2. Stanford, California. pp. 1–4. Available at http://palimpsest.stanford.edu/byorg/abbey/ap/ap02/ap02-2/ap02-202.html [viewed 11 June 2002].

Apsey, M., Laishley, D., Nordin, V., and Peillé, G. 2000. The perpetual forest: using lessons from the past to sustain Canada's forests in the future. For Chron. **76**: 29–53.

Armleder, H.M., and Stevenson, S.K. 1995. Silviculture systems to maintain caribou habitat in managed British Columbia forests. *In* Innovative silviculture systems in boreal forests. Proceedings, Symposium, 2–8 October 1994, Edmonton, Alberta. *Edited by* C.R. Bamsey. Clear Lake Ltd., Edmonton, Alberta. pp. 83–87.

Askins, R.A., Lynch, J.F., and Greenborg, R. 1990. Population declines in migratory birds in eastern North America. Curr. Ornithol. **7**: 1–57.

Attiwill, P.M. 1994. The disturbance of forest ecosystems: the ecological basis for conservative management. For. Ecol. Manage. **63**: 247–300.

Bailey, R.G. 1995. Description of the ecoregions of the United States. United States Forest Service, Fort Collins, Colorado. Available at http://www.fs.fed.us/land/ecosysmgmt/ecoreg1_home.html [viewed 25 June 2003].

Baskerville, G.L. 1992. Forest analysis: linking the stand and forest level. *In* The ecology and silviculture of mixed-species forests, a festschrift for David M. Smith. *Edited by* M.J. Kelty, B.C. Larson, and C.D. Oliver. Kluwer Academic Publishers, Dordrecht, Netherlands. pp. 257–277.

Baskerville, G.L. 1994. Gaelic poetry for deaf seagulls: encore. For. Chron. **70**: 562–564.

(BCMWLAP) British Columbia Ministry of Water, Land and Air Protection. 2001. Is British Columbia's rich biodiversity protected? Available at http://wlapwww.gov.bc.ca/soerpt/1protectedareas/ecosystems.html [viewed 23 June 2003].

Bergeron, Y., and Harvey, B. 1997. Basing silviculture on natural ecosystem dynamics: an approach applied to the southern boreal mixedwood forest of Quebec. For. Ecol. Manage. **92**: 235–242.

Bernadotte, B., and Gustafsson, U. (*Editors*). 1987. Swedish forest: facts about Swedish forestry and wood industries. Skogsstyrelsen (National Board of Forestry), Jönköping, Sweden. 111 p.

Bernsohn, K. 1981. Cutting up the north: the history of the forest industry in the northern interior. Hancock House Publishers, North Vancouver, British Columbia. 192 p.

(BFN) Boreal Forest Network. no date. The boreal forest: a global crisis in the making. Boreal Forest Network, Winnipeg, Manitoba. Available at http://www.borealnet.org/overview/index.html [viewed 18 September 2001]. 4 p.

Binkley, C.S. 1997. Preserving nature through intensive plantation forestry: the case for forestland allocation with illustrations from British Columbia. For. Chron. **73**: 553–559.

Binkley, C.S., and Forgacs, O.L. 1997. Status of forest sector research and development in Canada. Faculty of Forestry, University of British Columbia, Vancouver, British Columbia. 23 p.

Booth, D.L., Boulter, D.W.K., Neave, D.J., Rotherham, A.A., and Welsh, D.A. 1993. Natural forest landscape management: a strategy for Canada. For. Chron. **69**: 141–145.

Boyce, M.S., and Haney, A. (*Editors*). 1997. Ecosystem management: applications for sustainable forest and wildlife resources. Yale University Press, New Haven, Connecticut. 361 p.

Bunnell, F.L. 1992. De mo' beta blues: coping with the landscape. *In* Proceedings, Seminar on Integrated Resource Management, 7–8 April 1992, Fredericton, New Brunswick. *Compiled by* C.M. Simpson. Forestry Canada, Fredericton, New Brunswick. Inf. Rep. M-X-183E/F. pp. 45–58.

Burton, P.J. 1995. The Mendelian compromise: a vision for equitable land use allocation. Land Use Policy, **12**: 63–68.

(CBI) Canadian Boreal Initiative. 2003. A world treasure in our backyard. Canadian Boreal Initiative, Toronto, Ontario. Available at http://www.borealcanada.org/index_e.cfm [viewed 25 June 2003].

(CBWG) Canadian Biodiversity Working Group. 1994. Canadian biodiversity strategy: Canada's response to the convention on biological diversity. Environment Canada, Hull, Quebec. 52 p.

(CCEA) Canadian Council on Ecological Areas. 2002. Some quantitative environmental and socioeconomic characteristics of Canada's terrestrial ecozones. Available at http://www.ccea.org/ecozones/quant.html [viewed 16 May 2002].

(CCFM) Canadian Council of Forest Ministers. 1998. National forest strategy 1998–2003: sustainable forests, a Canadian commitment. Canadian Council of Forest Ministers, Ottawa, Ontario. 47 p.

(CCFM) Canadian Council of Forest Ministers. 2002. National forest database program. http://nfdp.ccfm.org/cp95/data_e/ [viewed 16 May 2002].

(CCI) Cortex Consultants Inc. 2001. Timber supply analysis technical workshop: course notes. Available at http://www.cortex.org/Course_Notes_v1.pdf [viewed 27 May 2002]. 57 p.

Clark, T.P., and Gilbert, F.F. 1982. Ecotones as a measure of deer habitat quality in central Ontario. J. Appl. Ecol. **19**: 751–758.

Costanza, R., Norton, B.G., and Hakell, B. (*Editors*) 1992. Ecosystem health: new goals for environmental management. Island Press, Washington, D.C. 269 p.

(CPAWS) Canadian Parks and Wilderness Society. 2002–2003. CPAWS national boreal campaign. Available at http://www.cpaws.org/boreal/ [viewed 25 June 2003].

Cretney, W. 1996. Dioxins and furans: stemming their flow in British Columbia's marine environment. Fact sheet. Institute of Ocean Sciences, Fisheries and Oceans Canada, Sidney, British Columbia. Available at http://www-sci.pac.dfo-mpo.gc.ca/mehsd/services/dioxin/dioxin-e.htm [viewed 11 June 2002].

(CSSP) Clayoquot Sound Scientific Panel. 1995. Sustainable ecosystem management in Clayoquot Sound. Report No. 5, Clayoquot Sound Scientific Panel. Cortex Consultants Inc., Victoria, British Columbia. 296 p.

De Jong, J., Larsson-Stern, M., and Liedholm, H. 1999. Greener forests. Skogsstyrelsen (National Board of Forestry), Jönköping, Sweden. 208 p.

Delgamuukw v. British Columbia.1997. 3 S.C.R. 1010, allowing in part appeal from (1993), 104 D.L.R. (4th) 470 (B.C. C.A.), varying (1991), 79 D.L.R. (4th) 185 (B.C.S.C.). Available at http://www.lexum.umontreal.ca/csc-scc/en/pub/1997/vol3/html/1997scr3_1010.html [viewed 26 May 2002].

Devall, B. (*Editor*). 1993. Clearcut: the tragedy of industrial forestry. Sierra Club Books and Earth Island Press, San Francisco, California. 291 p.

Drengson, A., and Taylor, D. 1997. Shifting values: seeding forests and not just tree$. *In* Canadian issues in environmental ethics. *Edited by* W. Cragg, A. Greenbaum, and A. Wellington. Broadview Press, Peterborough, Ontario. pp. 35–49.

Duffy, D.C., Boggs, K., Hagenstein, R.H., Lipkin, R., and Michaelson, J.A. 1999. A landscape assessment of the degree of protection of Alaska's terrestrial biodiversity. Conserv. Biol. **13**: 1332–1343. Available at http://www.uaa.alaska.edu/enri/aknhp_web/biodiversity/state_of_state/cons_biology/consbio.html [viewed 23 June 2003].

Duinker, P.N. 1994. Cumulative effects assessment: what's the big deal? *In* Cumulative effects assessment in Canada: from concept to practice, papers from the 15th Symposium of the Alberta Society of Professional Biologists. *Edited by* A.J. Kennedy. Alberta Society of Professional Biologists, Edmonton, Alberta. p. 11–24.

Dyer, S.J., O'Neil, J.P., Wasel, S.M., and Boutin, S. 2001. Avoidance of industrial development by woodland caribou. J. Wildlife Manage. **65**: 531–542.

Environment Québec. 2003. Protected areas in Québec: a pledge for the future. Available at http://www.menv.gouv.qc.ca/biodiversite/aires_protegees/orientation-en/appendix1c.htm [viewed 23 June 2003].

Epp, A.E. 2000. Ontario forests and forest policy before the era of sustainable forestry. *In* Ecology of a managed terrestrial landscape: patterns and processes of forest landscapes in Ontario. *Edited by* A.H. Perera, D.L. Euler, and I.D. Thompson. UBC Press, Vancouver, British Columbia. pp. 237–275.

Erdle, T.A. 1998. Progress toward sustainable forest management: insight from the New Brunswick experience. For. Chron. **74**: 378–384.

Erdle, T.A. 1999. The conflict in managing New Brunswick's forests for timber and other values. For. Chron. **75**: 945–954.

(ERGL) Environics Research Group Limited. 1992. 1991 national survey of Canadian public opinion on forestry issues. Environics Research Group Limited, Corporate Research Associates Inc., and CROP Inc. Prepared for Forestry Canada, Ottawa, Ontario.

(ESWG) Ecological Stratification Working Group. 1996. A national ecological framework for Canada. Agriculture and Environment Canada, State of the Environment Directorate, Ecozone Analysis Branch, Ottawa, Ontario.

(EWG) Ecoregions Working Group. 1989. Ecoclimatic Regions of Canada, First Approximation. Canada Committee on Ecological Land Classification. Ecological Land Classification Series, No. 23. Sustainable Development Branch, Canadian Wildlife Service, Environment Canada, Ottawa, Ontario. 119 p. and map at 1:7 500 000 scale. Available at http://sis.agr.gc.ca/cansis/references/1989ec_a.html [viewed 27 May 2002].

(FAO) Food and Agriculture Organization. 2003. State of the world's forests, Annex 2, data tables. Food and Agriculture Organization of the United Nations, Rome, Italy. Available at ftp://ftp.fao.org/docrep/fao/005/y7581e/y7581e11.pdf [viewed 25 June 2003]. 26 p.

(FEMAT) Forest Ecosystem Management Assessment Team. 1993. Forest ecosystem management: an ecological, economic, and social assessment. Report of the Forest Ecosystem Management Assessment Team, Government Printing Office, Washington, D.C.

Fengyou, W., and Jingwen, L. 1995. Silvicultural systems in boreal forests in east Asia. In Innovative silviculture systems in boreal forests. Proceedings, Symposium, 2–8 October 1994, Edmonton, Alberta. Edited by C.R. Bamsey. Clear Lake Ltd., Edmonton, Alberta. pp. 17–20.

Fernow, B.E. 1902. Economics of forestry, fifth edition. T.Y. Crowell Co., New York, New York. 520 p.

Fernow, B.E. 1913. A brief history of forestry in Europe, the United States and other countries, third edition. University of Toronto Press, Toronto, Ontario. 506 p.

(FOCS) Friends of Clayoquot Sound. 1998. Friends of Clayoquot Sound time-line. The Friends of Clayoquot Sound. Available at http://www.ancientrainforest.org/history_focs/timeline.html [viewed 10 June 2002].

Franklin, J.F. 1989. Toward a new forestry. Am. For. 11: 37–44.

Franklin, J.F., and Forman, R.T.T. 1987. Creating landscape patterns by cutting: ecological consequences and principles. Landscape Ecol. 1: 5–18.

Fritz, E.C. 1989. Clearcutting: a crime against nature. Eakin Press, Austin, Texas. 124 p.

(FSC) Forest Stewardship Council. 2001. Forest Stewardship Council of Canada. Available at http://www.fsccanada.org/about/history.shtml [viewed 10 June 2002].

Gauthier, D. (Editor). 1992. Ecological areas framework for developing a nation-wide system of ecological areas, Part 1: a strategy. Occasional Paper #12. Canadian Council on Ecological Areas, Ottawa, Ontario. 50 p.

Gillis, A.M. 1990. The new forestry: an ecosystem approach to land management. BioScience, 40: 558–562.

Gladstone, W.T., and Ledig, F.T. 1990. Reducing pressure on natural forests through high-yield forestry. For. Ecol. Manage. 35: 69–78.

Gosselink, J.G., Shaffer, G.P., Lee, L.C., Burdick, D.M., Childers, D.L., Leibowitz, N.C., Hamilton, S.C., Boumans, R., Cushman, D., Fields, S., Koch, M., and Visser, J.M. 1990. Landscape conservation in a forested wetland watershed: can we manage cumulative impacts? BioScience, 40: 588–600.

Grossnickle, S.C. 2000. Ecophysiology of northern spruce species: the performance of planted seedlings. NRC Research Press, Ottawa, Ontario. 407 p.

Grumbine, R.E. 1994. What is ecosystem management? Conserv. Biol. 8: 27–38.

Hammond, H. 1991. Seeing the forest among the trees: the case for wholistic forest use. Polestar Press, Vancouver, British Columbia. 309 p.

Hannelius, S., and Kuusela, K. 1995. Finland, the country of evergreen forest. Forssan Kirjapaino Oy, Helsinki, Finland. 192 p.

Hansen, A.J., Spies, T.A., Swanson, F.J., and Ohmann, J.L. 1991. Conservation of biodiversity in managed forests: lessons from natural forests. BioScience, **41**: 382–392.

Hare, F.K., and Ritchie, J.C. 1972. The boreal microclimates. Geogr. Rev. **62**: 334–365.

Henry, J.D. 2002. Canada's boreal forest. Smithsonian Institution Press, Washington, D.C. 176 p.

Higman, S., Bass, S., Judd, N., Mayers, J., and Nussbaum, R. 1999. The sustainable forestry handbook: a practical guide for tropical forest managers implementing new standards. Earthscan Publications Ltd., London, U.K. 289 p.

Hunter, M.L. 1993. Natural fire regimes as spatial models for managing boreal forests. Biol. Conserv. **65**: 115–120.

Hunter, M.L. (*Editor*). 1999. Maintaining biodiversity in forest ecosystems. Cambridge University Press, Cambridge, U.K. 698 p.

Hunter, M.L., and Calhoun, A. 1996. A triad approach to land-use allocation. *In* Biodiversity in managed landscapes. *Edited by* R.C. Szaro and D.W. Johnston. Oxford University Press, New York, New York. pp. 477–491.

(IGC) Institute for Global Communications. 1996. "Rio Cluster" of U.N. proceedings. Available at http://www.igc.org/habitat/un-proc/ [viewed 10 June 2002].

Isaev, A., and Shvidenko, A. 2002. Land resources of Russia: forest. International Institute for Applied Systems Analysis, Laxenburg, Austria. Available at http://www.iiasa.ac.at/Research/FOR/russia_cd/for_des.htm#top [viewed 23 June 2003].

James, C.R., Adamowicz, V., Hannnon, S., Kessler, W., Murphy, P., Prepas, E., Snyder, J., Weetman, G., and Wilson, M.A. 1997. Sustainable forest management and its major elements: advice to the Lands and Forest Service on timber supply and management. Alberta Forest Management Science Council, Edmonton, Alberta. Available at http://www.borealcentre.ca/reports/sfm.html [viewed 28 May 2002].

Jordan, G.A., and Baskent, E.Z. 1992. A case study in spatial wood supply analysis. For. Chron. **68**: 503–516.

Karr, J.R. 1991. Biological integrity: a long-neglected aspect of water resource management. Ecol. Appl. **1**: 66–84.

Kay, C.E. 1994. Aboriginal overkill: the role of Native Americans in structuring western ecosystems. Hum. Nat. **5**: 59–398.

Keenan, R.J., and Kimmins, J.P. 1993. The ecological effects of clear-cutting. Environ. Rev. **1**: 121–144.

Kessler, W.B., Salwasser, H., Cartwright, C.W., and Caplan, J.A. 1992. New perspectives for sustainable natural resources management. Ecol. Appl. **2**: 221–225.

Kimmins, J.P. 1991. The future of the forested landscapes of Canada. For. Chron. **67**: 14–18.

Kimmins, J.P. 1995. Sustainable development in Canadian forestry in the face of changing paradigms. For. Chron. **71**: 33–40.

Kohm, K.A., and Franklin, J.F. (*Editors*). 1997. Creating a forestry for the 21st century: the science of ecosystem management. Island Press, Washington, D.C. 475 p.

Kouki, J. (*Editor*). 1994. Biodiversity in the Fennoscandian boreal forests: natural variation and its management. Finnish Zoological and Botanical Publishing Board, Helsinki, Finland. Ann. Zool. Fenn. **31**(1). 217 p.

Krebs, C.J., Boutin, S., and Boonstra, R. (*Editors*). 2001. Ecosystem dynamics of the boreal forest: the Kluane Project. Oxford University Press, New York, New York. 544 p.

Kulshreshtha, S.N., Lac, S., Johnston, M., and Kinar, C. 2000. Carbon sequestration in protected areas of Canada: an economic valuation. Research report. Department of Agricultural Economics, University of Saskatchewan, Saskatoon, Saskatchewan. Available at http://www.cd.gov.ab.ca/preserving/parks/fppc/carbonsequestration.pdf [viewed 23 June 2003]. 142 p.

Lanken, D. 1996. Boreal forest. Can. Geogr. **116**(3): 26–33.

Lavender, D.P., Parish, R., Johnson, C.M., Montgomery, G., Vyse, A., Willis, R.A., and Winston, D. (*Editors*). 1990. Regenerating British Columbia's forests. UBC Press, Vancouver, British Columbia. 372 p.

Lewis, H.T., and Ferguson, T.A. 1988. Yards, corridors, and mosaics: how to burn a boreal forest. Hum. Ecol. **16**: 57–77.

Lindenmayer, D.B., and Franklin, J.F. 2003. Towards forest sustainability. CSIRO Publishing, Collingwood, Victoria, Australia. 231 p.

Lindsay, K.M., and Smith, D.W. 2001. Evaluation of British Columbia Ministry of Forests aboriginal rights and title-consultation guidelines — the Ditidaht case study. Environ. Eng. Policy, **2**: 191–201.

Lundmark, J.E. 1995. Forestry's green revolution in Sweden — AssiDomän's utilization of site-adapted forestry with functional conservation aspects. *In* Innovative silviculture systems in boreal forests. Proceedings, Symposium, 2–8 October 1994, Edmonton, Alberta. *Edited by* C.R. Bamsey. Clear Lake Ltd., Edmonton, Alberta. pp. 52–55.

MacKinnon, A., Meidinger, D., and Klinka, K. 1992. Use of the biogeoclimatic ecosystem classification system in British Columbia. For. Chron. **68**: 100–120.

Marchak, P. 1983. Green gold: the forest industry in British Columbia. UBC Press, Vancouver, British Columbia. 474 p.

Marles, R.J., Clavelle, C., Monteleone, L., Tays, N., and Burns, D. 2000. Aboriginal plant use in Canada's northwest boreal forest. UBC Press, Vancouver, British Columbia. 368 p.

Maser, C. 1994. Sustainable forestry: philosophy, science and economics. St. Lucie Press, Delray Beach, Florida. 373 p.

McCrory, C., Mahon, K., Carr, A., and May, E. 1994. Clearcut: the tragedy of industrial forestry — stop clearcutting Canada's forests. Press release, 8 February 1994. Canada's Future Forest Alliance, New Denver, British Columbia. Available at http://forests.org/archive/canada/bcgood.htm [viewed 29 September 2002].

(MFP) Model Forest Program. 1998–2002. About the Canadian Model Forest Network. Canadian Forest Service, Ottawa, Ontario. Available at http://www.modelforest.net/e/home_/aboute.html [viewed 28 September 2002].

(NRCan) Natural Resources Canada. 2002. Boreal forest — a global ecosystem. Available at http://atlas.gc.ca/site/english/learning_resources/borealforest/boreal_global.html [viewed 29 September 2002].

Nelson, J.D., and Finn, S.T. 1991. The influence of cut block size and adjacency rules on harvest levels and road networks. Can. J. For. Res. **21**: 595–600.

Newman, P.C. 1985. Company of adventurers, volume I. Penguin Books Canada, Markham, Ontario. 413 p.

Noss, R. 1990. Indicators for monitoring biodiversity: a hierarchical approach. Conserv. Biol. **4**: 355–363.

Oliver, C.D., and B.C. Larson. 1990. Forest stand dynamics. McGraw–Hill, Toronto, Ontario. 467 p.

Ottertooth. 2000. Temagami milestones in environmental preservation: a timeline of the dramatic events. Available at http://www.ottertooth.com/Temagami/Milestones/tem_miles_env.htm [cited 10 June 2002].

Pape & Salter. 1998. Delgamuukw: a summary of the Supreme Court of Canada decision. Prepared for the Carrier Sekani Tribal Council, Prince George, British Columbia. Available at http://www.cstc.bc.ca/pages/Treaty_delgmkwsmry.htm [viewed 26 May 2002].

Persky, S. 1998. Delgamuukw: the Supreme Court of Canada decision on aboriginal title. David Suzuki Foundation and Greystone Books, Vancouver, British Columbia. 137 p.

Pielou, E.C. 1991. After the ice age: the return of life to glaciated North America. University of Chicago Press, Chicago, Illinois. 364 p.

Pincett, S. 1993. Some origins of French environmentalism. For. Conserv. Hist. **37**: 80–89.

Pratt, L., and Urquhart, I. 1994. The last great forest: Japanese multinationals and Alberta's northern forests. NeWest Press, Edmonton, Alberta. 222 p.

Pruitt, W.O. 1978. Boreal ecology. Studies in Biology 91. Edward Arnold, London, U.K. 73 p.

Pyne, S.J. 1982. Fire in America: a cultural history of wildland and rural fire. Princeton University Press, Princeton, New Jersey. 654 p.

Reed & Associates. 1978. Forest management in Canada, 2 vols. Canadian Forest Service, Environment Canada, Ottawa, Ontario.

Reed, F.L.C. 2001. Two centuries of softwood lumber war between Canada and the United States: a chronicle of trade barriers viewed in the context of saw timber depletion. Prepared for The Free Trade Lumber Council, Montreal, Quebec. Available at http://www.ftlc.org/index.cfm?Section=2&DownloadID= 37 [viewed 28 September 2002]. 78 p.

Robertson, A.G. 2000. Digital remote sensing for forest management: hype or panacea? The 25 year experience of Blue Ridge Lumber. Paper presented at Geographic Information Systems and Remote Sensing for Sustainable Forest Management: Challenge and Innovation in the 21st Century, 23–25 February 2000, Edmonton, Alberta. Available at http://www.ualberta.ca/~gis1/wshop/paper.pdf [viewed 10 June 2002].

Robinson, D.W., and Hawley, A.W.L. 1997. Social indicators and management implications derived from the Canadian Forest Survey '96. McGregor Model Forest Association, Prince George, British Columbia. Available at http://www.mcgregor.bc.ca/publications/SocialIndicators.pdf [viewed 25 June 2003]. 40 p.

Robinson, S.K., Thompson, F.R., Donovan, T.M., Whitehead, D.R., and Faaborg, J. 1995. Regional forest fragmentation and the nesting success of migratory birds. Science, **267**: 1987–1990.

Rowe, J.S. 1972. Forest regions of Canada. Environment Canada, Ottawa, Ontario. Can. For. Serv. Pub. No. 1300.

Runesson, U.T. n.d. Boreal forests of the world. Available at http://www.lakeheadu.ca/~borfor/world.htm [viewed 23 June 2003].

Ryan, A.G. n.d. Protected areas in Newfoundland & Labrador. Forest Resources, Government of Newfoundland and Labrador, St. John's, Newfoundland. Available at http://www.gov.nl.ca/forestry/protection/areas.stm [viewed 23 June 2003].

Salwasser, H., MacCleery, D.W., and Snellgrove, T.H.A. 1993. An ecosystem perspective on sustainable forestry and new directions for the U.S. national forest system. *In* Defining sustainable forestry. *Edited by* G.H. Aplet, N. Honson, J.T. Olson, and V.A. Sample. Island Press, Washington, D.C.

Schindler, D.W. 1998. A dim future for boreal waters and landscapes. BioScience, **48**: 157–164.

Schlich, W. 1896–1911. A manual of forestry: Vol. 1. Introduction to forestry, 2nd ed. (1896); Vol. 2, Silviculture, 4th ed. (1910); Vol. 3, Forest management, 4th ed. (1911). Bradbury, Agnew & Co., London, U.K.

Schneider, R.R. 2002. Alternative futures: Alberta's boreal forest at the crossroads. Federation of Alberta Naturalists and Alberta Centre for Boreal Research, Edmonton, Alberta. 152 p.

Schneider, R.R., Stelfox, J.B., Boutin, S., and Wasel, S. 2003. Managing the cumulative impacts of land uses in the western Canadian sedimentary basin: a modeling approach. Conserv. Ecol. [online], 7(1): 8. Available at http://www.consecol.org/vol7/iss1/art8/main.html [viewed 24 June 2003].

(SERP) State of the Environment Reporting Program. 1999. Sustaining Canada's forests: timber harvesting. SOE Bulletin No. 99-4. Indicators and Assessment Office, Ecosystem Science Directorate, Environment Canada and Natural Resources Canada, Ottawa, Ontario. Available at http://www.ec.gc.ca/ind/English/Forest/default.cfm [viewed 26 May 2002].

(SFMN) Sustainable Forest Management Network. 2001. The next stage of growth, annual report 2001. Sustainable Forest Management Network, Edmonton, Alberta. Available at http://sfm-1.biology.ualberta.ca/english/pubs/index.htm [viewed 4 June 2002]. 33 p.

(SFMN) Sustainable Forest Management Network. 2003. The sustainable forest management network, home [web page]. Available at http://sfm-1.biology.ualberta.ca/english/home/index.htm [viewed 25 June 2003].

Shutov, I.V. 1995. Trends and use of silviculture systems in the boreal forests of Russia. *In* Innovative silviculture systems in boreal forests. Proceedings, Symposium, 2–8 October 1994, Edmonton, Alberta. *Edited by* C.R. Bamsey. Clear Lake Ltd., Edmonton, Alberta. pp. 21–25.

(SSCAF) Standing Senate Committee on Agriculture and Forestry. 1999. Competing realities: the boreal forest at risk. Report of the Sub-Committee on Boreal Forest of the Standing Senate Committee on Agriculture and Forestry. Parliament of Canada, Ottawa, Ontario. Available at http://www.parl.gc.ca/36/1/parlbus/commbus/senate/com-e/BORE-E/rep-e/rep09jun99-e.htm [viewed 25 May 2002]. 96 p.

Statistics Canada. 1992. Land cover by province/territory and ecozone. Statistics Canada, Ottawa, Ontario. Available at http://estat.statcan.ca/HAE/English/modules/module-3/mod-3c.htm [viewed 16 May 2002].

Stiell, W.M. 1976. White spruce: artificial regeneration in Canada. Canadian Forest Service, Sault Ste. Marie, Ontario. Inf. Rep. FMR-X-85. 275 p.

Swift, J. 1983. Cut and run. Between the Lines, Toronto, Ontario. 283 p.

(TBFRA) Temperate and Boreal Forest Resources Assessment. 2000. Forest resources of Europe, CIS, North America, Australia, Japan and New Zealand (industrialized temperate/boreal countries). United Nations, Geneva, Switzerland. Available at http://www.unece.org/trade/timber/fra/tbfra-ad.htm [viewed 25 June 2003]. 445 p.

Tittler, R., Messier, C., and Burton, P.J. 2001. Hierarchical forest management planning and sustainable forest management in the boreal forest. For. Chron. **77**: 998–1005.

Toumey, J.W., and C.F. Korstian. 1947. Foundations of silviculture upon an ecological basis, second edition. John Wiley & Sons, New York, New York. 468 p.

(TRN) Taiga Rescue Network. 2002. Taiga Rescue Network, home [web page]. Available at http://www.taigarescue.org/index.php [viewed 29 September 2002].

Troup, R.S. 1928. Silvicultural systems. Oxford University Press, Oxford, U.K. 199 p.

Utzig, G.F., and Macdonald, D.L. 2000. Citizen's guide to allowable annual cut determinations. British Columbia Environmental Network Educational Foundation, Vancouver, British Columbia. 100 p.

Velichko, A.A. 1995. The Pleistocene Termination in northern Eurasia. Quat. Int. **28**: 105–111.

Wagner, R.G., and Colombo, S.J. (*Editors*). 2001. Regenerating the Canadian forest: principles and practices for Ontario. Fitzhenry & Whiteside, Markham, Ontario. 650 p.

Waltner-Toews, D., Kay, J.J., Neudoerffer, C., and Gitau, T. 2003. Perspective changes everything: managing ecosystems from the inside out. Front. Ecol. Environ. **1**(1): 23–30.

Wang, C.-W. 1961. The forests of China, with a survey of grassland and desert vegetation., Harvard University, Cambridge, Massachusetts. Maria Moors Cabot Found. Pub. No. 5. 313 p.

(WCED) World Commission on Environment and Development. 1987. Our common future. Oxford University Press, Oxford, UK. 400 p.

Weetman, G. 1994. Design deficiencies in Canadian forestry practices — a Canadian perspective. For. Chron. **70**: 645–646.

Weetman, G.F. 1995. Silviculture systems in Canada's boreal forest. *In* Innovative silviculture systems in boreal forests. Proceedings, Symposium, 2–8 October 1994, Edmonton, Alberta. *Edited by* C.R. Bamsey. Clear Lake Ltd., Edmonton, Alberta. pp. 5–16.

Wein, R.W. 1999. ENCS/BOT 204 — Introduction to plant resources, Module III, Chapter 5: Plant resources in forest industry landscapes. Department of Renewable Resources, University of Alberta, Edmonton, Alberta. Available at http://www.rr.ualberta.ca/courses/encs204/content/3_ch5.htm [viewed 11 June 2002].

(WHA) World History Archives. n.d. The Lubican Lake confrontation. Available at http://www.hartford-hwp.com/archives/44/index-caa.html [viewed 10 June 2002].

Wheeler, E.E. n.d. Alaska paper birch (*Betula papyrifera* var. *humilis*). Alaska Department of Community and Economic Development, Juneau, Alaska. Available at http://www.dced.state.ak.us/cbd/wood/tree5e.htm [viewed 23 June 2003].

Wilcove, D.S., and Robinson, S.K. 1990. The impact of forest fragmentation on bird communities in eastern North America. *In* Biogeography and ecology of forest bird communities. *Edited by* A. Keast. SPB Academic Publishing, The Hague, Netherlands. pp. 319–331.

Williams, M. 2003. Deforesting the earth: from prehistory to global crisis. University of Chicago Press, Chicago, Illinois. 689 p.

Woodley, S.J., Francis, G., and Kay, J. (*Editors*). 1993. Ecosystem integrity and the management of ecosystems. St. Lucie Press, Delray Beach, Florida. 220 p.

Yaroshenko, A.Y., Potapov, P.V., and Turubanova, S.A. 2001. Last intact forest landscapes of northern European Russia. English version *edited by* L. Laestadius. Greenpeace Russia, Moscow, Russia. Available at http://pubs.wri.org/pubs_description.cfm?PubID=3170 [viewed 23 June 2003]. 74 p.

Zasada, J.C., Gordon, A.G., Slaughter, C.W., and Duchesne, L.C. 1997. Ecological considerations for the sustainable management of the North American boreal forests. International Institute for Applied Systems Analysis, Laxenburg, Austria. Interim Report IR-97-024. Available at http://www.iiasa.ac.at/Publications/Documents/IR-97-024.pdf [viewed 29 September 2002]. 67 p.

Zonneveld, I.S. 1990. Scope and concepts of landscape ecology as an emerging science. *In* Changing landscapes: an ecological perspective. *Edited by* I.S. Zonneveld and R.T.T. Forman. Springer-Verlag, New York, New York. pp. 5–20.

Chapter 2

Sustainability and sustainable forest management

Wiktor L. Adamowicz and Philip J. Burton

"The idea of sustainability is surprisingly simple: resource consumption cannot exceed resource production over time."

Donald W. Floyd (2002)

Introduction

Is sustainability really such a simple idea? Is this simple idea easy to implement?

Modern discussion of sustainability began with the Brundtland Commission's (WCED 1987) statement,

"Sustainable development is development that meets the needs of the present without compromising the ability of future generations to meet their own needs."

Since then, definitions of sustainability and sustainable development have proliferated, to the point where they are used in almost every forest management context. However, there is a realization that the concept of sustainable forest management (SFM), despite its wide appeal, is complex and not widely understood by the public, by forest policy and management professionals, or even by researchers (Sample et al. 1993). This chapter outlines the components of sustainability and sustainable development, describes various concepts of sustainable forest management, and discusses the role of research in the movement towards SFM in the boreal ecosystem. We further explain why certain themes have been emphasized in the first 7 years of research conducted by Canada's Sustainable Forest Management Network (SFMN), and hence why these themes are the subject of repeated attention in this book. For broader discussions of sustainability, both in concept and in application, the reader is directed to writings by John Pezzey (1992; Pezzey and Toman 2002), the International Institute for Sustainable Development http://www.iisd.org, and regular releases of the *The Sustainability Report* http://www.sustreport.org.

W.L. Adamowicz. Program Leader, Sustainable Forest Management Network, G208 Bio Sciences Building, University of Alberta, Edmonton, Alberta T6G 2E9, Canada. Telephone: 780-492-3625. e-mail: vic.adamowicz@ualberta.ca

P.J. Burton. Symbios Research & Restoration, Smithers, British Columbia, Canada.

Correct citation: Adamowicz, W.L., and Burton, P.J. 2003. Sustainability and sustainable forest management. Chapter 2. *In* Towards Sustainable Management of the Boreal Forest. *Edited by* P.J. Burton, C. Messier, D.W. Smith, and W.L. Adamowicz. NRC Research Press, Ottawa, Ontario, Canada. pp. 41–64.

Components of sustainability

Fundamentals: intergenerational equity, wealth, and substitutability

The fundamentals of sustainability can be broken into three relatively simple concepts. First, sustainability implies a form of equity over all future generations. One interpretation is that sustainability implies that future generations will be provided an "environment" that is no worse than the one we enjoy today. Interpreted in an ecological sense, this means that ecological conditions will not be degraded to the point that productive capacity is reduced, species are lost, or irreversible losses occur; i.e., that ecological integrity will be maintained. Social and economic sustainability have also been interpreted in a similar fashion: that is, the productive capacity of the economic system, and function of the social system in resolving conflict and developing institutions, will be no worse in the future than it is today. This notion of intergenerational equity or ensuring that generations in the future benefit from the natural endowments and human creations we enjoy today is the cornerstone of sustainability (Fig. 2.1).

A second component of sustainability is to consider global "wealth" as derived from intertwined natural, economic, and social systems. We have traditionally considered wealth to be material wealth embodied in goods and services. Sustainability challenges us to consider wealth in terms of environmental assets and services as well as community attributes that contribute to our well-being. In an economic sense we are attempting to consider market and non-market values as we strive for sustainability. Management that provides for intergenerational equity based on a broader concept of wealth is making a step toward sustainability.

An approach to assessing "wealth" is to consider the capital that generates wealth: human capital (knowledge), natural capital (forests, biodiversity), constructed capital (infrastructure), and social capital (ability to collectively solve problems). One form of sustainability (sometimes called "strong sustainability") suggests that one never allows any of the capital stocks to decline, and that one only lives off of the "interest" from the stocks. In other words, we should not let the capital depreciate, and in fact, we may wish to evaluate investments in capital stocks. But such an approach would not allow a temporary or permanent reduction in one of the capital stocks in order to improve overall "wealth". Thus, a weaker form of sustainability has been suggested as the maintenance of overall "wealth" from these capital stocks, without regard for the levels of the capital in each area. Much activity in our forests can be evaluated in light of these definitions. Conversion of primary forest to managed growth may involve reductions in the natural capital stock, or (implicitly) the substitution of investments in constructed capital or human capital for natural capital. Policies promoting this substitution have largely been implemented under the notion that it enhanced overall wealth by the production of material goods and services. Today a different conclusion regarding forest management might be reached when one uses a broader definition of wealth and the concept of intergenerational equity.

The discussion above introduces the third fundamental component of sustainability, substitutability. Can we substitute constructed capital for natural capital to enhance wealth? To what extent can such substitutions occur? Current discussions of sustainability raise concerns about the irreversible loss of species, cultures, and knowledge. Are

Fig. 2.1. Inter-generational equity is one of the foundations of sustainability. How do we make sure that these young people, and their children and grandchildren, will have access to the same quantity, quality, and diversity of forest values as we do today? (*a*) Guitar maker in La Patrie, Quebec; (*b*) campers near Smithers, British Columbia.

(*a*)   (*b*)

Canadian Press photo by Paul Chiasson   *Photo by Phil Burton*

there substitutes for these items? Will other components of the ecosystem substitute for ones that have been changed or removed, and will they maintain or improve our broad concept of wealth and well-being?

Box 2.1 provides an overview of sustainability for an entire nation, including the forest sector. Note the elements of intergenerational equity, well-being (through techno-logical change and avoiding negative impacts), and avoidance of irreversible losses (implying low substitutability) embodied in the principles. Sustainable forest manage-ment can be considered a subset of this overall view of sustainability presented by Canada's National Round Table on Environment and Economy. High-level objectives such as these can be applied to cases of individual forest regions or even industries or communities (see http://www.iisd.org/sd/principle.asp for examples, additional infor-mation, and discussion).

While the concept of sustainability can be identified using the components of equity, wealth, and substitutability, many challenges arise when we attempt to implement this conceptual model. First, we know little about the preferences of future generations nor can we predict the technology that they will have available to them. To a certain extent concepts of wealth depend on knowledge of preferences. Our idea of what is valuable in the forest has changed dramatically over the past century (Chap. 1). Technology has also allowed us to benefit from enhanced material wealth, in many cases with fewer impacts on the natural world than in the past. Weitzman (1997), for example, argues that technological change has been the greatest single source of wealth enhancement over the past century. However, technological innovations may bring with them new con-cerns about impacts on other values, and we do not know if technology will continue to make such strides in the future, or if it will perhaps make greater strides.

Of course the major challenge to sustainability is that we cannot (yet?!) measure such integrated concepts of wealth and therefore we don't know if we are on the path towards sustainability. We do not know if wealth is declining, nor do we accurately know the degree to which other forms of capital can substitute for natural capital. With-out knowledge of wealth and substitutability we also do not know if our processes for managing forests are providing the right targets, incentives, and frameworks for achieving sustainability. Efforts to address issues of equity, and the identification and weighing of multiple forest values constitute recurrent themes of much socioeconomic research in SFM (see Chaps. 3–7).

Because of the difficulties in developing an integrated measure of wealth, sustainability debates have opted for multiple measure approaches or a multiple accounts framework. This is the basis of the "criteria and indicator approach" to sustainability. In addition, the criteria-and-indicator approach is often incorporated as a component of "adaptive management". Adaptive management is a process to assess our scientific models, assumptions, and values, and an approach that emphasizes continuous improvement (Chap. 21). Thus, criteria and indicators form the basis for the assessment of our progress towards sustainability, and adaptive management examines the processes by which we make decisions about resource management. It is no surprise that these two elements play such a prominent role in discussions of sustainability and forest management. These concepts are explored in greater detail below.

Sustainable forest management: definitions and approaches

Models of sustainable forest management were developed in parallel with the development of national, multi-sector concepts of sustainable development. In the last decade of the 20th century, many conferences, discussion papers, and books explored the several dimensions of "sustainable development" as applied to forestry (e.g., Aplet et al. 1993; Maser 1994; Bouman and Brand 1997). The principle of sustainability was adopted in March of 1992 as part of Canada's National Forest Strategy (CFS 1998) by the Canadian Council of Forest Ministers (CCFM), those government ministers with national, provincial, or territorial jurisdiction over forestry. The CCFM subsequently charged a steering committee of 30 broadly based stakeholder representatives, supported by a science panel and a technical committee, with establishing an integrated set of criteria and indicators for evaluating progress in achieving sustainability. The resulting framework was released on 19 October 1995 and is periodically updated along with reports on the national status of indicators (CCFM 1995). This set of definitions summarizes what Canadian society expects of its forests and of its forest managers, and has provided international leadership in the pursuit of forest sustainability (Floyd 2002). Although four criteria for SFM are ecological or biophysical and only two criteria are socioeconomic, all six criteria are necessary and of equal importance. The CCFM framework (Box 2.2) has become the de facto starting point for developing sustainable forest management plans and reforming forest management institutions in much of Canada. The CCFM framework employs the criteria and indicators approach to outlining what has "value" in forest systems. It does not address process or principle so much as the components that should be sustained or enhanced.

Canada's National Forest Strategy (Box 2.3) approaches the concept of SFM from a set of principles that involve recognition of multiple values, sustaining value, and encouraging learning. Including a framework for action, it is reviewed and revised on a 5-year cycle. The Canadian Forest Accord (Box 2.4), endorsed by government ministers and the representatives of the forest products industry and environmental organizations

---

Box 2.1. The principles of sustainability, as adopted by Canada's National Round Table on the Environment and the Economy.

The National Round Table on the Environment and Economy (NRTEE) has produced the following objectives for sustainable development, with a preamble:

"The natural world and its component life forms, and the ability of that world to regenerate itself, through its own evolution has basic value. Within and among human societies, fairness, equality, diversity and self-reliance are pervasive characteristics of development that is sustainable.

1. **Stewardship.** We must preserve the capacity of the biosphere to evolve by managing our social and economic activities for the benefit of present and future generations.
2. **Shared responsibility.** Everyone shares the responsibility for a sustainable society. All sectors must work towards this common purpose, with each being accountable for its decisions and actions, in a spirit of partnership and open co-operation.
3. **Prevention and resilience.** We must try to anticipate and prevent future problems by avoiding the negative environmental, economic, social and cultural impacts of policy, programs, decisions and development activities. Recognizing that there will always be environmental and other events which we cannot anticipate, we should also strive to increase social, economic and environmental resilience in the face of change.
4. **Conservation.** We must maintain and enhance essential ecological processes, biological diversity and life support systems of our environment and natural resources.
5. **Energy and resource management.** Overall, we must reduce the energy and resource content of growth, harvest renewable resources on a sustainable basis, and make wise and efficient use of our non-renewable resources.
6. **Waste management.** We must first endeavour to reduce the production of waste, then re-use, recycle and recover waste by-products of our industrial and domestic activities.
7. **Rehabilitation and reclamation.** Our future policies, programs and development must endeavour to rehabilitate and reclaim damaged environments.
8. **Scientific and technological innovation.** We must support education, and research and development of technologies, goods and services essential to maintaining environmental quality, social and cultural values and economic growth.
9. **International responsibility.** We must think globally when we act locally. Global responsibility requires ecological interdependence among provinces and nations, and an obligation to accelerate the integration of environmental, social, cultural and economic goals. By working co-operatively within Canada and internationally, we can develop comprehensive and equitable solutions to problems.
10. **Global development.** Canada should support methods that are consistent with the preceding objectives when assisting developing nations."

Keating 1994

---

in 1998, provides a similar approach to defining SFM as the maintenance or enhancement of the ecological, social, and economic components of forested areas. These policy statements (Boxes 2.1–2.4) collectively constitute the foundations for sustainable forest management in Canada, broadly agreed upon by most of the participants in forest policy debates across the country.

In addition to government agency descriptions of the principles of SFM, industrial groups and nongovernmental organizations have also constructed guidelines for SFM. The Forest Stewardship Council's principles of sustainable forestry are largely based on the principles for "ecoforestry" outlined in Box 2.5. Most conservation biologists consider the sustainability of forest ecosystem integrity a more pressing issue than the sustainability of a forest products sector able to meet the world's fibre needs (Noss 1993). While one response to this concern has been campaigns to protect large wilderness areas from consumptive human use, another approach has been to advocate such light levels of timber harvesting that the human/industrial footprint is almost imperceptible on the forest landscape. This ecoforestry movement has gained considerable momentum in recent years, with its own society, journal, and textbooks appearing on the scene (see http://www.ecoforestry.ca).

In somewhat of a contrast, and not surprisingly, several wood products manufacturing organizations place more emphasis on the industrial, economic, and community development aspects of forest management. Examples of such approaches are provided in Box 2.6. The contrast between Boxes 2.5 and 2.6 illustrates an important aspect of SFM — the heterogeneity of preferences and widespread differences in values exhibited by different groups of people. Concepts of value are critical for implementation of SFM; however, the diversity of values arising in different sectors makes the problem extremely challenging.

Most definitions of sustainability embrace the notion of multiple values being maintained over generations. As described above, however, most approaches attempt to break down the major forest values into categories (social, economic, ecological) as if these elements can somehow be assessed independently. These three components are commonly referred to as the "three pillars" of sustainability (e.g., Goodland 1995; Yamasaki et al.[1]). While in practice it is likely that such compartmentalization will be necessary, the real challenge is associated with the integration of these values. There are not three separate "systems" of ecology, economy, and society that can somehow be examined independently to determine if they are "being sustained". In reality, the phrases ecological, economic, and social apply to complex sets of values that we (humans) have. We must assess how to evaluate outcomes or plan options that have different effects on social, economic, and ecological components of our world. We will find that individuals and groups hold different values over these components. Sustainability does not involve sustaining these three as separate components. It involves coming to terms with tradeoffs between values arising from our choices of management actions, or finding win–win solutions if possible, and recognizing the differences in values across individuals, regions, countries, and generations.

Another perspective on the fundamentals of sustainability explicitly identifies the role of technology in providing for human needs (now and in the future) without degrading environmental values (Pezzoli 1997; Castro et al. 1999). The broad discipline of environmental engineering is based on the cost-effective minimization of detri-

---

[1]Yamasaki, S.H., Kneeshaw, D.D., Bouthillier, L., Fortin, M.-J., Fall, A., Messier, C., and Leduc, A. Balancing on the three pillars of sustainable forest management: integrating social, economic, and ecological indicators into the public participation process. Submitted to *Conservation Ecology* for review.

> ### Box 2.2. A framework for sustainable forest management, as endorsed by the Canadian Council of Forest Ministers (CCFM).
>
> The CCFM criteria for sustainable forest management can be paraphrased as follows:
>
> 1. **Conservation of biological diversity**
> - forest management must maintain the variety and quality of the earth's ecosystems, must not allow any species to go extinct, and must conserve genetic diversity in species being managed;
>
> 2. **Maintenance and enhancement of forest ecosystem condition and productivity**
> - the health, vitality, and rates of biological production in forest ecosystems must be protected (and even increased in some places) by minimizing the incidence of biotic and abiotic stresses, enhancing ecosystem resilience, and maintaining the biomass of selected components;
>
> 3. **Conservation of soil and water resources**
> - soil and water quantity and quality must be maintained in order to guarantee long-term forest productivity, provide potable water for human and wildlife use, and to provide suitable habitats for many other organisms;
>
> 4. **Forest ecosystem contributions to global ecological cycles**
> - forest management should promote sustained utilization and rejuvenation of forest ecosystems and protect them from widespread destruction by fire, pests, and conversion to alternate land uses in order to maintain or enhance their role in sequestering carbon and regulating regional hydrological cycles; it should further promote the manufacture of products that can act as long-term carbon pools and that have a low fossil fuel demand in their production;
>
> 5. **Multiple benefits to society**
> - forests must continue to provide the flow of wood products, commercial and nonmarket goods and services, and environmental and option values over the long term;
>
> 6. **Accepting society's responsibility for sustainable development**
> - forest management must respect Aboriginal and treaty rights, encourage Aboriginal participation in forest-based economic opportunities, sustain forest communities, and incorporate fair, effective and informed decision making through public participation.
>
> These criteria are explored in various chapters of this book. Further background on the CCFM is available at http://www.ccfm.org, while specific indicators and annual reports on their status can be found at their *Criteria and Indicators of Sustainable Forest Management in Canada* worldwide web page, http://www.ccfm.org/3_e.html.

mental environmental impacts of human industrial and community activities (Fig. 2.2). Since the 1970s there has also been widespread practice of conducting environmental impact assessments in advance of the construction of major industrial projects (Beanlands and Duinker 1984; Erickson 1994). Some assessments result in industrial development being curtailed, but they usually result in design modifications to reduce and mitigate impacts. More recently, "life cycle assessments" have started evaluating the entire stream of supplier, manufacture, distribution, and disposal costs and impacts associated with particular products (Curran 1996). Modifications of this approach, and associated impact assessment methodologies, may hold promise in identifying weaknesses

---

Box 2.3. Principles associated with forest management and the subsequent framework for action as identified in Canada's National Forest Strategy for 1998–2003.

Strategic direction 2. Forest management: practicing stewardship

Principles:

- ethical conduct on the part of all those who direct, practice, or judge performance in forest management is essential;
- sustainable forest management recognizes a forest's potential to sustain a range of values and the needs and rights of all users, and strives to find the best balance of uses based on the relative benefits and impacts of management alternatives;
- sustainable forest management requires an adaptive management approach, following exemplary forest practice that is grounded on the best available scientific knowledge;
- coordinated direction, applied to objectives from broad land use plans to local site-specific goals, must guide all forest operations;
- forest land tenure systems must balance rights with responsibilities, encourage sound stewardship, sustain a supply a resources, and provide opportunities for a fair return on investments;
- forestry practices must be based on a sound understanding of ecological principles and of the goals established for the forest.

Framework for action:

- plan for a full range of environmental, social, economic, and cultural forest values, guided by goals defined at appropriate scales and for appropriate time horizons;
- review and improve our silvicultural systems and practices;
- ensure the prompt renewal of disturbed forests;
- manage forests with concern for the economic, social, and ecological impacts of fire, insects, disease, competing vegetation, and climate change;
- encourage forest stewardship and the use of the best forestry practices.

CFS 1998

---

and alternatives in the management of forests and the manufacture of forest products (Castro et al. 1999). Research in forest engineering, improved (i.e., largely more cost-, energy, and feedstock-efficient) processing facilities, pollution prevention and control, and recognizing the global role of forestry in terms of climate change and carbon balances (Chaps. 15–20) has been fundamental in charting the path to greater sustainability.

Sustainability requires the management of alternative human actions while understanding the implications of any choice. Some alternatives may improve the generation of material goods at the expense of delivering non-market goods and services, or by increasing the risk to elements of the ecosystem. Another set of alternatives may involve improvement in the environment for traditional use of forest resources by Aboriginal People at the expense of energy sector exploration. Other options may improve habitat for wildlife species but reduce the opportunities for individuals to remain in a small community. These types of tradeoffs provide the challenges. The implications of our

Box 2.4. Highlights of the Canada Forest Accord, endorsed by the Canadian governmental ministers responsible for forests and representatives of the Canadian forestry community on 1 May 1998.

We believe:

- Healthy forest ecosystems are essential to the health of all life on earth.
- Out forest heritage is part of our past, our present, and our future identity as a nation.
- It is important to maintain a rich tapestry of forests across the Canadian landscape that sustains biological diversity.
- Continued economic, environmental, and social benefits must be maintained for the communities, families, and individual Canadians who depend on the forest for their livelihood and way of life.
- The spiritual qualities and inherent beauty of our forests are essential to our physical and our mental well-being.
- As forest stewards, we must ensure the wise use of our forests for the environmental, economic, social, and cultural well-being of all.
- All Canadians are entitled to participate in determining how their forests are used and the purposes for which they are maintained.

Our vision:

- All measures within our means will be taken to ensure healthy forests are passed on to future generations.
- We will fulfill our global responsibilities in the care and use of forests, maintaining their contribution to the environment and the well-being of all living things.
- Our needs will be met through developing and applying the best available knowledge, and through cooperation.
- Our forests will be managed on an integrated basis, supporting a full range of uses and values, including timber production, habitat for wildlife, and parks and wilderness areas.
- We will participate in setting objectives and priorities for managing our forests, based on how we value them and using the best available knowledge of their environmental, economic, social, and cultural features.
- A strong economic base for varied forest products, tourism and recreation will be supported within a framework of sound ecological and social principles and practices.
- Advanced training, skills, and education will be provided to those employed in forest-related activities, and stable, fulfilling employment opportunities will add to the quality of life in their communities.
- Through consultation, mutual respect, sharing of information, and clear and harmonious relationships among all those involved with forests, trust and agreement will be brought about and the effectiveness of forest conservation, management, and industrial development will be improved.

Our goal: sustainable forests

- Our goal is to maintain and enhance the long-term health of our forest ecosystems, for the benefit of all living things both nationally and globally, while providing environmental, economic, social, and cultural opportunities for the benefit of present and future generations.

FSC 1998

Box 2.5. An ecoforestry perspective on sustainable forest management.

The following *Ten Elements of Sustainability* are proposed by Smith (1997); similar criteria for forest sustainability are widely advocated in the ecoforestry community (e.g., Hammond 1991, 1997).

1. Forest practices will protect, maintain and (or) restore fully functioning ecosystems at all scales in both the short and long terms.

2. Forest practices will maintain and (or) restore surface and groundwater quality, quantity, and timing of flow, including aquatic and riparian habitat.

3. Forest practices will maintain and (or) restore natural processes of soil fertility, productivity, and stability.

4. Forest practices will maintain and (or) restore a natural balance and diversity of native species of the area, including flora, fauna, fungi, and microbes, for purposes of the long-term health of ecosystems.

5. Forest practices will encourage a natural regeneration of native species to protect valuable native gene pools.

6. Forest practices will not include the use of artificial chemical fertilizers or synthetic chemical pesticides.

7. Forest practitioners will address the need for local employment and community stability and will respect workers' rights, including occupational safety, fair compensation, and the right of workers to bargain collectively.

8. Sites of archaeological, cultural, and historical significance will be protected and will receive special consideration.

9. Forest practices executed under a certified forest management plan will be of the appropriate size, scale, time frame, and technology for the parcel, and adopt the appropriate monitoring program, not only to avoid negative cumulative impacts, but also to promote beneficial cumulative effects on the forest.

10. Ancient forests will be subject to a moratorium on commercial logging, during which time research will be conducted on the ramifications of management in these areas.

Smith 1997

Some of these concepts have been incorporated in the principles and criteria of the Forest Stewardship Council, available at http://www.fscoax.org and discussed further in Chap. 21.

---

actions affect economic, ecological, and social dimensions, at various spatial and temporal scales. Moving towards sustainability will involve developing approaches, tools, strategies, and institutions that help us manage our actions so that this diverse and complex set of values is recognized and appreciated. Among these developments are attempts to better integrate environmental assets into market economies either through valuation and recognition of the importance of environmental assets for the economic system (Daily 1997; Heal 1998) or through the use of market institutions, like forest certification, to signal the value and importance of environmental assets.

Collectively, these public commitments to sustainability (as described in Boxes 2.1–2.6) constitute the milieu of goals and expectations in which boreal forest managers find themselves today. If these are the aims of SFM, against which individual companies, management areas, and provinces are to be evaluated in the public eye, how well

Box 2.6. Some industrial perspectives on sustainable forest management.

The pulp and paper industry of Canada issued the following environmental statement as part of the 1992 annual report of the Canadian Pulp and Paper Association (now the Forest Products Association of Canada).

"The Pulp and Paper Industry of Canada shares with all Canadians important responsibilities to the environment in which we live and work. It supports the responsible stewardship of resources, including forests, recyclable materials, fish and the aquatic habitat, wildlife, air, land and water. Responsible stewardship makes possible sustainable economic development. In this spirit, the industry believes that a set of principles should govern its attitude and action in environmental matters. As endorsed by the member companies of the Canadian Pulp and Paper Association, these are as follows:

- The companies commit themselves to excellence in sustainable yield forestry and environmental management, and will conduct their business in a responsible manner designed to protect the environment and the health and safety of employees, customers and the public;
- The companies will assess, plan, construct and operate facilities in compliance with all applicable regulations;
- The companies will manage and protect forest resources under their stewardship for multiple use and sustained yield;
- The companies, beyond or in the absence of regulatory requirements, will apply sound management practices to advance environmental protection and reduce environmental impact;
- The companies will promote environmental awareness amongst employees and the public, and train employees in their environmental responsibilities;
- The companies will report regularly to their Boards of Directors on their environmental status and performance;
- The industry will work with governments in the development of regulations and standards based on sound, economically achievable technologies, and the analysis of environmental impact; and
- The industry will continue to advance the frontiers of knowledge in environmental protection through the support of scientific research and, as appropriate, apply such knowledge at its facilities."

From IISD (n.d.)

In October 1994, the American Forest and Paper Association (AFPA), whose members comprise many of the United States' forest products companies, adopted a set of sustainable forestry principles. These principles call for members to:

- practice a land stewardship ethic that integrates the reforestation, management, growing, nurturing, and harvesting of trees for useful products with the conservation of soil, air, water quality, wildlife and fish habitat, and aesthetics;
- use sustainable forestry practices that are economically and environmentally responsible;
- protect forests and improve long-term forest health and productivity;
- manage unique qualities (protect unique sites); and
- continuously improve forest management practices.

From Vogt et al. (1997, p. 322)

These AFPA principles have since evolved into the Sustainable Forestry Initiative (SFI) certification program, as outlined at http://www.afandpa.org/forestry/sfi/ and in Chap. 21.

Fig. 2.2. Forest industries and their regulators are faced with the challenge of generating useful products and economic activity, while minimizing any negative impacts on the natural environment and nearby communities. Shown here are the Millar Western pulp mill, residential neighbourhoods, and the Athabasca River in Whitecourt, Alberta.

Photo courtesy of Millar Western Forest Products Ltd.

do they measure up? If it is safe to assume that we are not currently meeting all the lofty goals of SFM as articulated in these principles, what research approaches, tools and technologies, and perspectives are needed to more effectively steer boreal forestry towards sustainability?

Measuring progress towards sustainable forest management: boundaries, risk, and monitoring

System boundaries

In order to discuss sustainability one must describe the boundaries of the system. The global environment is the ultimate system; however, to describe sustainable forest management, the system must be defined more narrowly. Note that this narrowing comes at a cost. In examining sustainability in a broader context we recognize the interactions between industrial sectors (e.g., wood, steel, concrete) and identify tradeoffs between the uses of these materials for economic development. The development of institutions that send signals regarding the environmental impacts of alternative resources (renewable and non-renewable) must also be conducted if we are to move towards sustainability. However, in this volume we limit ourselves to discussing the sustainability of the boreal forest and boreal forestry, recognizing that ecological influences will arise from beyond these boundaries, and trade in products and services will generate linkages between the forest sector and other sectors and between the boreal region and other regions. While discussing boreal forest sustainability one must also recognize that sustainability in this sector may arise at the expense of sustainability in another region (in the management of tropical forests, for example) or may result in undesirable outcomes because of the byproducts from the use of other materials (e.g., increased production of steel with associated increases in air emissions).

System boundaries are related to the issue of spatial and temporal scale. Within the system boundaries, an evaluation of sustainability will be multi-scale, recognizing that this increases the complexity of the task significantly. Arguments can be made for national, provincial, and ecoregion sustainability; forest products companies may strive for sustainability within the corporation or within individual divisions; communities also expect sustainability, while environmentalists may expect sustained old-growth forest or wildlife populations in every watershed or landscape unit. When managing for a sustained yield of timber, one of the first steps in forest planning is the explicit identification of the "forest estate" or the "sustained yield unit" from which a perpetual supply of wood fibre is expected. This step remains a cornerstone of forest management as it extends the pioneering concept of sustained yield to other forest values.

Risk

The most significant, and probably most ignored, element of sustainability is the role of risk. One can almost never state with certainty that a certain practice or plan is sustainable, or is not sustainable. Due to natural fluctuations in ecological and socioeconomic systems, and limitations in our understanding of the systems, there is a degree of uncertainty regarding all processes. There are risks associated with any action. There are uncertainties about future preferences for forest products and ecological services. Ideally, one would like to assess actions in terms of the probability of outcomes. This would facilitate decision analysis under risk. There is a great need for improved knowledge of risks, and thresholds, to enable such an analysis. Adaptive management, involving learning and continuous improvement, provides a vehicle for this improvement (see Chap. 21).

The policies of forest products companies and government agencies typically encompass various strategies for risk minimization. Hazards threatening the sustainability of forest management plans can generally be categorized as either biophysical (e.g., the threat of wildfire, insect outbreaks, windthrow and climate change, or inadequacies in the prediction of forest growth or the effectiveness of forest regeneration) or socioeconomic changes in values (e.g., the demand for different forest products, the importance placed on wilderness protection, or changes in elected governments). Some common approaches to the management of biophysical risk can include:

- Conservative estimates and wide margins of error in setting sustainable rates of timber harvest, and implementation of methods like the probabilistic analysis of sustainability (Chap. 6);

- Widespread use of artificial regeneration for certain tree species, rather than counting on natural regeneration that is sometimes less reliable;
- Procedures for the early detection and rapid suppression of wildfires and insect epidemics;
- The designation of very large sustained yield units, so that the relative impact of timber being lost through any one disturbance event (e.g., a large fire or a bad fire year) is lessened; and
- The maintenance of genetic diversity, species diversity, and ecosystem diversity at appropriate scales.

Likewise, socioeconomic risks typically can be addressed by:

- Producing a broad mix of products, and selling into many markets;
- Avoiding irreversible losses to habitats or ecosystem services that may provide value in the future;
- Diversifying the economic base of communities and investing in human capital to improve resilience to economic and natural system shocks;
- Designing and building processing facilities that are flexible in their use of different species and sizes of feedstock materials, and (or) are flexible in the specifications to which they manufacture products;
- Designing facilities so that they can adapt to changing environmental standards and avoid outcomes that may expose a company or agency to legal liabilities;
- Designing regulatory institutions and policies so that they can adapt to changing information on environmental outcomes and values as well as changing industrial technologies; and
- Considering public opinion and input in the development of corporate or agency policies as part of an adaptive management process.

Although the above strategies employ a degree of common sense and the principles of diversity and flexibility long practiced in business and portfolio management, risk assessment and management has evolved into an increasingly quantitative exercise. Conferences and interdisciplinary teams have been organized to identify and propose management solutions to biophysical risks (e.g., Arbez et al. 2002), economic risks (e.g., Lohmander 2000), and their interaction (e.g., IUFRO 2001) in forestry. An important aspect of risk assessment and management requires the calculation of event probabilities, so we can foresee a day when statistical analysts and actuarial scientists become an integral component of forest management teams, much as they are in the insurance sector. Many of the principles and practices of SFM can be considered exercises in risk management, where the tradeoffs involve comparisons of different kinds of risks (e.g., financial versus ecological) and their effects on people with different levels of risk tolerance and different preferences over the outcomes.

Criteria and indicators, monitoring, and sustainability

Above we note that there is no single metric upon which we can measure "well-being". Therefore, it is very common to adopt several metrics as proxies for well-being, as illustrated in a criteria and indicators approach. The criteria are perspectives on value or importance of elements arising from the forest resource (e.g., the CCFM criteria in

Box 2.2). Some of these are biological in nature, but because of the inseparability of the forest resource from the economy and community, some criteria also include community and economic aspects. The criteria used in criteria and indicator systems are often not well defined, and in many cases the values expressed in these criteria may be in conflict. Nevertheless, they provide an expression of the range of values of interest.

Several high-level multilateral meetings in the 1990s hammered out international agreements on criteria and indicator frameworks for sustainable forest management. Two of these, the Helsinki Process, initiated in June, 1993 (with subsequent refinements agreed to in Lisbon in 1998), and the Montreal process, initiated in February, 1995, are described in Box 2.7. Note considerable overlap and agreement in these two processes; the Russian Federation is the only country which has signed both accords. It is also worth noting that the CCFM framework (Box 2.2) is somewhat more development-oriented than the Montreal Process endorsed internationally as well as by the Canadian government.

The measurement of progress associated with the criteria is the task of a series of indicators. Indicators also provide the basis for monitoring as a component of adaptive management (Chap. 21). The monitoring program must be relational; that is, there must be some basis for comparison of the system with some form of benchmark or baseline. Without such benchmarks it will not be known if the fluctuation in an indicator is due to inherent variation or due to a detrimental management action. The construction and identification of appropriate benchmarks for ecological, economic, and social monitoring schemes is a significant challenge, yet it is an integral component of sustainability. Monitoring, in this approach, will also involve an understanding of the degree of historical variation. The "natural disturbance model" (see Box 2.8 and Chap. 9) for guiding ecosystem management endorses the emulation of natural processes and patterns as guidelines for the conservation of biodiversity. It depends on the assumption that our activities will not have a long-term destabilizing influence on the system if our future activities can be conducted in a manner that maintains the forest within its historical range of variation (Morgan et al. 1994; Landres et al. 1999). The natural disturbance emulation approach is emerging as a risk minimization strategy, and the dominant model for understanding and managing boreal forest ecosystems (Fig. 2.3). It was the cornerstone of the initial research program of the Sustainable Forest Management Network, and many forestry firms have adopted it as a guiding forest management principle (e.g., APFI 2001). The natural disturbance model attempts to provide a measure of forest condition and a measure of the ability of a forest to conserve biodiversity. There has been much recent research exploring the mechanisms of forest disturbance and ecosystem responses to disturbance (Chaps. 8–10), and then designing and testing forest management approaches at strategic, tactical, and operations levels to apply what has been learned (Chaps. 11–15).

While criteria and indicators approaches have dominated the scene in terms of measurement of sustainability, other groups have attempted to develop metrics or indexes of sustainability that integrate the ecological, economic, and social components. Several countries and regions have developed forest resource accounts in which timber and non-timber resources are blended to form assessments of the sustainability of the natural capital stock (see Chap. 6). In addition, measures of sustainability for countries, not just

Box 2.7. Two processes on criteria and indicators for sustainable forest management.

The *Helsinki Process*, now known as the Pan-European Forest Process, focuses on the sustainable development and management of forests in Europe, and resulted in the development of the Pan-European Criteria and Indicators for SFM in Europe (see Chap. 21). The European countries and the European Community (41 signatories) have agreed on six common criteria, 20 quantitative indicators, and 84 descriptive indicators for SFM at the regional and national levels, with operational level guidelines also developed (MCPFE 1998). Criteria for SFM are:

1. Maintenance and appropriate enhancement of forest resources and their contribution to global carbon cycles;
2. Maintenance of forest ecosystem health and vitality;
3. Maintenance and encouragement of productive functions;
4. Maintenance, conservation, and appropriate enhancement of biological diversity in forest ecosystems;
5. Maintenance and appropriate enhancement of protective functions in forest management (notably soil and water); and
6. Maintenance of other socio-economic functions and conditions;

The *Montreal Process* was the basis for the Canadian Council of Forest Ministers' (CCFM) framework on sustainable forest management and Canada's National Forest Strategy (CCFM 1995; CFS 1998). The CCFM criteria subsequently have been adopted by the Canadian Standards Association (CSA) for forest certification (see Chap. 21). Dealing with sustainable forest management in temperate and boreal forests outside Europe, the Montreal Process enlisted 12 participating countries (including Canada) in agreeing to seven criteria and 67 indicators of SFM (MPWG 1995). Criteria for SFM are:

1. Conservation of biological diversity;
2. Maintenance of productive capacity of forest ecosystems;
3. Maintenance of forest ecosystem health and vitality;
4. Conservation and maintenance of soil and water resources;
5. Maintenance of forest contribution to global carbon cycles;
6. Maintenance and enhancement of long-term multiple socio-economic benefits to meet the needs of societies; and
7. Legal, institutional and economic frameworks for forest conservation and sustainable management.

Castañeda et al. 2001

the forest sector, have been developed by creating adjusted economic accounts (adjusted by the appreciation or depreciation of natural capital) or by creating measures of genuine savings (savings rates adjusted by changes in natural capital; see Hamilton and Clemens 1999). Using these measures in which natural capital is aggregated with other forms of capital by valuing its outputs, Canada appears to be doing quite well. Positive and relatively large genuine savings rates are reported for this country by the Organisation for Economic Co-operation and Development (OECD) and the World Bank.

Other metrics of sustainability that construct composite indexes have also been developed. Some of these focus only on the forest sector while others examine the entire economy. Anielski (2001) provides a description of a variety of such metrics including the Genuine Progress Indicator or GPI. The GPI is in the spirit of capital-based accounting methods like resource accounts or genuine savings measures; however, non-market elements including environmental, social, and cultural capital are factored into the index using an implicit weighting scheme. The GPI for Alberta, for example, is developed from a variety of accounts including the Forest Account (Anielski and Wilson 2001). Examining the forest account, Alberta performs well on some measures (economic outputs, forest stocks) but poorly on others (fragmentation). The overall picture painted by the Alberta GPI is one of declining well-being, once all social and environmental elements are factored in. It is necessary to note, however, that in all such index measures the elements chosen for inclusion in the index and the weights placed on these elements in aggregation are very influential in determining the final outcome. At a global scale, indexes like the Environmental Sustainability Index (ESI 2002) describe the performance of countries over time. Canada performs well on this index (in the top five), not surprisingly as this metric is highly correlated with aggregate income and measures of social institutional capacity. Regardless of the index or system of indicators chosen, it appears that the key question is how to integrate measures of economy, environment, and society into clear, meaningful representations of sustainability.

Sustainable forest management in practice: information, public involvement, and adaptive management

A fundamental problem arises when indicators of sustainability tell us that some elements of the system are being compromised for other elements. For example, if the forest condition is moving beyond the range of natural variation, is the risk of loss of ecological integrity sufficient to demand a change in management that may jeopardize economic or social components of the system? In such potential conflict cases there are few absolutes. The key to such cases is a process for the evaluation of the alternatives, assessment of risks, and implementation of an adaptive management plan.

The evaluation process will involve:

- Using the best scientific information available, the (quantitative) assessment of the ecological, economic, and social implications, or the impact on important values, of a defined set of alternatives, including descriptions of the "risks" and "limits" of the systems;
- Engagement of "the public" (including policymakers, stakeholders, etc.) in assessment of the alternatives, where this engagement may take on various forms;
- Articulation of a desired future forest (including the path to reach the future) from the set of management alternatives, and the management actions that will best be used to reach this future forest;
- Construction of an adaptive management process where management is designed to reveal new information on system function; and
- Return to the beginning of the process with a modification of the initial understanding of the system, followed by repeated and continual implementation of the process.

Box 2.8. Emulating nature, the natural disturbance model, and coarse-filter biodiversity management.

Designing and evaluating forest management practices that more closely emulate natural disturbances has been a theme of SFMN research. This natural disturbance model ("the NDM") has been advocated by most forest ecologists (e.g., Hansen et al. 1991; Hunter 1993; Bergeron et al. 1999) and is implicitly promoted in a number of corporate and governmental forest management policies (BCMOF and BCMELPPC 1995; APFI 2001; FWA 2001; OMNR 2001). Advocates of "close-to-nature forestry" have promoted this cause from one perspective or another for more than a century, although the debate has usually reduced to a preference for even-aged vs. uneven-aged silviculture (see Chap. 13). Social demands for near-natural forest management and continuous cover forestry continue to grow, especially in Europe (von Gadow et al. 2002).

If "nature for nature's sake" were the only motivation for using the NDM to guide forestry, we might be accused of committing the "naturalistic fallacy," a term coined by the philosopher G.E. Moore (1903) to refute the conclusion that there is inherent goodness in the state of nature. This can be considered an extension of David Hume's admonition that "what is" never provides ethical guidance on "what ought to be." Perhaps the environmentally-aware children of the 1960s have blithely extended Barry Commoner's "Third Law of Ecology" (i.e., that "Nature knows best;" Commoner 1971) too universally, when its intention was primarily to warn of the dangers of introducing chemicals having no natural precedent into the environment.

But widespread adoption of the NDM is not primarily an ethical or even aesthetic decision, although many of its advocates may be so motivated (e.g., Suzuki and McConnell 1997). Rather, its utility is an application of the "coarse filter" approach to the conservation of biodiversity (Hunter 1991; Burton et al. 1992). The principles of forest sustainability and international commitments to biodiversity protection mean that it is important to provide for viable populations of all species. But we can't even identify all forest-dependent species (when considering non-vascular plants, fungi, arthropods, unicellular organisms, etc.); we can't possibly develop hundreds of species-specific habitat management plans (the "fine filter" approach adopted for some species at risk), and policies promoting any one species typically have negative effects on others (Cumming et al. 1994). So the prudent strategy is to maintain the composition, age-class distribution, landscape pattern, and stand-level structures under which indigenous species are known to have persisted through history.

Habitats created through forest harvesting instead of through natural processes such as wildfire can never completely duplicate those that occur naturally (see Chap. 8), and we have no assurance that they are optimal for any species. Nevertheless, we know that natural habitats prevailing until the modern era were at least adequate for all the species left for us to manage today, so maintaining the closest possible similarity between wild and managed landscapes is an inherently conservative approach. The generation of forest cutblocks with sinuous boundaries and islands of green trees can also serve social, aesthetic benefits which should not be discounted, and it may allow us to think of a landscape as largely wild and untamed. But for every person who likes to maintain the illusion of wilderness, there is another person who prefers a landscape of orderly domestication. Not only is it invalid to assert that natural patterns are right and artificial patterns are wrong (or vice versa), but we can't automatically describe naturalistic patterns as "pleasing to the eye" either. Whether landscapes sculpted according to the NDM are "good" or "better" (for meeting human needs and sensibilities, protecting more biodiversity, or achieving sustainable forest management) remains unsubstantiated until research and monitoring informs us more fully of the tradeoffs and net effects associated with this approach.

With thanks and tribute to the late Gene Namkoong (1933–2002) for stimulating this discussion. PJB.

Fig. 2.3. One theme of many research projects during the early years of the Sustainable Forest Management Network consisted of testing how forest management practices might effectively emulate patterns of natural disturbance. (*a*) Residual trees after salvaging timber killed by a recent wildfire; (*b*) harvesting live trees while leaving a similar pattern of residual trees. Both photographs courtesy of Millar Western Forest Products Ltd.

(a)   *(b)*

Photos courtesy of Millar Western Forest Products Ltd.

This process indicates the importance of understanding the interaction between the ecological, economic, and social systems. It also requires that public involvement be implemented carefully, with the intent to construct a desired future forest, and with commitment to further strive for sustainable forest management.

Conclusions

Despite its simple premise, sustainability is a difficult concept and the implementation of sustainable forest management is a difficult task. Moving towards sustainability requires us to judge the impact of current actions on the future, and to assess the risks and tradeoffs inherent in these actions, versus others available to us. Since we do not know the preferences of future generations, a prudent approach is to avoid irreversible losses and maintain a suite of values and options for the future. Sustainability will be implemented by processes that define the desired future forest based on best available knowledge of the systems, by appropriate monitoring programs, by learning through adaptive management, and by continuous improvement. Sustainability will also be implemented by changing institutions so that individuals and firms are motivated to move towards sustainability, with sustainability as the objective of economic activity, rather than a constraint.

Strong sustainability concepts should be employed for unique environmental assets and in cases where ecological integrity is at risk of being irreversibly compromised, in order to safeguard the potential for an infinite stream of future benefits from assets that have few or no substitutes. In such cases, maintaining natural capital will surpass the value of any short-term gains that may be garnered from exploitative (non-sustainable) policies and practices. In other cases, however, we must continue to evaluate natural capital in a broad sense and assess the degree to which short-term depreciation or appreciation of natural capital enhances well-being and wealth. If our practices today mean that we destroy or degrade the resource so that it can no longer generate particular irre-placeable value streams in the future, then we are obviously doing the wrong thing.

The move to sustainable forest management should be considered evolutionary, not revolutionary. In many ways, it can be considered the melding of ecosystem manage-ment principles with those of traditional sustained-yield forestry. Although recent or suggested changes in forest policies and practices may be referred to as a "shift from sustained-yield forestry to SFM", it must be strongly emphasized that sustained-yield timber management remains a foundation of SFM today. The new SFM paradigm sim-ply means that a perpetual yield of wood fibre is no longer a sufficient description of sustainable forestry. Sustainable forest management is still based on "sustained yield", but this now means the sustained yield of a whole array of forest goods, services, attrib-utes, and values, not just wood fibre. Indeed, the entire framework for SFM rests on the shoulders of preceding forestry paradigms, and owes much to the advocates of regulated logging, forest conservation, sustained-yield management, and multiple-use forestry who faced political resistance and public skepticism in their day.

Sustainable forest management is more than a purely scientific process that outlines a prescription for management actions. Sustainable forest management involves the identification and selection of values as well as developing innovative institutions. The role of the SFM Network that has emerged in this context is:

- to help construct descriptions of the implications of management alternatives by providing the best science about the dynamics of the boreal forest ecosystems, com-munities, and economies;
- to help construct the best mechanisms for assessment of alternatives and decision making;
- to engage in adaptive management so that key uncertainties can be examined;
- to design monitoring strategies that can be used to evaluate sustainability; and
- to help develop technologies, strategies, policies, and institutions that can better achieve the objectives of sustainability.

Sustainable resource management requires not just a commitment to intergenera-tional equity, but also sufficient understanding of the impacts of management decisions on numerous system components. Likewise, any implementation of sustainable resource management requires reliable predictions of the impact of alternative strategies. Research in these areas has helped us to be able to describe the multiple sources of wealth arising from the boreal forest, to understand the complex relationships between human actions and the valued elements of the forest, and to implement changes in prac-tices and policy that will maintain or enhance this resource over time. Consequently, diverse programs of basic, applied, and operational research will be part of any agenda to implement sustainable forest management.

Acknowledgements

The content of this chapter originated in an attempt by the Sustainable Forest Manage-ment Network's management team to define sustainability as it applies to forest management. This exercise was undertaken to provide a multidisciplinary focus to sustainability and to help direct the Network's research based on this multidisciplinary approach. The members of the management team contributing to this effort were Adamowicz, Eric Hall, Daryll Hebert, Christian Messier, Dan Smith, Terry Veeman, and Dale Vitt. We hope this document reflects the views of these individuals. However, any errors should be attributed to the compilers of this information (Adamowicz and Burton) and not to the members of the management team.

References

Anielski, M. 2001. The Alberta GPI blueprint. Pembina Institute for Appropriate Development, Drayton Valley, Alberta. Available at http://www.pembina.org/pdf/publications/gpi_blueprint.pdf [viewed 18 March 2003]. 84 p.

Anielski, M., and Wilson, S. 2001. The Alberta GPI accounts: forests. Pembina Institute for Appropriate Development, Drayton Valley, Alberta. Available at http://www.pembina.org/pdf/publications/20_forests.pdf [viewed 18 March 2003]. 126 p.

(APFI) Alberta-Pacific Forest Industries. 2001. Ecosystem management. Alberta-Pacific Forest Industries, Inc. Boyle, Alberta. Available at http://www.alpac.ca/Forest_Management/EcosystemManagement.htm [viewed 10 March 2003].

Aplet, G.H., Johnson, N., Olson, J.T., and Sample, V. (*Editors*). 1993. Defining sustainable forestry. Island Press, Washington, D.C. 328 p.

Arbez, M., Birot, Y., and Carnus, J.-M. (*Editors*). 2002. Risk management and sustainable forestry. Proceedings 45. Proceedings of the Scientific Seminar, 8 Sept. 2001, Bordeaux, France. European Forest Institute, Joensuu, Finland. 94 p.

(BCMOF and BCMELP) British Columbia Ministry of Forests and British Columbia Ministry of Environment, Lands and Parks. 1995. Forest practices code biodiversity guidebook. B.C. Ministry of Forests and B.C. Ministry of Environment, Lands, and Parks. Victoria, British Columbia. 99 p.

Beanlands, G.E., and Duinker, P.N. 1984. An ecological framework for environmental impact assessment. J. Environ. Manage. **18**: 267–277.

Bergeron, Y., Harvey, B., Leduc, A., and Gauthier, S. 1999. Forest management guidelines based on natural disturbance dynamics: stand- and forest-level considerations. For. Chron. **75**: 49–54.

Bouman, O.T., and Brand, D.G. (*Editors*). 1997. Sustainable forests: global challenges and local solutions. Food Products Press. Binghamton, New York. 378 p.

Burton, P.J., Balisky, A.C., Coward, L.P., Cumming, S.G., and Kneeshaw, D.D. 1992. The value of managing for biodiversity. For. Chron. **68**: 225–237.

Castañeda, F., Palmberg-Lerche, C., and Vuorinen, P. 2001. Criteria and indicators for sustainable forest management: a compendium. Forest Management Working Paper 5. Forestry Department, Food and Agriculture Organization of the United Nations, Rome, Italy. Available at http://www.fao.org/forestry/fo/fra/main/pdf/chap5.pdf [viewed 26 May 2002].

Castro, A.P., Hall, E.R., and Adamowicz, W.L. 1999. Life cycle assessment for sustainable forest management. Working Paper 1999-23. Sustainable Forest Management Network. Edmonton, Alberta. 27 p. Available at http://sfm-1.biology.ualberta.ca/english/pubs/PDF/WP_1999-23.pdf [viewed 10 March 2003].

(CCFM) Canadian Council of Forest Ministers. 1995. Defining sustainable forest management: a Canadian approach to criteria and indicators. Canadian Council of Forest Ministers. Natural Resources Canada, Ottawa, Ontario. Available at http://www.ccfm.org/ci/framain_e.html [viewed 5 June 2003].

(CFS) Canadian Forest Service. 1998. National forest strategy, 1998–2003: sustainable forests, a Canadian commitment. Canadian Forest Service, Ottawa, Ontario. Available at http://www.nrcan.gc.ca/cfs/nfs/strateg/title_e.html [viewed 25 March 2002]. 47 p.

Commoner, B. 1971. The closing circle: nature, man and technology. *In* Thinking about the environment. *Edited by* M.A. Cahn and R. O'Brien. M.E. Sharpe, Inc., Armonk, New York. pp. 161–166.

Cumming, S.G., Burton, P.J., Mapili, M., and Prahacs, S. 1994. Potential conflicts between timber supply and habitat protection in the boreal mixedwood of Alberta, Canada: a simulation study. For. Ecol. Manage. **68**: 281–302.

Curran, M.A. (*Editor*). 1996. Environmental life-cycle assessment. McGraw–Hill, New York, New York. 432 p.

Daily, G.C. (*Editor*). 1997. Nature's services: societal dependence on natural ecosystems. Island Press, Washington, D.C. 412 p.

Erickson, P.A. 1994. A practical guide to environmental impact assessment. Academic Press, San Diego, California. 266 p.

(ESI) Environmental Sustainability Index. 2002. Environmental sustainability index: an initiative of the Global Leaders for Tomorrow Environmental Task Force, World Economic Forum. Center for International Earth Science Information Network, Columbia University, Palisades, New York. Available at http://www.ciesin.columbia.edu/indicators/ESI/index.html [viewed 18 March 2003].

Floyd, D.W. 2002. Forest sustainability: the history, the challenge, the promise. The Forest History Society, Durham, North Carolina. 83 p.

(FWA) Forest Watch Alberta. 2001. Planning and practices survey of FMA holders in Alberta. Forest Watch Alberta, Edmonton, Alberta. Available at http://www.forestwatchalberta.ca/forestry/survey.html [viewed 26 May 2002].

Goodland, R. 1995. The concept of environmental sustainability. Ann. Rev. Ecol. Syst. **26**: 1–24.

Hamilton, K., and Clemens, M. 1999. Genuine saving in developing countries. World Bank Econ. Rev. **13**(2): 33–56.

Hammond, H. 1991. Seeing the forest among the trees: the case for wholistic forest use. Polestar Book Publishers, Vancouver, British Columbia. 309 p.

Hammond, H. 1997. Standards for ecologically responsible forest use. *In* Ecoforestry: the art and science of sustainable forest use. *Edited by* A. Drengson and D. Taylor. New Society Publishers, Gabriola Island, British Columbia. pp. 204–210.

Hansen, A.J., Spies, T.A., Swanson, F.J., and Ohmann, J.L. 1991. Conservation of biodiversity in managed forests: lessons from natural forests. BioScience, **41**: 382–392.

Heal, G. 1998. Valuing the future: economic theory and sustainability. Columbia University Press, New York, New York. 224 p.

Hunter, M.L. 1991. Coping with ignorance: the coarse-filter strategy for maintaining biodiversity. *In* Balancing on the brink of extinction: the Endangered Species Act and lessons for the future. *Edited by* K. Kohm. Island Press, Washington, D.C. pp. 266–281.

Hunter, M.L. 1993. Natural fire regimes as spatial models for managing boreal forests. Biol. Conserv. **65**: 115–120.

(IISD) International Institute for Sustainable Development. n.d. Pulp and paper industry of Canada environmental statement. Sustainable development principles, International Institute for Sustainable Development, Winnipeg, Manitoba. Available at http://www.iisd.org/sd/principle.asp?pid=22&display=1 [viewed 18 September 2002].

(IUFRO) International Union of Forest Research Organizations. 2001. The economics of natural hazards in forestry. Provisional Proceedings of the IUFRO Symposium in Solsona, Catalonia, Spain, 7–10 June 2001. Padua University Press, Solsona, Catalonia, Spain. 157 p.

Keating, M. 1994. Media, fish and sustainability: a paper on sustainable development and the Canadian news media. Working Paper #22. National Round Table on the Environment and the Economy, Ottawa, Ontario. 52 p. Available at http://www.nrtee-trnee.ca/publications/ WP_22_E.PDF [viewed 18 September 2002].

Landres, P.B., Morgan, P., and Swanson, F.J. 1999. Evaluating the utility of natural variability concepts in managing ecological systems. Ecol. Applic. **9**: 1179–1188.

Lohmander, P. 2000. Economic risk management in forestry and forest industry and environmental effects in a turbulent world economy — a project family. Department of Forest Economics, Faculty of Forestry, Swedish University of Agricultural Sciences, Umeå, Sweden. Available at http://www.sekon.slu.se/~plo/ERM/proj99.htm [viewed 11 March 2003].

Maser, C. 1994. Sustainable forestry: philosophy, science and economics. St. Lucie Press, Delray Beach, Florida. 373 p.

(MCPFE) Ministerial Conferences on the Protection of Forests in Europe. 1998. General declarations and resolutions adopted at the Ministerial Conferences on the Protection of Forests in Europe: Strasbourg 1990 – Helsinki 1993 – Lisbon 1998. Ministerial Conferences on the Protection of Forests in Europe, Vienna, Austria. 88 p. Available at http://www.minconf-forests.net/Basic/FS-Publications.html [viewed 26 May 2002].

(MPWG) Montreal Process Working Group. 1995. The Montreal Process. Montreal Process Working Group, Ottawa, Ontario. Available at http://www.mpci.org [viewed 26 May 2002].

Moore, G.E. 1903. Principia ethica. Reprinted in 1988 by Prometheus Books, Amherst, New York. 237 p.

Morgan, P., Aplet, G.H., Haufler, J.B., Humphries, H.C., Moore, M.M., and Wilson, W.D. 1994. Historical range of variability: a useful tool for evaluating ecosystem change. *In* Assessing forest ecosystem health in the Inland West. *Edited by* R.N. Sampson and D. Adams. Haworth Press, Binghamton, New York. pp. 87–111.

(NFSC) National Forest Strategy Coalition. 1998. Canada forest accord. National Forest Strategy Coalition, Ottawa, Ontario. Available at http://www.nrcan.gc.ca/cfs/nfs/strateg/ acontrol_e.html [viewed 9 April 2003].

Noss, R.F. 1993. Sustainable forestry or sustainable forests? *In* Defining sustainable forestry. *Edited by* G.H. Aplet, N. Johnson, J.T. Olson, and V.A. Sample. Island Press, Washington, D.C. pp. 17–43.

Pezzoli, K. 1997. Sustainable development: a transdisciplinary overview of the literature. J. Environ. Planning Manage. **40**: 549–574.

Pezzey, J. 1992. Sustainability: an interdisciplinary guide. Environ. Values, **1**: 321–362.

Pezzey, J.C.V., and Toman, M.A. 2002. The economics of sustainability: a review of journal articles. Resources for the Future, Washington, D.C. Discussion Paper 02-03. Available at http://www.rff.org/disc_papers/PDF_files/0203.pdf [viewed 17 June 2002].

Sample, V.A., Johnson, N., Aplet, G.H., and Olson, J.T. 1993. Introduction: defining sustainable forestry. *In* Defining sustainable forestry. *Edited by* G.H. Aplet, N. Johnson, J.T. Olson, and V.A. Sample. Island Press, Washington, D.C. pp. 3–8.

Smith, W. 1997. The Pacific Certification Council. *In* Ecoforestry: the art and science of sustainable forest use. *Edited by* A. Drengson and D. Taylor. New Society Publishers, Gabriola Island, British Columbia. pp. 200–203.

Suzuki, D., and McConnell, A. 1997. The sacred balance: rediscovering our place in nature. Greystone Books, Vancouver, British Columbia. 259 p.

Vogt, K., Gordon, J., Wargo, J., Vogt, D., Asbjornsen, H., Palmiotto, P.A., Clark, H., O'Hara, J., Keeton, K., Patel-Weynand, T., and Witten, E. 1997. Ecosystems: balancing science with management. Springer-Verlag, New York, New York. 470 p.

von Gadow, K., Nagel, J., and Savorowski, J. (*Editors*). 2002. Continuous cover forestry: assessment, analysis, scenarios. Kluwer Academic Publishers, Dordrecht, Netherlands. 368 p.

(WCED) World Commission on Environment and Development. 1987. Our common future. Oxford University Press, Oxford, U.K. 400 p.

Weitzman, M.L. 1997. Sustainability and technical progress. Scand. J Econ. **99**(1): 1–13.

Chapter 3

Just another stakeholder? First Nations and sustainable forest management in Canada's boreal forest

Marc G. Stevenson and Jim Webb

The Onkwehonweh replied, "you (the Whiteman) pronounced yourself as my father, and with this I do not agree, because a father can tell his son what to do and can also punish Him. We will not be like Father and Son, but like Brothers. This friendship shall be everlasting and the younger generations will know it and the rising faces from Mother Earth will benefit by our agreement." The Whiteman said, "I understand, I confirm what you have said, that this will be everlasting as long as there is Mother Earth. We have confirmed this and our generation to come shall never forget what we have agreed. Now it is understood that we shall never interfere with one another's beliefs or laws for generations to come."

The Record of the Two Row Wampum Belt,
translated by Huron Miller (1980)

Introduction

Social, economic, health, and other pathologies stemming from chronic underemployment, welfare dependency, soaring birth rates, and other problems related to poverty are endemic to Canada's approximately 1.2 million Aboriginal Peoples. As Aboriginal Peoples struggle to hold on to their cultural values, traditions, and identities, and to address these problems, they remain, through a combination of factors, marginalized from Canadian society and economy.

In Canada, 80% of First Nations communities are situated in productive forest areas (NAFA 1995), and possess traditional cultures that reflect long-term reliance on forest resources. Perhaps more than any other industry, economies based on the sustainable

---

M.G. Stevenson.[1] Sustainable Forest Management Network, G208 Biological Sciences, University of Alberta, Edmonton, Alberta T6G 2E9, Canada.
J. Webb. Little Red River Cree Nation, High Level, Alberta, Canada.
[1]Author for correspondence. Telephone: 780-492-2476. e-mail: marc.stevenson@ualberta.ca.
Correct citation: Stevenson, M.G., and Webb, J. 2003. Just another stakeholder? First Nations and sustainable forest management in Canada's boreal forest. Chapter 3. *In* Towards Sustainable Management of the Boreal Forest. *Edited by* P.J. Burton, C. Messier, D.W. Smith, and W.L. Adamowicz. NRC Research Press, Ottawa, Ontario, Canada. pp. 65–112.

Box 3.1. First Nations' perspectives of key challenges and issues in sustainable forest management.

- *Protection of environment*: Government policy and industrial activities impacting ecosystem integrity and First Nations' relationships/responsibilities to the land;
- *Preservation of culture*: A host of forces undermining cultural traditions, values, identity, knowledge, social, and economic relationships, etc.;
- *Economic development*: Lack of skills, human capital, and resources to participate in economic development and land-use planning;
- *Social development*: Chronic unemployment, welfare dependency (>80%), social problems, rapidly growing population, lack of skills, human capital, and resources to affect change;
- *Aboriginal and treaty rights*: Frustrated by government intransigence to:
  - consult in a manner consistent with Supreme Court decisions;
  - adopt principles of treaty interpretation as laid out by the courts;
  - amend legislation and policy to accommodate Aboriginal/treaty rights; and
- *Self-determination and governance*: Limited capacity/ability to:
  - exert influence over resource development;
  - rebuild First Nations' systems of governance/management; and
  - have First Nations' knowledge and values count in decision making.

---

use of forest resources may be capable of addressing the current and future needs of Canada's non-urban Aboriginal Peoples. While many of these peoples still depend on the forest for nutritional, social, cultural, spiritual, and other needs, to date less than 5% of the on-reserve labour force is involved in forestry. Moreover, most First Nations lack the institutional frameworks, capacity, and resources to support an active dialogue with the provincial governments and forest industries to effect a change in this scenario.

Over 94% of the forest lands in Canada are publicly owned, and over 70% are administered under provincial law, regulations, and policies. Most provincial forestry regimes have encouraged development of timber resources through award of forest tenures to large industrial forest corporations (Chap. 1). These agreements emphasize the harvest and re-growth of timber, with the objective of insuring a sustained yield of fibre to support milling operations and employment opportunities. Over the last few decades, this policy has been modified to allow some consideration of other forest values. However, for the most part, this recognition amounts to little more than "tinkering" with timber-dominated approaches to forest management. In this context, Aboriginal Peoples are generally viewed as one of many "stakeholders", whose interests are not only perceived to conflict with sustained yield approaches, but which, if not properly managed, could become constraints on timber production through political interference and civil disobedience.

In this chapter, we argue that Aboriginal Peoples are "not just another stakeholder". They have rights and interests, recognized at national and international levels, which compel government and industry to work with Aboriginal governments to ensure that Canada's forests continue to sustain Aboriginal cultures and economies, while contributing to regional economic stability. In order to do this we must come to terms with Aboriginal and First Nations issues and challenges in the forest (Boxes 3.1 and 3.2).

There are very valid and practical reasons for government and the forest products industry to work with First Nations and Aboriginal communities to increase their contributions to sustainable forest management. Registered Indians have the lowest labour force participation rate of any Aboriginal group in Canada at 54%, and the highest unemployment rate at 27%. These figures are significantly greater on-reserve, especially in rural forested areas, where unemployment levels and welfare dependency rates exceeding 80% are not uncommon. At the same time, Canada's Aboriginal population is growing faster than any other group in Canada. Today, 50% of Canada's Aboriginal population is under the age of 25 (Fig. 3.1). Within 15 years, if current demographic trends continue, 400 000 more Aboriginal Peoples will be added to Canada's potential labour force (Statistics Canada 1996). If employment is not found to meet the needs of Canada's Aboriginal Peoples, we can expect the social and economic costs to Canada to be exorbitant, and can anticipate that rapidly growing on-reserve Aboriginal population will have little interest in the sustainability of Canada's forests or forest economy.

This is not to suggest that employment in the formal economy is the answer for all, or even, most Aboriginal communities. Clearly, traditional activities are nutritionally, socially, culturally, and spiritually important in achieving overall well-being. Nevertheless, one of the greatest obstacles in finding the desired balance between the formal and informal economies is the lack of employment and opportunity in the formal sector. In fact, it may even be argued that, for many Aboriginal communities, money generated through wage labour employment is critical to sustaining traditional values, activities, and the informal economy.

Because of the historical and legal position of First Nations in Canada, accommodating Aboriginal rights and interests must be considered a fundamental building block for sustainable forest management (SFM) in Canada's boreal forests. The challenge we, as a society, face is in implementing ways and means to significantly increase Aboriginal participation in the management of forest uses and in the development of forest

---

Box 3.2. Sustainable Aboriginal Communities (SAC) initiative: the Sustainable Forest Management Network's response to Aboriginal issues.

The SAC group, under the direction of Dr. Cliff Hickey at the University of Alberta, was created to address the issues and challenges in Box 3.1, and undertakes research in four key areas:

- Integrating Aboriginal Peoples, values, knowledge, and management systems into sustainable forest management (SFM);
- Accommodating Aboriginal and treaty rights into SFM;
- Developing Aboriginal capacity and economies in the context of SFM; and
- Developing Aboriginal criteria and indicators for SFM.

More recently, SAC, as part of a Network initiative, has begun to focus on developing:

- More effective institutions for Aboriginal Peoples in SFM;
- A model of forest tenure that reconciles treaty rights with forest policy; and
- Non-timber forest and value-added wood product opportunities for First Nations.

Fig. 3.1. Little Red River Cree Nation youth, Richard Durmaine and Dylan Seweepagaham (standing), preparing a sled for transport of moose meat after a kill.

Photo by Conroy Seweepagaham

resources. Meeting such challenges offers great promise not only to address the needs, rights, and interests of Canada's boreal forest-dependent Aboriginal Peoples, but to solve current economic and social problems in their communities. Decreased welfare dependency, social assistance, medical, and other costs associated with poverty are some of the benefits that would accrue to Canada. Other, perhaps not so obvious, benefits include an enhanced ability for Canada to live up to international and national commitments; increased nutritional, social, cultural, and spiritual wellness for Aboriginal Peoples; and the maintenance of biodiversity. Aboriginal Peoples traditionally played a major role in sustaining the ecological integrity and biodiversity of Canada's forests (Stevenson and Webb 2003). Opportunities exist in industrial forestry (logging, silviculture, management, etc.), as well as in the development of local non-timber forest goods and services, and associated value-added products and market opportunities. The latter remain largely unexplored in the context of boreal forest-dependent First Nations and Aboriginal communities, but may prove to be the most economically viable and sustainable industries over the long term.

Sustainable forest management

Sustainable forest management recasts the ongoing struggle between environment and the economy (and arguably politics) in a different light, creating a new framework and benchmark for Canada's forest products industry. Under the weight of growing environmental awareness and public pressure, the "goalposts" are shifting from maximum sustained yield approaches to SFM. However, the concept means different things to different people. To most forest products companies, government regulators, and the public at large, SFM has more to do with finding a balance between commercial forestry and the conservation of biodiversity while attempting to answer the question, "How much is enough?" In other words, how many trees can we remove without jeopardizing the timber resource and other values (environmental, recreational, etc.)? Alternatively, how much of the forest can be protected without damage to our economic sustainability?

For many people who live in and depend upon Canada's forests, especially Aboriginal Peoples, the issue is much greater than balancing the needs of industrial forestry and the conservation of biodiversity. To many Aboriginal Peoples the concept of SFM raises the question: "Sustainability of what, for what, and for whom?" While most Aboriginal Peoples are generally concerned about the ecology of their forests and how forests should be used to create community economic self-reliance, these concerns are embedded within a broader context. It is the sustainability of the nutritional, social, cultural, spiritual, and other values and needs that they derive from the forest that are at issue in any dialogue about SFM. Such values arise from long-term use and occupation of forests, and reflect unique systems of knowledge and management practice about the place and role of humans in forest ecosystems, and how best to manage human uses of forest resources in the interest of long-term sustainability. It is these values, needs, and knowledge and management systems that proponents of SFM must embrace and accommodate if society's use and dependence upon Canada's forests is ever going to provide equitable and sustainable benefits to all, both now and in the future. Such values and needs, and the knowledge and wisdom that informs them, constitute one of the most compelling reasons why Aboriginal Peoples ought to be actively involved in a dialogue, from which they have been largely excluded so far, leading to the development and implementation of sustainable management regimes within their traditional territories.

In the same vein, there are strong legal, scientific, and other arguments to be made for the significant and meaningful inclusion of Aboriginal Peoples in SFM. In fact, it might be reasonably argued that if we continue on our present course of treating Aboriginal Peoples as just another stakeholder, SFM will not be achievable. At the same time, the costs and consequences of maintaining the status quo (in regards to the involvement of Aboriginal Peoples in forestry, and in decisions taken with respect to their forests in the absence of informed consent) could be exorbitant for Canada.

This chapter describes the commitments that Canada has made both internationally and nationally to improve the roles of its Aboriginal Peoples in SFM. In Canada, Indigenous peoples are referred to as Aboriginal Peoples, and their rights and interests are referred to as Aboriginal, and (or) treaty rights and interests. The term "First Nations" refers to ethnically/geographically defined self-governing bodies of Indian peoples. This chapter also presents the legal and scientific rationales for greater Aboriginal participation in the development of forest resources. Serious consideration of these commitments by government and industry presents obvious challenges to the status quo. Meeting these challenges will require sustained commitment, and perhaps investment over the short term. However, the benefits of changing forest policies, regulations, and practices to accommodate the needs, rights, and interests of Canada's boreal forest-dependent Aboriginal Peoples ultimately will pay significant dividends to Canada far into the future. It is in this context that forest certification may provide a fruitful ground for, and a common path to, increased Aboriginal participation in managing resource uses on their traditional lands.

This chapter concludes with an analysis of the experience of the Little Red River Cree and Tallcree First Nations in the hope of providing direction and incentive for Canada's forest regulators and companies. This experience may also prove useful to Indigenous peoples elsewhere as they seek greater control over their traditional homelands.

Canada's commitments

International commitments

Canada is recognized internationally for its efforts to promote human rights, and is signatory to a number of multi-national agreements that support the increased involvement of Indigenous peoples in sustainable management and development (Box 3.3). Among other international instruments that address Indigenous rights, only the International Labour Organization Convention 169 (ILO 1989) has reached the stature of international law and is legally binding on those nations ratifying the agreement. ILO 169 imposes a duty of good faith consultation on state governments with Indigenous and tribal peoples regarding development projects and other activities affecting them, while laying out criteria for consultation. These include obtaining informed consent, protecting Indigenous rights and interests, and addressing the concerns of Indigenous peoples.

Box 3.3. Canada's international commitments to Indigenous peoples following the United Nations Council on the Environment and Development (UNCED) meeting in Rio de Janeiro in 1992.

Indigenous people and their communities, and other local communities, have a vital role in environmental management and development because of their knowledge and traditional practices. States should recognize and duly support their identity, culture and interests and enable their effective participation in the achievement of sustainable development.

Rio Declaration, Principle 22

In view of the interrelationship between the natural environment and its sustainable development and the cultural, social, economic and physical well-being of indigenous people, national and international efforts to implement environmentally sound and sustainable development should recognize, accommodate, promote and strengthen the role of indigenous people and their communities.

Agenda 21, Chapter 26, clause 26.1

Subject to its national legislation, respect, preserve and maintain knowledge, innovations, and practices of indigenous and local communities embodying traditional lifestyles relevant for the conservation and sustainable use of biological diversity and promote their wider application with the approval and involvement of the holders of such knowledge, innovations, and practices and encourage the equitable sharing of the benefits arising from the utilization of such knowledge, innovations and practices.

Convention on Biological Diversity, Article 8j

IGC 1996

> **Box 3.4. Gathering Strength, Canada's Aboriginal Action Plan *(Canada 1997)* commits Canada to:**
>
> • Supporting professional development initiatives in lands and environmental stewardship;
> • Supporting accelerated transfer of land management to First Nations;
> • Exploring with First Nations treaty issues and how historic treaties can be understood in contemporary terms;
> • Working to strengthen the co-management process and to provide increased access to land and resources;
> • Working to accelerate Aboriginal participation in resource-based development in and around Aboriginal communities and to improve the benefits communities receive from these developments; and
> • Increasing funding for resource initiatives so that First Nations can derive more benefits from resource development, co-management of resources, and harvesting and contracting opportunities related to resources.

Despite what may be honourable intentions to live up to commitments made in these international covenants and declarations, Canada has not been entirely successful in applying their spirit and intent at home. On the contrary, Canada has been reprimanded by the United Nations for failing to respect the rights of Aboriginal Peoples to adequate lands and resources as set out by the International Covenant on Civil and Political Rights (United Nations 1967). In addition, it has, to date, failed to take any specific federal, or provincial legislative steps towards implementing provisions in these agreements, such as Article 8j of the Biodiversity Convention (IGC 1996)

National commitments

Canada has made a number of national policy commitments to its Aboriginal Peoples over the years, intended to increase their roles in the sustainable management and development of forests. For example, Canada's response to the recommendations of the Royal Commission on Aboriginal Peoples, *Gathering Strength: Canada's Aboriginal Action Plan* (RCAP 1996), commits the nation to specifically addressing economic, political, and legal reforms for First Nations and Aboriginal communities (Box 3.4).

The Canadian Council of Forest Ministers (CCFM), a body of federal and provincial ministers, specifically addressed a range of Aboriginal Peoples' issues in Strategic Direction Seven of the *National Forest Strategy* (CCFM 1998). Benefiting directly from the input of the National Aboriginal Forestry Association (NAFA), Strategic Direction Seven is guided by principles that "recognize and make provision for Aboriginal and Treaty rights and responsibilities, and respect the values and traditions of Aboriginal Peoples regarding the forests for their livelihood, community and cultural identity" (CCFM 1998). These principles are clearly articulated in the commitments of the *National Forest Strategy's* "Framework for Action" (Box 3.5).

The CCFM has also endorsed a set of national criteria and indicators (CCFM 1995) that include direct reference to the recognition of Aboriginal and treaty rights (Criterion 6.1) and participation of Aboriginal communities in SFM (Criterion 6.2). However,

---

Box 3.5. The National Forest Strategy's Framework for Action (CCFM 1998) commits Canada to:

- Ensuring the involvement of Aboriginal Peoples in forest management and decision making;
- Increasing Aboriginal Peoples' access to forest resources for both traditional and economic development activities;
- Supporting Aboriginal employment and business development in the forest sector;
- Increasing the capacity of Aboriginal communities, organizations, and individuals to participate in and carry out sustainable forest management; and
- Achieving sustainable forest management on reserve lands.

---

much to the dismay of Aboriginal organizations (e.g., NAFA 2003), the CCFM has, to date, not supported the creation of a separate criterion to address Aboriginal issues, rights, or interests (CCFM 2003). Ross and Smith (2002) assert that Canada's constitutional structure, whereby the federal government assumes the responsibility for "Indians and lands reserved for Indians", while provinces assume the responsibility for the management of natural resources, has allowed provincial governments to virtually ignore any need to recognize and respect the rights of Canada's Aboriginal Peoples. Provincial behaviour, in turn, handcuffs Canada from making any significant progress towards its commitments, while rendering any statements that portend otherwise potentially hypocritical and without substance.

Perhaps it is too early to expect significant progress towards implementing Canada's international and national commitments to increase the role of Aboriginal Peoples in SFM. Despite what may be good intentions, existing programs, policies, legislation, and institutions at the federal and provincial levels are not particularly accommodating of Aboriginal rights, needs, and interests. By way of providing motivation and direction to government and industry, the following two sections present the legal and scientific bases for significant change and improvement in SFM policy and practice. The following section clarifies Canada's legal obligation to address Aboriginal Peoples involvement in the context of SFM, while encouraging provincial governments to reform policy and legislation.

Legal basis

The fiduciary relationship

Canada's Indian, Inuit, and Metis peoples have constitutionally protected rights beyond those normally enjoyed by most Canadians. Section 35.1 of the Canadian Constitution Act, 1982, recognizes and affirms "the existing aboriginal and treaty rights of the Aboriginal Peoples of Canada". This wording begs the question as to what these existing rights are, and what burdens of proof are necessary to demonstrate their existence. While the exact nature of these rights remains a subject of debate between the Crown (federal and provincial governments) and Canada's First Nations, one thing is certain: Canada and its provincial governments have a fiduciary relationship with Canada's

Fig. 3.2. The Two Row Wampum: the foundation for all treaties and agreements made with Europeans during the colonial history of North America.

Aboriginal Peoples, which extends back to the Royal Proclamation of 1763 after the British defeat of New France.

Although the Royal Proclamation of 1763 is not a constitutional document, it has the force of law in Canada and is referenced in Section 25 of the Canadian Charter of Rights and Freedoms (Schedule B to the Canada Act, 1982 (U.K.), c.11) and in numerous court decisions. In essence, the Royal Proclamation recognized the rights of Indians to lands under their possession, reserving such lands as their "hunting grounds". The Crown's desire to open up "Indian land" for settlement provided the *raison d'etre* for it to enter into treaties with Aboriginal Peoples. In those parts of Canada where treaties were not negotiated, "Aboriginal Title" is seen as a burden on Crown Title, and is the source of the fiduciary obligation. In those parts of Canada where treaties were made, the treaty relationship is the basis of the fiduciary obligation.

The metaphor of "Two Row Wampum", embodied in the 1664 Treaty of Albany between the Iroquois and the Dutch, is viewed by most First Nations people as equal in importance to the Royal Proclamation (Fig. 3.2). Both devices were the focus of the 1764 Niagara Conference between the imperial Crown's governor and the assembled Chiefs of Canada's Aboriginal Peoples. The proceedings of this conference provide the strongest grounding for interpreting the basis for the Crown's fiduciary obligation towards Aboriginal Peoples. According to the Mohawks of Kahnawake (see http://www.kahnawake.com), the concept of the Two Row Wampum was developed by their ancestors so they could peacefully co-exist, conduct trade and share resources with the Europeans. The Two Row Wampum symbolizes the principles of sharing, mutual recognition, respect, and partnership, and is based on a nation-to-nation relationship that respects the autonomy, authority, and jurisdiction of each nation. The two rows symbolize two paths or two vessels travelling down the same river of life together. One, a birch bark canoe, represents the Original peoples, their laws, their customs, and their ways. The other, a ship, is for the European peoples, their laws, their customs and their ways. They travel down the river together, side by side, each in their own boat, neither trying to steer the other's vessel.

Treaty rights

Aboriginal Peoples possess Aboriginal rights and, frequently, treaty rights. The Supreme Court of Canada has defined an Aboriginal right as "an element of a practice, custom or tradition integral to the distinctive culture of the Aboriginal group claiming the right", of which Aboriginal title is one kind of Aboriginal right. Treaty rights, on the other hand, derive from legally binding agreements made between the Crown and Aboriginal nations. Such agreements, from the perspective of the Crown, usually arise in the context of the extinguishment of certain rights enjoyed by the Aboriginal nations since time immemorial (e.g., exclusive title) in exchange for other rights and privileges (e.g., reserves, health care, housing). Federal and provincial governments have and continue to view the historic treaties (Fig. 3.3) as surrenders of land, taking a literal and restricted interpretation of the texts of the treaties. Alternatively, First Nations have traditionally looked upon these treaties as solemn promises between sovereign nations to share and co-exist in peace. Within this view, treaty rights can be characterized as either treaty-protected Aboriginal rights, which were recognized and affirmed by the Crown in treaty negotiations, or as unique treaty rights, which reflect commitments or inducements offered by the Crown, and accepted by First Nations in treaty negotiation. In this context, treaty commitments by the Crown related to respect for, non-interference with, and protection of the Indian way of life, are seen as the basis of the Crown's obligation to maintain the forest as an ecosystem that facilitates cultural sustainability.

A key feature of the numbered historic treaties, which served as an inducement for First Nations to sign treaties, was the guarantee that they could continue to live off the land as their forefathers had done. This treaty clause is focused, in part, on facilitating economic activities that arose after contact with non-aboriginal peoples and the assertion of Crown sovereignty.

> *The Indians shall have the right to pursue their usual vocations of hunting, trapping, and fishing throughout the territory surrendered . . ., subject to regulations . . . by the government, . . . and saving and excepting such tracts as may be required or as may be taken up from time to time for settlement, mining, lumbering, trading or other purposes.*
>
> From the text of Treaty 6 (1876)

A number of recent court decisions have addressed the "taking up" clause in the historic treaties, arguing, on the one hand, that forestry or National Parks are not incompatible with the exercise of treaty rights. On the other hand, it is reasoned that developments that could not have been reasonably contemplated or envisioned by either party at the time do not constitute a justifiable "taking up" of the land or interference of a treaty right in the absence of consultation or informed consent (e.g., Halfway River v. British Columbia 1997; Miskisew v. Regina 2001).

Differences in the meaning and significance of treaties have forced Canadian courts to adopt a novel perspective whereby treaties are viewed as neither simple contracts nor international-like agreements between nation states, but *sui generis* or unique. In a number of recent decisions by the Supreme Court of Canada, where Aboriginal parties have claimed infringement of treaty rights, or failure of the federal government to uphold its fiduciary obligation arising from a treaty, special principles of treaty interpretation have emerged, based, in part, on contract case law. Foremost amongst these is that Aboriginal treaties include not only the text of the treaty, but the understandings and oral promises of both parties at the time of signing (Box 3.6).

While these principles are intended to assist courts in rulings involving treaty interpretation, the practical results of such readings tend to be modest. The Supreme Court

Fig. 3.3. Post-Confederation numbered Treaty areas in Canada (from *The Atlas of Canada*, Natural Resources Canada).

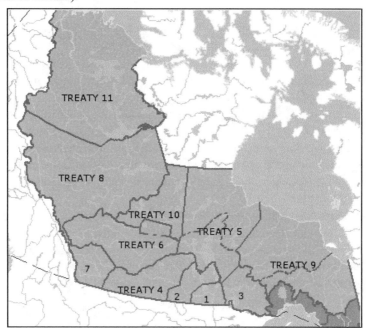

of Canada sees its role as defining parameters to assist First Nations and federal and provincial governments to negotiate agreements. Nevertheless, these principles also instruct federal and provincial governments to redress their literal interpretations of treaties, as well as those policies, legislation, and regulations that flow from such interpretations.

Natural Resource Transfer Agreements

Most of the Crown colonies that joined confederation brought with them their land base and negotiated articles of confederation with the federal Crown. In contrast, the provinces of Manitoba, Saskatchewan, and Alberta were created out of the Northwest Territories. In 1930 the federal government entered into agreements with each of these provinces for the primary purpose of transferring control of natural resources and Crown lands to the provinces. These agreements were confirmed by provincial legislation and given force of law by the Parliament of Canada through the enactment of the Constitution Act, 1930.

This transfer of jurisdiction has created a headache of large proportions for those First Nations who had signed historic treaties with the federal government. While the provinces assumed control over natural resources, the fiduciary obligation owed them was still vested with the federal government. The issues at stake for First Nations are whether treaty Indians are subject to provincial game laws of general application, and whether the provinces are required to act as fiduciaries should provincial legislation or resource development infringe upon a treaty right. Two sections of these agreements are germane to an understanding of Aboriginal and treaty rights within the numbered

> ***Box 3.6. Principles of treaty interpretation as summarized by the Supreme Court of Canada.***
>
> - Aboriginal treaties should be liberally construed and ambiguities or doubtful expressions should be resolved in favour of the Aboriginal signatories;
> - The goal of treaty interpretation is to choose from among the various possible interpretations of common intention the ones which best reconcile the interests of both parties at the time the treaty was signed;
> - In searching for the common intention of the parties, the integrity and honour of the Crown is presumed;
> - In determining the signatories' respective understanding and intentions, the court must be sensitive to the unique cultural and linguistic differences between the parties;
> - The words of the treaty must be given the sense which they would naturally have held for the parties at the time;
> - A technical or contractual interpretation of treaty wording should be avoided; and
> - Treaty rights of Aboriginal Peoples must not be interpreted in a static or rigid way. They are not frozen at the date of signature. The interpreting court must update treaty rights to provide for their modern exercise. This involves determining what modern practices are reasonably incidental to the core treaty right in its modern context.
>
> Extracted from such decisions as Regina v. Badger, Regina v. Regina v. Sioui, Regina v. Simon, Regina v. Horseman, and Regina v. Sundown, these principles were summarized by McLachlin (1999).

treaties areas. While Section 2 of the Natural Resource Transfer Agreements makes transfer of administrative responsibility over lands and resources subject to "existing trusts and interests", Section 12 affords specific protection for Indian support and sustenance, and impels provincial governments to address forestry legislation and natural resource development policy to accommodate the exercise of treaty rights (Box 3.7).

Aboriginal rights

For many First Nations, Aboriginal rights flow from the Creator and include the rights to lands, resources, culture, language, and self-determination. Aboriginal rights and title

> ***Box 3.7. Section 12 of the Saskatchewan Natural Resources Transfer Agreement.***
>
> *In order to secure to the Indians of the Province the continuance of the supply of game and fish for their support and sustenance, Canada agrees that the laws respecting game in force in the Province from time to time shall apply to the Indians within the boundaries thereof, provided, however, that the said Indians shall have the right, which the Province hereby assures them, of hunting, trapping and fishing game and fish for food at all seasons of the year on all unoccupied Crown lands and on any other lands to which the said Indians may have a right of access.*
>
> Natural Resources Transfer Agreement, Saskatchewan 1930

emerge out of prior occupation of the land. As Chief Justice Lamar stated in Regina v. Van der Peet (1996):

> *When Europeans arrived in North America, Aboriginal Peoples were already here, living in communities on the land, and participating in distinctive cultures, as they had done for centuries. It is this fact, and this fact above all others, which separates Aboriginal Peoples from all other minority groups in Canadian society and which mandates their special legal and now constitutional status.*
>
> Regina v. Van der Peet (1996)

To enjoy constitutional protection, an Aboriginal right must have existed prior to 1982, and cannot have been extinguished by the sovereign prior to this time. Moreover, to be recognized and affirmed by Section 35 of the Canadian Constitution, an Aboriginal right must satisfy the Van der Peet test. Namely, it must be part of a custom, practice, or tradition integral to an Aboriginal community's distinctive culture prior to European contact, and its modern practice cannot be substantially different from its precontact activity. Although Aboriginal rights are not defined in the Canadian Constitution, a number of judicial decisions, beginning in 1990, have begun to clarify what rights are protected under Section 35.

The Supreme Court has established what has become known as the "Sparrow Test" for the infringement, and justification thereof, of an Aboriginal right. If an Aboriginal right can be proven, the government must justify its regulation if it interferes with this right. However, it is up to the First Nation to show that the regulation is unreasonable, imposes undue hardship, and (or) denies it of its preferred means of exercising the right. If an infringement is demonstrated, the Crown must justify it, by showing that:
- there was a valid legislative objective;
- after conservation, resource allocations gave priority to the Aboriginal right;
- there has been as little infringement as possible to effect conservation; and
- the Aboriginal group has been consulted with respect to the conservation measure being implemented.

Aboriginal title

Aboriginal title is special kind of Aboriginal right that goes beyond those directly tied to historic practices before contact. In Delgamuukw v. Regina (1997), the Supreme Court of Canada found that Aboriginal title is the broad, sometimes exclusive, right to use land for a variety of purposes, whether historically precedented or not (Fig. 3.4). In other words, Aboriginal title flows from traditional use of land by an Aboriginal community, which, in itself, may be sufficient to establish an exclusive right of occupancy, and permits a First Nation to make modern uses of land that may have no direct analogues in historic practice. However, the exercise of Aboriginal title is subject to the limitation that "uses cannot destroy the ability of the land to sustain future generations of aboriginal peoples" (Delgamuukw v. Regina 1997). In the Supreme Court's decision in Delgamuukw, a direct link is made between cultural and ecological sustainability, which are at the fore of SFM.

Fig. 3.4. Gitxsan dancers in Ottawa, celebrating the Supreme Court of Canada's far-reaching Delgamuukw decision, 16 June 1997.

Canadian Press photo by Fred Chartrand

Rules of consultation and compensation

The infringement of Aboriginal rights or title can be justified if there is a valid and compelling legislative objective to do so, such as conservation or economic development (e.g., agriculture, forestry, mining, hydroelectricity). However, infringement is subject to the fiduciary obligation of the Crown to Aboriginal Peoples; the honour of the Crown is always at stake. As the Delgamuukw decision states, because Aboriginal title encompasses the right to choose the ends to which a piece of land can be put, and to the extent that Aboriginal rights amounting to less than title constitute an ongoing interest in the land ". . . the fiduciary relationship between the Crown and aboriginal peoples may be satisfied by the involvement of aboriginal peoples in decisions taken with respect to their lands" (Delgamuukw v. Regina 1997). The Supreme Court has further articulated additional principles of consultation based on Aboriginal rights:

> ***There is always a duty of consultation. Whether the aboriginal group has been consulted is relevant to determining whether the infringement of aboriginal title is justified****. . . . The nature and scope of the duty of consultation will vary with the circumstances. In occasional cases, when the breach is less serious or relatively minor, it will be more than a duty to discuss important decisions that will be taken with respect to the lands held pursuant to aboriginal title. Of course, even in these rare cases when the minimum acceptable standard is consultation, this* **consultation must be in good faith, and with the intention of substantially addressing the concerns of the aboriginal peoples** *whose lands are at issue. In most cases, it will be significantly deeper than mere consultation.* **Some cases may require the full consent of an aboriginal nation** *. . .*

> Delgamuukw v. Regina 1997, para. 168
> [emphasis added]

Additional principles of consultation have been articulated by other court decisions (Box 3.8).

In the case of forestry planning, the duty to consult is primarily that of the Crown's, not that of industry. However, in Haida v. British Columbia and Weyerhauser (2002), the court ruled that forest products companies obtaining provincial forestry licenses that infringe constitutionally protected Aboriginal rights must also consult with the affected First Nations in order to accommodate their rights and interests. In other words, a forest company that accepts a license knowing full well that its execution would violate an Aboriginal right could be held to be a constructive trustee (Woodward 1999). The fact that industry must now consult with Aboriginal Peoples, nevertheless, does not discharge the Crown of this duty. Most provincial governments and forest products companies have waited for courts of law to recognize Aboriginal rights and title before consulting with First Nations. However, in Haida v. British Columbia and Weyerhauser (2002) and Taku River Tlingit First Nation v. Ringstad et al. (2002), the courts ruled that First Nations do not have to go before a court of law to prove their Aboriginal rights and title. The fiduciary obligation owed to them by the Crown is sufficient to initiate consultation towards accommodating their rights and interests.

The economic aspect of Aboriginal title suggests that compensation is as relevant to the question of justification as consultation, and is part of the same justification test for infringement. Thus, not only is the Crown compelled to undertake adequate consultation, fair compensation will ordinarily be required when Aboriginal title is infringed. Under circumstances where government acts are likely to infringe Aboriginal and treaty rights and interests, including those that are less than title (e.g., traditional use rights and interests), the Crown has a fiduciary obligation to consider how best to accommodate Aboriginal interests. Compensation may be required where the action amounts to an expropriation. Other means of accommodation or mitigation may be required, depending on the circumstance.

Institutional inertia

In Canada, the regulation of forest management comes under the jurisdiction of the provinces. The gulf between existing provincial forestry policy and legislation and the ability of provinces to accommodate Aboriginal and treaty rights, however, has not grown any narrower, despite the increasing number of Supreme Court and other court rulings clarifying and defining these rights. Provincial forestry policy in Canada, for the most part, remains mired in timber management assumptions that grant significant rights to third party interests (Chap. 1). As Ross and Smith (2002) have recently pointed out in a synthesis report for the Sustainable Forest Management Network (SFMN), provincial systems of tenure are a structural and systemic impediment to the recognition and protection of Aboriginal and treaty rights in forest management in Canada. Here, clearly, is an area in need of institutional reform (Chap. 7); a number of prevailing forestry policies across Canada undermine the ability of Aboriginal Peoples and First Nations to continue traditional land-use practices and to translate their underlying forest values into a contemporary expression which is essential for the exercise of their constitutionally protected rights. For example, the determination of annual allowable cuts (AACs), the allocation of long-term tenures to third parties, and the requirement to

> **Box 3.8. Additional principles of consultation articulated by Canadian courts.**
>
> - It is the Crown, not the Aboriginal group, that must initiate consultation;
> - The Crown must fully inform itself of Aboriginal and treaty rights and the effect of a law or regulation on a First Nation;
> - Full information and disclosure must be provided to a First Nation, and this must be done in a timely manner and on a continuing basis, as required;
> - Consultation must be meaningful, i.e., the views of the Aboriginal party must be taken seriously, and the First Nation must be given the opportunity to make a reasonable assessment of what the Crown is doing, including receiving sufficient information so it can determine the effects of the Crown's actions;
> - The duty to consult implies rules of procedural fairness, whereby the Aboriginal party is entitled to know all the details of the action and to respond accordingly;
> - There is an obligation on First Nations to participate in consultation, once initiated by the Crown;
> - First Nations and Aboriginal Peoples are entitled to separate and distinct consultation processes; and
> - Industry must consult with the affected First Nation if it is in "knowing receipt" of a licence or permit that infringes Aboriginal rights or title.

build and operate mills as a condition of tenure allocation all serve to further reinforce the exclusion of Native peoples from the modern economy as forest resources (the basis for both traditional practices and for future economic self-sufficiency) are liquidated at rates which (arguably) are not ecologically, socially, or economically sustainable.

For First Nations and other Aboriginal Peoples, AACs do not consider Aboriginal non-timber forest uses, rights, interests, and values. Yet, these are critical to determining the land base available for commercial timber production (i.e., the supply of fibre for dimensional lumber and pulp and paper) and the sustainable rate of harvest (Ross and Smith 2002). The current practice of focusing on the extraction of maximum volumes of timber are especially detrimental to: (1) the exercise of constitutionally protected rights and traditional activities, which are critical to sustaining Aboriginal cultural values, sustenance requirements, and social relationships; and (2) Aboriginal efforts to develop economies based on non-timber forest and value-added products and markets, which have the potential to be more ecologically, socially, and economically sustainable than the industrial forestry model. The process of allocating long-term tenures, which in most provinces is discretionary, shielded from public view, and done in the absence of consultation (Ross and Smith 2002), serves to exclude Aboriginal Peoples from effectively participating in commercial forestry, while denying their rights to their traditional lands and resources. The requirement to build and operate a mill as a condition of tenure allocation also serves to exclude Aboriginal Peoples from commercial forestry, while precluding the use and development of non-timber forest products and of timber for value-added production, which might be more in keeping with Aboriginal needs and values (NAFA 1997).

However, as the preceding discussion underscores, there is a strong legal case to be made for tenure reform that accommodates Aboriginal and treaty rights into forestry policies and practices. Indeed, it might be argued that if the industrial forestry model is not modified to accommodate the rights and social, economic, and ecological requirements of boreal forest dependent communities, especially Aboriginal ones, SFM will exist only in theory and not as a practice. Equity is a key component of the sustainability and certainty that we all, as a society, seek. The sharing of knowledge, management authority, and responsibility is also critical to these objectives, a subject to which we now turn.

Scientific basis: alternative approaches to Aboriginal Peoples contributions to sustainable forest management

Sustained yield management approaches are arguably driven more by basic business principles and political decisions than by "science", unless people challenge this hegemony. To the extent that science comes into play at all, it is usually in the context of using excess revenues to commission scientists to fill in the details and to provide the numbers that might guide or somehow "tweak", but not undermine, existing practices. This serves as a systemic barrier not only to the meaningful inclusion of Aboriginal Peoples and their knowledge, but to the more general and open use of western science in informing existing forestry policies and practices (Chap. 23). The following section is concerned not so much with many and varied contributions of science to SFM; this is addressed in other chapters. Rather, it lays the groundwork for an alternative model for the application of both scientific knowledge and the knowledge of Aboriginal Peoples in planning and decision making. In so doing, it is hoped that in the future, both will be accorded greater consideration in setting the course for SFM than they presently enjoy.

Systemic barriers and challenges to the meaningful involvement of Aboriginal Peoples and their knowledge in SFM

Aboriginal Peoples confront many systemic barriers that restrict their participation in decisions taken with respect to forestry and industrial activities on their traditional lands. Even at the most basic level, where Aboriginal Peoples and forest managers and scientists sometimes come together to address mutual concerns and interests in the context of cooperative management, Aboriginal knowledge and management systems "take a back seat" to those of western science. Forest managers and scientists should be among the greatest champions for the meaningful inclusion of Aboriginal Peoples and their knowledge and management systems into SFM. Yet, less by design than parsimony and force of circumstance — for when one's voice is weak, one generally speaks only for oneself — these cultural barriers have become a significant impediment for Aboriginal Peoples in the practice of cooperative management and SFM. This section tackles some of these systemic barriers head-on.

Aboriginal Peoples may possess a wealth of ecological and related knowledge that can inform and assist the goals of SFM. Many elders and traditional land users, for example, have developed extensive knowledge bases about the behaviours, spatial and temporal distributions, health and conditions of specific plants and animals, and the factors that influence these phenomena. On the face of it, much of this "traditional ecological knowledge" (TEK) may be useful for making informed decisions to sustain various forest values. Most notably, being the product of both personal experience and cumulative knowledge passed down through the generations, TEK may reveal much about natural variation over time in valued ecosystems, species, and their interrelationships. Many Aboriginal Peoples have also been witness to the individual and combined impacts of natural (e.g., fires) and human (e.g., logging) disturbances and their effects on forest resources. It is these types of information and knowledge that forest managers and planners generally do not have access to, but often need (Chap. 9), in order to make informed decisions and plans over broad temporal and spatial scales.

In practice, however, and to date, TEK has made very little impact on the ways forests and forest resources are managed. Even though the integration of TEK into decision making and planning continues to be one of the key considerations around which cooperative assessment and management processes and regimes are constructed (e.g., see the Mistik Management case study in Chap. 22), Aboriginal Peoples remain generally frustrated by the lack of their meaningful involvement, and the failure to incorporate their knowledge and values in environmental resource management (ERM), planning, and decision making, particularly in forestry. The reasons for this are complex and multi-faceted, but must be explored in some detail, if (1) the "real" contributions of the knowledge of Aboriginal Peoples to SFM are ever going to be realized, and (2) effective policies, institutions, and practices are ever going to be developed to fully and equitably incorporate Aboriginal Peoples into SFM.

The information, knowledge, and procedures used to achieve the objectives of ERM are normally determined by the dominant culture, particularly those "cultured" in the western scientific and capitalist traditions, from which sustained-yield management and even the modern paradigm of SFM emerge. Under this management philosophy and system, cost/benefit determinations and technical information about specific resources (such as tree or animal species abundance, distribution, ages, condition, regeneration rates, natural mortality rates, harvest rates, etc.) are often used to the exclusion of other

---

Box 3.9. On power, decision making, and alternative knowledge systems.

For any particular domain [such as forest management] several knowledge systems exist, some of which, by consensus, come to carry more weight than others, either because they explain the state of the world better for the purposes at hand (efficacy) or because they are associated with a stronger power base (structural superiority). . . . A consequence of the legitimization [sic] of one kind of knowing as authoritative is the devaluation, often the dismissal, of all other kinds of knowing. Those who espouse alternative knowledge systems then tend to be seen as backward, ignorant, or naive trouble-makers. . . . Whatever they might think they have to say about the issues up for negotiation is judged irrelevant, unfounded, and not to the point. . . . The constitution of authoritative knowledge is an ongoing social process that both builds and reflects power relationships within a community of practice. . . .It does this in such a way that all participants come to see the current social order as a natural order, that is, the way things (obviously) are. The devaluation of non-authoritative knowledge systems is a general mechanism by which hierarchical knowledge structures are generated and displayed.

Jordan 1997

Fig. 3.5. Little Red River Cree elder, Malcom Auger, skinning a rabbit.

Photo by Tanja Schramm

types of knowledge. It thus becomes the "authoritative knowledge base" and the main currency used in decision making. This is not to suggest that other types of information and knowledge are not considered; forestry decisions, as noted above, to date have been based largely on economic and political considerations. Indeed, science is often used in such as way as to support economic and political objectives. As Jordan (1997) writes in an essay about an "authoritative knowledge system" far removed from that of forest management, "the power of authoritative knowledge is not that it is correct, but that it counts" (Box 3.9).

However, contrary to the popular belief of many of its practitioners, western science is not as objective, "value-free", or rigorous as they wish it be. It is, in fact, "value-laden", much like TEK. Western science and its derivatives such as forestry, engineering, and other technical fields, are imbued heavily with Judeo-Christian ideology, capitalist philosophy, agrarian tradition, and gender bias whereby, "Man", standing above and apart from nature, holds dominion over it. The types of knowledge that "count" under such management systems are usually only those that can be "counted". However, not only is much of the TEK held by Aboriginal Peoples not concerned with systematically and dispassionately collected quantitative information, it is qualitatively different from that approach to knowledge acquisition. The success or failure of Aboriginal subsistence activities (Fig. 3.5) is known to be more dependent on knowing the factors that influence the behaviours, availability, distributions, movements, interrelationships, etc., of resources rather than on quantitative information about the specific resources themselves (Stevenson 1996; Usher and Wenzel 1987). Even in some of the progressive attempts to develop "scientific" forest management guidelines (e.g., the Canadian Council of Forest Ministers criteria and indicators; CCFM 1995), the issues

Fig. 3.6. The incorporation of TEK into environmental resource management practice: the status quo.

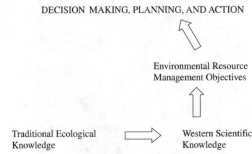

up for discussion and the knowledge sought to address them reflect the biases and practices of the dominant culture, whereby politics, economics, societal values, and science are tightly interwoven into a fabric that is virtually impossible to penetrate from an Aboriginal perspective. Battiste and Youngblood Henderson (2000) in their recent book, *Protecting Indigenous Knowledge and Heritage: A Global Challenge*, refer to this as the "Eurocentric Synthesis".

Under most cooperative management arrangements having a mandate to incorporate TEK, the common practice has been to take whatever aspects of this knowledge managers and natural resource scientists find useful, and then merge them with western scientific knowledge to inform environmental resource management objectives and, ultimately, planning and decision making (Fig. 3.6). There are, however, many problems and pitfalls associated with this common scenario. Aboriginal elders and TEK holders often do not accept (or even contemplate) such modular and incomplete uses of their knowledge.

First, the kinds of TEK considered by resource managers and planners constitute only a small component of the entire knowledge base that Aboriginal Peoples possess. It is granted that many Aboriginal Peoples possess "specific environmental knowledge" about resources that environmental resource managers may find useful for SFM as presently conceived and practised. This knowledge, or more precisely "information", may be actively sought by environmental resource managers and forest planners, either because they do not have ready access to it through their own experiential knowledge base/system, or because it is easily accommodated within their western scientific way of thinking. However, this practice fails to account for the fact that specific environmental knowledge held by Aboriginal Peoples is embedded within a broader knowledge base that includes both ecological and non-ecological knowledge, and traditional and non-traditional (contemporary) knowledge (Fig. 3.7). It serves no one's interests to isolate the specific environmental knowledge of TEK holders to the exclusion of related knowledge sets and its broader socio-cultural context (from which it derives its true meaning and value).

Moreover, many TEK holders find the process of accessing their environmental knowledge troubling. Such knowledge exists primarily in oral form. As such, it has complex social and cultural functions, meanings, and values not easily grasped by outsiders. Non-Aboriginal researchers typically interview TEK holders using local interpreters to translate often difficult to understand concepts and issues into a foreign language (i.e., English in most of North America). The interview is recorded on audio or video tapes and then transcribed, in whole or in part, onto paper and (or) maps. Specific aspects of this information are then singled out for their contribution to existing data sets and established procedures. At each step of the way there is a loss of information, knowledge, and context. Not only is knowledge increasingly divorced from the social and cultural contexts where it more properly resides, its owners are progressively separated from knowledge that they once owned and controlled. If this process were not problematic enough, the transcribed information often becomes the authoritative reference, effectively excluding the people who once owned and controlled the rights to this knowledge from decision making. Not only is there no accountability to the TEK holders in this process, it can also be an infringement of individual intellectual property rights if the proper protocols are not observed.

Through its systemic and ongoing decontextualization and "dumbing down" for the purposes of furthering the modern resource management agenda, the knowledge of Aboriginal Peoples becomes sanitized and made more palatable and useable to the dominant culture. In effect, knowledge becomes information (which can be easily transmitted and is the main currency accepted in ERM), and TEK assumes the role of hand-maiden to western scientific knowledge (see Fig. 3.6). In so doing, alternative ways of seeing, knowing, and relating to the natural world are devalued and dismissed, reflecting the authoritative knowledge system of ERM and strengthening the existing power relationships that support it.

The overall effect of the status quo described above has been that Aboriginal Peoples have effectively contributed little to ERM generally, or to forest management specifically. But this disappointing fact must be considered in the following light: the ecological knowledge of Aboriginal Peoples, even their specific environmental knowledge (Fig. 3.7), does not exist to inform western science or ERM. Indeed, contrary to the claims of many researchers and even some First Nations, TEK may have little to offer conventional resource management. The full contributions of Aboriginal Peoples and their knowledge to managing for sustainability in Canada's boreal forests will not be realized until they can be implemented in their entirety, or at least until their knowledge and management systems are considered equally with those of western scientific knowledge ERM perspectives in forest planning and decision making.

Systemic solutions

Perhaps the best way to conceptualize the cultural differences, and the potential contributions of each in the context of SFM, is to address the issue of what each management system attempts to manage. For those cultured in the conventional European resource management tradition, specific resources and spatially defined areas (forest stands or forest estates, or, more recently, mapped ecosystems or landscapes) are often arbitrarily designated as management units. In order to manage these units, specific forms of quantitative information (soil depths, timber volumes, furbearer population numbers, etc.) are sought for the management unit. The "human factor" is still, for the most part, treated as an outlier that obscures and distorts the "real" data needed for decision making. Institutional processes and procedures for ERM have not kept pace with a growing

Fig. 3.7. A model of the components of Indigenous knowledge and their inter-relationships (from Stevenson 1996).

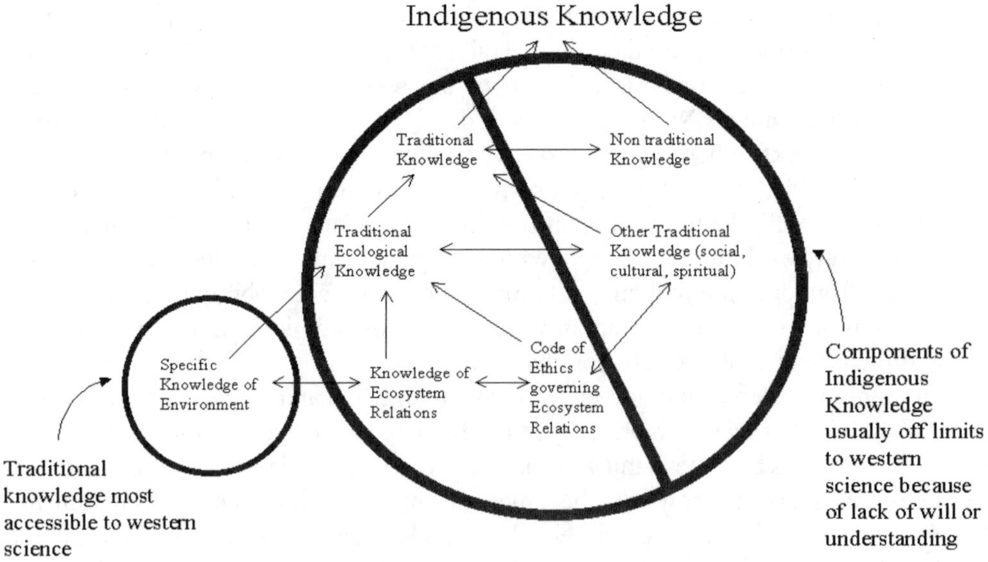

Indigenous Knowledge

consensus that human ecology is often a critical factor in understanding ecological relationships and managing human uses of a region's natural resources.

In many traditional Aboriginal communities the notion that humans are capable of managing species of plants or animals is unfathomable, even sacrilege. "Manage animals?" as one Inuit hunter once told the principal author, "Only God can do that!" While Aboriginal Peoples traditionally manipulated their environments for desired ends, the presumption that humankind could effectively manipulate animal populations is a rejection of Aboriginal Peoples' responsibility to the Creator, an abrogation of the reciprocal relationship with all things in the natural world. To the extent that Aboriginal Peoples manage anything, it is human activities and their relationships with or connections to the natural world that are managed. These they can do something about. Thus relationships, not specific resources or spatially bounded areas or ecosystems, become "the management unit", and the nexus around which Aboriginal Peoples traditionally constructed their knowledge base and world view. This perspective has been conveyed to the principal author and other researchers (e.g., Spak 2001) by elders from many different Aboriginal cultures (Inuit, Chipweyan Dene, Cree, Gwich'in, Shuswap, Nuu-cha-nulth). From the perspective of the dominant European culture, this world view might be rationalized as being a consequence of unscientific thinking, low populations, and primitive technologies. But such dismissal would greatly underestimate its ultimate veracity and potential power in guiding the quest for sustainability.

The differences between the Aboriginal management philosophy and that of conventional ERM are not trivial, and cannot be easily dismissed. Environmental resource managers will often admit that they do not really manage resources, but human activities. However, the types of information required for management decisions almost invariably focus on "resource specifics" to the exclusion of knowledge and understanding about resource "relationships". Although natural resource scientists and managers are becoming conceptually aware of the importance of understanding real-world ecosystem relationships, the snob-appeal of "pure" scientific research still means that most ecological research still excludes the ecological influences of humans or is conducted at scales of little use to resource managers (Baskerville 1997). In other words, humans are still viewed as extraneous to the types of information required to understand how the world works. To the extent that information relevant to human–ecosystem relationships is considered at all, it is usually in the rather simplistic (i.e., short-term, one-way, linear) context of measuring human impact on whatever resources or lands are being managed, whereby humans are considered extraneous and disruptive. The role of human beings in sustaining ecological integrity (ecosystem composition, structure, and functions) rarely enters into equation.

The real contribution of Aboriginal Peoples and their knowledge to sustainable management and economic development in Canada's boreal forests may be in the area of illuminating relationships, not only among non-human species (plants and animals), but between humans and the natural world. Acknowledging the fact that conventional ERM and traditional Aboriginal management systems and philosophies have fundamentally different origins, histories, and objectives provides both a challenge and an opportunity for those charged with implementing SFM. On the one hand, it represents a threat to the authoritative knowledge system in power; change will not come without much hand wringing and sacrifice. On the other hand, it opens the door to a paradigm shift in cooperative management and decision making, whereby the social, economic, cultural, ecological, and other sustainability requirements of Canada's boreal forest-dependent Aboriginal communities can truly be accommodated.

Figure 3.8 presents a model for the meaningful and equitable involvement of Aboriginal Peoples and their knowledge in managing for sustainability in Canada's boreal forest. This model was originally developed by the principal author in the early 1990s, and is now being implemented with SFMN funding by Fikret Berkes among the Cree of Shoal Lake, Manitoba. There are two key features of this model. First, just like an ecological food web, it has more strands (ecological relationships) than nodes (ecological components), making the management of relationships potentially a more powerful and effective tool than the management of components. Secondly, it utilizes the "web of life" model not to inform western science or conventional ERM, but to inform traditional Aboriginal management approaches, ways of knowing, and ways of relating to each other and the natural world. In this model, the focus is not on TEK per se or how it can serve the interests of the Eurocentric synthesis, but on the management approaches and ecological relationships that this knowledge was intended to inform. Here then, is a practical and mutually beneficial application of the Two Row Wampum (Miller 1980):

> *The Whiteman said, "What will happen if any of your people may someday want to have one foot in each of the boats we have placed parallel?" The Onkwe-honweh replied, "If this so happens that my people wish to have their feet in each of the two boats, there will be a high wind and the boats will separate and the person that has his feet in each of the boats shall fall between the boats; and there is not a living soul who will be able to bring him back to the right way given by the Creator, but only one: The Creator Himself."*

One caveat to the implementation of this model is that traditional management systems, and the lifestyles that they sustained, have been so decimated and marginalized through a host of forces originating in the dominant culture (e.g., imposition of state and church on the lives of Canada's Aboriginal Peoples) that, in many contexts, elements of these systems may need to be reconstituted. Ultimately, under this model, cooperative decisions and actions are taken based on information, knowledge, and wisdom from both management streams. In this scenario, neither management system is devalued at the expense of the other, and an opportunity is created to learn from each other and to truly develop practices and policies that balance the sustainability requirements of Canada's Aboriginal Peoples and with those of other users of the boreal forest.

One way the cooperative management model portrayed in Fig. 3.8 could potentially be operationalized is to utilize the "valued ecosystem component" (VEC) concept common to environmental impact assessment literature (e.g., Beanlands and Duinker 1984) as the unit of interest to environmental resource managers. With its emphasis on specific resource information, ERM and its practitioners are ideally positioned to provide information and knowledge regarding the assessment and management of VECs. Alternatively, the contributions of many TEK holders and traditional land-users may be best realized in the context of providing wisdom and knowledge relevant to managing "valued ecosystem relationships" (VERs), including, and perhaps most importantly, valued human–forest resource relationships (Fig. 3.9). Theoretically, VECs could include VERs, and in some environmental assessments Aboriginal values and relationships with key species have been identified as VECs. Nevertheless, management considerations almost always focus on information about the resource to the exclusion of knowledge of its relationships. At a minimum, both should be considered equally in decision making. In reality, however, each should be accorded consideration commensurate with their respective contributions to achieving ecological, social, cultural, and economic sustainability in Canada's forests. Setting management priorities and developing plans focused around sustaining VECs and VERs would be a process of negotiation among the First Nation/Aboriginal community, provincial government, forest products company, and other stakeholders.

For Aboriginal Peoples and First Nations to participate effectively in SFM, they must have the time, opportunity, and resources to undertake research and to develop

Fig 3.8. Model for the integration of Aboriginal Peoples and their knowledge and management systems into environmental management and decision making.

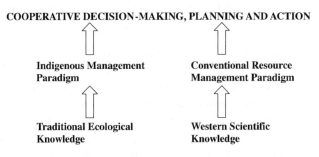

Fig. 3.9. "Components and relationships" in the web of life. Note that for any one component there are obviously many more relationships than depicted.

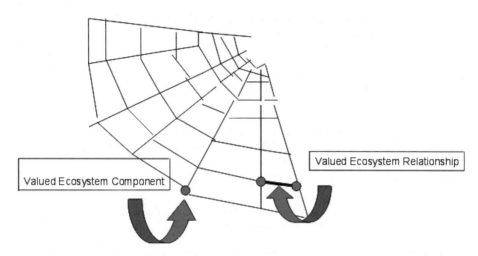

their own data sets and management plans before coming to the table with government and industry. This desire is often ignored by the latter in their rush to open up the boreal forest to economic development. This could be done in cooperation with government and industry at the start, or through the implementation of a research/planning model such as that developed by the SFMN with several Yukon First Nations, and informed by experiences with the SFMN partner, the Little Red River Cree Nation of northern Alberta (Fig. 3.10). In the first stage of this model, the First Nation or Aboriginal community documents, assesses, and prioritizes its needs, uses, and values with respect to the forest and forest resources (Step 1). Based on this work it develops its own land use and forestry objectives, plans, policies, and regulations (Step 2). Following this, the First Nation/Aboriginal community sits down with government and industry to negotiate effective policies, institutions, and strategies to meet its needs, rights, and interests in the context of SFM along with those of other stakeholders (Step 3). The final stage embraces an adaptive management approach, modifying existing policy and practice (in a structured manner; see Chap. 21) where appropriate and as required, to meet the sustainability requirements of First Nations/Aboriginal communities and other stakeholders (Step 4).

The alternative to considering potential solutions to the systemic barriers that continue to thwart meaningful involvement of Aboriginal Peoples and their knowledge in SFM is, arguably, not sustainable and could have significant negative repercussions for Canada. In practice, SFM must become more than conventional ERM, and must include, in a significant way, the perspectives, needs, rights, interests, policies, and practices of First Nations and other users of the forest in planning and decision making (see also Chap. 4). The institutional momentum and investment that drives the status quo must be overcome in order to develop policies and practices that are, from an Aboriginal perspective, ecologically, socially, and economically sustainable. This will take time and long-term commitment on the part of government, industry, and First Nations. Failure to accommodate the rights and interests of Aboriginal Peoples would scuttle any

Fig. 3.10. A model to address the needs, rights, and interests of Aboriginal Peoples in sustainable forest management.

1. Document, assess and prioritize needs, values, and uses of forests and forest resources

2. Develop First Nations' land use and forest plans, policies and regulations

3. Integrate with government policies and regulations through negotiation and development of new institutions

4. Monitor and assess effectiveness of management actions

movement towards sustainability — they are not about to go away, as their population, and need for economic and social stability, grows exponentially with each generation.

Yet, there are things that all of us can do right now to set the wheels in motion. Most of all, we can all be more critical of what we believe to be "true", and cognizant of the implications of maintaining the power relationships and institutional structures that facilitate timber-based approaches and the status quo. Environmental resource managers, for example, can be more critical of the assumptions and culturally rooted biases that orient and guide their thinking. They can also be more open to alternative ways of knowing and relating to the natural world. In turn, they must be more aware of the multiplicity of effects caused by the imposition of conventional ERM on Aboriginal Peoples who relate to the natural world in different ways.

Researchers and scientists who work with Aboriginal Peoples and TEK holders must be careful not to be co-opted into and influenced by the conventional ERM paradigm to the extent that the traditional management and knowledge systems of Aboriginal Peoples are muted, or not recognized at all. After all, the Aboriginal systems have a long and proven track record of sustainability. Granted, interactions with the dominant culture have upset and undermined these systems. Nevertheless, the facts remain that modern European models of resource management are the "new kid on the block", and that the management systems of Aboriginal Peoples may have much to offer SFM.

Researchers, because of their own cultural backgrounds and assumptions, frequently and unwittingly become advocates of (and agents for) the authoritative knowledge systems and power structures that dismiss Aboriginal management and knowledge systems. While scholars may, through research and education, assist First Nations to regain a rightful voice in the management of human activities on their traditional lands, it is the Aboriginal Peoples themselves who must find a way of reconciling existing ERM theory and practice with their own traditional management and knowledge systems.

Finally, Aboriginal Peoples who have been forced into "the ERM game" for marginal gains must weigh the advantages of doing so against rebuilding and re-instituting their own systems of management and knowledge in the context of asserting their constitutionally protected Aboriginal and treaty rights. Many Aboriginal Peoples who participate in cooperative management regimes are seen by traditionalists as having been coerced into accepting the reductionist language, concepts, procedures, and (most importantly) power relationships of ERM in exchange for an increase in management authority or participation in economic development. To the extent that these perceptions may be true, only the Aboriginal Peoples and governments involved can determine the relative cost-effectiveness of their actions. In order to do this, Aboriginal Peoples may wish to take the initiative to be just as critical of their own roles as those of environmental resource managers, scientists, and other stakeholders in supporting existing approaches, policies, and relationships. It is only through the critical examination of all our roles that Canada will develop policies, practices, and institutions that truly embrace SFM through the accommodation of the diverse and pressing needs, rights, and interests of its boreal forest-dependent First Nations and Aboriginal Peoples.

Allies and adversaries: cooperative management as an interim solution

The Little Red River Cree and Tallcree First Nations

The Little Red River Cree and Tallcree First Nations (LRRC/TC), are mentor partners in the SFMN. They occupy a large portion of the lower Peace River watershed in northwestern Alberta (Fig. 3.11; Box 3.10). Members of both bands signed Treaty 8 in 1899 on assurances that the treaty was not a land surrender, but an agreement to share lands and resources with settlers in a peaceful manner in exchange for a number of Crown commitments (Box 3.11). Historic accounts of the period provide unequivocal support for the Little Red River Cree Nation's interpretation of treaty. Over the last 100 years, the LRRC/TC have tried, with little success, to persuade the Crown and non-Aboriginal

Fig. 3.11. Map of LRRC/TC Special Management Areas and 1926 Reserve Area promised them by the Crown in 1926. The traditional lands of these two First Nations is considerably larger and overlaps with other First Nations in all directions.

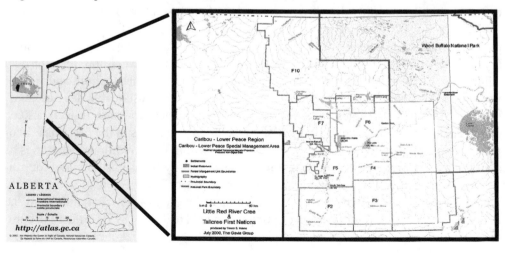

Box 3.10. The Little Red River Cree and Tallcree First Nations: some general facts.

- The population of these two Cree bands at the time of Treaty 8 was signed in 1898 was 168;
- Today, there are currently over 4000 people of these two First Nations, 3200 of which live on seven reserves;
- Unemployment within these communities is between 85% and 95%;
- Over 70% of households rely on social assistance for sustenance and support;
- Most households consume bush food on a regular basis, with over 80% relying on non-timber forest products;
- The general health of the on-reserve population, which constitutes 35% of the population base served by the Northwest Health Service, accounts for 60% of the region's hospital workload;
- Housing is substandard and overcrowded, with an average household occupancy of nine people;
- Two reserve communities have no permanent road access; and
- Social assistance recipients have to spend 98% of their monthly social assistance funds to purchase the food necessary to maintain their health.

Lawn 2001

---

interests that this sharing has to be maintained in a manner which does not undermine their uses of and relationships with the forest.

Some First Nations are willing to wait for validation and settlement of outstanding land claims as a vehicle for restructuring relationships with the Crown, the forest industry, and other users of forest resources. The LRRC/TC are not among them. Rather, they are looking for pragmatic interim processes capable of providing some immediate economic benefit and an effective voice in the dialogue leading to cultural, social, and economic sustainability. Cooperative management is viewed by the LRRC/TC as a promising approach to sustaining Cree cultural values and traditional activities, while re-establishing economic self-reliance prior to the final resolution of treaty relations.

The LRRC/TC have demonstrated a willingness to share their forests and its resources with non-Indigenous peoples, provided that this sharing does not destroy (1) the forest, (2) those things in the forest that have economic, cultural, and spiritual value, or (3) Cree relations with resources that are integral to their culture and way of life. These three caveats reflect the essence of what the term "sustainable development" means to LRRC/TC peoples. The history of this willingness to share reflects a keen understanding of the needs and benefits of cooperation. Today, as in the past, these peoples have undertaken to share renewable resources within their traditional territories as a means of supporting economic and cultural sustainability for themselves and for others who have settled on their lands or use its resources. The pre-confederation treaty relationships established under Two Row Wampum, the Royal Proclamation, and the treaty relationships committed to by the Crown under Treaty 8 (1899), reflect the commitment of these two First Nations peoples to act with honour, patience, and respect.

Unfortunately, most Canadians have failed to understand or recognize the true nature of the treaty relationship. In the face of commitments to Indian peoples to protect their ways of life, governments have incrementally destroyed their livelihoods and have attempted to assimilate them into the dominant culture. The history of treaty relations between LRRC/TC and the Crown establishes the centrality of "righting wrongs" to any final resolution involving SFM (Box 3.12).

For the LRRC/TC, the failure of the Crown to honour its treaty commitments has shaped and defined their approaches to the rights of their peoples and sustainable development in national and international policy arenas. Incrementally, the dialogue has disclosed a need to "restructure the relationships" between Canada and the LRRC/TC Nations. More recently, the federal Crown has come to accept that Canada's Indigenous peoples have inherent rights to take part in framing dialogues about sustainability. This change reflects a growing recognition of the relationship between ecological sustainability and cultural sustainability, whereby Canada's boreal forest-dependent Aboriginal Peoples are perceived to be so much a part of the forest, that, if industrial uses of the boreal forest destroys its ecological integrity, we will have committed cultural genocide (Cassidy and Dale 1988). Yet, despite this recognition, the federal and provincial Crowns appear to lack the political will to act toward Indian peoples in a manner that respects their constitutionally protected rights and upholds their honour and fiduciary obligations. This lack of will reflects a general perception within Canadian politics that we, as a society, ought not be accountable for things which happened in the past, and a desire that Indians should simply be treated like everyone else.

The LRRC/TC leadership understands that meaningful action to restructure treaty relationships between their Nations and Crown governments will probably not take place within their lifetimes. Reviews of court proceedings, negotiation processes, Aboriginal policy guidelines, and other government policy documents disclose the inclination of the Crown and its legal advisors to deny the existence of Aboriginal rights

---

Box 3.11. Crown treaty commitments most relevant to cultural sustainability.

(1) Other than taking up portions of these lands from time to time, the Crown would not interfere with the Indians' way of life, and that they would be as free to live as before;

(2) Settlers would not compete with the Indians in pursuit of their traditional livelihoods of hunting, trapping, and fishing, and the Crown would protect the Indians from such competition;

(3) The Crown would make no laws concerning fish and wildlife, except for those in the best interest of the Indian peoples;

(4) The Crown would provide instrumental support to Indians to sustain their traditional vocations of hunting, trapping, and fishing in the interest of self-reliance;

(5) The Crown would provide ". . . a fair portion of the lands surrendered . . .", to the Indians as settlement and development took place;

(6) If some Indians elected to undertake farming as a livelihood, the Crown would provide them with the instrumental support to do so; and

(7) The Crown would care for the old and the indigent in times of need.

> **Box 3.12. Broken promises: claims of the Little Red River Cree Nation against the Crown for breeches of fiduciary obligation arising from treaty.**
>
> • Treaty commissioners committed "constructive fraud" during negotiations by failing to provide information about resource values and to disclose the Crown's true intentions of forced assimilation;
>
> • The Crown failed to protect the Indian's traditional vocations of hunting, trapping, and fishing through various actions and inactions (e.g., not carrying through with its commitment to create "special reserves", setting aside Crown lands for exclusive use by Metis peoples, creating Wood Buffalo National Park and not recognizing the treaty rights of First Nations peoples within it, and transferring administrative authority over lands and resources to the provincial Crown (1930) in a manner which extinguished the Indian's commercial treaty right to hunt); and
>
> • The Crown neither provided the instrumental support for the Indians' traditional way of life, transition to agricultural livelihoods, nor a reserve land base to foster economic self-reliance.

until they are upheld by Canadian courts — a position no longer tenable in the context of recent court rulings in Haida and Taku (see legal basis section, above). When they can no longer avoid addressing Aboriginal and treaty rights, Crown governments generally impose unilateral policies and narrow technical definitions of these rights in order to limit their responsibility when these rights are infringed. Government intransigence to accommodate the rights and interests of its Aboriginal Peoples may be one of the greatest barriers to SFM in Canada, not only in terms of social/cultural sustainability, but because it threatens ecological sustainability; we argue elsewhere that the "traditional ecological footprint" of Aboriginal Peoples once played a critical role in maintaining boreal forest biodiversity (Stevenson and Webb 2003). But there are signs that things are changing.

Cooperative management: an interim strategy

In 1988, the chiefs and councils of LRRC/TC, in consultation with elders, made a fundamental policy decision that set the direction for the use of "cooperative management" as an "interim measure" to re-establish influence over renewable resource extraction within their traditional territories. This strategic alliance for joint action concerning a range of shared resource development issues was devised because it was recognized that cooperative management approaches could be beneficial, and to structure a strategic approach that would be mutually so. The two First Nations insisted on the establishment of agreements that acknowledge "government-to-government" relations grounded in existing international, federal, and provincial policies, and on explicit definitions of the concepts and terms used in agreements. Such agreements were part of a conscious strategy to develop a cooperative management-planning regime whereby the First Nation governments could establish a high level of influence over renewable resource use activities within their traditional territories. This was further enhanced by the establishment of institutional relations between governments and corporate interests, and a formal research planning process grounded in, and supported by, award of large First Nation provincial forest tenures.

In 1991, the LRRC/TC initiated a policy dialogue about "joint planning and shared decision making" with the federal and provincial ministries responsible for resource management. Subsequently, in 1995 they and the government of Alberta signed a memorandum of understanding (MOU) for cooperative management planning of all renewable natural resources within a 35 000 km^2 "Special Management Area" (SMA), situated to the west of Wood Buffalo National Park (Fig. 3.11). Under this MOU, a management board was mandated to undertake a "landscape assessment" and to develop a "resource management philosophy and goal statement". The MOU identifies the fundamental principles and concepts to be used in achieving the latter, including "sustainable development", "ecological management", and "adaptive management". Within its planning mandate, the Board is required to consider social, cultural, economic, and ecological aspects of the planning landscape, and to use this information to develop MOU-Board recommendations that will:

• establish resource-use priorities which are compatible with sustainable development and with traditional use of the SMA by First Nation peoples;
• establish guidelines and objectives for the management of renewable natural resources within the SMA;
• identify economic, employment, and training initiatives and opportunities for the First Nations within the SMA; and
• identify special initiatives to address First Nation concerns regarding management of wildlife and wildlife habitat within the SMA.

Between 1995 and 1999 the First Nations and two forest products companies (Tolko and Footner Forest Products) reached agreements that have resulted in provincial award of forest tenures within one-half of the SMA and the development of long-term timber supply agreements indexed to industry/First Nation commitments for cooperative management and joint actions to address First Nation employment, training, and business development initiatives, as well as wildlife and wildlife habitat considerations. Tolko Industries Ltd. operates a dimensional sawmill in the town of High Level that requires 1 million m^3 of coniferous timber per year, 20% of which is supplied by the First Nations. Footner Forest Products Ltd. is developing an oriented strandboard (OSB) mill at High Level that will require 1 million m^3 of deciduous timber a year, of which 35% will be provided by First Nation tenures.

As these forest companies, the government of Alberta, and the LRRC/TC are all members of the SFMN, the MOU established a formal research-planning relationship between the Board and the SFMN under the Sustainable Aboriginal Communities research group. In this context, some 15 research projects have been undertaken over the last 7 years with SFMN funding to support cooperative planning processes within the SMA. Current research-planning efforts are focused on a number of issues relevant to the ecological, cultural, and economic sustainability requirements of the two First Nations (Box 3.13). Resource use scenarios will be developed on the basis of this research, and will be used by the management board to establish a balanced set of renewable resource use recommendations as part of the resource management philosophy and goal statement for the SMA. These recommendations will then be submitted by

---

Box 3.13. Current research planning efforts of the LRRC/TC First Nations.

- "Traditional use" study in order to establish the "ecological footprint" of First Nations peoples within the special management area (SMA);
- Initiation of research on bison disease, habitat, and bison recovery potential within the SMA and the Wood Buffalo National Park eco-region;
- Research into plantation approaches to timber yield sustainability;
- Development of a carbon sequestration model for use in the SMA management planning process;
- The modification and use of a multiple/cumulative impact assessment model; and
- The use of research information compiled from previous research projects and resource inventories for development of a series of potential future renewable resource use scenarios within the SMA.

---

the Board for consideration by the First Nations and provincial governments. Upon approval of the "Resource Management Philosophy and Goal Statement" by the government of Alberta, the MOU Board will provide advice and recommendations about resource development and the management mechanisms and processes required to implement the integrated management regime, develop administrative or contractual relationships required for implementation, and amend existing regulations, policies, or laws as needed.

The LRRC/TC are using cooperative management to reframe the institutional regime required for SFM within their traditional territories. This strategic approach ought not be confused with the widely advocated public participation or "community-based" management approaches (Chap. 4). The constitutionally protected rights of Canada's Aboriginal Peoples force us to consider that they are "not just another stakeholder". The emphasis on "government-to-government" relations demanded by the First Nations signals their reliance on the "Two Row Wampum" model, which locates the nexus for policy dialogue between existing institutional frameworks for forest management (i.e., institutionalized relations between Crown governments and industry) and a nascent First Nations institutional framework for governance over matters which are integral to the identity and culture of these First Nations peoples.

The LRRC/TC reject the notions that they should be treated only as members of the public or that "community-based" approaches will allow for a more equitable consideration of their values, perspectives, interests, and objectives. Since 1992, these two First Nations have relied upon traditionally grounded processes to establish the governance mandates which chiefs and councils use to interact with Crown governments and industrial corporations. Within this strategic approach, cooperative management focuses on "external relations", while "internal" processes of dialogue and consensus building (i.e., nascent government processes) are infused with Indigenous values, perspectives, interests, and objectives. First Nations individuals may take part in "community-based" or public participation roundtables and initiatives, but this does not substitute for the role of their governments working directly with other governments.

This intentional separation of "external" and "internal" dialogues is not generally well understood by researchers working within First Nation communities. The

LRRC/TC governments, in providing access to their communities, depend on researchers to assist in their efforts to re-establish influence by "translating" their views, needs, and aspirations into language which is understandable to Crown governance institutions and the resource developers operating under Crown tenure and management policies. This role, which is grounded in an appreciation of "language competence" (i.e., they understand enough of the "internal dialogue" within First Nation communities to provide plausible accounts of these values, perspectives, interests, and objectives to external audiences), allows some researchers to play a boundary-spanning function between two cultures.

The LRRC/TC cooperative management strategy is broad-based, relying upon a large range of institutional relations within a set of loosely coupled policy arenas at various levels. Their entry into these arenas has been facilitated by membership in a large number of organizations, (e.g., North Peace Tribal Council, Treaty 8 First Nations of Alberta, the National Aboriginal Forestry Association, The Assembly of First Nations, the Centre for Indigenous Environmental Resources). Representatives of the LRRC/TC First Nations also participate on provincial and federal–provincial boards and committees focused on issues related to renewable natural resources and traditional use. They also work cooperatively with a number of university-based research institutions, most notably the SFMN and the University of Alberta, to define and inform aspects of sustainability which are of interest to them.

Within these arenas, the LRRC/TC are seeking coalitions with other policy actors in order to address common interests. This approach does not assume that any of these coalitions are grounded in a larger, shared, broad-based set of common perspectives, interests, and objectives. Within any policy arena, a potential "ally" for one issue could well be an "adversary" in relation to another issue within the same or another policy arena. LRRC/TC policy actions across these arenas is consistent and focused on fostering awareness of, and support for, their influence over resource use activities within their traditional territory through the cooperative management planning process. This "allies and adversaries" approach to cultural sustainability has emerged as the central dynamic within the cooperative management planning process.

Cooperative approaches to cultural sustainability

The notion that organizations may be "allies" in one context and "adversaries" in another is illustrated by the ongoing policy dialogue to define and operationalize the concept of "traditional use" in the cooperative planning process. The mandate of the MOU Board calls for the development of resource-use priorities within the SMA that are compatible with sustainable development and traditional use by the First Nations. By way of facilitating this objective the LRRC/TC developed a set of three discussion papers for the Board, government, and industry consideration in August 2000. The second of these, *A Cooperative Management Approach to Cultural Sustainability Within the Caribou – Lower Peace Special Management* Area (LRRCN and TCFN 2000), explored the relationship between Aboriginal and treaty rights and the concept of traditional use, as it has been used by the Canadian courts and in socio-cultural anthropology.

This paper argued that the purpose of Section 35.1 of the Canadian Constitution is to protect integral and defining features of Aboriginal cultures, and to enable Aboriginal Peoples to preserve their relationships with their lands. The paper also outlined the relationship between Aboriginal and treaty rights, traditional use, and the concept of "cultural sustainability" with the latter being defined as (Sharvit 1999):

". . . a development/resource use process that meets the cultural and material needs of the present, without compromising the ability of future generations to retain their cultural identity, social relationships and values, and the management over resource usage which is consistent with the cultural values of a peoples . . ."

Within this context, "traditional use" was defined as (LRRCN and TCFN 2000):

". . . all material and social patterns of occupation, land and resource use, their associated values, customs and traditions, and the economic, social, cultural, spiritual and political relationships within which these use patterns have developed, continued and evolved within aboriginal cultures and societies . . ."

From a Cree perspective, cultural sustainability and ecological sustainability are inseparable. Consequently, the LRRC/TC assert that, as a people whose culture evolved within a northern boreal ecosystem, their cultural survival is dependent upon the ecological sustainability of the boreal landscape. Conversely, the biological diversity and ecological integrity of the boreal forest may be dependent upon their cultural survival (Stevenson and Webb 2003).

To date, the governments of Canada and Alberta have not agreed with this definition, nor have they proposed alternative definitions. There have been discussions about the need for a mutually acceptable definition within the MOU process and within a separate federal/provincial–Aboriginal consultation process. These discussions focused on the identification of a "best practices" approach to the conduct of traditional use studies, but have ended up being adversarial. Provincial government reluctance to discuss the issue is, in large part, a function of their general strategy to delay action on issues related to Aboriginal and treaty rights. When pressed, government representatives have ventured that "traditional use" is somehow related to hunting, trapping, and fishing rights established under the treaty, and affirmed within Section 12 of the 1930 Natural Resource Transfer Agreement (NRTA). But they refuse to acknowledge the interconnectedness of culture and ecology, or that "traditional use" is related to a "bundle" of inherent rights and interests associated with constitutional protection of cultural sustainability under Treaty 8, Section 2 of the NRTA, and Section 35.1 of the Canadian Constitution.

In order to assist the MOU Board, and as a means of developing criteria and indicators for the concept, the LRRC/TC have provided a second operational definition of "traditional use". This definition, which has been adopted and supported by the MOU Board, identifies three criteria for establishing "traditional use" parameters, with measurable indicators for each:

- Cultural resources (cabins, trails, gravesites, spiritual sites, etc.);
- Traditional use commodities (moose meat, berries, medicinal plants, etc.); and

- Traditional use activities (use practices which require a specific ecological setting within a boreal forest landscape).

In proposing this approach to defining and measuring traditional use, the First Nations and SFMN-funded researchers are undertaking to document traditional use criteria and to demonstrate a transparent relationship among traditional use, cultural sustainability, and ecological sustainability. The First Nations use the term "ecological footprint" (in the same sense as Wackernagel and Rees 1996) to advocate that these three criteria, which inform historic and current traditional use parameters, be combined with demographic projections for First Nations populations to understand how much biodiversity must be conserved within a forest landscape in order to meet current and future traditional use requirements. This research may yet demonstrate that Indians, in their traditional use of the forest, may be one of the best "key indicator species" for development of biodiversity parameters (Stevenson and Webb 2003). To the extent that accommodation of these traditional use needs may require provincial establishment of management regimes for conserving biodiversity, it is easy to understand why government and industry leaders could well feel threatened by such approaches to SFM planning. The two First Nations are also keenly aware that this approach may solidify alliances with environmentally conscious stakeholders, who aspire to the establishment of high conservation value forests. At the same time, the LRRC/TC prefer to use this approach to the reconciliation of traditional uses with other uses, in a manner that allows for an acceptable balance between ecological and economic sustainability in order to enhance cultural sustainability.

Cooperative management approaches to First Nations economic self-reliance

Treaty 8 (1899) addressed First Nations economic self-reliance through a number of commitments (see Box 3.11). From the LRRC/TC perspective, all of these promises have been ignored, denied, or defined in such narrow technical terms by the Crown as to be without meaning or substance. Various specific claims and court actions provide the two First Nations with an opportunity to redress historic wrongs and ongoing breeches of fiduciary obligation. However, these problems will not be resolved quickly; at the rate the Crown is responding to these actions, most Cree children will be grandparents before they are settled. Given the appalling conditions under which the LRRC/TC peoples exist, and given their current and projected demographic profiles, few will be better off by waiting for treaty resolution as a means to further the economic sustainability of the communities. Economic solutions must be found *now* to address the extreme levels of welfare dependency and unemployment (and the social problems that such poverty has created) in the five LRRC/TC communities.

The MOU allows for the identification of ". . . economic development, employment and training opportunities and initiatives for the First Nations within the SMA". There is a parallel North Peace Tribal Council capacity-development initiative, which allows for coordination of federal, provincial, and First Nations resources to explore, develop, and implement economic initiatives. The long-term timber supply agreements between the two First Nation forest corporations and the two mill operators are indexed to parallel agreements between the First Nations. These mill operators have also agreed to transfer woodlands management responsibility to the First Nations forest corporations and institute capacity-development processes to facilitate First Nations capture of all woodlands-based business and employment opportunities within the SMA. Woodlands operations business projections (KPMG 1998) estimate that it may be possible for the LRRC/TC to use these agreements to capture some 300 jobs in woodlands operations and management within the SMA.

The LRRC/TC are currently implementing development initiatives under the auspices of these interrelated agreements. In 2000, Little Red River Forestry Ltd. assumed responsibility for woodlands management within the First Nations tenure areas. This corporation is currently undertaking a "pilot demonstration" woodlands management initiative within a portion of the joint-tenured area held by the two mill operators in order to demonstrate management capacity and capture more management opportunities in the future. Likewise, it has developed in-house capacity to undertake data management and computer support services through the establishment of an agency relationship with one of the larger forest management consulting corporations. Strategically, the capture of woodlands management within the SMA will provide the two First Nations with discretionary authority to award all contracts for woodlands operations, and to make such contracts contingent upon inclusion of reasonable training, employment, and subcontract provisions for its members.

Both First Nations are engaged in corporate joint-venture initiatives. The joint-venture agreements with existing woodlands operations corporations have been used to facilitate the training and employment of First Nations members, and will support future sub-contracted, owner-operator initiatives. This "bootstrap" approach to capacity-development, and the capture of woodlands-based jobs and business opportunities, allows the First Nations to use the existing capacity of cooperating woodlands contractors to develop "in-house" capacity among their own peoples. In addition to other cooperative development initiatives (e.g., a joint First Nations – federal and provincial government industry $18 million venture to construct all-weather roads into Garden River and Fox Lake), "sunset" and buy-out provisions within these corporate joint ventures may provide the certainty required to build a sustainable economic future based on the forest.

Barriers to First Nations cultural sustainability

In their relatively successful effort to regain some influence over resource use within their traditional territory, the LRRC/TC have encountered adversaries. Renewal of the MOU has been delayed because of perceptions held by various groups within adjacent communities, especially the Municipal District Council, that these agreements were giving the Indians too much influence over economic matters affecting non-Aboriginal interests. In the "Local Committee" process used by the province to establish resource-use parameters for the creation of a Caribou Mountains Wildland Park, the First Nations found support in the majority of local stakeholders, including industry and government. Support was grounded in committee review of a large body of scientific information, generated and presented in part by the First Nations, about unique ecological features of the proposed park. Although the recommendations of the Local Committee were adopted by the Province, the Municipal District Council and the Fort Vermilion local of the Metis Nations of Alberta voiced strong objections to the establishment of the park.

These perceptions have also emerged in the context of the ongoing dialogue about the need for action to reduce risk of brucellosis transmission from diseased bison (*Bison bison athabascae*) ranging within the SMA and Wood Buffalo National Park. Over the last 10 years, the two First Nations have invested a considerable amount of time, effort, and money to develop a wood bison disease eradication/recovery strategy. The LRRC/TC look to disease-free bison as a major source of nutritional and economic well-being for its peoples. The Wentzel River herd, a small herd of 110 free-ranging bison, has been the focus of research to determine whether this herd is suitable for inclusion in bison recovery initiatives based on genetic variability. These strategies are included as part of the national Wood Bison Recovery Strategy (Gates et al. 2001). However, in Alberta, free-ranging bison outside of identified "protection areas" are not protected by provincial hunting regulations. Two groups, the Alberta Professional Outfitters Association and the Fort Vermilion Local Metis Nation of Alberta, actively oppose proposals for the interim protection and research of these bison on the grounds that it would eliminate their current ability to hunt these animals. They argue that First Nations stewardship of wood bison would give the LRRC/TC rights that are adverse to the interests of other hunters.

Prospects for co-management

Government intransigence and local opposition from non-First Nation groups remain significant barriers to LRRC/TC efforts to resume control over the management and development of resource use on their traditional lands. Indeed, they have been major factors in the ongoing delay in renewing the MOU, which expired in March of 2001. The leadership of these First Nations made a thoughtful and strategic decision a decade ago to utilize "cooperative management" and "the industrial forestry model" (i.e., use of provincial forest tenures) as means of obtaining greater stakes in their traditional lands. But there were prices to be paid. In one study (Treseder 2000), certain community members expressed serious reservations about the impacts of logging practices on their ability to exercise their rights in the forest; the author notes the potential for conflict "between the objectives of employment in the forest industry and protection of First Nation land uses". Ross and Smith (2002) point out these findings underline a tension between the pursuit of economic development by means of industrial timber allocations, and the objective of regaining control over the traditional territory in order to promote compatibility between industrial resource uses and traditional uses.

While this tension is not unlike that found between development and conservation interests in other communities and in society as a whole, Natcher (2001) discovered that many community members felt that participation in commercial forestry would strengthen Cree values by providing the financial means to spend more time in the bush (Fig. 3.12). Ultimately, band members are being asked to tolerate an economic strategy that involves them in the very industry that poses one of the greatest threats to their homeland, and perhaps cultural survival, in exchange for economic benefits (Natcher

Fig. 3.12. Chief Johnsen Seweepagaham, Little Red River Cree Nation, at home in the boreal forest.

Photo by Conroy Seweepagaham

2001). This provides one of the rationales behind a joint LRRC/TC-SFMN initiative to assess the educational, training, cultural, and other needs of community members with respect to forestry (Natcher 2001). Another study that may provide a useful tool in the co-management process is that of Natcher and Hickey (2002). These researchers have developed a series of local-level criteria and indicators that will potentially assist local decision makers to balance the needs and values of community members with respect to the forest.

From the perspective of the LRRC/TC, cooperative management has been effective in relation to the gains made. When it ceases to be effective, or when it appears to be intolerably co-optive, these two First Nations will abandon it, and seek other means to achieve their directives. Given the number of mutual dependencies created under this cooperative management regime, each party has to constantly evaluate the cost/benefits of cooperative action, and determine whether the level of cooptation is tolerable in relation to the benefits that accrue to them as a result of cooperative processes.

In this context, the LRRC/TC strategy shares certain similarities with that of other First Nations engaged in forest management and development under existing provincial forest policies and regulations (Table 3.1). In most of these cases, these First Nations seek to be active partners in the development of forest management plans that, in addition to enabling sustainable timber harvesting, attempt to maintain ecosystem integrity, traditional uses, and cultural values (Ross and Smith 2002). Despite the fact that the adoption of the industrial forestry model is seen by the LRRC/TC as a means to regain control over activities on their traditional lands, provincial standards of forest management (which apply to all forest operations) have not yet been modified to ensure the harmonization of these operations with traditional activities of the First Nation communities (Ross and Smith 2002).

Is the cooperative management approach under the industrial forestry model capable of addressing the social/cultural, economic, and ecological sustainability requirements of the LRRC/TC First Nations? Although it may be too early to determine whether timber harvesting and traditional resource use are indeed compatible and sustainable in this context, Treseder and Krogman (2002) conclude that:

> *Co-management challenges industrial forestry to consider a wider range of viewpoints and values in forest management and incorporate alternative methods for conflict management. By giving a high priority to cultural sustainability criteria, First Nation participants in the co-management process in northern Alberta challenge the forest industry and government to re-think the pace of development, the rates of return required to be profitable, and measures to improve First Nation employment within the industry. In the context of larger societal demands for more sustainable forestry and more influence by local users over resource management decisions, co-management may indeed challenge the industrial model and affect the practice of industrial forestry in the northern boreal forest.*

Summary and conclusions

Canada has made commitments, both nationally and internationally, to increase the roles of its Aboriginal Peoples in the economic development of forest resources and the management of forest activities. However, the inability to implement these commitments locally and regionally has been the function of a number of systemic barriers, stemming principally from the limited participation of Aboriginal Peoples in Canada's policy arenas. At the same time, federally administered economic development programs are out of step with First Nations' abilities to build capacity and take advantage of such opportunities. Moreover, many forest-related programs are framed within the industrial forestry model and, more often than not, serve to undermine cultural sustainability, ecological integrity, and other values that many Aboriginal communities and other stakeholders wish to preserve. In order to address the sustainability requirements of Aboriginal communities and First Nations, these programs need to be restructured and refocused along the lines suggested by the model presented in Fig. 3.10, whereby research has a significant role to play in informing decisions and actions at all four stages.

There are strong and irrefutable legal arguments for greater Aboriginal participation in the management of Canada's forests. Forest products companies will face an increasingly uncertain and precarious future as more and more First Nations seek equity through litigation, international pressure, and civil disobedience (e.g., road blocks) until treaty rights and Aboriginal title issues are dealt with adequately. The creation of new treaties with First Nations may help us in this journey. But as the current impasse in treaty negotiations in British Columbia reminds us, we should not expect much progress on this front in our lifetimes. Amendment of provincial legislation may be a more effective route, but again, there is little political will to change the status quo, even though it may not be ecologically, economically, or socially sustainable. As this lack of sustainability is made more widely known, it may provide the impetus for governments to

Table 3.1. Selected First Nations initiatives in forest management and development.

Agreement	Key features	Perceived benefits	Current challenges
Algonquins of Barriere Lake, Trilateral Agreement with Quebec and Canada	• Lays groundwork for development of integrated resource management plan (IRMP) over 10 000 km^2 • Area subject to 36 existing CAFFs • 3 phases: (1) data collection/analysis; (2) development of IRMP; (3) development of implementation plan	• Goodwill fostered among province, First Nation, and forest companies • Phase 1 complete, Phase 2 near completion • Amount and size of area cut reduced in 7 key management areas to allow traditional use	• Federal government withdrawal of support for IRMP • Blockades, suspension of logging activities, and impending mill closures
Nuu-Chah-Nulth/ Iisaak Forest Resources, Clayoquot Sound, B.C.	• Site of industry, environmental groups and First Nations conflict over old-growth forestry in 1980s and early 1990s • Joint (51%) venture with Weyerhauser • Long-term area-based tenure • Co-management with province	• AAC and stakeholder conflicts reduced significantly • Equal representation of First Nations and non-First Nations members on Advisory Board • Province continues to accept Board's recommendations • Non-timber forest and cultural values predominant in decision making • FSC certified • MOU signed with environmental groups • Internationally recognized conservation achievement	• Value-added commitments unrealized • Iisaak does not generate enough revenue for self-sufficiency
Forestry Provisions of the Nisga'a Final Agreement, B.C.	• First modern treaty in B.C. • Fee simple title to 2000 km^2 • Recent AAC (165 000 m^3) not sustainable • 5-year transition to First Nation control over forestry	• Increased AAC and management authority to First Nation over 5 years • Reduction in AAC in Year 5 (135 000 m^3) • Ability for Nisga'a to make, monitor, and enforce own forestry laws • First Nation capture of additional forest tenures (<150 000 m^3) • Existing licensees to provide contracts to First Nation	• 10 year prohibition against Nisga'a building a mill • Limited ability of Nisga'a to reduce rate of logging over short term, or to work outside industrial model to sustain ecological and cultural values

Table 3.1. (*concluded*).

Agreement	Key features	Perceived benefits	Current challenges
Forestry Provisions of the Bilateral Agreement Between the Crees and the Province of Quebec	• Treaty settled disputes between Cree and province re: forestry, mining, and hydro development • Joint Cree/Quebec body determines appropriate levels of development to balance economic, ecological, and First Nations needs • Adoption of Cree traplines as forest management unit • Institution of Cree tallymen as managers • Cree knowledge and values used in establishment of many new policies and regulations in new forestry regime	• Greater Cree participation in and control over pace of development in Cree territory • Greater ability to balance Cree lifestyles with resource development • Adoption of specific logging practices to protect residual cover (e.g., mosaic cutting, <100 ha cutblocks, no logging on traplines with over 40% cover removed)	• Ability to develop Cree approaches to SFM constrained by adherence to existing provincial forest policy • Only 1% of traplines can be protected as areas/sites of cultural values • Special management standards applied to only 25% of traplines to enhance critical wildlife habitat
NorSask Forest Products Ltd., Meadow Lake Tribal Council, Saskatchewan	• Canada's largest First Nations owned forest company • Joint venture/management agreement with Millar Western (Mistik Management) • Produces both pulp and dimensional lumber • No effluent discharge from pulp mill	• <1% of forest cut annually • Co-management agreements with many Aboriginal communities • Forest management plan based on TEK • Co-management and advisory boards meet monthly with Mistik to discuss size and location of cuts, protection of wildlife, native values, etc. • Mistik employs 500–600 people seasonally, 200 permanently • Forest management plans extend >200 years	• Sustaining a balance (over the long term) of ecological, economic, and social/cultural needs and values • Sustaining "clear-cutting" as the predominant form of harvesting • Developing secondary manufacturing opportunities

change provincial forestry policy and legislation in ways that can accommodate Aboriginal rights and interests in the context of SFM.

We may have much to learn from Canada's Aboriginal Peoples about living sustainably; their knowledge and management systems have stood the test of time. Yet, the full contributions of Aboriginal Peoples and their knowledge and ways of managing their relationships with the forest are rarely realized in SFM. The "Two Row Wampum" provides a model, advocated by First Nations since the coming of Europeans, to incorporate Aboriginal Peoples and their systems of knowledge and management into SFM.

We honour the spirit and intent of our forefathers when we use the "Two Row Wampum" to guide us in our dealings with Canada's First Nations and Aboriginal Peoples.

The LRRC/TC gazed into the future and did not like what they saw. As a result, they have been forced to make some difficult decisions. They have chosen to work within the industrial forestry model, have been awarded tenures by the Province of Alberta and are seeking to meet their socio-cultural, economic, and ecological sustainability require-ments through the development and implementation of cooperative management plan-ning and joint venture agreements with industry. There are lessons to be learned for forest products companies, provincial governments, and other First Nations and forest-dependant communities in the LRRC/TC experience. However, time will only tell whether the systemic barriers they now face will be overcome, and whether the path they have chosen is the right one.

Meanwhile, a small number of other First Nations across Canada are exploring alter-native tenure, management, and economic development arrangements to achieve eco-nomic and social stability based upon sustainable use of the forest (see Ross and Smith 2002). For example, Aboriginal communities in northwestern Saskatchewan have developed their own version of co-management, as described in the Mistik Management case study in Chap. 22. These and other efforts at SFM through Aboriginal-based co-management (Table 3.1) may have lessons for other First Nations, other forest-dependent communities, and their industrial and government partners, allies, and adver-saries. In the interests of creating certainty for everyone, we (as researchers, civil servants, industry representatives, and environmentally and socially concerned citizens of Canada) must all take some responsibility and action to ensure that Aboriginal Peoples meaningfully participate in the economic development and management deci-sions undertaken with respect to their forests and traditional lands. The social, eco-nomic, and cultural costs of not doing so are unacceptable.

In regard to translating what we know into action, as noted above, we can all begin to be more critical of the cultural biases and assumptions, and their consequences, that we bring to the table when we deal with First Nations in the context of SFM. There are other things that we can be doing to facilitate the transition to social, cultural, economic, and ecological sustainability for Canada's boreal forest-dependent Aboriginal commu-nities, and (inseparably) economic and political certainty for Canada's forest products industry (Box 3.14). The stability and certainty we all seek for ourselves and our chil-dren cannot be accomplished in isolation from each other. Informed by research, we must develop the courage to paddle, not in "one boat", but together, side-by-side, into uncharted waters. We must explore ways to adapt the industrial forest model so that it can evolve into SFM and stand tall on the pillars of ecological, economic, social, and technological sustainability. Can existing practices be modified to meet the needs, rights and interests of Aboriginal Peoples, and other stakeholders? If so, how, and at what costs? What is the price of reform in provincial forestry policy and legislation? Alter-natively, what are the economic, political, ecological, social, cultural, and other costs of no reform?

At the same time, we need to develop alternatives to industrial forestry that are more compatible with the maintenance of boreal forest biodiversity, traditional uses and

---

Box 3.14. Some actions we can undertake to create certainty and sustainability in the boreal forest (FG, federal government; PG, provincial government; IN, forest industry; FN, First Nations/Aboriginal communities; AC, Academics and researchers).

1. Act as a fiduciary and uphold the honour of the Crown at all times (FG, PG).
2. Employ the "Two Row Wampum" model, or a "Nation-to-Nation" approach when dealing with First Nation/Aboriginal communities (FG, PG).
3. Inform oneself of Aboriginal and treaty rights (FG, PG, IN, FN, AC).
4. Work with government to accommodate Aboriginal and treaty rights and interests (IN, FN, AC).
5. Continue to exercise constitutionally protected Aboriginal and treaty rights, and traditional relationships with the forest (FN).
6. Lobby governments to reform existing forest policies and tenure regimes to create economic, political, and jurisdictional certainty in the forest (FN, IN).
7. Reform forestry and natural resource policies, tenure regimes, and regulations (especially annual allowable cut allocations, area-based tenure allocations, and mill appurtenancy requirements) to accommodate Aboriginal and treaty rights (PG).
8. Be more critical of own cultural biases/assumptions, aware of the effects of relying on the knowledge and management systems of the dominant culture to the exclusion of other systems, and reflect on one's own involvement in maintaining existing power relationships (FG, PG, IN, FN, AC).
9. Undertake research to inform the roles and contributions of Aboriginal Peoples, their knowledge, values, and management systems to decision making (FN, AC, FG, PG, IN).
10. Redesign Aboriginal economic development programs to accommodate the need for First Nations to document and prioritize forest uses and values, and to develop land-use plans based on the above research (FG, FN).
11. Implement commitments made to First Nations/Aboriginal groups in the National Forest Strategy and through the Canadian Council of Forest Ministers criteria and indicators (FG, PG).
12. Work with First Nations and Aboriginal groups to become certified (IN).
13. Do not wait for court decisions before initiating consultations with First Nations/Aboriginal communities (PG, FG, IN).
14. Develop agreements with First Nations/Aboriginal communities regarding resource access/use, the sharing of economic benefits and management authority/responsibility (PG, IN).
15. Work with First Nations/Aboriginal communities to diversify product lines and incorporate value-added wood manufacturing and non-timber forest product opportunities (PG, IN, AC).
16. Explore intensive forest management, triad zoning approaches, and other alternatives to existing industrial forestry practices to alleviate pressure on high conservation value forests (IN, FN, AC).

---

values, and the economic, social, and cultural sustainability of forest-dependent Aboriginal communities. The development of non-timber forest products (NTFPs) and markets may offer Aboriginal communities and First Nations opportunities to create businesses that potentially are not only more ecologically sustainable than existing industrial forestry practices, but more economically stable and socially viable. Similarly, the development of value-added wood products (VAWPs) has the potential to rebuild local Aboriginal economies and communities, while reducing the need to extract large volumes of timber to meet AAC allocations for dimensional lumber and pulp mills. The expansion of local economies based on NTFPs and VAWPs should not be viewed as a challenge to industrial forestry nor the traditional subsistence economy. On the one hand, the development of NTFPs and VAWPs present an opportunity that may contribute to the long-term sustainability of Canada's forest products industry and forests through product diversification and a reduction in the need for large, potentially unsustainable AACs. On the other hand, the development of such products and markets has the potential to reduce ecological impacts on the forest, while providing income to sustain traditional uses and their associated cultural values. However, economic, social, and ecological research and analyses relevant to the development of NTFPs and VAWPs, and the capture of markets that would sustain their production, is woefully lacking and needs to be urgently undertaken.

More efficient zoning of forest management, along the lines suggested by the Senate of Canada (SSCAF 1999), also has the potential to reconcile competing rights and interests in the name of sustainability. Viewed from an Aboriginal perspective, such an approach might partition Canada's boreal forest into intensively managed areas for timber/fibre production (e.g., tree plantations), extensively managed areas (balancing multiple uses with the conservation of biodiversity), and protected areas (where uses do not detract from ecological integrity, e.g., traditional hunting and trapping). It has been estimated that intensive management practices could result in an eight-fold increase in timber production, freeing up more existing forest for ecological protection and other uses (SSCAF 1999). Many First Nations in Canada's western provinces oppose the conversion of prime boreal forest habitat to agriculture, which is marginal at best (even with heavy government subsidies), on their traditional lands. The LRRC/TC view the conversion of these marginal agricultural lands on their traditional territories to tree plantations as a vehicle to protect existing forests, as well as their rights and interests in the forest, and to create employment for their peoples. However, processes, models, and frameworks that would be required to implement such "triad" zoning approaches in the context of First Nations interests need to be developed, tested, and demonstrated.

The expansion of industrial forestry into the central and northern reaches of Canada's boreal forest provides an opportunity for provincial governments, First Nations, and forest products companies to work together to design institutional arrangements and frameworks that are ecologically, socially, and economically sustainable. Imposition of existing regimes and regulations (which are geared towards the industrial forest model) on First Nations, in the absence of the recognition and accommodation of Aboriginal and treaty rights, is not sustainable. Provincial government forestry policy and legislation, in consultation with the First Nation governments, needs to be amended to address Aboriginal rights and interests in order to create the political certainty and economic stability required to sustain Canada's forest products industry and the people who live in the forest. In the interest of all, it is time for provincial governments to inform themselves of Aboriginal and treaty rights, to stop dealing with Aboriginal communities as "just another stakeholder", and to interact with Aboriginal governments on a nation-to-nation basis.

In an effort to enhance their own long-term viability, and not content on waiting for government to "step up to the plate", some forest companies, including a few SFMN industry partners, have taken significant steps towards creating sustainable relationships with Aboriginal Peoples and governments. These companies have, with their Aboriginal partners, developed progressive policies, joint-venture agreements, and co-management arrangements whereby Aboriginal governments and First Nations have assumed greater economic and management roles in forestry. In the context of forest products certification (see Chaps. 7 and 21), and the public's growing awareness of environmental and social issues, these companies are distancing themselves from the competition. In the future, it is conceivable that fibre or wood products that do not meet ecological, social, and cultural criteria established by forest product certifiers will not be marketable, and thus economically moribund. But industry must go further, and entertain and test ideas and solutions "outside the box", i.e., not grounded in the current industrial forestry model. Successful industry–Aboriginal collaborations may also lead the way for government to develop more progressive policies and legislation that will create the political stability, economic certainty, social equity, and ecological integrity we all seek.

Finally, in order to effect change in the status quo, Aboriginal and First Nation governments must not remain on the sidelines. While some First Nations in the boreal forest have developed their own specific consultation guidelines, systems of environmental governance, and land-use plans, most have not. Informed by research, these First Nations need to take stock of their forest resources and uses, and to develop use priorities based on these and the futures they envision for their peoples (i.e., Step 1 in the research model presented in Fig. 3.10). By way of facilitating their planning efforts (Step 2), Aboriginal governments could undertake studies aimed at assessing the economic, social, cultural, and other resource values in their territories, and the efficacy of developing non-timber forest and value-added wood products and markets, in addition to industrial timber-based approaches. This research, combined with studies relevant to determining the effectiveness of alternative policies, practices, and institutions (e.g., zoning approaches) will inform the development of joint-venture and cooperative management arrangements with government and industry that have the potential to accommodate everyone's rights and interests (Step 3). It is only through focused effort on everyone's part, in concert with one another and informed by decisions based on research, that we will create a future that is ecologically, economically, and socially sustainable. Non-action is not a sustainable option.

Acknowledgements

This chapter has benefited by the comments of an anonymous reviewer and Drs. Vic Adamowicz, Phil Burton, and Cliff Hickey. We would like to thank the Little Red River Cree and Tallcree peoples and all other Aboriginal Peoples with whom we have worked for assisting in the development of ideas and thoughts contained in this paper.

References

Baskerville, G.L. 1997. Advocacy, science, policy, and life in the real world. Conserv. Ecol. [online], **1**(1): 9. Available at http://www.consecol.org/vol1/iss1/art9 [viewed 4 June 2002].

Battiste, M., and Youngblood Henderson, J. 2000. Protecting indigenous knowledge and heritage: a global challenge. Purich Publishing Ltd., Saskatoon, Saskatchewan. 324 p.

Beanlands, G.E., and Duinker, P.N. 1984. An ecological framework for environmental impact assessment. J. Environ. Manage. **18**: 267–277.

Canada. 1997. Gathering strength: Canada's Aboriginal Action Plan. Minister of Indian Affairs and Northern Development, Ottawa, Ontario. 36 p.

Canadian Charter of Rights and Freedoms. Schedule B to the Canada Act, 1982 (U.K.) 1982, c.11.

Cassidy, F., and Dale, N. 1988. After native claims? The implications of comprehensive claims settlements for natural resources in British Columbia. Institute for Research on Public Policy, Halifax, Nova Scotia. 230 p.

(CCFM) Canadian Council of Forest Ministers. 1995. Defining sustainable forest management: a Canadian approach to criteria and indicators. Canadian Forest Service, Ottawa, Ontario. Available at http://www.nrcan.gc.ca/cfs/proj/ppiab/ci/framain_e.html [viewed 25 March 2002].

(CCFM) Canadian Council of Forest Ministers. 1998. National forest strategy 1998–2003: sustainable forests, a Canadian commitment. Canadian Council of Forest Ministers, Canadian Forest Service, Ottawa, Ontario. 47 p.

(CCFM) Canadian Council of Forest Ministers. 2003. CCFM C&I Review: Draft revised indicators for sustainable forest management. Canadian Council of Forest Ministers Criteria and Indicators Secretariat, May 2, 2002. Canadian Forest Service, Ottawa, Ontario. 20 p.

Delgamuukw v. Regina. 1997. Reasons for judgement. 3 SCR 1010. Supreme Court of Canada, Ottawa, Ontario. Available at http://lexum.umontreal.ca/doc/csc-scc/en/pub/1997/vol3/html/1997scr3_1010.html [viewed 3 July 2003].

Gates, C.C., Stephenson, R.O., Reynolds, H.W., van Zyll de Jong, C.G., Schwantje, H., Hoefs, M., Nishi, J., Cool, N., Chisholm, J., James, A., and Koonz, B. 2001. National recovery plan for the wood bison (*Bison bison athabascae*). National Recovery Plan No. 21. Canadian Wildlife Service, Environment Canada, Ottawa, Ontario. 50 p.

Haida v. British Columbia and Weyerhauser. 2002. BCAA 462. Court of Appeal for British Columbia, Vancouver, British Columbia. Available at http://www.courts.gov.bc.ca/jdb-txt/ca/02/04/2002bcca0462.htm [viewed 3 July 2003].

Halfway River v. British Columbia (Minister of Forests). 1997. British Columbia Supreme Court, Victoria, British Columbia. Can. Native Law Rev. **45**(4).

(IGC) Institute for Global Communications. 1996. "Rio Cluster" of U.N. proceedings. Available at http://www.igc.org/habitat/un-proc/ [viewed 10 June 2002].

(ILO) International Labour Organization. 1989. Convention on Indigenous and Tribal Peoples in Independent Countries. No. 169, 28 I.L.M. 1382.

Jordan, B. 1997. Authoritative knowledge and its construction. *In* Childbirth and authoritative knowledge: cross-cultural perspectives. *Edited by* R. Davis-Floyd and C. Sargent. University of California Press, Berkeley, California. pp. 55–79.

KPMG. 1998. Forest business incubator centre: business plan. Project report prepared for the Little Red River Cree Nation and Tallcree First Nation, John D'Or Prairie and Fort Vermilion, Alberta. 32 p.

Lawn, J. 2001. Food costs in Treaty 8 communities of northern Alberta: Northern River Basins Food Consumption Study. Prepared for the Alberta Treaty 8 Health Authority. 48 p.

LRRCN and TCFN (Little Red River Cree Nation and Tallcree First Nation) 2000. A cooperative management approach to cultural sustainability within the Caribou – Lower Peace Special Management Area.

August, 2000. Little Red River Cree Nation and Tallcree First Nation, John D'Or Prairie and Fort Vermilion, Alberta. 10 p.

McLachlin, J. 1999. Dissenting opinion in Regina v. Marshall. Supreme Court of Canada, Ottawa, Ontario.

Miller, H. 1980. Record of the Two Row Wampum Belt. Turtle Quarterly (Winter issue). Native American Center for the Living Arts, Niagara Falls, New York. Excerpts available at http://hometown.aol. com/mitetben/miketben.htm [viewed 3 July 2003].

Miskisew v. Regina (Minister of Canadian Heritage). 2001. FCT 1426, No. 1877. Federal Court of Canada, Ottawa, Ontario.

NAFA (National Aboriginal Forestry Association). 1995. Aboriginal forest land management guidelines: a community approach. National Aboriginal Forestry Association, Ottawa, Ontario. 150 p. plus appendices.

NAFA (National Aboriginal Forestry Association). 1997. Value-added forestry and Aboriginal communities: the perfect fit. National Aboriginal Forestry Association, Ottawa, Ontario. 150 p.

NAFA (National Aboriginal Forestry Association). 2003. Letter from Mr. Steve Ginnish, Chair of the National Aboriginal Forestry Association, to Honourable Herb Dhaliwal, Minister of Natural Resources, Government of Canada, 23 July 2003.

Natcher, D. 2001. Building capacity through forestry education: community and industry assessment (Phase One). Report prepared for the Sustainable Forest Management Network, Edmonton, Alberta.

Natcher, D., and Hickey, C. 2002. Putting the community back into community-based resource management: a criteria and indicators approach to sustainability. Human Org. **61**: 350–363.

Natural Resources Transfer Agreement, Saskatchewan. 1930. Government of Canada, Ottawa, Ontario, and Government of Saskatchewan, Regina, Saskatchewan.

RCAP (Royal Commission on Aboriginal Peoples) 1996. Report of the Royal Commission on Aboriginal Peoples, 5 volumes. Canada Communications Group, Ottawa, Ontario.

Regina v. Van der Peet. 1996. Reasons for judgement. 2 SCR 507. Supreme Court of Canada, Ottawa, Ontario. Can. Native Law Rev. **177**.

Ross, M., and Smith, P. 2002. Accommodation of Aboriginal rights: the need for an Aboriginal forest tenure (synthesis report). Sustainable Forest Management Network, Edmonton, Alberta. Available at http://sfm-1.biology.ualberta.ca/english/pubs/PDF/other_6.pdf [viewed 27 September 2002]. 49 p.

Royal Proclamation of 1763. London, England. R.S.C. 1985, Appendix II, no.1. Available at www.bloorstreet.com/2000block/rp1763.htm [viewed 27 June 2003].

Sharvit, C.Y. 1999. A sustainable co-existence? Aboriginal rights and resource management in Canada. Masters thesis, Faculty of Law, University of Calgary, Calgary, Alberta. 258 p.

Spak, S. 2001. Canadian resource co-management boards and their relationship to indigenous knowledge: two case studies. Doctoral thesis, Department of Anthropology, University of Toronto, Toronto, Ontario. 235 p.

(SSCAF) Standing Senate Committee on Agriculture and Forestry. 1999. Competing realities: the boreal forest at risk. Report of the Sub-Committee on Boreal Forest of the Standing Senate Committee on Agriculture and Forestry. Parliament of Canada, Ottawa, Ontario. Available at http://www.parl. gc.ca/36/1/parlbus/ commbus/senate/com-e/BORE-E/rep-e/rep09jun99-e.htm [viewed 25 May 2002]. 96 p.

Statistics Canada 1996. 1996 Census Aboriginal data. Statistics Canada, Ottawa, Ontario. Available at www.statcan.ca/Daily/English.

Stevenson, M.G. 1996. Indigenous knowledge in environmental assessment. Arctic, **49**(3): 276–291.

Stevenson, M.G., and Webb, J. 2003. First Nations: measures and monitors of biodiversity. Paper presented at the BorNet International Conference, 26–28 May 2002, Uppsala, Sweden. Ecol. Bull. In press.

Taku River Tlingit First Nation v. Ringstad et al. 2002. BCCA 59. Court of Appeal for British Columbia, Vancouver, British Columbia. Available at http://www.courts.gov.bc.ca/jdb-txt/ca/02/00/2002bcca 0059.htm [viewed 3 July 2003].

Treaty 6. 1876. Between Her Majesty the Queen and the Plains Wood Cree Indians and other Tribes of Indians at Fort Carleton, Fort Pitt and Battle River with Adhesions. Cat. No. R33-0664, 1962. Queen's Printer, Ottawa, Ontario.

Treaty 8. 1899. Made 21 June 1899. Reprinted from the 1899 edition by ©Roger Duhamel F.R.S.C. Queen's Printer and Controller of Stationery, Ottawa, Ontario. Cat. No. Ci-72-0866. IAND Pub. No. QS-0576-000-EE-A-16.

Treseder, L.C. 2000. Forest co-management in northern Alberta: conflict, sustainability and power. Masters thesis, Department of Renewable Resources, University of Alberta, Edmonton, Alberta. 159 p.

Treseder, L.C., and Krogman, N.Y. 2002. Forest co-management in northern Alberta: does it challenge the industrial model? J. Environ. Sustain. Dev. **1**: 210–223.

United Nations. 1967. International Covenant on Civil and Political Rights. U.N. Doc. A/6316 at 49.

Usher, P.J., and Wenzel, G. 1987. Native harvest surveys and statistics: a critique of their construction and use. Arctic, **40**: 145–160.

Wackernagel, M., and Rees, W. 1996. Our ecological footprint: reducing human impact on the earth. New Society Publishers, Gabriola Island, British Columbia. 160 p.

Woodward, J. 1999. No innocent bystanders: consequences of the failure to consult. Paper presented at "Law '99: Aboriginal Law in Canada '99" Conference, 18–20 November 1999, Vancouver, British Columbia.

Chapter 4

Public involvement in sustainable boreal forest management

Fiona Hamersley Chambers and Tom Beckley

> *"Public involvement is based on two-way communication (which includes listening) . . . In contrast, the main goal of public relations is to sell the preferred option . . ."*
>
> Tom Beierle[2]

Public involvement, sustainable development, and sustainable forest management

The purpose of this chapter is to explore the main issues that resource managers must consider when planning and undertaking public involvement. The demand for effective public involvement (PI) in the management of Canadian Crown lands is being expressed in a number of ways: lobbying for change of laws governing PI requirements, direct action against forest products companies and their contractors, and public opinion research. We discuss these features of increased demand and how this is being met (or not, as the case may be) through recent legal and policy incentives. This is followed by a discussion of the major benefits of and challenges to PI, such as the lack of consensus on what constitutes effective participation. Finally, we investigate the question of how best to achieve effective PI in forest management, and propose conclusions and recommendations to further this key objective of sustainable forest management (SFM).

Over the past three decades, a consensus has grown among managers, local resource users, the general public, and researchers that the public should be more involved in Canadian forest management. Historically, decisions regarding forest lands in Canada have been carried out by professional forest managers employed by timber tenure hold-

F. Hamersley Chambers.[1] School of Environmental Studies, University of Victoria, P.O. Box 1700, STN CSC, Victoria, British Columbia V8W 2Y2, Canada.
T. Beckley. Faculty of Forestry and Environmental Management, University of New Brunswick, Fredericton, New Brunswick, Canada.
[1]Author for correspondence. Telephone: 250-595-6465. e-mail: fiona.chambers@pobox.com

Correct citation: Hamersley Chambers, F., and Beckley, T. 2003. Public involvement in sustainable boreal forest management. Chapter 4. *In* Towards Sustainable Management of the Boreal Forest. *Edited by* P.J. Burton, C. Messier, D.W. Smith, and W.L. Adamowicz. NRC Research Press, Ottawa, Ontario, Canada. pp. 113–154.

[2]Tom Beierle, Resources for the Future, Washington, D.C., personal communication, 2003.

ers and by government staff, with limited opportunities for meaningful public input. In Canada and the United States, management of public forest lands has been based on a managerial model in which government administrators were entrusted to identify and pursue the common good of maximizing social welfare (Beierle and Cayford 2002, p. 2). Since the 1970s, however, this managerial model has given way to a pluralist approach which attempts to strike a balance between expertise and accountability, setting government administrators as arbiters among different interests within the public rather than a source of objective decision making (Beierle and Cayford 2002, p. 3). As in other areas of resource development, citizens everywhere are now playing an increasingly meaningful and direct role in forest planning and decision making (Duinker 1998).

The last two to four decades have seen gradual moves towards managing our public forests for a wider range of values rather than just timber production (Chap. 1). It turns out, however, that true multiple use forestry is more difficult to practice than originally expected. For example, what should be the tradeoff between different forest products such as timber and water? In addition to this, forests are now viewed as having intrinsic and spiritual value, which many people feel needs to be protected also (Devall and Sessions 1985). It is clear that in order to adequately understand and manage for these complex (and often competing) values, it is necessary to involve forest users and the general public in the forest management process. While citizens have historically been content so long as the public forest provided jobs (locally), fibre (corporately and perhaps nationally), and opportunities for recreation (regionally), today a broader spectrum of values is demanded of the forest, and forest managers are expected to deliver this full spectrum.

In this new vision of PI, government and industry are increasingly seen as the stewards and managers of these resources for the public owners, with citizen participation an accepted (and indeed necessary) element of forest decision making. Since the Canadian public owns the vast majority (94%) of the country's forests, it therefore has responsibilities and rights to participate in determining the future of this resource (CIF/IFC 1998). In theory, since the land is publicly owned, the demand for greater public involvement should be respected in a democratic society. Indeed, a fundamental challenge for governing bodies is "reconciling the need for expertise in managing administrative programs with the transparency and participation demanded by a democratic system" (Beierle and Cayford 2002, p. 3). This also assumes, however, that public involvement mechanisms are in place to permit citizens to question these authorities and to have access to specific information. So an equally important aspect of public involvement is the reporting back to the public in a relevant manner on the status of its forests. This has been the purpose behind annual reports on the "*State of Canada's Forests*" prepared by the Canadian Forest Service since 1995 (e.g., CFS 2001). Examples of reasonably exhaustive public consultations regarding issues around air protection, defining protected areas, and what conservation techniques to apply include the New Brunswick Protected Area Strategy, British Columbia's Commission on Resources and Environment and its Protected Areas Strategy, Ontario's Lands For Life program, and Alberta's Forest Conservation Strategy. The conviction that the public has a right to be involved in forest management is also supported in a number of international initiatives such as the Forest Stewardship Council (FSC n.d.) certification program, as well as in provincial and territorial forestry legislation.

This same period has also seen a myriad of attempts to define, and move towards, sustainable development and (by extension), SFM. Examples of Canadian initiatives in response to this trend include the Canadian Council of Forest Ministers' *National Forest Strategy* (see Chap. 2), the New Brunswick Vision Document (see http://www.gnb.ca/0078/vision.htm), and the British Columbia Commission on Resources and Environment (see http://www.luco.gov.bc.ca). Significant international initiatives have also been undertaken, such as the 1992 Rio Earth Summit where the issue of forest sustainability was heatedly discussed, the United Nations-sponsored Intergovernmental Panel on Forests, the Helsinki process in Europe, and the Montreal process for many of the boreal and temperate forest countries such as Canada (Chap. 2). Indeed, even a cursory investigation demonstrates the inexorable link between public involvement in forest management and achieving sustainable development (Box 4.1).

The Brundtland Report (WCED 1987) defined sustainable development as development that meets the needs of the present without compromising the ability of future generations to meet their own needs. In the context of forestry, one interpretation of this is the understanding that forest managers will consult with the public to ensure that forestry activities sustain both today's forests and their users in all their diversity, as well as those in the future (Box 4.2). Indeed, one of the major reasons for the failure of conventional top-down management to achieve SFM, as outlined historically in Chap. 1, is that it has not meaningfully included the public in the forest planning and management process. Achieving SFM implies moving beyond the conventional approach of

---

Box 4.1. The role of public involvement in sustainable forest management:

- To ensure that managers consider the widest range of the public's values in forest decision making, including economic, social, and ecological values through a process that is proactive, open, and fair;
- To assist managers in determining the socially desirable management directions for which they should be striving;
- To set the bounds for choices on forest management practices, strategies, and policies;
- To bring unique local knowledge and insights into decision making (i.e., traditional ecological knowledge of Aboriginal people, and local knowledge of other forest users such as recreationists and tourism outfitters);
- To engender flexibility, to permit local experimentation that fosters learning about promising approaches;
- To build trust amongst the participants and legitimacy for the process;
- To provide training programs for both resource professionals and lay citizens, aimed at raising skills and understanding of approaches and techniques for productive public participation in forest decision making;
- To define the institutional framework within which forest decisions shall be made;
- To provide strong forums for citizen learning about forest ecosystems, the views and values of other stakeholders, and management options and alternatives; and
- To provide a forum for addressing and resolving conflicts and attempting to develop consensus on forest management choices.

Adapted from Duinker 1998

sustained yield forestry where the focus was on a steady volume of fibre, often to the detriment of other forest values and resources. Achieving this switch from predominantly timber management to SFM therefore requires that the full range of desired forest values be identified. Since these values can be nationally, locally, and personally specific (like "beauty in the eye of the beholder"), it is prudent that "many eyes" be involved in not only setting SFM objectives, but also (to a degree) in their implementation and monitoring. As argued in Chaps. 2 and 5, modern views of SFM include a strong and meaningful PI process as well as support for the related issues of community health and sustainability.

To address these concerns, some of the work funded by Canada's Sustainable Forest Management Network (SFMN) is intended to quantify a number of the important qualitative dimensions of the PI process. This includes polling participants and sponsors of public participation regarding their satisfaction with the processes in which they are involved. In 2001 the SFMN formed a Public Participation Research Group (PPRG), with the goal of shedding light on key uncertainties still hampering strong progress in implementing good public participation in forest management and policy decision making. SFMN research sponsored through the PPRG is now exploring issues such as the effect of forest management knowledge transfer to the public (under the direction of Luc Bouthillier, Université Laval), public involvement in environmental assessment (John Sinclair, University of Manitoba), and public participation in sustainable forest management (Russ Waycott, Stora Enso Port Hawkesbury Ltd.).

Since its inception, a guiding principle of SFMN research has been that decision making in forest policy and management must reflect societal values related to the forest resource (Chap. 1). Indeed, during a SFMN Partners Retreat in Grande Prairie in 1999 "the partners identified public involvement as one of their highest priorities for Network research" (SFMN 1999, p. 1). Through their work, however, researchers have concluded that eliciting and measuring these values is often difficult (Bengston 1994). A goal of the Network has therefore been to investigate the effectiveness of available PI mechanisms in eliciting societal values. For example, current areas of SFMN research with a PI component include issues regarding alternative PI methods, indicators and monitoring protocols, trust and accountability mechanisms, the relative importance of participants' values in the process, and cross-cultural involvement, specifically eliciting and incorporating Aboriginal values. Research themes in the SFMN aim to investigate novel social and economic institutions and provide opportunities for the development and empowerment of First Nations, northern communities, industry, provincial governments, and other members of the Canadian public (Adamowicz 1999). The inclusion of a chapter on PI in forest management in this volume is a direct result of this focus and concern, both on the part of the SFMN, of forest stakeholders, and of greater civil society. This chapter draws heavily on a case study from northwestern Saskatchewan, since the authors are most familiar with this example, and it allows for direct quotes from "on-the-ground" participants as well as an academic analysis.

Different communities should be heard

Sociologists and others have grappled with the definition of community for decades. A review by Hillery (1955) of the different ways that the term is used in academic litera-

---

Box 4.2. The role of public involvement in sustainable development.

Sustainable development is an approach by which human activities account for long-term health, wealth and equity, through respect for the tolerance of the environment, of the economy and of social acceptance, including fairness toward future generations. Because such an encompassing and value-laden approach lends itself to interpretation and to varied applications, the adoption and evolution of sustainable forms of development hinge on public acceptance, and therefore on the public's trust in decision-making . . . It also brings a further requirement, to incorporate society's changing views and values into decisions about planning processes, research priorities, forest practices, manufacturing practices, marketing and consumption.

CCFM 1998, *National Forest Strategy 1998–2003*, p. ix

---

ture found that people use the term to refer to specific geographical places. But, in other instances, people use the term to refer to groups of people with shared values, networks, or experience. For example, we talk about the business community, the academic community, and the environmental community. These social groups are not necessarily geographically bounded.

These different concepts of community help make the point that we need multiple PI tools to meet the needs of multiple publics. There are at least three different groups that need to be targeted for PI, two of which are reflected in the distinction laid out above. First, we need tools to directly involve local, forest-dependent places in forest decision making, as these "communities of place" are often most affected (economically, aesthetically, etc.) by specific forest policies and management actions. Second, we need PI tools to address "communities of interest", which may or may not be locally based. These are the traditional stakeholders that have taken a direct and active interest in forest issues. They include industry representatives, recreationists, environmentalists, First Nations, unions, and countless others. They are scattered across the socio-political landscape of forestry just as communities of place are scattered across the physical landscape. Finally, there are the masses, or the general public. To extend the community metaphor, it is useful to think of this group as the "global village". This represents "everyone else", and is therefore an overwhelming majority, but a largely disinterested majority. Nevertheless, given that Canada's forests are mostly publicly owned, the forest should still be managed to reflect the values of this majority of its owners. The term global village also reminds us that in our globally linked, increasingly interdependent world, constituencies even outside our borders have an interest in how we manage our forests.

In practice, most PI exercises engage the local public with locally based resource management professionals (either industry or government employees, or both). Yet, at the local level, actors are constrained in the latitude with which they can act. Often centralized provincial governments set the broad policy objectives for resource management. And they rely on other hierarchically structured (and often non-local) institutions, namely corporations, to achieve those objectives on a given land base. The general public, and sometimes even the global village, is the group that should play a role in issues of larger (provincial, national, international) scope. The types of information solicited of that group should be general goals and objectives rather than the particulars of how to achieve them. This is because members of the global village generally lack the technical knowhow regarding the best means to achieve these desired goals. Survey research is one of the best methods for eliciting such broader values and objectives at larger scales of analysis.

In contrast, face-to-face workshops, town hall meetings, and advisory groups are best suited for micro-managing problems and issues related to turning broad-scale, general goals and objectives into practice. As well, these face-to-face processes can be useful for eliciting special local contextual issues and creating ideas for mitigating the negative effects of certain management strategies or practices on particular forest values. For example, a broad objective may specify that a certain percentage of a forest be accessible for fibre extraction. Part of the process that involves the global village may entail asking whether these interests feel that current harvest rates are too high, about right, or whether they could be higher. Thus, the general public could have input (to be combined with data from growth and yield models, rates of natural disturbance, etc.) as to the level of harvest at a large scale: in other words, *how much* to harvest. Advisory groups or stakeholder groups are better suited for determining *where* and *how* to cut. That is, the local or regional community may have important preferences and input regarding cutblock size, shape, residuals in variable retention harvests, areas of special significance for recreation or traditional uses, and so on.

Other issues, such as *who* should benefit from exploitation of public lands may require input from communities of place, communities of interest, and the global village. For example, the general public may desire some say as to whether a given province is divided up into 6 large licenses or 30 medium-sized ones. Once that decision is made, the local advisory bodies may suggest how the local benefits are distributed (e.g., which corporations are allowed access to Crown land, and what criteria they must meet in order to operate on Crown land). Some of these issues tread sensitive ground, because our society also values free-market principles, and many feel the market should sort out some of the *"who benefits"* questions.

Public involvement in forest management, planning, and policy entails a suite of tools, and some are better suited than others for a particular purpose. As a society, we need to engage all three types of "publics". Obtaining input from these different groups requires different tools and strategies.

Effectively incorporating PI into SFM represents a significant challenge to all forest managers and policymakers. In the closing session of a 1999 SFMN workshop on forest co-management held in Calgary, participants reiterated their challenge to communities and local resource users to ensure that PI processes such as co-management continue to be synonymous with a more sustainable way of using resources, creating healthy communities, and maintaining the natural environment. Participants also felt that Aboriginal and non-Aboriginal communities must demonstrate that they can practice forestry differently and sustainably. If they cannot, then it will be difficult to justify increased PI in forest management, as these new ways of doing business are often more expensive and time consuming (Chambers 1999).

The basis for public involvement in Canadian forest management

Increased opportunities for public involvement

The trend of increased PI in forest resources management (planning, policy, and decision making) is due to a complex mix of contemporary issues as well as differing motivations among the various parties involved. In the Canadian experience, this includes changing public expectations and values regarding forest resource ownership and management. In part, increased PI in forest management is a reflection of society's recognition that forests are not solely a source of timber, but of non-timber values ranging from recreational opportunities to supporting First Nations subsistence activities. Public opinion research shows that people want greater opportunities to participate in forest management (see Box 4.3). Interestingly, a general interest in seeing more PI occur does not always translate into more individuals participating. It is easy to respond positively in opinion research that the public should have a greater voice in resource management; it is quite another thing to personally commit to what are often long and detailed volunteer processes with very uncertain payoffs.

In Canada, a number of factors have contributed to the public's demand for an increased role in forest management decision making and planning processes. These include higher levels of education in the general population, improved communications networks, better organized interest groups, and a growing cynicism about the ability of a centralized government or industry to effectively manage natural resources in a sustainable manner. "Communities are demanding a greater say in decisions affecting them and citizens have begun to realize that science (including social science) is

---

Box 4.3. Perception is reality: the public feels left out.

In recent survey research in Newfoundland, there is strong evidence to suggest that there is a demand for greater public involvement in forest management and that status quo practices are not currently meeting that demand. A majority of survey respondents (56.8%) expressed that citizen participation is of great value. An additional 31.7% expressed that citizen participation in forest management has some value. Less than 7% thought that citizen involvement has little or no value. The survey also asked about whether government and industry are currently supplying ample opportunity for public input. Only 12.7% suggested the government provides some or a great deal of opportunity for public involvement, in contrast to 76.3% who believe there is little or no opportunity for public views to be heard. Furthermore, less than 9% felt industry provides a great deal or some opportunity, compared to 83.3% who believe that they provide little or no opportunity for public views to be heard. In actuality, both government and industry do have active public involvement programs, but that is clearly not the perception of the general public (Bonnell 2001).

National survey data have shown that 73.7% of the Canadian public feel that there is value in involving the public in setting forest management goals (Robinson et al. 1997).

unable to solve our social, economic and environmental problems" (Stauch 1997, p. 4). There is also a growing sense of disenfranchisement as politicians seem increasingly obligated to advance the interests of corporate globalization, labour unions, or the solidarity and dogma of political parties rather than the interests of their constituents. Recent protests at the meetings of international leaders (e.g., Vancouver in 1997, Seattle in 1999, Quebec City and Genoa in 2001, Kananaskis/Calgary in 2002) vividly portray this frustration by large segments of society who seek "a place at the table" of political decision making. While doing more PI will not necessarily decrease the level of conflict or the number of logging road blockades (Fig. 4.1), it may well increase the legitimacy of the process of forest management. In other words, through participation and education as well as being responsive to public input, an effective PI process can make forest management acceptable to a broader range of stakeholders and the public.

The case that increased PI has come solely through direct pressure from the public or stakeholders may be overstated. The forest products industry and provincial governments are affording more opportunities for public input, but in many cases these are responses to more indirect pressures. In contrast to vocal local or regional residents, these indirect demands take the form of voters and taxpayers, certifying bodies, and company shareholders. One can imagine a complex scenario in which a company with a forest tenure might increase PI activities to satisfy a certification audit, so that it main-

Fig. 4.1. These protestors are objecting to industrial forest management activities in Nova Scotia. Domestic protests can indicate a failure of effective public consultation on the part of forest managers and their leaders. Companies and governments often view consultative processes as a way to defuse high-profile protests, and may undertake lengthy public involvement programs with only token changes on the ground — essentially a "talk and log" strategy. In response, community activists have learned not to cease protest activities as a condition for negotiation and discussion.

Canadian Press photo

tains market share and thus keeps it shareholders happy. It is difficult to extract and isolate such a motivation from direct local pressure for more PI opportunities, or from the requirements of government regarding public input.

Much of the public appear to want the benefits of PI without having to do it themselves. Full citizen control of a given resource, or even co-management or joint management, is not always desired by stakeholders due to time commitments, possible liability issues, and the accountability that comes with having management authority. Often, communities or stakeholders want to give input, and they want that input considered, but they do not want the responsibility that citizen control or delegated power entails. Most of the public is content to let government manage the forest resource as long as the outcome is acceptable to them. The public wants the best result with the least amount of effort, and will only undertake PI that involves a high level of commitment if they are very distrustful of existing delegated managers. A lower level of involvement such as an advisory board or participating in a survey may be the desired end goal of the participants. This was the case with a proposed co-management board in Saskatchewan, when the stakeholder group opted to develop an advisory board rather than a co-management structure. The Divide Forest Advisory Council Corporation was formed, with the members formally adopting a constitution that specifically labels the board as advisory rather than aspiring to a higher level of involvement. The board insisted on this because they felt that within the existing management structure all government could allow them to be was advisory, and also because the members did not want to take on a larger management role or responsibilities (Hamersley Chambers 1999).

Centralized management systems have also failed to sufficiently incorporate the local knowledge held by both Aboriginal and non-Aboriginal resource users into forest management. Recent experience has shown that incorporating this knowledge can lead to better management decisions, reduced conflict, and greater compliance with rules as a result of buy-in from affected local resource users (Hamersley Chambers 1999, p. 24). The incorporation of local knowledge through a PI process (whether at the information gathering stage or at the management design stage as advocated in Chap. 3) therefore has the potential to result in a more effective and efficient forest management system. This can be achieved through a public process that involves the acceptance and incorporation of both western scientific and local knowledge in the management of the resource(s) in question (Berkes 1989; Pinkerton 1989). Public involvement processes can provide an effective means for local communities and residents to provide grassroots expertise and knowledge about the forest to both government and industry managers. This reciprocal flow of information has the potential to reduce the cost of research for government managers and, increasingly, the forest industry.

The meaningful involvement of local resource users in the forest management process is an important tool in integrating state-level and local-level resource management systems. In Berkes' (1994, p. 18) theoretical model of the continuum of resource management, state-level management is defined as management that is "carried out largely by some centralized authority such as a federal agency, is based on scientific data and analysis, and uses the authority of government laws and regulations for enforcement". Local-level management, on the other hand, is characterized by decentralized structures and often involves customary authority, such as that existing in traditional Aboriginal management systems. Local-level (or "self-management") systems, unlike state-level systems, do not rely on government legitimization for their existence (Feit 1988). In this way, PI can be appreciated as a management process that attempts to integrate conventional state-level centralized management with local-level management systems through a more participatory and decentralized process.

Industry can take the first step

Like government, the forest products industry is also responding to the call for greater PI that has been created by rising public consciousness, conflict over forest resources, and also by the lobbying efforts of major conservation groups. For example, there is growing international demand for environmentally benign or "green" wood products such as those certified by the Forest Stewardship Council, an independent certification organization based out of Oaxaca, Mexico. Indeed, the SFMN itself was developed partly in response to the concern that Canada needs to reassure the international community that it is serious in its pursuit of SFM if it is to continue to access many world markets (Adamowicz 1999). In response to these pressures, a number of forest products companies have acted in advance of governments in developing a wide spectrum of PI initiatives to facilitate citizen participation in their operations. For instance, a key component of a joint forest ecosystem management process undertaken between Daishowa–Marubeni International (DMI) and Canadian Forest Products Ltd. (Canfor) in the mixedwood boreal forest of northwestern Alberta was the creation of a public advisory group. Not only was this group designed to be a decision-making body and involved in every step of the planning process, but the resulting forest management strategy had to bear the approval of this committee before final submission to government. This commitment involved a substantial investment of resources and time by industry and those members of the public who accept the invitation to participate, including monthly educational workshops and meetings for 3 years to get the public advisory committee members up to speed (Barker et al. 1999).[3]

The Canadian forest industry has also recognized through recent experiences that long-term conflict and the resulting negative publicity can create significant political and economic costs. For example, in the NorSask Forest Management License Area (FMLA) in Saskatchewan, an 18-month logging blockade at Keeley Lake during 1992–1993 by the Canoe Lake First Nation ended when industry, through its own initiative and with the tacit support of government but no public funding, established community-based co-management[4] boards (FHI 1996). These boards, which are still

---

[3]This process is currently being evaluated by SFMN-funded research: "Alberta forest management in the public sphere: a province-wide case study of public advisory groups", led by Debra Davidson at the University of Alberta; project description available at http://sfm-1.biology.ualberta.ca/english/research/policy.htm#policy02 [viewed 9 July 2003].

[4]The debate regarding what constitutes "true" co-management is ongoing, with some researchers claiming that effective co-management can happen in the absence of government involvement (Hamersley Chambers 1999) and others claiming that it must be institutionalized in a partnership of equals such as a settled Aboriginal land claim (Notzke 1993). For the purposes of this chapter, each public involvement case study is referred to in the terms used by the participants, whether it is an advisory or co-management process. In the case of NorSask Forest Products and Mistik Management, participants refer to their process as "co-management".

operating 10 years later, represent a generally successful attempt at a participatory process to address the growing conflict between industry and local inhabitants over forest development (Hamersley Chambers 1999). Across Canada, managers are learning that involving stakeholders and the general public before conflict gets out of control is the most cost-effective way of doing business. For example, a senior industry manager in the NorSask FMLA stated that:

> *Before, if someone didn't like something, they would go straight to the Minister, and things would get out of perspective, out of control. It was time-consuming, expensive and often a waste of time and resources with a lot of bad feelings for practically the same management result. Now, people come to discuss issues with the co-management board before they go anywhere else . . . We have a forum now to discuss issues, to explain our point of view and knowledge. Now, a lot of issues are solved before they really become issues.*

(Hamersley Chambers 1999, p. 89)

The forestry community itself has made a number of efforts to include PI in their planning and operations. For example, the Canadian Council of Forest Ministers (CCFM 1995) has identified a number of PI indicators that are considered vital to the practice of SFM. While these indicators are unfortunately difficult to quantify and aggregate at regional, provincial, and especially national levels, they do capture the proper spirit of intent with PI. Examples of these indicators include:

Indicator 6.4.1: Degree of public participation in design of decision-making processes;
Indicator 6.4.2: Degree of public participation in decision-making processes; and
Indicator 6.4.3: Degree of public participation in implementation of decisions and
monitoring of progress toward sustainable forest management.

Over the past 10 years, steady reductions in government programs and funding for research and enforcement in the forest sector have resulted in the cancellation of many publicly funded activities in support of progressive forest management. Some of these responsibilities have been devolved to private forest products companies, while some are now being undertaken (either officially or behind the scenes) by non-governmental organizations (NGOs) such as Forest Watch volunteers, or by third-party auditors commissioned to support forest product certification. The argument could be made that government devolution of its forest management responsibilities does not extinguish the responsibility for public consultation in forest management, but merely transfers it to other sectors such as private corporations.

The increasing role and responsibilities of industry in PI, however, have not come without debate. For example, during the SFMN-sponsored Calgary workshop on co-management in 1999 (Chambers 1999), some participants felt that the forest industry should not be involved as voting members in some PI processes like co-management at all. Others were of the opinion that since government is increasingly devolving its management responsibilities (through staff and service cutbacks) to industry and that corporations are picking up the tab for this, it is necessary for this sector to be involved. While the shift of control from centralized government to the regional level may have many positive outcomes, it also raises a number of concerns such as who will monitor whether industry is doing an effective and fair job of PI or not. Many participants at the SFMN workshop felt that industry has a strong vested interest in the forest resource (i.e., timber) that is often in direct conflict with that of traditional resource users such as trappers and wild rice harvesters. Consequently, they argue that industry should not have voting privileges in PI processes. Some participants thought that since much of Canada's forest lands are publicly owned, it was the responsibility of government, and not industry, to ensure that a democratic management process is followed to manage these lands. Participants concluded that the success of industry in PI is based upon a company's willingness to obey the law, be honest, and to take a long-term and holistic view of resource development (Chambers 1999).

A number of the frustrations discussed above have also been expressed by industry. A meeting of the SFMN Public Involvement Task Force in August of 1999 concluded that requiring forest companies to implement strategies that include the public does not necessarily lead to effective PI, and that both forest companies and many public constituencies are frustrated by the generally poor success of PI mechanisms to date (SFMN 1999). However, it is also important to note here that PI is a fairly new responsibility of industry. Many companies are doing their best to meet these new requirements, including substantial investments of time and resources. Early efforts in this regard were typically undertaken with inappropriately educated staff and consultants, but in generally good faith. Their effectiveness should not necessarily be judged by past actions prior to corporate policy shifts or recent refinements. For its part, SFMN work should improve this process, as it has emphasized the education of highly trained personnel.

Legal basis for increased public involvement

There are an increasing number of legal reasons and incentives for government and industry to integrate PI into forest management. These include provincial and territorial Forest Acts such as the B.C. Forest Practices Code, forest tenure arrangements that increasingly include PI requirements, recent court decisions, and Aboriginal treaties. For example, a B.C. Commission on Resources and Environment report (BCCORE 1995, Vol. 3, p. i) states that: "The government should now recognize the general right of participation in law as a necessary element of the provincial land use strategy."

Due to differences in legal rights, it is useful to consider Aboriginal and non-Aboriginal groups separately in terms of their participation in forest management (Chap. 3). The practical reality of economic development and the reliance upon forest resources by non-Aboriginal citizens means that government managers must meet these other obligations to Canadians as a whole while protecting Aboriginal interests. In recent years, many decisions in the Canadian courts have directed the provinces to recognize and protect Aboriginal and treaty rights in resource development and planning. The federal government has recognized the active role of Aboriginal people in SFM in the *National Forest Strategy* (CCFM 1998) and in the Canadian Council of Forest Ministers *Criteria and Indicators for SFM* (CCFM 1995). Aboriginal people in Canada therefore participate in the public involvement process from a very different legal position than non-Aboriginal stakeholders (Smith 1995). For in-depth discussion of Aboriginal involvement in forest management, please see Chap. 3.

To ensure that forest products companies manage for a broader range of forest values than just timber supply, provincial and territorial governments are increasingly using legislation to require PI in the forest planning process. Examples of this in Ontario alone include the Crown Forest Sustainability Act of Ontario (1994) and the Forest Management Planning Manual (Duinker 1998). In Saskatchewan, Sections 11 ("Forest Management Committees") and 94 ("Delegation") of the 1997 provincial Forest Resources Management Act make specific provisions for public consultation in forest management. For example, Section 11 states that:

The minister, pursuant to the regulations:

(a) shall establish a Provincial Forest Policy Advisory Committee to advise the minister on matters relating to the management of forest resources, including the preparation, approval, implementation, amendment, revision or audit of any plan or Saskatchewan Forest Accord prepared pursuant to this Act; and

(b) may establish forest management committees for those areas that are designated by the minister to facilitate local involvement in the management of forest resources.

Section 94 of the Saskatchewan Act also pertains to the development of PI processes as it addresses the decentralization of certain decision-making powers and authority from the provincial to the local management level. For example, Section 94 states that:

The minister may delegate the exercise of any of the minister's powers, or the carrying out of any of the minister's responsibilities, pursuant to this Act or the regulations that are prescribed in the regulations for the purposes of this section to:

(a) any forest management committee; or

(b) any person or category of persons other than an officer or employee of the department.

Most forest tenure agreements in Canada require that management plans be reviewed by the general public every 5 or 10 years. Provincial forestry regulations require that forest products companies benefiting from timber rights on public lands initiate their planning process in collaboration with other forest users such as First Nations, recreationists, and municipal authorities. By law, once these plans are elaborated, they must undergo a consultation process that is open to the general public.

Policy basis for increased public involvement

National and provincial governments and industry-initiated corporate policies are providing an unprecedented opportunity for the Canadian public to participate in forest management. Indeed, "the purpose of participation has shifted from merely providing accountability to developing the substance of policy" (Beierle and Cayford 2002, p. 5). Examples of this include regional land-use strategies, such as the 1990s Commission on Resources and Environment in British Columbia, and provincial and national policy development processes. Even private forest owners are realizing that careful involvement of the local public in forest planning can improve public relations and possibly even increase the revenues generated from these private lands (Fullerton 2000). Indeed, forest managers and policy-makers have ample opportunity for creating and implementing PI exercises in forest decision making. Opportunities are commonly presented with environmental impact assessments, forest management plans, model forests, and community forests (CIF/IFC 1998). At the provincial level, examples of forest-related PI policies are Ontario's 1993 policy framework for sustainable forests, Manitoba's 1990 workbook approach to policy on sustainable forest development, and Yukon Renewable Resource's 1995 forest policy initiative (Duinker 1998).

At the federal level, examples of public involvement policies that facilitate citizen involvement include the 1998 *National Forest Strategy*, the National Roundtable on the Environment and Economy, and the Canadian Standards Association's (CSA's) Forest Certification Program. In 1992, the Canadian Council of Forest Ministers' report "*Sustainable Forests: A Canadian Commitment*" presented a number of Strategic Directions including the following three, which relate directly to increased public participation in forest management:

> *(1) The public is entitled to participate in forest policy and planning processes, recognizing that this carries with it obligations and responsibilities.*
> *(2) Effective public participation requires an open, fair and well-informed process, with generally accepted procedures and deadlines for decisions.*
> *(3) To participate effectively, the public must be aware and informed, with access to comprehensive and easy-to-understand information on forest resources.*

The *National Forest Strategy 1998–2003* also makes specific policy commitments to SFM and the right of the public to be involved in forest management: "all Canadians are entitled to participate in determining how their forests are used and the purposes for which they are managed" (CCFM 1998).

Government policy influences not only the development of public involvement processes, but whether and how the results from these initiatives are implemented. In summary:

> *The road from policymaking to policy implementation is long and complicated. A host of political, social, and legal influences come into play. In examining implementation, we see participation in its larger context, that is, how it fits into the complex web of public policy. From this perspective, public participation appears as only part of the machinery - perhaps only a small part - that turns ideas into action.*
>
> (Beierle and Cayford 2002, p. 56).

Public involvement is essential to achieving adaptive forest management

Another impetus for increased public participation is the recognition that this input is essential to achieving adaptive management, an increasingly popular approach in the lexicon of natural resource management (see Chap. 21 for a thorough discussion). Adaptive management is not a new concept, with the term appearing more than two decades ago (Holling 1978) and having since been popularized by a number of authors (Walters 1986; Lee 1993). Indeed, many formal adaptive management systems recommend or even require various forms of public participation.

Adaptive management has been mostly used in the management of ecological systems as opposed to social systems. However, the same characteristics of ecological systems that led to the widespread adoption of adaptive management also exist in the social and policy realm. Indeed, adaptive management can be considered the implementation of any management cycle that has continual improvement as one of its objectives. The ecological characteristics of forests and forestry which lead to the need for adaptive management are described by Kessler (1999) as follows:

- a poor understanding of how systems work;
- inaccuracies in measurements, inventories, models, and assumptions, and;
- a degree of unpredictability or chaos that exists in nature.

Social and policy systems are similarly fluid, emerging, interactive, and evolutionary. For example, things like comprehensive land claims or the post-World War II boom in recreation represent unpredictable changes in social values for forests. Many academics agree that we are at a very rudimentary stage in assessing or taking "inventory" of social values for natural resources. While conventional management has tended to focus on the forest's immediate economic or commodity value, SFM suggests that we need to think in terms of the whole spectrum of social and economic values that exist for forest resources. As a result, a strong connection exists between effective PI and implementing an SFM approach to forest management. For example, learning about PI and its incorporation into forest management is part of an adaptive management process. The use of PI in an agency can and should be subject to continuous evaluation and improvement.

Benefits of public involvement in forest management

A major reason for the increase in PI initiatives in Canadian forest management over the past two decades is the wide spectrum of potential benefits offered by these processes to key stakeholders. A number of these are summarized in Table 4.1, under the main headings of benefits to government, industry, communities of interest, communities of place, and to the global village. Just as the *motivations* for undertaking or participating in PI vary from player to player, so too do the *benefits* or the perceptions of benefits vary from person to person and from group to group. One may find a stakeholder group entering a process because they feel they will achieve greater power, an industry group agreeing to sponsor public involvement because they think that it will solidify a market, and a government agency participating in or sponsoring a process because they feel it will increase their legitimacy. Moreover, there are different benefits (actual and perceived) associated with different forms of PI. The benefits to all participating parties in a focus group or structured workshop are likely quite different than that of a co-management process. Table 4.1 presents a hypothetical framework for assessing what types of benefits might accrue to different groups participating in PI processes.

One of the most obvious and important benefits of PI is that it elicits public values and preferences toward resources and resource use. Often it is important to do this in a systematic way (e.g., with a representative random sample) so the results of a process

Table 4.1. Potential benefits of public involvement in forest management are different for everyone. Furthermore, the relative importance of each benefit varies with the problem at hand and among communities from place to place, hence the cells left blank in this generic table.

Benefit	Government	Industry license holders	Communities of interest	Communities of place	Global village
Reduction in forest resource conflicts					
Incorporating public values into planning and decision making					
Government and industry able to meet legal and policy obligations					
Educating and informing the public					
Improving the quality of forest management decisions					
Greater cooperation, trust, and improved relations between stakeholders					
Community empowerment, development, and education					
Mutual learning between stakeholder groups					
Development of sustainable forest management systems					

may be extrapolated to a larger population. Surveys are arguably the most effective tool for values elicitation. Many of the face-to-face forms of PI (open houses, field tours, town hall meetings, advisory groups, and the like; Fig. 4.2) are better suited for building trust and legitimacy among local players. These sorts of tools may reduce conflict

Fig. 4.2. Open houses (*a*) and field tours (*b*) hosted by forest products companies or forest management agencies are often used to inform the public of activities planned for public land,

(*a*)

(*b*)

Photo courtesy of Millar Western Forest Products Ltd.

Photo by Carol Murray,
courtesy of Mistik Management Ltd.

if resource managers are responsive to local public concerns regarding how industrial license holders achieve their management objectives.

Arguably one of the most important benefits of PI is the reduction in forest resource conflicts. By facilitating wider involvement in forest resource planning and management, these processes minimize resource conflicts that can lead to blockades and other civil protest (Witty 1994; Beckley and Korber 1995; Roberts 1996). Research also suggests that decision making that is collaborative in nature rather than adversarial results in decisions that are longer lasting and more satisfying (Susskind and Cruikshank 1987). Indeed, it appears that a great deal of conflict could be avoided by simply increasing communication between stakeholder groups so that initial misunderstandings or erroneous assumptions are cleared up before they spiral into genuine conflict (Hamersley Chambers 1999). Public involvement can also help differentiate value-based disagreements from data-based disagreements. For example, some conflict stems from different goals and normative views, while other conflict stems from disagreement over beliefs about impacts and outcomes.

While conflict most often arises among forest users, it can also develop between these users and urban-based environmental organizations and citizens. Although conflicts can sometimes be beneficial (for example, to shake management agencies or corporations out of inertia and complacency), for the most part these are destructive and harmful, especially when communities become divided into bitter factions. In facilitated processes it is not unheard of that opposing interests discover that they agree on 80% of the issues. However, a disproportionate amount of time, energy, and resources are devoted to contending the disputed 20% of issues, rather than on furthering the portion of the agenda for which there exists a consensus.

It is important to note that PI increases the chance (but by no means guarantees) that conflict will be reduced. Due to the value differences that will continue to exist by virtue of the diverse stakeholder groups involved, complete conflict reduction is not an achievable goal. Rather, PI is an important part of managing this issue. Through this process, conflict can be managed, and the value differences that underlay conflict channelled into creative solutions for achieving multiple benefits from our forests. Likewise, the fact that there is no obvious conflict at the end of a process does not mean that everything is well. It is possible that contentious issues have been avoided instead of being resolved by either not discussing them or by excluding stakeholders that are controversial (Beierle and Cayford 2002).

Public involvement in forest decision making has also been shown to channel existing and new conflicts into collaborative searches for accommodating solutions (CIF/IFC 1998). This input is also invaluable in understanding tradeoffs about timber management decisions, particularly those involving non-timber values (such as recreational opportunities or wildlife habitat), many of which are different for forest-based or urban constituents. For example, local and regional (forest-based) publics may be more concerned about fish and wildlife resources as sources of food and recreation, while urban or extra-regional publics may be concerned about a wider range of organisms (biodiversity) for their conservation and intrinsic values. In other words, public input helps improve decision making by identifying the heterogeneity of diversity in values — something that is important for managers and policymakers to know.

Involving the public in forest management can also lead to increased cooperation and improved relations between stakeholders, as diverse groups who often had no tradition or formalized way of meeting with each other sit down at the table together (Hamersley Chambers 1999). Meeting as a stakeholder group provides a mechanism for fostering the consent of local resource users and reducing conflict through a process of participatory democracy (Pinkerton 1989). As different parties get involved, they start to know and understand each other, and sometimes even trust each other (Buchy and Hoverman 2000; Beierle and Cayford 2002, p. 29).

The process of PI also helps to identify stakeholder groups that should be involved in managing the forest resource, thereby legitimizing the participation of interests that have historically been excluded through conventional management systems.

Higher quality decision making with greater stakeholder buy-in is a common result of PI with the result being "better, fairer and more stable decisions" (BCCORE 1995). These decisions should be more cost-effective or more satisfying to a range of interests, and improve the information or analytical foundations on which decisions are made (Beierle and Cayford 2002, p. 27). By including the public in the resource management process, the ability of managers and users to develop, successfully implement, and abide by enforcement regimes can be significantly improved. Local and regional governments can also benefit from PI by reduced challenges to their authority. This is because management power and responsibility are shared to varying degrees through the involvement process with other resource users, who are then more likely to accept management decisions, since they have had a say in forming them. Conventional centralized management, in contrast, has a history of producing decisions that local resource users either ignore or actively oppose, sometimes resulting in civil disobedience and negative publicity. Box 4.4 outlines some common ways of addressing the substantive quality of decisions. Although this is not an exhaustive typology and some points are arguably more important than others it does provide a useful checklist for managers and participants to review.

Box 4.4. Evaluating the quality of decision making.

- Cost-effectiveness — do the decisions or recommendations made by participants lead to actions that are more, or less, cost-effective than a probable alternative in solving a forest management problem?

- Joint gains — as a result of the agreement, are some participants better off without any participants being worse off?

- Opinion — do participants or case study authors feel that decisions are better than a probable alternative, not according to concrete criteria but in terms of general satisfaction with an outcome or in terms of a range of quality criteria?

- Added information — do participants add information to the analysis that is not otherwise available?

- Technical analysis — do participants engage in technical analysis to improve the foundations on which decisions are based?

- Innovative ideas — do participants come up with innovative ideas or creative solutions to problems?

- Holistic approach — do participants introduce a more holistic and integrated way of looking at a forest-related problem?

- Other measures — do participants improve the technical quality, the environmental benefits, or other aspects of a decision?

Adapted from Beierle and Cayford 2002, p. 27

Other commonly recognized benefits of PI for local resource users include gaining influence in management decisions, creating higher quality decisions, increasing the buy-in of stakeholders in management decisions, and a generally fairer management process (Witty 1994; SERM 1995*b*). On the other hand, local-level PI processes are also known to exclude certain groups. Managers must therefore be aware that strategies to make involvement inclusive should be developed and implemented.

Public involvement often leads to greater cooperation and trust between government managers, industry, and local forest users. A useful definition of "trust" in the context of PI is given by Beierle and Cayford (2002, p. 30) where "trust" is understood to involve two components: (1) competence, or the ability to do what is right; and (2) fiduciary duty, which is the will to do what is right. Trust in an institution means that the public has confidence that the lead agency involved is capable of understanding the public interest and is required to do so. Research suggests that good pre-existing relationships are conducive to the success of PI, but other research suggests that pre-existing conflict and mistrust have a greater impact with less intensive PI processes than with more intensive ones (Beierle and Cayford 2002).

By bringing different groups together to sit at the same table, a PI process can result in the creation of new relationships between resource users and managers. By working together, stakeholders learn to appreciate other points of view and to accommodate a diversity of interests in management decisions (Hamersley Chambers 1999). This process of mutual learning is particularly effective if the individual players involved remain consistent over the long term. In turn, this improved trust can create an increased willingness by users and government to explore management alternatives, to develop and implement enforcement plans, and to allow greater self-management for local resource users (Pinkerton 1989; Roberts 1996). It is important to note that a lack of trust often stems from a society-wide mistrust of institutions. Therefore, it is crucial in PI processes to determine how broadly the formation of trust extends beyond participants to the wider public (Beierle and Cayford 2002, p. 26).

From the point of view of local forest resource users, community empowerment and development are key benefits of increased PI. These can include economic development as well as the building of community capacity through training, educational opportunities, and the opportunity to develop general management and leadership skills. An exhaustive study of 239 public involvement cases in the U.S.A. concluded that, "When insufficient attention was devoted to educating and informing the public, participants remained largely powerless to engage effectively in decisionmaking" (Beierle and Cayford 2002, p. 32). While most of these case studies did a good job of educating direct participants, they did not generally provide educational outreach to the broader public nor incorporate their values into decision making (Beierle and Cayford 2002). Higher levels of PI such as co-management can lead to increased local control over resources as well as a greater degree of organization, credibility, cultural identity, and self-reliance among resource users (Witty 1994). Another benefit of PI is creating a greater sense of control over participants' working lives and their community's environment. Indeed, it "is difficult to appreciate the importance of what a sense of control can mean until one looks at situations in which it is lacking" (Pinkerton 1989, p. 26).

In more advanced PI processes such as advisory committees and management boards stakeholders can sometimes agree to share the costs of management with government and industry. Finally, including sectors of the public such as resource users in the management process helps to ensure that both local knowledge and scientific knowledge work together to create better resource management decisions that are accepted by a wider segment of society. The forest industry can also benefit from this by reduced conflict and through access to local and traditional knowledge about the resource(s) in question. The incorporation of local knowledge into forest management planning and operations can lead to better data gathering and analysis; for example, there is often improved communication about resources and species populations (Hamersley Chambers 1999).

What does good public involvement look like?

Desirable qualities for public involvement strategies

One of the common misperceptions regarding PI is that there is one correct way to do it. In fact, just as the public is diverse in its values and interests in the forest, so too must PI strategies be diverse to capture that variety of views. There is no one correct way to undertake and succeed in PI for the purposes of SFM. The diversity of situations, participants, and methods inherent in the Canadian landscape necessitates a deep toolbox and more of a "shotgun" approach. Indeed, "forest practitioners and participants in public involvement exercises would do well to design processes specifically to suit their own circumstances and tastes" (Duinker 1998). Despite this diversity, researchers and practitioners have come up with a number of specific qualities that they believe lead to successful PI strategies, as well as direct and indirect methods for incorporating public values into forest planning and policy. For example, the *National Forest Strategy* (CCFM 1998, p. 12) states that:

> *The key to maintaining effective public participation is the development of better models for public input, with processes and mechanisms that are clearly defined, fair and open, with deadlines for decisions and with a review of results that will ensure the accountability of those responsible for the welfare of forests.*

Canada's *National Forest Strategy* also describes principles and actions of effective PI in forest management (Box 4.5). Other CCFM criteria include a process that ensures a two-way flow of information, and the incorporation of traditional and local knowledge into management planning.

Beierle and Cayford (2002, p. 49) conclude that there is a direct and positive correlation between the process followed for PI and the success of the initiative. The following four criteria are strongly linked to success in PI, with number one the most important and number four the least (Beierle and Cayford 2002):

(1) Responsiveness of the lead agency;
(2) Motivation of the participants;
(3) Quality of deliberation; and
(4) Degree of public control.

Duinker (1998) recommends the inclusion of the following elements in any process that involves the public in forest management issues. While each process must be constructed to specifically suit the objectives and circumstances at hand, these principles

---

Box 4.5. Public participation principles and the subsequent framework for action as identified in Canada's National Forest Strategy (1998–2003).

Strategic Direction 3. Public participation: many voices

Principles:

- Public participation in forest policy and planning processes is essential, and carries with it obligations and responsibilities for all involved;
- Effective public participation in forest management and planning requires an open, fair, and well-defined process, with generally accepted procedures and timely deadlines for decisions;
- Effective public participation requires current information from a variety of sources, including publicly funded forest resource databases.

Framework for action:

- Heighten public awareness and knowledge of forests;
- Improve access to and provision of information on forests that meets the needs of the public;
- Ensure that the views of the public are considered in forest management planning and decision-making processes.

CCFM 1998, p. 47

(while not guaranteeing success) invariably characterize all successful PI endeavours in Duinker's experience:

(1)   Openness, fairness, and inclusiveness in selection of participants;
(2)   Clear mandate and purpose;
(3)   Professional design and implementation;
(4)   Informal but structured process;
(5)   Design for positive-outlook problem-solving to elicit collective solutions;
(6)   Variety of techniques for eliciting input;
(7)   Clear influence on decision making;
(8)   Sufficient time and supporting technical resources;
(9)   Keeping decision makers informed throughout; and
(10)   Reasonable and realistic expectations.

Direct and indirect methods for incorporating public values in forest planning and policy

A large and growing number of methods exist to involve the public in forest management and decision making. As already indicated, trying to find one "right" way to undertake and evaluate public involvement is neither likely nor even desirable (Beierle and Cayford 2002). The many different aspects of "community" involved in forest management help make the point that we need multiple PI methods to meet the needs of multiple publics, both directly and indirectly.

Direct methods are those that involve face-to-face interactions with the public, generally as small subsets. These methods directly solicit public attitudes, values, and beliefs. Commonly used direct methods for incorporating public values in forest planning and policy include town hall meetings, workshops, public information meetings, public hearings, surveys, open houses, and public advisory boards. Decision making in these methods is often consensus based, and can also include mediation and alternative dispute resolution protocols. The benefits of direct methods are that they tend to provide good, detailed information on field-level operations, build trust among stakeholders, and target those most interested and affected.

Indirect methods are those that are not face-to-face, and generally involve large, representative samples of the public as a whole rather than smaller groups. Examples of indirect methods include qualitative interviews (semi-structured surveys), quantitative interviews (mail, telephone surveys), and participant and non-participant observation. The main benefits of these methods are that they provide excellent representativeness, good general information on values and attitudes, and can allow for very frank, honest reporting.

Public involvement methods tend to fall under four main categories: public meetings and hearings, advisory committees not seeking consensus, advisory committees seeking consensus, and negotiations and mediations. Methods become more intensive as they progress from the first of these to the last, also becoming more oriented towards forging agreements among a small group of interests rather than gathering information from a wide range of stakeholders. Not surprisingly, more intensive methods correlate strongly with success at forging direction for management. While more intensive processes tend to produce participants who are effective in solving problems and getting decisions implemented, these processes require greater funding and staff support than less intensive ones. However, it is important to note that less intensive processes are more effective at involving the wider public, as the participants in intensive processes are not likely to reflect the diverse socio-economic characteristics of the wider public (Beierle and Cayford 2002, p. 46).

The three different groups outlined earlier in this chapter that should be targeted for PI are communities of place, communities of interest, and the global village. It is important to note that these three different target groups require very different methods of engagement. One of the appealing attributes of communities of place is that they are geographically bounded, so that it is possible to employ face-to-face methods with this type of public. There are a variety of methods that are useful, from public meetings, to focus groups, open houses, field trips, and workshops. One caution with these approaches, however, is that they must be structured in such a way as to facilitate two-way information flow. In the past, the tendency has too often been to use such tools to "tell and sell" information from the sponsoring institution, whether government or industry.

While face-to-face methods in communities of place create good will and allow for flexibility and responsiveness (e.g., you can have a dialogue), they should not be the only method employed to solicit local views. Opinion research in a defined local or regional area will provide added breadth to information collected through face-to-face methods. Opinion research is also less flexible. However, it is likely more representative of the broader public, at whatever scale such a method is employed. Opinion research is one of the only methods available to obtain views and values for the global village, that is, the broadest conception of the public. And despite this view's detachment from the issues and perhaps from the forest itself, this group is again by far the largest majority and on that basis alone deserves consideration.

Methods for obtaining public input from communities of interest will vary, depending upon the locale or distribution of the interest group. In some instances one will be dealing with a community of interest that is quite geographically bounded, such as a local all-terrain vehicle (ATV) or snowmobile club, or a local watershed protection organization. In such instances, face-to-face interaction is critical and the methods listed above are appropriate. Local interest groups may provide feedback through formal or informal individual sessions with decision makers, or through group processes that involve other communities of interest. These latter tools (roundtables, consensus processes, advisory boards with broad representation of stakeholders, etc.) provide the added benefit of allowing forest users to hear, and hopefully come to understand, the views and values of groups with which they may be competing. Many of the participants in a 1999 SFMN-sponsored workshop in Grande Prairie were not aware of the range of PI activities that had been used in other areas. A recommendation of this group was that a directory of PI mechanisms would be very useful, along with suggestions as to when and where these different methods could be effectively employed (Chambers 1999).

The distinction has been made between PI as an end in itself (an approach, an ideology for community development), or as a means to an end (a method, a set of guidelines and practices) (Buchy and Hoverman 2000). The purpose of this distinction is not to label different ends of the scale as better or worse, but to describe a distinction between different processes. Whether PI is intended to be an end in itself or a means to an end, therefore, has an impact on which methods of engagement are used.

Challenges to public involvement in Canadian forest management

There is no consensus on what public involvement is supposed to achieve or how it is best structured

Public involvement in Canadian forest management is evolving in a wide variety of situations that range from comprehensive land claims (where Aboriginal communities have a constitutionally protected right to be consulted) to informal industry consultations with non-Aboriginal communities and limited government involvement. As a result, great differences are emerging in structures and practices as well as expectations and goals of public involvement. While there is strength in diversity, this situation also makes defining PI as well as evaluating and comparing its successes difficult. In fact, much of the debate appears to centre on the different understandings that the parties involved have of what PI really means and what the goals of the process are. At a general level, academics and practitioners agree that PI refers to (CIF/IFC 1998):

> *. . . any situation where people other than resource-management professionals and tenure holders in forest decision-making are invited to give opinions on any matter in the decision process. There is a wide variety of degrees of public participation, and also of techniques and forms of participation. Public participation includes everything from surveys and open houses to full decision-making partnerships.*

Combined with the unique needs and expectations of each PI scenario, the diversity of models available assures that every attempt at enhancing SFM through these processes likewise evolves as a distinctive arrangement.

The fact that there is no consensus on what PI is supposed to achieve or how it is best structured can be seen as both a challenge and an opportunity. The challenge for resource managers is to choose a participation mechanism that matches the needs of their particular situation. This can be a daunting process, as there are a large number of different models and methods from which to choose, and participants themselves often come to the table with very different goals and expectations. The opportunity presented by this diversity of methods and motivations is that of great flexibility, and therefore good potential for success.

Power as central to public involvement: it changes dynamics, relationships

The issue of power is central to PI in forest management, as undertaking a participatory process ultimately changes existing relationship patterns and dynamics. Since people tend to participate in PI processes because they want greater control, their participation stops being meaningful when the expected transfer or share of decision-making power does not happen (Buchy and Hoverman 2000). Experience has shown that government officials "often jealously guard their authority against encroachment by other agencies, and they are not in the habit of sharing power with those they have the authority to regulate" (Osharenko 1988, p. 44). For example, civil servants may reject change on account of a perceived diminishing of influence and power (though this is rarely admitted). The goal of increased local involvement in resource management and decision making through public involvement can therefore conflict significantly with existing power structures. Other stakeholders may also fear the burden of responsibility that comes with decision-making power (Beckley and Korber 1995; SERM 1995a).

Who needs to be involved?

"The public" is not a uniform block; rather, it is made up of diverse sets of (often) competing and (somewhat) overlapping interests and values. A major challenge for managers, therefore, is to decide "who's values" should be included and what the role of each member or sector of the public should be. It is often difficult to involve all stakeholders within a PI process. Reasons for this include the diversity of stakeholders within each community, the absence of clear stakeholder organizations, the expense involved, and general apathy unless something affects an individual or interest group directly. However, it should be noted that "apathy" is not necessarily something which needs to be combated, as it could well be a symptom of satisfaction, poor leadership, over-consultation, or frustration due to ignored or poorly implemented recommendations (Chambers 1999).

The issue of how well the chosen groups represent the general public should be addressed early in the process (Box 4.6). Organized stakeholder groups tend to be the motivated and vested interests at the extremes of any debate. On the other hand, most PI processes do a poor job of integrating the views of the greater number of lightly effected and disinterested general public. The scale of involvement is a significant challenge, as focusing on the local community level does not capture the entire public, and may in fact be at odds with broader public opinion and interests (W.L. Adamowicz, University of Alberta, personal communication, May 2002). This situation is exacerbated by the trend of urbanization in our country: 79.4% of Canadians now live in urban areas (Statistics Canada 2003).

Participants at a 1999 SFMN Grande Prairie workshop on PI wondered how these processes could be structured to ensure that all relevant stakeholders would participate, including the "passive public". Representatives from industry, academia, government, and abroad felt that determining who should participate in each process was a great challenge, and were particularly interested in how to balance the interests of various publics such as urban and rural communities, active and passive participants, and the interests of customers versus those of the local residents (SFMN 1999). A clear definition of the rights, responsibilities, and roles of participants should be outlined early in each process (CIF/IFC 1998). For example, have stakeholder, government, and industry interests been clearly identified, and which of these need to be involved? Should participants be appointed by designated stakeholder groups (and if so, which ones?), elected from the public at large, or contracted by government or industry to represent particular perspectives?

Box 4.6. Issues of representivity.

In recent research that reviewed 239 public involvement case studies across the U.S.,

In nearly 60% of the 63 studies that were coded for socio-economic representative-ness participants were not at all representative of the wider public. In 58% of the 74 cases with information on interest group representation, participants or case study authors identified interests that were missing from the table. The fact that partici-pants were not necessarily a good reflection of the wider public might not be impor-tant if participants sought out input from the wider public directly. However, of 129 case studies with relevant information, participants consulted the wider public in 39% of the cases, consulted to some extent in 27% of the cases, and did not consult at all in 34% of the cases. Although participants had a high degree of influence in the majority of cases, important questions remain about how well those participants reflected the values of the public they were meant to represent.

Beierle and Cayford 2002, p. 24

Another difficult issue is how to involve people or organizations with radical opin-ions. Such people can make the work of PI difficult, disrupting the process and even making it unworkable or impossible. However, there are also advantages to having a wide range of views and ideas represented at the table (Chambers 1999). One proposed solution to these problems is to invite contentious groups to the table as presenters of information rather than as part of the consensus group. Often they will opt out of a process if they see they are going to be a clear minority in a consensus structure, though once they are out, their perspective is generally lost. An alternative could be for them to present their science, their logic, and their perspective for consideration by planning teams or subcommittees. This puts them on equal footing as government or industry sci-entists, who are often invited to provide "objective" and "scientific" information (though their facts and opinions may simply represent the value-laden questions they have asked; see Chap. 3). Research has demonstrated that if these sorts of questions are not asked and the responses clarified at the start of a process, the credibility of the process can be damaged and create the appearance that agency and community time and money have been squandered (Buchy and Hoverman 2000).

Another issue related to stakeholder involvement is that while government and industry representatives participate as part of their job, representatives of local commu-nities are usually interested volunteers or individuals already holding elected positions on Hunters and Trappers committees, Village Councils, or other existing governing bodies. This results in community representatives being *de facto* unpaid volunteers, which also impacts who is interested in being involved in the process, and how much time and dedication they bring to it (Chambers 1999). Fair reimbursement for travel expenses, and even for time, can be a key ingredient in recruiting and retaining com-mitted representatives of the public.

Consensus-based decision making

A significant amount of literature and processes have been dedicated to advocating consensus-based decision making (e.g., the B.C. Commission on Resources and the Environment (CORE), and the National Round Table on Environment and Economy, to name two Canadian examples). There is a general perception in society that consensus-based decisions are more creative, effectively address participants' concerns and interests, and produce decisions that are more widely accepted than those produced by conventional management processes (Susskind et al; 1999). Recent research suggests, however, that a consensus-based process may not always be more appropriate or successful than other PI methods such as surveys or an open house, depending on the specific goals and context of the exercise (Coglianese 1999; Gregory 2000).

Along with the potential benefits of a consensus approach come a number of concerns and restrictions. Sometimes contentious groups are purposefully avoided while others may walk away from the table or refuse to sign the final submission if it does not meet their needs. For example, "More-intensive public participation processes also demonstrate a strong tendency to reach consensus by leaving out participants or ignoring issues" (Beierle and Cayford 2002, p. 48). The constraint of having to reach consensus may encourage participants to avoid important but controversial issues, resulting in abstract principles and vague standards that mitigate the usefulness of the final report (Coglianese 1999, p. 6). For example, it is easy to produce vague recommendations, since everyone can agree to them and they can be widely interpreted (advocating "sustainable development" through "consensus by the lowest common denominator"?). Reports produced by consensus are often not analytical, innovative, or specific enough to be useful and do not necessarily ensure better decisions or reduce conflict (Coglianese 1999).

A consensus approach is therefore not appropriate for all PI situations. Participants at a town hall meeting or an open house, for example, are normally there to inform themselves and have their voices heard, not to develop consensus on appropriate forest management practices. In general (Coglianese 1999, p. 15),

> *Rather than viewing conflict as the problem and consensus as the solution, public managers should instead focus squarely on the substantive problems facing the environment and regulated firms. They should decide when and how to engage in public dialogue based foremost on what will serve the overall public interest, not on what will lead to a consensus among those inside the policy loop.*

For forest managers, it is interesting to note that "consensus in and of itself may not play much of a role in determining which processes will be successful . . . the data provide some support for recent challenges to the wisdom of seeking consensus as an explicit goal of public participation" (Beierle and Cayford 2002, p. 47). In summary, no consensus exists as to whether and when a consensus-based approach is the best method to choose! While consensus should be sought in some situations, other PI methods often produce many of the benefits claimed by consensus-based approaches but without much of the resources, information support, contention, and time required by this model.

Issues with government participation and coordination

Not only do stakeholder groups arrive with different expectations, but institutionally, government and industry are often at the PI table to "inform" stakeholders about "how things are done" rather than to share management authority (the "decide-announce-defend" approach). More often than not, government sees participation as a vehicle to inform and make the public aware of issues rather than to share or reform management power and responsibility. Public involvement processes can create expectations that government agencies may be unable or unwilling to respond to (SERM 1995b). This is particularly true when dealing with Aboriginal communities and natural resource issues, since many First Nations see co-management and other high levels of PI as "consultation", an interim measure on the road to eventual self-government and treaty or land claim settlements. Common concerns cited by PI participants as major sources of conflict and frustration with government representatives include slow decision making, sending low-level managers with little or no authority, poor communication, and a division of responsibility for overlapping issues within government (Chambers 1999).

A central dilemma for government is to maintain the balance between ministerial authority and the delegation of management responsibility and decision-making powers to the local level. In theory, government officials have the authority to make resource-management decisions, since these managers act on behalf of their superiors, who are democratically elected. That is, they are expected to uphold the public good. However, contemporary pressures pit government-as-resource-developer against government-as-environmental-trustee. At the same time, civil servants are starting to acknowledge that they need information and knowledge that can only be tapped through local, decentralized processes. As a central tenet of SFM, PI often comes in conflict with government's desire to maintain decision-making control. For example, these processes challenge conventional resource managers to integrate the concerns of other (often previously ignored, yet equally important) resource users into planning and operations. These other forest users may have completely different values, and thus opinions and priorities concerning resource management, than government managers.

The public must take on responsibilities as they accept these new rights that come with increased participation

It is important to note that with the new opportunities being granted to communities and civil society through PI processes comes the responsibility of citizens to become adequately prepared to make meaningful input (CIF/IFC 1998). For example, communities that are directly involved in, and benefiting from, increased decision-making powers regarding their local forest resources also have a responsibility to learn about related issues such as forest ecology, the economics of forest products, and other stakeholder values. Indeed, by accepting a role in forest management, the public assumes responsibility and seeks to be knowledgeable and informed (CCFM 1998). However, the reality is that the public often lacks the basic knowledge, capacity, and interest to effectively participate. For instance, the public may want good outcomes, but they are not willing to put in the time to achieve this by consistently attending meetings for weeks or months. Individuals are often unlikely to invest time or other resources in PI unless they feel that an issue affects them directly (Hamersley Chambers 1999). In a 1996 national survey, only 21% of the Canadian public sampled considered themselves to be well informed on forest issues in their local area, and only 13% of the national sample felt well informed on national forest issues (Robinson et al. 1997).

The limits of science in prescribing management practices

Participants at a 1999 Calgary SFMN conference on forest co-management felt that some PI processes have the potential to undermine Aboriginal rights, culture, language, and aspirations. For example, a lack of capacity (in areas such as management experience and institutional structure at the local level) often leads to regimes and knowledge from the dominant culture (i.e., driven by commercial concerns and reductionist science) being imposed and forming the centre of the process. As a result, local systems of knowledge and management are often invalidated and ignored (Chap. 3). In order to avoid this situation, participants must recognize the limits of science in prescribing management practices. Participants (especially the industry and government sponsors of participatory activities) must recognize that values issues have a legitimate place in forest management. Science can tell us what can happen by manipulating the system, but what we want from the system is a value judgement, not something that science can tell us. Science, and the information that scientific studies generate, can be useful for constructing scenarios and predicting responses to various actions. But science cannot tell us what the best course of action is. That depends upon our management objectives, and our management objectives derive from our values. A worthy pursuit in many face-to-face PI processes is to present the group with scenarios based on scientific findings (Chap.14), to have this form the basis for discussions about what ought to be done.

Other forms of knowledge must also be recognized and afforded some legitimacy. Most of traditional knowledge related to ecosystems is based on long-term observation (sometimes over generations). Scientific studies, in contrast, typically involve larger sample sizes or more observations, but over a shorter time horizon. Therefore, both types of knowledge may prove useful. Too often, western science is privileged at the management table and for both structural and cultural reasons, traditional knowledge is devalued.

Resource managers misunderstand public values and preferences

One of the reasons that it is crucial to solicit the views and values of the public (or of the various publics) is that these views and values have proven to be significantly different from those of professional resource managers. For a number of decades, resource managers rather cavalierly assumed that they knew what the public wanted. This assumption eventually resulted in management actions being implemented that were out of step with the mainstream. For example, Wagner et al. (1998) describe survey results from Ontario that compared professionals' value orientations and perceptions of risk with those of the general public:

Forestry professionals tended to be less supportive of some environmental values and forest management goals, perceive everyday and forestry activities to be less risky, be more trusting of science and government, and be more accepting of forestry activities than the general public.

This sort of finding has been confirmed several times over the past decade and a half (see Twight and Lyden 1989; Vining and Ebreo 1991; McFarlane and Boxall 2000). In fact, warnings about paying attention to value differences between resource professionals and the public go back several decades. In the 1960s, Behan (1966) suggested that professional foresters and the public differ substantially in their perception of risk, their acceptance of science over participatory democracy as an organizing principle for resource management decision making, and in their view of the proper role of resource professionals in society. Behan (1966) forcefully contended that the role should be to implement the public will (for public lands management, anyway) rather than to determine management objectives *for* the public.

Public involvement is critical precisely because of the demonstrated differences between the values of resource management professionals and those of the public. In Canada, most registered professional foresters either serve the public directly (as employees of provincial or federal governments) or they manage the public's land (as industrial foresters operating on Crown land). In either case, these professionals need to realize that they have been educated and socialized in a manner that has often created a gulf between them and the public. Their own codes of professional ethics suggest that they need to do everything they can to bridge that gulf.

Inter-cultural communication barriers

Since many public involvement processes involve Aboriginal communities and traditional resource users, intercultural communication must be addressed. For example, in much of boreal Canada it is not only basic courtesy but essential for effective participation to determine if English, French, or Cree is the preferred language, and if translators are required so that nobody is excluded from participating. As indicated in Chap. 3, cultural biases are pervasive and are expressed in more than language differences. In fact, "cultural" (or rather, "sub-cultural") differences exist among most communities and constituencies of the public, as reflected in different values, assumptions, jargons, and heroes of professional foresters, bureaucrats, environmentalists, and businesspeople.

Achieving effective public involvement in forest management

How do we undertake effective public involvement?

In order to advance and improve the practice of PI in forest management, a number of actors in society have key roles to play. In this section, we review some of elements that we believe are necessary to further the development of public involvement practice in Canada; some of these are summarized in Box 4.7. As mentioned previously, the changes required to improve current PI initiatives include both micro-process issues (tinkering and experimenting with specific mechanisms and techniques), but also macro-structural issues such as institutional reform, political will, and willingness to pay and to participate. Some of these problems are inter-related. For example, research has demonstrated that some people do not participate because they do not believe their involvement will have an impact (Crowfoot and Wondolleck 1990). If institutional

> **Box 4.7. Recommended options for effective public involvement (PI) in sustainable forest management (SFM):**
>
> - Reassess both micro-process and macro-structural issues;
> - Industry to embrace PI as a central tenet of SFM practice;
> - Industry to provide staff with appropriate PI training;
> - Industry to send fairly senior staff to PI processes;
> - Industry to communicate in non-technical language;
> - Stakeholders to continue participation even if not all goals are met;
> - Stakeholders to educate themselves about related forest issues;
> - Stakeholders to be clear, honest, and consistent about desired outcomes;
> - Stakeholders to contribute funding and build capacity;
> - The general public to increase their knowledge and capacity;
> - The general public to contribute funding to the process;
> - Academics to examine the issue of non-participation in PI;
> - Academics to develop new curriculum and innovative PI approaches;
> - Government to provide technical, administrative, policy, and financial support;
> - Government to incorporate PI input into planning, decisions, and operations;
> - Government to send fairly senior staff to PI processes with appropriate training; and
> - Government to support alternative structures such as shared management boards.

reform increased public confidence in PI mechanisms, it could be easier to recruit participants and respondents. Perpetuating ineffective processes reinforces the cycle of poor attendance at events, and allows industry and government to claim that ". . . everyone must be satisfied because no one comes to the public meetings".

As stated above, people participate in PI for a wide variety of reasons. Likewise, the definition of what constitutes "effective" public involvement differs significantly, depending on who is asking the question, what their values are, and what their goals for the process are. Public involvement can be seen as a necessary evil for all participants, but this does not mean that it should be entered into with a negative attitude.

There are very real incentives for different communities of interest to participate and make the PI process either effective or doomed to failure. For example, what incentives are there for government to give up management authority to a community, or for a forest products company to undertake an expensive process that will negatively impact their profit margin? Government might decide that it can save money by delegating management functions to the local level and the forest industry might conclude that the reduced production and poor publicity associated with a road blockade will cost them more in the long term than undertaking PI. In these cases, the benefits of public involvement outweigh the costs. A simple definition of effective public involvement therefore might be that the benefits to the parties involved are greater than the costs that would be incurred by their not participating. For example, Mistik Management, a forest company operating in northwestern Saskatchewan, first invested in PI in a significant way after an 18-month logging road blockade. This conflict generated negative publicity both nationally and internationally and significantly affected their operations in the area.

Although this was a negative incentive for undertaking PI, the resulting co-management boards (Fig. 4.3) and benefits have far exceeded most expectations at the start of the process.

Industry

The forest products industry clearly has a key role to play in PI. It operates at the front line of forestry; its actions directly impact the public. Industry has made significant progress in embracing PI as an important and necessary part of the practice of forest management in Canada but more needs to be done. The basis for increased PI (that is to say, the need and demand for it) has been demonstrated earlier in this chapter. For years, industry paid lip service to this idea, but in reality, many industry practitioners tried to pass off public relations as PI. In general, public relations involves a simple one- or two-way flow of information (getting the message out) where industry asks the public what their values and preferences are and takes these into account at their discretion. Public involvement is based on two-way communication that provides a role for the public in each of three policymaking stages: (1) defining the problem, (2) coming up with options to solve it, and (3) choosing among those options. In contrast, the main goal of public relations is to sell the preferred option presented in Stage 3 (Tom Beierle, Resources for the Future, Washinton, D.C., personal communication, 2003).

    Another responsibility for industry is to train staff members to effectively participate in PI processes. This may entail company-wide training on cultural awareness and respect for diversity, as well as training for specific staff on the various tools available

Fig. 4.3. The Canoe Lake Co-management Board in the NorSask Forest Management License Area of northwestern Saskatchewan. Government agencies and forest products companies are now experimenting with different models of co-management, often considered the zenith of public involvement in sustainable forest management.

Canadian Press/Halifax Daily News photo by Paul Darrow

to solicit public involvement. Further to this, companies might consider hiring people trained in mediation or conflict resolution, or send their own employees back to school to obtain these skills.

Industry must also demonstrate a commitment to public involvement by sending fairly senior staff to public involvement processes. Sending inexperienced or junior staff sends the wrong message. It suggests that the process is not important to the company. There are additional dangers to sending junior staff to such processes as official corporate representatives. For example, they may not have enough experience with the company to clearly articulate company positions, interests, or desired outcomes to the public.

Industry sponsors also need to be sensitive to the level of technical knowledge held by the public. Nothing can quiet a crowd of lay participants faster than strings of undefined acronyms and long-winded dissertations by company staff on certification standards, yield curves, and computer simulation of wood supply. Every effort must be made to convey this sort of information in understandable, non-technical language. Finally, industry needs to demonstrate leadership, creativity, and innovation. This is what market-oriented institutions are supposed to do best. There are many existing tools, such as opinion research, workshops, interactive field tours, and various other PI methods that involve direct contact with stakeholders and the public that could be improved upon with industry input.

Stakeholder groups

Stakeholder groups ("communities of place" and "communities of interest") are the "bread and butter" of PI in Canadian forest management. These groups have been some of the most articulate proponents of the need to integrate PI into SFM and have been the most consistently active participants in existing processes. Through this participation, many have learned a great deal about the broad spectrum of forest values that exist in society. Some have responded to this well, and others have responded by redoubling their efforts to promote their own positions. It is incumbent upon the organized stakeholder interests in forest management to recognize that they will not always get their way, but that this is not a sufficient reason to abandon the process of dialogue on desired forest outcomes. Stakeholders need to commit to continuing to work within an agreed-upon system for constructive dialogue, even if this does not always achieve their desired results.

We have already suggested that stakeholders and the general public involved in PI have a responsibility to educate themselves on forest issues. There is some self-interested motivation in doing this, as professional resource managers tend to place much more value on stakeholder opinions if they appear to be well-reasoned and informed. This is not to say that value-based input is less legitimate than input informed by western science. In fact, there is growing consensus that the dichotomy between "facts" and values is a false one. Values shape what scientists choose to study, how they formulate their hypotheses, and how they report their results. Stakeholders may choose to inform their opinions from a variety of sources. However, if participants can demonstrate that they have read and understand some of the scientific background material on a given topic, they may gain some legitimacy among their co-participants. That may make them more open to less science-based input. We believe that many of the desired outcomes for forests are shared by most stakeholders (e.g., healthy, productive, ecologically sustainable forests). Where differences often lie is in the perception of risk and in the preferred means to achieve specific outcomes.

Stakeholders also need to be clear, consistent, and honest about their ultimate desired outcomes. Industry and government have expressed some frustration with the environmental community for "always wanting more" whenever concessions are made or incremental reforms are introduced. On the one hand, this continual redefinition of goals and objectives is part of adaptive management. On the other, industry and government officials often feel that nothing they do will ever satisfy certain hardline stakeholders. If stakeholders honestly articulate what would ultimately satisfy them (even if those goals include the absence of industrial forestry in a given area), that offers a clear starting point for negotiation and compromise. The incremental approach often practiced gives industry and government false hope that conflict will "go away" once some minor tinkering with the status quo system of management occurs.

Secure, long-term financial support is particularly important to the success of PI. While it is important to have initial funding available to get community members properly involved and be able to match the resources of government or industry, a community should plan to start to generate its own income to support the process. During a 1999 SFMN workshop on forest co-management, some practitioners saw government funding as a ploy by government to control the process, an umbilical cord that keeps people disenfranchised and dependent. Based on this perception, the only way out of the situation would be for local resource users to contribute their own funding, since "those who pay the piper pick the tune". There are many ways to facilitate this financial independence for stakeholder tables, including revenue-sharing by resource users, seeking grants from non-profit foundations, and other alternatives such as casino revenue (Chambers 1999).

If some fundamentally different sort of management scenario with greater stakeholder or community authority is a desired outcome for certain stakeholder groups, then they must demonstrate a capacity to take on that role. This means demonstrating an understanding of forest ecology, the capacity to model treatments and responses, and the ability to evaluate the likely socio-economic consequences of alternative forest management scenarios (feasibility and benefit cost studies). This knowledge will help ensure that society has the necessary information to make informed tradeoffs. There are many potential sources and bases of knowledge, including science, direct experience, the media, cultural traditions and practices, religion, and others. The method chosen should help stakeholders to articulate from where their forest values emerge, on what basis their interests or positions are held, and how negotiable or non-negotiable their interests may be, depending upon their origin.

One of the most commonly identified constraints in PI is that of a lack of capacity (technical, logistical, and financial expertise) within communities. Community capacity refers to the ability of communities to meet PI needs such as experienced translators, secretaries, and leadership from within each community instead of relying on outside consultants (Chap. 5). Building the capacity of participants helps to develop an informed public. During a 1999 SFMN-sponsored forest co-management workshop, most practitioners cited an urgent need for local people with leadership, negotiating, secretarial, and organizational skills to undertake management responsibilities. For example, facilitators and leaders must have enough management skills to care for participants and prevent burnout. With so many different and often competing interests represented, there is a great need for leaders and participants with strong diplomacy and cross-cultural sensitivity skills. Practitioners also cited many problems with training and keeping these people involved in the process. As in industry and government, there is a problem that once an individual has received training and management experience, he/she often leaves for a better job somewhere else. Participants also need to gain something from their involvement. This might include stipends, desired skills, respect from their peers, or a sense of accomplishment through their participation (Chambers 1999).

To address these concerns, public involvement processes can provide on-the-job training to improve the capacity of members. When outside experts must be hired, a "replacement" policy can be built in to ensure that the community builds capacity. Successful examples of this include contracts whereby the consultant has a trainee who "job shadows" while the consultant "works himself out of a job". Finally, a goal of capacity building should be to give traditional resource users and community representatives the ability to articulate their concerns in a medium that government and industry understand and to educate these other players to acknowledge traditional knowledge and methods (Chambers 1999).

Training and education opportunities offered through PI have the potential to strengthen community health as people gain skills and feelings of pride and self-worth. Other related issues include the reluctance of many community members to leave the community to receive training, internal divisions, and jealousy by members who do not benefit directly from the process. If the community is suffering from unmet basic needs (like reliable employment and decent housing) and social issues such as family violence and drug abuse, then its members cannot be expected to focus their attentions on issues such as forest management.

The general public

It goes without saying that in order for PI to be effective, informative, and useful to resource managers, the public needs to be active in the process. This may not always mean a long-term commitment. Rather, it may be as simple as responding to public opinion surveys. Some members of the public must be willing to put in more time and to follow through with monitoring, but this does not have to include everyone.

Another key role of the public involves their willingness to share the costs for PI. Public involvement is an input into the management process. If we want more public involvement in forest management, and if we want better quality PI, it will cost money. How we pay can vary. We can pay higher taxes if the government is to be held accountable for improving PI, or we can pay higher prices for forest products if industry is to provide it. In either case, consumers or taxpayers must be willing to contribute more money to the cause, or redirect public monies from other priorities.

Government

The role of government in PI is to provide technical, administrative, financial support, and advice to the process, as well as to represent wider provincial and national interests. Public involvement will not likely be successful if the values elicited through such a process are not reflected in planning, decision making, and operations. In such instances, all that PI accomplishes is to raise expectations. So listening well is not enough. Acting upon public expectations and concerns by requiring that forestry be conducted in a way that is socially acceptable is a significant role of government in PI.

Since government currently holds the legal and management responsibilities for most natural resources in Canada, it is critical that bureaucrats be involved in the PI process. However, government representatives must be accountable and be senior enough to make decisions at the table. Junior bureaucrats who must constantly confer with more senior managers slow down the decision-making process and lead to frustration on both sides. Linking outside researchers and support services with community needs and concerns should also be a major government role.

Most importantly, government must be willing to fund public involvement in forest management. It can be expensive to do PI well. Such group processes make very high demands of time and support resources, especially when technically complex issues are involved and the process involved a consensus model (Duinker 1998). There are a number of potentially significant costs associated with undertaking PI, such as those associated with measuring, assessing, and incorporating the elicited public sentiment. For group processes (such as focus groups, open houses, and public hearings), common expenses include facility rentals, mileage for participants, professional facilitation, snacks, and distribution of information. Costs associated with social science methods such as surveys and participant observation include printing, graduate student stipends, field expenses, telephone or postage, data entry and analysis. Government organizers should also be willing to extend deadlines as well as ensure that sufficient technical resources and reliable information are available to participants in a timely fashion if educated decisions are to be made (Duinker 1998, p. 109).

Similar to industry, government needs to develop or hire more expertise in the area of PI, conflict resolution, and mediation. This too will cost money, or will require a shift in departmental priorities. The significant downsizing that has occurred in many natural resources departments across Canada has involved major structural change in how these agencies do business. Such institutional changes often present opportunities to invest human resources in new priority areas. For example, government downsizing presents an opportunity for those directly affected by forest management to be more directly involved in PI. Rather than serving as a go-between, government can support new institutional structures such as joint ventures and co-management that entail direct contact between stakeholders, community residents, and those that are most involved in on-the-ground management of forest ecosystems (i.e., the forest products industry). If government helps to set the rules of engagement and provides oversight, they could foster more and better quality PI with a modest amount of human resources. However, doing this means relinquishing some power and decision-making authority.

Government's role in PI varies, depending on the nature and purpose of a particular planning exercise. In order to gain public trust and to add legitimacy to any public process, government departments need to clarify their roles and consistently act within these. For example, in some processes, government may be the sponsor, while in other processes the government may be just another stakeholder. Or more likely, one government agency may be the sponsor of a planning process (e.g., the forestry branch), but experts from other government departments (environmental protection, fish and wildlife, etc.) may be at the table as stakeholders, representing stakeholders, or simply to provide information. There is clearly a need to identify whether government takes part in PI processes as a stakeholder or as a facilitator; often both roles are filled, though by different government agencies. For those people serving as facilitators (whether government or not), it is important that the responsible individuals are trained for this role in the areas of conflict resolution, mediation, and facilitation. When government or government agencies participate as a stakeholder, they must be prepared to take a place at the table as an equal participant (i.e., not one that holds the trump card of regulatory authority). Many outsiders view government as monolithic and uniform in its interests. Clarification of the various roles and mandates of the different government departments that may be involved will help clarify the situation for participants.

Academics

Public involvement is a new but important area in forest management. As such, academics have a key role to play in raising the profile and improving the quality of these processes. We need to learn more about what works well in PI, what does not seem to work, and why. Unfortunately, it is impossible to do controlled, laboratory-like experiments in this area. At best, we can treat the diversity of PI processes that exists in the world as a natural experiment. While we cannot control for certain variables, through observation and comparison among processes, we can advance our understanding on this topic.

A specific research contribution that academics can make is to examine the issue of non-participation in PI. Some research on this topic has recently been published (Diduck and Sinclair 2002). We need to understand more completely why some people do participate, but also why most do not. Is it because they feel the costs in terms of time or money (travel) are too high, or because their expectations of effecting any real change are too low? Is it simply a matter of apathy? Or have the sponsors of such activities done a poor job of recruiting? Are people reluctant to donate their time to help someone else do their job (i.e., work they get paid for)? As well, research on non-participants could reveal the degree to which this segment of the public has values that differ dramatically from those who do participate.

Academics might also contribute by proposing novel and innovative ways for including the views and values of members of the general public in forest management. At present our level of sophistication in how to undertake PI is low.

Academics can contribute to improved PI in forest management by incorporating information on the subject into their teaching curricula. There are two dimensions to this. They can instil in their students an understanding that conflict management and human relations are an important part of forest management. As well, they can offer some tools and resources to actually begin to develop a conflict management toolkit for resource professionals.

Promising future directions

Since the 1970s, PI has played an increasingly critical role in Canadian forest management planning, decision making, and operations. The fact that the public is now an accepted and important player in forest management planning exercises is reflected in national and provincial forest-related policy and regulations. While practitioners and academics agree that the theory and practice of PI is still maturing, the principle that the public has a right to participate is now embraced in virtually all jurisdictions for all levels of decision making. There is a growing appreciation that the absence of meaningful public participation in decisions has potentially serious consequences. Decisions that fail to balance public interests (through a lack of reliable information or as a result of relevant interests not being given sufficient attention) lead to instability, continuing conflict, lack of sustainability, and long-term inefficiencies in the use of government personnel and funding (BCCORE 1995).

Despite the increasing recognition and practice of PI, however, a number of important caveats still exist. Increased PI in the management of forest resources under conventional systems does not automatically lead to SFM. Despite progress, there is still a concern that mechanisms for public input in forest decision making require strengthening and further development (CIF/IFC 1998). There is also a need to establish minimum criteria for monitoring and measuring success so that groups can evaluate their experiences and learn from them under the umbrella of adaptive management. Since these are all barriers to SFM, these must be addressed and resolved in order for this goal to be achieved.

There may be a false assumption that just because local communities (Aboriginal and non-Aboriginal) are involved, forest practices will be acceptable to the public and more or less sustainable. It is extremely difficult for unhealthy communities to undertake the management responsibilities and functions required for successful PI. Local control and ownership of forest resources therefore does not always lead to SFM. The assumption that combining a PI process, stakeholder groups, and local control automatically leads to SFM does not bear up under scrutiny (Chambers 1999).

In Canada, it appears that we can look forward to a growing maturity of practice led by creative process designers and implementers who are experimenting with an increasing variety of methods. These people are supported by a sharing of successes and failures through active networks of practitioners and participants, such as the SFMN. It is safe to assume that some public advisory groups will demonstrate sufficient maturity and willingness to assume higher degrees of responsibility, and will eventually begin sharing decision-making roles with professionals and agency officials (CIF/IFC 1998).

At the same time, the trend of government cutbacks and decentralization opens up both opportunities and potential pitfalls for PI in forest management. In British Columbia, for example, "[Premier] Campbell's pledge to streamline land-use planning has reduced opportunities for public input . . . Twenty District Offices [of the Ministry of Forests] will be closed, reducing opportunities for public participation in managing our public forest resources" (Wareham and Stauffer 2002).

While there are obvious short-term gains, it appears that the most significant benefits of practicing forest PI are long term. It is interesting that in an industry that by its very nature is required to take the long-term view on account of the time required to grow a crop of trees, people often expect quick solutions in the sphere of PI. Those who do are often dissatisfied with experiences that do not yield tangible, immediate returns. Just like a marriage or a partnership, the establishment of the relationship is just the beginning. The harder work, and the more important work, is maintaining that relationship decade after decade. In conclusion, it is those people most affected — the local communities, resource users, and youth (who will inherit the land, the employment, and lifestyle opportunities) — by forest activities who should ultimately judge the success of a PI process (Chambers 1999). We will never have "social acceptability" from all members of the public, but we will, hopefully, continue to use PI to improve forest management and decision making. "Forest management is really just an attitude. It begins with wants and values, is driven by people and eventually policies and is fine tuned by science" (Hebert 1999). Public involvement is therefore a central part of this iterative process that aims to achieve sustainable management of Canada's forest heritage.

Acknowledgements

The authors would like to extend our sincere thanks to the Sustainable Forest Management Network for the funding to undertake this research, and SFMN partners for providing us with photos to illustrate our chapter. Anonymous reviewers provided us with excellent suggestions to improve the content and flow of our text — thank you!

References

Adamowicz, W. 1999. Preface and abstract. *In* Proceedings of the 1999 Sustainable Forest Management Network Conference: Science and Practice: Sustaining the Boreal Forest, 14–17 February 1999, Edmonton, Alberta. *Edited by* T. Veeman, D.W. Smith, B.G. Purdy, F.J. Salkie, and G. Larkin. Sustainable Forest Management Network, Edmonton Alberta. pp. v–viii and pp. 1–2.

Barker, T., Oberle, F., and Vinge, T. 1999. Forest ecosystem management planning: practical lessons from the P1/P2 pilot project. *In* Proceedings of the 1999 Sustainable Forest Management Network Conference: Science and Practice: Sustaining the Boreal Forest, 14–17 February 1999, Edmonton, Alberta. *Edited by* T. Veeman, D.W. Smith, B.G. Purdy, F.J. Salkie, and G. Larkin. Sustainable Forest Management Network, Edmonton, Alberta. pp. 205–213.

(BCCORE) British Columbia Commission on Resources and Environment. 1995. Public participation: rights and responsibilities, Community Resource Boards. Volume 3, Provincial Land Use Strategy. Commission on Resources and Environment, Victoria, British Columbia. 153 p. Available at http://www.luco.gov.bc.ca/lrmp/plus/vol3/vol3-01.htm#P55_3385 [viewed 4 June 2002].

Beckley, T. M., and Korber, D. 1995. Sociology's potential to improve forest management and inform forest policy. For. Chron. **71**: 712–719.

Behan, R.W. 1966. The myth of the omnipotent forester. J. For. **64**: 398–407.

Beierle, T., and Cayford J. 2002. Democracy in practice: public participation in environmental decisions. Resources for the Future Press, Washington, D.C. 149 p.

Bengston, D.N. 1994. Changing forest values and ecosystem management. Soc. Nat. Resour. **7**: 515–533.

Berkes, F. (*Editor*). 1989. Common property resources: ecology and community-based sustainable development. Bellhaven Press, London, U.K. 312 p.

Berkes, F. 1994. Bridging the two solitudes. Canadian Arctic Resources Committee, Ottawa, Ontario. North. Perspect. **22**(2–3): 18–20.

Bonnell, B. 2001. Assessing public opinion on sustainable forest management in western Newfoundland. Western Newfoundland Model Forest, Corner Brook, Newfoundland. 60 p.

Buchy, M., and Hoverman, S. 2000. Understanding public participation in forest planning: a review. For. Policy. Econ. **1**: 15–25.

(CCFM) Canadian Council of Forest Ministers. 1995. Defining sustainable forest management: a Canadian approach to criteria and indicators. Canadian Council of Forest Ministers, Canadian Forest Service, Ottawa, Ontario. Available at http://www.ccfm.org/ci/fra6_e.html [viewed 15 June 2003].

(CCFM) Canadian Council of Forest Ministers. 1998. National forest strategy 1998–2003: sustainable forests, a Canadian commitment. Canadian Council of Forest Ministers, Canadian Forest Service, Ottawa, Ontario. Available at http://nfsc.forest.ca/strategy4.html#contents [viewed 15 June 2003].

(CFS) Canadian Forest Service. 2001. The state of Canada's forests, 2000–2001; sustainable forestry: a reality in Canada. Available at http://www.nrcan.gc.ca/cfs-scf/national/what-quoi/sof/latest_e.html [viewed 9 June 2002].

Chambers, F. 1999. From co-management to co-jurisdiction of forest resources: a practitioners workshop. October 28–30, Calgary, Alberta. Sustainable Forest Management Network, Edmonton, Alberta. 25 p.

(CIF/IFC) Canadian Institute of Forestry/Institut forestier du Canada. 1998. Position paper: public partici- pation in decision-making about forests. Canadian Institute of Forestry, Ottawa, Ontario. Available at http://www.cif-ifc.org/publicparticipation.pdf [viewed 10 July 2003]. 4 p.

Coglianese, C. 1999. The limits of consensus. Environment, **41**: 28–33.

Crowfoot, J.E., and Wondolleck, J.M. (*Editors*). 1990. Environmental disputes: community involvement in conflict resolution. Island Press, Washington, D.C. 295 p.

Devall, B., and Sessions, G. 1985. Deep ecology: living as if nature mattered. Peregrine Smith, Layton, Utah. 267 p.

Diduck, A.P., and Sinclair, A.J. 2002. Public involvement in environmental assessment: the case of the non- participant. Environ. Manage. **29**: 578–588.

Duinker, P.N. 1998. Public participation's promising progress: advances in forest decision-making in Canada. Commonw. For. Rev. **77**(2): 107–112.

Feit, H. 1988. Self-management and state-management: forms of knowing and managing northern wildlife. *In* Traditional knowledge and renewable resource management in northern regions. *Edited by* M.M.R. Freeman and L.N. Carbyn. Boreal Institute for Northern Studies, Edmonton, Alberta. Occ. Pub. No. 23. pp. 72–91.

(FHI) Fraser Hamilton Inc. 1996. The NorSask Forest story: managing a forest in northwestern Saskatchewan. Fraser Hamilton Inc., Edmonton, Alberta. 172 p.

(FSC) Forest Stewardship Council. n.d. Principles and criteria. Available at http://www.fscoax.org/princi- pal.htm [viewed 2 May 2002].

Fullerton, G. 2000. Backyard forestry: Irving and worried neighbours reach a compromise. Atl. For. Rev. **6**(3): 56–58.

Gregory, R. 2000. Using stakeholder values to make smarter environmental decisions. Environment, **24**(5): 34–44.

Hamersley Chambers, F. 1999. Co-management of forest resources in the NorSask FMLA, Saskatchewan: a case study. Masters thesis, University of Calgary, Calgary, Alberta. 183 p.

Hebert, D. 1999. Embracing change. *In* Proceedings of the 1999 Sustainable Forest Management Network Conference: Science and Practice: Sustaining the Boreal Forest, 14–17 February 1999, Edmonton, Alberta. *Edited by* T. Veeman, D.W. Smith, B.G. Purdy, F.J. Salkie, and G. Larkin. Sustainable Forest Management Network, Edmonton, Alberta. pp. 3–5.

Hillery, G. 1955. Definitions of community: areas of agreement. Rural Soc. **20**(June): 111–123.

Holling, C.S. (*Editor*) 1978. Adaptive environmental assessment and management. John Wiley and Sons, New York, New York. 377 p.

Kessler, W.B. 1999. Sustainable forest management is adaptive management. *In* Proceedings of the 1999 Sustainable Forest Management Network Conference: Science and Practice: Sustaining the Boreal

Forest, 14–17 February 1999, Edmonton, Alberta. *Edited by* T. Veeman, D.W. Smith, B.G. Purdy, F.J. Salkie, and G. Larkin. Sustainable Forest Management Network, Edmonton, Alberta. pp. 16–22.

Lee, K.N. 1993. Compass and gyroscope: integrating science and politics for the environment. Island Press, Washington, D.C. 243 p.

McFarlane, B.L., and Boxall, P.C. 2000. Forest values and attitudes of the public, environmentalists, professional foresters, and members of public advisory groups in Alberta. Canadian Forest Service, Edmonton, Alberta, and Foothills Model Forest, Hinton, Alberta. Inf. Rep. NOR-X-374. Available at http://www.fmf.ab.ca/attitudes.html [viewed 15 June 2003].

Notzke, C. 1993. The Barriere Lake trilateral agreement. University of Lethbridge, Lethbridge, Alberta. Available at http://www.cnie.org/NAE/docs/notzke.html [viewed 9 July 2003]. 93 p.

Osharenko, G. 1988. Sharing power with Native users: co-management regimes for native wildlife. Canadian Arctic Resources Committee, Ottawa, Ontario.

Pinkerton, E. (*Editor*). 1989. Co-operative management of local fisheries: new directions for improved management and community development. UBC Press, Vancouver, British Columbia. 312 p.

Roberts, K. (*Editor*). 1996. Circumpolar Aboriginal people and co-management practice: current issues in co-management and environmental assessment. Proceedings of a conference, 20–24 November 1995, Inuvik, Northwest Territories. Arctic Institute of North America, Calgary, Alberta. 172 p.

Robinson, D., Robson, M., and Hawley, A. 1997. Social valuation of the McGregor Model Forest: assessing Canadian public opinion on forest values and forest management — results of the Canadian Forest Survey '96. Prepared for the McGregor Model Forest Association, Prince George, British Columbia. Available at http://www.mcgregor.bc.ca/publications/publications.html [viewed 15 June 2003].

(SERM) Saskatchewan Environment and Resources Management. 1995*a*. A framework for integrated land and resource management planning. Saskatchewan Environment and Resources Management, Sustainable Land Management Branch, Regina, Saskatchewan.

(SERM) Saskatchewan Environment and Resources Management. 1995*b*. Public involvement in the management of Saskatchewan's environment and natural resources: a policy framework. Saskatchewan Environment and Resource Management, Public Involvement Working Group, Regina, Saskatchewan.

(SFMN) Sustainable Forest Management Network. 1999. Discussion summary of the Public Involvement Taskforce. Meeting held 23 August 1999, Grande Prairie, Alberta. Sustainable Forest Management Network, Edmonton, Alberta. 4 p.

Shindler, B., and Neburka, J. 1997. Public participation in forest planning: 8 attributes of success. J. For. **95**(1): 17–19.

Smith, P. 1995. Aboriginal participation in forest management: not just another "stakeholder". National Aboriginal Forestry Association, Ottawa, Ontario. 11 p.

Statistics Canada. 2003. Highlights from the 2001 census of population. Available at http://geodepot.statcan.ca/Diss/Highlights/Page1/Page1_e.cfm [viewed 10 July 2003].

Stauch, J. 1997. Resident participation in non-profit housing: a case study. Master's thesis, Faculty of Environmental Design, University of Calgary, Calgary, Alberta. 116 p.

Susskind, L., and Cruikshank, J. 1987. Breaking the impasse: consensual approaches to resolving public disputes. Basic Books Inc., New York, New York. 288 p.

Susskind, L., Amundsen, O., Matsuura, M., Kaplan, M., and D. Lampe. 1999. Using assisted negotiation to settle land use disputes: a guidebook for public officials. Lincoln Institute for Land Policy, Cambridge, Massachusetts. 26 p.

Twight, B.W., and Lyden, F.J. 1989. Measuring Forest Service bias. J. For. **87**(5): 35–41.

Vining, J., and Ebreo, A. 1991. Are you thinking what I think you are? A study of actual and estimated goal priorities and decision preferences of resource managers, environmentalists and the public. Soc. Nat. Resour. **4**: 177–196.

Wagner, R., Flynn, J., Gregory, R., Mertz, C.K., and Slovic, P. 1998. Acceptable practices in Ontario's forests: differences between the public and forestry professionals. New For. **16**: 139–154.

Walters, C.J. 1986. Adaptive management of natural resources. McGraw–Hill, New York, New York. 374 p.

Wareham, B., and Stauffer, J. 2002. Attacking the foundations of sustainability: what provincial cuts mean to environmental quality in B.C. The Sierra Rep. **20**(2): 1.

(WCED) World Commission on Environment and Development. 1987. Our common future. Oxford University Press, Oxford, U.K. 400 p.

Witty, D. 1994. Co-management as a community development tool: the practice behind the theory. Plan Can. **34**(1): 22–27.

Chapter 5

Milltown revisited: strategies for assessing and enhancing forest-dependent community sustainability

Sara Teitelbaum, Tom Beckley, Solange Nadeau, and Chris Southcott

> *"The local community must understand itself finally as a community of interest — a common dependence on a common life and a common ground. And because a community is, by definition, placed, its success cannot be divided from the success of its place, its natural setting, and surroundings; its soils, forests, grasslands, plants, and animals, water, light and air."*

Wendell Berry (1987)

Introduction and theoretical underpinnings

The past decade has seen a rise in interest in the social aspects of forestry. A key reason for this interest is the much-quoted report *Our Common Future* by the World Commission on Environment and Development (WCED 1987), as reviewed in Chap. 2. This report calls for the creation of means to ensure that development is socially, economically, and environmentally sustainable. Following the WCED report, there has been a

S. Teitelbaum. Faculty of Forestry and Environmental Management, University of New Brunswick, Fredericton, New Brunswick, Canada.

T. Beckley.[1] Faculty of Forestry and Environmental Management, University of New Brunswick, Fredericton, New Brunswick E3B 5A3, Canada.

S. Nadeau. Natural Resources Canada, Canadian Forest Service, Fredericton, New Brunswick, Canada.

C. Southcott. Department of Sociology, Lakehead University, Thunder Bay, Ontario, Canada.

[1]Author for correspondence. Telephone: 506-453-4917. e-mail beckley@unb.ca

Correct citation: Teitelbaum, S., Beckley, T., Nadeau, S., and Southcott, C. 2003. Milltown revisited: strategies for assessing and enhancing forest-dependent community sustainability. Chapter 5. *In* Towards Sustainable Management of the Boreal Forest. *Edited by* P.J. Burton, C. Messier, D.W. Smith, and W.L. Adamowicz. NRC Research Press, Ottawa, Ontario, Canada. pp. 155–179.

concerted effort to develop social indicators of sustainability. Social indicators are standard measures or benchmarks of social and economic well-being.

In the forest products sector, government and industry officials as well as researchers have spent considerable effort trying to determine reliable indicators of sustainability for forest-dependent communities. The process has been extremely frustrating to many. The development of reliable social indicators is dependent on subjective interpretations of key concepts such as sustainability, community, and social pathologies. Determining objective measures that meaningfully allow one to compare the sustainability of one human community to another is challenging. Economic and ecological researchers face similar challenges, but they have had greater success in creating quantifiable indicators that reflect ecosystem and economic processes related to forestry. For sociologists these problems are neither new nor surprising. However, they are more evident than ever as the forest management sector has explicitly sought sociological input on the selection and design of indicators of sustainability. The man often cited as the "founder of sociology", Auguste Comte, made the point that sociology would be the last science to develop as a rigorous discipline because social phenomena are more complex and more difficult to understand than the relatively simple phenomena studied by the natural sciences. Max Weber, the early 20th century German sociologist, and probably the most universally respected sociologist today, always maintained that the social sciences and sociology in particular could never be used to develop reliably predictive laws (or indicators) that would be used to tell society what policy options it should follow. For the past three decades attempts to develop rigorous models of social behaviour that can be used to indicate to decision makers what they should do have often been criticized. More recently sociologists have questioned whether "doing sociology" does more harm than good (see O'Neill 1995).

Given this culture in sociology, it is perhaps surprising that any sociologists would be interested in attempting to develop social indicators of sustainability. There are some researchers, however, who are doing so. They are part of a counter-trend to those who question the practical value of sociology. These sociologists are aware of many of the criticisms leveled at them by others and indeed they accept many of them. The critics of the practical value of sociology would tell them that it is impossible to develop a perfect model of sustainability that can be used to develop social indicators. This may be so, but it is also true that it is possible to develop models that, while not perfect, are better than others — and that the development of a better model would be of some use to decision makers as long as they realize its limitations.

There is a long tradition of applied sociology in the United States (U.S.) that is particularly focused on rural communities. This substantial body of work spans the entire 20th century and looks explicitly at issues of community welfare and power, the impact of changing technologies, human–land relationships, and a wide range of other topics. The research was largely carried out through departments of Rural Sociology associated with land-grant universities. Ironically, the vast majority of this work deals only with agricultural rural communities. Despite the location of forestry communities in rural locales, U.S.-based rural sociologists rarely turned their analytical lenses in that direction until quite recently. Foresters, on the other hand, have long been concerned with sociological dimensions of community stability, which they associated with timber supply and employment.

Box 5.1. Challenges to sustainable forest communities:

- Maintaining healthy and productive forest ecosystems that provide a broad spectrum of values and benefits;
- Developing meaningful indicators of community well-being and sustainability:
  - processes and rates of change; and
  - intangible descriptors of cohesion, capacity, and resilience;
- Diversifying economic opportunities through management of the full spectrum of forest values; and
- Balancing local interests in publicly owned forests with the broader interests of an urbanizing population.

In Canada, forest-dependent communities have been a topic of research since the early 20th century. However, it wasn't until the 1950s that characterizing community sustainability or community well-being became an explicit research objective. Since then, social scientists have been working towards the development and refinement of more comprehensive and accurate models and measures of sustainability (Box 5.1). In this chapter, we trace the evolution of research on forest community sustainability in Canada from the 1950s onward. We discuss recent approaches used to characterize community sustainability and the methods used for collecting data. The chapter ends with a discussion of the management implications, particularly of recent work in the field for the forest products industry, government, and communities.

Early research on forest communities

Initially, research on forest-dependent communities revolved around those communities economically dependent on timber extraction or timber processing. In fact, the expression "forest-dependent community" was often synonymous with terms like "single-industry community" or "single-enterprise community". In the first half of the 20th century "company towns" were springing up in many parts of Canada, and private forest products companies and mining companies were playing a major role in their design, construction, service provision, and day-to-day operations. The little sociological research that existed at this time reflected this priority, and tended to be focussed on the planning and operation of forest industry communities (see, for example, Walker 1927, Grimmer 1934, Knight 1975).

The 1950s, 1960s, and 1970s saw a number of more descriptive studies looking at the phenomenon of the single-industry community in Canada and the associated social and economic characteristics typifying these places. In the study, *New Industrial Towns on Canada's Resource Frontier*, Robinson (1962) draws on four Canadian case studies of single-industry towns in order to generalize about the physical and social attributes of new industry towns. He describes their populations as young, male-dominated, and multi-ethnic. He sees high population mobility and volatile resource economies which detract from their overall sense of permanence and stability. However, they have strong participation in local recreation activities and a high number of recreational amenities. Other influential studies of a similar kind include *Single Enterprise Communities* by the Institute of Local Governance at Queen's University (1953) and Rex Lucas' (1971) *Minetown, Milltown, Railtown*. Blishen et al.'s (1979) *Socio-Economic Impact Model for Northern Development* was an early attempt to develop quantitative indicators for the assessment of communities. Robson (1986) provides a useful review of the materials published during this period.

Perhaps the most succinct description of the socio-economic characteristics of single-industry towns comes from Rex Lucas' co-worker, Alex Himelfarb. In an article written in 1977 and published in 1982, Himelfarb outlines the principal social, cultural, and economic characteristics shared by these communities (Himelfarb 1982). In this article, Himelfarb borrows heavily from Lucas but also integrates many of the findings of other research conducted in the 1970s, such as the work associated with the Centre for Settlement Studies at the University of Manitoba. Table 5.1 presents a summary of the main characteristics of single-industry towns. This summary is termed the Lucas/Himelfarb model but it also includes the observations of researchers such as Jackson and Pouchinsky (1971), Matthiasson (1970), and Riffel (1975).

In the United States, research on forest communities at this time was taking a slightly different direction. Population density, geography, and the pace of resource development created a greater management challenge than had yet been experienced in Canada. By the turn of the 20th century, large portions of the U.S. had been mined of the most valuable timber, and in the wake of that rapid development were scores of timber ghost towns (Robbins 1988). The main policy priority was the implementation of

Table 5.1. Characteristics typical of single-industry towns (the Lucas/Himelfarb model).

Main characteristics
1. New (at least in the 1960s)
2. Small (<30 000 people)
3. Isolated
4. Dominated by one resource-based industry
5. Planned
6. Economic dependence on outside forces
7. High degree of instability

Demographic characteristics
16. Highly mobile population
17. High degree of youth out-migration
18. Young population/fewer old people
19. More males than females
20. Larger families
21. Greater ethnic diversity

Economic characteristics
8. One dominant employer
9. Large industrial corporation based outside the region
10. Capital intensive and technology intensive
11. Primarily blue-collar unskilled or
12. Relatively high wages semi-skilled jobs
13. Few employment opportunities for women
14. Small service sector
15. Small retail sector

Cultural characteristics
22. High level of cultural dependence
23. Wage-earner culture
24. Male-dominated blue-collar culture
25. Lower levels of formal education
26. Negative environment for women

sustained yield management. This approach was designed to stabilize the flow of timber, thereby ensuring steady employment for local residents and promoting the long-term viability of forest-based communities (Nadeau et al. 1999). As a result, the notion of forest community *stability* and the measurement of stability became a major preoccupation for social scientists studying forest communities at this time. Because social stability was so closely linked with economic stability, early evaluations of community stability tended to revolve around the examination of economic indicators such as levels of employment, income in forest industries, and prices of forest products (for example, Waggener 1977, Hyde and Daniels 1986). Other research was oriented towards the evaluation of the sustained yield policy and its effect on community stability (Schallau 1974; Schallau 1989). One notable exception was the seminal work of Kaufman and Kaufman (1946), who already in 1946 looked beyond measures of economic success in the analysis of community well-being in the Libby-Troy area of Montana to include such considerations as public participation in the formation of forest policy, strength of local leadership, and training opportunities for youth.

The research undertaken during this time provided a strong foundation for later work on forest community sustainability. The early work clarified the concept of the single-industry community and distinguished it as a unique product of 20th century industrialism with unique sociological properties. However, recently researchers have begun to question the wisdom of relying too heavily on models and assumptions developed decades ago (Pharand 1988; Randall and Ironside 1996; Southcott 2000). Southcott (2000) criticizes earlier models because they are static, meaning that they just present a "snapshot" of a community at a particular point in time. He therefore advocates the development of more dynamic models which can account for the important social changes that are affecting communities, which he argues can be referred to as globalization, post-industrialism, or post-Fordism.

A parallel trend in forest social science in Canada is a more critical vein of scholarship. There are several examples of this. *Trouble in the Woods* (Sandberg 1992) is an edited collection of articles dealing with the allocation and use of the forest in Atlantic Canada. It documents how policy decisions consistently favoured certain interests, such as pulp and paper companies, over others such as woodlot owners and small independent mill owners. Similar themes, applied to a more contemporary case, are explored in Pratt and Urquhart's (1994) book, *The Last Great Forest*. Pratt and Urquhart (1994) describe the rapid expansion of the pulp industry in Alberta, and the role of government and other key policy actors in that process. Richardson et al. (1993) describe the same events from the inside. They focus on the deliberations over the allocation of forest pulp mill licenses in northern Alberta, arguing that community interests came second to those of non-local, corporate interests. While provincial governments invoke "the public interest" for major forest industrialization schemes, the negative local impacts are downplayed. This tradition of critical scholarship is further evidence of the diversity of theoretical traditions and perspectives within the social sciences. The remainder of this chapter focuses more on the efforts of applied social scientists interested in ways to improve forest management, policy, and (particularly) community sustainability. However, it is important to acknowledge a variety of parallel trends and traditions in forest social science scholarship.

Recent approaches to assessing forest community sustainability

Some researchers are taking up the challenge of developing new models for assessing community sustainability. These new models attempt to address the constantly evolving nature of communities and present a broader picture of the effects of forest dependency on communities. In doing so, they are contributing new insights and knowledge about the changes taking place in the social, economic, and cultural life of these communities.

A number of new concepts have been proposed to evaluate or assess the sustainability of forest communities. These concepts share a common emphasis on evaluating communities' *adaptability* to changing social and economic conditions rather than static measures of *stability*. Among these are the concepts of community capacity (see Box 5.2), community resiliency, and community well-being.

Nadeau (2002) uses the concept of community capacity and the associated determinants in her study of the Haut-St-Maurice, a forest industry region in Quebec. She was able to identify factors that contribute to and detract from community capacity over time based on data collected from local people through semi-structured interviews. For example, in the early days the community enjoyed plentiful forests, water, and agricultural land, strong leadership from within local institutions (such as church and mill), and strong community life. In those days, founders of the mills exerted a dominant influence on community planning, administration, and social organization. Over the years, a change of mill ownership led to a steady decline in the direct involvement of mill own-

---

Box 5.2. What is "community capacity?"

Definition of community capacity:

> "The collective ability of residents in a community to respond to external and internal stresses; to create and take advantage of opportunities; and to meet the needs of residents, diversely defined."

Kusel 1996

Determinants of community capacity (Kusel 1996; Doak and Kusel 1996; Nadeau 2002):

- Social capital (the will and ability of people to mobilize resources and work together);
- Human capital (the education, job experience, acquired skills, health, and mobility of individuals);
- Infrastructure (the "utilities" that support the economic and social activities of the community); and
- Natural resources (the goods and services delivered by nature, such as water, food, building materials, and recreational opportunities).

The four elements are also shaped by changes that emerge outside the community such as decisions of macro-level actors (multinational corporations, governments) or societal trends such as technological or demographic change.

Proponents of building community capacity stress the importance of paying attention to the interactions between these different constituents, since the collective effect creates a dynamic which might maintain, reinforce, or weaken the overall capacity of a community.

ers in community life, leaving room for local leaders to emerge. In recent years, industry downsizing has highlighted factors that contribute to weak local capacity, such as a long-standing lack of investment in human capital, a lack of entrepreneurship among the local population, and limited public participation in management of local natural resources. However, the community is evolving in social and political realms; the community is filling the niche left by industrial and religious leaders, and (as a consequence) its capacity is improving.

Community resilience is similar to capacity insofar as it emphasizes people's ability to alter their behavior and take charge of local institutions in order to work towards a positive outcome for the community. A team of social scientists (Quigley et al. 1996) that was part of the Interior Columbia Basin Ecosystem Management Project (ICBEMP) in the western U.S. recently applied this concept. They define community resilience as:

"The capacity of humans to change their behaviour, redefine economic relationships, and alter social institutions so that economic viability is maintained and social stresses are minimized" (Quigley et al. 1996).

In the ICBEMP, community resilience was measured at two different scales, the county level and the community level. In order to measure community resilience at the county level the team developed a socioeconomic resiliency index that integrated measures of economic resiliency, population density, and lifestyle diversity. Data were taken from different pre-existing databases. The authors of the report found that counties with high resilience tend to lie along major transportation corridors, or have high scenic amenities and quality of life (Quigley et al. 1996).

Another index was developed for the community level. This time the index was built from data collected specifically for the purpose of analyzing community resilience. The index is an aggregate measure of residents' perceptions of certain community characteristics and conditions, including aesthetic attractiveness, proximity to outdoor amenities, level of civic involvement, effectiveness of community leaders, economic diversity, and social cohesion among residents over time (Harris et al. 1998). Their results supported the findings at the county level regarding the importance of population size, economic diversity, and quality of life but provided new insights about other important factors such as local autonomy and experience coping with change.

Finally, another concept that is making its way into the community assessment literature is community well-being. It is described by Wilkinson (1991) as a concept meant to "recognize the social, cultural and psychological needs of people, their family, institutions, and communities". In other words, it is an attempt to incorporate a wide variety of factors into an assessment of the quality of life for a particular community. Due in part to the lack of an agreed-upon definition of the term, studies on community well-being have adopted very different approaches. Some studies have analyzed specific factors that influence well-being such as poverty or economic development (Cook 1995; Overdevest and Green 1995). Others have focussed on characteristics that determine the level of well-being in forest-dependent communities (Kusel and Fortmann 1991; Bliss et al. 1998). Bliss et al. (1998) compared two forest counties in the United States through the analysis of social structures, ownership patterns, forest sectors, and historical development patterns. This multi-prong approach allowed them to observe the interaction between economic and social well-being. For example, they observed that a high concentration of resource ownership and product specialization, though an important contributor to economic well-being, poses problems for social well-being because of the lack of commitment among mill owners to the social life of the community. It follows, therefore, that advice for corporations to develop products or services which confer global competitive advantage, does not apply equally to communities where diversity is more important for long-term success.

Marchak (1983) used another approach in her seminal study of forest-dependent communities in British Columbia. She identified factors that contribute to community instability. Marchak argues that forest industry towns in British Columbia are economically weak due to their dependence on large companies and volatile markets. Drawing on the example towns of Mackenzie, Terrace, and Ocean Falls, she highlighted a number of socio-economic traits she sees as detracting from the quality of life and stability of the community. For example, in Mackenzie, there were frequent layoffs, resulting in high employee turnover rates and high population mobility. The geographical isolation of the community meant that women co-habitating with forest workers (few women are employed in the primary forest sector) find little employment and feel socially isolated. Economic instability also creates a feeling of uncertainty about the future as families worry about the opportunities for their children as they grow up. Marchak (1983) concludes that those communities situated more centrally with a higher degree of economic diversification were in a better position to adapt to changes in the economy. She also suggests that planning for all rural areas must include efforts to diversify local economies away from such a strict dependence on industrial forestry and forest products manufacturing.

Barnes et al. (1999) situate recent forest industry restructuring and resulting community responses in coastal British Columbia in terms of the larger economic shifts from Fordism to post-Fordism. Fordism is characterized by standardized mass production, managerial culture, and stable single-industry communities; post-Fordism is based on flexible production, specialized products, and down-sized weakened communities. They argue that the shift to post-Fordist production has eroded jobs and their associated benefits in coastal communities, and has replaced the corporatist ethic with an entrepreneurial culture. Some communities have been more successful than others in developing this entrepreneurial culture and creating new opportunities for local development. In an article on Chemainus, B.C., Barnes and Hayter (1992) describe how that town avoided major economic decline through the development of grassroots tourism. The experience of other communities which have fared less well are also addressed by the authors (Barnes and Hayter 1994, Barnes et al. 1999)

In a recent national study, *Sustainability for Whom? Social Indicators for Forest-dependent Communities in Canada*, Beckley (2000) characterizes sustainability in terms of assets and liabilities. Assets refer to positive forces or capital stocks that contribute to community sustainability. They are the components upon which the community currently relies and upon which it can build to create long-term prosperity and sustainability. In the nine case study communities included in this study, recurring assets included things like the quality of the natural resource base and strong community leadership (Table 5.2). Liabilities, on the other hand, typically consist of challenges unique to each community. These may be related to human capital deficits, over-exploitation of resources, or low institutional capacity. Interestingly, however, some features were described as both assets and liabilities. For example, in one community, the leadership was described as an asset because the small cadre of experienced leaders was able to mobilize resources and effect positive change. Leadership was also identified as a liability in the same community because leadership was concentrated and decision making was the purview of a select few; new leadership was not being encouraged or nurtured. Community sustainability thus depends on conscious efforts to train new leaders, for all community organizations to plan for internal succession and capacity building.

Quantitative and qualitative approaches to assessing community sustainability

The process of developing more adaptive and dynamic models of community sustainability has also demanded a broadening of the methodological tools used to measure sustainability. Quantitative indicators are statistics that describe the social or economic conditions in a community, and have long been a favourite of social scientists and planners. A recent survey of 22 indicator studies by Beckley and Burkosky (1999) revealed that the most commonly used indicators were all easily quantifiable, including unemployment rates, poverty rates, education attainment, diversity of the economic base, real estate values, demographic stability, crime rates, and natural resource dependency. Their popularity stems from the fact that readily available data exist for these metrics. For example, Statistics Canada gathers information on these and other indicators and reports them at various geographic scales of analysis. In many instances, census subdivisions approximate community boundaries. These data provide a standardized and replicable source of baseline information about communities that is updated every 5 years.

Several Canadian researchers are using quantitative indicators to track the social and economic situation of Canadian forest-dependent communities. Southcott (2000), for

Table 5.2. Assets and liabilities identified for nine case study communities (Beckley 2000).

Assets	Liabilities
Natural resources	Youth out-migration
Diversified economic base	Dependence on forest products industry
Entrepreneurship amongst local population	Seasonal economy
Commitment to place	Vulnerability to service consolidation
Effective community organizations	High population mobility
Strong local political and economic leadership	Decline of retail sector associated with
An active subsistence economy	out-shopping
	Lack of renewal of leadership
	Concentration of leadership
	Gap between rich and poor
	Lack of formal human capital
	Job instability

example, explicitly set out to test some of the conclusions derived in the 1960s and 1970s. Using a suite of indicators that includes population change, industry structure, occupational structure, female employment, education levels, and degrees of self-employment, he compared northern forest-dependent communities with the descriptions characterized by the Lucas/Himelfarb model (Table 5.1). What he found is that the service sector in natural resource-dependent communities (NRDCs) has increased compared to the same communities in the past. At the same time, primary and manufacturing industries have decreased (Fig. 5.1). The occupational structure of these communities has also changed: traditional blue-collar jobs are decreasing in importance, the number of self-employed has increased quite dramatically (Fig. 5.2), and there are more women in the workforce. All these elements corroborate Barnes and Hayter's assertion the forest-dependent communities are entering a post-Fordist phase in their economic and social development. Yet when compared to national trends, northern forest-dependent communities nevertheless stand out as "islands of industrialism" with lower education levels, lower female participation in the workforce, and lower-wage service sector jobs. Table 5.3 and Box 5.3 display some of the changes taking place in forest- and resource-dependent communities (Southcott 2000).

In similar work, Randall and Ironside (1996) found that the non-resource economic sectors of resource-dependent communities are now more significant than had been indicated in the earlier work, and that women are benefiting from increased employment opportunities in these sectors. Ehrensaft and Beeman (1992) have indicated that change is occurring in the occupational and industrial structure of resource-dependent communities. They note the declining importance of jobs in the primary sectors and the increasing importance of employment in the public sector and administrative services.

Quantitative indicators also have some limitations. Firstly, aggregate statistics often mask important distributional inequalities within households, communities, or regions (Kusel 1996). Secondly, they have been criticized for missing important dimensions of community life — in particular those aspects that make each community unique. For

Fig. 5.1. Employment in primary/manufacturing industries as a percentage of all industries in natural resource-dependent communities (NRDCs), compared with Canada as a whole.

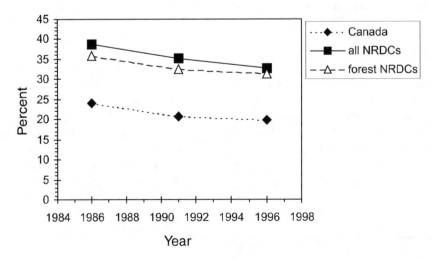

Fig. 5.2. Self-employed workers as percentage of all workers in natural resource-dependent communities (NRDCs), compared with Canada as a whole.

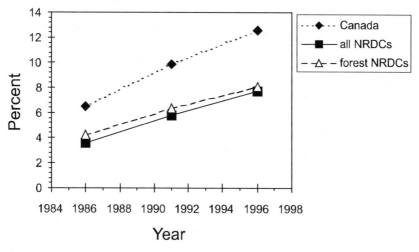

Table 5.3. The mixed blessing of single-industry forest dependence.

Advantages	Disadvantages
High wages for those in the resource sector	Uneven income distribution overall
Low unemployment	Limited local advancement opportunities
Stable population over time	High turnover of population
Access to resource-based recreation	Isolation
High-quality services and facilities	Low education attainment
Healthy local tax base due to corporate assessment	Vulnerable to global commodity price trends and to capital markets

example, factors such as access to decision-making processes or the strength of social support networks are easily overlooked (Stedman 1999). For this reason, interest in qualitative indicators is growing, especially in more local assessments of community sustainability. Qualitative indicators focus more on the perceptions of the local community towards things like social cohesion or the quality of local leadership. Their advantage is in their sensitivity to local circumstances and processes. For researchers interested in learning about sub-groups of a population based on ethnicity, gender, or age cohort, qualitative indicators can be particularly useful. However, the measurement of qualitative indicators is more energy intensive, most often requiring researchers to engage directly with community residents through participant observation, interviews, or surveys (see Parkins et al. 2001). It is also more difficult to replicate and compare qualitative indicators because of their subjectivity and the lack of uniform systems of valuation or measurement, making statistical analysis difficult. The adoption of techniques such as Rapid Rural Appraisal (RRA) and Participatory Rural Appraisal (PRA) have allowed for more time- and cost-efficient collection of qualitative indicator data (Greenwood 1998; Townsley 1996; Mukherjee 1995).

Qualitative indicators are also useful for describing causal phenomena. The traditional suite of quantitative, secondary data indicators from Statistics Canada are good for describing the "what" of community attributes, but are less useful for describing "how" and "why". We make the distinction here between profile indicators (the standard quantitative suite) and process indicators. The latter are indicators that measure or assess social processes, relationships between groups or individuals, perceptions, and behaviour based on perceptions. For example, population mobility is a profile indicator that describes how many people move in and out of a given community. It is a static measure in that it only tells you the numbers moving in and out; it provides no causal explanation as to why this is happening, nor whether it is favourable or not. A highly mobile population can be good from the perspective of start-up industries, but is bad for community cohesion and has been demonstrated to lead to other social pathologies (higher divorce rates, crime rates, etc.). So the value of a given index and how to interpret it clearly depends on its use and users.

Process indicators include such things as social cohesion, strength of social networks, and attachment to place. These describe social processes and causal variables that may explain high or low population mobility. Poverty may be partly explained by strong attachments to place. Beckley (2003) presents two sorts of place attachment that he describes as "anchors" and "magnets". There is currently much discussion about the amenity values for rural places and how environmental quality, a slower pace of life, and access to nature may serve as a draw to rural places (Power 1996; Beatley and Manning 1997). Beckley refers to these factors as magnets. However, there are also "anchoring" factors that keep people rooted in place, even though this may compromise their well-being as measured by quantitative indicators such as income and employment. Anchors are related to social structural factors, such as family dynamics, class position, and social networks. However, things like local ecological knowledge and the ability to derive a partial subsistence from the land may also be anchoring factors in place attachment. If people are unwilling to migrate from their ancestral homes, they may stay despite poor economic prospects. This is often the case in First Nations settlements (Chap. 3), and the degree to which it occurs can promote cultural cohesion and environmental conservation at the same as it constrains economic opportunity.

The Sustainable Forest Management Network (SFMN) is currently funding research that examines place attachment and the sense of place displayed by local residents in three diverse forested settings in Canada. Researchers hope to better understand what factors connect people to where they live, and how important some of these factors are relative to one another (Beckley and Stedman 2002). For example, are landscape characteristics more important than socio-cultural factors in forming meaning and attachment to place? Early results suggest that there are differences between rural places in the degree to which family, church, and community play an important role relative to the forest landscape, opportunity for recreation, and being close to nature.

Another example of a process indicator is a community's collective level of entrepreneurship, which in turn could help to explain employment levels in a given community. Forest-dependent communities have traditionally been characterized by large corporate employers paying high wages, and by underdeveloped service sectors. These forces (as well as social norms and values around wage work, unionization, and women's employment) historically have combined to keep levels of entrepreneurship low in single-industry towns.

Yet, process indicators are often more difficult to measure quantitatively. Entrepreneurship, as one component of community resilience, might be easily measured in the number of small business or small business start-ups. Other process indicators, such as the quality of leadership, the depth of the leadership base, social cohesion, and attachment to place more often require fieldwork, extensive interviews, and other methods of qualitative data collection. Further research is needed to determine methods and strategies for creating quantitative measures for process indicators so that they can be compared across communities (see Stedman 1999). Ultimately, we believe that process indicators will demonstrate far greater explanatory power regarding the sustainability of rural communities than profile indicators.

Future research, new directions

An important new direction in forest community research is an increasing focus on non-industrial forest-dependent communities. Until recently, very little attention was paid to forest-dependent communities not reliant on timber harvesting and processing. Activities such as the harvest of non-timber forest products, revenues from tourism and recreation, and subsistence practices were characterized as economically "fringe" or "marginal" especially in the field of forestry (some examples can be found in other disciplines such as anthropology and recreation studies). However, the transformations taking place in forest communities (most notably the decline of timber and other resource-extractive industries, and the growth associated with activities such as tourism, recreation, and retirement settlement) has prompted some researchers to broaden their definition of forest dependence and to concentrate more on learning about these types of communities (Power 1996).

One example is a study by Beckley and Hirsch (1997) on the importance of subsistence activities as a component of forest dependence. Their economic evaluation of subsistence activities in Aboriginal forest-dependent communities in the Northwest Territories revealed that approximately 30% of household income and income-in-kind was generated from subsistence activities such as hunting, trapping, berry picking, and craft making. Data from Fort Liard are presented as an example in Fig. 5.3. Other efforts of "natural resource accounting" are described in Chap. 6.

Other researchers have expanded their analysis of non-timber forest products to include not only economic benefits, but also the cultural and social motivations for engaging in certain forest activities. In a study of maple syrup producers in Vermont and Quebec, Hinrichs (1998) discovered that maple syrup production, while rarely a primary occupation, is an important complement to other seasonal activities such as farming or forest harvesting, and provides security through the diversification of economic activities. Furthermore, Hinrichs concluded that the motivations of syrup producers are not purely profit-oriented, revolving "less around absolute revenues than around the contribution such enterprises make in helping rural households to manage risk, cope with seasonality, define an identity and be part of a local community". Comte's suggestion that predictive modelling is difficult in social science is borne out in an activity such as syrup

Fig. 5.3. Mean household ($n = 71$) income and income-in-kind from "bush" (forest) sources in Fort Liard, Northwest Territories, in 1994 (Beckley and Hirsch 1997).

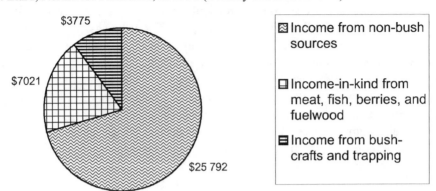

making. It involves a complex web of relations that is regionally bounded. The activity is partly social, partly economic; it may contribute significantly to attachment to place; and it may help build social networks and entrepreneurship.

While there are few non-timber forest products that comprise a substantial part of the economic base in many boreal communities, other forest-related activities such as hunting (for recreation and subsistence), balsam tipping for wreaths, mushroom harvesting, fiddlehead harvesting, etc., may be important anchors, particularly for the rural poor. Knowledge of the use and location of such resources may allow a person to remain in a rural community, embedded in dense social support networks, living semi-subsistence lifestyles. Comprehensive assessments of community resilience and sustainability and well-being should take into account these activities, but our methods for measuring this sort of forest dependence in all their complexity are just developing.

Many provincial and local governments are heavily promoting tourism as the saviour for declining rural communities. The wisdom of this approach has been thoroughly criticized elsewhere (Power 1996). Tourism employment tends to be low-wage and seasonal. It is also dependent upon discretionary income, inexpensive petroleum, and the absence of any prohibitive travel restrictions. Expenditures in tourism services are made after basic necessities are covered, and when regional and world economics and politics are stable. Another aspect of tourism that is often overlooked is the potential conflict that can arise when tourism activities or developments clash with local values or activities (e.g., undesirable aesthetics of industrial forestry to tourists, or undesirable changes to community infrastructure due to tourism development).

There are some significant obstacles to collecting data on these other types of forest dependence. As mentioned above, there is very little tradition of this type of research. Attempts to measure subsistence use of forests or amenity use of forests range dramatically in their methodological approaches. There are no large, quantitative databases to turn to, as Statistics Canada does not collect data on subsistence use of forests. Furthermore, Statistics Canada data on the tourism or visitor sector is collected in such a way that it is impossible to segregate employment attributable to business versus recreational travel, as well as local versus visitor expenditures for some services. As a result, it is currently impossible to create an index of tourism dependence using the same methodology traditionally employed for determining timber dependence. If tourism dependence is to be measured consistently and clearly, new methods will need to be developed to disaggregate service sector data.

Another emerging research direction is the study of how sub-populations such as youth, women, or ethnic minorities within communities are differentially affected by forest dependence. There has been a tendency to focus on the effect of economic restructuring on the male-dominated workforce, but this is changing. For example, research on the situation of women living in resource-dependent communities is moving beyond the dominant depictions of women as being socially isolated and disempowered. Gill (1990) illustrates the complexity of women's experience in her evaluation of how women and men evaluate and cognitively structure attributes of their quality of life in an isolated new resource town in British Columbia. A significant evaluative element for women was how well the environment met the needs of their children. On this point, women in Tumbler Ridge (a medium-sized coal-mining community incorporated in 1981) were less than satisfied with the quality of their children's educational experience but held favourable views of the level of safety for children. The social environment was also rated highly amongst women, although transience was identified as a problem. Another frequently mentioned drawback was the lack of shopping facilities. The existence of job opportunities for women was not a major concern, though the range of job choices was considered inadequate.

Beckley and Sprenger (1995), in their study of communities in the Lower Winnipeg Basin, explicitly set out to document the socio-cultural characteristics of all groups living in this geographic area and the different ways in which people interact with and make use of the forest. They observed an unequal distribution throughout the region of the benefits associated with the local newsprint mill. Mill workers mostly reside in one community, where they benefit from a secure environment, high amenities, and a good standard of living. However, other communities, including most of the First Nations population, experience very few tangible benefits from industrial forest development. This pattern is not unique to this area. Across Canada, industrial uses of forests and provincial grants of timber licenses often radically alter forest use patterns and practices among existing Aboriginal communities. In the case Beckley and Sprenger (1995) describe, Aboriginal employment in the newsprint mill is still modest because of a tight local labour market. But new owners of the newsprint mill have expanded their Aboriginal programs and have initiated joint ventures with First Nations in developing sawmills in the region. These developments are indicative of current directions in direct involvement of Aboriginal communities in forest resource management; other examples are provided in Chaps. 3 and 4.

Management implications

Research on forest communities has done a good job of identifying the challenges communities face in moving towards more sustainable social and economic systems. This chapter has touched on a number of these challenges, including human capital development, the need for economic diversification, and the need for better opportunities for women and minorities in resource communities. This research begs the question of how local and non-local actors can address these issues. While these are complex questions that require community-specific strategies, we have nonetheless included a series of suggestions. Recommendations are made for three categories: government, industry, and communities.

Management implications for government

It is widely recognized that economically diversified communities tend to be better prepared and fare better when dealing with economic downturns and economic instability. Federal, provincial, and state governments have many potential avenues for encouraging economic diversification. Unfortunately, such efforts are rarely coordinated and individual departments/ministries attack the problem in a piecemeal fashion. Natural resource agencies may be concerned with development and diversification of industries and products within the forest sector. Government tourism promotion functions are often paired with economic development or economic diversification functions in a single ministry. In contrast, human resources ministries may have their own unique approaches to local economic diversification. In Canada, there are only a few provinces that have ministries expressly devoted to rural or northern issues, and even these do not always take a coordinated approach to problems.

Piecemeal efforts to promote economic diversification are better than none at all. Often rural communities lack the information and networks that would enable them to capitalize on opportunities related to the changing global economy. Production is becoming increasingly decoupled from geography. That is, a growing proportion of the wealth generated in national or provincial economies is in information technology or business services that could really be located anywhere. This trend will likely continue and could offer opportunities to rural places, particularly those with high environmental amenity values, to attract new firms in the high-technology and information technology sectors (Power 1996). Provincial governments often serve as partners or promoters of new business development, and as such, they could direct economic growth toward areas that need employment.

One of the factors holding rural places back is the lack of skills and education required to seize opportunities and compete in the emerging economy. Those rural youth who receive post-secondary training usually move to cities because that is where the jobs are. Firms are often reluctant to locate in rural areas (except when a non-skilled, low-wage labour force is required) because of low levels of skills and training. Therefore, provincial governments could also play an important role in ensuring adequate levels of human capital development in rural regions. This refers to adult education as much as to youth education. If you train youth, they are likely to take their skills elsewhere. If you train or retrain middle-aged workers that are rooted in communities, they are more likely to stay and apply their new skills in the local economy. Distance learning is opening new doors to individuals and presenting them with opportunities to receive training that will facilitate their move into growing economic sectors as opposed to remaining mired in declining sectors. Provincial human resource and education departments should continue to make human capital development a priority.

With respect to forest management, some experimentation may be considered that would take sociological outcomes, rather than business- or environment-based outcomes as evaluation criteria. More explicitly, provincial, state, or federal natural resource ministries/departments may wish to experiment with alternative forms of forest tenures (see Chap. 7) on public land to see if there are different approaches to organizing the forestry sector that would lead to greater local economic benefit and greater community sustainability. There appears to be increasing demand for such experimentation, and some provinces such as British Columbia, Ontario, and Quebec are responding with pilot community forestry programs. Other experiments with encouraging results, such as the forest tenant farming project in the Bas St. Laurent Model Forest (Chap. 22), are also ongoing and deserve closer scrutiny and replication.

It is important to remember that policies of all sorts have implications for rural communities. Ministries or departments of natural resources and forestry tend to think that their policies primarily affect forests and natural resources, and perhaps specific stakeholder groups. While forest policy makers sometimes invoke the interest of communities in their policy proclamations, the focus is usually on jobs related to timber extraction. Community well-being is often viewed as the responsibility of some other department or ministry. There is a lack of coordinated and explicit policy toward rural communities and their development at both the federal and provincial levels. This means that rural communities have to deal with a myriad of government agencies to meet specific needs. For example, labour force training is typically handled by one department, natural resource management by another, tourism by another, and infrastructure by another. In some cases, multiple levels of government also complicate the policy issues that may dramatically affect quality of life in some locales (e.g., in First Nations communities). Some steps are being taken to coordinate efforts of various government departments. This type of strategy needs to be pursued more broadly and both industry and communities should lobby government departments to coordinate their efforts more effectively.

Management implications for industry

The organization of resource extractive industries in Canada has evolved into a partnership between government and industry. In this partnership private industry takes a great deal of responsibility for stewardship of public resources. This is true of mining and forestry in particular. While industry has formal responsibilities for forest stewardship and management under a wide array of Crown (public) land licenses and tenures, there are virtually no formal requirements for companies to take responsibility for the communities that depend upon the forest products industry. The same is true virtually worldwide, and trends in globalization are only reinforcing the right of corporations to exercise a great deal of mobility. The trends associated with global corporate mobility are not without consequences for the communities that are sometimes left behind.

In the past, many timber-dependent communities were operated almost as private fiefdoms of the companies who held the local Crown license. Most forest products companies got out of the business of "town management" in the 1960s and 1970s so that

> **Box 5.3. Taking responsibility: corporate capitalism with a social conscience.**
>
> Tembec represents an example of a company that takes an active interest in community affairs and community well-being, but does so with sensitivity. In the past, many companies took a very active interest in local affairs, to the extent that they helped set their own tax assessments, actively supported salaried staff in local politics, and helped decide major municipal budgetary allocations (see Beckley 1996; Nadeau 2002). Tembec has a different vision of corporate/community relations. This view has much to do with the fact that the company was born of a local employee buyout in the 1970s. While the company has grown tremendously since then (over 50 manufacturing units and 10 000 employees at the end of 2002), it has maintained its commitment to the communities in which it operates.
>
> Examples of Tembec's socially progressive policies include:
>
> (1) A return of 1% pre-tax profits from manufacturing facilities to their host communities. Allocation of these funds are decided by a local committee and are often used as matching funds — a practice that helps build local capacity;
>
> (2) Inclusion and respect for employees through profit sharing and also through opportunities for all levels of employees to participate in corporate policy and decision making;
>
> (3) Commitment to environmental quality at all levels; and
>
> (4) A local purchasing policy to support small business in the communities in which they operate.
>
> Beckley and Krogman (2002)

they could focus on forest management. The result of the widespread change in policy with respect to local management and operation of communities is that many of these places floundered. Factors that contribute to community sustainability, such as volunteerism, entrepreneurship, and strong local leadership, were lacking, a deficiency in part due to the legacy of paternalistic corporate control of timber communities in the early and middle parts of the last century.

From a broad perspective encompassing corporate policy and community policy, the challenge is to strike a balance between corporate contributions to community sustainability and what some might view as corporate "meddling in" or domination of community affairs. An extended discussion of suggestions for how to achieve this balance have been laid out elsewhere (Beckley and Reimer 1999). Briefly, industry can actively support and sponsor human capital development through broad adult education (not simply technical training) for the entire community, not just the industry's own labour force. A community with a diverse set of skills and abilities will be better equipped to deal with social and economic change in the future. Industry can also support leadership development in the form of team-building skills, self-actualization, and personal development. This will encourage people to solve their own problems, either on the job or in their communities, rather than looking for external solutions. This can be done through sponsoring workers to take leadership training courses, by bringing such seminars and workshops into the mills and making them available to other community members, by offering matching funds for starting up local community groups, and by sponsoring mentoring programs for potential leaders.

Industry should also encourage entrepreneurship. Historically, companies have been reluctant to develop business skills or leadership skills among their labour force because they do not want to lose good people. Today, modern mills require fewer and fewer workers, but wages in those industries remain quite high. This means that the capital accumulated by mill workers could be turned into small business start-up funds.

There is also a need to support those people in the community who are benefiting least from industry activities. There are often large inequalities between those employed in the mills and those employed in the service or other sectors. Traditionally, mills have contributed to the community in ways that have supported their workforce, which often has merely widened the gap. Companies should provide direct support to institutions in their host communities that benefit those most in need. For instance, rather than (or in addition to) supporting activities that benefit their employees, such as youth hockey or the local all-terrain vehicle or snowmobile club, industry should support women's shelters, food banks, employment re-training programs, meals-on-wheels, and other similar activities. These latter programs and charities benefit single moms, the elderly, the poor, and the unemployed. Improving conditions for the lowest segment of society makes a community a better place to live for all.

Finally, industry should provide all of the above services and support to the entire impact area of the mill. Throughout Canada, fibre-processing mills usually draw their wood from a wide area. In the past, forest products mills have focused their community development efforts in the communities in which their processing facilities existed. Once again, this concentrates wealth rather than disperses it. The result have been internal regional inequality, often between European and Native communities. Companies are increasingly realizing that their responsibility on the community side does not end at the town line in the community where their mill is located. Hopefully, this trend will continue.

Management implications for communities

There are a number of management implications for leadership, policy, and planning in rural communities. Changing global and local conditions are presenting serious challenges for natural resource-dependent communities but are also creating new opportunities. Sometimes communities feel helpless with respect to exogenous change that they view as inevitable. In this era of scaled-down governments and increasingly global industries, communities must be prepared to take on more responsibility for their own well-being.

Our research bears out the relative advantage of having a diversified local economy. While most communities realize this, many that are heavily vested in traditional resource sectors do not. Some community visions of possible futures are clouded by historical dependencies; their strategy for economic diversification involves attracting businesses that support or supply traditional commodity producers. The problem with this strategy is that when orders for pulp go down, the pulp mills' orders for chemicals, machinery, and business services go down as well. Communities need to identify economic diversification strategies that complement rather than compete with existing strengths. Examples of complementary economic activities would involve businesses that use a different resource base, or that do not compete directly with a similarly trained labour force (e.g., software designers rather than pipe-fitters). Many communities have found diversification opportunities through value-added manufacturing for niche markets. These may entail using waste materials from existing forest products processing facilities (see Chap. 19), or adding value to the products that come from existing mills. These are examples of complementary development strategies within the same general forest products sector, while other strategies can be developed to take advantage of a region's amenities, both natural and built.

There is a bit of a "Which comes first, the chicken or the egg?" problem with respect to the local skills development and economic diversification. It would not be wise for small communities in rural Nova Scotia or in the northern Prairie provinces to invest a great deal of effort in turning their youth into computer programmers in the hope that some information technology firm will locate in their community. The labour embodied in those individuals is more mobile than the industries; more often than not, those newly trained young people will go to where the jobs are rather than wait until the jobs come to their hometown. Jobs may come to smaller centres, though probably not to very remote places, if those communities invest in the proper infrastructure. In the new economy, that infrastructure includes things like fibre optics links for high-speed internet access, and cleaner and cheaper sources of energy (e.g., natural gas, thermal power). It also involves social and institutional infrastructure that will create a high quality of life and make a given rural locale a desirable place to reside — good pediatric care, a good school system, diverse recreational activities, scheduled airline service, distinctive homes, and the like. That type of infrastructure will likely attract young families to an area. Educators and community developers need to monitor how the global economy is developing in order to develop programs that will train youth and retrain adults in ways that will enable them to compete in future (rather than current) labour markets. Eventually, if people are sufficiently attached to where they grew up, and opportunities to apply their skills and knowledge emerge in such a place, more people will stay, or if they have left, some may return.

Another recommendation for communities is to lobby resource management agencies for access to and (or) control of publicly owned natural resources. The movement toward co-management in fisheries, forestry, and wildlife management is emerging slowly, but with positive results (Chaps. 4, 6, and 22). The demand for more local control is a function of perceived mismanagement of resources by government in the past (e.g., the Atlantic cod fishery) and by the perception that local places no longer benefit as much as they once did from their surrounding resource endowments. In the forest products sector, there are fewer jobs and they are located further away than they have been in the past. Today, many people watch the log trucks roll down the highway to large centralized processing facilities, and they wonder how the wealth represented in the load might benefit their community were it processed locally. In order for communities to make a strong case that they should have more access to or control over local Crown resources, they need to demonstrate both resource management capacity and business start-up/management capacity (often limited by access to capital). Once again, the issues of available local human capital, global competition for investment funds, and community development are linked.

Conclusions

Sociologists have been well aware of the problems associated with constructing rigid definitions around concepts such as community and sustainability. Sociologists have been struggling since the dawn of the discipline to define "community", and the early giants in the field warned of trying to impose mechanistic, predictive models or laws on phenomena as complex as human communities. Some sociologists still show disdain for attempts to categorize and predict outcomes for communities. However, despite these problems, other sociologists are forging ahead in an attempt to aptly portray and evaluate dimensions of life in forest communities. Forest communities are evolving, as is the paradigm of forest management in Canada. While the relationship between these phenomena remains unclear, it is important from a policy perspective to continue to assess the well-being and quality of life in those rural communities that rely on the surrounding forests for their livelihood.

The applied sociological researchers engaged in this line of enquiry are drawing on a wide array of methodological tools and approaches, several of which have been described in this chapter. As well, they are broadening the definition of what constitutes a forest-dependent community to include subsistence and non-timber goods, as well as amenity uses of forests. A purposive discourse is developing around the concept of community "sustainability". To a large degree the concept of community sustainability has been de-constructed into more workable concepts such as community capacity, community resilience, and community well-being.

To measure these concepts, both quantitative and qualitative indicators are being tested and applied and are providing a rich picture of the diversity of forest-dependent communities in Canada. They are also revealing that these communities are undergoing rapid change and cannot be evaluated on a "one-time basis". Just as inventories of forests must be updated periodically, continued attention to change in rural forest communities should be an ongoing endeavour. The indicators commonly used tell an interesting story, and one that has some clear management implications for governments, industry, and communities themselves (Box 5.4).

While forest community sustainability still offers many challenges to researchers, the real challenges are:

- to government policy makers who wish to see stable and productive communities with happy citizens;
- to industry managers, who desire a stable, healthy, capable labour force, and a nice locale in which to live and work; and
- (most importantly) to communities themselves.

The challenge communities face is to build their capacity to allow them to adapt to change. A major first step in this endeavour is to identify assets and liabilities. This is an area where researchers can help. Once this is done (and it should be periodically redone), communities must devise strategies to build on their strengths and to address their weaknesses.

Box 5.4. Options for enhancing community sustainability:

Governments can:

- encourage economic diversification in a co-ordinated manner;
- support or promote business services that are decoupled from geography, especially in rural places with high environmental amenity values;
- ensure adequate levels of human capital development in rural regions, especially in terms of adult education opportunities; and
- experiment with more diverse (including community-based) forms of forest tenure.

Industry can:

- strike a balance between contributing to the community and dominating it;
- support and sponsor human capital development through broad adult education;
- support leadership development in its employees and the community at large (through leadership training courses, bringing seminars and workshops to the community, sponsoring mentorship programs);
- offer matching funds for starting up or supporting local community groups;
- encourage entrepreneurship through training seminars and start-up funds available to the community;
- provide direct support to community institutions in the most need (not just those that benefit their employees); and
- expand community involvement and commitment to the entire millshed, not just the mill town.

Communities can:

- identify economic diversification strategies that complement rather than compete with the dominant employer;
- invest in sustainable and information-age infrastructure, such as cleaner and cheaper sources of energy and broadband communications technology;
- invest in social infrastructure attractive to families and all ages: good pediatric and geriatric care, a good school system, diverse recreational activities, scheduled airline service, pedestrian- and bicycle-friendly planning and zoning, etc.;
- lobby resource management agencies for access to and (or) control of public natural resources;
- develop and promote local resource management and entrepreneurial capabilities (human capital, investment capital) through adult training, education, and mentorship programs, and through the promotion of local purchasing and investment; and
- identify their assets and liabilities, and then devise community plans that build on their strengths and address their weaknesses.

Acknowledgements

We would like to thank the SFMN for providing the opportunity to prepare this chapter and for research funds to pursue studies of forest-dependent communitites in Canada. The reviewers made some astute observations and helped to improve the final product. We must thank the residents of forest-dependent communities across Canada who have helped us to understand their history and their current circumstances. Whatever insights we have gleaned about forest-dependent communities in Canada is largely due to the willingness of their residents to share their experience with us. Discussions and parallel research efforts with John Parkins, Rich Stedman, Bill Reimer, Eloise Murray, Naomi Krogman, Jeji Varghese, Tracy Burkosky, Audrey Sprenger, Dianne Korber, Luc Bouthillier, Bruce Shindler, and George Stankey have informed this work. We authors are solely responsible for its shortcomings.

References

Barnes, T.J., and Hayter, R. 1992. The little town that did: flexible accumulation and community response in Chemainus, British Columbia. Reg. Stud. **26**: 647–663.

Barnes, T.J., and Hayter, R. 1994. Economic restructuring, local development and resource towns: forest communities in coastal British Columbia. Can. J. Reg. Sci. **17**(3): 289–310.

Barnes, T., Hayter, R., and Hay, E. 1999. "Too young to retire, too bloody old to work": forest industry restructuring and community response in Port Alberni, British Columbia. For. Chron. **75**: 781–787.

Beatley, T., and Manning, K. 1997. The ecology of place: planning for environment, economy and community. Island Press, Washington, D.C.

Beckley, T.M. 1996. Pluralism by default: community power in a paper mill town. For. Sci. **42**(1): 35–45.

Beckley, T.M. 2000. Sustainability for whom? Social indicators for forest-dependent communities in Canada. Sustainable Forest Management Network, Edmonton, Alberta. Proj. Rep. 2000-34. Available at http://sfm-1.biology.ualberta.ca/english/pubs/PDF/PR_2000-34.pdf [viewed June 19, 2003]. 34 p.

Beckley, T.M. 2003. The relative importance of sociocultural and ecological factors in attachment to place. *In* Understanding community–forest relations. *Edited by* L.E. Kruger. USDA Forest Service, Portland, Oregon. Gen. Tech. Rep. PNW-GTR-566. pp. 105–126.

Beckley, T.M., and Burkosky, T.M. 1999. Social indicator approaches to assessing and monitoring forest community sustainability. Canadian Forest Service, Edmonton, Alberta. Inf. Rep. NOR-X-360. 13 p.

Beckley, T.M., and Hirsch, B. 1997. Subsistence and non-industrial forest use in the Lower Liard Valley. Canadian Forest Service, Edmonton, Alberta. Inf. Rep. NOR-X-352. 42 p.

Beckley, T.M., and Krogman, N.T. 2002. Social consequences of employee/management buyouts: two Canadian examples from the forest sector. Rural Sociol. **67**(2): 183–207.

Beckley, T.M., and Reimer, W. 1999. Helping communities help themselves: industry–community relations for sustainable timber-dependent communities. For. Chron. **75**: 805–810.

Beckley, T.M., and Sprenger, A. 1995. Social, cultural, and political dimensions of forest-dependence: the communities of the Lower Winnipeg Basin. Series 1995-1. Rural Development Institute, Brandon, Manitoba. pp. 22–61.

Beckley, T.M., and Stedman, R.C. 2002. Understanding forest users' sense of place: implications for forest management. *In* Advances in forest management: from knowledge to practice. Proceedings of the 2002 Sustainable Forest Management Network Conference, 13–15 November 2002, Edmonton, Alberta. *Edited by* T.A. Veeman, P.N. Duinker, B.J. Macnab, A.G. Coyne, K.M. Veeman, G.A. Binsted, and D. Korber. Sustainable Forest Management Network, Edmonton, Alberta. pp. 267–273.

Berry, W. 1987. Does community have a value? *In* Home economics: fourteen essays. North Point Press, San Francisco, California. pp. 179–192.

Blishen, B., Lockhart, A., Craib, P., and Lockhart, E. 1979. Socio-economic impact model for northern development. Department of Indian Northern Affairs, Ottawa, Ontario.

Bliss, J.C., Walkingstick, T.L., and Bailey, C. 1998. Development or dependency? Sustaining Alabama's forest communities. J. For. **96**(3): 24–30.

Cook, A.K. 1995. Increasing poverty in timber-dependent areas in western Washington. Soc. Nat. Resour. **8**: 97–109.

Doak, S.C., and Kusel, J. 1996. Well-being in forest dependent communities, part II, a social assessment focus. *In* Sierra–Nevada Ecosystem Project: final report to Congress, vol. II, assessments and scientific basis for management options. Centre for Water and Wildland Resources, University of California, Davis, California. pp. 375–401.

Ehrensaft, P., and Beeman, J. 1992. Distance and diversity in nonmetropolitan economies. *In* Rural and small town Canada. *Edited by* R. Bollman. Thompson Educational Publishing, Toronto, Ontario. pp. 193–224.

Gill, A. 1990. Women in isolated resource towns: an examination of gender differences in cognitive structures. Geoforum, **21**(3): 349–358.

Greenwood, D. 1998. Introduction to action research: social research for social change. Sage Publications, Thousand Oaks, California. 274 p.

Grimmer, A.K. 1934. The development and operation of a company owned industrial town. Eng. J. **17**(5): 221–223.

Harris, C.C., McLaughlin, W.J., and Brown, G. 1998. Rural communities in the Interior Columbia Basin: how resilient are they? J. For. **96**(3): 11–15.

Himelfarb, A. 1982. The social characteristics of one-industry towns in Canada. *In* Little communities and big industries. *Edited by* R.T. Bowles. Butterworths, Toronto, Ontario. pp. 16–43.

Hinrichs, C. 1998. Sideline and lifeline: the cultural economy of maple syrup production. Rural Sociol. **63**: 507–532.

Hyde, W.F., and Daniels, S.E. 1986. Below-cost timber sales and community stability. Proceedings, Below-cost Sales: A Conference on the Economics of National Forest Timber Sales, 18–19 February 1986, Spokane, Washington. *Edited by* B. Flamm, J. Hendee, and D. LeMaster. Wilderness Society, Washington, D.C. 266 pp.

Jackson, J.E.W., and Pouchinsky, N.W. 1971. Migration to northern mining communities: structural and social–psychological dimensions. Center for Settlement Studies, University of Manitoba, Winnipeg, Manitoba. 158 p.

Kaufman, H., and Kaufman, L.C. 1946. Toward the stabilization and enrichment of a forest community: the Montana Study. USDA Forest Service Region 1 and University of Montana, Missoula, Montana. 95 p.

Knight, R. 1975. Work camps and company towns in Canada and the U.S.: an annotated bibliography. New Star Books, Vancouver, British Columbia. 80 p.

Kusel, J. 1996. Well-being in forest-dependent communities, part I: a new approach. *In* Sierra Nevada Ecosystem Project: Final Report to Congress. Rep. No. 39. Wildland Resources Centre, University of California, Davis, California. pp. 361–374.

Kusel, J., and Fortmann, L.P. 1991. What is community well-being? *In* Well-being in forest-dependent communities (volume 1). *Edited by* J. Kusel and L. Fortmann. Forest and Rangeland Resources Assessment Program and California Department of Forestry and Fire Protection, Berkley, California. pp.1–45.

Lucas, R.A. 1971. Minetown, milltown, railtown: life in Canadian communities of single industry. University of Toronto Press, Toronto, Ontario. 433 p.

Marchak, P. 1983. Green gold: the forest industry in British Columbia. UBC Press, Vancouver, British Columbia. 474 p.

Matthiasson, J.S. 1970. Resident perceptions of quality of life in resource frontier communities. Center for Settlement Studies, University of Manitoba, Winnipeg, Manitoba. 41 p.

Mukherjee, N. 1995. Participatory rural appraisal and questionnaire survey: comparative field experience and methodology innovations. Concept Publishing Co., New Delhi, India. 163 p.

Nadeau, S. 2002. Characterization of community capacity in a forest-dependent community: the case of the Haut Ste-Maurice. Ph.D. dissertation. College of Forestry, Oregon State University, Corvallis, Oregon. 245 p.

Nadeau, S., Shindler, B., and Kakoyannis, C. 1999. Forest communities: new frameworks for assessing sustainability. For. Chron. **75**: 747–754.

O'Neill, J. 1995. The poverty of postmodernism. Rutledge, London, U.K. 205 p.

Overdevest, C., and Green, G.P. 1995. Forest dependency and community well-being: a segmented market approach. Soc. Nat. Resour. **8**: 113–134.

Parkins, J., Varghese, J., and Stedman, R. 2001. Locally defined indicators of community sustainability in the Prince Albert Model Forest. Canadian Forest Service, Edmonton, Alberta. Inf. Rep. NOR-X-379. 39 p.

Pharand, N.L. 1988. Forest sector-dependent communities in Canada: a demographic profile. Economics Branch, Canadian Forest Service, Ottawa, Ontario. Inf. Rep. E-X-39. 34 p.

Power, T.M. 1996. Lost landscapes and failed economies: the search for a value of place. Island Press, Washington, D.C. 304 p.

Pratt, L., and Urquhart, I. 1994. The last great forest: Japanese multinationals and Alberta's northern forests. NeWest Press, Edmonton, Alberta. 222 p.

Quigley, T.M., Haynes, R.W., and Graham, R.T. 1996. Integrated scientific assessment for ecosystem management in the Interior Colombia Basin and portions of the Klamath and Great Basins. USDA Forest Service, Portland, Oregon. Gen. Tech. Rep. PNW-GTR-382. 303 p.

Randall, J.E., and Ironside, R.G. 1996. Communities on the edge: an economic geography of resource-dependent communities in Canada. Can. Geogr. **40**(1): 17–35.

Richardson, M., Sherman, J., and Gismondi, M. 1993. Winning back the words: confronting experts in an environmental public hearing. Garamond Press, Toronto, Ontario. 190 p.

Riffel, J.A. 1975. Quality of life in resource towns. Center for Settlement Studies, University of Manitoba, Winnipeg, Manitoba. 107 p.

Robbins, W.G. 1988. Hard times in paradise: Coos Bay, Oregon 1850–1986. University of Washington Press, Seattle, Washington. 194 p.

Robinson, I. 1962. New industrial towns on Canada's resource frontier. University of Chicago, Chicago, Illinois. Dept. Geog. Res. Pap. No. 73. 190 p.

Robson, R. 1986. Canadian single industry communities: a literature review and annotated bibliography. Rural and Small Town Research and Studies Programme, Department of Geography, Mount Allison University, Sackville, New Brunswick. 137 p.

Sandberg, L.A. 1992. Trouble in the woods: forest policy and social conflict in Nova Scotia and New Brunswick. Acadiensis Press, Fredericton, New Brunswick. 275 p.

Schallau, C.H. 1974 Forest regulation II — can regulation contribute to economic stability? J. For. **72**: 214–216.

Schallau, C.H. 1989. Sustained yield versus community stability: an unfortunate wedding? J. For. **87**(9): 16–23.

Southcott, C. 2000. Social change in northern forest-dependent communities: for a dynamic model of community sustainability. Paper presented at the Forest Sustainability Beyond 2000 Conference, 14–18 May 2000, Thunder Bay, Ontario.

Stedman, R.C. 1999. Sense of place as an indicator of community sustainability. For. Chron. **75**: 765–770.

Townsley, P. 1996. Rapid rural appraisal, participatory rural appraisal and aquaculture. Food and Agriculture Organization, United Nations, Rome, Italy. Fish. Tech. Pap. 358. 109 p.

Waggener, T.R. 1977. Community stability as a forest management objective. J. For. **75**: 710–714.

Walker, J.A. 1927. Company towns. J. Town Plann. Inst. **6**(2): 147–148.

(WCED) World Commission on Environment and Development. 1987. Our common future. Oxford University Press, New York, New York. 400 p.

Wilkinson, K.P. 1991. The community in rural America. Greenwood Press, Westport, Connecticut. 141 p.

Chapter 6

The economics of boreal forest management

W.L. (Vic) Adamowicz, Glen W. Armstrong, and Mark J. Messmer

"One of the anomalies of modern ecology is the creation of two groups, each of which seems barely aware of the existence of the other. The one studies the human community, almost as if it a separate entity, and calls its findings sociology, economics, and history. The other studies the plant and animal community and comfortably relegates the hodge-podge of politics to the liberal arts. The inevitable fusion of these two lines of thought will, perhaps, constitute the outstanding advance of the present century."

Aldo Leopold (1935)
Unpublished essay

Introduction

Canada's boreal forest is a source of wealth for Canadians and the world. Traditionally this wealth has been thought of as employment in and industrial output from the forest products sector (Fig. 6.1). These are still important today, with the boreal forest being a major contributor to an important sector of Canada's economy. The recent Canadian Senate report on the boreal forest (SSCBF 1999) stated:

In 1997, there were just over 13,000 separate establishments involved in the forest industry in Canada. Of that number, 9,636 were involved in logging, 2,872 in the wood industry, and 686 establishments were paper and allied mills. British Columbia accounts for over 4,000 of these establishments, while the boreal provinces have 7,050. In addition, a small fraction of New Brunswick's 1,238 establishments handle wood from that province's small boreal forest

W.L. (Vic) Adamowicz.[1] Department of Rural Economy, University of Alberta, Edmonton, Alberta, T6G 2H1, Canada.

G.W. Armstrong. Department of Renewable Resources, University of Alberta, Edmonton, Alberta, Canada.

M.J. Messmer. Weyerhaeuser Company Ltd., Edmonton, Alberta, Canada.

[1]Author for correspondence. Telephone: 780-492-3625. e-mail: vic.adamowicz@ualberta.ca

Correct citation: Adamowicz, W.L., Armstrong, G.W., and Messmer, M.J. 2003. The economics of boreal forest management. Chapter 6. *In* Towards Sustainable Management of the Boreal Forest. *Edited by* P.J. Burton, C. Messier, D.W. Smith, and W.L. Adamowicz. NRC Research Press, Ottawa, Ontario, Canada. pp. 181–211.

Fig. 6.1. Forests are a source of economic activity and employment.

Photo by Roger Nesdoly, courtesy of Mistik Management Ltd.

area. In the 'boreal provinces,' employment in logging and forestry services, wood industries and paper and allied industries, totaled approximately 395,000 in 1997, out of a Canadian total of 830,000.

While forest management in the boreal forest has a significant economic footprint, the intensity of activity has been changing. Increasing substitution of capital for labour characterizes the industry, although increases in the volume of output in some regions means that the number of people employed in the industry has not decreased as dramatically as the labour to capital ratio. Again, quoting from the Canadian Senate report (SSCBF 1999):

Over the last two decades, the mechanization of logging operations and advances in processing technologies has combined to gradually decrease the number of jobs per unit of production. This is particularly evident in the pulp and paper sector where the jobs/1000 tonnes of pulp and paper dropped from about 3.4 in 1975 to about 1.6 by 1993. In lumber production the drop was less dramatic, going from 1.8 to 1.0 jobs/1000 cubic metres produced over the same time frame. The smallest change was seen in the harvesting sector, where the jobs/1000 m³ harvested fell from 0.5 to about 0.3 . . . . [S]ince 1989, employment in the forestry industry has decreased from representing 9.3 per cent of total Canadian employment to 7.3 per cent, while in absolute terms, there has been only a slight decline. This situation was made possible by a significant expansion of industrial capacity reflected indirectly in the value of sales by the industry and by an increase in the amount of timber harvested.

The boreal forest is clearly a major contributor to the wealth of Canada.

The boreal forest is home to a large number of Canada's Aboriginal People. Many Aboriginal communities are facing significant challenges with rapid population growth rates and high levels of unemployment. They are struggling to find a balance between maintenance of the land for traditional uses and industrial activities that provide employment for community members. They also face challenges related to the interpretation of their rights to forest resources and the duty of industrial users to consult with Aboriginal People before acting on the land (Ross and Smith 2002; see Chap. 3).

The boreal forest in Canada also provides a wide variety of environmental and other non-timber values, ranging from recreation and non-timber forest products to wilderness areas and wildlife habitat. These values are increasingly being recognized in planning decisions, policies and economic analyses. Ecosystem services (natural processes affecting water quantity and quality, carbon cycling, etc.) are also increasingly recognized as valuable to society. These important environmental values must be considered as part of the economics of boreal forest management; they constitute a component that is challenging to include in such analysis because of their largely non-market nature, but also a component that is probably growing rapidly in value. Given the multi-dimensional character of the boreal forest, how would one describe the economics of boreal forest management? Given the changing nature of the boreal forest, what policy and management alternatives might be considered best from an economic perspective? What tools of economic analysis could be used to help achieve sustainable forest management (Box 6.1)? Given the changing nature of the boreal forest, the forest products industry, and forest-based communities, what actions should be taken?

The Canadian Senate report quoted above ended with a host of recommendations for the management of the boreal forest. Most of these recommendations addressed protected areas and the implications of forest management to Aboriginal People and communities. Of the 35 recommendations and sub-recommendations, only two were directly aimed at economic aspects of boreal forest management. These recommendations suggest increased value-added activity to enhance employment, and retraining programs to allow displaced workers to remain in their communities (SSCBF 1999). In this chapter

---

Box 6.1. Challenges for sustainable forest management:

For industry:

- Incorporating timber and non-timber values into forest management planning;
- Identifying, through adaptive management, the complex linkages between management actions, ecological systems, and human values arising from the forest; and
- Implementing monitoring schemes that measure the value of the forest natural capital stock.

For government and society:

- Designing institutions (e.g., forest tenures) that provide incentives for sustainable management, technology development, and wise use of resources;
- Using, where appropriate, economic instruments to guide economic activity; and
- Investing in monitoring and reporting on economic indicators of sustainability.

we offer a number of additional recommendations based on the underlying economic elements of boreal forest management. We outline an economic approach to forest management and we provide a set of tools that can be used by managers and policy makers to enhance the economic sustainability of boreal forest management.

After briefly describing what economists do, we begin this chapter with a review of the basic elements of market goods in the boreal forest — timber values. Insights into the financial fundamentals of timber values will lead into a broader discussion of timber and non-timber values and the trade-offs implicit in forest management in the boreal forest. This is followed by a discussion of tenure and policy arrangements and their relationship to the economic aspects of forest management in the boreal forest. Finally, economic sustainability is discussed conceptually, followed by the presentation of a variety of tools, available to policy makers and managers, which may help achieve sustainability.

This chapter provides an overview of the economic aspects of forest management at the forest level. It does not examine in any detail issues of international trade or global economic structure, since the focus of this book is the management of "forests". It should also be recognized that economic aspects of management are difficult to untangle from institutions, public values, and social aspects of forest management. This chapter concentrates on economics but linkages with Chaps. 3–5 and 7 will arise throughout. Institutional structure, such as tenure, is a critical issue to understand when assessing economic aspects of forest resources. In this chapter we examine some limited aspects of institutions, while in Chap. 7 institutions are examined more broadly and the dynamics of institutions are outlined. Much of the value arising from the forest is derived from public goods, and public involvement processes are a key way of identifying and perhaps even quantifying those values. This chapter describes economic valuation of non-timber goods and services and their integration into forest management, while Chap. 4 outlines public involvement in general as a mechanism for identifying public values. Value arising from the forest varies over individuals or groups of individuals. In Canada, Aboriginal People have unique interests and rights to forest resources. The challenges of incorporating values of Aboriginal People into forest management are not addressed in any detail in this chapter, but are examined in Chap. 3. Finally, there is a linkage between forests and small communities. The economic aspects of employment and development in communities are discussed briefly in this chapter, but the larger concept of community sustainability and community dynamics is the focus of Chap. 5. It is somewhat artificial to separate economics from public involvement, community sustainability, Aboriginal People's values, and institutional design. Yet we hope that by separately examining these issues we will be better prepared to analyze the entire set of social and economic issues arising in sustainable forest management.

Economics and forestry

Economics is the study of the allocation of resources under situations of scarcity. Economists study social systems; more specifically, firms, consumers, and other economic agents as they make choices in economic contexts. Economic science is usually concerned with explaining the behaviour of agents (in markets or in other market-like situations) and employs theories of individual consumer choice, industry behaviour, and other theories of behaviour in forming explanations of what can be expected if markets, policies, or other factors change. This *positive* approach of economic science has been used to understand human choices in markets for consumer goods, recreation behaviour, industrial behaviour, and government behaviour.

In addition to a *positive* ("what will happen?") approach, economists are also involved in *normative* ("what should happen?") forms of analysis. These normative analyses employ the same underlying behavioural models of positive analysis but are used to develop policy alternatives to lead to the best outcomes for society. As social scientists, economists study social welfare, which refers to measures of the well-being of all individuals in a society. Economists are interested in finding policy or institutional approaches that provide the highest levels of aggregate social welfare.

There are several key points to be made before embarking on a discussion of the economics of boreal forest management. First, economists are not concerned solely with market or financial returns. Economists are concerned about *social welfare*, which includes all aspects of well-being, whether they are derived from market or non-market sources. Second, economists tend to define social welfare as the sum of welfare of all agents. In other words, welfare is defined as the sum over consumers and producers. The search for the use of resources that maximizes aggregate welfare is termed efficiency analysis. Efficiency analysis examines how to set policy or frame institutions to create the largest aggregate welfare but it does not generally weight welfare by social group or sector (e.g., placing higher weights on low-income groups or certain sectors of the economy). Examining who benefits and who loses from policy choices is referred to as equity analysis, and is an important input to policy decisions. However, equity issues can only be examined once the aggregate benefits and costs have been evaluated, and require a broader social context than can be handled by economic analysis alone. Thus, economists focus on efficiency, but typically describe the equity implications of choices so that these considerations can be considered in the policy domain.

Third, while much of economic analysis focuses on describing benefits and costs of various choices (policies, management strategies, institutions, etc.), a key role of economic science is the development of innovative approaches for managing market and social systems. Property rights and tenure systems, for example, are examples of institutions that generate different levels of social welfare depending on their structure and assignment. Suggesting new institutional arrangements and evaluating their impact on social welfare is a key role for economic science: this is considered further in Chap. 7.

Finally, economic efficiency has historically been defined as maximizing social welfare. In an intertemporal or dynamic context this has been interpreted as maximizing the present value of social welfare where less weight is implicitly placed on future generations through discounting future gains and losses. While discounting is consistent with the behaviour observed in market or financial contexts, and thus helps explain the behaviour of firms, it may not be consistent with notions of *sustainability* (as explored in Chap. 2). *Economic sustainability* may be thought of as choosing policies or strategies that result in non-declining economic welfare (where welfare includes market/financial as well as non-market considerations).

This last point raises a fundamental challenge in analyzing the economics of sustainable forest management. Market behaviour will be largely motivated by the present value of financial returns to forestry, and other industrial sectors operating in the forest. Economic sustainability requires assessment of market and non-market elements, and requires strategies that result in non-declining welfare. Geoffrey Heal (2001, p. 2) states,

> *There are two points that are central to sustainability: a concern for what happens in the long run, and a respect for the constraints that the natural world places on the dynamics of human societies and the well-being of their members.*

The conflict between Heal's points and traditional economic systems will be evident throughout this chapter in discussions of current policy, as well as in suggestions of new strategies to improve the prospects of achieving economic sustainability.

Economic fundamentals of the boreal forest

Much of the forest economics literature focuses on the value of forest land and immature stands for timber production, and (a related problem) the optimal age at which to harvest a stand of trees. Martin Faustmann, a German forester writing in 1849 (see Gane 1968), is usually credited with the first correct formulation for the determination of the value of bare forest land for timber production. For the purposes of this discussion imagine that a forest stand is being managed using an even-aged silvicultural system: the primary management activities consist of a final harvest of all trees in the stand, and reforestation after harvest. These reforestation efforts result in another sequence of final harvest and reforestation at some point in the future, and so on in perpetuity. The number of years between successive final harvests is the *rotation age* for the stand. Net harvest revenues at any potential harvest age, reforestation costs, and the discount rate are all known and fixed. The value of bare forest land for any given rotation age, would be the sum of discounted net revenue (i.e., net present value) associated with an infinite stream of reforestation and harvest cycles. The optimal forest rotation age would be the rotation age that yields the greatest net present value resulting from this perpetual cycle of management. One way of determining the optimal rotation is to compare the costs and benefits of harvesting the stand at a particular age versus delaying the harvest decision for at least one more year. The fundamental decision to be made is whether to harvest the stand immediately, or to postpone the harvest to a later date. The rotation age is the youngest age for which the optimal decision is immediate harvest, though greater timber yields and habitat values may still accrue at greater ages (see Chap. 13).

There are two financial forces pushing the decision maker to decide for an immediate harvest: the financial gains resulting from harvest can be invested in the best financial alternative to provide future returns, and immediate harvest allows for a quicker start on the next rotation, leading to a quicker realization of the benefits of subsequent harvests. If the returns to alternative investments are high (e.g., high returns in capital investment markets or interest bearing accounts) then the incentive for immediate harvest will be increased.

The forces working in favour of a delayed harvest are the growth rates of the trees (measured in incremental value per unit area), and the costs of reforestation. If the trees grow quickly, the rate of growth of the forest as an asset may outstrip the growth of financial assets in an interest bearing account. If reforestation requires an outlay of money, any delay in this expenditure has a benefit. Using this simple model, the optimal time for harvesting such a stand of trees occurs when the benefits of delaying harvest for one year just equal the benefits of harvesting and investing the proceeds.

How can such a simple model of optimal rotation be useful? How can a model that ignores harvest flow policies, future price uncertainties, and the interrelationships between stands be helpful in understanding the economics of forestry? The answer is that this simple Faustmann model outlines the fundamental *financial* issues associated with forest management. Note that the word *financial* is used here to refer to the monetary aspects or timber value aspects of forests — we have not yet considered the non-timber value of forests. Several insights are worth identifying:

(1) Because forestry as an industry competes in the same financial markets as all other industries, forest managers must make a return that is competitive with other industrial returns or face a removal of capital from the industry. Investors can place their funds in a variety of industries: if forestry cannot provide sufficient returns to attract investors, then capital will not be invested (or retained) in the sector. The financial benefits of harvest (including the returns to investments of proceeds) are determined largely by forces outside the control of the individual firm or the sector (housing demand, interest rates, etc.). These financial considerations are ever-present in the forest business manager's objectives.

(2) Standing harvestable timber has high value, while forest lands that have recently been harvested or contain young stands, have low value. The value of recently harvested forest land for timber production can be viewed as the present value of all future stands of trees that will grow on that land. Because alternative investments exist that allow forest products companies and (or) shareholders to invest in interest bearing or capital accounts, future financial benefits are worth less than current benefits to the firms. This concept, known as discounting, usually results in the *financial* value of future forests being very low, relative to standing harvestable timber. It is important to note that the value of a young stand, at any point in time, is determined by the current and future costs and benefits associated with the stand and land it occupies, and unrelated to the costs incurred in establishing the stand. These are sunk costs, and should not affect current or future decisions. There is much debate about discounting and the rates of discount that should be used when making social policy decisions, including environmental policy decisions. However, in the financial market, rates of return on capital are determined by market forces and are a critical element in decisions regarding investment and activity in a sector.

(3) In a relatively slow-growing forest, such as Canada's boreal forest, there is little *financial* incentive to put effort into reforestation. Once again, since future returns are discounted, the trade-off between today's reforestation expenditures and expected future returns makes reforestation financially unjustifiable. Table 6.1 illustrates the case for Alberta's boreal forest. If one examines the returns from investing in silviculture, Rodrigues (1998) shows that the net present value (present value of returns minus the costs) is negative for intensive silviculture, positive but small for extensive silviculture, and positive and relatively large for no silviculture. Rodrigues examined many variations of silvicultural options and yield projections

Table 6.1. The net present value of investment at the stand level (per hectare) (from Rodrigues 1998).

Investment	Silvicultural strategy		
	None	Extensive	Intensive
Regeneration lag	28 years	9 years	2 years
Silviculture costs			
Site preparation and planting at year 0			$920
Site preparation and aerial seeding at year 0		$200	
Glyphosate treatment at year 8		$241	$241
Present value of costs at year 0		$339.46	$1081.98
Present value of stumpage from year 100 at year 0	$271.46	$376.45	$550.18
Net present value (NPV)	$271.46	$ 36.99	–$531.80

and the same general conclusions arose. Similar findings have been presented by Benson (1988), McKenney et al. (1992, 1997), the British Columbia Ministry of Forests (BCMF 1999), and other authors. A stylized presentation of this phenomenon is presented in Box 6.2.

(4) If the benefits of growing trees for carbon sequestration begin to materialize (via the Kyoto protocol or other national or international agreement), then the incentive to practice intensive forestry will likely increase. The degree of intensive management that will be practiced under such systems is an empirical question that will depend on the "price" offered for sequestered carbon (see van Kooten and Hauer 2001; van Kooten et al. 2000; and Chap. 20).

(5) From a *financial* standpoint, silvicultural expenditures aimed at controlling the long-term development of future crops of trees in much of the slow-growing boreal forest are often bad investments. This is a fundamental insight arising from Faustmann logic. The financially optimal strategy may be to harvest the existing stand and leave it to regenerate naturally. The optimal *financial* strategy is essentially to treat the stand as if it were a nonrenewable resource (like a mine) and not invest in the future crops of trees. This point requires some clarification. In most of the boreal forest, trees will eventually re-establish on a harvested site, even without management intervention, so there is certainly a renewable aspect to the boreal forest. However, regeneration lags will likely be longer and realized growth rates will likely be lower than would be seen with reforestation efforts. Because current harvest levels are largely supported by stocks of mature timber (i.e., stands older than their optimal rotation age), it is unlikely that future harvests at or above current levels can be considered to be financially prudent.

(6) *Financial* aspects of forest management do not necessarily coincide with other concepts of sustainability, including economic sustainability. Financial considerations typically focus on optimizing returns relative to alternative investments (rates of return). Since this places little weight on ecological objectives and the generation of forest benefits to future generations, this approach may be viewed as inconsistent with concepts of sustainability. At a large scale (multi-industry, national, or international), operating under financial incentives may be consistent with economic

---

Box 6.2. *Financial aspects of boreal forests: an example.*

Assume that the yield curve below (Fig. 6.2) represents a forest stand. The logging costs are assumed to be $5000/ha, and the value of the wood (what the mill would be willing to pay for the wood delivered to the gate) is $60/m^3. The optimal forest rotation is 90 years at a 4% discount rate and 80 years at a 10% discount rate, under a "leave for natural" (LFN) reforestation prescription (no cost).

Fig. 6.2. Standard trend in merchantable volume of a boreal forest stand over time.

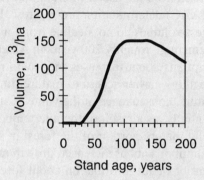

Fig. 6.3. Minimum yields at different stand ages required to recoup silvicultural investment made under different discount rates or no cost.

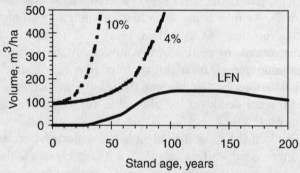

Figure 6.3 shows the yields required to recoup a $500/ha silivicultural investment at 4% and 10% discount rates compared to the assumed yields for natural regeneration. At a 4% discount rate, the yield at 70 years would have to be 2.5 times that under LFN in order to justify a $500/ha investment (i.e., MAI must increase from 1.2 to 3.1 m^3 ha^{-1} year^{-1}).

---

sustainability (non-declining human welfare) if returns from forest use are re-invested into other natural and (or) constructed capital. However, this is a complex issue that will not be easily resolved. In any event, it must be recognized that the financial forces driving firm behaviour are not necessarily consistent with ecological notions of sustainability. Thus policies that require ecological sustainability (even simple ecological notions like long-run maintenance of standing timber volume) will typically be at odds with financial incentives. An understanding of these financial forces is essential to the development of any sustainable forest management policy.

(7) Forest management legislation and regulations often link the allocation of timber harvest rights to a commitment for reforestation, and to past performance on those commitments. However, it is crucial to recognize that these regulations are contrary to the underlying financial forces driving firm behaviour. This phenomenon also helps one understand the desire of firms to take advantage of allowable cut effects (ACEs) associated with increased growth or intensively managed timber. The allowable cut effect arises from the use of intertemporal harvest flow constraints in annual allowable cut (AAC) determination procedures[2]. Harvest flow policies typically require that harvest volumes do not decline from one period to the next over a lengthy planning horizon (e.g., roughly 200 years in Alberta). If surplus mature timber exists, then assumptions about increases in growth rates (to be realized in the future) can be used to allow earlier access to mature timber. The desire for allowable cut effects is a natural consequence of the value of standing timber and harvest flow constraints. Essentially, as a result of the intertemporal flow constraint, a firm can mine the financially mature portion of the forest at an increased rate, by making and justifying assumptions about increased growth rates (perhaps as a result of silvicultural investment or better information about forest production). Reforestation can therefore be a rational financial decision for a firm operating in a policy environment prescribing intertemporal harvest flow policies.[3] However, ACE can only exist when there is a policy linkage between the allowable harvest levels in different time periods. An interesting side effect of intertemporal harvest volume constraints and the resulting ACE is that they tend to make volume-increasing silvicultural investments more financially attractive than value-increasing investments (e.g., pruning for clear sawlogs). Intertemporal volume flow constraints introduce market distortions that can lead to socially inefficient silvicultural expenditures. For additional detail on the allowable cut effect see Luckert and Haley (1995) and Luckert (1996).

(8) The "bottom line" on the fundamental financial aspects of boreal forest management is that from a social welfare perspective it is likely desirable in many parts of the boreal forest to operate using "light touch" approaches or natural regeneration. Interestingly, such practices may also be ecologically desirable as they may allow the regenerating stands to more closely follow natural successional pathways. While a financial perspective suggests minimizing the expenditure on silviculture, various social and ecological arguments have been used to suggest the opposite. We examine some of these issues below.

---

[2]Intertemporal harvest flow policies link the allowable harvest volume in one time period to that in at least one other. Common implementations of these policies include even flow (where the harvest volume in each period of a planning horizon must be equal) and non-declining yield (where the harvest volume in any period can be no lower than that in the previous period).

[3]Hegan and Luckert (2000) show that even in cases where allowable cut effects are available they may not be financially desirable in all cases. They identify specific conditions under which this strategy is financially advantageous to the firm.

A commonly stated goal of resource management is the creation or maintenance of employment. One of the economic realities of forest management is that in the foreseeable future, employment, particularly employment per unit of output, will continue to fall because of technical change. Unless there are unforeseen circumstances, capital will continue to substitute for labour within forest harvesting and management sectors.[4] Employment within the forest harvesting sector could be increased through increases in the amount of land allocated for forest management, but many jurisdictions across Canada have fully allocated the land available for industrial forestry. It has been demonstrated that it is primarily the factors that define the demand for timber that determine employment levels, not policies defining even flows of timber harvest or policies that would increase the supply of timber (McCandless and Messmer 1996). While economists are interested in employment (or unemployment) in the economy as a whole as an indicator of economic activity and social welfare, maintaining constant or increasing employment within any one sector (such as forestry) is not necessarily desirable. When examining the benefits and costs of a particular project, economists tend *not* to consider employment creation within a sector or region beneficial unless that region is expected to suffer from chronic unemployment. If the region will not suffer (or is not suffering) from chronic unemployment then additional employment creation is simply a transfer from one sector to another and will not generate any net economic benefits.

This overview of fundamental forest economics shows that in many parts of the boreal forest the key *financial* value is in the standing timber. The pure financial incentives are to remove mature standing timber and minimize effort on regeneration and related activities. Of course, regulations have been enacted to control such behaviour. We have attempted to demonstrate that a basic understanding of these key features is required to understand the economics of boreal forest management. In the following section, we address extensions to the simple financial model and bring in considerations of non-timber values and economic management of both timber and non-timber resources.

Economic aspects of boreal forests: timber and non-timber goods and services

Over the past few decades the significance of a wide range of non-timber goods and services in forest management objectives and outcomes has increased (Chap. 1). Various approaches have been employed to incorporate recognition of wildlife habitat, visual quality, and biodiversity into forest management objectives and outcomes. Many non-economists believe that economics is unable to make a significant contribution to the management of non-timber goods and services because such values are not priced in a marketplace. Private entrepreneurs typically do not have rights or responsibilities to

---

[4]Value-added enterprises present an opportunity to increase employment and economic returns to the sector. However, markets are required for value-added enterprises to be successful. Canada may have some comparative advantage in developing markets for value-added goods, but many other countries (especially those with lower labour costs; see Pearse 2001) actively and successfully compete on this front.

these non-priced goods and services and therefore do not have strong incentives for maintaining or supplying them. However, advances in the economic analysis of non-priced goods over the past three decades have been substantial. Once again a few fundamentals associated with non-timber goods and services arise from economic analysis.

(1) The first fundamental outcome of several forms of economic analysis is that the value (or importance of) non-timber goods and services relative to timber values has risen over time and will probably continue to rise. There are several reasons for this. First, in cases involving unique environmental resources (unique habitats, species, etc.; see Fig. 6.4) there are few substitutes, while there *are* substitutes for wood fibre resources and products generated by fibre resources. This lack of substitutes for environmental goods will result in increasing relative values for these goods over time. Second, if incomes continue to increase, it is likely that the demand for environmental goods will increase at a faster rate. A variety of analyses have shown that the demand for environmental quality tends to increase with increasing income levels. There has been much debate and discussion in economics about a phenomenon called the environmental Kuznets curve that shows an upside-down U-shaped relationship between income and the level of environmental degradation (see Kristrom 2001 for a summary). In other words, environmental degradation is low when countries are at low-income levels, high when incomes are moderate, and low again when income is high. This phenomenon has been observed for some pollutants and measures of environmental degradation, but not for others. Furthermore, it appears that this relationship likely arises from a combination of preferences for environmental quality as incomes rise, better technology in higher income countries, and (or) improved institutional frameworks in higher income countries. These factors will all likely continue to generate increasing demand for environmental quality.

Fig. 6.4. The boreal forest — a mosaic of timber and non-timber values.

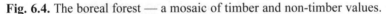

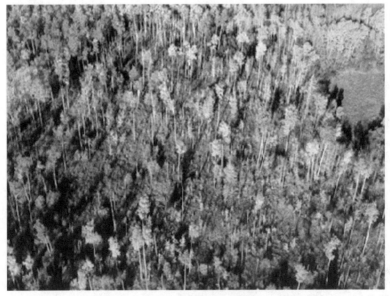

Photo by Roger Nesdoly, courtesy of Mistik Management Ltd.

(2) From an economic perspective, both biological and economic/social factors are involved in the value of non-timber goods and services. That is to say, biological factors alone will not be able to provide value measures of environmental goods. Value in economic analysis is determined by the trade-offs that individuals are willing to make, and by people's perceptions of what is scarce or abundant. If individuals are willing to forego forest sector development in order to maintain a unique environment untrammeled by human industrial activity, then the value of the environment is implicitly greater than the value of the timber resources. Of course such values differ over individuals, and groups of individuals. Owners of the forest products company may not be willing to forego the value of the timber extraction, and may therefore be willing to exploit timber resources under some form of social license (or rule of acceptable practice). The purpose of the social license contract is to establish a more definitive and meaningful role to include the recognition of society's values for non-timber resources. The challenge for economic analysis arises here. Economic analysis involves the evaluation of trade-offs or values over all individuals. Value, in this context, depends on human preferences and trade-offs, and not solely on ecological information. The sum of these individual values is assumed to reflect society's overall value. Naturally one cannot capture all values and some decisions about the "extent of the market" will have to be made. Nevertheless, such an approach hopes to capture the relevant values for those significantly affected, thereby identifying the importance of non-timber values.

(3) As human populations increase, the importance of environmental values will increase, but in complex ways. Aggregate values associated with recreational uses of forests will likely increase, even though value accruing to any particular individual may decrease (because of congestion, increased development, etc.). As people move into the urban–wildland interface, there will be increased pressure to reduce industrial activity, and to prevent human harm and injury from what would otherwise be considered natural processes and activities (e.g., wildfires and predation from large carnivores). As human populations grow, the impacts of industrial activity on air and water quality will have larger value impacts. Values associated with traditional uses by Aboriginal People (see Chap. 3) will increase as their populations increase and the resources on which they depend become scarcer as the lands that they use for traditional activities become more developed and fragmented (unless preferences change dramatically — which is possible given the dynamic interaction between preferences and institutions described in Chap. 7). There will likely be increasing conflict between non-timber resource users (e.g., motorized and non-motorized recreationists) and there will be increasing conflict between individuals interested in the use of forest resources and those interested in preservation of forests (as protected areas). Currently, boreal forest regions in Canada have low population densities. As population levels increase in concert with rising levels of personal wealth, the relative importance of environmental resources will also probably increase.

In markets, the prices of goods reflect their values. Many trade-off decisions are made in markets and the resulting prices are indicators of the overall value of a good or service. Environmental goods, however, are not typically traded in markets. There is no price mechanism that helps reveal their value. In order to evaluate environmental goods, economists have created many techniques to assess trade-offs and determine implicit values from these trade-offs (Adamowicz and Boxall 1998). In some cases behaviour in markets is observed and this reveals a value for traditionally non-priced resources. For example, if consumers are observed to pay higher prices for certified lumber or choose certified lumber over non-certified lumber (see Chaps. 7 and 21) when all other attributes of the lumber remain the same, this indicates a value associated with the sustainable practices required for certification. Models of recreation choice behaviour (where people go, how often they go, etc.) can be constructed to reveal the trade-offs individuals make between traveling further and visiting sites with different environmental quality levels. Recreation values illustrate that values arising from forests are complex combinations of biological attributes (wildlife, views, etc.) and economic/social attributes (location of populations, the price of gasoline, congestion at recreation sites and on roads, etc.). In addition, economists have developed a variety of techniques that attempt to elicit trade-offs people are willing to make through structured interviews and questionnaires (Adamowicz and Boxall 1998). Each of these techniques has its limitations, however: in each case a combination of biological information and human behaviour is employed to assess how individuals value environmental goods and services. We present some of these techniques for measuring the value of non-timber goods and services below.

If one knows the value of non-timber goods and services, how can they be incorporated into forest management planning? The simple Faustmann optimal rotation model described earlier relies on a description of how the value of a timber stand changes with age, assuming that the stand is harvested at any given age. This represents a "flow" value, or a value associated with the removal of goods (timber) from the site. Many environmental values are best thought of as "stock" values, or values that are realized at the site. Some of these (e.g., wildlife habitat) may be related to vegetation conditions and, therefore, to stand age. Hartman (1976) extended the Faustmann model to incorporate both the flow of timber values and the stock of environmental values. The optimal Hartman rotation can be either shorter or longer than the rotation age identified by the Faustmann model, depending on the nature of the environmental good, and could even suggest that no harvesting take place. Van Kooten et al. (1995) examine carbon values associated with stock changes in this framework.

As originally formulated, the Hartman rotation is based on a single stand or hectare. In reality, the value of environmental goods depends on the spatial and temporal distribution of a variety of forest attributes over a large area. It has also been shown that the age of trees in a forest stand is often not a good indicator of the stand's ability to provide non-timber amenities (Bunnell et al. 1998). However, several other practical methods have been developed to incorporate environmental values into forest management. Adamowicz et al. (2002), for example, develop a spatial economic model that measures the impact of forest harvesting on the hunting activities of Aboriginal People. This spatial information can be used by forest managers to examine different harvesting plans and to choose ones that reduce the impact on non-timber values (see Chap. 14). These models can also be used in an optimization framework to construct optimal forest harvesting plans that maximize timber plus non-timber values or maximize timber value subject to a non-declining non-timber value constraint. Nanang (2002) has developed such a framework. Further details on the theory and application of economic analysis with timber and non-timber objectives can be found in Bowes and Krutilla (1985) and Mendelsohn (1996).

Methods for valuing non-timber goods and services

Methods that identify the value of recreational activities in forests have been the focus of many economists. In part this is because of the direct linkage between the environment (habitat, access, etc.) and human activity. These methods include assessments of actual choices made by recreationists (revealed preference methods) or structured interview questions (stated preference methods) to develop models of recreation behavior. These models then provide the information on trade-offs, values, and predicted behavioral change associated with changing environmental conditions. These methods combine data collected through surveys or other techniques with econometric modelling of these survey data. Decision support tools have been developed (Adamowicz and Boxall 1998; Akabua et al. 2000) that indicate which regions have high recreation values and which regions would be most significantly affected by changes to forest attributes. Boxes 6.3 and 6.4 contain examples of such tools.

In addition to recreation values, researchers have examined the value of carbon sequestration (a non-market value that may soon become a market value), non-timber forest products (mushrooms, berries, etc.), values of subsistence resource use (Haener et al. 2001), values associated with visual quality, and values of ecosystem services (water flow, etc.). Each of these values is a component of the overall mix of values generated from the forest resource. Each of the values listed above are related to some activity or product arising from the forest (recreation, drinking water, etc.). The value associated with protected areas, or the value society places on removing industrial activities from some forest regions, is a more challenging value to capture. There is little doubt that the public is interested in wilderness and protected areas (see, for example, Adamowicz et al. 1998). Yet the degree to which the information necessary to make such complex trade-off decisions can be conveyed and digested remains an open question. Values associated with a forest condition (for example, preferences for protected areas) that have no link to actual human behaviour are called *passive use values*. Several researchers have attempted to quantify passive use values (Adamowicz et al. 1998). These measures are derived from structured questions about hypothetical trade-offs and they continue to be controversial. Nevertheless, these types of trade-offs will likely be some of the most important factors in future forest management.

Economic sustainability

In the introduction, we discussed the fact that economic behaviour observed in markets is not necessarily consistent with sustainability. Sustainability is a constraint or a condition (or even a "responsibility"!) that many sectors and jurisdictions are attempting to place on their economic activities. In some ways, intertemporal harvest flow constraints

Box 6.3. Measuring recreational values.

Figure 6.5 describes the preferences of hunters for forest attributes. The graph arises from survey-based research that identifies hunters' preferences in terms of hunting site selection. They respond to various factors in the forest environment including access, evidence of forest harvesting, and wildlife populations. The horizontal axis measures "utility" or satisfaction, while the vertical axis itemizes forest characteristics. The preferences for three groups of hunters (categorized by age) are presented. Forest sites can be characterized by the attributes on the horizontal axis and the attractiveness of each forest site can be quantified. Given behavioural relationships of this type, models predicting recreation activity in response to changes in the forest environment can be constructed. Box 6.4 provides an example of such a simulator.

Fig. 6.5. Factors affecting hunters' choices of forest regions (Dosman et al. 2002).

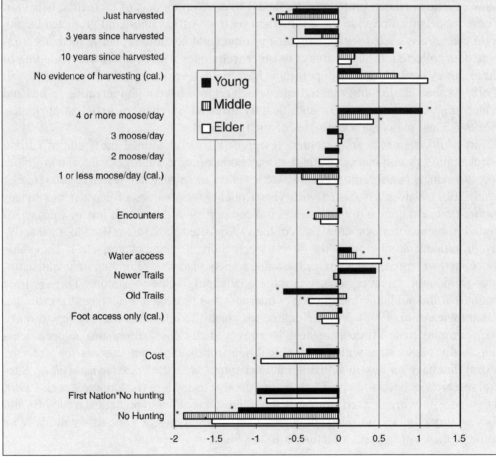

were early attempts to impose sustainability on market driven activity. Arguably, this could have been handled by more explicit definitions of property rights, yet these were hardly deemed necessary in the face of cheap and abundant natural resources. Concepts of sustainability have become more complex. This fact, coupled with increases in society's wealth and technical change in the forest products sector has resulted in significant shifts in the demand for non-timber resources, and has greatly affected the kind and

Box 6.4. *Integrating recreational values into forest management.*

The simulator below (Fig. 6.6) is an example of a computer model that predicts hunter site choice and identifies the economic value of changes in the landscape. The bars indicate the proportion of the hunters going to a given WMU (wildlife management unit). Each WMU is described by a number of forest attributes (similar to those indicated in Fig. 6.5). Changing attributes will result in a change in hunter visitation patterns and economic welfare. In the figure below, a change at WMU 346 is simulated. This change makes WMU 346 unattractive relative to the base case and thus fewer hunters are expected to travel there. Hunters instead reallocate to other sites or choose not to go at all (the "None" option). The welfare measures provide estimates of the per trip economic value of the change (negative in this case).

Fig. 6.6. Output from a spatial recreational choice simulator (Akabua et al. 2000).

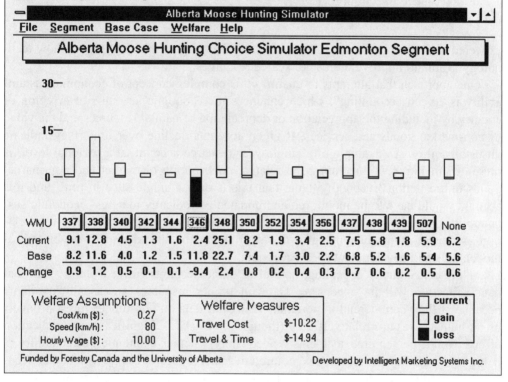

WMU	337	338	340	342	344	346	348	350	352	354	356	437	438	439	507	None
Current	9.1	12.8	4.5	1.3	1.6	2.4	25.1	8.2	1.9	3.4	2.5	7.5	5.8	1.8	5.9	6.2
Base	8.2	11.6	4.0	1.2	1.5	11.8	22.7	7.4	1.7	3.0	2.2	6.8	5.2	1.6	5.4	5.6
Change	0.9	1.2	0.5	0.1	0.1	-9.4	2.4	0.8	0.2	0.4	0.3	0.7	0.6	0.2	0.5	0.6

Welfare Assumptions		Welfare Measures	
Cost/km ($):	0.27		
Speed (km/h):	80	Travel Cost	$-10.22
Hourly Wage ($):	10.00	Travel & Time	$-14.94

☐ current ☐ gain ■ loss

Funded by Forestry Canada and the University of Alberta   Developed by Intelligent Marketing Systems Inc.

amount of forest that can be utilized for wood products. Within this constantly shifting scene, monitoring to determine the effects of such changes has also become very complex. As introduced in Chap. 2, there are various definitions of sustainability in the ecological, forestry, and economics literature. Once again, economic concepts of sustainability include both the natural aspects (biodiversity, scenery, etc.) as well as the human aspects (employment, output, products). An overview of economic sustainability concepts can be found in Mittelsteadt et al. (2001); more technical presentations are found in Heal (1998) and in Pezzey (1992). Economic sustainability relies on maintaining the welfare of present and future generations, where this welfare depends on market as well as non-market goods. It is recognized that natural systems have limits that must be accounted for in the way we manage our current economic activities. Welfare is also

Fig. 6.7. Example of a regional forest resource account (Haener and Adamowicz 2000*b*).

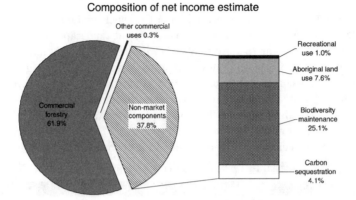

Composition of net income estimate

expected to depend on how resources are shared among people (equity) and how technology is able to improve our use of scarce resources.

One approach that attempts to quantify this complex concept of economic sustainability is "green accounting", which enhances the economic accounts of a region or country by including the appreciation or depreciation of natural resources and the value of non-market goods and services. If green accounts decline over time, they indicate unsustainability. An example of a single-year resource account, at a regional level, is presented in Fig. 6.7, illustrating the relative contribution of market and non-market goods to the regional economy. Note that this is only a single slice in time, and this account would have to be monitored and updated periodically to assess economic sustainability. Also, the value of the forest resources (including appreciation or depreciation associated with changing growth rates) should be linked directly to biological models of stand development and succession.

While conceptually appealing, natural resource accounts suffer from the fact that data collection will be expensive. Data on timber resources are usually available because of their commercial importance. However, data on non-timber forest products or on human uses are relatively rare (though see Fig 5.3). Significant data collection efforts would be required to identify adequately whether economic sustainability is being achieved. Further research is required to select the resources on which to concentrate such efforts. In addition, sources for funding such monitoring programs will need to be developed. Nevertheless, the importance of resource accounting is highlighted in the recent book *Nature's Numbers* published by the U.S. National Academy of Sciences, where it is indicated that the collection of data for non-market accounts should be a high priority (Nordhaus and Kokkelenberg 1999, p. 3).

The institutional foundations of the economics of boreal forest management

The foundation of boreal forest management is the set of property rights embodied in tenure systems for timber and non-timber resources. These tenure systems provide the rules within which economic agents (firms and individuals) operate. These tenure systems also provide insight into certain behaviours as illustrated by harvest flow policies and allowable cut effects discussed previously. A thorough description of the institutions underlying forest management is the main focus of Chap.7, but a short discussion of the implications of institutional arrangements on the economics of forest management is relevant here.

A key issue underlying the economic aspects of boreal forest management is the tenure assigned to timber and non-timber values. Because tenures are associated with timber resources, and economic agents are able to capture value from these resources, forests have been the focus of economic development projects throughout the boreal regions of the world. Non-timber products (including wildlife resources) have by and large been retained as public goods. These goods are not subject to ownership and management by individuals but are retained as property of the Crown. Rights for non-timber goods are allocated through administrative measures often involving rules of capture. Here a fundamental challenge arises in economic management of the forest resource. Tenure for forest resources provides the forest industry with some incentive to manage timber resources to maximize their economic value. But there is little incentive for the management of non-timber resources except to obtain approval to operate. Recreation resources, non-timber products, and protected areas are often provided as a cost to timber extraction, or are provided at public expense. In both cases, tools and methods designed to enhance the value of these non-market activities have attracted little interest. Additional regulation may be imposed to require forest managers to consider non-timber elements, but there remain no direct incentives for the maintenance or enhancement of non-timber goods. Framed in another way, forest managers capture few of the benefits of non-timber products but are required to manage for these resources (see Luckert and Haley 1995). This is likely to lead to a situation where forest products companies try to find the least costly ways to satisfy the requirements for the provision of these non-timber resources.

The current tenure regime commonly restricts trade between forest regions. In order to maintain local employment it is not uncommon for forest tenures to require that fibre harvested within a region be processed in that region or that rights to harvest timber be only given if a certain processing facility exists in a certain community. This is an explicit attempt to achieve economic development and employment objectives within regional economies, typically an important objective of elected government representatives (Pearse 2001). These linkages may also make it easier for processing facilities to search for capital as fibre supplies (or at least some percentage of the total demand) are linked to the mill. However, such restrictions often limit economic opportunities. Decoupling the business realities of forest harvesting and management activities from the business of running a wood processing facility would allow timber to flow to the facility with the highest demand. Disturbances from fire or insect outbreaks would have a lesser impact on local wood scarcities caused by inflexibilities in tenure. Wood processing facilities would be free to purchase wood from wherever they choose, which would result in a general increase in the efficiency of fibre use. At first glance, this would apparently be at the cost of less security in the fibre supply for each individual firm. However, a larger provincial or national "wood basket" from which all firms were free to purchase wood could actually increase the security in fibre supply, as a result of an "insurance effect" against catastrophic forest fires and other disturbances. Naturally, any such efforts would have to include considerations of non-timber values at the appropriate scales.

The government-mandated vertical integration of forest management and mill operations has obvious costs in Alberta (and other regions with similar ecological and tenure conditions). Much of Alberta's boreal forest falls in the boreal mixedwood section (Rowe 1972). The mesic sites in this part of the boreal forest often support both trembling aspen (*Populus tremuloides*) and white spruce (*Picea glauca*), often as part of a successional pathway with aspen as the pioneer species, followed by aspen and spruce coexisting in a mixedwood phase, potentially culminating in a climax of pure white spruce (see Chaps. 8 and 13). Until the early 1980s, aspen in Alberta had little or no commercial value. Forest tenures were allocated on land with a large white spruce component; the other land was considered unsuitable for commercial forestry. The construction of oriented strandboard (OSB) and pulp mills in the 1980s and 1990s suddenly changed the prevailing view of aspen as a "weed species" to aspen as a valuable source of fibre. As a result, much forest land that was previously non-commercial now had value, and new forest tenures were allocated to take advantage of this new-found wealth. Forested land is not divided into contiguous stands of white spruce and aspen; rather, aspen and spruce stands are interspersed, and the species are usually interspersed within any given stand (see Fig. 22.1). Tenures were (and still are) allocated based on the species used by the mill associated with the tenure. On a stand-by-stand basis, the forest is arbitrarily divided into a softwood land base and a hardwood land base with different agencies and companies responsible for management of the different land bases. This "divided land base" leads to a number of inefficiencies. Because of intertemporal harvest volume constraints, firms with harvest rights to white spruce stands will try to eradicate aspen on stands within "their" land base. Firms holding the harvest rights to aspen stands have little incentive to protect white spruce understories during harvest. In the former case, money is being spent fighting natural successional pathways in order to gain a benefit associated with the ACE. In the latter case, a resource with potential value is being ignored because the firm responsible for management does not have rights to that resource, which is temporarily an understory species. Other inefficiencies that may be introduced include the development of parallel road networks and the waste of "incidental" harvest volume. These inefficiencies have been evaluated in a detailed simulation study by Cumming and Armstrong (2001), and similar results were produced in an optimization approach by Hauer and Nanang (2001).

There are alternatives to this type of management regime: Mistik Management is a forest management company that supplies hardwood and softwood to mills in the boreal mixedwood of northwestern Saskatchewan (see Chap. 22 for a further description). Daishowa–Marubeni International Limited (primarily hardwood) and Canadian Forest Products Ltd. (exclusively softwood) have developed a joint forest management plan in the Peace River region of Alberta. Provincial forest policy in Alberta has begun to change in response to studies reported in Box 6.5 (sponsored by the Sustainable Forest Management Network, SFMN) and other research, and the subsequent development of effective alternatives.

Tenure systems guide the use of resources. In most of Canada's boreal forest these tenure systems tend to be relatively inflexible in terms of harvesting and processing relationships, trading fibre resources across boundaries, and assigning rights to harvest individual tree species in mixedwood forests. The tenure system usually gives the manager of a forest no control over or ability to capture benefits from other forest resources, even though in many cases the forest manager is required to include consideration of these other forest resources in management. Finally, other users of the land base (e.g., the energy sector in Alberta) can affect the resource but they operate under different tenure rules that are created without reference to tenure provided to the forest sector (see Fig. 22.6).

These tenure institutions are in place for specific reasons, but typically to capture a share of economic returns that satisfies local political and economic interests. Given the changing benefit structure of the forest, it is necessary to re-evaluate current tenure institutions. The economic activity generated within the current tenure system leads to "unnatural" (and likely unsustainable) ecological conditions in boreal mixedwoods, a system that provides incentives for intensive management only in the presence of intertemporal harvest volume flow policies and allowable cut effects, and a system that is regionally inefficient in providing resource values to industry and society because of inter-regional trade barriers. Intensive silvicultural investments may be economically rational for firms operating in a policy environment that leads to the allowable cut effect. From a social point of view, however, these investments should be considered in light of the rather discouraging results on the returns to intensive management. Allowable cut effects recognize the value associated with an accelerated liquidation of mature timber. Resources invested in intensive silviculture in such cases may be more wisely spent in other activities, perhaps in value-adding silviculture or other forest-bound investments, or public investments. Every dollar spent on government-mandated silviculture (or dollars spent as a result of the incentives of the ACE) is a dollar that could have been captured through stumpage fees, and allocated in the broader context of public policy. The problem of shortages in wood supply has several other potential policy-based solutions (enhancing trade between regions, decoupling the vertical integration between woodlands and mills, etc.) that may be attained at lower cost.

It is important to explicitly recognize that some degree of "timber mining" (i.e., unsustainable rates of harvesting older, natural forest) is occurring in the boreal forests of the world. But it is also important to consider that this may be a reasonable public policy, as long as management decisions are based on sound understanding of the public's wishes for the future conditions for their forest, rather than a sustained yield-driven approach to resource management. It is likely that the public cares at least as much about the state or *condition* of their forest as it does about the amount of *output* it is producing. A significant component of the value is in the stock or forest condition. This is tied to the increasing importance of existence or intrinsic values. The decoupling of today's harvest from tomorrow's assumed growth, and the reliance on multiple output (timber and non-timber) measures of sustainability (e.g., James et al. 1997) will provide for more flexible and efficient forest management.

Box 6.5. Economic analysis of alternative tenure arrangements.

Simulation of the economic implications of alternative tenure arrangements provides a vehicle for policy analysis. Figure 6.8 illustrates the implications of changing tenure regimes from one that effectively "divides the land base" by species composition into one that allows the land base to be treated as a single unit. The number of entries is much larger across the land base as a whole in the "divided" case, while the entries are more concentrated in the undivided case. The result in the undivided case is a type of natural "triad" (three-class land zoning) in which the southern area receives much more intensive activity, while the north remains relatively untouched. The annual cost savings under the undivided strategy is about $7 000 000 (Cumming and Armstrong 2001). Hauer and Nanang (2001) performed a similar analysis on a different land base; Fig. 6.9 summarizes their results. The graph contains indexes of "shadow prices" or the marginal costs of regenerating, harvesting, transporting the wood to the millgate, and milling for each mill. The marginal costs associated with the base run (divided land base) are higher than those associated with the non-divided land base.

Fig. 6.8. Simulated number of entries on the same land base managed by rules for (*a*) divided and (*b*) undivided management of coniferous and deciduous tree species (Cumming and Armstrong 2001).

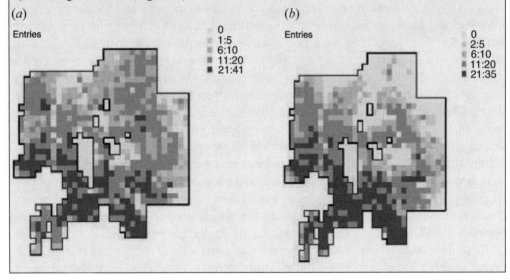

Where to from here? Economic tools for sustainable forest management

The above discussions highlight several fundamental aspects of forest resources in the boreal region. The combination of tenure systems, biological conditions, and human activity have led to an array of timber and non-timber values arising from forest resources and varying degrees of efficiency in capturing these values. What economic tools can be used to improve management in the boreal forest or enhance our attempts to achieve economic sustainability within the boreal region? The following subsections explore some approaches and methods that may help to achieve such goals.

Box 6.5. (concluded).

Fig. 6.9. Simulated index of marginal costs over several decades, comparing "Baserun" (divided) versus "Scen1" (undivided) scenarios (Hauer and Nanang 2001).

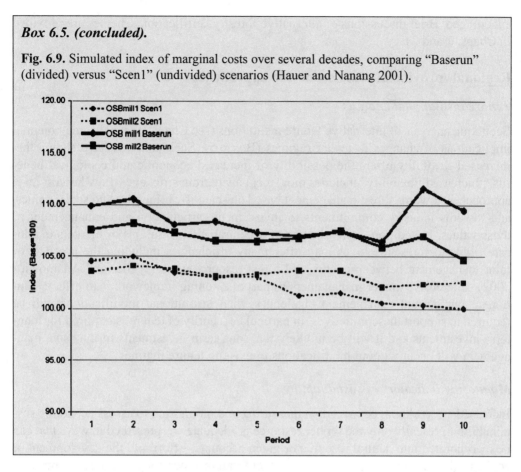

Global-scale economic approaches

Certification/market signals for sustainable forest management

Providing clear signals and incentives is an important element required for resource management. At a global scale, signals for timber values are provided through prices and sales of forest products. Signals for non-timber values or for values associated with sustainable forest management are not typically provided through markets, with perhaps the exception of forest certification. Certification provides a mechanism for signaling the demand for various aspects of sustainable forest management. Allowing forest products companies to compete not only on product price but also on the sustainability of their practices, their involvement with local communities, and their partnerships with Aboriginal People, could help to clarify the incentives for these non-timber aspects of forest management. Economic analysis of the price or market share response to certification has been limited, but increasingly it is being shown that certification schemes and eco-labels do provide ways to signal attributes of interest to buyers of forest products. As interest in environmental values increase, interest in certification and signaling schemes should also increase. Methods that allow the market to signal environmental performance and allow firms to capture the benefits of such performance, should be encouraged. Brief discussions of alternative forestry certification schemes are provided in Chaps. 7 and 21.

Regional-/provincial-scale economic approaches

Tenure/institutional analysis

Economic analysis of alternative tenure institutions (see Chap. 7) reveal some potential implications of changes to tenure regimes (Box 6.5). Such analyses have shown that increased flexibility offers the possibility of increased economic and ecological bene-fits. "Increased flexibility" includes movement away from strict even-flow harvest level approaches, towards forest management based on scientific principles and social values, and towards industry commitments to invest in the articulation and enhancement of those values. In addition, increased use of decentralized institutions or markets to allocate rights to forest resources may offer many efficiencies including the prospect of joint management between energy and forest resource users (Weber and Adamowicz 2002). If intensive timber management as part of a zoning framework is to offer a solution to our current management challenges, then institutional modifications will be required to support these initiatives. In particular, security of tenure is required for long-term investments and it will be unlikely that long-term investments in intensive management will occur without modifications to existing tenure regimes.

Monitoring indicators of sustainability

Increased effort should be placed on measuring and monitoring to assess economic sustainability. Presently only the timber resource is adequately represented in ways that can be incorporated into natural resource or green accounts. Effort into the development of conceptual and empirical methods for incorporating other forest resources and non-forestry human activities into these accounts is essential for the assessment of the sustainability of forest resources and our uses of them. The methods proposed by Vincent and Hartwick (1997), and implemented to some degree by Tamai (2000), would go a long way in evaluating the value of the capital stock (broadly defined) in forestry.

Forest-/stand-scale economic approaches

Probabilistic sustainability

The nature of the boreal forest includes significant disturbances by natural agents such as fire and insects (Chap. 8). This requires that economic analysis be done in a probabilistic fashion with due consideration of the risks of these disturbances. In the determination of timber supply (see Chap. 11), for example, harvest level projections should be evaluated in a risk framework where any chosen harvest level has the possibility of not being met because of natural disturbances (Armstrong 2000). Measures of economic sustainability, especially indicators of forest resource accounts, should also include consideration of the risk of disturbance (Haener and Adamowicz 2000a). While full consideration of the risk of catastrophic disturbance is challenging, even relatively simple evaluations of risks will be tremendous improvements over current practices. Economic analysis of this type can provide guidance on the appropriate size of the sustained yield unit(s) over which we can hope to balance risks, thereby being able to maintain sustained yields of timber or non-timber goods at a given level of probability.

Economics of silvicultural investments and the allowable cut effect

Economic analysis of silvicultural investments was discussed above. Our recommendation is that analyses of silvicultural investments recognize the true economic aspects associated with silviculture as well as those related to the ACE. The costs of silvicultural investments and the ACE must be evaluated against the benefits of these practices, benefits that in the current situation arise from regulatory approaches that are increasingly being questioned. In much of the boreal region, relatively modest investments or "light-touch" silviculture can be used to maintain the economic value of the timber stock, especially when the timber stock is evaluated over a larger scale. Such practices may also be beneficial for many non-timber values.

Economics of zoning

The triad principle of land-use zoning has been proposed as a solution to land-use challenges in Canadian boreal forests (see Chaps. 11–13). This approach suggests a partitioning of the land base into extensively managed, intensively managed, and protected areas. The concept of forest land-use specialization into these three sectors has merit from an economic perspective (Vincent and Binkley 1994), as well as from the perspectives of social equity and future adaptability (Burton 1995). However, a successful triad program will require that the movement to a particular zoning scheme improve the sum of timber and non-timber values. The discussion above casts some doubt on the merits of the intensive management component except in those cases where growing conditions are such that significant returns to initial investments are generated, or in cases where the silvicultural costs are relatively small. In other words, it is possible that the optimal intensively managed area is zero. The protected area component of the triad also requires investigation. What is the appropriate size, position, and temporal location of a protected area? Can economic methods be used to determine the optimal mix of protected area and harvested area? Economic methods have been employed to identify candidate areas based on minimizing the cost of achieving specific ecological objectives (where costs are defined as the opportunity costs associated with not being able to use the protected area for industrial purposes). These analyses reveal significant savings associated with the use of cost minimization approaches to protected areas planning (Ando et al. 1998). However, these methods do not help in determining the optimal economic combination of protected area and forest management area. Research efforts that use market and non-market analysis to identify such trade-offs are required. Without such trade-off analysis it will be difficult to identify economically optimal triad schemes.

Incorporating non-timber values

A variety of methods for quantifying non-timber values have been developed and tested. For non-timber values associated with recreation and tourism, for example, models of spatial, dynamic economic benefits can be integrated into forest management models

(Adamowicz and Boxall 1998). Similarly, methods for incorporating non-timber values into resource accounts have been developed and deployed in some jurisdictions (Nordhaus and Kokkelenberg 1999). In most cases the challenges associated with incorporating these models into forest management is not associated with the modelling tools, but relates to data availability or methodological problems associated with inventorying human uses and non-timber goods and services. A very significant question is, "Who should pay for the development of such inventories?"

Methods to assess timber and non-timber values over time and space

Economically efficient management of boreal forest resources requires methods to inventory, monitor, and project timber and non-timber values over large areas of space and long periods of time. A concept that has been suggested for this type of assessment is the "desired future forest" or DFF approach. The DFF concept can be thought of as a method that outlines the implications of alternative forest management plans in time and space. As such it is consistent with the notion of finding forest management options that consider timber and non-timber values and explicitly recognize the trade-offs involved in forest management alternatives. For example, the analysis presented in Armstrong et al. (2003) can be considered a partial DFF approach. Armstrong et al. (2003) illustrate the impacts of alternative management approaches on economic outputs and wildlife habitat within a natural disturbance management framework (see Chaps. 9 and 11). This is not to suggest that implementing a search for the DFF will be easy. It will be fraught with all of the challenges of evaluating uncertain timber and non-timber values under different management regimes. However, given the increases in computing power and visualization tools, such methods are within reach.

Conclusions

In this chapter we have examined economic aspects of boreal forest management. These issues have been examined at a relatively small spatial extent (from stands to landscapes or regions). We have specifically avoided discussion of provincial-scale economic issues (economic impact analysis, employment issues, etc.) or national economic issues (trade policy, balance of trade, competitiveness, etc.) associated with forestry. It is not that those issues are unimportant, only that we wish to provide a view that may aid forest managers and policy makers dealing with forest management policy at these smaller spatial extents (Box 6.6).

Our discussion of the economics of boreal forest management can be summarized in the following way. First, we have attempted to show that a "correct" economic approach to forest management will involve the "best" use of forest resources, where these may be market or non-market uses associated with the forest, or even "non-use" of forest resources. Conceptually, economically efficient forest resource use will involve an evaluation of market and non-market values. The economic signals regarding resource scarcity, for timber and non-timber resources, need to be able to reach the managers.

Economic analysis can be carried out on alternative management plans, or on alternative policy and institutional arrangements. In this chapter we have provided an illus-

---

Box. 6.6. Options for sustainable forest management:

- Incorporate relevant non-timber values (recreation, etc.) directly into forest management planning tools;

- Incorporate natural disturbance, through such methods as probabilistic sustainability assessment, into planning tools;

- Use current forest planning tools as mechanisms to collect environmental values, particularly non-use values, as part of a desired future forest (DFF) evaluation process;

- Evaluate investments in intensive forest management and zoning from a broader economic perspective than just sustained timber yield;

- Implement monitoring schemes to examine the value of the forest capital (in the broad sense) and link these to forest level planning;

- Examine alternatives to current tenure arrangements, particularly those that limit trade between regions and that constrain the flow of resources to their highest and best use;

- Begin pilot projects for schemes that remove divided land base policies, decouple facilities from fibre sources, and integrate multiple users on the land base; and

- Promote market and policy instruments that provide incentives for sustainable forest management, including certification and decentralized regulatory measures (economic instruments).

---

tration using tenure as an example of a policy analysis case. In addition, a number of tools for identifying and quantifying non-timber values have been presented. Non-timber valuation methods are still being developed, but they are at the point where they can be applied to forest management. These methods can be applied to identify the impact of alternative management plans on non-timber values, and in some cases they can be used to identify the expected behavioural response to alternative management plans (e.g., recreational participation rates).

We have outlined what we believe are the financial fundamentals of boreal forest management. Policies have been developed that require silvicultural investments, but the economic logic of some of these policies must be questioned. In the absence of allowable cut effects, silvicultural approaches that require large initial investments and provide returns in the relatively distant future are difficult to rationalize on a financial basis. Natural regeneration approaches using knowledge about advance regeneration or seedbed preparation and seed dispersal (see Chaps. 8 and 13) may be much more financially attractive.

While many silviculture expenditures may not be defensible from a financial standpoint, there are cases where silviculture for non-timber values may make good economic sense. Silviculture for non-timber purposes may include filling in critical age classes for wildlife habitat, or stand manipulation to reduce the risks of wildfire effects. However, the forest products industry will have difficulty in capturing the returns from these types of silvicultural investments and it will be difficult to clearly identify the value of such investments. A notable exception is silviculture for carbon credits (see Chap. 20), although even in this case there are several uncertainties in the strength of the incentives for such activity. Nevertheless, these types of silvicultural decisions will undoubtedly increase in their importance in the future.

An issue that the discussion above raises is the question of "who pays" for the generation of environmental services arising from forest lands. This is a complex issue that includes questions about the effect of institutions on benefit flows and beneficiaries. This discussion of institutions is the topic of Chap. 7. The link to this chapter should be relatively clear. Institutional frameworks that provide for private markets (wood, perhaps recreation, etc.) may function well in that beneficiaries of the products will pay for them. For ecological services such linkages are more difficult. They may arise through indirect schemes like forest certification. They may also arise through traditional approaches where the public sector concentrates its efforts on addressing public goods and externalities. The mechanisms through which the public sector provides incentives for public good "creation" and maintenance, or penalties for damage to public goods, are complex and multifaceted, especially in the Canadian context of public forest lands.

The main challenges to incorporating economic analysis into forest management are institutional inflexibilities and the limitations of inventory on human uses and non-timber goods and services. Institutions or policies that prevent economic signals for timber or non-timber values to be transmitted to the managing agents will not aid in achieving sustainability. The development of policies that provide such signals, whether they are planning-oriented (command and control approaches) or more incentive-based decentralized approaches, is essential if the Canadian boreal forest is to be managed sustainably. In a world with increasing competition among global timber suppliers (Songhen et al. 1999) and uncertainty from climate change, the forest products sector must increase its ability to respond to economic signals (see also Pearse 2001). Finally, the development of a broader system for monitoring economic sustainability, based on indicators that include natural capital depreciation and non-market values, is critical for the assessment of our forest management strategies and for the incorporation of economic

Fig. 6.10. The boreal forest, providing educational value.

Photo by Roger Nesdoly, courtesy of Mistik Management Ltd.

analysis into an adaptive management framework (Fig. 6.10). Some of these issues are tackled in the next chapter.

Acknowledgements

The authors would like to thank the Sustainable Forest Management Network for funding and supporting research on the economics of sustainable forest management. They also acknowledge NSERC, SSHRC, CFS, and Mistik Management Ltd. for research support. Thanks also to G.C. van Kooten and two anonymous reviewers for their constructive comments on this manuscript. Finally, they thank Roger Nesdoly of Mistik Management Ltd. for providing photographs. Adamowicz was a Gilbert White Visiting Fellow at Resources for the Future, Washington, D.C., when much of this chapter was written. He gratefully acknowledges their support.

References

Adamowicz, W., and Boxall, P. 1998. Forest management and non-timber values: opportunities and challenges. Proceedings of technical papers, Canadian Woodlands Forum, 25 March 1998, Montreal, Quebec. Canadian Pulp and Paper Association, Montreal, Quebec. pp. 95–99.

Adamowicz, W., Boxall, P., Williams, M., and Louviere, J. 1998. Stated preference approaches for measuring passive use values: choice experiments and contingent valuation. Am. J. Agric. Econ. **80**(1 Feb.): 64–75.

Adamowicz, W., Boxall, P., Haener, M., Zhang, Y., Dosman, D., and Marois, J. 2002. Assessing the impacts of forest management on Aboriginal hunters: evidence from stated and revealed preference data. Paper presented at the Second World Congress of Environmental and Resource Economics, June 2002, Monterey, California.

Akabua, K.M., Adamowicz, W.L., and Boxall, P.C. 2000. Spatial non-timber valuation decision support systems. For. Chron. **76**: 319–327.

Ando, A., Camm, J., Polasky, S., and Solow, A. 1998. Species distributions, land values and efficient conservation. Science, **279**: 2126–2128.

Armstrong, G.W. 2000. Probabilistic sustainability of timber supply considering the risk of wildfire. Working Paper 2000-10. Sustainable Forest Management Network, Edmonton, Alberta. Available at http://sfm-1.biology.ualberta.ca/english/pubs/PDF/WP_2000-10.pdf [viewed 6 November 2002]. 15 p.

Armstrong, G.W., Adamowicz, W.L., Beck, J.A., Cumming, S.G., and Schmiegelow, F.K.A. 2003. Coarse filter ecosystem management in a nonequilibrating forest. For. Sci. **49**: 209–223.

(BCMF) British Columbia Ministry of Forests. 1999. Guidelines for developing stand density management regimes. Forest Practices Branch, British Columbia Ministry of Forests, Victoria, British Columbia. 94 p.

Benson, C.A. 1988. A need for extensive forest management. For. Chron. **64**: 421–430.

Bowes, M.D., and Krutilla, J.V. 1985. Multiple use management of public forestlands. *In* Handbook in natural resource and energy economics. *Edited by* A.V. Kneese and J.L. Sweeney. Elsevier Science Publishers, New York, New York. pp. 531–569.

Bunnell, F., Kremsater, L, and Dunsworth, B.G. 1998. A forest management strategy for the 21st century: ecological rationale. University of British Columbia and MacMillan Bloedel Limited, Vancouver and Nanaimo, British Columbia. 286 p.

Burton, P.J. 1995. The Mendelian compromise: a vision for equitable land use allocation. Land Use Policy, **12**: 63–68.

Cumming, S.G., and Armstrong, G.W. 2001. Divided land base and overlapping forest tenure in Alberta, Canada: a simulation study exploring costs of forest policy. For. Chron. **77**: 501–508.

Dosman, D., Haener, M., Adamowicz, W., Marois, J., and Boxall, P. 2002. Assessing impacts of environmental change on Aboriginal People: an economic examination of subsistence resource use and value. Department of Rural Economy, University of Alberta, Edmonton, Alberta. Proj. Rep. 02-01. 33 p.

Gane, M. (*Editor*). 1968. Martin Faustmann and the evolution of discounted cash flow: two articles from the original German of 1849. Inst. Pap. No. 42. Commonwealth Forestry Institute, Oxford, U.K. Reprinted in J. For. Econ. **1**(1): 7–44. Available at http://www.urbanfischer.de/journals/jfe/cotent/1995/Faustmann.pdf [viewed 24 June 2003].

Haener, M.K., and Adamowicz, W.L. 2000*a*. The incorporation of fire and price risk in regional forest resource accounts. Ecol. Econ. **33**: 439–455.

Haener, M.K., and Adamowicz, W.L. 2000*b*. Regional forest resource accounting: a northern Alberta case study. Can. J. For. Res. **30**: 264–273.

Haener, M, Dosman, D., Adamowicz, W., and Boxall, P. 2001. Can stated preference methods be used to value attributes of subsistence hunting by Aboriginal people: a case study in northern Saskatchewan. Am. J. Agric. Econ. **83**: 1334–1340.

Hartman, R. 1976. The harvesting decision when a standing forest has value. Econ. Inquiry, **14**: 52–58.

Hauer, G.K., and Nanang, D.M. 2001. A decomposition approach to modelling overlapping tenures in Alberta: a case study. Sustainable Forest Management Network, Edmonton, Alberta. Proj. Rep. 2001-25. Available at http://sfm-1.biology.ualberta.ca/english/pubs/PDF/ PR_2001-25.pdf [viewed 6 November 2002]. 23 p.

Heal, G.M. 1998. Valuing the future: economic theory and sustainability. Columbia University Press, New York, New York. 226 p.

Heal, G.M. 2001. Optimality or sustainability? Paper presented at the European Association of Environmental and Resource Economics annual meeting, June 2001, Southampton, U.K.

Hegan R.L., and Luckert, M.K. 2000. An economic assessment of using the allowable cut effect for enhanced forest management policies: an Alberta case study. Can. J. For. Res. **30**: 1591–1600.

James, C.R., Adamowicz, W., Hannon, W., Kessler, W., Murphy, P., Prepas, E., Snyder, J., Weetman, G., and Wilson, M. 1997. Sustainable forest management and its major elements: advice to the Lands and Forest Service on timber supply and management. Alberta Forest Management Science Council, Edmonton, Alberta. Available at http://www.borealcentre.ca/ reports/sfm.html [viewed 28 May 2002]. 15 p.

Kristrom, B. 2001. Growth, employment and the environment. Swedish Econ. Policy Rev. **7**: 155–184.

Luckert, M.K. 1996. Welfare implications of the allowable cut effect in the context of sustained yield and sustainable development forestry. Working Paper. Department of Rural Economy, University of Alberta, Edmonton, Alberta. Available at http://www.re.ualberta.ca/STAFF-PAPERS/sp-96-03.PDF [viewed 21 June 2003]. 16 p.

Luckert, M.K., and Haley, D. 1995. The allowable cut effect (ACE) as a policy instrument in Canadian forestry. Can. J. For. Res. **25**: 1821–1829.

McKenney, D., Fox, G., and van Vuuren, W. 1992. An economic comparison of black spruce and jack pine tree improvement. For. Ecol. Manage. **50**: 85–101.

McKenney, D.W., Beke, N., Fox, G., and Groot, A. 1997. Does it pay to do silvicultural research on slow growing species? For. Ecol. Manage. **95**: 141–152.

McCandless, L., and Messmer, M. 1996. Employment and timber supply policies in British Columbia: a time-series analysis. Forest Practices Branch, British Columbia Ministry of Forests, Victoria, British Columbia. 64 p.

Mendelsohn, R. 1996. An economic–ecological model for ecosystem management. *In* Forestry, economics and the environment. *Edited by* W.L. Adamowicz, P.C. Boxall, M.K. Luckert, W.E. Phillips, and W.A. White. CAB International, Wallingford, U.K. pp. 213–221.

Mittelsteadt, N.L., Adamowicz, W.L., and Boxall, P.C. 2001. A review of economics sustainability indicators. Working paper 2001-11. Sustainable Forest Management Network, Edmonton, Alberta. Available at http://sfm-1.biology.ualberta.ca/english/pubs/PDF/WP_ 2001-11.pdf [viewed 6 November 2002]. 35 p.

Nanang, D. 2002. Optimization based planning tools for sustainable forest management. Ph.D. Dissertation. Department of Rural Economy, University of Alberta, Edmonton, Alberta. 152 p.

Nordhaus, W.D., and Kokkelenberg, E.C. (*Editors*). 1999. Nature's numbers: expanding the national economic accounts to include the environment. Executive Summary. Panel on Integrated Environmental and Economic Accounting, Committee on National Statistics Commission on Behavioral and Social Sciences and Education National Research Council. National Academy Press, Washington, D.C. 250 p.

Pearse, P.H. 2001. Ready for change: crisis and opportunity in the coast forest industry. A report to the Minister of Forests on British Columbia's coastal forest industry. British Columbia Ministry of Forests, Vancouver, British Columbia. Available at http://www.for.gov.bc.ca/ hfd/library/documents/phpreport/ [viewed 21 June 2003]. 36 p.

Pezzey, J. 1992. Sustainability: an interdisciplinary guide. Environ. Values, **1**: 321–362.

Rodrigues, P.M.J. 1998. Economic analysis of ecologically based mixedwood silviculture at the stand level. M.Sc. thesis. Department of Rural Economy, University of Alberta, Edmonton, Alberta. 114 p.

Ross, M., and Smith, P. 2002. Accommodation of Aboriginal rights: the need for an Aboriginal forest tenure (synthesis report). Sustainable Forest Management Network, Edmonton, Alberta. Available at http://sfm-1.biology.ualberta.ca/english/pubs/PDF/other_6.pdf [viewed 9 October 2002]. 51 p.

Rowe, J.S. 1972. Forest regions of Canada. Canadian Forest Service, Department of Fisheries and Environment, Ottawa, Ontario. Pub. No. 1300. 172 p.

Tamai, S. 2000. Sustainability of Canada's forest sector: estimating economic depreciation of Canada's timber resources. M.Sc. thesis. Department of Rural Economy, University of Alberta, Edmonton, Alberta. 123 p.

Songhen, B., Mendelsohn, R., and Sedjo, R. 1999. Forest management, conservation and global timber markets. Am. J. Agric. Econ. **81**: 1–13.

(SSCBF) Standing Senate Committee on Agriculture and Forestry. 1999. Competing realities: the boreal forest at risk. Report of the Sub-Committee on Boreal Forest of the Standing Senate Committee on Agriculture and Forestry. Parliament of Canada, Ottawa, Ontario. Available at http://www.parl.gc.ca/ 36/1/parlbus/commbus/senate/com-e/BORE-E/rep-e/rep09jun99-e.htm [viewed 25 May 2002].

van Kooten, G.C., and Hauer, G. 2001. Global climate change: Canadian policy and the role of terrestrial ecosystems. Can. Public Policy, **27**: 267–278.

van Kooten, G.C., Binkley, C.S., and Delcourt, G. 1995. Effect of carbon taxes and subsidies on optimal forest rotation age the supply of carbon services. Am. J. Agr. Econ. **77**: 365–374.

van Kooten, G.C., Stennes, B., Krcmar–Nozic, E., and van Gorkom, R. 2000. Economics of afforestation for carbon sequestration in western Canada. For. Chron. **76**: 165–172.

Vincent, J.R., and Binkley, C.S. 1994. Efficient multiple-use forestry may require land-use specialization. Land Econ. **69**: 370–376.

Vincent, J.R., and Hartwick, J.H. 1997. Accounting for the benefits of forest resources: concepts and experience. A report for the Forestry Department of the Food and Agriculture Organization, Rome, Italy. Harvard Institute for International Development, Harvard University, Cambridge, Massachusetts. Available at http://www.fao.org/DOCREP/005/AC272E/ AC272E00.HTM [viewed 21 June 2003].

Weber, M., and Adamowicz, W. 2002. Tradable land use rights for cumulative environmental effects management. Can. Public Policy, **28**: 581–595.

Chapter 7

Designing institutions for sustainable forest management

Harry Nelson, Ilan B. Vertinsky, M.K. Luckert, Monique Ross, and Bill Wilson

Introduction

Institutions are "the rules of the game in a society, or more formally, the humanly devised constraints that shape human interactions" (North 1998). They include governance structures and social arrangements (Williamson 1985). Differences in choices made in different regions of Canada and the world with respect to development paths or management regimes of forests reflect to a large extent the differences in institutional structures among these regions. These structures include the constitutional order (the fundamental rules), the institutional arrangements (the operational rules within the constitutional order and the organizations devised to support them), and cultural endowments (the behavioral norms and mental models shared by the society or groups within it; Clague 1997, pp. 368–369). Changing cultural endowments and the constitutional order are much slower processes than changing institutional arrangements. Thus, much of the debate about how to achieve sustainable forest management (SFM) largely focuses on institutional arrangements. Our objective in this chapter is to explore how institutional arrangements that exist in different regions in Canada lead to behavioral patterns that either enhance SFM or lead to unsustainable forest management. We also explore how changes in institutional arrangements can improve decision making and modify management practices. As a consequence of our analysis, we offer some suggestions for changes in institutional arrangements in Canada that may increase the sustainability of forest management.

H. Nelson and I.B. Vertinsky.[1] Forest Economics and Policy Analysis (FEPA) Research Unit, University of British Columbia, Vancouver, British Columbia V6T 1Z4, Canada.
B. Wilson. Pacific Forestry Centre, Canadian Forest Service, Victoria, British Columbia, Canada.
M.K. Luckert. University of Alberta, Edmonton, Alberta, Canada.
M. Ross. University of Calgary, Calgary, Alberta, Canada.
[1]Author for correspondence. Telephone: 604-822-3886. e-mail: fepa@interchange.ubc.ca
Correct citation: Nelson, H., Vertinsky, I.B., Luckert, M.K., Ross, M., and Wilson, B. 2003. Designing institutions for sustainable forest management. Chapter 7. *In* Towards Sustainable Management of the Boreal Forest. *Edited by* P.J. Burton, C. Messier, D.W. Smith, and W.L. Adamowicz. NRC Research Press, Ottawa, Ontario, Canada. pp. 213–259.
Note: The views expressed in this chapter do not represent the official position of the government of Canada.

We start the chapter with a brief analysis of the concept of sustainable forest management as it relates to functional attributes that institutional arrangements may need to satisfy. We proceed in the next section to examine two fundamental elements of institutional arrangements: market and government regulatory systems. We relate each mechanism to the functional requirements of SFM to define market and government failures. We then proceed to describe the Canadian regulatory system and the role of markets within it. We examine the contributions that alternative patterns of forest governance make to SFM and their vulnerability to various kinds of institutional failures. We end the chapter with examination of alternative options for institutional change that might enhance SFM.

Sustainable forestry and fundamental requirements

Sustainable forestry has been broadly defined as the management of forests for both environmental and economic benefits in a way that does not compromise the ability of forests to deliver those benefits in the future (Chap. 2). Within Canada, efforts have been made to identify a set of principles that can guide sustainable forest management and against which current practices can be evaluated. Using these principles, sustainable management practices should (CCFM 2000):

- conserve biological diversity;
- maintain and enhance the forest ecosystem and productive capacity;
- conserve soil and water resources;
- recognize contributions to global ecological cycles;
- recognize multiple benefits to society; and
- accept the social need for sustainable development.

The concept of sustainable forest management implies management for a wide variety of objectives and considers the interests of many diverse stakeholders. Conflict between objectives is almost unavoidable. The stakes are high because of the importance of forests to human (and much other) life. Sustainable forest management is thus unavoidably a highly political process. Institutions which are devised to promote SFM must accommodate a wide variety of values and seek to resolve conflicts in ways which are consistent not only with interests of current stakeholders but also with those of future stakeholders (i.e., unborn generations). The breadth of the objectives means that the management process involves a high degree of uncertainty and requires a large amount of information, some of which is not available. Thus institutions designed to promote SFM must promote and accommodate acquisition of information (e.g., scientific research), validation of information (e.g., to resolve conflicts about scientific knowledge), and continuous learning. Since decisions cannot be postponed until all knowledge is available (since making "no decision" is a decision), the system must provide sufficient flexibility so as to ensure adaptation of management as new circumstances arise or new knowledge is acquired. Learning by doing (or sometimes not doing) must be embedded as an integral part of the institutional design.

In the long run, sustainable institutional arrangements must influence values and preferences of stakeholders, since certain value and preference patterns cannot accom-

> **Box 7.1. Institutional challenges for sustainable forest management:**
>
> - Developing the technical and scientific knowledge required for SFM;
> - Identifying and resolving tradeoffs between different objectives; and
> - Aligning individual and social preferences.

modate long-term sustainability. In the short run, the institutional system must provide incentives so as to align preferences of stakeholders and decision makers to promote behavior consistent with SFM. To be economically viable, institutions must not involve prohibitively high transaction and information costs. Goodin (1996) has suggested that institutional designs must also be robust. He suggests that institutional change must respond only to significant changes in the environment, but while not responding too readily, they must not become ossified. Public scrutiny is important to ensure public support and acceptance (which is necessary to have legitimacy in a democratic society). He also suggests that institutional designs will include variety, i.e., encourage the emergence of a number of different forms to allow experimentation. He recognized, however, that offering variety does not create necessarily a social laboratory in which the best forms self-select, but may instead end up in a "race to the bottom".

Basic elements of institutional arrangements

There are two ideal types of institutional arrangements that form the endpoints of a spectrum of governance mechanisms: (1) regulatory and (2) market based (Loasby 1999). Regulatory structures dictate behavior and control behavior through sanctions. The pre-eminent regulatory organizations are governments. One view of (democratic) government is that it is an institutional mechanism designed to address the well-being of individuals through collective action that individuals would be unable to do by themselves (Rosen 1985). Other authors have taken less altruistic views of government, particularly with regard to government regulation, suggesting that the principal aim is to either satisfy bureaucratic preferences or certain interest groups (Magee et al. 1989). If we adopt the democratic view, governments are thereby concerned not only about the efficient allocation of resources but also with issues such as equity, income distribution, and the provision of a variety of public goods. In Canada, the constitution identifies the powers of the Federal and Provincial governments. Both levels of government have powers with respect to various dimensions of managing the forest system. Provincial governments, however, have dominant roles in regulating forest management, not only through the regulatory powers granted by the constitution, but also as the owners of the majority of forest lands in Canada (Fig. 7.1).

Apart from regulation, governments have a choice of other instruments to affect SFM (Stanbury and Vertinsky 1998). They can manage forests directly. They can use economic instruments that affect prices and values to provide incentives that encourage individuals or organizations to take specific actions (e.g., taxes, subsidies, and charges for formerly free goods). They can also try, through education and the provision of

Fig. 7.1. Forest ownership in Canada is predominantly by the provinces (*a*), making provincial governments (as represented by the Alberta Legislature, *b*) the dominant institutions in determining Canadian forest management policy.

(*a*)   (*b*)

Photo by M. Abugov

information and various promotional strategies, to change individual preferences and actions.

Governments are not unitary organizations. Different departments and agencies have different objectives and perspectives. Governments are also exposed to efforts by a variety of non-government organizations (NGOs) and groups to influence their decisions and policies. Indeed, the interaction within government organizations and between governments and NGOs shapes to a significant extent the articulation of government SFM policies and actions. In the process of policy articulation, governments respond to legitimate demands of stakeholders and try to resolve conflicts among stakeholders. The pursuit of SFM involves, therefore, the creation of new institutions through which stakeholders may participate in the policy articulation process and the management of public forests (Delbridge 1988; Beckley and Reimer 1999). In British Columbia (B.C.), efforts to settle land claims through treaty processes will result in agreements to turn over land, timber rights, and jurisdiction to the First Nations. Where governments have not responded in a satisfactory manner to First Nations' demands, such demands have been pursued through the courts (Chap. 3). An institutional arrangement that emerged to provide for regulatory and management power sharing is the co-management agreement. Co-management agreements were employed in the United States to incorporate Aboriginal concerns over public lands. Within Canada, such co-management agreements have taken place north of the 60th parallel, where the federal government negotiated comprehensive land claim agreements in the past 15 years (Gawthrop 1999). South of the 60th parallel, there are few examples of such agreements in a government-to-government context, although interest in this approach is growing (Chaps. 4 and 22). Instead, much of what has taken place within Canada has been the creation of opportunities for First Nations to participate in the forest sector through different types of partnerships within the existing regulatory system. These do not provide a fundamental change in the institution but fit the First Nations into a pre-existing institutional framework by treating the First Nations as another license holder, timber processor, contractor, or other pre-existing role, depending upon the specific partnership.

Market institutions are an important vehicle through which economic private actions are shaped and take place (Chap. 6). The key institutional arrangements that affect the functioning of the market are rules that govern contracts (or the term of exchanges between different parties) and the definition of property rights (the rules of ownership and control). Markets and market mechanisms rely on decentralized decision making and thus tend to result in lower transaction costs. If markets are competitive and prices reflect the full benefits and costs to society, markets theoretically provide the most efficient means of allocating resources and bringing about socially desirable outcomes. Markets for some goods and services do not exist or fail to achieve socially desirable outcomes (e.g., because of externalities or lack of competition). Governments may then use legislation to create markets where they do not exist (e.g., carbon trading, discussed in Chap. 20) or to correct failures of the market pricing system (e.g., using subsidies and taxes). Theoretically, market failures in many cases can be corrected through government interventions. Such interventions presume that governments can identify the socially optimal pricing system that achieves SFM and implement it. The major danger lies, however, in the possibility of government failure.

Paradigms of forest management

Historically, the main objective in forestry was an emphasis on timber production (Chap. 1). Concepts of sustainability referred specifically to sustaining wood yield, and the forest was managed to maximize profits subject to sustaining timber harvests. The recognition that the forest has other economic values (e.g., recreation) led to the modification of the paradigm to include management for several economic objectives. More recently, public concerns about the impact of large-scale harvesting on other forest values, such as biodiversity and the integrity of underlying ecological processes, have led to the emergence of the new paradigm of SFM (Pearse 1993; Adamowicz and Veeman 1998; Chaps. 1 and 2). This change in views is not exclusive to Canada; it appears to be a paradigm shift that has also occurred in other regions with substantial forest resources such as the United States, Scandinavia, and New Zealand (Wilson et al.1999). Wilson et al. (1999) describe the forest policy adjustments exercised in New Zealand, Sweden, and Finland, for example, in response to this paradigm shift. The shift was triggered and legitimized by the Brundtland Report (WCED 1987), which introduced the concept of sustainable development. The report formed the basis for the United Nations (U.N.) Conference on Environment and Development in Rio de Janeiro in 1992, where new principles concerning forest management were articulated. The conference issued an authoritative statement of "non-legally binding principles for a global consensus on the management, conservation, and sustainable development of all types of forests". This international dialogue coincided with growing public concern in most of the developed countries about the impact of economic activities on the environment and the public debate that was taking place over sustainable development. This concern has also influenced how we think about forestry by highlighting issues such as climate change and some of the roles forests play in global ecological cycles, such as carbon sequestration (Cubbage et al. 1993).

> **Box 7.2. Incentive and coercive instruments in forestry.**
>
> Government can use different approaches (which need not be mutually exclusive) to promote SFM. Market-based approaches rely on influencing the choices made by decision makers by either (1) changing the incentives they face or (2) providing additional information to correct information asymmetries. In either case, government relies on decentralized decision making (e.g., markets) to achieve optimal outcomes. Regulatory-based approaches rely on more coercive approaches by prescribing specific outcomes. These approaches differ in their flexibility, the certainty with which they may be able to accomplish the desired outcomes, comes, the time frame in which the outcomes are achieved, and the distribution of costs and benefits among participants (which influences the economic and political feasibility of the approach). Examples of the use of such instruments in forestry include:
>
> **Incentive-based instruments**
>
> *Price manipulation and taxes* — changing the prices to which agents respond through taxes and (or) subsides; this may involve government expenditures such as cost-sharing for reforestation programs or assistance to adopt new environmental technologies; alternatively, it may be taxes and charges for emissions.
>
> *Modification of rights* — changes in lease arrangements to encourage lease holders to invest in more intensive silviculture by letting them capture the benefits through the incremental volume associated through their investment with perhaps modified charges for that volume;
>
> *Disseminating information* — relying on moral suasion through public scrutiny of firms' actions such as holding public meetings during the development of harvesting plans or development of benchmarking for best management practices to help private landowners achieve SFM; and
>
> *Provision of information* — using certification so that people's preferences for sustainably managed forests can be expressed in the marketplace.
>
> **Regulatory approaches**
>
> *Direct public management* of forest land;
>
> Prescribing management behaviour in the *contractual terms of tenure arrangements* for licenses operating on public land; and
>
> *Command-and-control measures* — government regulation regarding harvesting rates, the spatial distribution of harvests (e.g., green-up requirements as a condition to harvesting adjacent cutblocks), and harvesting practices such as the B.C. Forest Practices Code.

Paradigm, government, and market failures

From a socio-economic perspective, failure to achieve SFM can stem from either (1) a forestry paradigm failure, and (or) (2) policy or government failure, and (or) (3) market failure. A *forestry paradigm failure* occurs when societal consensus about the appropriate management of forests is inconsistent with the principles of sustainability or is based on faulty biological or ecological information. It may occur because the consensus is based on ambiguous principles, and there is no agreement on interpretations of these principles. Failure of translation may also occur when there is a lack of understanding of how the principles are translated into actions and consequences.

Policy or *government failure* can originate from the lack of commitment by the government to the SFM paradigm. Governments, in seeking international legitimacy, today are likely to endorse SFM in principle. There is evidence, however, that the degree of commitment to SFM depends to a large extent on the short-term costs that implementation of SFM policies may incur. The interpretation of SFM principles is likely to be influenced by the specific economic interests of a country and its domestic political makeup. Even when commitment is made, lack of scientific and technical knowledge, lack of information, and lack of resources may undermine the efforts to achieve SFM. Poorly designed forest policies or extra-sectoral policies may discourage SFM (OECD 1992; Kaimowitz et al. 1998). Institutional failures occur either as a result of government failure to create appropriate formal institutions that promote SFM or the existence of traditional institutions which impose barriers on SFM.

Market failures occur where there are absent, distorted, or malfunctioning markets. These types of failure are especially prevalent for environmental goods and services because of externalities, missing markets for environmental services and the presence of open access goods, and a general lack of information and knowledge, which create uncertainty and frustrate market functions (Richards 2000).

Forestry paradigm failures

There are several pre-conditions to avoiding forestry paradigm failures, i.e., reaching a social consensus about forestry which is inconsistent with SFM. Perhaps the most important one is public and professional awareness of the key role forests play in the maintenance of human and other life and the vulnerability of the forest system to destruction through human activities. The study circles in Sweden provide an excellent example of a massive educational movement through which thousands of people gained intimate knowledge of forest ecology and SFM (Wilson et al. 1999). Continuing education services by professional societies, government extension services, and university training programs that incorporate SFM principles are other examples of information dissemination that support the emergence of the SFM paradigm (Fig. 7.2). The second pre-condition is the emergence of post-materialistic values. A commitment to an SFM paradigm requires members of the society (including direct stakeholders in the forest) to internalize social values. A third pre-condition is commitment to the generation of scientific and technical knowledge to ensure that the societal consensus about the appropriate management is based on sound biological and ecological knowledge and that this knowledge is translated appropriately (from principles to actions). In the absence of a wide social consensus about SFM, government can create the appropriate system of incentives for increasing awareness and internalization of SFM principles into individual and organizational decisions.

Paradigm failures may arise when there are irreconcilable differences in stakeholders' beliefs about what constitutes SFM. In such circumstances, only significant advances in science may lead to sufficient convergence of views necessary for a paradigm to obtain social legitimacy. It should be noted that in a democratic society, a diversity of opinions that stimulate debate about SFM is desirable, provided that all parties accept democratic means of conflict resolution.

Fig. 7.2. Educational institutions play a central role not only in the development and transmission of new knowledge but also in the conveyance of attitudes and world views to future forest managers. Shown here is the Forestry and Geology Building at the University of New Brunswick.

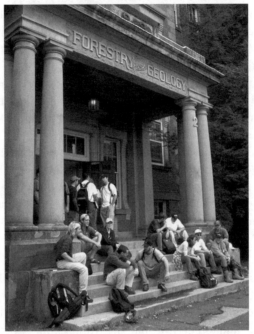

Photo by Joy Cummings

Government failure

Government failure may occur when government policies have unintended effects, since bureaucratic or political interventions impede incentives for SFM. For example, politicians may not properly evaluate the long-term impact of decisions, focusing instead on the short-term electoral benefits, or intervene to satisfy particular political constituencies and not to address SFM. Government agencies may have multiple objectives that conflict with one another, or are in conflict with objectives of other agencies concerned with forest management. Policies designed to address one problem may create perverse incentives that lead to undesirable outcomes, thus creating new problems. In tropical forestry, several authors have shown how unclear land tenure legislation, policies that favor one land use (such as agriculture) over another (such as sustainable forestry), and policies that systematically underprice forest products all contribute to deforestation and the resulting loss of biodiversity (Repetto 1988; Panayotou and Ashton 1992; Kaimowitz et al. 1998).

In general, governmental failure is part of a larger problem of institutional failure. We create organizations in part to overcome the cost of acquiring information through creating rules and operating procedures that guide decision making (Loasby 1999). The trade-off is that organizations tend to be less flexible to changing demands (relative to markets) and may not respond to changing preferences or new information because of these pre-set rules. For example, organizations by their nature select information that creates an incentive for action that will benefit the organization (Loasby 1999, p. 53).

This bias may ignore certain types of information, or lead to inappropriate actions (Goodin 1996, p. 10).

In the case of government organizations, institutional failure may create policies that are inappropriate, or the interaction between various policies and legislation may create perverse incentives and lead to undesirable and unanticipated outcomes. Bass et al. (1998) describe the difficulty public forestry institutions have in accommodating the new demands on forests and competing interests from multiple stakeholders and the possibility of "capacity collapse" in which existing institutions break down and cannot respond effectively.

Market failures

Economic analysis of market failures focuses on two elements that affect market functioning: (1) property rights, which relate to alignment of incentives; and (2) transaction costs, which relate to problems of costly information.

Property rights are defined as the right to the stream of economic benefits from the use of an asset where that right is enforceable and supported by society (Haley and Luckert 1990). The theory of property rights has been articulated most notably by Ronald Coase and Harold Demsetz, who argued that it is critical to ensure that the correct incentives are aligned with the ownership: where the two are not aligned, market failure can occur (Coase 1960; Demsetz 1967; Scott 1983).

In general, ownership of natural resources falls into one of four main categories: open access; state-owned; private property; and common-pool property. Under open access, no party exerts effective ownership and property rights are absent; classic examples are open seas fisheries or areas under state control where there is no effective enforcement. State-owned is when the government retains the property rights; private property is when property rights are held by individuals and private enterprises; and common-pool describes property rights owned collectively such as by a community, village, or ethnic group.

Property rights to use a resource are rarely absolute, as they are usually hedged by restrictions and obligations placed on the owner. Often they may be incomplete in that they are not fully defined, or only permit the rights-holder certain benefits. In some cases, social mores may restrict what are considered acceptable uses of the right. Scott (1983) has emphasized that property rights can be considered a bundle of different rights with differing degrees of strength. The most important characteristics of property rights are comprehensiveness, durability, transferability, divisibility, and enforceability. To the extent that any of these are weakened, incentives for stewardship of the asset are reduced and (or) absent, and may lead to market failures. Market failure to achieve SFM may also result from a misalllocation of property rights. Failure to recognize the rights of future generations may result in resource use that depletes the endowments left for these generations. This is the case if the current generation does not internalize SFM principles, and the government has awarded full property rights to the current generation. The property rights approach has been the principal analytic tool used to examine tenure agreements in Canada, as discussed further below.

New Institutional Economics (NIE) does not dispute the importance of property rights but raises the importance of information and the transaction costs it creates in understanding how markets function. Under the assumptions of NIE, it is costly to acquire information, and agents are weakly rational and weakly moral; therefore, they may withhold or distort information at times in order to better their position, an assumed behavior called "bounded rationality". Information may be incomplete, either because the information is absent, or it is asymmetric (where one party has less information than the other, and the one with more information cannot be expected to divulge it freely). Transaction costs emerge when it is costly to gather information or when exchanges involve transmission and processing of large quantities of information. Transaction costs can also be attributed to the other components of the exchange, although these are usually assumed to be less significant: the costs of bargaining, coordinating participants, and drawing up contracts, for example.

The presence of transaction costs leads to different forms of market institutions, such as different types of contracting and organizational forms (Williamson 1985). These transaction costs may be high enough to preclude a mutually beneficial exchange, and even if exchange does take place, it may be incomplete or limited (since the parties are not able to fully value the good or right that is being exchanged) and therefore lead to market failures (Williamson 1985, p. 29). Under these circumstances, institutions can be thought of as ways of dealing with the problem of information costs. Institutions are a response to uncertainty. As Loasby (1999, p. 46) points out,

> *"They are patterns acquired from others that guide individual actions, even when these actions are quite unconnected with any other person. They econo-mize on the scarce resource of cognition, by providing us with ready-made anchors of sense, ways of partitioning the space of representations, premises for decisions, and bounds within which we can be rational or imaginative. They constitute a capital stock of other people's reusable knowledge, although like all knowledge, this is fallible."*

Overlapping institutions in forestry

There are a number of institutions that are either directly or indirectly involved in the management of forest land in Canada. First, there are a number of different entities that own forest lands — private individuals and companies, provincial governments, the fed-eral government, and First Nations — although provinces hold the majority of forest land within Canada. Provincial governments have relied on the use of leases, or tenure agreements, with companies for the commercial utilization of Provincial Crown (pub-lic) lands. These companies sell the products they manufacture in domestic and foreign markets. This creates a linkage between government policies and their direct impact on companies' costs and indirect factors such as customer preferences and trade agree-ments, all of which influence companies' abilities to produce and sell forest products in these different markets.

The multiple environmental values and non-timber attributes of forest lands bring in other interests and institutions. Non-timber resources such as fish and wildlife may be the responsibility of different agencies within provincial governments or may fall under federal jurisdiction. First Nations may have treaty rights to off-reserve resources.

Environmental groups may be concerned about particular species or types of habitats found in different regions. Local communities may have concerns about the impact of forest management practices on local watersheds or recreational opportunities. The interactions between the different components of the institutional system that govern forests lead to the emergence of paradigms that guide forest management over the years. A paradigm represents basic beliefs and assumptions which are perceived to be legitimate and which represent a social "consensus". Such "consensus" is institutionalized and therefore is robust; those attacking it may be viewed as illegitimate. Persistent attacks may lead, however, to a paradigm shift and the possibility that there may emerge a new consensus about what constitutes a new paradigm (although it can also lead to persistent disagreement if stakeholders cannot reconcile their values).

Canada's current forest management system and SFM

The forest management system in Canada is vulnerable to the three types of failures just discussed. While there is almost universal acceptability of SFM as a paradigm, there is no consensus on the practical meaning of the concept (Chap. 2). There is a consensus on the dimensions of SFM (e.g., the CCFM Criteria and Indicators of SFM, Box 2.2) but not on the definition of standards. Note that standards in a dynamic conceptualization of SFM may refer to process criteria not to outcomes. Without a well-defined universally accepted paradigm of SFM, governments are more likely to yield to pressures of interest groups. Much of the leadership in forging practical definitions of SFM has been left to conflicting private certification systems. Where governments have attempted to prescribe tight standards for forest management, transaction costs have risen substantially without necessarily achieving the objectives of sustainability (e.g., the 1995 Forest Practices Code in B.C.). Despite the attempts to create markets for environmental goods through certification and labels, prices for such goods are inadequate. For example, premiums for wood products certified by the Forest Stewardship Council (FSC) are generally small or nonexistent and are therefore unlikely to cover the added costs of certification (Baldwin 2001; Kim and Carlton 2001; Kiekens 2000). Tenure systems and the property rights they confer do not create, in most cases, strong incentives for the practice of SFM.

Under the Canadian Constitution, provinces were granted the rights to forest resources in 1867.[2] Therefore, forest lands are owned predominantly by the provinces, although private ownership is significant in certain regions of the country. Minor amounts of forest land are held collectively (if one considers the timberlands held by the federal government on behalf of First Nations), and federal ownership of forest land is concentrated in the Yukon and Northwest Territories. First Nations ownership of forest lands is expected to increase with the settlement of pending treaty negotiations in British Columbia.

---

[2]The exception were the prairie provinces — Alberta, Saskatchewan, and Manitoba — that received jurisdiction over their provincial forests when the federal government returned control of natural resources to them in the 1930s (Myre 1998).

In the next subsection, we briefly describe formal efforts by provincial and federal governments to address SFM. More extensive description of these institutions and the evolution of provincial forest policies can be found in Howlett (2001). We then examine the possible role of certification as an alternative to public regulation as a way to achieve SFM.

Government efforts to address SFM

In the long run, the interaction between private and public institutions shapes the legal and regulatory system in place. That system provides, in the short run, the framework through which the influence of the various organizations is exerted on forest management. Within the framework of the law, governments have a large degree of discretion, guided by policies and strategies, which articulate government objectives and priorities. Governments can delegate some areas of their discretion to private organizations or individuals (e.g., leaving matters to the discretion of private professionals). The articulated policies and strategies of the government may influence the decisions of private actors who manage public forest lands as they consider the threat of future government interventions.

Changes and proposed changes to forest legislation provide a good indication of the commitment of a government to SFM. Three provinces (British Columbia, Ontario, and Saskatchewan) have adopted new forestry legislation in recent years that incorporate the principles of sustainable forest management into new Forest Acts, while Quebec has pending legislation to incorporate those principles, and Manitoba is contemplating changes to its legislation (CCFM 2000). Provinces have also passed a variety of legislation dealing with various environmental aspects of forestry, such as providing riparian buffers and the identification of protected areas. New Brunswick is using habitat supply models in establishing harvesting guidelines, while Saskatchewan and Manitoba require environmental impact assessments for forest management plans, the first two provinces to do so (SSCBF 1999).

The federal government has played a role in both financing research and supporting efforts to coordinate the development of SFM. It has been engaged in involving the provinces and territories in developing a strategic framework for Canada's forests, the National Forest Strategy (NFS). These efforts originally started in the early 1980s; they have steadily expanded, covering both more topic areas and including more participants from a wider spectrum. The Canadian Council of Forest Ministers (CCFM), a group including the provincial and federal ministers responsible for forestry, was formally created in 1985; it has overseen the NFS process since then (Young and Duinker 1998). One of the primary goals of the CCFM has been the development of the criteria and indicators of sustainable forest management in a Canadian context.

As part of that research effort, the federal government has sponsored the development of model forests across the different regions of Canada, which serve as opportunities to experiment with different approaches to forest management and to investigate the implementation of criteria and indicators at the local level. Today, there are 11 model forests found across Canada, their locations and links to each forest are provided by MFP (2002). The federal government has also funded the Network of Centres of Excellence (NCE) in sustainable forest management, a cooperative research network involving private industry, academics, and others (Chap. 1). It has also funded several other programs: the National Forest Science and Technology Course of Action to develop the means by which to identify and measure sustainable forest management; and the First Nations Forestry program designed to provide funding to help improve economic conditions for First Nations and to gain experience in the forestry sector.[3] The federal government has also ensured that Canada has been represented at the various international forums discussing sustainable forestry, including the U.N. Forum on Forests (UNFF) and its predecessor the International Forum on Forests (IFF), discussions on climate change under the Framework for Climate Change (FCCC), and the ongoing Montreal Process (Natural Resources Canada 2000). These international efforts are described in more detail in a subsequent section.

Certification

An increasing regulatory role is being assumed by private certification organizations. Through a system of buyers' clubs and direct pressures, forest product producers seeking market access to certain markets must attain certification (Fig. 7.3). Some producers seek to differentiate their product and obtain premiums over market prices, while others commit to certification to meet their convictions.

Competing organizations have been established by environmental and industry groups. There are three major systems of certification that influence forestry practices in Canada (see Chap. 21 for more detailed discussion). The Forest Stewardship Council (FSC) was established by a consortium of environmental non-governmental organizations (ENGOs) to promote sustainable forest management and has articulated principles it regards as guidelines for SFM globally (Box 21.6). These principles are adapted in each geographical region to meet local conditions. The Council certifies certifiers who agree to follow and apply its principles. If chain of custody is maintained, producers can use the FSC logo to identify their products as ones originating in an FSC-certified forest. The FSC through its certification process defines SFM and regulates the practice of those firms that commit to its certification process.

The ISO 14000 certification processes have been developed as environmental management process certification systems by the International Standards Organization. The certification system requires improvement in the environmental management system and continuous improvement in practices and does not prescribe standards or outcomes, as does the FSC system. The ISO does not directly certify forest land; however, the volume of forest land under companies management that have been certified under ISO is often used in comparison with other certification systems that do directly certify specific forest lands (Fig. 7.4).

The CSA forest certification system was developed by the Canadian Standards Association at the initiative of the forest products industry in Canada to offer an alternative to the FSC certification. The system combines process and product certification. The Forest Products Association of Canada (FPAC) has recently announced that certifi-

---

[3]Both of the programs were initially 5-year programs. The National Forest Science and Technology Course of Action was started in 1998 and the First Nations Forestry program in 1996. More information about the latter can be found at www.fnfp.gc.ca.

Fig. 7.3. International protests (*a*) and threats to boycott wood harvested from wild forests have prompted forest companies, governments, and ENGOs to devise certification systems (*b*) that assure various levels of sustainable forest management. All logos in (*b*) are registered trade marks or copyrighted.

(*a*)   (*b*)

Canadian Press/Associated Press photo by Anat Givon

cation will be mandatory for its members (McKinnon 2002). The corresponding U.S. industry association, the American Forest and Paper Association (AFPA), also requires its members to have their woodlands certified, although it does not specify the kind of certification, and its own Sustainable Forestry Initiative (SFI) functions much like a certification system (Box 2.6).

Each of the certification systems is in competition with the others to obtain acceptance by both buyers and producers. The acceptance of a significant proportion of buyers provides a certification system not only with legitimization but also with strong sanctions. Indeed the certification process in such circumstances becomes a private regulatory system with coercive powers (Wilson et al. 2001; Vertinsky and Zhou 2000).

Regulatory and policy systems, and the role of markets in Canada

The main economic benefit forest lands provide is typically from the commercial production of timber. There are other sources of economic activity, including recreation, and non-timber forest products, but timber values dominate when viewed from an economic perspective. Consequently, the main goal of provincial policy in the past has been to direct timber towards companies that then utilize the timber to make forest products such as lumber, pulp, and paper. This has generated both direct payments through stumpage and royalty payments, as well as employment (and the indirect taxes associated with income taxes on both employees and companies).

Governments face four fundamentally different options in deciding how to develop state-owned timber resources. These include having the state own and operate the timber resource and the processing facilities, selling timber (or the right to harvest the timber) directly, selling off the timber resources completely, or long-term leasing. Each of these alternatives differs significantly in terms of their reliance on external markets to organize the development and utilization of timber resources.

Fig. 7.4. Certified forest area (ha) in Canada, by certification system, 2000–2002. *ISO certifies management systems, not forest management areas, so the ha refer to the tenure area of certified companies.

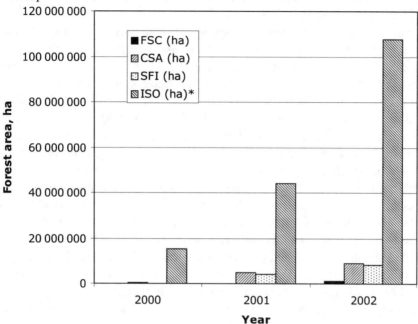

Provinces, with few exceptions, have retained ownership of the timber resource (forests) and have leased out the right to harvest those trees under a variety of tenure arrangements. The forest product companies are in the business of producing and selling into competitive markets. As such, they need to be able to cover their costs of production and earn a sufficient rate of return on their investments. Governments recognize the need and attempt to reconcile it with other government objectives when developing policy. The main focus of government policy for much of the post-war period has been on sustained yield — ensuring an even flow of timber over time (Chap. 1).

The main bodies for developing policy are created through the forest legislation that is unique to each province, and the set of tenure arrangements each province has created. Together, they specify the rights companies enjoy to the economic benefits derived from harvesting crown timber, and the obligations and responsibilities they assume. While various provincial governments have introduced different forms of tenure and are experimenting with new types, as discussed below, it should be emphasized that to date provincial governments retain full discretion over two key variables — the control of harvest levels and the assessment of charges for harvesting forest products.

Tenure

While there is a multitude of tenure arrangements which differ within and among provinces, these arrangements all share one common purpose: to grant the license holder the right to harvest a certain volume of Crown timber over time. Under these arrangements, the tenure holder gains the commercial benefits of utilizing Crown land for harvesting timber, but also agrees to certain obligations. These rights and obligations and the incentives they create vary under the various forms of tenure, and may lead to significant differences in behavior by tenure holders, depending on the characteristics of the type of tenure arrangement they face.

The most recent and comprehensive survey of tenure arrangements in Canada was carried out by Haley and Luckert (1990). They used the property rights approach to compare different types of tenure across the various provinces using six principal dimensions associated with defining a property right: comprehensiveness, duration, transferability, right to economic benefits, exclusiveness, and security. These six dimensions, and their impact upon the rights held by the tenure holder, are described below. For the most part, tenure arrangements have not changed substantially since their report, although there have been some minor modifications within the various provinces. From a company's perspective, tenure arrangements are a critical factor in the company's behavior. The more secure the rights, and the more assurance the company has over what its cutting rights will be in the future, the more certainty they provide, which is a precondition for long-term investment and stewardship. Secure and clearly defined tenure arrangements help a company determine the optimal size of its plant, what kind of products it should produce, and to what extent it should undertake future investments to maintain its productive capacity. Companies have also used these cutting rights as a basis for attracting long-term capital to build and maintain processing facilities.

Along with the cutting rights granted under tenure arrangements, companies are also sensitive to the system of charges and obligations these rights entail. To the extent that changes in these obligations impose costs, they can reduce the economic benefits to the company and may impair its financial viability. These changes can be substantial enough to reduce the value of these rights. Furthermore, to the extent such changes become unexpected and arbitrary, they create uncertainty, which also reduces the value of expected future economic benefits the company can expect to receive from its cutting rights. We now review the important attributes of the rights forest companies enjoy under their tenure arrangements and the obligations they incur, and additional discussion is provided in Box 7.3.

Comprehensiveness refers to what the property right covers. The more comprehensive the right and the greater the range of economic benefits it entitles one to, the more valuable the right becomes. Historically, all timber tenure agreements in Canada just granted rights to harvest timber. Recreation, wildlife, and other uses of the forest resource are specifically excluded from such agreements, although in Alberta, the Crown can enter into agreements with tenure holders to provide recreational opportunities.

Duration refers to how long the property right lasts. The longer it lasts, the longer one can enjoy the stream of economic benefits it provides, and the more valuable it becomes. The majority of tenure arrangements in Canada are long-term, renewable agreements held by the company with the Crown. Some tenure is held under short-term, non-renewable agreements.

Transferability refers to the ability of the property right holder to transfer or sell their right to others. All provinces place restrictions on the right of the tenure holder to transfer or sell tenure; it must be with the consent of the Minister or provincial government, and may entail changes in the original agreement. In addition, there may even be restrictions to the tenure holder on the sale or transfer of logs or wood residue derived from the tenure.

As noted, a property right is the *right to enjoy the economic benefits of an asset*, and to the extent that this right is unimpaired by any charges, fees, restriction, or other obligations it is worth more. Under tenure agreements, the tenure holder receives the economic benefits through harvesting and processing Crown timber. The provincial governments retain an interest in the economic benefits through fees they charge tenure holders. All provinces have some form of charges for the right to harvest timber. These charges cover a range of instruments; they may be based on the area of land or the volume of timber harvests; they may be fixed rates or adjusted on a regular basis; and they may be designed in a way that makes them sensitive (or insensitive) to differences in cost, wood quality, or end-product use.

These charges can be calculated in two fundamentally different ways: either through competitive bidding or through some administrative approach that utilizes information to determine appropriate charges. How these fees are calculated can therefore have an impact on the economic benefits that the tenure holder will retain after their payments to the Crown, as can other obligations (described below) that the tenure holder typically incurs.

Exclusiveness refers to the ability of the property right holder to prevent others from utilizing the resource. To the extent others cannot be excluded, and can appropriate the economic benefits, the right becomes correspondingly diminished. For the most part, tenure holders enjoy the exclusive right to harvest timber within a specific area; even under volume-based agreements, once a cutting area has been determined, the company will enjoy the exclusive rights to harvest within that area. Where mixed stands exist, companies may only have the exclusive right to specific tree species within an area.

Security refers to the degree of confidence the right holder has in the strength of its right. The stronger the right, the more valuable it becomes. Governments typically retain the right to alter the terms of the tenure arrangement upon periodic intervals coinciding with the renewal of long-term leases. In addition, most agreements allow the government to reduce the timber harvest if there is an overall reduction in the sustained yield, and allow government to withdraw land for other uses. Some tenure agreements allow for compensation if there is a reduction under certain cases, while others do not.

Tenure agreements commonly contain restrictions that influence the rights companies have under these different forms, and it is through these restrictions that governments address concerns about social and environmental issues. These *operational stipulations and controls* are the obligations the tenure holder agrees to and the ways in which governments monitor and enforce tenure agreements. These responsibilities typically fall under the areas of management, harvesting, processing, reforestation, and other silvicultural activities. While the government may share some of the responsibilities, especially for some of the smaller tenures, in general, governments have shifted these responsibilities onto industry in the last two decades.

Under management responsibilities, the company may be responsible for planning, road building, reforestation, and protection. The license holder may be required to solicit input from the public (Chap. 4) or other stakeholders as well as to take other government objectives regarding environmental goals into account (this is described more

---

Box 7.3. Property rights and tenure.

Tenure policy — the arrangements under which governments grant access to timber on Crown land — is the fundamental cornerstone of forest policy throughout Canada. Provincial governments often use tenure to satisfy a number of different objectives, ranging not only from the desire to promote economic activity and the economic benefits it generates but also to provide social objectives such as community stability (through requirements for such things as cut control and appurtenancy clauses that tie wood under lease to a specific mill).

In some cases, these objectives might conflict with other objectives; for example, the government seeks to ensure the best possible economic return for the Crown while at the same time placing restrictions upon the use of wood from a lease reduces its value. Pearse (2001) has argued that on the B.C. Coast the number of restrictions placed upon tenure, coupled with other provincial policies regarding stumpage and land use, have so reduced the value of the wood generated through these long-term leases that firms are unwilling to invest any more capital in the industry.

Paradoxically, then, policies meant to promote community stability in the short run may have the long-run effect of jeopardizing the economic viability of the B.C. coastal forest industry. Pearse advocates strengthening some of the property rights associated with tenure by reducing these operational stipulations and controls by giving companies more flexibility in how they choose to allocate their wood as well as lengthening the time frame in which they can recoup their investment. Coupled with other changes, Pearse argues that this modification of the property rights associated with tenure will provide the right incentives for the coastal industry to re-invest in its facilities and in better management practices.

---

fully under forest planning) in developing management plans. These plans can provide commitments the company makes to engage in certain activities (or carry them out in a specified manner) for which they may be held accountable.

In terms of harvesting, the government typically places some restriction on the amount that can be harvested both annually and over 5–10-year time frames. This cut control allows the company some limited flexibility to deviate to some degree from the harvest level over the course of the period in response to changing market conditions. At the same time, it ensures that the flow of wood through the mill remains relatively constant (since companies that sharply reduce harvests recognize that they must make it up at some future period and run the risk of doing so when economic conditions do not warrant it).

The government may also impose utilization standards, requiring the company to harvest certain types and sizes of wood (such as down to certain minimum sizes or stump height). Companies are also typically responsible for building and maintaining roads, as well as sometimes deactivating them after use. Provincial governments also require all companies under the terms of their license to follow regulations and standards over what are considered acceptable harvesting practices, designed to protect environmental values and maintain the long-run productivity of the forest. The government may also direct companies to harvest certain types of timber damaged by either insects or fire. In Quebec, the proposed legislation requires companies within a management area to harvest their share of damaged or diseased wood at the direction of the Minister, or risk losing their cutting rights.

In terms of processing timber, the government may attach other obligations, such as building and maintaining processing facilities (otherwise known as appurtenancy agreements), or restrictions on the disposition of harvested timber (requiring it to be processed within the company's facilities or not allowing it to be exported without processing). In B.C., some licenses require that a certain portion of the timber must be harvested by a firm other than the license holder, thereby providing economic opportunities for smaller logging contractors. The social objectives of appurtenancy and similar restrictions are the provision of local employment, the encouragement of secondary industries, or the promotion of community stability.

Companies are typically responsible for carrying out reforestation activities after they harvest (although for some smaller forms of tenure the government may assume that responsibility). These responsibilities may end with completion of replanting the harvested area or once forest regeneration has reached some pre-specified standard (Chap. 13). Companies may also be responsible for forest protection from wildfire.

Major types of provincial tenure arrangements in Canada

In general, most timber harvested in Canada is under large-scale, area-based renewable tenure arrangements held by companies (Ross 1995). Much of the timber harvested under these arrangements is dedicated to a company's own processing facilities, which is reinforced by appurtenancy conditions attached to the cutting rights. For the most part, the obligations increase with the duration and size of the tenure held by the firm; for example, area-based, large-scale renewable agreements typically carry the most responsibilities in terms of planning, management, and silviculture. By way of contrast, in programs such as the Small Business Forest Enterprise Program (SBFEP) used in B.C., timber sales are for limited terms and are non-replaceable. In addition, these sales often carry less management responsibility (the government may prepare management plans, build roads, and carry out reforestation). We illustrate these common characteristics below with examples from various provinces and note how they may influence the incentives for SFM.

Area- vs. volume-based rights

One way in which tenure types differ is that the cutting rights may be expressed as a timber volume commitment over time, rather than harvesting rights over a specific area of land. In this case, the license holder harvests from a certain land area and may not necessarily return to that area in the future. Consequently, the license holder is likely to have no incentive beyond doing what is required in terms of reforestation or taking measures to preserve the productivity of the site. This problem is less likely to arise when the cutting rights of a company are area based. In this case, the company knows that it has the exclusive rights within a certain geographic area, provided the government does not alter the terms of the lease over time.

For the most part, most tenures in Canada are area based, such as Tree Farm Licenses (TFLs) in British Columbia, Forest Management Agreements (FMAs) in Alberta and Saskatchewan, Sustainable Forest Licenses (SFLs) in Ontario, and Contrat d'Approvisionnement et d'Amenagement Forestier (CAAF) in Quebec. The only province where volume-based rights account for a significant percentage of the harvest is British Columbia, where Forest Licenses (FLs) provide the licensee with an annual volume drawn from a Timber Supply Area (which itself is a geographic unit). Attempts to convert these FLs into area-based licenses (which mainly occur in the interior of the province) in the early 1990s ran into opposition from the general public that felt this would increase company control over provincial forest lands.

Duration of cutting rights

All provinces make some distinction between long-term cutting rights – those that can be renewed on a periodic basis — and short-term cutting rights that are granted for a fixed period without any provision for renewal. Short-term cutting rights do not mean that the license holder, once a license expires, cannot seek new short-term licenses. The large-scale, area-based licenses that most large companies hold are (with little exception) renewable or replaceable, typically every 5–10 years into a new term, though the term of these renewable agreements is usually for 20 or 25 years (Chap. 1). At the agreed-upon intervals of review or renewal, government typically retains the ability to alter the harvest volume and may impose other obligations. Provinces also create forms of tenure, however, that allow for short-term timber sales that are for a fixed term (typically 5 years) and are non-replaceable. These non-renewable tenure forms are used for a number of different purposes: to address special circumstances (such as temporarily elevated levels of cutting to respond to disease or fire); as a source of fibre to other suppliers; and in some cases, to attract industry or investment.

Size of operators

The majority of tenure arrangements in each province are in the form of long-term renewable agreements with major forest product companies that use the timber in their own large, capital-intensive processing facilities. These may be large lumber mills, oriented strandboard (OSB) plants, or pulp and paper operations. However, most provinces also make some allowance in their tenure systems for smaller operations, either through the special allocation of cutting rights (which may specifically exclude larger operators) or through tenure agreements that have a different package of rights and obligations compared to those held by major operators. These licenses generally tend to be non-renewable and, as noted earlier, may impose fewer obligations on the tenure holder. Examples can be found in most provinces. In British Columbia, the province utilizes the Small Business Forest Enterprise Program (SBFEP) to make short-term non-replaceable timber sales to independent loggers, small companies without long-term tenure, and small-scale value-added manufacturers. In Alberta, the province uses timber permits for small non-integrated mills and independent loggers; in Saskatchewan, the province uses forest products permits to grant access for specified products; Ontario also makes similar provisions in its system; and in Quebec, the Convention d'Amenagment Forestier (CAF) is used for groups that do not have wood processing facilities.

Social objectives

Several provinces have established tenures that are meant to accomplish social objectives such as offering additional economic opportunities within rural communities and

First Nations. British Columbia has woodlot licenses, which are small, area-based licenses that can be held by individuals or small groups, and recently introduced community forest agreements, which are area-based tenures held by local communities including First Nations. In Quebec, the provincial government is proposing to introduce a new form of tenure under pending legislation that will be available to communities and First Nations. In some cases, research and experimentation are promoted or supported by special-use permits, woodlot licenses, or other special arrangements for the designation and management of research forests or experimental forests. Some innovative tenure arrangements are also being explored under Canada's Model Forest program; see Chap. 22 for details.

Stumpage

Provincial governments retain a financial interest in publicly owned forests through charging for the timber harvested under the various forms of tenure. Governments have two fundamentally different ways in which they can determine the appropriate charges: auctioning off the right or determining the residual value of the wood after a company has recovered its production costs. For the most part, provinces have chosen to use these residual pricing systems to calculate timber values, although a number of different systems have evolved. In Ontario, there are different charges for the wood that is harvested, depending upon whether it will be utilized for lumber, panelboard, or pulp. In Quebec, the provincial government utilizes sales of private timber in determining the level of charges, and then takes factors such as cost, species, and location into account in determining specific charges. In British Columbia, the government uses changes in lumber prices to determine the overall level of charges, and then a variety of factors to determine differences in the relative value across the two regions of the province to distribute charges for the major license holders. However, the province also uses competitive bidding for a portion of its timber sales under the SBFEP and is in the process of converting the program into a new program called BC Timber Sales where all sales would be made on a competitive basis. In the past, the program has operated several experimental log markets where it has contracted out harvesting and has sold logs directly through competitive bidding through log yards (the largest and longest running program, the Vernon Log Market, was recently discontinued). Other log yards operate in the province. The community of Revelstoke holds a small TFL and sells its logs through a log market it operates, while other private operators run similar ventures from time to time.

Often, a portion of stumpage may be allocated to a particular purpose such as reforestation or protection (many provinces assess charges that are earmarked for such uses). British Columbia dedicated a portion of the stumpage increase in 1994 (known as superstumpage) to forest renewal and research. In some cases, such as the Nisga'a, First Nations can determine and collect the stumpage on timber harvested within their treaty area.[4] On reserve lands, the federal government collects the fees harvested for First

---

[4]During an initial 5-year transition period, the provincial government will determine and collect stumpage, which will then be remitted to the Nisga'a (subject to a minimum payment of $6 per cubic metre and other minor adjustments). After the transition period, the Nisga'a will determine and collect any timber fees.

Nations. Under some joint ventures and cooperative arrangements (see an example in Chap. 22), First Nations or other local bodies may receive a fee in addition to any stumpage payments made to provincial governments (these are typically negotiated privately between the partners).

Addressing environmental and social objectives

Provincial governments respond to environmental and social concerns in two ways. First, provinces incorporate these concerns in the ongoing management of timber resources by developing procedures and standards in two main areas — forest practices and forest planning — and then requiring their use through regulation or conditions of tenure, or both. Secondly, governments also engage in higher level planning and specific polices to address SFM objectives. These policies help interpret regulations. Finally, many provincial governments are expanding the opportunity for the public to participate in the development of management plans, although participation, for the most part, consists mainly of identifying concerns that managers may respond to in developing the plans. In some provinces (e.g., Ontario) there is formal recourse to environmental assessment processes which may empower stakeholders. The full range of public involvement options are discussed in Chap. 4.

Forest planning

A tenure holder's responsibility may encompass the preparation of several different levels of plans (Table 1.2). First, a tenure holder may be required to prepare a long-term forest management plan, showing the general activities to be carried out over a time frame that can be up to 20 years in length (these are sometimes described as management plans or forest plans). Virtually all tenure holders are required to prepare more detailed tactical plans that describe the operations to be carried out over a shorter time frame (between 1 and 5 years), indicating where roads will be built and harvesting will take place (development plans). Tenure holders are also expected to prepare plans which show how a company intends to harvest a cutblock and the steps the company will take to reforest it (operating plans). The level of detail in these plans, and the aggregate number of plans a tenure holder may be required to prepare, differ from province to province. This hierarchy and provincial and national variations are briefly reviewed in Chap. 1, while Chap. 11 provides new perspectives on forest plans, Chap. 12 explores recent approaches in development planning, and Chap. 13 presents a range of options for site plans in the boreal forest.

Most provinces require companies to solicit the input of other stakeholders in the preparation of some of these plans. In British Columbia, the planning process specifically prescribes the procedure for public review and comment in the writing of development plans. In Alberta, the holder of an FMA is expected to solicit public input in the development of their management plans according to procedures outlined in a planning manual prepared by the government. In Saskatchewan, the holders of Forest Management Agreements must develop their management plans in consultation with the public and First Nations. As part of the consultation, the tenure holder must report issues and concerns raised and the steps taken to address the concerns. In Ontario, procedures are specified for establishing local citizens' groups that are involved in the preparation of management plans. A planning manual describes procedures to be followed, along with specified times for public comment and review. First Nations can opt to participate in the public process or pursue a separate consultation process. In Quebec, local governments, First Nations, and any party with a responsibility for managing wildlife or that holds a special permit for maple sugar production within the area is invited to participate in the development of management plans. The tenure holder is required to identify where they have responded to any concerns raised and how they will be addressed; in addition, the plan undergoes public review and comment for a fixed period of time before adoption at which point groups not initially consulted can apply, if they have standing, to be included in the planning process. In New Brunswick, there is no formal requirement for public or local participation.

These planning processes commonly require the tenure holder to take into account the impact of planned harvesting activities on other forest resources. At the forest management level, they may require the tenure holder to identify how certain forest values will be maintained, as is the case in Saskatchewan. In Ontario, the planning manual specifies other resources such as old-growth forest, traplines, and other non-timber values that have to be taken into account in developing the management plan. In development plans, the tenure holders identify the activities that they will undertake (for example, location of roads, road construction, watercourse crossings, scope and nature of harvesting, and reforestation). The description of these planned activities may become measurable standards against which their performance is assessed.

Approval of these plans rests with the government. Appeal of these plans is typically directed to the Minister or to the official that approves the plans, and no province has a statutory appeal process other than B.C., where the Forest Practices Board may appeal an approved development plan.

Provincial governments may also carry out lower level plans on behalf of small tenure holders. They may also develop strategic or higher level plans, which provide land use and resource emphasis direction that must be taken into account when lower level plans are developed. The procedures for carrying out these strategic plans may involve the solicitation of input from different stakeholders, or the development of specialized planning bodies, although it is the government that retains ultimate authority.

Forest practices

Forest practices refer to the activities carried out by tenure holders: timber harvesting; road construction, maintenance, and decommissioning activities; fire protection and management activities; and silviculture. As noted above, in the preparation of development and site plans, tenure holders establish the activities they will undertake and against which they may be held accountable. Provincial governments have typically addressed the environmental impacts of these activities through regulations and restrictions. Standard examples include requirements to leave riparian buffers, road construction standards to minimize impact on water quality, and limiting or restricting the type of harvesting activities that can take place in more sensitive areas such as steep hillsides subject to erosion.

The government may specify or prescribe how certain activities are to be carried out (this approach to regulation is process based). For example, road construction may have to meet pre-established standards in terms of how the road will be built (Chap. 15). Alternatively, the government may prescribe a certain result it hopes to achieve, such as no sedimentation from road construction. In this case, the regulation is result based, and the standard will be the outcome (such as a certain allowable level of sediment deposition in a stream bed).

For the most part, most provincial governments have used prescriptive or process-based rules to protect environmental values for forest practices. The level of specificity of these rules can vary; they may be province-wide, region- or ecosystem-specific. The ability to vary from pre-specified rules may differ. In some cases, there may be no alternative to following procedures spelled out in a manual. In other cases, there may be a process by which a tenure holder can pursue a different approach.

British Columbia has perhaps the best known example of a process-based regulatory system that relies on carefully described procedures in order to protect and enhance non-timber values. The B.C. Forest Practices Code specifies how activities are to be carried out, and published a series of guidebooks that provides the detail. The legislation also creates several independent bodies, the Forest Practices Board and the Forest Appeals Commission, to monitor practices and adjudicate disputes. In Alberta, some of the rules concerning forest practices are negotiated for each individual tenure holder to reflect local circumstances, and the province is working on developing regional and ecozone-based guidelines for forest management plans. In Ontario, activities and plan development have to be carried out in accordance with a manual that specifies certain procedures the company has to follow. Under proposed legislation in Quebec, the government may prescribe regional standards and procedures, although the government may allow for departure from such plans if it can be shown that they will achieve the same result.

Enforcement and monitoring

All provinces allow for monitoring of the tenure holder's performance. Penalties range from administrative action, such as stop-work orders, to fines (which are typically several thousand dollars but can range up to $1 000 000 in B.C.), or to a reduction or loss of cutting rights if the infraction is severe enough. Several provinces use past performance as a factor in current evaluations to provide an additional incentive to meet performance requirements. In B.C., a record of past contravention and fines may lead to increased penalties for future infractions and the potential non-replacement of a replaceable license. In addition, the Forest Practices Board may challenge the adequacy of monetary penalties leveled by the government. A licensee may appeal an administrative penalty to the Forest Appeals Commission, and the Forest Practices Board may become party to the process, thereby increasing the risk to an operator of not meeting its obligations. In Alberta, a company with a good track record will face reduced scrutiny and see its operational and site plans approved more quickly.

Fig. 7.5. International and bilateral trade agreements can have a tremendous impact on export volumes and tariffs, shown here (*a*) by a backlog of dimensional lumber awaiting export in Richmond, B.C. (*b*) Canadian International Minister of Trade Pierre Pettigrew and British Columbia Forest Minister Michael de Jong preparing for softwood lumber negotiations with the U.S.A.

(*a*)   (*b*)

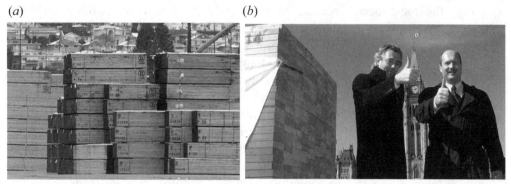

Canadian Press photos

International formal institutions

The last 30 years have seen an increasing focus on environmental issues at the international level. Countries have pursued bilateral agreements as well as broader global agreements governing many resource issues, often under the auspices of the United Nations. Canada has been an active participant in many of these, and several are applicable to sustainable forestry. International arrangements include a mix of "hard law" and "soft law". Hard law refers to binding commitments that countries may enter into through international or multilateral treaties or agreements, while soft law covers general principles that countries agree to pursue under conventions, protocols, and declarations.

Currently, there is no "hard law" that is specifically applicable to sustainable forestry. Rules applicable to international trade under the World Trade Organization (WTO) constitute hard law that may have an impact on SFM. In general, WTO rules are meant to facilitate trade and do not allow countries to create restrictive trade practices under the guise of other objectives. For example, environmental labeling of products is acceptable so long as it is not a requirement (Ruddell et al. 1998). While WTO rules allow countries to enact special restrictions in order to conserve or protect the environment or resources (see Box 7.4), attempts to limit trade because of environmental considerations have been contentious and, until recently, unsuccessful. The best known example of an attempt to limit imports is the U.S.–Mexico dolphin–tuna dispute in the early 1990s over the Mexican fishing practices under which the GATT/WRTO Panel found that the U.S. could not use domestic laws over foreign environmental practices to justify trade restrictions (Nordström and Vaughan 1999). More recently, the WTO heard a case in which the U.S. enacted regulations to restrict the import of shrimp from countries that did not employ devices to reduce the by-catch of sea turtles. The U.S. restrictions were originally found to be in conflict with WTO rules but were subsequently upheld in 2002 after the U.S. funded a series of conferences and reached an agreement with some of the affected countries on how to protect sea turtles. Several countries are appealing this most recent decision. More recently, the United States has renewed its trade actions regarding Canadian softwood lumber, arguing that certain forest management practices in Canada in effect subsidize Canadian forest product firms (Fig. 7.5). The U.S. claims that the government-set stumpage rates in Canada are lower than those found in the U.S. and that the difference constitutes a subsidy. The WTO dispute panel will rule on whether the U.S. method of calculation (which involves cross-border comparisons) is valid and whether or not the government determination of stumpage payments falls under WTO rules governing subsidies.

The North American Free Trade Agreement (NAFTA) governs the rules under which Canadian lumber is ex ported to the other two member countries, the U.S., and Mexico (Mexican imports are negligible). NAFTA governs trade in goods as well as investment, and also provides a process to resolve trade disputes. Unlike the WTO dispute process, the NAFTA dispute process simply examines whether each country complied with its own domestic law, although disputes under the investment provisions consider whether or not government actions or policies discriminate against foreign firms and constitute an expropriation.[5] The dispute over current U.S. countervailing and anti-dumping duties will be heard under the WTO dispute settlement process as well as the NAFTA process (although the NAFTA process is expected to focus more narrowly on procedural issues).

In addition, Canada, the U.S., and Mexico established the Commission for Environmental Cooperation (CEC) through an environmental side agreement to NAFTA. The organization is charged with monitoring and disseminating information on the environmental aspects of trade and economic development in North America. The CEC can also hear complaints as to whether countries are enforcing their own domestic environmental laws but has no enforcement powers. To date, the CEC has received a number of complaints regarding Canadian forestry practices from both First Nations and environmental groups.

There are several international efforts under way that have some impact upon sustainable forestry. The United Nations Conference on the Environment and Development (UNCED) in 1992 approved an action program, Agenda 21, in which governments would promote sustainable development. Significant progress with respect to forest management occurred in Chap. 11 of Agenda 21 and the non-legally binding "Forest Principles". Under the auspices of UNCED, the Rio Declaration, the Convention on Biological Diversity (CBD), and the Framework Convention on Climate Change (FCCC) were adopted, as well as a statement on the need to move towards sustainable

---

[5]Chapter 11 of NAFTA deals with the investment provisions and has already been the focus of one dispute in which an American lumber firm, Pope & Talbot, sued the Canadian government over the quota it received under the Softwood Lumber Agreement, arguing that it was unfairly treated and had received too little quota. The company was unsuccessful in its main complaint. Several Canadian companies (Canfor, Doman, and Tembec) have currently filed suits under Chap. 11 provisions over current U.S. countervailing duties while one U.S. firm with a Canadian subsidiary (Weyerhaeuser) is appealing.

forest management. It should be noted that many of these conventions also discuss the importance of indigenous involvement and participation. Efforts to develop a more formal convention on forests at UNCED failed, as participants were unable to reach a consensus. It did lead to an ongoing process at the United Nations, which has sponsored several intergovernmental panels on forests that culminated in the creation of the U.N. Forum on Forests (UNFF), which is attempting to establish a legal framework for forests (NRCan 2000). The most recent meeting of the UNFF (its second) again displayed the lack of consensus between members over defining what constitutes SFM and who should participate in those decisions (IISD 2002). Within the FCCC, the main focus is on developing a way to reduce greenhouse gas emissions. Countries are currently deciding on mechanisms and how to debit/credit abatement with significant implications for forestry practices, depending on how the debates are resolved. For example, if credits are given for afforestation, this would create an incentive to convert marginal agricultural land into forestry. It was recently agreed that certain forest management practices could be recognized as sinks for carbon sequestration (although they can also be sources, depending upon forest practices and forest losses associated with disease and fire; Chap. 20).

Canada helped initiate the Montreal Process (Box 2.7) in 1995, in which a group of non-European countries that have temperate and boreal forests are attempting to develop a consensus over a common set of criteria and indicators by which sustainable

---

Box 7.4. International trade and SFM.

Some people have argued that there is a potential conflict between free trade and SFM and that trade liberalizing measures, including WTO trade rules designed to prevent countries from restricting trade, may conflict with the greater need to protect environmental values. They believe that these efforts will prevent countries form establishing policies that will ensure environmental and resource sustainability and may inhibit the use of government policy to attain other social objectives such as community stability. There is also a concern that WTO trade rules may frustrate the use of other Multilateral Environmental Agreements (MEAs), which are typically designed to address a particular environmental issue such as ozone depletion or trade in endangered species, and may also reduce the effectiveness of approaches such as eco-labelling (which is currently allowed only if it is voluntary and not a government requirement).

The WTO does allow exceptions under Article XX, where countries take measures to conserve and protect renewable and non-renewable resources, while other parts of the agreement allow countries to enact measures to protect human, plant, and animal health. Ongoing negotiations under the WTO framework, known as the Doha round, are currently investigating the compatibility of WTO rules with MEAs in general and the use of eco-labelling.

While it is true that current MEAs do not have many teeth, the main danger in strengthening international environmental rules around trade is the increased likelihood that such rules will be used to disguise protectionist measures and that the major victims of such measures will be developing countries. In addition, such an expansion would open up new avenues of harassment for major exporters such as Canada. To protect forests, perhaps an international forest convention in which members can reach a consensus is the appropriate tool, similar to the approach used for other environmental issues with international dimensions.

forestry practices can be identified and evaluated (Chap. 2). Certification of sustainable forest management constitutes an example of private "soft law" (and is discussed in greater detail in Chap. 21). As it becomes more pervasive the sanctions associated with lack of certification become more effective. In addition to the systems discussed earlier (FSC, ISO, and CSA in Canada), other countries have introduced their own national or regional certification schemes. These include the Sustainable Forestry Initiative (SFI) being pursued by companies and industry associations in the U.S., and the Pan-European Forest Certification system (PEFC, which is actually a common framework for recognizing regional or country specific schemes) being pursued in Europe. Currently, forests in Canada are certified under ISO, CSA, SFI, and FSC (Fig. 7.4).

Within the European community, countries are trying to articulate SFM through the European Ministerial Conferences on the Protections of Forests in Europe (MCPFE); this is also known as the Helsinki Process, which started in 1990 (Box 2.7). PEFC is using criteria and indicators developed through this approach. In addition, a number of European countries have recently revised their forest policies to take into account biodiversity and other issues being raised in international dialogues, as well as processes to identify criteria and indicators of SFM. At the same time, some countries noted that there were concerns about the economic viability of forestry within their countries, especially coupled with the need to provide more non-timber benefits that could be expected to increase costs. Some countries had adopted programs to mitigate the costs to private forest owners of reduced harvests or increased planning requirements through providing financial assistance or technical support (United Nations 2001). As in Canada, the amount of certified forest land in Europe has grown substantially in recent years.

Institutional and policy impediments to SFM in Canada

Existing tenure and stumpage arrangements are not designed to respond to the issues raised by sustainable forest management

As described above, forest management in Canada is a complex mix of institutions. It is not well equipped to deal with some of the issues raised under SFM. Some of the conflicts that arise are due to the disincentives to forest management created under current tenure arrangements. Pearse (1990) notes that the tenure system in Canada raises several fundamental issues that influence both the competitiveness of firms and their responsiveness to other concerns. First, Pearse notes that the historical method of tenure allocation, under which firms received tenure in exchange for commitments to build and maintain processing facilities, tends to fix production patterns and reduce industry's flexibility in responding to new markets and products. Second, the more narrow the scope of the tenure agreement, which usually tends to be limited to granting rights and responsibilities solely associated with timber, the less opportunity there is to internalize externalities. Third, the security of the agreement is important because insecurity creates risk. Coupled with the security is the scope for intervention; if the rights and obligations under the tenure agreement can be freely modified on short notice, the tenure holder will perceive such agreements to be more risky. Fourth, the allocation of management responsibilities is also important: the more responsibilities the tenure holder incurs, the more costs they incur. Coupled again with the responsibilities is the distribution of resource rent or economic payments available to be captured by governments on behalf of the public. If too much is captured, firms may be unable to reinvest in the business, and over time the competitive position of the industry will weaken. The impacts of these issues on sustainable forest management in Canada are discussed below.

One of the fundamental problems of the tenure system is that the lack of ownership, coupled with the long time frame between harvests, reduces the likelihood a company will receive the benefits from its reforestation activities. This reduces a firm's willingness to invest in silviculture or activities that promote or maintain long-run productivity that they otherwise might, since they have no assurance they will be able to capture the future benefits from this investment. Consequently, silviculture performed on Crown lands tends to be the minimum required by government. This problem is compounded when tenure holders have volume- rather than area-based rights, since there is little assurance the company will return to harvest the same site in the future. This problem has been emphasized by Pearse (1990) and Haley and Luckert (1990) among others.

Problems have also been pointed out with restrictions on tenure such as appurtenancy clauses. By requiring firms to operate and maintain facilities as a condition of the license, firms may be unable to respond to changes in demand. As such, these conditions reduce firms' flexibility in responding to changing markets. They may also reduce the efficiency of operations by requiring firms to process logs internally rather than finding the best market for the fibre. On the other hand, appurtenancy has been a cornerstone of regional employment and community stability policies (Chap. 5).

Economists have also noted that some aspects of tenure may be perverse from an economic perspective; utilization or cut-control requirements may require firms to harvest or extract more timber volume per unit of land even though it is unprofitable to do so at the time. In some cases, utilization standards may conflict with requirements to leave a certain portion of fibre behind that is considered necessary for forest ecosystem functions. At the same time, some selective logging (partial cutting) practices run the risk of high-grading, extracting the most valuable timber and leaving less desirable trees behind; high-grading could be based on size or species or both. Over time, this may lead towards stands that are deficient in particular age classes or species and therefore no longer representative of that particular ecosystem.

Provincial governments currently use tenure arrangements to grant the right to harvest timber. They also grant the right to use other resources on the same forest lands, such as furs, minerals, and the right to conduct seismic exploration for oil and gas development. This often creates problems. First, different provincial agencies may be involved, and there may not be a comprehensive assessment of the environmental impacts resulting from different activities. Secondly, there may be an additional level of administrative hurdles for operations approval, depending upon how rights are defined and incorporated within the current tenure system. Conflicts may require the involvement of other agencies and the need to establish whose rights should prevail (or what weight they should be given and who should be the arbiter). In addition, some provinces, such as B.C. and Alberta, grant overlapping forest tenures; for example, different parties receive the right to hardwood and softwood, or pulpwood and sawtimber, within the same area. This usually complicates planning and operations, since the involvement of two different parties may lead to the creation of two separate plans, each with their own interest.

As noted earlier, most stumpage in Canada is determined by the government through administrative processes. If these procedures are unresponsive to the costs of obligations, they can make timber uncompetitive, and companies can be expected to resist costs and obligations for which they are not compensated (especially if they perceive they are put at a disadvantage to their competitors). Tenure obligations, such as development of forest plans, can become quite costly, depending on the requirements. Only large, well-capitalized firms might be able to undertake such obligations. Under these circumstances, small communities and First Nations might be effectively excluded from managing forests even if granted the opportunity.

Stumpage costs, the imposition of obligations, or restrictions on how the firm utilizes Crown timber can eventually be expected to have an impact on a firm's willingness to undertake additional investment. This, in turn, may eventually reduce their ability to pay stumpage and contribute to the revenue stream on which provincial governments depend. The same is true for any uncertainty the government creates for the firm, if the conditions under which the tenure holder has access can change quickly and with no redress.

Regulatory approach to environmental concerns raises costs

In terms of operations, governments have typically approached the provision (or protection) of environmental values through the specification of rules and standards (the "command and control" approach). Stanbury and Vertinsky (1998) point out that rules may provide little benefits for the cost imposed, and generally do not allow firms to seek better ways of accomplishing a given objective. This has been a complaint with the Forest Practices Code in B.C., which has led to substantial increases in the cost of harvesting (McIntosh et al. 1997). The province is currently embarked upon an effort to move towards a result-based code that can achieve the same objectives although at a lower cost (BCMOF 2002a).

In addition, regulatory schemes tend to be inflexible and may become less responsive over time to the issues or problems that they were originally designed to address (Stanbury and Vertinsky 1998). For these reasons, specifying results are generally preferred, since they allow companies to innovate and search for the best way (lowest cost) to provide these non-timber outputs. In B.C., the emphasis on guidelines and regulatory process has led to an unwillingness of government officials to sign off on plans that do not conform directly to the proposed guidelines at the risk of facing penalties. Furthermore, companies appeal any penalties that are imposed, no matter how minor, since they face the prospect of increased penalties or potentially reduced access to timber if they have a blemished record.

However, moving to a results-based system raises difficulties. First, the long time frames of industrial development, tree growth, and ecosystem recovery mean that unsatisfactory "results" may not become apparent for decades after unsustainable practices are employed and have become common practice; this makes the selection of scientifically rigorous indicators paramount. Other foundations of SFM, related to the protection of biodiversity and ecological functions, either require the collection of vast amounts of data or involve poorly understood interactions between timber harvesting and various ecosystem processes. Under the current system, companies have no incentive to undertake such research, since it is questionable as to what economic benefits they may provide, and furthermore, other companies (i.e., their competitors) may benefit from any research they undertake. Governments have little incentive to fund such research, preferring the apparently less costly approach (to the government) of simply imposing requirements and regulations that companies must follow.

Resource allocation in the presence of externalities and public goods

Economic theory suggests that the positive externalities associated with forests, and public goods such as wildlife, clean water, and carbon sequestration, will be underprovided if we rely on the private sector to supply such goods. At the same time, for those externalities and public goods that are national or international in scope (such as biodiversity or carbon sequestration), there is no reason to expect provincial governments to provide the right mixes of goods. In the case where the environmental benefits are widespread and accrue mainly to those outside a province, and the costs of complying are borne through reduced economic activity within a province, the province is unlikely to take action. In some cases, this dynamic may even be at work within certain regions in a province, if a particular area is perceived as having values whose maintenance or preservation require a significant reduction in harvesting.

In cases where there is no market for the good, the opportunity to utilize market prices is absent, and the typical response has been for government to place restrictions upon the firm. Firms have no incentive to incorporate non-timber values into planning beyond that required by the government. This creates the problem that non-timber values become a constraint in the planning process from the company's point of view if it is the one preparing the plan. Companies have no incentive to attempt to produce these non-timber outputs in a positive fashion and, given a choice, may not choose plans that offer substantial improvements in environmental benefits. Put another way, non-timber objectives become residual values to the harvesting decisions, given the imposition of external constraints to meet these objectives. This may also be true for government agencies if their effectiveness is solely measured by the degree to which they encourage economic activity or generate government revenues. To the extent that companies are required to exercise judgment or make assumptions, they have an incentive to relax the effect of such constraints. Ross (1995) describes two examples that illustrate this dynamic. In the first, a tenure holder in British Columbia argued that an annual allowable cut (AAC) determination improperly incorporated non-timber objectives in the calculation and appealed the AAC set by the Chief Forester. In the second, Ross quotes Gordon Baskerville, who conducted an assessment of Ontario's guidelines and concluded that non-timber values were treated as constraints and not as meaningful objectives, with the consequence that there was no real attempt to integrate these values into forest management and no measurement of results (Ross 1995, pp. 170–171).

Ecosystem management has been offered as a way to incorporate environmental concerns in forest management (Kohm and Franklin 1997), but it still leaves unresolved some of the fundamental questions. What environmental values should be maintained and at what level? How do managers translate landscape level attributes to the stand level? As Lackey (1999) shows, the conflict surrounding ecosystem management depends on the underlying assertions and values that differ significantly between those proponents that envision ecosystem management as a continuation (with some modification of approaches and emphasis on different goods) of existing multiple-use management practices with those that see it as a fundamental shift in the way that we approach society and current lifestyles with their emphasis on material goods.

Insufficient R & D

Research and development (R & D) in Canada in the forest industry is generally considered to be low (Binkley 1995; Binkley and Forgacs 1997; Globerman et al. 1999). Research and development is crucial for innovation and expected to be a critical component if Canada's forest industry is expected to remain competitive. We can distinguish between research that offers foreseeable commercial benefits and research that does not. Firms engage in research to discover production methods that lead to increased productivity and (or) reduced costs, or to develop new higher value products. The payoff to firms is the profits that flow from successful R & D. To the extent forest management practices required under sustainable management are more costly (without a corresponding increase in product prices), firms will find it more difficult to fund R & D.

In addition, there is also other research that will become increasingly important in the context of sustainable forestry. Understanding environmental processes and the impact of forest practices upon those processes are crucial to the decisions we make about how to manage forests. In this case, however, there is no commercial incentive to undertake this research. The problem is compounded by the fact that any new information, once discovered, becomes open to all, and this "spill-over" further reduces the value to any individual firm that undertakes such research. Such research tends to be considered "open science". Under open science, research results are publicly disseminated so that they can be tested by other researchers. This reliance on external critiquing, the process of peer review, is an integral part of validating research results.

These two problems take place in the context of declining public funding and privatization of many of the management costs associated with provincial forest lands. A critical question then becomes "Who will pay for such research?" (Byron and Sayer 1999).

Jurisdictional issues frustrate policy innovation

There are a number of jurisdictional issues that complicate efforts to address sustainable forestry. Authority split across several different provincial agencies can make decision making more time consuming and may lead to bureaucratic in-fighting, frustrating attempts to develop coordinated policies. For example, Carrow (1997) notes that in the Bras D'Or of Nova Scotia, 22 different government agencies have some authority within the 360 000 ha watershed and that they have been unable to come up with an integrated management plan in 20 years. At the same time, the traditional provincial responsibility for natural resources has meant that provinces resist attempts by others, most notably the federal government, to intrude in the environmental arena (Hoberg and Harrison 1994; Harrison 1996). This makes it more difficult for the federal government to enter into international commitments, since although the federal government makes the commitment, it is the provinces that have to implement it, and Canada has been critiqued for failing to act on its commitments in terms of SFM practice and reducing carbon emissions for just this reason (WRI 2000; Chap. 20). In addition, other authors have noted that the dynamics of the parliamentary system itself may tend to insulate it from environmental pressure. Because they enjoy a degree of autonomy, regulatory officials in parliamentary-based governments are likely to be shielded from direct pressures by environmental interest groups. Consequently, they may choose to represent diffuse interests, although at the same time they are also freer not to do so (Vogel 1993, p. 264).

The political system in Canada also makes policy innovation (the development of new policies) both more feasible as well as more subject to policy reversal (since new policies can just as easily be introduced to reverse the pre-existing policy; Weaver and Rockman 1993, p. 452). However, where jurisdiction is shared between the federal and provincial governments, such as in natural resources, such innovation becomes less likely in the absence of a shared consensus (Weaver and Rockman 1993). Hoberg and Harrison (1994) note that the provinces have been quite skeptical of federal efforts in the environmental arena and quick to assert provincial jurisdiction. If, however, the provincial and federal governments can reach a consensus, it appears that they are quite capable of substantive policy changes (Feigenbaum et al. 1993, pp. 73–74).

Existing institutions do not address social concerns

A common complaint has been that current tenure systems exclude communities and First Nations (Jaggi 1997; Higgins 1999). Fully allocated timber supplies make it difficult to accommodate new groups such as First Nations. Even where tenures have been made available, the argument has been that they do not permit management and utilization in a significantly different way than those used by companies — the agreements focus on timber production and extraction (M'Gonigle 1998).

Current planning processes, either at the strategic or management level, may frustrate companies, local communities, and First Nations. Where processes are not transparent, information is lacking, and the rights local or First Nations groups hold are not clear; groups may distrust the process and not participate fully. From a company's perspective, such processes may also create uncertainty if it is unclear how the government reaches decisions and the criteria it uses. To the extent that parties to the process can attempt end-runs and circumvent the process by appealing to government directly, such planning processes are likely to be empty exercises. Carrow (1997) and Howlett and Rayner (1998) note that forest policy development still takes place within a closed policy network, dominated by government and industry that does not readily incorporate other viewpoints.

Problem of the planning approach

There is the hope that some of the more contentious questions can be better addressed through the provision of additional information and more elaborate consultation guidelines. Evidence from the U.S. suggests that this be wishful thinking to the extent that additional information appears to do little to change people's preconceived notions (O'Toole 2000). The planning process also changes the roles of people and gives weight to some groups over others, which creates new sources of political conflict. Finally, the collaborative approach advocated as a solution to some issues may not work. In the U.S.

(and increasingly elsewhere), decisions reached by a collaborative process run the risk of being put aside owing to litigation by a group or party unhappy with the outcome.

Local communities also have concerns about the economic benefits offered by current forest management (Chaps. 3–5). Reliance on export markets, the substitution of capital for labour, and the increasing scale of manufacturing plants all have an impact upon local economies. Over time, mills have become larger while employing fewer people and drawing their wood from a larger area, typically in response to economic forces determined by distant markets. Unsurprisingly, communities in areas supplying the wood complain that they do not receive their share of the benefits and have little say in local land-use issues (Markey et al. 2000).

First Nations also have complaints about sharing in the economic benefits from the forestry activity, and strong concerns that the emphasis on timber production that ignores cultural values and other resources in which they have an interest and rights. In recent years, both environmental groups and First Nations have solicited support from groups in the U.S. over Canada's forest policy. This is part of a larger trend in which groups excluded from the policymaking process within Canada seek to achieve their goals by looking for support outside of Canada (see Box 7.5).

Current efforts to correct market and institutional failures

Below we describe the efforts that have been recently undertaken to address the issue of sustainable forestry. The focus in this section is on attempts by the primary actors — provincial governments, the federal government, and industry — to address issues

---

Box 7.5. Disempowered groups seeking political leverage outside of Canada.

"If Canada's relatively closed, government-dominated decision arena has acted to exclude nongovernmental interests, its proximity and close trade ties to the United States have created new channels of policy access and constraint in recent years. Provisions of the Canada–U.S. free trade agreement of 1988 bar many of the NEP [National Energy Program] policies that discriminated against American firms and consumers while guaranteeing that Canadian producers will not be subject to any import tax imposed by the United States. In addition, Canadian interests that find themselves excluded from the Canadian policymaking process increasingly find that they may be able to get leverage in the United States. The clearest case to date concerns the next phase of the James Bay hydro project, which will flood hundreds of thousands of acres that are the homeland of Cree Indians. Unable to block the development at the provincial level, the Cree took their lobbying to the New York State legislature. The governor of New York eventually canceled a huge contract to purchase power from Hydro-Québec, which the province had counted on to win financing for the project" (Feigenbaum et al. 1993, p. 73).

An example of this dynamic at work in the area of Canadian forest policy are recent announcements by U.S. environmental groups and First Nations in B.C. and Quebec, who intend to petition the U.S. government to impose duties on softwood lumber shipments from those two provinces, arguing that they have been effectively subsidizing production by permitting environmental degradation through lax forest practices (McKenna 2001).

Box 7.6. British Columbia's pilot projects and results-based regulation.

The Forest Practices Code in British Columbia prescribes how forest management activities are to be carried out by tenure holders. The responsibilities of these tenure holders (typically forest products companies) encompass the development of forest plans that show future harvesting, road-building, silvicultural practices, and other related forest activities. The Forest Practices Code relies heavily on a detailed procedural approach in which the plans must meet environmental and social objectives and where the license holder is required to obtain pre-approval by the government of these plans before operations can commence. The provincial government recently embarked upon a series of pilot projects to investigate whether it could move away from this model of regulation towards a results-based approach, in which the same objectives could be met but at a lower cost through reduced administrative requirements and by allowing more flexibility in meeting certain performance criteria.

One of the pilot projects is in the Ft. St. John Timber Supply Area, a 4 700 000 ha region in the northeastern corner of the province containing four biogeoclimactic zones, mostly boreal forest. The project involves the four major tenure holders within the region (including the provincial government through its Small Business program) and covers both area-based and volume-based tenures that are either coniferous or deciduous only. Under the draft legislation, the license holders would participate in a Forest Development Plan (FDP) that would show the activities to be carried out over a 5-year period within the TSA while actual operations would be identified on an annual basis. Both the 5-year and annual plans would have to meet the objectives identified in either a higher level plan or required under the Forest Practices Code. The license holders would no longer be required to seek pre-approval of site-specific plans; rather, so long as they complied with the FDP the license holder could conduct its on-site operations. Compliance would be met through the use of certification systems to audit both the procedures, the development of benchmarks against which performance can be measured, and strengthened sanctions through higher penalties for non-compliance.

This approach offers promise in terms of reducing administrative requirements and providing an opportunity for integrated resource development and planning across different tenure types and responsibilities. The government of B.C. is now proposing to extend this approach to the entire province (BCMOF 2002a). The success of such an approach will ultimately depend upon its ability to develop meaningful and publicly acceptable criteria governing environmental and social objectives, and demonstrate that they are being met.

---

raised by the challenge of sustainable forest management. These efforts involve changes in processes and the formal incorporation of previously excluded groups and other concerns into the existing way forests are managed; for the most part these efforts have maintained the existing institutional structure (policy framework, economic incentives, and focus on timber production). The subsequent section discusses more fundamental changes to the way our forests are currently being managed.

Achieving environmental objectives while reducing regulatory costs

Currently, several provinces are pursuing attempts to streamline the regulatory process.[6] Within B.C., the government has created pilot projects that will encourage different

---

[6]This is an illustrative list and not a comprehensive survey of such efforts across Canada.

approaches to meeting environmental and social objectives. A pilot project in Ft. St. John in the Northeastern corner of the province is designed to facilitate integrated management and coordinated planning within a Timber Supply Area (Box 7.6). This is to be accomplished by having multiple local licensees jointly prepare one plan together rather than having each of them prepare their own individual plans and build their own roads. Projects proposed for other parts of the province range from the development of a stewardship plan by the tenure holder to identify landscape level values that will be addressed through the forest development plan to experimentation with different ways of collecting stumpage from small woodlot owners (BCMOF 2002b).

Alberta is considering a scheme under which more discretion would be given to the judgment of professional foresters, who would sign off on plans prepared by companies. At the same time, the government would create a process by which the public could examine plans and raise objections (Anonymous 2001).

Accommodating outside concerns through new forms of tenure and encouraging new partnerships

Governments are also experimenting with new forms of tenure. B.C. has introduced community tenures in 1998, and Quebec has proposed to do so in its forest legislation. B.C. has also expanded its woodlot program, which offers individuals and communities, including First Nations, the opportunity to manage small areas of forest. The Bas St. Laurent Model Forest in Quebec is experimenting with a novel form of tenure, that of "tenant forest farmers" (Chap. 22).

In addition, First Nations are entering into a number of different institutional arrangements concerning forest resources (Chaps. 3 and 22). NAFA (2000) identified five principal forms: (1) joint venture agreements, (2) cooperative business arrangements, (3) forest services contracting, (4) socio-economic partnerships, and (5) forest management planning. Though not a comprehensive survey, the study found one form or another of these partnerships in all the provinces except the Atlantic Provinces, with B.C. having the most partnerships, and the most common type of partnership was a joint venture, followed by forest services contracting. In B.C., efforts to support such joint ventures have led to an increase from less than 1% of the AAC to 6% of the AAC held in aboriginal joint ventures today (T. Niemann, Senior Advisor, Sustainable Forest Management Standards (Criteria and Indicators), Forest Practices Brandh, B.C. Ministry of Forests, Victoria, B.C., personal communication, 2001).

Responding to environmental concerns through technological innovation

Industry has responded to some of the environmental concerns through technological innovation. In recent years, machines have been developed that have a lighter footprint and increased use of harvesting systems (such as cable logging) that help reduce impacts on the forest floor (Chap. 15). Provinces and companies are also experimenting with different forms of harvesting systems, such as variable retention, where a portion of the trees are retained in the timber stand (Chap. 13).

Strategic land-use planning

Several provinces have engaged in large-scale land-use planning to identify areas with high ecological values or environmental sensitivity. These lands are then zoned for non-development and set aside. Examples include the Protected Areas Strategy in B.C., Alberta's Special Places Program, Saskatchewan's Forest Accord, and Ontario's Lands for Life. Saskatchewan's Forest Accord not only provides a plan in how Saskatchewan's forests are to be used; it also makes express allowance for a new forest accord to be developed in 10 years time.

Box 7.7. Information needs and designing institutions.

In order for markets to function, it is necessary that all the information needed to make informed choices be available to all the participants. Yet this condition may not hold: information may be absent, asymmetric (not all decision makers will have the same information), or costly to acquire. This condition raises three different sets of questions related to the information requirements of SFM. The first set involves the identification of sustainable forest management:

- Should we rely on government to define SFM (e.g., using mandatory standards, criteria and indicators, etc.)?
- Should we rely on the expertise of scientists and (or) professional foresters?
- Should we rely on certification by third-party environmental or industry groups to define SFM?

The second set of questions relate to the institutional issues of how we generate information, pay for it, and use it:

- Does the "public good" nature of much of the information we think we need require government funding?
- If government funding is required, which level should provide it — provincial, national, multilateral — and how do we maintain funding commitments over time?
- How do we incorporate the dynamic nature of sustainable management — "always learning new things"— and how can we institutionalize responses into the existing institutional structure?
- How do we design systems for information when "we don't know what we don't know"?
- Will adaptive regulation — using results-based approaches — allow companies the opportunity for learning-by-doing?
- What information needs exist if we are to use results-based approaches (rather than regulatory approaches)?
- How can we minimize the associated enforcement costs with this approach (e.g., do we have independent auditors, do we provide people the right to litigate)?

Finally, there are the questions related to how to determine which outputs to produce and to what extent we can use market mechanisms to achieve the optimal outcome:

- Can we create information-generating processes and the appropriate incentives to use decentralized decision making, or do we need to rely on regulatory mechanisms?
- Will markets necessarily be more efficient? We can observe difficulty in integrated planning in jurisdictions with fragmented ownership and the problems of internalizing externalities — there may be circumstances in which the economies of scale or scope overcome the costs associated with government regulation.

Designing strategies for institutional change

In thinking about how to design institutions that will produce sustainable forest management, we return to the two ideal types of institutional arrangements — regulatory and market. We might also add a third consideration, the generation of reliable information (Box 7.7). But the first step is to make markets work better (i.e., smarter).

Redesigning property rights

One area that has received attention has been the redesign of current tenure arrangements (see Box 7.3). As identified earlier, these proposals in one form or another change the property rights associated with tenure. In some cases, the proposals are designed to more closely align ownership with the correct incentives. Luckert (1998) suggests using sharecropping arrangements to give companies the rights to a future interest in timber stands they plant to encourage more intensive silviculture. Several authors have suggested strengthening tenure rights by modifying them through various means: lengthening the term, requiring the government to provide compensation when rights are substantially modified, making rights more freely transferable, and removal of some of the operational controls placed upon rights (Pearse 1988; Haley and Luckert 1990). Some authors (Duinker 1994) have encouraged using existing institutional constraints over even-flow harvesting, such as the allowable cut effect[7], to encourage more intensive silviculture (as is case in B.C. with the Innovative Forestry Practices Agreement and similar agreements proposed in Alberta).

Pearse (1988) notes that property rights evolve in response to growing scarcity, and one government response has been to either create charges for previously free goods (such as the right to discharge pollution) or to create new property rights (such as exclusive areas for outfitters or rights to run trap lines). Typically, these new rights are not integrated into existing tenure arrangements but are instead granted separately, creating an overlapping series of rights that do not permit firms to internalize externalities (much as different timber companies may have the rights to different kinds of timber, as discussed earlier). Pearse (1988), among others, has noted that for some goods such as recreation, there may be a market for these goods, and the firm might provide the right level of outputs if it could respond to price signals much as they do in determining what mix of forest products to produce.

Another approach has been through diversifying ownership patterns in order to provide a mix of different perspectives and management objectives. As noted earlier, most forest land in Canada is provincially owned, and of the forest land that is commercially exploited, most of it is under long-term licenses to companies producing forestry products. Some proposals, as discussed above, have centred on diversifying tenure arrangements: providing different types of tenure (such as community forests) and broadening the representation of who holds it (communities, First Nations, and other groups; Beckley 1998).

---

[7]The allowable cut effect is the influence that any enhancement to forest growth and yield, usually through incremental silviculture, has on the overall timber supply and hence the amount that can be annually harvested.

More fundamental changes to current patterns might range from redefining either ownership or the role of the government. For example, public ownership could be retained, but rather than relying on long-term leases, the government would assume responsibility for managing forest land and supplying logs directly. Alternatively, the government could privatize some of its holdings, selling forest land to communities, corporations, and "forest-farming" individuals.

Binkley (1997) and others have suggested that overall timber production and environmental benefits can be increased through zoning, which permits for more intensive land use, through implicitly relaxing environmental constraints on a portion of the land base (relative to current management practices) while supplying those environmental goods on the remaining land base. Such a zoning system, while implicit in many historic land-use patterns, has served as the basis for much Sustainable Forest Management Network research in stand and landscape management (Chaps. 11–13).

One way in which current management practices and emphasis might change is through more complete recognition of Aboriginal rights. While there has been an acknowledgement of such rights across Canada, actual participation in decision making has been scant. Provincial governments sometimes refuse to recognize Aboriginal concerns in forest-level planning, arguing that they should be addressed at a higher level. However, the forest development plans are the ones that immediately impact Aboriginal communities, and there appears to be no formal process to address these concerns at the higher level other than consultative discussions that take place through the various land-use planning processes. This dynamic is most evident in B.C., where First Nations are frustrated in their attempts to participate in decision making over local resource issues associated with ongoing forest development while treaty discussions continue interminably.

To some degree there may be opportunities to utilize existing institutions through improving First Nations institutional capacities and increasing public participation. To the extent there are common values and ground that can be reached through better processes, such efforts will likely be fruitful. However, much of the current decision making takes place at the provincial level, and there may be substantial differences in values, depending on perspective — local, provincial, national, or international — on how a particular forest should be managed that cannot be addressed through procedural design. For example, Robson et al. (2000) show that there is a great deal of similarity regarding general principles of forest management between the residents of Prince George, B.C. (a forest-dependent community), and provincial and national attitudes. There are, however, significant differences in local attitude over actual forest practices (such as harvesting methods) compared to provincial and national attitudes.

Changing incentives

Another set of approaches centres around creating incentives. These can be positive in terms of monetary payments for attaining or satisfying certain performance criteria, and negative through financial penalties or sanctions. Indeed, internalization of the full value of environmental goods and services can be achieved if governments are ready to pay the full price of such services and goods. Certification can prevent market failures stemming from asymmetric information. While today premiums for certified goods are insignificant, certification is increasingly seen as necessary to sell into international markets, and the lack of certification may lead to the loss of markets.

Several authors have discussed mechanisms for financing (transfer payments, carbon trading, certification and "fair trade", marketable forest obligations) and incentive contracts (Richards 2000; Crossley 1997; Lippke and Fretwell 1997). In some cases, the financing is created through incorporating environmental goods, previously free goods, into the market system. Examples include paying for biodiversity (Sedjo and Simpson 1995) and carbon sequestration (Richards 2000; Profor 2000). These approaches raise many issues. Should these be incorporated into existing tenure rights? If new property rights are granted, who should receive them? How would such rights fit into the existing legal framework within Canada?

There is also the question of whether some of these proposals, designed to address problems like deforestation and biodiversity, are appropriate in a Canadian context.

From a Canadian perspective, it is not necessarily a lack of managerial capacity nor a lack of infrastructure (inadequate physical and social capital) that precludes sustainable management. Many SFM practices advocated within Canada will undoubtedly result in higher costs and (or) lower harvest volumes, and will therefore reduce economic margins. This needs to be recognized in the design of stumpage systems and has implications for industry structuring (e.g., fewer sawmills means monopsonistic markets for logs and consequently even lower prices). At the same time, recent trade disputes between the U.S. and Canada have focused on issues of cost and the allocation of timber. Any changes to stumpage payments, financial assistance to offset the cost of different harvesting or management practices might be avoided by government because they could lead to additional trade complaints. At the very least, the threat of such action and the possibility that managers' actions may be open to such scrutiny is likely to reduce flexibility and the ability to respond efficiently to changing circumstances and needs.

There are also limits to what markets can accomplish, and there is reluctance to employ them for the protection of environmental values (Box 7.8). For public goods such as biodiversity, even if we were able to design a rights-based system in which one could trade, we face the "free rider" problem, which suggests such goods will be underprovided. Creating taxes and subsidies linked to the output of particular goods to move towards the appropriate equilibrium also means we run the considerable risk of not choosing the appropriate level, since we do not know whether or not markets will necessarily respond in the manner we assume. This is especially true if we are concerned that there is some critical level or threshold over (or under) which damage will occur, and we cannot assume that the market will necessarily provide the optimal amount. Furthermore, many environmental concerns have to do with the localized nature of many resources (such as wildlife populations), and market mechanisms may not work as well here, since standard markets tend to allocate the supply of resources to the lowest cost source. We also know that the transaction costs of additional taxes, subsidies, or voluntary (i.e., local/regional) trade restrictions can be quite costly, depending on the need for measurement and enforcement. This is why standards are sometimes used for such things as tail pipe emissions from cars. While trading of emission permits is theoretically possible, the costs of establishing such a system would quite likely outweigh the benefits gained by moving away from the standards-based approach.

Because it is difficult to design tenure systems and prices that accurately target SFM we will continue to need to rely on a regulatory framework, and we need to think how we can make regulation work "smarter" as well. One approach examined earlier is the emphasis on results-based regulation, rather than procedure-based (we referred to the example in Alberta of specifying a particular desired outcome and letting companies innovate to find the least cost way to achieve that outcome). Somewhat paradoxically, however, the strength of the regulatory approach in designing an institutional framework will not be in specifying the desired outcome — but rather in establishing a dynamic, ongoing process in which those desired outcomes can be identified.

Certification is an alternative approach through *de facto* (though largely voluntary) regulation. Although it uses market mechanisms (improved access for sustainably produced wood, better prices if they should materialize, etc.) to encourage sustainable management, the most important part of the approach is the dynamic regulatory process. Most of the current certification systems are tied to a series of criteria and indicators, which are meant to generate information that then feeds back into the objectives and management practices. In this case, we rely on private markets to enforce and support sustainable forest management as it is defined and re-defined on an ongoing basis.

This "privatization" of regulation by a third party (other than the government or firm) will only accomplish SFM if it truly reflects the components of what are required to achieve SFM. Currently, there are several different certification systems competing in the marketplace, as we noted earlier. Some of them, such as ISO, rely more on self-

---

> ### Box 7.8. If market instruments are so good, why might they not be used?
>
> Market-based instruments, such as the use of emission permit trading systems, have demonstrated their effectiveness in their flexibility and ability to achieve desired outcomes at a substantially lower cost compared to the typical approach of government regulation. Why, then, might industry, government, and environmental groups prefer government regulation to achieve SFM? Some reasons include:
>
> *Uncertainty* — the "production function" for SFM is unknown owing to the large number of interdependencies so that "do's and don'ts" offer more certainty in terms of results;
>
> *Ideological* — for some environmental groups, market failures represent moral failure of profit-seeking firms, and they do not believe market instruments (such as taxes or revised rights) can rectify the problem; in addition, they believe it is not appropriate to grant the "right to harm the environment"; some may also prefer the regulatory approach if they believe it will make some logging uneconomic;
>
> *Rent-seeking* — from a government perspective, they may prefer regulation because it maintains/establishes government control, which may be beneficial in either providing for increased bureaucracy or a means of soliciting political support (through visible measures); industry may prefer it because it knows how the system works (whom to appeal to) and avoids the uncertainty created by introducing competitive forces that may alter the competitive balance among firms.
>
> For these reasons, and perhaps more, stakeholders may choose to seek government regulation even where there are more effective (and efficient) ways of achieving the desired outcome. Economic efficiency is not part of the political calculation.

regulation and improvement of internal processes rather than the achievement of certain outcomes. Other systems will be more oriented towards either industry or environmental groups reflecting the main supporters of each particular system.

Ultimately, however, the certification approach will only work to the extent that the ideas it embodies are supported by consumers. If there is a paradigm failure, or if the ideas embodied in the certification system are not shared by the public, such an approach will be ineffective. Alternatively, if consumers do not clearly understand the message embodied in certification systems and certified products, the approach will be less effective than it might otherwise be. The increasing number of certification systems now competing in the marketplace makes this a strong possibility. If consumers cannot distinguish between one system or another, or do not have some independent way to evaluate each on their own merits, then the certification approach will likely fall short of its goal of encouraging SFM.

In the long run, it is clear that an essential component of SFM will be a strong educational commitment to clearly articulate the requirements (and financial costs) of achieving SFM. Not only will it provide the political support to adopt necessary changes, it will also help ensure that the marketplace recognizes the environmental benefits associated with sustainably produced forest products. There are opportunities for institutional reform, evolution, and innovation on many fronts in meeting the SFM challenge (Box 7.9).

As a final note, it should be emphasized that for institutional change to be effective in bringing about SFM it must reflect a new commitment to the values underlying SFM by all key stakeholders. Without such a commitment the implementation of SFM will be in doubt, especially without commitment by the government and the public. Even with such commitment the costs of information, monitoring, and enforcement may be prohibitive, reducing the reach and effectiveness of the institutional system.

Box 7.9. Some options for improving the institutional support for sustainable forest management:

- Moving to results-based regulation; identifying and specifying outcomes and then creating incentives to ensure outcomes are met through increased penalties (see Box 7.6);
- Redesigning tenure so that firms have an incentive to invest in forestry through measures such as eliminating cut control requirements and operational stipulations that reduce the value of the wood produced and don't offer corresponding environmental or social benefits (see Box 7.3);
- Limited but bold experimentation — we should encourage diversity in experiments in how forests are managed and wood is produced (e.g., auctions, log yards, community forests, and different forms of tenure) and recognize that there is likely not just one appropriate model of forest management or policy applicable across Canada; and
- Greater reliance on scientific expertise and professional knowledge to monitor and enforce SFM practices; also matched by institutional arrangements to protect independence of such monitors and the use of sanctions (by the appropriate professional associations).

Acknowledgements

We would like to thank the SFMN and SSHRC for funding, and Phil Burton and Tony Lempriere for insightful suggestions.

References

Adamowicz, W.L., and Veeman, T.S. 1998. Forest policy and the environment: changing paradigms. Can. Pub. Pol. **24**(2): s51–s61.

Anonymous. 2001. The edge. Feb. 2001 issue. Mediawest, Sherwood Park, Alberta.

Baldwin, S. 2001. Sustainable or certified forestry? Timber Mart – South Market Newsletter, Vol. 6, No. 2, 2nd Quarter.

Bass, S., Balogun, P., Mayers, J., Dubois, O., Morrison, E., and Howard, W. 1998. Institutional change in public sector forestry: a review of the issues. IIED Forestry and Land Use Series No. 12. International Institute for Environment and Development, London, U.K. 12 p.

(BCMOF) British Columbia Ministry of Forests. 2002*a*. Consultation on a new framework for B.C. forest practices. British Columbia Ministry of Forests, Victoria, British Columbia. Available at http://www.resultsbasedcode.ca/ [cited 19 June 2002].

(BCMOF) British Columbia Ministry of Forests. 2002*b*. Results-based FPC pilot projects [online]. British Columbia Ministry of Forests, Victoria, British Columbia. Available at http://www.for.gov.bc.ca/hfp/rbpilot/ [cited 20 June 2002].

Beckley, T. 1998. Moving toward consensus-based forest management: a comparison of industrial, co-managed, community and small private forests in Canada. For. Chron. **74**: 736–744.

Beckley, T., and Reimer, W. 1999. Helping communities help themselves: industry–community relations for sustainable timber-dependent communities. For. Chron. **75**: 805–810.

Binkley, C. 1995. Designing an effective forest sector research strategy for Canada. For. Chron. **71**: 589–595.

Binkley, C. 1997. Preserving nature through intensive plantation management. For. Chron. **73**: 553–558.

Binkley, C., and Forgacs, O. 1997. Status of forest sector research and development in Canada. University of British Columbia, Faculty of Forestry, Vancouver, British Columbia. Available at http//forcast.forest.ca/pdf/bi [viewed 1 December 2001].

Byron, N., and Sayer, J. 1999. Organising forestry research to meet the challenges of the Information Age. Inter. For. Rev. **1**(**1**): 4–10.

Carrow, J.R. 1997. Canada's quest for forest sustainability: options, obstacles, and opportunities. For. Chron. **73**: 113–120.

(CCFM) Canadian Council of Forest Ministers. 2000. Criteria and indicators of sustainable forest management in Canada: national status 2000. Natural Resources Canada, Ottawa, Ontario. Available at http://www.ccfm.org/ci/2000_e.html.

Clague, C. 1997. Institutions and economic development: growth and governance in less-developed and post-socialist countries. John Hopkins Press, Baltimore, Maryland. 290 pp.

Coase, R. 1960. The problem of social cost. J. Law Econ. **3**: 1–44.

Crossley, R.A., Lent, T., Calllejon, P., and Sethare, C. 1997. Innovative financing for sustainable forestry. Unasylva, **188**: 23–31. Available at ftp://ftp.fao.org/fo/Unasylva/188e.pdf.

Cubbage, F.W., Regen, J.L., and Hodges, D.G. 1993. Climate change and the role of forest policy. *In* Emerging issues in forest policy. *Edited by* P. Nemetz. UBC Press, Vancouver, British Columbia. pp. 86–98.

Delbridge, P. 1988. The growth of community pressure. Can. For. Ind. (Dec.): 40–43.

Demsetz, H. 1967. Toward a theory of property rights. Amer. Econ. Rev. **57**: 347–359.

Duinker, P. 1994. What will it take? Policy needs of intensive silviculture. For. Chron. **70**: 134–136.

Feigenbaum, H., Samuels, R., and Weaver, R.K. 1993. Innovation, coordination, and implementation in energy policy. *In* Do institutions matter? Government capabilities in the United States and abroad. *Edited by* R.K. Weaver and B. Rockman. Brookings Institution, Washington, D.C. pp. 49–109.

Gawthrop, D. 1999. Vanishing halo: saving the boreal forest. Greystone Books, Vancouver, British Columbia. 225 p.

Globerman, S., Nakamura, M., Ruckman, K., Vertinsky, I., and Williamson, T. 1999. Technological progress and competitiveness in the Canadian forest products industry. Canadian Forest Service, Science Branch, Ottawa, Ontario.

Goodin, R. 1996. Institutions and their design. *In* The theory of institutional design. *Edited by* R. Goodin. Cambridge University Press, New York, New York. pp. 1–53.

Haley, D., and Luckert, M. 1990. Forest tenures in Canada: a framework for policy analysis. Forestry Canada. Ottawa, Ontario. Inf. Rep. E-X-43. 104 p.

Harrison, K. 1996. The regulator's dilemma: regulation of pulp mill effluents in the Canadian federal state. Canadian Journal of Political Science, **XXIX**(3): 469–496.

Higgins, C. 1999. Innovative forest practice agreements: what could be done that would be innovative. For. Chron. **75**: 939–942.

Hoberg, G., and Harrison, K. 1994. It's not easy being green: the politics of Canada's green plan. Can. Pub. Pol. **10**: 119–137.

Howlett, M. 2001. Canadian forest policy: adapting to change. University of Toronto Press, Toronto, Ontario. 446 p.

Howlett, M., and Rayner, J. 1998. Do ideas matter? Policy network configuration and resistance to policy change in the Canadian forest sector. Can. Pub. Admin. **38**: 382–410.

(IISD) International Institute for Sustainable Development. 2002. Summary of the second session of the United Nations forum on forests. International Institute for Sustainable Development, Winnipeg, Manitoba. Available at www.iisd.ca/linkages/forestry/unff/unff2 [viewed 1 June 2002].

Jaggi, M. 1997. Current developments in Aboriginal forestry: provincial forest policy and Aboriginal participation in forestry in Ontario, Canada. Int. J. Ecofor. **13**(1): 13–20.

Kaimowitz, D., Byron, N., and Sunderlin, W. 1998. Public policies to reduce inappropriate tropical deforestation. *In* Agriculture and the environment: perspectives on sustainable rural development. *Edited by* E. Lutz. World Bank, Washington, D.C. pp. 302–322.

Kiekens, J.P. 2000. Forest certification. Eng. Wood Prod. J. (Spring issue). Available at http://www.forestweb.com/APAweb/ewj/2000_spring/f_forestcertification.html.

Kim, Q.S., and Carlton, J. 2001. Timber industry goes to battle over rival seals for 'green' wood marketplace. The Wall Street Journal (23 May).

Kohm, K.A., and Franklin, J.F. (*Editors*). 1997. Creating a forestry for the 21st century: the science of ecosystem management. Island Press, Washington, D.C. p. 475.

Lackey, R. 1999. Radically contested assertions in ecosystem management. J. Sustain. For. **9**(1/2): 21–34.

Lippke, B., and Fretwell, H. 1997. The market incentive for biodiversity. J. For. **95**(1): 4–7.

Loasby, B.J. 1999. Knowledge, institutions and evolution in economics. Routledge, London, U.K.

Luckert, M. 1998. Efficiency implications of silvicultural expenditures from separating owner-ship and management on forest lands. For. Sci. **44**: 365–378.

Magee, S., Brock, W., and Young, L. 1989. Black hole tariffs and endogenous policy theory. Cambridge University Press, Cambridge, U.K. 438 p.

Markey, S., Vodden, K., and Ameyaw, S. 2000. Promoting community development for forest-based communities: final report. Community Economic Development Centre, Simon Fraser University, Burnaby, British Columbia. Available at http://www.sfu.ca/cedc/forestcomm/reports/cedcforestcommrpt31.pdf [viewed 13 May 2003]. 179 p.

McIntosh, R.A., Alexander, M.L., Bebb, D.C., Ridley-Thomas, C., Perrin, D., and Simons, T.A. 1997. Financial state of the forest industry and delivered wood cost drivers. Prepared for the Economics and Trade Branch, B.C. Ministry of Forests, Victoria, British Columbia by KPMG Perrin, Thorau & Associates Ltd. and Simons. Available at http://www.for.gov.bc.ca/het/costs/fin-toc.htm [viewed 13 May 2003].

McKenna, B. 2001. Environmental groups petition for surcharge. The Globe and Mail (10 May, Sect. B4). Toronto, Ontario.

McKinnon, J. 2002. Canadian Forest Association sets new forest management requirements. Dow Jones Newswire (28 January).

(MFP) Model Forest Program. 2002. About the Canadian Model Forest Network. Canadian Forest Service, Ottawa, Ontario. Available at http://www.modelforest.net/e/home_/aboute.html [viewed 28 September 2002].

M'Gonigle, M. 1998. Living communities in a living forest: towards an ecosystem-based struc-ture of local tenure and management. *In* The wealth of forests: markets, regulation, and sus-tainable forestry. *Edited by* C. Tollefson. UBC Press, Vancouver, British Columbia. pp. 152–185.

(NAFA) National Aboriginal Forestry Association. 2000. Aboriginal-forest sector partnerships: lessons for future collaboration. National Aboriginal Forestry Association and Institute of Governance. Ottawa, Ontario. Available at http://www.iog.ca/publications/forest_partner-ships.pdf .

Nordström, H., and Vaughan S. 1999. Special studies 4: trade and environment. Special Studies 4. World Trade Organization, Geneva, Switzerland. Available at http://www.wto.org/eng-lish/tratop_e/envir_e/stud99_e.htm [viewed 5 October 2002].

North, D. 1998. Where have we been and where are we going? *In* Economics, values, and organ-ization. *Edited by* Ben-Ner, Avner, and L. Putternam. Cambridge University Press, Cam-bridge, U.K. pp. 491–508.

(NRCan) Natural Resources Canada. 2000. Sustainable forest management: a continued com-mitment in Canada. Natural Resources Canada, Ottawa, Ontario.

OECD (Organization for Economic Co-operation and Development). 1992. Market and govern-ment failures in environmental management: wetlands and forests. Organization for Eco-nomic Co-operation and Development, Paris, France.

O'Toole, R. 1999. Reforming the Forest Service. Journal of Forestry (May): 34–36.

Panayotou, T., and Ashton, P.S. 1992. Not by timber alone: economics and ecology for sustaining tropical Forests. Island Press, Washington, D.C. 282 p.

Pearse, P. 1988. Property rights and the development of natural resource policies in Canada. Can. Pub. Pol. **14**(3): 307–320.

Pearse, P. 1990. Introduction to forestry economics. UBC Press, Vancouver, British Columbia. 226 p.

Pearse, P. 1993. Forest tenure, management incentives and the search for sustainable development policies. *In* Forestry and the environment: economic perspectives. *Editied by* W.L. Adamowicz, W. White, and W.E. Phillips. CAB International, Wallingford, U.K. pp. 77–96.

Pearse, P.H. 2001. Ready for change: crisis and opportunity in the coast forest industry. A report to the Minister of Forests on British Columbia's coastal forest industry. Vancouver, British Columbia. Available at http://www.for.gov.bc.ca/pab/news/phpreport/index.htm [viewed 9 October 2002]. 36 p.

Profor. 2000. Financing of sustainable forest management. Commission on Sustainable Development, Intergovernmental Forum on Forests. Fourth session, January 31 – February 11, 2000. Workshop report viewed on 26 April 2000 at www.un.org/eas/sustdev/ecn17iff2000-sprep.htm.

Repetto, R. 1988. The forest for the trees? Government policies and the misuse of forest resources. World Resources Institute, Washington, D.C. 105 p.

Richards, M. 2000. Can sustainable forestry be made profitable? The potential and limitations of innovative incentive mechanisms. World Dev. **28**: 1001–1016.

Robson, M., Hawley, A., and Robinson, D. 2000. Comparing the social values of forest-dependent, provincial and national publics for socially sustainable forest management. For. Chron. **76**: 615–622.

Rosen, H. 1985. Public finance. Irwin, Homewood, Illinois. 641 p.

Ross, M. 1995. Forest management in Canada. Canadian Institute of Resources Law, University of Calgary, Calgary, Alberta. 388 p.

Ruddell, S., Stevens, J., and Bourke, I.J. 1998. International market access issues for forest products. For. Prod. J. (Nov./Dec.): 20–26.

Scott, A. 1983. Property rights and property wrongs. Can. J. Econ. **16**: 555–573.

Sedjo, R., and Simpson, R.D. 1995. Property rights, externalities, and biodiversity. *In* The economics and ecology of biodiversity decline. *Edited by* T. Swanson. Cambridge University Press, Cambridge, U.K. pp. 79–88.

(SSCBF) Standing Senate Committee on Agriculture and Forestry. 1999. Competing realities: the boreal forest at risk. Report of the Sub-Committee on Boreal Forest of the Standing Senate Committee on Agriculture and Forestry. Parliament of Canada, Ottawa, Ontario. Available at http://www.parl.gc.ca/36/1/parlbus/commbus/senate/com-e/BORE-E/rep-e/rep09jun99-e.htm [viewed 25 May 2002].

Stanbury, W., and Vertinsky, I. 1998. Governing instruments for forest policy in British Columbia: positive and normative analysis. *In* The wealth of forests. *Edited by* C. Tollefson. UBC Press, Vancouver, British Columbia. pp. 42–77.

United Nations. 2001. Forest policies and institutions in Europe 1998–2000. Geneva Timber and Forest Study Papers ECE/TIM/SP/19. Geneva, Switzerland. 21 p.

Vertinsky, I., and Zhou, D. 2000. Product and process certification: systems, regulations and international marketing strategies. Inter. Mark. Rev. **17**: 231–252.

Vogel, D. 1993. Representing diffuse interests in environmental policy making. *In* Do institutions matter? Government capabilities in the United States and abroad. *Edited by* R. Weaver and B. Rockman. Brookings Institution, Washington, D.C. pp. 237–271.

(WCED) World Commission on Environment and Development. 1987. Our common future. Oxford University Press, Oxford, U.K. 400 p.

Weaver, R.K., and Rockman, B. 1993. When and how do institutions matter? *In* Do institutions matter? Government capabilities in the United States and abroad. *Edited by* R.K. Weaver and B. Rockman. Brookings Institution, Washington, D.C. pp. 445–461.

Williamson, O. 1985. The economic institutions of capitalism. Free Press, New York, N.Y. 450 p.

Wilson, W., Takahashi, T., and Vertinsky, I. 2001. The Canadian commercial forestry perspective on certification: national survey results. For. Chron. **77**: 309–313.

Wilson, W., van Kooten, G.C., Vertinsky, I., and Arthur, L. 1999. Forest policy: international case studies. CABI Publishing, Cambridge, U.K. 273 p.

(WRI) World Resources Institute. 2000. Canada's forests at a crossroads: an assessment in the year 2000. Available at www.globalforestwatch.org [viewed 1 December 2001].

Young, J., and Duinker, P. 1998. Canada's national forest strategies: a comparative analysis. For. Chron. **74**: 683–693.

Chapter 8

A process approach to understanding disturbance and forest dynamics for sustainable forestry

E.A. Johnson, H. Morin, K. Miyanishi, R. Gagnon, and D.F. Greene

The ideas of community structure and the expression of dominance, that of biological succession, and finally that of climax, are based largely upon the assumption of long-term stability in the physical habitat. Remove this assumption and the entire theoretical structure becomes a shambles.

. . . Disturbances have been so frequent and so generally effective that the expected "climaxes," or "equilibria," recede into pure speculation. Natural successions either do not occur at all or are limited to such incomplete fragments as can be accomplished between upheavals.

Hugh Raup (1956, p. 45)

Introduction

There has been a change in boreal forest management and silviculture from a focus on wood products to include other products and services of the forest (see Chap. 2). This has required a rethinking of forest management from only sustained-yield and regulated forests to the more inclusive ecosystem-based management that considers management for nontimber values as well (Slocombe 1993; Kaufmann et al. 1994). One of the several goals in ecosystem management has been to incorporate the understanding of

E.A. Johnson. Department of Biological Sciences, University of Calgary, Calgary, Alberta, Canada.

H. Morin and R. Gagnon. Départemente des Sciences fondamentales, Université du Québec à Chicoutimi, Chicoutimi, Quebec, Canada.

K. Miyanishi.[1] Department of Geography, University of Guelph, Guelph, Ontario N1G 2W1, Canada.

D.F. Greene. Department of Geography, Concordia University, Montreal, Quebec, Canada.

[1]Author for correspondence. Telephone: 519-824-4120, ext. 6720. email: kmiyanis@uoguelph.ca

Correct citation: Johnson, E.A., Morin, H., Miyanishi, K., Gagnon, R., and Greene, D.F. 2003. A process approach to understanding disturbance and forest dynamics for sustainable forestry. Chapter 8. *In* Towards Sustainable Management of the Boreal Forest. *Edited by* P.J. Burton, C. Messier, D.W. Smith, and W.L. Adamowicz. NRC Research Press, Ottawa, Ontario, Canada. pp. 261–306.

natural disturbance that has been gained in the last 30 years (Pickett and White 1985; Johnson 1992; Coutts and Grace 1995).

If there is one message from disturbance studies, it is that the dynamics of ecosystems cannot be understood without considering natural disturbances at different scales of time and space. As introduced in Chap. 2 (see Box 2.8) and discussed further in Chap. 9, one widely proposed approach to ecosystem management is to attempt to have management practices emulate natural disturbances, thereby retaining the range of natural variation in ecosystems in order to maintain resilience and sustainability (Hunter 1993; Swanson et al. 1994; Bunnell 1995; DeLong and Tanner 1996; Holling and Meffe 1996; Cissel et al. 1999; Landres et al. 1999). The objective in this approach is to determine the natural variation of the disturbance regime (Johnson and Van Wagner 1985) and then to maintain this variation through management practices. Swanson et al. (1994) suggested an example of a three-variable regime (Fig. 8.1) in which disturbance frequency is emulated by forest rotation age, disturbance size by cut-block size, and disturbance intensity by utilization standards. Certainly other variables can be imagined, but these descriptors are illustrative of the "regime approach" to describing regional disturbance ecology.

One can think of many potential problems with emulating the range of natural variation in disturbance (Parsons et al. 1999); we will discuss two that will help in understanding the process approach to disturbance ecology used in this chapter. The first and perhaps most obvious problem is that natural disturbances (e.g., wildfires) are seen, at least implicitly, as eventually being replaced by the emulated disturbance of management (Chap. 9). Usually little or no consideration is given of the fact that complete exclusion of natural disturbance may be impossible and that the landscape pattern will always be a product of both natural disturbance and forest management practices. Despite the technological advances made in firefighting, the boreal forest has experi-

Fig. 8.1. Three-variable (frequency, size, and intensity) disturbance regime proposed by Swanson et al. (1994), showing the range of variation produced by natural disturbances and by forestry practices.

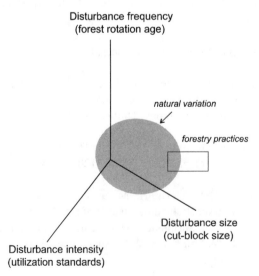

enced extremely large fires in the past two decades. Thus, despite the optimistic approach of managing biophysical risk through "early detection and rapid suppression of wildfires" (Chap. 2), there is little evidence that such activities have a significant impact on the landscape pattern (Bridge 2001).

A second and more fundamental problem is the descriptive nature of the goal in this approach, typically to replicate the variation in *pattern* of the disturbance (e.g., Bergeron and Harvey 1997; Cissel et al. 1999) or the remnant *structures* left by the disturbance (Hansen et al. 1991). Such templates for management do not consider how the disturbances work (i.e., the disturbance processes involved) and in any precise way what their effect is on ecosystem components and processes. However, without an understanding of the natural disturbance processes (e.g., smoldering combustion in forest fires), and the ecosystem processes (e.g., tree population dynamics) as well as how the disturbance causes the ecosystem effects, we do not know what disturbance variables to emulate and if those chosen will produce the ecosystem effects desired. This has been a chronic problem in prescribed burns that often use simplistic descriptions (e.g., hot or cool fires) of the natural fire regime for guidance (Johnson and Miyanishi 1996).

The problem in choice of variables when no processes are understood can be seen in the Swanson et al. (1994) model (Fig. 8.1) and in later modifications of this model (Chap. 9, Fig. 9.1). Disturbance frequency and size describe characteristics of the disturbance but not of specific ecosystem effects. Intensity has been used to characterize either the disturbance process or an ecosystem effect and has been used almost interchangeably with severity. Furthermore, although severity is generally defined in terms of ecosystem effects or responses (e.g., tree mortality, duff consumption, soil alteration, etc.), it is often given as a characteristic of the disturbance itself (e.g., "severe fire behaviour" in Chap. 9). One must start by indicating what ecosystem process one wishes to affect and work back to the specifics of the disturbance processes. The regime approach, as it is presently formulated, cannot lead to this. A more detailed discussion of the process–response approach can be found in Johnson (1985) and Johnson and Miyanishi (2001).

Underlying the approach of emulating natural disturbance may be the belief that, by mimicking the pattern, one can escape the necessity of understanding the disturbance processes and ecosystem responses involved (the coarse filter approach). This explains the use of largely descriptive indexes of patterns (Turner et al. 1989; Ripple et al. 1991; McGarigal and Marks 1995; Gustafson 1998). Unfortunately, different parameters of species populations respond differently to different processes in disturbances, and may or may not show any response to landscape pattern. The different scale-dependence of species (Chap. 9) is in part related to this different process–response relationship. A strictly empirical and clinical approach (rather than the process–response approach) to research in disturbance ecology does not provide any scientific basis for selection of the appropriate disturbance variables to be emulated or the ecosystem responses to be subsequently monitored. Accepting the belief of not needing to understand the disturbance processes and ecosystem responses reduces natural resource management to determining what natural disturbance variation is to be measured. Thus the emulation idea is being implemented in management policy (e.g., in British Columbia, Alberta, Ontario, etc.) *before* the hypothesis has been tested or a well-defined model has been developed. Even in the context of adaptive management (Chap. 21), there is a requirement for

---

Box 8.1. Challenges to sustainable forest management:

- To understand disturbance processes, not just emulate disturbance patterns;
- To sustain important ecosystem processes, including tree population dynamics;
- To characterize regional differences in disturbance regimes in a meaningful way;
- To recognize the constraints to forest management imposed by regional disturbance processes; and
- To consider the forest management opportunities offered by regional disturbance processes.

---

forecasting the results of management practices through the use of quantitative models and evaluating the forecasts through monitoring. Process-based models will lead to better predictions than strictly empirical models that may, at times, be based on spurious correlations between arbitrarily selected variables.

Perhaps part of the reason for this reversal of the hypothesis testing-application sequence is the emotive nature of emulating, mimicking, or faking the variation of natural disturbance. The unquestioning acceptance of the hypothesis and its implementation in policy made it seem unnecessary for researchers to question the underlying assumptions and to conduct research that would lead to a better understanding of the processes involved in natural disturbances as well as the connections between these processes and the responses of the various ecosystem components. Thus, from the beginning, the goals of the disturbance group in the Sustainable Forest Management Network included testing the hypothesis of emulating natural disturbance and building a better process–response understanding for the boreal forest ecosystem.

In this chapter, we will consider two important natural disturbances on tree populations in the North American boreal forest: wildfire; and outbreaks of the eastern spruce budworm (*Choristoneura fumiferana*). The discussion will use tree population dynamics as the ecosystem process of interest and explain disturbance processes as they affect recruitment and mortality. Foresters should be interested in understanding tree population dynamics (Box 8.1), even if they presently have solutions (e.g., using mechanical site preparation to address problems of regeneration). Such understanding would lead to more reasoned solutions to future problems that may arise (and less need for a trial-and-error approach to finding such solutions). While we have not used data or examples from outside North America, the approach taken here to study disturbance ecology is generalizable.

Forest tree populations and disturbance

A forested landscape consists of a hierarchy, starting at the bottom with individual trees, these trees being grouped together into local populations, and the local populations into regional populations. Most local tree populations have cohort structure; i.e., they are divided into groups of individuals born at more or less the same time and having similar mortality histories. Note that cohorts are not subpopulations because they are defined only by their birth and mortality patterns. Their offspring cannot be recruited into their cohort; instead, the offspring form new cohorts.

Within the regional population, the local populations or their cohorts act as individuals in the sense that whole local populations or certain cohorts within local populations can be exterminated. In this sense, the regional population has a regional mortality rate. An example of local population mortality is the occurrence of a large high-intensity wildfire that may destroy a number of local populations by killing all trees within these populations. Examples of cohort mortality in regional populations include:

- low-intensity surface fires that may kill only the younger age classes (cohorts) in one or more local populations, leaving the older age classes (cohorts) intact; and
- insect outbreaks such as spruce budworm that may kill only the older cohorts, leaving the younger cohorts unaffected.

Local populations and their cohorts also act as individuals within regional populations in the sense that they are recruited into the regional population. For example, following a high-intensity wildfire that destroys one or more local populations, new local populations will arise to replace those destroyed. These new local populations (i.e., recruits) arise from seeds dispersed from adjacent surviving local populations, except when the species has serotinous cones (e.g., jack pine, *Pinus banksiana*) or can resprout (e.g., trembling aspen, *Populus tremuloides*). In these latter cases, the new local populations will primarily arise from seeds or sprouts produced *in situ* by the previous local population. In the case of cohort (rather than local population) mortality due to disturbances such as low-intensity surface fires or spruce budworm outbreaks, the recruitment of new cohorts (even of nonserotinous and nonsprouting species) will be primarily from offspring of the surviving cohorts; thus, dispersal from adjacent local populations becomes less important. As a result, long-distance dispersal plays a significant role only when the whole local population (not just a cohort within the local population) is killed and when there are no *in situ* sources of propagules (seeds or sprouts). This means that, because of its serotinous cones, jack pine doesn't generally have a regional population structure, since after a fire it is not dependent on other local populations for its recruitment. The exception to this is when the local population is destroyed before its oldest cohort has reached reproductive maturity. On the other hand, white spruce (*Picea glauca*) does have a regional population structure, since, after being destroyed by a fire, a local population is dependent on dispersal from surviving adjacent populations for its recruitment.

Local populations can be subdivided into cohorts of individuals that recruit at similar times and have similar mortality schedules. Cohorts are recognized because different age classes (cohorts) of the population recruit in different numbers at different times and can find themselves in different mortality environments (e.g., the tree cohort starting soon after a fire lives its life in a different environment than subsequent cohorts which start in the shade of an established canopy).

Recruitment into a local population will come from individuals within that local population and also from individuals in adjacent local populations. If the local population is completely extirpated by a disturbance, then seeds will have to disperse from surviving adjacent populations, unless, of course, the species has serotinous cones or can resprout. Recruitment varies significantly between cohorts as a result of temporal variability in the availability of seeds and suitable seedbeds.

Mortality in the local population also varies between cohorts. For example, the cohort which recruits shortly after extermination of the whole local population (e.g., by a crown fire) will have a better survivorship in most cases than subsequent cohorts recruiting in the shade of the initial post-disturbance cohort (Johnson and Fryer 1989). If only some individuals or certain cohorts within a local population are killed by disturbance (e.g., by insects), the survivorship of other individuals or cohorts will be improved. For example, if the cohort comprising the canopy trees are killed, the suppressed seedlings and saplings in the understory (called the seedling bank or advanced regeneration) are released, improving both their growth and survivorship.

A stable population requires that each cohort have the same recruitment and mortality; i.e., all cohorts are really the same as every other. From the above discussion, it is obvious that local populations of most boreal forest trees are *not* stable, since the different cohorts within any local population have different recruitment and mortality rates. Disturbances may cause mortality of certain cohorts within a local population (e.g., surface fire or insect outbreaks) or mortality of the whole local population (e.g., crown fire). The dynamics of these local populations subject to such disturbances can only be understood by seeing the forested landscape as being made up of many local populations that are connected to each other through seed and pollen dispersal. Recent studies of such metapopulations (Pulliam 1988; Kareiva 1990; Hanski and Gilpin 1991) and conservation biology (Hunter 1996) have shown that examining only stand-level processes (stand development or forest succession) does not provide an adequate understanding of forest dynamics.

Wildfires and tree population dynamics

Wildfires consist of a large number of processes (see Johnson and Miyanishi 2001) of which only those tied to population processes of birth, mortality, dispersal, and regional population mortality are discussed here.

Regional population mortality

Within local populations, mortality is described by death of a few individuals at a time; on the other hand, regional population mortality involves death of large numbers of individuals (sometimes the whole local population) and also large numbers in adjacent populations. This section will explain the role of wildfires in determining regional population mortality of forested landscapes by causing catastrophic mortality of local tree populations over large areas. First, we will explain why wildfires cause catastrophic mortality in the boreal forest, and second, we will address how many local populations would be affected by an individual wildfire.

In the North American boreal forest, crown fires account for most of the area burned (Johnson 1992; Fig. 8.2). Crown fires occur when the surface fire intensity (heat output) in kW m^{-1} produces enough heat to ignite the tree crowns. In order for the fire to stay in the crowns two criteria must be met: (1) the surface fire intensity must be above a threshold value, and (2) the crown fire intensity and rate of spread must be sufficient to transfer enough heat forward to maintain flaming combustion within the forest canopy (Van Wagner 1977). If these criteria are not met, the crown fire may be sporadic, burning individual trees or groups of trees while burning only on the surface in other areas.

Fig. 8.2. Active crown fire in the Northwest Territories, Canada. Note that the flaming front is a wall of fire from the surface into the crown. Also evident in this photograph is the atmospheric convection generated by the heat, resulting in turbulent coherent structure within the fire.

Crown fires are predominant in the boreal forest for two general reasons: (1) the occurrence of certain mid-tropospheric weather conditions that cause both extreme fuel drying and convective storms that produce lightning, and (2) a fuel structure and arrangement (a continuous canopy of predominantly needle-leaves arranged on numerous small branches) that promote flame laddering and active crown-to-crown spread.

The mid-latitude atmosphere is dominated by the westerly flow of air. Occasionally this flow is deflected by high-pressure systems which persist for more than 8 days. These mid-tropospheric patterns are often called blocking systems because they block the normal alternation of west–east movement of high- and low-pressure cells over a region, diverting the flow north or south. Because the persistent highs result in clear weather and high temperatures in spring and summer, they are associated with rapid fuel drying. These blocking highs are the result of incompletely understood atmospheric processes, but some have very characteristic patterns. Two of these patterns, which are known to be associated with years in which large areas of forest are burned, are the Pacific North America (PNA) pattern (Johnson and Wowchuk 1993; Flannigan and Wotton 2001) and the Hudson Bay High (Schroeder 1950). The PNA pattern consists of a persistent high over western Canada and the U.S.A., a low in the Gulf of Alaska and another low over Ontario and adjacent U.S.A. The Hudson Bay High occurs over Manitoba, Ontario, and adjacent U.S.A. These highs often persist for several weeks and, at times, have convective thunderstorms that produce the lightning for wildfire ignition. While past occurrences of atmospheric patterns such as the El Niño – Southern Oscillation (ENSO) have been well studied (e.g., Horel and Wallace 1981; Rasmussen and Wallace 1983) and future occurrences can be reasonably predicted, at the present time it is not possible to determine past occurrences of the PNA pattern in the same way. If this were possible, we might be able to determine not only a pattern in large-fire years in the boreal forest but also tie them more closely to the characteristic weather patterns which cause them. The ability to forecast these patterns could lead to better predictions of large wildfire burn years in the boreal forest (Johnson and Wowchuk 1993; Skinner et al. 1999; Flannigan et al. 2000). Some progress has been made in using the surface water temperature anomalies in the North Pacific Ocean to predict the PNA pattern (Walsh and Richman 1981).

In order to judge the *number* of local tree populations affected by an individual fire we require an understanding of (1) the size distribution of fires, and (2) the distribution of burned local populations.

Size distributions of fires can be presented in two ways, one giving the probability distribution of fire sizes on the landscape and the other giving the proportion of total area burned by each fire size. The probability density distribution (frequency histogram) is usually heavily skewed (Fig. 8.3). The data are always truncated at smaller areas because of the difficulty and economics of finding small fires, particularly in regions with little or no fire suppression. However, it does seem that the distribution increases to a peak in the lower size classes and then decreases with a very long tail if the record is carefully constructed. Surprisingly few attempts have been made to fit a function to the fire size distributions. Reed and McKelvey (2002) have shown for a diversity of regions that the size distribution of fires does not fit the power law relationships proposed by Malamud et al. (1998), Ricotta et al. (1999), and Cumming (2001); instead, they obtained the best fit with a competing hazard model of extinguishment. The size distribution of fires raises the question of the peak (Fig. 8.3); i.e., why do a large number of fires only grow to a certain size before going out while fewer fires remain smaller and few fires get very large? This must tell us something about ignition, extinction, and spread of fires, but exactly what is not yet clear. It may be related to the fact that the distribution of persistence of pressure systems has a peak at about 5 days (see fig. 2.10 in Johnson 1992).

One must be careful not to interpret this peak in the distribution of fire sizes as an indication of which fires have the most impact on the landscape. This latter issue is better shown by a Lorenz graph (Fig. 8.4). These graphs show what proportion of the total area is burned by fires in different size classes. When presented as a cumulative distribution, it shows how the distributions deviate from an equitable distribution of the total area around fires of different sizes (i.e., an equitable distribution would result in a straight diagonal line). In fact, the Lorenz graphs of fire distributions are all right-

Fig. 8.3. Density distributions of fire sizes for (*a*) northeastern Alberta and (*b*) Northwest Territories (Reed and McKelvey 2002).

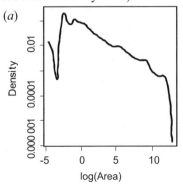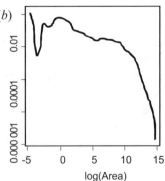

Fig. 8.4. Lorenz graphs for the fire size distributions from (*a*) northern Alberta and (*b*) north-western Ontario.

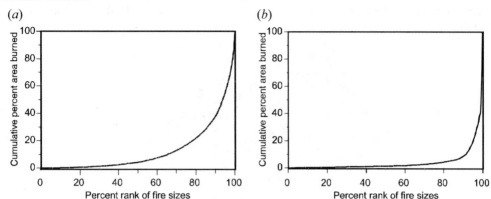

skewed, showing that large fires (although few in number) account for the majority of the total area burned.

The size distribution of fires gives the area covered by individual fires but not the number of local tree populations affected by fires. The ability to estimate the actual number of specific local populations (e.g., populations of white spruce) affected depends on an understanding of the structure of the landscape on which the species is distributed. All landscapes are made up of drainage channels and ridgelines between which hillslopes are hung. Hillslopes form the fundamental upland hydrological gradients and stream courses form the lowland hydrological gradients. Because the downward flow of water from the ridgelines to the hillslope bottoms also results in the downward movement of nutrients, hillslopes create the moisture and nutrient gradients that explain the important compositional gradients in vegetation (Bridge and Johnson 2000). Figure 8.5 gives an example of two such hillslope gradients in the mixedwood boreal forest on glaciofluvial and glacial till substrate. Note that both the steepness of the hillslope and the moisture and nutrient values depend in part on the surficial substrate.

Consequently, if there is a general pattern in hillslopes and stream courses, one should be able to predict some generalized distributions of species on the landscape. One method for doing this is to use digital elevation model (DEM) and surficial geology data in a hydrologic model to calculate a wetness index. The results of such studies (e.g., Sivapalan et al. 1987) suggest that most landscapes have distributions of wetness values similar to those in Fig. 8.6. Relatively dry parts of the landscape (ridge lines) are not very abundant and neither are very wet areas along stream courses and hillslope bottoms. The most abundant wetness values are those occurring just below the ridge line. This follows from the allometric relationship (Hack 1957) of basin length equals basin area$^{0.6}$. This relationship indicates that basins tend to get longer and narrower as they get larger and that the average distance from ridgeline to channel tends to remain relatively constant or to increase very slowly. Thus species populations which tolerate (and are more competitive at) the wetness values that are most abundant on the landscape will be more widespread. Remember that wetness values depend on surficial material, slope angle, catchment area, and the regional climate.

Fig. 8.5. Landscape patterns of canopy tree species composition in the mixedwood boreal forest in central Saskatchewan, based on the relationship between the stand positions on the moisture and nutrient gradients and the stand distances from the ridgeline (Bridge and Johnson 2000).

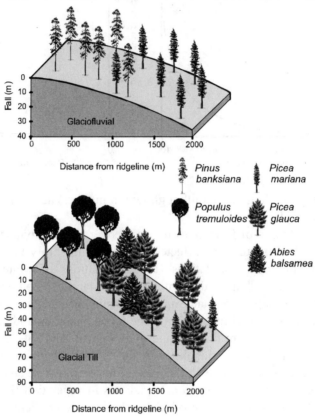

Fig. 8.6. Distribution of wetness values (topographic index) for a study site in central Saskatchewan (Tchir 2000).

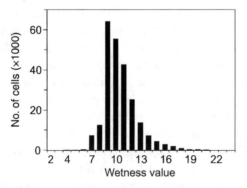

With the exception of stream courses and wetlands, fires burn landscapes with no apparent consistent reference to hillslopes. Although one can see specific parts of hillslopes not burning in a fire, rarely do we see a *statistically* consistent pattern. This tends to go against what most of us intuitively think. However, the conditions under which large fires occur in the boreal forest are more conducive to remnant unburned patches being explained by turbulent coherent structure in the fire (Jenkins et al. 2001) rather than by site factors. Thus it may be that the areas burned by fires above a certain size simply represent a sampling of the landscape wetness distribution (and hence the forest composition) as described above.

Now that we have some understanding of the number of local populations affected by individual fires, we can further ask what the catastrophic mortality is of these local populations. Interestingly, we know this catastrophic mortality because this is what fire frequency studies give (Johnson and Gutsell 1994). The time-since-last-fire distribution is, in fact, the survivorship of the regional population subject to crown fires. In the boreal forest, time-since-fire distributions are mixed negative exponentials (e.g., Fig. 8.7). The distributions are mixed because the fire frequency has changed more than once in the past 300 years. These changes were often at similar times over large areas (Johnson 1992), suggesting a large-scale climate mechanism. For example, changes in the mid-1700s and late 1800s are often associated with the Little Ice Age (Johnson and Larsen 1991; Bergeron and Archambault 1993). Between the changes in frequency, the distributions appear to be negative exponentials. This distribution suggests that there is no aging effect; i.e., as a forest gets older, there is no change in the instantaneous mortality (hazard) rate. Thus, old forests are not more subject to fire than younger ones, as has sometimes been suggested (Tymstra et al. 1998) although with little or no evidence presented.

There is some evidence that changes in fire frequency change the spatial mosaic of time since last fire. As currently understood in the mixedwood boreal forest, short fire cycles lead to younger, larger oblong-shaped polygons with irregular edges, while longer fire cycles lead to older, smaller, more circular polygons (Weir et al. 2000). This is because a short fire cycle results in most of the remnants of past burns (that are small, old, compact and circular polygons) being overburned; thus the polygons reflect the large oblong irregular shapes of these recent burns.

Fig. 8.7. Time-since-fire mixed negative exponential distribution for forest stands in Prince Albert National Park (Weir et al. 2000).

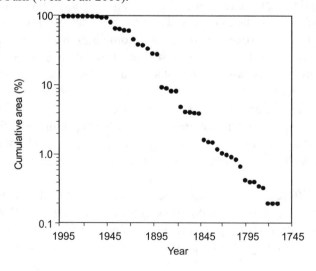

Post-fire recruitment

In the boreal forest where stand-replacing crown fire is the principal disturbance, most tree regeneration generally occurs within the first six years after a fire (St-Pierre et al. 1992; Charron and Greene 2002; Greene et al. 1999; Gutsell and Johnson 2002). This section will address the two factors determining tree recruitment in burned areas: availability of seed (source strength and distance) and availability of appropriate seedbeds for germination and seedling establishment. The spatial pattern of individual fires is of interest because dispersal and colonization are, in part, dependent on distance from surviving vegetation. Distance is particularly important for nonserotinous and nonsprouting species such as white spruce, balsam fir (*Abies balsamea*), and tamarack (*Larix laricina*) that require wind dispersal of their seeds from surviving individuals for recolonization of burned areas. Paper birch (*Betula papyrifera*), although capable of basal sprouting, cannot adequately colonize areas by vegetative means like trembling aspen or balsam poplar (*Populus balsamifera*) which can reproduce by root suckering.

Ignoring for the moment the surviving remnants within burns, the relationship between the perimeter of burns and burn area reveals that fires tend to become narrower as they get larger. This somewhat oblong shape results from wind-driven fires spreading faster in the direction of the wind than perpendicular to the wind. Therefore, the distance from the edge to any point within the burn will not increase greatly as the fire size increases. Interestingly, this allometric relationship is similar to that found for stream basins as discussed in the previous section. McAlpine and Wotton (1993) have also shown that the convolution of the burn perimeter is similar across several magnitudes of size. Thus, the mean dispersal distance into the burn for plants near the burn margin increases slowly with burn size (Fig. 8.8), primarily as a function of the relationship between burn area and burn length.

Besides seed sources from *outside* the burn, there are also sources *within* burns due to the areas of forest that escape burning: the residual stands. As mentioned in the previous section, these unburned patches within burns may be explained by turbulent coherent structure in the fire (Jenkins et al. 2001). Unlike what was originally believed, atmospheric physicists and fluid mechanicists have shown that turbulence is not without organization and that this organization forms "coherent structures" which are sometimes called hairpin or horseshoe vortices. The pattern of air movement associated with the development of these vortices seems to have something to do with survival of forest patches within fires but the process is still not clearly understood. The unburned residuals show similarities between different size burns. The amount of surviving forest per unit area of burn remains constant with increasing fire size, and the number of surviving forest polygons per unit area of burn decreases with burn area (Fig. 8.9). These two results suggest that, as the burns increase in size, the surviving polygons get bigger. The frequency distribution of sizes of surviving forest polygons in a burn fits an extreme value distribution with larger burns having larger modal remnant sizes than smaller burns.

Thus, despite differences in fire size, the maximum dispersal distance either from the burn perimeter or from surviving polygons typically is not greater than about 150 m (Greene and Johnson 2000) because of the oblong shape of fires, the convolution of the fire margins, and the ubiquity of residual stands in large fires. Not surprisingly, the dis-

Fig. 8.8. Mean distance to nearest edge as a function of nominal burn area, i.e., the total area inside the perimeter of the burn including any unburned remnants within the burn.

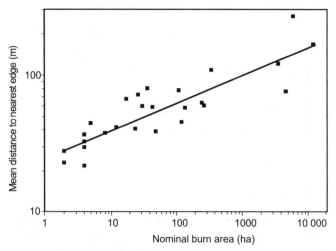

tribution of distances of any point in a burn to the *nearest* source of surviving trees fits an extreme value distribution.

The seemingly random dispersal of tree seeds and establishment of tree seedlings are, in fact, quite tractable. Most of these principles apply to natural regeneration in recently harvested cutblocks as well as in recent burns (Greene et al. 2002a). Seed dispersal of boreal trees is by wind; consequently, dispersal curves can be developed from a micrometeorological model of the flight of the seeds. Models based on this mechanism have been developed by Greene and Johnson (1989; 1992a, b; 1995; 1996) using heights at which seeds are released, terminal velocity (rate at which seeds fall in still air), and wind velocity (see the references listed above for details on specific values used in the discussion below). Below we give four examples of models. The first two are spatially explicit: a point source (single tree) model and an area source (collection of conspecific trees) model. The final two models are merely stand-scale mean recruitment densities for aerial seedbank species (e.g., jack pine and black spruce, *Picea mariana*) and for asexual recruitment of suckers (e.g., trembling aspen, balsam poplar).

The point source model results in a log normal distribution of recruited stems (F_{Dx}) as a function of distance (x) from an individual tree:

[1] $$F_{Dx} = (F/((2\pi)^{1.5}\sigma_x x^2)) \exp[(-0.5/\sigma_x^2)(\ln(x/x_{0.5}))^2]$$

The cumulative source strength F (seedlings produced per tree per T years) is given as $F = \overline{Q}S_J T$ where $\overline{Q}$ is the mean annual filled seed production, S_J is the juvenile survivorship of a cohort, and T is the number of years in which colonization can occur before the seed bed becomes so poor that colonization effectively ceases, here assumed to be $T = 4$. (One notes that masting is such a strong source of temporal variation that even 4-year sums will not render it unimportant.) According to Greene and Johnson (1994), $\overline{Q}$ is a function of seed mass (m in g) and basal area (B in m^2) of the seed-producing tree,

Fig. 8.9. (*a*) Amount of surviving forest per unit area of burn, and (*b*) number of surviving forest polygons per unit area of burn.

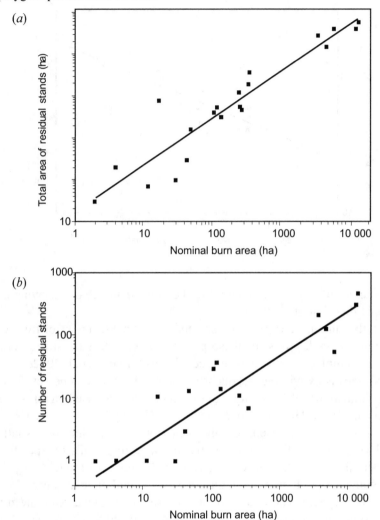

[2] $\bar{Q} = 3067m^{-0.58}B^{0.92}$

and $x_{0.5}$ is the median distance travelled by seeds of mass m:

[3] $x_{0.5} = z_a u_{0.5} / \bar{f}$

where f is the mean terminal velocity (m s^{-1}), $z_a = 0.75\, z_h$ (z_h is forest height and z_a is the abscission height), and $u_{0.5}$ is the median wind speed (around 2.5 m s^{-1} in recent burns and 0.7 m s^{-1} in intact forests; Greene and Johnson 1996).

The standard deviation of the logarithms of the horizontal distances travelled (σ_x) is a function of terminal velocity, and the vertical and horizontal turbulence. These values are typically around $\sigma_x = 1.0$ (Greene and Johnson 2000). From empirical evidence (Greene and Johnson 1998), juvenile survivorship (S_J) is proportionate to germinant size and seedbed type. Following the empirical literature, Greene and Johnson (1998) dichotomized upland seedbed types into "optimal" (mineral soil, thin humus, rotted logs) and "poor" (all others). The areally averaged survivorship will, of course, depend on the percentage allocated to these two seedbed categories (see below). If the optimal seedbeds comprise about 17% coverage (this value is typical of site-prepared clearcuts) then:

[4] $S_J = 0.28m^{0.52}$

for each cohort.

By rewriting equation [1] with the values given above, we get:

[5] $F = 3312B^{0.92}$

(The exponent on m becomes virtually zero so seed mass can be ignored.) Thus, essentially, source strength is a function of tree size. See Greene et al. (2002b) for a more thorough treatment of the relationship between light availability, tree size, and seed crop size.

Area sources are collections of contiguous point sources, and these are, familiarly, forests abutting clearcuts or burns, or perhaps trees adhering to the contours of hillslopes. The area source recruitment curve is simply the sum of individual recruitment curves of equation [1] assuming trees are more or less uniformly distributed within the area source. Further, where $x = 0$ is regarded as the edge of the area source. The density of recruits (F_{Dx} = number per m^2) is related to distance x from the edge as:

[6] $F_{Dx} = (F_D/2) \, \mathrm{e} \, \mathrm{xp}(-0.35(x\overline{f})^{0.4})$

The potential source strength is a density ($F_D = Q_D S_J \underline{T}$) where the seed density (Q_D = seeds m^{-2}) is estimated well within the source area as $\overline{Q}B_D$ (and B_D is the basal area per area). Meanwhile, S_J, juvenile survivorship, is based on the area of suitable seedbeds outside the area source (i.e., clearcut, burn, etc.).

Singly or in combination, the point source and area source models can be used to predict the sexual recruitment in burns, clearcuts, blowdowns, and other sorts of disturbances given any spatial configuration of sources. Greater realism can, of course, be gained by accounting for differences in prevailing winds (e.g., Greene and Johnson 1999; Greene and Johnson 1996), non-random distribution of seedbed types in Cartesian space, etc.

The foregoing models apply especially to species such as white spruce and balsam fir that are dependent upon sexual recruitment from a living source. These two common species do not seem to be well adapted to regenerating after fire because they lack serotinous cones and any vegetative form of regeneration (the occasional layering can be ignored). The typical distance from burn edges or unburned patches is generally greater than their median dispersal distance (150 m vs. around 30–50 m, depending on whether the stand is open or closed), and consequently, early colonization tends to be poor. Indeed, Greene and Johnson (2000) showed that these species seldom have adequate stocking beyond approximately 50 m from an area source in burns. Thus, these species are dispersal limited; their recolonization of even the sites in which they had been previously present may be questionable if no surviving individuals are nearby. Balsam fir, at least in the mixedwood boreal forest, has another problem in that the bottom hillslope position where it grows best is also limited in area, as previously pointed out. How then does it persist in the forest? One might speculate that it is the combination of its distribution on the landscape and the disturbance frequency that allows such persistence.

The ready supply of seeds in the aerial seedbanks of jack pine and black spruce, coupled with their tendency to form near-monocultures, lead to a simpler recruitment model. For a fire, their expected recruitment density becomes:

$$[7] \qquad F_D = Q_D S_J S_A$$

where Q_D is proportional to basal area/area (B_D); Greene and Johnson (1999) have found $Q_D = 35\,097 B_D^{0.86}$ for jack pine and $Q_D = 163\,400 B_D^{0.95}$ for black spruce. As above, S_J, denotes juvenile survivorship, and depends on the area of suitable seedbeds after a fire. Finally, S_A, the pre-abscission survivorship of the seeds during the passage of the flaming front, has been estimated as 1.0 for jack pine and 0.58 for black spruce.

The foregoing has made clear that a principal factor determining post-fire recruitment is the proportional coverage of seedbed types. Seedlings of boreal forest tree species generally have very poor juvenile survivorship on litter or on the thick organic layer overlaying the mineral soil (Chrosciewicz 1974, 1976; Zasada et al. 1983; Thomas and Wein 1985; Weber et al. 1987; Greene et al. 1999), especially at the germination stage, but also for the subsequent 2 years for a cohort (Charron and Greene 2002). Fire removes litter and the organic layer to varying degrees, exposing suitable seedbed (Fig. 8.10a). The area of the forest floor from which the F layer is removed thus determines the minimum area available for post-fire recruitment of tree seedlings. Obviously, in order to understand recruitment in burns, we require an understanding of the process by which the organic layer is removed by fire.

While the litter layer is largely removed by flaming combustion during passage of the fire front, the partially decomposed F and H organic layers, together known as duff, are generally consumed by smoldering combustion after the flaming front has passed (Wein 1981, 1983; Frandsen 1991; Johnson 1992; Hungerford et al. 1995; Fig. 8.10b). The duff is not consumed by flaming combustion because of its physical and chemical properties (see Miyanishi 2001). The decomposition of forest litter results in organic matter that is more compacted (with a higher packing ratio) and that has a lower volatile content and a higher lignin:cellulose ratio than the original litter. When heated, this compacted chemically altered organic matter pyrolyzes, producing char which oxidizes slowly as a solid rather than releasing sufficient combustible volatiles to sustain flaming combustion.

Smoldering differs from flaming combustion (rapid oxidation of gases) in being a much slower oxidation of a porous char-forming solid (Drysdale 1985). Smoldering involves the endothermic process of pyrolysis (thermal degradation), which releases some volatiles and forms char; the char then oxidizes exothermically. Smoldering stops when the heat generated by oxidation of the char or the transfer of heat to the virgin duff is too low to cause pyrolysis. The principal variables that influence this extinction process are duff moisture, density, and depth (Miyanishi and Johnson 2002). Moisture is important because latent heat of vaporization is a significant heat sink. Heat transfer between the oxidation zone and pyrolysis zone by conduction is determined by the tem-

Fig. 8.10. (*a*) Patchiness and patterning of duff consumption within burned stands following a stand- replacing crown fire in the boreal forest of central Saskatchewan. (*b*) The duff is consumed by smoldering combustion which occurs after passage of the flaming front. Note in (*a*) that regeneration of both herbaceous plants and trees occurs only within the burned out patches.

(*a*)   (*b*)

Photo by K. Miyanishi   *Photo by E. Johnson*

perature difference and thermal diffusivity of the duff. Thermal diffusivity is in part determined by density. Duff depth is important because it influences the heat transfer between the oxidation and pyrolysis zones. In thicker duff layers, the surface area over which oxidation is occurring is larger, thus generating more heat. Furthermore, thicker duff traps more of the heat and a lower proportion of the heat generated by oxidation is lost by convection from the bed. As the duff layer becomes thinner, not enough heat is transferred to the unaltered duff to sustain smoldering (Palmer 1957; Jones et al. 1994; Miyanishi and Johnson 2002).

Smoldering is patchy at different spatial scales (Dyrness and Norum 1983; St-Pierre et al. 1991; Miyanishi and Johnson 2002) because moisture, depth, and density of duff vary between canopy and gap locations within stands, down hillslopes, and regionally. Although different tree species intercept precipitation with different efficiencies, the canopies of all trees intercept to some degree, resulting in differences in moisture inputs to duff at the scale of metres or less. The tree canopy also intercepts both incoming solar and outgoing terrestrial radiation. The interception of solar radiation would decrease surface evaporation while interception of terrestrial radiation would decrease surface cooling at night, inhibiting dew formation. These opposing effects may tend to cancel each other's impacts on duff moisture. Greater litter accumulation beneath trees than between crowns would lead to differences in duff depths at this scale. Furthermore, differences in the chemical composition of litter from tree species and from herb and shrub species in gaps, as well as differences in moisture and radiation balances beneath tree crowns and in gaps, would affect decomposition rates. Thus, duff depths and densities would also vary at this scale. As a result, the patchiness of duff consumption within stands shows a pattern related to the location of fire-killed trees; i.e., the holes in the duff layer resulting from smoldering are spatially correlated with the trees (Miyanishi and Johnson 2002). As we have discussed before, moisture and nutrients change down hillslopes; this affects both duff moisture and canopy tree species composition. Both of these factors will also influence duff depth and density and hence smoldering propagation and extinction. Since precipitation and temperature affect the drying rate of duff, variation in these variables at the regional scale will influence the patterns of duff consumption. This very complex variation in duff consumption pattern is important, since almost all boreal tree recruitment from seeds after fire occurs only in patches in which most of the duff has been consumed (Charron 1998; Greene et al. 1999; Fig. 8.10*a*).

Above, we assumed that mineral soil and thin humus comprised about 20% of the immediate post-fire surface. Until recently, there were very few studies available that quantified this percentage. Miyanishi and Johnson (2002) reported about 40% coverage for the optimal seedbeds in two Saskatchewan fires, with D.F. Greene (unpublished data) finding a similar value for the 2000 Chisholm burn in Alberta. Charron and Greene (2002) found somewhat lower values in Saskatchewan, while Greene et al.[2] found much lower values (as low as 5% in lowland black spruce stands) for a Quebec fire. Clearly, there is still much to be learned about the combustion effects within and among fires.

Emulating the dynamics of fire?

Our present understanding of both wildfire and forest population dynamics is still relatively modest. Given our confusion and misunderstandings in the past of the role of disturbance in ecosystems (Raup 1956), it is not responsible to suggest applications that could be considered professional practice. However, we will try at this point to consider a selection of issues related to emulating wildfire effects.

Foresters have "learned by doing" what silviculture works in the boreal forest (see Chap. 13). Not surprisingly, many of these practices can be explained in terms of the tree population dynamics. In the boreal forest, as we have seen, there are two general types of tree dynamics: that of dispersal-limited species for which the regional dispersal and mortality (wildfire) are both important in local dynamics, and that of species with good on-site seed availability for which only regional mortality (wildfire) is of major importance.

Trees with serotinous cones (jack pine and black spruce) or vegetative reproduction (trembling aspen) do not, in general, depend on dispersal from adjacent populations for regeneration after fire. There are generally enough seeds or root suckers available to give adequate stocking if the local pre-fire population was dense enough and of reproductive age. Local population extinction would occur, on average, in 19% of the landscape if the fire cycle is 100 years and the age of first reproduction is 20 years (Fig. 8.11). Dispersal for serotinous populations is important only in this small part of the landscape. Aspen is more complicated, since its mortality will depend on killing not only the aboveground stems (fire intensity) but also the underground roots (smoldering combustion). Consequently, the regional mortality (wildfire) and pattern of duff consumption by smoldering combustion are the regional processes that are dynamically important in the local population.

---

[2]Greene, D.F., Noël, J., Bergeron, Y., Rousseau, M., and Gauthier, S. Tree recruitment across a burn severity gradient following wildfire in the southern boreal forest of Quebec. Submitted to Canadian *Journal of Forest Research* for review.

Tree harvesting by partial cutting (e.g., commercial thinning, shelterwood, or selection harvesting) is like a crown fire in that both processes generally kill most of the tree populations while leaving an *in situ* seed source. The size and shape of the cut is less important to serotinous species, since dispersal from outside the burn or cutblock is not as important. However, site preparation by prescribed burning or surface disturbance (scarification) is important (Chap. 13). Few comparisons have been made between the patterns created by smoldering in wildfires and the often more complete duff removal in prescribed slash burns or the pattern created by harvesting activities or mechanical site preparation methods such as disk trenching. Generally, site preparation has been successful at getting suitable densities and quality of recruits.

Trees with nonserotinous cones (white spruce and balsam fir) depend on both dispersal and regional mortality. The regional mortality issues are similar to those for serotinous species except that the age of first reproduction is typically older (30 years for white spruce and 35 years for balsam fir). This older age of first reproduction means that a slightly greater proportion of the landscape (20%) could potentially burn before becoming reproductively mature and sources of seeds for dispersal (Fig. 8.11). On the other hand, local populations of nonserotinous species also may become extinct on previously occupied sites if the sites are beyond the dispersal distance of surviving seed trees. Thus, in these species, fire and cutblock size and shape as they affect dispersal are important if natural regeneration is expected (Greene et al. 1999). However, if seedlings are planted, as is often done for white spruce, dispersal is not a problem. The successful establishment of white spruce plantations often depends on site preparation (to create plantable spots in deep duff, or to depress competition from shrubs and herbs; see Chap. 13), but as noted before, comparisons of forest floor patterns created by site preparation with those made by smoldering combustion remain unexplored at present.

Up to now, we have seen the effect of the wildfire processes primarily on the cohorts recruited immediately after and as a result of the fire-caused mortality and duff consumption. The mortality schedules of these trees seem to fall into roughly two types:

Fig. 8.11. The probability of regional mortality of a local tree population given the fire cycle and the age of first reproduction.

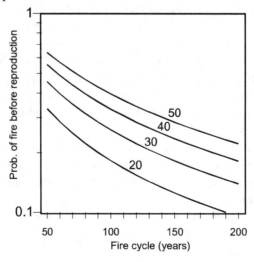

Fig. 8.12. Jack pine canopy and black spruce understory in the mixedwood boreal forest of central Saskatchewan. When accurately aged, these trees are found to be the same age. The differences in size are due to differences in growth rates between these two species and do not reflect differences in time of establishment.

Photo by E. Johnson

those for species such as jack pine and aspen with rapid individual height growth rates in the first 10 years, little shade tolerance and hence higher density-dependent mortality, and those for species such as white spruce, black spruce, and balsam fir with slower growth rates in the first 10 years, shade tolerance and hence lower density-dependent mortality. One result of the different growth rates in the first years after germination is widespread error in age determination at the base of the tree (as described in following sections of this chapter) and difference in tree height. Error in age determination is the result of not aging at the root collar but only near the base of the tree. If seedlings are fast growing, the error between root collar and tree basal age is small, but if seedlings are slow growing the error will be as much as 15 or more years (Gutsell and Johnson 2002). The result is that white spruce will appear to be coming in a decade or more after aspen and black spruce, often up to 20 years after jack pine. This apparent lag in recruitment suggests that the slower growing species were born into the stand some time after the fire when, in fact, they were born at the same time as the faster growing species. Further, because of the initial growth rate differences, the slower and faster species continue to be different in height until maturity (Gutsell and Johnson 2002; Fig. 8.12). The age error and height difference has led to the belief that one is seeing one species succeeding the next; however, they are often cohorts born at approximately the same time (Gutsell and Johnson 2002). This has implications for ecologically sustainable mixedwood management (Chap. 13).

What of the cohorts that are truly born after the fire cohort (i.e., not incorrectly interpreted as such because of aging error)? In most situations, these consist of slower growing, shade-tolerant species (e.g., white spruce, black spruce, balsam fir; Gutsell and Johnson 2002). They start in a shaded environment and generally on the rotted logs derived from the canopy cohort that burned decades before or, much later, canopy trees felled by more minor disturbances (wind, insects). These cohorts generally exhibit lower birth rates and higher mortality rates. Note that layering in black spruce and balsam fir is not included here; we are only talking about seed germination. What is the chance of these later cohorts reaching the canopy and becoming reproductive? Figure 8.13 shows the probability of reaching the canopy for a series of growth rates and fire cycles. The conclusion from this graph is that, even with growth rates that might occur in a gap caused by death of canopy tree(s), at the rate at which fires occur in the western boreal forest, there is little chance of understory trees reaching the canopy before the next fire occurs. This last conclusion has some significance to the forest if all wildfires cannot be suppressed.

It is often assumed that emulating the effects of fire through forest practices will replace wildfire (Chap. 9). If this does not occur, then what will happen to the age distribution of the forest when both cutting and wildfire compete for the wood supply? Several theoretical studies have been carried out (Martell 1980; Routledge 1980; Van Wagner 1983; Reed 1984; Reed and Errico 1985, 1986). In general, these studies show that if the goal of forest management is a regulated forest and if fire follows an exponential time-since-fire distribution, then the resulting forest age distribution will be a decreasing geometric distribution (Fig. 8.14). The actual rotation age and long-term yield will be reduced from that for a regulated forest alone. Interestingly, the shape of the forest age-class distribution is similar to that for fire alone although the age scaling is different; i.e., the old age tail isn't very old. The reason for this is that fire is continuing to burn all ages of the forest, while harvesting only removes older stands (above the rotation age).

Of course, one could ask what the forest age-class distribution actually produced by cutting and wildfire after 75 or more years is. Unfortunately, there are almost no documented examples of such forest age-class distributions in the literature. One example in eastern Ontario has a decreasing distribution not unlike the mixed time-since-fire distribution expected for this region (see Bridge 2001). The reason for this is not only continuing fire but also the fact that forest management areas are not completely productive

Fig. 8.13. Probability of reaching the canopy for different growth rates and fire cycles. Curves were calculated using growth rates of Picea glauca trees in the canopy and understory obtained from a 111-year-old stand in central Saskatchewan. The numbers 1, 3, and 5 indicate initial tree heights in metres. For example, a 1-m-tall tree growing in understory conditions has a 0.65 probability of being burned before reaching the canopy if the fire cycle is 100 years (data from Gutsell 2001).

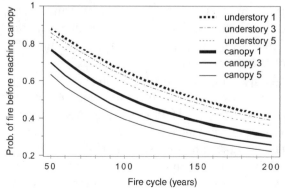

Fig. 8.14. Forest age distribution (decreasing geometric) resulting from logging and fire (Reed and Errico 1985).

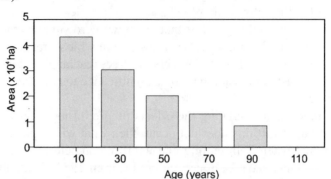

(i.e., non-productive areas are not harvested). Thus comparing the age distribution of a completely regulated forest to that of the whole landscape is not appropriate. Furthermore, changing ideas of forest practices, technological changes, and new and different markets have changed harvesting practices over time, and access to all parts of the management areas is not equal. Therefore, even if the whole landscape were productive, we should not expect to find a uniform age distribution after a century of harvesting.

Natural variation

Natural variation in a forested landscape (site differences aside) will here be considered in terms of the statistical distribution of regional and local tree mortality. Mortality distributions can be looked at in several ways, all of which are interrelated in a logical manner. $F(t)$ is the inverse of cumulative survival, usually called survivorship, $S(t)$; i.e., $F(t) = 1 - S(t)$. The cumulative survival is the chance of surviving (avoiding mortality) up to time t. Obviously, survival must start at 1.0 at $t = 0$ and decrease afterwards. Another way of looking at natural variation in mortality is the probability density distribution $f(t)$, that is, the probability of mortality in a time interval. Survival, $S(t)$, and the probability density distribution are related by $f(t) = 1 - S(t)dt$, which is to say that $1 - S(t)$ is the accumulation of $f(t)$. More importantly for our interest in natural variation, the probability density distribution $f(t)$ explains how mortality comes about, since $f(t) = q(t) \cdot S(t)$ where mortality $f(t)$ in the time interval dt is a result of the probability of surviving, $S(t)$, to the beginning of the interval t and then the instantaneous mortality $q(t)$ in the interval.

Now with this viewpoint, the distribution of instantaneous mortality of the regional and local population gives us the natural variation. For example, Fig. 8.15*a* gives the natural variation in instantaneous mortality $q(t)$ in a local population. The exact values and shape of this distribution will vary some by species, density, site, seedbed, and local weather, etc., as previously discussed. The natural variation in instantaneous mortality, $h(t)$, in regional populations (Fig. 8.15*b*) follows a mixed negative exponential or Weibull distribution (Johnson and Gutsell 1994). The distribution is constant between changes in the fire frequency (see previous sections).

Natural variation in the instantaneous mortality of both the local and regional mortality would have to incorporate fire (disturbance) into the local mortality. Local causes of mortality are important during the first 15–20 years, and fire becomes the most important cause of mortality after this. Silviculture is primarily used to manipulate local causes of mortality. However, if wildfires are still occurring (Johnson et al. 2001), then in later years this would be an important source of mortality.

Insect outbreaks and tree population dynamics

Spatial and temporal dynamics of outbreaks

Spruce budworm, *Choristoneura fumiferana* (Clem.), is the insect that causes the most disturbance to coniferous tree populations in eastern North America (Fig. 8.16). The last outbreak, which took place between 1974 and 1988, devastated more than 55 million ha of forest and destroyed between 139 and 238 million m^3 of spruce and fir, the equivalent of 10 years of intensive logging (QMER 1985).

Dendroecological analysis of growth reductions caused by budworm defoliation indicates that budworm populations have reached epidemic levels regularly over the past 300 years (Morin and Laprise 1990; Krause 1997; Fig. 8.17). Reliable defoliation surveys are available for the last outbreak and for part of the mid-20th century outbreak (Kettella 1983; Hardy et al. 1986; Boulet 1996; Candau et al. 1998). They have been used to develop models of the spatial and temporal dynamics of the budworm in order to predict the dynamics of the next outbreak (Candau et al. 1998; Gray et al. 1998, 2000; Williams and Liebhold 2000). However, even if these models are extremely well done, they can only reproduce the spatial and temporal patterns of the last outbreak. The advantage of dendroecological studies is that they can provide reliable information on past defoliations over periods up to 300 years, since trees maintain a record of defoliations in their ring width patterns (Fig. 8.17).

A recent detailed study of the Quebec vegetation zones south of the 50th parallel shows a very regular periodicity of outbreaks at the supra-regional scale. As indicated by O1–O6 in Fig. 8.18, there were three outbreaks per century during the 19th and 20th centuries (Jardon 2001). Based on historical data, it has been suggested that outbreaks during the 20th century became increasingly important and devastated larger areas (Blais 1983, 1984). This change has been attributed to human activities such as logging

Fig. 8.15. (*a*) Instantaneous mortality, *q*(*t*), in a local population, and (*b*) instantaneous mortality, *h*(*t*), in a regional population (fire frequency).

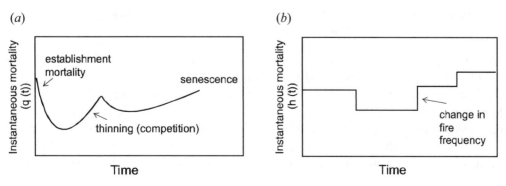

(*a*)

(*b*)

Fig. 8.16. Trees killed by spruce budworm in a balsam fir forest in Quebec. Note that the canopy trees are killed while the understory trees survive.

Photo by Natalie Perron

Fig. 8.17. Chronology of spruce budworm outbreaks in the boreal forest of Québec derived from dendroecological analysis from (*a*) Morin and Laprise (1990) and (*b*) Krause (1997).

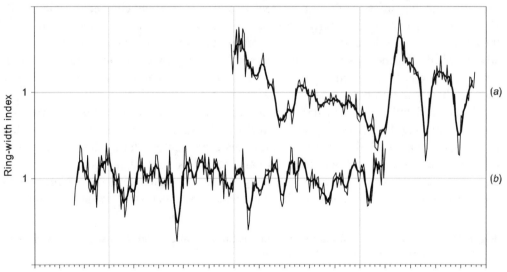

and fire suppression. However, until recently, data were scarce across eastern North America, making it difficult to test this hypothesis (Morin 1998).

The first attempt to compare histories of spruce budworm outbreaks in different locations across North America using published dendrochronological data showed that outbreaks in the 20th century occurred at a very regular interval (Morin 1998). Severe outbreaks, such as the ones that occurred at the beginning (O3) and the end (O1) of the century, were synchronized across locations, as indicated by the date of occurrence of

Fig. 8.18. Relative frequencies of sampled sites affected by the 6 last outbreaks (O1–O6) south of the 50th parallel in the province of Quebec (from Jardon 2001).

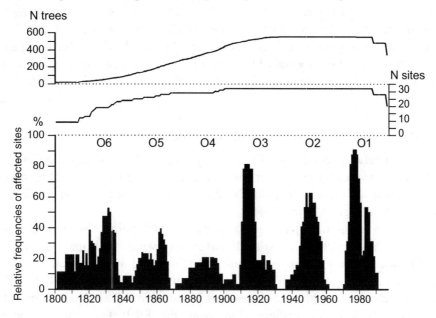

the smallest ring produced by defoliation. However, such synchronicity was not evident for the outbreak (O2) that occurred in the middle of the 20th century (Morin 1998). Evidence now suggests that there was a noticeable shift in outbreak patterns at the turn of the century. Even if the 19th century outbreaks covered the entire territory, these outbreaks showed a slow patchy pattern of dispersal within a stand as well as across the region. On the contrary, the severe 20th century outbreaks (O1 and O3) showed a rapid and synchronous dispersal pattern both within stands and across the entire area (Fig. 8.18). It is unlikely that human activities are the only cause of this shift because logging activities and fire suppression programs were not active in most of the area at the beginning of the 20th century when one of the most severe outbreaks (O3) occurred at almost every location at the same time, showing an explosive diffusion pattern (Jardon 2001). A change in forest age structure across the area from younger to older forests, containing more susceptible species such as balsam fir, may have contributed to this shift. This change to older forests may be linked to a lengthening of the fire cycle that occurred at the end of the Little Ice Age (Bergeron et al. 1995; Jardon 2001).

The spruce budworm diffusion pattern detected from dendroecological data for the last outbreak of the 20th century is in general agreement with the well-known diffusion pattern described by defoliation surveys for the same area (Kettela 1983; Hardy et al. 1986; Boulet 1996; Gray et al. 1998, 2000). However, dendroecological data based on four outbreaks showed that each outbreak had its own distinctive pattern of spread, some being detected first in the southwest and expanding toward the northeast (e.g., O1) and some showing the reverse pattern from the northeast to the southwest (e.g., O4) (Jardon 2001). The mechanisms by which the insects cover such large areas in such different patterns are not fully understood. The explosive pattern described for the O3 out-break suggests that the insect was present everywhere at an endemic level and that a large-scale variable such as climate may have acted as a synchronizing agent; this mechanism has been described as the Moran effect (Williams and Liebhold 2000). However, a diffusion pattern such as the one described for the O1 outbreak suggests that other variables (e.g., the dispersal of the insect from very dense populations) could have played a role in synchronization of the outbreak (Régnière and Lysk 1995; Jardon 2001).

Structure of mature boreal balsam fir populations

In the interval between destructive fires, the dynamics of natural boreal balsam fir pop-ulations are controlled by canopy openings caused by tree mortality. This mortality is often associated with spruce budworm outbreaks that sculpt the landscape in a patchy mosaic of variable age (Morin 1994).

A study of a boreal fir population situated at the 50th parallel north of Lac St.-Jean, Quebec, that was exposed to a major disturbance (a wind storm that resulted in large areas of blowdown) has been used to put forward a model for understanding the mech-anisms involved in regeneration of the population (Morin 1990). The defoliation and opening of the site by the blowdown resulted in growth release of the seedlings already established on the forest floor (the seedling bank; Fig. 8.19). Thus, the original popula-tion was replaced by a population of the same species with an apparent even-aged struc-ture. Such regeneration from a seedling bank (Thompson and Grime 1979; Morin 1990) is also representative of even-aged stands originating solely from a budworm outbreak (Morin 1990; Fig. 8.20). Using the dendroecological model presented for the blowdown (Morin 1990) and the chronology of spruce budworm outbreaks for the region (Morin and Laprise 1990), Morin (1994) evaluated the importance of recurrent spruce budworm outbreaks in the dynamics of these balsam fir forests.

Most of the studied populations showed the same dynamics as the blowdown site and could be classified as budworm-origin even-aged stands. Characteristic growth releases in the sampled trees generally occurred after exceptionally narrow rings around 1952 (Group 1), 1914 (Group 2), and 1860–1890 (Group 3). A characteristic example of the age structures and growth curves obtained for each of the groups is presented in Fig. 8.20. In general, this pattern of age structure and rapid growth release of seedlings suggests a rapid mortality of the canopy trees, possibly associated with a partial or total blowdown. This situation is in good agreement with spruce budworm dynamics, which have been very consistent in different regions and across different outbreaks (MacLean 1980). This is a good example of a cyclic change that was proposed to explain recent dynamics of the fir forests of New Brunswick (Morris 1963; Baskerville 1975, 1986; MacLean 1984), Nova Scotia (MacLean 1988), Quebec (Ghent et al. 1957), and Ontario (Blais 1954; Ghent et al. 1957).

The seedling bank

During an outbreak, the seed supply is limited because of the feeding of larvae on the staminate cones (Blais 1952; Powell 1973). Therefore, in order to achieve replacement

Fig. 8.19. The permanent seedling bank under a closed balsam fir canopy.

Photo by H. Morin

of the canopy by trees of the same species, a well-established seedling bank of this species should be present in the stand when a major disturbance such as a budworm outbreak, blowdown, or clearcut occurs. In this part of the chapter, we describe the structure of the seedling bank and show how determining the age structure of the seedlings by ring counts at the presumed collar level can lead to erroneous interpretation of the mechanisms that permit regeneration of the fir population. Detailed dendroecological analysis of seedlings allows us to describe the forest dynamics that permit the seedling bank to stay alive under the canopy while maintaining the same age and size structure for decades.

Structure of the seedling bank

Seedlings from the seedling bank were sampled within the same populations used for the age structure analysis of mature populations. Ages were determined at the presumed collar level, i.e., the location in the stem where the maximum number of growth rings was found under a microscope using fine cross sections. This count was compared to the bud scar count down to this location. The most striking result was that the seedling banks had very similar age and height structures despite the various ages of the mature tree populations (Morin and Laprise 1997). Seedling age structures had normal distributions centred on 20 years (Fig. 8.21). Two hypotheses were proposed to explain the similarity of the age and height structures of the seedling banks: (1) the seedling bank is always being renewed by a turnover process; or (2) the bank establishes and grows to a certain point corresponding generally to 20 growth rings or 20 terminal bud scars, then it stops growing. If the turnover process is the dominant one, then the age structure of the seedlings should fit a "reverse J" shaped distribution characteristic of a stable population. Additionally, older seedlings would be expected in older stands and some mortality should be observed among those old seedlings. As shown in Fig. 8.21, the age structures of the seedling bank fit a normal distribution, not a reverse J. Also, relatively old dead seedlings were virtually absent. The lack of young seedlings in the seedling

Fig. 8.20. Age structure (5-year age classes) and mean growth curves from three typical mature balsam fir stands in the boreal zone of Quebec (from Morin 1994).

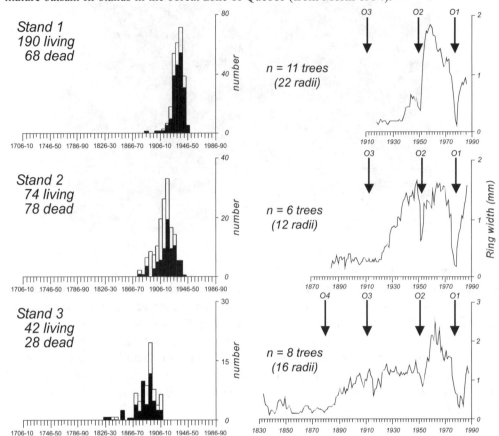

bank was probably due to the lack of seed production at the end of the last budworm outbreak, a consequence of the budworm feeding at the time of sampling (1985–1986). Results from this study and field observations did not support the turnover hypothesis and suggested that the seedling bank dynamics could, at least in part, be based on a strategy that enables the fir to survive under the canopy for long periods with a complete lack of detectable radial and apical growth.

Demography of the seedling bank

The best way to test the two hypotheses of seedling bank dynamics was to establish permanent plots in which we could follow all the steps of the regeneration process from the formation of the seeds to their dispersal, germination, seedling growth, and mortality. Four permanent plots were established in the autumn of 1994 in representative sites based on age structure analysis (Morin 1994). Duchesneau and Morin (1999) conducted detailed seedling demography surveys in order to characterize the ecological factors affecting germination and early establishment in understory balsam fir seedling banks and to determine if continued recruitment can potentially renew the seedling bank.

Fig. 8.21. Age structure of balsam fir seedlings from typical mature balsam fir stands of various ages (from Morin and Laprise 1997).

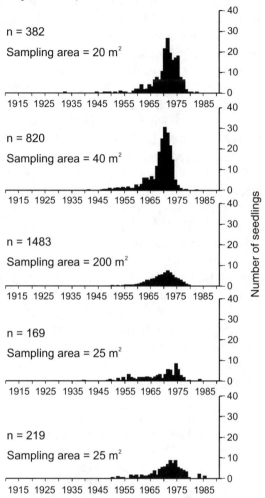

Seven years after the 1994–1995 mast year, very few seedlings from this cohort remained, and very few seedlings from subsequent years of seed production survived. On the other hand, the seedlings that were present in the bank at the beginning of the study were still present, had an unchanged height structure, and had experienced very little, if any, mortality. These results further supported our second hypothesis that a stagnation process played an important role in maintaining the seedling bank. Stagnation would imply that apical and radial growth must cease after a certain point (Morin and Laprise 1997). However, careful follow-up of the seedlings in the bank since 1994 demonstrated that more than 80% of the seedlings showed yearly apical growth, suggesting that the stagnation hypothesis was not supported. So, how can seedlings keep approximately the same age and height from year to year if they continue to grow every year?

Growth strategy of the seedling bank

As a balsam fir seedling grows in the forest understory, its stem is often buried in the litter and is sometimes characterized by prostrate growth (Fig. 8.22). These situations promote the formation of adventitious roots and reverse taper as the stem is buried, leading to a loss of growth rings (DesRochers and Gagnon 1997; Parent et al. 2000). Thus, measuring the aboveground portion of the seedling and calculating the age from a ring count at the presumed collar level (in many studies, at the level of the first adventitious root) leads to the erroneous age and height structure depicted in most studies (Ghent 1958; Batzer and Popp 1985; Côté and Bélanger 1991; Bélanger et al. 1993; Bergeron and Charron 1994; Osawa 1994; McLaren and Janke 1996; Galipeau et al. 1997; Morin and Laprise 1997; Sirois 1997; Kneeshaw and Bergeron 1998).

The formation of adventitious roots on a gradually buried stem represents an efficient mechanism that permits seedlings to maintain a balance between the photosynthetic and non-photosynthetic parts and to stay alive for long periods in a low-light environment (Parent et al. 2002). This is a good example of how an understanding of the mechanism involved allows us to understand the dynamics of regeneration, a phenomenon which previously had been misinterpreted in most cases.

From these results, it also becomes clear that age structures reported for mature balsam fir populations (Morin 1994) are not reliable but are artefacts of the aging method. These age estimates reflect the number of growth rings that we can count at the base of the trees after cross-dating, i.e., mainly the number of growth rings set down after the growth release, not the true ages of the trees. Although such age structures may allow us to study the importance of budworm outbreaks in the dynamics of balsam fir populations, they cannot be used to make any inferences about the regeneration processes.

Fig. 8.22. Typical balsam fir seedling from the seedling bank showing belowground portion with adventitious roots, terminal bud scars, hypocotyl region, and primary roots (from Parent et al. 2000).

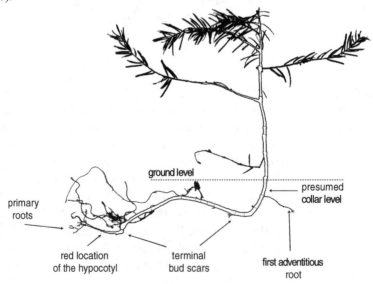

Post-outbreak regeneration of balsam fir forest

Regeneration following spruce budworm outbreaks has not been studied as intensively as post-fire regeneration in the boreal forest. Although both fire and insect outbreaks occur at the landscape scale and affect tree population structures, there are several important differences between these two catastrophic disturbances. For example, there are no time-since-last-outbreak maps for any large areas as there are time-since-fire maps and, therefore, no good estimates with confidence intervals of the disturbance cycle for budworm outbreaks. What we know is that, despite the 33-year recurrence of outbreaks at the landscape level, a particular stand will generally not be severely affected by two consecutive outbreaks. Unlike disturbance by fire which does not appear to be age-dependent (Johnson 1992), the impact of a budworm outbreak has been shown to be related to stand age (MacLean 1980). In general, spruce budworm will mostly attack mature trees (regardless of their size) except in the most severe years (Fig. 8.16). Stand structure and composition have also been recognized to be important for the occurrence of outbreaks as well as the impact of defoliation. Although outbreaks cover large areas and are relatively predictable at the landscape scale, we don't fully understand why a given mature stand will be more affected than a neighbouring one. The proportion of balsam fir, the preferred host of the budworm, in the stand is the only variable that has been found to be significantly related to the impact of budworm defoliation (Bergeron et al. 1995; MacLean and MacKinnon 1997).

The model of balsam fir populations affected by recurrent spruce budworm outbreaks that we present here differs from the classical successional model. The main reason for this difference is the dating technique we used that enabled us to show missing cohorts in the canopy and in the understory. Successional models based on static age structures (e.g., Morin 1994; Morin and Laprise 1997), should be re-examined in light of these new findings. We now know that it is probably impossible to date mature balsam fir exactly because of the difficulty of finding missing rings below the conventional shoot–root interface. However, new findings on the possibility of dating seedlings (as described above), as well as permanent plot monitoring, suggest a new approach to understanding the mechanisms involved in regeneration of balsam fir populations subjected to recurrent insect outbreaks.

Germination

Mature balsam fir trees produce seed almost every year, but with mast years of production every 2–8 years. The main requirement for seed germination is any seedbed with sufficient moisture. In natural mature fir populations, this type of seedbed is generally abundant and associated with mosses and fallen logs (Simard et al. 1998; Duchesneau and Morin 1999). Germination is thus a continuous process in mature fir populations. This situation is different during budworm outbreaks when there are no or few seeds produced because of larval feeding on the staminate cones (Blais 1952; Powell 1973). If the majority of canopy trees in a stand are killed during an outbreak, there will be some seed dispersal from nearby less affected stands as well as from the few survivors. This represents approximately 25% of the normal seed rain in a mature stand. If we consider that balsam fir starts to produce seeds 15 years after growth release and more efficiently after 20–30 years, and if there is no mortality among the germinants and no limitation in the seedbed availability, then the theoretical seedling bank age structure without mortality would reflect the pattern of viable seed dispersal in the stand (Fig. 8.21). Figure 8.21 is based on data from yearly monitoring of seed germination within thirty-two 60 × 60 cm permanent plots established in 1994 in four balsam fir stands of different ages (Duchesneau and Morin 1999). When an accurate dating of seedlings was done, we could detect in the seedling bank age structure the mast years of seed production. Of course, mortality also alters the age structure (Parent et al. 2002).

This theoretical pattern of seedling recruitment cannot always be seen in a study using conventional dating techniques such as ring counts at ground level or bud scale counting from ground level to the apex. One reason for this is, of course, mortality (a factor considered in more detail in the next sub-section), but a second reason is the dating technique. Using conventional techniques, the best figure one can obtain is a reverse-J age-class distribution, sometimes showing peaks corresponding to mast years of production (Morin and Laprise 1997). If the stand has been affected recently by an outbreak, as was the case with many stands we sampled, there will be a lack of recruits in the youngest ages. But sometimes the dating is simply not accurate enough to show mast years. Inaccurate dating techniques lead to missing cohorts in the age distribution, giving the impression that there are no or very few old seedlings, and that there are some recruits even during outbreak periods.

Mortality

Two conditions are required for seedlings to survive: a suitable seedbed (mainly mosses on rotten logs; Simard et al. 1998; Duchesneau and Morin 1999) and a minimum level of light to permit net photosynthesis. Within a mature fir stand with an established seedling bank, the principal limitation is the lack of available seedbeds. Thus, even during years of high seed production (e.g., 1994–1995 in the permanent plots described above), seedling establishment is limited. As shown in Fig. 8.23, initial seedling mortality within the first year is very high. Although mortality decreases exponentially during the first few years, after several seasons there are few, if any, seedlings left from this cohort (Morin and Laprise 1997). Thus these seedlings constitute a transient seedling bank whose relative cohort numbers reflect seed production.

When the two conditions of available seedbed and good light levels are available (e.g., following a budworm outbreak and (or) a blowdown event), higher seedling numbers establishing would result in a greater number of seedlings persisting beyond the first few years. Once well established, such seedlings experience very low mortality (Box 8.2a) and can thus survive a long time. The survivors of such cohorts form a persistent seedling bank. Consequently, balsam fir stands would simultaneously have two distinct seedling banks:

- a persistent one generally establishing shortly after an insect outbreak when seedbed and light availability are high; and
- a transient one establishing in all subsequent years between outbreaks when both conditions (and particularly seedbed availability) are in low supply.

Seedlings of the transient seedling bank will be younger and smaller and therefore subject to higher mortality than those of the persistent seedling bank. The seedlings of

Fig. 8.23. Survival curves of the 1995 balsam fir seedling cohorts from 4 permanent plots. Relative percentage of seedling survival is defined as the number of seedlings found at the time (weeks post-germination)/(total number of seedlings) × 100 (from Duchesneau and Morin 1999).

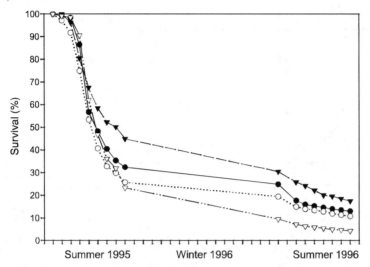

the persistent seedling bank that were 20–30 cm in height when the 1994–1995 mast year seedlings established are still alive in the permanent plots, while most of the seedlings born in 1994–1995 have since died.

For the transient seedling bank, the age-specific mortality rate (q_x *sensu* Johnson and Fryer 1989) is very high at germination and within the first year. This event can occur any time during the life span of the stand. Mortality stabilizes at a low level thereafter for the few remaining seedlings (Box 8.2*a*). At the time of site opening after an outbreak that kills the mature canopy trees, the q_x rises again during growth release of the older seedlings in the persistent seedling bank. Thus, the probability of seedlings from the transient seedling bank making it to the canopy during this period of canopy opening and growth release is very low. In contrast, the persistent seedling bank has a lower germinant mortality because of the availability of seedbed. Thus, a greater number of seedlings survive the initial high seedling mortality and the subsequent low mortality over the long period of suppression (Box 8.2*b*). These seedlings will be released by the canopy mortality due to a budworm outbreak. However, some thinning will gradually occur so that the q_x gradually rises, and the cycle starts again.

The three conditions necessary for the buildup of the persistent seedling bank (seeds, suitable seedbed, and sufficient light availability) are not always met very early after stand opening. For example, mature trees in one permanent plot were almost all killed by the last spruce budworm outbreak at the end of the 1980s. Fifteen years after the opening of the site, the persistent seedling bank had not yet established, as indicated by the high mortality of the small seedlings originating from nearby seed sources. Another permanent plot in which the release dates back to the 1952 outbreak still has only a transient seedling bank and a high mortality of the small seedlings because of a very low light environment under a very dense stand. Since we don't know exactly when the persistent seedling bank is building up, we don't know how long it takes to fully occupy

Box 8.2. *Mortality patterns of balsam fir seedlings.*

Fig. 8.24. The age-specific mortality rates, q_x, for balsam fir from (*a*) the transient seedling bank, and (*b*) the persistent seedling bank after the release of the stand by a spruce bud-worm outbreak.

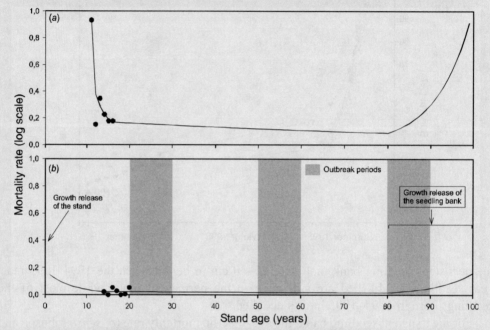

Closed circles indicate computed q_x from four permanent plots in mature balsam fir stands for both permanent and transient seedling banks (Morin 1994). The q_x pattern from the transient seedling bank represents any cohorts establishing in the understory anytime during the lifespan of the stand (100 years), whereas the q_x pattern for the persistent seedling bank represents the cohorts that establish when there are adequate seedbeds and light conditions at the beginning of the development of the stand.

the good seedbeds and exactly what the real range of ages in this bank would be. We can, however, presume that the window of opportunity during which the good conditions are met is quite narrow and that this event is crucial for the future of the stand. Precise counting of above- and below-ground bud scars indicates that seedlings can be as old as 60 years or more (Parent et al. 2001). Nevertheless, the persistent seedling bank constitutes a very discrete entity with a very low mortality. The theoretical figures of the age-specific mortality rate (q_x) given in Box 8.2 are based on data compiled from permanent plot monitoring for an understory cohort originating from a mast year in a mature stand (Box 8.2*a*) and from the persistent seedling bank (Box 8.2*b*). The rate corresponding to the constitution of the persistent seedling bank is based on the assumption that seeds germinated on a suitable seedbed in a favorable light environment.

Why is this regeneration phenomenon not detected in studies of static stand age structures? Without precise dating of the seedlings, no distinction can be made between the persistent and transient seedling banks. Inaccurate dating can lead to the impression that the age structure shows a typical reverse J shaped curve with a decreasing mortality rate, leading to the presence of one type of seedling bank, the transient seedling bank, and that there are no old seedlings. Permanent plot monitoring of the seedlings permits us to demonstrate that the persistent seedling bank has very low mortality in comparison to the transient seedling bank (Box 8.2).

Putting birth and death together

Putting together the birth and mortality models presented above, we can reconstruct the theoretical age distribution of a balsam fir stand after a spruce budworm outbreak (Box 8.3). The persistent seedling bank will rapidly start to grow following canopy mortality. Let us assume that the seedlings have an apparent 20 year range in ages and a structure like those suggested by the seedling bank age structures already presented (Morin and Laprise 1997). As these seedlings grow, seeds from nearby less-affected stands or survivors within the stand provide a continuous seed supply. Then, one of two scenarios can occur: (1) the density of the growing regeneration is very high; or (2) there are some open patches in the regeneration.

In the first scenario, there are a large number of very young seedlings but very few survive under the dense canopy of the growing seedlings. The seedlings grow up and some 15–20 years later they start to produce seeds themselves. There are still a few survivors, and this situation prevails until self-thinning of the stand (that can be triggered by another budworm outbreak some 30–40 years later) permits some seedlings to survive and occupy the best seedbeds left by the previous outbreaks, namely moss mats on old rotten logs.

If the second scenario prevails, the best seedbeds could be occupied earlier in the development of the stand. Since the stand is relatively young after release, it will be less affected than a mature one and the outbreak will have a thinning effect rather than causing extensive canopy death. The seedlings present at this time will constitute the persistent seedling bank. Mortality which was high at the beginning will stabilize and the few survivors will persist during the lifespan of the stand. During this life span, numerous seedlings will germinate during good seed years but very few will survive. If an outbreak kills the majority of the mature trees 30–40 years later, or, as is often the case, after two outbreaks 60–80 years later, the persistent seedling bank will take over the canopy after some 60–120 years of growth. This persistent seedling bank will play a crucial role in the dynamics of balsam fir stands. Its build-up determines the regenerative capacity of a stand subjected to spruce budworm outbreaks. Indeed, the transient seedling bank is very fragile, as it depends largely on mast years of seed production, and it does not persist long enough to permit adequate stand regeneration.

The theoretical figure in Box 8.3 is based on:
(1) data from permanent plots for the age structure of the growing seedlings 20–30, 50–60, and 80–90 years after release of the stand; they represent apparent ages and age structures because determining the exact age of mature balsam fir is not possible; however, the number of trees is realistic;
(2) the same distribution of seedlings/m^2 that was proposed for the recruits (Fig. 8.26); these are exact ages, since they come from permanent plot monitoring;

Box 8.3. *Balsam fir stand development under the influence of periodic spruce budworm outbreaks.*

Fig. 8.25. Reconstructed age distributions for a typical balsam fir stand after the growth release caused by a spruce budworm outbreak.

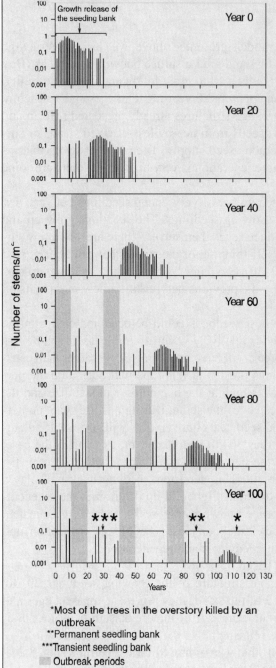

Year 0 seedling age structure is based on the apparent age structure of a typical balsam fir seedling bank (Morin and Laprise 1997).

The decreasing number of stems in this cohort is based on typical apparent age structure from stands 10, 20, 40, 50, 70, and 80 years after release (Morin 1994).

Age of the recruits is real age based on the recruitment pattern from permanent plot monitoring (Box 8.2).

Spruce budworm outbreaks occurred in this stand between 21–30, 51–60, and 81–90 years.

Age-specific mortality rates, as defined in Box 8.2, have been applied to permanent and transient seedling banks over the lifespan of the stand.

A high mortality rate was applied to the first cohorts of seedlings germinating under the dense seedling bank (year 0–20 years) until a spruce budworm outbreak thinned the stand between 21–30 years.

A lower mortality rate typical of the persistent seedling bank was then applied to these seedlings, and a high mortality rate typical of the transient seedling bank was applied to the other seedlings growing under the canopy on less favourable seedbeds.

The last outbreak killed the majority of the trees in the overstory.

*Most of the trees in the overstory killed by an outbreak
**Permanent seedling bank
***Transient seedling bank
Outbreak periods

Fig. 8.26. Theoretical seedling recruitment (without mortality) in a balsam fir stand after growth release of the seedling bank. The figure illustrates the lower recruitment at the beginning of the development of the stand because of seeding in from outside the stand, and the typical variability in seedling recruitment caused by recorded mast years. The estimates are made from permanent plot monitoring in stands after 10, 40, and 70 years of release.

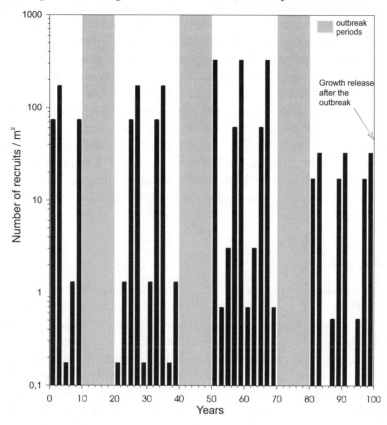

(3) the age-specific mortality rate applied to these recruits over time as determined from permanent plot data (Box 8.2); and

(4) the hypothesis that outbreaks occurred between 21 and 30 years, 51 and 60, and 81 and 90 years after the first opening, a typical cycle for the budworm (Jardon 2001).

We think that this model of balsam fir regeneration following a budworm outbreak is a more realistic representation of the dynamics of balsam fir populations. If we could date precisely all the cohorts in the population, we would expect this to be the result. It should be noted that the persistent seedling bank in this model is the result of the first cohorts to survive after the opening of the stand. These seedlings can be very old, almost as old as the developing stand itself. It must also be emphasized that this persistent seedling bank is the future of the stand and that it will form the next stand after the opening of the canopy. This interpretation is in contrast to many other studies that suggest that younger seedlings, and even seedlings that germinate during or after the outbreak, will form the next stand. In this model, younger seedlings are always smaller than the persistent seedlings and they have little chance, if any, to be dominant or even to reach the canopy.

Some management implications

It is usually thought that balsam fir regeneration is widespread and that it occurs readily after natural or anthropogenic disturbances. This is based on the assumption that there is a continuous supply of seedlings to the seedling bank and that all those seedlings can participate in the regeneration of the stand and reach the canopy. The model presented here is not in accordance with this assumption. Rather, regeneration of balsam fir stands relies heavily on a persistent seedling bank. If the persistent seedling bank is not present, the transient seedling bank is a very fragile one on which to count if a disturbance affects the stand. In this case, if there was a run of poor seed production years or if there was a budworm outbreak that stopped seed production for 10–15 years, the survivors of the transient seedling bank would be insufficient to adequately regenerate the stand. A good example of a situation where the persistent seedling bank is not present and where there are problems with regeneration is that of dense balsam fir populations managed on short 40-year rotations (Côté and Bélanger 1991). In these heavily stocked stands, there is not enough light on the ground to permit seedlings to survive. Also, these intensively managed stands have few logs (the best seedbeds for survival of balsam fir seedlings) on the ground.

Conclusions

The importance of natural disturbance to the dynamics of forest trees is well recognized. Also, forest dynamics are now recognized as occurring at the scale of both local and regional tree populations. Unfortunately, a precise understanding of how disturbances actually affect the dynamics of both regional and local tree populations is far from understood. Past disturbance studies have generally not attempted to develop this understanding but have focused mostly on simply describing patterns of disturbances and their effects. This descriptive approach does not lead to an understanding of the dynamics (i.e., dispersal, birth, recruitment, and mortality rates) of these tree populations, nor of how different disturbances affect these rates. We hope that some of the discussion in this chapter has revealed the need for a more precise understanding of the connection between the disturbance processes and population dynamics processes, a connection that is essential for sustainable management of forests (Box 8.4).

The studies of disturbances caused by wildfires and insect outbreaks discussed in this chapter have shown the importance of studying the cohort structure of local tree populations. They have also shown that only certain cohorts are responsible for regeneration of the canopy following disturbance, despite relatively continuous germination and seedling establishment. Following either wildfire or insect outbreaks, the cohort that will replace the dead canopy trees establishes under conditions of abundantly available seedbed and relatively high light conditions. All subsequent seedlings establishing under less optimal conditions have lower recruitment and higher mortality rates such that their chances of surviving long enough to grow into the canopy are virtually zero.

The situation described above differs from the traditional idea of succession in which the forest is viewed as a series of cohorts replacing each other. One of the aspects of the traditional notion of succession has been a rather linear view of how forest

---

Box 8.4. Options for improving sustainable forest management:

- Recognize that fire and insect disturbances in the boreal forest are part of the ecosystem and cannot and should not be completely excluded. Therefore, these disturbances and their effects on tree population dynamics (age-specific mortality, recruitment, cohort structure) should be taken into consideration in any forest management plans (e.g., in setting sustainable harvest levels).
- Recognize that, as a result of such disturbances, many tree populations operate on a regional scale with dispersal and local population extinctions as important processes. Thus, landscape-scale forest management should incorporate these regional-scale population processes and not focus solely on stand-level stages such as stand initiation, stem exclusion, understory regeneration, and old growth.
- Finally, recognize that an understanding of tree population dynamics at both the local and regional scales indicates that traditional succession arguments do not adequately explain the forest dynamics. Changes in forest composition and structure over time are a result of these population dynamic processes.

---

dynamics work. Even the more recent ideas of succession consisting of multiple pathways (Fastie 1995) do not seem to completely encompass the incredible flexibility of tree populations. Despite the constraints of the life history strategies of each species population (e.g., serotinous and nonserotinous cones, shade tolerance, growth rate, etc.), these species are capable of using these strategies in a very diverse manner, depending on the particular way disturbances affect their populations. The only way to understand this flexibility is to understand the disturbances and their effects on tree population dynamics (i.e., birth and death rates). Careful studies of these disturbances and the ways in which they affect population dynamics at both the local and regional scale may lead to some novel approaches to forest management.

Acknowledgements

We gratefully acknowledge funding support to all of the authors from the Sustainable Forest Management Network and the Natural Sciences and Engineering Research Council of Canada Discovery Grants Program, and to H.M. and R.J. from the Consortium de recherche sur la forêt boréale commerciale. Many thanks to François Gionest, Pierre-Yves Plourde, and Danielle Laprise for their help with field work and data processing, and Marie Puddister for producing all of the final figures for the chapter.

References

Baskerville, G. 1975. Spruce budworm — super silviculturist. For. Chron. **51**: 4–6.
Baskerville, G. 1986. Understanding forest management. For. Chron. **62**: 339–347.
Batzer, H.O., and Popp, M.P. 1985. Forest succession following a spruce budworm outbreak in Minnesota. For. Chron. **61**: 75–80.
Bélanger, L., Allard, D., and Meek, P. 1993. Dynamique d'établissement d'un peuplement bi-étagé de bouleau blanc et de sapin baumier en zone boréale. For. Chron. **69**: 173–177.

Bergeron, Y., and Archambault, S. 1993. Decreasing frequency of forest fires in the southern boreal zone of Quebec and its relation to global warming since the end of the 'Little Ice Age'. Holocene, **3**: 255–259.

Bergeron, Y., and Charron, D. 1994. Postfire stand dynamics in the southern boreal forest (Québec): a dendroecological approach. Ecoscience, **1**: 173–184.

Bergeron, Y., and Harvey, B. 1997. Basing silviculture on natural ecosystem dynamics: an approach applied to the southern boreal mixedwood forest of Quebec. For. Ecol. Manage. **92**: 235–242.

Bergeron, Y., Leduc, A., Morin, H., and Joyal, C. 1995. Balsam fir mortality following the last spruce budworm outbreak in northwestern Quebec. Can. J. For. Res. **25**: 1375–1384.

Blais, J.R. 1952. The relationship of the spruce budworm (*Choristoneura fumiferana* (Clem.)) to the flowering condition of balsam fir (*Abies balsamea* (L.) Mill.). Can. J. Zool. **30**: 1–29.

Blais, J.R. 1954. The recurrence of spruce budworm infestations in the past century in the Lac Seul area of northwestern Ontario. Ecology, **35**: 62–71.

Blais, J.R. 1983. Trends in the frequency, extent, and severity of spruce budworm outbreaks in eastern Canada. Can. J. For. Res. **13**: 539–547.

Blais, J.R. 1984. Réflexions sur l'épidémiologie de la tordeuse des bourgeons de l'épinette (*Choristoneura fumiferana* (Clem.)) suite à 40 années d'études. Rev. Entomol. Qué. **29**: 27–34.

Boulet, B. 1996. Le suivi à long terme des populations de la tordeuse des bourgeons de l'épinette Choristoneura fumiferana (Clem.), au moyen de pièges à phéromone. *In* Comptes rendus du Séminaire sur la TBE. En savons-nous assez pour lutter efficacement contre la tordeuse? Gouvernement du Québec, Ministère de ressouces naturelles, Direction de la recherche Forestière. Quebec, Quebec. pp. 49–57.

Bridge, S.R.J. 2001. Spatial and temporal variations in the fire cycle across Ontario. Northeast Science and Technology, Ontario Ministry of Natural Resources, South Porcupine, Ontario. NEST Tech. Rep. TR-043. Available at http://nesi.mnr.gov.on.ca/spectrasites/internet/nesi/media/documents/main/netr043.pdf [viewed 4 June 2003]. 36 p.

Bridge, S.R.J., and Johnson, E.A. 2000. Geomorphic principles of terrain organization and vegetation gradients. J. Veg. Sci. **11**: 57–70.

Bunnell, F.L. 1995. Forest-dwelling vertebrate faunas and natural fire regimes in British Columbia. Conserv. Biol. **9**: 636–644.

Candau, J.N., Fleming, R.A., and Hopkin, A. 1998. Spatiotemporal patterns of large-scale defoliation caused by the spruce budworm in Ontario since 1941. Can. J. For. Res. **28**: 1733–1741.

Charron, I. 1998. Sexual recruitment of trees following fire in the southern mixedwood boreal forest of Canada. M.Sc. thesis, Department of Biological Sciences, Concordia University, Montréal, Quebec. 109 p.

Charron, I., and Greene, D.F. 2002. Post-wildfire seedbeds and tree establishment in the southern mixedwoodboreal forest. Can. J. For. Res. **32**: 1607–1615.

Chrosciewicz, Z. 1974. Evaluation of fire-produced seedbeds for jack pine regeneration in central Ontario. Can. J. For. Res. **4**: 455–457.

Chrosciewicz, Z. 1976. Burning for black spruce regeneration on a lowland cutover site in southeastern Manitoba. Can. J. For. Res. **6**: 179–186.

Cissel, J.H., Swanson, F.J., and Weisberg, P.J. 1999. Landscape management using historical fire regimes: Blue River, Oregon. Ecol. Appl. **9**: 1217–1231.

Côté, S., and Bélanger, L. 1991. Variations de la régénération préétablie dans les sapinières boréales en fonction de leurs caractéristiques écologiques. Can. J. For. Res. **21**: 1179–1795.

Coutts, M.P., and Grace, J. (*Editors*) 1995. Wind and trees. Cambridge University Press, Cambridge, U.K. 485 p.

Cumming, S.G. 2001. Forest type and wildfire in the Alberta boreal mixedwood: what do fires burn? Ecol. Appl. **11**: 97–110.

DeLong, S.C., and Tanner, D. 1996. Managing the pattern of forest harvest: lessons from wildfire. Biodiv. Conserv. **5**: 1191–1205.

DesRochers, A., and Gagnon, R. 1997. Is ring count at ground level a good estimation of black spruce age? Can. J. For. Res. **27**: 1263–1267.

Drysdale, D. 1985. An introduction to fire dynamics. John Wiley and Sons, Chichester, U.K. 424 p.

Duchesneau, R., and Morin, H. 1999. Early seedling demography in balsam fir seedling banks. Can. J. For. Res. **29**: 1502–1509.

Dyrness, C.T., and Norum, R.A. 1983. The effects of experimental fires on black spruce forest floors in interior Alaska. Can. J. For. Res. **13**: 879–893.

Fastie, C.L. 1995. Causes and ecosystem consequences of multiple pathways of primary succession at Glacier Bay, Alaska. Ecology, **76**: 1899–1916.

Flannigan, M.D., and Wotton, B.M. 2001. Climate, weather, and area burned. *In* Forest fires: behavior and ecological effects. *Edited by* E.A. Johnson and K. Miyanishi. Academic Press, San Diego, California. pp. 351–373.

Flannigan, M.D., Todd, B., Wotton, M., Stocks, B., Skinner, W., and Martell, D. 2000. Pacific sea surface temperatures and their relation to area burned in Canada. *In* Third Symposium on Fire and Forest Meteorology. American Meteorological Society, Boston, Massachusetts. pp. 151–157.

Frandsen, W.H. 1991. Smoldering spread rate: a preliminary estimate. *In* Proceedings of the Eleventh Conference on Fire and Forest Meteorology. *Edited by* P.L. Andrews and D.F. Potts. Society of American Foresters, Bethesda, Maryland. pp. 168–172.

Galipeau, C., Kneeshaw, D., and Bergeron, Y. 1997. White spruce and balsam fir colonization of a site in the southeastern boreal forest as observed 68 years after fire. Can. J. For. Res. **27**: 139–147.

Ghent, A.W. 1958. Studies of regeneration in forest stands devastated by the spruce budworm II. age, height, growth and related studies of balsam fir seedlings. For. Sci. **4**: 135–146.

Ghent, A.W., Fraser, D.A., and Thomas, J.B. 1957. Studies of regeneration in forest stands devastated by the spruce budworm I. evidence of trends in forest succession during the first decade following budworm devastation. For. Sci. **3**: 184–208.

Gray, D., Régnière, J., and Boulet, B. 1998. Prédiction de la défoliation par la tordeuse des bourgeons de l'épinette au Québec. Service Canadien des forêts, Centre de foresterie des Laurentides, Sainte-Foy, Quebec. Notes de Recherche No 7. 8 p.

Gray, D., Régnière, J., and Boulet, B. 2000. Analysis and use of historical patterns of spruce budworm defiolation to forecast outbreak patterns in Quebec. For. Ecol. Manage. **127**: 217–231.

Greene, D.F., and Johnson, E.A. 1989. A model of wind dispersal of winged or plumed seeds. Ecology, **70**: 339–347.

Greene, D.F., and Johnson, E.A. 1992a. Fruit abscission in *Acer saccharum* with reference to seed dispersal. Can. J. Bot. **70**: 2277–2283.

Greene, D.F., and Johnson, E.A. 1992b. Can the variation in samara mass and terminal velocity on an individual plant affect the distribution of dispersal distances? Am. Nat. **129**: 825–838.

Greene, D.F., and Johnson, E.A. 1994. Estimating the mean annual seed production of trees. Ecology, **75**: 642–647.

Greene, D.F., and Johnson, E.A. 1995. Long-distance wind dispersal of tree seeds. Can. J. Bot. **73**: 1036–1045.

Greene, D.F., and Johnson, E.A. 1996. A model for the dispersal of seeds by wind from a forest to a clearing. Ecology, **77**: 595–609.

Greene, D.F., and Johnson, E.A. 1998. Seed mass and early survivorship of tree species in upland clearings and shelterwoods. Can. J. For. Res. **28**: 1307–1316.

Greene, D.F., and Johnson, E.A. 1999. Modelling recruitment of *Populus tremuloides*, *Pinus banksiana*, and *Picea mariana* following fire in the mixedwood boreal forest. Can. J. For. Res. **29**: 462–473.

Greene, D.F., and Johnson, E.A. 2000. Tree recruitment from burn edges. Can. J. For. Res. **30**: 1264–1274.

Greene, D.F., Zasada, J.C., Sirois, L., Kneeshaw, D., Morin, H., Charron, I., and Simard, M.J. 1999. A review of the regeneration dynamics of North American boreal forest tree species. Can. J. For. Res. **29**: 824–839.

Greene, D.F., Kneeshaw, D., Messier, C., Lieffers, V., Cormier, D., Doucet, R., Coates, K.D., Groot, A., Grover, G., and Calogeropoulos, C. 2002*a*. Modelling silvicultural alternatives for conifer regeneration in boreal mixedwood stands (aspen/white spruce/balsam fir). For. Chron. **78**: 281–295.

Greene, D.F., Messier, C., Asselin, H., and Fortin, M.-J. 2002*b*. The effect of light availability and basal area on cone production in *Abies balsamea* and *Picea glauca*. Can. J. Bot. **80**: 370–377.

Gustafson, E.J. 1998. Quantifying landscape spatial pattern: what is the state of the art? Ecosystems, **1**: 143–156.

Gutsell, S.L. 2001. Understanding forest dynamics incorporating both local and regional ecological processes. Ph.D. dissertation. University of Calgary, Calgary, Alberta. 228 p.

Gutsell, S.L., and Johnson, E.A. 2002. Accurately aging trees and examining their height growth rates: implications for interpreting forest dynamics. J. Ecol. **90**: 153–166.

Hack, J.T. 1957. Studies of longitudinal stream profiles in Virginia and Maryland. United States Geological Survey, Washington, D.C. Prof. Pap. 294-B. 97 p.

Hansen, A.J., Spies, T.A., Swanson, F.J., and Ohmann, J.L. 1991. Conserving biodiversity in managed forests — lessons from natural forests. BioScience, **41**: 382–392.

Hanski, I., and Gilpin, M.1991. Metapopulation dynamics: brief history and conceptual domain. Biol. J. Linn. Soc. **42**: 3–16.

Hardy, Y., Mainville, M., and Schmitt, D.M. 1986. An atlas of spruce budworm defoliation in eastern North America 1938–1980. USDA Forest Service, Cooperative State Research Service, Washington, D.C. Misc. Pub. No. 1449. 52 p.

Holling, C.S., and Meffe, G.K. 1996. Command and control and the pathology of natural resource management. Conserv. Biol. **10**: 328–337.

Horel, J.D., and Wallace, J.M. 1981. Planetary-scale atmospheric phenomena associated with the Southern Oscillation. Mon. Weath. Rev. **109**: 813–829.

Hungerford, R.D., Frandsen, W.H., and Ryan, K.C. 1995. Ignition and burning characteristics of organic soils. *In* Proceedings of the Tall Timbers Fire Ecology Conference on Wetlands: A Management Perspective. *Edited by* S.I. Cerulean and R.T. Engstrom. Tall Timbers Research Station, Tallahassee, Florida. pp. 78–91.

Hunter, M.L., Jr. 1993. Natural fire regimes as spatial models for managing boreal forests. Biol. Conserv. **65**: 115–120.

Hunter, M.L., Jr. 1996. Fundamentals of conservation biology. Blackwell Science, Cambridge, Massachusetts. 482 p.

Jardon, Y. 2001. Analyses temporelles et spatiales des épidémies de la tordeuse des bourgeons de l'épinette au Québec. Ph.D. thesis, Université du Québec à Chicoutimi, Chicoutimi, Quebec. 157 p.

Jenkins, M.A., Clark, T., and Coen, J. 2001. Coupling atmospheric and fire models. *In* Forest fires: behavior and ecological effects. *Edited by* E.A. Johnson and K. Miyanishi. Academic Press, San Diego, California. pp. 258–302.

Johnson, E.A. 1985. Disturbance: the process and the response — an epilogue. Can. J. For. Res. **15**: 292–293.

Johnson, E.A. 1992. Fire and vegetation dynamics: studies from the North American boreal forest. Cambridge University Press, Cambridge, U.K. 129 p.

Johnson, E.A., and Fryer, G.I. 1989. Population dynamics in lodgepole pine – Engelmann spruce forests. Ecology, **70**: 1335–1345.

Johnson, E.A., and Gutsell, S.L. 1994. Fire frequency models, methods, and interpretations. Adv. Ecol. Res. **25**: 239–287.

Johnson, E.A., and Larsen, C.P.S. 1991. Climatically induced change in fire frequency in the southern Canadian Rockies. Ecology, **72**: 194–201.

Johnson, E.A., and Miyanishi, K. 1996. The need for consideration of fire behavior and effects in prescribed burning. Restoration Ecol. **3**: 271–278.

Johnson, E.A., and Miyanishi, K. 2001. Strengthening fire ecology's roots. *In* Forest fires: behavior and ecological effects. *Edited by* E.A. Johnson and K. Miyanishi. Academic Press, San Diego, California. pp. 1–9.

Johnson, E.A., and Van Wagner, C.E. 1985. The theory and use of two fire history models. Can. J. For. Res. **15**: 214–220.

Johnson, E.A., and Wowchuk, D.R. 1993. Wildfires in the southern Canadian Rocky Mountains and their relationship to mid-tropospheric anomalies. Can. J. For. Res. **23**: 1213–1222.

Johnson, E.A., Miyanishi, K., and Bridge, S.R.J. 2001. Wildfire regime in the boreal forest and the idea of suppression and fuel build-up. Conserv. Biol. **15**: 1554–1557.

Jones, J.C., Goh, T.P.T., and Dijanosic, M.J. 1994. Smouldering and flaming combustion in packed beds of *Casuarina* needles. J. Fire Sci. **12**: 442–451.

Kareiva, P. 1990. Population dynamics in spatially complex environments: theory and data. Phil. Trans. Roy. Soc. London B. Biol. Sci. **330**: 175–190.

Kaufmann, M.R., Graham, R.T., Boyce, D.A., Moir, W.H., Perry, L., Reynolds, R.T., Bassett, R.L., Mehlhop, P., Edminster, C.B., Block, W.M., and Corn, P.S. 1994. An ecological basis for ecosystem management. USDA Forest Service, Fort Collins, Colorado. Gen. Tech. Rep. RM-246. 22 p.

Kettela, E.G. 1983. A cartographic history of spruce budworm defoliation from 1967 to 1981 in eastern North America. Canadian Forest Service, Fredericton, New Brunswick. Inf. Rep. DPC-X-14. 8 p.

Kneeshaw, D.D., and Bergeron, Y. 1998. Canopy gap characteristics and tree replacement in the southeastern boreal forest. Ecology, **79**: 783–794.

Krause, C. 1997. The use of dendrochronological material from buildings to get information about past spruce budworm outbreaks. Can. J. For. Res. **27**: 69–75.

Landres, P.B., Morgan, P., and Swanson, F.J. 1999. Overview of the use of natural variability concepts in managing ecological systems. Ecol. Appl. **9**: 1179–1188.

MacLean, D.A. 1980. Vulnerability of fir-spruce stands during uncontrolled spruce budworm outbreaks: a review and discussion. For. Chron. **56**: 213–221.

MacLean, D.A. 1984. Effects of spruce budworm outbreaks on the productivity and stability of balsam fir forests. For. Chron. **60**: 273–279.

MacLean, D.A. 1988. Effects of spruce budworm outbreaks on vegetation, structure, and succession of balsam fir forests on Cape Breton Island, Canada. *In* Plant form and vegetation structure. *Edited by* M.J.A. Werger, P.J.M. van der Aart, H.J. During, and J.T.A. Verhoeven. SPB Academic Publishing, The Hague, Netherlands. pp. 253–261.

MacLean, D.A., and MacKinnon, W.E. 1997. Effects of stand and site characteristics on susceptibility and vulnerability of balsam fir and spruce budworm in New Brunswick. Can. J. For. Res. **27**: 1859–1871.

Malamud, B.D., Morein, G., and Turcotte, D.L. 1998. Forest fires: an example of self-organized critical behavior. Science, **281**: 1840–1841.

Martell, D.L. 1980. The optimal rotation of a flammable forest stand. Can. J. For. Res. **10**: 30–34.

McAlpine, R.S., and Wotton, B.M. 1993. The use of fractal dimension to improve wildland fire perimeter predictions. Can. J. For. Res. **23**: 1073–1077.

McGarigal, K., and Marks, B.J. 1995. FRAGSTATS: spatial pattern analysis program for quantifying landscape structure. USDA Forest Service, Portland, Oregon. Gen. Tech. Rep. PNW-351. 122 p.

McLaren, B.E., and Janke, R.A. 1996. Seedbed and canopy cover effects on balsam fir seedling establishment in Isle Royale National Park. Can. J. For. Res. **26**: 782–793.

Miyanishi, K. 2001. Duff consumption. *In* Forest fires: behavior and ecological effects. *Edited by* E.A. Johnson and K. Miyanishi. Academic Press, San Diego, California. pp. 437–475.

Miyanishi, K., and Johnson, E.A. 2002. Process and patterns of duff consumption in the mixedwood boreal forest. Can. J. For. Res. **32**: 1285–1295.

Morin, H. 1990. Analyse dendroécologique d'une sapinière issue d'un chablis dans la zone boréale, Québec. Can. J. For. Res. **20**: 1753–1758.

Morin, H. 1994. Dynamics of balsam fir forests in relation to spruce budworm outbreaks in the boreal zone of Quebec. Can. J. For. Res. **24**: 730–741.

Morin, H. 1998. Importance et évolution des épidémies de la tordeuse des bourgeons de l'épinette dans l'est du Canada: l'apport de la dendrochronologie. Geogr. Phys. Quat. **52**: 237–244.

Morin, H., and Laprise, D. 1990. Histoire récente de la Tordeuse des bourgeons de l'épinette au nord du Lac Saint-Jean (Québec): une analyse dendrochronologique. Can. J. For. Res. **20**: 1–8.

Morin, H., and Laprise, D. 1997. Seedlings bank dynamics in boreal balsam fir forests. Can. J. For. Res. **27**: 1442–1451.

Morris, R.F. 1963. The dynamics of epidemic spruce budworm populations. Mem. Ent. Soc. Can. **31**: 7–27.

Osawa, A. 1994. Seedling responses to forest canopy disturbance following a spruce budworm outbreak in Maine. Can. J. For. Res. **24**: 850–859.

Palmer, K.N. 1957. Smouldering combustion in dusts and fibrous materials. Combust. Flame, **1**: 129–154.

Parent, S., Morin, H., and Messier, C. 2000. Effects of adventitious roots on age determination in balsam fir (*Abies balsamea* (L.) Mill.) regeneration. Can. J. For. Res. **30**: 513–518.

Parent, S., Morin, H., and Messier, C. 2001. Balsam fir (*Abies balsamea* (L.) Mill.) establishment dynamics during a spruce budworm (*Choristineura fumiferana* (Clem)) outbreak in the a south-boreal forest: an evaluation of the impact of aging techniques. Can. J. For. Res. **31**: 373–376.

Parent, S., Morin, H., and Messier, C. 2002. Missing growth rings at the trunk base in suppressed balsam fir saplings. Can. J. For. Res. **32**: 1776–1783.

Parsons, D.J., Swetnam, T.W., and Christensen, N.L. 1999. Uses and limitations of historical variability concepts in managing ecosystems. Ecol. Appl. **9**: 1177–1178.

Pickett, S.T.A., and White, P.S. (*Editors*). 1985. The ecology of natural disturbance and patch dynamics. Academic Press, Orlando, Florida. 472 p.

Powell, G.R. 1973. The spruce budworm and megasporangiate strobili of balsam fir. Can. J. For. Res. **3**: 424–429.

Pulliam, H.R. 1988. Sources, sinks, and population regulation. Am. Nat. **132**: 652–661.

(QMER) Quebec Ministère de l'Énergie et des Ressources. 1985. La nouvelle politique forestière. Ministère de l'Énergie et des Ressources, Gouvernement du Québec, Québec, Québec.

Rasmussen, E.M., and Wallace, J.M. 1983. Meteorological aspects of the El Niño/Southern Oscillation. Science, **222**: 1195–1202.

Raup, H.M. 1956. Vegetational adjustment to the instability of the site. *In* Proceedings of the 6th Technical Meeting of the International Union Conservation of Nature and Natural Resources, Edinburgh, 1956. International Union for the Conservation of Nature and Natural Resources, London, U.K. pp. 36–48.

Reed, W.J. 1984. The effects of the risk of fire one the optimal rotation of a forest. J. Env. Econ. Manage. **11**: 180–190.

Reed, W.J., and Errico, D. 1985. Assessing the long-run yield of a forest stand subject to the risk of fire. Can. J. For. Res. **15**: 680–687.

Reed, W.J., and Errico, D. 1986. Optimal harvest scheduling at the forest level in the presence of the risk of fire. Can. J. For. Res. **16**: 266–278.

Reed, W.J., and McKelvey, K.S. 2002. Power-law behaviour and parametric models for the size-distribution of forest fires. Ecol. Model. **150**: 239–254.

Régnière, J., and Lysk, T.J. 1995. Population dynamics of the spruce budworm, *Choristoneura fumiferana* (Clem). *In* Forest insect pests in Canada. *Edited by* J.A. Armstrong and W.G.H. Ives. Canadian Forest Service, Science and Sustainable Development Directorate, Ottawa, Ontario. pp.1–11.

Ricotta, C., Avena, G., and Marchetti, M. 1999. The flaming sandpile: self-organized criticality and wildfires. Ecol. Model. **119**: 73–78.

Ripple, W.J., Bradshaw, G.A., and Spies, T.A. 1991. Measuring forest landscape patterns in the Cascade Range of Oregon, USA. Biol. Conserv. **57**: 73–88.

Routledge, R.D. 1980. The effect of potential catastrophic mortality and other unpredictable events on optimal forest rotation policy. For. Sci. **26**: 389–399.

Schroeder, M.J. 1950. The Hudson Bay high and the spring fire season in the lake states. Bull. Am. Meteor. Soc. **31**: 111–118.

Simard, M.J., Bergeron, Y., and Sirois, L. 1998. Conifer seedling recruitment in a southeastern Canadian boreal forest: the importance of substrate. J. Veg. Sci. **9**: 575–582.

Sirois, L. 1997. Distribution and dynamics of balsam fir (*Abies balsamea* (L.) Mill.) at its northern limit in the James Bay area. Ecoscience, **4**: 340–352.

Sivapalan, M., Wood, E.F., and Beven, K.J. 1987. On hydrologic similarity. 2. a scaled model of storm runoff production. Water Resour. Res. **23**: 2266–2278.

Skinner, W.R., Stocks, B.J., Martell, D.L., Bonsal, B., and Shabbar, A. 1999. The association between circulation anomalies in the mid-troposphere and area burned by wildland fire in Canada. Theor. Appl. Clim. **63**: 89–105.

Slocombe, D.S. 1993. Implementing ecosystem-based management. BioScience, **43**: 612–622.

St-Pierre, H., Gagnon, R., and Bellefleur, P. 1991. Distribution spatiale de la régénération après feu de l'épinette noire (*Picea mariana*) et du pin gris (*Pinus banksiana*) dans la forêt boréale, Réserve faunique Ashuapmushuaj, Québec. Can. J. Bot. **69**: 717–721.

St-Pierre, H., Gagnon, R., and Bellefleur, P. 1992. Régénération après feu de l'épinette noire (*Picea mariana*) et du pin gris (*Pinus banksiana*) dans la forêt boréale, Québec. Can. J. For. Res. **22**: 474–481.

Swanson, F.J., Jones, J.A., Wallin, D.O., and Cissel, J.H. 1994. Natural variability implications for ecosystem management. *In* Eastside forest ecosystem health assessment. Vol. 2, ecosystem management: principles and applications. *Edited by* M.E. Jensen and P.S. Bourgeron. USDA Forest Service, Portland, Oregon. Gen. Tech. Rep. PNW-318. pp. 89–104.

Tchir, T.L. 2000. A model of fragmentation resulting from human settlement in the boreal forest of Saskatchewan. M.Sc. thesis, University of Calgary, Calgary, Alberta. 86 p.

Thomas, P.A., and Wein, R.W. 1985. The influence of shelter and the hypothetical effect of fire severity on the postfire establishment of conifers from seed. Can. J. For. Res. **15**: 148–155.

Thompson, K., and Grime, J.P. 1979. Seasonal variation in the seed banks of herbaceous species in ten contrasting habitats. J. Ecol. **67**: 893–921.

Turner, M.G., O'Neill, R.V., Gardner, R.H., and Milne, B.T. 1989. Effects of changing spatial scale on the analysis of landscape pattern. Landscape Ecol. **3**: 153–162.

Tymstra, C., McGregor, C., Quintilio, D., and O'Shea, K. 1998. Is fire a wildcard in Alberta's protected areas strategy for forest conservation. *In* Linking protected areas with working landscapes conserving biodiversity. Proceedings of the Third International Conference on Science and Management of Protected Areas. *Edited by* N. Munro and J.H.M. Willison. Science and Management of Protected Areas Association, Wolfville, Nova Scotia. pp. 542–551.

Van Wagner, C.E. 1977. Conditions for the start and spread of crown fires. Can. J. For. Res. **7**: 23–34.

Van Wagner, C.E. 1983. Fire behavior in northern conifer forests and shrublands. *In* The role of fire in northern circumpolar ecosystems. *Edited by* R.W. Wein and D.A. MacLean. John Wiley and Sons, New York, New York. pp. 65–80.

Walsh, J.E., and Richman, F.M.B. 1981. Seasonality in the associations between surface temperature over the United States and the North Pacific Ocean. Mon. Weath. Rev. **10**: 767–782.

Weber, M.G., Hummel, M., and Van Wagner, C.E. 1987. Selected parameters of fire behavior and *Pinus banksiana* Lamb. regeneration in eastern Ontario. For. Chron. **63**: 340–346.

Wein, R.W. 1981. Characteristics and suppression of fires in organic terrain in Australia. Austral. For. **44**: 162–169.

Wein, R.W. 1983. Fire behavior and ecological effects in organic terrain. *In* The role of fire in northern circumpolar ecosystems. *Edited by* R.W. Wein and D.A. MacLean. John Wiley and Sons, New York, New York. pp. 81–85.

Weir, J.M.H., Johnson, E.A., and Miyanishi, K. 2000. Fire frequency and the spatial age mosaic of the mixed-wood boreal forest in western Canada. Ecol. Appl. **10**: 1162–1177.

Williams, D.W., and Liebhold, A.M. 2000. Spatial synchrony of the spruce budworm outbreaks in eastern North America. Ecology, **81**: 2753–2766.

Zasada, J.C., Norum, R.A., Van Veldhuizen, R.M., and Teutsch, C.E. 1983. Artificial regeneration of trees and tall shrubs in experimentally burned upland black spruce/feather moss stands in Alaska. Can. J. For. Res. **13**: 903–913.

Chapter 9

Comparing forest management to natural processes

Sybille Haeussler and Dan Kneeshaw

> *". . . what we observe is not nature itself but nature*
> *exposed to our method of questioning."*

Werner Heisenberg (1958)

Introduction

Forest management can be defined as human intervention into the nature, extent, and timing of disturbance to forest ecosystems for the purpose of obtaining desired goods and services. Foresters and other land managers have long observed that many of the ecological consequences of logging resemble those of natural disturbances and have used these similarities to enhance the productivity and supply of important forest resources such as timber and game (Leopold 1933; Smith 1986). Since the mid-1980s, however, there has been a major reassessment of the relationship between forestry and the disturbance processes that operate in natural, unmanaged environments. Disturbance agents such as wildfire, wind, insects, and flooding, once viewed as devastating, wasteful, and a threat to human progress, are today increasingly seen as essential to the diversity and functioning of forest ecosystems and as a model or template for progressive forest management. With this change in perspective has come a need to go beyond the assumption that all disturbances have equivalent effects on forest ecosystems and to critically examine the long-held notion that clearcut logging and its variants are the most appropriate cutting practice for boreal forest ecosystems because boreal forests are adapted to stand-destroying fire.

Recent scientific interest in disturbance has fuelled an explosion of work in boreal, temperate, and tropical systems to characterize all types of natural disturbances and their effects on biological diversity and ecosystem processes (see Chap. 8). Since 1995, much of the research related to natural disturbance in Canadian boreal forests has been carried

S. Haeussler[1] **and D. Kneeshaw.** Groupe de recherche en écologie forestière interuniversitaire, Université du Québec à Montréal, C.P. 8888, Succ. Centre-ville, Montréal, Quebec H3C 3P8, Canada.
[1]Author for correspondence. Telephone: 250-847-6082. e-mail: skeena@bulkley.net

Correct citation: Haeussler, S., and Kneeshaw, D. 2003. Comparing forest management to natural processes. Chapter 9. *In* Towards Sustainable Management of the Boreal Forest. *Edited by* P.J. Burton, C. Messier, D.W. Smith, and W.L. Adamowicz. NRC Research Press, Ottawa, Ontario, Canada. pp. 307–368.

out under the umbrella of the Sustainable Forest Management Network (SFMN). A major component of Network research has been to characterize, in detail, the similarities and differences between logging and associated forest practices and natural disturbance processes, especially wildfire. Information garnered from these comparisons and from similar work done elsewhere has been used to develop and test innovative silvicultural approaches (see Chap. 13) that reduce discrepancies between natural processes and forest management (Veeman et al. 1999; Lindenmayer and McCarthy 2002).

This chapter reviews this work. We begin by introducing pertinent ecological concepts such as the natural disturbance model, ecosystem integrity, and natural variability; we then go on to show how these concepts apply to Canadian boreal forest management at regional, landscape, and stand scales of planning and operations. At each scale, we identify similarities and differences between logging or other silvicultural practices and natural disturbance processes, then outline recommendations or give examples of innovative forest management based on natural disturbance dynamics. Because Canadian experience in boreal forest management is mostly short term, we have also drawn on information from similar forests in the United States of America, Fennoscandia, and Russia to assess some of the longer term outcomes of boreal forest management.

This review concentrates on comparisons between forest management and wildfire. It has been accepted wisdom that fire is the dominant disturbance agent in unmanaged boreal forests. Thus, most of the natural disturbance research in boreal forests has been fire based, and most operational applications of a natural disturbance based approach to boreal forest management are built on a model of stand-destroying wildfire. However, as Bergeron et al. (1998) and Cumming et al. (2000) point out, the role of fire may have been overemphasized. New research is showing that gap or patch disturbances caused by insects, pathogens, large animals, wind, snow, and ice damage are more common in boreal forests than previously thought, and that in some regions and situations they are more important than wildfire (Kuuluvainen 1994; Kneeshaw and Burton 1997; Kneeshaw and Bergeron 1998; Cumming et al. 2000; McCarthy 2001). A comprehensive natural disturbance based approach to forest management cannot focus exclusively on wildfire dynamics, ignoring the role that smaller scale and less severe disturbances have played in structuring boreal stands and landscapes. But while it is intuitively obvious that as fire suppression reduces wildfire incidence, logging may be able to take its place in stand initiation, it is less clear how management practices affect the occurrence of insect, disease, and storm damage and what benefits might be gained by emulating these agents in a managed forest setting. Moreover, the interactions between wildfire, fire suppression, logging history, and these other agents of forest disturbance form an extremely tangled web of which forest research is just beginning to make sense. We describe some of these interconnections and their implications for forest management in this chapter. The challenge is to determine the degree to which the natural disturbance model can and cannot provide guidance for sustainable forest management (Box 9.1).

Our topic here is terrestrial forest ecosystems and the impacts of forest management on terrestrial organisms and forest soils. A complementary research program in aquatic ecology is providing important insights into the ecological consequences of accelerated forest harvesting on Canadian boreal waters. This research effort, summarized in Chap.

Box 9.1. Challenges to sustainable forest management:

- The natural disturbance model asserts that forest management must retain the variability and unpredictability to which species are adapted and under which ecosystems have developed if it is to maintain biological diversity and sustain ecosystem processes; in contrast, traditional forest management relies on command-and-control approaches in which variability and unpredictability must be reduced and controlled;

- The goal of maintaining ecological integrity challenges forest managers to look beyond short-term, stand-scale concerns and consider ecosystem function across a huge range of spatial and temporal scales;

- Disturbance rates and human impacts vary greatly across the Canadian boreal forest; at both national and regional scales, the extent to which forest management compensates for, or is additive to, natural disturbance processes remains highly uncertain;

- At landscape scales, the pattern of disturbances and the age and species structure of forests are important to biological diversity and ecosystem function; forest management tends to narrow the range of patterns observed and introduces elements (such as roads) that have no natural analogue;

- At stand scales, changes in the distribution and type of residual organic matter left behind further modify community composition and threaten the viability of both fire-dependent and late successional forest-dependent species and communities;

- Interactions between activities or events can produce unexpected or cascading effects at a variety of scales; one such example is an increased abundance of trembling aspen (*Populus tremuloides*) in many boreal landscapes;

- Substantial discrepancies exist between forest practices and natural disturbance processes at all scales; much can be done to reduce these discrepancies, but no matter how sensitive and intelligent our interventions may be, forest management will result in progressive and cumulative changes to boreal forests as we know them today; and

- A major challenge will lie in deciding to what degree boreal forestry should aim to counteract or offset the rapid rate of ecological change by striving to remain within the range of historic variability, or should adopt practices that embrace and accommodate change.

10, contains many parallel developments to the work done in terrestrial systems and expands our appreciation of system-wide links in ecological function.

The natural disturbance model for ecosystem management

During the 1990s, a consensus emerged among ecologists that the most promising approach for conserving biological diversity and ecosystem function in managed forests was to emulate the disturbance processes that drive forest succession and dynamics in natural, unmanaged forests (Hunter 1993; Attiwill 1994; Christiansen et al. 1996; Bergeron and Harvey 1997; Kohm and Franklin 1997). The natural disturbance model or paradigm has been rapidly adopted by land and resource management agencies as a means for achieving sustainable forest management goals which stress the maintenance of biological diversity and long-term sustainability of ecosystem processes in addition to traditional concerns for optimizing resource use and economic stability. This approach incorporates understanding of natural disturbance processes into the development and assessment of management alternatives (Gauthier et al. 1996; Coates and Burton 1997; Angelstam 1998; Bergeron et al. 1999*b*; Lundquist and Beatty 2002).

Although the natural disturbance model is being applied to ecosystem management and resource development in forest and non-forest biomes all over the world (Attiwill 1994; Michener et al. 1998), it is particularly appealing in boreal forests for two reasons. First, boreal forests include some of the largest remaining tracts of sparsely settled land in which natural disturbances continue to dominate. These areas can serve as reference or benchmark systems for study. Second, the prevalence of large, stand-destroying wildfire and insect outbreaks in boreal ecosystems holds out the promise that these forests may be one system in which industrial-scale logging can be carried out sustainably without deviating substantially from natural forest dynamics (Pastor et al. 1998; Burton et al. 1999). In Canada, governments, the forest products industry, and multi-agency research partnerships such as the SFMN have been relatively quick to incorporate elements of a natural disturbance approach into corporate policy and legislation. However, implementation of the natural disturbance model has lagged behind both scientific understanding (Schindler 1998*b*; see also Chap. 8) and policy rhetoric (Boutin 1999).

The conceptual or scientific basis for the natural disturbance model hinges on a single, compelling hypothesis, which has its basis in Darwin's theory of evolution by natural selection. The hypothesis is that because biota have evolved with and have adapted to events such as fire, storms, flooding, and herbivory, they are more likely to maintain viable populations if novel human interventions such as timber harvesting are as similar as possible to the environmental conditions that prevailed during their evolution (Heinselman 1971; Hunter 1990). Likewise, ecosystem processes such as carbon, nutrient and water cycling, and community succession are more apt to be maintained if the frequency, spatial extent, and severity of disturbance conform to those that shaped these processes in the past. The natural disturbance model is an extension of the "coarse filter" approach to biological conservation (Noss 1987; Hunter 1991), which assumes that conserving a representative array of undamaged ecosystems within a region will maintain viable populations of the vast majority of species in that region, many of which have habitat requirements (even identities) that are still unknown. The natural disturbance model also formally recognizes that ecosystems and landscapes are dynamic entities and that stabilizing outputs (for example, by establishing non-declining, even-flow targets for timber or wildlife production) is neither achievable nor desirable.

There is broad agreement that the challenge for forest managers is to develop techniques of resource extraction that do not cause local extinctions of species, reduce ecosystem productivity, or damage the processes that underlie the resiliency of forest ecosystems to disturbance (Pastor et al. 1998, Chap. 8). However, the means by which to accomplish this goal within highly dynamic boreal forest ecosystems are far from clear. As outlined by Chapin et al. (1996) and Landres et al. (1999), a sustainable forest economy is one in which management interventions do not cause the system to move beyond or outside the range of natural oscillations characteristic of the system (Fig. 9.1*a*, *b*). Holling and Meffe (1996) proposed a closely related but potentially conflicting hypothesis which asserts that the pathology of conventional "command and control"

Fig. 9.1. Range of variability in disturbance events and ecosystem traits in unmanaged and managed systems: (*a*) and (*c*) illustrate the range in disturbance size, severity, and frequency under unmanaged (lightly shaded sphere) and managed (heavily shaded sphere) conditions (adapted from Swanson et al. 1994); (*b*) and (*d*) represent temporal fluctuations in an ecosystem trait or indicator before and after management begins. In the upper diagrams (*a* and *b*), management interventions have caused the system to move outside the natural range of variability. In the lower diagrams (*c* and *d*), the system remains within the range of natural variability but management has constrained the amplitude of oscillations.

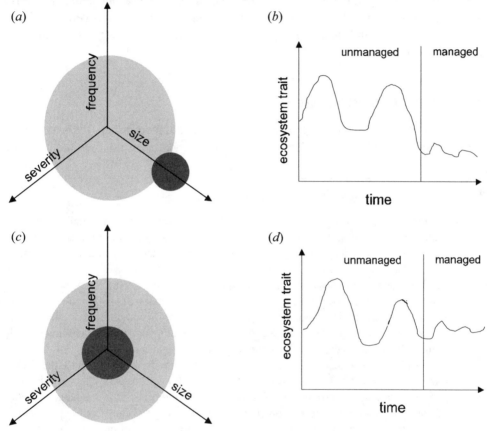

management is that it seeks to reduce the range of variability occurring in dynamic and unpredictable natural systems (Fig. 9.1c, d). They argue that enforced reduction in the variability of oscillations causes managed ecosystems to become less resilient to disturbance or stress, and eventually leads to a collapse in their ability to supply resources and services of value to humans. Holling and Meffe (1996) suggested a "Golden Rule Of Natural Resources Management" which states that natural resource management should strive to retain critical types and ranges of natural variability in ecosystems. While the rule itself would likely meet with little disagreement among ecologists and natural resource managers today, the difficulty lies in determining what types and ranges of natural variability are critical to ecosystem function, what deviations from natural disturbance regimes are acceptable, and how best to achieve these in managed systems.

Despite the intuitive or logical appeal of the natural disturbance model, it remains an unproven, working hypothesis. In Canadian boreal forests, a variety of short-term tests of the model have been completed by means of chronosequence studies (e.g., Lee et al. 1997), and several large-scale and long-term interdisciplinary projects have recently been initiated (Brais et al. 1999; Volney et al. 1999). At the long time scales and huge spatial scales that are meaningful for forestry operations, it is neither logistically possible nor ethically defensible to formally test the hypothesis that an agricultural model of forest management will drive many species to extinction and cause widespread collapse of ecosystems. History has, however, supplied us with several unplanned experiments on a grand scale that provide important lessons for Canadian boreal forestry (Box 9.2).

Defining ecological integrity

The concept of ecological or biological integrity was adopted in the 1990s as a means of broadening the goals of ecosystem management to encompass not only the living ele-

---

Box 9.2. Loss of ecological integrity, some examples.

- *In Fennoscandia*, centuries of settlement and forest exploitation caused major alteration to disturbance regimes. A regime of intensive plantation silviculture was established 80–120 years ago to address widespread deforestation and land degradation, resulting in a managed landscape with very high control over tree species composition, stocking density, and age-class structure, with minimal losses to wildfire and other disturbance agents and rigorous elimination of dead, dying, and non-commercial trees (Östlund et al. 1997). Fennoscandian forestry is a global success story in terms of wood production, but success in the forest products sector has come at a high cost to indigenous biological diversity. In Sweden, some 1500 species are threatened by forest practices (Berg et al. 1994); in Finland, Hanski (2000) predicts that past forest management will ultimately cause the extinction of more than 1000 forest-dwelling species. By contrast, in the neighbouring underdeveloped Russian province of Karelia, wildfire and other natural agents of disturbance continue to dominate boreal forest dynamics. This region retains viable populations of many of the large mammal, plant, fungal, and arthropod species that have become rare and endangered in Fennoscandia, and it is serving as a benchmark to guide the restoration of Finnish and Swedish forest ecosystems (Siitonen and Martikainen 1994; Kuuluvainen 2002).
- *In Canada's Maritime Provinces*, European settlement began in the 1600s. Forest fragmentation from agricultural settlement plus several centuries of progressive high-grading to remove the largest, most valuable timber resulted in gradual decline of the original forest. In the 1950s, a major expansion of the forest industry, coupled with an aggressive insecticide program to prevent large-scale timber losses to the eastern spruce budworm (*Choristoneura fumiferana*), caused sudden and drastic changes to system dynamics, stand age structure, and species composition (MacLean 1984; Baskerville 1988). The cumulative result was large tracts of highly degraded publicly owned forest with reduced timber values, chronic susceptibility to insect attack, and reduced resistance to other environmental stresses (Baskerville 1988).

ments of ecosystems but also the full suite of physical, chemical, and biological processes that give rise to their diversity, productivity, and resilience. In other words, it explicitly considers not only the composition or structure of ecosystems, but their function as well. The forestry examples described in Box 9.2 illustrate the two sides of the ecosystem integrity coin. In the first example of Fennoscandian forestry, there is no strong evidence that current forest practices are causing a reduction in the productive capability of the forest to sustain high rates of timber production (Örlander et al. 1996; Aamlid et al. 2000), but it is clear that intensive forest management has caused losses of biological diversity at multiple scales and for a wide variety of taxonomic groups. In the example of the Canadian Maritime Provinces, concerns have focused primarily on the decline in productivity and resilience of degraded forest ecosystems. But these two facets of ecosystem integrity are parts of a single interdependent whole. If the natural disturbance hypothesis and Holling and Meffe's (1996) corollary arguments concerning natural variability in ecosystems are correct, Fennoscandian forestry will not be able to sustain the productivity and health of its forests over the long term in the presence of massively simplified and tightly regulated ecological systems. By the same token, the biological diversity of Maritime forests (if not already greatly reduced) is unlikely to be maintained in forest ecosystems with an artificially maintained age and species structure.

Angermeier and Karr (1994) defined ecological integrity as "*a system's wholeness, including presence of all appropriate elements and occurrence of processes at appropriate rates*". Parks Canada, which has a policy goal of maintaining ecological integrity in Canada's national parks, defines it as "*a condition where the structure and function of an ecosystem are unimpaired by human-induced stresses*" (Bouchard 1997). While many other definitions exist, these two examples highlight the strengths and limitations of the concept. What is common to all definitions is that the degree of ecological integrity is measured in terms of the system's departure from an unimpacted benchmark system (see Box 9.3 for some definitions).

The strength of the ecological integrity concept is that it provides a more comprehensive, whole-system focus for natural resource management than the maintenance of biological diversity, which focuses exclusively on living things. It explicitly recognizes that biologically diverse ecosystems cannot exist without a complex air–water–earth life-support system (Rowe 1992). Assigning appropriate emphasis to the maintenance of ecosystem processes may also help to overcome a species-centred preoccupation with numbers of taxa and species extinctions that result from the mistaken impression that biodiversity is synonymous with species diversity (Angermeier and Karr 1994; Purvis and Hector 2000). Moreover, it helps to shift the focus of management from a reactive to a preventative mode. In a system with relatively few, highly versatile species and low functional redundancy, changes in the rates of ecosystem processes such as successional patterns or carbon and nutrient cycling may be detected well before local and regional species extinctions occur. The more holistic focus on ecosystem integrity seems particularly appropriate for boreal forest management because of the coarse scale of forest management and ecological monitoring operations over large areas and because most boreal forest taxa are widely dispersed, resilient to disturbance, and lack specialized

> **Box 9.3. Some definitions.**
>
> *Range of variability* is a non-statistical term that can refer to the standard deviation, a 95% confidence interval, a percentile range, or the full range of observations (minimum to maximum), depending on the particulars of the study (Landres et al. 1999). It is most often presented as a distribution (e.g., the stand age distributions in Boxes 9.4 and 9.7). What is important is that the oscillations of the system around its central tendency are of greater interest than the central tendency itself. For example, it is more useful to know the complete size distribution of wildfires in a region than it is to know the size of the average fire (Hunter 1993). The abbreviations *NRV* (natural range of variability) or *HRV* (historic range of variability) are often used to describe management approaches that use the range of conditions in unmanaged systems or prior to European settlement, respectively, as guidelines or targets for ecosystem management.
>
> *Benchmarks* — The range of variability is observed in reference systems that have had minimal human influence but are otherwise as similar as possible to the modified system(s). In the boreal forest, either past conditions (historic and prehistoric) within the same tract of land, or contemporary conditions within a nearby, unmanaged ecosystem or landscape can serve as benchmarks. They are analogous to experimental controls, though not necessarily in a statistically rigorous sense.
>
> *Non-linear or threshold responses* — Decisions about what degree of deviation from benchmark conditions or what level of human impact is deemed acceptable rely heavily on the identification of non-linear or threshold responses to disturbance parameters (Frelich and Reich 1998; Romme et al. 1998; Chapin et al. 2000). If an indicator variable displays a linear response to disturbance, it can be very difficult to decide objectively how much disturbance is acceptable, since less (or more) is always better. However, many ecological processes and systems exhibit either non-linear (curved) or threshold (stepped) responses to external phenomena. If such changes to ecosystem condition are irreversible, then acceptable limits of change may have to be set well back from the threshold to reduce the risk of a permanent loss of ecosystem integrity (Brown et al. 1999). An example of an ecological threshold from the North American boreal forest is the effect of frequent fires on conifer regeneration. Areas of burned-over black spruce (*Picea mariana*) or jack pine (*Pinus banksiana*) forest will fail to regenerate if a second fire occurs before the regenerating trees are old enough to produce a cone crop. In this case, the threshold fire frequency would be the number of years required for juvenile trees to produce a sizeable cone crop.

habitat requirements. Thus, they are unlikely to serve as highly sensitive, early warning indicators of impending ecosystem damage.

The major limitations of the ecological integrity concept relate to the difficulty of establishing appropriate benchmarks or limits to acceptable change (Haila 1995; Landres et al. 1999; Millar and Woolfenden 1999):

(1) Who decides what rates of ecological processes are acceptable? Some analysts argue that such decisions are technical matters that should be established through scientific investigation, while others believe these are primarily social questions for which all different value perspectives should be heard (see Chap. 4);

(2) If ecological integrity describes the condition of an ecosystem that has had no human impact, then by definition, all human impact decreases ecological integrity and is "bad" for ecosystems (see Box 2.8). In this view, "anthropogenic stress is a debilitating agent, not a revitalizing one" (Rapport and Whitford 1999), thus there is no need to define acceptable change initiated by humans nor acceptable management practice, as such a thing can not logically exist. Minimal human involvement or interaction with forest ecosystems becomes the de facto target for resource management;

(3) By extension, all "natural" change to forest ecosystems is deemed acceptable or "good". This view ignores evidence that severe natural disturbances such as repeated fires can cause long- term losses of diversity and forest productivity even in relatively pristine ecosystems and that deteriorations in ecosystem condition such as forest decline often result from the cumulative or simultaneous imposition of anthropogenic and natural stresses such as drought, insect, or disease outbreaks; and

(4) Finally, there is the difficulty of distinguishing between natural and human-caused environmental change, given the non-equilibrium conditions and constant change that prevail in all ecological systems. Past or even current conditions may not be useful benchmarks when weather systems, physiography, and biota are undergoing rapid change. Furthermore, global and long-distance transport of pollutants such as greenhouse gases, ozone-depleting chemicals, acid rain, toxic aerosols, and invasive species make it impossible to locate or judge truly non-impacted benchmark systems.

Assessing variability at multiple scales

Variability in ecosystems occurs in nested hierarchies of process and response occurring over a complete spectrum of spatial and temporal scales. There is no single scale that is appropriate for describing, examining, or understanding ecological phenomena (Levin 1992; Parker and Pickett 1998). In the case of wildfire, rates of disturbance are best observed at spatial scales several orders of magnitude larger than the largest disturbance event (approximately 100 000 ha) and temporal scales several orders of magnitude longer than the average interval between events (approximately 100 years). On the other hand, responses of organisms to the disturbance are best observed at spatial scales relevant to the body size of the organism and its home range or dispersal distance, and at time scales (typically greater than the life span of individuals) that allow demographic processes, such as reproduction and mortality, and the succession of plant and animal communities to be observed. Although evolutionary time scales (typically thousands to millions of years required for speciation) are relevant to natural selection for adaptations to disturbance, they are not considered in detail here.

Hunter (1993), in one of the first papers to explicitly consider the use of natural disturbance regimes as spatial models for boreal forest management, identified three different ways to use wildfires as models for logging:

(1) match the rate of harvest to the periodicity of wildfire;

(2) match the size and distribution of openings to patterns created by wildfire; and

(3) leave residual organic matter such as slash, logs, snags, live trees, seedlings, and seeds behind to more closely match the legacy left by fire.

Each of Hunter's recommendations involves a different scale of observation, planning, and execution, and corresponds to a level in the traditional hierarchy of forest management (Chap. 1; Tittler et al. 2001a):

(1) The periodicity of wildfire and the harvest rate are determined at broad geographic scales corresponding to administrative or ecoclimatic regions (see Chap. 11);

(2) The size and distribution of wildfires and cutblocks are delineated on maps and air photos at landscape or watershed scales (see Chap. 12); and

(3) Observations and decisions about the amount of residual plant life left behind are made in the field within forest stands (see Chap. 13).

This list of possibilities outlines the range of spatial scales to be considered when comparing forest management to wildfire. While Hunter's (1993) analysis was restricted to wildfire, the same approach can be applied to other disturbances that operate in unmanaged forests. Table 9.1 gives examples of bioindicators or indices that can be used to monitor biodiversity and ecosystem processes at regional, landscape, and stand scales. For each of these indicators, the natural disturbance model poses four questions about the variability found in managed versus unmanaged forests:

(1) Do forest practices create conditions outside the range of variability (see Box 9.3) observed in unmanaged forests?

(2) Do forest practices replicate the full range of variability observed in unmanaged forests?

(3) If significant differences exist in either 1 or 2 above, how do they affect the maintenance of biological diversity and long-term ecosystem function?

(4) How can forest practices be modified to address these differences and minimize the loss of ecological integrity?

Table 9.1. Framework for a multiscale comparison of forest management and natural disturbance processes with some example indicators of ecological integrity.

| | Example indicators of ecological integrity | | | |
| | Biodiversity | | | Ecological |
Scale	Structure	Composition	Diversity	processes
Region 10^8–10^5 ha	Land use	Ecological zones	Variety of distinct landscapes	Disturbance rate
Landscape 10^5–10^2 ha	Pattern of openings	Forest cover	Variety of stand types and ages	Succession
Stand 10^2–10^{-2} ha	Living and dead trees	Species composition	Within-community species richness	Destruction of organic matter

National and regional variability

The big picture

An important initial question is whether logging is indeed replacing natural disturbances when viewed from a nationwide perspective. Kurz and Apps (1999) compiled Canadian wildfire, logging, and insect mortality statistics dating back to the 1920s, the earliest date for which figures are available (Fig. 9.2a). From 1920 to 1970, the data show that the area of wildfires diminished, insect mortality first rose then fell, and the area logged climbed steadily (Barney and Stocks 1983), resulting in no net increase in the total area disturbed over this period. After 1970, the figure shows a sharp increase in both wildfire and insect mortality coinciding with a major expansion of the Canadian forest products industry, approximately doubling the total annual area disturbed in comparison to previous decades. The accuracy and impact of Fig. 9.2a is weakened by the fact that many forest fires and insect outbreaks were not detected or recorded prior to 1960 (NRCan 1997). However, Van Wagner (1988) and Kurz et al. (1995) believe that the large increase in annual area disturbed within Canada after the 1970s is real. They con-

Fig. 9.2. Canada-wide rates of wildfire, insect outbreaks, and logging disturbance at three time scales. (*a*) Five-year averages, 1920–2000 (updated from Kurz and Apps 1999); (*b*) annual variability, 1980–2000 (Canadian Forest Service data); (*c*) and (*d*) two contrasting hypothetical scenarios for the total area disturbed over the past millennium. In (*c*) the trend since 1980 is anomalously high; in (*d*) it is well within the historical range of variability. No Canada-wide data are available for this period. Insect disturbance rates in (*a*) and (*b*) use the multipliers of Kurz and Apps (1999) to convert area moderately or severely defoliated into area killed.

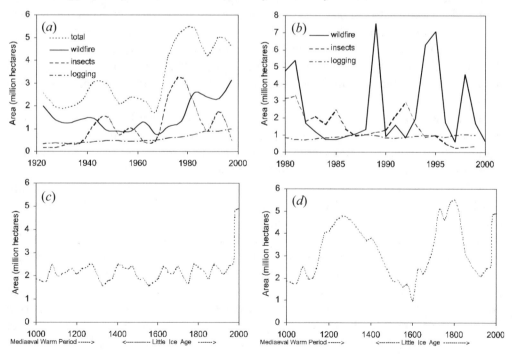

tend that this increase resulted from warmer and drier climatic conditions (the 1990s and 1980s were the warmest decades on record in Canada (Environment Canada 1999)). They also state that the increase was caused by a shift in the forest age-class distribution towards older, more susceptible stands during the preceding period of low disturbance frequency (but see contrary arguments by Bessie and Johnson 1995; Johnson et al. 1998).

Figure 9.2b–d show heuristically how a multi-scale analysis of natural variability can help to provide additional insight into the degree of ecological threat posed by recent fire, insect, and logging trends. In these examples, the time period was scaled down or up by one order of magnitude while the spatial scale, Canada, was unchanged.

Figure 9.2b shows inter-annual variability in wildfire, insect mortality, and logging rates for the period 1980–2000. At this scale of observation, wildfire rates fluctuate wildly at 1–3-year intervals according to fire weather conditions, insect outbreaks display 3–7-year oscillations, but logging rates are relatively invariable because of cut-control policies imposed to provide stability of employment. Logging alone clearly does not retain the natural range of interannual variability. Does this represent a critical type of natural variation in the sense of Holling and Meffe (1996) or is it trivial? Can we identify meaningful ecological processes, such as effects of smoke production and dispersion on regional precipitation patterns or uncoupling of cyclic predator–prey dynamics (e.g., Krebs et al. 2001), that might be affected if this interannual variability were eliminated in highly regulated forest systems? To date, no such critical ecological relationships have been identified or explored.

Figure 9.2c and d depict two hypothetical disturbance rate scenarios scaled up to the time period from 1000 AD to 2000 AD. According to the natural disturbance model, whether current high rates of fire, insect, and logging disturbance represent a major and unprecedented threat to boreal ecosystem integrity will depend on which of the two scenarios more closely represents the natural or historic range of variability. In the first case (Fig. 9.2c) the recent rise in disturbance rates is anomalously high; that is, it lies outside the range of variability experienced during the past millennium, and is probably environmentally risky and socially unacceptable (from the perspective of social values; see Chap. 4).

In the second scenario (Fig. 9.2d), the current peak in disturbance rates is just one in a series of peaks that have occurred in the recent past in response to exogenous climatic cycles and endogenous factors such as changes in forest age structure. With this scenario, the acceptability of recent rates of disturbance is less clear. The *Golden Rule* of Holling and Meffe (1996) asserts that maintaining such variability will help to maintain boreal ecosystem integrity, while attempts to artificially dampen the cycle with increased fire and insect suppression and reduced logging could cause ecosystem collapse on a grand scale. But the arguments in favour of curbing disturbance rates are also cogent. Schindler (1998a, b) has argued persuasively that boreal forest ecosystems are facing an unprecedented degree of threat from climate change, atmospheric deposition of acids and toxic trace metals, and stratospheric ozone deposition. In his view, current rates of clearcut logging constitute an unacceptable additional stress during a period when fire frequencies and other mortality factors are at historic highs. Carbon budget modelling of the Canadian boreal forest (Kurz et al. 1995; Kurz and Apps 1995, 1996,

1999) suggests that during the 1980s, boreal forests became a net source of atmospheric C, after acting as a carbon sink for the period 1920–1980. Moreover, disturbances such as forest fires and logging have a positive feedback effect that may accelerate the rate of global warming (Kurz et al. 1995; Weber and Stocks 1998; see Chap. 20). The need to consider multiple threats to ecosystem integrity at global scales may override concerns about maintaining historic and prehistoric rates of disturbance variability at national and lower scales.

Unfortunately, there are no Canada-wide data available from prior to 1920 to assess which of the two hypothetical scenarios more accurately reflects the past. There is a growing cross-continental network of boreal fire histories developed with dendrochronological, time-since-fire mapping techniques (Heinselman 1973; Yarie 1981; Foster 1982; Suffling et al. 1982; Cogbill 1985; Bergeron 1991; Bergeron and Archambault 1993; Dansereau and Bergeron 1993; Larsen 1997; AEM 1998a[2]; Lefort et al. 1999; Weir et al. 1999). These fire studies, together with a smaller number of palaeoecological reconstructions (Swain 1973; Carcaillet et al. 1999; Laird 1999; Laroque et al. 1999) suggest that large-scale changes in fire activity on the order of scenario Fig. 9.2d have occurred throughout the Holocene. There may have been a previous peak in fire activity near the end of the Little Ice Age (mid- to late 1800s), a period when many of the studies cited above reported a shift to lower fire frequencies. Disturbance chronologies for major insect pests such as the eastern spruce budworm (*Choristoneura fumiferana*) and forest tent caterpillar (*Malacosoma disstria*) have been reasonably well developed using dendrochronological and macrofossil evidence (Blais 1983; Morin et al. 1993), but other non-fire disturbances such as blowdown are poorly documented (but see Jonsson and Dynesius 1993; Ruel 1995; Ulanova 2000). Here again, available evidence suggests that large-scale oscillations in tree mortality rates are characteristic of boreal forest dynamics.

The studies cited above were independent regional analyses that have not been systematically pooled to assess continent-wide trends. It is possible that there was sufficient regional asynchrony in major disturbance years (e.g., Foster 1982) to dampen the range of variability at a national scale.

Regional variability

Canadians tend to think of their boreal forest as a vast, homogeneous biome that stretches unbroken from the Atlantic to the Pacific Coast. It is true that the characteristic plant and animal species of the boreal zone, such as black spruce (*Picea mariana*), white spruce (*P. glauca*), trembling aspen (*Populus tremuloides*), Labrador tea (*Ledum groenlandicum*), woodland caribou (*Rangifer tarandus caribou*), lynx (*Lynx canadensis*), and snowshoe hare (*Lepus americanus*), have continent-wide distributions, and that regional distinctions are much less pronounced than in the temperate zone. However, there is substantial regional variation within the boreal forest that affects ecosystem

---

[2]Applied Ecosystem Management Ltd. 1998a. Fire history of the Little Rancheria caribou herd winter range. Unpublished report for Yukon Renewable Resources and Indian and Northern Affairs Canada, Whitehorse, Yukon. 41 p.

Fig. 9.3. Rates and types of disturbance in the (*a*) western and (*b*) eastern boreal forest of Canada, 5-year averages from 1920 to 1990. Insect disturbances include a multiplier that converts area defoliated to total area killed (from Kurz and Apps 1999).

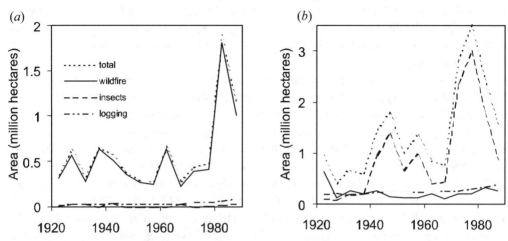

rates and processes, levels of biological diversity, and the degree to which the ecological integrity of the region is at risk. Two of the most important regional differences are the frequency and prevalence of wildfire versus other disturbances, and the degree of human intervention or land use.

Kurz and Apps (1999) separated the Canadian boreal forest into two regions, west and east of the Ontario–Manitoba boundary, and tallied rates of wildfire, logging, and insect disturbances for the period 1920–1989 (Fig. 9.3). What is evident is a stark contrast in the relative importance of wildfire and insects as disturbance factors. In the Boreal West, fire has been and continues to be the dominant disturbance factor and vastly outstrips the area logged, particularly since the 1970s when huge areas burned in the northern Prairie Provinces and adjacent territories in response to short periods of extreme fire weather (Weber and Stocks 1998). In the Boreal East, fires have been less important than logging for most of the past century while insect mortality, mainly spruce budworm damage, has affected a far greater area than either fire or logging combined. Here, as in the West, there was a large increase in total forest disturbance after 1970. The recent, coincident rise in disturbance rates is attributed to a combination of climatic warming, increases in average forest age since the end of the Little Ice Age (Bergeron and Flannigan 1995), and changes in stand structure and species composition induced by a history of logging and fire and insect suppression (Blais 1983; Baskerville 1988), all of which lead to increased susceptibility to insects, disease, windthrow, and possibly fire (Fryer and Johnson 1988; Kurz and Apps 1999).

Figure 9.4 gives a spatial overview of the causes and consequences of regional variability in disturbance regimes across Canada. The Canadian boreal forest biome has been subdivided into six broad ecozones based on geological features, climate and forest cover (Fig. 9.4*a*; EWG 1989). "Boreal" ecozones are dominantly forested and contain a mix of spruce, pine, fir and poplar–birch forest whereas the more northerly, colder "Taiga" ecozones are dominated by open black spruce – lichen taiga and wetland and contain productive closed forest only in favourable locations such as on alluvial flats.

Fig. 9.4. Regional variation in the Canadian boreal forest. (*a*) Ecozones based on geology, climate, and forest cover (EWG 1997); (*b*) past fire frequency (modified from Simard 1975 and Johnson 1992); (*c*) current land use (modified from Bird 1998); and (*d*) predicted decrease (–) or increase (+) in fire weather index (FWI) with a $2 \times CO_2$ climate (modified from Bergeron and Flannigan 1995).

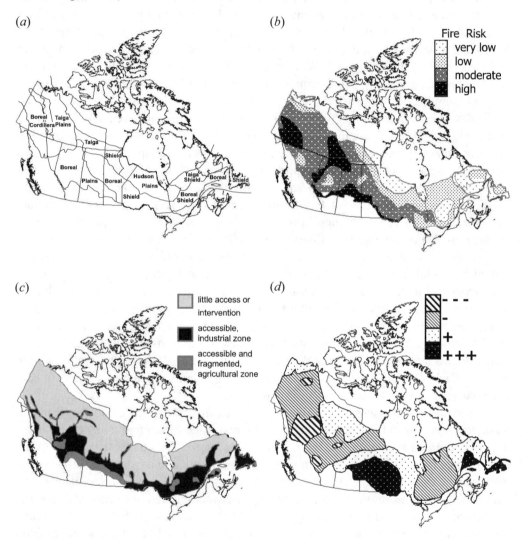

The Boreal Cordillera is a rugged, mountainous ecozone, with climate and vegetation that vary greatly depending on topographic position and elevation. It includes the driest boreal forests in Canada as well as areas of wet climate. The Taiga Plains, Boreal Plains, and Hudson Plains have mostly very flat relief with deep soils, whereas the Boreal Shield and Taiga Shield have hummocky relief with frequent outcrops of Precambrian granite interspersed with wetlands and lakes. East of the Cordillera, the amount of precipitation increases gradually from <300 mm/year in the foothills of the Rocky Mountains to 900–1600 mm/year in parts of Newfoundland (EWG 1989).

Under unmanaged conditions, the frequency and extent of summer fires (Fig. 9.4*b*) generally decreases from south to north as the fire season becomes shorter and the availability of continuous fuels decreases. Open taiga has longer fire cycles and generally smaller fires than occur in closed boreal forest (Payette et al. 1989). Within the Boreal Cordillera, fire frequency is extremely variable (50–300+ year fire cycle) but generally moderate on a region-wide basis because of the lack of large areas of homogeneous terrain. East of the Cordillera, the natural fire frequency is high over much of the Boreal Plains and western Shield, decreasing from a <50–100 year average fire cycle in Alberta and Saskatchewan (Larsen 1997; Weir et al. 1999) to 100–300 years in eastern Ontario and Quebec (Cogbill 1985; Bergeron 1991; Lefort and Leduc 1998[3]) to 500 years or more in humid Labrador (Foster 1982).

Today, the number of fires is highest in the south, but the total area burned is largest in the mid-boreal forest, south of the taiga, where there are extensive uniform tracts of fire-prone black spruce and jack pine (*Pinus banksiana*) or lodgepole pine (*P. contorta*). Human land-use patterns play an important role (Fig. 9.4*c*). Southern, settled areas of boreal forest are highly fragmented and accessible and are dominated by less flammable deciduous and mixedwood forest, interspersed with agricultural land (Tchir and Johnson 1999). There the "natural" fire regime is now almost inconsequential; humans are the most important source of fire ignition, and most fires remain small because there are few extensive and inaccessible tracts of fire-prone coniferous forest. The sparsely settled industrial forest zone, where large-scale clearcut logging takes place, contains more extensive tracts of coniferous forest and is therefore more fire-prone. Both human and lightning fire ignitions are important, but logging disturbance is now generally dominant, except when extremely dry fire years intervene. Lightning-caused fires continue to play the largest role in boreal forest dynamics in the northernmost sections of the Boreal Plains and Boreal Shield and southern Taiga where the tracts of coniferous forest are sufficiently continuous to carry a fire, but too remote and non-productive to have yet been accessed for logging and fire suppression.

A final factor in assessing the relative roles of fire and logging and the degree of threat to the ecological integrity of the boreal forest is the future frequency of fire under a global warming scenario (Fig. 9.4*d*). Although Chap. 20 discusses the effects of climate change in greater detail, we can note that both future climates and the effect of climate change on fire frequency are highly uncertain. Using a combination of historical data analysis that examines past relationships between climatic patterns and fire frequency and a simulation modelling approach that couples global climate models (GCMs) to fire weather index (FWI), Canadian researchers have made some preliminary predictions (Bergeron 1991; Flannigan and Van Wagner 1991; Wotton and Flannigan 1993; Bergeron and Flannigan 1995; Flannigan et al. 1998; Li et al. 2000). The simulations indicate that while overall fire frequency will likely increase under a doubled CO_2 scenario, there will be important regional variations in this trend. Increased CO_2 does not produce a uniform warming trend, but rather causes shifts in major meteorological phenomena such as the southern oscillation (El-Nino effect; Swetnam and

---

[3]Lefort, P., and Leduc, A. 1998. Les perturbation forestières au Québec et leurs implications dans la conservation des écosystèmes forestiers exceptionnels. Unpublished report for Ministère des Ressources naturelles du Québec, Québec, Quebec. 96 p.

Betancourt 1990) and northern persistent blocking high-pressure systems (Johnson and Wowchuk 1993) that affect seasonal temperature and precipitation patterns and the incidence of extreme fire weather at a subcontinental scale. Thus, while some portions of the Canadian boreal forest, notably the central Boreal Plains and adjacent Shield and Taiga ecoregions, are predicted to have longer, warmer, drier summer seasons with a significant increase in fire frequency, others such as the eastern Boreal Shield and Hudson Plains are predicted to have wetter, perhaps even slightly cooler fire seasons with fewer fires. Paradoxically, even an increase in fire frequency with global warming may not translate directly into greater overall fire activity because changes in climate and fire frequency bring concomitant changes in fuel accumulation and fire behaviour. In the Sierra Nevada of California, Swetnam (1993) found that periods of cool weather caused a decrease in fire frequency but an increase in fire size and severity. These examples illustrate the danger of assuming that global warming is a straightforward recipe for increased fire activity.

Past fire frequency, human land use and fire suppression effectiveness, and predicted future fire frequencies can be combined to produce a conceptual model of the degree to which logging threatens the future integrity of boreal forest ecosystems (Table 9.2). In this simple regional model, the degree of risk to boreal integrity is assumed to be highest where the future disturbance regime is most at variance with the past disturbance regime. Thus, if there is minimal human intervention and predicted climatic change is small, the risk is very low. If fire frequency in the past was low (e.g., as in the La Biche River valley of southeastern Yukon; Box 9.4), but both logging rates and future fire frequencies are predicted to be high, then the risk is very high. The reverse scenario, with high past disturbance rates and low future disturbance rates, also produces a high risk of ecological change.

Table 9.2 excludes natural disturbance agents other than fire, under the simplifying assumption that logging can replace fire but that forest management has less impact on other disturbances such as herbivory, pathogens, and storm damage. But as we have seen, these disturbances also exhibit regional variability, and they will certainly change along with the climate. Moreover, all of these agents both affect, and are affected by, rates of wildfire and logging in ways that may either counteract or compound our simple model (McCullough et al. 1998; see Box 9.5 for pertinent examples).

Table 9.2. Risk of loss of ecological integrity based on past and predicted future disturbance regimes.

Fire frequency		Risk to ecological integrity	
Past	Future[a]	Low rate of logging	High rate of logging
Infrequent	Increasing	Moderate–high risk	Very high risk
	No change	Low risk	High risk
	Decreasing	Low risk	High–moderate risk
Frequent	Increasing	Low–moderate risk	High risk
	No change	Low risk	Moderate risk
	Decreasing	High risk	Moderate risk

[a]Based on predicted climate change and degree of fire suppression.

Box 9.4. Old growth in the boreal?

The La Biche River drainage is tucked in the extreme southeast corner of Yukon Territory. It supports a lush, highly productive, and structurally diverse ecosystem with an exceptionally rich flora and fauna, including amphibian, bat, and plant species not recorded elsewhere in Yukon. The La Biche lies just to the east of the Boreal Cordillera, within the Taiga Plains ecozone but is still climatically influenced by nearby mountain ranges. Summer precipitation is high and there are few lightning strikes and very few ignitions. Fire history mapping has shown that in marked contrast with the rest of southern Yukon, less than 5% of the La Biche has been affected by fire within the past 100 years, and most of the area has not burned within the past 250 years (Fig. 9.5). As a consequence, multi-layered, all-aged, mixed stands of white spruce, trembling aspen, paper birch (*Betula papyrifera*), balsam poplar (*Populus balsamifera*), and subalpine fir (*Abies lasiocarpa*) predominate. Lodgepole pine stands are restricted to a few recently burned areas, and black spruce is mainly confined to wetlands. Many trees are able to achieve ages and sizes close to their biological maxima (Fig. 9.6), a rare phenomenon in boreal regions. Gap dynamics driven by eastern spruce budworm, western balsam bark beetle (*Dryocytes confusus*), decay fungi (*Fomitopsis pinicola, Inonotus* spp.), and flooding on the river floodplains are the major forest disturbances. This disturbance regime has given rise to a diverse, structurally complex forest with very different plant and animal communities than are found within the young, mostly even-aged fire-mosaic that dominates the southern Yukon.

Fig. 9.5. Forest age-class distribution of the La Biche River drainage. Approximately 65% of the area is comprised of a single large area of forest that has escaped burning for approximately 250 years (from AEM 1998*b*[4]).

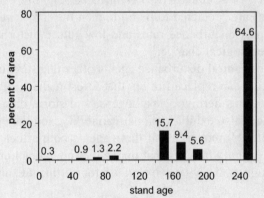

The La Biche River drainage demonstrates that fire may not be the most appropriate planning model for all western boreal forest types. Within this valley, fires occur too infrequently to serve as the basis for calculating allowable cuts, and an even-aged management regime dominated by clearcut logging would not maintain the area's unique old-growth values. The dominant pattern of low intensity, single-tree or small-group mortality would be more effectively reproduced with small patch cuts and selection logging. It is thought that the mixed species composition of La Biche forests prevents spruce budworm and other insects and disease agents from reaching the epidemic outbreak levels that might occur in purer stands.

Sources: AEM 1998*b*[4], Eckert 1998

---

[4]Applied Ecosystem Management Ltd. 1998*b*. La Biche River drainage forest dynamics assessment. Unpublished report for Yukon Renewable Resources and Indian and Northern Affairs Canada, Whitehorse, Yukon. 112 p.

> **Box 9.4. (concluded).**
>
> **Fig. 9.6.** Old-growth *Populus balsamifera – Abies lasiocarpa – Picea glauca* forest in the La Biche River valley with a lush understory that includes atypical species such as devil's club (*Oplopanax horridus*) and lady fern *(Athyrium filix–femina)*.
>
>
>
> *Photo by Bruce Bennett*

Landscape pattern and process

It is immediately apparent looking down from an airplane window or examining an aerial photograph that clearcut logging radically transforms forest landscapes and does so in ways that appear very different from the effects of wildfire, insects, and other natural disturbances. This change in landscape pattern is not just unsettling to look at, but may have important ecological consequences (Box 9.5). In the 1980s a new branch of science known as landscape ecology emerged to examine the impacts of human intervention on the spatial patterning and the functioning of landscapes. While much landscape ecology research is devoted to analyzing and describing spatial pattern itself, at the root of the discipline is the assumption that pattern and process are interrelated. In other words, human changes to landscape pattern will affect biological diversity and rates of critical ecosystem processes, and vice versa. This new science, with its findings on the effects of habitat fragmentation on wildlife ecology and biodiversity, has led to development of the new approaches to forest planning described in Chap. 12.

Two important ecological processes that can be examined at landscape scales are forest succession, and the colonization and population viability of many forest-dwelling species. Hydrological processes such as streamflow and soil erosion are also typically monitored at landscape scales and are discussed further in Chap. 10.

Box 9.5. Positive feedback loops in a changing boreal forest.

Changes in disturbance regimes can affect forest health in indirect and unexpected ways. During the past century, lengthening fire return intervals in Quebec and other regions of eastern Canada have increased the abundance of balsam fir (*Abies balsamea*), a species poorly adapted to fire and the preferred host of the eastern spruce budworm (*Choristoneura fumiferana*). This increase in balsam fir has been directly linked to an increase in the frequency and extent of budworm outbreaks. Logging exacerbates the problem because both current "careful" logging and pre-1970s "selective" logging favour the retention of advance regeneration, thereby increasing the proportion of shade-tolerant balsam fir in future stands.

In the Canadian Prairie Provinces, new research is showing how forest fragmentation and global warming may interact with insect populations to cause regional declines in the health of trembling aspen forests. As a consequence of land clearing and logging, aspen stands today tend to be fragmented into smaller parcels than under former wildfire-dominated regimes. Fragmented landscapes may experience longer forest tent caterpillar (*Malacasoma disstria*) outbreaks than unfragmented landscapes because predatory flies that help keep tent caterpillar populations in check are slower to migrate from one land parcel to the other than the caterpillar itself (Roland 1993). Elevated temperatures along the edges of forest fragments may be another contributing factor. Other forest insects such as wood-boring beetles further compound the drought stress experienced during recent Prairie summers, leading to widespread aspen mortality and crown dieback (Fig. 9.7).

Fig. 9.7. Crown dieback in trembling aspen in central Saskatchewan, likely resulting from cumulative effect of drought and wood-boring insects.

Photo by Ted Hogg, courtesy of Canadian Forest Service

Sources: Blais 1983, Roland and Taylor 1997, Rothman and Roland 1998, Hogg 1999

Forest succession and landscape diversity

The varied landscape mosaic of the boreal forest accounts for much of its aesthetic appeal and is fundamental to its biological diversity. This mosaic of ecosystems results from both physiographic variability and the history of disturbance and subsequent forest succession. A vital question for boreal forest management is how the diversity of the forest mosaic will be modified over time by forest practices. Differences in succession between managed and unmanaged forest landscapes can be assessed by examining the distribution of age classes and tree species composition (cover type) among forest stands (Gauthier et al. 1996; He and Mladenoff 1999). Indices of landscape diversity are essentially measures of the combined variability of these two parameters.

Suffling et al. (1988) used forest cover and stand age to determine landscape diversity across a gradient of forest fire frequency in unmanaged, unaccessed boreal forests of northwestern Ontario. In agreement with the intermediate disturbance hypothesis, they found that landscapes with intermediate fire frequencies (0.2–0.3% of land area burned per year) had a greater diversity of community types than landscapes with either lower or higher fire frequencies (Fig. 9.8). The implication of these findings is that fire suppression will increase the diversity of disturbance-prone landscapes by allowing more stands to reach older successional stages, but will decrease diversity in landscapes with intermediate to low disturbance frequency by eliminating early seral stages (Suffling et al. 1988). Extending this argument to include logging, intermediate disturbance theory suggests that in the short run, clearcutting can increase landscape diversity in areas of low fire frequency by increasing the proportion of young stands (right arrow, Fig. 9.8). This effect will diminish over time as harvesting over the entire landscape gradually eliminates old-growth stands. When combined with effective fire suppression, logging can also potentially increase the diversity of highly fire-prone landscapes where the natural fire cycle is shorter than the average logging rotation (left arrow), since it allows more stands to reach maturity. In high-diversity landscapes with intermediate fire cycles (middle arrow), logging and fire suppression are likely to reduce landscape diversity unless they succeed in closely mimicking the age-class and species mosaic found in unmanaged landscapes. While the dashed-line scenario of Fig. 9.8 oversimplifies the dynamics of real forest landscapes in which wildfire, insects, and other natural events continue to interact with forest management, it illustrates the point that implementation of any single management strategy risks homogenizing the forest at some scale (Box 9.6).

One of the most obvious and most studied differences between managed and unmanaged forest landscapes is the effect that a regulated forest rotation has on the age-class distribution of forest stands. Theory predicts that if stands burn more or less at random, their age structure will approximate a negative exponential distribution in which 37% of stands are older than the average fire cycle (Fig. 9.9a; Johnson and Van Wagner 1985). Although real forest landscapes rarely follow the negative exponential curve exactly, they do contain many stands older than the average fire cycle (Fig. 9.9b). With traditional even-aged forest management, by the time all commercially valuable stands were harvested, only non-commercial forest stands would be left to exceed the rotation age. Thus, a loss of old growth and associated old-forest species is inevitable in fully accessed forest landscapes. Various approaches to address this discrepancy have been proposed and implemented for boreal forest ecosystems including old-growth and ripar-

Fig. 9.8. The relationship between landscape diversity and wildfire disturbance rate in boreal landscapes of northern Ontario. Without fire suppression or logging (solid line), landscape diversity peaks at intermediate fire frequencies that maintain a mix of early, mid, and late seral stages (data from Suffling et al. 1988). With fire suppression and logging (dashed line), short-term diversity will likely increase in landscapes that had low or high fire frequencies ("up" arrows) and decrease in landscapes that had intermediate fire frequency ("down" arrow). Unless forest management regimes are varied, the net result will be reduced diversity among landscapes at the regional scale.

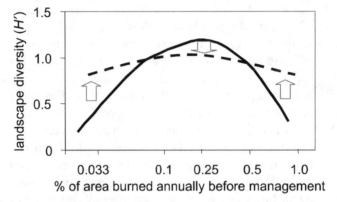

ian reserves, longer forest rotations, and partial cutting or uneven-aged forest management systems (Bergeron and Harvey 1997; Burton et al. 1999; Bergeron et al. 1999*b*; Harvey et al. 2002; see also Chap. 11). On some newer forest management tenures, annual harvest rates are calculated based on the length of the average fire cycle (Armstrong et al. 1999), but this approach only partially addresses age-class imbalances because it is the range of stand ages in the landscape rather than mean stand age that is important for biological diversity (Andison 1999*b*).

The extent to which the age-class structure of managed landscapes deviates from that in wildfire-dominated landscapes depends on several factors. As shown in Fig. 9.9*a*, the greater the discrepancy between the managed forest rotation and the average fire cycle, the greater the effect on the age-class structure. Where the average harvest

---

Box 9.6. Managing diversity at multiple scales.

Because disturbance histories have evolved continuously throughout recorded and prehistoric times, Harvey et al. (2002; and see Chap. 11) suggest basing the target forest age class and species mosaic on a disturbance cycle that allows for the maximum diversity of forest types (i.e., the top of the hump in Fig. 9.8) rather than rigorously adhering to a "natural" fire cycle that was calculated for a single period of time. Such a tactic should certainly help to prevent loss of diversity *within* landscapes, but there is a risk that uniformly applying a maximum diversity logging cycle to all forest landscapes or management units may lead to homogenization of the forest at the larger, regional scale, as shown in Fig. 9.8. Although there may be heterogeneity of silvicultural systems within each forest management unit, these same strategies will be repeated from one forest management unit to the next. Multi-scale analysis teaches us that what appears to be an optimum policy at one scale is unlikely to be optimal at all scales. In this sense, there is no single best approach or single best scale for optimizing forest diversity.

rotation exceeds the natural fire cycle, there will be a deficit of both young and old stands; where the rotation length is substantially shorter than the natural fire cycle there will be an excess of young stands and much less old growth. Further complicating the picture is the fact that within any forest landscape, not all ecosystems are equally likely to burn. Xeric ecosystems on ridge crests and south-facing slopes are typically more fire-prone than moist ecosystems on lower- and north-facing slopes (Zackrisson 1977; Foster 1982; Suffling et al. 1988). This spatial variability in disturbance cycles is generally not replicated in managed boreal forests. In fact, because rotation lengths are generally based on the culmination of mean annual increments, they tend to be longest on dry, slow-growing sites, and shortest on mesic and moist sites. The age-class mosaic can become further distorted if moist, lower slope and riparian forests are preferentially cut, while fire-prone xeric ecosystems are bypassed and allowed to become abnormally old. The ASIO forest planning model, implemented by some Swedish forest enterprises (Angelstam 1998) directly addresses this age-class distortion by adopting ecosystem-specific logging rotations.

Temporal variability in disturbance regimes is also very different in managed and unmanaged forest landscapes. Catastrophic disturbances such as insect epidemics, fires, or windstorms are episodic, occurring regionally every 30–40 years in the case of spruce budworm (Blais 1983; Royama 1984) or every 6–16 years for forest tent caterpillar (Sippell 1962; Ives 1973; Daniel and Myers 1995) and with considerable variability around these means, especially at stand scales (Jardon 2001). During outbreaks, peak destruction takes place over time scales ranging from a matter of minutes to a few years and is often season-specific. Logging, on the other hand, aims to be more or less continuous, with roughly equal annual harvests within a management unit and with activities spread throughout much of the year. Such discrepancies can create age-class distributions that are substantially at variance with those in unmanaged landscapes and

Fig. 9.9. Age distributions in unmanaged and managed boreal forests: (*a*) theoretical age distributions for 50-, 100-, and 300-year fire cycles (negative exponential distribution) and a 100-year logging rotation (rectangular distribution); adapted from Johnson (1992); (*b*) actual 1980 age-class structure of a central British Columbia sub-boreal forest with a disturbance cycle of ~133 years and its predicted age-class structure in 2080 if 80% of the forest is logged and 20% is non-commercial and left undisturbed (adapted from Burton et al. 1999).

(*a*)

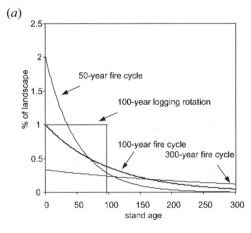

(*b*)

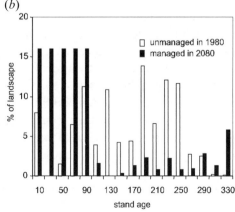

may leave no temporal refuge for disturbance-sensitive wildlife species. The extent to which harvest activities can be modified to more closely mimic the temporal patterns of nature depends on how the landscape is subdivided or amalgamated into forest management units or tenures. A large management unit should, in theory, allow more flexibility to schedule logging activities so that large-scale disturbances are episodic and portions of landscapes can be left to recover for decades at a time. However, such an approach is at odds with societal pressures for smaller scale, community- or privately owned tenures that provide continuity of local employment and income (Chaps. 4, 5, and 7).

After a disturbance, the species composition of boreal stands (including overstory trees, understory vegetation, canopy epiphytes, vertebrates, invertebrates, and microbes) changes as they age, most noticeably in the southern mixedwoods and in regions with longer fire cycles where pioneering, fire-adapted species are typically replaced over time by old-forest and gap-dependent species (Fig. 9.10a; Frelich and Reich 1995; Kneeshaw and Bergeron 1998). With conventional even-aged forest management there is limited opportunity for species replacement, particularly when there is strong silvicultural control over stocking density and species composition (Fig. 9.10b). Thus, even relatively minor reductions in the variability of the forest age-class structure may have complex and cascading effects on the composition and diversity of the spatial landscape mosaic (Gauthier et al. 1996). Moreover, most dendroecological and palaeoecological studies show that both disturbance frequency and post-disturbance successional patterns have changed during the Holocene (Johnson 1992; Millar and Woolfenden 1999). The current spatial landscape mosaic is thus a product not only of contemporary disturbance and climatic regimes, but includes legacies of past climates and past disturbance regimes (Bergeron et al. 1999a).

It will be a tall order to understand this complex forest history well enough to replicate it within a managed forest regime. While replication may be impossible, forest managers can do much to retain the complexity of the forest mosaic by adopting a variety of harvest methods and silvicultural regimes (Lieffers et al. 1996; Angelstam 1998), rotating crop species (DeLong 2002), and periodically allowing all portions of the landscape to undergo extended rotation lengths (Bergeron et al. 1999b).

Effects of spatial landscape patterns on forest-dwelling species

Research in temperate and tropical regions has shown that animal and plant communities can be highly sensitive to changes in spatial landscape pattern such as the degree of forest fragmentation or connectivity. Although many of the same principles apply, findings from these predominantly agricultural and urbanized landscapes may not translate directly to Canadian boreal forest conditions (Corkum et al. 1999; Bourque and Villard 2001). For example, some boreal bird communities may be less sensitive to edge and fragmentation effects than their counterparts in temperate deciduous forests because they are pre-adapted to landscape fragmentation by stand-destroying fire (Boulet et al. 1999; Brongo et al. 2000) or because brood parasites and nest predators are less abundant (Bourque and Villard 2001). Imbeau et al. (2001) compared eastern Canadian and Fennoscandian boreal bird communities, concluding that migrant birds like warblers

Fig. 9.10. Theoretical species and age-class structure of unmanaged and managed boreal land-scapes. (*a*) Unmanaged landscape in which each physiographic unit or soil type has a negative exponential age-class distribution determined by its fire cycle and a characteristic successional sequence of forest types (adapted from Gauthier et al. 1996). (*b*) Managed landscape; a highly regulated forest estate composed of single-species stands with a uniform age-class distribution up to the rotation age. Real landscapes are more irregular than either of these theoretical scenarios.

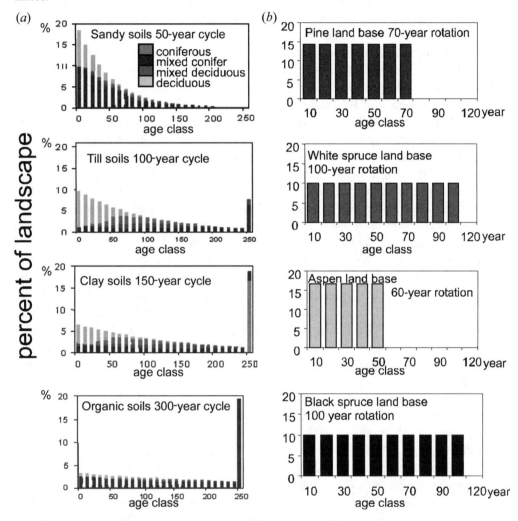

may be less sensitive than results from temperate forests suggest. Instead, they concluded that resident bird species such as woodpeckers and owls that are highly dependent on stand-scale features such as snags for nesting or foraging are likely to be at highest risk from logging.

An important difference between boreal and temperate forests is the degree to which landscapes have been permanently modified from their original condition. Songbirds in mixedwood forests of northern Alberta may be able to compensate for loss of deciduous habitat by moving into adjacent coniferous stands and older clearcuts (Norton et al.

2000), something they are less able to do in temperate areas where the surrounding matrix is agricultural or urban land. In New Brunswick, a series of studies is using one the most extensively altered industrial forest landscapes in Canada to test the hypothesis that vertebrate communities are resistant to spatial landscape change up to a critical threshold (see Box 9.3), after which their behaviour and reproductive success are substantially impaired (Bowman et al. 1999; Villard et al. 1999). Results from this research should help to answer important questions about how much habitat alteration is acceptable in newly accessed boreal forest. Preliminary results from a study of the ovenbird (*Seiurus aurocapillus*) and black-throated blue warbler (*Dendroiea caerulescens*), both Neotropical migrants, found no difference in reproductive success between moderately harvested (~70% mature forest remaining) and intensively harvested (~45% mature forest) landscapes (Bourque and Villard 2001). For these bird species, critical thresholds of forest retention may be 30% or less, as suggested by Andrén (1994) in his review of fragmentation studies worldwide. Another possibility is that effects of habitat loss may manifest themselves only after a lengthy delay, a phenomenon known as the "extinction debt". Tilman et al. (1994) suggested that populations may decrease slowly before reaching a threshold, after which local and species-wide extinctions occur at a high rate. With these cautions in mind, Potvin et al. (2000) have taken a relatively conservative stance on this issue in Quebec black spruce landscapes, for example, recommending for marten (*Martes americana*) that a minimum of 40–50% mature forest cover be retained with no more than 30% forest removal over a 20-year period.

The above-cited studies examined broad issues of how much mature forest to retain and how much cutting is permissible. The pattern or spatial arrangement of retained and cutover units, although less important than the total amount of habitat available (Andrén 1994; Drapeau et al. 2000), also affects the survival and reproductive success of forest-dependent organisms. The availability of new spatial analytical tools has spawned a huge number of studies that measure differences in s ize, shape, amount of edge, spatial complexity, dispersion, connectivity, and other landscape metrics between cutover and unmanaged forest landscapes (e.g., Vernier and Cumming 1999). These studies have shown that landscape metrics for clearcut- and wildfire-dominated landscapes are often substantially different. For some variables, notably opening size (Hunter 1993), logging decreases the range of variability found in systems dominated by wildfire and insect disturbances (McRae et al. 2001). For other metrics, such as the degree of dispersion produced by patchwork clearcuts (DeLong and Tanner 1996; Rempel et al. 1997), logging produces patterns well outside the extremes of pattern found in unlogged landscapes. Furthermore, hierarchies of nested pattern (i.e., smaller patches within larger patches) are typically less complex (Elkie and Rempel 1999), particularly when the small-scale patchiness produced by insects, pathogens, and storm events is overlain onto the complex mosaic of live and dead tree patches resulting from wildfire. Modification of harvest patterns and increased landscape-level planning are under way in many regions of Canada to reduce these discrepancies (Chap. 12).

It is fairly easy to demonstrate differences in spatial arrangement between logged and naturally disturbed landscapes, but it is more difficult to show how and why these differences in landscape pattern may be important to forest-dwelling species. Much current landscape ecology research focuses on identifying which of the many different landscape metrics are best related to ecosystem function and plant and animal behaviour (Vernier and Cumming 1999; Schmiegelow et al. 1999), which are essential to emulate (Andison 1999b), and again, what are critical thresholds for retention. Both seed dispersal modelling (Greene and Johnson 1999) and empirical evidence (Galipeau et al. 1997) indicate that small remnant forest islands (fire skips) play a relatively minor role in the recolonization patterns of coniferous trees except when the disturbed area is extremely large (Chap. 8). However, these same tree islands are very important for recolonization of disturbed habitats by forest-dwelling birds (Mackenzie and Steventon 1996; Tittler et al. 2001b), arthropods (Lindgren et al. 1999), lichens, liverworts, and mosses (Peck and McCune 1997). The size of residual tree patches influences their effectiveness. In northwestern Quebec, for example, abundance of *Usnea* spp., the most sensitive of the common epiphytic lichens, is negatively affected when retained patches are less than 1 ha (Rheault et al. 2003).

The landscape pattern or signature produced by logging is dictated by provincial harvesting policy which tends to be applied uniformly over large regions for up to a decade or more, then switches abruptly in response to political pressures and the fashion of the day. Responding to public concerns over the effects of large-scale progressive clearcutting on wildlife habitat, aesthetics, and stream hydrology, regulations in British Columbia and Ontario have in recent decades imposed a dispersed checkerboard pattern of cutblocks of 40–150 ha size, logged in a two- or three-pass system, with approximately 20–30 years anticipated between passes. This strategy produces a signature that differs radically from the landscape pattern produced by wildfire (Fig. 9.11a, b; DeLong and Tanner 1996) and requires an extensive active road network (Box 9.7), leading to calls for aggregated cutblocks that would more closely resemble wildfire patterns and would preserve large unfragmented tracts of closed forest for interior forest-dwelling species over the long term (Harris 1984). Simulation models have shown the long-term reductions in interior forest habitat that result from dispersed cutting practices and the difficulty of changing the pattern once it is established (Li et al. 1993). Meanwhile, in

Fig. 9.11. Landscape patterns created by (*a*) dispersed block cutting in British Columbia, (*b*) wildfire in British Columbia, and (*c*) aggregated block clearcutting in Quebec (adapted from Delong and Tanner 1996 and Potvin et al. 1999).

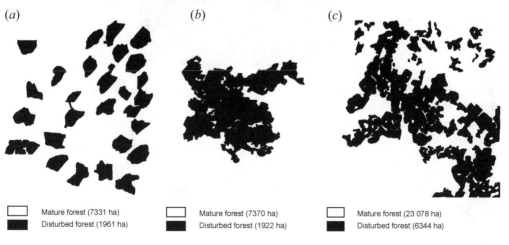

(*a*)	(*b*)	(*c*)

Mature forest (7331 ha)
Disturbed forest (1961 ha)

Mature forest (7370 ha)
Disturbed forest (1922 ha)

Mature forest (23 078 ha)
Disturbed forest (6344 ha)

Box 9.7. The special case of forest roads.

Roads are an element of managed forests with major environmental impacts across a full spectrum of spatial scales. They have no analogue in unmanaged systems, though they partially resemble streams in the way they control the flow of energy, matter, and biota through ecosystems, and they are similar to newly deglaciated till in terms of disturbance severity. At regional scales (Fig. 9.12*a*), road density dictates and indicates human land use, the direction of carbon fluxes, the extent of aspen dominance, and rates of species invasion and extinction. At landscape scales (Fig. 9.12*b*), roads supplant hydrological networks as the organizing framework around which landscapes are structured, and control flows of energy, materials, and species into, out of, and across watersheds. At stand and smaller scales (Fig. 9.12*c*), road surfaces represent the radical extreme of the disturbance–diversity spectrum. At all scales, the full ecological impact of roads greatly exceeds the percent of land surface occupied. It has been conservatively estimated that direct ecological impacts of roads affect 20 times the land area they cover (Forman 2000). Indirect effects, such as increased hunter access to remote areas, influence a much greater area.

Fig. 9.12. Forest roads viewed at three spatial scales.

(*a*)

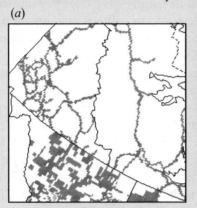

Map courtesy of Natural Resources Canada

(*b*)

Photo by Dave Coates

(*c*)

*Photo by Dave Cheyne,
courtesy of Alberta Pacific
Forest Products*

Roads affect terrestrial and aquatic ecosystems in seven general ways (Trombulak and Frissel 2000):

- Direct mortality of sessile and slow-moving organisms from road construction;
- Direct mortality from collision with vehicles;
- Modification of animal behaviour, movement, and reproductive success;
- Alteration of the physical environment (temperature, water, light, dust);
- Alteration of the chemical environment (heavy metal and salt deposition);
- Spread of exotic species; and
- Changes in human land and water use.

Box 9.7. (concluded).

Roads do not affect all ecosystems and species equally. In the mountainous Boreal Cordillera, for example, roads have greater effects on hydrology and slope stability than in the gentle terrain of the Boreal Shield and Plains. Reptiles and amphibians, at their northern range limits, suffer disproportionately from vehicle collisions.

The intensity of human land use is very strongly correlated with road networks. This makes it difficult to separate effects of roads from those of the accompanying land use. But it also means that roads serve as a very effective indicator of the cumulative impact of human use. Wolves (*Canis lupus*), for one, will not establish viable populations above a region-specific threshold density of roads (Jensen et al. 1986), and the distribution of the woodland caribou is closely correlated with the remaining unroaded territory in boreal North America. Conversely, the distribution of invasive, exotic plant species in the Canadian north is nearly perfectly correlated with the presence and density of road networks (Cody 1996).

Roads are an aspect of managed forests that is well outside the range of natural variability at all ecological scales. Their effects are universally negative for boreal ecosystem integrity, and in the context of a natural disturbance model, they are unequivocally not acceptable. Perhaps for this reason, ecologists have concluded they require no further study and have systematically excluded them from most experimental designs and sampling protocols. The responsibility for reducing environmental impacts has been left to forest managers, planners, and road engineers who have few alternatives with which to work and little incentive to pursue them. What is really needed is a comprehensive, multi-scale assessment that demonstrates to policymakers and the public the full environmental, social, and economic costs of roads in the boreal forest, and a well-funded, transdisciplinary scientific effort to find effective solutions.

Sources: Jensen et al. 1986, Cody 1996, Forman 2000, Trombulak and Frissel 2000, McRae et al. 2001

Quebec, harvest regulations have until now produced clustered patterns of 100–250 ha cutblocks (slightly smaller since 1996) separated by 60–100 m buffer strips (Fig. 9.11c; Potvin et al. 1999). While the aggregated pattern produces landscape metrics that can be quite similar to those left by wildfires (Perron et al. 2000; but see Gluck and Rempel 1996), it has become socially and politically unacceptable. Not only does the sudden loss of forest cover over very large areas have profound aesthetic impacts, it can have negative consequences for aquatic habitat (Garcia and Carignan 1999, Chap. 10), for terrestrial species with intermediate-sized home ranges, and for organisms dependent on these species. Potvin et al. (1999) concluded that in many cases a dispersed checkerboard pattern might better integrate the needs of multiple forest users in Quebec.

The lesson in these examples is that no single landscape pattern, including that generated by unconstrained wildfire, will be optimal for all forest-dwelling species and forest users at all points in time and at all scales. The short- and long-term effect of any landscape pattern on wildlife and other organisms depends on their specific habitat needs, home range size (Hunter 1990; Potvin et al. 1999), and dispersal characteristics (Dettki et al. 2000). Governments in all jurisdictions have tended towards a single "optimal" solution that eventually homogenizes the landscape, thereby undermining the objective of creating a landscape that satisfies both environmental and human needs. As a case in point, Rempel (2001) showed that in Ontario, consistent application of optimum moose habitat guidelines tended to reduce heterogeneity of the managed forest at the landscape scale.

Box 9.8. Fire history guides landscape planning in the Foothills Model Forest, Alberta.

The Weldwood Forest Management Agreement (FMA), located on the eastern slopes of the Rocky Mountains near Jasper National Park and Hinton, Alberta, is unique in Canada in having a detailed time-since-fire map prepared in 1960, by forester Jack Wright. When the Foothills Model Forest began its comprehensive natural disturbance research program in 1995, Wright's map proved to be just the ticket for getting the landscape-level planning effort off on a solid footing (Fig. 9.13).

Fig. 9.13. Forest age-class distribution in the Foothills Model Forest reconstructed to 1950 from Andison 2000).

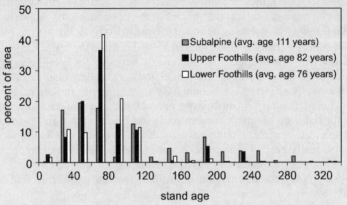

One of the early questions forest managers asked was how much old-growth spruce forest had historically been present within the FMA. Though the map itself was just a single snapshot in time, it was able to serve as the basis for a simulation model that estimated that during the past 200 years, the proportion of spruce forest in age classes over 200 years varied between 0 and 33%. The current level of 10% spruce in older age classes is well within that range. However, landscape plan projections consistently showed that in the future more than 33% of spruce stands would be over 200 years old, mainly because of riparian buffer restrictions put in place to protect streamside environments (Fig. 9.14).

Fig. 9.14. White spruce riparian buffer in the Foothills Model Forest, Alberta.

Photo courtesy of Foothills Model Forest

Box 9.8. (concluded).

This result challenges forest managers to either reconsider buffer restrictions or to reassess the usefulness of a natural-range-of-variability (NRV) approach for addressing contemporary social and environmental issues like streamside protection. Uniform implementation of lakeside and streamside buffers has been questioned elsewhere (e.g., Chap. 10) because a certain amount of canopy disturbance is required to maintain the integrity of hydro-riparian ecosystems. The spatial distribution of old spruce forests in unmanaged landscapes is also considerably different from the linear strips that result from streamside protection policies. Further research is underway to address these issues.

Sources: Burton 1998; Andison 1999a, 2000; McCleary and Andison 2001

Fortuitously, the changing history of Canadian logging standards has left behind a legacy of varied landscape patterns, each with its own set of environmental strengths and weaknesses, rather than a single pattern. As a result of the research summarized above, the current fashion is to recommend flexible regional and local harvest standards based on a detailed multi-scale understanding of natural disturbance regimes and their effects on landscape pattern, but adapted as necessary to take into account the needs of stakeholders and sensitive, at-risk species and ecosystems. This approach, properly applied (see an example in Box 9.8), should increase diversity within and among landscapes, and thus increase the diversity of organisms and human values that can be accommodated. However, it is unlikely to replicate the landscape pattern produced by boreal wildfire regimes.

Stand-scale comparisons

Disturbance severity

When a wildfire, windstorm, or insect outbreak passes through a boreal forest stand, its immediate effect on plant and animal communities and on forest soils depends on its severity. Disturbance severity refers to the degree to which living organisms are killed, the extent to which dead organic matter is consumed or removed, and the degree to which microclimate and soils are altered (Chap. 8). One simplified way to represent a gradient of disturbance severity is by means of the percent destruction and removal of organic matter. The ecological impact of a disturbance such as wildfire can then be quantified in terms of the distribution of the total burned area across this severity gradient (Fig. 9.15a).

The site-level impacts of logging can likewise be described by the amount of organic matter killed, removed, or displaced during harvesting and site preparation (Fig. 9.15b; Keenan and Kimmins 1993). Stand-scale discrepancies between natural and logging-related disturbance can then be identified by comparing the spectrum of disturbance severities across a typical forest site. For the example of wildfire versus logging, Fig. 9.15c illustrates three important areas of discrepancy and suggests potential solutions:

Fig. 9.15. Schematic portrayal of disturbance severity gradients for (*a*) wildfire and (*b*) logging. (*c*) Strategies for reducing major discrepancies between wildfire (solid line) and logging (dashed line).

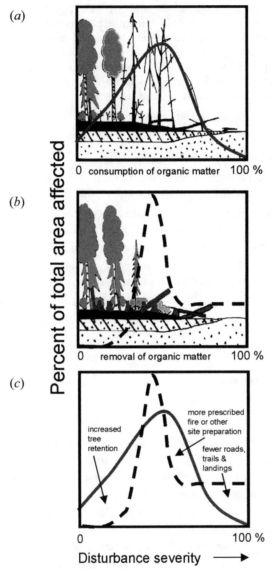

(1) *Snags and residual vegetation* — The crown fires that prevail in the Canadian boreal forest consume tree foliage, branches, and bark, leaving behind many blackened stems. Patches of live and partially killed trees and understory vegetation remain within fire skips (Eberhart and Woodward 1987) and along ragged fire edges (Andison 1999*b*). These unburned patches act as important refugia for forest-dwelling species of insects, orchids, lichens, bryophytes, fungi, and other organisms with limited dispersal capabilities (Gandhi et al. 1999; Dettki et al. 2000). The legacy of live and dead trees also creates structural heterogeneity in the regenerating stand, providing habitat for plant and

animal species that depend upon large old trees, snags, and decaying logs (McComb and Lindenmayer 1999). By contrast, traditional clearcut logging removes all merchantable stems from the site and fells unmerchantable stems, leaving few live or dead standing trees. One of the first and most widely adopted means to reduce differences between natural disturbance and clearcutting has been to modify logging techniques to increase the amount of merchantable and non-merchantable trees and shrubs retained within clearcut stands (Coates and Steventon 1995). Leaving behind small patches of trees and shrubs greatly increases the occurrence on clearcuts of a wide variety of animals and plants that do not tolerate fully open habitats (Lindgren et al. 1999; Potvin et al. 1999; Tittler et al. 2001*b*). Partial cutting approaches that leave behind even more residual material have also been advocated for boreal forests (Bergeron and Harvey 1997) but have not yet been widely implemented. Partial cutting may be particularly applicable for emulating fire behaviour in cool, moist patches of spruce forest that are often skipped or only lightly touched by fire (Buddle et al. 2000; Gandhi et al. 1999), or in open, irregular pine forests where ground fires were more common than intense crown fires (Bergeron and Brisson 1990; AEM 1998*a*[2]; Angelstam 1998).

(2) *Forest floor reduction* — Wildfires typically kill more forest understory vegetation and remove larger quantities of litter and duff than does clearcut logging, particularly when the logging is done on snow, as often is the case in the boreal forest (Crites 1999; Nguyen-Xuan et al. 2000; Reich et al. 2001). Duff consumption by fire is patchy and exposed mineral soil is often concentrated around the base of dense clumps of coniferous trees, where dry accumulations of coniferous needles, cones, and other litter are consumed through smouldering combustion (Miyanishi et al. 1999; see Fig. 8.10*a*). These well-dispersed patches of mineral soil and thin humus are vital seedbeds for coniferous tree regeneration as well as for many pioneering shrub, herb, bryophyte, and lichen species (Viereck 1973; Zasada 1986; Johnson 1992; Charron and Greene 1999; Nguyen-Xuan et al. 2000). Combustion of vegetation and organic layers also releases stored nutrients and stimulates nutrient cycling and root growth through increased soil warming (MacLean et al. 1983; Van Cleve et al. 1983). By contrast, current logging guidelines in Canada attempt to minimize machine damage to soils and protect existing coniferous regeneration and other vegetation (Canuel 1989; BCMOF 1996; CCFM 1997). Clearcutting results in a very different pattern of soil disturbance in which most of the land surface has minimally disturbed organic layers and understory vegetation, with exposed and compacted mineral soils concentrated along skid trails. While such practices reduce physical soil damage that impairs site productivity (Brais and Camiré 1998), it seems that in many boreal ecosystems, most notably the less productive northern and nutritionally poor black spruce-dominated or pine-dominated forests, lack of sufficient soil disturbance may prevent adequate conifer regeneration from seed and alter successional trajectories (Carleton and MacLellan 1994; Reich et al. 2001). Insufficient forest floor disturbance may also cause long-term declines in forest productivity through poor regeneration, reductions in nutrient availability, and increased dominance of ericaceous shrubs (Keenan and Kimmins 1993; Zackrisson et al. 1996; Horvath et al. 1999; Smith et al. 2000). As schematized in Fig. 9.15*c*, judicious use of prescribed fire and low-impact mechanical site preparation after logging may address some of these concerns by increasing the similarity of post-logging succession and nutrient cycling

processes to those prevailing after fire (Prévost 1996; Bulmer et al. 1998; Kranabetter and Yole 2000).

(3) *Extreme soil disturbance* — Very severe fires remove most organic debris and incinerate soil organic layers. Deep penetration of heat into the soil can kill roots, rhizomes, and buried seeds (Rowe 1983). Such severe fire behaviour is most likely after periods of extreme fire weather and often occurs when an earlier disturbance, such as blowdown, insect mortality, or a previous fire, has left behind large accumulations of dry, dead fuel. Severe fire patches often regenerate to atypical semi-permanent shrub or herbaceous vegetation. These communities may contain rare or unusual species or provide unique wildlife habitat and thus contribute significantly to stand- and landscape-level biodiversity. But severe wildfires also cause threshold-type disruptions (Box 9.3) to successional, soil, and hydrological processes that would be considered degrading if they occurred frequently, on a large scale, or were initiated by human activity. At both stand and landscape scales, severe burning of large areas occurs very rarely. With logging, on the other hand, some 3–30% of every disturbed area (CCFM 1997) is converted to roads, trails, landings, gravel pits, semi-permanent skidtrails, and other machine disturbances that do not revert to indigenous forest cover for lengthy periods of time and are prime sites for exotic species invasion. The percentage of such area varies with the method and season of logging but is unquestionably outside the range produced by fire in terms of its severity, its frequency, and most likely the total area affected. Reducing the extent of extreme soil disturbance and soil compaction, for which no post-glacial natural analogues exist, is one of the most obvious ways that forest management can be improved to better emulate natural forest disturbances.

The concept of a severity gradient can also be useful for identifying and addressing stand-level discrepancies between forest management practices and non-fire disturbances. Although such disturbances are ubiquitous, detailed comparisons with logging are not abundant in the boreal literature (but see Price et al. 1998 for coastal British Columbia). Non-fire disturbance events, particularly those caused by insects and disease, tend to be skewed towards the low-severity end of the disturbance spectrum (Fig. 9.15a), causing tree mortality in gaps and small patches with little understory or soil damage (Kneeshaw and Bergeron 1998, 1999). However, catastrophic windthrow involving full destruction of the tree canopy, significant soil disturbance, and damage to tree regeneration and understory layers can also occur (Ulanova 2000; Frelich 2002). Herbivores, pathogens, and storms are often selective agents of tree mortality, targeting not only particular species or species-groups, but often specific size and vigour classes as well. Baskerville (1975) and MacLean (1980) describe, for example, how all overstory fir trees may be killed in mature and overmature stands of balsam fir, whereas in immature stands the budworm may only thin trees and lead to a multi-layered stand. Thus, although their effect may be less catastrophic than either wildfire or clearcut logging, insects and storm events also profoundly affect stand age structure and the composition of forest communities.

Stand structure after clearcut logging bears little resemblance to the heterogeneous structure of old stands exposed to multiple small-scale and gap disturbances. This discrepancy plainly can be addressed through increased use of partial cutting. Despite the long fire return intervals in much of eastern Canada and portions of the western

Cordillera, and the preponderance of structurally complex stands in these regions, partial cutting is not a significant component of harvest operations in any jurisdiction in boreal Canada.

Clearcut logging with protection of advance regeneration and non-fire gap disturbances differ less in their effect on forest understory layers and forest floors than in their effect on the overstory canopy. In both instances, most of the understory is released from overstory competition with relatively minor disturbance or displacement of the forest floor. Recruitment of trees and other plants tends to come from well-developed seedling banks (advance regeneration) and from shade-tolerant species that survived the disturbance more so than from seed-regenerating pioneering species. This is particularly true after winter logging and during the first logging rotation (see Chap. 8 for problems associated with maintaining the seedling bank following short rotation logging). Insect- and disease-created gaps typically open up gradually over a period of several years (5–12 years for spruce budworm, depending on outbreak severity; MacLean 1980) while logging tends to be abrupt. The gradual increase in light availability may confer a competitive advantage on shade-tolerant and semi-tolerant species over shade-intolerant species in natural versus managed systems (Kneeshaw 2001).

Storm events such as ice and snow damage, windstorms, flooding, and mass wasting involve increasing degrees of organic matter displacement and mineral soil exposure. In Russian boreal forests, for example, pit and mound topography accounted for 7–12% of the surface area of uneven-aged spruce forests, increasing to 15–25% after catastrophic windthrow (Ulanova 2000). After such events, understory diversity is often substantially increased by early seral colonizers.

Low-severity disturbances that open up the tree canopy often result in increased abundance of understory species, particularly tall shrubs such as green alder (*Alnus crispa*), beaked hazelnut (*Corylus cornuta*), or mountain maple (*Acer spicatum*), and in extreme cases can cause a shift to non-forest vegetation (Batzer and Popp 1985; Kneeshaw and Bergeron 1999). Silvicultural techniques that attempt to emulate low-severity gap-scale disturbance must somehow replenish tree seedling banks and understory heterogeneity before subsequent crops are cut, or there will be gradual impoverishment of the understory over time (Chap. 8). However, indiscriminate partial cutting that allows tall understory shrubs to flourish can lead to low tree stocking levels. Planting, rather than relying on advance regeneration, in partially cut stands can help to overcome this dilemma, but at costs that may be too high for typical boreal forest operations (Greene et al. 2002). Mechanical site preparation techniques such as excavator mounding that emulate pit and mound disturbances, may be another effective, albeit costly, means for enhancing tree regeneration and understory heterogeneity following partial cutting.

Energy, water, and nutrient dynamics

Disturbances alter the energy, water, and nutrient balance of forest ecosystems. The degree to which these cycles are affected depends on disturbance severity; however, even at broadly equivalent severity levels, there are important differences between natural disturbances and logging. Forest destruction has a major effect on the radiation balance, leading to changes in the temperature and moisture of soils and air (Keenan and

Kimmins 1993). An abundance of snags, which act as absorptive and radiative black bodies, and the presence of live residual trees, means that extremes of air temperature, wind, and humidity are normally less pronounced after wildfires and insect kills than in clearcuts. Edge effects extending into adjacent undisturbed stands are also generally softer. In Sweden, Söderström (1988) concluded that the interior of managed stands was droughtier than that of unmanaged old-growth stands because repeated thinning enhanced air movement and penetration of sunlight. On the other hand, changes to soil and organic layer temperatures and moisture are typically more extreme after wildfire than after logging of similar severity because of the presence of charcoal, ash, and hydrophobic surface horizons (Viereck 1983; Smith et al. 2000). Because temperature and moisture availability regulate rates of growth and many other important physiological and biochemical processes, small differences in micro-environmental conditions may substantially alter community composition, biomass production and ecological succession. For instance, balsam fir and spruce tend to regenerate most abundantly at the southern, shady ends of spruce budworm gaps, presumably because of their intolerance of moisture stress (Kneeshaw and Bergeron 1999). The relative humidity within stands also profoundly affects the species composition of lichen, liverwort, and moss communities (Söderström 1988).

An important concern in forestry research for over a century has been to assess the nutrient budgets of forest stands to determine whether multiple forest rotations can be sustained without causing long-term losses of forest productivity (Kimmins 1987). Criticism of intensive forest management often stems from a concern that logging and silvicultural practices cause excessive nutrient export, soil erosion, and leaching and do not allow for the renewal of nutrient capital that occurs during natural succession. Although much empirical research and simulation modelling has been carried out to address these issues, relatively few studies have directly compared nutrient dynamics after wildfire or other natural disturbances with those in managed stands (but see Hamel and Paré 1999; Smith et al. 2000; Reich et al. 2001). It is evident that wildfire also causes nutrient export, soil erosion, and leaching, and that in the case of severe or multiple fires, these losses may be extreme, leading to long-term diminution of forest productivity and impaired forest regeneration (Leopold 1933; Zackrisson 1977). On the whole, however, boreal forest productivity is not substantially impaired by fire (Kimmins 1987). On the contrary, in many boreal ecosystems fire reverses a natural process of gradually declining soil nutrient availability, renewing ecosystem productivity by liberating nutrients that have accumulated in unavailable forms in soil organic layers and in dead and decaying wood (Van Cleve et al. 1991). Restoration of declining soil nitrogen levels has also been documented following severe spruce budworm outbreaks (Paré et al. 1993).

Comparative effects of logging and natural disturbances on forest nutrient dynamics cannot be easily generalized. Local soils and topography, dominant tree species, stand successional stage, understory vegetation composition, and disturbance severity are all extremely important sources of variation, and each of the major soil nutrients (N, P, K, Ca, etc.) responds to that variation in a slightly different manner. Because fires volatilize nitrogen, N losses tend to be higher after wildfire than after logging of comparable severity. The reverse is true for exchangeable cations and phosphorus, which are deposited in ash after fires, but can be exported from the site after logging (Paré and Munson 2000). Because most nutrients are found in foliage, their export is highest with whole tree logging and with deciduous trees logged during the dormant season after nutrients have been translocated back into the trunk.

Most experimental nutrient budget work in the boreal forest has produced only short-term results and simulation model development is in its early stages. Nonetheless, the accumulating body of research is suggesting that long-standing concerns about nutrient depletion in managed forests may have been overstated. Although sizeable nutrient losses do occur, the ability of most ecosystems to regain adequate levels of nutrients after even fairly severe fire, logging, or site preparation is higher than previously believed (Brais et al. 1995, 2002; Örlander et al. 1996; Bulmer et al. 1998; Paré and Rochon 1999; Kranabetter and Yole 2000). As stated earlier, insufficient disturbance during logging, rather than too much disturbance, may be the most important limiting factor in nutritionally poor black spruce ecosystems where rates of C and N cycling diminish over time. Recent results from a large chronosequence study in the southern boreal forests of Minnesota (Reich et al. 2001) back up these findings. Eighty young (25–40 years old) and mature (70–100 years old) jack pine, trembling aspen, and black spruce stands that regenerated after either wildfire or logging were compared. No significant differences in biomass, productivity, or mineral nutrient (N, K, Ca, Mg) availability were found between fire- and logging-origin stands of comparable age and forest type. While soil carbon levels tended to be higher after fire than after logging in aspen stands, the opposite trend was observed in pine stands. The authors concluded that there is no evidence that disturbance by logging limited productivity at stand scales as compared to fire.

Diversity and composition of post-fire and post-logged communities

Boreal plant and animal communities are well adapted to a range of disturbance severities. Many studies have found that after timber harvest, the richness and diversity of plant and animal communities are not reduced when compared to wildfire, and often, species richness or diversity is higher after logging than after wildfire because post-logging communities contain representation of a greater range of successional stages (Crites 1999; Hobson and Schieck 1999; Nguyen-Xuan et al. 2000; Reich et al. 2001). These observations appear to apply to a wide variety of taxa, site types, and geographic areas, though there are some exceptions (see Rees and Juday 2002). For certain highly specialized taxonomic groups such as ectomycorrhizal fungi (Dahlberg et al. 2001) or woodpeckers, diversity may be higher after fire than after clearcut logging. Moreover, these short-term findings apply to boreal forests that were unmanaged prior to logging and have not been subjected to repeated silvicultural interventions. Longer term results from Fennoscandian forests in their second or third logging rotation with intermediate cleaning, spacing, and thinning treatments have shown reductions in the diversity of a wide variety of taxonomic groups, particularly organisms associated with dead and decaying wood and large old aspen (*Populus tremula*) trees (Niemelä 1997; Kruys and Jonsson 1999; Box 9.2).

While few studies have shown overall reductions in the number and diversity of plant and animal species, virtually all fire versus logging comparisons have found that post-fire and post-logged plant and animal communities differ substantially in community composition or structure. Differences are particularly strong 1–3 years after disturbance, with early post-fire communities being distinct from all other successional stages (Fig. 9.16). This phenomenon has been observed in bird communities (Hobson and Schieck 1999), spider and beetle assemblages (Niemelä et al. 1993; Buddle et al. 1999, 2000), and in vascular and non-vascular plant communities in both aspen (Crites 1999) and coniferous (Nguyen-Xuan et al. 2000) forest types. Early post-fire communities are characteristically dominated by a few (pyrophilous) species strongly associated with fire. For example, post-fire bird communities contain an abundance of woodpeckers attracted to newly killed snags (Hobson and Schieck 1999). Vascular plants that are strongly associated with early post-fire succession include *Geranium bicknellii*, *Corydalis aurea*, *C. sempervirens*, *Arabis hirsuta*, *Polygonum cilinode*, and *Pinus contorta* var. *latifolia* (Rowe 1983). Post-fire cryptogams include the mosses *Ceratodon purpureus*, *Polytrichum juniperinum*, and *Leptobryum pyriforme* (Jonsson 1993; Rees and Juday 2002), the liverwort *Marchantia polymorpha* and several terrestrial *Cladonia* and *Stereocaulon* lichen species (Nguyen-Xuan et al. 2000). This pyrophilous flora characteristically produces long-lived seeds, cones, or diaspores that are stimulated by high temperatures and preferentially establish on sterilized, ash-covered mineral soil.

Because fire in the boreal forest occurs at long and unpredictable intervals, there are proportionally few species (compared with Mediterranean ecosystems, for example) that are strongly dependent upon fire. All of these species possess one or more adaptations that enable them to persist through long fire-free intervals. They either remain dormant for centuries at a time (e.g., *Geranium bicknellii*, *Ceratodon purpureus*), or are highly mobile and can quickly disperse to new fires (*Ceratodon purpureus*, *Picoides arcticus*). While most fire specialists are able to survive and reproduce in unburned habitats such as clearcut-logged areas or insect-killed forest gaps, they all have their highest densities and reproductive success in early post-fire habitats. Preliminary evidence from Fennoscandia suggests that at least some of pyrophilous species may be unable to sustain viable populations with long-term fire exclusion, and that prescribed burning after logging may not be a sufficient substitute for burned uncut forest habitat (Wikars and Schimmel 2001). While microbial and arthropod communities likely contain the largest numbers of truly pyrophilous species, these taxonomic groups are poorly studied, especially in Canada. A research priority in these little-known groups is to identify indicator species that may be able to act as surrogates for fire-specialist taxa as a whole (Spence et al. 1999). Although the functional importance of these organisms in post-fire ecosystems is not well known, it is believed that they play a role in nutrient capture (Marks 1974), charred wood decomposition, and nutrient and energy cycling.

Most fire specialists dominate their habitat for only a few years before becoming dormant or moving on to other newly burned habitats, and except in extreme cases there is rapid succession to later seral stages. Chronosequence studies indicate that 10–20 years after disturbance, differences between post-fire and post-logged communities are greatly diminished, although they never completely disappear within the timeframes studied (Hobson and Schieck 1999; Lee and Crites 1999; Reich et al. 2001). Moreover,

Fig. 9.16. Variability in the composition of plant and animal communities after wildfire and logging, portrayed as multivariate ordinations of sample plots in species-space. (*a*) Aspen plant community chronosequence, Alberta (from Crites 1999); (*b*) aspen beetle community chronosequence, Alberta (from Buddle et al. 1999); (*c*) black spruce plant communities, Quebec (from Nguyen-Xuan et al. 2000). Early or severely burned post-fire communities (open circles) are distinct from all other communities, and logging does not fully replicate the range of variability found after fire. In the Alberta chronosequences, communities partially converge with time (darker symbols are older).

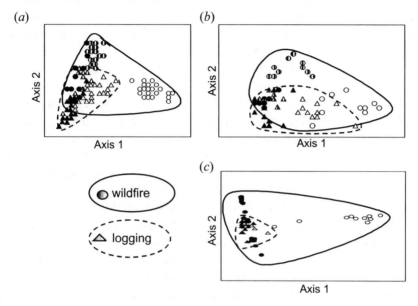

compositional changes after 10 years are much less dramatic than during the first decade. Figure 9.16 illustrates an important consequence of these findings for biodiversity: with logging, the range of variability in plant community composition across stands and landscapes is likely to be diminished.

Reich et al. (2001) drew attention to an important weakness of existing chronosequence studies that has major implications for the maintenance of ecological integrity. Because researchers have compared like stands only (e.g., post-fire aspen-dominated stands were compared with post-logging aspen-dominated stands) they were unable to detect changes in tree species composition that may arise from the disturbance agent. Most boreal conifers regenerate much better from seed on burned than on logged seedbeds (Johnson 1992), whereas aspen, birch, and other broadleaved tree species resprout readily after logging. Failure of coniferous regeneration, and a marked shift to aspen and other broadleaved tree species after logging (especially in large, unburned clearcuts) has been observed across boreal Canada (Carleton and McLellan 1994; McRae et al. 2001) as well as in the northern United States (Grigal and Bates 1997; Reich et al. 2001). In boreal regions where balsam fir or subalpine fir (*Abies lasiocarpa*) are abundant and low-impact logging techniques are used to protect advance regeneration, the proportion of these fire-sensitive species in regenerating stands is much higher after logging than after wildfire.

In summary, stand-scale comparisons of wildfire and logging suggest that logging does not generally cause reductions in species diversity or ecosystem productivity when compared to wildfire for extensively managed stands during the first post-logging rotation. However, there are significant shifts in species composition such as increases in deciduous and fir species that affect biological diversity at landscape and larger scales, and that will certainly change nutrient cycling and other processes in comparison with unmanaged benchmark systems. The viability of certain fire-specialist and sensitive forest-dwelling species may also be at risk. Extreme disturbances that cause long-term soil degradation and encourage exotic species invasion are more prevalent after logging.

Much less work has been done in boreal systems to examine differences between non-fire disturbance effects on plant and animal community diversity and succession with those of clearcut and partial logging. There is a large base of published literature, mainly from Fennoscandia, showing that mature managed stands subjected to multiple silvicultural treatments lack equivalent sizes and amounts of coarse woody debris, large living and dead trees, upturned root wads, and non-commercial tree species, and have structurally homogeneous understories compared with old unmanaged stands. These differences have negative consequences for many forest-dwelling species including vertebrates, invertebrates, epiphytic and epixylic plants, and fungi (Niemelä et al. 1988; Söderström 1988; Kuusinen 1994; Esseen et al. 1996; Dettki and Esseen 1998; Hanski 2000). However, few of these Fennoscandian studies explicitly compared natural and managed disturbance processes such as gap creation and soil dynamics. In Canada, large multidisciplinary experimental projects such as the EMEND project in Alberta (Spence et al. 1999), and the SAFE study at the Lac Duparquet Research and Teaching Forest in Quebec (Box 9.9; Brais et al. 1999) have begun testing whether proposed new natural dynamics based silvicultural techniques will better recreate the range of variability found in unmanaged stands and better sustain populations of sensitive forest-dwelling species (Haeussler et al. 2002).

Interactions among scales and processes

Some differences between natural disturbance processes and logging in the boreal forest are system-wide, while others are specific to a particular geographic region, landscape, or stand type. The increase in trembling aspen abundance following human settlement and clearcut logging, for example, appears to be extremely widespread (Carleton and MacLellan 1994; McRae et al. 2001). Over the same period of time, the northward retreat of viable woodland caribou populations into unsettled landscapes dominated by black spruce (Heinselman 1973; Cumming 1992) also seems to be a continent-wide phenomenon. On the other hand, concerns that logging causes long-term declines in nutrient availability appear to be prevalent in the far north, especially on black spruce – ericaceous ecosystems (Van Cleve et al. 1991; Horvath et al. 1999), Such concerns are less prevalent in southern boreal mixedwoods, particularly on aspen-dominated ecosystems where rates of nutrient cycling are higher (Paré et al. 1993; Reich et al. 2001).

Multi-scale analysis reveals that differences that are apparent at one scale of observation may disappear or reverse themselves when observed at another scale. In scaling

> **Box 9.9. Working with not against natural succession in Abitibi, Quebec.**
>
> One of the most perplexing issues for southern boreal mixedwood management has been the problem of species replacement. When a forest company harvests a valuable stand of pure white spruce, the expectation is that will replace it with vigorous young spruce of equal value, rather than a degraded stand of lower value poplar and scrub. Provincial regulations are mostly based on the principle of replacing conifers with conifers; however, these policies do not take into account that on finer textured soils where the fire cycle is long, white spruce-dominated stands tend to arise through a gradual process of forest succession in which spruce regenerates along with aspen and other pioneering tree species but does not become dominant until more than 100 years later when shade-intolerant species have largely dropped out of the stand. In attempting to replace spruce with spruce, silviculturists can become locked in a vicious cycle of hardwood removal that apparently can only be won through large-scale applications of herbicides.
>
> Not only is this battle expensive and socially unpopular, it also has ecological consequences. Deciduous-dominated early successional stages play important roles in forest nutrition and support diverse plant and animal communities. Canadian forestry statistics, meanwhile, suggest that the battle against the hardwoods is being lost: our boreal forests are increasingly dominated by hardwoods while pure conifer and mixedwood stands decline in abundance, and this too has major consequences for ecological integrity and sustainability.
>
> On mesic, clay-dominated ecosystems in the Lac Duparquet Research and Teaching Forest of Quebec, researchers are attempting to find a win–win solution that respects natural forest dynamics while not requiring silviculturists to wait 200 years to produce a harvestable stand of spruce. They are testing a combined regime that regenerates hardwood-dominated stands by clearcutting and uses small patch and selection cutting to accelerate succession towards mixedwood and conifer-dominated stands (Figs. 9.17 and 9.18). Ecological concerns are to be addressed by maintaining a balance of hardwood, mixedwood, and coniferous stands across the landscape and by allowing all stands to periodically cycle through each of the three successional stages. Variations on this theme are also being developed for black spruce and xeric pine ecosystems. See Chaps. 11 and 22 for further details on this approach.
>
> Sources: Bergeron and Harvey 1997, CCFM 1997, Bergeron et al. 1999*b*

up from stands to landscapes, we find, for example, that although logging often increases species diversity within forest stands as compared to wildfire, there may be less diversity across a landscape if the range of community types is diminished through loss of late successional conifer-dominated communities, and distinctive early post-burn communities (Fig. 9.16; Carleton and MacLellan 1994). As illustrated in Fig. 9.8, fire suppression and logging can increase diversity within landscapes with short or long pre-settlement fire cycles (i.e., alpha diversity); however, this may result in a region having lower overall diversity (i.e., gamma diversity) if all of its component landscapes then become similar in character. Scaling downwards from regions to landscapes and stands, it is apparent that while trembling aspen is becoming more abundant on a North America-wide scale, intensive conifer silviculture has reduced the aspen component in some industrial landscapes and in many individual stands to the point where concerns about long-term nutrient cycling and wildlife habitat have arisen.

Fig. 9.17. Forest dynamics on clay soils in southern boreal mixedwood forests of Quebec. Natural forest succession (white arrows) from deciduous-dominated to mixed to coniferous-dominated stands. Conventional silviculture, dominated by clearcutting (dashed arrows), returns all stands to a deciduous-dominated stage. Proposed natural dynamics based silviculture incorporating a mix of clearcutting (dashed arrows), partial, and selection cutting (solid arrows) to generate and maintain deciduous, mixedwood, and coniferous stands (adapted from Bergeron and Harvey 1997).

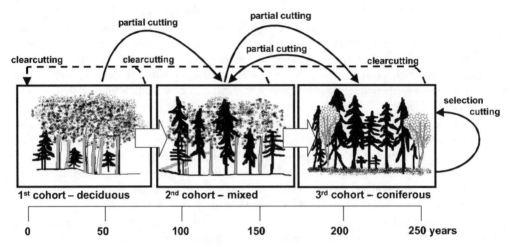

The trembling aspen example also illustrates how incongruities can arise across temporal scales. Even where aspen is sufficiently abundant in young stands to satisfy nutrient cycling needs, there may be long-term biodiversity concerns about the loss of large old aspen trees and their associated flora and fauna under a management strategy focused on short forest rotations (Esseen et al. 1992; Crites and Dale 1997; Boudreault et al. 2000).

Underlying the paradox of insufficient aspen during a period of apparent aspen super-abundance is a broader phenomenon that has been termed the "unmixing of the mixedwoods". The intricate intermixing of aspen, spruce, and other forest species that has arisen over thousands of years of natural disturbance dynamics operating at multiple scales is being progressively dismantled and reorganized along simpler lines (Fig. 9.10*b*). To simplify administrative and operational procedures, forest managers have attempted to artificially separate intimate mixtures of deciduous and coniferous trees into discrete rows, single-species stands, and divided forest land bases. Although recent policy- and field-level changes have been made in most Canadian provinces to implement better mixedwood forest management (Greene et al. 2002), much remains to be done (Chaps. 13 and 22). As described by Grover and Greenway (1999), this is an example of a multi-scale problem for which solutions can only be found by working simultaneously at multiple scales of forest policy, planning, and operations.

The issue of trembling aspen expansion shows how a problem can be perceived differently when it is observed at or within different temporal or spatial scales, making it difficult to find agreement on acceptable limits for change. The problem of forest roads

Fig. 9.18. Experimental gap-based partial cutting in a 130-year-old mixedwood stand in the SAFE project, near Lac Duparquet, Quebec, to accelerate conversion to a coniferous stand.

Photo by Dave Coates

illustrates how an ecological problem can be perfectly consistent regardless of the scale of observation, yet still defy rational resolution.

Conclusions

Can forest practices effectively mimic natural disturbance processes?

The natural disturbance model asserts that reducing differences between forest management activities and the disturbances that operate in wild, unmanaged forests is the most promising way to maintain the integrity of forest ecosystems. We began this chapter by suggesting the need for a closer look at the premise that a regime of large-scale clearcutting is the most appropriate management strategy for boreal forests because they are characterized by regular, catastrophic, or stand-destroying disturbance. We then compared clearcut logging with wildfire and other natural disturbance processes at several spatial scales and found substantial discrepancies at all scales. These differences affect rates of processes as well as the structure, composition, and diversity of forest stands, landscapes, and regions. In many instances, forest management creates conditions that are outside the range of variability observed in unmanaged wildfire- and insect-dominated systems. At the stand scale, for example, logging roads and trails are an extreme form of soil disturbance that remove a higher percentage of soil organic matter and modify physical soil processes to a greater degree and for a longer period than even the most severe wildfire. At landscape scales, regularly spaced patch clearcut openings (Fig. 9.11*a*) produce an overdispersed pattern of forest age classes that is well outside the range produced by natural disturbances. In boreal regions with long fire cycles such as the southeastern Yukon or southern Labrador, clearcutting on an approximately 100-

year rotation greatly exceeds the rate of forest disturbance in pre-industrial times. In each of these examples, logging creates novel ecological conditions to which the local or regional biota are not pre-adapted, and is likely to result in progressive change in the character or composition of boreal forest stands, landscapes, and regions.

We also found many instances where forest management activities create conditions within the range of variability observed in unmanaged systems, but their uniform implementation is reducing the range of variability in ecosystem conditions. For example, in a western boreal landscape with a historic average fire return interval of 100 years, setting a uniform rotation age of 100 years will not recreate the full variety of successional processes, stand types, and age classes observed in the unmanaged system. Fully one-third of the stands in the unmanaged landscape could be expected to exceed the average return interval while many other stands will burn more than once within a 100-year period.

Forest management as it is currently practiced in the boreal forest is likely to cause progressive change in the character of the boreal forest, and to reduce the variability or diversity below that found within natural or unmanaged systems. Because most boreal species are generalists with a variety of adaptations enabling them to recover from disturbance, logging per se does not initially reduce stand-scale diversity for most taxonomic groups studied. On the contrary, it can increase stand-scale diversity by retaining a broader representation of early, mid, and late seral stages and by creating a more variable pattern of soil and vegetation disturbance than occurs after most boreal wildfires. As longer term experience in Fennoscandia has shown, conventional forest practices can reduce stand-scale diversity through successive silvicultural interventions such as brushing, cleaning, and thinning that eliminate much of the fine-scale heterogeneity within stands. Successive rotations of a single crop species or species mix prevent species replacement and the gap-scale pattern that develops through natural succession in stands where the fire cycle is longer than the average lifespan of the dominant tree species. However, retention of coarse woody debris and wildlife trees, and creative application of a diverse range of silvicultural practices in time and space, should largely prevent this loss of stand-scale diversity from occurring.

Forest practices also have the potential to reduce diversity at larger spatial scales by:
(1) reducing or eliminating distinctive post-burn communities (Fig. 9.16) and sensitive late seral species;
(2) imposing a single, uniform pattern of cutblocks across the landscape (Fig. 9.11);
(3) reducing the range of tree species mixes, community types, and age classes present (Figs. 9.9 and 9.10); and
(4) imposing uniform forest policies and management regimes over broad geographic areas (Fig. 9.8b).

Awareness and willingness to adopt a variety of practices at all spatial scales will do much to avert this. In many regions — though certainly not all — boreal forest disturbance cycles are not hugely at variance with managed forest rotations, and since most boreal landscapes and regions also include substantial areas of commercially unproductive and unallocated forest, it should often be possible to adopt such approaches without drastic reductions in allowable cut.

It is unrealistic to expect that forest management can effectively mimic natural disturbance processes in ways that will prevent ecosystem change. We have shown that logging is distinct from wildfire and other disturbances in many large and small ways. Undoubtedly, many more differences exist that cannot yet be imagined. Each of these differences has potentially cascading implications for terrestrial ecosystem processes and terrestrial biological diversity. The same is true for boreal aquatic ecosystems, perhaps even more dramatically so (Schindler 1998*a*, *b*; Garcia and Carignan 1999; Chap. 10). Foresters can, and should, modify forest management at all scales (stand, landscape, and region) to reduce discrepancies with natural processes, while recognizing that no matter how sensitively and intelligently they are applied, forest management practices will result in substantial, progressive, and cumulative changes to boreal forests as we know them today.

Changes and diversity losses caused by forest practices are not the sole threat to boreal forest integrity. Forest management for timber production does not occur in isolation from many other radical changes taking place within the global boreal forest. Climate change, insect and disease dynamics, wildfire regimes and storm patterns, long range transport of pollutants and invasive exotic organisms, mining, oil and gas exploration, and hydroelectric developments all interact with and compound the effects of logging and silvicultural disturbance to forest stands and landscapes. A major challenge for forest ecosystem management in this century will lie in deciding whether forest practices should aim to counteract or offset the rapid rate of change by deliberately maintaining rates of ecological processes within their historic range of variability, or whether practices should be modified to accommodate and embrace change, for example, by favouring tree species that are better able to cope with recurring drought or novel forest insects and disease agents. Most likely, a mix of these two approaches should be adopted.

Adapting the natural disturbance model

Rigid application of a natural disturbance approach to forest management can lead to a variety of philosophical problems, as outlined earlier. The most troublesome of these is that, taken to its logical extreme, the natural disturbance model deems any human intervention inappropriate, while all non-human disturbances, no matter how damaging they may be to ecosystem productivity and diversity, are deemed acceptable. One potential solution to this dilemma is to adopt a sliding scale of strictness in applying the model. At one end of the scale will be large wilderness areas in which human intervention is kept to an absolute minimum, while at the other end of the scale will be intensively managed plantations in which maximizing timber production takes precedence over biodiversity and ecosystem dynamics. Where a particular forest management unit should sit within this spectrum is a matter of socio-political and geographical context, and, in the case of public forest land, will invariably be contentious. A triad zoning approach with three broad categories of land use (protection forest, ecologically managed forest, and intensively managed forest) has been proposed as appropriate for the Canadian boreal forest (Chaps. 11–13). Within a triad structure, it is possible to incorporate elements of a natural disturbance approach into all three land-use categories. And even within the ecologically managed portion of the land base, it will undoubtedly be necessary to adopt a sliding scale of strictness depending on the remoteness of a particular landscape, its balance of timber versus other forest values, and the specifics of its natural disturbance history (Box 9.10).

Several alternatives to, or variants of, the natural disturbance model, are possible. For example, a "conservation biology" approach (R.J. Mitchell[5]; Palik et al. 2002) has been advocated for the southeastern U.S. and is appropriate for boreal jurisdictions such as Fennoscandia and parts of Canada, where a true natural disturbance approach is not feasible because of the highly fragmented and modified nature of the residual forest land base. This approach can be viewed as either a refinement of the coarse-filter model or as an integration of coarse- and fine-filter approaches. Wherever feasible, management practices are brought more closely in line with natural forest dynamics. Major discrepancies that are too difficult or costly to address are identified. The next step is to identify elements of biological diversity (usually plant and vertebrate species) that are most at risk from these discrepancies. Focussed research identifies which elements of the natural disturbance regime are critical to the survival of endangered species, and then only those elements are either emulated in management practice or mitigated artificially. For example, small-scale and highly targeted prescribed burning may be carried out to maintain or restore populations of endangered pyrophilous plant or insect species (Kirkman et al. 1998; Ehnström 2001).

Another variant on the natural disturbance model that may be appropriate for North American boreal forest ecosystems can be termed a "biocomplexity" approach. This model is grounded in ecological theory but is fundamentally untested. It acknowledges that a critical flaw of the natural disturbance based approach is that it attempts to recreate historic or current conditions at a time when rapid changes to climate and global species distributions are occurring and human influences are pervasive and increasing. Under a biocomplexity model, research is carried out to increase understanding of natural variability at all spatial and temporal scales, but the knowledge is not used to reproduce or preserve past conditions. Rather, the intent is to create a future for boreal forests that will be more compatible with human needs, concepts of sustainable resource management and changing environments. For example, it may be both desirable and possible to increase the diversity and complexity of boreal forest ecosystems at multiple scales. Empirical evidence shows that boreal forests have lower species richness than many other forest biomes, and that this low diversity is both cause and consequence of the catastrophic disturbances it regularly undergoes (Gaston 2000; McCann 2000). Ecological theory tells us that an increase in the diversity of the boreal forest should reduce the scale and amplitude of these disturbances and increase the system's resilience to the stresses of global change (McCann 2000; Tilman 2000). For instance, a forest management strategy that does not replace large tracts of uniform, single-aged boreal forest with monospecific plantations but instead creates complex spatial mosaics that include mixed species and multi-layered stands could reduce the spread of wildfires and increase resistance to insects, disease, blowdown, and exotic species invasion.

---

[5]Forest ecologist, J.W. Jones Ecological Research Center, Newton, Georgia, personal communication, 2001.

Box 9.10. Options for sustainable forest management.

Major problems:

- Many forest practices result in attributes outside the natural range in variability (e.g., reductions in coarse woody debris, roads, changes to forest age-class structure);
- Forest management may also be within the range of variability but may homogenize the forest through uniform practices;
- Effects may not be observed immediately but may occur after substantial time lag;
- Forestry-caused problems are compounded by other environmental changes (e.g., climate change, other industrial developments, invasive species); and
- It is difficult to account for all differences, as some may be unknown.

Changes needed:

- Consider ecological impacts at small to large scales and over short to long time frames;
- At the stand scale, retain coarse woody debris and wildlife trees, reduce but do not eliminate organic matter at the soil surface, vary silvicultural applications, and plan for species succession;
- At the landscape scale, ensure a variety of practices, patterns, and species mixes; allow for temporal and spatial continuity of stands; and
- At regional and larger scales, consider cumulative effects of disturbance rates and human land use.

Alternative models for change:

- Zoning (such as the triad approach; see Chap. 12), including a sliding scale of adherence to the natural disturbance model;
- A conservation biology approach which combines some aspects of a natural disturbance model with a fine-filter strategy that identifies elements of biodiversity most at risk and proposes interventions to address them; and
- A biocomplexity approach that does not preserve past conditions but aims to increase diversity and complexity of ecosystems at all scales to enhance resilience in an uncertain future.

The established model for intensifying forest management has been to simplify and streamline ecosystems along the lines of a Fordian model of industrial mass production (Jacobs 2000). Even the triad approach accepts that an agro-industrial component is essential for a sustainable forest products industry. A "biocomplexity" model rejects that premise and suggests instead that the key to sustainable forest management is to use ecological principles found in nature to build more complexity into human-altered systems.

To suggest that our current understanding of the dynamics of boreal forest ecosystems is sufficient to emulate nature, let alone "improve on" nature, is immensely presumptuous. We are told, quite rightly, that *"Nature is not only more complex than we think; it is more complex than we **can** think"* (Egler 1970, p. 21). With the enormous pressures facing boreal forest ecosystems in coming decades, perhaps it is time that forest managers and forest scientists rose to the challenge.

Acknowledgements

Support provided by the SFMN, NSERC, and FQRNT is gratefully acknowledged. Thanks to Yves Bergeron, Dave Coates, and Jose Prades for helpful discussions during manuscript preparation, and to Phil Burton and Christian Messier for their editorial suggestions.

References

Aamlid, D., Tørseth, K., Venn, K., Stuanes, A.O., Solberg, S., Hylen, G., Christophersen, N., and Framstad, E. 2000. Changes of forest health in Norwegian boreal forests during 15 years. For. Ecol. Manage. **127**: 103–118.

Andison, D.W. 1999*a*. Assessing forest age data in foothills and mountain landscapes of Alberta. Foothills Model Forest, Hinton, Alberta. Available at http://www.fmf.ab.ca/pdf/forestage.pdf [viewed 16 September 2002]. Alberta Foothills Disturb. Ecol. Res. Ser. Rep. No. 1. 29 p.

Andison, D.W. 1999*b*. Natural pattern emulation for forest management: buyer beware. *In* Science and practice: sustaining the boreal forest. Proceedings of the 1999 Sustainable Forest Management Network Conference, 14–17 February 1999, Edmonton, Alberta. *Edited by* T.S. Veeman, D.W. Smith, B.G. Purdy, F.J. Salkie, and G.A. Larkin. Sustainable Forest Management Network, Edmonton, Alberta. pp. 591–596.

Andison, D.W. 2000. Landscape-level fire activity on foothills and mountain landscapes of Alberta. Foothills Model Forest, Hinton, Alberta. Alberta Foothills Disturb. Ecol. Res. Ser. Rep. No. 2. Available at http://www.fmf.ab.ca/pdf/fireactivity.pdf [viewed 16 September 2002]. 44 p.

Andrén, K. 1994. Effects of habitat fragmentation on birds and mammals in landscapes with different proportions of suitable habitat: a review. Oikos, **71**: 355–366.

Angelstam, P.K. 1998. Maintaining and restoring biodiversity in European boreal forests by developing natural disturbance regimes. J. Veg. Sci. **9**: 593–602.

Angermeier, P.L., and Karr, J.R. 1994. Biological integrity versus biological diversity as policy directions. BioScience, **44**: 690–697.

Armstrong, G.W., Cumming, S.G., and Adamowicz, W.L. 1999. Timber supply implications of natural disturbance management. *In* Science and practice: sustaining the boreal forest. Proceedings of the 1999 Sustainable Forest Management Network Conference, 14–17 February 1999, Edmonton, Alberta. *Edited by* T.S. Veeman, D.W. Smith, B.G. Purdy, F.J. Salkie, and G.A. Larkin. Sustainable Forest Management Network, Edmonton, Alberta. pp. 369–374.

Attiwill, P.M. 1994. The disturbance of forest ecosystems: the ecological basis for conservative management. For. Ecol. Manage. **63**: 247–300.

Barney, R.J., and Stocks, B.J. 1983. Fire frequencies during the suppression period. *In* The role of fire in northern circumpolar ecosystems. SCOPE Report 18. *Edited by* R.W. Wein and D.A. MacLean. Wiley, Chichester, U.K. pp. 45–62.

Baskerville, G.L. 1975. Spruce budworm: super silviculturist. For. Chron. **51**: 138–140.

Baskerville, G.L. 1988. Redevelopment of a degrading forest system. Ambio, **17**: 314–322.

Batzer, H.O., and Popp, M.P. 1985. Forest succession following a spruce budworm outbreak in Minnesota. For. Chron. **61**: 75–80.

(BCMOF) British Columbia Ministry of Forests. 1996. Forest practices code guidebooks. British Columbia Ministry of Forests and British Columbia Ministry of Environment, Lands and Parks. Victoria, British Columbia.

Berg, Å., Ehnström, B., Gustafsson, L., Hallingbäck, T., Jonsell, M., and Weslien, J. 1994. Threatened plant, animal, and fungus species in Swedish forests: distribution and habitat associations. Conserv. Biol. **9**: 718– 731.

Bergeron, Y. 1991. The influence of island and mainland lakeshore landscapes on boreal forest fire regimes. Ecology, **72**: 1980–1992.

Bergeron, Y., and Archambault, S. 1993. Decrease of forest fires in Quebec's southern boreal zone and its relation to global warming since the end of the Little Ice Age. The Holocene, **3**: 255–259.

Bergeron, Y., and Brisson, J. 1990. Fire regime in red pine stands at the northern limit of the species' range. Ecology, **71**: 1352– 1364.

Bergeron, Y., and Flannigan, M.D. 1995. Predicting the effects of climate change on fire frequency in the southeastern Canadian boreal forest. Water Soil Air Pollut. **82**: 437–444.

Bergeron, Y., and Harvey, B. 1997. Basing silviculture on natural ecosystem dynamics: an approach applied to the southern boreal mixedwood forest of Quebec. For. Ecol. Manage. **92**: 235–242.

Bergeron, Y., Engelmark, O., Harvey, B., Morin H., and Sirois, L. 1998. Key issues in disturbance dynamics in boreal forests: introduction. J. Veg. Sci. **9**: 464–468.

Bergeron, Y., Gauthier, S., Carcaillet, C., Flannigan, M., Prairie, Y., and Richard, P.J.H. 1999*a*. Variability in fire frequency and forest composition in Canada's southeastern boreal forest: a challenge for sustainable forest management. *In* Science and practice: sustaining the boreal forest. Proceedings of the 1999 Sustainable Forest Management Network Conference, 14–17 February 1999, Edmonton, Alberta. *Edited by* T.S. Veeman, D.W. Smith, B.G. Purdy, F.J. Salkie, and G.A. Larkin. Sustainable Forest Management Network, Edmonton, Alberta. pp. 74–80.

Bergeron, Y, Harvey, B., Leduc, A., and Gauthier, S. 1999*b*. Forest management guidelines based on natural disturbance dynamics: stand- and forest-level considerations. For. Chron. **75**: 49–54.

Bessie, W.C., and Johnson, E.A. 1995. The relative importance of weather and fuels on fire behaviour in subalpine forests. Ecology, **76**: 747–762.

Bird, R. 1998. The boreal forest. National atlas of Canada. Canadian Geographic and Canadian Forest Service, Geomatics Canada, Ottawa, Ontario. (Poster and maps.)

Blais, J.R. 1983. Trends in the frequency, extent and severity of spruce budworm outbreaks in eastern Canada. Can. J. For. Res. **13**: 539–547.

Bouchard, S. 1997. Terminology used by Parks Canada. Minister of Public Works and Government Services, Ottawa, Ontario. Terminol. Bull. 236. 456 p.

Boudreault, C., Gauthier, S., and Bergeron, Y. 2000. Epiphytic lichens and bryophytes on *Populus tremuloides* along a chronosequence in the southwestern boreal forest of Québec, Canada. The Bryologist, **103**: 725–738.

Boulet, M., Darveau, M., and Belanger, L. 1999. Is reproductive success of birds affected by forestry practices in a boreal black spruce forest [abstract]? *In* Science and practice: sustaining the boreal forest. Proceedings of the 1999 Sustainable Forest Management Network Conference, 14–17 February 1999, Edmonton, Alberta. *Edited by* T.S. Veeman, D.W. Smith, B.G. Purdy, F.J. Salkie, and G.A. Larkin. Sustainable Forest Management Network, Edmonton, Alberta. p. 733.

Bourque, J., and Villard, M.A. 2001. Effects of selection cutting and landscape-scale harvesting on the reproductive success of two neotropical migrant bird species. Conserv. Biol. **15**: 184–195.

Boutin, S. 1999. Adaptive ecosystem management: all talk and no action. *In* Science and practice: sustaining the boreal forest. Proceedings of the 1999 Sustainable Forest Management

Network Conference, 14–17 February 1999, Edmonton, Alberta. *Edited by* T.S. Veeman, D.W. Smith, B.G. Purdy, F.J. Salkie, and G.A. Larkin. Sustainable Forest Management Network, Edmonton, Alberta. pp. 22–23.

Bowman, J., Forbes, G., and Dilworth, T. 1999. The spatial structure of small mammal populations in a managed forest. *In* Science and practice: sustaining the boreal forest. Proceedings of the 1999 Sustainable Forest Management Network Conference, 14–17 February 1999, Edmonton, Alberta. *Edited by* T.S. Veeman, D.W. Smith, B.G. Purdy, F.J. Salkie, and G.A. Larkin. Sustainable Forest Management Network, Edmonton, Alberta. pp. 58–63.

Brais, S., and Camiré, C. 1998. Soil compaction induced by careful logging in the claybelt region of northwestern Quebec (Canada). Can. J. Soil Sci. **78**: 197–206.

Brais, S., Camiré, C., and Paré, D. 1995. Impacts of whole-tree harvesting and winter windrowing on soil pH and base status of clayey sites of northwestern Quebec. Can. J. For. Res. **25**: 997–1007.

Brais, S., Harvey, B., and Bergeron, Y. 1999. Testing ecosystem management in the southeastern boreal mixedwood forest [abstract]. *In* Science and practice: sustaining the boreal forest. Proceedings of the 1999 Sustainable Forest Management Network Conference, 14–17 February 1999, Edmonton, Alberta. *Edited by* T.S. Veeman, D.W. Smith, B.G. Purdy, F.J. Salkie, and G.A. Larkin. Sustainable Forest Management Network, Edmonton, Alberta. p. 736.

Brais, S., Paré, D., Camiré, C., Rochon, P., and Vasseur, C. 2002. Nitrogen net mineralization and dynamics following whole-tree harvesting and winter windrowing on clayey sites of northwestern Quebec. For. Ecol. Manage. **157**: 119–130.

Brongo, D., Drapeau, P., and Giroux, J.-F. 2000. Edge avoidance and preference of forest birds in boreal mixed-wood forests in Abitibi, Quebec (Canada) [abstract]. *In* Disturbance dynamics in boreal forests, restoration and management of biodiversity, 21–25 August 2000, Kuhmo, Finland. *Edited by* L. Karjalainen and T. Kuuluvainen. University of Helsinki, Helsinki, Finland. p. 62.

Brown, J.R., Herrick, J., and Price, D. 1999. Managing low-output agroecosystems sustainably: the importance of ecological thresholds. Can. J. For. Res. **29**: 1112–1119.

Buddle, C.M., Spence, J.R., Langor, D.W., and Pohl, G.R. 1999. The succession of litter-dwelling arthropod communities following disturbance by harvesting and wildfire. *In* Science and practice: sustaining the boreal forest. Proceedings of the 1999 Sustainable Forest Management Network Conference, 14–17 February 1999, Edmonton, Alberta. *Edited by* T.S. Veeman, D.W. Smith, B.G. Purdy, F.J. Salkie, and G.A. Larkin. Sustainable Forest Management Network, Edmonton, Alberta. pp. 476–481.

Buddle, C.M., Spence, J.R., and Langor, D.W. 2000. Succession of boreal forest spider assemblages following wildfire and harvesting. Ecography, **23**: 424–436.

Bulmer, C., Schmidt, M.G., Kishchuk, B., and Preston, C. 1998. Impacts of blading and burning site preparation on soil properties and site productivity in the sub-boreal spruce zone of central British Columbia. Victoria, British Columbia. Can. For. Serv. Inf. Rep. BC-X-377. 33 p.

Burton, P.J. 1998. Designing riparian buffers. Ecoforestry, **13**(3): 12–22.

Burton, P.J., Kneeshaw, D.D., and Coates, K.D. 1999. Managing forest harvesting to maintain old growth in boreal and sub-boreal forests. For. Chron. **75**: 623–629.

Canuel, B. 1989. Guide d'utilisation de la coupe avec protection de la régénération (abattage mécanique). Ministère de l'Énergie et des Ressources du Québec, Service des traitements sylvicoles, Québec, Quebec.

Carcaillet, C., Frichette, B., Richard, P.J.H., Bergeron, Y. Gauthier, S., and Prairie, Y.T. 1999. *In* Science and practice: sustaining the boreal forest. Proceedings of the 1999 Sustainable Forest Management Network Conference, 14–17 February 1999, Edmonton, Alberta. *Edited by* T.S. Veeman, D.W. Smith, B.G. Purdy, F.J. Salkie, and G.A. Larkin. Sustainable Forest Management Network, Edmonton, Alberta. pp. 87–91.

Carleton, T.J., and MacLellan, P. 1994. Woody vegetation responses to fire versus clear-cutting logging: a comparative survey in the central Canadian boreal forest. EcoScience, **1**: 141–152.

(CCFM) Canadian Council of Forest Ministers. 1997. Criteria and indicators of sustainable forest management in Canada. Technical Report. Canadian Forest Service, Natural Resources Canada, Ottawa, Ontario. 137 p.

Chapin, F.S., Torn, M.S., and Tateno, M. 1996. Principles of ecosystem sustainability. Amer. Nat. **148**: 1016–1037.

Chapin, F.S., Zavaleta, E.S., Eviner, V.T., Naylor, R.L., Vitousek, P.M., Reynolds, H.L., Hooper, D.U., Lavorel, S., Sala, O.E., Hobbie, S.E., Mack, M.C., and Diaz, S. 2000. Consequences of changing biodiversity. Nature (London), **405**: 235–242.

Charron, I., and Greene, D.F. 1999. The influence of seedbed type on regeneration of trees following fire in the mixedwood boreal forest of Saskatchewan [abstract]. *In* Science and practice: sustaining the boreal forest. Proceedings of the 1999 Sustainable Forest Management Network Conference, 14–17 February 1999, Edmonton, Alberta. *Edited by* T.S. Veeman, D.W. Smith, B.G. Purdy, F.J. Salkie, and G.A. Larkin. Sustainable Forest Management Network, Edmonton, Alberta. p. 738.

Christiansen, N.L., Bartuska, A.M., Brown, J.H., Carpenter, S., D'Antonio, C., Francis, R., Franklin, J.F., MacMahon, J.A., Noss, R.F., Parsons, D.J., Peterson, C.H., Turner, M.G., and Woodmansee, R.G. 1996. The report of the Ecological Society of America committee on the scientific basis for ecosystem management. Ecol. Applic. **6**: 665–691.

Coates, K.D., and Burton, P.J. 1997. A gap-based approach for development of silvicultural systems to address ecosystem management objectives. For. Ecol. Manage. **99**: 337–354.

Coates, K.D., and Steventon, J.D. 1995. Patch retention harvesting as a technique for maintaining stand level biodiversity in forests of north central British Columbia. *In* Innovative silvicultural systems in boreal forests. Symposium Proceedings, 4–5 October 1994, Edmonton, Alberta. *Edited by* C.R. Bamsey. Clear Lake Ltd., Edmonton, Alberta. pp. 102–106.

Cody, W.J. 1996. Flora of the Yukon Territory. NRC Research Press, Ottawa, Ontario. 643 p.

Cogbill, C.V. 1985. Dynamics of the boreal forests of the Laurentian Highlands, Canada. Can. J. For. Res. **15**: 252–261.

Corkum, C.V., Fisher, J.T., and Boutin, S. 1999. Investigating influences of landscape structure on small mammal abundance in Alberta's boreal mixed-wood forest. *In* Science and practice: sustaining the boreal forest. Proceedings of the 1999 Sustainable Forest Management Network Conference, 14–17 February 1999, Edmonton, Alberta. *Edited by* T.S. Veeman, D.W. Smith, B.G. Purdy, F.J. Salkie, and G.A. Larkin. Sustainable Forest Management Network, Edmonton, Alberta. pp. 29–35.

Crites, S. 1999. A test case for the natural disturbance model: understory vegetation communities following fire and harvesting in aspen mixedwood stands. *In* Science and practice: sustaining the boreal forest. Proceedings of the 1999 Sustainable Forest Management Network Conference, 14–17 February 1999, Edmonton, Alberta. *Edited by* T.S. Veeman, D.W. Smith, B.G. Purdy, F.J. Salkie, and G.A. Larkin. Sustainable Forest Management Network, Edmonton, Alberta. pp. 607–615.

Crites, S., and Dale, M.R.T. 1997. Diversity and abundance of bryophytes, lichens and fungi in relation to woody substrate and successional stage in aspen mixedwood boreal forests. Can. J. Bot. **76**:641–651.

Cumming, H.G. 1992. Woodland caribou: facts for forest managers. For. Chron. **68**: 481–491.

Cumming, S.G., Schmiegelow, F.K.A., and Burton, P.J. 2000. Gap dynamics in boreal aspen stands: is the forest older than we think? Ecol. Applic. **10**: 744–759.

Dahlberg, A., Schimmel, J., Taylor, A.F.S., and Johannesson, H. 2001. Post-fire legacy of ecto-mycorrhizal fungal communities in the Swedish boreal forest in relation to fire severity and logging intensity. Biol. Conserv. **100**: 151–161.

Daniel, C.J., and Myers, J.H. 1995. Climate and outbreaks of the forest tent caterpillar. Ecography, **18**: 353–362.

Dansereau, P.-R., and Bergeron, Y. 1993. Fire history in the southern boreal forest of north-western Quebec. Can. J. For. Res. **23**: 25–32.

DeLong, S.C. 2002. Using nature's template to best advantage in the Canadian boreal forest. Silva Fenn. **36**: 401–408.

DeLong, S.C., and Tanner, D. 1996. Managing the pattern of forest harvest: lessons from wild-fire. Biodivers. Conserv. **5**: 1191–1205.

Dettki, H., and Esseen, P.-A. 1998. Epiphytic macrolichens in managed and natural forest land-scapes a comparison at two spatial scales. Ecography, **21**: 613–624.

Dettki, H., Klintberg, P., and Esseen, P.-A. 2000. Are epiphytic lichens in young forests limited by local dispersal? Ecoscience, **7**: 317–325.

Drapeau, P., Leduc, A., Giroux, J.-F., Savard, J.-P.L., Bergeron, Y., and Vickery, W.L. 2000. Landscape-scale disturbances and changes in bird communities of boreal mixed-wood forests. Ecol. Monogr. **70**: 423–444.

Eberhart, K.E., and Woodward, P.M. 1987. Distribution of residual vegetation associated with large fires in Alberta. Can. J. For. Res. **17**: 1207–1212.

Egler, C.D. 1998. Known range of dusky flycatcher extended northeast to the Kotaneelee Range, Yukon. Birders J. **7**: 205–207.

Egler, F. 1970. The way of science: a philosophy of ecology for the layman. Hafner Publishing Company, New York, New York. 145 p.

Ehnström, B. 2001. Leaving dead wood for insects in boreal forests — suggestions for the future. Scand. J. For. Res. **3**(Suppl.): 91–98.

Elkie, P.C., and Rempel, R.S. 1999. Identifying hierarchical structure in Ontario landscapes as a component of emulating natural disturbance. *In* Science and practice: sustaining the boreal forest. Proceedings of the 1999 Sustainable Forest Management Network Conference, 14–17 February 1999, Edmonton, Alberta. *Edited by* T.S. Veeman, D.W. Smith, B.G. Purdy, F.J. Salkie, and G.A. Larkin. Sustainable Forest Management Network, Edmonton, Alberta. pp. 221–227.

Environment Canada. 1999. Climate trends and variations bulletin for Canada. Available at http://www.msc-smc.ec.gc.ca/ccrm/bulletin/index.html [viewed 12 September 2002].

Esseen, P.-A., Ehnström, B., Ericson, L., and Sjöberg, K. 1992. Boreal forests — the focal habi-tats of Fennoscandia. *In* Ecological principles of nature conservation. *Edited by* L. Hansson. Elsevier Applied Science, London, U.K. pp. 252–325.

Esseen, P.-A., Renhorn, K.-E., and Pettersson, R.B. 1996. Epiphyte lichen biomass in managed and old- growth forests: effect of branch quality. Ecol. Applic. **6**: 228–238.

(EWG) Ecoregions Working Group. 1989. Ecoclimatic regions of Canada, first approximation. Canada Committee on Ecological Land Classification, Canadian Wildlife Service, Environ-ment Canada, Ottawa, Ontario. Ecol. Land Class. Ser. No. 23. 119 p.

Flannigan, M.D., and Van Wagner, C.E. 1991. Climate change and wildfire in Canada. Can. J. For. Res. **21**: 66–72.

Flannigan, M.D., Bergeron, Y., Engelmark, O., and Wotton, B.M. 1998. Future wildfire in cir-cumboreal forests in relation to global warming. J. Veg. Sci. **9**: 469–476.

Forman, R.T.T. 2000. Estimate of the area affected ecologically by the road system in the United States. Conserv. Biol. **14**: 31–35.

Foster, D.R. 1982. The history and pattern of fire in the boreal forest of southeastern Labrador. Can. J. Bot. **61**: 24–59.

Frelich, L.E. 2002. Forest dynamics and disturbance regimes. Cambridge University Press, Cambridge, U.K. 276 p.

Frelich, L.E., and Reich, P.B. 1995. Spatial patterns and succession in a Minnesota southern-boreal forest. Ecol. Monogr. **65**: 325–356.

Frelich, L.E., and Reich, P.B. 1998. Disturbance severity and threshold responses in the boreal forest. Conserv. Ecol. [online], **2(2)**:7. Available at http://www.consecol.org/vol2/iss2/art7 [viewed 16 September 2002].

Fryer, G.I., and Johnson, E.A. 1988. Reconstructing fire behavior and effects in a subalpine forest. J. Appl. Ecol. **25**: 1063–1072.

Galipeau, C., Kneeshaw, D.D., and Bergeron, Y. 1997. White spruce and balsam fir colonization of a site in the southeastern boreal forest as observed 68 years after fire. Can. J. For. Res. **27**: 139–147.

Gandhi, K.J.K., Spence, J.R., Langor, D.W., and Morgantini, L. 1999. Fire-skips and carabid beetles of northern Rockies: 'natural lessons' for the forest industry. *In* Science and practice: sustaining the boreal forest. Proceedings of the 1999 Sustainable Forest Management Network Conference, 14–17 February 1999, Edmonton, Alberta. *Edited by* T.S. Veeman, D.W. Smith, B.G. Purdy, F.J. Salkie, and G.A. Larkin. Sustainable Forest Management Network, Edmonton, Alberta. pp. 466–471.

Garcia, E., and Carignan, R. 1999. Impact of wildfire and clearcutting in the boreal forest on methyl mercury in zooplankton. Can. J. Fish. Aquat. Sci. **56**: 339–345.

Gaston, K.J. 2000. Global patterns in biodiversity. Nature (London), **405**: 220–227.

Gauthier, S., Leduc, A., and Bergeron, Y. 1996. Forest dynamics modelling under a natural fire cycle: a tool to define natural mosaic diversity in forest management. Environ. Monit. Assess. **39**: 417–434.

Gluck, M.J., and Rempel, R.S. 1996. Structural characteristics of post-wildfire and clear-cut landscapes. Environ. Monit. Assess. **39**: 435–450.

Greene, D.F., and Johnson, E.A. 1999. Patterns of residual (unburnt) stands in boreal forest fires. *In* Science and practice: sustaining the boreal forest [abstract]. Proceedings of the 1999 Sustainable Forest Management Network Conference, 14–17 February 1999, Edmonton, Alberta. *Edited by* T.S. Veeman, D.W. Smith, B.G. Purdy, F.J. Salkie, and G.A. Larkin. Sustainable Forest Management Network, Edmonton, Alberta. p. 757.

Greene, D.F., Kneeshaw, D.D., Messier, C., Lieffers, V., Cormier, D., Doucet, R., Coates, K.D., Groot, A., Grover, G., and Calogeropoulos, C. 2002. A biological and economical analysis of silvicultural alternatives to the conventional clearcut/plantation prescription in boreal mixedwood stands (aspen/white spruce/balsam fir). For. Chron. **78**: 281–295.

Grigal, D.F., and Bates, P.C. 1997. Assessing impacts of forest harvesting — the Minnesota experience. Biomass Bioenergy, **13**: 213–222.

Grover, B.E., and Greenway, K.J. 1999. The ecological and economic basis for mixedwood management. *In* Science and practice: sustaining the boreal forest. Proceedings of the 1999 Sustainable Forest Management Network Conference, 14–17 February 1999, Edmonton, Alberta. *Edited by* T.S. Veeman, D.W. Smith, B.G. Purdy, F.J. Salkie, and G.A. Larkin. Sustainable Forest Management Network, Edmonton, Alberta. pp. 419–428.

Haeussler, S., Bedford, L., Leduc, A., Bergeron, Y., and Kranabetter, J.M. 2002. Silvicultural disturbance severity and plant communities of the southern Canadian boreal forest. Silva Fenn. **36**: 307–327.

Haila, Y. 1995. Natural dynamics as a model for management: is the analogue practicable? Arctic Centre Public. **7**: 9–26.

Hamel, B., and Paré, D. 1999. Interpretation of nutritional constraints of forest productivity in relation to disturbance, deposit and region with the use of multivariate foliar indices. *In* Science and practice: sustaining the boreal forest [abstract]. Proceedings of the 1999 Sustainable Forest Management Network Conference, 14–17 February 1999, Edmonton, Alberta. *Edited by* T.S. Veeman, D.W. Smith, B.G. Purdy, F.J. Salkie, and G.A. Larkin. Sustainable Forest Management Network, Edmonton, Alberta. p. 765.

Hanski, I. 2000. Extinction debt and species credit in boreal forests: modelling the consequences of different approaches to biodiversity conservation. Ann. Zool. Fenn. **37**: 271–280.

Harris, L.D. 1984. The fragmented forest: island biogeography theory and the preservation of biotic diversity. University of Chicago Press, Chicago, Illinois. 211 p.

Harvey, B.D., Bergeron, Y., Leduc, A., and Gauthier, S. 2002. Stand-landscape integration in natural disturbance-based management of the southern boreal forest. For. Ecol. Manage. **155**: 369–385.

He, H.S., and Mladenoff, D.J. 1999. Spatially explicit and stochastic simulation of forest-landscape fire disturbance and succession. Ecology, **80**: 81–99.

Heinselman, M.L. 1971. The natural role of fire in northern conifer forests. *In* Proceedings, Fire in the Northern Environment — A Symposium, 13–14 April 1971, Fairbanks, Alaska. *Edited by* C.W. Slaughter, R.J. Barney, and G.M. Hansen. USDA Forest Service, Portland, Oregon. pp. 61–72.

Heinselman, M.L. 1973. Fire in the virgin forests of the Boundary Waters Canoe Area, Minnesota. Quatern. Res. **3**: 329–382.

Heisenberg, W.K. 1958. Physics and philosophy: the revolution in modern science. Harper & Row, New York, New York. 206 p.

Hobson, K.A., and Shieck, J. 1999. Changes in bird communities in boreal mixedwood forest: harvest and wildfire effects over 30 years. Ecol. Applic. **9**: 849–863.

Hogg, E.H. 1999. Simulation of interannual responses of trembling aspen stands to climatic variation and insect defoliation in western Canada. Ecol. Model. **114**: 175–193.

Holling, C.S., and Meffe, G.K. 1996. Command and control and the pathology of natural resource management. Conserv. Biol. **10**: 328–337.

Horvath, R., Ruel, J.-C., Ung, C.H., and Munson, A. 1999. Effect of fire and careful logging on growth in black spruce stands. In Science and practice: sustaining the boreal forest. Proceedings of the 1999 Sustainable Forest Management Network Conference, 14–17 February 1999, Edmonton, Alberta. *Edited by* T.S. Veeman, D.W. Smith, B.G. Purdy, F.J. Salkie, and G.A. Larkin. Sustainable Forest Management Network, Edmonton, Alberta. pp. 442–445.

Hunter, M.L. 1990. Wildlife, forests and forestry: principles of managing forests for biological diversity. Prentice–Hall Inc., Englewood Cliffs, New Jersey. 370 p.

Hunter, M.L. 1991. Coping with ignorance: the coarse-filter strategy for maintaining biodiversity. *In* Balancing on the brink of extinction: the Endangered Species Act and lessons for the future. *Edited by* K.A. Kohm. Island Press, Washington, D.C. pp. 266–281.

Hunter, M.L. 1993. Natural disturbance regimes as spatial models for managing boreal forests. Biol. Conserv. **65**: 115–120.

Imbeau, L., Mönkkönen, M., and Desrochers, A. 2001. Long-term effects of forestry on birds of the eastern Canadian boreal forests: a comparison with Fennoscandia. Conserv. Biol. **15**: 1151–1162.

Ives, W.G.H. 1973. Heat units and outbreaks of the forest tent caterpillar, *Malacosoma disstria* (Lepidoptera: Lasiocampidae). Can. Entomol. **105**: 529–543.

Jacobs, J. 2000. The nature of economies. Modern Library, New York, New York. 190 p.

Jardon, Y. 2001. Analyse temporelle et spatiale des épidémies de la tordeuse des bourgeons de l'épinette au Québec. Ph.D. thesis, Université du Québec à Montréal, Montréal, Quebec. 171 p.

Jensen, W.F., Fuller, T.K., and Robinson, W.L. 1986. Wolf *Canis lupus* distribution on the Ontario- Michigan border near Sault Ste. Marie. Can. Field-Nat. **100**: 363–366.

Johnson, E.A. 1992. Fire and vegetation dynamics: studies from the North American boreal forest. Cambridge University Press, Cambridge, U.K. 149 p.

Johnson, E.A., and Van Wagner, C.E. 1985. The theory and use of two fire history models. Can. J. For. Res. **15**: 214–220.

Johnson, E.A., and Wowchuk, D.R. 1993. Wildfires in the southern Canadian Rocky Mountains and their relationship to mid-tropospheric anomalies. Can. J. For. Res. **23**: 1213–1222.

Johnson, E.A., Miyanishi, K., and Weir, J.M.H. 1998. Wildfires in the western Canadian boreal forest: landscape patterns and ecosystem management. J. Veg. Sci. **9**: 602–610.

Jonsson, B.G. 1993. The bryophyte diaspore bank and its role after small-scale disturbance in a boreal forest. J. Veg. Sci. **4**: 819–826.

Jonsson, B.G., and Dynesius, M. 1993. Uprooting in boreal spruce forests: long-term variation in disturbance rate. Can. J. For. Res. **23**: 2383–2388.

Keenan, R.J., and Kimmins, J.P. 1993. The ecological effects of clear-cutting. Environ. Rev. **1**: 121–144.

Kimmins, J.P. 1987. Forest ecology. Macmillan Publishing Company, New York, New York. 531 p.

Kirkman, L.K., Drew, M.B., and Edwards, D. 1998. Effects of experimental fire regimes on the population dynamics of *Schwalbea americana*. Plant Ecol. **137**: 115–137.

Kneeshaw, D.D. 2001. Are non-fire gap disturbances important to boreal forest dynamics? *In* Recent research developments in ecology. *Edited by* S.G. Pandalarai. Transworld Research Press, Trivandrum, India. pp. 43–58.

Kneeshaw, D.D., and Bergeron, Y. 1998. Canopy gap characteristics and tree replacement in the southern boreal forest. Ecology, **79**: 783–794.

Kneeshaw, D.D., and Bergeron, Y. 1999. Spatial and temporal patterns of seedling and sapling recruitment within canopy gaps caused by spruce budworm. Ecoscience, **6**: 214–222.

Kneeshaw, D.D., and Burton, P.J. 1997. Canopy age structures of some old sub-boreal *Picea* stands in British Columbia. J. Veg. Sci. **8**: 615–626.

Kohm, K.A., and Franklin, J.F. (*Editors*) 1997. Creating a forestry for the 21st century: the science of ecosystem management. Island Press, Washington, D.C. 475 p.

Kranabetter, J.M., and Yole, D. 2000. Alternatives to broadcast burning in the northern interior of British Columbia: short term tree results. For. Chron. **76**: 349–353.

Krebs, C.J., Boonstra R., Boutin, S., and. Sinclair, A.R.E. 2001. What drives the 10-year cycle of snowshoe hares? BioScience, **51**: 25–35.

Kruys, N., and Jonsson, B.G. 1999. Fine woody debris is important for species richness on logs in managed boreal spruce forests of northern Sweden. Can. J. For. Res. **29**: 1295–1299.

Kurz, W.A., and Apps, M.J. 1995. An analysis of future carbon budgets of Canadian boreal forests. Water Air Soil Pollut. **82**: 321–332.

Kurz, W.A., and Apps, M.J. 1996. Retrospective assessment of carbon flows in Canadian boreal forests. *In* Forest ecosystems, forest management and the global carbon cycle. NATO ASI Ser. 1, Vol. 40, Global Environmental Change. *Edited by* M.J. Apps and D.T. Price. Springer-Verlag, Heidelberg, Germany. pp. 173–182.

Kurz, W.A., and Apps, M.J. 1999. A 70-year retrospective analysis of carbon fluxes in the Canadian forest sector. Ecol. Applic. **9**: 526–547.

Kurz, W.A., Apps, M.J., Stocks, B.J., and Volney, W.J.A. 1995. Global climate change: disturbance regimes and biospheric feedbacks of temperate and boreal forests. *In* Biotic feedbacks in the global climate system: will the warming feed the warming? *Edited by* G.M. Woodwell and F.T. Mackenzie. Oxford University Press, New York, New York. pp. 119–133.

Kuuluvainen, T. 1994. Gap disturbance, ground microtopography and regeneration dynamics of boreal coniferous forests in Finland: a review. Ann. Zool. Fenn. **31**: 35–51.

Kuuluvainen, T. 2002. Disturbance dynamics in boreal forests: defining the ecological basis of restoration and management of biodiversity. Silva Fenn. **36**: 5–12.

Kuusinen, M. 1994. Epiphytic lichen flora and diversity on *Populus tremula* in old-growth and managed forests of southern and middle boreal Finland. Ann. Bot. Fenn. **31**: 245–260.

Laird, L.D. 1999. Changes in fire regime in response to climate over the last millennium [abstract]. *In* Science and practice: sustaining the boreal forest. Proceedings of the 1999 Sustainable Forest Management Network Conference, 14–17 February 1999, Edmonton, Alberta. *Edited by* T.S. Veeman, D.W. Smith, B.G. Purdy, F.J. Salkie, and G.A. Larkin. Sustainable Forest Management Network, Edmonton, Alberta. p. 775.

Landres, P.B., Morgan, P., and Swanson, F.J. 1999. Overview of the use of natural variability concepts in managing ecological systems. Ecol. Applic. **9**: 1179–1188.

Laroque, I., Bergeron, Y., Campbell, I.D., and Bradshaw, R.H.W. 1999. Influence of disturbance on the distribution and regeneration of boreal species [abstract]. *In* Science and practice: sustaining the boreal forest. Proceedings of the 1999 Sustainable Forest Management Network Conference, 14–17 February 1999, Edmonton, Alberta. *Edited by* T.S. Veeman, D.W. Smith, B.G. Purdy, F.J. Salkie, and G.A. Larkin. Sustainable Forest Management Network, Edmonton, Alberta. p. 776.

Larsen, C.P.S. 1997. Spatial and temporal variations in boreal fire frequency in northern Alberta. J. Biogeogr. **24**: 663–673.

Lee, P., and Crites, S. 1999. Early successional deadwood dynamics in wildfire and harvest stands. *In* Science and practice: sustaining the boreal forest. Proceedings of the 1999 Sustainable Forest Management Network Conference, 14–17 February 1999, Edmonton, Alberta. *Edited by* T.S. Veeman, D.W. Smith, B.G. Purdy, F.J. Salkie, and G.A. Larkin. Sustainable Forest Management Network, Edmonton, Alberta. pp. 601–606.

Lee, P.C., Crites, S., Neitfeld, M., Nguyen, H.V., and Stelfox, J.B. 1997. Characteristics and origins of deadwood material in aspen-dominated boreal forests. Ecol. Applic. **7**: 691–701.

Lefort, P., Gauthier, S., and Bergeron, Y. 1999. A 235 year boreal forest fire history in eastern Canada: the part of climate and land use [abstract]. *In* Science and practice: sustaining the boreal forest. Proceedings of the 1999 Sustainable Forest Management Network Conference, 14–17 February 1999, Edmonton, Alberta. *Edited by* T.S. Veeman, D.W. Smith, B.G. Purdy, F.J. Salkie, and G.A. Larkin. Sustainable Forest Management Network, Edmonton, Alberta. p. 778.

Leopold, A. 1933. Game management. Scribner, New York, New York. 481 p.

Levin, S.A. 1992. The problem of pattern and scale in ecology. Ecology, **73**: 1943–1967.

Li, H., Franklin, J.F., Swanson, F.J., and Spies, T.A. 1993. Developing alternative forest cutting patterns: a simulation approach. Landscape Ecol. **8**: 63–75.

Li, C., Flannigan, M.D., and Corns, I.G.W. 2000. Influence of potential climate change on forest landscape dynamics of west-central Alberta. Can. J. For. Res. **30**: 1905–1912.

Lieffers, V.J., Macmillan, R.B., MacPherson, D., Branter, K., and Stewart, J.D. 1996. Semi-natural and intensive silvicultural systems for the boreal mixedwood forest. For. Chron. **72**: 286–292.

Lindenmayer, D., and McCarthy, M.A. 2002. Congruence between natural and human forest disturbance: a case study from Australian montane ash forests. For. Ecol. Manage. **155**: 319–335.

Lindgren, B.S., Lemieux, J.P., and Steventon, D. 1999. Effects of silviculture systems on arthropod community structure: contrasting clearcut and patch retention harvests in high elevation forests. Prince Rupert Forest Region, British Columbia. Ministry of Forests, Smithers, British

Columbia. For. Sci. Ext. Note No. 37. Available at http://www.for.gov.bc.ca/prupert/ Research/Extension_notes/Enote37.pdf [viewed 28 May 2003]. 4 p.

Lundquist, J.E., and Beatty, J.S. 2002 A method for characterizing and mimicking forest canopy gaps caused by different disturbances. For. Sci. **48**: 582–594.

Mackenzie, K.L., and Steventon, J.D. 1996. Bird use of a patch retention treatment in SBSmc forests. Prince Rupert Forest Region, British Columbia. Ministry of Forests, Smithers, British Columbia. For. Sci. Ext. Note No. 16. Available at http://www.for.gov.bc.ca/prupert/ Research/Extension_notes/Enote16.pdf [viewed 28 May 2003]. 4 p.

MacLean, D.A. 1980. Vulnerability of spruce–fir during uncontrolled spruce budworm outbreaks: a review and discussion. For. Chron. **56**: 213–221.

MacLean, D.A. 1984. Effects of spruce budworm outbreaks on the productivity and stability of balsam fir forests. For. Chron. **60**: 273–279.

MacLean, D.A., Woodley, S.J., Weber, M.G., and Wein, V. 1983. Fire and nutrient cycling. *In* The role of fire in northern circumpolar ecosystems. *Edited by* R.W. Wein and D.A. MacLean. Wiley, Chichester, U.K. SCOPE Rep. 18. pp. 111–134.

Marks, P.L. 1974. The role of pin cherry (*Prunus pensylvanica* L.) in the maintenance of stability in northern hardwood ecosystems. Ecol. Monogr. **44**: 73–88.

McCann, K.S. 2000. Diversity–stability debate. Nature (London), **405**: 228–233.

McCarthy, J. 2001. Gap dynamics of forest trees: a review with particular attention to boreal forests. Environ. Rev. **9**: 1–59.

McCleary, K., and Andison, D.W. 2001. What we know about fire in riparian zones — so far. *In* Proceedings of the Natural Disturbance and Forest Management Symposium, 5–7 March 2001, Edmonton, Alberta. Sustainable Forest Management Network, Edmonton, Alberta. Available at http://www.fmf.ab.ca/pdf/symposium.pdf [viewed 16 September 2002]. p. 21.

McComb, W., and Lindenmayer, D. 1999. Dying, dead and down trees. *In* Maintaining biodiversity in forest ecosystems. *Edited by* M.L. Hunter, Jr. Cambridge University Press, Cambridge, U.K. pp. 335–372.

McCullough, D.G., Werner, R.A., and Neumann, D. 1998. Fire and insects in northern and boreal forest ecosystems of North America. Ann. Rev. Entomol. **43**: 107–127.

McRae, D.J., Duchesne, L.C., Freedman, B., Lynham, T.J., and Woodley, S. 2001. Comparisons between wildfire and forest harvesting and their implications in forest management. Environ. Rev. **9**: 223–260.

Michener, W.K., Blood, E.R., Box, J.B., Couch, C.A., Golloday, S.W., Hippe, D.J., Mitchell, R.J., and Palik, B.J. 1998. Tropical storm flooding of a coastal plain landscape. BioScience, **48**: 696–705.

Millar, C.I., and Woolfenden, W.B. 1999. The role of climate change in interpreting historical variability. Ecol. Applic. **9**: 1207–1216.

Miyanishi, K., Bajtala, M.J., and Johnson, E.A. 1999. Patterns of duff consumption in *Pinus banksiana* and *Picea mariana* stands. *In* Science and practice: sustaining the boreal forest. Proceedings of the 1999 Sustainable Forest Management Network Conference, 14–17 February 1999, Edmonton, Alberta. *Edited by* T.S. Veeman, D.W. Smith, B.G. Purdy, F.J. Salkie, and G.A. Larkin. Sustainable Forest Management Network, Edmonton, Alberta. pp. 112–115.

Morin, H., Laprise, D., and Bergeron, Y. 1993. Chronology of spruce budworm outbreaks near Lake Duparquet, Abitibi Region, Quebec. Can. J. For. Res. **23**: 1497–1506.

Nguyen-Xuan, T., Bergeron, Y., Simard, D., Fyles, J., and Paré, D. 2000. The importance of forest floor disturbance in the early regeneration patterns of the boreal forest of western and central Québec: a wildfire vs logging comparison. Can. J. For. Res. **30**: 1353–1364.

Niemelä, J. 1997. Invertebrates and boreal forest management. Conserv. Biol. **11**: 601–611.

Niemelä, J., Haila, Y., Halme, E., Lahti, T., Pajunen, T., and Punttila, P. 1988. The distribution of carabid beetles in fragments of old coniferous taiga and adjacent managed forest. Ann. Zool. Fenn. **25**: 107–119.

Niemelä, J., Langor, D., and Spence, J.R. 1993. Effects of clear-cut harvesting on boreal ground-beetle assemblages (Coleoptera: Carabidae) in western Canada. Conserv. Biol. **7**: 551–561.

Norton, M.R., Hannon, S.J., and Schmiegelow, F.K.A. 2000. Fragments are not islands: patch vs. landscape perspectives on songbird presence and abundance in a harvested boreal forest. Ecography, **23**: 209–223.

Noss, R. 1987. From plant communities to landscapes in conservative inventories: a look at the Nature Conservancy (USA). Biol. Conserv. **41**: 11–37.

(NRCan) Natural Resources Canada. 1997. The state of Canada's forests 1996–1997. Canadian Forest Service, Ottawa, Ontario. 128 p.

Örlander, G., Egnell, G., and Albrekston, A. 1996. Long-term effects of site preparation on growth in Scots pine. For. Ecol. Manage. **86**: 27–37.

Östlund, L., Zackrisson, O., and Axelsson, A.-L. 1997. The history and transformation of a Scandinavian boreal forest landscape since the 19th century. Can. J. For. Res. **27**: 1198–1206.

Palik, B.J., Mitchell, R.J., and Hiers, J.K. 2002. Modeling silviculture after natural disturbance to sustain biodiversity in the longleaf pine (*Pinus palustris*) ecosystem: balancing complexity and implementation. For. Ecol. Manage. **155**: 347–356.

Paré, D., and Munson, A. 2000. Maintenance of soil fertility in the boreal forest. *In* Towards ecological forestry: a proposal for indicators of SFM inspired by natural disturbances. *By* D.D. Kneeshaw, C. Messier, A. Leduc, P. Drapeau, R. Carignan, D. Paré, J.-P. Ricard, S. Gauthier, R. Doucet, and D. Greene. Sustainable Forest Management Network, Edmonton, Alberta. Available at http://sfm-1.biology.ualberta.ca/english/pubs/PDF/SP_kneeshaw_en.pdf [viewed 28 May 2003]. pp. 29–35.

Paré, D., and Rochon, P. 1999. Toward guidelines for a sustainable management of boreal forest floor nutrients. *In* Science and practice: sustaining the boreal forest. Proceedings of the 1999 Sustainable Forest Management Network Conference, 14–17 February 1999, Edmonton, Alberta. *Edited by* T.S. Veeman, D.W. Smith, B.G. Purdy, F.J. Salkie, and G.A. Larkin. Sustainable Forest Management Network, Edmonton, Alberta. pp. 237–240.

Paré, D., Bergeron, Y., and Camiré, C. 1993. Changes in the forest floor of Canadian southern boreal forest after disturbance. J. Veg. Sci. **4**: 811–818.

Parker, V.T., and Pickett, S.T.A. 1998. Historical contingency and multiple scales of dynamics within plant communities. *In* D.L. Peterson and V.T. Parker. Ecological scale: theory and applications. Columbia University Press, New York, New York. pp. 171–192.

Pastor, J., Light, S., and Sovell, L. 1998. Sustainability and resilience in boreal regions: sources and consequences of variability. Conserv. Ecol. [online], **2**: 16. Available at http://www.consecol./org/vol2/iss2/art16 [viewed 16 September 2002].

Payette, S., Morneau, C., Sirois, L., and Desponts, M. 1989. Recent fire history of the northern Quebec biomes. Ecology, **70**: 656–673.

Peck, J.E., and McCune, B. 1997. Remnant trees and canopy lichen communities in western Oregon: a retrospective approach. Ecol. Applic. **7**: 1181–1187.

Perron, N., Bélanger, L., and Gagnon, R. 2000. Forest management almost emulates recent fires at the landscape level in the black spruce forest in Quebec (Canada) [abstract]. *In* Disturbance dynamics in boreal forests, restoration of management and biodiversity, 21–25 August 2000, Kuhmo, Finland. *Edited by* L. Karjalainen and T. Kuuluvainen. University of Helsinki, Helsinki, Finland. p. 82.

Potvin, F., Curtois, R., and Belanger, L. 1999. Short-term response of wildlife to clear-cutting in Quebec boreal forest. Can. J. For. Res. **29**: 1120–1127.

Potvin, F., Bélanger, L., and Lowell, K. 2000. Marten habitat selection in a clearcut boreal landscape. Conserv. Biol. **14**: 844–857.

Prévost, M. 1996. Effets du scarifage sur les propriétés du sol et l'ensemencement naturel dans une pessière noire à mousses de la forêt boréale québecoise. Can J. For. Res. **26**: 72–86.

Price, K., Pojar, J., Roburn, A., Brewer, L., and Poirier, N. 1998. Clearcut, windthrow — what's the difference? Northwest Sci. **72**: 30–33.

Purvis, A., and Hector, A. 2000. Getting the measure of biodiversity. Nature (London), **405**: 212–219.

Rapport, D.J., and Whitford, W.G. 1999. How ecosystems respond to stress. BioScience, **49**: 193–203.

Rees, D.C., and Juday, G.P. 2002. Plant species diversity on logged versus burned sites in central Alaska. For. Ecol. Manage. **155**: 291–302.

Reich, P.B., Bakken, P., Carlson, D., Frelich, L.E., Friedman, S.K., and Grigal, D.F. 2001. Influence of logging, fire and forest type on biodiversity and productivity in southern boreal forests. Ecology, **82**: 2731–2748.

Rempel, R. 2001. Signatures of sustainability: a multiscale landscape approach to assessing habitat suitability [abstract]. Presented at a conference on Natural Disturbance & Forest Management: What's Happening and Where It's Going, 5–7 March 2001, Edmonton, Alberta. Foothills Model Forest and Sustainable Forest Management Network, Edmonton, Alberta. Available at http://sfm-1.biology.ualberta.ca/english/pubs/PDF/WS_2001-3.pdf [viewed 28 May 2003]. p. 17.

Rempel, R.S., Elkie, P.C., Rodgers, A.R., and Gluck, M.J. 1997. Timber-management and natural-disturbance effects on moose habitat: landscape evaluation. J. Wild. Manage. **61**: 517–524.

Rheault, H., Drapeau, P., Bergeron, Y., and Anders-Esseen, P.-A. 2003. Edge effects on epiphytic lichens in managed black spruce forests of eastern North America. Can. J. For. Res. **33**: 23–32.

Roland, J. 1993. Large-scale forest fragmentation increases the duration of tent caterpillar outbreak. Oecologia, **93**: 25–30.

Roland, J., and Taylor, P.D. 1997. Insect parasitoid species respond to forest structure at different spatial scales. Nature (London), **386**: 710–713.

Romme, W.H., Everham, E.H., Frelich, L.E., Moritz, M.A., and Sparks, R.E. 1998. Are large infrequent disturbances qualitatively different from small, frequent disturbances? Ecosystems, **1**: 524–534.

Rothman, L.D., and Roland, J. 1998. Forest fragmentation and colony performance of forest tent caterpillar. Ecography, **21**: 383–391.

Rowe, J.S. 1983. Concepts of fire effects on plant individuals and species. *In* The role of fire in northern circumpolar ecosystems. *Edited by* R.W. Wein and D.A. MacLean. Wiley, Chichester, U.K. SCOPE Rep. 18. pp. 135–154.

Rowe, J.S. 1992. The ecosystem approach to forestland management. For. Chron. **68**: 222–237.

Royama, T. 1984. Population dynamics of the spruce budworm *Choristoneura fumiferana*. Ecol. Monogr. **54**: 429–462.

Ruel, J.-C. 1995. Understanding windthrow: silvicultural implications. For. Chron. **71**: 434–445.

Schindler, D.W. 1998*a*. A dim future for boreal waters and landscapes. BioScience, **48**: 157–164.

Schindler, D.W. 1998*b*. Sustaining aquatic ecosystems in boreal regions. Conserv. Ecol. [online], **2(2)**: 18. Available at http://www.consecol.org/vol2/iss2/art18 [viewed 12 September 2002].

Schmiegelow, F., Vernier, P., Demarchi, D., and Cumming, S.G. 1999. Seeing the forest beyond the trees: a cross-scale approach to wildlife habitat assessment. *In* Science and practice: sustaining the boreal forest. Proceedings of the 1999 Sustainable Forest Management Network Conference, 14–17 February 1999, Edmonton, Alberta. *Edited by* T.S. Veeman, D.W. Smith, B.G. Purdy, F.J. Salkie, and G.A. Larkin. Sustainable Forest Management Network, Edmonton, Alberta. pp. 232–236.

Siitonen, J., and Martikainen, P. 1994. Occurrence of rare and threatened insects living on decaying *Populus tremula*: a comparison between Finnish and Russian Karelia. Scand. J. For. Res. **9**: 185–191.

Simard, A.J. 1975. Forest fire weather zones of Canada. Canadian Forest Service, Environment Canada, Ottawa, Ontario. (Map.)

Sippell, L. 1962. Outbreaks of the forest tent caterpillar, *Malacosoma disstria* Hbn., a periodic defoliator of broad-leaved trees in Ontario. Can. Entomol. **94**: 408–416.

Smith, C.K., Coyea, M.R., and Munson, A.D. 2000. Soil carbon, nitrogen and phosphorus stocks and dynamics under disturbed black spruce forests. Ecol. Applic. **10**: 775–788.

Smith, D.M. 1986. The practice of silviculture, 8th edition. Wiley & Sons, New York, New York. 578 p.

Söderström, L. 1988. The occurrence of epixylic bryophyte and lichen species in an old natural and a managed forest stand in northeast Sweden. Biol. Conserv. **45**: 169–178.

Spence, J.R., Volney, W.J.A., Lieffers, V.J., Weber, M.G. Luchkow, S.A., and Vinge, T.W. 1999. The Alberta EMEND project: recipe and cooks' argument. *In* Science and practice: sustaining the boreal forest. Proceedings of the 1999 Sustainable Forest Management Network Conference, 14–17 February 1999, Edmonton, Alberta. *Edited by* T.S. Veeman, D.W. Smith, B.G. Purdy, F.J. Salkie, and G.A. Larkin. Sustainable Forest Management Network, Edmonton, Alberta. pp. 583–590.

Suffling, R., Smith, B., and Dal Molin, J. 1982. Estimating past forest disturbance and disturbance rates in northwestern Ontario: a demographic approach. J. Environ. Manage. **14**: 45–56.

Suffling, R., Lihou, C., and Morand, Y. 1988. Control of landscape diversity by catastrophic disturbance: a theory and a case study of fire in a Canadian boreal forest. Environ. Manage. **12**: 73–78.

Swain, A.M. 1973. A history of fire and vegetation in north-eastern Minnesota as recorded in lake sediments. Quatern. Res. **3**: 383–396.

Swanson, F.J., Jones, J.A., Wallin, D.O., and Cissel, J.H. 1994. Natural variability — implications for ecosystem management. *In* Eastside forest ecosystem health assessment, vol. II, ecosystem management: principles and applications. *Edited by* M.E. Jensen and P.S. Bourgeron. USDA Forest Service, Portland, Oregon. Gen. Tech. Rep. PNW-GTR-318. pp. 80–94.

Swetnam, T.W. 1993. Fire history and climate change in giant sequoia groves. Science, **262**: 885–889.

Swetnam, T.W., and Betancourt, J.L. 1990. Fire-southern oscillation relations in the southwestern United States. Science, **249**: 1017–1020.

Tchir, T.L., and Johnson, E.A. 1999. A model of fragmentation resulting from human settlement in the mixedwood boreal forest of Saskatchewan. *In* Science and practice: sustaining the boreal forest. Proceedings of the 1999 Sustainable Forest Management Network Conference, 14–17 February 1999, Edmonton, Alberta. *Edited by* T.S. Veeman, D.W. Smith, B.G. Purdy, F.J. Salkie, and G.A. Larkin. Sustainable Forest Management Network, Edmonton, Alberta. pp. 456–460.

Tilman, D. 2000. Causes, consequences and ethics of biodiversity. Nature (London), **405**: 208–219.

Tilman, D., May, R.M., Lehman, C.L., and Nowak, M.A. 1994. Habitat destruction and the extinction debt. Nature (London), **371**: 65–66.

Tittler, R., Messier, C., and Burton, P.J. 2001*a*. Hierarchical forest management planning and sustainable forest management in the boreal forest. For. Chron. **77**: 988–1005.

Tittler, R., Hannon, S.J., and Norton, M.R. 2001*b*. Residual tree retention ameliorates short-term effects of clear-cutting on some boreal songbirds. Ecol. Applic. **11**: 1656–1666.

Trombulak, S.C., and Frissell, C.A. 2000. Review of ecological effects of roads on terrestrial and aquatic communities. Conserv. Biol. **14**: 18–29.

Ulanova, N.G. 2000. The effects of windthrow on forests at different spatial scales: a review. For. Ecol. Manage. **135**: 155–167.

Van Cleve, K., Chapin, F.S., Dyrness, C.T., and Viereck, L.A. 1991. Element cycling in taiga forest: state-factor control. BioScience, **41**: 78–88.

Van Cleve, K., Oliver, L., Schlentner, R., Viereck, L.A., and Dyrness, C.T. 1983. Productivity and nutrient cycling in taiga forest ecosystems. Can. J. For. Res. **13**: 747–766.

Van Wagner, C.E. 1988. The historical pattern of annual burned area in Canada. For. Chron. **64**: 182–185.

Veeman, T.S., Smith, D.W., Purdy, B.G., Salkie, F.J., and Larkin, G.A. (*Editors*) 1999. Science and practice: sustaining the boreal forest. Proceedings of the 1999 Sustainable Forest Management Network Conference, 14–17 February 1999, Edmonton, Alberta. 816 p.

Vernier, P., and Cumming, S.G. 1999. Predicting landscape patterns from stand attribute data in the Alberta boreal mixedwood. *In* Science and practice: sustaining the boreal forest. Proceedings of the 1999 Sustainable Forest Management Network Conference, 14–17 February 1999, Edmonton, Alberta. *Edited by* T.S. Veeman, D.W. Smith, B.G. Purdy, F.J. Salkie, and G.A. Larkin. Sustainable Forest Management Network, Edmonton, Alberta. pp. 689–694.

Viereck, L. 1973. Wildfire in the taiga of Alaska. Quatern. Res. **3**: 465–495.

Viereck, L.A. 1983. The effects of fire in black spruce ecosystems of Alaska and northern Canada. *In* The role of fire in northern circumpolar ecosystems. *Edited by* R.W. Wein and D.A. MacLean. Wiley, Chichester, U.K. SCOPE Rep. 18. pp. 201–220.

Villard, M.-A., Bourque, J., and Gunn, J.S. 1999. Harvesting intensity and reproductive success of forest birds: a landscape perspective. *In* Science and practice: sustaining the boreal forest. Proceedings of the 1999 Sustainable Forest Management Network Conference, 14–17 February 1999, Edmonton, Alberta. *Edited by* T.S. Veeman, D.W. Smith, B.G. Purdy, F.J. Salkie, and G.A. Larkin. Sustainable Forest Management Network, Edmonton, Alberta. pp. 64–68.

Volney, W.J.A., Spence, J.R., Weber, M.G., Langor, D.W., Mallet, K.I., Johnson, J.D., Edwards, I.K., Hillman, G.R., and Kishchuk, B.E. 1999. Assessing components of ecosystem integrity in the EMEND experiment. *In* Science and practice: sustaining the boreal forest. Proceedings of the 1999 Sustainable Forest Management Network Conference, 14–17 February 1999, Edmonton, Alberta. *Edited by* T.S. Veeman, D.W. Smith, B.G. Purdy, F.J. Salkie, and G.A. Larkin. Sustainable Forest Management Network, Edmonton, Alberta. pp. 244–249.

Weber, M.G., and Stocks, B.J. 1998. Forest fires and sustainability in the boreal forests of Canada. Ambio, **27**: 545–550.

Weir, J.M.H., Johnson, E.A., and Miyanishi, K. 1999. Fire frequency and the spatial age mosaic of the mixedwood boreal forest of Saskatchewan. *In* Science and practice: sustaining the boreal forest. Proceedings of the 1999 Sustainable Forest Management Network Conference, 14–17 February 1999, Edmonton, Alberta. *Edited by* T.S. Veeman, D.W. Smith, B.G. Purdy, F.J. Salkie, and G.A. Larkin. Sustainable Forest Management Network, Edmonton, Alberta. pp. 81–86.

Wikars, L.-O., and Schimmel, J. 2001. Immediate effects of fire-severity on soil invertebrates in cut and uncut pine forests. For. Ecol. Manage. **141**: 189–200.

Wotton, B.M., and Flannigan, M.D. 1993. Length of the fire season in a changing climate. For. Chron. **69**: 187–192.

Yarie, J. 1981. Forest fire cycles and life tables: a case study from interior Alaska. Can. J. For. Res. **11**: 554–562.

Zackrisson, O. 1977. Influence of forest fires on the North Swedish boreal forest. Oikos, **29**: 23–32.

Zackrisson, O., Nilsson, M., and Wardle, D.A. 1996. Key ecological function of charcoal from wildfire in the boreal forest. Oikos, **77**: 10–19.

Zasada, J. 1986. Natural regeneration of trees and tall shrubs on forest sites in interior Alaska. *In* Forest ecosystems in the Alaskan taiga: a synthesis of structure and function. *Edited by* K. Van Cleve, F.S. Chapin, P.W. Flanagan, L.A. Viereck, and C.T. Dyrness. Springer-Verlag, New York, New York. pp. 44–73.

Chapter 10

Impacts of forest disturbance on boreal surface waters in Canada

E.E. Prepas, B. Pinel-Alloul, R.J. Steedman, D. Planas, and T. Charette

"Water always has a source, and trees always have roots."

Chinese proverb

Background

Introduction

Boreal forests contain just over half of Canada's surface fresh water (CCEA 2002). The boreal landscape can be thought of as a mosaic of watersheds, where every square metre of forest drains into some form of water body. Despite the linkage between surface waters and the land area that drains into them, forest management strategies tend to focus on terrestrial components of the watershed.

Until recently, there has been a lack of information on the effects of watershed disturbance on water quality in boreal surface waters. Current strategies for sustainable management are based on the assumption that the impact of wildfire must be understood before we can intelligently manage forests for timber harvesting (Chaps. 1, 8, and 9).

E.E. Prepas.[1] Faculty of Forestry, Lakehead University, Thunder Bay, Ontario P7B 5E1, Canada.

B. Pinel-Alloul. Groupe de Recherche Interuniversitaire en Limnologie et en Environement Aquatique (GRIL), Département des Sciences Biologiques, Université de Montréal, Montréal, Quebec, Canada.

R.J. Steedman. Professional Leader, Environment, National Energy Board, Calgary, Alberta, Canada.

D. Planas. Département des Sciences Biologiques, Université du Québec à Montréal, Montréal, Quebec, Canada.

T. Charette. Department of Biological Sciences, University of Alberta, Edmonton, Alberta, Canada.

[1]Author for correspondence. Telephone: 807-343-8623. e-mail: ellie.prepas@lakeheadu.ca

Correct citation: Prepas, E.E., Pinel-Alloul, B., Steedman, R.J., Planas, D., and Charette, T. 2003. Impacts of forest disturbance on boreal surface waters in Canada. Chapter 10. *In* Towards Sustainable Management of the Boreal Forest. *Edited by* P.J. Burton, C. Messier, D.W. Smith, and W.L. Adamowicz. NRC Research Press, Ottawa, Ontario, Canada. pp. 369–393.

Box 10.1. The challenge of including aquatic values in forest management plans.

- Determine how surface water responds to disturbance in the watershed;
- Identify regions and water bodies particularly sensitive to disturbance;
- Evaluate the effectiveness of protective practices such as the retention of vegetated buffers and controlling the proportion of catchment logged;
- Develop tools and guidelines that assist management of forests in natural landscapes; and
- Plan forest development, harvest rates, and logging operations accordingly in order to reduce impacts on water quality and aquatic ecosystems.

Surface water studies, begun with the establishment of the Sustainable Forest Management Network (SFMN) in 1995, focused on boreal aquatic system response to logging and wildfire in the watershed (Carignan and Steedman 2000).

The objective of this chapter is to distill results from these studies into a representative picture of watershed disturbance effects in the Canadian boreal forest, and to show how forest management can be refined to incorporate these concepts (Box 10.1). The focus is lakes, but information on streams is also presented where available.

Surface water health

Surface water health can be quantified by a number of physical (e.g., temperature, oxygen, water clarity), chemical (e.g., ions, nutrients, buffering capacity), and biological (e.g., algae, algal toxins, zooplankton, fish) indicators. More difficult to determine are "acceptable" standards for ecosystem health and sustainability. Provincial and federal water quality guidelines for the protection of aquatic life exist for a number of water quality parameters (Box 10.2). But should surface water be managed to prevent deterioration or to preserve current aquatic health? The following sections highlight the important indicators of surface water health (from Mitchell and Prepas 1990).

Algae and nutrients

Lake health is often defined by changes in nutrient and dissolved oxygen concentrations and lake greenness (see Box 10.3 for an explanation of several terms used in this chap-

---

Box 10.2. How do we determine if surface waters are in good or poor health?

Good quality surface waters meet certain measurable standards:

Alberta (Alberta Environment 1999)
- total phosphorus <50 µg/L;
- dissolved oxygen >6.5 mg/L (7-day mean).

Ontario/Quebec (OMEE 1994; MEQ 1992)
- total phosphorus (lakes) <10 µg/L;
- dissolved oxygen >5–8 mg/L (dependent on water temperature).

Canada, proposed for drinking water (CCME 2002)
- microcystin-LR <1.5 µg/L (a toxin produced by some blue–green algae).

ter). Lake greenness (algal biomass) is determined directly by measuring the concentrations of the green photosynthesizing pigment chlorophyll *a* in a water sample. A more indirect method is to measure water clarity. This can be simply and quickly done with a Secchi disk, a black and white disk that is lowered down through the water column until it can no longer be seen. The "Secchi disk depth" is the midpoint between the depth at which the disk disappears from view when lowered and reappears when pulled up

---

Box 10.3. Glossary of technical terms.

algae — The photosynthesizing portion of the plankton; see also plankton.

biomass — Weight of living matter.

catchment — The land area that contributes surface runoff to a water body.

chlorophyll a — One of the green photosynthetic pigments; the concentration of chlorophyll *a* in water is a measure of greenness in the water column, and an indicator of algal biomass; see biomass.

blue–green algae — Prokaryotic organisms in the algae community; blue–green algae (also called cyanobacteria) are typified by cells without a nucleus or organelles and with photo sensitive pigments dispersed throughout the cell; some species are capable of obtaining nitrogen for metabolism from atmospheric nitrogen.

epilimnion — The warm uppermost layer in a thermally stratified lake that is subject to mixing by wind.

fresh water — Water with concentration of total dissolved solids below 500 mg/L.

greenness — Abundance of algae in the water column; see chlorophyll *a*.

hypolimnion — The cool bottom layer in a thermally stratified lake that is separated from surface influences by a thermocline; see thermocline.

MCLR — Abbreviation for Microcystin-LR, a liver toxin (hepatotoxin) produced by certain blue–green algae.

plankton — The community of small and microscopic organisms living in open water.

runoff — The water reaching a lake, stream, or ocean after flowing over land or through the surficial layers of the land.

Secchi disk depth — The depth in water to which a Secchi disk (a 20-cm-diameter disc with alternating black and white quadrants) can be seen from the surface; Secchi disk depth is an easy measurement of water clarity.

stratified — Divided into layers; in stratified lakes, there may be mixing within a layer but little mixing occurs between layers; layers have different densities, which may be determined either by temperature and (or) salinity.

thermocline — The layer of water in a lake between the epilimnion and hypolimnion in which the temperature gradient is greatest and exceeds a change of 1°C per metre of depth; see also epilimnion, hypolimnion, and stratified.

water renewal time — The average time required to refill a basin with new water if it were to be emptied; it is usually calculated by dividing the volume of the waterbody by the average annual outflow.

watershed — The total area of the catchment and lake; see catchment.

zooplankton — The non-photosynthesizing or animal portion of the plankton.

Based on Ruttner 1974, Mitchell and Prepas 1990

again. Water clarity can also be measured with an electronic sensor that determines light intensity through the water column. Turbidity and colour affect water clarity: turbidity is a measure of suspended particles (e.g., mud, silt, algae) in the water column; colour is a measure of the quantity of humic material in water.

Carbon (C), phosphorus (P), and nitrogen (N) are essential nutrients for algae and aquatic plant growth and hence influence the biomass and productivity of the entire aquatic food chain (illustrated in Fig. 10.1). Carbon concentrations in water usually meet or exceed requirements for algal and aquatic plant growth because of an abundant atmospheric supply (CO_2). There is no atmospheric source of P. It is often considered the growth-limiting nutrient in lakes; the P concentration will often determine algal biomass and growth. Although N has an atmospheric reserve, only a subset of algae can access this N source. These N-fixing algae convert gaseous N into a useful (bioavailable) form. Worldwide, P and N are the most common culprits for nutrient-stimulated water quality concerns. Phosphorus and nitrogen enter surface waters via runoff, groundwater, wet atmospheric deposition, and recycling from bottom sediments.

Most algae and aquatic plants can only use bioavailable forms of nutrients, for example: nitrate (NO_3^-) nitrite (NO_2^-), ammonium (NH_4^+), and phosphate (PO_4^{3-}). Nearly all the organic C in natural inland waters is in the form of dissolved organic C. Organic C content in lake and stream waters can be influenced by natural attributes (e.g., the presence and type of wetlands), natural disturbance (e.g., wildfire), or human disturbance (e.g., logging) in the watershed (D'Arcy and Carignan 1997; Carignan et al. 2000).

Fig. 10.1. Lake food-web interactions. Nutrients mainly enter lakes from the catchment, atmosphere, and lake bottom decomposition. Nutrients usually limit the productivity of algae. Grazers consume algae. Fish are the top predators in lake ecosystems and feed on various components of the aquatic community.

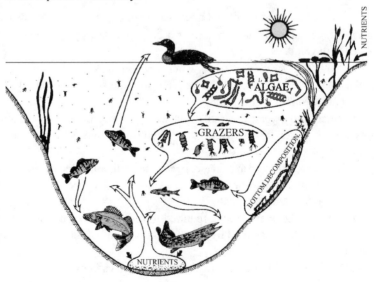

Temperature and lake stratification

Air temperature, wind, sun exposure, and physical features of the watercourse or basin govern thermal patterns in lakes and streams. Sunlight heats lake water more near the surface and thus forms a layer of less dense warm water overlying a denser cool zone. These layers are referred to as the epilimnion (top of the lake) and hypolimnion (bottom of the lake) (Fig. 10.2). When these two layers are formed, the lake is classified as thermally stratified. Throughout summer, the depth of the surface layer can increase because of strong winds and warmer air temperatures. With the onset of cooler air temperatures in the fall, the surface layer becomes thinner until eventually the water column becomes uniform in temperature. Generally, thermally stratified lakes tend to be deep and lakes that mix or are only weakly stratified tend to be shallow, since heat from the sun is more likely to reach the bottom of shallower lakes.

Oxygen

Most aquatic organisms require dissolved oxygen for respiration. The atmosphere is a good supply of oxygen to surface waters. Dissolved oxygen concentrations tend to be high in moving waters because turbulence aerates the water. In still waters, dissolved oxygen concentration can vary with depth. In unproductive deep lakes, the dissolved oxygen content of the surface layer is generally lower than in the deep-water layer because oxygen is less soluble at high water temperatures. In nutrient-rich deep lakes, bacterial decomposition at the lake bottom consumes oxygen, and can lead to dissolved oxygen depletion in deep water. In shallow lakes, the profile of dissolved oxygen can be variable during the ice-free season, with uniform concentrations when the water column is well mixed and low concentrations near the lake bottom when the water column is weakly thermally stratified. Ice and snow cover on lakes not only cut off the atmos-

Fig. 10.2. Visual interpretation of thermal stratification of a lake's water column. Stratification (left) occurs when sunlight heats lake water more near the surface and thus forms a layer of less dense warm water overlying a denser cool zone. Generally, thermally stratified lakes tend to be deeper than those that mix or stratify only weakly (right).

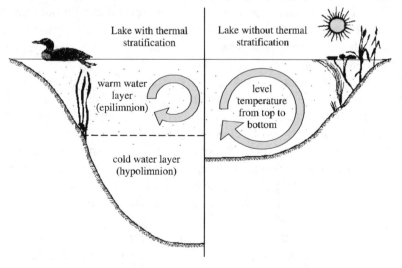

pheric supply of oxygen to surface water, but can limit light penetration and therefore reduce oxygen generation by photosynthetic aquatic plants and algae.

Ions and buffering capacity

Natural waters contain chemicals that separate into positively (cations) and negatively (anions) charged particles called ions. Cations common to freshwaters include calcium (Ca^{2+}), magnesium (Mg^{2+}), sodium (Na^+), and potassium (K^+); common anions include bicarbonate (HCO_3^-), carbonate (CO_3^{2-}), sulphate (SO_4^{2-}), and chloride (Cl^-). These ions may combine to form salts, such as calcium carbonate. Salinity in fresh water is best expressed as the sum of the major cations and anions. The salinity of water bodies is variable and depends on ionic contributions from subsurface flow (groundwater) and direct deposition from the atmosphere.

Concentrations of magnesium, sodium, potassium, and chloride are considered relatively stable and undergo only minor spatial and temporal fluctuations within a lake or stream in response to biological changes. Alternatively, calcium, bicarbonate, carbonate, and sulphate are dynamic, and their concentrations are strongly influenced by microbial metabolism. Changes in concentrations of these major cations can influence the metabolism of many aquatic organisms.

The capacity of water to neutralize acidity is termed the buffering capacity, measured as alkalinity. Alkalinity mainly relates to concentration of common anions such as calcium bicarbonate ($CaHCO_3^-$) and carbonate ($CaCO_3^{2-}$). Total alkalinity is measured as the quantity of strong acid required to lower the pH of a water sample to a specific level.

Aquatic organisms

Algae are small (often microscopic) organisms capable of photosynthesis and are generally found in water. Cell shape, mode of locomotion (e.g., whip-like flagella), biochemical characteristics (i.e., pigment composition), and physiological traits (e.g., mode of nutrition) are some characteristics used to classify algae. Some algae are really bacteria (called cyanobacteria or blue–green algae), and it is these organisms that can proliferate into surface scums and even produce toxins. Several algal groups are common in boreal lakes: blue–greens (Cyanophyta), chlorophytes (Chlorophyta), chrysophytes (Chrysophyta), diatoms (Bacillariophyta), dinoflagellates (Dinophyceae and Pyrrophyta), and euglenoid flagellates (Euglenophyta).

Zooplankton (henceforth referred to as grazers) are mainly small herbivores that eat algae and in turn are eaten by predacious invertebrates or fish (Fig. 10.1). There are three main groups of grazers: (1) Rotifera, (2) Cladocera, and (3) Copepoda (this latter group includes the Calanoida and Cyclopoida). Grazers are grouped by feeding mode, habitat, and physiological differences. Populations of various species of grazers change over space and seasons in response to changes in light conditions, water temperature, dissolved oxygen concentrations, food supply, predation, growth, and behavioral factors (Pinel-Alloul 1995). For example, calanoid copepods are commonly found in the open water of clear, nutrient-poor lakes.

Fish are the top aquatic carnivores in many freshwater ecosystems, where they feed on various components of the aquatic community: other fish, zooplankton, algae, rooted plants, bottom dwellers, and dead plant and animal matter (Fig. 10.1). Therefore, fish abundance and diversity can be affected by the productivity of lower levels of the aquatic food chain. Also, because they respire through gills, they are sensitive to factors like dissolved oxygen and turbidity. Nutrient-rich systems can support a greater abundance of fish food; however, dissolved oxygen depletion in these systems can reduce the biodiversity of all levels of the food chain and restrict fish habitat to warm surface waters. Nutrient enrichment can cause a shift of fish species from those requiring cool temperatures and high dissolved oxygen (e.g., lake whitefish, *Coregonus clupeaformis)* to warm-water species (e.g., white sucker, *Catostomus commersoni).*

Basic differences between surface waters on the Boreal Plain and Shield

Canadian boreal forests are divided into six major geological and climactic subregions (Chap. 1), with the closed-forest boreal zones being under the most pressure from industrial forestry (Fig. 10.3):
(1) Boreal Cordillera of the northwestern Canada;
(2) Boreal Plain of western and central Canada; and
(3) Boreal Shield of central and eastern Canada.
Because of a lack of information on response of lakes in the Boreal Cordillera to watershed disturbance, we will focus on the Boreal Plain and the Boreal Shield.

Generally, surface waters on the Boreal Plain are situated on low relief, often poorly drained sedimentary and glacial materials compared to those of the more steeply sloped Boreal Shield. As a consequence, watersheds on the Plain have, on average, more than twice the wetland coverage of those on the Shield (NWWG 1998). Boreal Plain lakes tend to be relatively rich in nutrients (Box 10.4; Mitchell and Prepas 1990). Recycling

Fig. 10.3. Conceptual representation of impacts of wildfire and logging on watershed and lake nutrient balance.

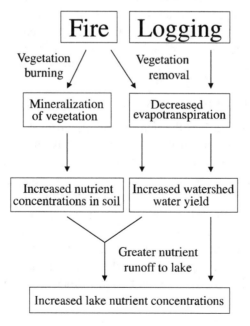

Box 10.4. Water quality: Boreal Plain versus Boreal Shield.

On average, compared to the Boreal Shield, Boreal Plain lakes are:

- 8 times greener;
- 37 times more alkaline;
- more concentrated in nutrients:
- 6 times more total phosphorus;
- 5 times more total nitrogen.
- more concentrated in dissolved substances:
- 3 times more dissolved organic carbon;
- 6–8 times more anions;
- 17–28 times more cations.

Data from Prepas et al. 2001*b* (upland lakes only),
D'Arcy and Carignan 1997, Carignan et al. 2000, Steedman 2000

---

of nutrients from decomposition of dead matter at the lake bottom (Fig. 10.1) can be an important source of nutrients in lakes on the Boreal Plain (Shaw and Prepas 1990). Also, glacial materials are less resistant to dissolution and erosion processes than the hard rock of the Boreal Shield. Forested watersheds underlain with hard rock formations (granite, basalt) export fewer nutrients than sedimentary watersheds (sandstone, slate), which in turn export fewer nutrients than glacial materials (Duarte and Kalff 1989). Therefore, lake water on the Boreal Shield is relatively dilute compared with that of the Plain. Surface waters on the Plain have higher buffering capacity and are generally less acidic (Prepas et al. 2001*b*) than on the Shield (D'Arcy and Carignan 1997).

Lake algal communities on the Boreal Plain are composed mainly of blue–greens (Prepas et al. 2001*b*), which flourish in nutrient-rich systems. In contrast, Shield lakes are dominated mostly by flagellated algae, such as small chrysophytes, diatoms, dinoflagellates, and cryptophytes (Planas et al. 2000), which often predominate in nutrient-poor systems. Grazer communities in both boreal subregions are typically dominated by rotifers based on numbers of animals, while cladocerans and copepods are the largest contributors based on biomass (Pinel-Alloul 1993; Patoine et al. 2000).

Catchment disturbance effects on surface water

Catchment disturbance can affect the quality and quantity of receiving waters (Likens and Bormann 1974). The extent of these effects depends on the nature and severity of the disturbance. Since trees take up water through the process of transpiration, their removal results in excess soil water. Soil saturation can cause greater export of water and nutrients (through both subsurface and overland flow) from the catchment after snowmelt and rainstorms (Fig. 10.3; Roby and Azuma 1995).

Wildfire disturbance can increase the availability of water-soluble nutrients, since fire mineralizes organic nutrients contained in vegetation. Severe fires can burn off the surface organic layer of soils, exposing the underlying mineral soil (Bayley et al. 1992). Such an increase in soil nutrient availability can further increase nutrient export from catchments (Lamontagne et al. 2000). Nutrients can even be directly deposited into aquatic systems through smoke and ash (Spencer and Hauer 1991).

Algae are very sensitive to nutrient fluxes in their environment. Greenness typically rises with increased nutrient loading to lake systems (Wetzel 2001). Changes in nutrient status can lead to a complex series of biological changes that can affect both the structure and function of entire aquatic ecosystems, including plankton, bottom-dwelling invertebrates, and fish.

Improved forest harvesting methods may minimize these impacts on aquatic systems. Buffer strips, unharvested areas left around lakes and streams, may moderate water quality changes in disturbed watersheds by retaining runoff particulates and (or) water-soluble nutrients. The efficiency of buffer strips on boreal lakes has only recently been investigated under the auspices of several collaborative research initiatives (Carignan and Steedman 2000; Steedman 2000; Prepas et al. 2001a). These studies, described in the next section, include lakes in the Canadian Boreal Plain and Boreal Shield.

Surface waters on the Boreal Plain

In experimental studies done to date throughout northern Alberta, lake catchments ranged in size from 0.1 to 82 km^2 and were relatively flat (less than 10% slope). Lakes were relatively small (0.1–11 km^2) and ranged in maximum depth from 2 to 38 m. Drainage ratios (catchment area divided by lake volume) were large and water renewal times were moderate (mean ~6 years). Study sites could be categorized as upland (less than 50% wetland cover) or lowland (over 50% wetland cover; Figs. 10.4 and 10.5). Upland catchments contained both deciduous (e.g., trembling aspen [*Populus tremuloides*], balsam poplar [*P. balsamifera*]) and coniferous (e.g., white spruce [*Picea glauca*]) tree species, desirable from a forest-harvesting perspective. Lowlands were characterized by either open areas or tree species currently of lesser commercial value (e.g., black spruce [*Picea mariana*], balsam poplar, paper birch [*Betula papyrifera*], and tamarack [*Larix laricina*]). Geology in the area consisted of soft sedimentary bedrock overlain by deep glacial materials. Precipitation across the region varied from 300 to 625 mm (ESWG 1996), and 70% of the annual precipitation fell during summer. Box 10.5 outlines the factors examined in each set of lakes.

Catchment disturbance and water quality in uplands

Undisturbed

Landscape characteristics influencing nutrients and algal communities in surface waters were investigated on the Boreal Plain. In undisturbed upland-dominated lakes, nutrient concentrations were mainly controlled by (1) drainage ratio and (2) percent rich fen coverage (see Box 10.6). Drainage ratio was most closely related to P concentrations and less so to N and dissolved organic C concentrations. Lakes with relatively large drainage ratios (i.e., large catchment area relative to lake volume) had the highest P concentrations.

Fig. 10.4. Location of lakes and streams studied in the Boreal Plain of northern Alberta, Canada.

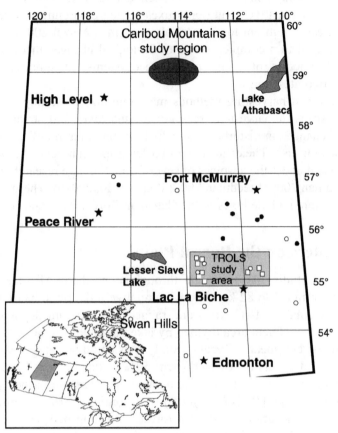

The drainage ratio is often used as a surrogate for "water renewal time", the average time required to refill a basin with new water if it were emptied. A lake that receives a relatively large amount of water from its catchment, as compared to its volume (high drainage ratio, short renewal time), can replace basin volume more quickly than a lake with little catchment input. Therefore, aquatic systems with high drainage ratios are closely connected to their catchments, rather than atmospheric or within-lake processes.

Wetlands can be classified based on peat formation (Box 10.6). Peat consists of a permanent layer of mosses that accumulate owing to stable seasonal water levels and low water flow. Nutrients become locked in the moss layer, unavailable to decomposers. Therefore, peat accumulates because of slow decomposition relative to plant production. Peat-forming wetlands include bogs and fens, and non-peat-forming wetlands include marshes and swamps. In upland catchments of boreal Alberta, type of wetland was related to lake water nutrient concentrations: rich fens retained total P and N (see Box 10.7 on identifying and classifying fens). The amount or type of wetland coverage did not appear to affect dissolved organic C concentrations in lake water.

In general, shallow lakes had higher N concentrations, likely due to the recycling of lake-bottom nutrients from decomposition (Fig. 10.1). Shallow lakes were also greener than deep lakes. Lake greenness and algal community structure were related to N and

Fig. 10.5. Oblique aerial photograph of a Boreal Plain lake (SPH20) with 28% of its catchment logged. Note the checkered pattern of cutblocks and the 20 m vegetated buffer strip on the shoreline.

Photo by Ellie Prepas

Box 10.5. Projects on the Boreal Plain that provided information for this review.

Uplands

- Nineteen undisturbed lakes from two studies (SFMN and TROLS) were combined to provide background information, specifically, the relationship between watershed characteristics and water quality (Prepas et al. 2001*b*);

- Eleven lakes from the TROLS study that adopted a "Before and After Control Impact" design were used to evaluate the effects of forest harvesting on water quality (Prepas et al. 2001*a*); and

- Four streams from the Forest Watershed and Riparian Disturbance (FORWARD) study were used to evaluate the effects of fire on water quality (Prepas et al. 2003; Prepas, unpublished data).

Lowlands (bog-dominated watersheds)

- Seven undisturbed lakes from an SFMN study were used to provide background information, specifically, the relationship between watershed characteristics and water quality (Prepas et al. 2001*b*); and

- Twenty-nine lakes and three streams from the Caribou Mountains Research Partnership study were used to evaluate the effects of fire on water quality (McEachern et al. 2000).

Box 10.6. Characteristics of wetlands in Canada's boreal forests.

Non-peat-forming wetlands

- Marshes: - open wetlands with relatively high water flow;
           - dominated by sedges (*Carex* spp.), cattails (*Typha* spp.), and (or) rushes
             (*Scirpus* spp.).
- Swamps: - thickly or sparsely treed or shrubby;
           - vegetated by tamarack (*Larix laricina)*, black spruce (*Picea mariana*),
             birch (*Betula* spp.), and (or) willow (*Salix* spp.).

Peat-forming wetlands

- Fens: - water levels at or near the peat surface; water flows through them;
           (1) open, dominated by sedges, rushes, cottongrasses (*Eriophorum* spp.); or
           (2) shrubby, dominated by birch and willow; or
           (3) wooded, dominated by black spruce, tamarack, birch, and (or) willow.
- Bogs: - water levels are 40–60 cm below the peat surface; low water flow;
           - dominated by peat mosses (*Sphagnum* spp.), feather mosses (*Pleurozium
             schreberi, Hylocomium splendens*), and lichens *(Cladonia* spp., *Cladina* spp.);
           - open or treed (black spruce only).

Based on Vitt et al. 1996

(or) P concentrations. Shallow lakes with relatively large watersheds and low rich fen cover (less than 8% of catchment area) were the most nutrient rich, and may be more susceptible to nutrient enrichment from watershed disturbance.

Logging

The effects of forest harvest (in which a mean of 15% of the area in each catchment was logged) were investigated 2 years before and 2 years after harvesting in 11 upland-dominated lakes in the southern portion of Alberta's boreal mixedwood ecoregion (Fig. 10.4). Drainage ratios were large (mean 14), rich fens covered less than 8% of catchment areas, and water renewal times were relatively short (~6 years). The lakes were divided into two categories:

(1) deep with cold-water fish habitat (maximum depth over 6 m); and
(2) shallow with no or intermittent cold-water fish habitat (maximum depth under 6 m);
     see Fig. 10.2 for illustration of how cold-water habitat is related to lake depth.

One year after logging, mean total P concentrations in the surface layer of the lake increased 40% in shallow lakes and rose only slightly in deep lakes. Lakes with large increases in total P concentrations also had large drainage ratios, implying the importance of surface runoff following watershed disturbance. In response to drier conditions in the second post-treatment year, there was a 50% decrease in mean total P concentrations in deep lakes, and a more modest decline (15%) in mean total P concentrations in shallow lakes. Watershed characteristics and treatments did not affect the concentration of other nutrients (N, dissolved organic C, pH, ions). These results are consistent with a harvest-enhanced wet year response.

Box 10.7. Identifying fens in Alberta.

Fens are readily distinguished from other wetland types with aerial photos. They are peat-lands with water levels at or near the surface, and therefore show flow patterns. A treed rich fen is dominated by tamarack, *Larix laricina*; by area, it has over 20% tamarack and under 2% black spruce (*Picea mariana*; Beckingham and Archibald 1996). In the case of a non-forested fen, rich fens can be identified from the ground by vegetation; they are dominated by brown mosses (species of the genera *Drepanocladus*, *Brachythecium*, and (or) *Callier-gonella*), whereas poor fens are covered with species of *Sphagnum* (Beckingham and Archibald 1996; L.A. Halsey, Department of Biological Sciences, University of Alberta, Edmonton, Alberta, personal communication).

Buffer strip width (about 20, 100, or 200 m) was not related to changes in nutrient concentration in lake water; lakes surrounded by catchments with wide buffers had the largest P increases after harvest. Also, the amount of catchment harvested was the second strongest predictor of changes in lakewater P concentration, but this parameter was not statistically significant, probably because of the overwhelming influence of drainage ratio on post-harvest P concentrations and the narrow range of watershed harvested (0–35%).

Algal response to logging depended on lake depth. Deep lakes were about 20% less green after harvesting due to a decrease in the biomass of algae other than blue–greens. Shallow lakes were at least 75% greener after harvesting due to a threefold increase in the biomass of inedible blue–green algae. Blue–greens are adapted to high P conditions and are the most likely algal group to increase in biomass following P enrichment of lakes on the Boreal Plain (Trimbee and Prepas 1987). Additionally, the concentration of microcystin-LR (MCLR), a common liver toxin produced by certain blue–green algae (Kotak et al. 1995), increased almost 10 times (from 2.9 to 23 µg/g) the first year following harvesting and remained high in the second post-treatment year. The MCLR increase was magnified in shallow lakes, where blue–greens increased the most. This study was the first to monitor cyanotoxin concentrations before and after watershed disturbance.

Grazer abundance followed variation in edible algal abundance and decreased by 45% overall after harvesting, mostly owing to decreased abundance of large grazers (cladocerans and calanoids).

Wildfire

The water quality in four 3rd and 4th order streams before and after wildfire (two undisturbed streams, two with watersheds 84–90% burnt) was measured in the Swan Hills, Alberta (Fig. 10.4). After fire, water export from the two burnt watersheds increased about two times, possibly owing to reduced uptake of water by vegetation and (or) reduced infiltration of water into soils. The increase in water export occurred during summer storm events. Wildfire increased P export in the streams, and most of this was attributed to a sevenfold increase in P bound to particles. Therefore, fire may have enhanced erosion in the watershed and (or) in the stream channel. The effects of fire on stream export were strongest in the last year (Year 2) after disturbance.

Bioavailable N (in the form of nitrate) exports were over two times higher following wildfire, possibly due to enhanced leaching from soils. Increases in bioavailable N were recorded recently in nearby Delorme Lake, the same year as a severe burn (in which 100% of the forest cover in the catchment was consumed by fire; E.E. Prepas, unpublished data). Following fire, there were no detectable differences in dissolved organic C concentrations and colour in the burnt streams relative to reference streams.

Catchment disturbance and water quality in lowlands

Undisturbed

Seven lowland-dominated lakes (mean wetland coverage 76% of catchment area) were studied to evaluate landscape variables influencing water quality in central Alberta's boreal mixedwood ecoregion. Percent wetland cover explained much of the differences in nutrient concentrations among lakes: 78% of total P, 40% of ammonium, and 74% of dissolved organic C. Most (80%) bioavailable N was ammonium, which was attributed to greater wetland cover and water-logged (oxygen-poor) soils.

By separating peatland cover into three basic categories (rich fens, poor fens, and bogs; see Boxes 10.6 and 10.7), rich fens were found to retain total P and total N, whereas poor fens played a neutral role. Bogs were the major wetland type and in contrast to rich fens, they exported N and P. Unlike upland-dominated lakes, amount and type of wetland cover were the dominant factors controlling nutrient concentrations, colour, and lake greenness in the surface water of lowland lakes.

Wildfire

Water quality of 15 burnt headwater lakes and one burnt stream in permafrost-laden, bog-dominated watersheds was evaluated for 2 years after a severe wildfire (mean 78% of the catchment burnt) in the Caribou Mountains of northern Alberta. Water colour was more than two times higher in surface water with burnt watersheds compared to reference systems and lakes in the boreal mixedwood ecoregion to the south (Prepas et al. 2001*b*). Water colour was strongly linked to dissolved organic C content in all study lakes, indicating a strong humic fraction in the peatland-derived dissolved organic C pool. Mean dissolved organic C concentration was 60% higher in burn-impacted lakes than in reference lakes, suggesting catchment sources of dissolved organic C were important.

Surface waters in burnt watersheds situated in permafrost had elevated P and N concentrations compared to reference systems. Mean bioavailable P was almost seven times higher in burnt relative to reference lakes. Also, total P concentrations were over five times greater in burnt streams than in reference streams. Concentrations of bioavailable N (nitrate) were also greater (three times, a lesser degree than P) in burnt than in reference systems. Larger increase in exports of P than N following fire has been previously reported in other wetland-dominated systems (Bayley et al. 1992). Nitrogen retention of undisturbed Alberta peatlands exceeds 98% (Li and Vitt 1997), similar to N-retention values of burnt peatlands in Ontario (Bayley et al. 1992). Therefore, burnt peatlands may lose their ability for P-retention more than N-retention. Elevated nutrient concentrations and reduced light cancelled each other out in burn-impacted lakes, which had similar greenness to reference lakes.

Fire also affected lake acidity and ion balances in these permafrost dominated sites. Lakes in burnt watersheds were more acidic (and less buffered) relative to reference lakes (mean pH = 6.9 and 7.6, respectively). Therefore, the large proportion of inundated peatlands in the Caribou Mountains exported bog water and dissolved ions after wildfire. The underlying permafrost layer may have enhanced the transport of dissolved substances to lakes by impeding percolation of water through the soil. Few relationships were observed between nutrients and physical parameters of the lakes and their catchments. In the permafrost dominated watersheds, drainage ratio explained up to 70% of the change in dissolved organic C concentrations and colour following fire. In general, lakes with small catchments were more likely to be impacted by fire than were larger lakes with extensive catchments.

Surface waters on the Boreal Shield

In experimental studies in northwestern Ontario (Fig. 10.6) and northern Quebec (Figs. 10.7 and 10.8), lake catchments were small (1–5 km^2) and had moderate slopes (under 10%). Catchments were composed of 75- to 100-year-old black spruce, jack pine (*Pinus banksiana*), trembling aspen, with some paper birch, balsam fir (*Abies balsamea*), eastern white cedar (*Thuja occidentalis*), eastern white pine (*Pinus strobus*), and red pine (*P. resinosa*). Wetlands covered about 10% of the landscape, which annually received

Fig. 10.6. Photograph of snags on the burnt shoreline of a Boreal Plain lake (Delorme Lake, Alberta).

Photo by Shawn Pinder

Fig. 10.7. Oblique aerial photograph of a Boreal Shield lake (L42) in northwestern Ontario with 74% of its catchment logged and no vegetated buffer strip left on the shoreline.

Photo by Robert Steedman

about 1000 mm of precipitation (40% snow). Surface waters were situated on bedrock (a mixture of gneiss and granite) covered by glacial deposits that are thin to nonexistent (0–2 m). The lakes were small (0.15–0.81 km^2) and ranged in maximum depth from 5 to 37 m. The focus lakes in Ontario had relatively small drainage ratios (mean ~2) whereas those in Quebec had ones more similar to the Boreal Plain (mean ~6). The factors studied in each set of lakes are outlined in Box 10.8.

Catchment disturbance and water quality in Ontario

Logging (CLEW project)

Three upland-dominated lakes in northwestern Ontario, about 200 km NW of the city of Thunder Bay, were investigated 5 years before and 2 years after harvesting (Fig. 10.6). Harvesting removed on average 65% of catchment forest, either with or without shoreline logging. Clearcutting was associated with a small increase (5% or less) in mid-lake wind speed, an effect prevented by leaving a forested buffer strip along shorelines. No detectable impacts of harvesting were associated with lakewater nutrient and ion concentrations and greenness. Adult lake trout habitat, measured as the lake volume containing cold water, remained intact after logging.

High precipitation magnified the increase in dissolved organic C after logging. This climate effect is consistent with the strong harvesting impacts on Boreal Plain lakes during the periods of higher runoff (see previous section). Shoreline buffer strips did not appear to protect lake water quality and adult lake trout habitat from the effects of catchment clearcutting.

Fig. 10.8. Oblique aerial photograph of Boreal Shield lakes in northern Quebec (C48 and C29) with 64–73% of catchments logged and 20 m vegetated buffer strips left on shorelines.

Photo by Richard Carignan

Wildfire (ELA project)

Three streams in northwestern Ontario (one lowland, two upland) were investigated 9 years before and 9 years after fire burnt 50–100% of the catchments (Fig. 10.6). Nitrogen concentrations increased at least twofold after fire in all streams. Phosphorus concentrations only increased (over threefold) in the stream draining bogs.

Box 10.8. Projects on the Boreal Shield that provided information for this review.

Northwestern Ontario

Coldwater Lakes Experimental Watersheds (CLEW)
- effects of experimental clearcut logging on water quality and trout habitat were evaluated in three lakes (Steedman 2000; Steedman and Kushneriuk 2000)

Experimental Lakes Area (ELA)
- impacts of wildfire on phosphorus and nitrogen concentrations were evaluated in three streams (Bayley et al. 1992)

Quebec

- 47 undisturbed lakes in two studies were used to provide background information on Boreal Shield lakes, specifically, the relationship between watershed characteristics and water quality (D'Arcy and Carignan 1997; Carignan et al. 2000)
- effects of wildfire and logging on lake water quality and biota were evaluated in 22 lakes from SFMN-Q (Carignan et al. 2000; Patoine et al. 2000; Planas et al. 2000; St-Onge and Magnan 2000; Pinel-Alloul, Planas, Carignan et al. 2002)

Catchment disturbance and water quality in Quebec

Undisturbed

Characteristics of 47 undisturbed watersheds in northern Quebec were related to lake nutrients and algal communities (Fig. 10.7 and reference lakes in Fig. 10.8). Nutrient concentrations (total P, bioavailable N, dissolved organic C), water clarity, and greenness were mostly controlled by (1) drainage ratio and (2) catchment slope. Flat topography created waterlogged conditions that increased soil exposure to weathering. As a consequence, concentrations of total P and certain ions (Ca^{2+}, Mg^{2+}) were higher in surface waters of flat watersheds. However, bioavailable N, a nutrient readily leached from soils, was lower in surface waters with steep watersheds. Lakes with relatively large watersheds and flat catchments were the most nutrient rich, and may be more susceptible to nutrient enrichment from watershed disturbance.

Logging

Water quality, plankton, and fish communities of 13 headwater lakes were compared to 17 undisturbed lakes for 3 years after clearcutting 40% of catchment areas (Fig. 10.8). Forested buffer strips (20 m) were left along lake shorelines, permanent streams, and wetlands. Lakes in the undisturbed and logged watersheds were similar: they were small (0.6 km^2), moderately deep (maximum depth 14 m), and had short water renewal times (mean 1.5 year).

Loss of nutrients (P, N, potassium, dissolved organic C) from catchments increased after cutting (Lamontagne et al. 2000). Although these exports represented small nutrient losses from the forest, they impacted water quality. Cut lakes had up to two times higher nutrient concentrations (total N and P) and over two times greater concentrations of certain ions (potassium and chloride) than reference lakes (Carignan et al. 2000). The strongest response of Boreal Shield lakes to logging was increased dissolved organic C concentration, and it persisted for at least 3 years. Higher dissolved organic C concentration reduced the transmission of light and limited photosynthesis. Furthermore, higher dissolved organic C concentrations in cut lakes has been linked to increased mercury contamination in the food web, from zooplankton to predatory fish (Garcia and Carignan 1999, 2000).

Post-harvest changes in water quality were strongly related to (1) the extent of catchment clearcutting (the ratio of catchment area harvested to lake volume) and (2) the drainage ratio. On the Boreal Shield, harvesting may impact lake water quality when the drainage ratio exceeds 4, and when more than 30% of the catchment is harvested (Carignan et al. 2000).

Post-logging nutrient enrichment did not transfer up the food chain. Only a small and short (first year post-harvest) increase in greenness was noted in logged lakes compared to reference lakes (Planas et al. 2000). This uncoupling between the nutrient pulse and algal production in logged watersheds may be explained by light limitation of algal growth due to higher dissolved organic C concentrations. This is reflected by the dominance of algae that can obtain C by means other than photosynthesis (cryptophytes and chrysophytes); they can use bacteria and organic matter to survive in light-limited environments (Tranvik et al. 1989). Also, grazers (calanoid copepods) often associated with clear, nutrient-poor lakes decreased by 43% in logged lakes. This grazer group may be a good bioindicator of logging impacts on higher levels of the food web. In addition, logging reduced the abundance of small yellow perch (*Perca flavescens*) and white sucker (*Catostomus commersoni*) in Quebec Boreal Shield lakes (St-Onge and Magnan 2000).

Wildfire

Water quality, plankton, and fish communities of nine headwater lakes were compared to 17 undisturbed lakes during 3 years after an intensive wildfire burned 50–100% of their catchments (Fig. 10.8). Concentrations of nutrients (P, N) and several major ions (Ca^{2+}, Mg^{2+}, K^+, Cl^-, SO_4^{2-}) in impacted lakes were two to seven times higher than in undisturbed lakes. In contrast, no change was observed in water colour and dissolved organic C concentrations (Carignan et al. 2000). Increases in P, Ca^{2+}, and Mg^{2+} were longer lived (at least 3 years) than nitrate and K^+ increases (2 years). Post-fire nutrient enrichment was proportional to the relative burn area (i.e., the amount of catchment burned relative to the lake area or volume) and the drainage ratio.

In contrast to logging, nutrient enrichment effects after fire transferred up the food web. Offshore greenness was two to three times higher and biomass of small plankton (algae and rotifers) was 59% higher in burned lakes as compared to reference lakes. The greatest response was observed the second year after wildfire, when production of algae (diatoms) and small grazers (rotifers) characteristic of enriched lakes on the Boreal Shield were at their highest (Patoine et al. 2000; Planas et al. 2000). Similar to logging, the relative abundance of small yellow perch and white sucker decreased in response to fire (St-Onge and Magnan 2000).

Recommendations

Historically, the focus of forest management in Canada has been to minimize the direct impacts of activities associated with logging (roads and stream crossings) at the stand level. Management is now moving towards minimizing impacts at the watershed scale (M. Donnelly)[2]. A watershed-level approach to sustainable forest management involves ensuring that the needs of present and future forest users will be met (CCFM 1995; see Chap. 2). There is a need to identify the values in aquatic systems that industry, communities, and regulators want to protect before entering a planning process (Chap. 4). For instance, should aquatic systems be protected from change beyond an established threshold or should they be managed according to provincial/federal guidelines that protect surface waters from negative impacts on aquatic life? This chapter presents important advances in identifying water quality parameters that can serve as tools in the watershed management process (see Box 10.9). The next challenge is to establish thresholds for these parameters.

---

[2]Donnelly, M. 2003. Ecologically based forest planning and management for aquatic ecosystems in the Duck Mountains, Manitoba. Draft manuscript prepared for submission to J. Env. Eng. Sci.

Box 10.9. Summary of disturbance impacts on boreal surface waters.

Influence of geology on surface water disturbance

(*a*) Similarities between granitic (i.e., Shield) and sedimentary (i.e., Plain) basins:

- ↑ total P concentrations;
- drainage ratios >4, presence of bogs, and high precipitation enhance impacts; and
- no evidence that vegetated buffer strips around water bodies protect water quality or cold-water fish habitat.

(*b*) Differences between granitic and sedimentary basins:

- ↑ blue–green algae and cyanotoxins in lakes on logged sedimentary plain; and
- shallow lakes more vulnerable on sedimentary Plain;

Influence of disturbance type on surface water

(*a*) Similarities between impacts of fire and logging:

- ↑ total P concentrations; and
- ↓ abundance of young fish in granitic basins; no data for sedimentary Plain.

(*b*) Differences between impacts of fire and logging:

Overall:

- ↑ bioavailable N in burnt but not necessarily in logged systems; and
- ↓ large grazer biomass (calanoids) in logged lakes.

On granite bedrock:

- ↑ dissolved organic C and water colour in logged lakes;
- algae limited by decreased light in logged lakes but growth increased (mostly diatoms) in burnt lakes;
- ↑ mercury in grazers and fish in logged lakes; and
- ↑ small grazer biomass (rotifers) in burnt lakes.

On sedimentary plain:

- ↑ dissolved organic C in burnt lowland lakes underlain by permafrost but not in logged lakes to the south; and
- growth of algae limited by decreased light in burnt lowland lakes underlain by permafrost but increased in logged lakes to the south.

Across Canada's boreal forests, total P concentration in water was the most consistent indicator of disturbance in first order watersheds, irrespective of the type of perturbation. There is at present little information on the length of time required for recovery from disturbance in boreal surface waters; most studies were short term (<3 years). Lakes on granite and sedimentary basins were affected throughout the length of time studied after disturbance (up to 3 years). However, there is some evidence that impacts may be long-lasting under some circumstances: lowland lakes underlain by permafrost recovered slowly over decades after disturbance.

Across boreal landscapes, the drainage ratio is a key indicator of the sensitivity of surface waters to watershed disturbance. Upland aquatic systems with a drainage ratio of at least 4 are most vulnerable to changes in water chemistry following watershed disturbance (Yamasaki et al. 2001; see Box 10.10 on how to delineate watershed boundaries), based on the primarily short-term (<3 year) studies to date. In general, streams have very large drainage ratios (i.e., short water renewal times). Lakes with drainage ratios of at least 4 and streams should be managed as systems strongly connected to their catchment. In the limited number of studies to date, there is no evidence that the quality of surface waters is sensitive to watershed disturbance when drainage ratios are below 4.

Wetland type also affected the response of surface waters to disturbance. In general, bogs enhanced and fens reduced nutrient release to aquatic systems. The influence of wetlands was strong in lowland watersheds. In upland systems, the influence of wetlands may be relatively unimportant as compared to drainage ratio.

Some indicators of disturbance differed across geological gradients. Biological indicators appeared most sensitive on the sedimentary plain, where nutrient concentrations can easily be perturbed beyond critical thresholds. Because blue–green algae are adapted to high P environments, increased blue–green algal biomass and cyanotoxin production are a danger in lakes underlain by glacial deposits. On the sedimentary plain, shallow lakes appeared most susceptible to disturbance over a short (<3 years) time scale: P concentrations, blue–green algal biomass, and their associated toxins were higher in shallow lakes, as compared to deep lakes with cold-water habitat. Contrasts in geology accounted for differences in aquatic indicators of watershed disturbance.

In general, the nature of aquatic indicators depended on type of disturbance. Across boreal landscapes, bioavailable N fractions increased in burnt but not in logged watersheds. Conversely, the biomass of large zooplankton can be useful biological indicators of logging disturbance but not fire. Dissolved organic C concentrations increased in logged but not burnt upland lakes on granite. On the sedimentary plain, dissolved organic C increased in burnt lowland lakes underlain by permafrost, whereas no such response was found on upland logged lakes to the south. The response of dissolved organic C to disturbance seems to be determined by a combination of soil disturbance intensity (fire versus logging) and substrate type (lowlands underlain by permafrost versus uplands on glacial till).

Box 10.10. Methods to delineate watershed boundaries.

1. Stereoscopic examination of aerial photos allows a three-dimensional bird's eye view of the landscape and as such shows surface slopes and drainage directions.

2. Geographic Information Systems (GIS) offer an automated method of watershed boundary delineation. An experienced technician can rapidly delineate watershed boundaries without prior knowledge of landscape drainage patterns. GIS techniques require the use of digital representations of topography, or Digital Elevation Models (DEMs), a series of grid networks that vary in resolution.

The skills of the technician will determine the efficiency of the approach. Both methods work well in upland regions that have substantial relief, but ground truthing is needed in areas of low topography in most watersheds. Watershed boundaries delineated with the GIS method can be easily integrated with other digital information, such as composition of vegetation and soils in watersheds. Perhaps the biggest failing of the GIS method is that the information generated is only as good as the resolution of the database and the skills of the user: errors or lack of precision in DEMs are translated into the watershed boundary delimitation.

Box 10.11. Guidelines to preserve surface water quality in Canada's boreal forests.

- Different types of disturbance (fire, logging) will produce distinct effects on surface waters; neither has exclusively acceptable or unacceptable impacts.
- Impacts of forest disturbance on water quality are more detectable when the drainage ratio (catchment area divided by lake volume) exceeds 4.
- Effects of forest disturbance on water quality are observable when 30% of trees are removed in a catchment. However, in the case of shallow lakes (maximum depth <5 m), impacts have been recorded when less than 30% of catchment has been perturbed.
- Alternative methods of logging (e.g., various forms of partial cutting, variable retention; see Chap. 13) offer opportunities for reduced impacts of watershed disturbance on surface waters.
- Rich fens in the catchment can dampen nutrient export to surface waters, so catchments containing rich fens may be able to support a higher rate of cut.
- Adopt slower rates of clearcut harvesting or a higher proportion of within-stand retention for watersheds where the drainage ratio is >4, maximum lake depth is <5 m, or rich fens are uncommon.
- There is no evidence that vegetated buffer strips around water bodies protect water quality.

Variation in dissolved organic C had important biological consequences. Increased dissolved organic C reduced water transparency, and effectively limited algal growth. Greenness is a good indicator of fire but not of logging in upland lakes in granite catchments. In addition, mercury concentrations in zooplankton and fish increased with dissolved organic C in granite catchments. Because its impacts can be widespread, dissolved organic C concentration may prove an informative indicator of watershed disturbance.

Thresholds for aquatic indicators of disturbance must be construed to preserve aquatic system values established by public consultation. Furthermore, there is a need to ascertain how much land can be disturbed before water quality reaches set thresholds. Based on studies in first order watersheds, we suggest that clearcutting should not exceed 30% of the catchment area (Box 10.11). However, disturbance thresholds for water quality may vary among forests and watersheds that differ in order (headwater versus 2nd order plus). A challenge in using a watershed-level approach to forest management is determining the level or order of watersheds to be managed. Based on regional differences in water quality indicators, the order of watersheds managed likely also depends on regional factors. Each forest is unique in its geological, biological, and climatic form.

To determine disturbance response of aquatic systems for a forest, industrial users could establish permanent aquatic sampling sites, similar to permanent sample plots used in forest stand monitoring, where changes in surface water can be tracked over decades. Key water quality indicators that could be measured in both lakes and streams include water clarity, dissolved organic C, total P, and bioavailable N. The first two indicators were recently proposed for inclusion in water quality monitoring components of forestry management on the Boreal Shield (Yamasaki et al. 2001). Biological indicators include: contaminant concentration in invertebrates and vertebrates, algal biomass and composition, and calanoid biomass (in lakes). Such biological indicators require expertise in taxonomy and chemistry. Despite a large and variable landscape, indicators of watershed disturbance are emerging for surface waters across the Canada's boreal forest.

References

Alberta Environment. 1999. Surface water quality guidelines for use in Alberta. Environmental Sciences Division, Alberta Environment, Edmonton, Alberta. Available at http://www3.gov. ab.ca/env/protenf/publications/surfwtrqual-nov99.pdf [viewed 5 June 2003]. 20 p.

Bayley, S.E., Schindler, D.W., Beaty, K.G., Parker, B.R., and Stainton, M.P. 1992. Effects of multiple fires on nutrient yields from streams draining boreal forest and fen catchments: nitrogen and phosphorus. Can. J. Fish. Aquat. Sci. **49**: 584–596.

Beckingham, J.D., and Archibald, J.H. 1996. Field guide to ecosites of northern Alberta. Canadian Forest Service, Edmonton, Alberta. 5 Spec. Rep. 5. 28 p.

Carignan, R., and Steedman, R.J. 2000. Impact of major watershed perturbations on aquatic ecosystems. Can. J. Fish. Aquat. Sci. **57**(Suppl. 2): 1–4.

Carignan, R., D'Arcy, P., and Lamontagne, S. 2000. Comparative impacts of fire and forest harvesting on water quality in boreal shield lakes. Can. J. Fish. Aquat. Sci. **57**(Suppl. 2): 105–117.

(CCEA) Canadian Council on Ecological Areas. 2002. Some quantitative environmental and socioeconomic characteristics of Canada's terrestrial ecozones. Environment Canada, Ottawa, Ontario. Available at http://www.ccea.org/ecozones/stats/quant.html [viewed 3 June 2003]. 1 p.

(CCFM) Canadian Council of Forest Ministers. 1995. Defining sustainable forest management: a Canadian approach to criteria and indicators. Natural Resources Canada, Ottawa, Ontario. Available at http://www.ccfm.org/ci/framain_e.html [viewed 3 June 2003].

(CCME) Canadian Council of Environment Ministers. 2002. Canadian environmental quality guidelines, summary table updated 2002. Canadian Council of Environment Ministers, Winnipeg, Manitoba. Available at http://www.ccme.ca/assets/pdf/e1_06.pdf [viewed 5 June 2003]. 12 p.

D'Arcy, P., and Carignan, P. 1997. Influence of catchment topography on water chemistry in southeastern Quebec Shield lakes. Can. J. Fish. Aquat. Sci. **54**: 2215–2227.

Duarte, C.M., and Kalff, J. 1989. The influence of catchment geology and lake depth on phytoplankton biomass. Arch. Hydrobiol. **115**: 27–40.

(ESWG) Ecological Stratification Working Group. 1996. A national ecological framework for Canada. Centre for Land and Biological Resources Research, Research Branch, Agriculture and Agri-food Canada, Ottawa, Ontario. Available at http://sis.agr.gc.ca/cansis/nsdb/ecostrat/intro.html [viewed 3 June 2003].

Garcia, E., and Carignan, R. 1999. Impact of wildfire and clear-cutting in the boreal forest on methyl mercury in zooplankton. Can. J. Fish. Aquat. Sci. **56**: 339–345.

Garcia, E., and Carignan, R. 2000. Mercury concentrations in northern pike (*Esox lucius*) from boreal lakes with logged, burned, or undisturbed catchments. Can. J. Fish. Aquat. Sci. **57**(Suppl. 2): 129–135.

Kotak, B.G., Lam, A.K.-Y., Prepas, E.E., Kenefick, S.L., and Hrudey, S.E. 1995. Variability of the hepatotoxin, microcystin-LR, in hypereutrophic drinking water lakes. J. Phycol. **31**: 248–263.

Lamontagne, S., Carignan, R., and D'Arcy, P. 2000. Element export in runoff from eastern Canadian Boreal Shield catchments following forest harvesting and wildfires. Can. J. Fish. Aquat. Sci. **57**(Suppl. 2): 118–128.

Li, Y., and Vitt, D.H. 1997. Patterns of retention and utilization of aerially deposited nitrogen in boreal peatlands. Ecoscience, **4**: 106–116.

Likens, G.E., and Bormann, F.H. 1974. Linkages between terrestrial and aquatic ecosystems. BioScience, **24**: 447–456.

McEachern, P., Prepas E.E., Gibson J.J., and Dinsmore, W.P. 2000. Forest fire induced impacts on phosphorus, nitrogen, and chlorophyll *a* concentrations in boreal subarctic lakes of northern Alberta. Can. J. Fish. Aquat. Sci. **57**: 73–81.

(MEQ) Ministère de l'Environnement du Québec. 1992. Critères de qualité de l'eau. Service d'évaluation des rejets toxiques et Direction de la qualité des cours d'eau, Ministère de l'Environnement du Québec, Québec, Quebec. Available at http://www.menv.gouv.qc.ca/eau/criteres_eau/ [viewed 3 June 2003]. 425 p.

Mitchell, P.A., and Prepas, E.E. (*Editors*). 1990. Atlas of Alberta Lakes. The University of Alberta Press, Edmonton, Alberta. 690 p.

(NWWG) National Wetlands Working Group. 1998. Wetlands of Canada. Sustainable Development Branch, Canadian Wildlife Service, Environment Canada, Polyscience Publications Inc., Montreal, Quebec. Ecol. Land Class. Ser. No. 24. 452 p.

(OMEE) Ontario Ministry of the Environment and Energy. 1994. Water management policies, guidelines, provincial water quality objectives of the Ministry of the Environment and Energy. Queen's Printer for Ontario, Toronto, Ontario. Available at http://www.ene.gov.on.ca/envision/gp/3303e.pdf [viewed 3 June 2003]. 32 p.

Patoine, A., Pinel-Alloul, B., Prepas, E.E., and Carignan, R. 2000. Do logging and forest fires influence zooplankton biomass in Canadian Boreal Shield lakes? Can. J. Fish. Aquat. Sci. **57**(Suppl. 2): 155–164.

Pinel-Alloul, B. 1993. Zooplankton community structure in hardwater hypertrophic lakes in Alberta. Water Sci. Technol. **27**: 353–361.

Pinel-Alloul, B. 1995. Spatial heterogeneity as a multiscale characteristic of zooplankton community. Hydrobiologia, **300/301**: 17–42.

Pinel-Alloul, B., Planas, D., Carignan, R., and Magnan, P. 2002. Synthèse des impacts écologiques des feux et des coupes forestières sur les lacs de l'écozone boréale du Québec. Rev. Sci. Eau, **15**(1): 371–395.

Planas, D., Desrosiers, M., Groulx, S.R., Paquet, S., and Carignan, R. 2000. Pelagic and benthic algal responses in eastern Canadian Boreal Shield lakes following harvesting and wildfires. Can. J. Fish. Aquat. Sci. **57**(Suppl. 2): 136–145.

Prepas, E.E, Pinel-Alloul, B., Planas, D., Methot, G, Paquet, S, and Reedyk, S. 2001*a*. Forest harvest impacts on water quality and aquatic biota on the Boreal Plain: introduction to the TROLS Lake Program. Can. J. Fish. Aquat. Sci. **58**: 421–436.

Prepas, E.E., Planas, D., Gibson, J.J., Vitt, D.H., Prowse, T.D., Dinsmore, W.P., Halsey, L.S., McEachern, P.M., Paquet, S., Scrimgeour, G.J., Tonn, W.M., Paszkowski, C.A., and Wolfstein, K. 2001*b*. Landscape variables influencing nutrients and phytoplankton communities in Boreal Plain lakes of northern Alberta: a comparison of wetland- and upland-dominated catchments. Can. J. Fish. Aquat. Sci. **58**: 1286–1299.

Prepas, E.E., Burke, J.M., Chanasyk, D.S., Smith, D.W., Putz, G., Gabos, S., Chen, W., Millions, D., and Serediak, M. 2003. Impact of wildfire on discharge and phosphorus export from the

Sakwatamau watershed in the Swan Hills, Alberta during the first two years. J. Environ. Eng. Sci. **2**. In press.

Roby, K.B., and Azuma, D.L. 1995. Changes in a reach of a northern California stream following wildfire. Environ. Manage. **19**: 591–600.

Ruttner, F. 1974. Fundamentals of limnology. *Translated by* D.G. Frey and F.E.J. Fry. University of Toronto Press, Toronto, Ontario. 307 p.

Shaw, R.D., and Prepas, E.E. 1990. Groundwater-lake interactions: II. nearshore seepage patterns and the contribution of ground water to lakes in Central Alberta. J. Hydrol. **119**: 121–136.

Spencer, C.N., and Hauer, F.R. 1991. Phosphorus and nitrogen dynamics in streams during a wildfire. J. N. Amer. Benthol. Soc. **10**: 24–30.

Steedman, R.J. 2000. Effects of experimental clearcut logging on water quality in three small boreal forest lake trout lakes. Can. J. Fish. Aquat. Sci. **57**(Suppl. 2): 92–96.

Steedman, R.J., and Kushneriuk, R.S. 2000. Effect of experimental clearcut logging on thermal stratification, dissolved oxygen and lake trout habitat volume in three small boreal forest lakes. Can. J. Fish. Aquat. Sci. **57**(Suppl. 2): 82–91.

St.-Onge, I., and Magnan, P. 2000. Impact of logging and natural fires on fish communities of Laurentian Shield lakes. Can. J. Fish. Aquat. Sci. **57**(Suppl. 2): 165–174.

Tranvik, L.J., Porter, K.G., and Sieburth, J.M. 1989. Occurrence of bacterivory in *Cryptomonas*, a common freshwater phytoplankton. Oecologia, **78**: 473–476.

Trimbee, A.M., and Prepas, E.E. 1987. Evaluation of total phosphorus as a predictor of the relative biomass of blue–green algae with emphasis on Alberta lakes. Can. J. Fish. Aquat. Sci. **44**: 1337–1342.

Vitt, D.H., Halsey, L.A., Thormann, M.N., and Martin, T. 1996. Peatland inventory of Alberta: phase 1: overview of peatland resources in the natural regions and subregions of the province. Prepared for the Alberta Peat Task Force. Sustainable Forest Management Network, Edmonton, Alberta. 117 p.

Wetzel, R.G. 2001. Limnology. Academic Press, San Diego, California, 1006 p.

Yamasaki, S.H., Côté, M.-A., Kneeshaw, D.D., Fortin, M.-J., Fall, A., Messier, C., Bouthillier, L., Leduc, A., Drapeau, P., Gauthier, S., Paré, D., Greene, D., and Carignan, R. 2001. Integration of ecological knowledge landscape modelling, and public participation for the development of sustainable forest management. Sustainable Forest Management Network, Edmonton, Alberta. Final Proj. Rep. 2001-27. Available at http://sfm-1.biology.ualberta.ca/english/pubs/PDF/PR_2001-27.pdf [viewed 3 June 2003]. 27 p.

Chapter 11

Forest management planning based on natural disturbance and forest dynamics

Brian D. Harvey, Thuy Nguyen-Xuan, Yves Bergeron, Sylvie Gauthier, and Alain Leduc

Introduction

Forest management in the boreal forests of Canada has evolved with changes in resource availability, in forest harvesting and information technology, in markets for forest products, in social values, and in our basic understanding of the boreal ecosystem. The proverbial shifting paradigm in forestry (see Chap. 1; Kimmins 1995, 2002) has led us to the current prevailing concepts of sustainable forest management and forest ecosystem management. Without a doubt, the paradigm of forest ecosystem management is also shifty and somewhat intractable, to the extent that what it does and does not constitute is the subject of considerable debate (Lackey 1998, 1999).

Although sustainable forest management (SFM) also has many interpretations (Chap. 2), it has gradually replaced the more monolithic objective of sustained wood yield as the basis of forest management policies across Canada. In most provinces, SFM is now written into legislative acts regulating the management of forests. Most provincial forestry legislation stipulates that forest management must be conducted according to a general or strategic management plan that is usually prepared by industry and approved by government. In this regard, strategic management plans represent a legal commitment to sustainable forest management. The need for a strategic or general forest plan is a common feature of forest management planning around the world (Chap.

B.D. Harvey,[1,2] **T. Nguyen-Xuan,**[2] **Y. Bergeron,**[2] **S. Gauthier,**[3] **and A. Leduc.**[2] Groupe de recherche en écologie forestière – interuniversitaire (GREFi), Université du Québec à Montréal, Montréal, Quebec, Canada.

[1]Author for correspondence. Université du Québec en Abitibi-Témiscamingue, 445 boulevard de l'Université, Rouyn-Noranda, Quebec J9X 5E4, Canada. Telephone: 819-762-0971, ext. 2361. e-mail: Brian.Harvey@uqat.ca

[2]Chaire Industrielle CRSNG–UQAT–UQAM en aménagement forestier durable, Université du Québec en Abitibi-Témiscamingue, Rouyn-Noranda, Quebec, Canada.

[3]Ressources naturelles Canada – Service canadien des forêts, Centre de foresterie des Laurentides, Sainte-Foy, Quebec, Canada.

Correct citation: Harvey, B.D., Nguyen-Xuan, T., Bergeron, Y., Gauthier, S., and Leduc, A. 2003. Forest management planning based on natural disturbance and forest dynamics. Chapter 11. *In* Towards Sustainable Management of the Boreal Forest. *Edited by* P.J. Burton, C. Messier, D.W. Smith, and W.L. Adamowicz. NRC Research Press, Ottawa, Ontario, Canada. pp. 395–432.

1), and is also central to forest certification requirements (explored further in Chap. 21). It is important to keep in mind, however, that because forest management activities must be planned on temporal scales ranging from several months to one or more forest rotations, and on spatial scales ranging from single cutblocks to entire forest management areas, planning is by definition hierarchical (Tittler et al. 2001). This hierarchical structure is introduced in Chap. 1 (see Box 1.3 and Table 1.2) and is the organizational basis for this chapter and the two that follow it. In the next chapter, Table 12.1 provides some good examples of how some common forest management concerns that might be addressed across the three levels — strategic, tactical, and operational — of planning. Chapter 12 focuses on tactical management planning, and Chap. 13 on stand-level silviculture applied at the operational level. This chapter is primarily concerned with long-term, strategic management planning and the preparation of a "forest plan". In particular, it explores how one might take what we have learned about the dynamics of natural disturbances in the boreal forest (and how they may or may not be emulated by logging activities; Chaps. 8–10), and use it for guidance in forest-level planning.

According to the Canadian Council of Forest Ministers (CCFM 1995), sustainable forest management aims for the multiple use of forest resources by present and future generations while maintaining the integrity of the forest ecosystems. The "coarse-filter approach" and natural disturbance based management have been promoted as sustainable approaches to forest management (Franklin 1993; Attiwill 1994; Bergeron and Harvey 1997) and to restoring biodiversity in intensively managed landscapes (Angelstam 1998; Kuuluvainen 2002a, b). In Ontario and Alberta, natural disturbance templates have been adopted as an ecological basis for provincial forest management policies. Exploring the applicability of the natural disturbance model to SFM was one of the main research themes of the Sustainable Forest Management Network (SFMN; Chap. 1).

Figure 11.1 provides a schematic framework of the strategic forest management planning process. Weetman (2000) quotes Gordon Baskerville in outlining nine steps for implementing this process (Box 11.1). Perhaps the most important points here are:
(1) the need to manage a forest in the context of the dynamic pattern of stand types;
(2) the importance of explicitly stating values for which the forest is to be managed; and
(3) tracking these values through space and time.

This fundamental aspect of scale in forest management has been evoked earlier by Baskerville (1986) and has been reiterated by others in New Brunswick (Erdle 1998; Erdle and Sullivan 1998).

The Forest Stewardship Council (FSC 2000) provides another list of the required elements of a sustainable forest management plan (Box 11.2); note the need to describe a management system "based on the ecology of the forest in question". Forest management planning based on natural disturbance and forest dynamics requires consideration of ecological information in most steps of the forest management planning process. The nature, precision, and extent of the information used will vary, depending on the planning phase and, of course, the availability of relevant information. Furthermore, as in any planning exercise, this process is not a closed and single process. Depending on the regulatory framework and institutional culture, Aboriginal input (Chap. 3), public involvement (Chap. 4), economic considerations (Chap. 5), and political and corporate policies (Chap. 7) will influence one or several steps of the process. And because plan-

Fig. 11.1. The general strategic management process in forest planning.

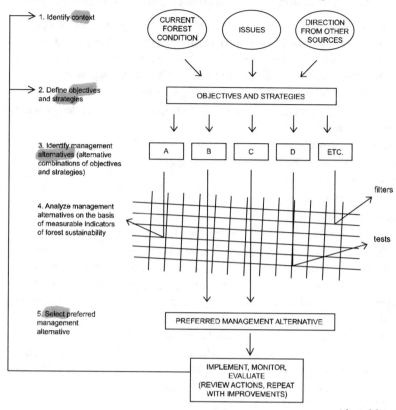

Adapted from OMNR 1996

ning is spatially and temporally hierarchical, implementation of the strategic plan must be linked to the other — tactical and operational — levels of forest management planning. Though spanning long periods (from approximately 20 years to one or two full timber rotations), the strategic planning framework typically has built-in opportunities for recurrent evaluation of the initial plan (generally every 5 years), allowing for modification of management strategies based on feedback from the field indicating changes in the forest condition and changes in the socio-economic milieu. Good strategic planning, therefore, is not rigid, but rather is responsive and adaptable (Chap. 21).

The objective of this chapter is to provide forest managers with some insights into, and guidelines (in a non-regulatory sense) for, integrating the concept of natural disturbance based forest management into strategic planning. Several of the major challenges to this approach are touched on in Box 11.3. The following sections show how ecological information can be used in the various steps of the strategic forest management planning process. This approach is illustrated with some of the SFMN work being conducted in northwestern Quebec. The examples given also illustrate the importance of policy, public involvement, and an adaptive approach to implementation of natural disturbance based forest management. This chapter emphasizes the steps undertaken in devising and articulating a novel forest management strategy, but in so doing, it also illustrates the basic requirements for all strategic forest planning.

Box 11.1. Nine steps required for a sustainable forest management plan.

According to G. Baskerville, former Head of the Department of Forest Resources Management at the University of British Columbia and former Dean of Forestry at the University of New Brunswick, a forest plan must have the following nine components in order to be credible, implementable, auditable, and to lend itself to adaptive management.

1. Define a specific forest area for management (order of 200 000 ha);
2. Designate a forest manager (an actual person) responsible for the design and implementation of management of the defined forest;
3. Create a general description of values sought from the forest;
4. List the target values to be obtained from the specific forest:
- what level for each value?
- what spatial distribution?
- what temporal distribution?
- what rules for tradeoffs in event of conflict among values?
5. Simulate 100-year forest development without harvesting or management intervention; show availability of timber and other values by decade and by broad geographic location;
6. Simulate 100-year forest development with harvesting and management interventions, employing three or four alternative strategies; show availability of timber and other values by decade and by broad geographic location;
7. Build an implementation guide for the strategy of choice:
- a harvest schedule: what will be cut, where, when, with first 20-year detail, 100-year broad picture, all of it stand-type specific, age-class specific, and location specific;
- a treatment schedule: stand conditions to be treated, timing of a treatment, location specific, and related to harvest schedule;
- availability schedule: show values (timber/non-timber) that will be available by amount/quality, at what locations, with detail for the first 20 years;
8. Implementation over time:
- follow harvest schedule or record/justify deviations;
- follow treatment schedule or record/justify deviations;
- use a forest practices code to ensure quality of implementation;
- assess actual availability of all values and compare to forecasts;
9. Periodic review: compare values actually achieved to those forecast, determine the basis for any differences, review goals, revise the forest plan as outlined in steps 1 through 8.

Steps 1 through 4, plus 9, provide opportunities for policy input and require public involvement and stakeholder consensus; steps 5 through 8 must be technically rigorous and must be done by forestry professionals.

From Weetman 2000

Description of the current forest condition — *where are we now?*

One of the first steps in the preparation of any forest management plan is the assessment of the current forest condition, including forest age structure and composition, and a host of other attributes. A forest plan generally requires the description of the management unit land base, the different land uses attributed to it, and the socio-economic con-

---

Box 11.2. The preparation of a forest management plan is one of 10 principles required for Forest Stewardship Council (FSC) certification.

The Forest Stewardship Council provides a concise summary of the rationale for preparing and documenting a strategic forest plan.

Principle 7: Management Plan

A management plan — appropriate to the scale and intensity of operations — shall be written, implemented, and kept up to date. The long term objectives of management, and the means of achieving them, shall be clearly stated.

7.1 The management plan and supporting documents shall provide:

- management objectives;
- description of the forest resources to be managed, environmental limitations, land use and ownership status, socio-economic conditions, and a profile of adjacent lands;
- description of silvicultural and (or) other management system, based on the ecology of the forest in question and information gathered through resource inventories;
- rationale for rate of annual harvest and species selection;
- provisions for monitoring of forest growth and dynamics;
- environmental safeguards based on environmental assessments;
- plans for identification and protection of rare, threatened, and endangered species;
- maps describing the forest resource base, including protected areas, planned management activities, and land ownership; and
- description and justification of harvesting techniques and equipment to be used.

7.2 The management plan shall be periodically revised to incorporate the results of monitoring or new scientific and technical information, as well as to respond to changing environmental, social, and economic circumstances.

7.3 Forest workers shall receive adequate training and supervision to ensure proper implementation of the management plan.

7.4 While respecting the confidentiality of information, forest managers shall make publicly available a summary of the primary elements of the management plan, including those listed in Criterion 7.1.

FSC 2000

---

text. In contrast to forested land in the United States and Fennoscandian countries, the boreal forest of Canada is almost entirely publicly owned and under provincial jurisdiction. This feature may, in many respects, facilitate development of a clear portrait of a forest management area or region in that most forest inventory, forest ecosystem classification information, wildlife inventory and so on, can be provided by one or more provincial resource agencies. In addition, most of the boreal forest has never been previously logged or cultivated (Fig. 11.2), in some ways making the underlying ecological relationships easier to discern and making planning simpler, but also conveying an important responsibility as we try to bring some of the world's last wild forest under human control. Consequently, implementation of forest-level management strategies in boreal Canada may be less complicated than in other regions or countries characterized by a patchwork of land tenures and land-use histories. While much of the country's forests may be considered remote compared to major forest regions in the United States

Box 11.3. Challenges for sustainable forest management plans:

As indicated in Boxes 11.1 and 11.2, all forest management plans must provide a framework and rationale for:

- delivering a sustainable volume of timber and other forest products;
- projecting the shifting mosaic of forest conditions over space and time;
- minimizing or mitigating environmental impacts; and
- identifying all other forest values and modelling their evolution under different management strategies.

Strategic forest planning has addressed the first concern for half a century in North America, while solutions to the second objective have largely depended on the development of computer technology. But the sincere adoption of the final two objectives has required a change of mind set that has not yet expressed itself in most Canadian forest managers. Where these objectives do show up in forest plans, they typically receive low priority for implementation.

Over the last three decades habitat protection has become a growing concern in forest management. Tools such as habitat suitability models and fine-filter management guidelines for particular valued wildlife species have been developed and applied in many provinces. However, it is only more recently that maintaining biological diversity beyond a few species of commercial value has been recognized as a worthy objective of forest management in Canada. While maintaining biodiversity is a concept that may remain somewhat vague in the minds of many people, it has become an accepted social value and the object of commitment of both government and the private sector. Natural disturbance based management attempts to build in biodiversity maintenance through a coarse-filter approach.

What problems has traditional forest management encountered in attempting to manage biodiversity?

Limited historic data and site-specific information concerning disturbance ecology;

- Little experience or interest in establishing forest-level objectives for maintaining ecosystem diversity;
- Inadequate interpretation of relationships between climate, site factors, natural disturbance regime (explicative variables) and forest dynamics and mosaic patterns;
- Few long-term, large-scale monitoring programs in place to evaluate the effects of natural disturbance and different management regimes on various wildlife species;
- Relatively little concern for effects of modifying forest age structure and composition on maintenance of ecosystem diversity (particularly of forests in over-mature age classes); and
- Slow development in integrating natural stand dynamics into silviculture as a means of maintaining stands with attributes similar to those of older stands on the landscape.

(for example), forest management in Canada is coming under increasing scrutiny from local communities, Indigenous people, and national and international environmental groups (Shindler 1998; Weetman 2001).

The role of ecological classification

Ecological information is a key component in developing a natural disturbance based management strategy, and provides a functional level of understanding of the system to

Fig. 11.2. Much of the boreal forest in Canada, Alaska, and Russia is unroaded wilderness, never before logged, in which natural disturbance regimes still prevail, and current forest age-class structures are not necessarily "equilibrial" or "sustainable" at all scales.

Photo courtesy of Millar Western Forest Products Ltd.

be managed. Permanent site characteristics such as slope position, aspect, surficial deposit, and soil moisture regime may interact with local climatic conditions in influencing forest dynamics (Bonan and Shugart 1989; Angelstam 1998). Thus, an understanding of the influence of permanent site variables on forest dynamics must be included in the development of a forest management strategy based on natural forest dynamics (Rülcker et al. 1994). Because forest ecosystem classification (FEC) is an essential tool for forest management, all provinces have developed hierarchical FEC programs that organize ecological information at various scales from the climatic- and physiographic-based supra-regional scales down to the fine-scale site type with its internal vegetation dynamics. Natural disturbance based management attempts to preserve the ecological integrity (see Woodley 1993) of the forests by perpetuating some of the processes historically driven by natural disturbances through the use of anthropogenic disturbances such as forest harvesting and regeneration systems. Thus an essential element in the development of a natural disturbance based management strategy is an understanding of the relationships that exist between natural disturbances and forest dynamics (Chap. 8). Some aspects of gathering and organizing information on regional natural disturbance and forest dynamics are presented in Box 11.4.

The importance of inventory

In order to establish an area-based portrait of the territory to be managed, the spatial representation of the different forest types (and of the disturbance regime that generated them) needs to be quantified. Inventory has always been a big part of forest management because it provides the information necessary for developing a portrait of current forest conditions and the basis on which future conditions are projected. Forest inventories may take on a variety of forms but generally include a mix of remote sensing tech-

---

Box 11.4. How to characterize the natural disturbance regime of a forest.

Forest disturbances can generally be classified into those that operate at the stand or landscape scale ("catastrophic disturbances") and those that operate within stands ("gap disturbances"). The importance of each kind of disturbance, and their selectivity with respect to species and location, provide useful information for the identification of feasible forest management objectives and the selection of silvicultural regimes (Chap. 13). So start out by compiling the disturbance history of your region and seek patterns (associated with terrain, dominant species, and stand age) as to when catastrophic and gap disturbances each prevail:

- Gather all available data on past disturbances and transcribe it to a map; seek out local knowledge and archival records regarding individual disturbance events; obtain all the mapped forest inventory data available;
- Classify mapped ecosystems or forest cover types into even-aged and uneven-aged stands;

Then different procedures are required for describing the parameters of disturbance (see Chaps. 8 and 9) in such a manner that silvicultural interventions can maintain similar patterns, processes, and habitat conditions.

Stand-replacing disturbances

- Emphasizing even-aged stands first to focus on catastrophic disturbances, determine the year of origin (time since disturbance) of each stand, or obtain a random sample of all forest stand types and ages. Time since disturbance is often not the same as stand age (Weir et al. 2000), so scars on living trees and the ages of shade-intolerant species should be used for establishing dates for stand-replacing disturbances.
- Identify the disturbance agent that initiated each stand (e.g., wildfire, insect outbreak, windthrow);
- Aggregate contiguous stands originating from the same disturbance agent at the same time to portray the spatial extent of individual disturbance events (e.g., individual forest fires);
- Characterize each disturbance event in terms of its:
  - causative agent;
  - % survival and distribution of pre-disturbance trees;
  - evidence of understory or soil disturbance;
  - total area;
  - some index of its shape complexity (perimeter to area ratio, sinuosity index, etc.);
  - any species selectivity in the action of the disturbance;
  - any topographic or other landscape selectivity in the action of the disturbance.
- Compile this information in terms of event-return intervals, age-class distributions, and size-class distributions of stand-initiating disturbances, with different distributions for different tree species or topographic locations if such differences are evident.

Box 11.4. (concluded).

Gap disturbances

- Characterize the extent of gap dynamics in a representative sample of all stand types and ages, starting with uneven-aged stands but including even-aged stands of different ages; note mortality agent, gap-making species, gap-filling species, gap dimensions and area (Runkle 1992); and

- Compile this information in terms of gap size distributions (e.g., Coates and Burton 1997) by forest type and age, and by mortality agent or topographic location if such distinctions are warranted.

Collectively, the attributes of stand-replacing and gap disturbance patterns then provide a template for design of a forest management strategy and associated silvicultural systems based on the emulation of natural disturbance.

---

nologies — from aerial photography to satellite imagery — and a range of field inventory techniques. Moreover, while the resolution of many remote sensing technologies is remarkable, field inventory is absolutely essential for validating or "ground-truthing" the interpretation of remote images of ecological units within a landscape or region. Typically mapped as "stands" or "forest cover polygons", these units are generally mapped at a scale appropriate to a particular level of management planning — strategic, tactical, or operational. While not always providing the ideal resolution for forest managers, any given scale of mapping is a compromise among factors, including the costs of the mapping program, data and information management constraints, and the inherent spatial variability of the biophysical attributes that define the mapping units themselves.

Field inventory is both essential to the exercise of mapping and complementary to forest cover maps themselves. Standard provincial forest inventory maps, even if they integrate topographic, site, and hydrological information, do not typically include information on all the resources or attributes of interest. Traditional forest inventory was essentially timber inventory; it was concerned with commercial tree species, stem-diameter distribution, stand density, wood volume, and tree ages. In many instances, it even ignored the utility of taking stock of advance regeneration. In the last two decades, however, there has been considerable development both in enhanced forest resources inventory (Bissonnette et al. 1997) and in habitat suitability or supply models based on inventory data of forest attributes (Verner et al. 1986; Thompson et al 2003; see also Chap. 14). As well, the now universal use of global positioning systems (GPS) and geographic information systems (GIS) has greatly facilitated inventory and the accurate mapping of all resources. Potentially, inventory programs could incorporate, locate, and reference a wide range of attributes important to non-timber forest values too:

- berry and mushroom patches;
- water bodies and sources;
- important foraging, migration, and wintering grounds for ungulates;
- beaver dams and lodges;
- known bear dens and raptor nests;
- endangered or rare species or communities; and
- cultural sites and areas of value for traditional land uses.

When integrated into a structured geo-referenced database, the location and associated documentation of these resources could then easily be accessed and displayed. However, any initiative to develop an enhanced forest resource inventory will invariably be compared with the cost of conducting basic timber cruise inventory. The key is in developing a system through which *useful data* can be rapidly collected in the field and compiled.

Characterizing natural disturbance

The different types of natural disturbances that historically have occurred in a region need to be documented early in the planning process, along with their effects on (and role in) forest dynamics. Box 11.4 provides some guidance in how to undertake this task. Disturbances may range in scale from the death of individual trees to catastrophic landscape-scale wildfires. They may also involve a variety of disturbance agents such as fire, windstorms, flooding, insects, and disease outbreaks. The natural disturbance regime is generally described in terms of each agent's relative importance, both temporally and spatially, and historical magnitude observed in a region (Pickett and White 1985). As well, synergistic aspects of disturbance, for example between different types of disturbance such as insect outbreaks and fire (Bergeron and Leduc 1998) or outbreaks and logging (Radeloff et al. 2000), and the specificity of disturbances to particular species, forest age classes, or landforms (White et al. 1999), should be characterized as best possible. When setting the stage for forest-level management, it is preferable to characterize the natural disturbance regime at a regional level because topographic and climatic conditions often result in distinct disturbance attributes across different regions (Bergeron et al. 2001). The spatio-temporal distribution of different disturbance types and events within a region also needs to be documented in order to provide an explanation for distinctive patterns within the regional mosaic (Rülcker et al. 1994; Angelstam 1998). Furthermore, single disturbance events are rarely uniform in their severity across the affected area, creating a stand mosaic at a finer scale. Studies of specific fire events provide a means of characterizing relative severity within burns and can help to establish relationships between site and landscape features and fire severity (Chap. 8). Finally, disturbance regimes also vary temporally, so they must also be considered in the context of the historic natural range of variability (Bergeron et al 1998; Landres et al. 1999; White et al. 1999). Developing a forest management regime in which silvicultural treatments are designed to reproduce effects similar to or within the natural variability of historic disturbances is a cornerstone of natural disturbance based management.

Natural disturbances influence forest dynamics through direct physical impacts on vegetation and the environmental changes that they induce. Disturbance affects such forest processes as stand initiation or rejuvenation, forest succession, and nutrient cycling. The relationship between the natural disturbance regime and forest dynamics generates a variety of forest types in time and space within a given region. Each of these forest types can be characterized by certain attributes to which the forest flora and fauna will respond. Knowledge of these attributes, particularly those related to propagation and establishment, and the processes that generate them is key to the development of any management strategy that aims at maintaining the integrity of the forest ecosystem (see Chap. 9.) An understanding of the mechanisms through which disturbances affect these attributes and processes is essential to the successful development of a natural disturbance based forest management strategy and in developing models to predict the effects of these strategies on timber and other values (Chaps. 12 and 14). These concepts are fleshed out in terms of a management strategy for black spruce (*Picea mariana*) – feathermoss forests in Box 11.5.

As mentioned, description of the current forest condition for the management unit (Fig. 11.1) includes documentation of the natural disturbances and forest dynamics that characterize it. It is widely recognized, however, that virtually no part of the planet remains entirely unaffected by anthropogenic factors, and Canada's boreal forests are no exception, despite their image as untouched wilderness. Some human-related disturbances may be largely local or regional in nature and relatively easy to document: statistics generally exist on forest harvesting, renewal, protection, road construction and other management activities, number and area of fires ignited by people, and hunting and trapping harvests. The consumption of other non-timber resources (wild fruits, mushrooms, other plant resources) that may affect stand dynamics are probably more difficult to document. Other human-related influences, including emissions of atmospheric pollutants such as CO_2, CH_4, and other greenhouse gases, affect forest dynamics indirectly through climate change on a planetary scale (Chap. 20). Certain aerial-borne pollutants, including sulphur and nitrogen compounds and heavy metals, may adversely affect productivity of both aquatic and forest ecosystems on regional or supra-regional scales (Zasada et al. 1997), particularly in areas with low buffering capacity (McLaughlin 1998). Moreover, Weber and Flannigan (1997) anticipate the impacts of climate change on fire regime, including effects on annual area burned, fire frequency and severity, to be more significant than their direct effects on species distributions, migrations, and extinction. Given that climate change is affecting the dynamics of the very forest ecosystems that we are managing, anticipating future climate effects on forest development is probably as important as understanding the natural variability of past disturbances, and it underscores the necessity of incorporating flexibility in establishing strategic objectives.

Strategic forest management objectives — *where do we want to go?*

Describing the desired future forest and the values we expect to be sustained in it constitutes one of the most crucial steps of the strategic forest management planning process. This step establishes the basis upon which all subsequent decisions are made and upon which all forest management actions are designed. Ultimately — or ideally — it should correspond to the society's definition of the roles forest ecosystems currently serve, and the roles desired for those ecosystems in the future. As such, this step is not limited to the strategic planning process of a given forest management unit, but usually starts at higher levels of decision making. For example, "sustainable forest management" is a strategic objective set in the forest legislation and policies of most provinces as well as in the National Forest Strategy (CFS 1998). This policy contrasts with policies of forest liquidation or conversion to farmland, which have (at various times and places) reflected the alternative priorities of governing bodies (Chap. 1). The six general

Box 11.5. A natural disturbance based forest management strategy for the black spruce – feathermoss forests of northwestern Quebec.

This strategy aims to implement the forest management concept proposed by Bergeron et al. (1999). The approach serves as the basis for the management plan for the Lac Duparquet Research and Teaching Forest (as described in Chap. 22), and for a pilot project by forest products companies in the region. The assessment of its implementation is carried out in collaboration with Sustainable Forest Management Network partners (Québec Ministère des Ressources naturelles and Tembec Forest Products) in a pilot territory located about 150 km north of La Sarre, Quebec, and roughly covering 4750 km^2 (49°37′30″–50°22′30″N and 78°30′00″–79°30′00″W).

Studies examining the regional natural disturbance regime (Bergeron et al. 2001, 2002) have demonstrated that wildfire frequency, wildfire size, and wildfire severity are all characterized by a certain range in variability. For example, fire frequency within the study area has changed over the last 300 years, resulting in an elongation of the fire return interval: 80 years before 1850, 150 years between 1850 and 1920, and 325 years since 1920. As for fire size, historic records indicate that fires ranging in size from 950 to 20 000 ha could be considered characteristic for the black spruce zone of western Quebec. However, it is important to note that although they account for less than 10% of the fire events, fires of 1000 ha or more generate close to 90% of the burned area. Finally, fire severity maps have shown that although the proportion of unburned forest (residual "islands") is relatively constant (5%), the proportion of partially burned forest varies from 30 to 50%, while 45–65% of the forest is severely burned. The unexpected prevalence of partially burned forest in the natural landscape suggests excellent opportunities for the successful use of silvicultural systems based on partial cutting.

Studies conducted in the area have documented the natural forest dynamics resulting from the previously described disturbance regime (Harper et al.[4]). For example, the mean stand age of 139 years calculated for the region suggests that, at least theoretically, 52% of its area is composed of forests that have burned more than 100 years ago, of which 22% consists of forests that are older than 200 years. In the extended absence of fire, stands that initially regenerate to relatively short-lived and shade-intolerant species (jack pine, *Pinus banksiana,* and trembling aspen, *Populus tremuloides*) generally exhibit compositional changes as these species are replaced by black spruce (Fig. 11.3). In stands that initially regenerate to black spruce after fire, composition tends to remain stable, but structural changes such as decreasing canopy height and reduced stem density occur through time (Fig. 11.4). The onset of paludification in old-growth forests is also an example of the influence of the altered regional disturbance regime on ecosystem processes. Finally, study results have shown that the forest flora and fauna tend to respond (mostly in abundance) to changes in stand structure, although relatively few non-vascular plant or bird species were found to be exclusive to old-growth stands (Boudreault et al. 2002; Drapeau et al. 2003).

---

[4]Harper, K.A., Boudreault, C., DeGrandpré, L., Drapeau, P., Gauthier, S., and Bergeron, Y. 2003. Structure, composition and diversity of old-growth black spruce boreal forest of the Clay Belt region in Québec and Ontario. Environ. Rev. In press.

Fig. 11.3. Natural disturbance and forest dynamics model for the black spruce – feathermoss forest of northwestern Quebec. BS, black spruce; JP, jack pine; TA, trembling aspen; cc, clear cutting; pc, partial cutting; p, plantation; s, scarification.

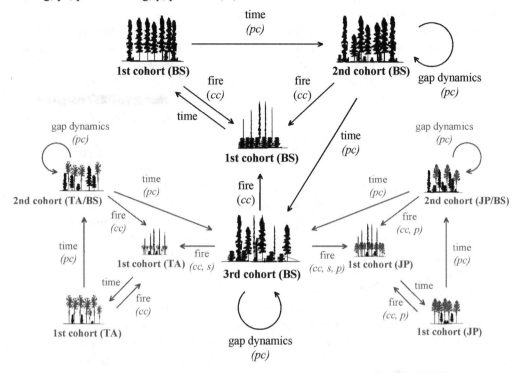

Fig. 11.4. Photographs portraying stands at each stage of the three-cohort model of natural stand development in black spruce – feathermoss forests. (*a*) First-cohort, consisting of an even-aged population of trees; (*b*) second-cohort, with an irregular stand structure as the first cohort dies off; and (*c*) third-cohort, self-maintaining old-growth forest dominated by gap dynamics.

SFM criteria established by the Canadian Council of Forest Ministers (CCFM 1995; see Box 2.2) constitute broad strategic objectives set by society — or at least are considered to be endorsed by current social values — which must be integrated into the strategic management objectives of a given forest management unit.

The preparation of the strategic plan for a given forest management unit requires that specific objectives (see Box 11.6) be set for that particular territorial unit. When considering all of the benefits society obtains from forest ecosystems, the initial list of potential management objectives may be extensive. For example, here are some of the subject areas that need to be considered when defining the management objectives for a given forest management unit (FMU) as provided by the *Alberta Forest Management Planning Manual* in its guidelines (AEP 1998):

- timber objectives (coniferous/deciduous/mixedwood management, sustainability, harvest priorities);
- silvicultural objectives (reforestation, growth rates, yields);
- forest protection objectives (fire and fuels, insects and diseases);
- ecological objectives (biodiversity, forest connectivity, ecological integrity);
- watershed management objectives (water quality and quantity, erosion, siltation, flooding, riparian zone management);
- fish and wildlife objectives (habitat, access);
- aesthetics and recreational objectives; and
- socio-economic objectives (commercial opportunities, social benefits).

Another list of land uses and land management objectives, and their associated requirements, impacts, and possible conflicts, is provided in Table 11.1, using the example of the Lac Duparquet Research and Teaching Forest in northwestern Quebec. Note that some uses of forest land (e.g., mineral exploration and mining) may be outside the jurisdiction of forest managers and a forest management plan. Yet these uses too must be addressed in a manner analogous to higher level policy direction, since overlapping tenures and resource access rights exist in many forms.

Some of these management objectives are complementary while others may be conflicting (as touched on briefly in Chap. 22). Hence, to facilitate the development of a congruent management strategy for the FMU, specific priorities must be assigned to the different objectives. It is useful to compile a matrix scoring the compatibility of different forest values. A simple example of this exercise from the Lac Duparquet Forest is presented in Table 11.2. Dealing with the incompatibilities then poses a challenge. One solution may be to reflect local or regional priorities and promote particular resources accordingly, but implementation of this approach typically compromises sustainability of the less tangible forest values. Another strategy is to zone the management of some

---

Box 11.6. Objective (definition):

"a concise, measurable statement of a desirable future condition for a resource or resource use which is attainable through management action".

British Columbia Ministry of Forests
"Guide to Writing Resource Objectives and Strategies"
http://www.for.gov.bc.ca/hfp/planning/writing/page45.html

Table 11.1. Identification of forest management objectives and values: an example describing the uses (values) of the Lac Duparquet Research and Teaching Forest, Quebec.

Use (value)	Extent of current use	Particular needs	Environmental impacts	Constraints on other uses	Constraints caused by other uses	Solutions to potential conflicts
Research in ecology, forestry	• Very important • 30–40 people/ summer • 12+ participating institutions	• Access to the area • Short- to long-term protection of study sites • Minimal infrastructure for lodging and work	• Generally minimal • Destructive sampling of trees, vegetation, soil removal • Flagging of sites sometimes abundant; occasional abandonment of research equipment (seed/ litter traps, posts, etc.) • Concentration of activities in some sites could have an environmental impact (trampling, destructive sampling)	• Could negatively influence potential for tourism activities if strict site protection is required • Scientific activities could increase eco-tourism potential • Possible conflict with archaeological research activities in the conservation zone	• Forest harvesting and other management activities can conflict or be complementary • Cottage development could be incompatible • Uncontrolled access could destroy or damage research sites • Mining activities (drilling exploration, and development) generally incompatible	• Assure planning and integration of research and forest management activities • Restrict cottage development in the conservation zone • Encourage all-terrain vehicle (ATV) and snowmobile use on marked, designated trails • Signage on research sites • Provide GPS locations and decriptions of all research sites to QMNR, Lands & Mines Sections • Seek protection status for Conservation Zone
Archaeological research	• Important • ca. 50 sites located in the research forest and on islands of Lake Duparquet	• Access to potential sites • Location and integral protection of sites	• Impacts of digs are important but restricted to relatively small areas	• Any disturbance (forest) harvesting, drilling, cottage develoment) can potentially reduce the integrity of sites • Potential for tourism development • Archaeological activities support the justification for the Conservation Zone	• Little compatibility with acitivites that disturb sites	• Precisely locate all known sites • Moratorium on forestry and mining activities and recreational development in proximity to sites • Public awareness campaign concerning protection of archaeological resources

Table 11.1. (*continued*).

Use (value)	Extent of current use	Particular needs	Environmental impacts	Constraints on other uses	Constraints caused by other uses	Solutions to potential conflicts
University- and college-level teaching	• Increasingly important but seasonal	• Access to the forest • Diversity of ecological and forest conditions in natural and managed forests • Minimal infrastructure for teaching/laboratory activities	• Minimal	• See research	• Mining acitivites (especially exploitation)	• No problem anticipated
Environmental monitoring and assessment	• Important in the Conservation Zone and on some islands on Lake Duparquet	• Long-term protection of sites • Minimal local human disturbance	• Minimal	• Incompatible with all uses that disturb the local environment • Possible conflict with archaeological research in Conservation Zone	• Forest harvesting • Mining activities • Recreational development • Wild berry, mushroom harvesting, etc. • Toursim activities if access restrictions are not respected	• Provide GPS locations and descriptions of all monitoring sites to QMNR, Forest, Lands and Mines sections • Exclude incompatible activities • Site signage • Seek protection status for Conservation Zone
Forest harvesting and silvicultural activities	• Important since 1998 • Annual harvest of about 75 ha	• Access to forest • Productive (or potentially productive) forests	• Multiple (habitat loss and fragmentation; modification of forest age structure and composition; possible impacts on soils and hydrology) • Forest regulations to be respected	• Ecological monitoring and assessment • Archaeological research • Influences on hunting, trapping, tourism potential • Potential for harvesting certain wild plants may diminish while wild blueberries and raspberries tend to increase	• The importance of maintaining site integrity for monitoring, basic research, and archaeological work make forestry activities incompatible on certain sites	• Restrict forestry activities to the Management Zone and concentrate the other conflicting acitivites in the Conservation Zone • Minimize dispersion of harvesting zones to reduce fragmentation of the forest mosaic • GPS — locate traplines • Respect environmental regulations

Table 11.1. (*continued*).

Use (value)	Extent of current use	Particular needs	Environmental impacts	Constraints on other uses	Constraints caused by other uses	Solutions to potential conflicts
Hunting	• Very important • ca 30 hunting camps in the Lake Duparquet Forest • Itinerant hunting is also important	• Abundant wildlife habitat • Healthy populations of wildlife (moose and ruffed grouse in particular • Minimum density of hunters (isolation)	• Minimum • Wildlife kill • Some camps in disrepair • Some illegal harvesting of fire wood	• Other activities in the forest are not recommended during moose hunting season • Potential to use camps off-season for recreational purposes	• New road construction increases hunter densities and hunting pressure and reduces the quality of the hunting experience for resident hunters • Forest harvesting can temporarily reduce visual landscape quality, reduce or improve wildlife habitat quality, and affect wildlife behaviour and hunting potential • Mining activities (especially mine development) destroys habitat	• See forest harvesting • Consider location of hunting camps when planning roads and harvest blocks • Respect regulations or increase the perimeter of forest cover around camps • Explore models for minimizing environmental impacts of mining activities
Trapping	• Important • 6 traplines included in the Lake Duparquet Forest	• Maintenance of healthy populations and good habitat for beaver, muskrat, mink, weasel, pine marten, fisher, otter, lynx, wolves • Exclusivity of trapping rights in the area	• Direct impacts on fur-bearing animal populations • Control of beaver populations (keystone species) could limit the influence of this species on habitat, surface drainage, and other species • Possibility of killing non-target species	• Possibility of creating a monitoring network of trapping harvests to evaluate impacts of forestry activities on wildlife populations	• Habitat loss or fragmentation by road construction, forest harvesting • Bear hunt and trapping compete for the same resource • Mining activities (especially mine development) destroy habitat	• Increase buffer strip widths on streams, rivers, and lakes • Minimize harvesting dispersion to reduce fragmentation • GPS — locate traplines • Respect regulations

Table 11.1. (*continued*).

Use (value)	Extent of current use	Particular needs	Environmental impacts	Constraints on other uses	Constraints caused by other uses	Solutions to potential conflicts
Recreational (cottage) development	• Restricted • Some cottages on Lake Hébécourt and the western shores of Lac Duparquet	• Solid, well-drained sites • Access by car or boat	• Generally local • Possibility of illegal cutting of wood, trail-making, modification of shoreline habitat, inadequate septic facilities	• Can negatively impact research and monitoring activities, hunting, and trapping and create conflicts with forestry and mining	• Incompatible with mining activities	• Restrict cottage development in the Lac Duparquet Forest • Encourage responsible activities associated with remote cottages • Possibility of modifying forest interventions (partial cutting, slash cleaning, etc.) close to cottage zones
Fishing and pleasure boating	• Very important on Lakes Duparquet and Hébécourt	• Healthy fish populations • Clean environment • Unrestricted water access • Road access to boat launch sites	• Pressure on fish populations • Water pollution • Noise • Illegal cutting of firewood during ice-fishing season	• May influence potential of non-motorized activities, particularly canoe-camping	• Mining can destroy habitat • Forest harvesting, road construction can potentially impact fish habitat and spawning sites	• Public awareness campaign concerning illegal wood harvesting and respect for the environment
Snowmobiles and ATV's ("quads")	• Moderately to very important	• Groomed or ungroomed trails	• Possible disturbance of wildlife populations • Noise and air pollution	• Non-motorized outdoor activities (eco-tourism) • Can damage research sites and experimental installations	• Forest harvesting can reduce quality of the visual landscape, but this group of forest users is generally less sensitive to cutovers than other users	• GPS — locate trails and encourage their use to reduce free-roaming movement
Eco-tourism, hiking, mountain biking, back-country snow-shoeing, skiing	• Currently limited but good potential	• Attractive natural or semi-natural environment • Tranquility • Minimum of infra-structures	• Minimum • Possible erosion on frequently used or poorly maintained trails	• May present conflicts with other uses (mining, forestry)	• User group most sensitive to forest harvesting and associated development	• Promote these activities especially in selected parts of the Conservation Zone and sectors that will not be harvested in the near future

Table 11.1. (*concluded*).

Use (value)	Extent of current use	Particular needs	Environmental impacts	Constraints on other uses	Constraints caused by other uses	Solutions to potential conflicts
Mineral exploration and development	• Important in the past • Some potential for new activity (gold, zinc, copper)	• Exploitable ore bodies	• Very important • Habitat loss and contamination, soil and water contamination, air pollution, noise • Mine spoils require reclamation, often bear toxic concentrations of heavy metals, and can generate acidic runoff or leachate	• Generally incompatible with other land uses	• Designation of Conservation Zone could limit mining development in this zone	• Explore models to minimize environmental impacts of mineral exploration and development • Request nullification of claims in part of the research forest • GPS — locate all important sites of archaeological interest, wildlife habitat, ecological research, and monitoring sites and provide this information to QMNR, Lands, Mines and Forests, and Wildlife sections
Harvesting of wild mushrooms, plants	• Limited	• Knowledge of potential habitat • Access to the forest	• Minimal (high potential could be over-exploited)	• Minimal	• Forest harvesting can be favourable or unfavourable • Mining activities are incompatibles	• No problem anticipated

bundles of compatible values into particular resource emphasis areas or management intensity zones. A compatibility matrix such as that presented in Table 11.2 helps assess the need for a zoning strategy, and suggests which land uses, resources, and management approaches can be conducted in shared zones. Zoning is further discussed in Chap. 12.

The priority assigned to each objective within the management strategy for the FMU will be influenced by provincial and regional forest management strategies, which have often addressed similar issues at a larger spatial scale. For example, in Ontario the "Lands for Life" land-use planning process has resulted in a land-use strategy that prioritizes different management objectives for different land-use zones. Ecological management objectives are prioritized in territorial units assigned various degrees of protection, specific socio-economic objectives are prioritized in enhanced management areas, whereas integrated forest resource use is prioritized in general-use areas. Forest management units generally consist of a combination of the latter two land-use designations, although they may also include or are often adjacent to protected areas too. Thus, this provincial land-use strategy provides a basis for the clarification, geographical limitation, and prioritization of the management objectives of each FMU, that is, of each geographical area in which one or more forest products companies must plan and manage for sustained timber yield and other values. Nonetheless, this approach still relies on zoning and mapping of management priorities among zones within a FMU. Similar provincial and regional land-use planning processes and strategies influence the preparation of forest management plans in several Canadian provinces (e.g., Land Use Plans and Land and Resource Management Plans in British Columbia). Although traditionally based on political, socio-economic, and administrative considerations, provincial and regional land-use policies and strategies increasingly recognize the importance of a sound ecological basis (such as the need for protecting representative and distinctive landscapes, with spatial connectivity among them, or managing on a watershed basis) and the need for public participation in the provincial, regional, and local planning processes (see Chap. 4).

Sustainable use of the timber resource is usually the first objective enunciated by forest management regulations. Ostensibly, any consumptive use of forest resources then requires that ecological processes and the integrity of forest ecosystems remain intact. Questions concerning ecological monitoring (which processes? which indicators of integrity?) are not addressed here, but it is clear that this requirement represents a global constraint on strategic management. As uncertainty and risk are inherent realities of long-term forest management planning, formulation of strategic objectives involves making numerous assumptions concerning stand and forest dynamics and silvicultural treatment effects on these dynamics. The imperative of adopting an adaptive management approach (Holling 1978) which integrates a design for learning and modifying management within a monitoring framework is treated in Chap. 21. The conservation of biological diversity, the maintenance and enhancement of forest ecosystem condition and productivity, and the conservation of water and soil resources are criteria of sustainable forest management (CCFM 1995) that can serve as general forest management objectives. Specified value ranges for some of the indicators associated with these criteria can serve as more detailed management objectives. For example, a desired proportion and area occupied by specific forest types or (older) age classes, a specified

Table 11.2. Assessing the relative compatibility of different land uses (values): an example from the area of the Lac Duparquet Forest, Quebec. ✓, generally compatible; ≈, may require measures to harmonize activities; ∅, incompatible (requires zoning).

	Forestry	Mining	Research, teaching, monitoring	Archaeol. field research	Hunting, fishing	Trapping	Cottage development	Snowmobile, ATV use	Eco-tourism	Wild plants, berry harvesting
Forest harvesting, silviculture		∅	≈	≈	≈	≈	≈	≈	≈	≈
Mining exploration & development			∅	∅	∅	∅	∅	∅	∅	∅
Research, teaching, environmental monitoring				≈	✓	✓	∅	≈	✓	✓
Archaeological field research					✓	✓	∅	✓	✓	✓
Hunting, fishing						✓	≈	✓	✓	✓
Trapping							≈	✓	✓	✓
Recreational (cottage) development								✓	≈	✓
Snowmobile, ATV use									≈	✓
Eco-tourism										✓

Box 11.7. Setting objectives for the black spruce – feathermoss forest pilot territory.

A pilot project for natural disturbance based forest management described in Box 11.5 includes portions of the two Forest Management Units managed by Tembec Forest Products and Nexfor-Norbord Industries. This pilot project represents a "scaling up" of an experimental forest management approach to an operational scale in a forest industry context. One of the primary management objectives for the pilot territory is a sustained timber supply. Maintaining forest productivity is also an important strategic objective, given the distinctive nutrient dynamics (paludification and the abundance of black spruce [*Picea mariana*] bog forests) of the Northern Claybelt region of northwestern Quebec and northeastern Ontario (Boudreault et al. 2002). In the context of sustainable forest management, these objectives also imply maintaining major forest ecosystem processes such as stand regeneration, succession, and nutrient cycling.

Non-industrial uses of the forest resources in the pilot territory mostly include recreational uses by local residents or traditional activities by First Nations communities (angling, hunting, trapping, berry picking). Outfitting operations, tourist-based recreational activities, and cottages are relatively rare. Nonetheless, social acceptance of the forest management practices is particularly important at the local level. Public support is probably more dependant on strategic objectives set in terms of enhancing or maintaining sustained levels of these other extractable forest resources rather than considerations for aesthetics or national/international concern for biodiversity or wilderness protection.

In the region, there is currently very little land area assigned any degree of protection, although the Quebec government has recently initiated a program to set aside at least 8% of the provincial land base for protection by 2005. Conservation of regional biodiversity is considered a high-priority strategic objective for the pilot territory. This entails maintaining the full array of forest habitats necessary to the flora and fauna of the region. Ecological studies conducted in the area indicate that special attention should be provided to recently burned (but unsalvaged) stands, over-mature forests, and true old-growth forests. The first two stand types constitute important habitat for specific bird communities (Drapeau et al. 1999, 2001), while the latter may be important reproductive habitat for woodland caribou (*Rangifer tarandus caribou*), a species listed as "of special concern" in Quebec. The absence of large tracts of protected forests in the region as well as the precarious status of certain forest species (e.g., American marten [*Martes americana*], woodland caribou) also underscore the need for such strategic objectives as limiting habitat fragmentation and access (by humans, natural predators, and competitors) to certain areas of the pilot territory.

Given that this area was established to assess the implementation of a management strategy based on natural disturbance and forest stand dynamics, a major strategic objective is to maintain, through forest management, a forest mosaic similar to one that would result from the regional natural disturbance regime (Bergeron et al. 2002). More specifically, this implies:

(1) maintaining the proportion of the area occupied by different stand types at levels similar to those that would be observed under a natural disturbance regime;

(2) establishing spatial and temporal distribution patterns for anthropogenic disturbances (mostly timber harvesting) over the area that would resemble those that would be observed under a natural disturbance regime; and

(3) maintaining the heterogeneous nature of disturbance severity observed within single natural disturbance events.

Box 11.7. (concluded).

The actual values assigned as management objectives in regards to the different indicators associated with these objectives are derived from the natural disturbance regime characterizing the region, as described in Box 11.5.

As an example, the desired proportion of the pilot territory occupied by each of the 10 stand types represented in the forest dynamics model (Fig. 11.3) can be based on the fire cycle observed in the region and the transition age between the different cohorts, as suggested by Bergeron et al. (1999). Figure 11.5a represents the expected age-based representation of the different successional series in the pilot territory. It was derived from the negative exponential curve (Van Wagner 1978) that would characterize the age representation expected for a region where all stands are equally susceptible to fire and the mean stand age is of 139 years. It also incorporates the expected representation of the different successional series (Gauthier et al. 1996). Figure 11.5b portrays an age-based representation of the 10 stand types of Fig. 11.3 (early and late cohort 1 are grouped) that would also produce a mean stand age of 139 years for the pilot territory. The transition age between the first cohort and the second cohort was established at 150 years, and between the second and third cohort at 275 years. The age distribution of each cohort was normalized over the life span of each successional stage.

Bergeron et al. (1999) and Harvey et al. (2002) suggest that this distribution be used as the strategic objective for the desired representation of the different stand types, as it approximates the distribution that would be observed under a natural disturbance regime. Thus, one of the strategic objectives for the pilot territory is to maintain the following area-based proportions for the different stand types:

- 3% of 1st cohort — TA (trembling aspen),
- 1% of 2nd cohort — TA (trembling aspen),
- 6% of 1st cohort — JP (jack pine),
- 2% of 2nd cohort — JP (jack pine),
- 59% of 1st cohort — BS (black spruce),
- 16% of 2nd cohort — BS (black spruce), and
- 13% of 3rd cohort — BS (black spruce).

---

level of fragmentation and connectivity of forest ecosystem components, minimal population levels and changes over time of selected species and species guilds, all constitute management objectives that can be set at the level of the FMU. A natural disturbance based management regime should aim at establishing the desired value ranges for these criteria within the natural historical range of variation. Hence, information on regional forest ecosystem classification and historic patterns (including spatial distribution and severity) of natural disturbance, particularly at the regional level, should provide the disturbance and ecological framework for targeting ecosystem diversity as a forest-level management objective.

Notwithstanding the effects of atmospheric pollution and climate change, an adequate representation of protected areas is considered essential for measuring the effects of forest management interventions. The creation of a network of natural benchmarks is intended to serve as a reference with which to evaluate management objectives and the effects of forest management practices. These reference areas may be established on a regional level as part of a regional strategy or they may need to be set within the forest management unit itself, especially in the case where they are absent (or particular forest types are not represented) at the regional level. Strategic management objectives set at the local level (i.e., for the FMU) are influenced by objectives established at the provincial and regional levels. In some instances, provincial or regional policies actually provide guidelines on the strategic management objectives to be set within the FMU. For example, the Ontario Ministry of Natural Resource's *Policy Framework for Sustainable Forests* states that:

> *"The long term health and vigour of Crown forest should be provided for by using forest practices that, within the limits of silvicultural requirements, emulate natural disturbances and landscape patterns while minimizing adverse effects on plant life, animal life, water, soil, air and social and economic values, including recreational values and heritage values."*

Section 1(3)2, Ontario Crown Forest Sustainability Act (1995)

Translating these general "motherhood" principles into concrete management objectives can be difficult. The government of B.C. has provided guidelines (available at http://www.for.gov.bc.ca/hfp/planning/writing/pagei.html and at http://www.gov.bc.ca/srm/down/sustainable_resource_management_planning.pdf) to assist in setting land management objectives. Box 11.7 walks through the process of evaluating the sorts of considerations explored above, and setting them in the context of a natural disturbance model, to devise specific landscape objectives for a pilot territory in northeastern Ontario and northwestern Quebec.

Development of forest management strategies — *what are some different ways of getting there?*

Once strategic management objectives have been enunciated, the next step in the planning process is the development of specific forest management strategies to attain these objectives. This step basically requires the definition and documentation of forest interventions, the definition of silvicultural systems to be applied to different forest types, and finally the definition of forest management strategies. There exists a hierarchical relationship between these elements, as a silvicultural system can be viewed as the temporal organization of different interventions within a stand (as elaborated upon in Chap. 13), and management tactics involve the spatial arrangement and scheduling through time of different silvicultural systems within a management unit (as per Chap. 12). But a management strategy consists of an overall approach for manipulating the forest landscape, for its inventory, forest health/protection, ecological preservation, and the identification of particular resources to be managed, enhanced, and protected.

While natural disturbance and forest dynamics act at scales from single organisms to entire landscapes, many of the dynamic processes involved in disturbance (such as stand initiation or rejuvenation, succession, and nutrient cycling) are often perceived at the stand level. Scaling up from the stand- to the landscape-level is an important part of developing a forest management strategy in that it involves understanding and managing the landscape dynamics that will ultimately result from stand-level interventions.

Fig. 11.5. Determining the desired age-based representation of stand types to be set as a strategic management objective. (*a*) Negative exponential; the theoretical age-class distribution of stands subject to random burning with a mean stand age (MSA) of 139 years. (*b*) Cohort-based age structure; a three- cohort forest management model, also having a MSA of 139 years. BS, black spruce; JP, jack pine; TA, trembling aspen.

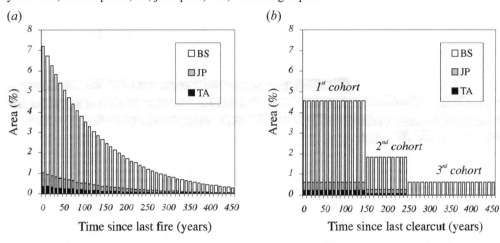

Andison and Kimmins (1999) provide a good example of the need for scaling up for the boreal mixedwoods in British Columbia. Although stand-level interventions need to be planned in detail at the operational level (Chap. 13), they also need to be considered at the strategic level because they are the means of implementing the strategy. In other words, strategic planning mainly establishes the management objectives and the course of actions necessary to reach them. Given that these actions are ultimately translated into stand-level interventions, it is necessary to have a general knowledge and understanding of all possible and available interventions in order to set a proper course of action. More specifically, it is necessary to identify which strategic objectives can be met with the use of different intervention types and to plan where, when, and how the various interventions should be applied. In the context of a management strategy based on natural disturbance and forest dynamics that aims to maintain ecological integrity, this requires ecological information about the effects of forest interventions on natural processes and stand dynamics (Lindenmayer and McCarthy 2002). Chapters 9 and 10 represent such an exercise as it relates to the impacts of forest harvesting on the terrestrial and aquatic components of the boreal forest ecosystem. Chapter 13 provides information on the effects of forest interventions on stand dynamics and how different interventions can be used to build different silvicultural systems. The deployment of these interventions in space and time is the topic of Chap. 12.

A silvicultural system is the combination and application of the different silvicultural interventions necessary to achieve specific management goals. Although traditionally defined on the basis of timber management objectives, they can also be defined on the basis of ecological objectives or any of the other forest values previously mentioned. Examples of the timber-based systems are composition-based silvicultural systems (e.g., coniferous, hardwood, or mixedwood management), age-based silvicultural systems (even-aged or uneven-aged management), and regeneration-based silvicultural systems (single-pass, two-pass, and multiple-pass management, with natural or artificial regeneration). Examples of ecologically motivated systems are structure retention based systems (full/green retention, variable retention, and no retention management systems), or the process-based systems described in Chap. 13 (natural, semi-natural, plantation, and intensive management systems).

As stated above, although silvicultural systems are generally defined on the basis of stand-level management objectives, they must also be considered in strategic planning as they represent the means through which the strategic objectives are reached. This involves identifying which of the strategic objectives are to be met by applying different silvicultural systems. In the context of a natural disturbance based management strategy that aims to maintain general forest processes, this implies comparing the forest dynamics resulting from different silvicultural systems to those observed with natural disturbance dynamics (Engstrom et al. 1999). If ecosystem patterns rather than processes are the focus and maintaining structure and composition of older forest types on a portion of the managed landscape is the objective, testing partial cutting or selective silvicultural systems that allow some stem extraction while leaving those structural and composition attributes that are considered to be of importance in old-growth forests has been proposed. Acquired knowledge on natural disturbance and forest dynamics can actually be used as an inspiration for the development and definition of silvicultural systems (Lieffers et al.1996; Bergeron and Harvey 1997; Bergeron et al. 1999, 2002).

Obviously, more than one management strategy may exist for attaining the same objective, in which case the implications of different alternatives for other resources or management activities will have to be evaluated. Because forest management inevitably involves making decisions about trade-offs, forest-level modelling is an essential tool in building strategic scenarios and comparing how different options may play out for different resource values (see the section on strategic planning models in Chap. 14). Meeting minimum area objectives of old-growth forest may be achieved by designing longer forest rotations for a portion of the land base or by establishing permanent protected areas. In the latter case, the "SLOSS debate" — single *l*arge or *s*everal *s*mall reserves — constitutes a further variation on "How to get there?" Or if increasing wood production is a priority, a forest products company might choose to increase silvicultural investment generally, or to zone a specific portion of its FMU for intensive, high fibre production management.

The definition and documentation of forest interventions and silvicultural systems provides the basic toolbox with which the forest management strategy can be developed. As advocated in this chapter, the strategy mainly consists in choosing or developing silvicultural systems that will facilitate attainment of the strategic objectives set in the previous step of the planning process. However, because silvicultural systems vary in their potential to achieve different strategic objectives, several silvicultural systems will inevitably be required to achieve strategic objectives related to maintaining forest type diversity or a mosaic of stand structures and age classes within the forest landscape. Using a variety of silvicultural systems creates a variety of treatment outcomes and is thus apt to produce more of the variability that is characteristic of regions driven by natural disturbance (Bergeron et al. 2002). The definition of the strategy is not only based on the nature of the silvicultural systems to be used, but also must consider the temporal and spatial distribution of these systems over the management unit. These considerations are necessary for the evaluation of the strategy in terms of temporally and spatially explicit strategic management objectives, e.g., establishing a managed forest landscape configuration that maintains critical features of landscapes created by natural disturbance dynamics, maintaining continuous areas of mature, over-mature, and old-growth forests over the landscape, limiting general and (or) permanent access in all areas of the management unit, protecting areas of special concern, etc. The inclusion of spatial considerations in the development of the management strategy is also necessary for management units that do not cover a homogeneous landscape, as the potential for different silvicultural systems may vary geographically within such units. For this reason zoning may constitute one possible approach to developing a strategic plan. Many of the spatial and temporal considerations of strategic planning are linked to aspects of tactical planning. Chapter 12 provides a more detailed discussion of the spatial aspects that need to be considered at both the forest and landscape levels.

The variety of defined silvicultural systems and their many possible combinations in terms of nature, space, and time allow for the elaboration of a multitude of potential management strategies for a given forest management unit. These different strategies will demonstrate the varying capacity to meet the objectives established for the forest management unit. The choice of the most appropriate strategy is the topic of the next section.

Assessment of forest management strategies — *what is the best way of getting there?*

The previous steps of the strategic planning process have established the strategic objectives for the management unit and devised alternative strategies describing different potential courses of action to fulfil them. The next step is to decide which strategy will actually be implemented in the forest management unit or in different portions of the management unit. The selection of planning indicators and the development and use of strategy assessment tools (Chap. 14) will facilitate the choice of the strategy that is best suited for the specified strategic objectives.

As previously discussed, the definition of strategic objectives in the planning process generally starts with broad goals that are then translated into more specific objectives that can be targeted and measured. The maintenance of minimum population levels of selected species or species guilds is a common objective of this sort. In the process of selecting the strategy (combination of silvicultural systems and how they are allocated in time and space in the forest) to be implemented in the forest management unit, the expected success at achieving each stated management objective must be assessed. However, as it is mostly based on silvicultural considerations, the definition of the management strategy provides little direct information on the status of other ecological elements such as changes in animal population levels. Thus the link between the forest management strategy and other strategic objectives must be made indirectly through the forest components on which the strategy directly acts. In this case, these may include such elements as the quantity and quality of available habitat (or residual habitat in extensively harvested areas), or habitat connectivity, attributes which are often tabulated at the landscape level (Chap. 12). The impact of alternative management strategies on indicators identified in the FMU objectives can then be directly measured, and can serve as the basis for selecting the preferred option. Kneeshaw et al. (2000) and Yamasaki et al. (2001) have called such elements "planning indicators", as they can be readily used in the strategic planning process. Examples of planning indicators provided by Kneeshaw et al. (2000) and Yamasaki et al. (2001) include:

- age-class structure of the forest;
- tree species composition of the forest;
- configuration of the forest; and
- road density.

The selection of "indirect" planning indicators requires a sound knowledge and understanding of the links that exist between the different elements of the forest ecosystem. This ecological information is also useful for the definition of "direct" planning indicators in forest management strategies based on natural disturbance and forest dynamics. The definition of direct and (or) indirect planning indicators for each of the strategic objectives set for the management unit is a prerequisite for the adequate assessment of all potential strategies.

Planning indicators provide a basis upon which alternative management strategies can be assessed and compared. Portraying the anticipated spatial and temporal distribution of forest conditions, articulated in both descriptive and quantitative terms, is a fundamental step in the selection of management strategies. As an example, the dispersion of different silvicultural approaches for management of the black spruce – feathermoss forests is explored in Box 11.8. However, the analysis of the potential impact of each strategy on the different indicators often requires the consideration of a large number of variables and analysis and synthesis of a considerable amount of information. In this respect, computerized decision support systems are invaluable strategic planning tools that allow the integration of a large number of variables in the representation and projection of the system to be managed (Chap. 14). Forest simulation models that are of particular interest in strategic planning are those that can cover the temporal and spatial horizons associated with full implementation of the management strategy, as well as those that can simultaneously integrate many of the planning indicators associated with the strategic objectives set for the management unit; see Box 11.9 for an example. In general, more than one model is used in order to cover all of the planning indicators necessary to assess the alternative strategies. These models are used to evaluate the capability of each alternative strategy to address the different strategic objectives (Fig. 11.5). In the case where more than one alternative strategy meets the selection criteria, preferred strategies can be ranked according to the priorities previously assigned to the different strategic objectives.

The development of forest management strategies based on natural disturbance and forest dynamics requires a good knowledge of the components of the forest ecosystem and understanding of the links that exist among them, links that represent the fundamental processes of ecosystem disturbance, recovery, and development (Chap. 8). Given that this information is not always complete when alternative management strate-

Box 11.8. Assembling stand intervention options into some alternative strategies.

Figure 11.3 illustrates the natural disturbance and forest dynamics model for the black spruce – feathermoss forest of northwestern Quebec. As suggested by Bergeron et al. (1999), this natural dynamics model can also serve as a basis for the elaboration of a forest management strategy for the pilot territory. In the case of the black spruce – feathermoss forest of northwestern Quebec, it is proposed that the 10 main stand types be managed for by the use of several types of silvicultural interventions (see Chap. 13 for more information on silvicultural options). Thus, the proposed management strategy for the pilot territory uses complete (clear cut) harvesting techniques (cc) in order to initiate stand regeneration as well as partial cutting techniques (pc) to maintain or establish the structural and compositional characteristics of later successional stages. The strategy incorporates both even-aged and uneven-aged silvicultural systems. The potential for uneven-aged management to maintain the structural characteristics of later successional stages of black spruce forests has been documented by MacDonell and Groot (1996) in the nearby Ontario Claybelt.

Figure 11.6 proposes a spatial and temporal dispersion pattern for the different silvicultural systems that is inspired by the distribution patterns of natural disturbances documented for the region. The dispersion strategy includes two main elements: (1) the dispersion of management zones; and (2) the dispersion of silvicultural systems within each of these zones. The dispersion of the management zones attempts to emulate the mosaic of large forest tracts historically created over the landscape by individual natural disturbance events, generally fire in our case. It is principally the result of an analysis of the potential for the agglomeration of the different silvicultural systems in different sections of the pilot territory. This analysis led to the definition of three types of management zones: (1) even-aged management zones (E) where primarily clear-cut harvesting (cc) will be used; (2) uneven-aged management zones (U) where predominantly partial harvesting (pc) will be used; and (3) light management zones (L) where non-productive forests predominate and where harvesting activities will be very limited. These zones range from 5000 to 40000 ha in size.

Within each management zone, the three main silvicultural/harvesting systems (cc, pc, and no intervention (nc)) are dispersed in an attempt to recreate the heterogeneity of disturbance severity spatially observed within single disturbance events. Figure 11.6 illustrates the potential internal dispersion pattern of these three harvesting systems within each type of management zone. In the examples provided, the relative proportions of cc, pc, and nc of the commercial forests in the even-aged management zone (E), are 51%, 17%, and 28%, respectively. In the uneven-aged management zone (U), this ratio shifts to 21%, 56%, and 22%, and in the light management zone (L) to 29%, 27%, and 50%. Each even-aged management zone (E) should be harvested over a period of 5–10 years while uneven-age management zones could be actively managed over much longer time periods, as they are mainly subjected to multiple entries. Harvesting in light management zones will only be partial. Over the entire pilot territory, the potential area of each of the management zones (even-aged, uneven-aged and light) is 63%, 9%, and 28%, respectively. To avoid the contiguity of regenerating management zones, even-aged management blocks should be dispersed in space and time.

The forest management strategy described above constitutes one possible combination of silvicultural systems and their temporal and spatial distribution over the pilot territory. Variations in the size as well as the spatial and temporal dispersion patterns of the management zones would provide other possible natural disturbance based strategies for the pilot territory, as would variations in the different types of management zones defined.

Fig. 11.6. Dispersion strategy for management zones and the harvesting systems within each zone. Zones: emz, even-aged management; umz, uneven-aged management; lmz, light management. Harvesting systems: cc, clear cutting; pc, partial cutting; nc, no cutting intervention.

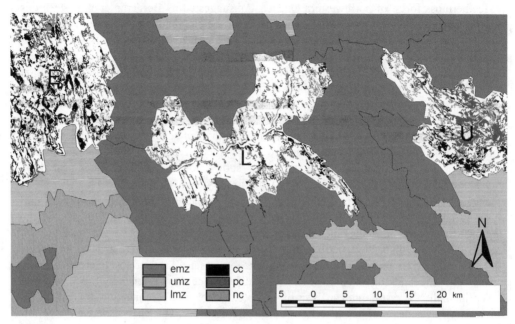

gies are being considered, a lot of management is based on assumptions. Thus, assessment of the management strategy also represents an opportunity in the planning process to identify knowledge gaps and to develop experimental designs and operational trials that will test different silvicultural systems or management strategies. This includes the design of monitoring programs that should be an integral part of the implementation of the preferred strategy and serve in its evaluation (Chap. 21). The FERNS (Forest Ecosystem Research Network of Sites) Network, established by the Canadian Forest Service (see http://www.pfc.forestry.ca/ecology/ferns/index_e.html), provides excellent examples of forest ecosystem research sites and silvicultural trials underway across Canada. The ongoing maintenance, monitoring, and evaluation of such long-term experiments and monitoring sites are essential for improved decision-making in the selection of strategies for sustainable forest management.

Implementation and evaluation of the best strategy — *did we get there?*

The process of strategic forest management planning broadly aims at establishing the general course of action to be carried out in order to attain the management objectives that have been set for the forest management unit. It provides the rationale and general direction of the management strategy to be implemented. It determines the types of silvicultural systems and forest interventions to be used. However, these silvicultural

Box 11.9. Model-based analysis of alternative strategies.

Two strategic management objectives previously defined for the black spruce – feathermoss pilot territory are a sustainable timber supply and the maintenance or conservation of regional biodiversity (Box 11.7). Planning indicators that can be associated with these objectives are the annual allowable cut (AAC) and the area-based proportion of the pilot territory occupied by the different stand types (i.e., the three broad temporal cohorts portrayed in Figs. 11.4 and 11.5b). In order to assess the possible implementation of the natural disturbance based forest management strategy suggested by Bergeron et al. (1999) for the black spruce-dominated forests of northwestern Quebec, a timber supply analysis was performed to compare the effects of three different management strategies on the two aforementioned planning indicators (Nguyen-Xuan 2000).

The three alternative strategies consisted of:

(1) an even-aged management strategy (clear-cut only, CPRS);

(2) an initial mixed management strategy (clear-cut and partial-cut to a 14-cm diameter limit, CPPTM14); and

(3) an alternative mixed management strategy (clear-cut and partial-cut with a 16-cm diameter limit, CPPTM16).

The timber supply analysis model used was SYLVA II (Version 1.3.2, Ministère des Ressources Naturelles du Québec [MRNQ] 1997, Québec, Quebec), the model used for the preparation of all general forest management plans in Quebec (MRNQ 1998). Results indicate that at the end of the 150-year simulation period, the mixed management strategies were closer to meeting the "cohort" area-based objectives than the even-aged management strategy (Fig. 11.7). The analysis also suggests that the use of a mixed management strategy does not significantly affect the AAC (Table 11.3). One should note, however, that in the mixed management strategies, only 15–20% of the managed forests were deemed suitable for partial harvesting.

In this analysis, only three alternative strategies were defined. However, as mentioned in the text, other possible alternatives that could also be examined include a strategy based on even-aged management with prolonged rotation ages for a portion of the forest area. The effects on allowable cut of such a strategy need to be quantified, however. In addition, more spatially explicit analyses are also required to assess the effects of the alternative strategies on spatially based planning indicators. These analyses will require the use of additional indicators and planning tools (Chaps. 12 and 14).

Table 11.3. Annual allowable cut (AAC) calculated for the pilot-territory for three alternative strategies.

Strategy	AAC (m³/year)	% volume from clear cutting	% volume from partial cutting	% of the area suitable for partial cutting
CPRS[a]	130000	100	0	0
CPPTM14[b]	132000	76	24	21
CPPTM16[b]	132000	74	26	17

[a]CPRS, harvesting with protection of regeneration and soil.

[b]CPPTM, harvesting with protection of small merchantable stems (up to 14 or 16 cm dbh).

Fig. 11.7. Area-based proportion of the commercial forests of the pilot territory occupied by each broad cohort at the initial (year 0) and final (year 150) points in the SYLVA II simulation for the three alternative management strategies. The three successional series (black spruce, jack pine, trembling aspen), as well as the early and late phases of cohort 1, are grouped.

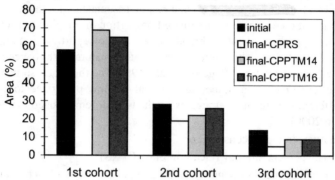

systems are actually implemented at the stand level. Hence, there still remain some analyses and decisions to be made at finer temporal and spatial scales, and these are usually performed at the tactical and operational planning levels (discussed in Chaps. 12 and 13, respectively). All three planning levels (strategic, tactical, and operational) are linked and dependant on each other (Tittler et al. 2001). Through the direction it provides, strategic planning will influence the development of the tactical and operational plans, and it is at these levels that implementation actually occurs over the management unit. While strategic planning and identification of management objectives and strategies is typically redone at 20-year intervals, the tactical and operational planning processes are usually repeatedly reinitiated on a 5-year and 1-year cycle, respectively. This periodic reinitiation provides the opportunity to evaluate the ability of forest interventions and the overall management strategy to meet the management objectives set during the strategic planning phase.

Similar to planning indicators, which provide the means during the planning phase to evaluate the capacity of management strategies to meet management objectives, monitoring indicators constitute a means of verifying these predictions following the partial or complete implementation of a strategy (Kneeshaw et al. 2000; Yamasaki et al. 2001). In contrast to planning indicators, monitoring indicators also provide a means to directly measure the effects of forest interventions and the management strategy on various components of the forest ecosystem. For example, the density at which disturbed sites become restocked with trees, the concentration of dissolved organic carbon in streams and lakes, or the structure and abundance of the avian community all provide for the direct evaluation of different strategic objectives (Yamasaki et al. 2001). Results from monitoring activities also permit the reassessment and improvement of silvicultural systems and management strategies, and as such close the loop of the strategic planning process (Box 11.10). They should be viewed as an integral part of experimental designs that aim to test management strategies, and all such strategies should be openly considered experiments in management, as more fully discussed in Chap. 21.

This chapter and the other chapters in the "Forest Ecology and Management" section of the book provide the scientific and operational basis for the development of management strategies based on natural disturbance and forest dynamics. The ecological knowledge and forest management tools described in these chapters are essential for the successful development and implementation of such management strategies. We believe that the natural disturbance emulation paradigm is one of several approaches that improve our ability to protect biodiversity and sustain the full range of forest values (Box 11.11). However, successful implementation also depends on the other dimensions that have been discussed in the previous section of this book. Favourable social, economic, and institutional contexts are also necessary for the development of sustainable

---

Box 11.10. Monitoring and ongoing work in the black spruce – feathermoss forest.

The natural disturbance based management strategy proposed for the pilot territory is based on several management hypotheses. One key hypothesis that is currently being tested is that forest stands subjected to partial harvesting methods will demonstrate similar ecological characteristics, including structural and compositional attributes, as those observed in natural forests of later successional stages. Partial harvesting is not a recent silvicultural innovation; however, its effects on stand dynamics and biodiversity are not well documented for the boreal forests of Quebec. Mensurational, floristic, and wildlife monitoring of partial-cutting trials performed in the area has been integrated into the management strategy. Preliminary results are encouraging, and as additional ecological data are collected and analyzed, they will serve to refine the natural disturbance based strategy and the partial-cutting silvicultural systems.

The work involved in the development of a natural disturbance based forest management strategy for the black spruce – feathermoss forests of northwestern Quebec has been ongoing for almost 10 years now. Its possible implementation has been assessed in the pilot territory over the last 3 years. During this period, it has been demonstrated that the scientific context (i.e., knowledge of the natural ecosystem dynamics) is satisfactory for the successful implementation of the strategy. Operational aspects (silvicultural knowledge and planning tools) are being developed and promoted through on-the-ground harvesting trials. The economic context may also be favourable to applying the natural disturbance based management approach developed here. Tembec Forest Products is currently working with the Forest Stewardship Council to acquire certification and gain the associated market advantage.

Other social and institutional aspects related to natural disturbance based management will be critical to its successful implementation. The overall approach and the use of mixed management (even-aged and uneven-aged) systems in the boreal forest are not clearly supported by current Quebec forest policies and regulations. Moreover, social acceptability (within Aboriginal communities and at local, regional, and provincial levels) of natural disturbance based forest management still remains to be assessed. While increased emphasis has been placed recently on more public input in forest management planning (see Chaps. 4 and 22) and on integrated resource management, neither of these aspects of the current paradigm of sustainable forest management preclude putting disturbance-based management into practice. Certainly, an effort must be made to explain to the public, to policy makers, and to forest managers what natural disturbance based management is, what the rationale for applying this approach is, and what the implications are for various resource values. Not a panacea to all forest sustainability issues (see Chap. 9), the natural disturbance based management model offers one manner in which forest management strategies can be diversified and creatively modified to better achieve the goals of SFM.

Box 11.11. Options for improved forest management.

The shift from conventional to more natural disturbance based management in the boreal forest may not be as difficult or as radical a move as many consider, but it should be done with the understanding that (as emphasized by Szaro et al. 1999):

- knowledge of natural disturbances will always be incomplete;
- interactions exist between scales of disturbance, between disturbance agents and events, and between the factors of climate, vegetation, landscape, site, and disturbance; and
- monitoring of both management and natural disturbance is essential.

The coarse-filter approach essentially aims to maintain ecosystem diversity and ecological processes, and in particular, to protect and maintain a certain portion of older forests on the landscape. These objectives may be at least partially accommodated by expanding or allowing greater flexibility in current regulations aimed at protecting rare or characteristic landscape features or specific habitat types, and buffering around aquatic features. Furthermore, we know that fire, insect outbreaks, and other natural disturbances in the boreal forest produce more variable outcomes than has traditionally been portrayed in popular treatments on the subject. Consequently, management strategies that attempt to *incorporate greater variety* in terms of:

- silvicultural systems;
- the retention of live and dead trees;
- the nature and degree of soil disturbance;
- the size of disturbance events; and
- the spatial configuration of disturbances

will help to move managed forests closer toward the natural disturbance template.

An *understanding of the regional natural disturbance regime and stand dynamics* provides the basis for evaluating current management strategies and developing new ones. At the level of the forest or the forest management unit, the objective is to create or maintain a natural landscape composition and forest mosaic and an age structure that includes the historic range of forest age classes, while still permitting a sustained timber supply. Management and silviculture at the stand level are the tools used to meet these objectives. A mix of treatments, including use of partial or selective cutting (to create openings similar to those created by low-intensity or patchy fire, small pockets of insect or disease outbreak, local stand breakup, canopy replacement, and gap dynamics) will have to become part of the boreal forest manager's toolbox. Fortunately, partial cutting is not a new form of intervention, so the ever-changing forest products industry is well positioned to adapt existing technologies and techniques for sustaining the full range of forest values in the boreal context.

---

forest management plans. Significant participation by all forest stakeholders throughout the strategic planning process is considered the best way to recognize, and subsequently plan for and sustain, all forest values.

Acknowledgements

Funding for work presented in this chapter was provided by the Natural Sciences and Engineering Research Council of Canada (NSERC), the Sustainable Forest Management Network, the Ministry of Education of Quebec (Fonds pour la formation des chercheurs et à l'aide à la recherche — FCAR) and Tembec Forest Products Group and Norbord (Nexfor) Industries. We would like to acknowledge our appreciation to the researchers of the GREFi (Groupe de recherche en écologie forestière interuniversitaire) and the SFM industrial Chair (Chaire industrielle CRSNG–UQAT–UQAM en aménagement forestier durable). We are also grateful to co-editor Phil Burton for his contribution to the chapter.

References

(AEP) Alberta Environmental Protection. 1998. Interim forest management planning manual — guidelines to plan development. Alberta Environmental Protection, Land and Forest Service, Edmonton, Alberta. 46 p.

Andison, D.W., and Kimmins, J.P. 1999. Scaling up to understand British Columbia's boreal mixedwoods. Environ. Rev. **7**: 19–30.

Angelstam, P.K. 1998. Maintaining and restoring biodiversity in European boreal forests by developing natural disturbance regimes. J. Veg. Sci. **9**: 593–602.

Attiwill, P.M. 1994. The disturbance of forest ecosystems: the ecological basis for conservation management. For. Ecol. Manage. **63**: 247–300.

Baskerville, G. 1986. Understanding forest management. For. Chron. **62**: 339–347.

Bergeron, Y., and Harvey, B. 1997. Basing silviculture on natural ecosystem dynamics: an approach applied to the southern boreal mixedwood forest of Quebec. For. Ecol. Manage. **92**: 235–242.

Bergeron, Y., and Leduc, A. 1998. Relationships between changes in fire frequency and mortality due to spruce budworm outbreak in the southeastern Canadian boreal forest. J. Veg. Sci. **9**: 492–500.

Bergeron, Y., Richard, P.J.H., Carcaillet, C., Gauthier, S., Flannigan, M., and Prairie, Y.T. 1998. Variability in fire frequency and forest composition in Canada's southeastern boreal forest: a challenge for sustainable forest management. Conserv. Ecol. [online], **2**(2): 6. Available at http://www.consecol.org/vol2/iss2/art6/ [viewed 11 June 2003].

Bergeron, Y., Harvey, B., Leduc, A., and Gauthier, S. 1999. Forest management guidelines based on natural disturbance dynamics: stand- and forest-level considerations. For. Chron. **75**: 49–54.

Bergeron, Y., Gauthier, S., Kafka, V., Lefort, P., and Lesieur, D. 2001. Natural fire frequency for the eastern Canadian boreal forest: consequences for sustainable forestry. Can. J. For. Res. **31**: 384–391.

Bergeron, Y., Leduc, A., Harvey, B.D., and Gauthier, S. 2002. Natural fire regime: a guide for sustainable forest management of the Canadian boreal forest. Silva Fenn. **36**: 81–95.

Bissonnette, J., Bélanger, L., Larue, P., Marchand, S., and Huot, J. 1997. L'inventaire forestier multiressource: les variables critiques de l'habitat faunique. For. Chron. **73**: 241–247.

Bonan, G.B., and Shugart, H.H. 1989. Environmental factors and ecological processes in boreal forests. Ann. Rev. Ecol. Syst. **20**: 1–28.

Boudreault, C., Bergeron, Y., Gauthier, Y., and Drapeau, P. 2002. Bryophyte and lichen communities in mature to old-growth stands in eastern boreal forests of Canada. Can. J. For. Res. **32**: 1080–1093.

(CCFM) Canadian Council of Forest Ministers. 1995. Defining sustainable forest management: a Canadian approach to criteria and indicators. Canadian Council of Forest Ministers, Natural Resources Canada, Ottawa, Ontario. Available at http://www.ccfm.org/ci/framain_e.html [viewed 5 June 2003]. 22 p.

(CFS) Canadian Forest Service. 1998. National forest strategy, 1998–2003: sustainable forests, a Canadian commitment. Canadian Forest Service, Natural Resources Canada, Ottawa, Ontario. Available at http://www.nrcan.gc.ca/cfs/nfs/strateg/title_e.html [viewed 25 March 2002]. 47 p.

Coates, K.D., and Burton, P.J. 1997. A gap-based approach for development of silvicultural systems to address ecosystem management objectives. For. Ecol. Manage. **99**: 337–354.

Drapeau, P., Bergeron, Y., and Harvey, B. 1999. Key factors in the maintenance of biodiversity in the boreal forest. Sustainable Forest Management Network, Edmonton, Alberta. Proj. Rep. 1999-08. Available at http://sfm-1.biology.ualberta.ca/english/pubs/PDF/PR_1999-08.pdf [viewed 2 November 2002]. 13 p.

Drapeau, P., Leduc, A., Savard, J.-P., and Bergeron, B. 2001. Les oiseaux forestiers, des indicateurs des changements des mosaïques forestières boréales. Nat. Can. **125**: 41–46.

Drapeau, P., Leduc, A., Bergeron, Y., Gauthier, S., and Savard, J.-P. 2003. Bird communities in old growth black spruce forests of the Claybelt region: expected problems and management solutions. *In* Old-growth forests in Canada — a science perspective. *Edited by* A. Mosseler. Canadian Forest Service, Ottawa, Ontario. In press.

Engstrom, R.T., Gilbert, S., Hunter, M.L., Merriwether, D., Nowacki, G.J., Spencer, P. 1999. Practical applications of disturbance ecology to natural resource management. *In* Ecological stewardship: a common reference for ecosystem management, vol. II, a practical reference for scientists and resource managers. *Edited by* R. Szaro, N.C. Johnson, W.T. Sexton, and A.J. Malk. Elsevier Science, Oxford, U.K. pp. 313–330.

Erdle, T. 1998. Progress toward sustainable forest management: insight from the New Brunswick experience. For. Chron. **74**: 378–384.

Erdle, T., and Sullivan, M. 1998. Forest management design for contemporary forestry. For Chron. **74**: 83–90.

Franklin, J.F. 1993. Preserving biodiversity: species, ecosystems or landscapes. Ecol. Appl. **3**: 202–205.

(FSC) Forest Stewardship Council. 2000. FSC principles and criteria, document 1.2. Forest Stewardship Council, Oaxaca, Mexico. Available at http://www.fscoax.org/html/1-2.html [viewed 12 June 2002].

Gauthier, S., Leduc, A., and Bergeron, Y. 1996. Forest dynamics modelling under natural fire cycles: a tool to define natural mosaic diversity for forest management. Environ. Monitor. Assess. **39**: 417–434.

Harvey, B.D., Leduc, A., Gauthier, S. and Bergeron, Y. 2002. Stand-landscape integration in natural disturbance-based management of the southern boreal forest. For. Ecol. Manage. **155**: 369–385.

Holling, C.S. (*Editor*). 1978. Adaptive environmental assessment and management. John Wiley & Sons Ltd., Toronto, Ontario. 377 p.

Kimmins, J.P. 1995. Sustainable development in Canadian forestry in the face of changing paradigms. For. Chron. **71**: 33–40.

Kimmins, J.P. 2002. Future shock in forestry. Where have we come from; where are we going; is there a "right way" to manage forests? Lessons from Thoreau, Leopold, Toffler, Botkin and Nature. For. Chron. **78**: 263–271.

Kneeshaw, D., Leduc, A., Drapeau, P., Gauthier, S., Paré, D., Carignan, R., Doucet, R., Bouthillier, L., and Messier, C. 2000. Development of integrated ecological standards of sustainable forest management at an operational scale. For. Chron. **76**: 481–493.

Kuuluvainen, T. 2002*a*. Disturbance dynamics in boreal forests: defining the ecological basis of restoration and management of biodiversity. Silva Fenn. **36**: 5–11.

Kuuluvainen, T. 2002*b*. Natural variability of forests as a reference for restoring and managing biological diversity in boreal Fennoscandia. Silva Fenn. **36**: 97–125.

Lackey, R.T. 1998. Seven pillars of ecosystem management. For Ecol. Manage. **40**: 21–30.

Lackey, R.T. 1999. Radically contested assertions in ecosystem management. *In* Contested issues of ecosystem management. *Edited by* P. Corona and B. Zeide. Food Products Press, Binghamton, New York. pp. 21–34.

Landres, P., Morgan, P., and Swanson, F. 1999. Overview of the use of natural variability concepts in managing ecological systems. Ecol. Appl. **9**: 1179–1188.

Lieffers, V.J., Macmillan, R.B., MacPherson, D., Branter, K., and Stewart, J.D. 1996. Semi-natural and intensive silvicultural systems for the boreal mixedwood forest. For. Chron. **72**: 286–292.

Lindenmayer, D., and McCarthy, M.A. 2002. Congruence between natural and human forest disturbance: a case study from Australian montane ash forests. For. Ecol. Manage. **155**: 319–335.

MacDonell, M.R., and Groot, A. 1996. Uneven-aged silviculture for peatland second-growth black spruce: biological feasibility. Canadian Forest Service, Natural Resources Canada, Sault Ste. Marie, Ontario. NODA/NFP Tech. Rep. TR-36.

McLaughlin, D. 1998. A decade of forest tree monitoring in Canada: evidence of air pollution effects. Environ. Rev. **6**: 151–171.

(MRNQ) Ministère des Ressources Naturelles du Québec. 1998. Manuel d'aménagement forestier. Ministère des Ressources Naturelles du Québec, Québec, Quebec. 122 p.

Nguyen-Xuan, T. 2000. Développement d'une stratégie d'aménagement forestier s'inspirant de la dynamique des perturbations naturelles pour la région Nord de l'Abitibi (année 1). Rapport de recherche effectuée dans le cadre du Volet 1 du programme de mise en valeur des ressources du milieu forestier. Ministère des Ressources Naturelles du Québec, Rouyn-Noranda, Quebec. Available at http://web2.uqat.uquebec.ca/cafd/pdf/nguyen1.pdf [viewed 12 June 2003]. 67 p.

(OMNR) Ontario Ministry of Natural Resources. 1996. Forest management planning manual for Ontario's Crown forests. Queen's Printer, Toronto, Ontario. 452 p.

Pickett, S.T.A., and White, P.S. (*Editors*). 1985. The ecology of natural disturbance and patch dynamics. Academic Press, San Diego, California. 472 p.

Radeloff, V.C., Mladenoff, D.J., and Boyce, M.S. 2000. Effects of interacting disturbances on landscape patterns: budworm defoliation and salvage logging. Ecol. Appl. **10**: 233–247

Rülcker, C., Angelstam, P., and Rosenberg, P. 1994. Natural forest-fire dynamics can guide conservation and silviculture in boreal forests. Skog Forsk Results No. 2, Uppsala, Sweden. 4 p.

Runkle, J.R. 1992. Guidelines and sample protocol for sampling forest gaps. USDA Forest Service, Portland, Oregon. Gen. Tech. Rep. PNW-GTR-283. Available at http://old.lternet.edu/research/pubs/forgap/ [viewed 12 June 2003]. 44 p.

Shindler, B. 1998. Does the public have a role in forest management? Canadian and U.S. perspectives. For Chron. **74**: 700–702.

Szaro, R., White, P.S., Engstrom, R.T., Harrod, J., Aplet, G., and Collins, B. 1999. Disturbance and temporal dynamics: why disturbance and temporal dynamics are important to ecological stewardship. *In* Ecological stewardship: a common reference for ecosystem management, vol. I, key findings. *Edited by* N.C. Johnson, A.J. Malk, R.C. Szaro, and W.T. Sexton. Elsevier Science, Oxford, U.K. pp. 39–45.

Thompson, I.D., Baker, J.A., and Ter-Mikaelian, M. 2003. A review of the long-term effects of post-harvest silviculture on vertebrate wildlife and predictive models, with an emphasis on boreal forests in Ontario, Canada. For Ecol. Manage. **177**: 441–469.

Tittler, R., Messier, C., and Burton, P.J. 2001. Hierarchical forest management planning and sustainable forest management in the boreal forest. For. Chron. **77**: 998–1005.

Van Wagner, C.E. 1978. Age-class distribution and forest fire cycle. Can. J. For. Res. **8**: 220–227.

Verner, J., Morrsion, M.L., and Ralph, C.J. (*Editors*). 1986. Wildlife 2000: modeling habitat relationships of terrestrial vertebrates. University of Wisconsin Press, Madison, Wisconsin. 470 p.

Weber, M.G., and Flannigan, M.D. 1997. Canadian boreal forest ecosystem structure and function in a changing climate: impact on fire regimes. Environ. Rev. **5**: 145–166.

Weetman, G.F. 2000. Worrying issues about forest landscape management plans in British Columbia. *In* Ecosystem management of forested landscapes: directions and implementation. Symposium proceedings, 26–28 October 1998, Nelson, British Columbia. *Edited by* R.G. D'Eon, J.F. Johnson, and E.A. Ferguson. UBC Press, Vancouver, British Columbia. pp. 319–324.

Weetman, G.F. 2001 Distinctly Canadian silviculture and forest management. For. Chron. **77**: 441–445.

Weir, J.M.H., Johnson, E.A., and Miyanishi, K. 2000. Fire frequency and the spatial age mosaic of the mixedwood boreal forest in western Canada. Ecol. Appl. **10**: 1162–1177.

White, P.S., Harrod, J., Romme, W.H., and Betancourt, J. 1999. Disturbance and temporal dynamics. *In* Ecological stewardship: a common reference for ecosystem management, vol. I, key findings. *Edited by* N.C. Johnson, A.J. Malk, R.C. Szaro, and W.T. Sexton. Elsevier Science, Oxford, U.K. pp. 281–312.

Woodley, S. 1993. Monitoring and measuring ecosystem integrity in Canadian National Parks. *In* Ecological integrity and the management of ecosystems. *Edited by* S. Woodley, J. Kay, and G. Francis. St. Lucie Press, Boca Raton, Florida. pp. 155–176.

Yamasaki, S.H., Côté, M.-A., Kneeshaw, D.D., Fortin, M.-J., Fall, A., Messier, C., Bouthillier, L., Leduc, A., Drapeau, P., Gauthier, S., Paré, D., Greene, D., and Carignan, R. 2001. Integration of ecological knowledge, landscape modelling, and public participation for the development of sustainable forest management. Sustainable Forest Management Network, Edmonton, Alberta. Proj. Rep. 2001-27. Available at http://sfm-1.biology.ualberta.ca/english/pubs/PDF/PR_2001-27.pdf [viewed 2 November 2002]. 27 p.

Zasada, J.C., Gordon, A.G., Slaughter, C.W., and Duchesne, L.C. 1997. Ecological considerations for the sustainable management of the North American boreal forest. International Institute for Applied Systems Analysis, Laxenburg, Austria. Interim Rep. IR-97-024/July. Available at http://www.iiasa.ac.at/Publications/Documents/IR-97-024.pdf [viewed 12 June 2003]. 62 p.

Chapter 12

Tactical forest planning and landscape design

David W. Andison

*"The future is not just what lies ahead:
it is something that we and Nature create together."*

Richard Forman (2003)[1]

Introduction

The planning of forest management activities takes place on several different levels. In general (Gunn 1996), planning begins with strategic policies, goals, and objectives (such as those explored in Chap. 11), which are translated into tactical activity schedules, which in turn are rendered into specific operational actions on the ground (such as silvicultural activities, as described in Chap. 13). Table 12.1 includes some typical examples of the planning and decision making hierarchy in forest management.

This chapter focuses on tactical, or intermediate, levels of planning. Tactical planning has to understand, interpret, and implement strategic goals within the realities of local-level constraints and opportunities. For instance, setting the level of the annual allowable cut (AAC), the amount of old-growth forest, and recreational opportunity priorities are all strategic decisions. Tactical planning takes these objectives and creates a schedule of management activities in space and over time that is most likely to achieve these goals. Such a schedule will typically project road building, and the sizes, locations, and timing of harvesting activities 2–10 years into the future. The schedule of tactical activities is then used as the foundation from which operational planners create more specific plans involving the choice of harvesting and hauling options, and the design of silvicultural operations.

Of the three levels of planning, tactical planning has undergone (and is still experiencing) the most dramatic changes over the past two decades. While both strategic and

---

D.W. Andison. Bandaloop Landscape-Ecosystem Services, 3426 Main Ave., Belcarra, British Columbia V3H 4R3, Canada. Telephone: 604-939-0830.
e-mail: andison@bandaloop.ca

Correct citation: Andison, D.W. 2003. Tactical forest planning and landscape design. Chapter 12. *In* Towards Sustainable Management of the Boreal Forest. *Edited by* P.J. Burton, C. Messier, D.W. Smith, and W.L. Adamowicz. NRC Research Press, Ottawa, Ontario, Canada. pp. 433–480.
[1]Graduate School of Design, Harvard University, Cambridge, Massachusetts, personal communication, 13 June 2003.

Table 12.1. Some examples of hierarchical forest management planning.

	Level of responsibility	
Strategic (Chap. 11)	Tactical (Chap. 12)	Operational (Chaps. 13, 15)
Define forest-wide harvest levels non-spatially	Optimize harvest sequence and road networks in time and space	Schedule annual resources to obtain desired transportation services and forest product mix
Establish desired levels and types of recreational activity	Design and maintain access and aesthetic considerations	Cooperatively build and maintain roads and trails, apply partial cutting techniques (Chap. 15)
Maintain historic levels of old-growth forest	Define and protect a sample of old-growth areas, identify future old growth potential, establish guidelines for residual stands and trees	Manage selected stands on extended rotations, apply extended rotations, apply careful logging techniques
Sustain current population levels of caribou	Model future landscape scenarios based on defined habitat needs, and devise plans that protect, maintain, and create habitat as required	Apply appropriate silvacultural and road network management options to create suitable habitat mix (Chap. 13)
Maintain genetic diversity of native tree species	Design and maintain regionally integrated seed collection and tree nursery program	Schedule appropriate site preparation and planting operations, encourage natural regeneration (Chap. 13)
Maintain productive capacity of the land	Apply appropriate rotation lengths, minimize netdowns for roads and landings, map all erosion-prone sites	Choose appropriate season of harvest and equipment, reclaim landings and roads wherever possible, avoid operating on steep slopes (Chap. 15)

tactical planning have evolved in response to changes in societal views of forest management in general, tactical planning has especially benefited from the evolution of technical tools for forest landscape data presentation and modelling. Good tactical planning becomes essential in sustaining an acceptable (and usually complex) balance of forest values.

This chapter first discusses some background material on the history and current challenges of tactical planning, and then covers a variety of elements, tools, and techniques used in tactical planning today. Figure 12.1 arranges these elements, tools, and techniques as a roadmap for the chapter sections. It also functions as an idealized system of tactical planning, starting at the bottom, and working up towards a planning solution.

Why tactical planning?

Historically, tactical planning was a relatively small part of the overall forestry planning system. Until only a few decades ago, timber was the only value actively being managed for. Little or no strategic emphasis was given to other (non-timber) objectives such as wildlife, biodiversity, or recreation. Allowable harvest level estimates were often the only strategic goal, and how and where harvesting occurred and roads were located was largely an exercise in cost efficiency. In other words, unless there was a compelling reason to do otherwise, the only tactical planning involved logging the most valuable wood closest to the processing mill.

The importance of tactical planning has increased significantly over the past few years for a number of reasons, including changing values, planning complexity, and recognition of the advantages of considering needs and wants in spatial terms.

Fig. 12.1. Chapter 12 outline: an idealized tactical planning system.

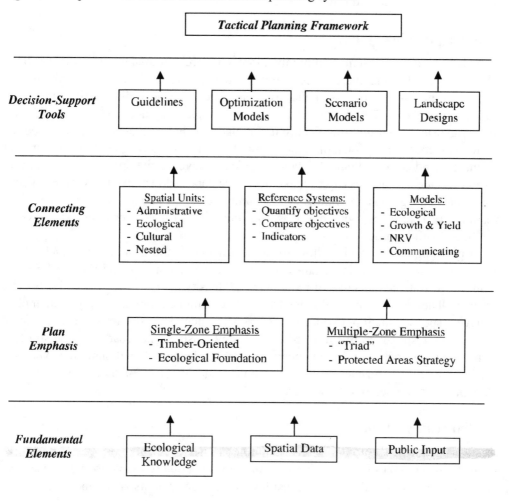

Growing and changing values

As developed in the chapters constituting the second part of this volume, forest management goals more or less reflect the values of society (see also Gregory 1999). Initial demands by society of its forests used to be few and simple. Prior to 1950, providing wood products at the lowest price was the primary objective for management of most boreal forests. In the 1940s, the first ideas of "sustained yield" emerged as we began to question the long-term continuity of our wood supply (Apsey et al. 2000; Chap. 1). As the end of the twentieth century approached, there was growing recognition that forests provided values other than wood products and that sustainable timber management did not necessarily safeguard all of those values. Today, the list of forest values includes water, livestock grazing, wild meat, furs, camping, aesthetics, hunting, fishing, hiking, biodiversity, wildlife habitat, and even spirituality (Weintraub and Davis 1996; Yoshimoto 2001). As the number of goals increases, the potential cost of such goals and how they might interact or conflict over time and space multiplies many-fold. It is simply not possible to resolve these problems without the ability to consider and compare management choices in space and time.

A better planning system

Complex planning problems commonly adopt a hierarchy to make them more tractable (Watt 1966). Planning hierarchies allow us to partition data, decision making, and responsibilities into more reasonable, solvable dimensions according to time and space scales (Baskent et al. 2000). Attempting to make all planning decisions at a single level in a single step is not only illogical, but also onerous. Forest management adopted the strategic–tactical–operational planning hierarchy many decades ago for this reason. For example, stand-scale ecological data are needed to develop site-specific treatment prescriptions, but are unnecessary to make decisions about allowable cut levels. Similarly, public advisory groups are valuable for making decisions about when and where harvesting will take place, but usually are not qualified to make decisions about which equipment will be used to do so.

A planning hierarchy also allows us to deal with risk and uncertainty, which generally increase with longer time frames and larger areas (Weintraub and Davis 1996). For example, a strategic objective of reducing road density over a large area and a 10-year period will not necessarily be affected by the successful petitioning of a group of anglers to have a specific small network of roads remain open. The (shorter term) operational plan simply responds to this unforeseen event by closing other roads. To the strategic plan, the group's actions were an uncertainty, but to the operational plan, they were just another variable to consider. Thus, it is the inherent flexibility of short-term plans that allow long-term objectives to be met.

Spatial reality check

Tactical plans are the point at which planning shifts from being non-spatial to spatial. Considering our goals in space imposes limitations not otherwise obvious. For instance, strategic estimates of allowable harvest levels are usually non-spatial, and thus assume equal access to any eligible wood. Thus, distant, small patches of eligible forest will be included in strategic calculations of allowable harvest. However, in reality, many such areas may never be harvested because of high logging and transportation costs. Similarly, today there are social, economic, and ecological considerations that prevent equal access to all areas of the landscape. Issues of habitat needs, patch size, and connectivity for sensitive wildlife, aesthetics, or aquatic protection can and do limit how much, where, and in what way wood may be harvested on a given landscape.

The impact of these spatial limitations is usually a net reduction in the (non-spatial estimate of) allowable harvest, and a net increase in total wood costs. One of the roles of tactical planning is to optimize wood costs and volume within the bounds of related spatial issues and constraints. In other words, it pays to plan in space.

What are the challenges?

While every planning situation is unique, there are some challenges common to most tactical plans (Box 12.1). These issues include the translation of objectives into spatial terms, balancing a large number of values, considering biological implications, dealing with risk, and the role of public input.

Moving the plan to space

Tactical plans shift the management context from general, non-spatial goals to specific, measurable, spatial objectives. When allowable harvest level estimates were the only strategic goal, this transition involved easily quantitative terms such as annual wood volume or revenue (Buttoud 2000). Today, that translation is much more challenging. For example, how do more contemporary goals such as "maintain water quality", or "allow for pleasing aesthetics" translate to spatial reality? In the first instance, we have incomplete knowledge of the dynamic biological processes, and we are likely to have differences of opinion on how to accomplish this goal. In the second instance, the attainment of superior aesthetics is a qualitative and subjective attribute, and thus difficult to capture and compare as a measure of success.

The shift to thinking and planning in spatial terms is perhaps the most difficult transition in the entire planning hierarchy, and the problem has likely not been given its due respect. Yet, if planning objectives cannot be simply stated and universally understood in spatial terms, tactical planning will almost surely fail.

Box 12.1. Challenges for tactical forest planning:

- Shifting the planning context from general, non-spatial goals to specific, measurable, spatial objectives;
- Understanding and balancing a large number of disparate values;
- Accounting for key biological patterns and functions at intermediate scales;
- Identifying and dealing with key risks and uncertainties; and
- Finding ways of sharing the decision-making responsibility.

Balancing a large number of values

While it often seems as if forest management goals are compatible and independent, conflicts are frequent when they are translated into spatial reality. For example, "*maximizing access for camping and hiking*" is likely to conflict with "*maintaining natural levels of large ungulate biodiversity*" because road density (a likely tactical measure of both goals) might have an optimal intermediate level for camping and hiking, but should be minimized (to reduce hunting pressure) for biodiversity values. Similarly, we already know that allowing for biodiversity and habitat requirements usually reduces harvest levels (Armstrong et al. 1999), increases costs (Ohman 2000), and can reduce forest industry employment (Toivonen 2000). Other tradeoffs are less obvious; for example, what are the true costs of increasing aesthetic appeal or recreational opportunities?

Tactical plans are responsible for finding a solution that generates an acceptable balance of tactical objectives representing our various values. At the very least we need a mechanism for understanding the consequences of value tradeoffs and exploring the feasibility of achieving certain goals through time and over space.

Considering biological implications

The growing importance of biodiversity as an issue is indicative of our increasing unwillingness to risk compromising the long-term capacity of forest landscapes to provide the full range of values (see recent volumes by Voller and Harrison 1998; Gustafsson and Weslien 1999; Hunter 1999; Larsson and Danell 2001). Thus, biological considerations are important to consider at all levels of planning. However, until recently, biological considerations were limited to either regional (strategic) or stand (operational) scales of planning. At regional scales, biological considerations have included the designation of protected areas and ecological reserves, the level of allowable harvest, and minimum level(s) of old forest. From there, biological considerations moved directly to the impacts of harvesting at the stand scale. There was widespread belief that by minimizing logging impacts on every cutblock, and then fully regenerating every cutblock, we would minimize the ecological impacts of timber harvesting and maintain overall landscape structural integrity. Our legacy of that planning logic is vast areas of small, regularly shaped blocks dispersed at regular "checkerboard" intervals (i.e., fragmentation; Franklin and Forman 1987).

We now know that the maintenance of biological integrity involves maintaining patterns and processes at a wide range of spatial and temporal scales (Noss 1990). For example, fragmented landscape patterns have been associated with negative impacts on wildlife (Schmiegelow and Hannon 1999), water and nutrient cycling, erosion potential, and microorganism, pathogen, and insect activity (Robinson et al. 1992; Li et al. 1993). We have also learned that patch composition and configuration are important for bird habitat (Hannon 1999; Drapeau et al. 2000; Schmiegelow and Beck 2001), and the presence of very large tracts of undisturbed old forest patches on the landscape is important for many large ungulates such as caribou. More generally, we have come to recognize that dynamics of the landscape mosaic over large areas and long time periods is a fundamental aspect of natural landscapes, and must be accounted for by planning at the appropriate scale. Our historical tendency of trying to stabilize or "balance" natural systems has been both unsuccessful and harmful.

By focusing our ecological efforts at specific or convenient scales, and not accounting for intermediate-scale ecological dynamics, not only are we unlikely to be able to manage for species that respond to those scales, but also we risk inadvertently degrading the entire ecosystem. There are very real limits to the capacity of even boreal forest systems to absorb and adapt to change. This means we must deal with ecological tradeoffs very carefully — perhaps differently than we do other tradeoffs. For instance, we would gain a considerable volume of valuable wood by harvesting through all riparian zones, but at a cost of denying some streams and rivers large woody debris, which is important for streambed composition, organic matter retention, and channel morphology, and thus fish and invertebrate communities (Harvey 1998; Gomi et al. 2001). So by taking advantage of the immediate, short-term objective of increasing wood volume, we risk the long-term decline of several other objectives over larger areas specified in terms of fish population levels, recreational opportunities, bank erosion, and flood risk.

Dealing with risk and uncertainty

There is risk and uncertainty at all levels of planning. Uncertainty exists in our models (O'Neill and Rust 1979), natural disturbance (Weintraub and Davis 1996), scientific understanding (Ludwig et al. 1993), inventories, societal values, markets (Marshall 1987), and the effects of climate change (Ludwig et al. 1993). Planning cannot reduce risk, but it should identify and quantify it, and then decide on a course of action that has the greatest chance of success (Colberg 1996). Unfortunately, we are not in the habit of either identifying or communicating the sources of uncertainties with stakeholders and decision makers (Marshall 1986), let alone dealing with it in tactical plans. For example, one of the most formidable forest management planning risks is natural disturbance. Despite our best control efforts, forest fires continue to burn, wind continues to blow down trees, and insects and disease continue to infest forests (Chap. 8). At the strategic level, it is relatively straightforward to include rates of natural disturbance (and probability of stand succession) in non-spatial projections (e.g., an average of x% of land base is burned every year). However, at the tactical level of planning, natural disturbance events are unpredictable both in time and space. This is a significant challenge that is historically dealt with by ignoring the risk at tactical scales, and revising plans if and when such events occur. Similarly, planning output is usually presented as a forgone conclusion, as opposed to one of many likely possibilities.

Sharing decision making

Tactical planners are increasingly looking for ways of effectively involving the public in the process of tactical-level decision making. This trend is a relatively recent phenomenon (see Chap. 4). Historically, public input into forest management decisions were centralized, accomplished at strategic levels of planning through public authorities to reflect the common goals of society (Cortner and Schweitzer 1983). Decisions about allocations or regulations for industrial forests were particularly centralized, since their ownership and markets are international in nature (Beckley 1998). Aside from presentations made to local communities about the particulars of a chosen management option, little or no consideration of local interests or needs was given (Buttoud 2000).

The growing and shifting values of society have changed standard practices by which foresters interact with the public. On one hand, the public wants to be more involved in how objectives are translated into actions on the land. Trust and credibility of forest management are serious concerns today (Shindler 2000), due in part to sometimes well-intended, but unilateral decisions being made on behalf of the public by forest management agencies or governments. On the other hand, greater involvement by the public can be a tremendous asset to forest management agencies. Many values are spatially defined, meaning it is important to make sure local-level needs and knowledge are being integrated into decision making (Schreier et al. 1996). Thus, having the public more directly involved in the tactical planning process potentially leads to better, more defendable plan solutions. Having the public involved can also help build objectivity and credibility. The diverse methods available for incorporating public input in forest planning and management are discussed in Chap. 4.

Fundamental elements of tactical planning

Today there is a wide range of sophisticated tools and techniques available to assist with, facilitate, and direct tactical planning, and many ways of approaching the problem. However, in a perfect world, regardless of which tools are used and which philosophies adopted, there are three fundamental building blocks that all tactical plans ideally have in common: ecological knowledge, spatial data, and public input.

Knowledge of the ecological system

Tactical planning requires knowledge of all parts of the forest and forest management system: ecological, social, and economic. Social and economic issues are dealt with in the second part of this book, and will not be discussed further here. For the purposes of this chapter, the focus of the discussion is on ecological knowledge.

Our understanding of how forest ecosystems function has taken place largely at finer scales, and in fragments. For instance, we have considerable understanding of how tree species grow and respond to light, competition, nutrients, water, and even climate change. We also have a substantial knowledge base of selected large fur-bearing vertebrates such as wolves (*Canis lupus*), deer (*Odocoileus* spp.), caribou (*Rangifer* spp.), and bears (*Ursus* spp.), and our understanding of the ecology and response to forestry of small mammals has developed more recently (Corkum 1999; Fisher 1999; Bowman et al. 2000; Moses and Boutin 2001; Sullivan et al. 2001). Considerable research has also been done on the habitat needs of most of the terrestrial bird species in the boreal forest (Olsen 1999; Hannon 2000; Hoyt 2000; Villard 2000; Schmiegelow and Beck 2001) and their responses to forest fragmentation and edge creation (Cotterill and Hannon 1999; Schmiegelow and Hannon 1999; Drapeau et al. 2000).

Highly specific understanding of species and processes is often referred to as "fine-filter" knowledge. In tactical planning, fine-filter ecological knowledge is represented as the amount, type, and (or) spatial arrangement of vegetation (Railsback 2001), as it relates to species habitat, forage, or cover requirements of particular species. This gives us the ability to relate specific management activities to probable impacts on those selected species.

There are two risks associated with using fine-filter knowledge alone as the basis for evaluating ecological tradeoffs in tactical planning. First, we may not be allowing for species or functions that are critical to the health of the whole system. We only have detailed fine-filter knowledge of a fraction of forest species. Most of those for which we have no knowledge are either invertebrates, below ground, or in the forest canopy, but we should not assume they are any less important than more obvious species. The second risk is that habitat is only one of many biological values that our forests provide. Forest landscapes regulate water and nutrient flows, fix atmospheric carbon, influence climatic patterns, and absorb and transform atmospheric and water-borne toxins (Davis 1993). These functions are believed to be fundamental to the long-term sustainability of all aspects of forest systems, including the production of timber. Tactical planning which considers only habitat values for a selected number of species is unlikely to sustain all levels and types of biological values.

In response to these shortfalls, scientists have begun taking a more comprehensive view of understanding ecological systems. This type of research and management is often referred to as a "coarse filter" approach (Hunter 1990; Oliver 2001), since it is concerned with how the forest system functions as a whole, as opposed to the individual components therein. For example, by studying the patterns of vegetation change across a landscape over several decades and thousands of hectares, one can observe the dynamics of habitat availability and locations, and develop and test hypotheses about the regulation, interaction and movement of nutrients, water, and species, and thus be better equipped to sustain all these elements of biodiversity. Coarse-filter ecological knowledge thus provides the context (e.g., understanding of vegetation pattern dynamics) within which fine-filter knowledge (e.g., habitat requirements for species) is considered and applied.

The appeal of the coarse-filter approach to planners and managers is such that the field of landscape ecology has expanded considerably over the last two decades. Much of this research focuses on studying the types, sizes, frequencies, timing and severity of natural disturbance events, since disturbance is the process most responsible for long-term vegetation dynamics. Collectively, these attributes comprise a "disturbance regime" (White 1979).

Disturbance regime attributes and disturbance processes of the boreal forest are generally well understood (Chap. 8). Wind, insects, disease, floods, and even browsing by large mammals are all active disturbance agents, but the dominant disturbance vector in most parts of the boreal is stand-replacing forest fires that created patches of even-aged forest. On average, in Canada it takes 60–80 years to burn an area equivalent to the area of a given landscape (the "fire cycle"), but this can range between 20 and 500 years depending on the geographic location (Ward and Tithecott 1993; Bergeron et al. 2001). Annual variation in fire activity can be even greater due to both climatic shifts (Bergeron and Archambault 1993) and prolonged periods of drought (Bessie and Johnson 1995). For example, there is evidence to suggest that the fires of 1919 consumed over 10 million ha in Alberta and Saskatchewan.

There is also tremendous variation in the sizes of fires. The vast majority of fires are quite small. Very large fires (i.e., over 100 000 ha) are extremely rare events, but are responsible for the vast majority of the area consumed by fire in boreal-type landscapes (Ward and Tithecott 1993; Bergeron et al. 2000; Cumming 2001; Andison 2003).

We also know that although boreal wildfires are often referred to as stand replacing (Johnson 1992), there is evidence that severity varies considerably within any given fire (Malanson 1985). In the sub-boreal spruce landscapes of British Columbia, Clark (1994) found that both large and small individual trees survived fires in 28% of sampled stands, and in Saskatchewan, Andison (1999) found evidence of multiple stand ages in 42% of 550 sample plots. Islands of unburned residual trees in boreal landscapes (see Box 12.2) have been quantified by Eberhart and Woodard (1987), DeLong and Tanner (1996), and Bergeron et al. (2000).

Finally, there is also evidence to suggest that fire regimes can change dramatically over short distances. Differential fire sizes have been associated with topographic complexity, and fire susceptibility has been correlated with topographic position (Clark 1990). Several studies describe distinct fire regimes for different geophysical areas of landscapes (Suffling et al. 1982; Barrett and Arno 1991).

While such details on disturbance regimes are certainly valuable, of perhaps greater importance is simply the message that forest landscapes are highly dynamic. Considering forest landscapes as truly dynamic entities is a powerful planning concept, but a relatively recent one to most. Over several years a given piece of forest appears static to us because tree growth and forest succession generate gradual, often imperceptible, changes. One can return to the same hunting area every year and recognize the same patterns and landmarks. Based on these experiences, we tend to equate sustainability with constancy. This perception influences how we make forest management choices. However, if we consider a forest landscape mosaic over millions of hectares and several decades, a highly dynamic system is evident. For example, fire has consumed an average of 0.6% per year across a 7.5 million ha area in west-central Alberta since 1961 (Andison et al. 2003). In other words, even with fire control efforts, almost one-third of the area has been burnt in the last 40 years. Such fire activity can render a landscape mosaic virtually unrecognizable, unlike the more local example above.

The concept of the shifting landscape mosaic is particularly valuable for tactical planning. In landscape ecology terms, tactical planning is the process of developing, comparing, and choosing between various future landscape mosaics. Thus, the type, timing, sizes, and severity (i.e., what and where to leave as residual forest elements) of harvesting activities are all tactical-level decisions. Furthermore, having the benefit of "natural" patterns as a template for tactical planning creates an understandable and ecologically defendable foundation for decision making. Linked with fine-filter knowledge, it provides a holistic framework with which to consider biodiversity values. The attraction of this framework is such that many of the techniques currently being adopted for tactical planning involve natural pattern foundations to some degree.

In the end, ecological knowledge is perhaps the most powerful decision aid there is for tactical planning, particularly with the rise of landscape ecology. It allows uncertainties to be clarified, risks and thresholds to be identified, informed tradeoffs to be made, and it reinforces how reliant we are on the continued health of the entire forest

Box 12.2. *Morphology of a forest fire.*

It is easy to think of a forest fire as a continuously disturbed area scattered with a few residual individuals and clumps of unburnt material. Reality is much more complex and variable. Fire events usually involve numerous individual burnt patches. The type, number, size, and spatial arrangement of unburnt patches within each disturbance event are equally important "natural pattern" considerations. In fact, on average, almost one-third of each fire event in the Alberta foothills is at least partially unburnt (Fig. 12.2).

Fig. 12.2. Smith Creek fire from 1956, which burned a gross area of 9117 ha near Hinton, Alberta.

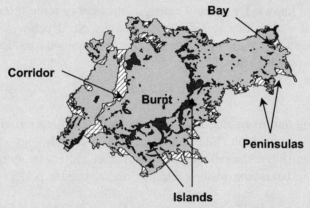

There are two major types of residual material within a fire: "islands" and "matrix". Residual matrix consists of peninsulas and corridors within the greater event area, but are still physically connected to the surrounding forest matrix. Matrix residuals account for 0–50% of the total area of a fire event, and averages 22% in the Alberta Foothills. Matrix residuals include both forested and non-forested areas.

Island residual patches are physically disconnected from the matrix, and thus completely surrounded by disturbed forest. Island residuals account for 0–20% of the area of a fire event, averaging about 9% in the Foothills.

The distinction between matrix and island residual material is a subtle one. For instance, if the fire shown above burned for even one more day, it is not difficult to see how the corridor, or any number of peninsulas could lead to the creation of islands. Depending on the importance of size and isolation to the survival of particular organisms, the distinction between islands and interior matrix residual patches may be functionally important or largely an artefact of classification. Nevertheless, all forms of residual vegetation and structure contribute internal heterogeneity in the wake of a natural disturbance event. This conceptual model of a forest fire is quite valuable. For instance, restricting ourselves to the question of island residuals would significantly underestimate the actual area of residual material in a fire event. Islands account for only 9% of the event area, while the total area in event residuals is almost 32%. The close relationship between island and matrix residual material also suggests that all forms of "residuals" should be considered as a package when planning cultural disturbance events. Finally, this model demonstrates highly variable mortality within so-called "stand-replacing" fire events, further supporting the notion of age complexity in the boreal forest.

Based on Quicknote #10 of the Foothills Model Forest Natural Disturbance Program, September 2001, by D.W. Andison system. Knowledge of disturbance regimes is also proving to be a valuable means by which we can convey the dynamic nature of landscape ecosystems. However, we must also be wary of the fact that science is by definition truth-in-progress. Our knowledge base will always be full of uncertainties, gaps, and assumptions. Despite these uncertainties, scientific conclusions are often presented as truth. If uncertainties, assumptions, and limitations are not disclosed, there will inevitably be a loss of credibility and confidence in the planning system because the public believed in the plan, its foundations, and its predictions. Furthermore, the most challenging part of science is not necessary *answering* questions, but *asking* them. Too often the relationship between the manager (who asks questions) and scientist (who answers them) is disconnected. Perhaps we do not yet have enough knowledge to truly manage forest ecosystems sustainably (Hobbs 1998; Chap. 8), but we are managing forests anyways. So it behoves us to be more responsible in understanding and applying science, asking good questions, and stating the limits of our knowledge (Box 12.3).

Data

The task of planning and managing landscapes is made much easier with reliable and plentiful data on the vegetation, soil, habitat, terrain, and hydrology (Pregitzer et al. 2001). Reliable, plentiful, and available spatial data can address many questions about the current status of a landscape, historical trends, and be used to project possible future conditions.

Unfortunately, spatial data in forestry are not always reliable or plentiful. Historical forestry spatial databases were designed for other uses at resolutions and extents often unsuitable for tactical planning, and those that are available are often incompatible with each other. For instance, newer forest inventories using cartographic projections (NAD) from 1983 cannot be overlain on older terrain data because it used NAD projections from 1927. Many of the older maps have not even been digitized, and of those that have, map scales, classes, and accuracy levels vary tremendously. Until we can resolve these issues, the suitability of historical data sets for spatial planning is often marginal at best (Waaub et al. 2000), and we must adjust our expectations and qualify any output using such data.

Fortunately, technology has risen to the challenge in the case of spatial data collection today. Modern tools, including data loggers, satellite imagery (Benton and Petterson 1999), and high-resolution digital photography (Pitt et al. 2000), can provide us with a dizzying array of data at low cost. Perhaps the most significant technological advancement in the past few decades has been the development of highly sophisticated hardware and software systems for displaying and analyzing spatial data (Weintraub and Davis 1996; Brown 2000). Geographic Information Systems (GIS), the Global Positioning System (GPS), and remote sensing have all transformed planning by the ease with which complex spatial relationships can be documented, studied, and communicated (Lachowski et al. 2000). Thanks to powerful integrated software packages, harvest plans can be overlaid onto soils and terrain maps to help identify sensitive sites and possible road locations (Brown 2000), fires can be mapped in detail, and the impact of proposed management activities on hiking trails can be compared visually.

Box 12.3. *Reliable ecological knowledge for tactical-level planning.*

Planning is one step in achieving the larger goal of sustainable forest management (SFM), and the successful outcome of any planning exercise relies on access to information of good quality. In the case of tactical planning, the availability of reliable ecological information for most forest systems in Canada has not kept pace with the advancement of new concepts in forest management. This is not due to a lack of financial and intellectual investment, but rather results from the challenge of acquiring such information at relevant spatial and temporal scales. Given the breadth of values we now expect our managed forests to provide, and the resultant rapid changes in forest management strategies, the lag in ecological research is not surprising. The following example serves to illustrate these issues for the boreal forest of western Canada.

When allocation of land to forest products companies expanded in the 1980s in the western boreal, many concerns over the conservation of biodiversity were raised. This precipitated a flurry of ecological research, but the initial focus was driven primarily by operating guidelines of the day (e.g., two-pass clearcut logging), and conservation paradigms imported from other systems (e.g., an emphasis on fragmentation and ecological response issues). While these studies have yielded valuable information, their application to tactical planning is constrained due to the limited spatial and temporal scales at which questions were posed: the emphasis was on stand or patch-level responses, independent of landscape context, and results were short-term.

The next wave of research was stimulated by promotion of a natural disturbance-based model for forest management. Fire-origin stands were directly compared to harvest-origin stands to test the assumption that clearcut harvesting could act as a surrogate for fire as a disturbance agent, and efforts were made to scale up investigations to address landscape-level issues. Again, some interesting findings resulted from these studies, not least of which was that type of disturbance and landscape context influence biotic response. But concrete guidelines for tactical planning were still not forthcoming (see Chap. 9).

Current research projects have responded in a somewhat divergent fashion to evolving issues. A growing emphasis is being placed on studying stand-level options for maintaining residual structure post-harvest, and on the exploration of intensive forestry (Chap. 13). A major challenge with these initiatives will be to quantify spatially explicit tradeoffs between stand, landscape, and forest-level objectives. Increased awareness that multiple land uses result in cumulative effects that influence forest management planning has resulted in ambitious attempts to apply broad-scale survey techniques, and measure responses at relevant scales. As well, the identification of meaningful indicators has received greater research attention.

What should be apparent from this example is that just as the priorities of, and planning tools available to, forest managers have evolved, so too have ecological research questions and approaches. As the two move forward in concert, there is a growing appreciation of the dynamic nature of the systems in which we operate, and the need for novel perspectives on the associated issues. Far from being crippled by a lack of reliable knowledge, we should make the best possible use of existing data through incorporation in models that test our assumptions about, and highlight uncertainties associated with, various forest management scenarios. Further, we should use this information to guide decisions about management options in order to further our understanding of system response and reduce key uncertainties. Such carefully designed management experiments will yield the most reliable knowledge, through active adaptive management (see Chap. 21).

While the solution sounds deceptively simple, translating this concept to reality is no small challenge. It requires a commitment from both managers and scientists to work together to ask better questions, explore innovative alternatives, embrace uncertainty, rally the public and institutional support necessary to overcome barriers to implementation, and to invest in long-term monitoring of outcomes. Tactical-level planning can provide the catalyst for many of these elements, if we are willing to think "outside the box".

By F.K.A. Schmiegelow (University of Alberta,
Department of Renewable Resources, Edmonton, Alberta)

The resolution and precision of currently available data is far beyond what we have had in the past, and is having its effect on management decisions. For instance, satellite imagery from the right time of year can determine understory species presence (Hall et al. 2000). In the boreal forest, this is nothing short of revolutionary, given the amount of understory white spruce (*Picea glauca*) suspected to exist where inventories have historically identified pure stands of trembling aspen, *Populus tremuloides* (Lieffers et al. 1996).

However, like all technological advancements, improved data comes with high costs and high risks. Until perhaps 20 years ago, the only necessary spatial data were forest inventories, which were done on 10- to 20-year cycles. Not only are we now demanding more data layers at higher levels of resolution, but also at higher frequencies. While one can never have enough knowledge, there is definitely a point at which one can have too much data. The cost of the various types of spatial inventories is a significant capital expenditure, and in some cases a questionable investment. We thus have to be careful to discriminate between data collection for its own sake (i.e., conducted because *we can*) versus data collection towards addressing specific questions, models, or decisions. For instance, it is not necessary to have highly detailed vegetation or topography data across an entire forest management area in order to make wood supply or recreation decisions. On the other hand, detailed topographic and vegetation overlays are invaluable for planning roads, silvicultural activities, and viewscape (or aesthetic) corridors. We also have to keep in mind that all data are designed for specific requirements, and extrapolating to other uses is dangerous. Forest inventories today are highly detailed and precise, but only poorly represent species habitat or biodiversity (Dussault et al. 2001; Nilsson and Gluck 2001) or fire history patterns (Andison 1999). As vital as raw spatial data are to the tactical planner, it is just as important that the data are of the right type, and at the right resolution to be useful.

Public input

Since tactical planning involves decisions about manipulating forests in time and space, it is often the point at which the public wants to be most involved. Given the problems forest managers face in translating often vague objectives into spatial reality, getting the public to help make tactical decisions is a valuable tool for the forest manager. Public input is thus not only a requirement of, but also a benefit to, tactical planning.

The issue of how and where to garner public input has been discussed at length elsewhere in this book (Chap. 4), and thus only an overview will be given here. Public input for tactical planning can take one of two forms: either enveloped within a larger public input strategy (which usually has the broader goal of identifying forest values), or focused specifically on the problem of making tactical decisions.

Public input as part of a holistic process

The simplest approach to gaining public input is to use permanent advisory boards, whose rotating membership is meant to represent a cross-section of society. Although advisory boards are used for a wide range of forest management issues, they can be a tremendous asset for picking tactical plan scenarios. Simply having advisory board involvement during the tactical planning process can alleviate the discomfort of others that decision making is biased.

A similar, but slightly more formal on-going relationship with the public can be generated through "co-management" boards where both local and high-level stakeholders are involved directly in decision making (Beckley 1998). Local-level stakeholders are solicited as part of the strategic planning cycle, and then given the chance to respond to, and participate in the development of tactical planning scenarios, and even operational forestry activities (examples are discussed in Chaps. 4 and 22).

An even more decentralized decision-making strategy is known as "community forestry". The idea is to allow local residents, stakeholders, and regulators to work together to not only come up with, but also to implement management solutions. This is especially important when it comes to translating strategic directions into tactical plans because the local public typically has greater familiarity with the land base and its many values and peculiarities than does any one planner or technical specialist. Community forestry has been hailed by some as the "key to sustainability" (Gray et al. 2001) where it corresponds to the appropriate scale to manage natural systems (Luckert 1999). However, it has been criticized for resulting in local-level special-interest compromises (Buttoud 2000), not representing the needs of non-locals who may have a legitimate stake in the process, and for being inefficient (Luckert 1999).

Public input as a discrete part of tactical planning

Public input for tactical planning can also be arranged directly. The simplest and most common technique is the use of focused local meetings or open houses. This usually involves creating several fully developed tactical plan scenarios and explaining the implications of each in lay terms. Feedback from such exercises is used to help select the final plan option, although in most cases there are a limited number of options from which to choose. The public presentation of tactical planning scenarios is now a minimum requirement for most forest management licence agreements in Canada.

Although ad-hoc meetings are the easiest way to involve the public, they are also the most limited, since they only play a part in planning near the end of the process. Plan scenarios, criteria for success, and decision variables of greatest interest to local participants are all predetermined prior to public consultation. In response to these shortfalls, it is possible to engage a small group of public representatives earlier to help define tactical planning options. Using spatial data, computer projection models, and visualization tools (see Chap. 14), and knowledge and understanding of habitat and other biodiversity needs, various plan options can be developed, explored, and compared using relevant criteria. Understandably, such a process is highly complex and requires an overarching framework, which includes the development of many other elements and data sets to be successful and effective. It may also require several such meetings or workshops to work through the range of issues and to enable participants to digest and reflect on what has been learned (Checkland and Scholes 1990). However, this is potentially one of the most effective ways of using public input for tactical planning. Such tools will be discussed in more detail later in this chapter.

How and where public input is used ultimately depends on the land base in question, and the local issues and culture. However, it is safe to say that the reliance of tactical planning on public input is greater than ever before, and continues to grow.

Plan emphasis

Most tactical planning exercises today are extensions of the historical forest planning philosophy, which attempts to identify, quantify, and satisfy a wide range of values against the ultimate objective of maximizing harvested timber volume or value. In other words, the primary emphasis is still usually on timber values, although the importance of other values has risen dramatically. The fact that allowable harvest levels are not normally a part of tactical planning exercises substantiates this observation.

However, that trend is rapidly changing, and will likely continue to do so. Timber-driven management emphasis has widely given way to multiple-use philosophies, and some are even moving towards a much stronger ecological focus. Furthermore, there is a growing trend towards partitioning land into areas of different emphases to avoid the problem of trying to satisfy all needs and values with a single plan, in a single area.

While a discussion of deciding on the management emphasis of a landscape is beyond the scope of this chapter, it is relevant to the choice of tools and techniques used to develop tactical plans. For example, one would presumably not choose optimization models (see below) to develop a tactical plan for a community-based plan heavily oriented towards non-traditional resource values.

Single-zone emphasis: ecological foundation

There is now an awareness that the ecological health of forests is necessary for those forests to continue to provide a wide array of values. More specifically, there is mounting concern that the idea of trying to maximize fibre production against other values such as recreation or biodiversity will not produce the desired objectives (Gilmore 1997). This has led to the concept of "ecosystem management". As proposed by Franklin (1989), ecosystem management supposes that ecological, social, and economic values are equally important. While there are many different definitions of ecosystem management, all would agree that it includes a scientific, ecological foundation for making management decisions and that it supports ecological integrity and sustainability (Franklin 1997; Gilmore 1997; Gordon and Lyons 1997).

The scientific foundation many have adopted for planning today is that of emulating natural disturbance patterns. Certainly the theory is attractive. By using natural patterns as a template on which to design our own disturbance activities, the risk of losing biological function and diversity at all scales is minimized, since the disturbance regime is evolutionarily familiar (Hansen et al. 1991; Franklin 1993; Noss 1998; Cissel et al. 1999). The key to using a natural pattern emulation strategy is thus to maintain the rates, intensities, and magnitudes of disturbances within their natural range of variation (NRV) (Landres et al. 1999; Chaps. 8 and 9). For example, Mistik Management in Saskatchewan (further discussed in Chaps. 4 and 22) has been sponsoring natural pattern research for several years with the intention of integrating this knowledge into plans. Their first attempt at an NRV-based tactical-scale plan borrowed heavily from the concept of a disturbance event, and is discussed in more detail in Box 12.4.

The NRV management strategy has attracted much attention and caused a good deal of debate over the last several years. While few argue with the principle of using NRV as a tool for biodiversity conservation, the question of how it is used in forest planning is another matter (see Chap. 9 for further discussion). For example, by committing to a natural disturbance emulation as the basis for planning, must all related pattern metrics (such as patch size, disturbance frequency, and so on) stay within the natural range? There are obvious problems with this all-or-nothing approach. For instance, harvest blocks hundreds of thousands of hectares in area are not likely to be socially acceptable. Similarly, although wildfires burn through forests of all ages and types, stands in the boreal forest younger than approximately 60 years of age are not merchantable. Other problems are that "natural" disturbance regimes are known to fluctuate in response to climatic and anthropogenic pressures (Weir et al. 2000).

A more reasonable use of an NRV foundation is less extreme, in which critical NRV parameters are established as a preliminary step to tactical planning and used as a baseline, or base-case scenario. The decision regarding the degree and manner in which such a natural baseline is used depends on other (ecological, social, and economic) objectives. A plan may deliberately stray beyond NRV for other reasons (such as mitigation costs, or risk of property loss). For example, while it is not reasonable to expect the elimination of roads in the boreal forest, there is no natural analogue. This suggests that the thoughtful deployment of roads should be one of the more critical aspects of tactical planning (see Box 9.7; Forman et al. 2003). The point is that we now have the capacity to identify and evaluate options, and make decisions in a more objective and conscious manner, because we have an ecological starting point.

As a starting point, trying to emulate the pattern of natural disturbances through forest harvesting is a very simple, but powerful concept. As we learn more about how natural patterns relate to other tactical objectives, we are finding more harmonies than conflicts. For example, in Scenario A of Box 12.4, it is not difficult to see how this pattern might align well with many fine-filter habitat objectives for interior-dependent forest species, since it means less edge, greater opportunities to allow for large undisturbed old forest areas elsewhere, and fewer roads. Furthermore, the concentration of management activities in one area (as opposed to the typical dispersed pattern) is cost effective, efficient, and can be more aesthetically pleasing. These discoveries are likely to lead to more extensive use of NRV knowledge in future applications.

On the other hand, we must be careful about how disturbance emulation models are promoted and applied. Because of (*a*) the removal of logs, (*b*) the use of heavy machinery, and (*c*) the construction of forest access roads, modern timber harvesting cannot closely mimic natural disturbances such as fire (Chap. 9). Furthermore, by simply increasing the average size of harvest blocks, one cannot claim to be emulating natural disturbance patterns. There are many types and scales of natural patterns, and they should be thought of as a package. While not all types and scales of natural patterns can be sustained, it is possible to commit to the *principle* of NRV, and to implement a strategy that acknowledges that the integration of consumptive and protective planning will evolve over time. For example, the Hinton Division of Weldwood of Canada has

Box 12.4. Emulating a natural disturbance event.

Mistik Management Ltd. manages a 1.8 million ha boreal mixedwood Forest Management Area in west-central Saskatchewan (see Chap. 22). With the cooperation of the Saskatchewan government, Mistik planned and has completed harvesting a single 11 000 ha operating area as a single disturbance "event" during 2001 and 2002. The design was based on research on the patterns of natural fire events in the area. An outline of the final plan (scenario A in Fig. 12.3) is compared with a more traditional first pass design in scenario B. Included in the comparison is the minimum network of semi-permanent roads necessary to maintain access for future treatments and (or) harvesting. Light grey represents forest >90 years old (i.e., mature); dark grey represents forest that is younger, and non-forested areas are in white.

Fig. 12.3. The deployment of roads and cutblocks is one of the most visible results of tactical forest planning and landscape design. Scenario modelling can demonstrate the visual, habitat, and economic consequences of following alternative development templates, such as (*a*) emulation of a wildfire burn pattern, and (*b*) a dispersed cutblock pattern.

(*a*) 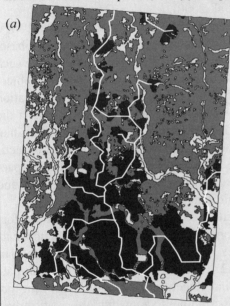   (*b*)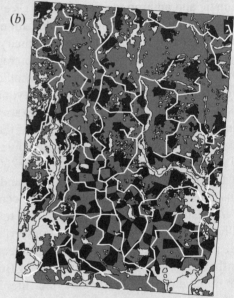

(Emulation) HARVESTING SCENARIO A
2678 ha harvested in 31 blocks
Ave. patch size = 84 ha (range 1–1104 ha)
Total disturbance edge = 167 km
50 km of roads

(Traditional) HARVESTING SCENARIO B
2680 ha harvested in 129 blocks
Ave. patch size = 21 ha (range 3–65 ha)
Total disturbance edge = 326 km
122 km of roads

The differences in the two scenarios are striking. Although the total area harvested is virtually identical, adjacency and block size limits typical of harvesting regulations today (scenario B) create almost twice as much disturbance edge, and 2.5 times as many roads as the natural pattern emulation design in scenario A. Adjacency limitations in scenario B also force harvesting to take place across the entire compartment, effectively fragmenting all of the unharvested mature forest. Looking ahead, scenario A gives Mistik the opportunity to recover roadbeds and "walk away" from this compartment for a number of years, or even decades, until the (less mature) wood in the north is ready for harvesting. In scenario B, activity in the compartment will be continuous for at least 20 years until all of the available wood is removed, during which time all of the roads must be maintained. Mistik expects reasonable cost savings in road construction and maintenance alone.

> **Box 12.4. (concluded).**
>
> The impacts of the two scenarios on wildlife and habitat are not difficult to envision. More roads certainly create travel corridors for predators, but far more important is that they allow greater access by humans, which means greater hunting pressure. Increased forest edge means less interior habitat, which means a shift in use among bird and mammal species. It is visually obvious how interior mature forest declines sharply from scenario A to scenario B. Finally, while many large ungulates may not mind sharing the landscape with us, generally the quicker timber harvesters can be in and out of an area, the more likely we are to maintain a greater variety of species.

invested considerable resources in understanding the natural disturbance regime of their management area through the Foothills Model Forest (see FMF n.d.) and generated a list of 78 natural pattern metrics, of which 20 have been integrated (R. Bonar, Chief Biologist, Weldwood of Canada Ltd., Hinton Division, Hinton, Alberta, personal communication, 2002). They have also assembled a natural pattern integration team to facilitate the process of defining and integrating these and future metrics over the next several years. So, without committing to absolute mimicry (of potentially infeasible and socially unacceptable natural patterns), Weldwood *has* committed to understanding and testing NRV as a package, and implementing those pieces that are achievable over time.

It should also be noted here that while not discussed, there is no reason why the primary plan emphasis on a given landscape could not be cultural (as opposed to timber or ecological). In fact, in many northern Canadian communities, this may very well already be the case.

Multiple-zone emphasis

Another byproduct of the increasing complexity of planning is trying to manage for all values across all landscapes. As discussed above, many of our objectives are conflicting, and there is increasing concern that ecological objectives are being unduly compromised to the point where we risk system degradation. Conversely, there are concerns over compromising critical economic or social values to facilitate ecological requirements. Regardless of whether the emphasis is timber, biodiversity, or cultural, it is not uncommon for many participants of a planning exercise to feel as if they have compromised their values in an attempt to satisfy all needs.

In response to this predicament, land bases may be partitioned into two or more areas of different resource emphasis. For instance, one zone might define a protected area in which all cultural activity (including harvesting) is excluded or kept to a minimum (Norton 1999). Another intensively managed zone might focus mainly on the production of wood, and a third zone perhaps managed extensively to include a mix of ecological, social, and economic values. This three-sided, or "triad" model of zoning is the one most commonly cited (Seymour and Hunter 1992; Burton 1995), but there is no reason why different zones or a different number of zone classes could not be used. The idea is simply to allow for different levels and types of management philosophies on different parts of the landscape, and for management of certain values or products to be concentrated where they can be produced most efficiently.

Adoption of a zoned approach to forest management is more of a strategic decision, so this topic is appropriately discussed in Chap. 11 as well. However, many of the issues cross over into tactical planning responsibilities. For instance, questions of the size and location of protected zones involve moderately detailed spatial knowledge of the distribution of organisms, habitat types, and disturbance regimes. Past efforts to set aside a certain percentage of land for conservation purposes have been criticized for leaving small, poorly designed, unconnected pieces of land that are not representative of the region, state, or province (Pressey and Logan 1998; Li et al. 1999; Norton 1999). In other words, the links to tactical reality were weak. Which forest values can still be accommodated in intensively and extensively managed zones, and which can only be protected in reserves? Can all non-timber, non-wilderness forest values be accommodated in extensively managed forest land, or will we need different zones for berry production, recreation, and wildlife? Perhaps the most important unresolved tactical issue with respect to adopting a zoning approach in the boreal forest is where and how to select land for intensive forestry, given the questionable reality of productivity gains in much of the North (but see Chap. 13).

In summary, the decision to adopt a zoning or triad approach to management is clearly a strategic one. However, there are many critical tactical issues that must be taken into account during and after this strategic choice. If and when zoning becomes reality, many of the tools and techniques discussed in this chapter would apply to defining the locations, sizes, uses of, and the relationships between different zones.

Connecting elements

Armed with ecological knowledge, spatial data, public input, and a plan emphasis philosophy, there is a wide range of tools and techniques available to the tactical planner, including defining appropriate spatial units, reference points for planning and monitoring, and models. These are referred to here as "connecting elements" because while none are *necessary* to conduct tactical planning, they are all valuable tools by which it is facilitated.

Defining tactical spatial units

Our spatial vocabulary in forest planning was historically limited to "stands", defined by forest inventories using structural and compositional characteristics, and very large "forests", defined by administrative boundaries of responsibility or ownership. Operational planning corresponds to the stand level, and strategic planning corresponds to the forest level. Consistent with the concept of the planning hierarchy, it would be valuable to have an intermediate spatial scale that corresponds to the objectives of tactical planning. For the sake of argument, we refer to these entities here as "sub-landscapes". Boreal "landscapes" are generally thought of as larger, more inclusive ecological zones, and generally associated with strategic planning.

The appropriate size of sub-landscapes, and the means by which they are defined is highly debatable. The simplest approach is to adopt existing intermediate-scale spatial entities where they exist. For example, "operating areas" or "compartments" already exist within most managed forests across Canada. Operating areas were originally designed for the purposes of distributing and sequencing road building and harvesting activities for tactical planning purposes and thus were mainly an administrative delineation, albeit a convenient one.

Others argue that the boundaries of tactical planning spatial entities should be redefined to correspond to ecological zones, since species and processes do not respect administrative boundaries (Waaub et al. 2000). For example, Alberta has been classified into eco-districts, sub-regions, and regions according to dominant vegetation, land form, elevation, and climate (Beckingham et al. 1996). Most provinces in Canada have similar land classification systems. In some cases, disturbance regimes have even been found to be associated with such ecological zones (Andison 1998, 2003), which adds further credibility to their use. On the other hand, since tactical planning is largely about communication and understanding the impact of decisions on the ground, pure ecological boundaries may not be entirely useful, since the average person cannot relate to them (Oliver et al. 1999). For example, eco-regions were not found to be useful for parks, conservation, or land management needs (Wright et al. 1999).

Another possibility is to adopt watersheds as intermediate spatial units. As the name suggests, watersheds are catchment areas defined by the direction of the flow of water. Conceptually, the use of watersheds is attractive, since they have both terrestrial and aquatic ecological meaning, but (*a*) they may be irrelevant from a social or economic point of view, and (*b*) our ability to define watersheds using existing terrain data on very flat areas of the boreal forest is limited (though see Box 10.10).

It is also possible to define spatial planning units purely from a cultural perspective. For example, where Aboriginal concerns and non-timber forest products are prevalent, registered trap lines or traditional First Nations clan territories might form convenient units for intermediate-scale planning (see Chap. 3 and the Mistik Management case study in Chap. 22).

Other factors that must be considered in the definition of spatial units is their size, and their relationship to each other. At 5–10 000 ha, operating areas may be too small for effective tactical planning purposes, and would require many dozens to cover a single defined forest area under management. However, the planning problem necessarily becomes more complex and more difficult for people to relate to at larger spatial units (Anderson et al. 2001). Boundary location is equally important because, regardless of the criteria used to delineate them, water, seeds, animals, and nutrients will continue to flow through and between any spatial entities we choose to define. Essentially, any spatial units we define are nothing more than cultural constructs identified for planning convenience.

In the end, the size and boundaries of sub-landscape units for tactical planning can be defined in many different ways, and the method must suit the needs (Babu and Reidhead 2000). Most agencies today rely on a combination of ecological, socio-economic, and administrative criteria to define sub-landscape entities for tactical planning purposes. The British Columbia Ministry of Forests defined "landscape units" across the entire province specifically for the purposes of tactical planning. The size of their landscape units varies up to 130 000 ha in the boreal region of the province, defined using a combination of ecological, physical, and access criteria (BCMOF and BCMELP

1999). Similarly, the Southern Rockies landscape planning pilot project in Alberta defined landscape units for tactical planning based on a combination of ecological, physical, practical, and administrative factors (AE and OOPDCI 2000).

Weldwood of Canada in Hinton, Alberta, defined "natural disturbance units" (NDUs) of between 10 000 and 30 000 ha based on ecological and practical attributes as the foundation for multi-scale analysis of seral-stage patterns across the forest management agreement area (see Box 12.5). This allows them to develop a nested hierarchy of spatial units to suit specific planning and management purposes (Bryce and Clarke 1996; Paredes 1996; Weintraub and Davis 1996). Thus, Weldwood essentially dealt with the problems of defining sub-landscapes by not defining them, but rather by identifying minimum-sized spatial units that can be grouped for either tactical, strategic, or operational purposes as the case may be.

Developing reference systems

A substantial challenge of tactical planning is to balance a wide range of values, some of which are mutually exclusive. For example, how does one make a reasonable trade-off between "*sustaining biodiversity values*" and "*increased recreational opportunities*"? The problem is that different objectives are not always quantifiable or logically linked, and when they are, the criteria for evaluating success are disparate. Tactical planning requires equitable methods of quantifying and comparing objectives, as well as defining objective measures of success.

Quantifiable objectives

The decision-making exercise is ultimately defenceless unless tactical objectives can be compared and contrasted in quantitative terms. Unfortunately, not all goals are easily translated. For example, aesthetics are an important value for most people, but highly subjective to define, let alone quantify. Some have suggested using "landscape professionals" to assess visual qualities, as they are more objective and consistent at rating landscapes than the general public. Others have explored ways of removing people from the process completely by quantifying scenes digitally using a standardized set of rules such as the percentage of tree or water cover, length of view, relative elevation change, and edge contrast (Palmer 2000).

It has also been suggested that for less tangible values such as aesthetics or recreational opportunities, we use opportunity costs to judge the value of *not* having certain landscape attributes (Gregory 1999; Buttoud 2000). However, even opportunity costs are notoriously subjective (Buttoud 2000). The true cost of a hiking trail that does not cross through harvest blocks will be different for different people.

Even ecological objectives can be challenging to quantify. Despite the depth of our fine-filter ecological knowledge, the choice of how many individuals of any particular species to sustain relies on our understanding of the habitat needs of that species. Even if we are right, the decision of how many individuals are "sufficient" or "sustainable" is often a judgement call.

One solution to this dilemma has been to adopt an NRV approach to quantifying biodiversity values. As previously discussed, the historical range of vegetation pattern

Box 12.5. Multi-scale planning.

Weldwood of Canada Ltd., Hinton Division, manages a 1 million ha area of foothills forest in Alberta. Concerned with finding ways to integrate natural range of variability (NRV) concepts into spatial reality, Weldwood sponsored a landscape simulation exercise that generated 100 historical possible landscape "snapshots" based on the historical rates, sizes, and shapes of fires generated from fire history research conducted by the Foothills Model Forest. The percent of each of four seral stages in four cover types was captured in 20-year intervals at five different spatial scales from 30 000 to 480 000 ha using a stochastic landscape disturbance simulation model, LANDMINE (Andison 1998), as shown in Fig. 12.4. The 30 000 ha areas are Weldwood's "natural disturbance units" (NDUs).

Fig. 12.4. Abundance of old (>180 years) spruce-dominated forest in landscapes of five different scales (1–16 Natural Disturbance Units) in the Upper Foothills natural subregion of the Weldwood Forest Management Area in Alberta, simulated using the stochastic model, LANDMINE. The natural range of variability (NRV) in the abundance and probability of old-growth spruce-dominated forest will depend on the scale under consideration.

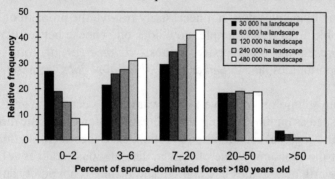

The output was used as an NRV baseline with which to compare the vegetation patterns of management options at a range of scales in time and space. For example, the simulation output suggested that there is no single best scale at which to manage old forest. At 30 000 ha, the dynamics of long-term fire behaviour are such that small amounts (<2%) of old spruce occurred 27% of the time. However, on 480 000 ha landscapes, very small amounts of old spruce occurred only 6% of the time. This pattern is not unexpected; large, infrequent fires can easily burn through most of one or more 30 000 ha NDUs in any given year (effectively eliminating old forest from that area), but it is less likely that fire will consume a 480 000 ha landscape. Thus there is no single spatial scale at which it is more appropriate to manage or monitor old forest.

Weldwood translated this understanding of old forest dynamics into an integrated set of long-range target ranges for large landscapes, combined with a strategy of planning for variable percentages of "potential" old-growth areas between NDUs, which allows for the dynamic nature of old forest.

The introduction of a nested hierarchy of spatial scales, anchored by the NDU concept and NRV baseline delineation, provides several benefits. First, it allows Weldwood to move beyond the artificial scales of "stand" and "landscape" to a more ecologically meaningful continuum. The scales at which planning should consider seral-stage percentages, cover-type percentages, and disturbance frequency are likely slightly different. Secondly, it allows for the logical integration of fine-filter and coarse-filter knowledge. For instance, Weldwood used the NRV pattern output from the simulation results as input for habitat supply models to determine NRV baselines for various habitat values. Finally, tracking and comparing patterns at a range of scales will help Weldwood avoid NRV "cherry picking". In our haste to integrate NRV concepts, it is too easy to adopt only the most obvious or convenient pattern changes.

metrics such as patch sizes and disturbance frequency are more objective, represent natural phenomena, and require no translation to be used in tactical planning. However, there is still a difference of opinion about which pattern measurements are important. For instance, the importance of connectivity, adjacency, and isolation for wildlife has been elegantly demonstrated theoretically (Gardner et al. 1989) and in selected instances empirically (Machtans et al. 1996; Desrochers and Hannon 1997). However, the value of forest connectivity was first recognized in the American Midwest where most of the natural vegetation has been replaced by intensive agriculture. The value of such attributes for the extensively managed boreal forest is debatable (Bunnell 1998; Hobbs 1998). Similarly, edge density is a commonly cited landscape attribute, yet its universal ecological relevance is not definitive (Cotterill and Hannon 1999; Ibarzabal and Desrochers 2001).

Comparing objectives

Quantifying plan objectives does not necessarily resolve the problem of how to reconcile disparate objectives. For example, how does one choose between "*patch-size distribution*" representing biodiversity values, "*kilometres of road*" representing recreational opportunities, and "*number of local full-time jobs*" representing local economic growth?

One possible solution is to convert other values into currency consistent with timber. Activities such as hunting, fishing, and recreation often contribute directly and indirectly to local and regional economies, and estimates of these contributions can be used directly against the economic value of timber. The obvious weakness of this approach is that the true worth of activities like hunting and fishing is not purely an economic one (Schreier et al. 1996; see Chap. 6).

Others argue that since all values are ultimately derived from the forest, we should be describing those values in terms of forest composition, structure, and pattern (Erdle and Sullivan 1998). For example, preference surveys formulated in such terms have been used to solicit choices of alternative management outcomes (Boxall and Macnab 2000). More powerful, interactive methods generate and present a range of future management scenarios that produce a set of possible outcomes in visual, graphical, and tabular form (Schreier et al. 1996; Waaub et al. 2000). The same idea can also be applied directly to a specific planning solution where the most desirable future forest landscape condition is chosen from among several options as the one most likely to achieve a set of objectives (Yamasaki et al. 2001; Andison et al. 2002). This has been referred to as "objective-based" or "target-oriented" forestry (Liu et al. 2000). The logic of target-oriented planning is that if a desired future state or condition of the forest can be defined, then the details of how to achieve it can be left to the technical planners (Nilsson and Gluck 2001).

There is considerable merit in this concept. For one thing, it forces us to focus on the relationship between our values and actual outcomes or products (Erdle and Sullivan 1998). It also allows the potential to move away from timber value as the primary focus of decision making. The "future forest condition" (FFC) concept also provides an objective set of decision variable outputs that have far less chance of being misinterpreted. Lastly, it also explicitly defines forest management activities as an agent of change of forest conditions, and only of values indirectly. This is a powerful and necessary message.

On the other hand, adopting the FFC concept is not a panacea. For instance, the method still assumes that the relationship between actions and outcomes is known and that we can differentiate between forest structures, patterns, and conditions that are "good" and "bad". The link to an NRV foundation is usually made for this reason. Lastly, defining a FFC also relies on high quality, complete, and suitable databases, and assumes that stakeholders understand enough about their values and the forest ecosystem to translate objectives into forest structure, composition, or pattern. Both are rare and expensive to develop.

Indicators of success

The third and last reference point valuable for tactical planning is the establishment of the criteria for measuring plan success. It is one thing to make informed decisions about how to best manage a piece of land, but quite another to be successful in achieving goals through those choices. There are many reasons why the intended future forest may not occur, or may not provide the values we had hoped. Recall that uncertainty and risk exist in many places, and to varying degrees. We thus need to establish a set of indicators that allow us to compare our plan predictions against actual outcomes.

To be truly effective, indicators must be understood, measurable, cost efficient, sensitive (to change), and achievable. The development of indicators (see Chap. 21) has challenged virtually all jurisdictions across Canada. At least part of the problem has been the separation of the planning process from the monitoring process. Indicators should be considered during tactical planning, not afterwards. The link is not as onerous as one might think. If our tactical planning objectives are phrased in predictive, quantitative, meaningful terms, then logically those objectives become the indicators. In other words, the output from tactical planning should ideally be a set of indicators and targets that are ultimately used to evaluate the intentions of the tactical plan. For example, if a plan objective is "*At least* X *kilometres of new roads will be built in the next year*", then the resulting indicator is logically "*kilometres of roads built annually*". Monitoring programs such as the Saskatchewan Forest Effect Monitoring Program (Andison et al. 2002) have adopted this direct link, to ensure that indicators are effective and that planning objectives relate to the real world.

Models

Models are by definition simplifications of reality (Sands 1988), and thus a valuable tool for the tactical planner who must understand and communicate the complexities of tradeoffs and options in a dynamic forest landscape. Models are found throughout the forest planning and management process, and are more thoroughly explored in Chap. 14. The discussion of models here will be limited to those specifically used in understanding and communicating elements of the tactical planning process. Decision-support models will be discussed in the next section.

Many types of models for the tactical planner are available. Ecological and forest dynamics models help to understand the dynamic needs of species and biological functions. For instance, fine-filter ecological habitat knowledge is captured and used in habitat suitability models (Railsback 2001). Such models state species requirements in terms that can be quantified, projected, compared, and contrasted to other habitat needs and coarse-filter descriptions of the FFC. Growth and yield model are capable of projecting stand volume growth of different tree species decades into the future in response to various site treatments, soil conditions, and even climate change.

There are also models for understanding and communicating the dynamic range of coarse-filter ecological patterns. The problem faced in the study and application of natural disturbance patterns is that disturbance activity takes place over such long time periods, making it difficult to generate these relationships empirically. As a result, stochastic simulation models such as LANDIS (Mladenoff et al. 1996), SELES (Fall and Fall 2001), and LANDMINE (Andison 1998) are used to generate dozens, or even hundreds of landscape possibilities using probabilities of disturbance sizes, shapes, frequencies, and severity based on the natural fire regime. This type of modelling can be used to generate NRV for many salient pattern features as well as compare them to the current or projected range of variation. Chapter 14 provides more detail on these types of models.

Some of the more powerful models at our disposal for communicating and understanding are not computer-based. Tables and graphs are precise and efficient ways of summarizing large amounts of data. However, they are not necessarily in a form understandable to everyone, and are not spatially explicit. For instance, a graph comparing deer habitat and hunting opportunities against different levels of local harvesting is a compelling way to demonstrate relationships between objectives, but does not take into account the spatial arrangement of harvesting, and may oversimplify some complex habitat relationships.

Maps and photographs are simple and understandable ways of summarizing large amounts of spatial data, but contain information that must be carefully interpreted (Wilson and McGaughey 2000). For example, the same landscape can be represented in very different ways depending on the classification scheme. A map of old-growth forest defined by anything older than 100 years will look very different than one defined by anything older than 140 years.

Maps and photos also suffer from being two-dimensional static "snapshots" in time, unable to show landscape dynamics, which is important for tactical planning. Technological advances have helped overcome this problem through the use of 3D visualization and computer-aided video techniques (Wilson and McGaughey 2000). It is now possible to not only "see" the landscapes that would develop from various planning scenarios virtually, but also to see how they change over time (Scott 2001). Furthermore, photo-realistic computer graphic techniques have the advantage of less interpretation, while providing a realistic image, and will undoubtedly see greater use in the future.

Tactical decision-support tools

So far, the discussion has been limited to the different elements that comprise tactical planning. But knowledge, data, public input, models, indicators, and an overarching philosophy must be brought together at some point towards making some decisions. A specialized type of model that combines many of the elements described above is called a "decision-support system" or DSS. Several different types of DSS models are available, but they are not necessarily mutually exclusive. In most cases, a combination of the systems discussed below is used.

Guidelines

A common approach to tactical planning is the adoption of rules, regulations, and guidelines. Guidelines for planning take many forms. For instance, almost every forest management company must comply with standard operating "ground rules", which dictate tactical-level decisions such as harvest block size, the number of years before adjacent areas may be cut, and which areas of the land base are eligible for harvesting. Guidelines also exist for protecting the habitat of particular species, aesthetics, water, and for preventing soil erosion and compaction. Many such guidelines include tactical-level specifications for a range of, or limits to, harvest block size, locations, connections, and amount or dispersion of residual trees. There are also guidelines to protect specific values, such as riparian corridors or unique ecological and cultural features.

The most recent additions to the guideline family are those which dictate the design of harvesting activities based on natural disturbance patterns. British Columbia was the first in Canada to produce landscape design guidelines, which establish seral-stage and patch-size targets based on five broad natural disturbance regimes (BCMOF and BCE 1995), which then must be integrated with a number of other management objectives in individual landscape units (BCMOF and BCMELP 1999). Manitoba has developed similar natural pattern emulation guidelines (Ehnes and Sidders 2002).

Guidelines are advantageous for several reasons. First, they are the best way of establishing and enforcing uniform, minimum standards. A higher level (strategic) planning exercise may identify certain thresholds (such as road density or percent of old forest) that are non-negotiable at the tactical planning level. In the same way, guidelines are often the only way to ensure the protection of important or special resources or values on landscapes (e.g., salt licks, breeding grounds, etc). Guidelines are also easy to implement and monitor, since they tend to be presented in a simplified form. Guidelines also can facilitate the adoption of a range of management philosophies or emphases. For example, different minimum environmental standards can be established for protected, intensive, and extensively managed zones of a triad. Lastly, guidelines are an effective conservative surrogate in the absence of knowledge or data. In other words, when we have insufficient knowledge or data with which to model a response to various management inputs, the next best option is to apply an educated rule-of-thumb that errs on the side of caution. For example, our concern over (and ignorance of the full range of) impacts of disturbance in riparian zones has led to simple no-harvest buffer zones.

On the other hand, guidelines are not a panacea, and the decision of when and to what degree to impose them on tactical planning should be made carefully. Guidelines have a long history in the management of the boreal forest in Canada, but that history began at a time when we knew far less, had limited access to spatial data and computer technology, and did not involve the public in decision making. Now, we can not only project probable vegetation patterns far into the future on a given landscape, but we can also make some predictions of the probable ecological impacts of those patterns. This changes our needs, and should change how we use guidelines.

There is no shortage of examples of forest regulations that have failed to achieve their stated goals. The British Columbia biodiversity guidelines (BCMOF and BCE 1995) were a laudable attempt to embody the greater goal of maintaining biodiversity. But a simulation exercise clearly demonstrated that it may not achieve its stated goals based on several key indicators of coarse-scale biodiversity at intermediate scales (Andison and Marshall 1999). Similarly, a recent review of 35 forestry guidelines from Ontario concluded that the system was cumbersome, inconsistent, and has resulted in a "short-circuiting" of the intended planning process (AVESL et al. 2000). Even the endangered species regulations in the United States are highly criticized for being narrowly focused on individuals instead of a range of scales and measurements (Hobbs 1998).

There are many reasons for guideline failure at the tactical scale:

(1) Our scientific knowledge tells us that landscapes have unique physiographic and biotic fingerprints (Spies and Turner 1999). However, strategic regulation typically specifies a uniform set of rules that assumes that all landscapes and all sites within them are homogeneous and equal (Oliver et al. 1999). When followed unquestioningly, guidelines do not allow us to respond to what we know about natural, local, ecological variation. The danger is that we may be inadvertently simplifying, or even misdirecting, a complex ecological system.

(2) Landscapes have a unique set of social and economic values and needs. Good intentions aside, universal rules conflict with the efforts of gaining local-level public input. In other words, the perceived values of greater society may be well represented with guidelines, but those of local stakeholders may not. Furthermore, guidelines to some degree usurp the public consultation process because certain decisions are taken out of their hands.

(3) Strategically established, general-level rules severely limit the opportunities for creative, locally optimal, and adaptive solutions to changing conditions (Gilmore 1997; Oliver et al. 1999). Guidelines result in undue restrictions, encourage lowest common denominator thinking, and in the end may produce a less thoughtful, sustainable balance (Baskent et al. 2000).

(4) A guideline insinuates a high level of certainty and finality with respect to the scientific basis for its rules. However, as discussed, scientific knowledge is always in a state of development and is never absolute.

(5) Guidelines represent a set of values, not plans. Even if we had complete scientific knowledge, it would be a mistake to assume that the scientific answer or a science-inspired direction is the most appropriate. Building forest management guidelines based on natural disturbance regimes has already become a reality (i.e., Ehnes and Sidders 2002). However, such models are narrowly focused on a particular set of values (biodiversity), at the expense of others. They assume not only that the sci-

ence behind the guides is right, but that the science and the guides are universally applicable. There is the real danger that the guide will become the plan rather than just informing it.

(6) When guidelines fail to achieve either higher level or local objectives, the perception is often that tactical planning failed, since this is the most obvious level of planning for most people. Rather, the failure of guidelines to achieve the desired goals is a strategic failure, since this is the level of decision making (Chap 11) that imposes guidelines.

(7) Guidelines are created to address individual issues, and tend not to be integrated. The potential for conflict and inconsistency among guidelines is very real (e.g., AVESL et al. 2000).

Guidelines have their place in forest planning, but the time has come to seriously examine the circumstances under which they are appropriate. The worst-case scenario is that guidelines not only become surrogate tactical planners, but also provide an inconsistent, ineffective framework, which results in over-regulated and less effective, defendable, and sustainable planning. Many feel that time has already come (Oliver et al. 1999; AVESL et al. 2000).

Optimization models

Tactical planning must ultimately find a means of projecting the impact of different decisions on the landscape patterns over time and space. In other words, we want to know what the impact of decisions A and B will be on values X and Y 20, 50, and 100 years from now. This is often dealt with through some form of "optimization model". As the name suggests, optimization models are designed to solve complex planning problems in space and time such that objective functions are either maximized or minimized.

The original optimization model in forestry is the "harvest scheduler", which attempts to maximize wood volume and (or) minimize wood costs from a specific land base subject to spatial, distance, and (or) inventory limitations (Merzenich 1991). The problem of allocating harvesting activities in time and space is well suited to mathematical formulations such as linear programming (LP). Model inputs include data on tree species, volumes, distances, growth and yield, and any physical restrictions or limitations that might prevent or restrict access. Output includes a schedule of harvesting activities for one or more decades. It is even possible for such models to deal with uncertainty directly using one of several sophisticated techniques such as fuzzy logic and stochastic programming (Marshall 1986; Mandelbaum and Martell 1996).

As planning objectives extend beyond timber value or costs, the problem to be solved by such a model becomes much more complex. Through dramatic advances in computer power, and some of the elements and tools discussed above (such as the use of common currency, NRV foundations, or indicators) optimization models today can handle dozens of objective functions relating to timber, habitat for multiple species, biodiversity, and even recreation.

Including other (non-commodity) values in optimization models can be done in one of two ways. The most common approach is to convey other values in terms of constraints on management activities (Liu et al. 2000). For instance, protective buffers or minimum areas of old forest are established as physical, spatial limits to any solution from within the model. There is still only one model objective function (maximizing timber volume or value, for instance), but now there are essentially "black-out" areas of the landscape where operations or certain features or activities cannot occur. However, one still has to make a subjective decision before running the model about when, or to what degree the constraints will be active.

The alternative way of dealing with other values in optimization models is to define them as parallel objective functions (instead of constraints). In other words, by stating competing objectives in spatial terms, we can now search for optimal levels of those attributes together with optimal levels of wood volume output (Roloff and Haufler 1996; Liu et al. 2000). This approach is more in keeping with the spirit of sustainable forest management. For instance, Rempel and Kaufmann (2003) have evaluated trade-offs between biodiversity and economic objectives using an LP optimization model linked to a tactical level block allocation model (STANLEY). The landscape created by a set of cutblock size and adjacency scenarios was evaluated as the total area of suitable habitat for marten (*Martes americana*), moose (*Alces alces*), caribou, and ovenbird (*Seiurus aurocapillus*). These variables were plotted through 50 years against timber flow. This allowed them to compare and contrast the effect of trends in cutblock size and adjacency on the ability to achieve the expect AAC, and on habitat values for different species (Rempel and Kaufmann 2003).

In this case, Rempel and Kaufmann (2003) were merely establishing the relationships embedded in trading off various values in spatial terms as a step towards choosing a desired forest future. More sophisticated optimization models can even help suggest the DFF by assigning a relative weighting factors to each objective. For instance, two adjacent landscapes may include the same set of objectives, but maximizing caribou habitat supply may be weighted the highest on landscape X, and maximizing wood flow may be weighted the highest on landscape Y.

There are several concerns over the application of optimization models to solve today's more complex forest management problems (Box 12.6). However, this is not to say that optimization models are either valueless or obsolete. On the contrary, they are still the best way to allocate effort to maximize output and (or) reduce costs for a well-structured, tractable problem. The problem is that our expectations of a decision-support tool have rapidly expanded to the point where we have begun to seek alternatives.

Near-optimization models

In recognition of the drawbacks of optimization models for tactical decision making, a new genre of tactical planning tools has appeared known as "near-optimization" models. Near-optimization models are also LP-based, but allow for a wider range of feasible solutions, and are specifically designed to deal with multiple objectives and complex problems. Alternative modelling techniques (such as simulated annealing and tabu searches) explore model outcomes within a more realistic, and larger near-optimal solution space (Duckstein et al. 1988; Baskent et al. 2000; Liu et al. 2000). Old-growth characteristics (Ohman 2000), habitat supply (Hof and Joyce 1992), or sediment loads (Hof and Bevers 2000) resulting from different management scenarios can be explored and

---

Box 12.6. Concerns with using optimization models as tactical planning DSS tools:

(1) *Model size*. As the model problem involves more objectives, constraints, area, and more data layers, it takes longer for the computer to find a solution. It is not unusual to wait days for a solution for a moderately complex problem, and the largest problems are, for all intents and purposes, unsolvable (Yoshimoto 2001).

(2) *Inability to quantify objectives*. For objectives to be compared, we must be able to quantify them. Despite our best efforts, the "worth" of things like aesthetics, hunting opportunities, or biodiversity may be quantifiable, but in a different currency than that of timber (Hof and Joyce 1992; Gilmore 1997).

(3) *Subjective weighting*. For the model to compare and contrast different combinations of objectives, we must assign relative weights to each. This exercise becomes a value-laden exercise — how much edge density or caribou habitat is "good"? Furthermore, relationships between such measurements are rarely linear in nature (Weintraub and Davis 1996). For instance, creating 10% less road may in fact create 50% more habitat of a certain type. This is difficult to capture with competing objective functions in a linear model.

(4) *Lack of transparency*. By creating very large modelling problems, one quickly loses an intuitive feel for what is happening. In other words, the model is responding to too many sets of interacting rules for its output to be explained logically (Weintraub and Davis 1996). If the model cannot be easily explained, it loses credibility, and thus its effectiveness as a tool.

(5) *Dealing poorly with uncertainty*. A large, complex model will necessarily involve a large number of assumptions and estimates, typically exercised over approximately 200 years (Puttock et al. 1998). Even very simple models are presented in such a manner that presumes parameters and relationships are known (Marshall 1986), but the larger the model, the longer the time frame, and the greater the number of parameters, the greater the uncertainty (Gunn 1996). "Bigger models" are expected to be "better models", when in fact their uncertainties and errors tend to be multiplicative.

---

spatially optimized in conjunction with fixed timber extraction levels. Management options can then be ranked based on other objectives such as habitat development (Drechsler 2000).

Near-optimization models reduce computer time dramatically, and the loss of solution precision of near-optimality versus perfect optimality is minimal (Liu et al. 2000; Rowse 2000). In other words, near-optimal models are almost as precise, and far faster than optimal modelling techniques. On the other hand, near-optimization models are still optimization models at heart and share many of the same drawbacks. They do not resolve issues of quantifying objectives and are no more transparent than optimization models. In fact, one could argue that the issues listed in Box 12.6 are exacerbated by near-optimization techniques.

Scenario models

Any decision aid designed to provide spatial problem solutions is subject to many assumptions and limitations, not the least of which is direct or indirect focus on fibre supply. An alternative approach to tactical planning is to begin at the end: define the conditions (which may include the composition, structure, diversity, function, and so on) of a desired future forest, and then work *backwards* in identifying the nature, location, and timing of management actions most likely to achieve those conditions. This is known as "scenario modelling".

Scenario models share many of the same technical elements as optimization models, but they are applied in a different way. Scenario modelling involves bringing the data, models, knowledge, and uncertainties to a larger group of stakeholders to play "*what if*" games that create a common understanding of how objective tradeoffs work, and how changing model objectives, weights, or constraints affects the outcome (Duckstein et al. 1988; Welker and Kollmyer 1996). For example, one could deal with a local concern over caribou habitat by imposing several different levels of habitat protection in a variety of model runs, and comparing the impacts in each case on other objectives such as timber harvest levels and recreational opportunities. If there was any uncertainty in the habitat needs of caribou, the habitat parameters could be altered to test several alternatives against the outcome.

It is not difficult to see how scenario models have the potential to overcome most of the disadvantages of decision-support models discussed earlier:

*   As the number of computer runs increases, and inputs are tested and compared to outputs, the greater the intuitive understanding of the model becomes. This alleviates the "black box" syndrome.
*   If simulation is being used, it quickly becomes obvious that the model is just suggesting future possibilities as opposed to a single predicted future, which conveys the idea of uncertainty and the existence of real options.
*   The impact of different levels of risk can be evaluated. For example, future estimates of forest fire activity can be tested against various biodiversity, recreation, and commodity measures to assess the risks and impacts of not achieving the desired future condition.
*   Scenario modelling facilitates direct stakeholder involvement in planning and decision-making process.
*   Finally, since there is no direct focus on optimizing one or more objectives, the ultimate objective becomes choosing a desired future forest, which is far easier to implement and monitor operationally, as opposed to dozens of disconnected objectives. In other words, decision making has matured from asking, "*Which outcome is best?*" to asking, "*Which set of inputs is most likely to result in a desirable outcome?*"

Scenario planning models hold tremendous promise to serve as an effective tactical planning platform. However, this represents a significant conceptual cultural shift, and it is still unclear how it will integrate with strategic planning and policy requirements (see ahead). It is also an extraordinary technical exercise, requiring time, expertise, resources, and technological and data capacities that remain beyond the reach of many agencies. For this reason, while examples of scenario planning pilot studies exist in Canada (Scott 2001; Yamasaki et al. 2001), most tactical planning exercises for the next few years will likely be based on a blend of optimal or near-optimal decision-support models and stakeholder-directed FFC visualization.

Designing landscapes

A final alternative for tactical planning is to adopt a *design system* for landscapes (as opposed to what I have been calling *planning systems*). Design systems focus on the arrangement of different elements in space and time to help create planning solutions for landscapes that are sensitive to both ecological and cultural needs. Drawing upon fields as diverse as hydrology, civil engineering, wildlife biology, and landscape architecture, it uses the landscape as a canvas on which plans are drawn for meeting diverse economic, social, and ecological goals (Diaz and Apostol 1992; Pojar et al. 1994; Diaz and Bell 1997). The resulting blueprints for forest development may still consist primarily of plans for road building and cutblock scheduling, but with considerable attention paid to resource emphasis zoning and other methods to avoid conflicts among land tenures and resource users. With appropriate consideration of flows and adjacencies, forest development planning at the tactical level thus becomes an exercise in applied landscape ecology, but borrows heavily on landscape architecture as well.

Landscape design systems are in many ways more holistic planning tools than either optimization or scenario modelling exercises, and use different reference points. For example, the process is less specific and scientific, and more abstract and conceptual. Some suggest that all landscapes can be effectively described and designed through universal design concepts such as the size and number of the largest patches, corridor integrity, stepping stones, and aggregating land-use activities (Dramstead et al. 1996; Forman and Collinge 1996). Diaz and Bell (1997) similarly use design "narratives" to help understand the flow of water, animals, and disturbance agents across landscapes. The attraction of the landscape design system is in its pragmatism. It allows us to create a common, visual language for understanding landscapes, it fully recognizes that we often make decisions based on preferences, and it reduces many of the scientific details down to a few simple, elegant rules. These are all powerful lessons.

However, there are also some associated dangers, not the least of which is moving away from a science-based decision-making process. A common theme to many chapters in this book is the enumeration of risks associated with oversimplifying natural processes and patterns (see discussion in Chaps. 8, 9, and 23, for example). On one hand, the idea of reducing spatial configuration issues down to a few simple rules is very attractive, and consistent with what we already know about NRV (such as the importance of large patches). In this respect, the proponents are well ahead of their time, and one would hope that we can ultimately move towards exactly this type of model. On the other hand, the design system allows for considerable subjectivity, and offers no means by which to verify the validity or importance of the rules. Simplifying must include elegance and conciseness, which requires a depth of understanding (of landscapes) that we do not yet possess. For example, recall that the concept of "corridors" is a debatable ecological attribute. In the end, we have to be aware of increasing calls for accountability and defensibility of decision making that may not tolerate this level of subjectivity.

Another problem is that this is largely a "soft" system of decision making that has many intangibles and is largely subjective in nature. It relies to a much greater extent on the knowledge and opinions of the persons participating in the process, and on the veracity and type of data they have available at the time. It also exploits techniques more familiar to social scientists for problem solving and conflict resolution. This makes it ideally suited to situations of conflict resolution, or where significant land development is a possibility, but it also moves away from the scientific foundations of ecosystem management.

In fact, landscape design systems are not (and were not necessarily meant to be) tactical planning options that replace any of the systems described above. They may not always result in the development of feasible planning solutions, and do not necessarily resolve all of the spatial issues facing a tactical forest planner today (such as a schedule of resources and activities over time required by operational plans). Rather, they represent an alternative system of developing land-use objectives and strategies for a given landscape through mutual understanding of issues and opportunities. In other words, they are another decision-making tool. However, the concepts of landscape design offer some powerful and novel ideas that will likely be part of the future of tactical planning. For instance, it is not difficult to see the value of stripping down complex spatio-temporal relationships into a handful of fundamental "design" elements for management and monitoring. The adoption of more generalized abstract concepts such as "flow" and "narrative" may also help bridge a challenging communication and education gap and allow stakeholders to appreciate landscapes as dynamic, complex systems.

One of the best examples of how a landscape design system may be modified and applied to a more rigorous need was recently completed in the southern Rockies area of Alberta (AE and OOPDCI 2000). Although the exercise was largely strategic, the project did effectively demonstrate the use of alternative planning design tools and techniques, while including many other planning elements discussed in this chapter. In the end, systems of planning are a reflection of not only how we want to manage resources, but how we perceive the resources being managed. The shift to "designing" landscapes is one of the strongest indications that our view of natural resources is evolving.

Obstacles to tactical planning

In a perfect and deterministic world, tactical planning would always be successful if the tools and techniques discussed above were fully developed and implemented. In reality, there are several impediments to tactical planning that are beyond the control of the participants.

Lack of coordination

Our current forest planning system responds to rapidly changing and increasingly complex conditions by being additive and inclusive. Individual issues are typically dealt with by regulations or guidelines designed only with that single issue in mind. It is not surprising to find goals and objectives from government programs and policies that conflict with each other (Cortner and Schweitzer 1983). For instance, Ontario now has 20 separate guidelines for biological issues alone, including woodland caribou, moose, fish, bats, eagles, and marten, plus another 15 guides for other issues. Although no direct testing for conflicts between guides has been done, a recent review of the guidelines concluded that the complete set of guidelines represented a confusing mixture of philosophies, approaches, and levels of detail. Worse, they are largely being applied without question by planners and managers (AVESL et al. 2000). Nor is there evidence that strategic policies consider consistency between levels of planning. For instance, policies that dictate that landscapes are expected to contribute equivalent portions of the overall allowable cut (Yoshimoto 2001) do not take into account local-level limitations, the existing inventory, or ground-level guidelines such as adjacency. In particular, this policy is likely to conflict with NRV guidelines about clustering harvesting activities and creating larger patch openings.

Similarly, the risk of imposing guidelines or rules at the wrong scale of planning (or monitoring) is considerable. We may inadvertently be making it impossible to meet higher level goals. For instance, enforcing rules about minimum areas of old growth at operational scales alone may make it impossible to meet strategic goals of old-growth patch-size or interior forest targets (because the opportunity for leaving single very large patches is diminished). In general, we have historically tended to consider the more important issues, but at planning scales that are too fine.

Forest tenure and cumulative effects

The ultimate political tool for achieving sustainability of public forests is the tenure system. For instance, under the current system in Alberta, forest companies are issued Forest Management Agreements (FMAs), which grant rights to manage a defined land base (Hauer and Nanang 2001). Similar agreements exist across Canada under different names (Tree Farm Licences, Forest Management Licence, etc.; see Chaps. 1 and 7; see also Beckley 1998). However, these rights are far from universal or exclusive. Several forest products companies may have the rights to different tree species on the same land (Cumming and Armstrong 2001). Similarly, small operators may obtain rights to small, selected areas through a government permit process (Hauer and Nanang 2001). Beyond the forest products sector are many other land uses that potentially affect sustainability in the boreal forest, including oil and gas development, fur trapping, agriculture, coal and peat mining, and human settlements (MacKendrick et al. 2001).

This is relevant to tactical planning because it isolates several different, unconnected tactical planning exercises, and there are no formal mechanisms in place that deal with the cumulative effects of what amounts to multiple levels of tenure. For instance, the planning and review process for forestry is 10–20 years, compared to only several weeks for the oil and gas industry (MacKendrick et al. 2001). Needless to say, this difference can be detrimental to meeting overall landscape objectives. Even a well-balanced, well-supported tactical design scenario by a forest management company may fail because of seismic line and oil well site installation that adversely affect levels of biodiversity and accessibility (see Fig. 22.6).

Similarly, a tactical planning objective to maintain dynamic landscape-level species distributions through a spatially flexible silvicultural regime will be in conflict with the objectives of individual operators not willing to have land converted from one kind of forest cover to another. Even the best tactical plans cannot achieve the most cost-efficient, optimal supply of forest outputs when planning cannot be coordinated (Cumming and Armstrong 2001; see Box 6.4). Some recent examples of improved coordination and innovative tenure arrangements are discussed in Chap. 22.

Blurring the lines between planning levels

As discussed earlier, the complexity of the forest system is such that partitioning the problem into more manageable parts based on a hierarchy of data, decision making, and responsibilities is required (Weintraub and Davis 1996; Baskent et al. 2000). However, until this point, we have been assuming that (*a*) the flow of information and decisions between planning levels is appropriate and plentiful, and (*b*) tactical planning is a single, well-defined level of planning. Neither is necessarily true.

Given the complexities of tactical planning today, and our technological capabilities, it is not difficult to see how we might be blurring the lines of responsibility and decision making between planning levels. For instance, some optimization models are capable of dealing with dozens of objectives over hundreds of thousands of hectares, and generate a 20-year spatial harvesting plan complete with habitat projections for dozens of species (Colberg 1996; Puttock et al. 1998; Liu et al. 2000; Nurullah 2000; Nelson 2001). The scales involved strongly suggest that strategic-level decisions are being made. Furthermore, the very process of learning about goal tradeoffs in space and time (while we develop a tactical plan) is likely to result in re-evaluation of our values and goals (which are strategic), and our choice of silvicultural or logging systems (which are operational choices).

On one hand, the integration of different levels of planning into a "meta-model" is an improvement in efficiency. By linking goals to objectives directly in a scenario planning exercise, for instance, we are more likely to develop a plan that represents our original values. On the other hand, the original advantages of having a distinct hierarchy of planning and decision-making levels still hold. The danger we face by creating and relying on meta-models is that as we continue to add new objectives, more detailed information, more levels of data, and more assumptions, we are ultimately creating a much larger problem to solve. Furthermore, as the functionality of these models increases, the temptation increases to override strategic goals and to dictate operational details.

The tendency to plan using meta-models may be a sign of our inability to develop strong links between planning levels. In a true planning hierarchy, location-specific needs, issues, conditions, and limitations are defined at low levels and communicated upwards, and objectives are set at the highest possible level and communicated down (Oliver et al. 1999). For example, for decades, many jurisdictions in boreal Canada had a policy of returning each harvested block to the current species composition. Yet there is no convincing ecological reason to do so, since boreal stand-types commonly alternate species dominance (Weir and Johnson 1998). The more appropriate scale at which to set species composition is the landscape — or higher. By mistakenly defining the objective of "maintaining cover types" as an operational issue as opposed to a strategic one, we are being inefficient and ecologically unsound.

Decisions being made at the lowest possible levels are also a form of risk management (see Chap. 2). By imposing rules at lower levels, it allows us to be more flexible (Paredes 1996) and to better deal with unforeseen changes to higher level values or objectives (Mandelbaum and Martell 1996). In the example provided above, setting the tree species composition goal at the landscape level rather than at the stand level now allows operational planners to take advantage of local site conditions, mast years, and harvesting techniques to make more appropriate regeneration decisions that are more likely to be successful (Greene et al. 2002).

A more prominent example of confusing the role of the different planning level decisions has already been discussed. The over-regulation of tactical-level decisions through guidelines and rules suggests that we may be trying to deal with the increasing complexities of forest management by indiscriminately regulating tactical decisions at strategic levels (Nilsson and Gluck 2001). The overabundance of guidelines suggests that many of the values represented by those guidelines could have been dealt with at a lower level of planning, if driven by more precise higher level goals.

A plea to think critically

One might presume from the contents of this and other chapters that the key to successful planning and management is the use of sophisticated models, extensive high-resolution data, cutting-edge problem-solving techniques, and a wide variety of experts. However, while it may be advantageous to have all of these things at one's disposal, they do not guarantee success, and in some instances may even detract from the real issues and problems to be resolved. Optimization models and GIS technology in particular have already led many to a sort of "technological paralyses". For example, many complex optimization-type models must be supported by types and amounts of data currently beyond the ability of most agencies to collect, and require extensive verification procedures that are often ignored (Nelson 2003). Similarly, the resolution of satellite sensors far exceeds our capacity for data storage and spatial manipulation, as well as our need for detail.

One of the most dangerous planning precedents being set today is the inclusion of literally dozens of habitat types in planning models (whether scenario or optimization) as output. The ease and confidence with which the models of today allow this to occur (complete with colour-coded, real-time graphical and tabular output) belies the inadequacy of our understanding of many species and their needs. The "*planning problem*" is not that we lack complete understanding of the ecological system (which we must ultimately learn to live with), but rather the misrepresentation of that fact within the decision-making process. Our efforts to be inclusive and technically adept do not deal with the real problem (i.e., knowledge gaps), and ultimately create another problem which we must ultimately face (i.e., the implicit assumption that we have the answers). If such assumptions are not dealt with during the planning stage (but instead after the fact), it will almost certainly lead to a failure to deliver the desired future condition, a failure to meet defined objectives and targets, and ultimately further erosion of forest management confidence by the public and stakeholders.

Good planning includes an understanding that (*a*) this chapter is mainly about tools (including even knowledge and data), and (*b*) not all of these tools will always be available or advantageous to use. The merits of whether, or in what way, each tool should be used should be carefully evaluated for each planning situation. For instance, if there is little or no information on habitat relationships, then it would be best to avoid tools that focus on habitat supply tradeoffs. If the planning and land-use issues are many and complex, avoid optimization modelling in favour of landscape scenario or design options.

The tools discussed in this chapter (indeed, this entire book) might be considered as parts to a planning "kit". The manner in which the different parts fit together to form a system or framework must suit the individual planning needs, and is therefore a different discussion altogether. However, it is worth spending some effort on this topic. One should not assume that the current planning framework (even for tactical purposes) is the most desirable or efficient. For the most part, we rely heavily on the existing planning framework, and simply include new tools and techniques as "add-ons". This usually amounts to more planning time and effort, but not necessarily better decisions or outcomes. At some point the question of designing new planning frameworks should be addressed. For example, the McGregor Model Forest developed an innovative system of planning centred on the scenario planning concept, involving public consultation, data, and knowledge (Scott 2001). While their solution potentially blurs the distinction between tactical and strategic levels of planning, it provides an excellent example of not only how various planning elements fit together, but also how adaptive feedback is used directly in the planning process.

The MacGregor experience also raises an important point about whole systems being greater than the sum of their parts. Thoughtful choice and integration of planning tools and elements can facilitate the development of key SFM principles. For example, despite the fact that virtually all forest management agencies claim to be practicing adaptive management, many are using the term synonymously with "learning from our mistakes", which is a much more passive activity. Being an *adaptive* forest manager means more than dealing with risks, uncertainties, and changing expectations (see Chap. 21). Planning must constantly be accounting for changes in markets, social values, knowledge, natural events, and modelling tools, realizing that not all changes in the system are due to management decisions.

Good planning must thus be a dynamic, resilient, adaptive activity (Holling 1978). For example, Gunn (1996) and Paredes (1996) advocate a "rolling plan" concept that deals with uncertainties in real time through the use of a closely nested hierarchy of models. The logic is that uncertainties increase with longer time scales, and so planning should focus on present conditions and decisions.

One of the more obvious ways of developing an adaptive, flexible planning framework is to link it directly with monitoring activities. After all, forest management can arguably be reduced to three main components: (1) planning appropriate actions; (2) implementing activities that are most likely to result in some desired forest condition; and then (3) monitoring the outcomes to check on our assumptions, predictions, and effectiveness. Historically, these activities have been too isolated from each other, with inadequate transfer of information among the three routine activities.

Saskatchewan has developed a Forest Impacts Monitoring Framework that couples monitoring activities directly with the planning process (Andison et al. 2002). Similarly, a Sustainable Forest Management Network initiative has created much the same sort of integrated framework in eastern Canada (Yamasaki et al. 2001), but approached the problem from the planning perspective. In both cases, the framework was built using many of the tools and techniques discussed in this chapter, including the use of an NRV baseline, desired future forest condition, and scenario modelling. They also took maximum advantage of data, knowledge, and public input, and involved guidelines only where required by law. Also in both cases, direct links were created between the desired future forest, indicator target predictions, and measurable outcomes that can be used to refine the planning process (Yamasaki et al. 2001; Andison et al. 2002).

These initiatives demonstrate how the success of planning can extend beyond simply comparing expectations against outcomes, to evaluating the knowledge, cooperation, trust, and tools gained through the process. Both are also excellent examples of how stepping away from the existing planning frameworks helped develop more effective and relevant planning systems.

Conclusions

Tactical planning is responsible for converting a large number of potentially conflicting, vague, and ill-constructed strategic-level goals into spatial and temporal reality in the face of considerable knowledge gaps, constant change, and economic, ecological, and social uncertainties. It is easily one of the greatest challenges of forest management. Understandably, there has been a tremendous growth in the number and sophistication of the tools and techniques available to aid the tactical planning process. The variety of tools available is a great advantage, since each planning situation can use a different combination or emphasis of tools. Not only is it necessary to have the proper tool for the job, but it is also important that tools are employed in the proper order and that their inherent weaknesses are recognized.

The sections of this chapter were deliberately presented as an idealized model of tactical planning as a hierarchy, beginning with the most basic elements, moving on to management philosophies, then to connecting elements, and finally to integrative decision-support systems. Figure 12.1 represents a map of this chapter, and an outline of this idealized model of tactical planning. This hierarchy also represents a logical order for tactical planning itself. Consider that there is limited value in adopting a natural pattern emulation strategy without first the knowledge of at least some basic natural disturbance patterns. Similarly, it is misleading to adopt sophisticated DSS systems when public input has been minimal. Note that at the top of Figure 12.1, the development of the planning framework necessarily includes consideration of all tools, elements, and philosophies. The exact nature of that system can take many forms, and the previous section briefly described some examples.

However, planning rarely takes place under ideal situations. There are almost always very real shortfalls and gaps in the tools and techniques discussed here, and tactical planning must proceed despite them. This dilemma is one that land management planners must deal with every day, and the answers are also the means of ultimately doing tactical planning better.

First, it must be understood and accepted by everyone involved (e.g., the public, scientists, regulators, and managers) that tactical planning is a long-term commitment by everyone. Scientists and managers must work together to ask good questions, and scientists must provide meaningful and understandable answers to stakeholders. Regulators must proceed on the assumption that the process is a dynamic one, and not restrict the learning process by inappropriate over-regulation. Managers and scientists must educate the public about knowledge and data and their limitations. The public must educate managers and scientists about what is important to them, and about location-specific forest values; the public must commit to being a part of a longer term process, as opposed to only participating in a single workshop or meeting.

Second, it must also be understood and communicated that tactical planning is not any one or two of the elements discussed in this chapter, but virtually all of them combined. Evolution is steady, but there may not always be any outward signs that progress is being made. Database development, for example, is expensive and largely goes unnoticed and unappreciated. Public involvement can take years to fully achieve. While planning workshops and open house meetings may seem to be tactical planning to many people, these are only one of many steps.

Third, we must admit, accept, allow for, and even plan on some of our decisions being wrong. Consistent with the theory of adaptive management, we must all accept that the only way to learn about how to make better management decisions is to adopt more of an experimental attitude. The boreal forest ecosystem is a relatively resilient system, and we can learn a lot from "pushing" it in selected instances. This is particularly applicable at tactical scales.

Fourth, the use and sophistication of the decision-making tools must take into account the level of development of the basic and connecting elements. For example, with poor data, knowledge, and public input, it would be misleading to adopt a highly sophisticated scenario planning decision-support methodology, since we make certain assumptions about the integrity of those elements.

Box 12.7. Options in tactical planning for sustainable forest management.

There is no single answer for tactical planning, and no set of rules that will provide a single best answer. However, there are options available to the tactical planner that will improve the process of tactical planning in any situation:

- *Stay current with "cutting edge" tools and techniques*. Do not assume that new tools are always better, but staying informed is the first step to upgrading your toolbox. If you are hopeful of the potential of certain tools, thoroughly evaluate the costs and benefits of them beforehand, and if necessary use test-cases or pilot studies. This applies equally to data, techniques, and models.

- *Constantly re-evaluate the plan emphasis*. This decision is likely beyond the scope of tactical planning, but the answer informs the planning system and type, number, and function of the tools required.

- *Participate in asking the questions*. As important as knowledge is as one foundation of tactical planning, some questions are worth answering more than others. Scientists need your help to address your critical gaps.

- *Challenge the system, when it needs it*. There are many obstacles to good tactical planning that have nothing to do with tactical planning.

- *Keep the big picture in mind*. Tactical planning is the "middle child" in many ways. It links through scale, policies, and objectives to several other management and monitoring initiatives. Good tactical planning is sometimes little more than making good connections.

- *Think critically*. You are, and always will be, the most valuable planning tool.

Finally, despite the many obstacles, all of the tools and techniques discussed here are valuable, and are making tactical planning more tractable (Box 12.7). Considering the status of our knowledge, technology, and management philosophies even 20 years ago, we have progressed tremendously on virtually all fronts in tactically planning for truly sustainable forest management.

Acknowledgements

Many thanks to Phil Burton for his patience and indispensable direction on designing and assembling this chapter. Thanks also to Fiona Schmiegelow for editorial help and for providing a sidebar discussion, and to Sue Hannon, Christian Messier, Rob Rempel, Greg Branton, and Alain Leduc for their valuable comments, additions, and editorial assistance throughout. Finally, I am grateful to the Sustainable Forest Management Network for the opportunity and support to write this chapter.

References

(AE and OOPDCI) Alberta Environment and Olson and Olson Planning and Design Consultants Inc. 2000. The Southern Rockies Landscape Planning Pilot Study, summary report. Alberta Environment, Edmonton, Alberta. 371 p.

Anderson, D.G., Wishart, R., Murray, A., and Honeyman, D. 2001. Sustainable forestry in the Gwich'in settlement area: ethnographic and ethnohistoric perspectives. Sustainable Forest Management Network, Edmonton, Alberta. Proj. Rep. 2001-9. Available at http://sfm-1.biology.ualberta.ca/english/pubs/PDF/PR_2000-9.pdf [viewed 13 June 2003]. 27 p.

Andison, D.W. 1998. Patterns of temporal variability and age-class distributions on a Foothills landscape in Alberta. Ecography, 21: 543–550.

Andison, D.W. 1999. Validating age data on the Mistik FMLA: laying the groundwork for natural disturbance research. Prepared for Mistik Management Ltd., Meadow Lake, Saskatchewan. Bandaloop Landscape-Ecosystem Services, Belcarra, British Columbia. 28 p.

Andison, D.W. 2003. Patch and event sizes on foothills and mountain landscapes of Alberta. Foothills Model Forest, Hinton, Alberta. Alberta Foothills Disturb. Ecol. Res. Ser. Rep. No. 4. 60 p.

Andison, D.W., and Marshall, P.L. 1999. Simulating the impact of landscape-level biodiversity guidelines: a case study. For. Chron. 75: 655–665.

Andison, D.W., Kimmins, J.P., Van Rees, K., Rempel, R., Hobson, K., Granger, R., McCallum, F., Melville, G., Chambers, P., Beatty, J., Greenwood, H., and Adamowicz, V. 2002. Saskatchewan forest ecosystem impacts monitoring framework, Part 1: Rationale and strategy. V1.6. Saskatchewan Environment and Resource Management, Prince Albert, Saskatchewan. 31 p.

Andison, D.W., Mueller, E., and Freehill, K. 2003. Changes to natural disturbance regimes. In Northeast Region integrated resource management situational assessment. Edited by D. Olson and H. Stelfox. Alberta Sustainable Resource Development, Edmonton, Alberta. Chapter 6.3.4. In press.

Apsey, M., Laishley, D., Nordin, V., and Paille, G. 2000. The perpetual forest: using lessons from the past to sustain Canada's forest in the future. For. Chron. 76: 29–53.

Armstrong, G.W., Cumming, S.G., and Adamowicz, W.L. 1999. Timber supply implications of natural disturbance management. For. Chron. 75(3): 497–504.

(AVESL) ArborVitae Environmental Services Ltd., CMC Ecological Consulting, and Callaghan and Associates Inc. 2000. Review of forest management guides. Prepared for the Ontario Ministry of Natural Resources, Toronto, Ontario. 309 p.

Babu, S.C., and Reidhead, W. 2000. Monitoring natural resources for policy interventions: a conceptual framework, issues, and challenges. Land Use Policy, **17**: 1–11.

Barrett, S.W., and Arno, S.F. 1991. Classifying fire regimes and defining their topographic controls in the Selway–Bitterroot wilderness. *In* Proceedings of the 11th Conference on Fire and Forest Meteorology, 16–19 April 1991, Missoula, Montana. *Edited by* P.L. Andrews and D.F. Potts. Society of American Foresters, Bethseda, Maryland. pp. 299–307.

Baskent, E.Z., Jordan, G.A., and Nurullah, A.M.M. 2000. Designing forest landscape management. For. Chron. **76**: 739–742.

(BCMOF and BCE) British Columbia Ministry of Forests and British Columbia Environment. 1995. Forest practices code: biodiversity guidebook. British Columbia Ministry of Forests and British Columbia Environment, Victoria, British Columbia. 99 p.

(BCMOF and BCMELP) British Columbia Ministry of Forests and British Columbia Ministry of Environment, Lands and Parks. 1999. Forest practices code: landscape unit planning guide. British Columbia Ministry of Forests and British Columbia Ministry of Environment, Lands and Parks, Victoria, British Columbia. 101 p.

Beckingham, J.D., Corns, I.G.W., and Archibald, J.H. 1996. Field guide to ecosites of West–Central Alberta. Canadian Forest Service, Edmonton, Alberta. Spec. Rep. 9. 380 p.

Beckley, T.M. 1998. Moving toward consensus-based forest management: a comparison of industrial, co-managed, community, and small private forests in Canada. For. Chron. **74**: 736–744.

Benton, R.A., and Petterson, K.M. 1999. Automated environmental monitoring and database management: fantasies and realities. For. Chron. **75**: 483–486.

Bergeron, Y., and Archambault, S. 1993. Decrease of forest fires in Quebec's southern boreal zone and its relation to global warming since the end of the Little Ice Age. The Holocene, **3**: 255–259.

Bergeron, Y., Gauthier, S., Carcaillet, C., Flannigan, M., and Richard, P.J.H. 2000. Reconstruction of recent and Holocene fire chronologies and associated changes in forest composition: a basis for forest landscape management. Sustainable Forest Management Network, Edmonton, Alberta. Proj. Rep. 2000-39. Available at http://sfm-1.biology.ualberta.ca/english/pubs/PDF/PR_2000-39.pdf [viewed 24 October 2002]. 20 p.

Bergeron, Y., Gauthier, S., Kafka, V., Lefort, P., and Lessieur, D. 2001. Natural fire frequency for the eastern Canadian boreal forest: consequences for sustainable forestry. Can. J. For. Res. **31**: 384–391.

Bessie, W.C., and Johnson, E.A. 1995. The relative importance of fuels and weather on fire behaviour in subalpine forests. Ecology, **73**: 747–762.

Bowman, J.C., Sleep, D., Forbes, G.J., and Edwards, M. 2000. The association of small mammals with coarse woody debris at log and stand scales. For. Ecol. Manage. **129**: 119–124.

Boxall, P.C., and Macnab, B. 2000. Exploring the preferences of wildlife recreationists for features of boreal forest management: a choice experiment approach. Can. J. For. Res. **30**: 1931–1941.

Brown, D.G. 2000. Image and spatial analysis software tools. J. For. **98**(6): 53–57.

Bryce, S.A., and Clarke, S.E. 1996. Landscape-level ecological regions: linking state-level ecoregion frameworks with stream habitat classifications. Environ. Manage. **20**(3): 297–311.

Bunnell, F.L. 1998. Managing forests to sustain biodiversity: substituting accomplishment for motion. For. Chron. **74**: 822–827.

Burton, P.J. 1995. The Mendelian compromise: a vision for equitable land use allocation. Land Use Policy, **12**: 63–68.

Buttoud, G. 2000. How can policy take into account the "full value" of forests? Land Use Policy, **17**: 169–175.

Checkland, P, and Scholes, J. 1990. Soft systems methodology in action. John Wiley and Sons, Chichester, U.K. 418 p.

Cissel, J.H., Swanson, F.J., and Weisberg, P.J. 1999. Landscape management using historical fire regimes: Blue River, Oregon. Ecol. Appl. **9**: 1217–1231.

Clark, D.F. 1994. Post-fire succession in the sub-boreal spruce forests of the Nechako Plateau, Central British Columbia. M.Sc. thesis. University of Victoria, Victoria, British Columbia. 124 p.

Clark, J.S. 1990. Landscape interactions among nitrogen mineralization, species composition, and long-term fire frequency. Biogeochemistry, **11**: 1–22.

Colberg, R.E. 1996. Hierarchical planning in the forest products industry. *In* Hierarchical approaches to forest management in public and private organizations, proceedings, 25–29 May 1992, Toronto, Ontario. *Edited by* D.L. Martell, L.S. Davis, and A. Weintraub. Canadian Forest Service, Petawawa, Ontario. Inf. Rep. PI-X-124. pp. 16–20.

Corkum, C.V. 1999. Response of small mammals to landscape structure at multiple spatial scales. M.Sc. thesis. University of Alberta, Edmonton, Alberta. 87 p.

Cortner, H.J., and Schweitzer, D.L. 1983. Limits to hierarchical planning and budgeting systems: the case of public forestry. J. Environ. Manage. **17**: 191–205.

Cotterill, S.E., and Hannon, S.J. 1999. No evidence of short-term effects of clear-cutting on artificial nest predation in boreal mixedwood forests. Can. J. For. Res. **29**: 1900–1910.

Cumming, S.G. 2001. A parametric model of the fire-size distribution. Can. J. For Res. **31**: 1297–1303.

Cumming, S.G., and Armstrong, G.W. 2001. Divided land base and overlapping forest tenure in Alberta, Canada: a simulation study exploring costs of forest policy. For. Chron. **77**: 501–508.

Davis, W. 1993. The global implications of biodiversity. *In* Our Living Legacy, Proceedings, Symposium on Biological Diversity, Spring 1991, Victoria, British Columbia. *Edited by* M.A. Fenger, E.H. Miller, J.A. Johnson, and E.J.R. Williams. Royal British Columbia Museum, Victoria, British Columbia. pp. 23–46.

DeLong, S.C., and Tanner, D. 1996. Managing the pattern of forest harvest: lessons from wildfire. Biodiv. Conserv. **5**: 1191–1205.

Desrochers, A., and Hannon, S.J. 1997. Gap crossing decisions by forest songbirds during the post-fledging period. Conserv. Biol. **11**: 1204–1210.

Diaz, N., and Apostol, D. 1992. Forest landscape analysis and design: a process for developing and implementing land management objectives for landscape patterns. USDA Forest Service, Portland, Oregon. R6 ECO-TP-043-92. Available at http://www.srs.fs.fed.us/pubs/viewpub.jsp?index=3048 [viewed 24 October 2002]. 64 p.

Diaz, N.M., and Bell, S. 1997. Landscape analysis and design. *In* Creating a forestry for the 21st century: the science of ecosystem management. *Edited by* K.A. Kohm and J.F. Franklin. Island Press, Washington, D.C. pp. 255–269.

Dramstead, W.E., Olson, J.D., and Forman, R.T.T. 1996. Landscape ecology principles in landscape architecture and land-use planning. Harvard University Graduate School of Design, Island Press, and American Society of Landscape Architects, Washington, D.C. 80 p.

Drapeau, P., Leduc, A., Giroux, J.-F., Savard, J.-P.L., Bergeron, Y., and Vickery, W.L. 2000. Landscape-scale disturbances and changes in bird communities of boreal mixed-wood forests. Ecol. Monogr. **70**: 423–444.

Drechsler, M. 2000. A model-based decision aid for species protection under uncertainty. Biol. Conserv. **94**: 23–30.

Duckstein, L., Korhonen, P., and Tecle, A. 1988. Multiobjective forest management: a visual interactive and fuzzy approach. *In* Symposium on systems analysis in forest resources. USDA Forest Service, Fort Collins, Colorado. GTR-RM-161. pp. 68–74.

Dussault, C., Rehaume, C., Huot, J., and Ouellet, J. 2001. The use of forest maps for the description of wildlife habitats: limits and recommendations. Can. J. For. Res. **31**: 1227–1234.

Eberhart, K.E., and Woodard, P.M. 1987. Distribution of residual vegetation associated with large fires in Alberta. Vegetatio, **4**: 412–417.

Ehnes, J., and Sidders, D. 2002. A guide to harvesting practices to regenerate a natural forest. Manitoba Model Forest, Pine Falls, Manitoba. Available at http://www.manitobamodelforest.net/publications/ Operator%20Guidelines.pdf [viewed 13 June 2003]. 29 p.

Erdle, T., and Sullivan, M. 1998. Forest management design for contemporary forestry. For. Chron. **74**: 83–90.

Fall, A., and Fall, J. 2001. A domain-specific language for models of landscape dynamics. Ecol. Model. **137**: 1–21.

Fisher, J.T. 1999. The influence of landscape structure on the distribution of the North American red squirrel, *Tamiasciurus hudsonicus*, in a heterogeneous boreal mosaic. M.Sc. thesis. University of Alberta, Edmonton, Alberta. 101 p.

Forman, R.T.T., and Collinge, S.K. 1996. The "spatial solution" to conserving biodiversity in landscapes and regions. *In* Conservation of faunal diversity in forested landscapes. *Edited by* R.M. DeGraaf and R.I. Miller. Chapman & Hall, London, U.K. pp. 537–568.

Forman, R.T.T., Sperling, D., Bissonette, J.A., Clevenger, A.P., Cutshall, C.D., Dale, V.H., Fahrig, L., France, R., Goldman, C.R., Heanue, K., Jones, J.A., Swanson, F.J., Turrentine, T., and Winter, T.C. 2003. Road ecology: science and solutions. Island Press, Washington, D.C. 481 p.

(FMF) Foothills Model Forest. n.d. Foothills Model Forest: a growing understanding. Available at http://www.fmf.ab.ca/ [viewed 24 October 2002].

Franklin, J.F. 1989. Towards a new forestry. Am. For. **95**: 37–44.

Franklin, J.F. 1993. Preserving biodiversity: species, ecosystems, or landscapes? Ecol. Appl. **3**: 202–205.

Franklin, J.F. 1997. Ecosystem management: an overview. *In* Ecosystem management: applications for sustainable forest and wildlife resources. *Edited by* M.S. Boyce and A. Haney. Yale University Press, New Haven, Connecticut. pp. 21–53.

Franklin, J.F., and Forman, R.T.T. 1987. Creating landscape patterns by forest cutting: ecological consequences and principles. Landscape Ecol. **1**: 5–18.

Gardner, R.H., O'Neill, R.V., and Dale, V.H. 1989. Quantifying scale-dependent effects of animal movement with simple percolation models. Landscape Ecol. **3**: 217–227.

Gilmore, D.W. 1997. Ecosystem management — a needs driven, resource-use philosophy. For. Chron. **73**: 560–564.

Gomi, T., Sidle, R.C., Bryant, M.D., and Woodsmith, R.D. 2001. The characteristics of woody debris and sediment distribution in headwater streams, southeastern Alaska. Can. J. For Res. **31**: 1386–1399.

Gordon, J.C., and Lyons, J. 1997. The emerging role of science and scientists in ecosystem management. *In* Creating a forestry for the 21st century: the science of ecosystem management. *Edited by* K.A. Kohm and J.F. Franklin. Island Press, Washington, D.C. pp. 447–453.

Gray, G.J., Enzer, M.J., and Kusel, J. 2001. Understanding community-based forest ecosystem management: an editorial synthesis. J. Sust. For. **12**(3/4): 1–23.

Greene, D.F., Kneeshaw, D.D., Messier, C., Lieffers, V., Cormier, D., Doucet, R., Grover, G., Coates, K.D., Groot, A., and Calogeropoulos, C. 2002. An analysis of silvicultural alternatives to the conventional clearcut/plantation prescription in boreal mixedwood stands. For Chron. **78**: 1–15.

Gregory, R. 1999. Identifying environmental values. *In* Tools to aid environmental decision-making. *Edited by* V.H. Dale and M.R. English. Springer-Verlag, New York, New York. pp. 32–58.

Gunn, E.A. 1996. Hierarchical planning processes in forestry: a stochastic programming decision analytic perspective. *In* Hierarchical approaches to forest management in public and private organizations, proceedings, 25–29 May 1992, Toronto, Ontario. *Edited by* D.L. Martell, L.S. Davis, and A. Weintraub. Canadian Forest Service, Petawawa, Ontario. Inf. Rep. PI-X-124. pp. 85–97.

Gustafsson, L., and Weslien, J. (*Editors*) 1999. Special issue: biodiversity in managed forests — concepts and solutions. For. Ecol. Manage. **115**(2–3). Elsevier Science, Oxford, U.K. 196 p.

Hall, R.J., Peddle, D.R., and Klita, D.L. 2000. Mapping conifer understory within boreal mixedwoods from Landsat TM satellite imagery and forest inventory information. For. Chron. **76**: 887–902.

Hannon, S.J. 1999. Avian response to stand and landscape structure in the boreal mixedwood forest in Alberta. Sustainable Forest Management Network, Edmonton, Alberta. Proj. Rep. 1999-6. Available at http://sfm-1.biology.ualberta.ca/english/pubs/PDF/PR_1999-06.pdf [viewed 24 October 2002]. 12 p.

Hannon, S. 2000. Avian response to stand and landscape structure in the boreal mixedwood forest in Alberta. Sustainable Forest Management Network, Edmonton, Alberta. Proj. Rep. 2000-37. Available at http://sfm-1.biology.ualberta.ca/english/pubs/PDF/PR_2000-37.pdf [viewed 24 October 2002]. 26 p.

Hansen, A.J., Spies, T.A., Swanson, F.J., and Ohmann, J.L. 1991. Conserving biodiversity in managed forests: lessons from wildfire. BioScience, **41**: 382–392.

Harvey, B.C. 1998. Influence of large woody debris on retention, immigration, and growth of coastal cut-throat trout (*Oncorhynchus clarki clarki*) in stream pools. Can. J. Fish. Aquat. Sci. **55**: 1902–1908.

Hauer, G., and Nanang, D.M. 2001. A decomposition approach to modelling overlapping tenures in Alberta: a case study. Sustainable Forest Management Network, Edmonton, Alberta. Proj. Rep. 2001-25. Available at http://sfm-1.biology.ualberta.ca/english/pubs/PDF/PR_2001-25.pdf [viewed 24 October 2002]. 23 p.

Hobbs, R.J. 199 8. Managing ecological systems and processes. *In* Ecological scale: theory and application. *Edited by* D.L. Peterson and V.T. Parker. Columbia University Press, New York, New York. pp. 459–484.

Hof, J., and Bevers, M. 2000. Optimal timber harvest scheduling with spatially defined sediment objectives. Can. J. For. Res. **30**: 1494–1500.

Hof, J.G., and Joyce, L.A. 1992. Spatial optimization for wildlife and timber in managed forest ecosystems. For. Sci. **23**: 489–508.

Holling, C.S. (*Editor*). 1978. Adaptive environment assessment and management. Wiley and Sons, Toronto, Ontario. 377 p.

Hoyt, J.S. 2000. Habitat associations of Black-backed *Picoides arcticus* and Three-toed *P. tridactylus* Woodpeckers in the northeastern boreal forest of Alberta. M.Sc. thesis. University of Alberta, Edmonton, Alberta. 96 p.

Hunter, M.L., Jr. 1990. Wildlife, forests, and forestry: principles of managing forests for biological diversity. Prentice–Hall, Englewood Cliffs, New Jersey. 370 p.

Hunter, M.L., Jr. (*Editor*). 1999. Maintaining biodiversity in forest ecosystems. Cambridge University Press, Cambridge, U.K. 698 p.

Ibarzabal, J., and Desrochers, A. 2001. Lack of relationship between forest edge proximity and nest predator activity in an eastern Canadian boreal forest. Can. J. For Res. **31**: 117–122.

Johnson, E.A. 1992. Fire and vegetation dynamics: studies from the North American boreal forest. Cambridge University Press, Cambridge, U.K. 129 p.

Lachowski, H., Maus, P., and Roller, N. 2000. From pixels to decisions: digital remote sensing technologies for public land managers. J. For. **98**(6): 13–15.

Landres, P.B., Morgan, P., and Swanson, F.J. 1999. Overview of the use of natural variability concepts in managing ecological systems. Ecol. Appl. **9**: 1179–1188.

Larsson, S., and Danell, K. (*Editors*). 2001. Science and the management of boreal forest biodiversity. Scand. J. For. Res. Suppl. 3. Taylor & Francis, Oslo, Norway. 123 p.

Leiffers, V.J., Stadt, K.J., and Navratil, S. 1996. Age structure and growth of understory white spruce under aspen. Can. J. For. Res. **26**: 1002–1007.

Li, H., Franklin, J.F., Swanson, F.J., and Spies, T.A. 1993. Developing alternative forest cutting patterns: a simulation approach. Landscape Ecol. **8**: 63–75.

Li, W., Wang, A., Ma, A., and Tang, H. 1999. Designing the core zone in a biosphere reserve based on suitable habitats: Yancheng Biosphere Reserve and the red crowned crane (*Grus japonensis*). Biol. Conserv. **90**: 167–173.

Liu, G., Nelson, J.D., and Wardman, C.D. 2000. A target-oriented approach to forest ecosystem design — changing the rules of forest planning. Ecol. Model. **127**: 269–281.

Luckert, M.K. 1999. Are community forests the key to sustainable forest management? Some economic considerations. For. Chron. **75**: 789–792.

Ludwig, D., Hilborn, R., and Walters, C. 1993. Uncertainty, resource exploitation, and conservation: lessons from history. Science, **260**(2): 35–36.

Machtans, C.S., Villard, M.-A., and Hannon, S.J. 1996. Use of riparian buffer strips as movement corridors by forest birds. Conserv. Biol. **10**: 1366–1379.

MacKendrick, N., Fluet, C., Davidson, D.J., Krogman, N., and Ross, M. 2001. Integrated resource management in Alberta's boreal forest: issues and opportunities. Sustainable Forest Management Network. Edmonton, Alberta. Proj. Rep. 2001-22. Available at http://sfm-1.biology.ualberta.ca/english/pubs/PDF/PR_2001-22.pdf [viewed 24 October 2002]. 29 p.

Malanson, G.P. 1985. Simulation of competition between alternative shrub life history strategies through recurrent fires. Ecol. Model. **27**: 271–283.

Mandelbaum, M., and Martell, D.L. 1996. Flexibility in forest management planning. *In* Hierarchical approaches to forest management in public and private organizations, proceedings, 25–29 May 1992, Toronto, Ontario. *Edited by* D.L. Martell, L.S. Davis, and A. Weintraub. Canadian Forest Service, Petawawa, Ontario. Inf. Rep. PI-X-124. pp. 98–106.

Marshall, P.L. 1986. A decision context for timber supply modelling. For. Chron. **61**: 533–536.

Marshall, P.L. 1987. Sources of uncertainty in timber supply projections. For. Chron. **62**: 165–168.

Merzenich, J.P. 1991. Spatial disaggregation process: distributing forest plant harvest schedules to subareas. *In* Systems analysis in forest resources. USDA Forest Service, Charleston, South Carolina. GTR-SE-74. pp. 250–254.

Mladenoff, D.J., Host, G.E., Boeder, J., and Crow, T.R. 1996. LANDIS: a spatial model of forest landscape disturbance, succession, and management. *In* GIS and environmental modelling: progress and research issues. *Edited by* M. Goodchild, L. Steyaert, B. Parks, C. Johnston, D. Maidment, M. Crane, and S. Glendinning. GIS World Books, Fort Collins, Colorado. pp. 175–179.

Moses, R.A., and Boutin, S. 2001. The influence of clear-cut logging and residual leave material on small mammal populations in aspen-dominated boreal mixedwoods. Can. J. For. Res. **31**: 483–495.

Nelson, J. 2001. Assessment of harvest blocks generated from operational polygons and forest-cover polygons in tactical and strategic planning. Can. J. For Res. **31**: 682–693.

Nelson, J. 2003. Forest-level models and challenges for their successful application. Can. J. For. Res. **33**: 422–429.

Nilsson, S., and Gluck, M. 2001. Sustainability and the Canadian forest sector. For. Chron. **77**: 39–47.

Norton, D. 1999. Forest reserves. *In* Maintaining biodiversity in forest ecosystems. *Edited by* M.L. Hunter, Jr. Cambridge University Press, Cambridge, U.K. pp. 525–555.

Noss, R.F. 1990. Indicators for monitoring biodiversity: a hierarchical approach. Conserv. Biol. **4**: 355–364.

Noss, R.F. 1998. At what scale should we manage biodiversity? *In* The living dance: policy and practices for biodiversity in managed forests. *Edited by* F.L. Bunnell and J.F. Johnson. UBC Press, Vancouver, British Columbia. pp. 96–116.

Nurullah, A.M.M. 2000. Spatial stratification in forest modelling. For. Chron. **76**: 311–317.

Ohman, K. 2000. Creating continuous areas of old forest in long-term forest planning. Can. J. For. Res. **30**: 1817–1823.

Oliver, C.D. 2001. Policies and practices: options for pursuing forest sustainability. For. Chron. **77**: 49–60.

Oliver, C.D., Boydak, M., Segura, G., and Bare, B.B. 1999. Forest organization, management, and policy. *In* Maintaining biodiversity in forest ecosystems. *Edited by* M.L. Hunter, Jr. Cambridge University Press, Cambridge, U.K. pp. 556–596.

Olsen, B.T. 1999. Breeding habitat ecology of the Barred Owl (*Strix varia*) at three spatial scales in the boreal mixedwood forest of north-central Alberta. M.Sc. thesis. University of Alberta, Edmonton, Alberta. 77 p.

O'Neill, R.V., and Rust, B. 1979. Aggregation error in ecological models. Ecol. Model. **7**: 91–105.

Palmer, J.E. 2000. Reliability of rating visible landscape qualities. Landscape J. **19**(1/2): 166–178.

Paredes, G.L. 1996. Design of a resource allocation mechanism for multiple use forest planning. *In* Hierarchical approaches to forest management in public and private organizations, proceedings, 25–29 May 1992, Toronto, Ontario. *Edited by* D.L. Martell, L.S. Davis, and A. Weintraub. Canadian Forest Service, Petawawa, Ontario. Inf. Rep. PI-X-124. pp. 67–82.

Pitt, D.G., Runesson, U., and Bell, F.W. 2000. Application of large-and medium-scale aerial photographs to forest vegetation management: a case study. For. Chron. **76**: 903–913.

Pojar, J., Diaz, N., Steventon, D., Apostol D., and Mellen, K. 1994. Biodiversity planning and forest management at the landscape scale. *In* Applications of ecosystem management: proceedings of the Third Habitat Futures Workshop, October 1992, Vernon, British Columbia. *Tech. coord. by* M.H. Huff, S.E. McDonald, and H. Gucinski. USDA Forest Service, Portland, Oregon. PNW-GTR-336. pp. 55–70.

Pregitzer, K.S., Goebel, P.C., and Wigley, T.B. 2001. Evaluating forestland classification schemes as tools for maintaining biodiversity. J. For. **99**(2): 33–40.

Pressey, R.L., and Logan, V.S. 1998. Size of selection units for future reserves and its influence on actual vs targeted representation of features: a case study in western New South Wales. Biol. Conserv. **85**: 305–319.

Puttock, G.D., Timossi, I., and Davis, L.S. 1998. BOREAL: a tactical planning system for forest ecosystem management. For. Chron. **74**: 413–420.

Railsback, S.F. 2001. Concepts from complex adaptive systems as a framework for individual-based modelling. Ecol. Model. **139**: 47–62.

Rempel, R.S., and Kaufmann, C.K. 2003. Spatial modelling of harvest constraints on wood supply versus wildlife habitat objectives. Environ. Manage. **32**. In press.

Robinson, G.R., Holt, R.D., Gaines, M.S., Hamburg, S.P., Johnson, M.L., Fitch, H.S., and Martinko, E.A. 1992. Diverse and contrasting effects of habitat fragmentation. Science, **257**: 524–526.

Roloff, G.J., and Haufler, J.B. 1996. Incorporating wildlife objectives into forest planning. *In* Hierarchical approaches to forest management in public and private organizations, proceedings, 25–29 May 1992, Toronto, Ontario. *Edited by* D.L. Martell, L.S. Davis, and A. Weintraub. Canadian Forest Service, Petawawa, Ontario. Inf. Rep. PI-X-124. pp. 125–130.

Rowse, J. 2000. Does a renewable natural resource usually have many near-optimal allocation paths? Nat. Res. Model. **13**(4): 503–533.

Schmiegelow, F.K.A., and Beck, J.A. 2001. Wildlife modeling and biomonitoring. Sustainable Forest Management Network, Edmonton, Alberta. Proj. Rep. 2001-3. Available at http://sfm-1.biology.ualberta.ca/english/pubs/PDF/PR_2001-03.pdf [viewed 24 October 2002]. 22 p.

Schmiegelow, F.K.A., and Hannon, S.J. 1999. Forest-level effects of management on boreal songbirds: the Calling Lake Fragmentation Studies. *In* Forest fragmentation, wildlife and management implications. *Edited by* J.A. Rochelle, L.A. Lehmann, and J. Wisniewski. Brill Press, Boston, Massachusetts. pp. 201–221.

Schreier, H., Thompson, W.A., van Kooten, G.C., and Vertinsky, I. 1996. A decomposed hierarchical system for forest land use allocation decisions with public participation. *In* Hierarchical approaches to forest management in public and private organizations, proceedings, 25–29 May 1992, Toronto, Ontario. *Edited by* D.L. Martell, L.S. Davis, and A. Weintraub. Canadian Forest Service Petawawa, Ontario. Inf. Rep. PI-X-124. pp. 136–147.

Scott, A. (*Editor*) 2001. The McGregor story: pioneering approaches to sustainable forest management. McGregor Model Forest Association, Prince George, British Columbia. Available at http://www.mcgregor.bc.ca/publications/McStory.pdf [viewed 24 October 2002]. 162 p.

Seymour, R.S., and Hunter, M.L., Jr. 1992. New forestry in eastern spruce–fir forests: principles and applications to Maine. Maine Agricultural Experiment Station, Orono, Maine. Misc. Publ. 716. 36 p.

Shindler, B. 2000. Landscape-level management: it's all about context. J. For. **98**(12): 10–14.

Spies, T.A., and Turner, M.G. 1999. Dynamic forest mosaics. *In* Maintaining biodiversity in forest ecosystems. *Edited by* M.L. Hunter, Jr. Cambridge University Press, Cambridge, U.K. pp. 95–160.

Suffling, R., Smith, B., and Dal Molin, J. 1982. Estimating past forest age distributions and disturbance rates in North-western Ontario: a demographic approach. J. Environ. Manage. **14**: 45–56.

Sullivan, T.P., Sullivan, D.S., and Lindgren, P.M.F. 2001. Stand structure and small mammals in young lodgepole pine forest: 10-year results after thinning. Ecol. Appl. **11**: 1151–1173.

Toivonen, H. 2000. Integrating forestry and conservation in boreal forests: ecological, legal and socio-economic aspects. Forestry, **73**(2): 129–135.

Villard, M.-A. 2000. Reducing long-term effects of forest harvesting on indicator species of closed-canopy mature forests. Sustainable Forest Management Network, Edmonton, Alberta. Proj. Rep. 2000-20. Available at http://sfm-1.biology.ualberta.ca/english/pubs/PDF/PR_2000-20.pdf [viewed 24 October 2002]. 13 p.

Voller, J., and Harrison, S. (*Editors*). 1998. Conservation biology principles for forested landscapes. UBC Press, Vancouver, British Columbia. 243 p.

Waaub, J., St.-Onge, B., Bergeron, Y., and Belanger, L. 2000. Integrated tools for decision aid in sustainable forest management. Sustainable Forest Management Network, Edmonton, Alberta. Proj. Rep. 2000-3. Available at http://sfm-1.biology.ualberta.ca/english/pubs/PDF/PR_2000-03.pdf [viewed 24 October 2002]. 19 p.

Ward, P.C., and Tithecott, A.G. 1993. The impact of fire management on the boreal landscape of Ontario. Ontario Ministry of Natural Resources, Toronto, Ontario. Aviat. Flood Fire Manag. Branch Pub. No. 305. 12 p.

Watt, K.E.F. 1966. The nature of systems analysis. *In* Systems analysis in ecology. *Edited by* K.E.F. Watt. Academic Press, New York, New York. pp. 1–14.

Weintraub, A, and Davis, L.S. 1996. Hierarchical planning in forest resource management: defining the dimensions of the subject area. *In* Hierarchical approaches to forest management in public and private organizations, proceedings, 25–29 May 1992, Toronto, Ontario. *Edited by* D.L. Martell, L.S. Davis, and A. Weintraub. Canadian Forest Service Petawawa, Ontario. Inf. Rep. PI-X-124. pp. 2–14.

Weir, J.M.H., and Johnson, E.A. 1998. Effects of escaped settlement fires and logging on forest composition in the mixedwood boreal forest. Can. J. For. Res. **28**: 459–467.

Weir, J.M.H., Johnson, E.A., and Miyanishi, K. 2000. Fire frequency and the spatial age mosaic of the mixed-wood boreal forest in western Canada. Ecol. Appl. **10**: 1162–1177.

Welker, J.C., and Kollmyer, C. 1996. Tactical level harvest scheduling based on strategic level woodlands planning. *In* Hierarchical approaches to forest management in public and private organizations, proceedings, 25–29 May 1992, Toronto, Ontario. *Edited by* D.L. Martell, L.S. Davis, and A. Weintraub. Canadian Forest Service, Petawawa, Ontario. Inf. Rep. PI-X-124. pp. 21–26.

White, P.S. 1979. Pattern, process, and natural disturbance in vegetation. Bot. Rev. **45**(3): 229–299.

Wilson, J.S., and McGaughey, R.J. 2000. Landscape-scale forest information. J. For. **98**(12): 21–27.

Wright, R.G., Murray, M.P., and Merrill, T. 1999. Ecoregions as a level of ecological analysis. Biol. Conserv. **86**: 207–213.

Yamasaki, S.H., Cote, M.-A., Kneeshaw, D.D., Fortin, M.-J., Fall, A., Messier, C., Bouthillier, L., Leduc, A., Drapeau, P., Gauthier, S., Pare, D., Greene, D., and Carignan, R. 2001. Integration of ecological knowledge, landscape modelling, and public participation for the development of sustainable forest management. Sustainable Forest Management Network, Edmonton, Alberta. Proj. Rep. 2001-27. Available at http://sfm-1.biology.ualberta.ca/english/pubs/PDF/PR_2001-27.pdf [viewed 24 October 2002]. 27 p.

Yoshimoto, A. 2001. Potential use of a spatially constrained harvest scheduling model for biodiversity concerns: exclusion periods to create heterogeneity in forest structure. J. For. Res. **6**(1): 21–30.

Chapter 13

Nature-based silviculture for sustaining a variety of boreal forest values

V.J. Lieffers, C. Messier, P.J. Burton, J.-C. Ruel, and B.E. Grover

"In other parts of the world, under conditions differing widely from those pre-vailing in Europe, there is great scope not only for the intelligent application of existing systems but also for the elaboration of new systems."

R.S. Troup (1928)

Introduction

Definitions and objectives

Silviculture is the theory and practice of controlling the establishment, composition, growth, and quality of forest stands to achieve the objectives of management (Sauvageau 1995). Historically emphasizing the "growing of trees", silviculture can now be said to encompass a wide range of forest stand manipulations for the implementation of ecosystem management. Until about 1990, silvicultural activities in Canadian boreal forests dealt almost exclusively on re-establishment of stands following logging in order to sustain the flow of wood products from forest lands (Chap. 1). The physical and intellectual energy expended on silviculture was focused on developing treatments that shortened conifer establishment times, reduced their competitors, and increased conifer productivity; goals were to meet the stocking and free-to-grow

V.J. Lieffers.[1] Dept. of Renewable Resources, 442 Earth Sciences Bldg., University of Alberta, Edmonton, Alberta T6G 2E3, Canada.
C. Messier. Dep. des Sciences Biologiques, Université du Québec à Montréal, Montréal, Quebec, Canada.
P.J. Burton. Symbios Research & Restoration, Smithers, British Columbia, Canada.
J.-C. Ruel. Faculté de Foresterie et Géomatique, Université Laval, Sainte-Foy, Quebec, Canada.
B.E. Grover. Alberta-Pacific Forest Industries Inc., Boyle, Alberta, Canada.
[1]Author for correspondence. Telephone: 780-492-2852. e-mail: vic.lieffers@ualberta.ca
Correct citation: Lieffers, V.J., Messier, C., Burton, P.J., Ruel, J.-C., and Grover, B.E. 2003. Nature-based silviculture for sustaining a variety of boreal forest values. Chapter 13. *In* Towards Sustainable Management of the Boreal Forest. *Edited by* P.J. Burton, C. Messier, D.W. Smith, and W.L. Adamowicz. NRC Research Press, Ottawa, Ontario, Canada. pp. 481–530.

standards of a particular province. This was particularly so because of the near-full allocation of forest lands to industry and the high rate of cutting of these lands.

In recent years, however, silviculturists have been caught in the middle of conflicting social and management objectives for the forest. Firstly, while they have been forced to meet the provincial standards/rules which focused on sustaining conifer wood production, now landscape management and biodiversity values may carry equal weight to wood production on public lands (OMNR 2001; see Chap. 11). Indeed, biodiversity and timber objectives for the management of a stand are often in direct conflict (Hunter 1990). Secondly, provincial policies (Pratt and Urquhart 1994), coupled with advances in pulping and panel board technology using hardwoods (mostly trembling aspen, *Populus tremuloides*), have created an industrial demand for these previously unexploited species. This has made silviculture more complex, as many jurisdictions now desire production of both hardwood and conifers from the same forest stand. A harsh environment, mutiple resource management goals, and a lack of knowledge make boreal silviculture challenging (Box 13.1).

Objectives for this chapter are to:

(1) describe the biological issues of forest stand dynamics, in particular those pertinent to boreal forests;

(2) discuss the technological, environmental, economic, and social issues related to various silvicultural practices; and

(3) describe and discuss examples of silvicultural systems (with an emphasis on novel approaches) for various sections of the boreal forest across Canada. These include treatments and systems that would be acceptable for the various resource emphasis options associated with the triad approach to zoned land management (as introduced in Chaps. 11 and 12).

Silviculture in forest management hierarchy

Forest Management can be divided into different hierarchical levels of planning (see Box 1.3 and Table 12.1). Strategic planning (as explored in Chap. 11) defines broad goals and objectives for a forest or large management unit. The tactical level (as developed in Chap. 12) defines broad levels processes to achieve these goals, often at the

---

Box 13.1 Silvicultural challenges for sustaining boreal forests:

- Low productivity forests (i.e., short growing season, cold soils);
- Long-rotation forests;
- High treatment costs;
- Uncertain biological and economical feasibility of intensive forestry;
- Low economic return on investment;
- Low tree species diversity;
- Understanding natural development of forest stands as models for stand management;
- Different stand management goals for public and industrial users of the forest;
- Conflict between fibre production and biodiversity conservation; and
- Inflexible government regulations.

Table 13.1. Considerations for development of silvicultural systems.

Biological	Technological	Social
Soils	Harvesting	Society/owner values
Macroclimate	Roading	Resources/values
Tree and plant establishment	Tools/equipment	Rules/regulations
Tree and plant growth/mortality	Schedules/planning	Economic feasibility
Insects	Safety	Costs/benefits
Fungi		
Wildlife		
Ecology/succession		
Natural disturbance regimes		

landscape level. The operational level (the subject of this chapter) defines many of the detailed issues that are applied to individual stands making up the forest. While some stand level treatments are tactical in approach (e.g., a decision to plan for natural or artificial regeneration), there are many operational details of how to treat a stand to control its development, thereby meeting economic, ecological, or social goals. The operational decisions such as types of logging or site preparation equipment, type of seedling to plant, number of seed trees and type of seedbeds for natural regeneration are often made by local foresters or forestry technicians. In recent years, however, there has been an increasing reliance on field workers and contractors to make many of the on-the-ground decisions. After training, contractors and operators often decide which trees to leave in partial harvesting systems, select skid trail locations, identify snags for retention, and resolve many other operational issues.

Silviculture as a tool for controlling forest stand development

The development of silvicultural systems can be hierarchically divided into three different components: biological, technological, and social (Table 13.1). First are the biological factors controlling forest stand establishment and stand development. These include species composition and limitations to the establishment, growth, and survival of trees related to: (*a*) site characteristics such as soils and macroclimate, (*b*) important historical events such as fire, frost, or impacts of insects (McCune and Allen 1985; see Chap. 8), and (*c*) competition over time among plants and between tree species. Collectively these factors interact to influence stand composition, productivity, and dynamics. Second, technological issues constrain the application of various silvicultural treatments. Availability of machines, pesticides, or limitations on their application (e.g., access, frost in soil, steepness of slope) all control the silviculture practiced. Third, social concerns (including the balance to be placed on timber and non-timber values), costs and benefits of treatments interact with the other two components to influence the type of silvicultural activities selected. While there are textbook examples of silvicultural systems (Matthews 1989; Smith et al. 1997), these have by no means exhausted the range of useful systems that might be applied to or developed for the boreal forests, given the specific biological and social circumstances of this region.

Site classification

The type of forest and the species typically found on a site are partially dependent upon the climate, soils, and biotic relationships (Krajina 1965) and partially on the biogeography and the restructuring of ecosystems since the retreat of the glaciers 10 000 years ago (Pielou 1991). Throughout Canada, the site factors and the forest compositions of a given area have been used to develop site classifications for forests (e.g., Pojar et al. 1987; Sims et al. 1989; Meades and Moores 1989; Beckingham and Archibald 1996; Cauboue and Tremblay 1992; and others), which have proved to be valuable tools for assisting foresters to develop better, site-specific silvicultural prescriptions. Though the relative emphasis on climate and geomorphology differs from region to region, all of these classifications recognize that the various forest sites (subject to different weather, soil moisture, soil nutrients, and disturbance regimes) are inherently different in their productivity, successional trajectories, and climax forest composition and structure. Such site differences bound the potential set of stand types that might occupy a given piece of ground. British Columbia has a formal system of site evaluation that must be followed in the preparation of Preharvest Silviculture Prescriptions (PHSPs) on Crown lands (BCMF 2000). Other provinces have not yet implemented such requirements, although many forest products companies or their consulting foresters do formal PHSPs without this requirement.

Stand dynamics and succession

Successful silviculture and stand-level treatments need to be based upon an understanding of stand dynamics, but must also be innovative to produce the necessary forest composition and structure required to sustain the timber and other social and biodiversity values desired in the forest management area. Given the training of silviculturists in the areas of stand initiation and stand dynamics, they should be well placed to develop forests of a wide range of structures that suit a variety of ecological and social needs, including wildlife, biodiversity, and non-timber forest products (Oliver 1992; Gautam and Watanabe 2002). Knowledge and skills related to stand dynamics and silviculture can be applied to any forest structure problem. As individual stands are the building blocks of the forest landscape, silvicultural manipulations must play a basic role in the development of any managed forest landscape.

Silviculture manipulates a forest stand to affect the direction of succession by adding or removing species/individuals during key times of stand development, or by changing conditions for growth of the stand (Wagner and Zasada 1991). It has long been a goal of silviculturists to manage forests by emulating and working with natural processes, though this approach has always had to compete with the agricultural model of wood crop production (Smith et al. 1997; Lähde et al. 1999). Whichever paradigm is followed, one key to the successful practice of silviculture is understanding stand dynamics so that stands can be better manipulated for a given objective.

Short-term dynamics within a rotation

Over the rotation of a forest, stand dynamics are driven by recruitment of trees, resource competition between trees and shrub/herbs, and mortality factors. These concepts are imbedded in the following discussions and in most models of forest stand dynamics (Shugart 1984; Botkin 1993; Chap. 14). Oliver and Larson (1996) describe four stages of redevelopment of a temperate forest following disturbance: *stand initiation, stem exclusion, understory re-initiation, and old growth*. While not a perfect fit, these phases are also useful to describe the stages of boreal forest development where mortality agents ("stand-replacing disturbance") such as windthrow, insects, diseases, and fire kill all the trees. In boreal forests fire leaves the biggest imprint on forest landscapes, however, windthrow or defoliation events might also initiate even-aged stand development (Chap. 8). Agents killing individual canopy trees also play an important role in the gap dynamics of the understory re-initiation and old-growth phases (Kneeshaw and Bergeron 1998) and in areas with low fire frequency. Within these canopy gaps, the same stages described below occur at a finer scale.

Stand initiation stage

This stage follows some major stand-replacing disturbance, such as fire, wind throw, insect outbreaks, or pathogen attacks. The forest environment at this stage is characterized by high levels of light, nutrients, water, and available growing space. The most successful species in this stage are often clonal, sprouting from roots and stumps of the previous generation (Hauessler et al. 1990; Peterson and Peterson 1992). Once sprouted, juvenile growth rates of clonal species are fast (Lieffers et al. 1996a).

Species establishing from seed and with slow juvenile growth must first have a seed source on or near the site, have seedbed conditions that allow successful germination, plus sufficient light for establishment (Messier et al. 1998; Greene et al. 2002). On drier and more nutrient-poor sites, jack pine (*Pinus banksiana*) or lodgepole pine (*P. contorta*) usually establish from seed cast from serotinous cones. On sites with organic soils, black spruce (*Picea mariana*) may be the first species to establish following fire. Species like white spruce (*Picea glauca*) may not easily regenerate following fire because it does not bank its seeds either in serotinous cones or in the soil. White spruce seed trees are easily killed by fires and seed is only produced in reasonable quantities during mast years (Nienstadt and Zasada 1990). Seed germination and germinant survival appear to be sensitive to seedbed conditions in inverse proportion to seed size (Burton 1997; Greene and Johnson 1998). White spruce germinants, for example, only survive in high frequency on mineral soil, H humus layers, or rotting wood seedbeds, and have very slow juvenile growth rates (DeLong et al. 1997). Mineral soils remain exposed and receptive to tree seed germination for only a few years after wildfire or logging (Stewart et al. 2000), after which seedlings establish preferentially on rotten logs (Lieffers et al. 1996b). Mortality related to competition, summer frost, overheating, flooding/anaerobic soils, frost heaving, insects and diseases, or hare browsing may prevent successful establishment. As a result, white spruce is notoriously difficult to regenerate in clearcut sites (Nienstadt and Zasada 1990). While it may sometimes recruit massively immediately following fires (Dix and Swan 1971), in other situations it may recruit into stands slowly over decades (Lieffers et al. 1996*b*). Balsam fir (*Abies bal-samea*) and subalpine fir (*A. lasiocarpa*) also do not store seeds in a seedbank or in serotinous cones. Their cones break up during late fall or winter, with the seeds often falling to the ground with the scales. Secondary dispersal by saltation over crusted snow can move the seeds 160 m or more from the tree (Frank 1990; Galipeau et al. 1997). As balsam fir is highly tolerant of shade, however, immediate establishment after distur-bance is not as critical for maintenance of its presence in stands (see Chap. 8 for a more thorough discussion). Also, harvested stands may re-establish from residual understory trees (advance growth) not destroyed by logging.

Stem exclusion stage

Eventually, the crowns of the developing trees fill in all of the growing space. Depend-ing on stocking levels, this period of crown closure might take place in 5–15 years in aspen or pine stands, but may not be achieved for 40 or more years if spruce constitutes the canopy. After crown closure, there is usually insufficient light at ground level to allow further recruitment of seedlings. This period of stand development is character-ized by intense intraspecific competition, rapid mortality, and self-thinning of the canopy trees. A theoretical basis for this self-thinning is described as the −3/2 power law (Harper 1977; Drew and Flewelling 1979). In essence, as canopy trees expand in height and volume, available resources are insufficient to support all of the trees and trees from the lower crown classes usually die.

Understory re-initiation stage

As trees become older and taller there is a marked reduction in leaf area index. This is also related to a physical separation of the crowns ("crown shyness", as described by Long and Smith 1992). This results in greater light transmission to lower levels, which allows release of suppressed trees below or recruitment of new individuals. In some cir-cumstances there is recruitment associated with the death of single trees or small groups of overstory trees related to insects, competition, and blowdown creating canopy gaps (Kneeshaw and Burton 1997; Kneeshaw and Bergeron 1998; Cumming et al. 2000; McCarthy 2001). This provides recruitment opportunities in the zone of increased light below, provided that there are suitable seedbeds for establishment of seedlings. As new germinants need sites with consistent moisture supply and low amounts of hardwood leaf litter (DeLong et al. 1997), rotten logs often are the most successful sites for regen-eration of conifer species (Lieffers et al. 1996*b*). In some circumstances root suckering of aspen may also occur in older stands (Paré and Bergeron 1995; Kelly et al.[2]), but this is a poorly understood process. The understory re-initiation phase allows development of pockets of a new cohort of trees within a previously even-aged stand, and can lead to the development of truly uneven-aged stands.

---

[2]Kelly, C., Messier, C., and Bergeron, Y. Mechanisms of shade intolerant deciduous tree maintenance between catastrophic disturbances in the Southern Boreal Mixed-Wood forest. Unpublished.

Old-growth stage

In very old stands, with the gradual loss of overstory trees and their replacement by recruitment, an uneven-aged stand may eventually develop; this is especially likely for stands dominated by the more shade-tolerant species. By some definitions this is old-growth forest (Oliver and Larson 1996), but in practice, old growth is often described as any stand with trees much older than typical for a region or older than the region's average natural disturbance interval (Achuff 1989; Hunter 1989). While truly uneven-aged stands are relatively rare in the boreal forests of western Canada because of frequent stand-replacing fires, there is evidence that gap dynamics may play a more important role in boreal forests than previously acknowledged (Cumming et al. 2000; McCarthy 2001). Truly ancient or "antique" forests that have escaped stand-level disturbance for a period longer than the age of the oldest trees within them (Achuff and La Roi 1977; Goward 1994; Burton et al. 1999) are rare in the western boreal region.

In the eastern boreal forest, both black spruce and birch/aspen-balsam fir forests often have uneven-aged structures when fire intervals are long enough (Bergeron and Harvey 1997; Bergeron et al. 1999), although very old forests tend to be dominated by balsam fir and eastern white cedar (*Thuja occidentalis*; Bergeron and Dubuc 1989). The prediction of forest composition for long fire cycles is complicated by interactions among site factors, post-fire composition, and stand vulnerability to spruce budworm (*Choristoneura fumiferana*), which feeds selectively on spruce and fir (Bergeron and Dansereau 1993; see Chap. 8). Batzer and Popp (1985) confirmed that, following a spruce budworm epidemic in northern Minnesota, conifer-dominated overstories changed to earlier successional stages in which trembling aspen and white birch (*Betula papyrifera*) predominated. Bergeron and Dansereau (1993) state that although the proportion of pure deciduous stands decreases with fire cycles exceeding 200 years, the proportion of mixed stands remains constant, suggesting that other disturbance factors (such as defoliating insects and stand breakup at maturity) must be interrupting the regional successsional trend from intolerant deciduous species to tolerant coniferous species. The shift to coniferous domination is gradual, and dependent on propagule sources, disturbance size, and disturbance intensity; presumably small and low intensity disturbances will favour shade-tolerant species such as balsam fir and white spruce. Kuuluvainen and Juntunen (1998) and Kneeshaw and Bergeron (1998) have shown that small-scale gap disturbances and gap regeneration can be common in boreal forests with long fire cycles. This evidence suggests that in various parts of the eastern boreal forest, the fire return interval is so long that other disturbances such as spruce budworm outbreaks or senescence are probably more important than fire in terms of tree and stand replacement.

Computer tools to simulate and visualize forest stand development, growth and yield, and succession have been developed to model stand dynamics over time (Chap. 14). Most of these models also have options for exploring the implications of management interventions based on the selective removal or addition of trees at one stage or another, sometimes in a spatially explicit manner (e.g., SORTIE, Coates et al. 2001).

Long-term dynamics and succession

Multiple successional pathways

Variation in the factors that affect short-term dynamics can lead to multiple pathways of stand development and succession (e.g., McCune and Allen 1985). Those interested in manipulating stand development for a particular goal should have an intimate knowledge of the possible natural trajectories of a particular forest site. Careful inspection of stand types and the reconstruction of stand history across a given forest region will usually reveal a wide number of "natural experiments" that will provide guidance for the purposeful manipulation of stand development. It is useful for silviculturists and forest ecologists to develop their skills at reading stand history. In general, one wishes to determine the temporal and spatial pattern of catastrophic disturbance agents, any selectivity in tree mortality or survival (in terms of species, size, etc.), the degree of disturbance to understory vegetation and the forest floor, the patterns of tree regeneration, the nature of intermediate disturbances, and how these phenomena are affected by site and microsite. It is the variation in these features which layer on each other to produce a myriad of different stand types, varying in their composition, structure, and rates of development. These are also the attributes which, in turn, are manipulated by silvicultural techniques to promote specific stand attributes or processes (see below).

Climate change

The boreal forest is projected to have large increases in temperature over the next century. Indeed, if model simulations are correct, the southern boreal forest may have a temperature regime more suited to temperate deciduous forests, grasslands, or scrub woodlands in the future (Hogg 1994; Hogg and Hurdle 1995; see Chap. 20). This will certainly put physiological stress on species adapted to cooler/moister conditions. Also, at the northern extreme, warming may allow massive invasion of the boreal forest species in the Arctic tundra, which may result in a net gain in area of boreal forests in North America (Bachelet and Neilson 2000). As species migrations and changes in soil properties may be slower than the change in climate, there are many uncertainties about how forests may respond to these changes. This chapter will not specifically address this problem, but it is clear that if the projected climate changes happen, silviculturists must react by selecting and moving the appropriate species and genotypes to locations at the southern and northern zones of the current boreal forest.

Examples of boreal forest dynamics

The stages of stand development found in boreal forests are illustrated by the following examples.

Dynamics of western boreal mixedwoods

In the western mixedwood boreal forest (east of the Rocky Mountains), the mesic or zonal forest sites are dominated by stands of trembling aspen and white spruce with some intermixing of balsam poplar (*Populus balsamifera*), white birch, black spruce, balsam fir, and pines (*Pinus banksiana, P. contorta*, and their hybrids).

Most natural mixedwood stands originated after fire. Deep-burning fires consume the duff (see Fig. 8.10), kill shallow roots and rhizomes of sprouting species, and prepare a mineral soil seedbed (Heinselman 1981). Light surface fires give sprouting species such as trembling aspen (Peterson and Peterson 1992) or bluejoint reedgrass (*Calamagrostis canadensis*) a distinct advantage over species establishing from seed (Lieffers et al. 1993). *Calamagrostis*, particularly on the moister sites, may dominate for a long period if it is not overtopped by woody species such as aspen. Thick beds of *Calamagrostis* shade and insulate the soil and are poor seedbeds for white spruce (Lieffers et al. 1993). White spruce may establish immediately after the fire (Dix and Swan 1971) provided that seed sources and seedbeds are developed, or may gradually recruit under the aspen (DeLong 1991; Lieffers et al. 1996b).

By 10 years after suckering, very dense aspen stands may develop leaf area indices of 6 or more (Pinno et al. 2001). This may result in light levels of <5% of above canopy irradiance. At these levels, dense aspen stands can exclude recruitment of species such as white spruce, which needs nearly 10% full light for establishment (Lieffers and Stadt 1994). The leaf area of aspen stands appears to peak by 15–25 years (Lieffers et al. 2002) and light transmission increases with stand age (Constabel and Lieffers 1996). By 40 or more years after stand initiation, aspen stands transmit sufficient light to support reasonably height growth of white spruce in the understory (Lieffers and Stadt 1994). The increased light to the understory may allow thick development of shrubs (Rowe 1961) or in some circumstances recruitment by aspen suckers (Carleton and Maycock 1978; Cumming et al. 2000). The aspen continues to dominate but usually loses vigour after 70 years. In most aspen-dominated stands, the stem initiation phase for white spruce continues until spruce plays a dominant role in the canopy. Spruce recruitment is most likely limited more by availability of seed sources (Stewart et al. 1998) and seedbeds (DeLong et al. 1997; Stewart et al. 2000) than by light. Stands may become dominated by spruce after 70 or more years, but some stands remain aspen-dominated (Fig. 13.1a).

Spruce-dominated stands persist longer but eventually start to undergo dieback, usually after 120 years. While insects or diseases are usually the final cause of death, loss of vigour may be related to litter build-up and reduction in nutrient cycling (Pastor et al. 1987), or perhaps increases in stem respiration and hydraulic resistance (Ryan et al. 1997). If there are seed sources available, the very shade-tolerant balsam fir will establish in the understory of spruce stands. Stands become all-aged as the canopy spruce are replaced. Stands become low-density and low-productivity mixtures of balsam fir and spruce in western Canada (Achuff and La Roi 1977). Tree productivity is low because recruitment is difficult due to understory dominance by dense layers of mosses and shrubs, and the rooting zone becomes dominated by accumulated organic materials with low nutrient availability. This is the old-growth phase for mesic sites in much of the mixedwood forest zone. In most circumstances fire will return stands to an early successional stage before they reach this stage (Rowe 1961; Achuff and La Roi 1977; Cogbill 1985).

Fig. 13.1. (*a*) Boreal mixedwood stand in northern Alberta. (*b*) Black spruce stand in northern Quebec. (*c*) Balsam fir stand 10 years after the spruce budworm epidemic in Abitibi, Quebec. (*d*) Forty-year-old aspen stand in Abitibi, Quebec. (*e*) View of the line between an old fire (on the left) now dominated by balsam fir and eastern white cedar and a young fire with the stand dominated by trembling aspen in Abitibi, Quebec.

Dynamics of eastern boreal black spruce forests

Black spruce is found on a wide range of sites and soil types in Eastern Canada (Arnup et al. 1988; Burns and Honkala 1990; Viereck and Johnston 1990). In the southern part of its range, it is mostly found in pure stands on extreme sites such as wet organic soils. In the middle portions of the boreal forest it is found in pure stands on nutrient-poor peatlands. It is often mixed with eastern white cedar and tamarack (*Larix laricina*) on richer peatland sites (Arnup et al. 1988), balsam fir on mesic sites, trembling aspen and white birch in mixedwood forests, and with jack pine on sandy soils. Upland black spruce stands are usually even-aged, but uneven-aged stands occur on wet soils (Fig. 13.1*b*).

Historically, fire has been the major disturbance initiating and terminating succession in black spruce ecosystems (Black and Bliss 1980). Black spruce is easily killed by fire and generally rates high in fire hazard (Viereck and Johnston 1990). Fire return intervals in the order of 75–150 years are common, but can reach 500 years in Labrador (Foster 1983) and the eastern part of Quebec (De Grandpré et al. 2000) where precipitation is more abundant and thus other types of disturbance become more important.

Black spruce cone production starts at 30 years and can continue up to 200 years of age (Viereck and Johnston 1990). The semi-serotinous cones remain partially closed and disperse their seeds over several years. The compact mass of cones on mature trees further enables some of them to retain viable seeds after the passage of fire. An aerial seed bank, therefore, is available in the crowns of mature trees. Even though successful black spruce regeneration usually occurs after fire, failures occur when the stand is too young to bear cones or when the fire is so intense that it destroys most of the cones (Payette et al. 2000). Fires that remove the surface organic layer usually provide good seedbeds for rapid establishment of black spruce (Viereck and Johnston 1990; St.-Pierre et al. 1992). A short period of recruitment reflects rapid seed dispersal after fire, a short interval of seed survival after dispersal, and decreasing receptivity of seedbeds over time (Black and Bliss 1980).

The stem exclusion stage in black spruce stands lasts approximately 10–50 years. The understory re-initiation stage generally occurs between ages of 50 and 120 years. Senescence and the development of old-growth attributes may begin between ages of 70 and 150 years, and occurs earlier on the most productive sites. Once canopy breakup starts, it proceeds rapidly (Arnup et al. 1988) through windthrow. Smith et al. (1987) have shown that the probability of windthrow in these aging black spruce stands is proportional to the height of dominant trees. The early breakup of productive stands may relate to the fact that trees attain taller, susceptible sizes earlier than on poor sites.

During stand development, advance regeneration of black spruce of predominantly vegetative ("layered") origin often becomes established (Doucet 1987; Ruel 1989; Archibald and Arnup 1993). In situations where fire does not occur before stand breakup, layers are able to release in the gaps that are created by windthrow (Doucet and Boily 1986). As a consequence, black spruce stands become increasingly uneven-aged with time since fire. This situation is quite common on wet soils or in parts of the species' range where rainfall is more abundant. Here, rapid moss growth facilitates the initiation of adventitious roots on lower branches, and fire return intervals are longer.

Dynamics of eastern boreal mixedwoods

The eastern boreal mixedwoods are strongly affected by a latitudinal gradient of temperature, which limits the northward expansion of the maple forest in the south and the southward expansion of balsam fir in the north. There is also a longitudinal gradient of precipitation (increasing from west to east), which affects the fire frequency and consequently the zones composition and dynamics. We can divide the eastern boreal mixedwoods into two large domains: (1) the yellow birch – balsam fir domain found in the south, and (2) the white birch – balsam fir domain that extends up to the black spruce forest in the north. The white birch – balsam fir domain can also be further divided into two sub-domains, a western one where precipitation is low (800–1000 mm/year) and the fire return interval is short (roughly <150 years), and an eastern one where precipitation is higher (900–1300 mm/year) and the fire return interval is >300 years (Grondin et al. 1996).

Most stands in the yellow birch – balsam fir domain are renewed by gap dynamics created by the mortality of older trees (mainly yellow birch, *Betula alleghaniensis*, and spruce) and the spruce budworm epidemic that kills most of the balsam fir population every 30–50 years (Blais 1983; Fig. 13.1*c*). Fire is also believed to be important, but few studies have documented its role; this forest also has a long history of logging. Regeneration and growth of both balsam fir and yellow birch are well suited to gap dynamics. Balsam fir is very shade tolerant and establishes on a variety of seedbeds. Because of its ability to survive in shade by limiting its growth (Parent and Messier 1995), it maintains a dense seedling bank positioned to respond to any sudden opening created by windthrow, mortality of larger trees, or the periodic recurrence of spruce budworm. In contrast, yellow birch is intermediate in shade tolerance, and requires mineral soil or decaying wood in order to germinate. Once established, it can grow much faster than most of its more shade-tolerant competitors, but its long-term survival depends on the presence and maintenance of mid- to large-sized gaps (at least 200 m^2) that provide more than 20% full sunlight. Different variants of the partial logging method seem to be successful in maintaining the naturally mixed composition of this forest type, whereas large scale clearcutting will tend to favour balsam fir and shade-intolerant species such as white birch and trembling aspen. On some mesic sites, a strong understory of mountain maple (*Acer spicatum*) may be present and can become a severe competitor with regenerating crop trees after harvesting (Archambault et al. 1998; Laflèche et al. 2000).

The differences in precipitation and fire regimes between the western and eastern sub-domains of the eastern boreal mixedwoods are responsible for associated differences in vegetation. In the western sub-domain, the proportion of conifers is lower, and because of rich clay soils, the understory tends to be denser and richer than its eastern counterpart. The dynamics of this forest are dominated by relatively frequent fires, strong gap dynamics influenced by the mortality of trembling aspen after 100 years, and the 30–50 year recurrence of spruce budworm epidemics. On mesic sites, aspen dominates after disturbance (Fig. 13.1*d*). Stands then shift to a mixture of aspen, balsam fir, white birch, and white spruce and finally to balsam fir, eastern white cedar, and white birch after 250 years (Fig. 13.1*e*). This diversity of disturbance events, the low precipitation and greater understory competition by mountain maple could explain the relatively lower abundance of balsam fir in the western sub-domain compared to the eastern sub-domain. In the latter, spruce budworm infestations dominate the natural disturbance dynamics. Balsam fir usually comes back when the budworm damages mature stands, owing to its abundant seedling bank, its ability to survive deep shade for long periods of time and still respond to release (Parent et al. 2000; Chap. 8). However, if the insect destroys a stand before the seedling bank is constituted, a pioneer phase would take place before balsam fir comes back.

Technological and social issues related to stand manipulation

This section discusses the technological, economic, and social concerns related to various silvicultural treatments used in boreal forests. Given the wide scope of this chapter we cannot discuss all of the available information on each of these treatments. Technical discussions on some of these practices can be found in Lavender et al. (1990) and in Wagner and Columbo (2001). One of the difficulties of applying sophisticated stand-level treatments in the boreal forest is that the low revenues generated from cutting a stand in this region may not justify the high costs of some treatments a silviculturist may wish to use. If simple financial analysis is applied to boreal forest siliviculture, many stand treatments cannot be justified on the basis of future wood products. In the boreal regions, stand-level treatments are usually justified on the basis of meeting provincial regulations for regeneration or stocking. While a thorough analysis of financial costs and social benefits is also not possible in this chapter, we acknowledge that financial considerations are often the over-riding factor in determining which systems forest managers eventually implement (Chap. 6). For example, cost savings associated with harvesting or vegetation control may be the strongest reasons for the widespread use of clearcut logging or the aerial application of herbicides to control competing vegetation. In contrast, however, environmental concerns about these techniques may eventually limit their use despite their efficacy and cost-efficiencies. Indeed, the social concerns related to the use of various treatments has been less discussed but may actually be one of the most important factors affecting choice of silvicultural treatment. The great conundrum facing silviculturists working on public lands in Canada is that social concerns (Table 13.2) may not allow them to use the most biologically or economically efficient silvicultural tool to achieve a silvicultural goal. In fact, many silviculturists feel threatened by the fact that "the public" does not like the treatments that they have selected for the forest.

In the following sub-sections we briefly describe various silvicultural procedures and discuss them in terms of several factors: (1) efficacy in achieving a silvicultural objective, and (2) the relative costs of treatments. We also discuss each one in relation to: (3) short-term or long-lasting impacts on forest soils/nutrient cycling, forest structure, and changes to biodiversity, and (4) social acceptability of the treatment and the type of forest that it produces.

Table 13.2. Silvicultural processes/activities arranged along a continuum from those viewed as highly natural (on the left) to those viewed as highly artificial or intensive (on the right). The relative position of these treatments on the gradient is based upon the authors' perception of social concerns regarding their use.

Natural →				Intensive
Regeneration				
Natural seeding	Aerial seed[1a]	Plant Local popn's[1a]	Plant improved stock[1a,3a]	Plant exotics[1a,3b,5b] Genetically modified organism[1a,2b,3b,5b]
Site preparation				
None	Slash burn[1a,2a]	Light scalp/mix[1a]	Disc trench/small mound[1a]	Deep plow or large mound[1b,5b]
Vegetation management				
None	Grazing[4a]	Mechanical[1a,2a]	Spot application herbicides[2a,4a]	Aerial spray herbicides[1a,2b,3b,4b]
Density control				
None		Juvenile spacing	Single comm. thin[3a]	Multiple comm. thin[3b]
Nutrition				
Natural		Spot fertilization[3a,4a]	Aerial fertilization[2a,3a,4b]	Sewage/biosolids/manure[2b,3b,4b]
Rotation length				
Long rotation		Normal[1a]		Short rotation plantation[1a,3b]
Harvest system				
Selection	Shelterwood[1a]	Nat. shelterwood CPRS[1a]	Green tree ret. seed tree[1a,3a]	Clearcut[1b,3b,5a]

Social concerns (*a*, moderate; *b*, severe):

[1]Aesthetics.

[2]Perceived risks to human or animal health.

[3]Perceived risks to biodiversity.

[4]Perceived risks of pollution.

[5]Persistence of treatment.

Forest regeneration

Forest regeneration systems are important for several reasons: regeneration impacts the species composition and structure of the stand, and is critical in assuring the sustainability of timber and forests. Environmentalists view naturally regenerated forests more positively than planted forests (Hammond 1991) because there is often higher diversity of species, and simply because trees are not in rows. Planting is often chosen by industrial foresters, however, because it reduces regeneration lags and the stochastic nature of tree establishment typical of natural regeneration systems. Because regeneration costs are often high and this investment must be carried through the life of the stand, regeneration systems are usually the biggest silvicultural costs in stand management (Brace and Bella 1988).

Site preparation

Seedbed and planting spot preparation are commonly practised before regeneration. This is done to reduce slash loads, reduce fire hazards, increase radiation transfer to the soil surface, increase mineralization rates, improve the soil water relations for establishing seedlings (draining water from or to roots), or to allow planters better access to planting spots. Also, vegetation competing with crop trees is often removed directly, or is reduced indirectly by removing the litter and thereby (1) damaging roots and rhizomes of competitors, (2) decreasing the availability of nutrients within the small patch of a future planting spot, or (3) producing a surface substrate that is unsuitable for establishment of competitors from seed. While these latter two justifications might also be negative for optimum tree growth in the absence of competitors, the ability to control competitors in the immediate vicinity of a seedling without herbicide use, however, can sometimes override concerns related to nutrition of the immediate microsite.

Fire — Wildfire has been shown to be a critical factor in reducing duff, increasing nutrient release, and increasing heat transfer into soil, thereby increasing biological activity (Feller 1982). Depending upon fire intensity and depth of burn (Granström 2001), fire has been important in maintaining many boreal species and in moulding their evolution. Fire has been shown to be critical in promoting the natural regeneration of boreal forests (Chap. 8). Use of prescribed fires has been useful for controlling unwanted vegetation, stimulating nutrient release, and changing soil thermal properties. It also produces similar seedbeds or conditions for clonal sprouting (in terms of organic matter reduction and chemical conditions) as would occur following wildfire, and has been used successfully for vegetation control and slash reduction in boreal conditions (Chrosciewicz 1988).

Surprisingly, however, prescribed broadcast burning ("slash burning") for site preparation has not been widely applied across Canada, and its use is decreasing. There are several reasons for this: smoke from fires poses a health problem and risk for traffic safety on highways; managers are always fearful of fires spreading out of control, thereby risking property and life; and escapes may destroy intact forests or valuable plantations. In boreal forests these escapes are more problematic than in other fire-prone forests, for two reasons: boreal forest tree species do not have thick bark that protects them from killing temperatures during ground fires; and boreal trees usually have their structural roots located in the thick organic matter of the forest floor where roots are vulnerable to smouldering ground fires (Fig. 8.10). As a consequence of these risks, expensive precautions such as ground crews, bulldozers, and aircraft must be on standby at the time of burning. This usually means that larger areas are burned at one time to justify the precautionary protection costs.

Given the expertise in fire management available across Canada, we should be encouraging moderate slash burns in boreal forests, despite the risks of escapes. Though our ability to mimic the behaviour and effects of natural wildfires is limited (Chap. 9), forest certification practices such as prescribed in Sweden and Finland require that some of the land area be burned during a planning period (Granström 2001). The timing of burns during the appropriate "burning window" minimizes risks of escape. Burning during periods of strong updraft conditions with wide dispersal of smoke, and quick mop-up of smouldering ground fires, is helpful in minimizing the amount of smoke at ground level. A way to minimize burning risks in boreal forests is to incorporate burn planning into landscape level planning of forest stands and their protection from fire (Chap. 12). Some features to consider are that fire escapes into mature timber may not be a serious problem if the burned areas are moderate in size and the wood can be salvaged. Recently established and mid-rotation stands may have had large inputs of investments, yet they are too young to have significant merchantable volume; a slashburn escape into these stands would be an economic disaster. Slash burning adjacent to these younger stands, therefore, poses a higher risk than beside mature timber.

Patch burning of seedbeds might be done using a propane burner mounted on a tractor, which allows burning under a wider range of conditions. The goal is to remove litter and blacken the soil to create warmer microsites with the chemical characteristics of naturally burned sites. Prescribed fire might also be used to renew stands where timber value is low because of insect infestation, rot or disease, as this may be cheaper than logging stands with little value.

Mechanical — There is a wide variety of treatments/equipment that can create microsites suitable for the regeneration of forest trees (Table 13.3). This equipment typically displaces the duff layer to expose the A horizon of mineral soil ("scalping"), mixes the forest floor with the mineral soil ("mixing"), or inverts the combined strata of duff and A horizon (typically in a mound or berm; Örlander et al. 1990). Trees planted on mechanically prepared microsties usually have large increases in initial growth rates compared to untreated microsites (Örlander et al. 1990); by a decade after planting, however, there may be little difference in height increment between trees from prepared and untreated microsites (Bedford and Sutton 2000). One of the major concerns related to mechanical treatments of boreal forest sites is access for machines during unfrozen conditions. Most row mounders and mixing treatments are limited to unfrozen or thin frost conditions. This often limits treatments to ripper plows, large backhoes (excavators), and bulldozer blading, which are clearly low-precision methods of site preparation. Over the past 50 years there has been continuous development of different types of equipment for this work (see McMinn and Hedin 1990 and Sutherland and Foreman 1995 for descriptions). Proper selection of the appropriate equipment and technique for given site conditions (moisture and nutrients) and type of competing vegetation could be tied to ecological site classification systems (e.g., Beckingham and Archibald 1996). In reality, however, there is usually only coarse matching of mechanical treatments with the ecological site type and its limiting factors, for tree growth identified in pretreatment assessments is multi-factorial and often not clearly understood. Machine availability, costs of operation, and access often override biological limitations for tree growth in the selection of the treatment.

Beside the above classification of treatments, site preparation methods can also be categorized by how much they vary from natural conditions and the permanence of their

Table 13.3. Microsite effects of inverted mounds, scalping, and mixing treatments. An attribute is expected to increase (+) or decrease (–) as result of the treatment, as indicated.

Site preparation treatment	Attribute	Light treatment	Moderate/heavy treatment
Inverted mound		Small volume (<40 L) or low berm height (<25 cm)	Large volume (≥40 L) or high berm height (≥25 cm)
	Machine type	Row mounder, disc trencher	Excavator mounder, deep plows
	Soil temp.	+ (Diurnal swings)	+ (Diurnal swings)
	Nutrient avail.	– (Mineral cap), ++ (org. layer)	– (Mineral cap), ++ (org. layer)
	Drainage	+	++
	Veg. control	+ (Inhibits seedlings)	+ (Inhibits seedlings and sprouts)
Scalp (scarify)		Shallow (to H layer of soil)	Deep (remove some mineral soil)
	Machine	Many different machines	Blading
	Soil temp.	– (At surface), + (at depth)	– (At surface), + (at depth)
	Nutrient avail.	+	– –
	Drainage	Neutral	– (Reduced infiltration, ponding)
	Veg. control	+	++ (Removes seeds/sprouts)
Mixing		Coarse (slow speed)	Fine (high speed)
	Machine	Disc trenching, drags	Merri crusher, rotory tillers
	Soil temp.	– (At surface), + (at depth)	– (At surface), + (at depth)
	Nutrient avail.	– (Short term), + (over 10 years)	– (Short term), + (over 10 years)
	Drainage	+ (Increased infiltration)	+ (Increased infiltration)
	Veg. control	– (Rhizomes/roots still sprout)	+ (Rhizomes/roots shredded)

impacts. Some treatments show little visible sign of having been implemented 5 or more years after treatment: light drags (anchor chains, "shark-fin" barrels), shallow and small scalps, and small mounds (<40 L in volume). Moderate treatments include heavy drags, disc trenching, shallow blading to mineral soil, and large mounds. Several long-lasting and severe treatments are deep plowing, backhoe mounding (Fig. 13.2a), and deep

Fig. 13.2. (*a*) Backhoe mounding on wet mineral soil, to prepare planting spots for white spruce, north of Peace River, Alberta. (*b*) Trembling aspen sucker regeneration 1 year after logging north of Peace River, Alberta. (*c*) Layering of black spruce. The ribbon is tied between the branch of the parent tree and the stem of the layer (regenerated tree). (*d*) Blue-joint reedgrass (*Calamagrostis canadensis*) competition 3 years after logging in northern Alberta.

blading using a bulldozer. Deep plowing produces a deep furrow and high berms that could be maintained for an entire rotation or more. Deep blading removes the surface litter and part of the A horizon, pushing these resources considerable distance from their original position, thereby lowering the site nutrient reserves. All of these treatments have social/environmental concerns that may limit their use (Table 13.2).

The generally superior early growth of many tree species on site-prepared microsites has long been recognized, especially in northern climates (Sutton 1993). However, concerns have been raised about site degradation (soil rutting, compaction and loss through erosion) caused by site preparation (Curran et al. 1990), and the growth form of crop tree roots in prepared microsites (Balisky et al. 1995). In recent years there has been a movement away from using heavy mechanical site preparation.

Chemical — In some regions, when there is a need to control the future expansion of shrubs, forbs, and grasses that are already well established at the time of logging, sites are treated with glyphosate, a systemic, broad-spectrum, foliar-active herbicide. As this treatment only controls the growth of living plants, chemical treatments are only suited to sites with no drainage, thermal, and nutritional problems that would have to be addressed by mechanical treatments. There have also been some mechanical treatments paired with the application of the soil-active herbicide hexazinone. These treatments have not been widely applied, as this herbicide persists in soil much longer than glyphosate (Michael et al. 1999). While these treatments are very cost effective, there are perceived and (or) real environmental and biodiversity concerns related to herbicide use (Reynolds et al. 1993).

Re-establishment of trees

Natural regeneration from seed — Regeneration from seed naturally cast on the site is often considered to be the best method of regeneration by people interested in natural forestry. Successful establishment requires that seed be available, that there is an appropriate seedbed, and that the environmental conditions be appropriate for germination and seedling establishment (Chap. 8). The first requirement is that seed be deposited on the site either from nearby seed trees or from cone-bearing slash. Many species, such as white spruce, produce large quantities of seed only sporadically in "mast" years. As the seedfall must coincide with the relatively short window of time in which seedbeds are receptive, waiting for natural seedfall is not always successful. There have been attempts to treat seed trees with giberellic acid or girdling to induce cone crops (Pharis et al. 1987; R.P. Pharis and B.E. Grover, unpublished data).

Spruce seed is reliably spread up to 150 m downwind of seed trees and will produce fully stocked stands provided that the understory vegetation is not too dense (Stewart et al. 1998; see also Chap. 8). Similar relationships can be devised for other species, dependent upon seed size and tree height (Greene and Johnson 1996; Leadem et al. 1997). But if microsites for seed germination have a heavy cover of hazelnut (*Corylus cornuta*), alders (*Alnus* sp.), or bluejoint reedgrass there will likely be little recruitment. For serotinous species, there is usually a large pulse of seed dispersal following wildfire as the aerial seedbank is released by the heat of the fire.

Spreading the serotinous cones in the slash of logged jack pine, lodgepole pine, or black spruce near the soil surface using anchor chains and sharkfin barrels (drag scarification) usually promotes regeneration of these species (Vyse and Navratil 1985; Béland et al. 1999) in a manner similar to regeneration following wildfires. In dragging, the cone-bearing slash is repositioned near the ground. The greater heating under fully open conditions near the soil surface (Oke 1987) will soften the resin bonds of the serotinous cones, thereby releasing the seed in a manner similar to the heating of these cones in wildfires. The drag scarification also exposes mineral soil, which is similar to the seedbed created by consumption of the duff in fires. Provided that the cone-bearing slash is spread uniformly on the cutover, drag scarification is an inexpensive and effective technique for regenerating sites using natural seed sources. Formulas to determine the amount of scarification are available in Vyse and Navratil (1985). Fire-origin stands can be distinguished from drag-scarified stands by a clumped distribution of seedlings, since the seeds dispersed from the tops of fire-killed standing trees are more uniformly distributed.

Mineral soil is usually the best seedbed for most tree species; however, rotten logs are noted to be the best seedbeds for white spruce in the understory of established stands (DeLong et al. 1997; Burton et al. 2000). For survival and good growth of seedlings there must be sufficient light (Lieffers and Stadt 1994; Wright et al. 1998), moisture, and nutrients; seedlings must also be able to withstand the site's temperature extremes and not be damaged by insects, diseases, and grazers. As there are many variables in natural regeneration (Greene et al. 2002), success is not assured and there may be delays before seedlings establish. A more complete analysis of the factors controlling natural regeneration of conifers and an estimate of the probability of successful establishment given seed sources, distances from seed source, seedbed conditions or survival of advance growth is provided in Greene et al. (2002). As boreal forests are very slow growing and the wood is often of relatively low value at time of harvest, it is logical to pursue cheaper strategies of regeneration through the examination and understanding of natural seeding.

Artificial seeding — In many circumstances, seedbeds may be available but seed is not. Seed trees may not be in close proximity if all mature trees have been harvested in a large area or if large fires have killed all trees. Also, mast years may be delayed until after the seedbed is no longer receptive. In such cases, seeding can be done either from aircraft or from ground dispersal equipment. Seed may cost more than $1 per thousand seeds. Spreading seed from aircraft is least efficient in terms of seed use, as most of the seed will land on inappropriate microsites. Recent trends in direct seeding have focused on the targeted delivery of seeds to selected microsites, sometimes with protective shelters to aid the survival of seeds and young seedlings. Direct seeding, however, has some drawbacks: (1) given possibilities of drought following application or rodent populations consuming the seed prior to it germination, establishment is not assured; (2) germinants are particularly vulnerable to vegetative competition, and often do not survive; and (3) seed is relatively expensive. Aerial seeding of pines or spruce has had some success in boreal forest sites (e.g., Alberta Lands and Forest Service), but given the uncertainties of the system noted above, it is unlikely to be widely applied to sites where there is an expectation of full stocking.

Planting — While regeneration costs of natural regeneration are low, the regeneration lag and the element of risk associated with natural seeding results in some organizations adopting the more costly step of planting seedlings. Planting of seedlings is considered to be a more assured means of regeneration than regeneration from seed (Grossnickle 2000), and promotes more uniform stocking. The regeneration liability associated with cutover stands on public land in most jurisdictions creates an urgency for rapid regeneration. For most organizations, the economic incentive of meeting regeneration standards, thereby maintaining the cutting rights on the mature timber in their management unit, outweighs the added costs of expensive artificial regeneration techniques. As planted stock does not always survive or grow to expectations, however, the assumption of assured regeneration using planting may sometimes prove to be wrong.

The exposure of mineral soil traditionally has been considered essential for the successful establishment of planted conifer seedlings. Recent research trials and field observations have noted spruce seedling root proliferation at the interface of organic and mineral horizons (Letchford et al. 1996), and even superior growth of spruce seedlings planted in some organic materials (Heineman 1991). It is speculated that greater availability of mineralized nitrogen in the upper and organic layers of intact soils are responsible for better spruce performance than on many mechanically prepared planting spots having a large subsoil component. Shorter root plugs on growing stock, Jiffy^R containers and copper-treated (root pruned) seedlings have been developed in order to promote a more natural, shallow rooting pattern. As a result, there is growing attention paid to the careful selection of planting spots to have optimal thermal properties and a shallow layer of surface organic matter (Balisky et al. 1995; Lloyd and Elder 1997). This "microsite planting" approach can often be done without expensive site preparation, so long as sufficient latitude in the spacing or species of crop trees is allowed. Also, in some regions tree planters now often do "mixed bag" planting on upland sites, establishing pine seedlings where mineral soil is exposed or the microsite is prone to dehydration, and planting spruce where a moderate duff layer or rich soil exists and where the microsite is unlikely to dry out during the growing season.

There is a large technology and information base on conifer seedling production in Canada (Lavender et al. 1990; Grossnickle 2000). Much of the seedling production systems are aimed at production of container and "plug" stock in greenhouses, and spruce is the most common type of planting stock in boreal forests (Grossnickle 2000). Stock production, coupled with the costs of site preparation and planting, is expensive relative to the normal rotation length and value of boreal forest stands. Most of this stock is derived from seed collected in the general vicinity of the planting sites, although increasingly seed is coming from seed orchards established from "plus tree" stock. Seedlings are usually planted onto sites that have been clearcut, but in recent years there has been some planting in shelterwood and in the understory of aspen stands (Stewart et al. 1998). "Plus tree" selections have been used to develop faster growing or disease resistant planting stock. Efforts are being made to clone individuals with extraordinary characteristics, but these remain largely in the research and demonstration stage. In most areas in the boreal zone, so long as harvested areas are not too extensive, there is usually some degree of unplanned natural regeneration that supplements any artificial regeneration efforts. Managers frequently plan for a combination of planting and natural regeneration to achieve the desired stocking. Thus local genotypes will also be maintained.

Intensively managed plantations often use exotic species; in the boreal forests of Canada these might be hybrid poplars (*Populus* × spp.), Scots pine (*Pinus sylvestris*), Norway spruce (*Picea abies*), or Siberian larch (*Larix sibirica*). These species are imported because of their fast growth rates or wood characteristics. Use of exotics, however, is often questioned because they displace native species and perhaps the flora and fauna that their native analogues support. There also are concerns that they will interbreed with native species of the same genera, thereby contaminating local gene pools.

There is ongoing research into development of planting stock by transferring genes into trees from unrelated organisms, thereby producing "genetically modified organisms" (GMOs) to attain greater growth or specialized characteristics such as herbicide-or insect-resistance (Pullman et al. 1998). To date such material has not yet been deployed in boreal forests. Given public resistance to use of GMOs, there will likely be need for careful regulation of this material (Mullin and Bertrand 1998) in natural settings and on public land. It is unlikely that GMO tree stock will be widely planted on public lands except in specially zoned areas.

Coppice regeneration — Boreal forest aspen and balsam poplar may have as much as 40 t/ha of root biomass (Peterson and Peterson 1992). At the time of fire or logging of a stand, the aboveground portion of the stand is killed and the auxin produced by the aboveground portion of the trees no longer flows to the root system (Schier 1972), thereby releasing buds on the root system from auxin-induced suppression. Suckering and stump sprouting (in the case of birch and other hardwoods) allows the next generation to capture much of the stored reserves and part of the structural root biomass of the previous generation (DesRochers and Lieffers 2001). The rapid rebuilding of stand leaf area and early development of high stand productivity (Pinno et al. 2001) is a reflection of this belowground legacy.

While aspen has the potential to sucker vigorously following logging (Fig. 13.2*b*), there are many sites where sucker density and growth do not meet expectations (Bates et al. 1993). There are several possible reasons for this: (1) root damage during felling and skidding may break the roots into small pieces and allow fungal and bacterial infections to decompose much of the root system before it can sucker; (2) soils compacted by heavy equipment may not allow sufficient aeration (McNabb et al. 2001) for root survival or sucker development; good recruitment of aspen suckers is found on summer-logged cutblocks with sandy loam and reduced moisture (so soils have not been compacted), but with sufficient disturbance during felling and skidding operations to have removed some of the organic layers covering the soil; (3) in lab studies, Zasada and Schier (1973) and Hungerford (1988) have noted increased suckering with soil temperature, but it is apparent that massive increases in suckering can be achieved with only slight increases in soil temperature if the organic matter is disturbed (Lavertu et al. 1994; Fraser et al. 2002), suggesting that increases in soil temperature following logging and soil disturbance are not the only factors stimulating suckering. Chilled soils, however, result in poor water relations (Wan et al. 2000) and slow growth of aspen (Landhäusser and Lieffers 1998). As wildfire would typically burn off some of these insulating layers, aspen regeneration should be more vigorous after moderate wildfires and slash burns compared to after logging (especially winter logging). Indeed, we speculate that on many sites in the mid- and upper boreal forest, aspen needs fire or soil disturbance to stimulate suckering and promote growth.

Black spruce is often regenerated from layering (Fig. 13.2*c*). Lower branches bend down with increasing size, touch the forest floor and become covered with mosses. They usually root at the point of contact and are capable of eventually becoming independent trees. With careful logging practices, these rooted branches can become the next generation of trees. Paquin et al. (1999) have shown that layers and seedlings of black spruce have quite similar physiological responses when they become acclimated to the removal of the canopy.

Vegetation control

Competition from vegetation is one of the more significant reasons for slow growth and mortality of planted conifer seedlings. Grasses such as bluejoint reedgrass (Fig. 13.2*d*), shrubs such as red raspberry (*Rubus idaeous*), alder, mountain maple, and hazelnut or undesirable trees (usually hardwoods) may quickly overtop regeneration established from seed or planting. While growth of seedlings is often suppressed with low levels of competitors, survival of planted seedlings is not affected until a high level of competition is experienced (Radosevich and Osteryoung 1987). Some of the most controversial aspects of forestry relate to the control of vegetation competing with crop trees. Recent trends in vegetation management include recognition of the ecosystem services and beneficial effects of some non-crop vegetation (even in conifer plantations), and the wider use of selective spot treatments rather than broadcast treatments.

Herbicide — Herbicides such as glyphosate, trichlopyr, 2,4-D, or hexazinone are often used to control herbaceous, shrubby, and hardwood species in establishing conifer stands (Campbell 1990). Aerial application of glyphosate offers a cheap and effective control of the species listed above, as it is a systemic poison to most green plants, killing both the above- and below-ground portions of the plant (Liu et al. 1996). While research indicates that glyphosate has little long-lasting impacts on mammals, birds, or insects (Lautenschlager 1993; Reynolds et al. 1993), the above herbicides are broad-spectrum and also kill non-target vegetation. Many people in society are uncomfortable with aerial application of herbicides on public forest lands, and herbicide use will likely remain an issue of contention between environmental groups and industrial foresters (Campbell 1990; Buse et al. 1995). As a result, herbicide use will likely be zoned relative to various land uses and forest resource values.

Herbicides can also be applied around individual crop trees using backpack sprayers and blowers. Application of basal bark treatments of trichlopyr or stem injections of glyphosate have been successful at controlling unwanted hardwoods. As these treatments target only the unwanted species/individuals, they are often considered more benign than broadcast applications. Unfortunately, the ground application of herbicide is at least twice as expensive as aerial application.

Mechanical treatments — Manual cutting of competing vegetation offers short-term control of competitors (Hart and Comeau 1992). As most of the problem species, however, are capable of resprouting from roots or rhizomes, they tend to quickly redevelop following treatment. As a result, there is usually a need for multiple treatments over several years to allow the conifer seedlings to grow above the usual height of the competitors. Indeed, there is evidence for increased density of suckers following brushing treatments, though there may be increased rot in sprouts regenerating from brushed stumps (B.E. Grover, unpublished data). Manual brushing is physically strenuous, working with motorized brush saws is dangerous on rough and slash-covered terrain, and the need for repeated treatments makes this approach expensive. Nevertheless, mechanical treatments are likely to be the most socially accepted treatment for vegetation control, providing seasonal employment opportunities (Chap. 5) as well as avoiding the use of widely mistrusted chemicals.

Density control and thinning

Many stands that are regenerated naturally have more trees than can be grown to maturity without some stem mortality. In anticipation of this mortality, some trees can instead be killed either in the early stages of stand development (pre-commercial thinning) without removing the cut stems, or cut in the middle or late years of stand development and removed for use as a commercial product (commercial thinning). Thinning can be used to change the rate of stand development. Low thinning, where the smaller (suppressed) individuals are removed, leaving the large trees to continue development (Smith et al. 1997), is the most common method of thinning in Canada. Low thinning accelerates stand development. It can also be used to accelerate development of old-growth characteristics such as open stand structures with relatively large trees (Oliver 1992). Conversely, stand rotations could be lengthened by removing some of the larger trees in mid-rotation, thereby releasing the smaller trees from suppression. This could be used to delay clearcutting operations. Of course, this could only be done with tree species such as white spruce and balsam fir that are capable of release from suppression (Urban et al. 1994).

In intensively managed plantations, stands are often established at relatively high density so that full leaf area and productivity can be reached at an early time. Trees are planted with regular square or hexagonal spacing so that the crowns can quickly and completely occupy all of the available space. This tends to maximize stand productivity and shorten rotation age.

Most studies of precommercial thinning have shown reduced growth at the stand level in the early stages, compared to unthinned stands. By reducing the total amount of growing stock, stand-level productivity is usually compromised in order to accelerate the growth of the remaining individual trees. Carefully maintaining the stands at moderate density, however may result in increased stand volumes (Day 1998). This latter view is not shared by Mitchell and Goudie (1998) who, using TASS simulations, report reduced final yield from all thinning systems. Nor could an analysis of thinning and yield in Quebec find any evidence for increased stand volumes (CCSMAF 2002). Stands that are commercially thinned, however, may result in greater total fibre productivity over the life of the stand as the trees that would have died and decomposed before final harvest would be captured in the thinning (Nyland 1996).

Thinning may be of considerable economic importance to a forest products company, if: (1) it can shorten rotations; (2) it could help bridge age-class gaps in their wood supply; (3) increased stem size at final harvest could increase revenue, depending upon the premium paid for larger-sized timber; and (4) increases in merchantable wood productivity may be rewarded by allowing more rapid timber harvest of mature stands (commonly known as the "allowable cut effect"). See the above discussion on thinning and total yield, however.

There are various methods of planning the thinning of a stand, all based on the premise that a stand could be fully occupied either by many small trees or fewer large ones, in a relationship known as the "–3/2 self-thinning law", a universal rule of plant population biology (Harper 1977). A spacing factor (Wilson 1946), stand density index (Reinecke 1933), or stand density management diagram (Drew and Flewelling 1979) all have been used to guide foresters as to how many trees to remove from the stands at any thinning entry. Stand density management diagrams are available for black spruce (Newton 1997) and lodgepole pine, white spruce, and Douglas-fir (*Pseudotsuga menziesii*; Farnden 1996). Indeed, in intensively managed systems, crop planning using multiple thinning entries (Day 1998) is expected to yield the highest volume gains. Widescale practice of such systems, however, may result in loss of snags and coarse woody debris from stands, the elimination of mixed species composition, and the biodiversity that these features support (Berg et al. 1995; Larsson and Danell 2001). Another difficulty of heavy precommercial thinning is the increase in knot size and the high proportion of juvenile wood that is found when stands are started at low density. A decline in wood quality could have a large impact on the value of the wood.

Fertilization

Many boreal forest sites have limited productivity because of deficiencies in nutrients, particularly nitrogen (Munson and Timmer 1989). There are relatively few N-fixing organisms in boreal forests, but low levels of N fixation associated with feather mosses may provide much of the boreal forest N (Deluca et al. 2002). Also, as there is significant amounts of N uptake in organic form (Lipson and Nasholm 2001), the cycling of N might be very tight in these systems. Nevertheless, forest fertilization has been demonstrated to produce remarkable increases in productivity on severely deficient sites (Prescott et al. 1996; Morrison 2001). Fertilizers could be applied early in stand development or late in the life of the stand to increase log size immediately before harvest. Fertilization can also be used to accelerate stand development of certain age classes to fill in age-class gaps. The challenge is to promote stand establishment and tree growth without encouraging the growth of competing vegetation, and at a stage in stand development when crop trees will respond. Commercial granular fertilizer (NH_4NO_3 or urea) is usually applied by aircraft, though application of municipal sewage slurries is also feasible near populated centres. In all of these treatments there is concern over nutrients leaching into ground or surface water (Pettersson 2001), and disease transfer in the case of sewage slurries.

Rotation length

In even-aged management systems the length of time that a stand is allowed to grow is often related to the age at which the mean annual increment in volume reaches its peak. This has been described as the technical or biological rotation age for the forest (Table 13.4). Rotation age, however, is often adjusted, depending upon the type of product demanded (pulp for paper vs. sawlogs for dimensional lumber vs. peelers for plywood), and the economic considerations of the investments made in the development of the stand. Therefore, the economic rotation is often different than the technical culmination of mean annual increment. Intensive management treatments such as fertilization or density management can be used to shorten rotations. The potential longevity of individual trees is usually much longer than the rotation age. In old stands, some trees may be more than twice the age of the silvicultural rotation (Table 13.4); indeed, one of the definitions of old-growth forest is that trees are much older than the local rotation age

Table 13.4. Potential longevity and age (years) at technical culmination of mean annual increment for typical boreal forest species in Quebec. Data from Pothier and Savard (1998).

Species	Potential longevity	Age at technical culmination[a]		
		Good sites	Medium sites	Poor sites
Abies balsamea	200	38	58	86
Betula papyrifera	140	54	69	74
Picea glauca	300	37	54	80
Picea mariana	200	38	68	119
Pinus banksiana	185	47	76	88
Populus tremuloides	200	46	66	77
Thuja occidentalis	400	55	82	109

[a]Age at technical culmination is the average age of fully stocked stands at which mean annual increment (MAI) is maximized.

(Hunter 1990). Managers therefore have some flexibility in their choice of rotation age. Older stands have large trees, and there is sufficient time for reinvasion of "old-growth" flora and fauna. They are therefore highly valued by those interested in biodiversity (Table 13.2). Depending on the tree species and climate, extended rotations may not necessarily reduce long-term yields, and represent a conceptually simple way of accommodating non-timber forest values (Curtis 1997; Burton et al. 1999).

Method of harvest and silvicultural systems

Harvesting can be done so that all of a forest stand or "cutblock" is removed at one time (clearcutting) or only part of the forest is removed at any harvest. Forests can also be even-aged (where 80% or more of the trees are recruited within a 20-year time span) or uneven-aged (with three or more age classes of trees). Different harvesting systems and methods for promoting cohorts of different age classes within stands have been used to increase biodiversity of forests (Kerr 1999).

Selection systems

Selection systems perpetuate an uneven-aged instead of even-aged stand structure (Mathews 1989; Smith et al. 1997). Selection systems retain some old-growth stand attributes while still allowing some timber extraction (Burton et al. 1999). Unlike natural old-growth stands, however, stands under selection management are internally dissected with logging trails and are periodically re-disturbed by logging. Selection systems have not been widely used in boreal forests but have been suggested as a mechanism for maintaining old-growth structures of relatively open canopies with some large trees (Bergeron et al. 1999).

Single-tree selection — This system involves the harvest of individual trees, leaving behind a stand with a wide range of tree sizes (ages) evenly dispersed throughout the stand. There are formal rules for targeting the number of trees in each size class (i.e., the q-factor) based on differences in the achievement of full stocking according to tree size (Smith et al. 1997). The stand might be re-entered for cutting every 10–25 years, which is known as the cutting cycle. This requires careful planning and highly skilled loggers, and is difficult to do with large machines such as feller bunchers. This system is well suited only to the perpetuation of shade-tolerant species, though lower stocking levels on dry sites also permit its application to trees generally considered intermediate in shade tolerance.

Group selection — In group selection, small gaps (up to two tree heights in diameter) are periodically cut out of the stand, thereby eventually cutting and regenerating the entire area, to create a fine-scale mosaic of patches of a range of ages. As the canopy gap is large enough to provide full light for several hours of the day, even shade intolerant species might be regenerated with this system (Coates and Burton 1997). Single tree selection mimics natural processes of canopy gap formation in older stands, and group selection initiates fine scale, even-aged cohorts of trees (as might naturally arise where small pockets of trees have died from root rot or insect attack; Bradshaw 1992; Coates and Burton 1997).

To reliably maintain an uneven-aged structure and produce a new cohort of regeneration after each cutting entry, however, formal control of basal area retention and (or) light estimation should be practiced to ensure regeneration. Recently developed computer tools such as MIXLIGHT (Stadt and Lieffers 2000) and SORTIE (Coates et al. 2001) could be used for such a purpose. Stands should be re-entered for harvesting at set intervals (i.e., the cutting cycle). Trees must be removed from various size classes so that the size–frequency distribution after cutting somewhat conforms to a reverse-J curve (Smith et al. 1997). This also requires that the cutting intensity is not too high and the cutting cycle is sufficiently short to ensure that a sufficient number of cohorts are maintained.

Diameter-limit cutting — Though typically not implemented with forest regeneration in mind, and hence not strictly a silvicultural system, diameter-limit cutting can be considered a crude form of the selection system. Here only those trees above some minimum size are cut, after which it is hoped that the growth of all remaining trees will increase (MacDonell and Groot 1996; Légère and Gingras 1998). This had been practiced widely in Canada prior to the development of mechanized logging, and continues today on many privately held woodlots that are not professionally managed. In many cases, this practice led to "high-grading", regeneration failures and invasion by shrubs or intolerant species. This has been the case following diameter-limit cutting in yellow birch – balsam fir stands, for example (Fig. 13.3*a*). Older birch–fir stands are often characterized by a sparse overstory of poor quality trees with a dense understory of mountain maple and pin cherry (*Prunus pensylvanica*) shrubs. If only the larger trees are removed, they are typically replaced only by the shrubs, not by smaller crop trees. Intense efforts are then required to bring these degraded stands into tree production using techniques such as scarification after strip cleaning (Fig. 13.3*b*).

Shelterwood systems

Shelterwood systems remove the existing mature trees in a series of cuts that extend over a small fraction of the rotation. The objective is to regenerate an even-aged stand

Fig. 13.3. (*a*) Degraded stand as a result of diameter-limit cutting in a yellow birch – balsam fir stand. The sparse overstory is yellow birch, and the understory is mountain maple. (*b*) Treatment of a degraded stand originating from diameter-limit cutting in a yellow birch – balsam fir stand. (*c*) Mixedwood shelterwood seed cut, near Edson, Alberta. (*d*) Planting of white spruce under aspen, near Edson, Alberta.

under part of the existing canopy (Wurtz and Zasada 2001). It works well with shade-tolerant species which require protection in early stages, but it can be applied to intolerant species in some situations. The *preparatory cut*, often considered optional, is applied to a dense even-aged stand some years prior to the time planned for establishment of regeneration. Objectives of the preparatory cut are to remove trees from the smaller size-classes to increase the vigour, seed-bearing capacity, and windfirmness of the remaining trees. It may also speed the breakdown of forest floor material and provide a better seedbed. Stands that need a preparatory cut include those with trees characterized by a high slenderness coefficient (height/diameter ratio), making them vulnerable to windthrow if too many trees are removed at one time. The *seed cut* (establishment cut) removes trees from the smaller size classes to open up enough growing space and light to allow establishment of the desired tree species (Fig. 13.3*c*). There may be 100 trees/ha retained, with up to 20 m^2/ha of total basal area. Trees from the dominant size categories are usually most wind- firm and likely the best seedbearers. As there are large numbers of trees left in the forest during the years of seedling establishment, this system is aesthetically pleasing relative to clearcutting. The sheltering trees are taken out in the *removal cut* once the next generation of trees is well established. Special care is required at this stage in order to avoid damaging the seedlings. As in the seed tree system, some overstory trees may be retained to meet wildlife and aesthetic objectives.

   Shelterwood systems are distinguished from seed tree systems (see below) in that a sufficient density of overstory trees is retained to provide seed but also protection to establishing seedlings (from full sunlight or radiation frost; Matthews 1989; Smith et al. 1997; Man and Lieffers 1999*b*). Shelterwoods could be used to simulate the natural die-off of the first cohort of trees in stands (e.g., from an insect outbreak, such as described in Chap. 8) and its replacement by a new generation of individuals. In maturing spruce–fir stands, a portion of the overstory trees could be carefully harvested to allow sufficient light transmission to the understory to allow regeneration of a new cohort of conifers; enough overstory, however, must be maintained to provide sufficient shade so that intolerant shrubs and grasses fail to develop. New computer models such as MIX-LIGHT (Stadt and Lieffers 2000) and SORTIE (Coates et al. 2001) use tree size, crown characteristics and stem densities to predict light in such partially cut stands, helping managers identify some of the consequences of canopy manipulation. Because of thick layers of organic matter, understory scarification is usually needed on most boreal forest sites. Given the sporadic nature of seed crops of these conifers, mechanisms to promote seed crops or direct seeding may be used to deliver seed to the sites to match the short "window of opportunity" (Lieffers et al. 1999) before the shrub/herb layer develops.

   Sometimes it is difficult to establish natural seedlings using shelterwoods. These systems, however, may still be valuable for control of summer frosts, avoiding a rise of the water table, protecting visual quality, or maintaining wildlife habitat. In such cases, planting can be used to establish the next generation of trees in a shelterwood understory. This would allow the environmental and structural benefits of the shelterwood system, but would increase the probability of successful regeneration and minimize the regeneration lag. Given the short time window before shrubs and herbs dominate the site after canopy reduction, there are distinct benefits to early establishment of seedlings that may outweigh the added planting costs.

   Irregular shelterwood — A transition between the natural shelterwood (see below) and selection systems is the irregular shelterwood. In this system, a full range of understory trees (from seedlings to saplings and small merchantable trees) are protected during logging, which can lead to stands of irregular (multi-cohort) structure (Groot and Horton 1994; Lieffers et al. 1996*b*). This method is more suited to open, irregular stands that have escaped fire for extended periods. In such conditions, saplings and small merchantable trees have large live crown ratios that permit them to survive and respond to release. In these stands, no attempt is made to reach a balanced diameter distribution as in single-tree selection, though (because smaller trees are often clumped) it may resemble group selection in residual stand structure and pattern.

   Natural shelterwood — Natural shelterwoods in mixedwood stands (mature aspen with understory spruce) can be used to simulate the gradual dieback of maturing aspen, thereby releasing the understory spruce from the overtopping aspen. In this version of the shelterwood system (Box 13.2), only the removal cut is applied. With a planned network of trails for harvesters and skidders, the aspen can be carefully cut to expose the spruce to greater light. As understory spruce usually has a large height/diameter ratio, it can be susceptible to windthrow when the overstory is removed. Leaving up to 20% (Navratil et al. 1994) or as little as 8% (B.E. Grover, unpublished data) of the aspen standing, either dispersed in the stand or in rows between the access trails, results in

Box 13.2. A natural shelterwood cycle for mixedwood silviculture.

Fig. 13.4. Schematic presentation of the major pathways and actions for mixedwood shelterwood management. Steps in process (labelled on the diagram) are as follows:
1. Clearcut a mature aspen dominated stand; 2. Let the stand regenerate by suckering;
3. Underplant the aspen stand at year 40 with white spruce if needed; 4. Harvest the aspen at year 60 with careful logging techniques, leaving much of the understory spruce intact;
5. Allow aspen suckering to occur on the skid trails; 6. Allow the stand to grow up to be spruce dominated, maturing when the spruce is 80+ years old; and 7. Allow the spruce-dominated stand to be clear cut and to sucker back to an aspen dominated stand. 8. Allow this stand to grow to maturity as an aspen dominated stand (long cycle) or return to Step 3 to underplant the aspen stand with spruce at year 40 (short cycle).

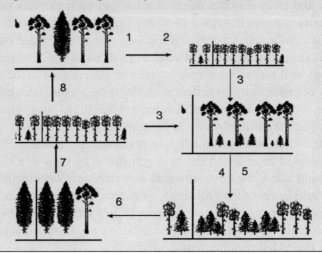

minimal windthrow of the spruce. In situations where the understory spruce is already established (i.e., typically taller than other understory vegetation), the natural shelterwood is an inexpensive system to develop partial stocking of white spruce. Given that the spruce is already part way through its rotation, any spruce that is saved during the logging of the aspen is very valuable (Brace and Bella 1988). This approach further ensures that spruce remains in mixedwood systems where aspen is harvested. The system is called "understory protection" in Alberta, "careful logging" in Ontario, and has similarities to the CPRS ("*coupe avec protection de la régénération et des sols*") system in Quebec (see below). In Ontario and Quebec, the size of regeneration is usually smaller and the windthrow problem is generally less important than in Alberta.

Underplanting spruce in natural shelterwood — Many mixedwood sites may be treed only by aspen and other hardwoods. These sites could support spruce, but it may not have recruited because of inadequate seed source or establishment conditions immediately after disturbance. Stands of aspen can be underplanted with white spruce seedlings several decades after establishment or, alternatively, 1–20 years prior to maturity of the aspen (Fig. 13.3*d*). This will result in the development of mixedwood stands similar to those targeted for the natural shelterwood described above. Planting could be laid out in patterns that would plan for harvester and forwarding access trails in the stand. Prior to harvesting the aspen overstory, the developing spruce trees should be tall enough to be visible to the loggers and no longer susceptible to competition from shrubs and grass following logging. Tall conifers, however, are more likely to be damaged during the careful logging or be blown over. This system is best suited to sites that will be troubled with aggressive grasses or hardwoods following clearcut logging; as the conifers will already be at least a meter tall at the time of logging, non-crop species will have less impact following logging. One of the biggest difficulties with the system can be damage from varying hares (*Lepus americanus*) immediately after planting at times when hare populations are high (Rodgers et al. 1993).

CPRS (*coupe avec protection de la régénération et des sols*) — CPRS has become a standard harvesting and silvicultural practice in mature conifer stands of Quebec and Ontario. Logging operations have been modified to reduce unnecessary destruction of advance regeneration of black spruce and balsam fir (Ruel 1989; MRNQ 1994). Using directional felling and widely spaced skidding trails (Archibald and Arnup 1993; Canuel 1987, 1989) have been shown to be effective at keeping near-full stocking of stands (Ruel 1990; Archibald and Arnup 1993). This understory regeneration usually survives and responds well to release (Hatcher 1964; Doucet and Boily 1986; Ruel et al. 2000) and can give yields similar to, or even greater than, fire-origin stands if tall advance regeneration is present (Paquin and Doucet 1992; Pothier et al. 1995).

The release of advance regeneration by CPRS logging is somewhat similar to that caused by the spruce budworm defoliation in balsam fir forests (Baskerville 1975; MacLean 1984), though there are some clear differences, such as the density of standing snags and fallen logs on the ground. In black spruce stands fire usually triggers stand initiation from seed origin. Use of CPRS in these stands deviates substantially from natural disturbance, as regeneration will be derived from layering, not seed. The structure of the layer-regenerated stand is usually more irregular than that of fire-origin stands (Groot and Horton 1994). However, regeneration by layering should not affect the genetic diversity of the future forest (Perry and Bousquet 2001).

Seed tree systems

Seed tree silvicultural systems are an even-aged system where most of the original stand is harvested at one time, leaving behind 10–50 well-spaced mature trees/ha to spread seed over the cut area. Once the seedlings are established, these trees can be carefully removed or, alternatively, permanently left to fulfil green-tree retention criteria for forest certification (Vanha-Majamaa and Jalonen 2001). This system can be used to simulate the seed cast from some residual trees left alive with patchy fire occurrence. This system has been successfully used in Scandinavia for regenerating Scots pine, and it is occasionally used in larch (*Larix* spp.), jack pine, and lodgepole pine forests. Provided that the site has some mineral soil exposure resulting from the action of logging machines or mechanical site preparation, seedlings should establish after the first seed crop. It is usually not successful on rich sites that are quickly invaded with competitive vegetation.

In boreal mixedwood forests, white spruce trees can be left in or near the edge of cutovers to act as a seed source for regenerating the stand (Greene et al. 2002). In larger cutovers, seed trees will need to be left in the interior of the block, usually in clumps to minimize blowdown. While this is likely to be an inexpensive regeneration system relative to clearcutting, it is also likely to result in regeneration delays and (or) produce less than full stocking of conifers.

Clearcutting

Clearcutting is the most widely applied silvicultural and harvesting system in Canada. Clearcutting was widely adopted after development of the powered skidder and feller buncher harvester. It offers an efficient and inexpensive logging system (Keenan and Kimmins 1993). With the widespread use of clearcutting, regeneration systems have tended to focus on planting of trees, and most provincial standards for regeneration of forests have been designed for use of clearcutting. Indeed, these standards tend to lock forest managers and silviculturists into a standard pattern of clearcut and plant. Only in recent years have there been efforts to promote a wider diversity of silvicultural systems to achieve various goals for a stand.

In terms of the silvics of the focal tree species, clearcutting is used when there is a desire to remove all of the stand in a single cut. Clearcutting is best suited to regeneration of shade intolerant species. Indeed, with species such as jack pine or lodgepole pine, it is difficult to regenerate them any other way. Complete exposure of site, however, makes sites vulnerable to nighttime frost, high surface heating, and rapid growth of shade-intolerant competitors.

Clearcutting lends itself to intensive management systems because the completely open sites allow intensive site preparation, species control, mechanized treatments, and a simple management system. The public perception of clearcutting, however, is generally negative and it is associated with low aesthetic appeal, loss of biodiversity, and environmental concerns such as nutrient loss or heavy runoff. To address some of these concerns green tree retention or variable tree retention (Franklin et al. 1997; Vanha-Majamaa and Jalonen 2001) cuts were developed. Here 10% of trees (ranging up to 30%) are left standing on cutting units. Residual trees are often undersized, wolf trees, or species of reduced commercial value. The main silvicultural value of these residual trees is to provide some of the habitat elements necessary for wildlife and biodiversity. Some of these residual trees of these stands often die within a decade, but will continue to provide habitat structure and coarse woody debris for the entire rotation.

Typical clearcut and plant conifers — Development of fully stocked spruce following clearcutting of boreal mixedwoods can be achieved by planting (Fig. 13.5a). Collections from wild stands or traditional tree breeding programs are frequently employed to provide the seed for these plantations. Commonly 1-year-old container-grown white spruce or black spruce is planted, either after mechanical site preparation (Table 13.3) or no site preparation if there are no limiting microsite factors that need to be addressed. The "clearcut and plant" system is well suited to meet the short regeneration times demanded by the regulations of most provincial governments. Thus, it is widely applied across Canada. Some of the difficulties with the system are: (1) high costs of seedlings, site preparation, and planting; cost may be greater than $2000/ha (Navratil et al. 1991; B.E. Grover, unpublished data); (2) heavy competition from non-crop species such as

Fig. 13.5. (*a*) White spruce plantation, site prepared and tended, located north of Peace River, Alberta. (*b*) Hybrid poplar plantation established by Alberta-Pacific Forest Industries 7 years earlier.

bluejoint reedgrass (Lieffers et al. 1993), shrubs such as raspberry, or tall shrubs may prevent establishment of the spruce (Jobidon et al 1999); (3) dense suckering and vigorous growth of the aspen may result in low light levels in some stands, killing or greatly suppressing growth of the spruce (Pinno et al. 2001); and (4) establishment of spruce in relatively exposed conditions results in stress to seedlings (Man and Lieffers 1999*b*; Grossnickle 2000), so the success of planting operations has not always met expectations (Navratil et al. 1991). Provided that the stand is initiated with sufficiently high densities of trees (perhaps 2000 stems/ha), multiple thinning entries can be planned. High initial densities reduce the time to crown closure, thereby reaching the time of full leaf area (and production) early in stand development.

If followed by relatively intensive vegetation control using either manual or chemical techniques, mesic sites can be converted to relatively pure stands of conifer at an early age. Indeed, many of the free-to-grow rules instituted by provincial regulators across Canada force managers in this direction. This conversion of mixedwood forests has been controversial; it is supported by those focusing on conifer production and criticized by supporters of management based upon emulating natural processes. There are theoretical reasons, however, to expect greater total productivity in mixed species stands (Man and Lieffers 1999*a*), and it has been demonstrated that aspen stands with a spruce understory have greater total productivity than aspen without spruce (MacPherson et al. 2001). While growth of tended, monospecific conifer plantations have yielded impressive volumes of the wood at an early age (McClain et al. 1994), as yet there has not been a careful comparison of yields in mixed vs. pure stands under similarly tended conditions in Canadian boreal forests (Man and Lieffers 1999*a*). Such data will be developed from the Western Boreal Growth and Yield Cooperative (see http://www.rr.ualberta. ca/Wesbogy/index.asp) program of permanent plot monitoring currently under way in the Prairie Provinces.

Choosing silvicultural systems for boreal forests

Silvicultural systems have traditionally been defined by:
- the method of harvest;
- the regeneration system for establishment of appropriate species (including soil treatments);
- methods of competition control;
- nutritional supplementation; and
- stand density/composition manipulations during development of the stand.

For the most part, silvicultural treatments have been applied to alleviate a biological constraint for tree establishment or growth. Silvicultural procedures, however, are also determined by the social acceptance of the treatment (see Table 13.2); efficacy of a treatment to meet a goal for stand development may be unimportant if society does not accept that treatment as a responsible procedure for forest management. This has been demonstrated by continued protests against clearcutting and herbicide use, despite scientific evidence that these procedures might be used safely and effectively in certain situations.

Many regions of Canada are in the process of adopting some form of priority-use zoning or a "triad" approach to forest management (Harris 1984; Rowe 1992; Hunter and Calhoun 1996). Under the triad vision of land-use emphasis, forest lands are placed into three categories: (1) intensively managed, (2) extensively managed, and (3) lands left as reserves (see Chaps. 11 and 12). Messier and Poulin (2001) advocates four levels of forest land use, separating intensive management into 2 categories: traditional intensive forestry with rotations greater than 40 years; and super-intensive forestry, often using hybrid species, with rotations of less than 40 years. While the triad concept has been well developed for landscape goals (Chap. 12), at this time there have not been good descriptions of the stand-level treatments suitable for each of the above categories. Indeed, with good planning and integration of stand-level treatments, managers may be able to develop forests that serve a wide variety of goals within any intensity of fibre production.

Below we propose four categories of silvicultural treatment systems to provide managers with a range of stand management options that would likely be acceptable in various land-use zones: (1) systems emulating natural processes, (2) semi-natural systems, (3) standard intensive systems, and (4) super-intensive systems. These systems cover the extremes from ecosystem restoration to intensive plantation culture (Burton 1999). This is an arbitrary number of categories, as in reality the categories listed fit on a gradient from no intervention to very intensive management (Table 13.2). While we may wish to create various landscape zones to meet timber, social, and biodiversity objectives, managers should be given some flexibility to create the desired range of stand types by different silvicultural procedures. For example, it might be desirable (both economically and environmentally) to maintain the use of intensive management procedures listed in Table 13.2 to create specialized habitats within a landscape zone emulating nature (Messier and Poulin 2001). Purists, however, may not agree with this sentiment.

Emulating natural processes

This is a management system in which natural regeneration systems are used exclusively. The philosophy is that the logging and subsequent "gentle" treatments should leave the forest, as much as possible, with a composition, structure, genetic pool, and presence of decaying wood similar to that which might be found after natural disturbance processes (tree death from fire, insect, diseases, or windthrow). Control of competitive vegetation and subsequent stand structure is not a priority in this approach. Since few (if any) natural disturbances in the boreal forest remove all standing boles (Chap. 8), nature-emulating logging systems would leave some of the original live and dead trees in the stand to influence the microclimate and the biota during the establishment of the next stand. Forest managers must, however, go beyond simply cutting the forest in the spatial patterns that might occur from natural disturbances such as wildfire. Given the continuous development in our understanding of conditions needed for seedling establishment, natural regeneration is likely to play a larger role in future stand management.

As many of the seedling regeneration processes are dependent upon seedbed conditions, treatments to the forest floor and shrub/herb layers also play a critical role in early stand establishment. For tree species regenerating from seed the following must be present: adequate seed source within an acceptable distance to regeneration sites; appropriate soil disturbance for seedbed receptivity; and appropriate light and moisture conditions for developing seedlings. Adopting this management system could also mean leaving the soil and duff with the appropriate level of disturbance for sprouting species such as aspen. Natural regeneration may not always have immediate success for all species. In some cases this will also demand that government regulators allow more latitude in stocking standards, and lower height growth standards for regenerating trees. Predictions of regeneration will usually be expressed in terms of probabilities of success given certain stand conditions (Greene et al. 2002).

Example of systems that somewhat emulate natural processes are variable retention clearcutting followed by coppice regneration (suckering) of poplars or with CPRS, selection, shelterwood, natural shelterwood, seed tree, and green tree retention systems, all described above. The effectiveness of these management systems in retaining mature tree cover and habitat values is well assured, but their effectiveness in stand regeneration depends on propagule availability, and may still require treatments to the forest floor that increase the probability of tree regeneration.

Semi-natural management

In many circumstances it is difficult to establish the desired tree species by natural means and the native seed rain from local populations. This is particularly true in situations where seed trees are widely scattered or seed crops are periodic. Artificial regeneration (direct seeding or the planting of seedlings) of desirable species immediately after logging on naturally suitable or artificially prepared microsites may quickly establish near-natural forest composition and structure. This will require, however, some movement of tree populations from related sites. Most provincial governments have seed transfer rules that restrict the distance or elevation of transfer of genetic material. In most cases, artificial regeneration also requires site preparation, which together create a more regular distribution of trees than might be found in fully natural regeneration systems. Examples of systems that fit in this category are site preparation and aerial seeding, shelterwood systems with planting, planting spruce under mature aspen, and planting spruce in boreal mixedwood stands (thereby creating even-aged mixtures of spruce and sucker-origin aspen). These practices are briefly described in earlier subsections.

Standard intensive management

Even-aged plantations with the primary goal of wood production can be considered the basic or standard management system on most industrial forestry lands. Currently, they also seem to be the target of most provincial silviculture regulations, unless there are overriding site or habitat constraints to timber production. Whether this approach should continue to be considered "the norm" for a forest management area depends on the balance of values being managed for (Chap. 11) and how those values are to be deployed in space and time (Chap. 12).

There may be economically important forest structures that landowners and society wish to promote. For example, in the mixedwood boreal forest relatively pure stands of large white spruce have highly desirable wood fibre but are slow to establish by natural processes. Such stands can be promoted and established earlier than would be usual through natural processes if the forester is prepared to prescribe mechanical site preparation, planting with genetically selected stock, vegetation management, fertilization, and commercial thinning. Wood production is a high priority in such stands, but this type of stand may be important for other values, especially if strategically located in the landscape. Description of such a system is provided above.

The concept of "intensive management" could also be applied to promote specific non-timber forest values (such as berries or ungulate browse), but there is little experience in doing so in the boreal forests of Canada. One example might be the ongoing use of prescribed fire by wildlife managers in the boreal cordillera of northern British Columbia to maintain open habitat and productive browse for big game (Churchill and Keller 1988; D. Hebert, Encompass Strategic Resources Inc., Creston, British Columbia, personal communication, 2001).

Super-intensive management

Intensive forest management methods represent the application of agronomic techniques to tree crops, with management for a single product (wood, in this case), with practices and impacts similar to those of modern agriculture (Burton 1999; Messier and Poulin 2001). Termed *"ligniculture"* in Quebec, such a management system usually means clearcutting, followed by intensive site preparation, planting a single uniform species of genetically selected trees, in rigid hexagonal spacing (for densest packing), followed by strict vegetation control, and possibly by thinning (Fig. 13.5*b*). Super-intensive management could use native species, exotic species such as hybrid poplar or larch (which can be managed on rotations of less than 20 years), or native and exotic species of spruce, fir, and pine (which can be harvested on a 40-year rotation). In the future, this system is also likely to use genetically modified tree species. These "tree farms" will, therefore, use the fastest growing and most valuable species or genotype that is biologically suited to a particular site, or does not generate excessive concerns related to biodiversity or aesthetics. Super-intensive management systems allow intervention using any of the stand management techniques (described in the previous major section), subject to the same sorts of environmental constraints and stewardship (such as streamside protection) that would usually apply to agricultural lands.

Dense, regular spacing of trees is central to a super-intensive system, where elevated levels of productivity are achieved because most of the site resources (light, moisture, nutrients, growing space) are channelled into crop plants. Intensive management also achieves productivity gains over wild or extensively managed stands by harvesting at a relatively young age, just after the culmination of mean annual increment (Table 13.4). Careful establishment and density control of conifer plantations in Ontario (McClain et al. 1994) and Quebec (J. Beaulieu, Laurentian Forestry Centre, Canadian Forestry Service, Sainte-Foy, Quebec, personal communication) have yielded nearly 300 m^3/ha by year 40. In Canada, hybrid poplar plantations have been the focus of intensive plantation

---

Box 13.3. Intensive poplar farm program at Alberta-Pacific Forest Industries Inc.

Alberta-Pacific Forest Industries has embarked upon a large-scale poplar farm program. These farms are designed to produce 15–17% of the fibre needs of the Alberta-Pacific pulp mill in Boyle, Alberta, when first harvest commences in the early 2020s. When fully implemented, there will be approximately 23 000 ha of private land under cultivation, with a 15–20-year rotation and an annual cut of 400 000 m^3. Realized growth rates will ultimately determine the exact rotation, cut volume, and total area if land is required to meet this goal.

This poplar program is based on research that began in 1993 (Fig. 13.5*b*) when the first collection of poplar hybrids were put into a trial at the Alberta-Pacific mill site. Since that time, several thousand different clones of hybrid poplars and aspens have gone through various stages of screening and testing. Both poplar and aspen (all of the genus *Populus*) hybrid breeding programs have also been started, and trials using these crosses were installed in 2002.

The primary selection traits in these programs include adaptability, growth, form, and wood quality. Combined with the tree improvement work is a dedicated silviculture program. This program covers research on site preparation, mounding, fertilizer rates and combinations, biosolid utilization, nursery stock and condition, and herbicides.

The aim of these research programs is to provide the best genetic material growing at the fastest rate possible in order to achieve the goal of an alternative fiber supply to the mill, grown on an economical short rotation. Where required, Alberta-Pacific also will support research that is relevant to reducing risk of crop failure (related to insects and disease, pollen flow, etc.).

Table 13.5. Projected fibre production possible in the boreal forests of Canada under a regime of 15% protected areas, 70% semi-natural management, 12% plantation management, and 3% super-intensive management (gross volume). The area of boreal forest lands (ha × 1000) is broken down into four productivity classes, according to estimated site index (SI) based on tree height at 50 years. It is assumed that each management system and each site classification will have the indicated levels of productivity, with more highly productive lands allocated to the more intensive management systems, and no super-intensive management applied to sites with SI <10.

Management regime	Productivity class, site index at 50 years				
	<10 m	10–15 m	15–20 m	>20 m	Total
Total area (1000s of ha)	60644	47972	30976	945	140537
Protected areas (~15%) (× 1000 ha)	9096.6	7195.8	4646.4	141.8	21080.6
Semi-natural management					
Area (~70%) (× 1000 ha)	47483.4	35979.2	19302.7	303.2	103068.5
Productivity (m^3 ha^{-1} $year^{-1}$)	0.7	1.0	1.2	1.5	
Production (× 1000 m^3)	33238.4	35979.2	23163.2	454.8	92835.6
Plantation management					
Area (~12%) (× 1000 ha)	4064	4797	4026.9	200	13087
Productivity (m^3 ha^{-1} $year^{-1}$)	1.5	3.0	3.5	5.0	
Production (× 1000 m^3)	6096	14391	14094.1	1000	35581.1
Super-intensive management					
Area (~3%) (× 1000 ha)	Nil	916.1	3000	300	4216.1
Productivity (m^3 ha^{-1} $year^{-1}$)	Nil	6.0	9.0	15.0	
Production (× 1000 m^3)	Nil	5496.6	27000	4500	36996.6
Total production (× 1000 m^3)					165413.3

Notes: The productive boreal forest of Canada (140.5×10^6 ha, as reported by Lowe et al. 1996) was divided into the four site index classes found in boreal zones (CFS 1999). The estimate of the area of each SI class was taken from site index determinations made from different permanent sample plots (PSPs) across Canada, and is based upon the dominant species in each PSP.

In these calculations we amalgamated the "Emulating Natural Processes" systems into the "Semi-Natural Management" category; there will likely to be only slight differences in productivity between these management systems, perhaps with a greater regeneration lag in the "Emulating Natural Processes" system. To account for this regeneration lag, we intentionally assigned a low estimate of productivity to these systems. The productivity estimates were scaled up for the "Plantation and Super-Intensive" systems based on the overall values in the literature (e.g., Van Oosten 2000).

operations. They may yield up to 37 m^3 ha^{-1} year^{-1} in coastal British Columbia, but the yields are likely to be much less than this on sites with colder winters and seasons of water stress (i.e., most of the boreal zone). There are relatively large poplar plantations managed by Domtar in Ontario and Quebec, by Pacifica Paper and Scott Paper in southern British Columbia, and by Alberta-Pacific Forest Industries in Alberta (Box 13.3). The current area in hybrid poplar plantations is ~7000 ha, nationwide (Van Oosten 2000). European larch (*Larix decidua*) plantations in Quebec have produced 5–8 Mg/ha in 5–10 years, or sawlog timber in 20–25 years. The Flakaliden project in central Sweden has demonstrated fibre production levels of 18 m^3 ha^{-1} year^{-1} when all water and nutrient limitations were removed (Bergh et al. 1999). While the inputs of resources into these types of plantations may be very high, these studies give us an understanding of the potential for gain in productivity with intensive culture.

In many regions of Canada, more intensive management has been used to achieve allowable cut effects and economic gain by cutting mature natural forests at a faster rate than would be allowed given estimates of natural yield. This has stimulated great interest in enhancing the yields of regenerated stands. Also, government regulators have promoted more intensive silvicultural practices by allowing/promoting relatively short rotations in AAC calculations. This implicitly applies pressure on silviculturists to find more intensive procedures to meet the high expectations for growth of regenerated stands.

Increased levels of management intensity, however, can also play an integral part in maintaining wood supply in forested landscapes managed to maintain biological diversity (Binkley 1997). In the calculations presented in Table 13.5 we demonstrate that with intensive management on a relatively small area, the same level of wood production can be maintained, even if 15% of the land base is set aside as reserves and a large part of the area (70%) is managed semi-naturally. Using the "State of Canada's Forests" report for 1998/1999 (CFS 1999), we have estimated that the current annual productivity of the boreal forest, and hence a sustainable rate of harvest, is about 141 × 10^6 m^3/year, with an actual cut of 108 × 10^6 m^3/year. (For this calculation, we have assumed that 20–35% of the AAC coming out of Nova Scotia, New Brunswick, Quebec, and Ontario was from hardwood forests, not boreal forest; we also assumed that only 20% of the AAC in British Columbia comes from the boreal forest.) The value of 165 × 10^6 m^3/year of wood projected as being available for harvest developed in the calculations of Table 13.5 is well above the natural productivity despite the reserves and the low productivity from a large part of the landscape.

Conclusions

Many silvicultural approaches and techniques are available to support a wide range of forest management objectives, even under the harsh conditions of boreal forestry (Box 13.4). Inspired by natural stand dynamics, silviculturists can manipulate stand structures and processes to various degrees in a continuum from nature emulation to super-intensive plantation culture. A diverse range of stand types is possible through such manipulations, making silviculture an increasingly important agent of landscape diversity in the

---

Box 13.4. Steps for sustainable boreal forest silviculture:

1. Identify site characteristics and factors limiting tree growth and stand development;
2. Understand succession and the critical stages where successional paths might be redirected or facilitated;
3. Zone forest land into different intensities of stand management;
4. Have clear management objectives for the stand, including forest and landscape level goals for future wood and habitat supply;
5. Select appropriate silvicultural tools/procedures that are efficacious but meet economic and social considerations;
6. Try to achieve your objectives at the landscape level, leaving room for failure at the stand level; and
7. Update government policies and regulations to ensure that appropriate fibre and biodiversity goals are achieved on the various land zones.

---

world's boreal regions. Implementation of a particular silvicultural regime depends on the attributes of the site and stand, and on the objectives (forest values to be emphasized) for its future. Foresters and provincial regulations throughout Canada have come to recognize that a single silvicultural approach is not appropriate in all ecosystems; now they need to recognize that no single silvicultural approach will not sustain all forest values. Silvicultural diversity will be one of the keys to sustainable forest management.

The fact that knowledge is central to wise and effective forest stand management points the way to ongoing research and education needs. Future research should focus on development of novel silvicultural systems that couple understanding of forest stand dynamics with environmentally appropriate treatments that match social objectives. The need for forest stand managers (silviculturists) to match ecological understanding and silvicultural policies with society's objectives for fibre production, biodiversity protection, aesthetics, and sustainability will demand ever-increasing levels of broad education and experience.

Acknowledgements

We thank the SFMN for funding and our colleagues (notably David Greene, Simon Landhäusser, and John Zasada) for discussions on boreal forest silviculture.

References

Achuff, P.L. 1989. Old-growth forests in the Canadian Rocky Mountain national parks. Nat. Areas J. **9**: 12–26.

Achuff, P.L., and La Roi, G.H. 1977. *Picea–Abies* forests in the highlands of northern Alberta. Vegetatio, **33**: 127–146.

Archambault, L., Morisette, J., and Bernier-Cardou, M. 1998. Forest succession over a 20-year period following clearcutting in balsam fir – yellow birch ecosystems of Eastern Québec, Canada. For. Ecol. Manage. **102**: 61–74.

Archibald, D.J., and Arnup, R.W. 1993. The management of black spruce advance growth in Northeastern Ontario. Ontario Ministry of Natural Resources, Timmins, Ontario. Northeast Sci. Technol. Tech. Rep. 008. 32 p.

Arnup, R.W., Campbell, B.A., Raper, R.A., Squires, M.F., Virgo, K.D., Wearn, V.H., and White, R.G. 1988. A silvicultural guide for the spruce working group in Ontario. Science and Technology Series, Vol. 4. Ontario Ministry of Natural Resources, Toronto, Ontario. 100 p.

Bachelet, D., and Neilson, R.P. 2000. Biome redistribution under climate change. *In* The impact of climate change on America's forest: a technical document supporting the 2000 USDA Forest Service RPA Assessment. USDA Forest Service, Fort Collins, Colorado. Gen. Tech. Rep. RMRS-GTR-59. pp. 18–39.

Balisky, A.C., Salonius, P., Walli, C., and Brinkman, D. 1995. Seedling roots and forest floor: misplaced and neglected aspects of British Columbia's reforestation effort? For. Chron. **71**: 59–65.

Baskerville, G.L. 1975. Spruce budworm: super silviculturist. For. Chron. **51**: 138–140.

Bates, P.C., Blinn, C.R., and Alm, A.A. 1993. Harvesting impacts on quaking aspen regeneration in northern Minnesota. Can. J. For. Res. **23**: 2403–2412.

Batzer, H.O., and Popp, M.P. 1985. Forest succession following a spruce budworm outbreak in Minnesota. For. Chron. **61**: 75–80.

(BCMF) British Columbia Ministry of Forests. 2000. Silviculture prescription guidebook, February 2000. Forest Practices Code of British Columbia. British Columbia Ministry of Forests, Victoria, British Columbia. Available at http://www.for.gov.bc.ca/tasb/legsregs/fpc/fpcguide/PRE/ [viewed 15 June 2003]. 53 p.

Beckingham, J.D., and Archibald, J.H. 1996. Field guide to ecosites of northern Alberta. Canadian Forest Service, Edmonton, Alberta. Spec. Rep. 5. 528 p.

Bedford, L., and Sutton, R.F. 2000. Site preparation for establishing lodgepole pine in the Sub-Boreal Spruce zone of interior British Columbia: the Bednesti trial, 10 year results. For. Ecol. Manage. **126**: 227–238.

Béland M., Bergeron, Y., and Zarnovican, R. 1999. Natural regeneration of jack pine following harvesting and site preparation in the Clay Belt of northwestern Quebec. For. Chron. **75**: 821–831.

Berg, Å., Ehnström, B., Gustafsson, L., and Hallingbäck, T. 1995. Threat levels to red-listed species in Swedish forests. Conserv. Biol. **9**: 1629–1633.

Bergeron, Y., and Dansereau, P.R. 1993. Predicting the composition of Canadian southern boreal forest in different fire cycles. J. Veg. Sci. **4**: 827–832.

Bergeron, Y., and Dubuc, M. 1989. Succession in the southern part of the boreal forest. Vegetatio, **79**: 51–63.

Bergeron, Y., and Harvey, B. 1997. Basing silviculture on natural ecosystem dynamics: an approach applied to the southern boreal mixedwood forest of Quebec. For. Ecol. Manage. **92**: 235–242.

Bergeron, Y., Harvey, B., Leduc, A., and Gauthier, S. 1999. Forest management guidelines based on natural disturbance dynamics: stand- and forest-level considerations. For. Chron. **75**: 49–54.

Bergh, J., Linder, S., Lundmark, T., and Elfving, B. 1999. The effect of water and nutrient availability on the on the productivity of Norway spruce in northern and southern Sweden. For. Ecol. Manage. **119**: 51–62.

Binkley, C.S. 1997. Preserving nature through intensive plantation forestry: the case of forestland allocation with illustrations from British Columbia. For. Chron. **73**: 553–559.

Black, R.A., and Bliss, L.C. 1980. Reproductive ecology of *Picea mariana* (Mill) at treeline, Northwest Territories, Canada. Ecol. Monogr. **50**: 331–354.

Blais, J.R. 1983. Trends in the frequency, extent, and severity of spruce budworm outbreaks in eastern Canada. Can. J. For. Res. **13**: 539–547.

Botkin, D.B. 1993. Forest dynamics: an ecological model. Oxford University Press, New York, New York. 309 p.

Brace, L.G., and Bella, I.E.1988. Understanding the understory: dilemma and opportunity. *In* Management and utilization of northern mixedwoods. Canadian Forest Service, Edmonton, Alberta. Inf. Rep. NOR-X-296. pp. 69–86.

Bradshaw, F.J. 1992. Quantifying edge effect and patch size for multiple-use silviculture — a discussion paper. For. Ecol. Manage. **48**: 249–264.

Burns, R.M., and Honkala, B.H. (*Editors*) 1990. Silvics of North America, vol. 1, conifers. USDA Forest Service, Washington, D.C. Agric. Handb. 654. 675 p.

Burton, P.J. 1997. Conifer germination on different seedbeds influenced by partially-cut canopies. Final Contract Rep., SSWG Proj. SS0049, for Silvicultural Practices Branch, British Columbia Ministry of Forests, Symbios Research & Restoration, Smithers, British Columbia. 66 p.

Burton, P.J. 1999. An assessment of silvicultural practices and forest policy in British Columbia from the perspective of restoration ecology. *In* Helping the land heal. Conference proceedings, 5–8 November 1998, Victoria, British Columbia. *Edited by* B. Egan. British Columbia Environmental Network Educational Foundation, Vancouver, British Columbia. pp. 173–178.

Burton, P.J., Kneeshaw, D.D., and Coates, K.D. 1999. Managing forest harvesting to maintain old growth in boreal and sub-boreal forests. For. Chron. **75**: 623–631.

Burton, P.J., Sutherland, D.C., Daintith, N.M., Waterhouse, M.J., and Newsome, T.A. 2000. Factors influencing the density of natural regeneration in uniform shelterwoods dominated by Douglas-fir in the Sub-Boreal Spruce zone. Working Paper 47. Research Program, British Columbia Ministry of Forests, Victoria, British Columbia. Available at http://www.for.gov.bc.ca/hfd/pubs/Docs/Wp/Wp47.htm [viewed 15 June 2003]. 65 p.

Buse, L.J., Wagner, R.G., and Perrin, B. 1995. Public attitudes towards forest herbicide use and the implications for public involvement. For. Chron. **71**: 596–600.

Canuel, B. 1987. Guide d'utilisation de la coupe avec protection de la régénération (abattage manuel). Ministère de l'Énergie et des Ressources, Ste-Foy, Quebec. 18 p.

Canuel, B., 1989. Guide d'utilisation de la coupe avec protection de la régénération (abattage mécanisé). Ministère de l'Énergie et des Ressources, Ste-Foy, Quebec. 30 p.

Campbell, R.A. 1990. Herbicide use for forest management in Canada: where are we and where are we going. For. Chron. **66**: 355–359.

Carleton, T.J., and Maycock, P.F. 1978. Dynamics of the boreal forest south of James Bay. Can. J. Bot. **56**: 1176–1173.

Cauboue, M., and Tremblay, J. 1992. Les stations forestières de la haute Côte-Nord: Méthodologie et synthèse générale. Centre de recherche et d'enseignement en foresterie de Sainte-Foy, Sainte-Foy, Quebec. 24 p.

(CCSMAF) Comité consultatif scientifique du Manuel d'aménagement forestier. 2002. Le traitement d'éclaircie précommerciale pour le groupe de production prioritaire SEPM. Avis scientifique. Forêt Québec, Québec, Quebec. 126 p.

(CFS) Canadian Forest Service. 1999. The state of Canada's forests, 1998–1999. Canadian Forest Service, Natural Resources Canada, Ottawa, Ontario. 112 p.

Chrosciewicz, Z. 1988. Forest regeneration on burned, planted and seeded clear-cuts in central Saskatchewan. Canadian Forest Service, Edmonton, Alberta. Inf. Rep. NOR-X-293. 16 p.

Churchill, B., and Keller, D. 1988. Prescribed burning for wildlife in the Peace River sub-region. *In* Wildlife and range prescribed burning workshop proceedings, 27–28 October 1987, Richmond, British Columbia. *Edited by* M.C. Feller and S.M. Thomson. Faculty of Forestry, University of British Columbia, Vancouver, British Columbia. pp. 87–100.

Coates, K.D., and Burton, P.J. 1997. A gap-based approach for development of silvicultural systems to address ecosystem management objectives. For. Ecol. Manage. **99**: 337–354.

Coates, K.D., Messier, C., Beaudet, M., Sachs, D.L., and Canham, C.D. 2001. SORTIE: a resource mediated, spatially-explicit and individual-tree model that simulates stand dynamics in both natural and managed forest ecosystems. Working Paper 2001-10. Sustainable Forest Management Network. Edmonton, Alberta. Available at http://sfm-1.biology.ualberta.ca/english/pubs/PDF/WP_2001-10.pdf [viewed 29 September 2002]. 29 p.

Cogbill, C.V. 1985. Dynamics of the boreal forests of the Laurentian highlands. Can. J. For. Res. **15**: 252–261.

Constabel, A.J., and Lieffers, V.J. 1996. Seasonal patterns of light transmission through boreal mixedwood canopies. Can. J. For. Res. **26**: 1008–1014.

Cumming, S.G., Schmiegelow, F.K.A., and Burton, P.J. 2000. Gap dynamics in boreal aspen stands: is the forest older than we think? Ecol. Appl. **10**: 744–759.

Curran, M., Fraser, B., Bedford, L., Osbert, M., and Mitchell, B. 1990. Site Preparation strategies to manage soil disturbance — interior sites. British Columbia Ministry of Forests, Victoria, British Columbia. Land Manage. Handb. Field Guide Insert No. 2.

Curtis, R.O. 1997. The role of extended rotations. *In* Creating a forestry for the 21st century: the science of ecosystem management. *Edited by* K.K. Kohm and J.F. Franklin. Island Press, Washington, D.C. pp. 165–170.

Day, R.J. 1998. The ancient and orderly European discipline of thinning is now a reality in North America. *In* Stand density management: planning and implementation. *Edited by* C. Bamsey. Clear Lake Ltd., Edmonton, Alberta. pp. 24–33.

De Grandpré, L., Morissette, J., and Gauthier, S. 2000. Long term post-fire changes in the northeastern boreal forest of Quebec. J. Veg. Sci. **11**: 791–800.

DeLong, C. 1991. Dynamics of boreal mixedwood ecosystems. *In* Northern Mixedwood '89. Proceedings, Symposium, 12–14 September 1989, Fort St. John, British Columbia. *Edited by* A. Shortreid. Forestry Canada and British Columbia Ministry of Forests, Victoria, British Columbia. FRDA Rep. 164. pp. 30–31.

DeLong, H.B., Lieffers, V.J., and Blenis, P.V. 1997. Microsite effects on first year establishment and over-winter survival of white spruce in aspen-dominated boreal mixedwoods. Can. J. For. Res. **27**: 1452–1457.

Deluca, T.H., Zachrisson, O., Nilsson, M.-C., and Sellstedt, A. 2002.Quantifying nitrogen-fixation in feather moss carpets of boreal forests. Nature, **419**: 917–920.

DesRochers, A., and Lieffers, V.J. 2001. Root biomass of regenerating aspen *(Populus tremuloides)* stands of different densities. Can. J. For. Res. **31**: 1012–1018.

Dix, R.L., and Swan, J.J.A. 1971. The role of disturbance and succession in upland forest at Candle Lake, Saskatchewan. Can. J. Bot. **49**: 657–676.

Doucet, R., 1987. La régénération préétablie dans des peuplements résineux et mixtes du Québec. L'Aubelle, Ordre des ingénieurs forestiers du Québec **57**(Suppl.): 2–5.

Doucet, R., and Boily, J. 1986. Croissance en hauteur comparée de marcottes et de plants à racines nues d'épinette noire. Can. J. For. Res. **16**: 1365–1368.

Drew, T.J., and Flewelling, J.W. 1979. Stand density management: an alternative approach and its application to Douglas-fir plantations. For. Sci. **25**: 518–532.

Farnden, C. 1996. Stand density management diagrams for lodgepole pine, white spruce and interior Douglas-fir. Canadian Forest Service, Victoria, British Columbia. Inf. Rep. 360. 36 p.

Feller, M.C. 1982. The ecological effects of slashburning with particular reference to British Columbia: a literature review. British Columbia Ministry of Forests, Victoria, British Columbia. Land Manage. Handb. 13. 60 p.

Foster, D.R. 1985. Vegetation development following fire in *Picea mariana – Pleurozium* forests of south eastern Labrador, Canada. Can. J. Ecol. **73**: 517–534.

Frank, R.M. 1990. Balsam fir. *In* Silvics of North America, vol. 1, conifers. *Edited by* R.M. Burns and B.H. Honkala. USDA Forest Service, Washington, D.C. Agric. Handb. 654. pp. 26–35.

Franklin, J.F., Berg, D.R., Thornburgh, D.A., and Tappeiner, J.C. 1997. Alternative silvicultural approaches to timber harvesting: variable retention harvest systems. *In* Creating a forestry for the 21st century: the science of ecosystem management. *Edited by* K.K. Kohm and J.F. Franklin. Island Press, Washington, D.C. pp. 111–139.

Fraser, E., Lieffers, V.J., and Landhäusser, S.M. 2002. Soil temperature and nutrition as drivers of root suckering in aspen. Can. J. For. Res. **32**: 1685–1691.

Galipeau, C., Kneeshaw D., and Bergeron, Y. 1997. White spruce and balsam fir colonization of a site in the southeastern boreal forest as observed 68 years after fire. Can. J. For. Res. **27**: 139–147.

Gautam, K.H. and Watanabe, T. 2002. Silviculture for non-timber forest product management: challenges and opportunities for sustainable forest management. For. Chron. **78**: 830–833.

Goward, T. 1994. Notes on old-growth-dependent epiphytic macrolichens in the humid oldgrowth forests in inland British Columbia, Canada. Acta Bot. Fenn. **150**: 31–38.

Granström, A. 2001. Fire management for biodiversity in the European boreal. Scand. J. For. Res. **3**(Suppl.): 62–69.

Greene, D.F., and Johnson, E.A. 1996. Wind dispersal of seeds from a forest into a clearing. Ecology, **77**: 595–609.

Greene, D.F., and Johnson, E.A. 1998. Seed mass and early survivorship of tree species in upland clearings and shelterwoods. Can. J. For. Res. **28**: 1307–1316.

Greene, D.F., Kneeshaw, D.D., Messier, C., Lieffers, V.J., Cormier, D., Doucet, R., Grover, G, Coates, K.D., Groot, A., and Calogeropoulos, C. 2002. An analysis of silvicultural alternatives to the conventional clearcut/plantation prescription in boreal mixedwood stands (aspen/white spruce/balsam fir). For Chron. **78**: 281–295.

Grondin, P., Ansseau, C., Bélanger, L., Bergeron, J.-F., Bergeron, Y., Bouchard, A., Brisson, J., DeGrandpré, L., Gagnon, G., Lavoie, C., Lessard, G., Payette, S., Richard, P.J.H., Saucier, J.-P., Sirois, L., and Vasseur, L. 1996. Écologie forestière. *In* Manuel de foresterie. *Edited by* J. Bérard. Ordre des ingénieurs forestiers/presses de l'Université Laval, Sainte-Foy, Quebec. pp. 133–279.

Groot, A., and Horton, B.J. 1994. Age and size structure of natural and second-growth peatland *Picea mariana* stands. Can. J. For. Res. **24**: 225–233.

Grossnickle, S.C. 2000. Ecophysiology of northern spruce species: the performance of planted seedlings. NRC Research Press, National Research Council of Canada, Ottawa, Ontario. 407 p.

Hammond, H. 1991. Seeing the forest among the trees: the case for wholistic forest use. Polestar Press, Vancouver, British Columbia. 309 p.

Harper, J.L. 1977. Population biology of plants. Academic Press, London, U.K. 892 p.

Harris, L.D. 1984. The fragmented forest, island biogeography and theory and the preservation of biotic diversity. University of Chicago Press, Chicago, Illinois. 211 p.

Hart, D., and Comeau, P.G. 1992. Manual brushing for forest vegetation management in British Columbia: a review of current knowledge and information needs. British Columbia Ministry of Forests, Victoria, British Columbia. Land Manage. Rep. 77. 36 p.

Hatcher, R.J. 1964. Balsam fir advance growth after cutting in Quebec. For. Chron. **40**: 86–92.

Hauessler, S., Coates, D., and Mather, J. 1990. Autecology of common plants in British Columbia: a literature review. Forestry Canada and British Columbia Ministry of Forests, Victoria, British Columbia. FRDA Rep. 158. 269 p.

Heineman, J.L. 1991. Growth of interior spruce seedlings on forest floor materials. M.Sc. thesis. University of British Columbia, Vancouver, British Columbia. 131 p.

Heinselman, M.L. 1981. Fire and succession in the conifer forests of northern North America. *In* Forest succession: concepts and application. *Edited by* D.C. West, H.H. Shugart, and D.B. Botkin. Springer-Verlag, New York. pp. 374–405.

Hogg, E.H. 1994. Climate and the southern limit of the western Canadian boreal forest. Can. J. For. Res. **24**: 1835–1845.

Hogg, E.H., and Hurdle, P.A. 1995. The aspen parkland in western Canada: a dry-climate analogue for the future boreal forest? Water Air Soil Pollut. **82**: 391–400.

Hungerford, R.D. 1988. Soil temperatures and suckering in burned and unburned aspen stands in Idaho. USDA Forest Service, Logan, Utah. Res. Note INT-378. 6 p.

Hunter, M.L. 1989. What constitutes an old-growth stand? J. For. **87**: 33–35.

Hunter, M.L. 1990. Wildlife, forests and forestry: principles of managing forests for biological diversity. Prentice Hall, Englewood Cliffs, New Jersey. 370 p.

Hunter, M.L., and Calhoun, A. 1996. A triad approach to land-use allocation. *In* Biodiversity in managed landscapes. *Edited by* R.C. Szaro and D.W. Johnston. Oxford University Press, New York, New York. pp. 477–491.

Jobidon, R., Trottier, F., and Charette, L. 1999. Chemical and manual release in black spruce plantations: case studies in balsam fire – white birch ecosystems in Quebec. For. Chron. **75**: 973–979.

Keenan, R.J., and Kimmins, J.P. 1993. The ecological effects of clear-cutting. Environ. Rev. **1**: 121–144.

Kerr, G. 1999. Use of silviculture systems to enhance biological diversity of plantations forests in Britain. Forestry, **72**: 191–205.

Kneeshaw, D.D., and Bergeron, Y. 1998. Canopy gap characteristics and tree replacement in the southern boreal forest. Ecology, **79**: 783–794.

Kneeshaw D.D., and Burton P.J. 1997. Canopy and age structures of some old sub-boreal *Picea* stands in British Columbia. J. Veg. Sci. **8**: 615–626.

Krajina, V.J. 1965. Biogeoclimatic zones and biogeocoenoses of British Columbia. University of British Columbia, Department of Botany, Vancouver, British Columbia. Ecol. Western North Am. **1**: 1–17.

Kuuluvainen, T., and Juntunen, P. 1998. Seedling establishment in relation to microhabitat variation in a windthrow gap in a boreal *Pinus sylvestris* forest. J. Veg. Sci. **9**: 551–562.

Laflèche, V., Ruel, J.-C., and Archambault, L. 2000. Évaluation de la coupe avec protection de la régénération et des sols comme méthode de régénération de peuplements mélangés du domaine bioclimatique de la sapinière à bouleau jaune de l'est du Québec, Canada. For. Chron. **76**: 653–663.

Lähde, E., Laiho, O., and Norukorpi, Y. 1999. Diversity-oriented silviculture in the boreal zone of Europe. For. Ecol. Manage. **118**: 223–243.

Landhäusser, S.M., and Lieffers, V.J. 1998. Growth of *Populus tremuloides* in association with *Calamagrostis canadensis*. Can. J. For. Res. **28**: 396–401.

Larsson, S., and Danell, K. 2001. Science and management of boreal forest biodiversity. Scand. J. For. Res. **3**(Suppl.): 5–9.

Lautenschlager, R.A. 1993. Effects of conifer release with herbicides on wildlife. Ontario Ministry of Natural Resources, Sault Ste. Marie, Ontario. For. Res. Inf. Pap. No. 111. 23 p.

Lavender, D.P., Parish, R., Johnson, C.M., Montgomery, G., Vyse, A., Wallis, R.A., and Winston, D. (*Editors*). 1990. Regenerating British Columbia's forests. UBC Press, Vancouver, British Columbia. 372 p.

Lavertu, D., Mauffette, Y., and Bergeron, Y. 1994. Effects of stand age and litter removal on regeneration of *Populus tremuloides*. J. Veg. Sci. **5**: 501–568.

Leadem, C.L., Gillies, S.L., Yearsley, H.K., Sit, V., Spittlehouse, D.L., and Burton, P.J. 1997. Field studies of seed biology. British Columbia Ministry of Forestry Research Program, Victoria, British Columbia. Land Manage. Handb. 40. 196 p.

Légère, G., and Gingras, J.-F. 1998. Évaluation de méthodes de coupe avec protection des petites tiges marchandes. FERIC, Montréal, Quebec. Rapp. Tech. RT-124. 12 p.

Letchford, T., Spittlehouse, D., and Draper, D. 1996. Mounding on a SBSwk1 site near Prince George. Forestry Canada and British Columbia Ministry of Forests, Victoria, British Columbia. FRDA Res. Memo 228. 5 p.

Lieffers, V.J., and Stadt, K.J. 1994. Growth of understory *Picea glauca, Calamagrostis canadensis* and *Epilobium angustifolium* in relation to overstory light. Can. J. For. Res. **24**: 1193–1198.

Lieffers, V.J., Macdonald, S.E., and Hogg, E.H. 1993. Ecology of and control strategies for *Calamagrostis canadensis* in boreal forest sites. Can. J. For. Res. **23**: 2070–2077.

Lieffers, V.J., Macmillan, R.B., MacPherson, D., Branter, K., and Stewart, J.D. 1996a. Semi-natural and intensive silvicultural systems for the boreal mixedwood forest. For. Chron. **72**: 286–292.

Lieffers, V.J., Stadt, K.J., and Navratil, S. 1996b. Age structure and growth of understory white spruce under aspen. Can. J. For. Res. **26**: 1002–1007.

Lieffers, V.J., Messier, C., Stadt, K.J., Gendron, F., and Comeau, P. 1999. Predicting and managing light in the understory of boreal forests. Can. J. For. Res. **29**: 796–811.

Lieffers, V.J., Pinno, B.D., and Stadt, K.J. 2002. Light dynamics and free-to-grow standards in aspen-dominated mixedwood forests. For. Chron. **78**: 137–145 .

Lipson, D., and Nasholm, T. 2001. The unexpected versatility of plants: organic nitrogen use and availability in terrestrial systems. Oecologia, **128**: 305–316.

Liu, S.H., Campbell, R.A., Studens, J.A., and Wagner, R.G. 1996. Absorption and translocation of glyphosate in aspen (*Populus tremuloides*) as influenced by droplet size, droplet number and herbicide concentration. Weed Sci. **44**: 482–488.

Lloyd, D., and Elder, R. 1997. Choosing a site that will get seedlings off to the best start possible. Can. Silv. Mag. **5**(2): 15–20.

Long, J.N., and Smith, F.W. 1992. Volume increment in *Pinus contorta* var. *latifolia*: the influence of stand development and crown dynamics. For. Ecol. Manage. **53**: 53–64.

Lowe, J.J., Power, K., Marsan, M.W. 1996. Canada's forest inventory 1991: summary by terrestrial ecozones and ecoregions. Canadian Forest Service, Victoria, British Columbia. Inf. Rep. BC-X-364E. Available at http://www.pfc.forestry.ca/cgi-bin/bstore/catalog_e.pl?catalog=4725 [viewed 16 June 2003]. 56 p.

MacDonell, M.R., and Groot, A. 1996. Uneven-aged silviculture for peatland second-growth black spruce: biological feasibility. Canadian Forest Service and Ontario Ministry of Natural Resources, Sault Ste. Marie, Ontario. NODA/NFP Tech. Rep. TR-36. 14 p.

MacLean, D.A. 1984. Effects of spruce budworm outbreaks on the productivity and stability of balsam fir forests. For. Chron. **60**: 273–279.

MacPherson, D.M., Lieffers, V.J., and Blenis, P.J. 2001. Productivity of aspen stands with and without a spruce understory in Alberta's boreal mixedwood forests. For. Chron. **77**: 351–356.

Man, R.Z., and Lieffers, V.J. 1999a. Are mixtures of aspen and white spruce more productive than single species stands? For. Chron. **75**: 505–513.

Man, R.Z., and Lieffers, V.J. 1999b. Effects of shelterwood and site preparation on microclimate and establishment of white spruce seedlings in a boreal mixedwood forest. For. Chron. **75**: 837–844.

Matthews, J.D. 1989. Silvicultural systems. Clarendon Press, Oxford, U.K. 284 p.

McCarthy, J. 2001. Gap dynamics of forest trees: a review with particular attention to boreal forests. Environ. Rev. **9**: 1–59.

McClain, K.M., Morris, D.M., Hills, S.C., and Buse, L.J. 1994. The effects of initial spacing on growth and crown development for planted northern conifers: 37 year results. For. Chron. **70**: 174–182.

McCune, B., and Allen, T.F.H. 1985. Will similar forests develop on similar sites? Can. J. Bot. **63**: 367–376.

McMinn, R.G., and Hedin, I.B. 1990. Site preparation: mechanical and manual. *In* Regenerating British Columbia's forests. *Edited by* D.P. Lavender, R. Parish, C.M. Johnson, G. Montgomery, A. Vyse, R.A. Wallis, and D. Winston. UBC Press, Vancouver, British Columbia. pp. 150–163.

McNabb, D.H., Startsev, A.D., and Nguyen, H. 2001. Soil wetness and traffic level effects on bulk density and air-filled porosity of compacted boreal forest soils. Soil Sci. Soc. Am. **65**: 1238–1247.

Meades, W.J., and Moores, L. 1989. Forest site classification manual: a field guide to the Damman forest types of Newfoundland. St. John's and Corner Brook, Newfoundland. Nfld./Can. FRDA Rep. 003. 366 p.

Messier, C., and Poulin, J. 2001. L'aménagement intensif des forêts: est-ce compatible avec sa conservation? Proceedings of the workshop on "l'aménagement des forêts feuillus", March 2001. Jardin Botanique de Montréal, Montréal, Quebec. 7 p.

Messier, C., Parent, S., and Bergeron, Y. 1998. Characterization of understory light environment in closed mixed boreal forests: effects of overstory and understory vegetation. J. Veg. Sci. **9**: 511–520.

Michael, J.L., Webber, E.C., Jr., Bayne, R.R., Fischer, J.B., Gibbs, H.L., and Seescok, W.C. 1999. Hexazinone dissipation in forest ecosystems and impacts on aquatic communities. Can. J. For. Res. **29**: 1170–1181.

Mitchell, K.J., and Goudie, J.W. 1998. The emperor's new clothes. *In* Stand density management: planning and implementation. *Edited by* C. Bamsey. Clear Lake Ltd., Edmonton, Alberta. pp. 45–58.

Morrison, I.K. 2001. Fertilization of conifers in Eastern Canada. *In* Enhanced forest management: fertilization and economics. Conference proceedings, 1–2 March 2001, Edmonton, Alberta. *Edited by* C. Bamsey. Clear Lake Ltd., Edmonton, Alberta. pp. 32–38.

(MRNQ) Ministère des Ressources naturelles du Québec. 1994. Une stratégie: aménager pour mieux protéger les forêts. FQ94-3051. Ministère des Ressources naturelles du Québec, Québec, Quebec. 197 p.

Mullin, T.J., and Bertrand, S. 1998. Environmental release of transgenic trees in Canada — potential benefits and assessment of biosafety. For. Chron. **74**: 203–217.

Munson, A.D., and Timmer, V.R. 1989. Site-specific growth and nutrition of planted *Picea mariana* in the Ontario Clay Belt. I. Early performance. Can. J. For. Res. **19**: 162–170.

Navratil, S., Branter, K., and Zasada, J. 1991. Regeneration in the mixedwoods. *In* Northern Mixedwood '89. Proceedings, Symposium, 12–14 September 1989, Fort St. John, British Columbia. *Edited by* A. Shortreid. Forestry Canada and British Columbia Ministry of Forests, Victoria, British Columbia. FRDA Rep. 164. pp. 32–48.

Navratil, S., Brace, L.G., Sauder, E.A., and Lux, S. 1994. Silvicultural and harvesting options to favor immature white spruce and aspen regeneration in boreal mixedwoods. Canadian Forest Service, Edmonton, Alberta. Inf. Rep. NOR-X-337. 78 p.

Newton, P.F. 1997. Algorithmic versions of black spruce stand density management diagrams. For. Chron. **73**: 257–265.

Nienstadt, H., and Zasada, J.C. 1990. *Picea glauca* (Moench) Voss White spruce. *In* Silvics of North America, vol. 1, conifers. *Edited by* R.M. Burns and B.H. Honkala. USDA Forest Service, Washington, D.C. Agric. Handb. 654. pp. 204–226.

Nyland, R.D. 1996. Silviculture: concepts and applications. McGraw–Hill, New York, New York. 633 p.

Oke, T.R. 1987. Boundary layer climates. Methuen, London, U.K. 435 p.

Oliver, C.D. 1992. Achieving and maintaining biodiversity and economic productivity. J. For. **90**(9): 20–25.

Oliver, C.D., and Larson, B.C. 1996. Forest stand dynamics, second edition. McGraw–Hill, New York, New York. 325 p.

(OMNR) Ontario Ministry of Natural Resources. 2001. Forest management guide for natural disturbance pattern emulation. Version 3.1. Ontario Ministry of Natural Resources, Queens Printer, Toronto, Ontario. 40 p.

Örlander, G., Gemmel, P., and Hunt, J. 1990. Site preparation: a Swedish overview. Forestry Canada and British Columbia Ministry of Forests, Victoria, British Columbia. FRDA Rep. 105. 61 p.

Paquin, R., and Doucet, R., 1992. Productivité de pessières noires boréales régénérées par marcottage à la suite de vieilles coupes totales au Québec. Can. J. For. Res. **22**: 601–612.

Paquin, R., Margolis, H.A., Doucet, R., and Coyea, M.R. 1999. Comparaison of growth and physiology of layers and naturally established seedlings of black spruce in a boreal cutover in Quebec. Can. J. For. Res. **29**: 1–8.

Paré, D., and Bergeron, Y. 1995. Above-ground biomass accumulations along a 230-year chronosequence in the southern portion of the Canadian Boreal forest. J. Ecol. **83**: 1001–1007.

Parent, S., and Messier, C. 1995. Effets d'un gradient de lumière sur la croissance et l'architecture du sapin baumier (*Abies balsamea*). Can. J. For. Res. **25**: 878–885.

Parent, S., Morin, H., and Messier, C. 2000. Effects of adventitious roots on age determination in balsam fir regeneration. Can. J. For. Res. **30**: 513–518.

Pastor, J., Gardner, R.H., Dale, V.H., and Post, W.M. 1987. Successional changes in nitrogen availability as a factor contributing to spruce decline in boreal North America. Can. J. For. Res. **17**: 1394–1400.

Payette, S., Bhiry, N., Delwaide, A., and Simard, M. 2000. Origin of the lichen woodland at its southern range limit in eastern Canada: the catastrophic impact of insect defoliators and fire on the spruce–moss forest. Can. J. For. Res. **30**: 288–305.

Perry, D.J., and Bousquet, J. 2001. Genetic diversity and mating systems of post-fire and post-harvest black spruce: an investigation using co-dominant sequence-tagged-site (STS) markers. Can. J. For. Res. **31**: 32–40.

Peterson, E.B., and Peterson, N.M. 1992. Ecology, management, and use of aspen and balsam poplar in the prairie provinces. Forestry Canada, Edmonton, Alberta. Spec. Rep. 1. 252 p.

Pettersson, F. 2001. Experience of fertilization in mineral soils in coniferous stands in Sweden. *In* Enhanced forest management: fertilization and economics. Conference proceedings, 1–2 March 2001, Edmonton, Alberta. *Edited by* C. Bamsey. Clear Lake Ltd., Edmonton, Alberta. pp. 14–26.

Pharis, R.P., Webber, J.E., and Ross, S.D. 1987. The promotion of flowering in forest trees by gibberellin A4/7 and cultural treatments: a review of the possible mechanisms. For. Ecol. Manage. **19**: 65–84.

Pielou, E.C. 1991. After the ice age: the return of life to glaciated North America. University of Chicago Press, Chicago, Illinois. 366 p.

Pinno, B.D., Lieffers, V.J., and Stadt, K.J. 2001. Measuring and modeling the crown and light transmission characteristics of juvenile aspen. Can. J. For. Res. **31**: 1930–1939.

Pojar, J., Klinka, K., and Meidinger, D.V. 1987. Biogeoclimatic ecosystem classification in British Columbia. For. Ecol. Manage. **22**: 119–154.

Pothier, D., and Savard, J. 1998. Actualisation des tables de production pour les principales espèces forestières du Québec. Ministère des Ressources naturelles du Québec, Québec, Quebec. 183 p.

Pothier, D., Doucet, R., and Boily, J. 1995. The effect of advance regeneration height on future yield of black spruce stands. Can. J. For. Res. **25**: 536–544.

Pratt, L., and Urquhart, I. 1994 The last great forest: Japanese multinationals and Alberta's northern forests. NeWest Press, Edmonton, Alberta. 222 p.

Prescott, C.E., Weetman, G.F., and Barker, J.E. 1996. Causes and amelioration of nutrient deficiencies in cutovers of cedar–hemlock forests in coastal British Columbia. For. Chron. **72**: 293–302.

Pullman, G.S., Cairney, J., and Peter, G. 1998. Clonal forestry and genetic engineering: where we stand, future prospects, and the potential impacts on mill operations. TAPPI J. **81**: 57–64.

Radosevich, S.R., and Osteryoung, K. 1987. Principles governing plant–environment interactions. *In* Forest vegetation management for conifer production. *Edited by* J.D. Walstad and P.J. Kuch. Wiley and Sons, New York, New York. pp. 105–156.

Reinecke, L.H. 1933. Perfecting a stand-density index for even-aged forests. J. Agric. Res. **46**: 627–638.

Reynolds, P.E., Scrivener, J.C., Holtby, L.B., and Kingsbury, P.D. 1993. Review and synthesis of Carnation Creek herbicide research. For. Chron. **69**: 423–430.

Rodgers, A.R., Williams, D., Sinclair, A.R.E., Sullivan, T.P., and Andersen, R.J. 1993. Does nursery production reduce antiherbiory defenses of white spruce? Evidence from feeding experiments with snowshoe hares. Can. J. For. Res. **23**: 2358–2361.

Rowe, J.S. 1961. Critique of some vegetational concepts as applied to the forests of northwestern Alberta. Can. J. Bot. **39**: 1007–1015.

Rowe, J.S. 1992. The ecosystem approach to forestland management. For. Chron. **68**: 222–224.

Ruel, J.-C. 1989. Importance de la régénération préexistante dans les forêts publiques du Québec. Ann. Sci. For. **46**: 345–359.

Ruel, J.-C. 1990. Advance growth abundance and regeneration after commercial clearcutting in Québec. *In* Conference on natural regeneration management, Fredericton, New Brunswick. *Edited by* C.M. Simpson. Forestry Canada, Fredricton, New Brunswick. pp. 115–132.

Ruel, J.-C., Messier, C., Doucet, R., Claveau, Y., and Comeau, P. 2000. Morphological indicators of growth response of coniferous advance regeneration to overstory removal in the boreal forest. For. Chron. **76**: 633–642.

Ryan, M.G., Binkley, D., and Fownes, J.H. 1997. Age-related decline in forest productivity: pattern and process. Adv. Ecol. Res. **27**: 213–262.

Sauvageau, F. 1995. Silvicultural terms in Canada. Canadian Forest Service, Hull, Quebec. 109 p.

Schier, G.A. 1972. Apical dominance in multishoot cultures from aspen roots. For. Sci. **18**: 147–149.

Shugart, H.H. 1984. A theory of forest dynamics: the ecological implications of forest succession models. Springer-Verlag, New York. 278 p.

Sims, R.A., Toweill, W.D., Baldwin, K.A., and Wickware, G.M. 1989. Field guide to the forest ecosystem classification for northwestern Ontario. Ontario/Canada FRDA, Ontario Ministry of Natural Resources, Thunder Bay, Ontario. 191 p.

Smith, D.M., Larson, B.C., Kelty, M.J., and Ashton, P.M.S. 1997. The practice of silviculture: applied forest ecology. Wiley and Sons, New York, New York. 537 p.

Smith, V.G., Watts, M., and James, D.F., 1987. Mechanical stability of black spruce in the Clay Belt of northern Ontario. Can. J. For. Res. **17**: 1080–1091.

Stadt, K.J., and Lieffers, V.J. 2000. MIXLIGHT: a flexible light transmission model for mixed species stands. Agric. For. Meteorol. **102**: 235–252.

Stewart, J.D., Hogg, E.H., Hurdle, P.A., Stadt, K.J., Tollestrup, P., and Lieffers, V.J. 1998. Dispersal of white spruce seed in mature aspen stands. Can. J. Bot. **76**: 181–188.

Stewart, J.D., Landhäusser, S.M., Stadt, K.J., and Lieffers, V.J. 2000. Regeneration of white spruce under aspen canopies: seeding, planting and site preparation. W. J. Applied For. **15**: 177–182.

St.-Pierre, H., Gagnon, R., and Bellefleur, P. 1992. Régénération après feu de l'épinette noire (*Picea mariana*) et du pin gris (*Pinus banksiana*) dans la forêt boréale, Québec. Can. J. For. Res. **22**: 474–481.

Sutton, R.F. 1993. Mounding site preparation: a review of European and North American experience. New For. **7**: 151–192.

Sutherland, B.J., and Foreman, F.F. 1995. Guide to the use of mechanical site preparation equipment in northwestern Ontario. Canadian Forest Service, Sault Ste. Marie, Ontario. NWST Tech. Rep. TR-87. 186 p.

Troup, R.S. 1928. Silvicultural systems. Oxford University Press, Oxford, U.K. 199 p.

Urban, S., Lieffers, V.J., and Macdonald, S.E. 1994. Radial growth of roots and stems of white spruce following release, measured with dendrochronology techniques. Can. J. For. Res. **24**: 1550–1556.

Vanha-Majamaa, I., and Jalonen, J. 2001. Green tree retention in Fennoscandian forestry. Scand. J. For. Res. **3**(Suppl.): 79–90.

Van Oosten, C. 2000. Activities related to poplar and willow cultivation and utilization in Canada 1996–1999. Report to 21st Session of the International Poplar Commission, Portland, Oregon, 24–28 September 2000. SilviConsult Woody Crops Tech. Inc., Nanaimo, British Columbia. 130 p.

Viereck, L.A., and Johnston, W.F. 1990. *Picea mariana* (Mill.) B.S.P. Black Spruce. *In* Silvics of North America, vol. 1, conifers. *Edited by* R.M. Burns and B.H. Honkala. USDA Forest Service, Washington, D.C. Agric. Handb. 654. pp. 227–237.

Vyse, A., and Navratil, S. 1985. Advances in lodgepole pine regeneration. *In* Lodgepole pine, the species and its management. *Edited by* D.M. Baumgartner, R.G. Krebill, J.T. Arnott, and G.F. Weetman. Washington State University, Pullman, Washington. pp. 173–186.

Wagner, R.G., and Columbo, S.J. (*Editors*). 2001. Regenerating the Canadian forest: principles and practice for Ontario. Fitzhenry & Whiteside, Markham, Ontario. 650 p.

Wagner, R.G., and Zasada, J.C. 1991. Integrating plant autecology and silvicultural activities to prevent forest vegetation management problems. For. Chron. **67**: 506–513.

Wan, X., Lanhäusser, S.M., Zwiazek, J., and Lieffers, V.J. 1999. Root water flow and growth of aspen (*Populus tremuloides*) at low root temperatures. Tree Physiol. **19**: 879–884.

Wan, X., Zwiazek, J., Lieffers, V.J., and Landhäusser, S.M. 2000. Hydraulic conductance in aspen seedlings exposed to low root temperatures. Tree Physiol. **21**: 691–696.

Wilson, F.G. 1946. Numerical expression of stocking in terms of height. J. For. **44**: 758–561.

Wright, E.F, Coates, K.D., Canham, C.D., and Bartemucci, P. 1998. Species variability in growth response to light across climatic regions in northwestern British Columbia. Can. J. For. Res. **28**: 871–886.

Wurtz, T., and Zasada, J.C. 2001. An alternative to clear-cutting in the boreal forest of Alaska: a 27-year study of regeneration after shelterwood harvesting. Can. J. For. Res. **31**: 999–1011.

Zasada, J.C., and Schier, G.A. 1973. Aspen root suckering in Alaska: effect of clone, collection date and temperature. Northwest Sci. **47**: 100–104.

Chapter 14

Modelling tools to assess the sustainability of forest management scenarios

C. Messier, M.-J. Fortin, F. Schmiegelow, F. Doyon,
S.G. Cumming, J.P. Kimmins, B. Seely, C. Welham,
and J. Nelson

Introduction

Today's challenge in forest management is to plan for sustainable, non-declining supplies of timber and non-timber values at various scales. This post-modern paradigm calls for a shift in our management strategies away from a focus on managing the individual trees or stands alone, toward managing whole landscapes. Hence, true ecosystem management (or ecological forestry, designed to utilize forest resources only to the extent that ecosystem composition, function, and structure are not threatened) tends to operate at spatial and temporal scales much larger than traditional forestry management practices, although some form of ecosystem practices have been advocated at the stand level. As stated by Korzukhin et al. (1996), ecosystem management requires: (1) assessing management options across a wide range of spatial scales, (2) predicting long-term effects of management actions, (3) understanding management effects on biological diversity, (4) predicting influences of specific components (e.g., biological legacies) on

C. Messier.[1] Groupe de recherche en écologie forestière interuniversitaire, Dép. des sciences biologiques, Université de Quebéc à Montréal, C.P. 8888, Succ. Centre-Ville, Montréal, Québec H3C 3P8, Canada.
M.-J. Fortin. Department of Zoology, University of Toronto, Toronto, Ontario, Canada.
F. Schmiegelow. Department of Renewable Resources, University of Alberta, Edmonton, Alberta, Canada.
F. Doyon. Institut Québécois d'Aménagement de la Forêt Feuillue, St-André-Avellin, Québec, Canada.
S.G. Cumming. Boreal Ecosystems Research Limited, Edmonton, Alberta, Canada.
J.P. Kimmins, B. Seely, C. Welham, and J. Nelson. Faculty of Forestry, University of British Columbia, Vancouver, British Columbia, Canada.
[1]Author for correspondence. Telephone: 514-987-3000, ext. 4009.
e-mail: messier.christian@uqam.ca

Correct citation: Messier, C., Fortin, M.-J., Schmiegelow, F., Doyon, F., Cumming, S.G., Kimmins, J.P. , Seely, B., Welham, C., and Nelson, J. 2003. Modelling tools to assess the sustainability of forest management scenarios. Chapter 14. *In* Towards Sustainable Management of the Boreal Forest. *Edited by* P.J. Burton, C. Messier, D.W. Smith, and W.L. Adamowicz. NRC Research Press, Ottawa, Ontario, Canada. pp. 531–580.

the larger system, (5) projecting population dynamics for a wide range of species, (6) comparing natural vs. human-induced disturbance patterns on forest landscapes, and (7) assessing global climatic influences on particular forests. All of these demands are characterized, however, by extraordinary complexity, limited availability of mechanistic hypotheses, and a paucity of data (Galindo-Leal and Bunnell 1995).

Biodiversity and its conservation indeed has become a major issue with respect to forest land management in Canada and elsewhere. The concern comes from landscape fragmentation, degradation, and elimination of habitats. So, the original concern about local species extinction has now been transformed into one of global concern, stewardship, and land ethics. We are now asking ourselves what stand and landscape attributes need to be maintained to preserve biodiversity. The large-scale forest harvesting of today creates considerable landscape variation, and without an understanding of its impact on species responses, it cannot be claimed that biodiversity is being maintained.

Traditional forestry calls for managing what was seen as relatively simple systems at the tree and stand levels, so simple tools were developed to address the questions arising from such a perception. Ecosystem management calls for managing a more complex system, so we need different and more complex tools at both the smaller and larger spatial and temporal scales to deal with this added complexity. Simulation models can organize the complexity of information and data into a coherent tool for analysing systems at these various scales. The effective use of simulation modelling in the analysis of living systems requires, however, knowledge of the biological and ecological processes, a comprehension of the relationships among these processes by means of system analysis, and the operational translation of these relationships using statistics and mathematics. Hence, modelling is a powerful tool available to the manager, to be used in concert with observation, personal experience, and experimentation to improve the understanding of forest dynamics and to assist in the development of sustainable management over a greater area and longer time period. Models complement empirical work on ecological processes and give insight into the dynamics of complex ecological systems. As Bossel (1991) states, forest modelling allows us to move from description toward explanation. There has been a steady increase in the development and use of models in the forestry research community over the last several decades (Box 14.1).

This chapter presents an overview of some conceptual and modelling approaches currently used at various scales in ecological forestry. These models may have their origins in the study of timber growth and yield (or forest mensuration), forest succession, timber harvest scheduling, or econometrics. But their application in the evaluation of sustainable forest management (SFM) scenarios increasingly requires a blending of predictors, indicators, and approaches from all of these various disciplines.

The role and importance of modelling and simulation models in forestry

What are modelling and simulation models?

Models are simplified representations of some aspects of reality that are developed to describe, analyse, understand, or predict the behaviour of something (Grant et al. 1997).

> **Box 14.1. The development and use of modelling tools and simulation models in forestry is increasing all over the world.**
>
> The development of modelling and simulation models in forestry is increasing quickly all over the world. A recent literature survey of articles from 1973 to 1999 in three leading international forestry journals (Canadian Journal of Forest Research, Forest Ecology and Management, and Forest Science) indicates that the proportion of all published papers which deals directly with modelling and (or) simulation models is increasing steadily, from less than 5% in the early 1970s to more than 15% today (Fig. 14.1).
>
> This increase is due to:
>
> - increased computer power;
> - increased participation of mathematicians and physicists in forestry issues;
> - increased general knowledge about forest ecosystem functioning; and
> - increased recognition of the complexity we are managing.
>
> **Fig. 14.1.** The steady increased use of models and modelling in papers published in forestry research journals.
>
>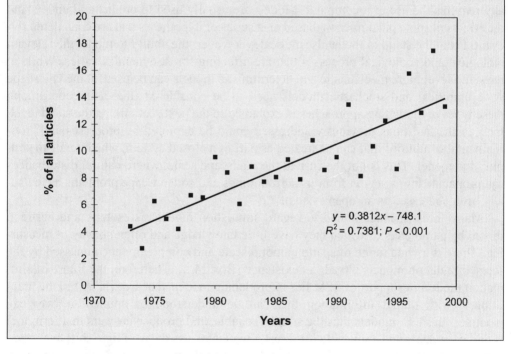
>
> $y = 0.3812x - 748.1$
> $R^2 = 0.7381; P < 0.001$

Models are useful for various kinds of representation, from a very qualitative or conceptual description of a phenomenon to a very detailed mathematical simulation of its dynamics. Models are widespread in everyday and technical usage. A mathematical model is like a verbal model, but using more concise, precise, and deductive language. The term model refers to any "analytical" or "mathematical" representation that quantifies response variables using mathematical equations, while the term "simulation model" refers to the generation of response variable values specifically by numerical approximation using a computer program (Haefner 1996).

It is important to distinguish between models used for understanding and those used for prediction (Bunnell 1989). Models for understanding are most useful as a research tool to help comprehend, collect, and link previously isolated bits of knowledge and to identify gaps where more work is needed. Such mechanistic models often fall far short of making realistic predictions, but their benefits often come from development of the model, a practice which forces one to organize and articulate the relationships among processes. Then there are models (based on our best understanding of the various processes or relationships) that are geared toward making predictions regarding some concrete ecosystem or forestry elements. Predictive models are often purely statistical rather than mechanistic, and may or may not help explain the system under study; this is not their goal. No matter what type of model, all good modelling processes begin and end with a good set of questions. The question(s) should direct the type of model or modelling approaches to be used. The real quest of dynamic modelling is to discover some (hopefully few) rather simple underlying principles that together derive understanding or insight from complexity (Hannon and Ruth 1997).

In forestry, as in many technical disciplines, models are usually made of mathematical equations. The complexity of these equations and of the data being processed usually requires the use of a computer. Models are usually used to predict and understand the effects of some phenomenon, based on a series of hypotheses that are considered relevant. From the stand to the landscape scale, however, the ability to model the relevant biological and ecological processes that are affecting forest dynamics varies. While at the stand level more mechanistic and deterministic models can be used, at the landscape scale empirical and stochastic models need to be considered (the distinction among these types of modelling approaches is explained in the next section). At the stand level, most of the questions previously addressed could be explored by processes and environmental conditions that can be treated as part of a closed system, which tends to simplify the model. This is not the case at the landscape scale, where natural disturbances can propagate into a region from its surroundings and so landscape problems have usually been addressed as an open system.

Many forest managers tend to regard simulation models as esoteric and abstract, driven by parameters of which they have little knowledge and often no way of measuring. There is a wide range of spatiotemporal scale and complexity encompassed by the forest simulation models already in existence (Box 14.2). Therefore, the future of simulation models in support of SFM lies in producing models that are relevant to the practitioner, with output information that can be understood or applied to everyday practices. In short, models must be simple, reliable, and produce answers in a form that is understandable and applicable for forester managers. As the paradigm shift to ecosystem management takes hold, foresters will need to deal with values other than wood production, such as species composition, stand structure, landscape fragmentation, soil nutrient reserves, water quality, and aesthetics. With this growing expectation for protecting, tracking, and managing diverse and complex forest values, the need for modelling, simulation models, and other sophisticated management tools is on the increase.

> ### Box 14.2. Examples of the scales and complexity of some models used today in forestry.
>
> Models in forestry can vary from very simple regression models to very complex process simulation models, and from simulation at the level of the leaf to that of a whole landscape. Such models can operate at various temporal (from seconds to decades) and spatial (from mm^2 to km^2) scales, depending on the phenomenon being simulated (Fig. 14.2). Like most things in life, models are never totally wrong or totally right. What is important is the ability to determine, control, and report the error associated with their calibration and use. They must therefore be evaluated in relation to specified objectives and their capacity to realistically simulate the main processes of interest.
>
> **Fig. 14.2.** Schematic representation of the various scales of complexity and time steps used in some of the models discussed in this chapter (adapted from Landsberg and Gower 1997).
>
>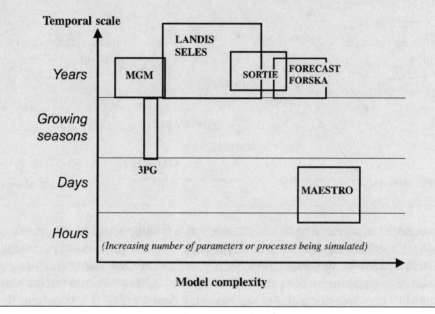

What are the different kinds of models and approaches used?

Most previous discussions about approaches to forest modelling have been divided between empirical vs. process-based and deterministic vs. stochastic modelling. Although these terms have been used to categorize models and modelling approaches, in reality, a continuum exists between these two dichotomies (Box 14.3). Process-based models using current understanding of key mechanisms or processes attempt to capture the system's internal structure, rules, and behaviour. In contrast, empirical models seek to describe statistical relationships among the data with limited regard to the system's internal structure, rules, or behaviour. Hence, empirical models describe only the relationships among variables without regard for the processes that generate these relationships. A review of the advantages and disadvantages of process-based vs. empirical modelling by various authors seems to indicate that empirical models are generally

Box 14.3. Comparison of models used in forestry in terms of their structural and functional realism.

Recently, models have been described or classified along two continuous axes: (1) increasing functional realism and (2) increasing structural realism (see figure below). Traditional forest growth and yield models, for example, are at the top of the triangle, since they have very little functional or structural realism. The current tendency in modelling is to have models that are both more functionally and more structurally explicit.

Fig. 14.3. A comparison of some stand and landscape models used in the boreal forest in terms of their structural and functional realism (adapted from Kurth 1994; see text for references for each of these models).

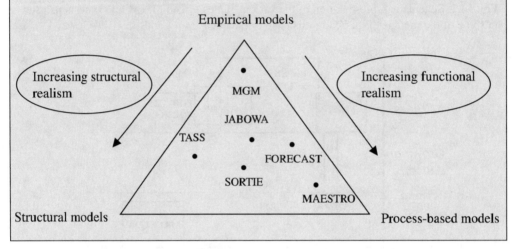

more precise when addressing phenomena that fall within the range of conditions for which the empirical relationship was developed, whereas process-based models are much better at addressing complex questions for which we have little previous experience, such as simulating multi-species stand dynamics, growth under a changing climate, and forest management under the emerging forest ecosystem paradigm. In this case, mechanistic models are preferable, since they more adequately predict system behaviour under new circumstances.

Both empirical and mechanistic models can be deterministic, meaning that only one outcome is predicted for a given set of starting conditions. Given the inherent role of chance involved in natural disturbances, landscape modelling of the boreal forest calls for stochastic models that either include a random error term, or select the value for a parameter from a probability distribution. These models produce a different result each time they are run, presenting a range of possible outcomes. Hence, they more accurately reflect the natural variability in the system and the uncertainty in expecting any particular outcome. Furthermore, as demonstrated by Hansen et al. (1995), many of the response variables can show non-linear behaviour across various levels of treatment, so stochastistic models, or even deterministic simulation models, can help us deal with some of this complexity.

Finally, another dichotomy that has developed recently is that of aspatial vs. spatial representation of model elements. Spatially explicit stand and landscape models and spatial decision support systems need to be developed to address the multiple ecological, economic, and social response variables that are sensitive to their spatial context. Such tools can help government agencies and forest companies achieve their sustainable management objectives, and can be an especially important tool in translating overall forest management objectives into mappable plans of action (Chap. 12).

In this chapter, we adopt the view of Korzukhin et al. (1996) that process-based models are more likely to meet the information needs presented by the complexity of implementing SFM practices in the boreal forest. Furthermore, our knowledge of many basic processes in forest dynamics is increasing quickly and there now exists a good, reliable, and growing bank of models and modelling approaches from which to choose when developing new simulation models, so that it is now possible to achieve a fairly high level of precision with process-based models. The lack of rigorous testing of the results of process-based models has often been criticized (Korzukhin et al. 1996). But Sievänen and Burk (1993) argue that most processes being simulated are introduced in the form of equations derived from well tested theories and hence it can be claimed that such models have already been tested, in part at least, during their construction. Process-based models are ideally suited to situations where fundamental mechanisms are known, but the system is so complex that it is difficult to develop a complete picture or make reasonable predictions. Most process-based models use some kind of empirically derived relationships at one level or another. For example, most forest stand dynamics models use empirically derived diameter vs. height relationships because they are fairly stable over a wide gradient of environmental conditions. For some specific applications, such as the prediction of tree cover or volume as a function of stand age in fully stocked plantations of a single species, empirical relationships alone may suffice.

The major problem when developing a spatially explicit simulation model, especially landscape models, is often the lack of data, because most of our knowledge and empirical data are at the level of individual trees and stands. Furthermore, when modelling at the landscape scale, researchers are faced with the trade-off of using precise, quantitative data over small areas, versus using coarse, less accurate data over larger areas. The fundamental problem is therefore to determine which processes should, and can, be translated from stand to landscape scales. While site factors may be critical determinants of productivity at the stand scale, at the landscape scale, however, disturbances are the most critical processes affecting forest dynamics. Hence, according to the spatial scale under study, some processes are more or less important than others. Simplicity in modelling is important to allow us to explore its capability and several models could be simplified without any loss to their predictive power (Bugmann 1996; Deutschman et al. 1999).

A brief history of modelling in forestry

Computer modelling has been with us for nearly 40 years (Hannon and Ruth 1997). Models have increased in complexity (both in terms of number of processes and the spatial scale being considered) along with the capacity of the computers available for running them. The need for forecasting systems has paralleled the overall evolution of forestry. Exploitative forest use in the pre-forestry stage of human-forest interactions had no need for forecasting tools, since there was no explicit effort to maintain the future supply of particular values. When forest exploitation led to resource depletion, forestry developed as an administrative, regulation-based social institution to achieve sustainability. This was mainly to conserve and sustain future supplies of domestic, industrial, and military wood, although in some cases early forestry sought to maintain game species for hunting (see Chap. 1). As this "administrative forestry" stage became more quantitative and industrial forestry more organized, the need arose for accurate forecasts of future harvestable timber volumes. This led to the development and widespread use of yield tables as predictive models. Historical records from Germany indicate that in western society, yield tables were available in the late 1700s. Earlier examples of such forecasting tools undoubtedly exist in Japan, China, and other highly organized cultures whose origins pre-date the developments in Europe.

The extensive accumulation of forest growth data in Germany led to the widespread use of empirical forecasting tools by the late eighteenth century. However, it was soon recognized that production forecasts based solely on technical and economic definitions of timber yield result in projections that are insensitive to both changing growth conditions (e.g., soil, climate, competition from non-crop vegetation) and social values (e.g., economics, wood value and uses). Recognition that these "historical bioassay" models were better mirrors of past growth than crystal balls of possible future growth led to the development of a branch of forest growth and yield research that focused on biomass production processes and the factors that regulate them.

One of the first major ecological studies of the factors limiting tree growth and yield was the work of Ebermayer (1876) in north German pine forests. Established on infertile sandy soils, these forests were subject to repeated litter and firewood collection, and exhibited growth and yield decline. Ebermayer was able to establish a relationship between soil fertility, nutrient cycling and ecosystem nutrient budgets, and the harvestable production of timber. This early work was an important cornerstone of the development of modern forest science. Later studies made an important contribution to our understanding of forest ecosystem function, and led to work by Möller (1945) and associates (Möller et al. 1956a, b) which laid the foundation for the photosynthesis – carbon balance approach to forest production research. These investigations led to studies of the relationship between foliage biomass and forest growth (e.g., Fujimori and Whitehead 1986), and the development of physiologically based process simulation models. The earlier forest models tended to deal with questions of tree size and volume of wood in monospecific stands and often included some physiological and structural details (Kellomäki and Kurttio 1991). Computer models that simulate the dynamics of a forest by following the fates of each tree were developed in the mid 1960s (Shugart and Prenctice 1992; Shugart 1998).

Increasingly, various approaches of modelling have been linked to produce models that can capture the strength of these various approaches (see Fig. 14.4). This is the case, for example, with SORTIE (Pacala et al. 1993, 1996), a spatially explicit stand dynamics model that mixes mechanistically and empirically derived relationships with some

Fig. 14.4. Schematic diagram of the most important processes entering into the makeup of process-based stand models. The circle indicates processes that could be modelled at the landscape level; S indicates stand level processes; L indicates landscape-level processes.

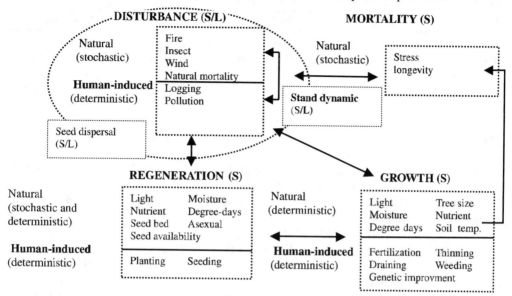

structural and functional realism in terms of stand canopy and light attenuation with the canopy.

An overview of simulation models used in the boreal forest

Numerous models of all kinds have been developed in forestry and forest ecology over the last 40 years. This section is an attempt to review the specificity and applicability of some representative examples of models that are currently used in the boreal forest. Table 14.1 compares some of them in terms of their structure, level and applications.

Tree-level models

There exist two general types of tree-level models, the process-based and structural tree models. The first type tries to emulate the basic physiological processes that are known to occur within trees, whereas the second type focuses entirely on the spatial position and orientation of the structural components of the tree. Recently, major advances have been made in both types and we are now starting to see the merging of these two approaches in what Sievänen et al. (1999) call the functional-structural models (FSM). ECOPHYS is a good example of a well-known ecophysiological process-based model that has been used to simulate the growth of poplar (*Populus* sp.; Rauscher et al. 1990). The model AMAP, developed by a group of French scientists, is, on the other hand, probably one of the best examples of a tree structural model (Reffye et al. 1997). The AMAP model has been used in various ornamental, agricultural, and forestry applications (Reffye et al. 1995).

Table 14.1. Examples of some single models used or developed for the boreal forest of Canada and Fennoscandia and their main characteristics.

Name of model	Type of model			
	Structure	Prediction	Level	Application
	Tree			
LIGNUM	Process based	Deterministic	Leaf/tree	Tree growth dynamics/ wood quality/ vigour
	Stand			
MGM	Empirical/ non-spatial	Deterministic	Tree/stand	Yield curves
SORTIE	Process based/ empirical/spatial	Deterministic/ stochastic	Tree/stand	Stand dynamics
FORSKA, LOKI	Process based/ empirical/ partially spatial	Deterministic	Tree/stand	Stand dynamics
FORECAST	Process based/ empirical/non- spatial	Deterministic	Tree/stand	Stand dynamics/ yield
MAESTRO	Process based/ non-spatial	Deterministic	Shoot/tree/ stand	Carbon fixation
	Landscape			
LANDIS	Process based/ empirical/ spatial	Stochastic	Landscape	Landscape dynamics
SELES	Process based/ empirical/ spatial	Stochastic	Landscape	Landscape dynamics
FEENIX	Process based/ empirical/ spatial	Stochastic	Landscape	Landscape dynamics
TARDIS	Hybrid, low resolution	Stochastic	Regional	Regional-scale Analogs of landscape dynamics
	Harvest scheduling			
FORMAN2000	Empirical/spatial	Deterministic	Landscape	Wood supply
FPS-ATLAS	Empirical/spatial	Deterministic	Landscape	Wood supply Landscape pattern
COMPLAN	Empirical/spatial	Deterministic	Landscape	Wood supply Landscape pattern
ECHO	Empirical/spatial	Deterministic	Landscape	Wood supply Landscape pattern

Table 14.1. (*concluded*).

| Name | Type of model | | | |
of model	Structure	Prediction	Level	Application
FSOS ECHO	Empirical/spatial	Deterministic	Landscape	Wood supply
				Landscape pattern
Woodstock/ Stanley	Empirical/spatial	Deterministic	Landscape	Wood supply Landscape pattern

Note: See text for references for each of these models.

The model LIGNUM (Perttunen et al. 1996, 1998) is part of the new generation of hybrid models that considers both metabolism (process-based relationships) and tree architectural relationships. LIGNUM grows a simulated tree as a population of shoots that connect to previously formed woody structure. The woody structure results from the secondary thickening of the previously formed shoots. In the model the shoots are called tree segments. A crown of the model tree consists of tree segments, branching points, and buds. Each pair of tree segments is separated by a branching point, as are the buds and tree segments. A tree segment may contain sapwood, heartwood, bark, and foliage. Growth of the tree is driven by the quantity of available photosynthate. The annual photosynthetic production by the crown is linearly proportional to the crown's light interception. The model calculates annual light interception for each shoot for all of the different directions of the sky hemisphere by considering the possible shading of other shoots in those directions. The crown-level annual light interception is the sum of all shoot interceptions (see Perttunen et al. 1998 for a more detailed explanation). The carbon consumption due to maintenance respiration is subtracted from the photosynthetic production to yield the photosynthate available for growth. Maintenance respiration is the sum of maintenance respiration of all tree organs. Root growth, for the time being, is calculated using the functional balance principle (Valentine 1985).

LIGNUM can be used to evaluate how tree structure interacts with light availability to ensure the growth and survival of individual trees. In practice, LIGNUM is being used to better understand the effects of various silvicultural systems on the growth and survival of various tree species through the interacting effects of the changing light dynamics and tree size and architecture (Perttunen et al. 2001; Lo et al. 2001; Nikinmaa et al. 2003). So far, it has been calibrated for Scots pine (*Pinus sylvestris*), jack pine (*P. banksiana*), and sugar maple (*Acer saccharum*). Such a model could also be used by geneticists to evaluate how various morphological and architectural traits might affect the long-term growth of various species under various light conditions.

Stand-level models

Stand models have been traditionally divided into three broad types: the growth and yield historical bioassay models, the process-based models, and the more population-

and community-type models that often mix empirically derived relationships with the explicit simulation of some key processes. Growth and yield historical bioassay models are well known and used by foresters, while population/community and process-based models were developed and used mostly by forest ecologists and ecophysiologists. Within the scope of these various types of stand-level models, there are numerous models that have been proposed, developed, and applied to various degrees. Here only a few models are highlighted and the readers is referred to Oliver and Larson (1990) for more detailed discussion of the growth and yield models, to Landsberg and Gower (1997) for the process-based models, and to Botkin (1993), Kimmins (1997), and Shugart (1998) for the population/community models. Of course, there has been convergence among the various approaches over the years, so that boundaries among these three categories are getting blurred. This convergence is a good thing and one suggestion is to continue development of constructs that combine traditional (i.e., historical bioassay) and process modelling (Monserud 2002). Another approach is to link modular, sub-ecosystem component models within ecosystem level meta-models (Mäkelä 2001), and use relationships based on smaller scale (tree and stand) models to drive higher scale models (stand and landscape, respectively; Stage 2001). Furthermore, Occam's razor is still a fundamental tenet of model construction, and the combination of both empirical and process-based relationships in simulation modelling appears to be the most likely to satisfy its conditions: that a model should be as simple as possible, but as complex as necessary to achieve its desired forecasting capabilities.

Growth and yield historical bioassay models

Some growth and yield (G&Y) models are directed more at the stand level, predicting wood fibre yield at a given time, and so utilize average stand data such as stand volume per hectare. Some other G&Y models predict tree growth and use mean or individual tree data such as stem diameter and height. Several good examples of G&Y models have been developed recently (MGM, TASS, TIPSY, VDYP, STIM, SDMDs, Prognosis BC) by the British Columbia Ministry of Forests (see http://www.for.gov.bc.ca/research/gymodels). These models differ considerably in their complexity, assumptions, and data requirements.

One of the G&Y models most applicable to boreal forests is MGM, a "mixedwood growth model". MGM is a deterministic, non-spatial, individual tree-based stand growth simulator (Huang and Titus 1994). The model was calibrated for pure or mixed stands of white spruce (*Picea glauca*), trembling aspen (*Popoulus tremuloides*), lodgepole pine (*Pinus contorta*), and black spruce (*Picea mariana*) in Alberta, British Columbia, Saskatchewan, and Manitoba. MGM simulates tree growth and mortality relationships based on data derived from permanent sampling points. It then predicts stand growth according to the basic composition and density of the stands and the site index. It is a purely empirically derived model, since no attempts are made to simulate any given biological process. Outputs are stand volume per hectare that can be used directly by forest managers.

Physiologically based models

Numerous process models have been developed over the last 30 years. These models compartmentalize stand development into a set of processes that represent the relationship between growth and environmental conditions. Physiologically based models have the ability to forecast growth under a wide variety of future conditions, so long as the physiological processes that are expected to change are represented within the model. Two important limitations of these models are the degree of complexity that can be developed as more processes are included, and the lack of ability to forecast future growth under changing ecosystem conditions if all the key processes that may change are not included. There has been little use of such models in applied forestry decision support applications to date.

The strength of these models is in predicting stand physiological processes. In fact, some of the gap models (e.g., LINKAGES; Pastor and Post 1986) have incorporated so many physiological processes that they could also be classified as physiologically based as well. Other well known stand physiologically based models are MAESTRO (Wang and Jarvis 1990*a*), FOREST-BGC (Running and Gower 1991), and BIOMASS (McMurthie and Landsberg 1992). All of these models somehow simulate the carbon flow (entry and exit) within stands and uses various levels of detail and time scales to achieve this. A good review of some of these models can be found in the book by Landsberg and Gower (1997).

MAESTRO is probably one of the best known forest physiologically based models. It simulates net photosynthesis and transpiration at the leaf level within various sections of the crown of individual trees within a stand. It also incorporates many details about the structure and arrangement of the leaf and distribution of light within the crown; as such, it can be considered somewhat spatially explicit. MAESTRO has been used mainly as a research tool to investigate the interaction between the climate and various canopy exchanges processes (Wang and Jarvis 1990*b*). But, as stated by Landsberg and Gower (1997), it could also be used to evaluate the carbon gain consequences of various silvicultural treatments in structurally simple stands.

Population/community models

The so-called gap models — Several forest models have been developed to model forest stand succession at the stand level. These models are notable in that they were some of the first to deal with multiple species and natural forest regeneration and mortality, requiring remarkably little site and silvics data for calibration and model initiation. The first to be developed, and the model which has had the most impact, was JABOWA (Botkin et al. 1972). This model attempted to capture the gap dynamics of a forest by modelling the differential vertical use of light by trees growing within the zone of influence of, or the space occupied by, a single potential canopy tree, hence a "gap" model. JABOWA models forest dynamics in representative plots of a fixed size (e.g., 10 m × 10 m) using three processes: a deterministic logistic growth subroutine predicting annual growth increment of each tree; a stochastic birth subroutine which adds new saplings; and a stochastic death subroutine which randomly selects which trees will die.

Numerous other gap models were developed from JABOWA, focusing on different processes and evolving from being non-spatial to spatially explicit models. Some of the more widely used gap models include FORET (Shugart 1984), LINKAGES (Post and Pastor 1996), ZELIG (Smith and Urban 1988), and FORSKA (Prentice and Leemans 1990). The reader is directed to Shugart (1998) for a comprehensive review. The family of gap models originating from JABOWA has provided countless forestry applications (Botkin 1993). In recent years, several models and modelling approaches have been linked together to address various issues. For example, FORSKA is a gap model that simulates assimilation and respiration rates, with annual growth of each individual tree then calculated as a function of net assimilation in its vertical leaf area profile scaled by a species-specific growth rate factor (Leemans 1992). To evaluate the past and possible future impact of management on C sequestration in forest (see Chap. 20), FORSKA and carbon budget models have been also linked (Price et al. 1997). Finally, to evaluate alternative silvicultural regimes in the Pacific Northwest, ZELIG has been used with economic models and habitat suitability regressions to explore the impacts of alternative forest harvesting patterns on logging revenues and bird density (Hansen et al. 1995).

Competition-based models — In these models, representations of competitive interactions between trees are used to modify empirical, historical bioassay growth projections. Examples of such tools include the distance-dependent, individual tree and stand model, TASS (Mitchell 1975; Mitchell and Cameron 1985), the distance-independent stand model, PROGNOSIS (Wykoff et al. 1982), and in graphical form, the Stand Density Management Diagram Model (Woods 1999). These are mainly plant population models for even-aged monoculture tree crop management, although extensions of these models to include non-tree vegetation, uneven-aged stand structures, and mixed species stands are under development. A more recent and somewhat different model in this class is SORTIE — a distance dependent, individual tree model based on empirical relationships between light and radial growth, which simulates light competition (Pacala et al. 1993, 1996). Several of these models are being used to aid forest stand management.

Using an individual-based approach, SORTIE (Pacala et al. 1993) was designed as a spatially explicit model of the successional dynamics of mixed species forests where the x, y coordinates of each individual tree are modelled within a plot. The model was originally developed for mixed forests in eastern North America, and can be viewed as a spatially explicit and empirically calibrated version of the same processes simulated in gap models. One of the most distinctive features of SORTIE is that the model was designed to be calibrated and initiated using direct input of data from rigorous field studies. In 1995, scientists from the B.C. Forest Service in Smithers, British Columbia, began a collaboration with the original developers of SORTIE to calibrate the model for the conditions found in the interior cedar–hemlock forests of northwestern B.C. Significant changes have been made to the original SORTIE model to take into account unique features of western conifer-dominated forests, and unique challenges posed by the use of the model in testing and developing new silvicultural systems. SORTIE/BC is currently being calibrated and further modified for the mixedwood and black spruce boreal forests of Canada (SORTIE/Boreal; Coates et al. 2003).

SORTIE now consists of four submodels: (1) seedling recruitment — a function of parent tree proximity and seedbed (substrate) suitability, (2) resource availability — understory light dynamics as a function of species-specific light extinction coefficients, (3) subcanopy tree growth — a function of light availability, and (4) tree mortality — a function of recent growth rates. Additionally, the model has a very flexible user interface to allow incorporation of a wide range of partial cutting strategies (e.g., understory protection, diameter limit harvesting, shelterwood, patch retention, and clearcutting). In SORTIE/Boreal, work is under way to add an understory vegetation submodel and to improve our prediction of growth and mortality of mature canopy trees. These attributes will make the model ideal for examining stand dynamics and succession in structurally complex mixed or single species stands, and for exploring the implications of alternative stand management strategies. Current model predictions include: (1) spatial distribution and sizes of all individuals in a simulated stand; (2) DBH and height distributions, by species; (3) changes in basal area and density, by species, over time; (4) tables of basal area and densities of both adult and juvenile trees; and (5) distribution of subcanopy light levels.

Net primary productivity models — The FORCYTE–FORECAST model series use empirical historical bioassay data to derive estimates of the rates of key growth and ecosystem processes to calculate the net primary productivity of a forest. This is a "backcasting" approach to process modelling that roots process simulation firmly in the empirical field data. These rates, together with empirical data on processes that cannot be backcasted, are then used in the ecosystem simulation model. This approach was initially developed in the FORCYTE model series (Kimmins and Scoullar 1979; Kimmins 1985, 1988, 1990, 1993) and is now the basis for the ecosystem simulation model, FORECAST (Kimmins et al. 1999; Seely et al. 1999). A full account of FORECAST's internal structure, components, and processes is provided in Kimmins et al. (1999) and Seely et al. (1999). The overall approach was to develop a modelling framework rather than a single multipurpose model, a strategy consistent with suggestions for forest stand modelling by Stage (2001) and Mäkelä (2001). An important difference between this type of hybrid simulation modelling and the gap models is that the limitations on growth imposed by light and nutrients are simulated mechanistically (as in a process model) rather than using a growth modifier approach (see Kimmins et al. 1999 for further details). Moisture and temperature are represented implicitly in the present version of FORECAST, whereas they are modelled explicitly in the gap models.

Net primary productivity (NPP) in FORECAST is simulated as a function of foliage biomass and light interception. As part of an internal calibration procedure the model calculates a measure of foliage nitrogen efficiency (net primary production per unit of foliage nitrogen content) by comparing estimates of total biomass production with measures of foliage nitrogen and simulated levels of light interception within the canopy. This estimate of foliar nitrogen efficiency under varying light profiles provides the driving function for FORECAST (see Kimmins 1986; Kimmins et al. 1999). Foliar nitrogen content has been shown to provide a good measure of the photosynthetic capacity of foliage (Brix 1971; Ågren 1983a, b; Yarie 1997). Its application as a general model-driving function is also consistent with the suggestion of Landsberg (2001) that foliar nitrogen efficiency is a useful surrogate for the complex processes underlying primary production.

The FORECAST user has also the option to simulate plant growth using light competition alone, or with both light and available nutrients. When the nutrient option is added, potential growth derived from light-related NPP (as described above) is moderated according to whether expected nutrient demand for this new growth can be satisfied by the nutrient supply. Nutrient demand is calculated on the basis of this new production, and empirical data on nutrient concentrations in each biomass component. Nutrient supply takes account of the availability of nutrients within the plant from internal cycling, availability of nutrients in the soil (from mineralization of litter and humus, and other sources), nutrient uptake from the soil (a function of the proportion of soil volume occupied by fine roots, and of nutrients captured by competitor species), and in the case of nitrogen-fixing plants, biological N-fixation. Nutrient input from precipitation and throughfall can also be simulated if desired and calibration data are available.

Landscape-level models

Natural (insect epidemics, wildfires) and man-made (logging activities) disturbances operate at more than one spatial and temporal scale, generating a complex forested landscape mosaic and influencing forest regeneration. The only way to evaluate the long-term effects of different disturbance regimes on forest regeneration and dynamics at the landscape scale is by means of model simulations (Shugart 1998). Spatial modelling of landscape ecology draws on forestry and geographic information science. The advances in technical tools such as the computational capability and lower cost of new computers, geographic information system (GIS) software, and remote sensing provide the foundation for the spatial modelling approaches presented here. Landscape disturbance models have more and more applications, as reviewed in the recent book by Mladenoff and Baker (1999).

LANDIS (landscape disturbance and succession model)

LANDIS is a spatially explicit and stochastic model that simulates forest landscape change (disturbance and succession) over long time domains and large, heterogeneous landscapes (Mladenoff and He 2000). The aim of this model is not to predict timber volumes but to investigate the impacts of natural and man-induced disturbances at large spatial and temporal scales. The model is raster based, meaning that it operates on each equally sized grid cell or pixel that collectively describe a landscape. Model start-up requires GIS maps describing initial conditions of the forests (e.g., species presence/absence, stand age) and a synthetic land-type map that classifies the initial environmental conditions (e.g., soil type, topography, and moisture). While the initial forest maps of cohort age and species composition are updated throughout the simulation, the environmental maps remained static. The simulation time step is 10 years, and the cell resolution is given by the GIS data. LANDIS models spatially explicit disturbances such as fire, wind, and harvesting (Gustafson et al. 2000) as well as ecological processes such as seed dispersal. Other ecological processes such as succession and species ageing have only a temporal dimension. Landscape disturbance processes are stochastic. Species response to these spatial processes is modelled using a fixed number of species' vital attributes according to the different land-types such as species regeneration and longevity, seed dispersal (effective and maximum), shade tolerance, and fire tolerance. LANDIS produces time series output and maps of forest composition (presence/absence data) and age as well as time-since-disturbance maps.

SELES (spatial explicit landscape event simulator)

SELES (Fall and Fall 2001) is a declarative modelling language that can be used to model specified key processes at the landscape scale. It is a discrete-event simulation engine that interprets and executes such models. SELES language is sufficiently general to facilitate the construction of models and sub-models that are quasi-continuous, periodic, or episodic; that are deterministic, process-oriented, or stochastic; and in which processes may operate locally, regionally, or globally and be either spreading or non-spreading. The declarative nature of the language supports comparison and modification, allows rapid model prototyping, and provides a structured framework to guide model development. SELES is useful as a research tool as well as a decision-support tool for exploring problems related to conservation and resource management. SELES is raster based, like LANDIS, so the basic spatial unit is a pixel, a cell. Time in SELES is a continuous variable; the engine takes care of processing events in their chronological order. Thus, time steps may vary in size, and processes with different temporal resolutions can be integrated. Processes modelled using SELES have included forest succession and encroachment, natural disturbances such as fires, insect outbreaks and flooding, and management activities such as timber harvesting and livestock grazing.

All SELES models are spatially explicit, with the spatial plane consisting of one or more raster layers (variable maps) of an arbitrary, but common, extent or resolution. A SELES model consists of two components. First, a set of global variables (raster layers) together define the landscape state. These include layers or maps that may change during the simulation (e.g., forest age, species composition), and layers that are static (e.g., slope and elevation). The initial state for each layer can be defined by a digital map from a GIS, a classified digital image from remote sensing, or a synthetic map from one of the SELES static landscape models. Secondly, a set of landscape events defines the model behaviour. SELES models landscape events that can incorporate aspects of cellular automata, percolation models, discrete-event simulation, and spatio-temporal Markov chains. A landscape event is a generic framework for describing the characteristics of an agent of change, and forms a semi-independent sub-model that usually reflects a single key process (e.g., seed dispersal, fire, timber harvesting). A landscape event may specify global behaviour (e.g., the interval between fire on the landscape), local behaviour (e.g., stand dynamics), and a set of consequent effects or state changes.

FEENIX (a forest ecosystem emulator)

FEENIX (Cumming et al. 1998) is a spatial dynamic simulation model of the ecology and management of forest landscapes. As with LANDIS and models specified with SELES, FEENIX is raster based, stochastic, and usually initialized from GIS data. However, FEENIX is somewhat specialized as regards its target ecosystem and the processes it simulates. Although its precursors have been used to model both terrestrial and marine ecosystems, FEENIX is designed for boreal forests, especially the mixedwood forests of western Canada. Its principal ecological submodels of wildfire, stand dynamics, and forest bird distributions are intended to utilize the empirical research of the Sustainable Forest Management Network (SFMN) Boreal Ecology and Economics Synthesis Team. For example, the spatial resolution of 3 ha was chosen to match the effective sampling radius of annual forest bird point count surveys, and the basic simulation time step is 1 year. The forest management submodel belongs to the class of deterministic harvest schedulers described below, with some relatively novel capabilities related to current developments in mixedwood forest management (see Chap 13). These include internal calculation of periodic cut levels, a very general blocking algorithm, a range of spatial sequencing options, and a limited ability to optimize harvest sequencing through time. As FEENIX is reasonably fast (typical execution times for a 300 000 ha landscape are roughly 1 s per model time-step), it can be used for interactive scenario analysis and in workshop settings. A graphical user interface controls the display of dynamic map layers and data graphs, and the settings of key parameters and simulation options. FEENIX can also perform replicated multifactorial simulation experiments automatically. An enhanced random number library is provided to ensure statistical independence of model runs.

TARDIS

TARDIS (Cumming and Armstrong 2001; Cumming and Vernier 2002) is a low spatial-resolution simulation model of regional forest dynamics. It represents forest regions as a regular grid of loosely interacting cells or landscapes, connected by a dynamic network of primary roads. This grid would typically be defined by a standard mapping system, such as 10 km by 10 km Universal Transverse Mercator (UTM) grid, or the township land survey grid used in much of western Canada. Thus, model cells would represent areas of roughly 10 000 ha. TARDIS was prototyped on a 74 000 km^2 study region of 825 cells with mean size 9000 ha, but can be applied to much larger regions. For each cell, TARDIS maintains a list of polygons representing mapped forest stands and other planar features, and static attributes such as mean elevation and interpolated climate normals. TARDIS would typically be initialized from the polygon attribute tables of a GIS coverage of 1:20 000 forest inventory maps. The underlying topology, or spatial configuration of polygons within cells, is not maintained. To model ecological processes that are sensitive to landscape configuration, TARDIS relies on statistical models (Cumming and Vernier 2002). This abstraction markedly reduces computational complexity. As one result, Monte-Carlo simulation experiments require only a few hours of computer time. As a scenario analysis tool, TARDIS is designed to evaluate strategic or policy alternatives in forest management under the risk of large fires. As a research tool, TARDIS is used to evaluate and apply methods for scaling some capabilities of high-resolution landscape models up to regional extents.

Harvest scheduling models

Harvest scheduling models are used to forecast timber supply (harvest flows, growing stock, revenues, and costs) according to inventories, growth and yield data, plus management assumptions related to economic, environmental, and social objectives. Harvest forecasts are commonly made for several rotations at which time the harvests and growing stock usually stabilize. Linear programming (LP) models and deterministic simulation models were first applied to harvest scheduling in the 1960s and they are still in use. As the need for spatial resolution in harvest scheduling became more pressing, new models capable of handling the integer and mixed-integer problems associated with spatial planning evolved. These models fall into three board categories: (1) deterministic simulation, (2) metaheuristics, and (3) hierarchical models.

Deterministic simulation models usually operate on time steps of 5 or 10 years, although this can vary depending on the level of detail desired. They are deterministic in that all inputs and events are assumed to be known with certainty. Sensitivity analysis is used to explore these assumptions. The objective function is usually to maximize harvest volume or net present value of the harvest. Periodic harvest flow constraints, such as an even flow of timber, are added to smooth harvests over time. The simulation model first ages the forest one time period, then sorts eligible stands according to a harvest priority (e.g., oldest first, closest to mill, or to minimize mortality) and finally proceeds to harvest stands. As each stand is harvested, the periodic harvest volume is incremented and the queue of eligible stands is updated. Periodic harvesting stops when the volume target is reached or no more eligible stands exist, either due to constraints or a lack of stands at or above the minimum harvest age. The forest is then aged another time period, and the process is repeated. This continues until the desired number of periods has been modelled. Simulation models require frequent intervention from the analyst to adjust harvest targets, constraints, and management assumptions until the desired outcome is achieved. The main advantages of deterministic simulation models are: (1) they are simple to use and understand, (2) they are able to handle large problems and carry many stand attributes, and (3) they have fast run times. Examples of time-step, deterministic models are FPS-ATLAS (Nelson 2000) developed at the University of British Columbia, FORMAN2000 (Jordan and Wightman 1993) developed at the University of New Brunswick, and COMPLAN (ORM 2002) developed in the private sector.

The meta-heuristic models use algorithms such as simulated annealing, tabu search, and genetic algorithms (see Davis et al. 2001 for more information about these algorithms) to schedule harvests. Lockwood and Moore (1993) pioneered this field with the introduction of simulated annealing to harvest scheduling. Because there is no direct solution technique for optimally solving very large spatial problems, all these algorithms use random moves and probabilistic acceptance/rejection criteria. Inputs and events are still considered deterministic, only the solution technique is random. In contrast to time-step simulation models, the meta-heuristics work with the entire solution (all time periods) simultaneously. They begin with an initial random solution. A common move entails randomly swapping the harvest of a stand in one period with a stand in another time period. If the move improves the objective function, then the move is accepted; otherwise it is rejected according to a probability function based on the number of moves already made. In the early stages, there is a high probability that an inferior move is accepted, but in the late stages this probability approaches zero. The purpose of accepting inferior moves is to avoid local optima.

The meta-heuristic models offer several advantages. First, any mathematical problem, regardless of its structure and characteristics, can be formulated and solved with these methods. Second, multiple objectives are routinely incorporated into these models. For example, targets can be set for timber flows, seral stages, patch-size distributions, visual objectives, and so on. The objective function of the algorithm then tries to minimize deviations from each specified target by using the random moves described above. Third, these models produce very good solutions that are superior to time-step simulation models. The disadvantages of meta-heuristics are their complexity, the need for extensive parameter tuning, difficulty in replicating solutions and longer solution times relative to simulation. Extensive sensitivity analysis is needed to explore model parameters plus different penalty functions and goal weights for each forest modelled. Examples of meta-heuristics that use simulated annealing are the ECHO model (MMF 2002), the Forest Simulation and Optimization System (FSOS), and the Swedish core area model by Ohman (2000). For further reading on simulated annealing we recommend van Deusen (1999) and Nelson (2001). Richards and Gunn (2000) and Sessions and Bettinger (2001) provide further information on the formulation and application of tabu search.

Hierarchical models typically combine a non-spatial LP model at the strategic level with a spatial model at the tactical level. The Woodstock (LP) and Stanley (spatial) models (Feunekes and Cogswell 1997) fit into this category and are widely used on very large forest estates. The Woodstock model operates without spatial detail and then passes its generalized solution to Stanley where it is mapped into harvest units according to standard constraints like opening size, seral stage targets, and so on. The main advantages of hierarchical systems are simplicity, speed, and minimal spatial data. This makes it possible to model very large forest estates quickly with minimal computing effort and spatial detail. The disadvantage is that the spatial model is usually only run for a period of less than one rotation, which makes it difficult to assess changes in forest structure and pattern that may only become evident in the long-term.

All the spatial harvest scheduling models tend to use vector polygons rather than raster mode to minimize the number of stands and therefore the number of decision variables. This does not exclude the use of raster approaches, however, but for the above reason they are rarely used. The harvest scheduling models all rely on growth and yield data provided by stand-level models. This information is stored as volume/age libraries for each stand type and stand volumes are retrieved when a stand is scheduled for harvest. In addition to timber supply statistics, harvest scheduling models also produce projected inventory maps that can be used by other models to assess habitat, hydrology, and visual impacts. Random events to simulate natural disturbances can also be included in the harvest scheduling models. These can be relatively simple routines within the model itself, or results can be imported from a more complex disturbance model.

Some examples of approaches and applications

This section does not attempt to be complete in dealing with all possible topics concerning models of the boreal forest, but merely reports on the status of current research and some examples of current approaches trying to link various models or approaches to help in decision making for sustainable forest management. Emphasis is directed toward the management implications of various simulation models or approaches. Scaling up models and processes to the landscape to understand the effects of the various processes acting on different scales is becoming important.

Quebec's "Integration" project

Introduction

Forest stakeholders are demanding that forest management guarantee the conservation of biodiversity and ecosystem productivity while maintaining resources, including the production of wood at a reasonable cost, for all users (Chap. 1). To achieve sustainable forest management, key criteria and indicators need to be identified that take into account the social and economic, as well as the ecological, context of forestry in order to develop practices that will ensure viable forests for future generations (Yamasaki et al.)[2]. To facilitate such a shift in the forest management paradigm and decision making for sustainable forest management, a conceptual framework needs to be developed for integrating sustainability indicators and simulation tools predicting the state of these key indicators after different management scenarios over large regions and long periods of time. This is the primary goal of Quebec's so-called "Integration" project.

The Integration project aims to develop a conceptual framework and decision tools facilitating the understanding of the ecological impacts of management decisions for all the participants involved in forest management (ecologists, public, stakeholders). This is achieved through the development of local- and landscape-level indicators integrating scientific knowledge and expertise from a team of scientists from various fields working closely together with various stakeholders (Kneeshaw et al. 2000; Côté et al. 2001). Specifically, the steps involved in this project are to: (1) develop and integrate indicators of sustainability at an operational level, (2) answer the scientific needs of the forest products industry and forest communities with respect to better understanding and evaluation of the forest resource, (3) develop simulation tools to model the spatial forest dynamics according to various harvesting management scenarios, and (4) test the validity of these decision-making tools for the Mauricie region of Quebec (located between $47°57'-49°08'$ N and $74°52'-73°45'$ W). Financial support from the forest products industry, its involvement in data acquisition, and its combined involvement with government agencies in consultation with the Integration research group ensured the development of operationally useful indicators and decision tools.

---

[2]Yamasaki, S.H., Kneeshaw, D.D., Bouthillier, L., Fortin, M.-J., Fall, A., Messier, C., and Leduc, A. 2002. Integration of indicators, landscape modeling, and public participation for sustainable forest management. Unpublished manuscript submitted to *Conservation Ecology*.

Biophysical criteria

The first phase of this project consisted of the development of biophysical criteria and indicators for sustainable forest management. Based on research conducted in several SFMN projects in Abitibi and Mauricie regions, Kneeshaw et al. (2000) and Yamasaki et al. (see footnote 2) developed sustainability indicators according to five criteria: biodiversity, aquatic environments, soils, forest productivity, and socio-economics.

For illustration purposes, we will present here only the biodiversity indicators aiming to maintain forest ecosystem integrity (see Kneeshaw et al. 2000 for more details about the other indicators). The conceptual framework used to develop the biodiversity indicators was the coarse-filter approach aiming to monitor indicators that favour the maintenance of forest ecosystem diversity. Specifically, we focus on indicators related to the maintenance of the diversity of forest habitat types required by forest species throughout succession. Hence, as proposed by Bergeron et al. (1999), age-class structure of the forest is used as a forest biodiversity indicator. Indeed, natural forest age-class structure issued from the regional wildfire regime can be used as a target forest mosaic composition at the landscape scale (Bergeron et al. 1999), rather than imposing strict composition objectives on each stand. This landscape-level management approach is more appropriate in the boreal forest where large fires continue to occur and where the synergistic impact of fire and logging need to be integrated (Chap. 9). Hence, forest planning over a larger region makes it possible to maintain a targeted forest age-class structure. To do so, alternative silvicultural methods need to be promoted. Several harvesting strategies were implemented into a spatially explicit model: "SELES/Boreal" (status quo), "Bergeron", "Burton", and "triad"; here we present the results of three of them. The first harvesting strategy is the industrial status quo that is driven by a target annual percentage of landscape to harvest (here 0.65% of the productive area) and where the minimum harvest age is 100 years. We chose a harvest level of 0.65% per year, resulting in a rotation of about 150 years, as this is approximately the level suggested by Bergeron et al. (1999) and Burton et al. (1999) for areas with a fire cycle of 100 years and where no partial logging is done. Investigation of the impact of varying this harvest level in synergy with various fire regimes is still in progress (but see Fortin et al. 2002). Then the two other harvesting scenarios, "Bergeron" and "Burton", both specify target age-class structures that attempt to balance forestry and biodiversity objectives (see Bergeron et al. 1999; Burton et al. 1999). The targeted forest age-class distributions are illustrated in Fig. 14.5, where the percentage (%) indicates the proportion of the forest in given age-classes according to each harvesting strategy. Both alternatives to the status quo require that 35% of the land area be managed under rotation lengths longer than the current industrial norm, but only the Bergeron scenario explicitly preserves very old forests (up to 300 years).

Simulation tool

These three harvesting strategies were implemented into a spatially explicit model of boreal forest dynamics at the landscape scale that we developed using the simulation language SELES (Fall and Fall 2001, as described above). SELES is a raster-based simulation tool that can model the interactive effects of several processes and events, such

Fig. 14.5. Target forest age-class structures under three alternative harvesting scenarios.

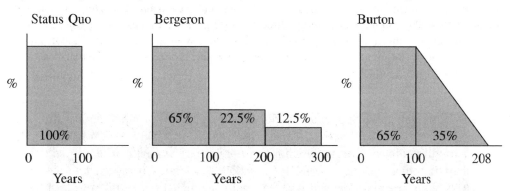

as natural disturbances (fires of various size, shape, and severity) and the proposed harvesting strategies (Status Quo, Bergeron, and Burton) that affect both stand structure and composition at the landscape level. Given that forest spatial configuration influences the behaviour of certain ecological phenomena (such as wildfire and regeneration by seed) and is crucial for the evaluation of certain biodiversity or sustainability indicators (forest fragmentation, for example), the model must be spatially explicit.

Results

For the present study, Integration researchers have tested these three scenarios with and without wildfire (Fig. 14.6) where each scenario (i.e., various combinations of harvesting strategy and fire regime) was run 30 times over a period of 500 simulated years. Results indicate that in the absence of fire, both mature (100–200 years old) and old (>200 years old) forests are well represented over the entire landscape. As soon as the effects of fire are added, however, the old forests virtually disappeared with the Status Quo and Burton harvesting strategies, but some proportion is maintained under the Bergeron harvesting strategy.

This simple example illustrates well that at the theoretical harvest level tested here (i.e., 0.65% per year, which is lower than what is allocated in reality), the Status Quo harvesting strategy puts a high risk to growing stock and future harvest levels when the synergistic effects of fire events are taken into account (Fig. 14.6). Sustainable forest management planning should therefore favour alternative harvesting types (such as proposed by Bergeron et al. 1999) that incorporate some flexibility to absorb the effects of natural disturbances, here fire, on forest age structure.

Future development

The landscape model currently produces outputs on the forest stands (composition and age) and disturbances (size, shape, severity of natural disturbances, time since fire, time since logging). The Integration project team is using this spatially explicit model to investigate the impacts of modifying the disturbance regimes and harvesting scenarios. In the next phase of this project, insect outbreaks will be added as another natural disturbance. Furthermore, wildlife indicators aiming to evaluate forest connectivity at the

Fig. 14.6. Age-class (0–100, 101–200, and >200 year) maps after 100 years of simulation and average total area (in 100 000 ha) with standard deviation of each age class after 500 years of simulation based on 30 replicates for the three harvesting strategies: (*a*) "Status Quo", (*b*) "Status Quo" with fire, (*c*) "Bergeron", (*d*) "Bergeron" with fire, (*e*) "Burton", and (*f*) "Burton" with fire. Black areas indicate bodies of water; progressively lighter shades indicate progressively older stands.

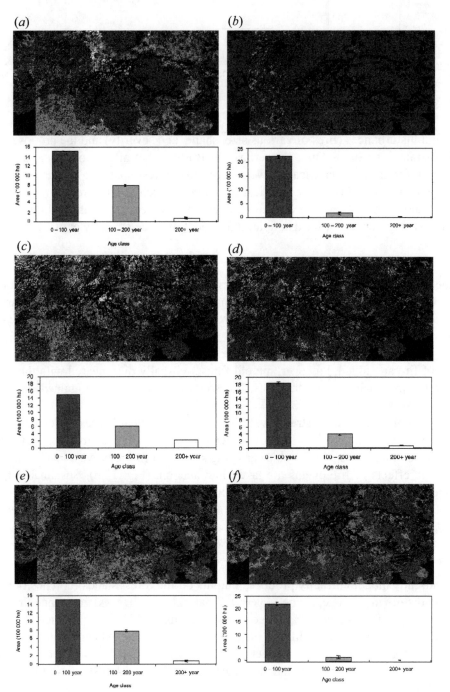

stand, landscape, and watershed levels will be implemented. Finally, to improve our ability to investigate tree succession and stand volume, a formal linkage with the stand-level model SORTIE (described above) will be made. With these new features, the project team is hoping to investigate the relationships between processes governing forest regeneration to: (1) understand the feedback mechanisms between forest dynamics and disturbances, given time-lagged responses from one spatio-temporal scale to another (stand to landscape); (2) investigate the consequences of changes in the fire regime and logging rotation; and (3) identify critical thresholds and (or) combinations of disturbances that are more likely to affect forest sustainability, according to key indicators.

These investigations will generate several outputs and the next methodological challenge will be to compare and identify significant differences among the results from the various scenarios studied. This implies three types of analyses: aspatial (comparison of stand frequency distributions), spatial (comparison of stand temporal and spatial patterns), and temporal (comparison of stand temporal patterns). First, the simulated stand age and composition distributions will be compared to the observed data using goodness-of-fit analyses. Second, cohesive spatial pattern of forest structure and disturbance will be tested using existing spatial statistics (Fortin 1999) and landscape indices (Gustafson 1998). New spatio-temporal statistics will also be developed to test whether or not patterns are cohesive in space through time (Dubé et al. 2001).

The results from these simulations and their spatial and ecological interpretations will then be used in the final phase of this project, which is the development of a social learning process and tools, integrating the interests of different stakeholders, for use in forest management decision making. The ecological indicators need to be interpretable by stakeholders, and they need to address their socio-economic concerns. The final selection of indicators must therefore be made as part of the process of deliberation among the different participants involved in the sustainable management of the forest, as they need to acknowledge and understand the long-term implications of management plans. Based on the results of a pilot study involving a round-table in the Mauricie, the Integration team is presently developing socio-economic indicators and technology transfer methods that will help participants understand the links between biophysical and socio-economic indicators (Côté et al. 2001; Yamasaki et al. [see footnote 2]).

The Biodiversity Assessment Project (BAP)

Introduction

In adopting the principles of SFM, forest administrators are committed to maintaining sustainable ecosystems while using the forest for human profit (Grumbine 1994). Biodiversity conservation in managed forests has been recognized by scientists as a cornerstone for ensuring forest sustainability (Gustafsson and Weslien 1999). Forest managers are thus in need of strategic planning analytical tools for assessing alternative management strategies in terms of biodiversity values. The Biodiversity Assessment Project (BAP) has addressed this problem and has been applied to a publicly owned forest managed by Millar Western Forest Products (MWFP) in Alberta (Duinker et al. 2000).

The objectives of this project were to:

(1) Develop models of bio-indicators appropriate to central-west Alberta conditions for evaluating landscape patterns, distribution of ecosystems, and distribution of quality habitats of selected vertebrate species;

(2) Analyze and compare bio-indicator performance over the long term according to forest projections obtained under alternative forest management scenarios and a natural disturbance regime (NDR) scenario; and

(3) According to the results, design promising new management strategies and advise on the implementation of a biodiversity monitoring plan.

BAP structure

BAP has been conceived to be embedded in the adaptive management loop (Duinker and Baskerville 1986; Walters 1986; Chap. 21). Specifically, it addresses the "forecasting" phase, where potential responses of the forest to proposed management actions are simulated using projection tools, and relevant indicator models are applied to the projections to track changes in abundance or quality of a valuable resource. The analysis of the indicator model outcomes leads to a reformulation and retesting of the management strategies until an acceptable management strategy is achieved.

The indicators

As biodiversity covers multiple aspects, BAP researchers developed indicator models for landscape patterns, ecosystem diversity, and habitat supply of some vertebrate species (Duinker et al. 2000; Doyon and Duinker 2000; Doyon et al. 2000; Table 14.2). Each of these three levels of biodiversity forms an independent analytical module of a

Table 14.2. Indicator models used in the Biodiversity Assessement Project.

Landscape pattern	Ecosystem diversity	Wildlife habitat supply
Patch area	Area-weighted mean age	Birds
Patch shape	Tree species dominance	Barred Owl
Mean edge contrast index	Tree species distribution	Brown Creeper
Contrast-weighted edge length	Habitat type proportion	Least Flycatcher
Core area proportion	Habitat diversity	Northern Goshawk
Length of different adjacency		Pileated Woodpecker
Mean nearest-neighbour distance		Ruffed Grouse
Contagion		Spruce Grouse
		Three-Toed Woodpecker
		Varied Thrush
		Mammals
		American Marten
		Elk; Moose; Canada Lynx
		Northern Flying Squirrel
		Snowshoe Hare
		Southern Red-Backed Vole
		Woodland Cariboo

suite of bio-indicator models. Such a strategy was inspired by the coarse- and fine-filter approach in conservation biology (Hunter 1990) where landscape pattern and ecosystem diversity indicators serve as coarse filters while the habitat supply models (HSMs) serve as fine filters.

The forest projections

Forest management scenarios were defined to explore contrasting conditions resulting from different guidelines in terms of silviculture and clearcut spatial layout (Fig. 14.7). GIS-Complan (ORM 2002) and Woodstock/Stanley (described above), two spatially explicit harvest schedulers, were used to project the long-term sustainability of various cut levels and the spatio-temporal allocation of harvest and silvicultural treatments according to the forest management rules specified for each scenario over 200 years. After exploring the results of a first round of scenarios (BAU, ASP, I2P, and ETP), three new scenarios (CST, BSI, and LFC) were developed to reflect more balanced strategies and were analysed in a second round (Fig. 14.7).

The BAP team recognizes that forests are naturally dynamic and that impacts of human interventions should be assessed in the context of the range of variation of the biodiversity indicators under the natural disturbance regime (NDR). They used LAN-DIS (He and Mladenoff 1999; see description above), a spatially explicit landscape simulator, to explore the baseline NDR scenario and generate the envelope of natural variation for each bio-indicator. LANDIS was parameterized specifically to reflect the regional vegetation-fire regime dynamics of the MWFP forest (Doyon 2000).

Fig. 14.7. Forest scenarios compared in the first round (BAU, ASP, I2P, ETP) of the Biodiversity Assessment Project (BAP) simulations were organized according to a 2 × 2 (silviculture by spatial layout) factorial design resulting in four combinations. A second round of comparisons involved more realistic scenarios (BSI, LFC, CST).

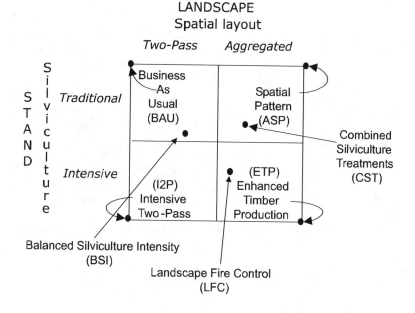

The analysis

The BAP research group used the "red flag" technique to compare outputs from the different forest management scenarios. This technique scrutinizes the behaviour of each bio-indicator one by one, all along the simulation horizon, in each scenario, and determines whether the values fall inside the envelope of natural variation. The envelope of natural variation is defined by the 95% confidence interval of the indicator range of variation in the NDR scenario. Each time a bio-indicator shows a value outside the envelope of natural variation, a red flag is raised. The scenario which raises the most red flags for a given bio-indicator requires the most attention in impact mitigation.

Results

Example of landscape pattern indicator — All scenarios, except the business-as-usual, produce a less contrasted forest than wildfire does (Fig. 14.8). Burnt patch borders were often highly complex and green patches remaining in the burnt areas increase the abundance of highly contrasted edges. Moreover, cutblocks tend to be larger, on average, than burned areas. The five scenarios with some silvicultural intensification consistently have lower contrast-weighted edge length values than the two using the traditional silviculture strategy. Silviculture intensification tends to set apart forest sections maintained under short rotation from other sections under long rotation regimes, resulting in lower edge contrasts. This result has important implications in terms of edge dynamics and fragmentation (Chap. 12).

Fig. 14.8. Contrast-weighted stand edge length in the Millar Western Forest Product forest management area under seven forest management scenarios simulated for 200 years. Horizontal dotted lines define the 95% confidence interval of the indicator under the Natural Disturbance Regime scenario. Scenarios are described in Fig. 14.7.

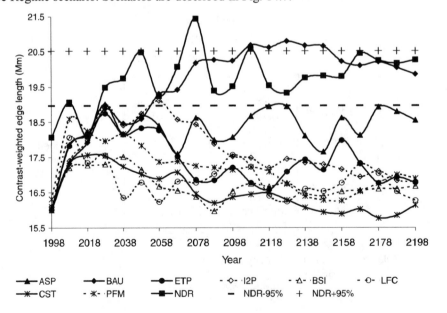

Example of ecosystem indicator — Intensifying the silviculture led to an increase in the area-weighted age beyond what is naturally observed according to the NDR simulation (Fig. 14.9). Such a counter-intuitive result can be explained by the even-flow constraint that was applied in the timber supply analysis. In fact, this is the "triad" effect, where silvicultural intensification allows one to focus timber production on more productive sites, relaxing the pressure on extensively managed stands that can attain ages older than otherwise observed.

Example of wildlife habitat supply indicator: the Least Flycatcher — Forest management scenarios did not differ much in terms of foraging habitat for the Least Flycatcher (*Empidonax minimus*; Fig. 14.10). Whatever the scenario, there is an overall reduction in habitat suitability values right at the start of the simulation, which then proceed to level off after 40 years. Least Flycatcher foraging habitat is related to the hardwood component of the forest and the BAU scenario is the one that preserves the most of it. Consequently, this scenario performs the best for most of the simulation horizon. "Hot spots" of good Least Flycatcher foraging habitat are detectable in the southeast of MWFP's forest management area under the CST scenario (Fig. 14.11).

Usefulness of the approach

The approach and tools developed in BAP have been fully integrated with the planning of Millar Western Forest Plan's forest management area. Analysis of the outputs allowed the BAP team to define a preferred forest management scenario with realistic objectives. Indeed, by exploring the extreme of the forest management decision space, trade-off functions were detected. For example, it appeared clear that intensive silvicul-

Fig. 14.9. Area-weighted age of Millar Western Forest Product forest management area under seven forest management scenarios simulated for 200 years. Horizontal dotted lines define the 95% confidence interval of the indicator under the Natural Disturbance Regime scenario. Scenarios are described in Fig. 14.7.

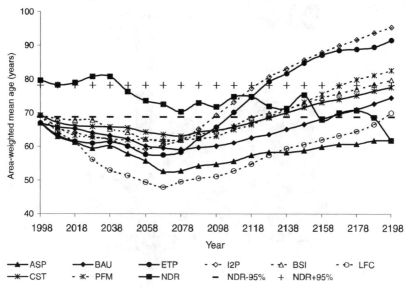

Fig. 14.10. Overall mean of the Least Flycatcher foraging suitability index values of Millar Western Forest Product forest management area under six forest management scenarios simulated for 200 years. Scenarios are described in Fig. 14.7.

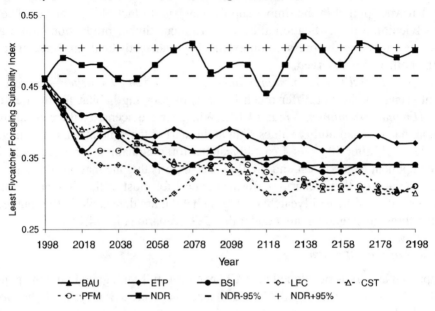

Fig. 14.11. Map of the mean by pixel of Least Flycatcher foraging suitability index values for a simulation of 200 years of the CST (combined silviculture treatment) forest management scenario.

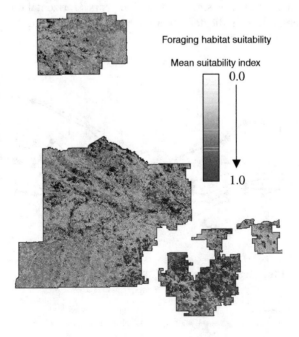

ture, with its multiple-entry practices, reduces the downed woody debris beyond a critical threshold. Species depending on this resource (e.g., marten [*Martes Americana*], northern flying squirrel [*Glaucomys sabrinus*], southern red-backed vole [*Clethrionomys gapperi*], and spruce grouse [*Dendragapus canadensis*]) would clearly experience a reduction in the quality of their habitat under intensive silviculture scenarios. Such knowledge has been useful to define an acceptable level of silvicultural intensity. Detecting critical forest conditions with biodiversity indicators has also served to design so-called "Best Management Practices" (BMPs). Biodiversity-sensitive practices like snag retention, green-patch retention, and alternative timber harvesting systems will be implemented according to defined conditions in the next management cycle. Also, as required in any SFM system, a monitoring program had to be defined by MWFP for the next management cycle. BAP analysis of indicator model outcomes has helped in selecting appropriate indicators deemed critical for maintaining biological values and in designing data collection protocols. Finally, BAP has served as a tool for engaging the public in the decision-making process (see further discussion in Chap. 21).

The Boreal Ecology and Economics Synthesis Team (BEEST)

Forest managers in Alberta recognize that the principles of adaptive management can help define effective mechanisms for achieving sustainable forest management. This recognition is expressed in numerous government documents, and is reflected in the management plans of many forestry companies. Implementation of adaptive management requires that existing interdisciplinary experience and scientific knowledge be integrated into predictive models, in order to evaluate the potential outcomes of alternative policies (Walters 1997; Chap. 21).

The Boreal Ecology and Economics Synthesis Team (BEEST) was established through the SFMN in 1997. It is an integrated, multi-disciplinary research group, with strengths in forest ecology, forest and resource economics, conservation biology, forest management, and simulation- and optimization-based modelling approaches. Team members are based at the Universities of Alberta and British Columbia. Representatives from government departments and forest products companies are active participants. Their objective is to develop an integrated suite of models of natural forest dynamics and human activities, and apply these models to evaluate management scenarios in terms of ecological and socio-economic indicators, in an adaptive management framework. Their research focuses on three broad areas:

(1) Improved understanding of the relationships between natural processes and human activities;

(2) Development of mathematics and simulation tools incorporating these interactions; and

(3) Analysis of policy alternatives.

Their modelling efforts seek to bridge multiple spatial scales, from stands to landscapes to forest regions.

Spatial-dynamic models can be very effective tools for scenario evaluation and policy analysis (e.g., Walters 1997). They can also help to generate and test hypotheses about complex landscape dynamics that are otherwise hard to visualize or think about.

However, they are most useful when at least some model components are well parameterized. Researchers have found that many of the basic relationships that govern interactions between natural processes and human activities in the boreal forest are poorly understood, at least for the purposes of predictive modelling. Therefore, the BEEST group has invested considerable effort in establishing the empirical foundation of their program. In particular, it has conducted basic research in the areas of fire ecology and dynamics (Armstrong 1999; Cumming 2000; 2001*a*, *b*) and in modelling wildlife response to changes in forest composition and structure (e.g., Schmiegelow et al. 1999; Cumming and Schmiegelow 2001; Vernier et al. 2001).

Modelling approach

The BEEST project team has developed two modelling platforms, FEENIX and TARDIS, designed for simulating wildfire, stand dynamics, forest harvesting, and wildlife habitat and population dynamics in boreal forests. FEENIX (see Cumming et al. 1998, and the description above) operates at a relatively fine resolution (3 ha) and at spatial extents of up to 500 000 ha. It can deal with patch-level questions such as fire ignition and spread, the dispersal of seeds or individual birds and mammals, landscape pattern, and some tactical aspects of timber harvest such as cutblock layout and spatial operability constraints. TARDIS (Cumming and Armstrong 1999, 2001; see description above) works at a much lower spatial resolution (100 km^2), relying on the aggregate statistical properties of compartment-level forest inventory data. However, it can simulate entire ecological regions or large forest estates. As such, TARDIS is intended primarily as a strategic level, policy analysis tool. These two modelling frameworks are not linked explicitly; rather statistical or process models, appropriate to their spatial scale and resolution, are developed from the same underlying data sets.

Fire — Fire is modelled as a three-stage stochastic process. The first stage models the number of detected fires (or fire arrivals) per unit area and time as a Poisson process, influenced by fuel type or stand composition. The second stage models fire "escape", or the probability that a given fire will grow larger than some small threshold size, which is related to fire suppression effort and effectiveness. The third stage is fire growth (modelled in FEENIX) or final size (modelled in TARDIS), and determines how big a fire will become, and what it will burn. All three stages depend on local fuel type, and on landscape context. Data used to generate these models include provincial fire history records (fire location, size, and cause), low-resolution forest inventories, and mapped fires linked to forest inventory maps. The resultant integrated fire models simulate patterns of burning that vary in time and space, and that can respond to forest management.

Stand dynamics — Forest stand development can be either deterministic or stochastic. Deterministic models are based on growth and yield tables and prescribed silvicultural treatments. BEEST researchers have also developed a spatially explicit stochastic model of mixedwood stand dynamics (Cumming et al. 1999). This model was designed to represent the complex dynamics of aspen and white spruce using as few parameters as possible, while respecting competing hypotheses regarding the mechanism and timing of white spruce recruitment (Chaps. 8 and 13). After fire or harvest, stands follow a volume-based growth trajectory defined by the initial stocking of spruce and an intrinsic rate of growth, which can vary with site index. Seed production (with masting), seed dispersal, regeneration substrate, and recruitment submodels determine the initial stocking of disturbed stands, and any later recruitment into mature stands. Parameter estimates for the stochastic model are based on the literature, expert opinion, and simulation experiments.

Forest harvesting — Timber harvest scheduling can be simulated based on a number of economic and operational criteria. Periodic volume requirements may be either given (e.g., specified annual allowable cuts) or computed from the models' forest state and yield curves, using (for example) the Hanzlik formula (Davis et al. 2001). The latter option allows the BEEST to simulate dynamic replanning, as after a large fire.

In TARDIS, a multi-pass compartment-level scheduler attempts to minimize periodic delivered wood costs to one or more mills, while respecting various operational and spatial constraints. TARDIS can accommodate multiple land-bases and forest tenures. FEENIX has similar capabilities, but can also simulate a range of harvest block layout and cut sequencing options, and designs optimal road networks that minimize haul costs. As well, FEENIX can read and execute exogenous harvest schedules generated by TARDIS, or by optimizing tools such as Woodstock/Stanley. Thus, it can evaluate the effects of real schedules on ecological indicators, and assess their sensitivity to fire. Data requirements for both models include spatially organized inventories, yield curves, operability constraints, and estimated costs of road building, harvesting (tree-to-truck), silviculture, and transportation, which determine delivered wood costs.

Wildlife response — BEEST has used forest birds as a focal species group for evaluating aspects of the ecological consequences of alternative management scenarios, though the modelling methods it has developed are generic and could be applied to other groups of species. Their focus on forest birds results from the availability of high quality data, and known sensitivities of this group to forest condition. They model bird species distribution and abundance at several spatial scales, commensurate with their landscape simulators, using bird survey and forest inventory data. Fine-scale models are based on patterns of bird abundance from long-term studies near Calling Lake, Alberta (e.g., Schmiegelow et al. 1997), modelled using local (3 ha) and neighbourhood-scale (81 ha) habitat variables that characterize forest structure and composition, including anthropogenic disturbances (i.e., these models are parameterized with robust empirical data from the region to which they are applied). In FEENIX, these models predict the expected abundances of a species at a 3 ha patch, assuming standardized sampling. The coarse-scale models used in TARDIS are based on statistical models of presence/absence data from the Alberta Breeding Bird Atlas (Semenchuk 1992), using aggregate habitat measures for 100 km^2 landscapes, derived from forest inventory data (Cumming and Schmiegelow 2001). Data requirements include distribution and abundance data for focal species or species groups, and appropriately scaled forest inventory information.

Model applications

BEEST landscape simulation tools are designed to evaluate, and to facilitate decisions related to, many issues associated with SFM. In particular, its work is focussed on policies regarding emulation of the natural range of variability as a management guideline (Chap. 9), the economic and ecological costs and benefits of strategies to reduce losses to large fires (e.g., fire suppression, landscape fuels management, and salvage logging policies), the costs and benefits of alternative landbase design strategies (e.g., overlapping forest tenure arrangements and elements of reserve design; e.g., Box 6.4), and analysis of incentives and regulations to control the impact of the energy sector on the forest. Here, we present two case studies of applications.

Fire, harvesting and forest birds: an evaluation of the natural disturbance paradigm — Forest companies are increasingly required to develop management plans that provide for the maintenance of non-timber values (Chap. 1). In recognition of the complexity of managing for single species, many forest products companies are adopting a landscape approach to biodiversity management, based on characterization and emulation of the natural disturbance regime for their management areas. The underlying assumption is that by maintaining indices of forest structure (stand ages and tree species compositions) within the range of natural variability, habitat values and hence biological diversity, will be conserved (Chaps. 1 and 9). Beyond evaluating the tradeoffs associated with achieving and maintaining target forest structures, there is also a need to understand how spatial aspects of habitat distribution, such as patch size and juxtaposition, affect suitability, and how animals respond to the habitat mosaics created by forest harvesting. Furthermore, it is necessary to identify measurable indicators to check whether the biodiversity objectives encapsulated in a landscape management approach are, in fact, being realized.

BEEST used FEENIX to model the effect of three management scenarios on the distribution and abundance of forest birds on a 314 000 ha study region in the vicinity of Calling Lake in northeastern Alberta. The first scenario is a baseline for the natural disturbance model, as the only landscape disturbance agent considered is fire. There is no forest harvesting in the baseline scenario, and the effect of fire suppression is low. The second scenario projects future landscapes under an approximation of long-run non-declining sustained-yield harvesting in the absence of fire (i.e., a very optimistic outcome that assumes complete fire suppression). The third scenario incorporates both harvest and fire at current levels of suppression effectiveness. Bird responses were predicted by fine-scale abundance models, which include habitat variables affected by the pattern and rate of disturbance. For ease of interpretation, we portray the results with respect to the distribution of "good" habitat for the Swainson's Thrush (*Catharus ustulatus*), a migratory forest songbird which breeds in older mixedwood forest (Fig. 14.12). Habitat goodness is defined relative to long term empirical abundance data from control sites in large contiguous areas of old deciduous and mixed stands. FEENIX aggregates good cells into contiguous patches and computes summary statistics describing the total area of these patches, and the mean and standard deviation of their sizes. Together, these three statistics describe many aspects of patch configuration, such as edge density, core area, and patch isolation (Vernier and Cumming 1998).

Clearly, the total amount of good habitat declines markedly under both harvesting scenarios, relative to what is currently present, and to what is expected under a natural fire regime (Fig. 14.12). Furthermore, the simulated time series of patch configuration metrics (Fig. 14.13) quantify the visual impression that patches of good habitat become not only less abundant, but also smaller and more isolated, in both harvesting scenarios.

Fig. 14.12. The distribution and abundance of good Swainson's Thrush habitat (cells where predicted abundances are relatively high (black); (*a*) present landscape; after 200 years under (*b*) natural fire only, (*c*) harvest only, and (*d*) harvest with 1990s fire.

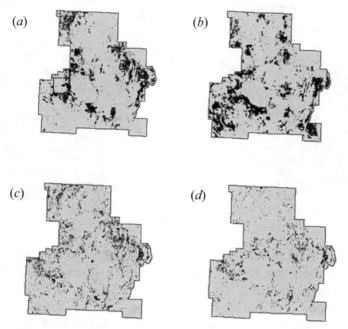

Fig. 14.13. Effect of management on the configuration of good habitat patches for the Swainson's Thrush; (*a*) mean patch size; (*b*) patch-size standard deviation. Scenarios simulated were natural fire regime only (case 1), current harvest rates without fire (case 2), and current harvest rates with 1990s fire levels (case 3).

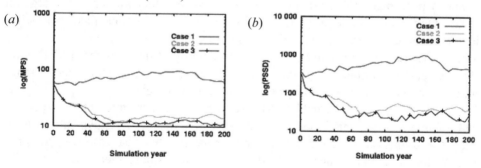

TARDIS can be used to conduct similar studies, at regional scales, using BEEST's coarse-scale habitat models. Figure 14.14 illustrates the predicted changes in the distribution of one woodpecker species (the Downy Woodpecker, *Picoides pubescens*) after 200 years of simulated forest harvesting at current rates of cut. In this case, the study area was a large (74 000 km^2) region of boreal mixedwood forest in northeast Alberta. BEEST researchers conclude from these simulations that current forest management will result in future landscapes that differ substantively from landscapes created by a natural fire regime. For species specialized in their association with older forest, such

Fig. 14.14. The predicted regional distribution of the Downy Woodpecker; (*a*) in 2000;
(*b*) in 2200, after 200 years of simulated forest harvesting in the absence of fire, as predicted
by TARDUS. Map values are the predicted probabilities of detecting the species within a
100 km² cell, under a standardized sampling protocol.

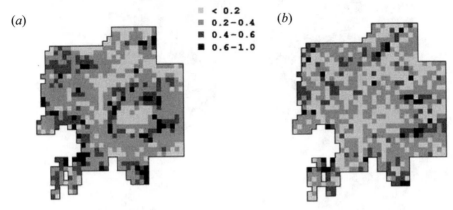

changes will result in both habitat loss and habitat fragmentation, which may affect
regional population persistence. These effects could be mitigated through manipulation
of the rate and pattern of harvest.

*Divided landbases and overlapping tenures in Alberta's mixedwood forests: an
evaluation of policy alternatives* — Sustainable forest management requires integrated
planning of all land-use activities within the region of interest. However, in Alberta, historic allocations of deciduous and coniferous rights within the mixedwood forest have
divided the landbase between separate firms, with different, but overlapping, tenure
arrangements. This complicates the forest planning environment considerably. Cumming and Armstrong (1999, 2001) modelled an instance of this problem, on a large forest estate supplying a single large pulp mill using mostly deciduous feedstock, within
which is embedded more than 17 volume-based coniferous quotas supplying sawmill
operations. These firms currently operate independently of one another. TARDIS was
used to evaluate the sustainability and cost effectiveness of the current tenure arrangement, relative to an alternative, integrated management strategy.

Under the "business as usual" (BAU) scenario, each firm attempts to satisfy its
annual volume requirements in the most cost-effective manner possible, subject to the
existing constraints on what and where it may harvest. This base case was compared
with an "integrated management" (IM) strategy, where a single agent was responsible
for all forest management actions, and allocated timber to individual mills through a
near-optimal annual harvest schedule.

The simulations suggest that the current softwood harvest is not sustainable under
BAU (Fig. 14.15*a*). The model predicts that, during many time periods, some quota
holders will experience significant shortfalls in wood supply. However, under IM there
are no shortfalls in wood supply, and substantial cost savings are predicted (Fig.
14.15*b*). Furthermore, under IM, many more large contiguous areas of forest (11% of
the total area) were never entered during the 200 year planning horizon. This has ecological implications for landscape management that are being explored further. This

Fig. 14.15. Implications of independent planning (BAU) vs. integrated forest management (IM) for (*a*) total softwood quota volumes achieved in the absence of fire, and (*b*) delivered wood costs for a large pulp mill operated by Alberta-Pacific Forest Industries Inc. (APFI) and mean costs for all softwood quota holders (Quotas).

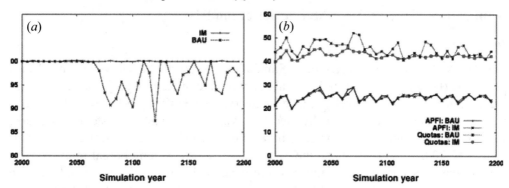

exercise demonstrated some of the inefficiencies inherent in the current landbase designation and tenure allocation policy in Alberta, and estimated some of the financial costs. It has led to calls for a review of forest tenure arrangements in Alberta (Cumming and Armstrong 2001). More recent work has shown that when the effects of wildfire are included in the analysis, the expected shortfalls in conifer supply become much larger, and occur much sooner than shown in Fig. 14.15*a*.

Future developments

In the western boreal forest, the energy sector is a major agent of disturbance (see Chap. 22). For example, the Environment Council of Alberta reported that more than 234 700 ha had been disturbed by petroleum exploration activities, an area almost equal to the 255 692 ha cut by forest product companies between 1956 and 1976 (Pratt and Urquhart 1994, p. 18). Until recently, data related to the location and intensity of energy sector activity were largely unavailable, and were therefore not represented in BEEST modelling or simulation exercises. However, these coverages are now becoming available, facilitating a number of new initiatives within the research group. Relationships between landscape attributes, resource use, and spatial behaviour of various economic agents (e.g., interactions between forest products and energy sector developments, and forestry, wildlife, and recreationists) are being modelled for inclusion in FEENIX and TARDIS. Applications of this work include econometric modelling of exploration and drilling activities in order to generate a spatially explicit set of land values for use in a reserve selection model, and to quantify impacts on landscape attributes that affect wildlife habitat and forestry values. FEENIX can generate maps of the probability of forest wildfire. Combined with the output of an optimizing timber supply model, these generate maps of expected financial loss, which the team is using to quantify the effect of fire on forest management plans, and the effect of large fires on long-run harvest rates. BEEST is embarking on a major comparative study of the affects of natural and artificial barriers to stream flow (beavers vs. roads) on riparian communities and the adjacent forest. New submodels of these processes will be incorporated into their simulation tools, so that large-scale consequences of greatly expanded road construction can be predicted. They are also exploring optimization approaches to many of the issues outlined here (e.g., Hauer and Hoganson 1995).

Application of the FORECAST series of models

Introduction

The Forest Ecosystem Management Simulation Group at the University of British Columbia (UBC) has established several collaborative modelling projects with government agencies and the forest products industry. The FORECAST model (described above), for example, has been calibrated for a variety of forest ecosystems in Canada, including in Saskatchewan to examine boreal mixedwood management strategies (Welham et al. 2002), and to evaluate the potential for re-establishing boreal forest ecosystems in the oil sands region of northeastern Alberta. Furthermore, the model has also been applied in an analysis of the Montane Alternative Silviculture Systems (MASS) project on Vancouver Island (in cooperation with the Canadian Forest Service, the B.C. Ministry of Forests, and Weyerhauser Canada). Its biomass-based approach has also proven useful for assessing the consequences of forest management on carbon storage (Seely et al. 2002). Other applications of FORECAST are currently underway in Norway, China, and the United Kingdom.

In addition to its stand-level applications, FORECAST has been linked to various landscape-level models. Scaling up from a stand to the landscape level introduces a new dimension to forest modelling, and its description is the focus of this sub-section.

Incorporating FORECAST output within existing landscape-level models using static linkages

Certain issues in forest resource management can be addressed only at large spatial scales. In this respect, their resolution depends, at least in part, on landscape-level models that can serve in decision support and in scenario/values trade-off analysis. Until recently, most landscape-level timber supply and habitat suitability models were driven by conventional historical bioassay growth and yield models (as described above), or used very simple assumptions about age-dependent stand development. Unfortunately, oversimplifying stand-level processes undermined the ability of these models to address the multi-scale and multi-value issues of stewardship, certification, and sustainability. To address these shortcomings and to facilitate a stronger link between stand-level ecology and multi-criteria landscape analysis, the UBC team has initiated a series of multi-disciplinary modelling projects. These projects include a multi-criteria analysis of alternative management options in a 40 000 ha landscape unit of the Arrow Lakes Timber Supply Area in southeastern B.C., and a similar project conducted in association with Canadian Forest Products Ltd. for TFL 48 (~500 000 ha in size) in the boreal region of northeastern B.C. In both cases, FORECAST was linked to the timber supply model, ATLAS (Nelson 2000), and a habitat supply model, SIMFOR (Daust and Sutherland 1997; Fig. 14.16). The project also included a visualization component to provide images to stakeholders involved in the development of the SFM plans for the respective projects.

Fig. 14.16. A schematic representation of the multi-scale modelling framework developed at the University of British Columbia.

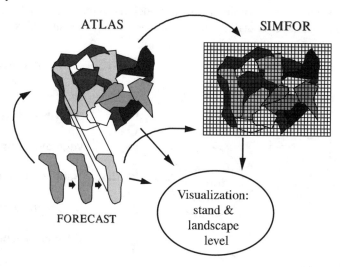

Each of the models in Fig. 14.16 was linked through a database approach whereby output from one model was used to provide input to another. Spatial ecological classification data (e.g., following B.C.'s Terrestrial Ecosystem Mapping or TEM protocol) was used in conjunction with vegetation inventory data to develop a stand-level database of the major forest types within the target forest area (Fig. 14.17). Each forest type, in conjunction with a set of silvicultural options, was considered a discrete forest unit. FORECAST was then used to create a series of stand attribute curves for each forest unit, including merchantable volume, species composition, stand structure, and carbon storage (Fig. 14.17). These unique forest units were subsequently assigned to individual polygons within the landscape models using a set of criteria including TEM site types, indices of site productivity, and current forest cover.

The above combination of models represents a powerful decision support tool to help the forest products industry and forest stakeholders understand the consequences of alternative ecosystem management scenarios for a broad range of indicators of SFM.

Developing landscape-level models with dynamic links to FORECAST

HORIZON and Possible Forest Futures (PPF) — The landscape-level model, HORIZON, is designed for watershed-scale scenario analyses. In contrast to the modelling framework described above, however, HORIZON will be dynamically linked to FORECAST using a cell-based spatial structure. This will permit the simulation of spatially dependent events between polygons, in particular seed dispersal. Hence, the consequences for natural regeneration of different patterns of timber harvesting can be explored, as suggested in Chap. 8.

With most forest scenario models, the "learning curve" is sufficiently steep to render them unsuitable for many non-technical users. A user-friendly landscape management game, "Possible Forest Futures" (PFF), has therefore been developed, based on the

Fig. 14.17. A diagrammatic representation of the methodology used to develop a set of discrete forest units to be modelled within FORECAST. A series of stand attribute curves is produced for each forest unit for use within the larger modelling framework. "TEM" refers to a terrestrial ecosystem map layer or classification; "SI classes" refers to site index classes of site productivity.

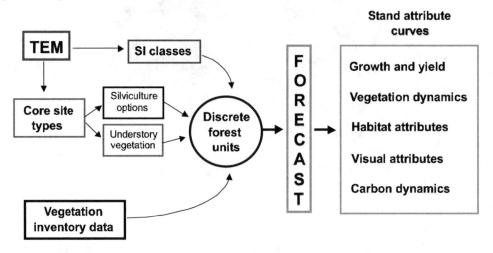

HORIZON model. During the game, the user is presented with a variety of stand management options (Fig. 14.18). These options are embedded within a landscape context, and illustrate many of the central issues surrounding forestry practice.

Local Landscape Ecosystem Management Simulator (LLEMS)

Variable retention (VR) is a system applied to even-aged stands in which scattered individual trees or tree clumps are retained at the time of final harvest (Arnott and Beese 1997; Franklin et al. 1997; Chap. 13). An important difference between VR and shelterwood systems is that the retained trees are not removed during subsequent entries. The main objective of VR is often to satisfy public aesthetic preference, but the system also has potential biodiversity and wildlife habitat benefits because it ensures a sustained supply of habitat features within a cutblock that is being managed for timber on shorter rotation. Variable retention may also have the added benefit of emulating some types of the stand structure that results from natural disturbance events. Critics of VR often claim it is nothing more than "clearcutting with reserves". Technically, if more than 50% of the harvested area is within the zone of influence of the cutblock edge and the reserved trees, such a layout is not considered a clearcut (Kimmins 1997).

One limitation of VR is a lack of growth and yield models and ecosystem management decision support tools that are capable of forecasting stand development under this silvicultural system. Other landscape models can represent complex patterns of polygons across a landscape and larger areas, but usually the polygons are internally homogeneous and lack any edge effects. To address some of the issues surrounding VR, the UBC Forest Ecosystem Management Simulation Group is developing LLEMS, a complex cutblock model with a spatial representation of edge effects, including light competition, seed production and dispersal, seedling establishment, and aspects of

Fig. 14.18. An example of a management choices screen (upper right) from the landscape management game, Possible Forest Futures (PPF), which is a user-friendly interface for HORIZON, which, in turn, is a landscape-level simulator dynamically linked to FORECAST.

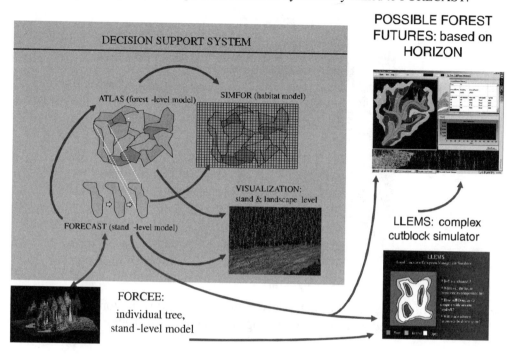

belowground soil and root processes. The model will be based on the pixel data structure developed for the HORIZON model, and possess a minimum resolution of 10 m × 10 m (Fig. 14.19). Individual pixels or groups of adjacent homogenous pixels will be represented by a version of FORECAST modified to facilitate spatial interactions. LLEMS will operate at the scale of an intermediate- to large-sized cutblock or group of cutblocks (10–1000 ha). Joint funding from International Forest Products Ltd. and the National Science and Engineering Research Council (NSERC) supports the development of LLEMS. The model will explore the long-term consequences of cutblock shape, orientation, and variable retention strategies on the economic, ecological, and social (i.e., visual) indicators of SFM.

Conclusions

As forestry evolves from the exploitative, administrative, and ecologically based, timber-focused stages to the social forestry stage (see Kimmins 2000; Chap. 1), new and different forest models and decision support tools are required. The traditional timber management paradigm is evolving into combinations of ecosystem management, adaptive management, and emulation of natural disturbances and the natural range of variation, and from regulation-based to results-based forestry. Multi-value ecosystem management models at various spatial scales are required. The FORECAST line of hybrid simulation models offers a range of decision support, communication, scenario analysis, and heuristic tools to serve this developing need. It marries the empirical

Fig. 14.19. An example of the pixel-based data structure in the model LLEMS, under development for representing the spatially explicit processes associated with edge ("ecotone") effects in small cutblocks and cutblocks with variable or patch retention of mature trees.

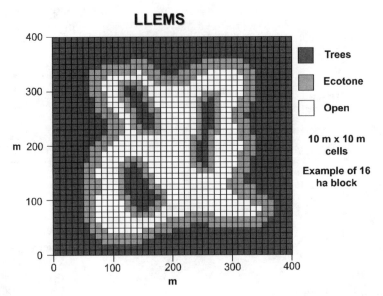

believability of traditional forest models with the flexibility to forecast alternative futures that is provided by process modelling (Korzukhin et al. 1996).

Concluding remarks

We have seen in this chapter that there is a wide range of modelling approaches aimed to address different goals. No one approach is inherently better than the other: they all provide insight and put emphasis on different deterministic mechanisms (e.g., tree growth), stochastic processes (e.g., wildfire), and domains of application (e.g., stand, landscape). Recent modelling developments include much convergence in the structure of many stand development models (based on regeneration, growth, and mortality modules), and of many landscape models (which, of necessity, include components of stand development and disturbance in a spatially sensitive framework). More and more models are able to incorporate non-timber considerations and outputs. We also see a vigorous lending and borrowing of ideas among assorted empirical, process, and stochastic modelling approaches. Finally, it is clear that hierarchical modelling and the passing of information among several spatial or temporal scales represents a fruitful area for improved forest management simulation.

These more complex modelling approaches, executed at various scales, are needed to address the current and future challenges in forestry (Box 14.4). This will be facilitated by the trend that models are being built to be generic and modular in design. The internal duties of a module can then be separated from the internal dynamics of the other modules. One can envision the collection of modules or a toolbox approach, which can be selected from and joined, depending on need (Reynolds and Acock 1997). Progress toward modular and generic models may make it possible to reach the stage where

---

Box 14.4. Why use modelling tools to assess the sustainability of our forest management?

As our view of the functioning forest and public demand for forest products is increasing in complexity, so is the need for tools that can evaluate and consider this complexity. As seen in the first 13 chapters of this book, the complexity of ecological and social aspects to consider is mind boggling. As Fred Bunnell (Faculty of Forestry, University of British Columbia, Vancouver, B.C.) is fond of saying about sustainable forest management, "It's not rocket science; it's much more complicated. . . ."

Among other things, we need to:

- Assess management options and social demands across a wide range of spatial and temporal scales;
- Predict influences of specific components on the larger system, and vice versa;
- Project population dynamics for a wide range of species and cohorts;
- Assess global climatic influences on particular forests; and
- Present the effects of a management plan, over a wide range of spatial and temporal scales, to the public in an understandable way.

For all of these reasons, and many more, simulation models and modelling tools will be used more and more in the future to help us manage the world's forests in a sustainable manner.

---

model development is through incremental improvements in process algorithms rather than new model construction. Addition of meaningful output variables (e.g., fraction of old-growth forest, habitat suitability indices) will also increase the utility of our models to forest practitioners. A pluralism of emphases, from individual-based to regional/global models, will continue to be useful for addressing problems at multiple scales, with meta-modelling used when linkage is needed (Baker and Mladenoff 1999), as seen within the four examples presented in this chapter.

Since the ultimate intended users of the above-mentioned models and modelling tools are forest managers, it is important that the flow of information between scientists building the models and forest managers that need them be improved in order to build better, more goal-orientated models. We also need to provide managers not only with quantitative predictions, but also with information about the certainty of our prediction. A precise prediction that is very uncertain might not be better than a vague prediction that is more certain.

In this regard, model development is mirroring the maturation of our own views regarding the roles of chance and variability in the maintenance of ecosystem integrity. Empirical research indicates that most ecosystems and landscapes can exhibit a wide range or stable or semi-stable states that we would consider more or less healthy; various methods of historical reconstruction and monitoring are now being used to describe a natural range of variability (NRV) for numerous bio-indicators. Likewise, models are becoming more stochastic, and (even when run on spatial databases) now are often designed to portray an array of possible outcomes. Expectations regarding mankind's ability to control boreal landscapes are maturing as well, with modelling tools helping forest managers make decisions in terms of probabilities and ranges of potential future states.

References

Ågren, G.I. 1983*a*. Nitrogen productivity of some conifers. Can. J. For. Res. **13**: 494–500.

Ågren, G.I. 1983*b*. The concept of nitrogen productivity in forest growth modeling. Mitt. Forstl. Bundes-Versuchsanst, **147**: 199–210.

Armstrong, G.W. 1999. A stochastic characterisation of the natural disturbance regime of the boreal mixed-wood forest with implications for sustainable forest management. Can. J. For. Res. **29**: 423–433.

Arnott, J.T., and Beese, W.J. 1997. Alternatives to clearcutting in B.C. coastal montane forests. For. Chron. **73**: 670–678.

Baker, W.L., and Mladenoff, D.J. 1999. Progress and future directions in spatial modeling of forest land-scapes. *In* Spatial modeling of forest landscape change: approaches and applications. *Edited by* D.J. Mladenoff and W.L. Baker. Cambridge University Press, Cambridge, U.K. pp. 333–350.

Bergeron, Y., Harvey, B., Leduc, A., and Gauthier, S. 1999. Forest management guidelines based on natu-ral disturbance dynamics: stand- and forest-level considerations. For. Chron. **75**: 49–54.

Bossel, H. 1991. Modelling forest dynamics: moving from description to explanation. For. Ecol. Manage. **42**: 129–142.

Botkin, D. 1993. Forest dynamics: an ecological model. Oxford University Press, Oxford, U.K. 309 p.

Botkin, D.B., Janak. J.F., and Wallis, J.R. 1972. Rationale, limitations, and assumptions of a northeastern forest simulator. IBM J. Res. Div. **16**: 106–116.

Brix, H. 1971. Effects of nitrogen fertilization on photosynthesis and respiration in Douglas-fir. For. Sci. **17**: 407–414.

Bugmann, H.K.M. 1996. A simplified forest model to study species composition along climate gradients. Ecology, **77**: 2055–2974.

Bunnell, F.L. 1989. Alchemy and uncertainty: what good are models? USDA Forest Service, Portland, Ore-gon. Gen. Tech. Rep. PNW-GTR-232. 27 p.

Burton, P.J., Kneeshaw, D.D., and Coates, K.D. 1999. Managing forest harvesting to maintain old growth in boreal and sub-boreal forests. For. Chron. **75**: 623–631.

Coates, K.D., Messier, C., Beaudet, M., and Canham, C.D. 2003. SORTIE: a resource mediated, spatially-explicit and individual-tree model that simulates stand dynamics in forest ecosystems. For. Ecol. Man-age. In press.

Côté. M.A., Kneeshaw, D.D., Messier, C., and Bouthillier, L. 2001. Increasing partnerships between sci-entists and forest managers. For. Chron. **77**: 85–89.

Cumming, S.G. 2000. Forest type, fire ignition, and fire frequency in boreal mixedwood forests. Working Paper 2000-11. Sustainable Forest Management Network, Edmonton, Alberta. Available at http://sfm-1.biology.ualberta.ca/english/pubs/PDF/WP_2000-11.pdf [23 October 2002]. 24 p.

Cumming, S.G. 2001*a*. Forest type and wildfire in the Alberta boreal mixedwood: what do fires burn? Ecol. Appl. **11**: 97–110.

Cumming, S.G. 2001*b*. A parametric model of the fire-size distribution. Can. J. For. Res. **31**: 1297–1303.

Cumming, S.G, and Armstrong, G.W. 1999. Divided land bases and overlapping tenures in Alberta's mixedwood forests: a simulation study of policy alternatives. Working Paper 1999-3. Sustainable For-est Management Network. Edmonton, Alberta. Available at http://sfm-1.biology.ualberta.ca/english/pubs/PDF/WP_1998-08.pdf [viewed 23 October 2002]. 30 p.

Cumming, S.G., and Armstrong, G.W. 2001. Divided land base and overlapping forest tenure in Alberta, Canada: a simulation study exploring costs of forest policy. For. Chron. **77**: 501–508.

Cumming, S.G., and Schmiegelow, F.K.A. 2001. Effects of habitat abundance and configuration, and the forest matrix, on distributional patterns of boreal birds. Working Paper 2001-1. Sustainable Forest Man-agement Network. Edmonton, Alberta. Available at http://sfm-1.biology.ualberta.ca/english/pubs/PDF/WP_2001-1.pdf [viewed 23 October 2002]. 27 p.

Cumming, S.G., and Vernier, P. 2002. Statistical models of landscape pattern metrics, with applications to regional scale dynamic forest simulations. Landscape Ecol. **17**: 433–444.

Cumming, S.G., Demarchi, D, and Walters, C. 1998. A grid-based spatial model of forest dynamics applied to the boreal mixedwood region. Working Paper 1998-8. Sustainable Forest Management Network, Edmonton, Alberta. Available at http://sfm-1.biology. ualberta.ca/english/pubs/PDF/WP_1999-03.pdf [viewed 23 October 2002]. 40 p.

Cumming, S.G., Demarchi, D., and Walters, C.J. 1999. A landscape model of white spruce dynamics for the boreal mixedwood. *In* Proceedings of the 1999 Sustainable Forest Management Network Conference, 14–17 February 1999, Edmonton, Alberta. *Edited by* T.S. Veeman, D.W. Smith, B.G. Purdy, F.J. Salkie, and G.A. Larkin. Sustainable Forest Management Network, Edmonton, Alberta. pp. 695–700.

Daust, D.K., and Sutherland, G.D. 1997. SIMFOR: software for simulating forest management and assessing effects on biodiversity. *In* The status of forestry/wildlife decision support systems in Canada: proceedings of a symposium, Toronto, Ontario, 1994. *Edited by* I.D. Thompson. Canadian Forest Service, Sault St. Marie, Ontario. pp. 15–29.

Davis, L.S., Johnson, K.N., Bettinger, P.S., and Howard, T.E. 2001. Forest management to sustain ecological, economic and social values, 4th edition. McGraw–Hill, New York, New York. 804 p.

Deutschman, D.H., Levin, S.A., and Pacala, S.W. 1999. Error propagation in a forest succession model: the role of fine-scale heterogeneity in light. Ecology, **80**: 1927–1943.

Doyon, F. 2000. Fire regime simulation of the Whitecourt forest using LANDIS. Biodiversity Assessment Project for Millar Western Forest Products. BAP Report 4. Millar Western Forest Products Ltd., Whitecourt, Alberta. 28 p.

Doyon, F., and Duinker, P.N. 2000. Species selection procedure. Biodiversity Assessment Project for Millar Western Forest Products. BAP Report 2. Millar Western Forest Products Ltd., Whitecourt, Alberta. 11 p.

Doyon, F., Higgelke, P., and MacLeod, H.L. 2000. Habitat supply models. Biodiversity Assessment Project for Millar Western Forest Products. BAP Report 6. Millar Western Forest Products Ltd., Whitecourt, Alberta. 11 p.

Dubé, P., Fortin, M.-J., Canham, C.D., and Marceau, D.J. 2001. Quantifying global gap dynamics and spatio-temporal structures in spatially-explicit models of temperate forest ecosystems. Ecol. Model. **142**: 39–60.

Duinker, P.N., and Baskerville, G.L. 1986. A systematic approach to forecasting in environmental impact assessment. J. Environ. Manage. **23**: 271–290.

Duinker, P.N., Doyon, F., Morash, R., Van Damme, L., MacLeod, H.L., and Rudy, A. 2000. Background and structure: Biodiversity Assessment Project for Millar Western Forest Products. BAP Report 1. Millar Western Forest Products Ltd., Whitecourt, Alberta. 20 p.

Ebermayer, E. 1876. Die gesammte Lehre der Waldstien mit Rücksicht auf die chemische Statik des Waldbaues. Springer, Berlin, Germany.

Fall, A., and Fall, J. 2001. A domain-specific language for models of landscape dynamics. Ecol. Model. **141**(1–3): 1–18.

Feunekes, U., and Cogswell, A. 1997. A hierarchical approach to spatial forest planning. *In* Proceedings of the International Symposium on System Analysis in Forestry, 28–30 May 1997, Traverse City, Michigan. Available from Remsoft Inc., Fredericton, New Brunswick.

Fortin, M.-J. 1999. Spatial statistics in landscape ecology. *In* Landscape ecological analysis: issues and applications. *Edited by* J.M. Klopatek and R.H. Gardner. Springer-Verlag, New York, New York. pp. 253–279.

Fortin, M.-J., Fall, A., and Didion, M. 2002. Intégration de la problématique des feux dans la planification forestière. L'Aménagement forestier et le feu, Actes de Colloque tenu à Chicoutimi les 9, 10 et 11 Avril, 2002. Ressources naturelles Québec, Quebec City. pp. 73–75.

Franklin, J.F., Berg, D.R., Thornburgh, D.A., and Tappeiner, J.C. 1997. Alternative silvicultural approaches to timber harvesting: variable retention harvest systems. *In* Creating a forestry for the 21st century: the science of ecosystem management. *Edited by* K.A. Kohm and J.F. Franklin. Island Press, Washington, D.C. pp. 111–139.

Fujimori, T., and Whitehead, D. (*Editors*). 1986. Crown and canopy structure in relation to productivity. Proceedings of the IUFRO Workshop, 14–20 October 1985, Japan. Forestry and Forest Products Research Institute, Ibaraki, Japan.

Galindo-Leal, C., and Bunnell, F.L. 1995. Ecosystem management: implications and opportunities of a new paradigm. For. Chron. **71**: 601–606.

Grant, W.E., Pedersen, E.K., and Marín, S.L. 1997. Ecology and natural resource management: systems analysis and simulation. John Wiley and Sons, New York, New York. 373 p.

Grumbine, R.E. 1994. What is ecosystem management? Conserv. Biol. **8**: 27–39.

Gustafson, E.J. 1998. Quantifying landscape spatial pattern: what is the state of the art? Ecosystems, **1**: 143–156.

Gustafson, E.J., Shifley, S.R., Mladenoff, D.J., He, H.S., and Nimerfro, K.K. 2000. Spatial simulation of forest succession and timber harvesting using LANDIS. Can. J. For. Res. **30**: 32–43.

Gustafsson, L., and Weslien, J. (*Editors*) 1999. Special issue: biodiversity in managed forests — concepts and solutions. Elsevier Science, Oxford, U.K. For. Ecol. Manage. **115**(2–3). 196 p.

Haefner, J.W. 1996. Modeling biological systems: principles and applications. Chapman & Hall, New York, New York. 473 p.

Hannon, B., and Ruth, M. 1997. Modeling dynamic biological systems. Springer-Verlag, New York, New York. 399 p.

Hansen, A.J., Garman, S.L., Weigand, J.F., Urban, D.L., McComb, W.C., and Raphael, M.G. 1995. Alternative silvicultural regimes in the Pacific Northwest: simulations of ecological and economic effects. Ecol. Appl. **5**: 535–554.

Hauer, G.K., and Hoganson, H.M. 1995. Tailoring a decomposition method to a large forest management scheduling problem in northern Ontario. INFOR **34**(3): 209–231.

He, H.S., and Mladenoff, D.J. 1999. Dynamics of fire disturbance and succession on a heterogeneous forest landscape: a spatially explicit and stochastic simulation approach. Ecology, **80**: 80–99.

Huang, S., and Titus, S.J. 1994. An age-dependent individual tree height prediction model for boreal spruce–aspen stands in Alberta. Can. J. For. Res. **24**: 1295–1301.

Hunter, M.L., Jr. 1990. Wildlife, forests and forestry: principles of managing forests for biological diversity. Prentice Hall Inc., Englewood Cliffs, New Jersey. 370 p.

Jordan, G.A., and Wightman, R. 1993. The FORMAN2000 forest management decision support system. *In* Proceedings of the 7th Annual Symposium on Geographic Information Systems in Forestry, Environment and Natural Resources Management, 15–18 February 1993, Vancouver, British Columbia. Polaris Conferences, Vancouver, British Columbia. pp. 89–92.

Kellomäki, S., and Kurttio, O. 1991. A model for the structural development of a Scots pine crown based on modular growth. For. Ecol. Manage. **43**: 103–123.

Kimmins, J.P. 1985. Future shock in forest yield forecasting: the need for a new approach. For. Chron. **61**: 503–513.

Kimmins, J.P. 1986. Predicting consequences of intensive forest harvesting on long-term productivity: the need for a hybrid model such as FORCYTE-11. *In* Predicting consequences of intensive forest harvesting on long-term productivity. *Edited by* G.I. Ågren. Swedish University of Agricultural Sciences, Department of Ecology and Environmental Research. Rep. No. 26. pp. 31–84.

Kimmins, J.P. 1988. Community organization: methods of study and prediction of the productivity and yield of forest ecosystems. Can. J. Bot. **66**: 2645–2672.

Kimmins, J.P. 1990. Modeling the sustainability of forest production and yield for a changing and uncertain future. For. Chron. **66**: 271–280.

Kimmins, J.P. 1993. Scientific foundations for the simulation of ecosystem function and management in FORCYTE-11. Forestry Canada, Edmonton, Alberta. Info. Rep. NOR-X-328. 88 p.

Kimmins, J.P. 1997. Models and their role in ecology and resource management. Chap. 17. *In* Forest ecology, second edition. Prentice Hall Inc., Englewood Cliffs, New Jersey. pp. 475–494.

Kimmins, J.P. 2000. Respect for nature: an essential foundation for sustainable forest management. *In* Ecosystem management of forested landscapes: directions and implementation. *Edited by* R.G. D'Eon, J.F. Johnson, and E.A. Ferguson. UBC Press, Vancouver, British Columbia. pp. 3–26.

Kimmins, J.P., and Scoullar, K. 1979. FORCYTE: a computer simulation approach to evaluating the effect of whole-tree harvesting on the nutrient budgets of Northwest forests. *In* Proceedings, Conference on Forest Fertilization. Contrib. No. 40. *Edited by* S.P.Gessel, R.M. Kenady, and W.A. Atkinson. Institute of Forest Resources, University of Washington, Seattle, Washington. pp. 266–273.

Kimmins, J.P., Mailly, D., and Seely, B. 1999. Modelling forest ecosystem net primary production: the hybrid simulation approach used in FORECAST. Ecol. Model. **122**: 195–224.

Kneeshaw, D., Leduc, A., Drapeau, P., Gauthier, S., Paré, D., Doucet, R., Carignan, R., Bouthillier, L., and Messier, C. 2000. Development of integrated ecological standards of sustainable forest management at an operational scale. For. Chron. **76**: 481–493.

Korzukhin, M.D., Ter-Mikaelian, M.T., and Wagner, R.G. 1996. Process versus empirical models: which approach for forest ecosystem management? Can. J. For. Res. **26**: 879–887.

Kurth, W. 1994. Morphological models of plant growth: possibilities and ecological relevance. Ecol. Model. **75/76**: 299–308.

Landsberg, J. 2001. Modelling forest ecosystems: state-of-the-art, challenges and future directions. *In* Proceedings of Forest Modelling for Ecosystem Management, Forest Certification, and Sustainable Management Conference, 12–17 August 2001, Vancouver, British Columbia. *Edited by* V. LeMay and P. Marshall. University of British Columbia, Vancouver, British Columbia. pp 3–21.

Landsberg, J.J., and Gower, S.T. 1997. Applications of physiological ecology to forest management. Academic Press, San Diego, California. 354 p.

Leemans, R. 1992. The biological component of the simulation model for boreal forest dynamics. *In* A systems analysis of the global boreal forest. *Edited by* H.H. Shugart, R. Leemans, and G.B. Bonan. Cambridge University Press, Cambridge, U.K. pp. 428–445.

Lo, E., Ming, W., Lechowicz, M., Messier, C., Nikinmaa, E., and Sievänen, R. 2001. Evaluation of LIGNUM for jack pine. For. Ecol. Manage. **150**: 279–291.

Lockwood, C., and Moore, T. 1993. Harvest scheduling with spatial constraints: a simulated annealing approach. Can. J. For. Res. **23**: 468–478.

Mäkelä, A. 2001. Modelling tree and stand growth: towards a hierarchical treatment of multi-scale processes. *In* Proceedings of Forest Modelling for Ecosystem Management, Forest Certification, and Sustainable Management Conference, 12–17 August 2001, Vancouver, British Columbia. *Edited by* V. LeMay and P. Marshall. University of British Columbia, Vancouver, British Columbia. pp. 27–44.

(MMF) McGregor Model Forest. 2002. ECHO planning system. Available at http://www.mcgregor. bc.ca/publications/publications.html [viewed 23 October 2002]. See also http://www. modelforest. net/e/home_/loca_/case_/scenarie.html [viewed 23 October 2002].

McMurthie, R.E., and Landsberg, J.J. 1992. Using a simulation model to evaluate the effects of water and nutrients on the growth and carbon partitioning of *Pinus radiata*. For. Ecol. Manag. **52**: 243–260.

Mitchell, K.J. 1975. Dynamics and simulated yield of Douglas-fir. For. Sci. Monogr. **17**: 1–39.

Mitchell, K.J., and Cameron, I.R. 1985. Managed stand yield tables for coastal Douglas-fir: initial density and pre-commercial thinning. British Columbia Ministry of Forests, Victoria, British Columbia. Land Manage. Rep. No. 31. 69 p.

Mladenoff, D.J., and Baker, W.L. (*Editors*) 1999. Spatial modelling of forest landscape change: approaches and applications. Cambridge University Press, Cambridge, U.K. 352 p.

Mladenoff, D.J., and He, H.S. 2000. Design, behavior and application of LANDIS, an object-oriented model of forest landscape disturbance and succession. *In* Spatial modeling of forest landscape change: approaches and applications. *Edited by* D.J. Mladenoff and W.L. Baker. Cambridge University Press, Cambridge, U.K. pp. 125–162.

Möller, C.M. 1945. Untersuchnuger über Laubmerge, Stoffnerlust und Stoffproducktion des Waldes. Forstl. Forsoegsuaes. Dan. **17**: 1–287.

Möller, C.M., Muller, D., and Nielsen, J. 1956*a*. Respiration in stem and branches of beech. Forstl. Forsoegsuaes. Dan. **21**: 273–301.

Möller, C.M., Muller, D., and Nielsen, J. 1956*b*. Graphic presentation of dry matter production of European beech. Forstl. Forsoegsuaes. Dan. **21**: 327–335.

Monserud, R.A. 2002. Large-scale management experiments in the moist maritime forests of the Pacific Northwest. Landscape Urban Plann. **59**: 159–180.

Nelson, J.D. 2000. Reference manual for FPS-ATLAS. University of British Columbia, Vancouver, British Columbia. 78 p.

Nelson, J.D. 2001. Forest and landscape models: powerful tools or dangerous weapons? *In* Proceedings of Forest Modelling for Ecosystem Management, Forest Certification, and Sustainable Management Conference, 12–17 August 2001, Vancouver, British Columbia. *Edited by* V. LeMay and P. Marshall. University of British Columbia, Vancouver, British Columbia. pp. 45–55.

Nikinmaa, E., Messier, C., Sievänen, R., Perttunen, J., and Lehtonen, M. 2003. Shoot growth and crown development: effect of crown position in three-dimensional simulations. Tree Physiol. **23**: 129–136.

Ohman, K. 2000. Creating continuous areas of old forest in long-term forest planning. Can. J. For. Res. **30**: 1032–1039.

Oliver, C.D., and Larson, B.C. 1990. Forest stand dynamics. McGraw–Hill Inc., New York, New York. 467 p.

(ORM) Olympic Resource Management. 2002. COMPLAN spatial forest modelling — harvest scheduler: software. Olympic Resource Management, Vancouver, British Columbia.

Pacala, S.W., Canham, C.D., and Silander, J.A., Jr. 1993. Forest models defined by field measurements: I. the design of a northern forest simulator. Can. J. For. Res. **23**: 1980–1988.

Pacala, S.W., Canham, C.D., Saponara, J., Silander, J.A., Jr., Kobe, R.K., and Ribbens, E. 1996. Forest models defined by field measurements: II. estimation, error analysis and dynamics. Ecol. Monogr. **66**: 1–43.

Pastor, J., and Post, W.M. 1986. Influence of climate, soil mostiure, and succession on forest carbon and nitrogen cycles. Biogeochemistry, **2**: 3–27.

Perttunen, J., Sievänen, R., Nikinmaa, E., Salminen, H., Saarenmaa, H., and Väkevä, J. 1996. LIGNUM: a tree model based on simple structural units. Ann. Bot. **77**: 87-98.

Perttunen, J., Sievänen, R., and Nikinmaa, E. 1998. LIGNUM: a model combining the structure and functioning of trees. Ecol. Model. **108**: 189–198.

Perttunen, J., Nikinmaa, E., Messier, C., and Sievänen, R. 2001. Adapting sugar maple (*Acer saccharum* Marsh) to LIGNUM. Ann. Bot. **88**: 471–481.

Post, W.M., and Pastor, J. 1996. Linkages — an individual-based forest ecosystem model. Climatic Change, **34**: 253–261.

Pratt, L., and Urquhart, I. 1994. The last great forest: Japanese multinationals and Alberta's northern forests. NeWest Publishers, Edmonton, Alberta. 222 p.

Prentice, I.C., and Leemans, R. 1990: Pattern and process and the dynamics of forest structure: a simulation approach. J. Ecol. **78**: 340–355.

Price, D.T., Halliwell, D.H., Apps, M.J., Kurz, W.A., and Curry, S.R. 1997. Comprehensive assessment of carbon stocks and fluxes in a Boreal–Cordilleran forest management unit. Can. J. For. Res. **27**: 2005–2016.

Rauscher, H.M., Isebrands, J.G., Host, G.E., Dickson, R.E., Dickmann, D.I., Crow, T.R., and Michael, D.A. 1990. An ecophysiological growth process model for juvenile poplar. Tree Physiol. **7**: 255–281.

Reffye, P.D., Houllier, F., Blaise, F., Barthélémy, D., Dauzat, J., and Auclair, D. 1995. A model simulating above- and below-ground tree architecture with agroforestry applications. Agroforestry Syst. **30**: 175–197.

Reffye, P.D, Fourcaud, T., Blaise, F., Barthélémy, D., and Houllier, F. 1997. A functional model of tree growth and tree architecture. Silva Fenn. **31**: 297–311.

Reynolds, J.F., and Acock, B. 1997. Modularity and genericness in plant and ecosystem models. Ecol. Model. **94**: 7–16.

Richards, E.W., and Gunn, E.A. 2000. A model and tabu search method to optimize stand harvest and road construction schedules. For. Sci. **46**: 188–203.

Running, S.W., and Gower, S.T. 1991. FOREST-BGC, a general model of forest ecosystem processes for regional applications. II dynamic carbon allocation and nitrogen budgets. Tree Physiol. **9**: 147–160.

Schmiegelow, F.K.A., Machtans, C.S., and Hannon, S.J. 1997. Are boreal birds resilient to fragmentation? An experimental study of short-term community responses. Ecology, **78**: 1914–1932.

Schmiegelow, F.K.A., Vernier, P., Demarchi, D., and Cumming, S.G. 1999. Seeing the forest beyond the trees: a cross-scale approach to wildlife habitat assessment. *In* Proceedings of the 1999 Sustainable Forest Management Network Conference, 14–17 February 1999, Edmonton, Alberta. *Edited by* T.S. Veeman, D.W. Smith, B.G. Purdy, F.J. Salkie, and G.A. Larkin. Sustainable Forest Management Network, Edmonton, Alberta. pp. 232–236.

Seely, B., Kimmins, J.P., Welham, C., and Scoullar, K. 1999. Defining stand-level sustainability, exploring stand-level stewardship. J. For. **97**: 4–11.

Seely, B., Welham, C., and Kimmins, J.P. 2002. Carbon sequestration in a boreal forest ecosystem: results from the ecosystem simulation model, FORECAST. For. Ecol. Manage. **169**: 123–135.

Semenchuck, G.P. (*Editor*). 1992. The atlas of breeding birds of Alberta. Federation of Alberta Naturalists, Edmonton, Alberta. 391 p.

Sessions, J., and Bettinger, P. 2001. Hierarchal planning: pathway to the future? *In* Proceedings of the First International Precision Forestry Cooperative Symposium, 17–20 June 2001, Seattle, Washington. College of Forest Resources, University of Washington, Seattle. pp. 185– 190.

Shugart, H.H. 1984. A theory of forest dynamics: the ecological implications of forest succession models. Springer-Verlag, New York, New York. 278 p.

Shugart, H.H. 1998. Terrestrial ecosystems in changing environments. Cambridge University Press, Cambridge, U.K. 537 p.

Shugart, H.H., and Prentice, I.C. 1992. Individual-tree based models of forest dynamics and their application in global change research. *In* A systems analysis of the global boreal forest. *Edited by* H.H. Shugart, R. Leemans, and G.B. Bonan. Cambridge University Press, Cambridge, U.K. pp. 313–334.

Sievänen, R., and Burk, T.E. 1993. Adjusting a process-based growth model for varying site conditions through parameter estimation. Can. J. For. Res. **23**: 1837–1851.

Sievänen, R., Nikinmaa, E., Nygren, P., Ozier-Lafontaine, H., Perttunen, J., and Hakula, H. 1999. Components of functional-structural tree models. Ann. For. Sci. **57**: 399–412.

Smith, T.M., and Urban, D.L. 1988. Scale and resolution of forest structural pattern. Vegetatio, **74**: 143–150.

Stage, A.R. 2001. How models are connected to reality — evaluation criteria for their use in decision support. *In* Proceedings of Forest Modelling for Ecosystem Management, Forest Certification, and Sustainable Management Conference, 12–17 August 2001, Vancouver, British Columbia. *Edited by* V. LeMay and P. Marshall. University of British Columbia, Vancouver, British Columbia. pp. 57–72.

Valentine, H.T. 1985. Tree-growth models: derivations employing the pipe model theory. J. Theor. Biol. **117**: 579–584.

van Deusen, P.C. 1999. Multiple solution harvest scheduling. Silva Fenn. **33**: 207–216.

Vernier, P., and Cumming, S.G. 1998. Predicting landscape patterns from stand attribute data in the Alberta boreal mixedwood. Working Paper 1998-7. Sustainable Forest Management Network, Edmonton, Alberta. Available at http://sfm-1.biology.ualberta.ca/english/pubs/PDF/ WP_1998-7.pdf [viewed 18 June 2003]. 21 p.

Vernier, P., Schmiegelow, F.K.A., and Cumming, S.G. 2001. Modeling bird abundance from forest inventory data in the boreal mixedwood forests of Alberta. *In* Predicting species occurrences: issues of accuracy and scale. *Edited by* J.M. Scott, P.J. Heglund, M.L. Morrison, J.B. Haufler, M.G. Raphael, W.A. Wall, and F.B. Samson. Island Press, Washington, D.C. pp. 559–571.

Walters, C.J. 1986. Adaptive management of renewable resources. MacMillan Publishing Company, New York, New York. 374 p.

Walters, C.J. 1997. Challenges in adaptive management of riparian and coastal ecosystems. Conserv. Ecol. [online], 1(2): 1. Available at http://www.consecol.org/vol1/iss2/art1 [viewed 17 June 2003].

Wang, Y.-P., and Jarvis, P.G. 1990a. Description and validation of an array model — MAESTRO. Agric. For. Meteorol. **51**: 257–280.

Wang, Y.-P., and Jarvis, P.G. 1990b. Effect of incident beam and diffuse radiation on PAR absorption, photosynthesis and transpiration of Sitka spruce: a simulation study. Silva Carelica, **15**: 167–180.

Welham, C., Seely, B., and Kimmins, J.P. 2002. The utility of the two-pass harvesting system: an analysis using the ecosystem simulation model, FORECAST. Can. J. For. Res. **32**: 1071–1079.

Woods, M.E. 1999. Density management diagrams . . . tools and uses. *In* Proceedings of Stand Density Management: Using the Planning Tools, 23–23 November 1998, Edmonton, Alberta. *Edited by* C.R. Bamsey. Clear Lake Ltd., Edmonton, Alberta. pp. 27–33.

Wykoff, W.R., Crookston, N.L., and Stage, A.R. 1982. User's guide to the Stand Prognosis Model. USDA Forest Service, Ogden, Utah. Gen. Tech. Rep. INT-133.

Yarie, J. 1997. Nitrogen productivity of Alaskan tree species at an individual tree and landscape level. Ecology, **78**: 2351–2358.

Chapter 15

Minimizing negative environmental impacts of forest harvesting operations

Reino Pulkki

Introduction

Harvesting is an integral part of forestry, making wood available for use by society while also being the first step in the managed renewal or rejuvenation of a forest. Wackerman et al. (1966) express the prevailing utilitarian realities in stating,

"The harvesting of timber crops according to correct forestry methods may logically be considered the principal operation in forestry in that the culmination of all forestry effort is a merchantable tree for service to mankind. Harvesting therefore is one of the most important operations in forestry, entirely aside from the fact that it provides the income necessary to carry on the whole forestry business."

More recently, though, the focus of harvesting may be to develop forest conditions for purposes other than timber production (e.g., production of wildlife habitat, recreation, aesthetics, maintaining forest health).

The objective of this chapter is to broadly outline potential negative environmental impacts of forest harvesting operations (Box 15.1) and, in more detail, outline how good advance planning can be used to minimize them. Some recent developments in equipment design and tools and techniques to mitigate negative environmental impacts will also be introduced. An interesting area of development is in off-road walking forest machines (Fig. 15.1).

Most jurisdictions have legal regulations/requirements, codes of practice, certification programs, guidelines, and best management practices (BMP) dealing with maximum negative environmental impact levels allowed, and rules, regulations, and techniques to minimize them and mitigate their impact. It is not possible to present all of this material in this chapter, so the reader is directed to the pertinent legislation and literature for their area. The reader is also directed to consult an excellent handbook (*Harvesting Systems and Equipment in British Columbia*) by MacDonald (1999) which

---

R. Pulkki. Faculty of Forestry and the Forest Environment, 955 Oliver Road, Lakehead University, Thunder Bay, Ontario P7B 5E1, Canada. Telephone: 807-343-8564. e-mail: Reino.Pulkki@Lakeheadu.ca
Correct citation: Pulkki, R. 2003. Minimizing negative environmental impacts of forest harvesting operations. Chapter 15. *In* Towards Sustainable Management of the Boreal Forest. *Edited by* P.J. Burton, C. Messier, D.W. Smith, and W.L. Adamowicz. NRC Research Press, Ottawa, Ontario, Canada. pp. 581–628.

Box 15.1. *Forest harvesting operations include all phases required to:*

- access the compartments[1] to be logged (road planning, lay out, construction, and maintenance);
- plan off-road transportation;
- plan and lay out the compartments to be logged;
- conduct the actual logging work (felling, in-stand and roadside processing, and extraction to roadside);
- transport the wood from roadside to the point of utilization; and
- any remedial work required to "fix" negative environmental impacts (e.g., fixing rutting damage, road deactivation).

[1]A compartment is defined as a part of a stand, a stand, or a group of stands delineated for harvesting by a particular logging method and system at a given time of the year.

Fig. 15.1. Prototype of an off-road walking forest machine (one-grip harvester).

Photo by Reino Pulkki

outlines risk analysis in harvesting equipment selection and describes in detail most logging equipment and systems applicable to the world's boreal forest.

Environmental issues in forest harvesting operations

The potential effects, both positive and negative, of forest harvesting operations can occur in eight broad categories:

- biological (terrestrial and aquatic) (e.g., harvested or rare species, or general biodiversity);
- water quality and quantity;
- soils;
- air;
- forest (tree) health;
- recreation and aesthetics;
- cultural, historic and societal values; and
- economic well-being.

Environmental impacts from harvesting operations can occur directly from the removal of the trees from the site (e.g., loss of nutrients, snags/cavity trees, large wood debris and shading, and change of microclimate), the building of roads required for access, and through physical impacts to the soil during harvesting (e.g., soil compaction, rutting, puddling, or displacement) and after harvesting (e.g., erosion). Ancillary impacts can result through fuel/oil leaks and spills, emissions, and noise.

The single largest environmental impact associated with harvesting operations is from the access roads, especially at water crossings. The potential environmental impacts of forest roads are:

- long-term or permanent removal of area from actual forest cover;
- soil erosion and sedimentation;
- landslips;
- actual physical impact at the stream/water crossing (e.g., presence of a culvert restricting fish migration);
- impact on water quality and quantity;
- altered water flow (lateral movement);
- compromised aesthetics;
- noise;
- dust;
- calcium chloride if used as dust suppressant;
- wildlife mortality from collision with vehicles, and greater hunting and poaching pressure;
- barrier to or corridor for wildlife movement; and
- opening up wilderness areas by providing access.

The effects of large areas being harvested and (or) reset to earlier stages of succession and stand development are also important, but are addressed in Chap. 12.

Planning and control to minimize negative environmental impacts

The key to minimizing environmental impacts of forest harvesting is good advance planning, implementation, and control of operations. This requires conducting onsite inventories of the compartments to be logged, supported by aerial photographs, GIS information, and other site-specific data that may be available. Field surveys must be conducted to obtain sufficient information to make an informed decision with regard to the most appropriate logging methods and systems to use, and the timing of the actual operations. At the same time, any risks associated with the harvesting operations can be minimized. The professional competence of the planner is key to minimizing environmental impacts.

Harvesting entails a chain of phases, the timing of which means either success or failure of the operation. From an industrial point of view the aim of harvesting planning is to supply a continuous, even flow of wood to the mill of the highest and most appropriate quality, and at the least cost with minimal environmental impact. To obtain the benefits of integrated planning, the same person or planning team deals with both silvicultural and harvesting planning. A *criterion* is a standard, rule, or test by which a judgement or decision can be made during the processes of planning and implementing objective(s). The plan objectives could, for example, be wood volume maximization, profit maximization, cost minimization, risk minimization, or a combination of these. *Control* is the process of ensuring that the objectives of the plan are achieved.

To plan successfully a forest harvesting planner/manager requires considerable data and information to make sound decisions. These data would include the logging chance factors (described in subsequent sections and sidebars) and the following:

- stand/compartment data
    - forest inventory (species, age, area, size, height, quality, volume)
        - siteclassification
        - terrain conditions (operability — slope, soil-bearing capacity, rockiness)
        - susceptibility to insect and fungal attack
        - location with regard to mill and road network
- infrastructure data
    - road network (current and planned)
    - railway lines, spurs, and sidings
    - rivers, lakes, jetties
- constraint data
    - environmental constraints (e.g., buffer strips, parks, terrain limitations, nature reserves)
    - terrain features
        - slopes, ridges
        - rivers, potential crossing points, bridge sites
        - non-passable sites, rocks, wetlands, deep gullies
        - location of road building materials
    - temporal (season of harvest, and fish spawning, tourism, and fire restrictions)
    - spring break-up period
    - labour availability, cost, and skill

  - mill requirements (volume, size, species, quality, delivery schedule)
  - legal/contractual (forest management agreements, union agreements)
  - cash flow constraints
  - capital availability
- technology data (harvesting, hauling, loading)
  - quantity of equipment
  - equipment age
  - operating and capital costs
  - productivity (in various operating conditions)
  - operating limitations (slope, size of landing required, road quality required)
  - implications for silviculture (quantity and distribution of slash, compaction,
  - effect on regeneration, site damage)

MacDonald (1999) presents decision keys for selecting the most appropriate harvesting equipment based on a number of the above factors. Historically, there was a tendency for a single harvesting system to be used throughout a company or division due to familiarity/experience with the system, and the desire to use existing equipment and contractors. Today, harvesting operations are beginning to rely on a greater diversity of logging systems that are used where most appropriate.

Harvesting planning process

As for most forestry activities (as reviewed in Chap. 1), there are three or four levels at which harvesting operations are planned: strategic (10–20 years), tactical (3–5 years), annual plan/budget (1 year), and operational (immediate to 1 year) (Brink et al. 1995).

Strategic planning (Chap. 11) forms an integral part of the long-term management strategy for a particular forest. The focus of strategic planning is in articulating forest-level management objectives. From a timber harvesting point of view, it identifies wood products based on markets and market trends, and matches the market with fibre resources available, while addressing political, economic, social, and environmental concerns.

In tactical planning (Chap. 12) the focus is on the macro-environment, and the goal is to balance the environmental and silvicultural requirements with harvesting and transport personnel and equipment. In terms of forest operations, the aim for the tactical plan is to match the most appropriate equipment with the silvicultural systems prescribed and the logging chance factors for particular areas designated for harvest: i.e., it matches the planned activities to spatially explicit constraints and opportunities on the land base. In the tactical plan a schedule of compartments to be harvested over the next 5 years is developed based on the forest management plan. The tactical plan provides the major inputs into the annual plan/budget.

The annual plan/budget ensures financial requirements are provided for in the operations scheduled for the year. The final choice of silvicultural system and its harvesting system is usually made at this stage (Chap. 13). The operational plan is a refinement of the tactical plan and deals with the actual implementation of the tactical plan. Quite often the operational plan is combined with the annual plan and budget as the basic working unit of harvesting operations and stand management. The focus of the opera-

---

Box 15.2. A fully written harvesting plan includes at least the following information:

- map in the appropriate scale (e.g., 1:10 000), indicating
  - roads;
  - cut block and compartment boundaries;
  - sensitive and difficult areas;
  - possible machine trails; and
  - landings;
- logging methods and systems used;
- compartment volumes by species and assortments;
- time schedule and budget for road construction;
- time schedule for compartments and logging phases;
- person-days required;
- machines required (number and hours);
- cost calculations by phases for each compartment; and
- detailed budget for the planning period.

---

tional plan is the compartment level, with definition of exact boundaries for equipment application and the detailed scheduling of all harvesting phases (Box 15.2).

When determining cutblock and compartment boundaries there are certain aspects related to the cutting and transport operations that need attention. Depending on the uniformity of the area, one aim would be to follow the natural boundaries of the landscape (e.g., rivers, ridges), while allowing easy access to the road network. In previously established plantations the detailed delineation generally would follow compartment boundaries as close as possible.

The detailed cut layout is also influenced by silvicultural considerations, harvesting methods, and systems used, and all the logging chance factors listed below in the section on factors influencing harvesting operations. There may also be higher level directions or government regulations on the size and shape of clearcuts, the need to leave coarse wood debris, residual trees, corridors, buffer strips and areas adjacent to previously clearfelled areas, and on the use of different harvesting methods or systems.

At the same time the planner must identify the possible alternatives applicable to the planning area and to each compartment: e.g., clearcutting versus various retention alternatives, labour intensive versus mechanized systems, reach-to-tree feller-buncher (Fig. 15.2) versus drive-to-tree feller-buncher, ground transport versus cable yarding, forwarding versus skidding, trucking versus rail, etc. An analysis of the implications of each alternative in relation to meeting the objectives and resource constraints is then required. Also, potential production and estimated costs in relation to manpower and machinery must be calculated and a detailed budget developed. The need for seasonal breaks (e.g., spring break-up) or seasonal logging must also be accounted for. Following the analysis a choice is made of the most suitable alternatives; however, they must be technically feasible, fit into the silvicultural prescriptions, be financially sound, be socially worthwhile, and fit within the environmental guidelines. Any number of differ-

Fig. 15.2. Reach-to-tree feller-buncher.

Photo by Reino Pulkki

ent alternatives could be chosen based on the characteristics of each individual compartment. The decisions made during this phase are also used in the detailed delineation of the area.

The choice of the primary transport equipment, and the type, standard, spacing, and layout of access roads must be selected. The sequence of the road construction phases must be determined and a budget developed. Also, at least a rough outline for transport directions and landing (wood storage/centralized processing area) locations are needed before a complete map of the logging site can be made.

The above phases are done before the final preparation and writing of the harvesting plan. The following steps are generally not included in the annual harvesting plan, but are done prior to the commencement of the harvesting operations.

Depending on the compartment size it can be divided into strips or other work units for management purposes. This helps staff in the supervision of the operation. These divisions are also necessary when piece work or any other production-based payment system is used. In any case all cut block and compartment boundaries, access roads, landings, and machine trails must be located and clearly marked using an established colour code for the organization (i.e., special flagging tape colours for side lines, end lines, landings, roads, reserves, corridors, etc.).

In many areas, and especially in selection logging, identification of the trees to be removed is commonly done in advance. Basal area is often used as the basis for future development of the stand and as a criterion for its thinning intensity (Chap. 13). When marking the trees to be removed, selection will also be made for sawlogs, veneer logs, pulpwood, etc. An axe, ribbon, or paint can be used for marking.

The exact locations of landings must be determined before logging operations begin. The purpose of a landing is to allow for effective handling, storage, and loading of wood. Therefore, there must be sufficient space to allow efficient wood piling and maneuvering of equipment. The best location for landings is on firm ground that is slightly elevated to allow good drainage, and optimally spaced in relation to skidding distance. Turn-around for trucks must also be constructed close by if the road is not looped. Skid trails or machine trails (these are not access roads for haul trucks but developed trails for off-road transport) are also often located within the logging area. This is done to increase production (e.g., especially when using forwarders, Fig. 15.3, operating at a maximum off-road transport distance of 600 m), to help avoid damage to the remaining stand, and to speed up and aid the cutting operations. Proper design and location of machine trails (Box 15.3) can greatly enhance operational efficiencies and minimize negative environmental impacts.

In many cases machine trails must cross small drainages or wet areas with poor bearing capacity. At small drainages temporary bridges and PVC (polyvinyl chloride plastic) pipe bundles (Mason and Moll 1995) are available to minimize the impact of equipment crossing at these sites. In wet areas special load distributing mats or gratings (e.g., Mason 1992) can be placed to minimize soil rutting and compaction. However, the

Fig. 15.3. Example of a forwarder (note the topped stem to produce a perch for raptors).

Photo by Reino Pulkki

Box 15.3. When locating machine trails the following points are considered:

- correct spacing;
- start planning from landing sites;
- condition in relation to transport volume (i.e., do not over-build);
- following easy routes through the terrain;
- alignment in direction of slope on steep terrain;
- use natural openings in the stand, if any;
- avoid soft spots, side slopes, swamps, rock outcrops, etc.;
- make short cuts for long trails;
- make wider trails if necessary to build over bad spots;
- make trails in loops to allow easier transport (especially in thinnings and selection cutting);
- avoid sharp bends in trails; and
- avoid adverse grades and if necessary plan so that you drive downhill when loaded (unless there is a risk of the load sliding forward, in which case you would back down steep hills).

---

best solution is to avoid machine travel in these areas when there is the potential for site damage (e.g., schedule the harvest during winter when the ground is frozen).

Role of the workers and supervisors

You can have the best plan developed, but it is the final implementation of the plan by the worker on the ground that will make or break it. The skill and motivation of the operator to economically implement the plan with minimum environmental impacts are crucial. The skill and motivation of the supervisor is critical in the layout of the roads and harvest areas, and this requirement will become more intense as special forms of harvesting are implemented in the field (e.g., partial cutting, thinning, special cutting patterns emulating natural disturbance patterns).

Operators and supervisors also require continuous training in proper operating techniques, especially in partial cutting, and in the basics of silviculture and ecology. As in all situations post-harvesting audits to determine damage levels to residual trees and the site, and to determine if the plan objectives have been met, are required as part of the control function and are one of the roles of the supervisor. Such "compliance monitoring" is best integrated in an overall system of adaptive management and continuous improvement (Chap. 21).

Forest harvesting operations

Road engineering

General principles

An adequate forest road network, composed of primary, secondary, and feeder roads, is necessary. Logging and hauling are generally year-round operations with truck hauling being suspended only during spring break-up and very wet periods when the bearing capacity of roads is insufficient to support heavy loads without damage to the road structure. Also, with the increase of intensive forest management, a permanent, well-planned network of geometrically designed forest roads is required to meet the demands of the hauling vehicles, and to ensure road safety and minimal environmental impact. In addition, forest roads are not "everlasting" structures and require periodic maintenance work. If the road is designed for temporary access for harvesting and regeneration, and (or) regular maintenance is not performed, then road deactivation is required.

The basic requirements when constructing forest roads are to build them as economically as possible, while minimizing any environmental impacts and maximizing traffic safety. This is especially true when constructing feeder roads because of the large number required and since they are often deactivated after operations in an area have ceased. Even if the feeder road forms part of a permanent network it still needs to be economically built, since the use of the road by heavy vehicles is infrequent.

Obtaining the most economical road possible requires good road planning, geometric design and locating, the wise use of road construction materials, the maximum use of materials found directly at the road construction site, and the use of drainage and compaction to increase and maintain the load bearing capacity of the road structure. It also requires that lower standard forest roads are not exposed to maximum loads during periods of reduced bearing capacity (e.g., spring break-up). Most forest roads (i.e., secondary and tertiary roads) are only single lane, thus meeting points (turn-outs) must be strategically located to allow vehicle passing. Also, turn-around points must be located on non-looping roads.

To minimize road construction costs and environmental impacts, roads must be planned well in advance of construction, and construction preferably occurs at least 1 year before harvesting operations begin within an area. This is because sufficient time must be reserved for the road construction work to allow:

- various work phases to be carried out during the most economical time of the year and when negative site impacts will be the least;
- construction work at various work sites to be linked; and
- roadbeds to dry and (or) stabilize before being subjected to heavy traffic.

In this way the same machines can be offered a maximum amount of work and the use of labour and supervisory staff optimized.

The forest road

General — A forest road can be defined as an engineered structure constructed to provide access to the area for the various forestry-related operations and forest uses. The forestry related operations, in addition to harvesting, could include cruising, pre-harvesting inspections and prescriptions, marketing, reforestation or afforestation, stand tending, stand protection (including fire control), and stand improvements. The various forest uses could include wood fibre production, rangelands, water control and production, wildlife management, recreation, and the production of other natural products (e.g., mushrooms and berries). To provide access to the immediate stand area whenever desired and in times of crisis (e.g., forest fires, insect outbreaks, or disease epidemics), in an intensively managed and multiple-use forest, a well-developed forest road network

> **Box 15.4. There are many factors that influence the layout of and the spacing between forest roads. Among others, some of these factors are:**
>
> - terrain conditions;
> - stand density or volume to be removed per hectare;
> - primary (off-road) transport cost;
> - secondary (truck) transport cost and demands;
> - road construction cost;
> - road maintenance cost;
> - forest area covered by roads and thus taken out of production (the opportunity cost of the land not growing wood);
> - level of forest management practiced (extensive versus intensive);
> - demands of other potential road users; and
> - environmental protection factors.

is essential. On the other hand, some resource management priorities (e.g., for wilderness values, or for maintaining healthy populations of edge-adverse species; see Chap. 12) demand that road development be minimized. Furthermore, all the above-mentioned forest activities do not require the same standard of forest road. The spacing and thus the density of the road network required for the various forest uses also varies.

Road network layout and optimum road spacing — There are a considerable number of factors that affect road location and the road network density required. However, one large source of error in road construction is building roads with no overall planned layout for a region: i.e., building one road this year, then building another road next year where it fits best, etc., without thinking about the area as a whole. As a result a lot of

Fig. 15.4. Relationships between total road and off-road transport cost (Ct), variable off-road transport cost (Cv), fixed off-road transport cost (Cf), and road construction and maintenance cost (Cr) when employing a grapple skidder.

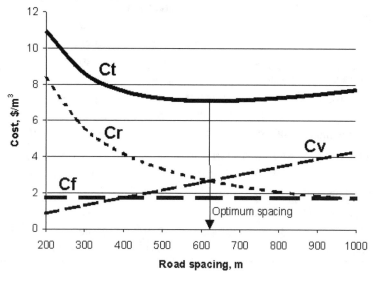

redundancy in road service areas occurs, and often the wrong standard of road is built (e.g., Smith 1989).

The road spacing that minimizes the total cost function for Box 15.4 is the best to employ. However, many of these factors are difficult to quantify and most often only the costs of off-road transport, and road construction and maintenance are used to determine optimum road spacing (Fig. 15.4).

Forest road standards — Of all the activities occurring in the forest, the logging operations impose the greatest demand on the road system. It is for this reason that forest roads are more or less constructed to take into account the needs of the logging operations, while also meeting environmental protection and safety standards. The side benefits of a high standard road network then spin off to the other forest activities and users. In cases where afforestation is the first activity, the forest roads are usually constructed to a low standard and are improved when thinning and (or) final harvests commence in the area. Intensive forestry in plantations and the other forest users require continuous access to the forest and thus a permanent forest road network. In areas of more extensive management, as is the case in many parts of the boreal forest, only temporary forest roads are required. In any case, no matter if the road is permanent or temporary, it has to be able to support the expected maximum loads and allow the desired speeds over its intended life.

There are three major international categories of forest roads: primary, secondary, and feeder (= tertiary). Most organizations use similar classes, and some also introduce a fourth class. When working with four road classes they can be referred to as primary, main, area, and branch forest roads. In this classification, secondary and feeder roads are broken into three classes. Winter roads are a special class of forest road not included in the above categories. Table 15.1 presents design characteristics for the three international road classes, while Table 15.2 presents design characteristics for main, area, and branch forest roads; the primary road standards are the same whether using three or four road classes.

Table 15.1. International design characteristics for forest roads constructed in different terrain types (McNally 1977).

		Characteristic					
		Design speed,	Min. radius of curvature,	Max. gradient,	Max. length of grade,	Road width,	Travel surface width,
Road type	Terrain	km/h	m	%	m	m	m
Primary	Flat–rolling	80–100	190–360	4	None	10–13	6–7.5
	Hilly	55–80	90–190	5–7	600 >4%	10–13	6–7.5
	Mountainous	40–55	50–90	7–9	400 >6%	10–13	6–7.5
Secondary	Flat–rolling	60–80	110–190	5	None	10–12	6–6.8
	Hilly	50–60	75–110	5–7	None	10–12	6–6.8
	Mountainous	35–50	35–75	7–9	750 >6%	8–9	6–6.8
Feeder/ tertiary	Flat–rolling	50–60	75–110	7	None	7.5–8.5	5.5–6
	Hilly	35–50	35–75	7–9	None	7.5–8	5.5–6
	Mountainous	25–35	30–35	9–12	1000 >9%	7.5–8	5.5–6

Table 15.2. Design characteristics for the lower three classes when using four forest road classes (i.e., the primary road class is the same as in Table 1) (Metsähallitus 1970).

Road characteristic	Road class		
	Main	Area	Branch
Design speed, km/h	30–60	30	30
Min. sight distance, m	70–150	70	70
Min. radius of curvature, m	50–200	50	50
Max. adverse loaded gradient, %	8	10	10
Max. length of loaded adverse grade, m	60 >12%	60 >12%	60 >12%
Road width, m	4	3.6	3.6
Pavement thickness,[1] cm	5–60	5–40	0–20

[1]Dependent upon the bearing capacity of the subgrade (pavement refers to sub-base, base-coarse, and surface layers).

Forest road engineering guidebooks (e.g., the B.C. Forest Practices Code, BCMOF and BCMELP 1995) also contain very detailed information on the geometric design of forest roads.

The standard of forest road to build depends upon many factors. The expected maximum axle loads to travel over the road play a significant role. One major function of the road pavement (that portion of the road over the natural underlying soil) is to distribute the applied axle loads onto the subgrade so the maximum stress on the subgrade does not exceed its bearing strength. Therefore, the higher the expected axle loads, the higher has to be the load distributing quality of the pavement. This is usually achieved by increasing the thickness of the pavement or by using higher quality road construction materials.

The *desired speed* below which the loaded vehicles should not have to travel, other than for traffic congestion, is another factor determining the standard required. The higher the desired speed, the wider have to be road right-of-way, shoulders, travel lane(s), and the horizontal curves. Also, the vertical curves must be reduced and the minimum sight distance increased. Thus desired speed is a major factor in determining the scale and footprint of all roads. Reducing transport speeds in an effort to reduce the necessary widths and lines of sight can be used to reduce road impacts in special management areas, but may have consequences in terms of truck turn-around times and overall productivity. Reduced road speeds also have a major impact on improving road safety.

The periods throughout the year when the forest road can be normally used at its designed load capacity is known as its *availability*. According to their availability, forest roads can be divided into year-round, all-weather (except during spring break-up and perhaps fall freeze-up), dry weather, and winter roads.

The road surface also tends to wear with use, and with repeated loading construction material, fatigue may develop in the pavement and subgrade. Therefore, roads with high *traffic densities* must be constructed to higher standards to resist the effects of traffic wear and fatigue. The *expected life* of the road also has a major influence on the number of load repetitions applied to the road. Also, if a permanent road is desired a high standard road is required to minimize the future maintenance costs. However, if a temporary road is required, future maintenance costs are irrelevant and a road suitable for the period in question is constructed.

In an intensively managed forest the *feeder forest roads* form the fine infrastructure providing access to the compartments. Feeder roads are the lowest standard of the forest roads and are generally single-lane, with vehicle meeting occurring at turn-outs; they have no or little shoulder; traffic intensity is light; and the periods of use are usually brief. It is important to note that feeder and branch roads can have lower design characteristics than presented in Tables 15.1 and 15.2. The major criterion in feeder road construction is that they have sufficient supportive strength to carry the maximum axle loads expected to travel over them without structural damage occurring to the road. In feeder road design, vehicle speed is not as important as the road construction costs. The distance required to travel along a feeder road, before it hooks up to a secondary road, is generally not too great: in most cases, less than 4 km. However, the total length of feeder roads required to service an area can be considerable: e.g., for off-road transport using skidders a road density of 1.7 km/km^2 (16.7 m/ha) is generally recommended. Since feeder roads are used sporadically and at low intensities, it is important to construct them as economically as possible. Therefore, if construction conditions are poor they may be constructed winding and narrow, as long as the road can be negotiated by fully loaded hauling vehicles. In certain situations it may be necessary to construct feeder roads that are capable of supporting only partial loads. This may be necessary if the natural underlying soils have poor bearing capacity, the road is short, the wood volume to travel over the road is small, and (or) road construction materials are scarce.

Secondary forest roads are intermediate in standard between the primary and feeder forest roads. The major design criterion for secondary roads is to obtain higher speeds. Vehicles can meet with priority being given to the loaded vehicle. Light and unloaded vehicles must slow down or even stop to allow loaded vehicles to pass at the design speed. The secondary roads connect the operating areas together and join up with the *primary forest roads* that generally link together the operating or management units. The primary forest roads are typically designed for high travel speeds and high traffic densities. A primary road has sufficient lane width to allow vehicles to meet at the design speed along any stretch of the road.

In countries that have a well-developed rural public road network, primary forest roads are generally lacking. In many instances there is no need for secondary forest roads either and thus the majority of the forest road construction effort is concentrated on feeder roads: i.e., area and branch roads according to finer classification presented in Table 15.2. In most of Canada, however, because of the large unaccessed areas, all forest road classes are required, and great care must be placed on planning the overall forest road network.

Structure of a forest road — A typical forest road cross-section is presented in Fig. 15.5. The two major structures of a road are the *subgrade* and the *pavement*. The subgrade is composed of the parent material of the roadbed and forms the underlying structural layer of the road. It may be in an undisturbed or disturbed state. The subgrade is in a disturbed state if it has been moved across to the roadbed from an embankment or ditch, or along the length of the roadbed from a cut section, and placed on top of the underlying undisturbed subgrade. The pavement is composed of all the material placed

Fig. 15.5. A typical road cross-section.

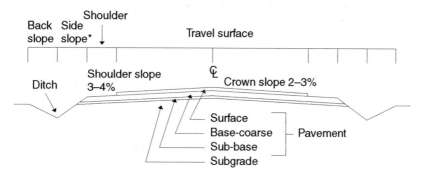

*From 2:1 to 4:1 depending on road class (horizontal:vertical)

on top of the subgrade and can usually be subdivided into a *surface layer*, a *base-course layer*, and a *sub-base layer*. However, in feeder roads the pavement is usually composed of a single layer of pit-run gravel. If there is a differentiation of layers, the quality of the material increases from the subgrade to the surface. Each overlying layer must have sufficient strength to resist the loads imposed on it and be thick enough to distribute the load on the underlying layer so it is not over-stressed. The pavement and the subgrade must be well compacted and together be able to support the maximum axle loads expected.

The sub-base layer is used to separate the upper layers of the pavement from the subgrade and is usually composed of fine sand. This breaks the capillary flow of water from the subgrade into the pavement, and thus prevents weakening of the pavement and frost heaving. The base-coarse layer is generally a coarse gravel material that gives the road structure its strength. For it to function fully it must also retain its strength irrespective of the amount of water in the road structure. In general, the base-coarse layer can be held as the most important layer in the road pavement. In forest roads the surface layer is composed of finer, high-quality gravel or crushed rock. It forms the wearing layer of the road and must resist the wearing forces imposed on it and provide a smooth running surface. Also, the surface material used must resist longitudinal (e.g., washboarding) and perpendicular (e.g., surface material thrown off to the side of the road) movement, minimize the amount of dust, and provide a hard surface with proper crowning to allow quick drainage of water to the ditches (i.e., water cannot be allowed to seep into the road structure).

Ashworth (1966) lists six requirements that a road pavement should fulfil:
- the pavement should be of sufficient thickness to spread the surface loading to a pressure intensity which the subgrade is capable of withstanding;
- the pavement itself should be sufficiently strong to carry the stresses imposed upon it;
- where necessary, the pavement should be sufficiently strong to prevent damage to a frost-susceptible subgrade;
- the pavement should be composed of frost-resistant materials;

Table 15.3. Maximum degrees of horizontal curvature subtended by a 30.48 m arc at various safe design speeds and super-elevations (Paterson et al. 1976).

Safe speed, km/h	Super-elevation, %					
	0	1	2	3	4	5
96	3.10	3.35	3.55	3.80	4.05	4.30
80	4.80	5.15	5.50	5.85	6.20	6.50
64	8.20	8.75	9.25	9.80	10.35	10.90
48	14.60	15.50	16.50	17.40	18.40	19.35
32	32.80	34.90	37.10	39.20	41.40	43.50

- the pavement should be impervious to the penetration of surface water which would weaken both the pavement and the subgrade; and
- the pavement surface should have a skid-resistant texture.

An additional requirement for a pavement surface is to provide a smooth running surface. A smooth surface imposes less strain on the hauling equipment, since the wheels roll with a minimum of resistance. A smooth surface also allows good surface drainage of water away from the pavement. Road surfaces generally have a 2–3% crown slope to allow run-off and thus prevent water from building up on the surface. On corners, it is recommended that the road contain a 1–5% super-elevation ("banked curves") to offset centrifugal forces (Table 15.3). The super-elevation must be over the entire width of the road, and the change from regular crown slope to super-elevation must gradually occur over a 30 m (at 48 km/h) to 53 m (at 96 km/h) distance (Paterson et al. 1976).

Forest road locating procedures

The following is a general procedure for locating a forest road network or road:
- obtain from higher level plans the general area to be accessed by the new road, its development objectives, and the construction and design standards intended for it;
- study aerial photographs of the area stereoscopically; locate the key areas (control points) through which the road must pass: e.g., possible bridge sites, saddles in hills, ridges between swamps, gravel pits, etc.;
- upon location of key points, join the points in such a manner as to give the shortest practical distance, while minimizing environmental impact risks and high construction cost risk;
- transfer these proposed try-lines to a base map and determine the bearing of each tangent; several routes or variations are usually selected using aerial photographs and stereoscope;
- wet and rocky areas are preferably avoided when making your initial route selection; since the soil along the road route usually forms the subgrade, it is preferable to select routes with gravel and sand;
- in general, highland types are followed in order to avoid wetlands and other drainage problems; good stands of pine, or mixed pine–spruce stands usually indi-

cate a sandy or gravel soil; the presence of fir and (or) aspen, mixed with pine, may indicate a loam, clay, or silty soil;

- gravel is often found where birch and pine are found growing together, or in esker ridges; often eskers form an excellent roadbed;
- drainage is rarely a problem when the roadbed is on a sandy outwash plain because of the porous subgrade;
- when the soil is clay or silty in nature, locate the road on a gentle side hill so that there is good drainage; fill can also be dug from the upper slope; southern slopes are preferred to northern slopes (i.e., faster drying because of the sun);
- to allow water drainage away from the road, long stretches of flat ground are avoided;
- where practical a take-off grade for side ditches of 2–4% is recommended;
- a curve must not appear on or at the bottom of a steep grade, or at the approach to a bridge crossing if it can be avoided;
- grades in excess of 200 m should not exceed 4% because of difficulty experienced by loaded and empty trucks in certain weather conditions;
- on single lane roads vehicle meeting points are located every 300–400 m and allow two haul trucks to pass with no problem (i.e., road travel surface with 8 m width and 50 m length);
- "walk" the various route alternatives; use of a hand compass and measuring distance by pacing or with a string-box is sufficiently accurate at this point; record the presence of various soil types, grades, and indications of gravel deposits; to allow re-alignment of the road blazing is kept to an minimum; stakes or flagging tape may be used in lieu of blazing;
- soundings taken in swamps and at bridge crossings are very useful in selecting the best location; fix your location on the air photo in relation to the "try-line" tangents;
- after all "try-lines" have been walked, select the best location; the final location may be a combination of several of the initial "try-lines";
- ribbon out the final location selected; if there are any questions or problem areas have your supervisor "walk" the location with you;
- frequently road location crews put numerous small curves into a road location to avoid small rock ledges, wet areas, etc.; where the cost is not great, it is best that the road pass over or through these obstacles (i.e., less road has to be built and hauling is cheaper and safer); too frequently construction crews must return to such sites to straighten out a curve or cut a rock to improve visibility; at critical points, as above, detailed cost calculations are beneficial;
- a three-person crew is usually the most efficient in locating a road (compass man, ribboner, combination chainman – ribboner); and
- for winter roads, stay on flat ground and avoid side-hills.

Forest road construction methods

Work method choice — Various methods are used in forest road construction. The choice of work method affects the machine types applicable and the final cost. However, it is quite clear that the choice of the most suitable method depends upon the terrain conditions, the size of the work site, and the availability of the various types of machinery.

The two basic methods employed are the crawler tractor method and the excavator method.

The crawler tractor method is most suitable on ridges and in hilly terrain where there is gravel or sandy soil, and the cut-and-fill work follow each other. In these situations the fill material is pushed along the roadbed. Grubbing, cut-and-fill work, and compacting of the subgrade material is best done during the summer in dry soil conditions. The crawler tractor method is also best for spreading the granular fill required for the sub-base, base-coarse, and surface layers, and results in the best compaction of the road structure.

The excavator method is most suitable on low-lying and frost-susceptible soils, as well as on peatlands, where it is important to raise the road structure and to lower the ground water table. With the excavator method the fill material is moved laterally to the roadbed. Grubbing, ditching, culvert work, and subgrade formation can begin in spring or early summer, since the method can be employed efficiently in wet soil conditions. In the spreading of granular fill and compaction of the road structure the excavator method is not as good as the crawler tractor method.

On steep slopes there are additional advantages of the excavator method. This is especially true if the cut material cannot be placed onto the lower slope (i.e., side-cast) on account of the risk of overloading the slope and causing failure. In these cases the cut material must be hauled to a "waste" site that is more stable. Kardoz (1993) states that side-cast failures usually occur on slopes greater than 70%, but can occur in wet morainal soils with slopes as low as 45%. Other advantages of the excavator method in steep terrain are less site disturbance, better sorting of materials, selective handling of material, better ditching, and the ability to remove stumps and rocks from the subgrade (Kardoz 1993).

The above are not competing methods and should be used in harmony so that the method most suitable for the situation at hand is employed. On large road construction projects both types of equipment may be used. In this case each machine type can be employed in the most beneficial way. However, the popularity of the excavator method has increased considerably in recent years. The increased use of excavators for forest road construction has resulted in more carefully built roads and reduced environmental footprints.

Right-of-way clearing — Once the road has been located in the terrain (i.e., centre-line or right-of-way (ROW) edges marked) the standing timber must be removed. The cleared ROW width depends on the road class, design speed, road horizontal curvature, environmental considerations (e.g., narrow ROWs where bank stabilization on slopes is required), and the logging system used (i.e., full tree and tree-length systems require a wider ROW). A wide ROW increases the line of sight along the road (especially around curves), and allows sunlight and wind to dry the road structure. However, a narrow ROW may be desired to shade the road to keep the road structure moist and therefore reduce dust problems. Another consideration is the area removed from productive forest or biodiversity habitat as ROW width increases. In general, on straight forest roads, ROW width varies from 15 to 40 m for secondary and primary roads (Table 15.4). On feeder roads the ROW width can vary from 12 to 20 m; however, standing timber must be removed to at least 2 m of the outside edges of the roadside ditches. In most cases a

Table 15.4. Recommended cleared right-of-way widths for secondary and primary forest roads for visibility and traffic safety (McNally 1977).

Design speed, km/h	Cleared right-of-way width, m
30	20
40	22
50	25
60	27
70	30
80	33
90	35
100	38

minimum standard of 20 m is used when not near environmentally sensitive areas. An important consideration is that minimizing ROW width wherever possible reduces the area of forest land required for purposes of access.

During the marking of the road location, the ROW width at road widenings (e.g., meeting points), turn-around spurs, and landings are increased accordingly. Also, the locations of roadside borrow pits and gravel pits must also be marked. These areas must be cleared of standing timber at the same time as the regular ROW clearing work. Also, the ROW must be widened on corners and at intersections to increase the line of sight and thus traffic safety. Near stream crossings (within 100 m of sensitive water crossings or water bodies) it is recommended that ROW width be decreased to a minimum (OMNR 1988).

Landing locations for the timber being removed from the right-of-way must also be marked to ensure they do not interfere with the grubbing, and cut-and-fill work. If landing locations are not marked the piles often end up too close to the centre-line or on good roadside fill sources.

Right-of-way clearing work could occur during any season, as long as rutting does not occur. However, since the off-road distances are often great and the terrain traversed variable, off-road mobility and access factors must be accounted for in the proper timing. Right-of-way clearing in the winter allows cutting on all soil conditions with minimal rutting and access problems. Cutting in the winter also minimizes the period of wood storage during the summer period and thus reduces the extent of wood deterioration and the risk of forming insect breeding sites (e.g., bark beetles and sawyer beetles). It is also best that the ROW clearing work occurs well in advance of the grubbing work to allow for initial drying of the soil and efficient scheduling of the subsequent work phases.

After the ROW has been cleared, the road centre-line, vehicle meeting points, turn-around spurs, and intersections are restaked. Ditch locations must also be staked, and in some cases, the area to be grubbed is also indicated (i.e., to minimize the extent of area grubbed).

Box 15.5. Good practices when working at water crossings (OMNR 1988; NBD-NRE 1994; MDNR 1995; WDNR 1995):

- Minimize amount of bare/exposed soil; no grubbing within 100 m of a stream crossing;
- Ensure proper structure type is chosen and that bridge opening or culvert size is appropriate for expected peak flow (e.g., 25–100-year flood);
- If building a permanent road, use permanent structures;
- On temporary roads, temporary bridges, culverts, or fords can be used;
- Install the structure at the driest time of the year and never during fish spawning/egg incubation periods;
- Cofferdams can be used in bridge construction but must not block more than one third of the waterway at a time;
- Install the structure as quickly as possible and minimize the number of machine crossings;
- A foundation of compacted good gravel is required under culverts to prevent sagging and sinking;
- The culvert length must be sufficient to prevent embankment material from plugging it;
- The culvert must be placed so that water flow and fish passage are not impeded (i.e., perched culvert). Depending on local regulations a culvert in a natural watercourse is placed into the stream bed to a depth of 10% of its diameter or at least 15 cm);
- In fish-populated natural watercourses the maximum slope of culvert should not exceed 0.5% to allow for fish passage;
- Use silt fences to prevent water contamination and sedimentation;
- Armour culvert ends and embankments, and use geotextiles and geogrids where appropriate to prevent bank/embankment erosion; and
- Follow good housekeeping standards and clean up the site once finished.

Water crossings — Since the largest impact of forest roads is at water crossings, great care must be taken during the actual selection of the crossing location, during the construction period, during the life of the road through proper road maintenance, and during removal of water crossing structures if the road is deactivated. Best practices when working at water crossings are given by a number of "best management practices" (BMP) manuals (e.g., OMNR 1988; NBDNRE 1994; MDNR 1995; WDNR 1995; Box 15.5).

Watershed run-off calculations are required before deciding on a culvert size or if a bridge is needed. Depending on the legislation for an area and intended life of the road the peak run-off is usually determined based on data or projections for 25–100 years. That is, design specifications for stream crossings must accommodate peak water flows that may only occur once in 25–100 years. Site surveys for bridges and major culverts are also required in most jurisdictions.

Based on the culvert calculations and the road plan, the proper number and sizes of culverts for the road must be obtained and transported to the work site. If a bridge is necessary a detailed schedule for the bridge construction is required. This plan would include a schedule for procuring bridge construction materials.

Since culverts should ideally be placed at the same time as ditching and subgrade preparation work, their transport out to the site may be difficult. In general, light culverts are the easiest to install. Also, if good non-frost susceptible fill must be hauled to the culvert location, to form the culvert bed and to fill in around the culvert, the final culvert placement must be delayed until after gravel hauling for the road base-coarse has reached the culvert placement location. For proper culvert placement and bridge construction procedures refer to appropriate manuals and guidelines such as OMNR (1988) and BCMOF and BCMELP (1995). Where unhindered fish migration is a concern, bridges or open-bottom arch culverts (Gillies 2002) are generally considered more appropriate than culverts; at the very least, flat-bottomed culverts with baffles are recommended for such situations.

Grubbing — Grubbing of the roadway surface is the first phase of actual road construction and is generally termed as preparatory work. It entails the removal of stumps, slash, small trees, brush, and rocks from the roadway, and minor subgrade levelling (i.e., cutting and filling up to a depth of 0.5 m; Antola 1988). The surface layer of vegetation and roots in peatlands and swamps supply a natural road base strengthening layer. To keep this layer intact it must not be grubbed; however, the stumps must be cut as low as possible. When ditches are to be dug the width of grubbing extends to the outside edges of the ditches. Grubbing does not usually include cut-and-fill work (i.e., subgrade levelling where cutting and filling depth is greater than 0.5 m), which is a distinct work phase and generally costed separately. The grubbing work can be done with either an excavator or crawler tractor. The choice of equipment depends on the conditions outlined earlier in this section.

In feeder (area and branch) forest road construction, grubbing can be as light as possible so that the surface organic layer is left as much as possible intact. This is especially important on frost susceptible and poor bearing soils, since the organic layer forms a good separation and bearing layer. Light surface grubbing also reduces environmental impacts and the risk of erosion.

In secondary and primary forest road construction, grubbing includes removal of the surface organic layer. This is required on account of higher speeds and the larger number of loads over the road. Because of the larger demands imposed on these road classes, the presence of organic material in the road structure may cause premature road failure. However, the area of roadway grubbing must be kept to a minimum to reduce environmental impacts and the risk of erosion.

Grubbing of all accessory areas, such as meeting points, turn-around spurs, intersections, and wood storage landings, occurs at the same time as the actual roadbed grubbing. Also, the surface organic material must be stripped from all roadside borrow pits and gravel pits. If there is a considerable volume of stumps, debris and rocks, roadside push-outs may be required. This material must not be placed so that water drainage or digging of the ditches is impeded. Also, push-outs must be planned ahead and cleared during the cutting of the road ROW. When the grubbed material is piled in windrows it can be piled at the sides of the ROW beyond the ditches, where ditching work and drainage are not impeded. Also, if large amounts of material are involved, openings in the windrows must be cleared to allow easy access to the forest from the road. If possible, the best alternative is to spread the grubbing debris as evenly as possible.

When using the excavator method in grubbing work, the loosening and lifting power of the machine are important factors. The excavator must be able to lift or move the largest stumps and rocks expected. In some cases a rock or stump may be too large and must be fragmented (e.g., by blasting, special impact hammer, or stump splitter) or covered with fill. Otherwise, the road alignment must be adjusted. The boom reach of excavators is another important factor; a long boom reach allows the excavator to obtain fill from further off the road, to move debris off to the side of the road, and to spread material from ditches and cut sections easier. The ability for the excavator to rotate 360° is another benefit. This allows some movement of fill along the road structure. To avoid excessive costs, large 30-tonne (Mg) excavators are only required in difficult terrain and (or) where there are large rocks and stumps. In most situations on fairly flat to rolling terrain, a medium-sized excavator (e.g., 20-tonne gross vehicle weight class) is most applicable.

As mentioned earlier grubbing with a crawler tractor is best done under dry soil conditions. It is the most applicable method for grubbing in rolling terrain with coarse-grained mineral soils (Antola 1988). When subgrade and fill materials must be transported along the roadbed the crawler tractor method is the most applicable. In many situations a crawler tractor is required for the cut-and-fill work. Therefore, the extra cost for transporting an excavator to the worksite for the grubbing and ditching work may not be warranted. The compaction of the subgrade is also superior with the crawler tractor method. The size of tractor required depends on the terrain conditions. Since forest road construction usually occurs in fairly difficult terrain, a 25–35-tonne class (e.g., Caterpillar D7-D8) crawler tractor is most often required. For very light grubbing and cut-and-fill work, and spreading of fill material, smaller 15–20-tonne class (e.g., Caterpillar D5-D6) crawler tractors are applicable. For proper placement of fill and ditching, the road centre-line and ditch locations must be re-staked after grubbing.

Road base strengthening — Road base strengthening through the use of brush mats, corduroy layers, or geotextiles (fibre fabrics) is often required on deep peat soils, and on wet frost-susceptible clay and clay-loam soils. Brush mats (overlapping branches and slash) and corduroy structures (logs laid parallel) are most applicable on poor bearing and deep peatlands. Geotextiles are applicable on both peatlands, and poor bearing and frost-susceptible clays and clay-loams. Brush mats and corduroy structures are used to "float" a road, raising the structure of the road to allow better drying, and to distribute loads more evenly and over a wider area. They also help to separate and prevent mixing of the road pavement material and the underlying soils.

Brush mats are formed from logging residues and brush, while corduroy structures are built from larger diameter tree-lengths and (or) logs. The two techniques can also be combined by placing tree-lengths along the road structure and the brush crosswise on top. Brush mats and corduroy structures must be built before digging the ditches or placing fill on the subgrade. This allows the fill to be placed on top and totally cover it. This reduces the rate of decomposition, since oxygen is not readily available within the structure. Nevertheless, brush mats and corduroy have a limited lifespan owing to decomposition and hence are generally suitable only for feeder roads.

The main functions of geotextiles (Fig. 15.6) in forest roads are to act as a separating layer to prevent mixing of road pavement and subgrade materials, and to reinforce and improve drainage of the road structure (Provencher 1992; Fannin 2001). By reinforcing the road structure, geotextiles help to distribute loads more uniformly over the

Fig. 15.6. End-dumping gravel onto a geotextile.

Photo by Tony Saint

subgrade and thus potentially increase the bearing capacity of the road. Geotextiles and geogrids are also used to help stabilize and reinforce roadside slopes and embankments, as well as their drainage (i.e., transmission of water through the geotextiles) (Wiest 1998). Geotextiles have a special role at water crossings to prevent erosion and deposition of sediments into the water through run-off. The type of textile chosen will depend on:

- the foundation (surface on which the textile is placed); and
- desired functions and thus the textile characteristics required (Steward et al. 1977; DOA 1992; Fannin 2001; Table 15.5), i.e.,
  - drainage (thickness, permeability),
  - filtration (permeability, porometry),
  - separation (porometry, abrasive resistance, tensile strength), and (or)
  - reinforcement (tensile strength, roughness).

When used, geotextiles are placed on the subgrade fill material to separate it from the road base-coarse layer. When geotextiles are used on peatlands they are placed over the fill material from the ditches. If no ditches are dug the fabric is placed directly on the peat surface and trucked fill is placed over top of it.

Whether using brush mats, corduroy structures, or geotextiles, the stumps on the road base are best removed. If stump removal will cause excessive surface disturbance in feeder road construction it is recommended that they be cut as low as possible because the organic top layer forms a good separation and bearing layer. When placing geotextiles, the separate sections must be overlapped to ensure full coverage after road settling. Sewing the sections together or at least staking the geotextiles to the subgrade

Table 15.5. Geotextile selection guide (Terrafix Geosynthetic, 1991).

Type of application	Required geotextile functions	Recommended geotextiles	Property/ characteristics
Roadway stabilization	Drainage	Thick non-woven	Thickness and permeability
Logging roads	Filtration	Thick non-woven	Pore size and permeability
	Separation	Non-woven/slit film woven	Small pore size
	Reinforcement	Non-woven/slit film woven	Abrasive tensile resistance
		Geogrid - different patterns	Aperature size, % open area
		Uniaxil - primary geogrid	Rib thickness, rigidity
		Biaxial - secondary geogrid	Tensile modules
Erosion control	Filtration	Biodegradable fibre fabric	Straw, coir fibre
	Drainage/separation		Synthetic fibre
Silt barriers for deposition	Filtration	Silt fences (geosynthetic)	Pore size, permeability
		Woven polypropylene	U.V. resistant

will prevent movement of the geotextiles under road use. Sewing geotextiles sections together is always recommended on poor bearing subgrades (Armtec 2002). When hauling fill and pavement material, end dumping (Fig. 15.6) is recommended to prevent damage to the road structure and the geotextile through the formation of ruts. If ruts do occur it is recommended they be filled with additional material, and not with material bladed from adjacent parts of the road pavement (Armtec 2002).

Care must be taken not to load too much weight on deep peatlands. This is because the road structure will gradually sink into the peat. To build up the road structure more gravel is required. This will add additional weight onto the peat, causing additional sinkage or total failure of the road. When dealing with difficult to access areas, due to peatlands, the option to access these areas with winter roads must be seriously considered.

Ditching — Ditching is another roadway preparatory work phase. It usually occurs directly after or in conjunction with grubbing. If suitable fill material can be obtained from the ditches it is generally placed on the grubbed roadbed. Rocks and stumps are separated from this material and placed to the side of the right-of-way. This work phase is best done with an excavator with a special profiled ditching shovel. However, a crawler tractor may be employed if the soil conditions are dry. When using a crawler tractor for ditching work, operator skill is an important factor.

Ditching is probably one of the most poorly done phases in forest road construction. The objective in forest road construction is to build a suitable road at minimal cost. The largest factor influencing road strength and thus road failure is water in the road structure (Pulkki 1982). The purpose of ditches is to lower the ground water table in the vicinity of the road and to direct water away from the road structure. If drainage is effective, poorer quality and less road construction material can be used.

Shallow "V"-ditches from 0.6 to 1.0 m in depth are generally sufficient. The depth depends on soil bearing strength and road class. If more fill is required the ditches can be dug deeper. However, the back and side slopes must be shaped so they do not exceed a slope of 2:1 (50%). On slopes a ditch is usually only required on the upper slope side of the road. The ditch should not be dug closer than three times its depth from the edge of the road surface (Antola 1988). In general, on mineral soils the minimum distance between the edge of the ditch and the road pavement material is 0.5 m; on peatlands this distance is at least 2.0 m (Antola 1988). When ditches are required on both sides of the road, the distance between them generally varies from 8 to 15 m, depending on the road class and soil conditions (Antola 1988). The ditch locations must be staked before ditching occurs. This ensures the ditches are straight and in the right place. At vehicle meeting points the ditch separation must be increased. At turn-around spurs and road intersections the ditches need to lead off into the forest, or a culvert must be placed if the water flow in the ditch cannot be broken. The ditch bottom must be as level as possible to ensure unhindered water flow. Any rocks that cannot be moved to the side of the road ROW can be sunk into the back slope so water movement is not blocked. Access points to the forest are located where damage to the ditch, or possible rutting or erosion, is minimized. These often can be located on heights of land where there is a natural drainage break. Otherwise, culvert placement may be required. Forest tractors must not be allowed to drive across ditches without the use of special portable bridges or the placement of a culvert.

The Ontario Ministry of Natural Resources (OMNR 1988) presents further techniques and considerations in ditching (summarized in Box 15.6), as well as techniques for mitigating environmental impacts. Some of these techniques involve the use of water bars, earth berms, sediment traps, check dams, silt fences, and diversion channels.

Cut-and-fill work — Cut-and-fill work always follows the roadway grubbing, roadbed strengthening, and ditching work. In this way the cut fill material is placed on top of the fill obtained from the ditches and on the prepared roadbed. Generally, cut-and-fill work is always required in rolling terrain. Cut-and-fill work also includes the formation of the ditches in the cut sections.

To minimize the amount of cut-and-fill work the objective is for the volume of material cut to be sufficient to fill the adjacent fill sections. In some cases deeper cutting may be required to reduce a grade to an acceptable slope. The volumes of cut-and-fill material are calculated and balanced when making the longitudinal road profile during the road planning phase. The volume cut would also include material from roadside ditches in the cut section. The cost for the cut-and-fill work is based on the volume which must be moved, the distance to the adjacent fill section, and whether the fill material must be hauled to a disposal site. Elevation stakes can be used to control the depth of cutting and filling work.

Box 15.6. Good practices when ditching (OMNR 1988):

- Ditching done during the subgrade preparation work encourages good drainage before gravel hauling begins;
- Dig ditches while working uphill to avoid trapping water;
- Use angled cross-culverts to direct water across the road structure (spacing depends on slope and soil type);
- Do not discharge ditches directly into waterways, but into the forest where the vegetation acts as a natural filter;
- Locate side ditches into the forest at regular intervals (spacing depends on slope and soil type) to prevent excessive flow and erosion potential;
- Check dams may be necessary to retard flow velocity on long slopes >3%, where side ditches are not possible;
- Interceptor ditches located off to the side of the road can be used to divert water away from steep slopes or cut sections;
- Locate landings and access points to the forest on high ground or install culverts to prevent blocking of the ditch; and
- Dig ditches as straight as possible, with an even/uniform ditch bottom.

A crawler tractor is best suited for cut-and-fill work in rolling terrain. However, the movement of fill by a crawler tractor is only economical up to distances of 100–130 m (Antola 1988). When pushing fill over long distances by a crawler tractor a "cut groove" can be used. Wheeled front-end loaders have been used economically for the transport of cut material over longer distances (130–300 m) (Antola 1988). Where the cut material must be hauled to a disposal site, as on steep slopes with a risk of side-cast overloading, the use of an excavator and dump trucks is best.

Roadside borrow pits and fill sources — To minimize forest road building costs, as much material as possible must be utilized from the immediate vicinity of the road. Where the volume from cut sections is not sufficient for the fill sections, material can be obtained from roadside borrow pits and from along the roadway (i.e., equivalent to making a very wide shallow ditch). When using borrow pits gravel trucks may be required for moving the material to the fill section. When obtaining material from along the roadway a crawler tractor can be used to push the material into the road centre and then along the roadway. A long-reach excavator can also be used to move fill into the road centre, while a crawler tractor is used to move the fill longitudinally along the road.

Fill and gravel hauling — When sufficient fill is not available from the ditches, cut sections, or from directly beside the road, it must be hauled in by truck or tractor. The closer the borrow pits are located to the road the lower the hauling costs will be. Since material used for fill is generally of low quality, it often contains many rocks. If the diameter of the rocks is larger than one-half the thickness of the fill layer, the rocks can be screened out or passed through a primary crusher. In difficult borrow pits excavators are required. When the working conditions are easy, a wheeled front-end loader is most applicable. When hauling fill, end-dumping is most often used. This requires the trucks to back up to the end of the fill section, at which point they dump. Sufficient turn-around spurs are required along the road to minimize the amount of backing and thus maximize hauling production. A small 15–20-tonne class (e.g., Caterpillar D5-D6) crawler tractor is the best machine for spreading the fill and compacting the subgrade.

Hauling gravel for the pavement layers does not begin until after the subgrade has been levelled and is dry, and preferably after it has been compacted. Often the formed subgrade is left to dry over an extended period during the summer before beginning gravel hauling. In this way the amount of high-quality gravel required for the pavement layers can be minimized. Occasionally, if the situation allows, gravel hauling may occur during the winter when the subgrade is frozen. This allows the trucks to run along the subgrade, with no risk of rutting, and continuously spread even layers of gravel for the pavement. When spreading the road pavement materials during the winter, all snow must be removed from the subgrade surface, and the pavement materials must be sufficiently dry or unfrozen to allow even spreading. On dry, good quality subgrades trucks can also drive over them while dumping with little risk of rutting. After each pavement layer is spread, it preferably is levelled and compacted. This ensures a good quality road at minimum cost.

The use of bottom-dumping trucks ("belly dumpers") is very applicable when the trucks can be driven along the road while dumping. Otherwise, gravel spreading can also occur by opening the box gate partially and tipping while driving along the road. End-dumping is required in situations where it is not possible to drive on the subgrade. This is also true if hauling lower quality material to fill in sections of the road. On some very large road construction sites requiring considerable hauling, scrapers may also be used.

Finishing work — Once the basic construction work is complete the finishing work must be done. This phase includes final grading and compaction of the road surface, and the levelling and tidying of roadside embankments and borrow pits. To prevent erosion, roadside areas can be seeded. Where there is a high risk of erosion, rip-rap, geo-grids, geotextiles, or other erosion control techniques are available. The finalizing work also includes the placing of traffic signs, kilometre markers and bridge markers. To assist in culvert location and maintenance their ends must be marked. Where beaver activity is evident, covering the up-stream end of the culvert with a steel grill or grid will prevent the beavers from plugging the culvert.

Road maintenance — Since forest roads degrade with use and over time, continual maintenance is required to ensure a good running surface and minimal environmental impact through erosion and disruption of water flow. The maintenance includes periodic grading of the road surface, ensuring culverts are not obstructing water flow, checking signage, roadside brush control, placement of dust inhibitors, and replacement of road surface material (resurfacing operations). To ensure that ditches are properly functioning they must also be periodically maintained and cleaned. When road failure occurs, e.g., owing to frost heaving, a section of the road may need to be rebuilt.

Road deactivation — In most jurisdictions a road must be deactivated if it will not be actively maintained. The purpose of deactivation is to minimize any potential environmental impact through erosion, loss of productive forest area, or undesired access into an area.

With road deactivation the road is permanently closed. All bridges, culverts, and related structures need to be removed, and the banks shaped and sloped. All channels at water crossings need to be reshaped to their previous shape. Where there is a risk of water flowing down the old road, water bars need to be built. Even where roads are subject to temporary closure, it is prudent to make them "self-maintaining" by removing culverts that could become plugged, installing cross-drains where water accumulates in ditches, and minimizing erosion and sediment generation. Areas with exposed soil need to be revegetated, and grass seeding or erosion control blankets may be necessary in these cases.

If the road is to be planted a sub-soil tiller (or ripper) can be used to loosen heavily compacted areas. Brush and slash can be distributed over the road. In some situations coarse woody debris is distributed over the roadway, while large timber and stumps have been buried right into the roadway. Finally, the road needs to be posted as closed and a barrier may be required.

Cutting, conversion, and extraction

Description of harvesting methods, systems, equipment, and their applicability to boreal forest

Harvesting methods — Harvesting method is defined by the form in which wood is delivered to the logging access road, and depends on the amount of processing (e.g., delimbing, bucking, barking, chipping) which occurs in the cutover. The different harvesting methods are cut-to-length (shortwood) harvesting, full tree harvesting, whole tree harvesting, and complete tree harvesting.

Fig. 15.7. Cut-to-length logging system, using a (1) one-grip harvester and (2) forwarder.

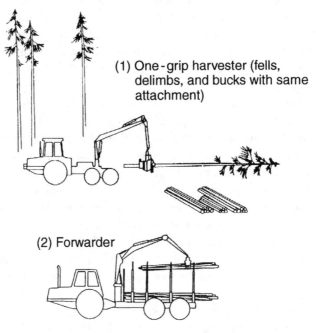

(1) One-grip harvester (fells, delimbs, and bucks with same attachment)

(2) Forwarder

Line drawings courtesy of FERIC, Western Division

Fig. 15.8. Tree-length logging with (1) motor-manual (chain saw) felling, (2) chain saw delimbing and topping, (3) cable skidding of tree-lengths, and (4) slashing (mechanical bucking) with a portable slasher.

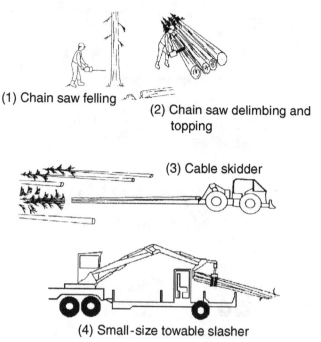

(1) Chain saw felling

(2) Chain saw delimbing and topping

(3) Cable skidder

(4) Small-size towable slasher

Line drawings courtesy of FERIC, Western Division

Under *cut-to-length (shortwood) harvesting*, trees are felled (cut off above the stump with stump height preferably less than one-half stem butt diameter), delimbed, and bucked to various assortments (pulpwood, sawlog, veneer bolt, etc.) directly in the stump area (Fig. 15.7). Trees can be topped down to a 5 cm top diameter, and limbs and tops can be left in windrows or spread over the cutover. Logging can be fully mechanized or motor-manual (i.e., using chain saws). Off-road transport is usually by forwarding (i.e., wood is carried off the ground), although cable skidders (which drag the wood at least partially on the ground, Fig. 15.8) are sometimes used. The cut-to-length method can be utilized in all silvicultural systems (e.g., clearfelling, thinning, individual tree selection). Roadside landings are minimal, since all processing is done in the cutover and high roadside piles can be made. The method also allows for efficient sorting and storage of various wood assortments. The method be used efficiently even when in-woods inventory levels are minimal, i.e., hot-logging (wood flows directly through the logging phases with no delay or storage) is very applicable. This method is re-establishing itself in North America owing to its "softer" environmental impact and flexibility.

In contrast *tree-length harvesting* (Fig. 15.8) involves felling, delimbing, and topping trees in the cutover. Delimbing and topping can occur in the stump area or at a point before the roadside. In softwoods, trees can be topped down to a 5 cm top; how-

Fig. 15.9. Conventional mechanized full tree logging system with (1) tracked, boom type feller buncher, (2) grapple skidder, (3) stroke delimber and (4) portable slasher.

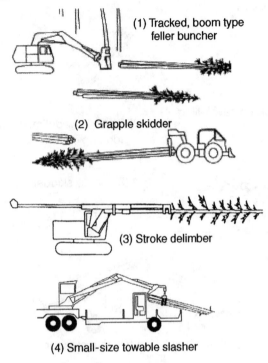

(1) Tracked, boom type feller buncher

(2) Grapple skidder

(3) Stroke delimber

(4) Small-size towable slasher

Line drawings courtesy of FERIC, Western Division

ever, topping generally occurs at a 7–10 cm top. In hardwoods, trees are often topped at a larger diameter (e.g., 15 cm) on account of their branching habit. Trees are mainly skidded to roadside with cable or grapple skidders. Crawler tractors and clam-bunk skidders are also used to some extent. The tree-lengths are bucked to pulpwood and logs at roadside, or can be left as tree-lengths for tree-length hauling to the mill. The tree-length method is most applicable to clearfelling, and can be used in row thinning. Landing requirements at roadside are much greater than for the cut-to-length method.

Full tree harvesting (Fig. 15.9) means trees are felled and transported to roadside with branches and top intact. Transport to roadside is mainly by cable or grapple skidders (Fig. 15.9). The full trees are processed at roadside or hauled as full trees to central processing yards or the mill. Roadside processing of full trees can include:

- full tree chipping and hauling of full tree chips to the mill;
- delimbing and topping to produce tree-lengths for hauling to the mill;
- delimbing, topping, and bucking to produce wood assortments for hauling as pulpwood to pulpmills or pulpwood using panel mills, and logs to sawmills or veneer mills; and
- chain flail delimbing–debarking prior to chipping to produce clean chips for transport to pulp and paper, or panel mills.

With the full tree method the limbs, tops, and wood residue (and, in the case of the chain flail delimber–debarker chippers, also the bark) are left in piles at roadside and

Fig. 15.10. Logging method use trends for Canada east of Alberta (data from the Canadian Pulp and Paper Association and Godin 2001).

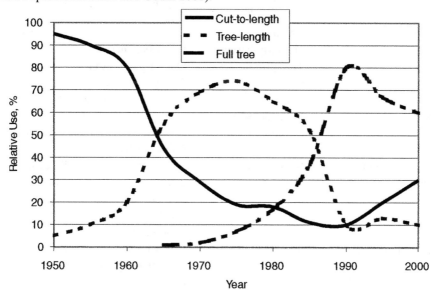

must be disposed of. The slash could be raked into piles and burned, or left as is for natural breakdown. Another alternative is to spread the slash or delimber–debarker mulch back into the cutover; such retention of biomass in the stand may have important implications to nutrient cycling and stand nutrition (Chap. 13). The full tree method is most applicable to clearfelling operations, and in some cases to first commercial thinnings where the material is chipped directly in the stump area or transported to roadside by forwarder. The landing requirement is the highest with this method. The full tree method is currently the most widely used method in Canada east of Alberta.

"Whole tree" harvesting can be an ambiguous term. For example, in the U.S. the term whole tree logging is equated to full tree logging. However, in this chapter a broader international definition of the whole tree method is used. In the whole tree method, full trees, including a large portion of the stump, are removed to roadside for processing and utilization. This method is seldom used in the boreal forests worldwide. *Complete tree harvesting* means that full trees, including the stump and major roots, are removed to roadside for processing and utilization. This method is seldom used.

Figure 15.10 presents the development in the use of the major logging methods in Canada east of Alberta. An important development has been the shift back towards in-stand processing and the use of the cut-to-length method. In 2001, cut-to-length logging accounted for 30% of the volume harvested in Canada east of Alberta (Godin 2001). This shift has been largely due to objectives to reduce the negative impact of logging on the environment, and the need for more flexible systems capable of working in smaller and more widely dispersed cutblocks. On a world-wide scale, Harstela (1999) states that the share of cut-to-length logging is also increasing, and is by far the dominant method used in the Nordic countries. Tree-length systems do not have the inherent operational

Table 15.6. Characteristics of the major logging methods.

Characteristic	Cut-to-length	Tree-length	Full tree
Felling equipment	Chain saw One-grip harvester Two-grip harvester	Chain saw Feller buncher One-grip harvester Two-grip harvester	Chain saw Feller buncher
Off-road transport equipment	Forwarder Cable skidder (limited use)	Cable skidder Grapple skidder Clam-bunk skidder Cable yarder	
Delimbing and topping location	Stump area	Stump area Cutover concentrated within cutover)	Roadside Not delimbed
Bucking location	Stump area	Roadside Centralized yard Mill Not bucked	Roadside Centralized yard Mill Not bucked
Slash distribution	Evenly spread Windrows	Evenly spread Small piles	Roadside piles No slash left
Roadside landing requirements and impact	Small	Large	Largest
Maximum effective off-road transport distance (straight-line)	600 m	Cable & grapple skidders, 300 m Clam-bunk skidder, 600 m	
Access road requirement	8.3 m/ha	Cable & grapple skidders, 16.7 m/ha Clam-bunk skidder, 8.3 m/ha	
Area with vehicular traffic	Low	Cable & grapple skidders, heavy Clam-bunk skidder, moderate	
Ground disturbance, dry	Low	Moderate	Heavy
Ground disturbance, frozen	Minimal	Low	Low
Ground disturbance, wet	Moderate	Heavy	Heavy
Protection of residual trees and regeneration	Good	Moderate	Poor

advantages of either full tree or cut-to-length systems (Pulkki 1997), and in 2001 accounted for 10% of the volume harvested in Canada east of Alberta (Godin 2001).

Harvesting system — A harvesting system is defined by the combination of equipment and procedures used to harvest an area. The individual components of the system can be changed without changing the harvesting method (i.e., the form in which the wood is delivered to roadside). A typical cut-to-length logging system could employ a one-grip harvester which fells, delimbs, and bucks the trees right in the stump area, and a forwarder to carry the pulpwood and logs to roadside (Fig. 15.7). With the tree-length

Table 15.7. Applicability of harvesting methods to silvicultural systems and operations.

Operation	Cut-to-length	Tree-length	Full tree
Even-aged			
Clearcutting	Good	Good	Good
Clear-cutting with standing snags and live trees	Good	Good	Good
Patch cutting	Good	Good	Good
Alternate strip cutting	Good	Good	Good
Progressive strip cutting	Good	Good	Good
Seed tree cutting	Good	Good	Moderate
Shelterwood cutting	Good	Moderate	Poor
Uneven-aged			
Group selection cutting	Good	Moderate	Poor
Individual tree selection cutting	Good	Poor	Poor
Other			
Overstory removal (shelterwood and seed tree)	Good	Moderate	Poor
Row thinning	Good	Moderate	Poor
Selection thinning	Good	Poor	Poor

method a common system would include motor-manual (chain saw) felling, delimbing, and topping, tree-length skidding to roadside, and roadside slashing (Fig. 15.8). A typical harvesting system used in full tree harvesting would include a feller buncher, grapple skidder, stroke delimber, and slasher (Fig. 15.9). Tables 15.6 and 15.7 summarize the characteristics of the cut-to-length, tree-length, and full tree harvesting methods, and their applicability to the various silvicultural systems.

Factors influencing harvesting operations: choice, planning, layout, scheduling

There are a considerable number of methods and systems available for harvesting in all silvicultural systems and stand conditions (Fig. 15.11). When choosing a harvesting method and (or) system, and when specifying access restrictions, logging area size, exclusion period length for adjacent areas, and whether two, three, or more entries into an area are required, the logging chance factors (Box 15.7) for the compartment must be considered. MacDonald (1999) presents useful decision keys for forest managers to determine the suitability of various logging equipment, depending on terrain conditions, tree size, and product produced.

The particular combination of logging chance factors will influence logging costs and potential impacts on the site and residual trees. Each logging system is best suited to specific conditions, with the full tree mechanized systems most suited to large concentrated harvesting operations, and small tree-length or cut-to-length systems more suited to small widely dispersed logging operations. The choice of the logging system will also influence the amount of access roads required in the area.

Fig. 15.11. Wood flow, quality, and logging options in wood procurement (Pulkki 1991).

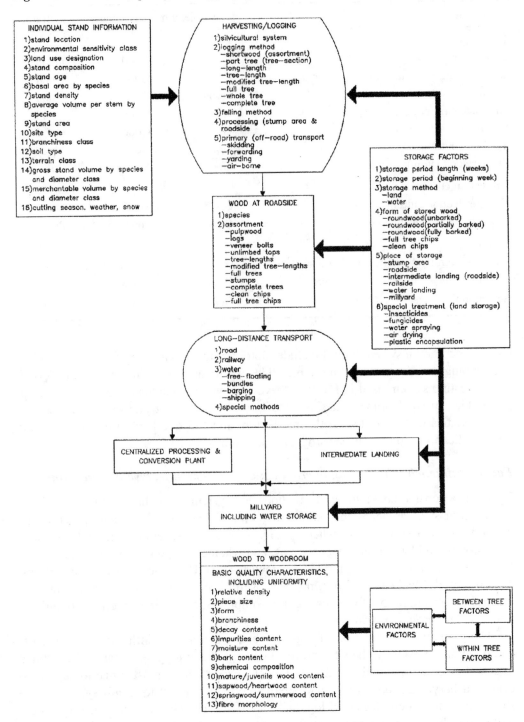

> **Box 15.7. The logging chance factors describing a compartment to be harvested can include the following information:**
>
> **Stand factors:**
> - age (e.g., over-mature, or need for thinning or partial cutting);
> - species of trees cut and silvics of the next crop;
> - tree size (volume, diameter, length), branchiness, and form;
> - volume per hectare and ratio of merchantable to non-merchantable trees;
> - wood quantity;
> - brush/undergrowth conditions; and
> - hygiene (i.e., presence of insects or disease, fire damage) and the necessity to salvage or contain insects/disease, or utilize wood before it is degraded beyond an economic limit
>
> **Site factors:**
> - environmental sensitivity class (e.g., soil depth/nutrients, riparian zones, rare or threatened species), special visual concerns, historically or culturally sensitive areas, or other special forest conditions (e.g., old growth);
> - terrain class (ground conditions — soil type and strength —, obstacles, slope — percent, shape, position on slope, and aspect);
> - land-use designation;
> - location in regard to distance to point of utilization, and source of workers and support services;
> - logging area size and extent to which compartments are scattered;
> - off-road transport distance (average and maximum);
> - ownership (private, forest industry, public);
> - recreational demands in area;
> - road network and other infrastructure available;
> - season to be harvested in (e.g., ground frozen in winter, snow depth);
> - visual management considerations;
> - wildlife management needs; and
> - weather conditions (rain, wind, low temperature, high temperature).

Harvesting mechanization has had a very positive impact on improving the working environment and safety for forest workers. Proper choice and use of the equipment will not result in increased site impacts when compared to the old manual logging systems. Many of the new machines available will result in less impact on the site than the narrow-tired skidders used in the 1960s and 1970s. However, proper planning of logging operations and use of equipment, following accepted BMPs, and good training of workers, are required to minimize negative environmental impacts of harvesting operations, while still meeting economic constraints.

A single harvesting and silvicultural policy which would take into account all the above factors is not practical, since many of the above considerations conflict, while others are not applicable to all situations. There is even conflict within a single factor: e.g., certain species of wildlife benefit from stand edges and small irregularly shaped patches, while other species require a large forest interior and thus larger clearcuts to provide large forest interiors in the future (Chap. 12). In other situations (e.g., in a remote primitive access area where only winter access may be allowed) larger clearcuts may be desired, with a minimum number of entries into the area.

A forest-level analysis of the above factors (e.g., species diversity, animal habitat, recreational opportunities, etc.) must be the focus. Hence, the mix of harvesting systems should be considered in forest-level planning (Chap. 11). However, stand-level factors are important to minimize the effects on water quality and soil, to maintain a wide variety of age classes and stand conditions in the forest, and to achieve the forest-level objectives. So the selection and deployment of alternative harvesting methods must be refined at each successive level of tactical (Chap. 12) and site-level (Chap. 13) planning.

Partial cutting operations

Introduction

Large clearcuts with minimal residual structure have traditionally constituted the "norm" for timber harvesting operations in most parts of the boreal forest for the last few decades. Recent expectations concerning biodiversity conservation, aesthetics, and ecosystem sustainability have prompted evaluation of a variety of alternatives.

"Alternative" forest operations aimed at boreal forests can be grouped into six broad categories:
*   strip cutting techniques to promote regeneration in the removed strips;
*   variable retention or patch retention techniques;
*   removal of most of the "short-lived" species (e.g., aspen, birch, and balsam fir) component of the stand, leaving the remaining softwoods, usually spruce;
*   removal of the majority of the overstory while protecting advance regeneration and any residual trees which may be left;
*   harvesting of small irregular-shaped patches (e.g., 0.1–0.5 ha in size) to promote regeneration of stands too susceptible to windthrow and snow damage, or which are of poor quality; and
*   thinning operations in 30–50-year-old stands to utilize volumes which would be lost to mortality, concentrate growth on the remaining trees, and make the remaining stand more wind firm and resistant to snow damage.

This section outlines some of the techniques, systems, and equipment available for "alternative" harvesting in boreal forests. The operational considerations and requirements in planning are discussed, as well as limitations of some of the "standard" current equipment. Requirements for equipment, minimal damage to residual trees and site, and prerequisites for success when partial cutting in boreal stands are also presented. There is some repetition of harvesting method and system recommendations covered in previous sections, while the associated silvicultural issues are addressed in Chap. 13.

Operating techniques

There are many techniques where conventional full tree harvesting systems have been adapted to partial cutting in boreal mixedwoods. Most of these techniques strive to remove most of the aspen, birch, and balsam fir component, while favouring the remaining softwoods (usually spruce). Cutting generally occurs as one pass over the entire cut area, two or three pass strip cutting, or progressive strip cutting starting on the leeward side of the stand. The cut and leave strips can have varying widths, and there can be edge areas where some selection cutting can occur. Navratil et al. (1994) present good descriptions of the many techniques available. Similar operating techniques using full tree, tree-length, or cut-to-length systems can be used when harvesting areas while protecting the advance regeneration: e.g., removal of overstory black spruce while protecting the advance growth.

Shelterwood cutting and selection thinning can also be employed. In general, cut-to-length systems employing forwarders for off-road transport are recommended. A cut-to-length operation would normally have 4 m wide machine trails, at a spacing of 20–30 m, with ghost trails within the leave corridors, if required. Motor-manual cut-to-length logging, with a small-size skidder or forwarder, is also applicable.

When a stand is opened up and is too susceptible to windthrow or snow damage, or is of poor quality or low volume, small (e.g., 0.1–0.5 ha) irregular-shaped patch cuts may be required. In this situation, cut-to-length systems employing forwarders are the most suited. This is due to the widespread nature of the patches, small size of patches, long off-road transport distances, and the need for minimal roadside landings.

Harvesting systems applied to partial cutting

Full tree — Full tree systems have been developed mainly for clearcutting. When used in partial or small patch cutting situations, full tree logging has the following disadvantages:

* large roadside landings;
* removal of limbs and tops from site, resulting in removal of nutrients from the stump area and concentrations of debris at roadside;
* wide-cut corridors;
* excessive damage to residuals (>20% damage to residuals is not uncommon), especially along corridors;
* limited skidding distance (usually limited to a maximum of 300 m); and
* considerable off-road vehicle traffic concentrated near the roadside landings.

Although there are some exceptions, feller-bunchers have been designed to work in clearcuts, and generally in a zone in front of the machine. A minimum cut strip width of 8 m is usually required. However, this depends on the feller-buncher employed, and is influenced by the operating procedure:

* the operator generally crowds the right side of the strip because the boom causes a blind spot);
* width of the machine (about 3 m or more);
* tail swing (0–3 m);
* area required to bunch wood; and
* maximum and minimum working reach (generally, maximum 6–7 m and minimum 3–5 m).

Tree-length — Some typical tree-length systems applied to partial cutting are as follows:

- conventional cut and skid, with motor-manual felling, limbing, and topping, cable skidding, and possibly roadside slashing; and
- one-grip harvester, and cable or grapple skidder.

The use of feller-bunchers, with in-stand delimbing by stroke delimber, is not possible in partial cutting situations when large standing residuals are left. This is because the delimber's boom will strike residual trees. Otherwise, the tree-length method is applicable to corridor and small patch cutting.

When compared to full tree systems, tree-length systems generally result in more or less the same amount of damage to residual trees, require slightly smaller roadside landings, and the tops and branches are left in the stump area. However, since the boles are skidded along the ground they pick up more dirt.

Cut-to-length method — Current mechanized cut-to-length systems have been designed specifically for widely dispersed, small patch, and partial cutting situations (e.g., Finnish/Swedish conditions). When used properly, cut-to-length systems can result in minimal residual stand damage. For example, in Finland the average damage to residuals is generally <5% (Sirén 1998; Sirén 2000), although in practice forest owners are demanding 0% damage. Damage levels of >20% to residuals (Ostrofsky et al. 1986; Nichols et al. 1994) and large landings typical to full tree systems used in Canada must not be tolerated. When removing large-size overstory trees (e.g., aspen) fixed-head

Fig. 15.12. Long-reach one-grip harvester and forwarder system (adapted from Metsäteho 1983). *Note*: processing can be done on machine trail so tops and limbs form a brush mat to minimize ground disturbance.

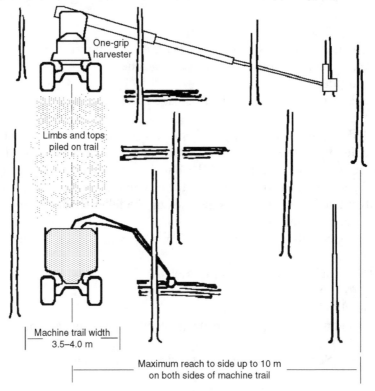

one-grip harvesters can be used to lift the trees through the stand to an open area for processing, thus minimizing damage to residuals.

Owing to the larger load size of forwarders (15 m^3 vs. 3–4 m^3 for skidders), the maximum off-road extraction distance is much further than with skidders (600 m vs. 300 m). As a result, a cut-to-length system employing a forwarder requires less road (8.3 m/ha versus 16.7 m/ha) and can be efficiently used when cutting small patches widely dispersed over the terrain. Since no processing occurs at roadside, and bucked logs and pulpwood are of short length and can be piled high, the cut-to-length method requires minimal roadside landings. As a result, more area can be left in a partially cut condition and site impacts are reduced. Since the wood is carried and cut inventory levels are low, site degradation is reduced, and wood quality is also maintained.

Specially designed one-grip harvesters can fell trees in front of and on both sides of the machine. Generally, the maximum reach for a one-grip harvester is about 10 m, and it can fell trees directly beside the machine. Thus the operating zone is much larger than for a feller-buncher. Because of its design the harvester can work in a 20 m wide swath through the stand, while travelling on one 4 m wide trail (Fig. 15.12). Usually the trees are limbed in front of the harvester, thus forming a brush mat on which both the harvester and forwarder can travel.

If the machine trail spacing is greater than 20 m, a "ghost trail" through the leave corridor is generally required. The ghost trail is about 3 m wide, being the path the harvester follows as it fells and directs the fall of the trees toward the machine trail. The harvester then returns down the machine trail, grabbing the tops of the trees, and processes them. When employing a small-size one-grip harvester and a 20 m machine trail spacing, the harvester can work within the stand from a ghost trail and pile the wood at the side of the machine trail.

The most appropriate operating techniques depend on the age of the stand, tree size, the spacing between the residual trees, and the terrain conditions.

Considerations/requirements

Economic — One of the most important factors to consider when choosing the most appropriate operating technique and system is economics. For a one-grip harvester and forwarder operation to be economically feasible the following are generally required (Pulkki 1996):

- greater than 50 m^3/ha removed;
- reasonably priced equipment required;
- minimum operation of two shifts per day;
- minimum production greater than 30 000 m^3/year, but preferably greater than 40 000 m^3/year; and
- realization that partial cutting costs more than clearcutting, and this has to be offset by lower stumpage fees or higher millgate prices.

Stand condition — When choosing the most appropriate operating technique and system, the stand condition must be considered. This includes the ability of the remaining trees to respond to the opening up of the stand (e.g., amount of live crown), and their susceptibility to windthrow or snow damage. Also, the presence of rot and insects and

Fig. 15.13. Effect of various factors on relative cost (excluding roads) of cut-to-length logging (Metsäteho 1983).

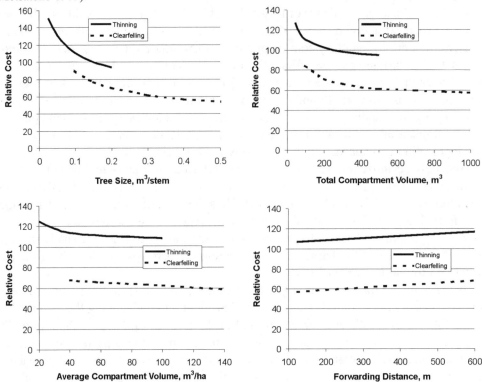

the quality of the remaining trees are important, unless non-timber values are paramount. There is no point in thinning or partially cutting stands that have little or declining value, or have low potential for future growth.

In addition to volume removed per hectare, off-road transport distance, harvesting area size, and other stand conditions that affect the harvesting operation also need to be considered. Among others, these factors include:

- size of the trees removed and retained;
- species removed and retained;
- branchiness of the trees;
- presence of undergrowth;
- ground conditions/terrain class; and
- amount of dead and unmerchantable trees.

Charts developed by Metsateho (1983) show the effect of tree size, forwarding distance, volume removed per hectare, and volume removed per logging chance on the relative cost of cut-to-length logging in clearfellings and thinnings (Fig. 15.13).

Equipment design — In partial cutting situations, equipment used must result in minimal residual tree and site impact. To minimize negative impacts the equipment must be narrow (<3 m), have low ground pressure (maximum 40–50 kPa, 6–7 psi), and have low-impact running gear (rubber tires or low track grousers).

Visibility from the cab should be excellent. In many cases the visibility from conventional feller-bunchers is limited to the right or left, depending on the location of the cab, boom, and engine cowling. Since individual trees are selected in partial cutting situations, good visibility of the crown is also necessary, unless trees are pre-marked (to cut or to leave). This requires a skylight or a large sloping front window on the cab.

On a fully operational scale, logging must also occur at night. This is especially true during winter when the daylight period is short. This requires good diffuse lighting systems, which light up the entire surrounding forest to allow good visibility and selection of trees.

The harvesting machines must have a wide operating zone in front and to the sides. This allows for minimal machine trails, machine movement within the stand, and stand damage. It also allows the harvester to reach trees from various positions along the machine trail. Working from both sides of the machine trail results in more time spent harvesting trees than moving within the stand. Similarly, a lot of time is spent positioning the machine so it is stable. Thus, minimizing equipment moves, while maximizing the number of trees felled at each position, will increase production.

Since most feller-bunchers and some excavators equipped with harvester heads pick the tree up to move it closer to the machine, they must be equipped with a counterweight or be specially designed (e.g., Timbco T420 and John Deere JD653E). This counterweight, or overhang in many cases, on the feller-buncher can result in a tail swing in excess of 3 m (i.e., the rear of the feller-buncher extends over tracks when turned 90°). When operating in partial cutting situations this is an important factor in regard to damage of residuals and cut strip width.

Often in partial cutting situations the protection of advance regeneration is a priority (see Chap. 13). In these situations the use of continuous saw heads is discouraged, unless the head is lifted over the advance growth before being dropped to the tree to be felled. Otherwise, when the head is moved horizontally toward a tree to be felled, any advance regeneration in its path will also be felled.

Stand damage — Damage levels to residuals of >20% typical of full tree operations are not acceptable. As mentioned above, the average damage to residuals in Finland is generally <5%, and the norm is to strive for 0% damage. This goal also pertains to the damage to tree roots and soil, or alteration of surface or sub-surface water flow.

Good planning and layout are prerequisites for minimizing stand damage. The layout must be done with the partial cutting in mind (i.e., typical clearcutting layout does not work). A minimum of machine trails must be ribboned in the stand (i.e., 20 m spacing at least). A trail width of 4 m is best, except on curves where it can be increased to 5 m. If the trail width is too narrow more residual trees will be damaged and thinning productivity will suffer. If the trail width is wider then the growing capacity of the site will not be used to its optimum. The machine trails must have no sharp curves, and must be in loops or have turn-arounds. Machine trails must also run straight up or down hills (maximum side slope 10%), and around obstacles and soft spots.

Proper operating technique and equipment selection are critical. It is important to understand that the terrain transport component, especially if skidders are used, generally causes the most damage to residual trees and the site. Some general rules to follow are:

- use cut-to-length systems;
- avoid full tree systems;
- use small, narrow, low ground pressure machines with long-reach capability;
- preferably use rubber-tired machines;
- if tracked machines are used, use low grousers;
- if a skidder is used, minimize turning around in the stand and use natural openings, or back in;
- limb the trees in front of the harvester to form a brush mat;
- use bumper trees to absorb most damage where necessary; and
- do not bring in excessively large equipment to handle a few large trees; get out with a chain saw to take care of them instead.

In addition to proper planning, layout, and selection of equipment, choosing the correct season to operate in is also important. Preferably, partial cutting operations in boreal mixedwood stands occur during the winter. This is because boreal mixedwood stands are generally found on fine soils (i.e., clays, silts, or loams) that are susceptible to compaction and rutting when wet and unfrozen. Spruce and aspen are very susceptible to attack from root-rotting pathogens because of their fine surface root systems, which are easily damaged by vehicle traffic on unfrozen fine soils. Also, bark adhesion to the wood is highest during winter, thus bark abrasions due to vehicle impacts are much less.

Perhaps the most important component for minimizing damage in partial cuts is the operator. In partial cutting operations it is imperative that highly skilled and motivated operators are chosen. These operators must be trained in both the operation of the machine and in the basics of forestry. They must have a general knowledge of silviculture, ecology, and mensuration. To make the operations profitable, it is the operator who must select the trees and ensure the correct basal area is maintained in the stand. Good supervision is also required to monitor the harvesting production results, damage to residuals and site, and give feedback to the operators.

Prerequisites for success

In summary the following prerequisites are required to ensure success in "alternative" or partial cutting operations in boreal forests:

- proper selection of harvesting method, equipment, system, and technique to meet the conditions;
- good planning, layout, and control;
- good operator training and commitment;
- good planner/supervisor training and commitment;
- company management commitment; and
- government commitment.

No single harvesting method or system is a panacea for all situations. Feller-buncher-based, full tree systems may be required in certain situations, while in others the cut-to-length systems may be more applicable. The key to choosing the right system for most situations is knowing the operating conditions, and good planning. The requirement for good training of operators, supervisors, and planners involved in partial cutting cannot be stressed enough.

Fig. 15.14. Large-size multi-axle low ground pressure forwarder.

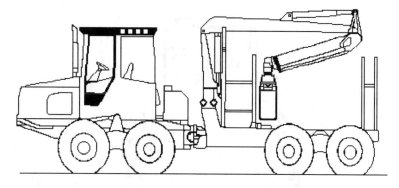

Fig. 15.15. Small-size multi-axle low ground pressure forwarder.

Some new equipment design features

Multi-axle low ground pressure forwarders

To minimize the impact of forwarders on forest soils and to minimize damage to residual trees and their root systems during thinning and partial cutting, small-size, multi-axle, low ground pressure forwarders have been developed (Figs. 15.14 and 15.15). Newly designed forest transport machines result in considerably less impact on the site than forwarders designed in the 1970s and early 1980s (Sirén et al.1987; Hakkila 1989). The use of these types of forwarders, as well as good operator training, has resulted in less than 2% of residual trees being damaged during thinning (Sirén 1998). Loaded ground pressures for these machines can be less than 21 kPa.

High flotation tires

To minimize detrimental ground disturbance on low-strength soils (bogs, and wet clays and silts) the easiest solution would be to harvest only during the winter when the ground is frozen. This is not always possible or feasible. Skidding wood from these areas with conventional logging tires (i.e., 46–61 cm in width) will most often result in unacceptable levels of ground disturbance, especially on the major skid trails and near

Fig. 15.16. High flotation and conventional tires for a skidder.

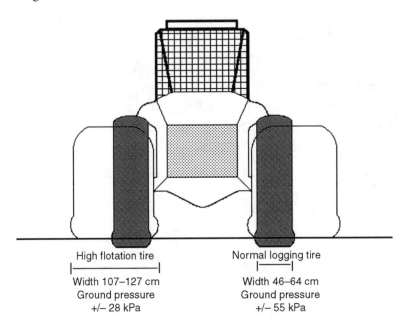

High flotation tire	Normal logging tire
Width 107–127 cm	Width 46–64 cm
Ground pressure	Ground pressure
+/– 28 kPa	+/– 55 kPa

the landings. To reduce the negative impact of skidding or forwarding, special high flotation tires have been developed (Fig. 15.16). These tires range in width from 107 to 127 cm, although tires up to 173 cm in width are available. Ground pressures with the high flotation tires can range from 10 to 24 kPa, depending on the tire width, and skidder or forwarder size.

High flotation tires have been shown to reduce site disturbance and damage to advance regeneration. On slopes the wide tires increase stability and mobility, and the reduced site disturbance results in less erosion. However, the tires are expensive, increase stress on the equipment (e.g., axles), are subject to puncture and sidewall wear, and severely limit the maneuverability and mobility of the equipment. A second set of narrower tires is required for the winter, since traction on snow is poor. Another problem with the wide tires is that they must be removed to be able to transport the skidder between work sites, and most garage doors have insufficient width for a skidder equipped with wide tires to pass. Also, the increased width of the equipment limits their applicability to logging applications where narrow machine trails are required (e.g., thinning, selection logging, shelterwood logging).

Environmentally friendly lubricants (bio-oils/lubricants)

In regard to fluid (e.g., engine oil, hydraulic oil, fuel, antifreeze) spills/leaks the best option is to use modern, well-maintained equipment with a good preventive maintenance program in place, and proper work procedures when changing and handling fluids. Most modern forest machines no longer have oil drain plugs, and oil (engine and hydraulic) is sucked out of the oil reservoir when changed. In this way all oil is captured and accounted for in environmental accounting systems (i.e., volume of oil disposed of at a licensed disposal site is at least a specified percent of total volume brought to the job site). Makkonen (1994*a*) presents a number of techniques that can be used to recover fluids from older machinery while minimizing the potential for a spill. If a spill does occur there are many options available for containing and recovering spilled oils (e.g., booms, loose particulates, pillows, mats, or pads) (Hamilton 1998). In the most severe case the contaminated soil must be dug up and disposed of at an approved site. Machines with double "hulls" have also been proposed to ensure fluids are captured within the body of the machine. Machines must also be washed at approved sites where any contaminants can be collected and disposed of accordingly.

In regard to environmentally friendly lubricants (EFL) a major area of concern is that there are few regulations or standards defining what they are (Beitelman 1996). The area of environmentally friendly lubricants is very new and much development is currently occurring in this field (Jokai 1995; Beitelman 1996; NFU 2002). The first major use of bio-lubricants was with chain saw oil in Europe and has expanded to hydraulic systems and transmissions (Makkonen 1994*b*; Beitelman 1996; Makkonen 1997*a, b*; NFU 2002). However, at this point users need to consult very carefully with their equipment suppliers to ensure warranty requirements are not violated through the use of EFLs.

Care must also be taken to prevent fuel spills during transport, at fuel storage areas, and while filling and operating machines. A fuel spill kit and an emergency response protocol in case of a spill are required for each operation.

GPS and GIS

Global Positioning Systems (GPS) and Geographic Information Systems (GIS) in themselves do not prevent or minimize environmental impacts. However, they do provide valuable tools for planners, managers, supervisors, and operators to help to minimize the impact of harvesting operations on the environment. Geographic information systems provide a valuable tool for planners when locating roads and in planning logging operations. Guimier (1999) envisages harvesting and stand-tending equipment equipped with GPS technology to help in implementing and controlling the operations. Haul trucks are already equipped with GIS and GPS systems to allow for more efficient hauling operations. Global Positioning Systems installed on harvesting and site preparation equipment are also used to help update harvested compartment GIS information.

Conclusion

The objective of this chapter was to broadly outline potential negative environmental impacts of forest harvesting operations and, in more detail, outline how good advance planning can be used to minimize them. Modern equipment is "friendlier" on the environment and when properly applied can have minimal impact on the site. New tools such as GPS and GIS allow the better planning, layout, and control of road construction and harvesting activities, while advances in geotextiles and geogrids give new alternatives for minimizing impacts of forest roads.

Environmental impacts from harvesting operations can occur directly from the removal of the trees from the site, the building of roads required for access, and through physical impacts to the soil during and after harvesting. Ancillary impacts can result through fuel/oil leaks and spills, emissions, and noise. However, the single largest environmental impact associated with harvesting operations is from the access roads, especially at water crossings.

There is no single road construction or harvesting method or system that is a panacea for all situations. The key to minimizing negative environmental impacts is to know your operating environment, and the methods, equipment, and systems available to you, and good advance planning, implementation, and control of operations.

To plan successfully, a forest harvesting planner/manager requires considerable data and information to make sound decisions. There are a considerable number of methods and systems available for road construction and forest harvesting in all silvicultural systems and stand conditions. However, the most appropriate method and system to employ needs to be technically feasible, fit into the silvicultural prescriptions, be financially sound and socially worthwhile, and meet environmental guidelines. The skill and commitment of the operators, supervisors, and planners cannot be overlooked, since they are critical in the proper implementation of any forest road construction or harvesting plan. The control and auditing of work results ensures that the plan has been successfully implemented, the objectives met, and negative environmental impacts minimized.

Also, when developing and implementing forest harvesting guidelines it is important to note that a single harvesting or silvicultural policy that would take into account the myriad of forest uses and values is not practical, since many values/uses conflict, while others are not applicable to all situations. A forest-level analysis of forest values (e.g., wood supply, species diversity, animal habitat, recreational opportunities, etc.) must be the focus. Hence, a mix of harvesting methods and systems should be considered in forest-level planning. However, stand-level factors are important to minimize the effects on water quality and soil, to maintain a wide variety of age classes and stand conditions in the forest, and to achieve forest-level objectives.

In the end it is the front-line supervisors and forest workers who implement the plans and are operating out in the terrain. As with managers and planners, constant training and skills upgrading, and continuing education, are necessary to keep supervisors and workers up-to-date on current standards in ecology, silviculture, harvesting equipment and techniques, environmental protection, guidelines, regulations, and legislation. A well-motivated workforce will also help ensure minimal adverse environmental impacts of forest harvesting operations.

References

Antola, A. 1988. Forest road construction techniques. University of Helsinki, Department of Logging and Utilization of Forest Products, Helsinki, Finland. Res. Notes 50. 104 p.

Armtec. 2002. Geotextile technical manual. Armtec Limited, Guelph, Ontario. 17 p.

Ashworth, R. 1966. Highway Engineering. Heinemann Educational Books Ltd., London, U.K. 284 p.

(BCMOF and BCMELP) British Columbia Ministry of Forests and British Columbia Ministry of Environment, Lands and Parks. 1995. Forest Practices Code Forest Road Engineering Guidebook. British

Columbia Ministry of Forests and British Columbia Ministry of Environment, Lands and Parks, Victoria, British Columbia. 153 p.

Beitelman, A.D. 1996. Environmentally friendly fluids. The REMR Bull. 13(2). 4 p.

Brink, M.P., Kellogg, L.D., and Warkotsch, P.W. 1995. Harvesting and transport planning — a holistic approach. South Afr. For. J. **172**: 41–47.

(DOA) Department of the Army. 1992. Military soils engineering. Department of the Army, Washington, D.C. Field Man. 5-410.

Fannin, J. 2001. Basic geosynthetics: a guide to best practices in forest engineering. *In* Proceedings of the International Mountain Logging and 11th Pacific Northwest Skyline Symposium, 10–12 December 2001. University of Washington, College of Forest Resources, Seattle, Washington. pp. 145–151.

Gillies, C. 2002. Installation of an aluminium open-bottom arch culvert. Forest Engineering Research Institute of Canada, Pointe Claire, Quebec. Advantage 3(16). 7 p.

Godin, A.E. 2001. Logging equipment database: 2001 update. Forest Engineering Research Institute of Canada, Pointe Claire, Quebec. Advantage 2(59). 2 p.

Guimier, D.Y. 1999. Canadian forestry operations in the next century. J. For. Eng. **10**(1): 7–12.

Hakkila, P. 1989. Logging in Finland. Acta For. Fenn. 207. 39 p.

Hamilton, P.S. 1998. Materials for containing and recovering spilled oils. Forest Engineering Research Institute of Canada, Pointe Claire, Quebec. Field Note Gen. 62. 2 p.

Harstela, P. 1999. The future of timber harvesting in Finland. J. For. Eng. **10**(2): 33–36.

Jokai, R. 1995. Use of vegetable-based hydraulic oil in forestry operations: an example. Forest Engineering Research Institute of Canada, Vancouver, British Columbia. Spec. Rep. 103. 38 p.

Kardoz, L. 1993. Good road design and construction techniques. Can. For. Ind. (Jan./Feb.): 24–26.

MacDonald, J. 1999. Harvesting systems and equipment in British Columbia. British Columia Ministry of Forests, Forest Practices Branch, Victoria, British Columbia. 197 p.

Makkonen, I. 1994*a*. Collection of waste oil from forestry machines. Forest Engineering Research Institute of Canada, Point Claire, Quebec. Field Note Gen. 37. 2 p.

Makkonen, I. 1994*b*. Environmentally compatible oils. Forest Engineering Research Institute of Canada, Point Claire, Quebec. Field Note Gen. 39. 2 p.

Makkonen, I. 1997*a*. A trial of bio hy-gard environmentally compatible hydraulic fluid in a feller-buncher. Forest Engineering Research Institute of Canada, Point Claire, Quebec. Field Note Gen. 54. 2 p.

Makkonen, I. 1997*b*. Performance of Raisio Biosafe 32 NE hydraulic oil in a Silvana selective cleaning machine. Forest Engineering Research Institute of Canada, Point Claire, Quebec. Field Note Gen. 60. 2 p.

Mason, L. 1992. Gratings with geotextiles at wetland crossings. USDA Forest Service, Technology Development Program, Washington, D.C. Timber Tech Tips (Dec.). 4 p.

Mason, L., and Moll, J.E. 1995. Pipe bundles and pipe mat stream crossings. USDA Forest Service, Technology Development Program, Washington, D.C. Timber Tech Tips (Feb.). 4 p.

McNally, J.A. 1977. Planning forest roads and harvesting systems. Food and Agriculture Organization of the United Nations, Rome, Italy. FAO For. Pap. 2. 148 p.

(MDNR) Minnesota Department of Natural Resources. 1995. Protecting water quality and wetlands in forest management: best management practices in Minnesota. Minnesota Department of Natural Resources, Division of Forestry, St. Paul, Minnesota. 140 p.

Metsähallitus. 1970. Metsätienormit. National Board of Forestry, Helsinki, Finland. 174 p.

Metsäteho. 1983. Metsätehon opas: leimikon suunnittelu. Painovalmiste Ky, Helsinki, Finland. 36 p.

Navratil, S., Brace, L.G., Sauder, E.A., and Lux, S. 1994. Silvicultural and harvesting options to favor immature white spruce and aspen regeneration in boreal mixedwoods. Can. For. Serv. Inf. Rep. NOR-X-337. 78 p.

(NBDNRE) New Brunswick Department of Natural Resources and Energy. 1994. Forest Management Manual for Crown Lands. New Brunswick Department of Natural Resources and Energy, Fredericton, New Brunswick. 35 p.

(NFU) National Farmers' Union. 2002. NFU President says government bump start needed for bio-oils. NFU, England, and Wales, Press Release 22 May 2002. 2 p.

Nichols, M.T., Lemin, R.C., and Ostrofsky, W.D. 1994. The impact of two harvesting systems on residual stems in a partially cut stand of northern hardwoods. Can. J. For. Res. **24**: 350–357.

(OMNR) Ontario Ministry of Natural Resources. 1988. Environmental guidelines for access roads and water crossings. Queen's Printer for Ontario, Toronto, Ontario. 64 p.

Ostrofsky, W.D., Seymour, R.S., and Lemin, R.C. 1986. Damage to northern hardwoods from thinning using whole-tree harvesting technology. Can. J. For. Res. **16**: 1238–1244.

Paterson, W.G., MacFarlane, H.W., and Dohaney, W.J. 1976. Standard classification for forest roads. Forest Engineering Research Institute of Canada, Vancouver, British Columbia. Tech. Rep. 9. 28 p.

Provencher, Y. 1992. Geotextiles: another road construction option. Forest Engineering Research Institute of Canada, Pointe Claire, Quebec. Tech. Note 182. 6 p.

Pulkki, R. 1982. The development of an economical method for measuring the bearing capacity of forest roads. University of Helsinki, Department of Logging and Utilization of Forest Products, Helsinki, Finland. Res. Notes 42. 85 p.

Pulkki, R. 1991. A literature synthesis on the effects of wood quality in the manufacture of pulp and paper. Forest Engineering Research Institute of Canada, Pointe Claire, Quebec. Tech. Note 171. 8 p.

Pulkki, R. 1996. Management tools and strategies: forest operations. *In* Advancing Boreal Mixedwood Management in Ontario: Proceedings of a Workshop, 17–19 October 1995. *Edited by* V.F. Haavisto, C.R. Smith, G.W. Crook, and C. Mason. Canadian Forest Service, Sault Ste. Marie, Ontario. pp. 61–65.

Pulkki, R. 1997. Cut-to-length, tree-length or full tree harvesting? Cent. Woodlands, **1**(3): 22– 27, 37.

Smith, R.T. 1989. Determining road class according to wood volume transported. H.B.Sc.F. Thesis. Lakehead University, Thunder Bay, Ontario. 47 p.

Sirén, M. 1998. Hakkuukonetyö, sen korjuujälki ja puustovaurioden ennustaminen. Metsäntutkimuslaitosken Tiedonantoja 694. 179 p.

Sirén, M. 2000. Silviculture result of one-grip harvester operation. J. For. Eng. **11**(2): 7–14.

Sirén, M., Ala-Ilomäki, J., and Högnäs, T. 1987. Harvennuksiin soveltuvan metsäkuljetuskaluston maastokelpoisuus. Folia For. 692. 60 p.

Steward, J., Williamson, R., and Mohney, J. 1977. Guidelines for the use of fabrics in construction and maintenance of low-volume roads. Portland, Oregon. USDA For. Ser. Rep. No. FHWR-TS-78-205. 176 p.

Terrafix Geosynthetic. 1991. Geotextile sales and information binder. Terrafix Geosynthetic, Toronto, Ontario. 8 p.

Wackerman, A.E., Hagenstein, W.D., and Mitchell, A.S. 1966. Harvesting timber crops. McGraw– Hill, New York, New York. 540 p.

(WDNR) Wisconsin Department of Natural Resources. 1995. Wisconsin's forestry best management practices for water quality. Wisconsin Department of Natural Resources, Bureau of For-estry, Madison, Wisconsin. 76 p.

Wiest, R.L. 1998. A landowner's guide to building forest access roads. USDA For. Serv. NA-TP-06-98. 45 p.

Chapter 16

Residues generated by the forest products industry

Clark P. Svrcek and D.W. Smith

Introduction

The categorization and quantification of outputs to air, water, and land are important for monitoring and regulatory purposes in most industries. Comprehensive analysis of discharges to these three media is essential in terms of addressing cumulative assessment of impacts to the environment and the creation of environmental policies and management plans. Regulatory agencies embrace multi-statute solutions to facility planning and permitting, compliance guarantee, education and outreach, research, and regulatory development issues (USEPA 1995b). The central idea that is driving this policy direction is that pollutant releases to each environmental medium affect each other and that strategies must proactively identify and address these inter-relationships by designing policies for an entire facility, entire sectors of the industry, and entire regions that are directly affected.

This chapter presents an overview of the solid, liquid, and atmospheric residual emissions related to the forest products industry. It should act as a primer for those individuals that are new or need a refresher in the residual waste streams that are output from wood processing operations. It is necessary to understand these industrial processes and the major categories of effluents, residuals, and byproducts associated with each processing stream in order to effectively reduce or control this pollution (Box 16.1). In describing this backgrounder information and the challenges associated with forest products industry residues, this chapter acts mainly as an introduction to the following three chapters, which will delve more deeply into the solutions and emerging technologies that are enabling us to better deal with these residual streams. While the bulk of this chapter focuses on residue streams from pulp processing and paper mills (owing to their diversity, volumes, and the severity of associated environmental and health concerns), there is also some discussion of residue streams from other aspects of the forest products industry, such as logyards and wood preserving. The first part of the chapter briefly outlines some of the waste streams and waste products characteristic of

C.P. Svrcek and D.W. Smith.[1] Environmental Engineering and Science Program, Department of Civil and Environmental Engineering, 304 Environmental Engineering, University of Alberta, Edmonton, Alberta T6G 2M8, Canada.
[1]Author for correspondence. Telephone: 780-492-4138. e-mail: dwsmith@civil.ualberta.ca
Correct citation: Svrcek, C.P., and Smith, D.W. 2003. Residues generated by the forest products industry. Chapter 16. *In* Towards Sustainable Management of the Boreal Forest. *Edited by* P.J. Burton, C. Messier, D.W. Smith, and W.L. Adamowicz. NRC Research Press, Ottawa, Ontario, Canada. pp. 629–668.

Box 16.1. Residue management challenges for sustainable forest management:

- To realize and manage the impact of atmospheric, solid, and aqueous emissions from processing operations as an important element in the development of a sustainable forest products industry;
- To keep a local, national, and international perspective and dialogue on issues such as cumulative effects monitoring and management, process modification, and residual treatment technology transfer;
- To identify issues of concern regarding forest industry residual streams that are as yet unre-solved; and
- To understand some of the interrelationships between the residual discharges to the air, water, and land.

---

wood fibre processing facilities. The chapter then provides some quantitative indication of residue emissions to all three environmental media, and concludes with a brief discussion on what is happening and changing in the industry to manage these residues.

Residue (or waste) streams from the forest products industry are complicated by the broadly diverse industries involved in wood fibre processing, from the actual harvesting of trees to the storage of wood, to lumber and plywood mills, and finally to pulp and paper mills. Even within the pulp and paper industry, varied processes, chemical inputs, and outputs that are used in pulp manufacture make difficult the categorization and quantification of residue streams (Fig. 16.1a). Chemical recovery systems reuse many process chemicals and water for some of these pulpmaking systems, but overall the chemical-intensive nature of pulp mill processes has been the focus of past and ongoing rulemaking (USEPA 1995b). The processes used to manufacture pulp (which is later converted into paper) are the major sources of environmental concerns for this industry, far more so than the residue streams from raw timber storage or plywood plants, for example. Pulpmaking processes are also the major sources of air and water pollutant outputs. Although a variety of processes are used in Canada and around the world to produce pulp, the vast majority of wood pulp is manufactured by the kraft chemical pulping process, which often releases nuisance odours and particulates to the air (Smook 1992; USEPA 1995b; de Choudens and Lachenal 1999). Bleaching processes, primarily used to whiten and brighten pulps for high-quality paper manufacture, may produce wastewaters containing chlorinated compounds such as dioxins. Overall, the pulp and paper making process is water-intensive: the pulp and paper industry is the largest industrial process water user in the United States, with a typical mill using 67 000–71 000 L of water per tonne of pulp produced, translating into a total industry discharge of 16 million m^3/day of water (USEPA 1995b). In Canada, a typical large pulp mill can produce 50 000–150 000 m^3 of wastewater effluent per day (Robinson et al. 1994). Pulp and paper mills usually operate wastewater treatment plants to remove materials responsible for biochemical oxygen demand (BOD), total suspended solids (TSS), and other pollutants before discharging wastewaters to a receiving water body, and likewise must have certain control measures in place for atmospheric emissions.

Almost all pulp mills in the world must comply with some form of regulatory control on residual discharges to the environment; most commonly, mills are subject to site-specific operation permits (or approvals) that are issued for a pre-defined number of years. These approvals, as well as an environmental impact assessment procedure, are more frequently being used to minimize the cumulative effects of mills in similar receiving water bodies and airsheds by dictating the types of residual management and pollution control technologies that must be employed at that particular mill. Siting of facilities also becomes an important application of residual management, as public and environmental concern about cumulative forest industry impacts can make a large difference in the long-term sustainability and acceptance of these industries, an excellent example of which was the siting of the Alberta Pacific Forest Industries pulp mill in the late 1980s and early 1990s along the Athabasca River in Alberta. The mill's location was debated because of the cumulative aspect of residual emissions to the environment from the other numerous mills in the Athabasca River basin. The Alberta Pacific mill ultimately went ahead with construction after a lengthy public consultation program and a change in the bleaching process to be employed at the mill that adequately satisfied regulators (Fig. 16.1b). To sum up the complicated interplay that led to this conflict and resolution is a quote from the first Sustainable Forest Management Network (SFMN) conference proceedings:

> *"The challenge of identifying and meeting the many and diverse 'wants' and 'needs' of forest stakeholders is now recognized, but the knowledge, understanding, and ability to predict, protect, and manage the ecological, social, economic and industrial interactions of forested regions is not yet in place. We are learning."*

(Smith and Veeman 1999)

An overview of processes and pollutants

There is no doubt that the forest industry in Canada is large: for 2001, the Forest Products Association of Canada (FPAC; formerly the Canadian Pulp and Paper Association) reports that 197 million m^3 of timber was harvested (PWC 2002). The Pricewaterhouse-Coopers study commissioned by FPAC also reports that in 2001 the total sales in forest-related industries was \$54.6 billion, with export sales at \$44.1 billion accounting for 11.8% of total Canadian exports (PWC 2002). The forest industry also employed, either directly or indirectly, approximately 984 000 people.

The forest products industry itself may be broken down into two main categories: the primary lumber and wood products industry, and the pulp and paper industry. The lumber and wood products industry includes facilities involved in (USEPA 1995a):
- cutting timber and producing pulpwood;
- storing solid wood materials;
- sawmills, lath mills, shingle mills;
- cooperage stock mills (cutting materials for use in wooden casks or tubs);
- planing mills;
- plywood mills; and
- facilities engaged in manufacturing finished articles made entirely or mainly of wood products.

Fig. 16.1. (*a*) Residues from pulp mills are made up of liquids, gases, and solids and are typically released into surface waters, the air, and disposed of in landfills. Mills are situated next to water bodies because of high water use and liquid waste disposal requirements. Note that most atmospheric emissions here are steam. (*b*) The wastewater facility is required to process a large volume of liquid as shown at this facility.

(*a*)

Photo courtesy of Millar Western Forest Products Ltd., Whitecourt, Alberta

(*b*)

Photo courtesy of Alberta Pacific Forest Industries Inc.

The lumber and wood products industry can often be further subdivided into five general categories:

* logging timber;
* sawing lumber;
* manufacturing panel products (including veneer and plywood, as well as reconstituted wood panels such as particleboard, hardboard, medium-density fibreboard, and oriented strandboard);
* value-added manufacturing of end-use products, such as doors, window frames, and furniture; and
* preserving wood.

The pulp and paper industry, on the other hand, produces commercial grades of wood pulp, primary paper products, and paperboard products. The practice of paper-making can be summarized neatly into three simple steps: pulpmaking; pulp processing; and paper/paperboard production. At the pulping stage, the prepared wood stock is digested into its fibrous constituents by breaking the bonds between fibres by different techniques. Box 16.2 details some of the basic differences between the various pulping techniques. Pulp processing removes impurities such as uncooked chips, cleans and recycles any residual cooking liquor via a series of washing steps, and increases pulp whiteness and brightness by adding bleaching chemicals, if necessary. Finally, the processed pulp is pressed into paper sheets and is dried on a paper machine as steam-heated rollers compress the sheets. While the harvesting of logs for subsequent lumber or paper production is a necessary step in the whole forest products industry, its wastes (e.g., sawdust, exhaust from harvesting equipment) are dispersed in the forest environment (see Chap. 15). On the other hand, most of the forest products residuals requiring special handling come directly from the pulping process, and as such these procedures are the focus of much of this chapter.

Kraft chemical pulping and traditional chlorine-based bleaching are both commonly used and may generate significant pollutant outputs. Kraft pulping processes produced approximately 80% of total U.S. pulp tonnage during 1993, and percentages in Canada were similar (Smook 1992; USEPA 1995b). While the use of traditional chlorine bleaching is in decline, a significant proportion of kraft mills currently use the process. Pollutant outputs from mechanical, semi-mechanical, and secondary fibre pulping are small when compared to kraft chemical pulping. In the pulp and paper industry, the kraft pulping process is the most significant source of air pollution. As well, pollutant outputs from chlorine bleaching — the chlorinated byproducts chloroform, dioxins, and furans — are problematic owing to their persistence in the environment, non-biodegradability, and toxicity. Table 16.1 outlines the process steps, the material inputs, and the major pollutant outputs by media for a pulp and paper mill using kraft chemical pulping and traditional chlorine-based bleaching. Some of the pollutant terms within the table are briefly defined in the footnotes below the table; however, a more rigorous description of some of the more pertinent pollutants is given in Boxes 16.3 and 16.4.

Pulp and paper mills use and generate materials that can be harmful to living organisms in the air, water, and on land. Pulp and paper processes generate large volume of wastewaters that might adversely affect freshwater and marine ecosystems. Furthermore, residual solid wastes from wastewater treatment processes may contribute to existing local and regional disposal problems, and air emissions from pulping processes and power generation facilities may release odours, particulates, or other pollutants. Kettunen (1999) has noted that water and air issues are most important to countries in Scandinavia, Eastern and Central Europe, and North America, whereas soil issues are more important to China and Western Europe. Of the sectors involved in the forest products industry, pulp and paper making contribute the largest amounts of residuals that adversely affect the environment. Even within the pulp and paper industry, pulping processes are the major source of environmental impacts, though each pulping process has its own set of process inputs, outputs, and resultant environmental concerns (USEPA 1995*b*). Papermaking activities have not been associated with significant environmental problems and are not specifically addressed by ongoing regulatory and non-regulatory initiatives of the United States Environmental Protection Agency (U.S. EPA) or Environment Canada. As such, non-integrated paper and paperboard mills (those mills without pulping facilities onsite) are not significant environmental concerns when compared to integrated mills or pulp mills (USEPA 1995*b*).

National Pollutant Release Inventory (NPRI) substances are reported to Environment Canada annually since the program's inception in 1992. The substances included under the NPRI are contained in the report "An Analysis of Pulp and Paper Mill Reports to the NPRI for 1999" (Environment Canada 2002). One hundred and fifteen facilities reported these substances under the Canadian Industrial Classification for pulp and paper mills in 1999. Table 16.2 provides a summary of the releases to the three media of concern for the 115 mills under consideration. Releases to air composed 78% of the total releases by mass, followed by 18% into water and 4% onto land (Environment Canada 2002). Analogous statistics from the U.S. are compiled for "toxic release inven-

---

Box 16.2. Overview of the pulping process.

Pulping refers to taking the prepared wood stock and digesting it into its fibrous constituents by breaking the bonds between the wood fibres. Different techniques can be employed to break these bonds, all requiring various chemicals and forms of energy. The different pulping techniques are necessary to obtain various grades of paper or paperboard. The three main techniques for creating pulp are chemical, semi-chemical, and mechanical pulp processes.

Chemical pulps

Chemical pulps are manufactured into products that have high-quality standards or require special properties. Chemical pulping degrades wood by dissolving lignin bonds holding the cellulose fibres together. This also involves cooking (or digesting) woodchips in aqueous chemical solutions at elevated temperatures and pressures.

A specific form of chemical pulping is kraft or bleached kraft pulping (also called sulphate pulping). Kraft cooking chemicals are selective in their attack on wood constituents, and the resulting pulp is notably stronger than those from other processes. The kraft process is flexible and can accommodate both hard and soft woods, and can tolerate contaminants frequently found in wood, such as resin. Lignin removal is high, up to 90%, allowing high levels of bleaching without pulp degradation owing to delignification. The chemicals used in kraft pulping can be recovered, making it economical and reducing some of the environmental releases.

Box 16.2. (concluded).

The bleached kraft pulping method uses a sodium-based alkaline pulping solution (called a liquor) consisting of sodium sulphide (Na$_2$S) and sodium hydroxide (NaOH). This liquor (called white liquor because of its colour) is mixed with woodchips in a digester. The output products are separated wood fibres (pulp) and a liquid that contains the dissolved lignin solids in a solution of reacted and unreacted pulping chemicals (called black liquor). The black liquor undergoes a chemical recovery process to regenerate white liquor for reuse in the first step. These processes convert approximately 50% of input wood into pulp, a relatively small yield when compared to the next two pulping processes.

Brown stock is another term of interest in chemical pulping. Brown stock washing refers to the recovery of spent cooking liquor (or weak black liquor) for reuse in the pulping process. There are a variety of brown stock washing technologies, including rotary vacuum washers, diffusion washers, rotary pressure washers, horizontal belt filters, wash presses, and dilution/extraction washers.

Another process term is the white water. White water is the drainage from wet stock (regardless of its colour) in pulping and papermaking operations. The white water will contain fibre as well as a variety of other furnish-derived materials. The white water system fulfills a number of important objectives; for further information the reader is directed to Smook (1992).

Semi-chemical pulps

Semi-chemical pulping involves partial digestion of raw materials (generally hardwood) in a weak chemical solution followed by mechanical refining for fibre separation. At most, the digestion step consists of heating the pulp in sodium sulphite (Na$_2$SO$_3$) and sodium carbonate (Na$_2$CO$_3$). The yield of semi-chemical pulping ranges from 55 to 90%, depending on the process variation used, but pulp residual lignin content is also high so bleaching is more difficult.

Mechanical pulps

Mechanically produced pulp is of low strength and quality and is often used for newsprint and other non-permanent paper products. Mechanical pulping uses physical pressure instead of chemicals to separate raw wood fibres. Process variations include stone-ground wood, refiner mechanical, thermomechanical, chemimechanical, and chemithermomechanical (sometimes called bleached chemithermomechanical pulp, or BCTMP). Pulp yields are quite high, up to 95%, but energy usage is also high. To offset its weakness, mechanical pulp is often blended with chemical pulp.

Kraft and mechanical pulp mills are the two processing trains most common in the Canadian pulp and paper industry.

Sources: Smook (1992), USEPA (1995*b*), de Choudens and Lachenal (1999)

tory" (TRI) chemical releases. The total amount of TRI releases generated by the pulp and paper industry in the U.S. provides a gross profile of the types and relative amounts of chemical outputs from mill processes, of which there is considerable overlap with the Canadian NPRI chemicals (USEPA 1995*b*; Environment Canada 2002). The U.S. pulp and paper industry releases 87% of its total TRI mass to the air, approximately 10% to

Table 16.1. Pollutants associated with kraft chemical-pulped, chlorine-bleached paper production (adapted from USEPA 1995*b*).

Process step	Material inputs	Process outputs	Major pollutant outputs[a]	Pollutant media
Fibre furnish (blend of fibrous materials used to make pulp) preparation	Wood logs Chips Sawdust	Furnish chips	Soil, grit, fibre, bark BOD[b] TSS[c]	Solid Water
Chemical pulping kraft process	Furnish chips	Black liquor (to chemical recovery system), pulp (to bleaching/ processing)	Resins, fatty acids Colour BOD COD[d] AOX[e] VOCs[f] (terpenes, alcohols, phenols, methanol, acetone, chloroform, methyl ethyl ketone) VOCs (terpenes, alcohols, phenols, methanol, acetone, chloroform, methyl ethyl ketone)	Solid Water Air
	Cooking chemicals: sodium sulphide (Na_2S), NaOH, white liquor (from chemical recovery)		Total reduced sulphur (TRS) compounds Organo-chlorine compounds (e.g., 3,4,5-trichloro-guaiacol)	
Bleaching	Chemical pulp	Bleached pulp	Dissolved lignin and carbohydrates Colour COD AOX Inorganic chlorine compounds (e.g., chlorate)[g]	Water
	Elemental chlorine (Cl_2), chlorine containing compounds		Organo-chlorine compounds (e.g., dioxins, furans, chlorophenols)	
	Hypochlorite (HOCl, NaOCl, $Ca(OCl)_2$) Chlorine dioxide (ClO_2)		VOCs (acetone, methylene chloride, chloroform, methyl ethyl ketone, carbon disulfide, chloromethane, trichloromethane)	Air/ water

Table 16.1. (*continued*).

Process step	Material inputs	Process outputs	Major pollutant outputs[a]	Pollutant media
Papermaking	Additives, bleached/ unbleached pulp	Paper/paperboard product	Particulate wastes Organic compounds Inorganic dyes COD Acetone	Water
Wastewater treatment facilities	Process wastewaters	Treated effluent	Sludge	Solid
			VOCs (terpenes, alcohols, phenols, methanol, acetone, chloroform, methyl ethyl ketone)	Air
			BOD TSS COD Colour Chlorophenolics Carbon disulfide VOCs (terpenes, alcohols, phenols, methanol, acetone, chloroform, methyl ethyl ketone)	Water
Power boiler	Coal, wood, unused furnish	Energy	Bottom ash: incombustible fibres	Solid
			SO_2, NO_x, fly ash, coarse particulates	Air
Chemical recovery system				
Evaporators	Black liquor	Strong black liquor	Evaporator noncondensibles (TRS, VOCs: alcohols, terpenes, phenols)	Air
			Evaporator condensates (BOD, suspended solids)	Water
Recovery furnace	Strong black liquor	Smelt Energy	Fine particulates, TRS, sulphur dioxide	Air

Table 16.1. (*concluded*).

Process step	Material inputs	Process outputs	Major pollutant outputs[a]	Pollutant media
Recausticizing	Smelt	Regenerated white liquor	Dregs	Solids
		Lime mud	Waste mud solids	Water
Calcining (lime kiln)	Lime mud	Lime	Fine and coarse particulates	Air

[a]Pollutant outputs may differ significantly based on mill processes and material inputs (e.g., woodchip resin content).
[b]BOD, biochemical oxygen demand.
[c]TSS, total suspended solids.
[d]COD, chemical oxygen demand.
[e]AOX, adsorbable organic halides.
[f]VOC, volatile organic compounds.
[g]Chlorate only significantly produced in mills with high rates of chlorine dioxide substitution to reduce dioxin and furan production.

Table 16.2. Summary of NPRI 1999 substance releases to various media from Canadian pulp and paper mills (Environment Canada 2002).

Media	Mass (Mg)[a]	Percent of total release
Air	25173	78
Water	6115	18
Land	1204	4
Total	32492	100

[a]1 Mg = 1 tonne = 1000 kg.

water and publicly owned wastewater treatment works (POTWs), and 2% is transferred offsite or disposed of on land. Other TRI industries average approximately 93% to air, 1% to water, and 6% to land. The larger proportion of water releases in the pulp and paper industry correlates with the water-intensive processes of this particular industry.

Methanol (CH_3OH) was the substance released in by far the greatest quantity in the Canadian NPRI, accounting for almost 54% of the total releases. Other substances released in relatively large quantities were ammonia (NH_3, 12% of total), hydrogen sulphide (H_2S, 9% of total), and sulphuric acid (H_2SO_4, 5% of total). Collectively, these four chemicals account for almost 81% of the residues released from mills. However, the composition of releases differs substantially between media, as presented in Table 16.3. Box 16.3 provides toxicity data on some of these substances. The following sections provide a more rigorous breakdown of information on the residuals discharged to each medium for pulp mills, and subsequent chapters will deal with the treatment of chemical contaminants in each medium.

Box 16.3. Health data for top pollutant releases.

The following is a list of some of the health-related data, in particular the lethal concentration or dose values, as well as some environmental persistence and partitioning information. Terms are explained following the table.

Substance	Physical state of residue	LC_{50}/LD_{50}	Level	Type of toxicity test	Water solubility	Vapour pressure	Partition coefficient log K_{ow}	Half-life	Bioconcentration factor (BCF)
Ammonia	Aqueous	LC_{50}	0.53 mg/L[a]	Rainbow trout (96 h)	531 g/L @ 20°C[a]	7600 mmHg @ 25.7°C[a]		0.16 days via volatilization[b]	
		LD_{50}	350 mg/kg[a]	Oral rat					
		LC_{50}	3360–7050 mg/m^{3}[a]	Inhalation rat (1 h)					
		LC_{50}	1.7 mg/L[b]	Freshwater toxicity					
Hydrogen sulphide	Gas	LC_{50}	0.007–0.55 mg/L[a]	Fathead minnow (96 h)	4.1 g/L @ 20°C[a]	20 mmHg @ 25.5°C[a]			No bioconcentration potential
		LC_{50}	673 ppm[a]	Inhalation mouse (1 h)					
		LC_{Lo}	600 ppm[a]	Inhalation man (30 min.)					
Methanol	Liquid	LC_{50}	8000 mg/L[a]	Trout (48 h)	Miscible[a]	100 mmHg @ 21.2°C[a]	–0.77[a]	7 days via volatilization[b]	
		LD_{50}	5600/7300 mg/kg[a]	Oral rat/ mouse		125 mmHg @ 25°C[c]		7 days via biodegradation[b]	For golden ide <10
		LC_{50}	64000 ppm[a]	Inhalation rat (4 h)					

Substance	Physical state of residue	LC_{50}/LD_{50}	Level	Type of toxicity test	Water solubility	Vapour pressure	Partition coefficient $\log K_{ow}$	Half-life	Bioconcentration factor (BCF)
Nonylphenol[d]	Liquid	LC_{50}	0.14 mg/L[a]	Fathead minnow (96 h)	Nearly insoluble[a]	9.2×10^{-5} mmHg @ 25°C[c]			For bay mussel 8–12[a]
		LD_{50}	1230 mg/kg[a]	Oral rat					
Sulphuric acid	Liquid	LC_{50}	100–330 mg/L	Flounder (48 h)	Miscible[a]	1 mmHg @ 145.8°C[a]			
		LD_{50}	2.14 g/kg	Oral rat					
		LC_{50}	320/510 mg/m^{3a}	Inhalation mouse/rat (2 h)					
Zinc	Solid	LC_{50}	<0.14 mg/L in soft water pH 8; 3.20 mg/L in hard water pH 5[a]	Brown trout (96 h)		1 mmHg @ 487°C[a]			
			124 mg/m^{3a}	Inhalation man, lowest toxic dose					

Sources: [a]Gangolli 1999, [b]USEPA 1996, [c]Yaws 1999.
[d]Note that nonylphenol chemical and toxicological properties are listed in place of nonylphenol polyethylene glycol ether, as it was not possible to find information specifically about nonylphenol polyethylene glycol ether. Instead, information for the parent compound is provided for purposes of comparison.

Methanol, the largest release by mass from the Canadian pulp and paper industry (15 607 Mg to air and 1835 Mg to water) is a miscible chemical and therefore very mobile in the environment; it also has a relatively high vapour pressure, suggesting that some of the liquid or aqueous methanol may vaporize and enter the atmosphere after being discharged as a liquid. The LD_{50}/LC_{50} values for methanol are quite a bit higher than some of the other major chemical releases. Ammonia, the major release to water, is one of the more toxic substances with fairly low LC_{50} values in freshwater trout toxicity tests. Hydrogen sulphide, the second largest contributor to residual air emissions at 3020 Mg in 1999, has LC_{50} values for inhalation in the parts per million magnitude; in the air residuals section of this chapter is some further discussion about threshold limits for the total reduced sulphur compounds, of which hydrogen sulphide is one.

Glossary of terms:

LC_{50}/LD_{50} — The lethal concentration (LC) or lethal dose (LD) administered to a test population, yielding a 50% survival rate in the test subjects.

Water solubility — A critical property that affects the fate and transport of chemicals in the aqueous environment. Refers to the ability of a contaminant to dissolve in liquid, most importantly a chemical's solubility in water, as we are most concerned with surface and groundwater contamination as a means of contaminant transport and an important exposure route. High water solubility translates into high mobility in the environment.

Vapour pressure — The pressure exerted by a vapour in equilibrium with its solid or liquid phase; this is the relative measure of volatility of a chemical in its pure state. A high vapour pressure will result in rapid vaporization to the atmosphere.

Octanol–water partition coefficient (K_{ow} or log K_{ow}) — The ratio of a chemical's concentration in an octanol phase [$CH_3(CH_2)_7OH$], an organic phase, to its concentration in water for a two-phase (octanol–water) system. It varies from 10^{-3} (hydrophilic) to 10^7 (hydrophobic) for many organic chemicals (log$_{10}$ values of -3–7). It also gives an indication of the tendency of a chemical to partition itself between fish muscle tissue (octanol) and water.

Half-life — A relative measure of chemical persistence in the environment, this parameter relates to the combined removal processes in an environmental medium (phase transfer, chemical transformation, biochemical transformation). The time it takes for the chemical concentration to halve in magnitude. Chemicals with shorter half-lives are not as persistent in the environment.

Fish bioconcentration factor (BCF) — The relative tendency for a chemical to partition into aquatic organisms (usually fish) coincident with the concentration of the chemical in the surrounding water. Given as the ratio of concentration in fish tissue to the concentration in water; typical ranges are 1–10^6, with 1 being low uptake by fish.

Table 16.3. Summary of onsite pulp and paper mill releases of specific residues to various media (Environment Canada 2002).

| Total mass released (Mg) | Medium | | |
	Air	Water	Land
Top three pollutants released	25173	6115	1204
	Methanol (62%)	Ammonia (37%)	Magnesium (68%)
	Hydrogen sulphide (12%)	Methanol (30%)	Zinc (12%)
	Sulphuric acid (7%)	Nitrate ion (20%)	Nonylphenol polyethylene glycol ether (11%)

Air releases can be traced to a variety of sources within a mill. In the United States, approximately 50% of air emissions are methanol, a byproduct of the pulpmaking process. The other major toxic chemicals released into the air are chlorinated compounds, sulfuric acid, and the chelator methyl ethyl ketone (MEK), all of which originate in the bleaching stage (USEPA 1995b). Methanol also accounts for approximately 40% of the water releases by pulp and paper facilities. Overall, methanol represents more than 49% of the pulp and paper industry's total TRI releases and transfers. The diversity of processes in the pulp and paper industry can be seen in the diversity of chemicals found in the annual Canadian NPRI and the U.S. TRI reports.

Water-borne residues

The most significant aqueous discharges in the forest products industry are associated with Canada's approximately 123 direct-discharging pulp and paper mills. This does not mean, however, that there are no significant residues left by the primary wood products industries; the industry's woodlands operations may indirectly contaminate water by changing the quality and quantity of water in natural water bodies. Harvesting practices often cause discharges of materials into surrounding waters, owing mostly to precipitation runoff carrying away contaminants like bark, sawdust, or woodchips that contribute to the total suspended solids and organic content of the resulting run-off (see Chap. 10). Once water is exposed to wood or logging debris, a leachate or run-off may be generated that contains many contaminating chemical compounds. In a few locations, mostly on the west coast of Canada, water is actually used to transport cut logs to processing facilities for the manufacture of wood products.

Occasionally, trees are sorted, handled, and stored at logyards and dryland sorts before making their way to a sawmill (USEPA 1995a). If stored on land, the logs are usually sprayed with water to keep them moist, to prevent cracking and reduce the risk of fire. More often than not this wood is stored without cover and is exposed to precipitation; when this water comes into contact with wood or associated debris (Fig. 17.26), a run-off leachate is generated that contains many contaminating materials. Most of the

Box 16.4. Summary of some pertinent aggregate aqueous pollution parameters.

Biochemical oxygen demand

Of the parameters contributing to the strength of pulp mill effluent, biochemical oxygen demand (BOD) is perhaps one of the most important of the regulated parameters, since it impacts the dissolved oxygen levels of the receiving water. BOD is an overall measure of the biodegradable organic strength of a wastewater, where the strength is measured indirectly via microbial oxygen consumption. Higher BOD increases the natural level of microorganism activity, which lowers dissolved oxygen concentration. In extreme cases, fish and other more complex forms of life can suffer from lack of oxygen. The standard BOD test consists of filling an airtight bottle of a specified size with the test fluid, a mixture of the waste sample and fully saturated dilution water, and incubating it at a specified temperature for 5 days (Eaton et al. 1995). The BOD_5 is then calculated as the difference between the initial and final dissolved oxygen concentration after a period of 5 days.

Chemical oxygen demand

The chemical oxygen demand (COD) test is widely used as a means to measure the total organic strength of a wastewater (Sawyer et al. 1994). It allows the measurement of a waste in terms of the total quantity of oxygen required for the oxidation of organic pollutants. As such, the COD is always greater than the BOD_5, as not all organic constituents of a waste are available for microorganism metabolism (Eaton et al. 1995). In fact, COD values can be much greater than BOD_5 values, especially when there are significant amounts of biologically resistant organic matter present in the waste stream, such as lignins in a wood-pulping waste (Sawyer et al. 1994).

Total suspended solids

Total suspended solids (TSS) refers to the solid material in the waste effluent that does not pass through a 2 μm filter (Eaton et al. 1995). TSS will cloud the receiving water, impacting aquatic life in the deeper reaches of the water body if they are dependent on sunlight. This could also be important to the exertion of BOD_5 in terms of how the organic material is presented to the microorganisms; dissolved solid material may be more accessible to the microbes, but over time the solid organic compounds could oxidize or dissolve into the water owing to other chemical reactions occurring. This ultimately results in different types of BOD — specifically, dissolved or suspended — where dissolved BOD refers to the organic compounds that are available to microbes as food (Chapra 1997). The test for TSS takes about an hour and a half and requires only a filter apparatus with filter paper and is very straightforward.

Colour

Colour in a water sample (e.g., Fig. 16.2a) is generally indicative of contact with organic debris, such as the lignins in a pulp mill effluent (Sawyer et al. 1994). Some of these organic compounds will be available to microorganisms as food, and will therefore contribute to the BOD of a wastewater. The test for colour is a relatively simple test: the most common ones consist of visual or spectrophotometric comparison to colourimetric standards.

Adsorbable organic halides

AOX, or adsorbable organic halides, is a bulk measure of the chlorinated material in a wastewater that has been in contact, as some point in the manufacturing process, with chlorine. Free chlorine and some other chlorinated bleaching agents will react with various organic wood constituents to form toxic chlorinated organic compounds. Among them, the most important are dioxins and furans because of their high toxicity and ubiquitous occurrence in processing wastewaters. The AOX method measures the total amounts of most, but not all, of the chlorinated contaminants known to be present in bleached chemical pulp mill wastewater.

Fig. 16.2. (*a*) Treated liquid effluents can contain high concentrations of colour-causing materials. Effluent plume from Weyerhaueser. (*b*) Surface aeration of an aerobic stabilization pond providing biological waste treatment.

(*a*)

Photo by Dan Smith

(*b*)

Photo by Dan Smith

contaminants, such as wood and bark-derived organics and particulates, originate from the wood itself, though run-off may also contain sediment from logyard erosion and greases and oils from logyard vehicles and other machinery used at the site. Table 16.4 contains a summary of some of the aqueous residuals from wood processing and storage industries. More discussion about these contaminants and their environmental health effects are found in Chap. 17.

Table 16.4. Major water pollutants from the wood processing industry (adapted from USEPA 1995*a*).

Process	Material input	Effluent characteristics
Logging	Trees, diesel, gasoline	BOD, COD, TSS, oils/gasoline from fuel spillage
Logyards/dryland sorts	Wood logs	BOD, COD, TSS
Sawing	Wood logs, diesel, gasoline	Mostly TSS, some hydrocarbon fuel in leachate
Surface protection	Wood, 3-iodo-2-propynyl butyl carbamate (IPBC), didecyl dimethyl ammonium chloride (DDAC)	Dripped formulation mixed with rainwater and facility washdown water, chemical spills of formulation
Plywood and veneer	Veneer, phenol-formaldehyde resins, urea-formaldehyde resins, melamine-formaldehyde resins, sodium hydroxide, ammonium sulphate, acids, ammonia	Not applicable
Reconstituted wood products	Wood particles, strands, fibre, same resins as plywood and veneer, methylenediphenyl diisocyanate resins	Not applicable
Wood preserving	Wood, pentachlorophenol, creosote, borates, ammonium compounds, inorganic formulations of chromium, copper, and arsenic, carrier oils	Dripped formulation mixed with rainwater and facility washdown water, chemical spills of formulation

Other than contaminated run-off from raw material storage, many of the plywood and panelboard plants do not have extensive aqueous residual outputs, as these plants try to reuse much of the woody material for product manufacture. Besides the pulp and paper industry, wood preserving has had some of the more hazardous residual outputs. Wood is treated with preservatives to protect it from mechanical, physical, and chemical influences. The most common preservatives, until recently, included water-borne inorganic compounds like chromated copper arsenate (CCA) and ammoniacal copper zinc arsenate (ACZA), and oil-borne organic chemicals like pentachlorophenol (PCP) and creosote (USEPA 1995*a*). However, a 9-year-old working group consisting of Health Canada's Pest Management Regulatory Agency (PMRA) and the U.S. EPA has been re-evaluating these products and actively seeking alternatives for their use in wood preservation (NAFTA TWGP 2002).

Before preservation, wood is conditioned in the open air (non-pressurized) or in a pressurized cylinder. Conditioning can be a major source of wastewater in the wood-preserving industry (USEPA 1995a). The open-air method involves the repeated use of preservative in a treatment tank with fresh preservative solution added to replace consumptive loss. The continual reuse of preservative leads to an accumulation of wood-chips, sand, stones, and other particulate debris contaminated with hazardous constituents in the bottom of the wood preservative treatment tanks. The contaminated debris is a major source of process waste for non-pressure processes. After pressurized wood preservative treatment, the treated wood is removed from the cylinder and placed on a drip pad where it remains until dripping has ceased. Preservative solution, wash-down water, and rainwater are collected on the drip pad and maintained in the process (USEPA 1995a). At water-borne preservative plants, these materials are transferred to a dilution water tank where they are blended with additional concentrate to make fresh treating solution. At oil-borne plants, these materials are processed to recover preservative and usable process solution. Excess waste liquid is treated onsite in a wastewater treatment plant or offsite at a POTW. After both pressure and non-pressure treatment, some unabsorbed preservative formulation adheres to the treated wood surface. Eventually this liquid drips from the wood or is washed off by precipitation. If the wood has been pressure treated, excess preservative will also exude slowly from the wood as it gradually returns to atmospheric pressure — this is known as kickback. If this kickback is not collected properly over a drip pad it may cause residual contamination on the ground or in surface water at the lumber's point of use.

The chemicals used in the wood-preserving process and the drip pads used to collect preservative drippage after treatment of wood have been the subject of considerable regulatory action (USEPA 1995a). The Environmental Protection Agency of the United States has issued final regulations regarding wood-preserving wastewater, process residuals, preservative drippage, and spent preservatives from wood-preserving processes at facilities that use chlorophenolic formulations, creosote formulations, and inorganic preservatives containing arsenic or chromium. These regulations are listed in the U.S. EPA document "Sector Notebook Project — Profile of the Lumber and Wood Products Industry" (USEPA 1995a). More recently, the EPA and Health Canada have both issued statements that manufacturers of wood treatment chemicals have agreed to make a transition away from the use of chromated copper arsenate (CCA) in treated lumber destined for the non-industrial market by December 31, 2003 (Health Canada 2002a, b; USEPA 2003). Industrial uses of CCA, such as on utility poles and for highway construction, are still acceptable. Alternatives to CCA being examined include alkaline copper quarternary (ACQ) and copper azole. This may result in decreased concentrations of CCA in some preservation drippage during the wood-preserving process, but the alternative preservatives may also lead to new and as-of-yet unforeseen water-quality issues.

General water pollution concerns for pulp and paper mills are effluent solids, biochemical oxygen demand, toxicity, and colour (USEPA 1995b). Effluent colour is often one of the most obvious indications of failure somewhere in the water treatment process (Fig. 16.2a). It is important to note, however, that most of the problematic contaminants in forest industry aqueous discharges are not formed in the manufacturing process, but rather they are unavoidably extracted from the wood that serves as the raw material for production. Additional toxicity concerns arise from the presence of chlorinated organic compounds such as dioxins, furans, and others (collectively referred to as adsorbable organic halides, or AOX) in wastewater after the chlorination bleaching/extraction process. These various pollution parameters are explained further in Box 16.4. Because of the large volumes of water used in pulp and paper processes, almost all mills have primary and (or) secondary wastewater treatment systems installed to remove particulates and BOD produced in the manufacturing process (Figs. 16.1*b* and 16.2*b*). These systems also provide significant removals (30–70%) of other important parameters, such as AOX and chemical oxygen demand (COD). Major sources of effluent pollution from pulp and paper mills are presented in Table 16.5.

Screening and cleaning operations during the pulp processing stage are usually sources of large volumes of wastewater. This effluent stream, called white water because of its characteristic colour, may contain significant BOD and is always a source of suspended solids from wood particles. Similar wastewater streams are also produced during the papermaking process. Fibre and liquor spills can also be a source of mill effluent, though they are typically captured and pumped to holding areas to reduce chemical usage through spill reuse and to avoid the necessity to treat them with the wastewater treatment facility. The condensates from chip digesters and chemical recovery evaporators are a low-volume, high-BOD effluent source, and some of the condensates contain reduced sulphur compounds (USEPA 1995*b*).

Suspended solids from a mechanical pulp mill are usually higher than from chemical pulping plants, generally because of the higher fine-particle content (Smook 1992). Thermomechanical pulps are known to contain longer fibres than other mechanical pulps (like stoneground wood) and on account of the relative absence of fines will generate less suspended solids in the effluent. Effluents from a mechanical pulp mill are toxic, owing mostly to acid extracts from the wood. These are predominantly resin acids and some unsaturated fatty acids like oleic and linileic acids. The primary factor governing effluent residual composition from a mechanical mill will be the composition of the raw material (i.e., the species of wood used; Smook 1992). Most northern softwoods such as spruce (*Picea* spp.), hemlock (*Tsuga* spp.), and western red cedar (*Thuja plicata*) are low in resins and fatty acids, and as a result create less pollution than high

Table 16.5. Major water pollutants from the pulp and paper industry (adapted from Smook 1992).

Source	Effluent characteristics
Water used in wood handling/debarking and chip washing	Solids, BOD, colour, acids (oleic and linileic)
Chip digester and liquor evaporator condensate	Concentrated BOD, may contain reduce sulphur
"White waters" from pulp screening, thickening, and cleaning	Large volume of water with suspended solids, may contain significant BOD
Bleach plant washer filtrates	BOD, colour, chlorinated organic compounds (AOX)
Paper machine water flows	Solids, often precipitated for reuse
Fibre and liquor spills	Solids, BOD, colour

resin-containing species, like pines (*Pinus* spp.). The age of the tree will also dictate the level of resin acids in the wastewater effluent.

The major source of pollution in the mechanical pulp mill will be the wood room, if the logs are debarked and chipped on site. These effluents are very heavy in suspended solids and acids, caused mostly by the bark. The second-largest source of pollution comes from the chips washing system, where the effluent again contains suspended solids and acids.

Table 16.6 provides a summary of the 20 substances released to receiving water bodies by the pulp and paper sector in Canada (based on 1999 NPRI data). For the sake of comparison, the 1995 TRI data from the U.S. is also provided.

Ammonia was released to water in the greatest quantity in Canada and was the most widely reported substance with 59% of all mills reporting releases (Environment Canada 2002). This is understandable as ammonia, along with other nutrients, is used in secondary wastewater treatment systems to promote the microbial degradation of organic contaminants in mill effluent before it is released to a receiving water body. If these nutrients are not completely utilized during microbial metabolism, they too will be released into the environment. It is interesting to note that the nitrate ion was present in the highest amount in American mill effluents; this can probably be explained by the fact that nitrates are produced through the biological oxidation of ammonia, and it is possible that the U.S. mills have longer detention times in their wastewater treatment plants, allowing for greater conversion of ammonia than in Canada. This observation could also be explained by variations in the points at which effluent was sampled from the secondary wastewater treatment facilities. Methanol accounted for the second-greatest release to water in both countries. Although methanol reaches the secondary treatment system via a waste stream, a portion of it is not converted and is released with the plant effluent. It is possible, too, that methanol is a degradation product of longer chain hydrocarbons and is therefore produced during the treatment process. Wastewater treatment systems can also be a significant source of cross-media pollutant transfer. As an example, waterborne particulate and some chlorinated compounds settle or absorb onto treatment sludge, and other compounds may volatilize during the wastewater treatment process.

In the past, releases to water have been greater in Canada. These have decreased substantially because of the 1992 Pulp and Paper Effluent Regulations that were passed under the Fisheries Act (Environment Canada 2002). For comparative purposes, total residual chemical releases to water totalled 15 190 Mg in 1994, approximately 2.5 times greater than in 1999. The widespread installation of secondary biological wastewater treatment facilities contributed greatly to the reduction in chemical releases, and have greatly benefited receiving water ecosystems by reducing sublethal effects on aquatic organisms and improving the vitality of benthic communities (Riebel 2000*a*). In fact, the installation of secondary treatment facilities at most mills has been credited with the current state of compliance, in both Canada and abroad.

Table 16.7 outlines some aggregate effluent characteristics of kraft and bleached chemithermomechanical pulp (BCTMP) mills in Alberta. The pollutant loadings are normalized to the amount of pulp product produced at the respective mills as air-dried tonnes (Mg). It should be noted that some of the effluent characteristics vary widely in

Table 16.6. Releases to water by the pulp and paper sector per year (adapted from USEPA 1998, Environment Canada 2002).

Substance	Canada (1999 data)					U.S. (1995 data)				
	Mg released	% of total releases to water	% cumulative	No. of mills reporting	% of mills reporting	Mg released	% of total releases to water	% cumulative	No. of mills reporting	% of mills reporting
Ammonia (total)	2248	36.8	36.8	68	59	1210	14	14	197	65
Methanol	1812	29.6	66.4	44	38	3045	36	50	175	57
Nitrate ion in solution at pH ≥ 6.0	1200	19.6	86	18	16	3614	42	92.4	54	18
Manganese (and its compounds)	674	11	97	33	29	31.7	0.4	92.8	2	1
Formaldehyde	35	0.6	97.6	6	5	37.5	0.4	93.2	60	20
Hydrochloric acid	21	0.3	97.9	3	3	0.3	0.004	93.2	161	53
Zinc (and its compounds)	22	0.4	98.3	11	10	159	1.9	95.1	50	16
Acetaldehyde	19	0.3	98.6	19	17	93	1.1	96.2	124	41
Nonylphenol polyethylene glycol ether	19	0.3	98.9	10	9	12.5	0.1	96.3	19	6
Hydrogen sulfide	18	0.3	99.2	18	16	NR[a]	—	96.3	0	0
Ethylene glycol	16	0.3	99.5	9	8	16.7	0.2	96.5	16	5
Phenol (and its salts)	14	0.2	99.7	5	4	6.87	0.08	96.6	99	32
Chromium (and its compounds)	13	0.2	99.9	1	1	24.6	0.3	96.9	6	2
Chloroform	3	0.04	100	6	5	143	1.7	98.6	81	27
Cresol (mixed isomers and their salts)	1	0.02	100	5	4	4.6	0.05	98.6	40	13
Isopropyl alcohol	1	0.01	100	1	1	NR	—	98.6	0	0
Sulphuric acid	0.1	0.002	100	1	1	0.5	0.006	98.6	148	49
Acrylamide	0.02	0.0004	100	1	1	NR	—	98.6	0	0
Methyl ethyl ketone	0.01	0.0002	100	1	1	19	0.2	98.9	58	19
Mercury (and its compounds)	0.003	0.00005	100	1	1	0.05	0.0006	98.9	1	0.3
Total	6115	100	—	115	—	8516	98.9	—	305	—

[a]NR, not reported.

Table 16.7. Aggregate effluent characteristics of kraft and BCTMP pulp mills in Alberta (McNamara 2003).

Parameter	Kraft mills[a]				BCTMP mills[b]			
	Avg.	SD	5%-ile	95%-ile	Avg.	SD	5%-ile	95%-ile
BOD (kg/ADt[c])	1.2	0.98	0.21	2.2	0.63	0.47	0.19	1.0
COD (kg/ADt)	27.1	17.6	11.1	47.3	12.1	7.7	4.5	16.8
Colour (kg/ADt)	40.8	29.5	18.1	75.9	20.4	12.4	8.3	29.8
TSS (kg/ADt)	2.1	1.0	1.0	3.1	1.4	0.91	0.50	2.1
AOX (kg/ADt)	0.25	0.22	0.08	0.51	—	—	—	—
Production (ADt/day)	1217	364	842	1609	655	109	551	746

[a]Number of kraft mills: 4. Based on 2001 data.

[b]Number of BCTMP (bleached chemithermomechanical pulp) mills: 3. Based on 2000 data.

[c]ADt, air-dried tonne (contains 10% moisture).

the respective mill categories, and as such, some percentile values are provided as well. All of the average aggregate pollution values are lower in the BCTMP mills, which is expected because of the difference in pulping processes and chemical requirements for the production of pulp. The difference can also be indicative of the different approval levels for each mill, which will vary depending on the era in which they were licensed, the geographic region, the receiving watershed, and local concerns. For a comparison of these same aggregate parameters in Canadian and Nordic mills the reader is directed to Chap. 17.

It is also interesting to note that the lowest AOX-discharging mills in Canada are located in Alberta, where 2000 AOX discharges were comparable to those from Nordic mills. The performance of the Alberta mills is probably driven by the relatively modern processes employed and treatment technology used in the jurisdiction.

In a similar comparison, de Choudens and Lachanel (1999) compared the residual aqueous pollution loads from three major groups of pulp manufacturing processes: bleached kraft, thermomechanical pulp (TMP), and deinked newsprint. The data presented in Table 16.8 appear contrary to the common perception that TMP mills pollute less than bleached kraft mills; indeed, it would appear as though all the pulp manufacturing processes available today are almost equally polluting in terms of their impact on water. It also appears that the environmental problems associated with the recycling of fibres are worse if one considers the huge amount of sludge produced. There is also the potential problem of leakage of hazardous components like heavy metals from deinking sludges. As this information is contrary to common beliefs about the difference in pollution loads from different mill types, a closer examination of the source data would be prudent before making any hard conclusions. However, the data do present a somewhat varied perspective of the difference between aqueous residuals from different mill types, suggesting that water pollution aspects alone do not appear to be an incentive to develop high yield or waste paper pulps (de Choudens and Lachenal 1999).

The largest challenge facing pulp producers in recent years has been the need to reduce chlorinated organics both in pulp and pulp mill effluents. Concentrations of chlo-

Table 16.8. Generic aqueous residual load for modern pulp manufacturing operations (de Choudens and Lachenal 1999).

	Bleached kraft	TMP	Deinked newsprint
Water consumption $(m^3/ADt^a)^b$	35	20	15
Untreated effluent			
BOD_7 (kg/ADt)	15	20	15
COD (kg/ADt)	50	55	30
TSS (kg/ADt)	10	10	90
AOX (kg/ADt)	1	0	0.01
Treated effluent			
BOD_7 (kg/ADt)	1	1	1
COD (kg/ADt)	20	15	10
TSS (kg/ADt)	1	0.5	0.5
AOX (kg/ADt)	0.4	0	0.005
Sludge production			
Total kg/ADt	15	20	100

[a] ADt, air-dried tonne (contains 10% moisture).

[b] These water consumption values are smaller than values quoted earlier in the chapter. They are reflective of "modern" mills, as in water consumption values for new mills or mills that are moving towards systems closure in 1999, rather than the earlier quoted data which was from the early 1990s. This should give the reader some indication of the speed at which residual management is changing in the pulp and paper industry.

rinated organic material depend on the quantity of chlorine-containing bleaching material used in the bleaching process, the type of chlorine-containing bleach, and the point in the process where the bleaching chemical is added. Up until a few years ago, the two chlorinated organics generally known about were dioxins and furans. The specific chemicals of concern were 2,3,7,8-tetrachlorodibenzo-*p*-dioxin (2,3,7,8-TCDD) and 2,3,7,8-tetrachlorodibenzofuran (2,3,7,8-TCDF); quantities were typically in the order of parts per trillion. These compounds can usually be controlled by good bleaching and pre-bleaching practices, particularly the substitution of chlorine gas with chlorine dioxide.

Now that dioxins and furans have essentially been eliminated by stopping the use of elemental chlorine, the challenge of eliminating all chlorinated organic material in effluent, as measured by AOX, has presented itself. In 1992, the British Columbia (B.C.) government amended the Pulp and Paper Mill Liquid Effluent Control Regulation to immediately reduce the allowable AOX discharges from 2.5 kilograms per air-dried tonne of pulp produced (kg/ADt) to 1.5 kg/ADt. Consequently, the AOX annual average discharge rate from B.C. pulp mills was less than 0.5 kg/ADt in 2000, compared to 6.5 kg/ADt in 1988 (Carey et al. 2002). This change is due mainly to advancements in the bleaching/pulping process, implementation of new technologies, and closure of the water loop in pulp screen systems.

The B.C. government has recently promulgated a review of the modified requirement for bleach plants to eliminate discharge of AOX by December 31, 2002 (Carey et al. 2002). A scientific review panel ultimately found that no evidence was available to indicate that further reductions of effluent AOX (beyond that already achieved) would result in any demonstrable environmental benefit. The panel found that a number of studies linked kraft black liquor (also called spent cooking liquor, the liquor separated from the wood fibre after the cooking reaction) to sub-lethal toxicity in aquatic organisms. The panel decided that ensuring these discharges were minimized could yield significant environmental benefits (Carey et al. 2002).

There is some debate as to whether there is any statistical correlation between discharges of AOX from kraft pulp mills and the discharges of 2,3,7,8-TCDD and 2,3,7,8-TCDF (Carey et al. 2002). If this is the case, there is also some question as to the wisdom of continuing to focus regulatory efforts on AOX. The aforementioned panel decided not to support the use of AOX as a regulatory surrogate for toxicity or environmental impact, but was of the opinion that AOX would be a useful indication of the chlorinating conditions at a bleach plant and would therefore be a useful production parameter. However, the U.S. EPA still chooses to regulate AOX discharges from bleached chemical pulp mills because it considers daily testing of the whole mill AOX discharge to be the most cost-effective means of ensuring compliance with regulations limiting the discharge of dioxins and furans (Carey et al. 2002).

The environmental effects monitoring (EEM) program was added to Canada's national pulp and paper mill effluent regulations to allow a science-based, iterative approach to regulation that would include monitoring, assessment, and management. In general, the EEM methodology is still undergoing refinement, as a number of scientific and technical concerns have arisen out of the first two cycles. For further discussion of the EEM review cycle process the reader is directed to Chap. 17.

The development of pulp mill effluent-induced flocculation has been extensively examined in rivers (Young 2001). This SFMN-sponsored research ultimately revealed that effluent-induced flocculation could contribute to anoxic conditions at the riverbed if its assimilative capacity was low or during the winter months when the river was under ice cover (see Chap. 17). More recent studies in this area have shown that microorganisms present in the river water and in the floc were sometimes capable of enhancing flocculation, though it was concluded that further testing was necessary to consider the effects of changing effluent and river water characteristics on this phenomenon (Joyce and Smith 2003).

Water use reduction is fast becoming the next looming environmental issue. There is a burgeoning expectation among the public that industries, which use up to 30% of the planet's freshwater, should be more vigilant about their consumption in the manufacturing process (Greenbaum 2002). Also known as progressive system closure, implementation of a program of water use reduction is a complicated and time-consuming process that can form a large part of a company's environmental management system. Decreasing water intake results in less wastewater to treat, as well as reduced chemical costs. However, moving to a closed system will also result in higher concentrations of residuals such as BOD, COD, TSS, and AOX, and may actually cause non-compliance of toxicity levels set out by regulatory agencies (Greenbaum 2002). As such, complete elimination of effluent may not be the most prudent choice, but a life-cycle analysis with a water balance will best determine how to manage system closure and the associated residual issues. Other minimum-impact mill concepts, such as minimizing water consumption and residual streams while creating sustainable value, are outlined in Reeve (1999). Some of these systems are currently being implemented in North America, with most of them focusing on minimizing losses from spent cooking liquor and other wood extract-containing liquors, as well as improving washing to minimize carryover to the bleach plant rather than just eliminating bleaching effluent.

More detailed discussion of some of these topics are described in Chap. 17 along with emerging abatement technologies in the treatment of boreal pulp and paper mill effluents released as part of the wastewater stream. Chapter 17 also offers some excellent comparisons with the regulatory efforts in countries abroad, such as Finland and Sweden.

Air-borne residues

Air pollutants are an important residual generated by industrial facilities because they can be transported long distances and can have detrimental health effects on living organisms, as well as being contained in an exposure medium that all terrestrial vertebrates require — air. While atmospheric emissions of some sort are inevitable in any industrial process, it is important to understand what the emissions are, their sources and their fate in the environment in order to delineate those emissions that have the most significant impact on environmental and human health. The common air pollutants emitted by the forest products industry include particulate matter (PM), sulphur compounds, nitrogen oxides, greenhouse gases (such as carbon dioxide, carbon monoxide, and, to a minor degree, methane), volatile organic compounds (VOCs), and hazardous air pollutants (HAPs). Chapter 18 goes into more detail and definition of all of these pollutants. As a brief introduction, though, particulate matter (PM) includes total suspended particulate matter as well as coarse and fine particulate matter. Sulphur oxides (SO_2, SO_3) are of concern from the standpoint of secondary particulates and acid-rain formation. Nitrogen oxides include the two principal oxides of nitrogen, nitrogen oxide (NO) and nitrogen dioxide (NO_2), formed from combustion; these too are of concern with respect to secondary particulates and acid-rain formation, but also as ozone precursors. For HAPs, although many of these compounds are emitted by the forest products industry and many are VOCs, the main compounds of interest include methanol, formaldehyde, and acetaldehyde.

In comparison with the pulp and paper sector, there is much less information on residual types and quantities discharged from the wood products sector (USEPA 1995a). Atmospheric emissions from this sector are highly variable by facility and depend on the type of product (sawn lumber, oriented strandboard, medium-density fibreboard, particleboard, plywood, or veneer), the wood species, and the type of equipment and materials used in the process (wood dryers, press vents, and pollution control equipment). Many of the emissions from the wood products industry come from the wood itself, either as particulates during handling and chipping or organic compounds emitted during heating and drying processes. Table 16.9 summarizes some of the atmospheric emissions for the wood processing industries. Each of these different facilities has been addressing the air quality issues aggressively over the past 10 years.

Table 16.9. Major air pollutants from primary wood processing industries (adapted from USEPA 1995*a*, CCME 2001).

Process	Material input	Air emissions
Logging	Trees, diesel, gasoline	PM_{10}, VOCs, CO, NO_x
Sawing	Wood logs, diesel, gasoline	PM_{10}, VOCs, CO, NO_x
Surface protection	Wood, 3-Iodo-2-Propynyl Butyl Carbamate (IPBC), Didecyl Dimethyl Ammonium Chloride (DDAC)	IPBC, DDAC, ethyl alcohol, petroleum naphtha
Plywood and veneer	Veneer, phenol-formaldehyde resins, urea-formaldehyde resins, melamine-formaldehyde resins, sodium hydroxide, ammonium sulphate, acids, ammonia	PM_{10}, VOCs, CO, CO_2, NO_x, formaldehyde, phenol, wood dust, condensable hydrocarbons, terpenes, methanol, acetic acid, ethanol, furfural
Reconstituted wood products	Wood particles, strands, fibre, same resins as plywood and veneer, methylenediphenyl diisocyanate resins	PM_{10}, VOCs, CO, CO_2, NO_x, formaldehyde, phenol, wood dust, condensable hydrocarbons, terpenes, methanol, acetic acid, ethanol, furfural
Wood preserving	Wood, pentachlorophenol, creosote, borates, ammonium compounds, inorganic formulations of chromium, copper, and arsenic, carrier oils	Pentachlorophenol, polycyclic organics, creosote, ammonia, boiler emissions, airborne arsenics, VOCs
Wood waste incinerators and conical wood waste incinerators (beehive burners)	Waste wood particles, some salt-laden wood	PM_{10}, $PM_{2.5}$, dioxins, furans

The reader is directed to Tables 18.1 and 18.2 in the chapter on air emissions, which give an excellent and more detailed account of the types and sources of atmospheric pollutants from pulp mills and other forest product sectors. Chapter 18 also presents detailed management and treatment options for atmospheric emissions.

In the sawn lumber industry, some residual wood from sawn lumber production is burned to produce steam or electricity (USEPA 1995*a*). A major emission of concern from wood boilers is PM, while carbon monoxide (CO) and organic compounds may be emitted in significant quantities if operating conditions are poor. Boilers that burn wood waste produce fly ash, CO, and VOCs. New boilers must meet new source performance standards for air pollutants. In addition, these sawn lumber mills are potential sources of toxic manganese air emissions.

Reconstituted wood products use all parts of the sawn log, and as such very little solid waste is generated. In mills where wood or other furnish (the blend of fibrous materials that is used to make pulp) are generated onsite, operations such as debarking, sanding, chipping, grinding, and fibre separation generate PM emissions in the form of sawdust and wood particulates. However, in the production of both panel products and reconstituted wood products, the principle environmental concerns stem from air emissions from dryers and presses.

Veneer impingement dryer emissions are characterized by organic aerosols and gaseous organic compounds, along with a small amount of wood fibre (USEPA 1995a). A mixture of organic compounds is driven from the green wood veneer as its water content is converted to steam in the drying process. These aerosols form visible emissions called blue haze. Emissions from the rotating-drum woodchip dryers used in reconstituted wood panel plants are composed of wood dust, condensable hydrocarbons, fly ash, organic compounds evaporated from the extractable portion of the wood, and may include products of combustion such as carbon monoxide (CO), carbon dioxide (CO_2), and nitrogen oxides (NO_x) if direct-fired units are used. The organic portion of industry emissions includes terpenes, resin and fatty acids, and combustion and pyrolysis products such as methanol, acetic acid, ethanol, formaldehyde, and furfural. The condensable hydrocarbons and a portion of the VOCs leave the dryer stack as vapour but condense at normal atmospheric temperatures to form liquid particles that create a blue haze (USEPA 1995a). Both the VOCs and the liquid organic mist are combustion products and compounds evaporated from the wood. Quantities emitted are dependent on wood species, dryer temperature, and fuel used.

The type of wood species burned also affects the composition of the effluent from rotary drum dryers. In a National Council of the Paper Industry for Air and Steam Improvement (NCASI) study cited by the U.S. EPA (USEPA 1995a), it was concluded that high total gaseous non-methane organic emission rates from the dryers occurred when the wood species processed had high turpentine contents, such as found in southern pines (e.g., loblolly pine, *Pinus taeda*). In a separate study on formaldehyde emissions, NCASI showed that dryers processing hardwood or a mixture of hardwood and softwood species had a moderate to dramatic increase in formaldehyde emissions at dryer inlet gas temperatures greater than 425°C, but dryers processing only softwood species had only a slight increase in formaldehyde emissions with increasing temperatures.

Emissions from board presses are dependent upon the type of resin used to bind the wood furnish together (USEPA 1995a). Emissions from hot presses consist primarily of condensable organics. When the press opens, vapours that exit may include resin ingredients such as formaldehyde, phenol, methylenediphenyl diisocyanate (MDI), and other organic compounds which are released to the atmosphere through the ventilation system. Formaldehyde emitted through press vents during pressing and board cooling operations is a function of the amount of excess formaldehyde in the resin as well as press temperature and cycle time. The formaldehyde to urea (F:U) mole ratio is a design parameter used for the principle adhesive in particleboard (PB) and medium-density fibreboard (MDF), and is one of several variables that can affect formaldehyde emissions. The higher the F:U mole ratio, the higher the board emissions of formaldehyde, therefore lowering this ratio is one way to control the press and board emissions of formaldehyde. Other variables include application rates, process rates, and the nature of the specific resin formations. Higher press temperatures generally result in higher formaldehyde emissions. In a NCASI study, emissions of formaldehyde and phenol from phenol-formaldehyde (PF) resins, used mainly for oriented strandboard (OSB), and structural plywood were not found to be related to any operating procedures, but were affected by different resin compositions (USEPA 1995*b*). The types of resins used can affect the amount of emissions. There was little information on emissions from the curing of MDI resins (used for OSB along with PF resins).

During wood preservation, aerosols and vapours may be released into the ambient air during chemical storage and mixing, solution storage, and during pressure treatment (once the treatment cylinder is opened). During the inorganic treatment process, additional vapours such as arsenic may be released to ambient air during the pressure-treating process, such as from the process tank or work vent during the initial vacuum stage, the flooding via vacuum, pressure relief and blow back, and the final vacuum (USEPA 1995*b*). Other typical air emission sources are volatilization of organic chemicals during wood preservation wastewater evaporation.

Conical wood waste incinerators, more commonly known as beehive burners, are used primarily in British Columbia to incinerate large quantities of wood waste produced by lumber mills each year (Fig. 16.3). These beehive burners are significant sources of airborne particulate matter (especially that less than 10 μm in diameter, PM_{10}), as well as dioxins and furans, both on a local and regional scale. However, there are no available data for actual dioxin and furan emissions from beehive burners on account of the practical impossibility of source testing for these crude incineration devices. They are archaic devices because they are only capable of burning wood residue at low temperatures owing to their overly simple construction — basically just a steel box — and it is not possible to raise the temperature of the incinerator to the point of efficient combustion, nor is it possible to fit the burners with any pollution control devices. A particular case of interest was the Bulkley Valley in northwest British Columbia, where meteorological conditions led to atmospheric inversions that exposed residents of the area to the thick wood smoke. The B.C. government attempted a provincial phase-out program of these burners in 1995, but it has been unsuccessful. However, the B.C. Environmental Appeals Board ruled in April of 2002 that several companies operating the crude incinerators in the Bulkley Valley must eliminate or reduce the emissions based on the legally proven fact that the burners will cause adverse effects on environmental and human health (SLDF 2002). Current attempts are being made to include them in the Canada-wide Standards (CWS) program for dioxins and furans as a means to eliminate their usage (CCME 2001).

Table 16.10 showcases some of the major air pollutant categories and their sources from the wood processing and pulp and paper making industries. Water vapours are the most visible air emission from a pulp and paper mill (Figs. 16.1 and 18.2), but are not usually regulated unless they cause significant visual impairment or prove to be a climate modifier (USEPA 1995*b*). Air pollution from a mechanical pulp mill consists of steam that may contain volatile resin components and turpentine (Smook 1992). Turpentine will not be a problem where heat recovery is practiced, though. In this case, the turpentine can be condensed and captured as a valuable byproduct and could be marketed or used as fuel in a boiler. Pulp and paper mill power boilers and chip digesters

Fig. 16.3. Conical wood waste incinerators (beehive burners) have been a concern with respect to the particulate matter contained in large volumes of wood smoke. These units are being phased out in favour of more environmentally acceptable methods of bark and sawdust disposal.

Photo by Tom Beckley

are generic sources of air pollutants such as particulates and nitrogen oxides (USEPA 1995*b*). Chip digesters and chemical recovery evaporators are the most concentrated sources of VOCs. The chemical recovery furnace is a source of fine-particulate emissions and sulphur oxides. In the kraft process, sulphur oxides are a minor issue in comparison to the odour problems created by four reduced sulphur gases, collectively referred to as total reduced sulphur (TRS): hydrogen sulphide (H_2S), methyl mercaptan (CH_3SH), dimethyl sulphide (CH_3SCH_3), and dimethyl disulfide ($CH_3S_2CH_3$). The TRS emissions are primarily from woodchip digestion, black liquor evaporation, and chemical recovery boiler processes. TRS compounds create odour nuisance problems at lower concentrations than sulphur oxides: odour thresholds for TRS compounds are approximately 1000 times lower than that for sulphur dioxide. Humans can detect some

Table 16.10. Major air pollutants from pulp and paper mill processes (adapted from Smook 1992).

Source	Pollutant type
Kraft recovery furnace	Fine particulates
Fly ash from hog fuel and coal-fired burners	Coarse particulates
Sulphite mill operations	Sulphur oxides
Kraft pulping and recovery processes	Reduced sulphur gases
Chip digesters and liquor evaporation	Volatile organic compounds
All combustion processes	Nitrogen oxides

TRS compounds in the air as a "rotten egg" odour at concentrations as low as 1 ppb (USEPA 1995b). Scrubber system particulate baghouses or electrostatic precipitators (ESPs) are common air pollution control technologies used in pulp mills.

The statistics presented in Table 16.11 indicate that the most significant contaminant released to the air in both Canada and the U.S.A. is methanol, which accounts for 62% of the atmospheric emissions reported by the pulp and paper sector, and is released by over half of the mills. This widespread reporting of methanol is not surprising, because it is produced in the pulping processes and is quite volatile. The second-greatest release to Canadian air is hydrogen sulphide; when compared to methanol, a much smaller quantity of hydrogen sulphide is released. It accounts for approximately 12% of air releases and is reported by 27% of the mills. Similarly to methanol, this substance is produced in wood transformation processes and also is quite volatile. One should note that methanol and hydrogen sulphide account for 73% of all releases to air by the pulp and paper sector, while the remaining 27% is accounted for by 18 substances. In the U.S., methanol and hydrochloric acid account for 74% of releases to the air, with the rest accounted for by 52 substances.

The Aerometric Information Retrieval System in the U.S. contains a variety of information related to stationary sources of air pollution from particular industries. With the exception of VOCs, there is little overlap with the TRI chemicals. Table 16.12 summarizes annual releases of carbon monoxide (CO), nitrogen dioxide (NO_2), particulate matter with an aerodynamic diameter of 10 m or less (PM_{10}), total particulates (PT), sulphur dioxide (SO_2), and volatile organic compounds (VOCs).

Recent progress in the treatment and control of forest products industry air pollution is described in Chap. 18.

Solid residues

The forest products industry generates large amounts of solid residues that can be recovered as usable raw materials or processed in to value-added products. Among them are the unused wood and trimming residues from logging operations, and various residues from wood processing such as waste chips and bark, hog fuel, sawdust, or effluent treatment sludge. Most of the reuse options in the forest products industry arise in the wood processing sector itself, as it is more economical to use the entire tree than deal with solid waste streams. It is also advantageous to recover some of the energy from some solid residues, as they easily can be incinerated and used directly as fuel or for heating water boilers.

A variety of solid residues is generated in the forest industry because of different wood processing operations employed, different input raw materials, and a wide range of finished products (from veneer to OSB, to the various grades of paper and paperboard). Based on the sources and characteristics of these wastes, they can be categorized into the following: (1) wood residues from logging, lumber production, wood paneling and pulp stock preparation; (2) wastewater treatment sludges; (3) wood and fly ashes from energy recovery boilers; and (4) miscellaneous solid wastes (USEPA 1995a). While all of these solid residues are described in detail in Chap. 19, a brief overview of them will be provided here.

Table 16.11. Releases to air by the pulp and paper sector per year (includes both fugitive and point source emissions; adapted from USEPA 1998, Environment Canada 2002).

Substance	Canada (1999 data)					U.S. (1995 data)				
	Mg released	% of total releases to air	% cumulative	No. of mills reporting	% of mills reporting	Mg released	% of total releases to air	% cumulative	No. of mills reporting	% of mills reporting
Methanol	15523	61.7	61.7	74	64	59101	62	62	175	57
Hydrogen sulfide	2931	11.6	73.3	31	27	NR[a]	—	62	0	0
Sulphuric acid	1727	6.9	80.2	58	50	5870	6	68.1	148	49
Ammonia (total)	1690	6.7	86.9	49	43	5427	5.7	73.8	197	65
Hydrochloric acid	1231	4.9	91.8	30	26	11035	11.6	85.4	161	53
Chlorine dioxide	947	3.8	95.5	41	36	582	0.6	86.0	95	31
Acetaldehyde	395	1.6	97.1	20	17	3996	4.2	90.2	124	41
Chlorine	348	1.4	98.5	26	23	637	0.7	90.8	154	50
Chloroform	111	0.4	98.9	6	5	4324	4.5	95.4	81	27
Cresol (mixed isomers and their salts)	76	0.3	99.2	5	4	406	0.4	95.8	40	13
Phenol (and its salts)	65	0.3	99.5	4	3	432	0.5	96.2	99	32
Formaldehyde	39	0.2	99.6	7	6	789	0.8	97.1	60	20
Isopropyl alcohol	37	0.1	99.8	3	3	NR	—	97.1	0	0
Manganese (and its compounds)	14	0.1	99.8	35	30	0.5	0.001	97.1	2	1
Chloromethane	12	0.05	99.9	1	1	256	0.3	97.3	16	5
Methyl ethyl ketone	11	0.04	99.9	1	1	691	0.7	98.1	58	19
Ethylene glycol	8	0.03	100	8	7	19	0.02	98.1	16	5
Zinc and its compounds	7	0.03	100	9	8	276	0.3	98.4	49	16
Phthalic anhydride	0.44	0.002	100	1	1	NR	—	98.4	0	0
Nitrate ion in solution at pH ≥ 6.0	0.2	0.001	100	1	1	0	0	98.4	54	18
Total	25173	100	—	115	—	95406	98.4	—	305	—

[a]NR, not reported.

Table 16.12. Pulp and paper mill air pollutant releases in the United States (adapted from USEPA 1995b, 1998).

Pollutant	1995 (tonnes/year)	1997 (tonnes/year)
CO	566232	530429
NO_2	357764	331872
PM_{10}	32270	34347
PT	103009	485891
SO_2	490689	161389
VOC	87866	97662

Table 16.13. Major solid waste residues from wood processing industries (adapted from USEPA 1995a).

Process	Material input	Main solid wastes
Logging	Trees, diesel, gasoline	Wood particles, bark
Lumber production		
Sawing	Logs, diesel, gasoline	Sawdust, bark
Surface protection	Wood, surface protectants [wood, 3-Iodo-2-Propynyl Butyl Carbamate (IPBC), Didecyl Dimethyl Ammonium Chloride (DDAC)]	Sawdust, chips, sand, stone, dirt, tar, emulsified or polymerized oils, tank sludge
Panelling		
Veneer and plywood	Veneer, phenol-formaldehyde resins, urea-formaldehyde resins, melamine-formaldehyde resins, sodium hydroxide, ammonium sulphate, acids, ammonia	Trim, sawdust, adhesive residues
Reconstituted wood products	Wood particles, strands, fibre, same resins as plywood and veneer, methylenediphenyl diisocyanate resins	Wood particles, adhesive residues
Wood preserving	Wood, creosote, borates, ammonium compounds, inorganic formulations of chromium, copper, and arsenic, carrier oils	Bottom sediment sludges

Tables 16.13 and 16.14 give a summary of the solid residues from primary wood processing operations and pulp processes, respectively. Wood residues are all the wood materials that are not suitable for manufacturing primary wood products. Most of the residual wood from sawn lumber production is reused as mulch, pulp, and furnish for some types of reconstituted wood panels; approximately 70% of a sawn log is utilized for lumber and other parts are used for co-products (USEPA 1995a). Some of the small

Table 16.14. Major solid waste residues from pulping processes (adapted from USEPA 1995*b*).

Process	Solid waste residue
Stock preparation	Bark, dirt, grit, wood wastes
Chemical pulping and bleaching	Resins, fatty acids
Chemical recovery system	Lime mud, lime slaker grits, green liquor dregs
Power and steam boiler	Fly ash, bottom ash, scrubber sludge
Wastewater treatment plant	Wastewater treatment sludge

residuals such as undersized chips, sawdust, and sawmill wood edges are gathered with pneumatic systems for combination with larger amounts destined for use in other products. While there is virtually no waste from the manufacturing process because all parts of the log are used for one product or another, wood residuals are high in organic matter and can impact surface waters if improperly handled. The woodchips that are prepared from these residues for combustion in wood-fired boilers are referred to as hog fuel. Table 19.5, in the chapter on solid wastes, summarizes the elemental composition along with heating values for selected wood species and waste residues.

Typical process residuals from treatments undertaken to protect wood surfaces are tank sludges that accumulate in the dip tank and (or) mix tank because of continuous reuse of the protectant (USEPA 1995*a*). Some plants use spray systems that generate sludge when the recovered formulation is filtered. Periodically, the accumulated sludge must be removed, and is typically placed on sawdust or woodchip piles onsite. The ultimate destination of the sludge is dependent upon the management of the sawdust piles. Plants have reported burning sawdust onsite or shipping it offsite for use as boiler feed for energy recovery. Depending on the particle size, some woodchips may be shipped to a pulp or paper mill. In the wood-preserving sector, sludges result if filters are used prior to solution reuse from wastewater treatment, and from the collection sumps at the facility.

Most of the environmental and health impacts associated with solid residues associated with the forest products industry can be attributed to four types of contaminants (see Chap. 19):
- heavy metals;
- toxic organic compounds, particularly chlorinated compounds such as dioxins and furans;
- pathogenic microorganisms; and
- nitrates.

The major sources of these contaminants have been identified in the waste streams of pulping processes. Of them, the greatest concern is associated with wood ashes and wastewater treatment sludges because of their presence in high quantities.

The significant residual solid waste streams from pulp and paper mills include bark, wastewater treatment sludges, lime mud, lime slaker grits, green liquor dregs, boiler and furnace ash, scrubber sludges, and wood processing residuals (USEPA 1995*b*). Aside from bark, wastewater treatment sludge is the largest volume residual waste stream generated by the pulp and paper industry. Sludge generation rates vary widely among mills, anywhere from 15 to 150 kg of sludge per air-dried tonne of pulp produced (USEPA 1995*b*). Pulpmaking operations are responsible for the bulk of sludge wastes, although treatment of papermaking effluents also generates significant sludge volumes. The organic component in pulp sludge is mainly wood fibre left behind in wastewater that cannot be biodegraded by the microorganisms in the wastewater treatment process (USEPA 1995*a*). For the majority of pulp and integrated mills that operate their own wastewater systems, sludge is generated onsite. A small number of pulp mills in the U.S., and a much larger proportion of papermaking establishments, discharge effluents to POTWs (USEPA 1995*b*). As chlorinated organic compounds (including dioxins, AOX) have a tendency to partition from effluent onto solids, wastewater treatment sludge has generated the most significant solid residue concerns for the pulp and paper industry (McNamara 2003). The U.S. EPA (USEPA 1995*b*) states that AOXs were present in bleached pulp mill sludges, resulting in calls to regulate both landfill disposal and land application of such sludges. To a lesser extent, concern has also been raised over whether chlorinated organics partition into pulp and paper products, to which the consuming public is then exposed, and a large portion of which ultimately becomes post-consumer residual solid waste.

Table 16.15 summarizes the release of 13 NPRI substances to Canadian land (based on 1999 data), and related TRI substances released to American soil (based on 1995 data).

Manganese and zinc accounted for the greatest masses of reportable chemical releases to land (68% and 12%, respectively) in Canada, whereas zinc and methanol accounted for the largest releases to land in the U.S. (61% and 33%, respectively). Wood wastes (i.e., bark, shavings, saw dust, chip rejects) are used as fuel in biomass burners to generate steam, which is in turn used in various process operations (Environment Canada 2002). Manganese and zinc are natural elements present in raw wood and are not destroyed upon incineration, and thus remain in wood ash. The wood ash is then landfilled at an onsite or offsite facility.

Residues from the papermaking portion of the process are usually generated from the ragger, the pulper, and the screens (Muratore 2000). These solid residues consist of contaminants present in waste paper bales, and of fibres that are entrained during the decontamination process. The ragger removes wires and large pieces of contaminants such as plastics. The pulper decontamination system yields rejects consisting of medium-sized plastics, non-pulpable paper, and adhesives. Screening devices and cleaners give small-sized rejects including plastics, polystyrene foam, wax, and hot melt adhesives. Muratore (2000) suggests incineration as a viable alternative to landfilling for many of these residues in an attempt to reclaim their energy value.

Just as is practiced in municipal and industrial secondary wastewater treatment plants, the idea of using sludges that accumulate during the wastewater treatment process (also called biosolids) as a value-added product for soil amendment is catching on in the pulp and paper industry (Riebel 2001). Riebel has also compiled a list of 17 World Wide Web resources for solid waste reduction from pulp and paper mills, including paper-mill residue compost usage, biosolids utilization, and ash reuse (Riebel 2002). In some SFMN-sponsored research, Zhou et al. (2000) examined the usage of fly ash and lime byproducts as beneficial road construction amendment materials based on

Table 16.15. Releases to land by the pulp and paper sector per year (adapted from USEPA 1998, Environment Canada 2002).

Substance	Canada (1999 data)					U.S. (1995 data)				
	Mg released	% of total releases to land	% cumulative	No. of mills reporting	% of mills reporting	Mg released	% of total releases to land	% cumulative	No. of mills reporting	% of mills reporting
Manganese (and its compounds)	819	68.1	68.1	26	23	32.3	1.8	1.8	2	1
Zinc (and its compounds)	142	11.8	79.9	6	5	1096	61	62.8	50	16
Nonylphenol polyethylene glycol ether	128	10.6	90.5	4	3	0.3	0.02	62.8	19	6
Methanol	73	6.1	96.6	18	16	586	32.6	95.4	175	57
Asbestos (friable form)	17	1.4	98	2	2	0	0	95.4	1	0.3
Chromium (and its compounds)	12	1	99	1	1	18	1	96.4	6	2
Ethylene glycol	9	0.7	99.7	3	3	0.66	0.04	96.5	16	5
Ammonia (total)	2	0.2	99.9	5	4	13.4	0.75	97.2	197	65
Formaldehyde	1	0.1	100	3	3	3.32	0.18	97.4	60	20
Nitrate ion in solution at pH $\geq$ 6.0	0.3	0.02	100	2	2	1.2	0.07	97.5	54	18
Acetaldehyde	0.1	0.01	100	5	4	7.9	0.44	97.9	124	41
Phosphoric acid	0.1	0.01	100	1	1	0.2	0.01	97.9	123	40
Cresol (mixed isomers and their salts)	0.002	0.0002	100	2	2	0.5	0.03	97.9	40	13
Total	1204	100	—	115	—	1797	97.9	—	305	—

technical, economical, and environmental considerations. Results were promising, as they showed that pulp mill wastes would have little adverse environmental impact and could be used as road construction amendments to improve soil strength and reduce deformation. The approach was put into operation at Alberta Pacific's facility in Alberta. Further options for the treatment and recycling, reuse, and reduction of solid wastes emanating from the forest products industry are explored in Chap. 19.

Conclusions

The residues released from pulp and paper mills into water, air, and soil have decreased over the last decade owing to increasingly stringent regulations, public pressure, and improved industrial attitudes about protecting the environment. Both Paprican (the Pulp and Paper Research Institute of Canada) and the SFMN are major players in organizing research to minimize residual impacts to the environment. Paprican initiatives include promotion of recycling to reduce bark and wood waste production, management of wastewater treatment plant residuals, and identifying additives that hold potential to increase dewatered sludge cake consistency, and subsequently decrease chemical use (Paprican 2002). With support from Environment Canada and Natural Resources Canada, Paprican is also addressing the gap in knowledge concerning $PM_{2.5}$ and PM_{10} emissions from pulp and paper mills, and mill contributions to ambient fine-particulate levels in neighbouring communities. A key aspect of this work is the use of a state-of-the-art dilution tunnel sampling system built at Paprican (Paprican 2002). Paprican scientists and technicians have also evaluated a new laboratory test developed in the U.S. and Europe for its potential as a monitoring tool for a future industry-wide effluent survey. As well, since BOD load is an important variable in treatment costs, Paprican has begun to characterize the major BOD components of typical newsprint and kraft mill effluents. Initial results with a newsprint effluent indicate that a rapid colorimetric assay can account for much of the BOD load, as opposed to waiting for the typical 5-day BOD result. As mills reduce water consumption, there is a potential for contravention of acute toxicity regulations, since toxicity is a concentration-based measurement (Paprican 2002). Laboratory results indicate that a TMP newsprint mill effluent, concentrated to simulate water use as low as 5 m^3/tonne, can still be successfully treated by the activated sludge process.

The last major compilation of SFMN-sponsored research (Veeman et al. 1999) included two sections on residue management options and technologies, one entitled "Residue liquid quality improvement and re-use" and the other "Minimizing impacts of gaseous emissions from forestry products manufacturing." The liquid residue management section included topics on advanced oxidation treatment processes for pulp mill wastewaters (Rehmat et al. 1999; Vaisman et al. 1999; Zhou and Smith 1999), various freezing techniques for treatment of pulp mill wastewater (Gao et al. 1999) and pulp mill membrane concentrates (Facey and Smith 1999), and biological treatment and process optimization for specific wastewater constituents such as methanol (Berube and Hall 1999) and resin acids (Mohn et al. 1999; Werker and Hall 1999; Yu and Mohn 1999), to name a few. Several papers on the usage of biofilters for treating gaseous emissions covered topics ranging from removal of TRS constituents to testing different

> **Box 16.5. Meeting the challenge for a sustainable pulp and paper industry:**
>
> - Further understanding and quantification of the environmental significance of residuals from the primary wood processing industries, such as runoff from logyards and dryland sorts and atmospheric emissions from beehive burners, would be of assistance to industry regulators and managers in helping them manage these risks for the protection of human and environmental health;
> - The technical and economic feasibility of novel pollution control technologies developed in Canada and internationally needs to be demonstrated at the pilot scale or mill scale, thus allowing engineers and plant managers to determine the appropriate roles of such technologies in future pulp and paper mill capital projects;
> - Replacement of archaic or nonexistent pollution control technologies, as is the case with conical wood waste incinerators (beehive burners), to ensure maximum protection of human health now and in the future; and
> - More scrutiny of the forest harvesting industry is required in terms of qualitatively describing the magnitude and adverse health effects of residues from this sector. To date, little to no hard data have been amassed for logging residues, and this should be a priority in the near future.

biofilter bed materials (Budwill and Coleman 1999; Dirk-Faitakis and Allen 1999; Wani et al. 1999, 2001). This research requires testing at both the pilot and mill scale in order to have the data available for designers and managers to make decisions for treatment of these various residual streams. For further examination of some of these residue treatment technologies and residue management practices, the reader is directed both to the aforementioned papers in the conference proceedings as well as the three subsequent chapters in this volume. Chapters 17 and 18, in particular, offer some detail on emerging abatement technologies to handle water and air residues, respectively.

Much good work has been done in the area of residue management from the forest products industry, from treatment of pollution emissions to administrative decisions on recycling and reuse options — but more good work is needed (Box 16.5). Constant vigilance from the staff of fibre processing facilities is required to ensure continual monitoring and quantification of harmful residuals so that the importance of these waste streams is kept in perspective. Corporate environmental reporting has been and will continue to be an important part of an environmental management system of a facility in terms of communicating the status of residues and environmental compliance to the regulators, employees, and the public (Riebel 2000*b*). This type of reporting adds credibility to a pulp and paper mill, and many opportunities exist to demonstrate environmental improvements and management of residuals from the pulp and papermaking processes.

References

Berube, P.R., and Hall, E. 1999. Determination of optimal operating temperature for the biological removal of methanol from synthetic kraft mill evaporator condensates. *In* Proceedings of the 1999 Sustainable Forest Management Network Conference — Science and Practice: Sustaining the Boreal Forest, 14–17 February 1999, Edmonton, Alberta. *Edited by* T.S. Veeman, D.W. Smith, B.G. Purdy, F.J. Salkie, and G.A. Larkin. Sustainable Forest Management Network, Edmonton, Alberta. pp. 263–269.

Budwill, K., and Coleman, R.N. 1999. Comparison of peat and waste wood as biofilter bed material. *In* Proceedings of the 1999 Sustainable Forest Management Network Conference — Science and Practice: Sustaining the Boreal Forest, 14–17 February 1999, Edmonton, Alberta. *Edited by* T.S. Veeman, D.W. Smith, B.G. Purdy, F.J. Salkie, and G.A. Larkin. Sustainable Forest Management Network, Edmonton, Alberta. pp. 623–628.

Carey, J., Hall, E., and McCubbin, N. 2002. Review of scientific basis for AOX effluent standards in British Columbia. Scientific Advisory Panel on AOX Effluent Standards in British Columbia. Prepared for the British Columbia Minister of Water, Land and Air Protection, Victoria, British Columbia.

(CCME) Canadian Council of Ministers of the Environment. 2001. Canada-wide standards for dioxins and furans: emissions from waste incinerators and coastal pulp and paper boilers. Canadian Council of Ministers of the Environment, Ottawa, Ontario. Available at http://www.ccme.ca/assets/pdf/d_and_f_backgrounder_e.pdf [viewed May 7, 2003]. 2 p.

Chapra, S. 1997. Surface water-quality modeling. McGraw–Hill, Boston, Massachusetts. 784 p.

de Choudens, C., and Lachenal, D. 1999. Perspectives in pulping and bleaching processes at the crossing of the millennium frontier. Water Sci. Technol. **40**(11–12): 11–19.

Dirk-Faitakis, C.B., and Allen, D.G. 1999. Treating cyclic air emissions from the forest products industry using a woodchip based biofilter. *In* Proceedings of the 1999 Sustainable Forest Management Network Conference — Science and Practice: Sustaining the Boreal Forest, 14–17 February 1999, Edmonton, Alberta. *Edited by* T.S. Veeman, D.W. Smith, B.G. Purdy, F.J. Salkie, and G.A. Larkin. Sustainable Forest Management Network, Edmonton, Alberta. pp. 629–635.

Eaton, A.D., Clesceri, L.S., and Greenberg, A.E. (*Editors*). 1995. Standard methods for the examination of water and wastewater, 19th edition. American Public Health Association, American Water Works Association, Water Environment Federation, Washington, D.C. 1325 p.

Environment Canada. 2002. An analysis of pulp and paper mill reports to the NPRI for 1999. National Office of Pollution Prevention, Environment Canada, Ottawa, Ontario. Available at http://www.ec.gc.ca/nopp/docs/rpt/pp1999/ [viewed May 6, 2003].

Facey, R.M., and Smith, D.W. 1999. Treatment of pulp mill membrane concentrates by freeze thaw. *In* Proceedings of the 1999 Sustainable Forest Management Network Conference — Science and Practice: Sustaining the Boreal Forest, 14–17 February 1999, Edmonton, Alberta. *Edited by* T.S. Veeman, D.W. Smith, B.G. Purdy, F.J. Salkie, and G.A. Larkin. Sustainable Forest Management Network, Edmonton, Alberta. pp. 313–319.

Gangolli, S. (*Editor*). 1999. The dictionary of substances and their effects, second edition. The Royal Society of Chemistry, Cambridge, U.K.

Gao, W., Smith, D.W., and Sego, D.C. 1999. Treatment of pulp mill membrane concentrates by spray freezing. *In* Proceedings of the 1999 Sustainable Forest Management Network Conference — Science and Practice: Sustaining the Boreal Forest, 14–17 February 1999, Edmonton, Alberta. *Edited by* T.S. Veeman, D.W. Smith, B.G. Purdy, F.J. Salkie, and G.A. Larkin. Sustainable Forest Management Network, Edmonton, Alberta. pp. 307–312.

Greenbaum, P.J. 2002. Making progress in water reduction. Pulp & Paper Can. **103**(2): 10–13.

Health Canada. 2002*a*. Chromated copper arsenate (CCA). Pesticide Management Regulatory Agency (PCMA), Health Canada, Ottawa, Ontario. Available at http://www.hc-sc.gc.ca/pmra-arla/english/pdf/rev/rev2002-03-e.pdf [viewed May 7, 2003]. 2 p.

Health Canada. 2002*b*. Update on the re-evaluation of copper chromated arsenate (CCA) treated wood in Canada. REV2002-01. Pesticide Management Regulatory Agency (PCMA), Health Canada, Ottawa, Ontario. Available at http://www.hc-sc.gc.ca/pmra-arla/english/pdf/rev/ rev_2002-01-e.pdf [viewed May 7, 2003]. 2 p.

Joyce, S., and Smith, D.W. 2003. The effect of microorganisms on pulp mill effluent induced flocculation in receiving waters. J. Environ. Eng. Sci. **2**(1): 39–46.

Kettunen, J. 1999. The forest cluster and society. Water Sci. Technol. **40**(11–12): 1–3.

McNamara, S. 2003. Determination of AOX removal efficiency in the wastewater treatment system at the Alberta-Pacific Forest Industries kraft pulp mill. M.Sc. Thesis. Department of Civil and Environmental Engineering, University of Alberta, Edmonton, Alberta.

Mohn, W.W., Martin, V.J.J., Muttray, A.F., and Yu, Z. 1999. Peeking into the black box: resin acid-degrading microbial populations in treatment systems for pulp mill effluent. *In* Proceedings of the 1999 Sustainable Forest Management Network Conference — Science and Practice: Sustaining the Boreal Forest, 14–17 February 1999, Edmonton, Alberta. *Edited by* T.S. Veeman, D.W. Smith, B.G. Purdy, F.J. Salkie, and G.A. Larkin. Sustainable Forest Management Network, Edmonton, Alberta. pp. 276–281.

Muratore, E. 2000. Energy value from waste paper mill rejects. Pulp & Paper Can. **101**(11): 57–60.

(NAFTA TWGP) North America Free Trade Agreement, Technical Working Group on Pesticides. 2002. Co-operative re-evaluation/re-registration of heavy duty wood preservatives (HDWP). Subcommittee: joint review of chemical pesticides. Health Canada, Ottawa, Ontario. Available at http://www.hc-sc.gc.ca/pmra-arla/english/pdf/nafta/review/jr0400fe-e.pdf [viewed May 7, 2003]. 8 p.

Paprican. 2002. Research program. Pulp and Paper Research Institute of Canada, Pointe-Claire, Quebec, and Vancouver, British Columbia. Available at http://www.paprican.ca/engl/ research/program/index. htm [viewed May 7, 2003] 9 p.

(PWC) PricewaterhouseCoopers. 2002. The forest industry in Canada, 2001. Forest Products Association of Canada, Ottawa, Ontario. Available at http://www.fpac.ca/Pricewat.pdf [viewed May 7, 2003]. 26 p.

Reeve, D.W. 1999. System closure, free enterprise and the environment. Water Sci. Technol. **40**(11–12): 5–10.

Rehmat, T., Branion, R.M.R., Rogak, S., Filopovic, D., Teshima, P., Hauptmann, E., Gairns, S., and Lota, J. 1999. Supercritical water oxidation for waste disposal. *In* Proceedings of the 1999 Sustainable Forest Management Network Conference — Science and Practice: Sustaining the Boreal Forest, 14–17 February 1999, Edmonton, Alberta. *Edited by* T.S. Veeman, D.W. Smith, B.G. Purdy, F.J. Salkie, and G.A. Larkin. Sustainable Forest Management Network, Edmonton, Alberta. pp. 270–275.

Riebel, P. 2000*a*. EEM Cycle Two reports are in: what next? Pulp & Paper Can. **101**(5): 58.

Riebel, P. 2000*b*. Corporate environmental reporting. Pulp & Paper Can. **101**(12): 140.

Riebel, P. 2001. Biosolids recycling. Pulp & Paper Can. **102**(11): 51.

Riebel, P. 2002. Solid waste reduction tips on the internet. Pulp & Paper Can. **103**(1): 11.

Robinson, R.D., Carey, J.H., Solomon, K.R., Smith, I.R., Servos, M.R., and Munkittrick, K.R. 1994. Survey of receiving-water environmental impacts associated with discharges from pulp mills. 1. Mill characteristics, receiving-water chemical profiles and lab toxicity tests. Environ. Toxicol. Chem. **13**(7): 1075–1088.

Sawyer, C., McCarty, P., and Parkin, G. 1994. Chemistry for environmental engineering, fourth edition. McGraw–Hill, New York, New York. 532 p.

(SLDF) Sierra Legal Defence Fund. 2002. Beehive burner victory. Newsletter, **31**: 3.

Smith, D.W., and Veeman, T.S. 1999. Foreword. *In* Proceedings of the 1999 Sustainable Forest Management Network Conference —Science and Practice: Sustaining the Boreal Forest, 14–17 February 1999, Edmonton, Alberta. *Edited by* T.S. Veeman, D.W. Smith, B.G. Purdy, F.J. Salkie, and G.A. Larkin. Sustainable Forest Management Network, Edmonton, Alberta. p. iii.

Smook, G.A. 1992. Handbook for pulp & paper technologists. second edition. Angus Wilde Publications, Vancouver, British Columbia. 395 p.

(USEPA) United States Environmental Protection Agency. 1995*a*. EPA Office of Compliance Sector Notebook Project — Profile of the lumber and wood products industry. Office of Compliance, Office of Enforcement and Compliance Assurance, United States Environmental Protection Agency, Washington, D.C. EPA/310-R-95-006.

(USEPA) United States Environmental Protection Agency. 1995*b*. EPA Office of Compliance Sector Notebook Project — Profile of the pulp and paper industry. Office of Compliance, Office of Enforcement and Compliance Assurance, United States Environmental Protection Agency, Washington, D.C. EPA/310-R-95-015.

(USEPA) United States Environmental Protection Agency. 1996. Superfund chemical data matrix (SCDM). United States Environmental Protection Agency, Washington, D.C. EPA/540/R-96/028. Available at www.epa.gov/superfund/resources/scdm/ [viewed May 7, 2003].

(USEPA) United States Environmental Protection Agency. 1998. EPA Office of Compliance Sector Notebook Project: Sector notebook data refresh — 1997. Office of Compliance, Office of Enforcement and Compliance Assurance, United States Environmental Protection Agency, Washington, D.C. EPA/310-R-97-010.

(USEPA) United States Environmental Protection Agency. 2003. Press advisory: EPA finalizes voluntary cancellation of virtually all residential uses of CCA-treated wood. Press release, 20 March 2003. United States Environmental Protection Agency, Washington, D.C. Available at http://yosemite.epa.gov/opa/admpress.nsf/ [viewed May 7, 2003].

Vaisman, E., Starosud, A., and Langford, C.H. 1999. Application of photocatalysis for oxidation of organics and decolorization of pulp mill effluents. *In* Proceedings of the 1999 Sustainable Forest Management Network Conference — Science and Practice: Sustaining the Boreal Forest, 14–17 February 1999, Edmonton, Alberta. *Edited by* T.S. Veeman, D.W. Smith, B.G. Purdy, F.J. Salkie, and G.A. Larkin. Sustainable Forest Management Network, Edmonton, Alberta. pp. 301–306.

Veeman, T.S., Smith, D.W., Purdy, B.G., Salkie, F.J., and Larkin, G.A. (*Editors*). 1999. Proceedings of the 1999 Sustainable Forest Management Network Conference — Science and Practice: Sustaining the Boreal Forest, 14–17 February 1999, Edmonton, Alberta. Sustainable Forest Management Network, Edmonton, Alberta.

Wani, A.H., Branion, R.M.R., and Lau, A.K. 1999. Biofiltration using compost and hog fuel as a means of removing total reduced sulfur gases from air emissions. *In* Proceedings of the 1999 Sustainable Forest Management Network Conference — Science and Practice: Sustaining the Boreal Forest, 14–17 February 1999, Edmonton, Alberta. *Edited by* T.S. Veeman, D.W. Smith, B.G. Purdy, F.J. Salkie, and G.A. Larkin. Sustainable Forest Management Network, Edmonton, Alberta. pp. 617–622.

Wani, A.H., Branion, R.M.R., and Lau, A.K. 2001. Biofiltration using compost and hog fuel as a means of removing reduced sulphur gases from air emissions. Pulp & Paper Can. **102**(5): 27–32.

Werker, A.G., and Hall, E. 1999. Limitations for biological removal of resin acids from pulp mill effluent. Water Sci. Technol. **40**(11–12): 281–288.

Yaws, C.L. 1999. Chemical properties handbook: physical, thermodynamic, environmental, transport, safety, and health related properties for organic and inorganic chemicals. McGraw–Hill, New York, New York. 784 p.

Young, S. 2001. Pulp mill effluent induced coagulation and flocculation in receiving waters. Ph.D. Thesis. Department of Civil and Environmental Engineering, University of Alberta, Edmonton, Alberta. 281 p.

Yu, Z., and Mohn, W.W. 1999. Microbial population dynamics of resin acid-degrading bacteria in biotreatment systems of pulp mill effluents. *In* Proceedings of the 1999 Sustainable Forest Management Network Conference — Science and Practice: Sustaining the Boreal Forest, 14–17 February 1999, Edmonton, Alberta. *Edited by* T.S. Veeman, D.W. Smith, B.G. Purdy, F.J. Salkie, and G.A. Larkin. Sustainable Forest Management Network, Edmonton, Alberta. pp. 282–287.

Zhou, H., and Smith, D.W. 1999. Ozone mass transfer in pulp mill ASB effluents. *In* Proceedings of the 1999 Sustainable Forest Management Network Conference – Science and Practice: Sustaining the Boreal Forest, 14–17 February 1999, Edmonton, Alberta. *Edited by* T.S. Veeman, D.W. Smith, B.G. Purdy, F.J. Salkie, and G.A. Larkin. Sustainable Forest Management Network, Edmonton, Alberta. pp. 327–332.

Zhou, H., Smith, D.W., and Sego, D.C. 2000. Characterization and use of pulp mill fly ash and lime byproducts as road construction amendments. Can. J. Civil Eng. **27**(3): 581–593.

Chapter 17

Forest industry aqueous effluents and the aquatic environment

Eric R. Hall

"All material entering a pulp mill must exit somewhere and therefore, zero emission is not achievable. The four potential modes of emission are in the product, in the sewer, in solid waste, or in gaseous emissions (the four S options: Sell, Sewer, Solid waste, Sky)."

John Carey et al. (2002)

Introduction

Many industrial sectors and manufacturing enterprises might be considered to be significant "users" of water. If a broad definition of water "use" can be adopted, it is easy to conclude that, perhaps after the hydroelectricity generation industry, the forest products sector may be the second largest user of water in Canada. The term "water use" has been defined to include both "withdrawal" and "in-stream" uses of water from surface or sub-surface sources (BCMELPEC 1993). Withdrawal refers to the removal of water from lakes, streams, and aquifers for agricultural, municipal, or industrial activity. In-stream water use would include navigation, fish and wildlife habitat provision, recreation, and hydroelectricity generation.

The water use patterns of the forest products industry include both withdrawal and in-stream aspects. The industry's woodlands operations may indirectly "use" water by contributing to impacts or changes in the quality and quantity of water in natural bodies such as lakes, rivers, and aquifers. Log processing sites and landfills operated by the industry may inadvertently use water by providing opportunities for precipitation to contact logs, bark, and wood chips before running off to surface water bodies. During this contact with various wood components, naturally occurring wood extractives may be leached from the solid material and conveyed to receiving water courses. In some locations in Canada, most notably on the west coast, water is still used for the transport of cut logs to processing facilities for the manufacture of wood products. The pulp and paper sector of the forest products industry is a large user of withdrawn water. With-

E.R. Hall. Department of Civil Engineering, University of British Columbia, Vancouver, British Columbia V6T 1Z4, Canada. Telephone: 604-822-2707. e-mail: ehall@civil.ubc.ca

Correct citation: Hall, E.R. 2003. Forest industry aqueous effluents and the aquatic environment. Chapter 17. *In* Towards Sustainable Management of the Boreal Forest. *Edited by* P.J. Burton, C. Messier, D.W. Smith, and W.L. Adamowicz. NRC Research Press, Ottawa, Ontario, Canada. pp. 669–712.

drawn water is used in pulp and paper mills for a variety of purposes before it is returned to the environment, sometimes with an increased heat content and often with higher concentrations of contaminants. The basic technologies of wood pulping and paper making and their associated waste streams and waste stream terminology (white water, brown stock, black liquor, etc.) are introduced in Chap. 16.

The opportunities for deleterious environmental impacts from water use by the Canadian forest products industry are magnified by the size of the industry and the geographical diversity of its operations (FPAC 2002). When all modes of water use are considered, the forest products industry can be seen as a significant contributor to potential impacts on water quality and quantity in Canada. Recent research into the relationships between forestry practice and natural aquatic systems is reviewed in Chap. 10. These relationships may be considered as emanating from land-based watershed impacts by the forest products industry. In the current chapter, I review issues related to water withdrawals by the forest products industry in boreal forest jurisdictions. In particular, I report on current regulations and industry performance with respect to the return of withdrawn water to the receiving environment — that is, forest industry aqueous discharges. Several charts are presented which compare regulatory limits to actual discharges in different boreal forest jurisdictions. I then go beyond current regulations by reviewing unresolved or emerging issues related to aqueous discharges and by discussing environmental control technology developments in Canada and abroad that may be of interest to managers and operators of forest products manufacturing operations (Box 17.1).

The regulatory environment in Canada and abroad

Depending upon the source cited, there are between 12 000 (CFS 2002) and 13 000 (SSCBF 1999) establishments operated by the Canadian forest products industry. Although the total number of forest industry sites is large, the most significant aqueous discharges may be those associated with Canada's approximately 123 direct-discharging pulp and paper mills (Fig. 17.1). For some time now, point source aqueous discharges from pulp and paper mills in Canada have been subject to regulations promulgated by both the federal and the provincial levels of government. In fact, pulp

---

Box 17.1. Challenges for sustainable pulp and paper manufacturing:

- To understand the current regulatory controls on forest industry aqueous discharges and the regulatory performance of the Canadian industry relative to those in other boreal forest jurisdictions;
- To evaluate recent regulatory and technology developments that may assist designers and managers in making continuing progress in the reduction of contaminant emissions in aqueous discharges from forest industry operations; and
- To identify issues of concern regarding forest industry aqueous discharges that are unresolved as yet.

Fig. 17.1. Pulp and paper mill effluent discharges have been regulated to protect human and environmental health in Canada since 1972.

Photo by Gordon J. Fisher

and paper mill effluent discharges are probably the most important aqueous discharges from the forest industry that are regulated for receiving-water protection.

There are several environmental protection objectives targeted by the variety of Canadian pulp and paper mill effluent regulations that have been implemented. Limits on the deposition of material that contributes to BOD (biochemical oxygen demand) are intended to protect receiving waters from the depletion of dissolved oxygen due to the natural decay of biodegradable organic matter in discharged effluents. Limits on TSS (total suspended solids) are designed to minimize aesthetic impacts in receiving waters and to limit physical deposition of wood fibre and other particulate material on lake and river bottoms. Uncontrolled deposition of suspended solids can result in significant loss of habitat for fish and bottom-dwelling aquatic organisms.

Governments in Canada have also been interested in limiting the discharge of persistent toxic substances from the bleached chemical pulp mill sector. The regulations developed to control such contaminants are of two types. A number of provincial governments have decided that this is best done by limiting the quantities of chlorinated organic matter that can be discharged by bleached chemical pulp mills. The chlorinated material is measured using a bulk, or lumped, parameter method known as "adsorbable organic halogen" (AOX). The AOX method measures the total amounts of most, but not all, of the chlorinated contaminants known to be present in bleached chemical pulp mill wastewater (Chap. 16). In contrast, the Government of Canada has declined to adopt AOX as a regulatory parameter (Halliburton et al. 1991) and instead has opted for a series of regulations under the Fisheries Act and the Canadian Environmental Protection Act that are designed to effectively eliminate the production and discharge of chlorinated dioxins and furans from bleached chemical pulp mills.

Canada is unique among the regulatory jurisdictions of the international boreal forest products industry in its requirement for acute toxicity testing of pulp and paper mill effluents. The 96-hour acute lethality test involves subjecting rainbow trout (*Oncorhynchus mykiss*) fingerlings to 100% pulp mill effluent for a period of 96 hours (McLeay and Sprague 1990). The tested effluent is judged to be non-toxic if more than 50% of the fish survive under the standard test conditions. Recent data show that more than 95% of the acute lethality tests carried out by the industry indicated compliance with the toxicity regulation (Kovacs et al. 2002). In fact, it is now commonly found that most, if not all, of the test fish survive during acute lethality testing (Carey et al. 2002).

Even though most Canadian pulp and paper mill wastewaters are not acutely toxic as judged by the 96-hour rainbow trout toxicity test, sub-lethal effects have been observed in a number of studies (e.g., Munkittrick et al. 1994; Parrott et al. 2000). Such observations have led the Government of Canada to require mills to conduct environmental effects monitoring (EEM) as a means of judging the adequacy of the environmental protection measures embodied in the regulated parameter limits above (Walker et al. 2002). Each direct-discharging mill in Canada is required to conduct an EEM study on a 3- or 4-year cycle and to report the results and supporting data to Environment Canada. The required components of an EEM program may include the elements listed in Box 17.2. In an EEM study, an effect is defined as a statistically significant dif-

---

Box 17.2. Key elements of an environmental effects monitoring (EEM) study to address pulp mill effluents:

- A fish survey to assess effects on fish;
- A benthic invertebrate community survey to assess effects on fish habitat;
- A fish-tainting study;
- Analyses of dioxins and furans in edible fish tissue to assess usability of the fisheries resource;
- Sub-lethal toxicity testing of effluent; and
- Chemical tracer study to delineate effluent dispersion characteristics.

Adapted from Walker et al. (2002)

ference in a measured endpoint between an effluent-influenced area and a reference area that is not exposed to effluent.

With the exception of the acute toxicity testing requirement, pulp and paper mill effluent regulations in other boreal forest jurisdictions are similar in nature to those of the Government of Canada. It is difficult to draw a completely accurate comparison of pulp and paper mill regulatory limits in different jurisdictions, as local requirements add considerable complexity to the picture. Figures 17.2–17.4 illustrate selected regulatory limits that are believed to have been in effect in 2000 for a number of boreal forest jurisdictions, plus those of the United States federal government. Since (in many cases) the applicable limits vary according to local circumstances and by the type of manufacturing technology used at individual mill sites, some of the regulatory limits are shown as bars which encompass this variation. For example, Fig. 17.2 summarizes regulatory limits for BOD discharges on a production-weighted basis (kg BOD_5/ADt, or kg of 5-day BOD per air-dried tonne [Mg] of product) for Canada and the United States. The Canadian federal regulatory limits of 12.5 kg/ADt on a daily basis, and 7.5 kg/ADt on a monthly average basis, apply to the vast majority of pulp and paper mills in Canada. The values at the lower end of the federal range apply only to a few mills, principally in Alberta, that are subject to more stringent limits. Similarly, the federal BOD_5 limits in the United States cover a wide range of values that encompasses limits for different pulp and paper manufacturing technologies. The actual regulatory limits in place at a given

Fig. 17.2. Ranges of BOD_5 (kg of 5-day biochemical oxygen demand per air-dried tonne of product) discharge permit limits for paper mills in several boreal forest jurisdictions, with averaging periods shown in parentheses (Swedish values as BOD_7, 7-day values).

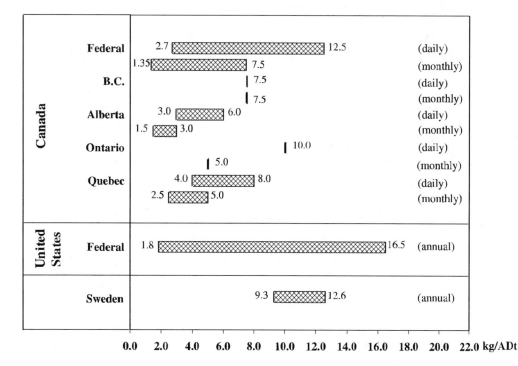

site in the United States could be much lower than the federal values, due to the imposition of additional requirements from state or local governments. For Sweden, a limited number of mills had BOD_7 limits included in their permits. However, it has been reported (Halliburton et al. 1999) that Sweden is moving away from BOD_7 limits by adopting COD (chemical oxygen demand) for regulatory purposes. For simple comparative purposes, a ratio of BOD_7/BOD_5 of about 1.3 may be assumed (Simons Ltd. Consulting Engineers and ÅF-IP 1995, as cited in Halliburton et al. 1999).

Figure 17.2 indicates that among the Canadian boreal forest jurisdictions that regulate pulp and paper mill effluent BOD, the tightest regulatory limits appear to have been imposed by Alberta and Quebec. The same is true for total suspended solids (TSS) discharges (Fig. 17.3). A number of Swedish mills have been permitted with TSS limits, but the TSS parameter is being phased out and replaced by COD in Finland.

Long-term AOX discharge limits vary from relatively low values of 0.1 to 0.4 kg/ADt in the Nordic countries, to 1.5 kg/ADt in Alberta. The values shown in Fig. 17.4 for the Nordic countries indicate target values proposed by the Nordic Council of Ministers (Hynninen 1998) for consideration in the local permitting process. The ranges of actual permit values for AOX in Sweden and Finland are shown for comparison.

Fig. 17.3. Ranges of TSS (total suspended solids) discharge permit limits for paper mills in several boreal forest jurisdictions (averaging periods shown in parentheses).

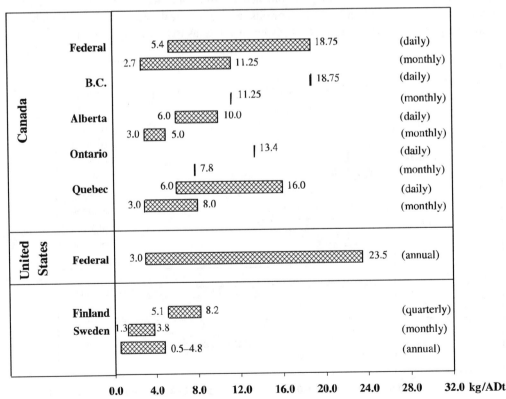

Fig. 17.4. Ranges of AOX (adsorbable organic halogen) discharge permit limits and targets for pulp and paper mills in several boreal forest jurisdictions (averaging periods shown in parentheses).

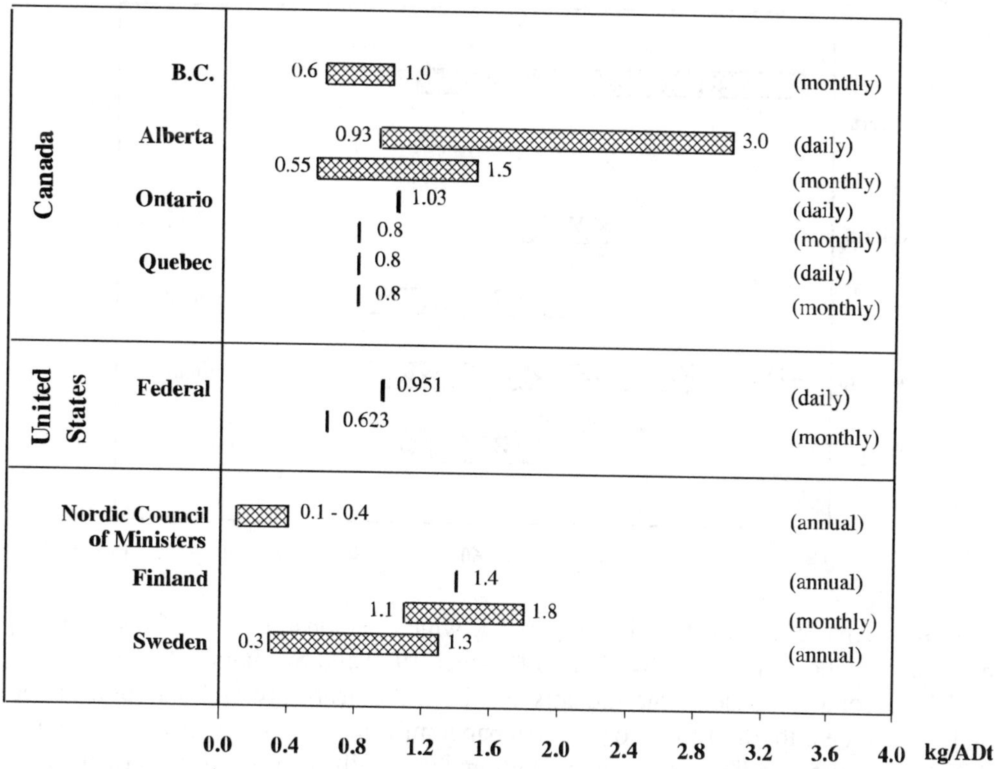

In addition to the parameters discussed above, Sweden and Finland utilize regulatory controls on a number of other parameters. For example, COD has been adopted in place of BOD for control of organic matter discharges in Sweden and Finland. The Nordic Council of Ministers target values for COD (Hynninen 1998) are shown in Fig. 17.5 along with known permit values (Halliburton et al. 1999; Carey et al. 2002). A unique requirement in Scandinavia is the imposition of limits for nitrogen and phosphorus discharges in some jurisdictions. Figure 17.6 summarizes the Nordic Council of Ministers target values for total phosphorus (total-P) and total nitrogen (total-N; Hynninen 1998).

Treatment technologies in use in Canada

The implementation of federal and provincial effluent regulations for Canadian pulp mills has resulted in a double-faceted response from the industry. The first involves a commitment to make future capital investment decisions in a manner that will meet productivity goals, while also moving the industry toward reduced water use and, therefore, toward reduced aqueous discharge volumes. In Canada, this investment philosophy has become known as "progressive systems closure" (Browne 2001). The gains made to

Fig. 17.5. Ranges of COD (chemical oxygen demand) discharge permit limits and targets for pulp and paper mills in several boreal forest jurisdictions (averaging periods shown in parentheses).

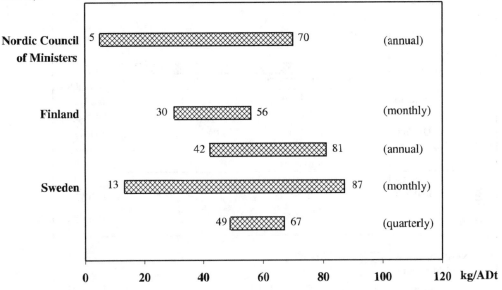

date in water use are illustrated in Fig. 17.7, which demonstrates that Canadian pulp mills have reduced water use by about 30% since 1989 and by about 60% since 1975.

The second response of the industry has been the near-universal adoption of secondary treatment for the 123 or so pulp and paper mills in Canada that discharge directly to receiving water bodies. There are 5 additional pulp mills that practice "zero effluent" operation (Fig. 17.8), through which all liquid effluent is recovered and reused internally (Hardman and Manolescu 1998). Of the mills employing secondary treatment (see Box 17.3), approximately one-third use aerated stabilization basin (ASB) systems, while the others utilize some form of the activated sludge process (AS).

Aerated stabilization basin technology was the most common form of pulp mill wastewater secondary treatment in Canada prior to the mid-1990s. In general, ASBs are

Fig. 17.6. Ranges of total phosphorus and total nitrogen discharge targets for pulp and paper mills in Nordic countries specified by Nordic Council of Ministers (averaging periods shown in parentheses).

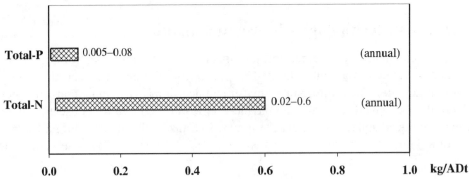

Fig. 17.7. Canadian industry average pulp and paper mill water use (data from Hearn and Vice 2001).

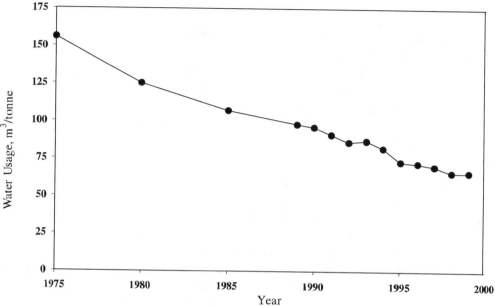

large, excavated basins that provide a minimum of 7 days retention time for BOD, TSS, and acute toxicity removal through the action of aerobic and facultative anaerobic micro-organisms (Fig. 17.9). The basin is usually partially mixed and aerated using floating mechanical aerators that maintain aerobic conditions near the water surface, while allowing anaerobic conditions to develop at the bottom of the basin. Figure 17.9 indicates a spill basin located beside the first zone of an ASB. Spill basins are used for diversion and retention of unusually contaminated effluents, which might otherwise overload an ASB. These diverted streams subsequently can be slowly introduced back into the treatment system at a rate that can be accommodated without process upset.

Fig. 17.8. Zero-discharge pulp mills require a large footprint for effluent treatment and water reuse.

Photo by Gordon J. Fisher

Box 17.3. Secondary wastewater treatment systems in use at Canadian pulp and paper mills.

- Total aerated stabilization basins: 34 mills
- Total activated sludge processes: 70 mills
  - Conventional: 48 mills
  - Pure oxygen: 12 mills
  - Sequencing batch reactor: 10 mills

(Based on personal communication with M. Paice,
Paprican, Pointe Claire, Quebec, 2002)

Microbial biomass produced during wastewater treatment in an ASB is allowed to settle to the bottom of the basin where it accumulates in an anaerobic benthic layer. Much of the accumulated biomass undergoes anaerobic digestion, whereby nutrients are released for reuse in the overlying aerobic layer. After a number of years of operation, an excessive accumulation of inert particulate material in the benthic layer may result in reduced hydraulic retention times in the ASB and a corresponding reduction in wastewater treatment efficiency. Available retention time can be restored by dredging and disposing of the solids from the benthic layer. Aerated stabilization basin systems are often preceded by a primary clarifier for initial removal of pulp fibre and other suspended material. Similarly, a final clarifier or a quiescent zone may be used to reduce TSS concentrations in the treated effluent prior to discharge to the receiving environment.

The main alternative to ASB technology used in Canada is now the activated sludge process, or one of the many variants of the process. Figure 17.10 illustrates a complete secondary system for the treatment of mechanical newsprint mill wastewater (McCubbin et al. 1991). This particular activated sludge system consists of a primary clarifier

Fig. 17.9. Schematic of a pulp and paper mill wastewater aerated stabilization basin treatment system.

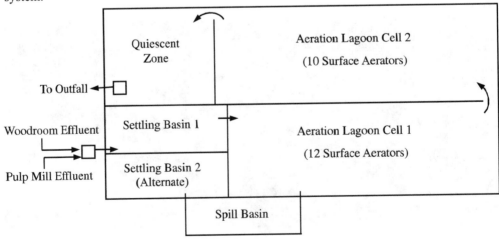

Fig. 17.10. Schematic of a mechanical newsprint mill activated sludge secondary treatment system (modified from McCubbin et al. 1991).

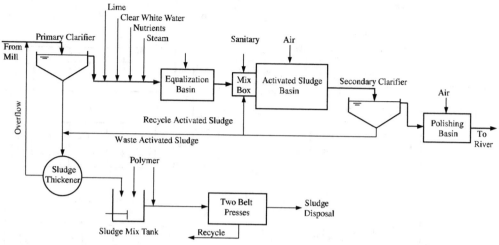

for initial TSS removal, a basin for flow and loading equalization, and a biological treatment stage (activated sludge) operated under aerobic conditions in a highly aerated, well-mixed tank. In contrast to ASB systems, the retention time of the biological treatment stage in the activated sludge process is usually less than 1 day. Microbial biomass and inert suspended solids are separated from the treated wastewater in a secondary clarifier. Most of the settled biomass is recycled to the aeration tank for reuse in the treatment process. A small amount of settled sludge (consisting of microbial biomass and accumulated inert material) is removed (wasted) from the process and then is dewatered and treated for disposal as a solid (see Chap. 19).

There are two variants of the conventional activated sludge process in common use in the Canadian pulp and paper industry. The first of these is the pure oxygen activated sludge process (Fig. 17.11). The mills that have installed pure oxygen systems have most likely done so to achieve one or more of the claimed advantages indicated in Box 17.4 (Strang 1992).

The main difference between conventional and pure oxygen activated sludge processes is the use of an enclosed bioreactor in the pure oxygen variant. Instead of air, pure oxygen is added at the upstream end of the bioreactor and is distributed throughout the liquid using submerged turbines or surface aerators. Gasses and liquids flow concurrently through the process. As treatment progresses along the length of the system, liquid-phase BOD and gas-phase oxygen are depleted and gaseous carbon dioxide

---

Box 17.4. Reported advantages of pure oxygen activated sludge systems:

- Smaller system size due to higher organic loading tolerance;
- Accommodation of foul condensates without pre-stripping;
- Low odour emissions;
- Better accommodation of transient loadings through control of oxygen partial pressure; and
- Lower sludge production due to more extensive endogenous respiration.

Fig. 17.11. Schematic of pure oxygen activated sludge system (modified from Strang 1992).

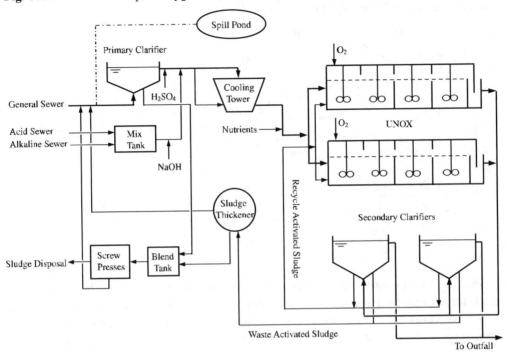

accumulates. In a recent review of 10 Canadian pure oxygen plants, it was noted that nominal hydraulic retention times employed in the biological treatment stages varied from 1.8 to 9.8 hours (Paice et al. 2002).

The survey by Paice et al. (2002) provides some interesting insights into the performance of pure oxygen activated sludge systems. It was concluded that, generally, the performance of the Canadian facilities compares favourably with that of the conventional activated sludge systems. However, one noteworthy difference between pure oxygen systems and conventional air systems is the much higher concentration of carbon dioxide observed in the oxygen-treated effluents. Whereas ASB and conventional activated sludge treated effluents may contain 20–30 mg/L of dissolved carbon dioxide (CO_2), measured concentrations in the pure oxygen system effluents ranged from 48 to 251 mg/L. At such high dissolved CO_2 levels, there is an increased risk of distress and mortality in the rainbow trout toxicity test. Paice et al. (2002) also report that limited data suggest that pure oxygen systems may not effectively remove some components of the reduced sulphur compound family that are sewered with foul condensates. Toxicity testing of one pure oxygen system effluent indicated that residual dimethyl sulphide (DMS) and dimethyl disulphide (DMDS) contributed to distress in the fish during the initial portion of the 96-hour test. However, when either CO_2 or DMS/DMDS are present in problem amounts, these volatile compounds are stripped from the wastewater by the aeration employed in the standard toxicity testing protocol. If the fish do not die early in the test, further mortality may be avoided by toxicant stripping as the testing proceeds.

The second activated sludge process variant that has been adopted in Canada is known as sequencing batch reactor (SBR) technology. The SBR approach usually follows primary clarification and always utilizes the same elements as a conventional activated sludge system including aeration, biomass settling, and sludge recirculation, but all of these take place in a single tank. The SBR tank operates in a batch mode with a cycle duration of 4–8 hours (Cocci et al. 1998). At the beginning of an SBR cycle (Fig. 17.12), wastewater is pumped to the aerated reactor until it is full (the "fill period"). The contents are then continually aerated until biological treatment is complete (the "aeration" or "react period"). The aeration system is then shut down and biomass is allowed to settle (the "settling period") until a clarified supernatant can be decanted from the surface (the "decant period"). Some of the settled biomass may be removed from the bottom of the basin during the decant period, or during a subsequent idle period. The batch cycle then begins again with the introduction of untreated mill effluent to the remaining biomass. Most pulp and paper mill SBR systems utilize multiple reactors (Fig. 17.13), with influent flow directed to only one of the reactors at a time. Sequencing batch reactor systems are thought to produce biomass with good settling characteristics and to be simpler to operate than conventional activated sludge processes (Courtemanche et al. 1997).

The discussion of pulp and paper wastewater treatment technologies in use in Canada would not be complete without mentioning two other approaches that have been adopted, albeit on a less extensive basis. Anaerobic pre-treatment followed by activated sludge polishing is practiced by a small number of mills that discharge high-strength mechanical pulping or neutral sulphite semi-chemical pulping effluents (MacLean et al. 1990; Garvie 1991). The anaerobic–aerobic approach was thought to provide an eco-

Fig. 17.12. Schematic of sequencing batch reactor operating cycle (from Courtemanche et al. 1997).

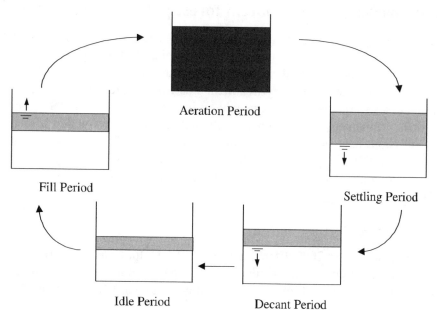

Fig. 17.13. Layout of a typical full-scale pulp and paper sequencing batch reactor wastewater treatment facility (from Cocci et al. 1998).

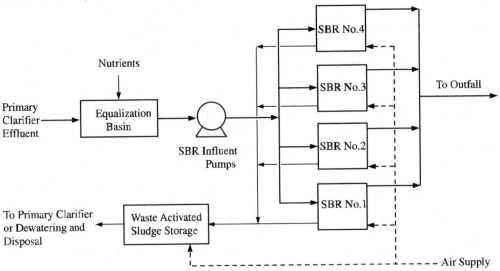

nomical means of treating high wastewater BOD concentrations while also minimizing the cost of processing and disposing of excess sludge. Finally, one bleached kraft mill has deployed a reverse osmosis system for treatment of selected evaporator condensates, as part of an extensive modernization program. No secondary treatment facilities have been installed at this location to date. Although BOD and TSS compliance data could not be obtained, published reports indicate that the final mill effluent was non-toxic to rainbow trout (Dube et al. 2000).

Current discharges from boreal forest pulp and paper mills

The widespread implementation of secondary treatment in the western boreal forest countries has greatly reduced the quantities of contaminants discharged by the pulp and paper sector. The situation in Finland is typical of many other jurisdictions (Fig. 17.14). Prior to the 1970s, when secondary treatment was first adopted at Finnish mills, production increases were always accompanied by increases in BOD and TSS discharges. Full secondary treatment has permitted continued growth in production levels, while aqueous emissions have decreased to historically low values. Similar observations have been reported in Canada, particularly for the period beginning in 1996, at which time most mills had completed the installation of secondary treatment systems (Hearn and Vice 2001).

BOD, TSS, AOX, dioxins, and furans

Figure 17.15 summarizes 2000 discharge levels of BOD_5 for Canada and BOD_7 for Finland and Sweden, relative to the regulatory limits that were in place at the time. There are two bars assigned to most jurisdictions portrayed in Figs. 17.15 through 17.17. The

Fig. 17.14. Historical data on pulp and paper production and effluent discharges from Finnish mills (from FFIF 2002).

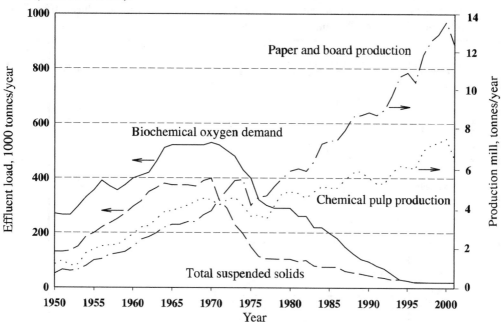

top bar indicates the range of regulatory limits or targets that applied in 2000. As noted above, the use of a bar allows the full range of applicable limits or targets to be illustrated. The second bar, in the form of a box plot, presents a summary of the available measured discharge data from all mills for 2000. The vertical lines in the box plots indicate the values of the 10th, 25th, 50th, 75th, and 90th percentiles of the data distributions. In most cases, the data underlying the box plots are annual averages for each mill in the jurisdiction, with all mill types pooled. Since the box plots utilize annual average data, they do not correspond directly to the BOD_5 regulatory values that are usually expressed as monthly averages. The available annualized data would tend to minimize the variation that might be seen on a monthly average basis. Nonetheless, the median or 50th percentile value should be similar in both cases. For a more complete discussion of the significance of the length of the averaging periods used in the calculation of effluent discharge values, the reader is referred to Carey et al. (2002).

Data from Canada indicate that pulp and paper mills are operating well within the regulatory limits established for BOD_5, with the Alberta pulp mill group demonstrating the lowest BOD_5 discharges. There are no regulatory targets for BOD specified under the Nordic Council of Ministers that would apply in Sweden and Finland. Data for BOD_7 (rather than the BOD_5 used in Canada) indicate that BOD discharges from Finnish mills are somewhat lower than in Alberta and substantially lower than many of the Swedish pulp and paper mill discharges.

The available TSS data for Canada, Finland, and Sweden indicate a similar situation (Fig. 17.16). Interestingly, the lowest TSS discharges appear to be produced in Finland,

Fig. 17.15. BOD regulatory limits in force in 2000 for selected Canadian jurisdictions (ranges of limits indicated by shaded bars) and reported 2000 annual average BOD discharges for all reporting mills in Canada, Finland, and Sweden (vertical lines on box plots indicate 10th, 25th, 50th, 75th, and 90th percentile values).

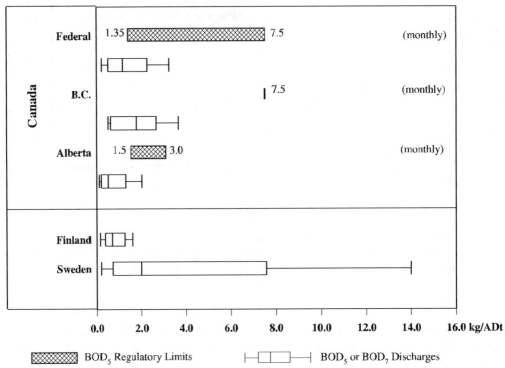

where complete secondary treatment and nutrient control has been required for some time.

The AOX discharge data in Fig. 17.17 indicate that mills in all jurisdictions are currently operating below the applicable limits or targets. The lowest AOX-discharging mills in Canada are located in Alberta, where 2000 AOX discharges were comparable to those from mills in Finland and Sweden. The performance of the low-AOX mills is driven largely by the relatively modern technology used in those locations, rather than by regulatory control, since the permit AOX limits in Alberta, Sweden, and Finland were substantially higher than the measured and reported AOX discharges.

Canada and the United States are two of the few jurisdictions that regulate the discharge of chlorinated dioxins and furans from bleached chemical pulp mill effluents. As outlined in Chap. 16, the term "dioxins and furans" represents a complex family of 210 congeners of polychlorinated dibenzo-*p*-dioxins (PCDDs) and polychlorinated dibenzofurans (PCDFs). Of this total, 17 congeners have chlorine substitutions at the 2, 3, 7, and 8 positions, and are considered to be toxic. When total concentrations of "dioxins and furans" are reported, they are usually expressed in terms of toxicity equivalents to the most toxic congener, 2,3,7,8-tetrachlorodibenzo-*p*-dioxin (TCDD; Luthe and Wrist

Fig. 17.16. TSS regulatory limits in force in 2000 for selected Canadian jurisdictions (ranges of limits indicated by shaded bars) and reported 2000 annual average TSS discharges for all reporting mills in Canada, Finland, and Sweden (vertical lines on box plots indicate 10th, 25th, 50th, 75th, and 90th percentile values).

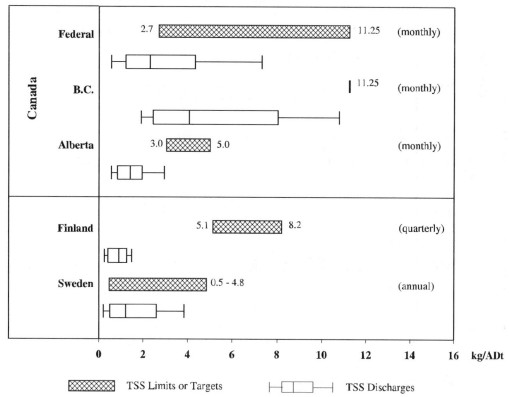

1994). The "toxic equivalent quantities" are then summed to give a total value (TEQ of TCDD).

The intent of the Canadian federal dioxins and furans regulations is to essentially eliminate the discharge of measurable levels of dioxins and furans in bleached chemical pulp mill effluents. A simplified illustration of the progress made is presented in Fig. 17.18, in which industry-wide data for Canada are summarized (Hearn and Vice 2001). Primarily due to mill modernization and technology changes, mill effluents are now routinely below the applicable detection limits for dioxins and furans TEQ. However, in some jurisdictions, TCDD and TCDF are occasionally still above detection limits (Carey et al. 2002).

Acute toxicity

In general, the boreal pulp and paper industry is operating well within the discharge limits established by regulatory authorities. In Canada, the one exception to this generalization may be the continuing occurrence of periodic failures of the acute lethality test at some mills. Kovacs et al. (2002), reporting the results of a recent survey of 74 Cana-

Fig. 17.17. AOX regulatory limits or targets, applicable in 2000 (ranges of limits or targets indicated by shaded bars), plus reported 2000 annual average AOX discharges for reporting mills in Canada, Finland, and Sweden (vertical lines on box plots indicate 10th, 25th, 50th, 75th, and 90th percentile values).

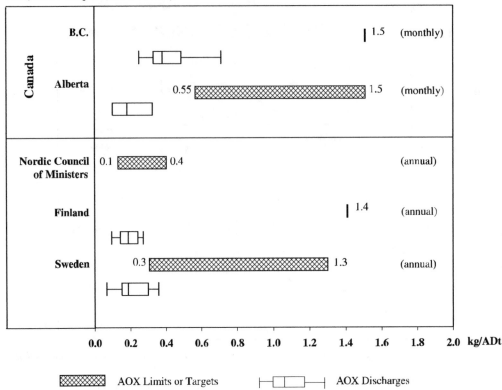

dian mills, indicated that in 1999, 19% of the mills experienced at least 1 rainbow trout toxicity test failure. One mill experienced 18 test failures during the survey period.

Several of the mills reporting toxicity failures also indicated that they were able to identify the cause. The most frequently cited cause appeared to be high ammonia levels in treated effluents, most likely due to excess supplementation of nitrogen for biological treatment. Other causes identified included black liquor losses (soap, slimicides, reduced sulphur compounds), low dissolved oxygen, and, in one case, poor toxicity removal associated with the start-up and optimization of a biotreatment process. Toxicity failures associated with cooling water discharges were most commonly attributed to toxic levels of residual chlorine. Chlorine is commonly used by pulp and paper mills to prevent biological growth in cooling water circuits.

Although failures of the federal acute lethality test are still occurring at Canadian mills, the frequency of failure appears to have decreased significantly between 1996 and 1999. Table 17.1 summarizes the toxicity compliance levels at 52 mills that participated in a series of annual surveys carried out by Paprican between 1996 and 1999. In 1996, when experience with secondary treatment was relatively limited in Canada, only 73%

Fig. 17.18. Historical dioxins and furans discharges from Canadian bleached chemical pulp mills (data from Hearn and Vice 2001).

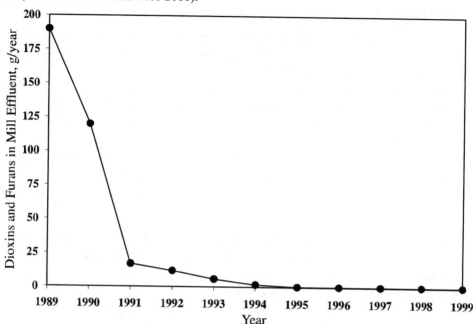

of the mills in this group operated without a toxicity failure. By 1999, compliance had levelled off at about 90%.

Environmental effects monitoring (EEM)

The environmental effects monitoring (EEM) program was added to Canada's national pulp and paper mill effluent regulations, to allow a science-based, iterative approach to regulation that would include monitoring, assessment, and management. The EEM program represents a sector-wide implementation of an adaptive management approach to continual improvement, as advocated in Chap. 21. The first pulp and paper EEM cycle was completed in 1996, and Cycle 2 was finished in 2000. Results from Cycle 3 are expected in 2004. For an overview of the EEM program and its results for Cycles 1 and 2, the reader is referred to Munkittrick et al. (2002*b*). In general, the EEM methodology is still undergoing refinement, as a number of scientific and technical concerns have arisen out of the first two cycles (Walker et al. 2002). In spite of the problems identi-

Table 17.1. Comparison of regulatory toxicity performance for 52 Canadian pulp mills between 1996 and 1999 (adapted from Kovacs et al. 2002).

Test organism	Mills with no toxic episodes			
	1996	1997	1998	1999
Rainbow trout, *Oncorhynchus mykiss*	38 (73%)	37 (71%)	46 (88%)	47 (90%)

> ***Box 17.5. Main response patterns observed from Cycle 2 EEM (environmental effects monitoring) freshwater fish survey.***
>
> 1. An increase in condition factor, increase in liver size, and increase in gonad size; this response pattern reflects an increase in food availability downstream of a mill, consistent with nutrient enrichment;
> 2. An increase in condition and (or) liver size, but a decrease in gonad size; this pattern represents a metabolic disruption, since the fish appear to have adequate food resources (increased growth, condition factor, and (or) liver size) but decreases in gonad size; and
> 3. A decrease in condition factor, liver size, and gonad size; this pattern is reflective of decreased food availability relative to the reference site.
>
> Modified from Munkittrick et al. (2002*a*)

fied, a number of significant observations have been reported. Cycle 2 freshwater fish survey data indicate that the installation of secondary treatment systems has reduced measurable responses in the fish survey portion of the program, from those observed during Cycle 1. The main responses evident after a preliminary review of Cycle 2 data are outlined in Box 17.5 and indicate that most mills discharging to freshwater reported statistically significant differences in condition factor, liver size, and gonad size between fish from the survey and reference sites (Munkittrick et al. 2002*a*). For mills with marine discharges, the fish survey (Fig. 17.19) methodology used did not indicate any direct effects on fish reproduction (Courtenay et al. 2002). For future EEM cycles,

Fig. 17.19. The second cycle of environmental effects monitoring indicated that the installation of secondary treatment at Canadian pulp mills has significantly reduced measurable effects on wild fish. The response of rainbow trout (*Oncorhynchus mykiss*, shown here) is used as one of the most common indicators of aquatic effluent toxicity.

there are still a number of methodological issues that need to be resolved in order to delineate direct and confounding impacts on wild fish in marine receiving waters.

The EEM sub-lethal toxicity testing results from Cycle 2 also indicated a general improvement in effluent quality due to the adoption of secondary treatment (Scroggins et al. 2002). Only two mills were required to conduct fish-tainting tests in Cycle 2. Although tainting was evident at both sites, in only one case was tainting related to the mill itself. Similarly, most mills were exempted from dioxins and furans analyses in edible fish tissue, since measured levels were low in Cycle 1. Of the 10 mills that conducted Cycle 2 dioxins and furans testing, 6 reported levels in fish tissue that exceeded Health Canada consumption guidelines (Environment Canada 2001).

Chemical oxygen demand (COD)

Many regulatory jurisdictions outside of North America have regulated the dissolved and suspended organic matter content of pulp and paper wastewaters through limits on COD discharges. Figure 17.20 presents the Nordic Council of Ministers target COD discharge levels for all mill types and the reported 2000 annual average discharges of COD from mills in Finland and Sweden. Very few data from North America are available for comparison to Fig. 17.20. McCubbin (2000) reported that annual average COD discharges from seven Ontario kraft mills fell in the range of 50–70 kg/ADt. Similar data for a broader range of mill types in British Columbia indicated COD discharges of 15–66 kg/ADt (Carey et al. 2002). In general, the Canadian COD discharges are higher than the 2000 discharges from Finland and Sweden.

Unresolved and emerging aqueous discharge issues

With a small number of exceptions, aqueous discharges from the boreal forest pulp and paper industry appear to be in compliance with existing regulations. The installation of

Fig. 17.20. Nordic Council of Ministers COD discharge targets (ranges of targets indicated by shaded bars) and 2000 annual average COD discharges from reporting pulp and paper mills in Finland and Sweden (vertical lines on box plots indicate 10th, 25th, 50th, 75th, and 90th percentile values).

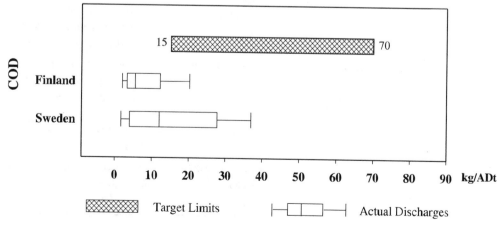

secondary treatment facilities at most mills has been credited with the current state of compliance, both in Canada and abroad. The Canadian compliance failures are related primarily to periodic breakthrough of rainbow trout acute toxicity at a number of sites. The unique Canadian regulatory requirement for EEM studies in 3-year cycles may provide the means by which any remaining significant impacts of pulp and paper mill effluents on fish and fish habitat can be identified. However, independently of the EEM program, a number of other issues related to aqueous discharges from the forest products industry can be presented.

Canadian pulp and paper mill wastewater issues

Sub-lethal toxicity

The Canadian Pulp and Paper Effluent Regulations (PPER) have required each direct-discharging pulp and paper mill to undertake two cycles of EEM testing. Since the PPER have been established under the authority of the federal Fisheries Act, the primary intent of the EEM testing is to determine whether the existing national regulations are providing adequate protection of fish, fish habitat, and human use of the fisheries resource in water bodies which receive pulp and paper mill effluent. As noted above, the measurement of sub-lethal toxicity is one monitoring requirement of the EEM program. In the Cycle 2 EEM, three sub-lethal toxicity assessment methods were applied twice per year for each of the 3 years of the cycle (Table 17.2).

In Canada, there are no regulations that limit the sub-lethal toxicity of pulp and paper mill effluents. Nonetheless, the results from the first and second EEM cycles indi-

Table 17.2. Approved sub-lethal toxicity test methods for EEM Cycle 2 (adapted from Scroggins et al. 2002).

Test	Receiving environment	Test organism
Early life stage development of fish	Marine or estuarine	Inland silverslide (*Menidia beryllina*) or topsmelt (*Athermops affinis*)
	Fresh water	Fathead minnow (*Pimephales promelas*) or salmonid (rainbow trout, *Oncorhynchus mykiss*)
Reproduction of an invertebrate	Marine or estuarine	Echinoid (sea urchin, generally *Strongylocentrotus* sp., or sand dollar, generally *Dendraster* sp.)
	Fresh water	*Ceriodaphnia dubia* (water flea)
Toxicity to an aquatic plant	Marine or estuarine	*Champia parvula* (red alga)
	Fresh water	*Selenastrum capricornutum* (green alga)

cate that many mills in Canada are discharging effluents that produce toxic responses in one or more of the sub-lethal tests. Environment Canada staff report that the sub-lethal toxicity results are useful to indicate temporal changes in effluents that reflect in-mill process changes or the start-up of effluent treatment systems (Scroggins et al. 2002). Even if there is no immediate intent to regulate sub-lethal toxicity, it would be helpful to understand the source of the toxic response(s) in current pulp and paper mill secondary effluents, so that sub-lethal toxicity could be avoided altogether.

In short, the cause of the sub-lethal toxicity identified in the EEM program is not known. Carey et al. (2002) attempted to correlate Cycle 2 sub-lethal toxicity data to AOX from mills in Ontario and British Columbia, without success. However, a strong correlation was noted between *Ceriodaphnia* sub-lethal toxicity and estimated black liquor losses for five bleached kraft mills in Ontario (Fig. 17.21). A similar analysis for selected mills in British Columbia demonstrated a weak relationship between sub-lethal toxicity and estimated black liquor losses (Fig. 17.22). In general, the British Columbia data indicated a general trend to increasing toxicity as the estimated black liquor content of wastewater increased. The observed correlations were very weak for the test results with fish and algae, but were higher for *Ceriodaphnia*. The authors of the report stated that they would have liked to have recommended a regulatory parameter that could be used to indicate and control black liquor losses. The use of chemical oxygen demand (COD) in some jurisdictions was noted, but it was felt that COD alone is not sufficiently specific to black liquor losses for use as a regulatory parameter at this time.

Fig. 17.21. Ceriodaphnia sub-lethal toxicity vs. estimated wastewater black liquor content from Ontario kraft mills (Carey et al. 2002). IC25 is the wastewater concentration causing 25% inhibition or effect.

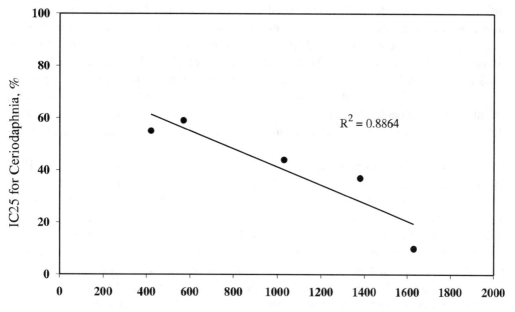

Fig. 17.22. Sub-lethal toxicity to fish, Ceriodaphnia, and algae vs. estimated black liquor solids in effluents from selected British Columbia kraft mills (Carey et al. 2002). IC25 is the wastewater concentration causing 25% inhibition or effect.

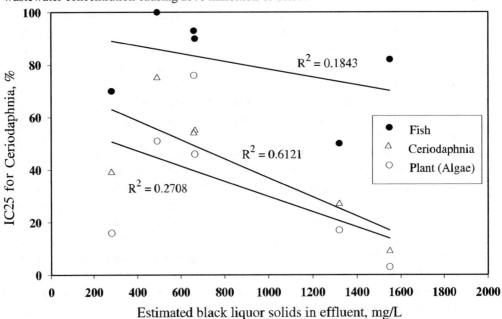

Sub-lethal impacts on wild fish

A considerable body of research from Canada, Scandinavia, the United States, and elsewhere has confirmed that pulp mill effluents are associated with impacts on wild fish in effluent-receiving waters. The most significant responses noted are effects on the livers of exposed fish and changes in a number of whole-organism reproductive parameters. The liver effects include increased liver size and the induction (stimulation) of mixed function oxygenase enzymes (MFOs; Munkittrick et al. 1998; Karels et al. 1999; McMaster 2001). Mixed function oxygenase enzyme induction is of interest because it can be used as an indicator of the *potential* for toxicity from substances that interact with an intracellular receptor known as the aryl hydrocarbon receptor, or AhR (Carey et al. 2002). If MFO induction is absent, it can be assumed that such substances are not present at toxic levels.

Several internal pulp mill process streams are known to be potent inducers of MFO. Spent pulping liquors, including weak black liquor, are one such source (Hodson et al. 1997), while bleach plant effluents are another (Burnison et al. 1999). The latter study also tentatively identified chlorinated stilbenes as a significant contributor to MFO induction. The stilbenes are a group of naturally occurring aromatic compounds that are present in the heartwood of pine trees. Subsequent studies by Burnison and colleagues have demonstrated that the non-chlorinated, native stilbenes are also MFO inducers (Carey et al. 2002). These observations suggest that the discharge of natural wood constituents in any of several internal process streams could be associated with the MFO induction observed in wild fish.

Of the several physiological responses of fish to pulp mill effluent exposure that have been noted in field and laboratory studies, the most serious are related to alterations in several reproductive parameters. Munkittrick et al. (1998) recently reviewed the literature on this topic and indicated that these responses have not been observed at all pulp mill sites. Further, it has not been possible to develop a mechanistic link between these effects and a particular chemical component of pulp mill effluent (Carey et al. 2002). A growing body of evidence is suggesting that non-chlorinated compounds may be important in this regard. One study has reported a link between plant sterols that occur naturally in wood and reproductive effects on fish (Tremblay and Van Der Kraak 1999). Another study has suggested that plant sterols could be converted to a known hormone in the receiving environment after effluent discharge (Jenkins et al. 2001).

The conclusion from the discussion of MFO induction and reproductive impacts on fish is that the causative agents may be non-chlorinated and that they may be discharged from many locations in a pulp mill. Any future regulations or technology developments designed to limit such impacts in the receiving environment may, therefore, need to address the general reduction of organic material in pulp and paper mill effluents. As noted previously, in some jurisdictions outside of Canada, regulations are in place to limit the discharge of all organic matter measurable as COD.

Nutrients

Secondary treatment of pulp and paper mill wastewater makes use of aerobic and facultative microorganisms that consume organic matter in the wastewater. Nitrogen and phosphorus are required nutrients for wastewater treatment microorganisms. If they are not present in the wastewater in adequate amounts, this deficiency will limit the ability of secondary treatment systems to remove BOD and acute toxicity. Pulp and paper wastewaters usually contain insufficient concentrations of nitrogen and phosphorus to support secondary treatment, so most mills practice nutrient supplementation by adding urea or ammonium and phosphate (Hynninen and Viljakainen 1995). Since in practice it is difficult to exactly match the addition of nutrients to the concentration of BOD in the wastewater, a conservative management approach would call for nutrients to be added in excess, so that BOD and toxicity removal can be maintained for compliance purposes. Any excess nitrogen and phosphorus would then be discharged in the treated effluent.

Even small concentrations of nitrogen and phosphorus in pulp mill effluents may be environmentally important, however, owing to the large volumes of the discharges associated with some pulp and paper mills. As noted previously, excess ammonia has been identified as a cause of some of the acute toxicity breakthrough events noted at Canadian mills (Kovacs et al. 2002). Both nitrogen and phosphorus may also be of concern if a receiving water is susceptible to nutrient-induced eutrophication (Bothwell 1992; Priha and Langi 2000; also see Chap. 10).

Canadian regulatory jurisdictions have not yet opted to regulate pulp and paper effluent nutrient discharges across the sector. However, the United States Environmental Protection Agency is actively working to develop criteria for nutrient concentrations in surface water bodies in several ecoregions of the United States (USEPA 1998). If

Fig. 17.23. Nordic Council of Ministers total phosphorus and total nitrogen discharge targets (ranges of targets indicated by shaded bars) and 2000 annual average total-P and total-N discharges from reporting pulp and paper mills in Finland and Sweden (vertical lines on box plots indicate 10th, 25th, 50th, 75th, and 90th percentile values).

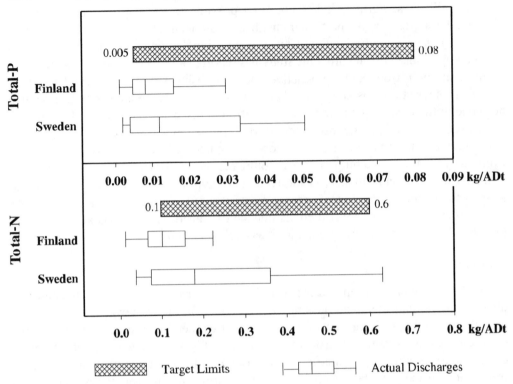

implemented, the criteria could influence the allowable levels of nutrient discharge in pulp and paper mill wastewaters. In Canada, at least one province is monitoring the issue of pulp and paper mill wastewater nutrients (P. Valupadas, Alberta Environment, Edmonton, Alberta, personal communication, 2002), although no regulatory action is planned at this time.

Nutrient discharges have been of regulatory interest in Scandinavia for a number of years (Junna and Ruonala 1991). Until recently, the nutrient guidelines set by the Nordic Council of Ministers have served as the targets for the development of permit decisions. Figure 17.23 summarizes the Nordic Council of Minister nutrient targets and the 2000 pulp and paper effluent discharges of total nitrogen and total phosphorus from mills in Sweden and Finland. Further discussions are underway in Europe that may further reduce the target values for nitrogen and phosphorus. Nutrient control may become an issue for future regulation in Canada, but probably only on a site-specific basis if receiving water eutrophication problems are identified through monitoring activities such as the EEM program.

Pulp mill effluent-induced coagulation and flocculation in rivers

Field studies on the Fraser River in British Columbia have provided evidence that suspended solids downstream of a pulp mill discharge were undergoing coagulation and flocculation that may have been induced by the pulp mill effluent (Krishnappan 2002). Related observations by other researchers have indicated that aqueous concentrations of pulp mill effluent contaminants decreased downstream of an outfall, while the concentrations in the sediment increased (Martinsen 1994; Judd et al. 1995, 1996). Although this observation could be explained by physical sedimentation of contaminants associated with effluent suspended solids, coagulation and flocculation could accelerate the phenomenon.

If organic material contained in pulp mill effluents undergoes coagulation and settling out on river bottoms, it may be of environmental concern for several reasons. First, the accumulated sediment may reduce the availability of suitable fish habitat. Second, the sediments may become a source of contamination for sediment invertebrates and fish. Finally, if the flocculated material is at all biodegradable, its deposition could increase sediment oxygen demand substantially and, in extreme cases, could result in sediment oxygen depletion.

In a University of Alberta study support by the Sustainable Forest Management Network (SFMN), researchers were able to confirm the occurrence of effluent-induced coagulation and flocculation in the Athabasca and Wapiti Rivers (Young and Smith 2001). Particles of less than 10 µm seemed to compose the size fraction that was most susceptible to flocculation. Several hydraulic, environmental, and chemical factors appeared to be important contributing factors in the process. The critical chemical substances included Na^+, Ca^{2+}, pulp fibre, hydrolytic lignins, proteins, carbohydrates, and microbial extracellular polymeric substances. Of most concern was the observation that the carbonaceous biochemical oxygen demand of deposited flocs was nine times higher than a suspension with the same substrate at similar concentrations. It was concluded that effluent-induced flocculation could indeed contribute to anoxic conditions at river bottoms if the river assimilative capacity is low, or in winter conditions under ice cover.

Best available techniques (BAT) specification in the European Union

In December 2001, the European Commission published a reference document on best available techniques (BAT) in the pulp and paper industry (EIPPCB 2001) that signals an intention by member states to consider BAT in the development of future permits for pulp and paper mill effluent discharges. In general, the European Commission definition of BAT for the reduction of emissions and the improvement of economic performance involves (EIPPCB 2001) the "implementation of the best available process and abatement technology in combination with the following:
- training, education, and motivation of staff and operators;
- process control optimization;
- sufficient maintenance of the technical units and the associated abatement techniques;
- environmental management system which optimizes management, increases awareness, and includes goals and measures, process and job instructions, etc."

Box 17.6. European Commission best available techniques (BATs) for kraft pulp mills:

- Dry debarking of wood;
- Increased delignification before the bleach plant by extended or modified cooking and additional oxygen stages;
- Highly efficient brown-stock washing and closed-cycle brown-stock screening;
- Elemental chlorine free (ECF) bleaching with low AOX (adsorbable organic halides) or totally chlorine free (TCF) bleaching;
- Recycling of some (mainly alkaline process) water from the bleach plant;
- Effective spill containment and recovery system;
- Stripping and reuse of condensates from the evaporation plant;
- Sufficient capacity of the black liquor evaporation plant and the recovery boiler to cope with the additional liquor and dry solids load;
- Collection and reuse of clean cooling waters;
- Provision of sufficiently large buffer tanks for storage of spilled cooking and recovery liquors and dirty condensates to prevent sudden peaks of loading and occasional upsets in the external effluent treatment plant; and
- In addition to process-integrated measures, primary treatment and biological treatment of the liquid waste stream is considered BAT for kraft mills.

From EIPPCB (2001)

The reference document itself provides definitions of the best available process and abatement technologies for five different pulp and paper manufacturing processes. For Canadian readers it is instructive to examine the BAT technology specifications for kraft and mechanical pulp mills, two processing trains that are common in the Canadian pulp and paper industry.

The European Union BAT technologies specified for kraft pulp mills (see Box 17.6) include many of the processing technologies and water use reduction measures being adopted by the Canadian industry under the philosophy of progressive systems closure (Browne 2001). One of the characteristics of European BAT for kraft mills is the use of steam stripping for condensate treatment and reuse, usually for pulp washing. For some mills, the reuse of stripped condensate may not be entirely straightforward. Unless the recycled condensate contains very low levels of contaminants, pulp quality may be affected (EIPPCB 2001) and volatile contaminants such as methanol and reduced sulphur compounds may be released inside the mill. To achieve methanol removal efficiencies of greater than 90% it may be necessary to use uneconomically high ratios of steam to condensate (Zuncich et al. 1993). Furthermore, removal efficiencies of total organic carbon achieved by steam stripping are typically lower than those achieved for methanol, if the condensate contains non-volatile or semi-volatile contaminants (Danielsson and Hakansson 1996). These considerations may necessitate a more advanced treatment approach if a substantial fraction of the kraft foul condensate is to be reused rather than sewered.

Box 17.7. European Commission best available techniques (BATs) for mechanical pulp mills:

- Dry debarking of wood;
- Minimization of reject losses by using efficient reject handling stages;
- Water recirculation in the mechanical pulping department;
- Effective separation of the water systems of the pulp and paper mill by the use of thickeners;
- Counter-current white water system from paper mill to pulp mill, depending on the degree of integration;
- Use of sufficiently large buffer tanks for storage of concentrated wastewater streams from the process; and
- Primary and biological treatment of the effluents and, in some cases, also flocculation or chemical precipitation.

From EIPPCB (2001)

The European BAT specifications for mechanical pulping operations (Box 17.7) do not explicitly include the zero-discharge technology that has been installed at a number of Canadian mills (Hardman and Manolescu 1998), although evaporation of concentrated process streams plus activated sludge treatment for the less contaminated wastewaters is referred to as "an interesting solution for upgrading mills". European mechanical pulp mill BAT describes effluent flows in the range of 12–20 m^3/ADt. It may be useful to know whether other technologies could improve the economy of the BAT train by increasing water recirculation rates and reducing the capacity requirements for evaporation and external biological treatment systems.

The discussion above is limited only to the kraft and mechanical pulping sectors of the pulp and paper industry. The European Commission reference document also provides BAT definitions for sulphite pulping, recycled fibre processing and papermaking, and related processes. The intention is that all European member states will consider the BAT target discharge values when existing pulp mill effluent discharge permits are renewed. When the recommended BAT emission levels to water for all mill types are pooled and the ranges of values are plotted, the magnitude of the potential changes in emissions targets for Finland and Sweden can be seen in Figs. 17.24 and 17.25. It is clear that the BAT emissions targets are substantially lower than the prevailing regulatory limits in Canada. It is also readily apparent that the BAT targets represent a reduction above and beyond the current Nordic Council of Ministers targets. If the BAT targets influence the regulated emission levels in renewed discharge permits, some of the mills in Finland and Sweden will need to reduce effluent emissions further.

Logyard/dryland sort run-off

The Canadian forest products industry processes enormous volumes of wood each year. Much of this wood is stored, handled, and sorted at logyards and dryland sorts located at thousands of sites throughout the country. These sites contribute to the generation of large quantities of woody debris, perhaps as much as 3–6% (by volume) of the wood processed (McWilliams 1992).

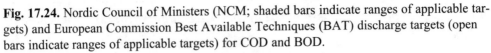

Fig. 17.24. Nordic Council of Ministers (NCM; shaded bars indicate ranges of applicable targets) and European Commission Best Available Techniques (BAT) discharge targets (open bars indicate ranges of applicable targets) for COD and BOD.

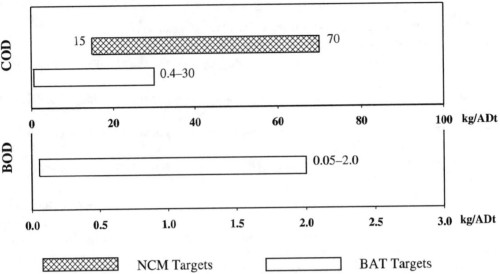

When wood is stored without cover, it is exposed to water in the form of rainfall or snow melt. Logyard operations may also utilize water for cleaning, fire prevention, and dust control. When water comes into contact with wood or woody debris, a leachate or run-off may be generated that contains many contaminating constituents (Fig. 17.26). Some of the contaminants originate from the wood itself. Run-off may also contain contaminants that originate from the equipment and the many materials used at the site (Orban et al. 2002). Orban (2000) reported that there are three groups of contaminants in logyard run-off that are of environmental concern: (1) wood and bark-derived organics, particulates, and foam; (2) sediment from logyard erosion; and (3) greases and oils from logyard vehicles and associated machinery.

Logyard run-off not only contains many of the wood-derived contaminants that are present in pulp and paper mill effluents, but also many other constituents that may exert deleterious effects in receiving environments. In general, the environmental concerns associated with logyard run-off include physical effects of particulate matter on fish habitat, acute and chronic toxic effects on fish and invertebrates, receiving water oxygen depletion, and bioaccumulation of toxic constituents in fish and other higher organisms (Liu et al. 2002). In light of these potential environmental impacts, regulatory agencies and researchers in boreal forest regions have been monitoring the logyard and sawmill run-off issue with interest (Zirnhelt 1989; McDougall 1996; Bailey et al. 1999; Borga et al. 1999; Liu et al. 2002; Orban et al. 2002).

Surface run-off from logyards, dryland sorts, and sawmills appears to be an overlooked aqueous emission from forest industry operations. Orban (2000) reviewed the regulatory and permitting procedures for logyard run-off in British Columbia, Alberta,

Fig. 17.25. Nordic Council of Ministers (NCM; shaded bars indicate ranges of applicable targets) and European Commission Best Available Techniques (BAT) discharge targets (open bars indicate ranges of applicable targets) for TSS, AOX, total-P, and total-N.

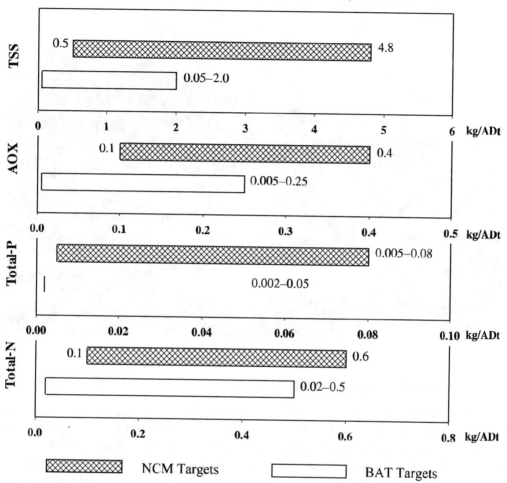

Ontario, New Brunswick, and Canada. None of the jurisdictions cited appeared to have regulations or formal permits for logyard run-off. The federal government (Samis et al. 1999) and most of the provinces have developed guidelines and best management practices recommendations for logyard run-off. The Province of Ontario requires logyard operators to conduct a storm water control study once every 3 years, the goal of which is to reduce the run-off contaminant loadings to the extent possible and to ensure that run-off is not acutely toxic. In Canada as a whole, the general provisions of the federal Fisheries Act would be applicable if logyard run-off is concluded to be harmful to fish or fish habitat. In contrast to the Pulp and Paper Effluent Regulations that apply to pulp and paper mill aqueous discharges, there have been no comparable regulations for logyard or sawmill run-off developed under the Fisheries Act.

Fig. 17.26. The contamination of leachate and runoff from log storage yards remains a largely unquantified and unregulated aqueous impact of the forest products industry.

Photo by Dan Smith

Emerging abatement technologies

The current status of aqueous discharges from the forest products industry has been achieved through capital-intensive investment in new manufacturing process technologies and pollution abatement systems. Further gains in environmental performance are not likely to come from the adoption of novel pollution abatement technology alone. As is clearly described in the literature emanating from ongoing research and development projects in Sweden (Axegård et al. 2002), Europe (EIPPCB 2001), Canada (Browne 2001), and elsewhere, future technology investments in the sector will focus primarily on the reduction of water use and on increasing the efficiency of resource and energy utilization. The indirect benefits to be gained from upgraded manufacturing technologies will include lower emissions of contaminants to the environment. A portion of the emissions reduction will accrue from process changes that reduce the formation of contaminants (pollution prevention). However, as discussed above, many of the most problematic contaminants in forest industry aqueous discharges are not formed in the manufacturing process, but rather they are unavoidably extracted from the wood that serves as the raw material for production. Interestingly, although the environmental focus in the industry has now shifted from external treatment to internal process modification, it has been recognized that novel pollution abatement technologies will still be needed if the goals of further reducing water use and improving process efficiency are to be successfully pursued. The following discussion reviews a selection of recent wastewater pollution abatement technology developments that may assist the forest products industry in further reducing its environmental footprint. The emphasis in the following material has been placed on research supported by the Sustainable Forest Management Network, although complementary developments from outside Canada are also discussed.

Chemical precipitation technologies

Chemical precipitation of wastewater

For some pulp and paper applications, chemical precipitation can be considered as an alternative to biological processes for secondary treatment of wastewater. Chemical precipitation is used in this fashion at a number of non-integrated paper mills in Sweden, either with or without a biological treatment stage (EIPPCB 2001). The chemical flocculants used are usually aluminum salts (alum) or ferric iron salts, with a polyelectrolyte supplement. The precipitated and flocculated material is removed by filtration or sedimentation, and the recovered sludge requires extensive downstream processing for dewatering and disposal. Data from the Swedish mills indicate removal efficiencies of 97–99% for TSS and 70% for COD when treating raw mill wastewater. The material removed is predominantly colloidal in nature, and little change in soluble biodegradable organic matter is reported. In jurisdictions in which even low-level nutrient discharges from pulp and paper mills are of concern, pre-treatment by chemical precipitation will reduce the BOD and biodegradable COD concentrations prior to biological treatment. This will result in lower nutrient supplementation requirements in the biological stage and lower overall discharges of nitrogen and phosphorus in the final effluent.

Recovery of chelants in TCF bleach plant filtrates

Totally chlorine free (TCF) bleaching of kraft pulp requires the addition of metal chelating agents prior to the hydrogen peroxide bleaching stage. These chelants are normally discharged with the sewered bleach plant filtrates. The most common chelants used for this purpose are ethylenediaminetetraacetic acid (EDTA) and diethylenetriaminepentaacetic acid (DTPA). The bleaching stage that involves chelant addition is usually designated as the "Q" stage. Although these chelants are essentially non-toxic in receiving waters, there is concern for their persistence in the environment and the possibility that these compounds may mobilize toxic metals from sediments into the overlying water. The containment of chelating agents in the kraft bleaching process would improve process economics and avoid any possibility of negative environmental impacts.

Recent studies have shown that chelating agents are not easily degraded in biological treatment systems, although degradation seems to be improved at moderately alkaline pH (Virtapohja and Alen 1998). An alternative approach under trial in Sweden is to dose the Q-stage filtrate with carbonate to raise the pH. Chelated metals are then released and precipitated. The precipitation process is aided by the addition of a polyethylene oxide flocculent. The insoluble sludge, containing metals, flocculating agents, and wood extractives can be separated by dissolved air flotation. The treated underflow containing the recovered chelants is then reused in the Q-stage of the bleaching process.

The process, known as the Kemira NetFloc system, has been developed to a mill-scale application for the recovery of EDTA (EIPPCB 2001). No operational or performance data were available at the time of writing.

Membrane separation technologies

Membrane filtration systems

Membrane filtration systems are widely used in industrial processing, water treatment, and wastewater treatment applications. If a membrane with the correct pore size and surface characteristics is chosen for an application, the resulting removal efficiencies for target contaminants are essentially 100%, and no undesirable substances are introduced into the treated water. It is now recognized that membrane filtration technologies may be useful in promoting water use reduction in pulp and paper mills. In general, as mill water use is reduced, the concentrations of dissolved and colloidal substances in mill water circuits increase. A membrane filtration system that can purge these undesirable contaminants would permit greater reductions in water use than could be achieved otherwise.

One such application is the in-line treatment of white water, in which membrane filtration is used to remove suspended solids, bacteria, colloidal matter, anionic trash, and other high molecular weight material from a portion of the excess white water (Elefsiniotis et al. 1995; Nuortila-Jokinen et al. 1999). The membrane permeate could then be returned to the white water system for reuse. The concentrate produced by membrane filtration is usually directed to an existing biological treatment facility, or is disposed of by incineration after further concentration.

The EIPPCB (2001) report indicates that there have been three full-scale applications of membrane filtration for white water treatment. The report is careful to point out that the capital and operating costs of membrane systems, which are substantial, reduce the applicability of the technology to special cases in which water supplies are scarce, or in which a particularly sensitive receiving water must be protected from the effects of a significant increase in mill production capacity. As was noted above in the section on secondary treatment technologies in use in Canada, one New Brunswick bleached kraft mill is currently using a full-scale reverse osmosis system for condensate treatment. The membrane unit is the only treatment system in use for organics removal at this location, as a secondary treatment process has not been installed (Dube and MacLatchy 2000).

Membrane bioreactor treatment

Some of the disadvantages of membrane filtration units can be overcome by combining membrane filtration technology with biological treatment in one integrated process — a membrane bioreactor (MBR). Wastewater is first introduced to a bioreactor, which can be operated under aerobic or anaerobic conditions, in which biodegradation of the organic material proceeds (Fig. 17.27). The bioreactor mixed liquor is then separated into a "solids" and "liquid" stream using a membrane filtration system that may be internal or external to the bioreactor. The suspended solids are retained in, or returned to, the

Fig. 17.27. Schematic of laboratory-scale high-temperature membrane bioreactor for in-mill process streams (Ragona and Hall 1998).

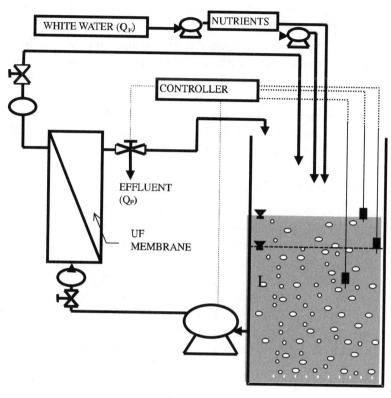

bioreactor for continued use in the process. The membrane-filtered effluent, or permeate, is discharged from the treatment process with negligibly small concentrations of suspended material (Ragona and Hall 1998).

The European Commission reference document identifies a number of potential pulp and paper applications of MBR involving either in-mill process stream treatment or external end-of-pipe treatment. One existing full-scale paper mill application is cited (EIPPCB 2001) in which excess MBR sludge is reinjected in the pulper, and a portion of the treated water is recycled to the mill. Other proposed applications are: (1) for reuse of condensates from black liquor evaporation; (2) for treatment of alkaline bleach plant filtrates; and (3) for treatment and reuse of white water.

As also noted for the biological treatment of atmospheric emissions (Chap. 18), one difficulty with most biological treatment approaches for pulp mill water reuse is the need to pre-cool high-temperature process streams to approximately 35°C prior to biotreatment. At higher temperatures, biomass settling characteristics become impaired, and eventually the treatment system will fail (Johnson 1995). Research sponsored by the SFMN at the University of British Columbia (UBC) has demonstrated that MBR technology can be used for white water treatment without substantial pre-cooling. The absolute retention of biomass by the membrane in an MBR eliminates the need to consider biomass settling characteristics in the process design. As long as the process biomass remains active, an MBR can be operated at temperatures at least as high as 55°C (Tardif and Hall 1997).

The UBC research program has shown that high-temperature aerobic MBR systems produce high-quality treated water when applied to both mechanical newsprint white water (Ragona and Hall 1998) and kraft mill evaporator condensates (Bérubé and Hall 1999). In the latter study, the capital and operating costs of MBR treatment were estimated to be less than one-half of those for conventional steam stripping (Bérubé and Hall 2001). Further, it was noted that the treated water quality from the MBR was substantially higher than that expected from steam stripping.

Advanced oxidation technologies

Pulp and paper mill effluents that have been treated to secondary standards still contain significant concentrations of recalcitrant and persistent organic material that is measurable as COD, AOX, or colour. For mills located on sensitive receiving waters, it may be necessary to adopt advanced or tertiary treatment techniques for further removal of the non-biodegradable residual contaminants. Laboratory (Oller 1997a, b, as cited in EIPPCB 2001) and pilot-scale studies have been completed on a system which combines ozonation and fixed-bed aerobic biofilm reactors for removal of recalcitrant organic matter. The ozonation stage oxidizes persistent organic compounds into more readily biodegradable material that can be removed in the biological treatment stage. Trial results from Europe with paper mill wastewater indicated that 50% COD removal efficiencies could be attained in a single-stage ozonation-biofilter. A two-stage process achieved up to 90% COD removal efficiency (EIPPCB 2001).

The main drawback to this system is the high capital and operating cost of the ozonation stage. Sustainable Forest Management Network researchers at the University of Alberta have demonstrated a new ozone bubble column that promises to significantly reduce treatment costs by improving the efficiency of ozone mass transfer to wastewater (Gamal El-Din and Smith 2001, 2002). In testing with bleached kraft mill wastewater, similar treatment efficiencies were noted with both a conventional fine-bubble contactor and the new impinging-jet contactor. However, the impinging-jet contactor required only 15% of the volume needed for the conventional ozonator.

In another SFMN research project at the University of Calgary, a supported TiO_2 photocatalyst promoted with small quantities of hydrogen peroxide achieved 93% removal of the carbon content of a pulp mill white water. The oxidation efficiency of the process appeared to be equally good for both dissolved and colloidal carbon components. A particularly interesting conclusion of the work is the suggestion that this technology could permit the installation of low-cost solar photocatalytic units at existing secondary treatment plants for enhanced removal of recalcitrant organic material. The exploitation of solar radiation would eliminate the expense of generating ultraviolet light by electrical means (Langford 2001).

Fig. 17.28. Schematic of potential mill-scale fungal/enzyme treatment system for mechanical newsprint white water.

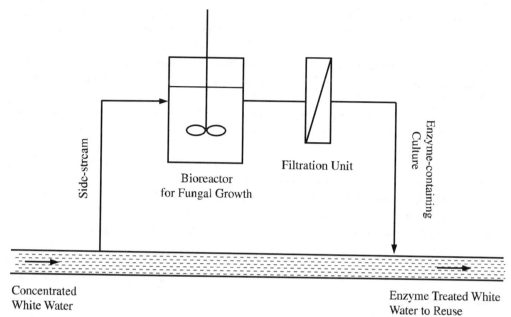

Novel biotechnologies

Bioaugmentation

Secondary biological treatment systems are periodically subjected to transient operating conditions that may result in incomplete acute toxicity removal. One occurrence of this type was reported at a Canadian mill during a period in which the biological treatment facility was undergoing start-up and optimization (Kovacs et al. 2002). During such periods of transient operation, many acutely toxic components of pulp mill wastewater may be insufficiently removed to ensure that the treated effluent is not acutely toxic. Resin acids are one such class of toxic compounds.

In an SFMN research program at the University of British Columbia, resin acid-degrading bacteria were isolated from biotreatment plant sludge, were characterized, and were tested for their suitability as inocula for bioaugmentation of biological treatment systems (Yu and Mohn 1999). In all, 18 unique isolates were characterized. When the resin acid-degrading capability of a laboratory-scale biotreatment system was inhibited by an extreme pH shock, selected isolates were used to inoculate the inhibited microbial communities (Yu and Mohn 2002). Three of the isolates were able to restore resin acid removal activity in the inhibited treatment systems. Further studies indicated that the isolates grew and persisted in the treatment systems, without adversely affecting the general treatment capability of the process. Although bioaugmentation for restoring resin acid biodegradation appears to be feasible, additional testing at a larger scale will be needed to confirm the practicality of the strategy.

Box 17.8. Meeting the challenge.

- Continuing effort is needed to bring the Canadian pulp and paper sector into complete compliance with existing effluent regulations, specifically the federal acute toxicity regulation;
- The impacts of stricter regulatory controls based on BAT (best available techniques) in other jurisdictions need to be related to future requirements for Canadian pulp and paper operations;
- Continuing research into unresolved pulp and paper effluent issues, such as sub-lethal toxicity and reproductive effects on fish, would enable the best design and technology choices to be made in the future;
- Further understanding of the environmental significance of run-off from logyards and dry-land sorts would be of assistance to industry regulators and managers; and
- The feasibility of a number of novel abatement technologies developed in Canada needs to be demonstrated at the pilot scale or mill scale, to permit designers and managers to determine the optimal roles of such technologies in future pulp and paper mill capital investment programs.

Rapid screening for resin acids

Another SFMN project at the University of British Columbia has developed an immunoassay for relatively rapid measurement of the resin acid levels in pulp mill wastewaters (Li et al. 1997; Serreqi et al. 2000). The technique was found to be very sensitive to dehydroabietic acid (DHA) and some DHA-like resin acids. Although it was reported that the method is subject to interference from other resin acids and some unknown contaminants, it would be best to consider the assay as a rapid screening tool for high resin acid levels in pulp mill process waters and effluents.

Enzyme treatment of white water

As an alternative to white water treatment in a membrane bioreactor as discussed above, SFMN-sponsored research has developed the concept of enzymatic treatment of concentrated mechanical pulping/newsprint mill process waters. In the proposed system (Fig. 17.28), a portion of the circulating white water is cooled and introduced to a bioreactor in which the fungus *Trametes versicolor* is grown on the dissolved and colloidal substances (DCS) present in the white water (Zhang et al. 2002). The enzymes produced and released by the fungus lead to extensive degradation of the original DCS components. The effluent from the bioreactor would then be filtered and the filtrate, containing the fungal enzymes, would be reintroduced into the white water system. Degradation of DCS components would then proceed in the circulating white water system itself. The degraded DCS components were found to have little or no impact on paper properties, in contrast to the undegraded DCS components that are present in untreated white water (Zhang et al. 2000).

Summary

Pulp and paper mills located in western boreal forest countries are generally in compliance with existing regulatory limits on aqueous discharges. In Canada, the one exception to this observation may be the occurrence of periodic failures of the acute toxicity test at a number of mill sites. European pulp and paper mill effluent discharge targets could be further reduced as a result of a best available techniques assessment completed for the European Union. In the future, some mills in Finland and Sweden may be required to reduce emissions from current levels to comply with the BAT limits.

The present high level of compliance has been attributed by many observers to the widespread adoption of secondary treatment by the pulp and paper industry. However, even with near-universal secondary treatment now in place, there still appear to be a number of unresolved or emerging environmental issues related to pulp and paper mill effluents. In addition, the issue of contaminated run-off from logyards and dryland sorting operations appears to be an overlooked, but potentially significant, aqueous emission from the forest products industry (Box 17.8).

Future technology developments that will have a major impact on aqueous emissions from the industry are expected to stem from changes to internal manufacturing processes. Nonetheless, there are several opportunities for coupling these new manufacturing technologies to novel pollution abatement technologies for additional gain. A number of research programs sponsored by the Sustainable Forest Management Network have generated excellent examples of the potential utility of novel pollution abatement technologies in the forest products industry.

Acknowledgements

The contribution made to this review by the following individuals and organizations is gratefully acknowledged: The Sustainable Forest Management Network, Zuohong Geng of the University of British Columbia, Anders Widell of the Swedish Environmental Protection Agency, The Finnish Forest Industries Federation, Prasad Valupadas of Alberta Environment, Yousri Hamdy of the Ontario Ministry of Environment, Frank Witthoeft of the B.C. Ministry of Water, Land and Air Protection, and Linda Maddison of Environment Canada.

References

Axegård, P., Backlund, B., and Warnqvist, B. 2002. The eco-cyclic pulp mill: focus on closure, energy-efficiency and chemical recovery development. Pulp & Paper Can. **103**: 26–29.

Bailey, H.C., Elphick, J.R., Potter, A., Chao, E., Konasewich, D., and Zak, B. 1999. Causes of toxicity in stormwater run-off from sawmills. Environ. Toxicol. Chem. **18**: 1485–1491.

(BCMELPEC) British Columbia Ministry of Environment, Lands and Parks and Environment Canada. 1993. State of the environment report for British Columbia. British Columbia Ministry of Environment, Lands and Parks and Environment Canada, Victoria, British Columbia. 127 p.

Bérubé, P.R., and Hall, E.R. 1999. Effects of kraft evaporator condensate matrix on methanol removal in a high temperature membrane bioreactor. Water Sci. Technol. **40**: 327–335.

Bérubé, P.R., and Hall, E.R. 2001. Cost comparison between a high temperature membrane bioreactor and a steam stripper for the treatment of foul evaporator condensate for reuse. TAPPI J. **84**: 62. Available at http://www.tappi.org/content%5Cpdf%5Cbookstore%5C01jun toc.pdf [viewed 27 April 2003].

Borga, P., Elowson, T., and Liukko, K. 1999. Environmental loads from water-sprinkled softwood timber. 2. Influence of tree species and water characteristics on wastewater discharges. Environ. Toxicol. Chem. **15**: 1445–1454.

Bothwell, M.L. 1992. Eutrophication of rivers by nutrients in treated bleached kraft mill effluent. Water Pollut. Res. J. Can. **27(3)**: 447–472.

Browne, T.C. (*Editor*). 2001. Water use reduction in the pulp and paper industry, edition 2. Paprican, Pointe Claire, Quebec. 172 p.

Burnison, B.K., Comba, M.E., Carey, J.H., and Parrott, J. 1999. Isolation and tentative identification of the compound responsible for fish MFO induction in a BKME. Environ. Toxicol. Chem. **18**: 2882–2887.

(CFS) Canadian Forestry Service. 2002. The state of Canada's forests 2001. Natural Resources Canada, Canadian Forestry Service, Ottawa, Ontario. Available at http://www.nrcan.gc.ca/cfs-scf/national/what-quoi/sof/sof01/profiles_e.html [viewed 10 July 2002].

Carey, J., Hall, E., and McCubbin, N. 2002. Review of the scientific basis for AOX effluent standards in British Columbia. Report to the British Columbia Minister of Water, Land and Air Protection. British Columbia Ministry of Water, Land and Air Protection, Victoria, British Columbia. Available at http://www.aoxpanel.ca [viewed 28 June 2002].

Cocci, A.A., Almost, S., and McCarthy, P.J. 1998. Sequencing batch reactors in the pulp and paper industry a benchmarking study. *In* Proceedings of the 1998 International Environmental Conference & Exhibit, TAPPI, **3**: 1203–1208.

Courtemanche, M., Tremblay, H., and Villeneuve, F. 1997. Sequencing batch reactor technology: an old technique successfully applied to the treatment of pulp and paper mill effluents containing high residuals of sulphite. *In* Proceedings of the 1997 International Environmental Conference & Exhibit, TAPPI, **2**: 853–863.

Courtenay, S.C., Munkittrick, K.R., Dupuis, H.M.C., Parker, R., and Boyd, J. 2002. Quantifying impacts of pulp mill effluent on fish in Canadian marine and estuarine environments: problems and progress. Water Qual. Res. J. Can. **37**: 79–99.

Danielsson, G., and Hakansson, M. 1996. Steam stripping of contaminated condensates from the black liquor evaporation plants. *In* Proceedings of the Minimum Effluent Mills Symposium, TAPPI, pp. 273–276.

Dube, M., and MacLatchy, D. 2000. Reverse osmosis treatment of condensates from a bleached kraft pulp mill: effects on acute and chronic toxicity of process streams and final effluent. *In* Proceedings of the 4th International Conference on Environmental Impacts of the Pulp and Paper Industry, 12 June 2000, Helsinki, Finland. *Edited by* M. Ruoppa, J. Paasivirta, K.-J. Lehtinen, and S. Ruonala. Finnish Environment Institute, Helsinki, Finland. pp. 270–276.

Dube, M., McLean, R., MacLatchy, D., and Savage, P. 2000. Reverse osmosis treatment: effects on effluent quality. Pulp & Paper Can. **101**: 42–45.

(EIPPCB) European Integrated Pollution Prevention and Control Bureau. 2001. Integrated Pollution Prevention and Control (IPPC): reference document on best available techniques in the pulp and paper industry. European Commission, Seville, Spain. Available at http://eippcb.jrc.es/pages/FActivities.htm [viewed 27 April 2003].

Elefsiniotis, T., Hall, E.R., and Johnson, R.M. 1995. Contaminant removal from recirculated white water by ultrafiltration and/or biological treatment. *In* Proceedings of the 1995 International Environmental Conference, TAPPI, **2**: 861–867.

Environment Canada. 2001. Environmental effects monitoring: cycle 2 reports are in! National EEM Office, Environment Canada, Hull, Quebec. Available at http://www.ec.gc.ca/eem [viewed 28 June 2002].

(FFIF) Finnish Forest Industries Federation. 2002. Production of the pulp and paper industry in Finland and waste water load 1950–2001. Finnish Forest Industries Federation, Helsinki, Finland. Available at http://english.forestindustries.fi/figures/figures.html?lang=en&pic= kuormit2 [viewed April 27, 2003].

(FPAC) Forest Products Association of Canada. 2002. 2001 annual review: paper & wood. Forest Products Association of Canada, Montreal, Quebec. Available at http://www.fpac.ca/english/news/public.htm [viewed 28 June 2002].

Gamal El-Din, M., and Smith, D.W. 2001. Maximizing the enhanced ozone oxidation of kraft pulp mill effluents in an impinging-jet bubble column. Ozone Sci. Eng. **23**: 479–493.

Gamal El-Din, M., and Smith, D.W. 2002. Theoretical analysis and experimental verification of ozone mass transfer in bubble columns. Environ. Technol. **23**: 135–147.

Garvie, R. 1991. Anaerobic/aerobic treatment of NSSC/CTMP effluent and biogas utilization. Preprints A. 77th Annual Meeting, Technical Section, Canadian Pulp & Paper Association, Montreal, Quebec. pp. 321–326.

Halliburton, D., Jones, S.A., Carey, J.H., Carlisle, D.B., Colodey, A.G., Myres, A., Lockhart, W.L., and Rogers, I.H. 1991. Effluents from pulp mills using bleaching. Environment Canada, Ottawa, Ontario. 60 p.

Halliburton, D., Maddison, L., and Simpson, D. 1999. Environmental requirements for industrial permitting case study on the pulp and paper industry — part one OECD, Paris, France. ENV/EPOC/PPC(99)8/FINAL/PART1.

Hardman, D., and Manolescu, D.R. 1998. Mill closure: the continuing challenge. *In* Proceedings of the 84th Annual Meeting, Technical Section, Canadian Pulp & Paper Association, Montreal, Quebec. pp. A403–A406.

Hearn, J., and Vice, K. 2001. Environmental progress report 2000–2001. Forest Products Association of Canada, Montreal, Quebec. Available at http://www.fpac.ca/english/news/ public.htm [viewed 28 June 2002].

Hodson, P.V., Maj, M.K., Efler, S., Burnison, B.K., Van Heiningen, A.R.P., Girard, R., and Carey, J.H. 1997. MFO induction in fish by spent cooking liquors from kraft pulp mills. Environ. Toxicol. Chem. **16**: 908–916.

Hynninen, P. 1998. Environmental control. *In* Papermaking science and technology. *Edited by* J. Gullichsen and H. Paulapuro. Fapet Oy, Helsinki, Finland. 234 p.

Hynninen, P., and Viljakainen, E. 1995. Nutrient dosage in biological treatment of wastewaters. TAPPI J. **78**: 105–108.

Jenkins, R., Angus, R.A., McNatt, H., Howell, W.M., Kempainnen, J.A., Kirk, M., and Wilson, E.A. 2001. Identification of androstenedione in a river containing paper mill effluent. Environ. Toxicol. Chem. **20**: 1325–1331.

Johnson, R. 1995. The performance of a sequencing batch reactor for the treatment of whitewater at high temperatures. M.A.Sc. Thesis, Department of Civil Engineering, University of British Columbia, Vancouver, British Columbia. 234 p.

Judd, M.C., Stuthridge, T.R., Tavendale, M.H., McFarlane, P.N., Mackie, K.L., Buckland, S.J., Randall, C.J., Hickey, C.W., Anderson, S.M., and Steward, D. 1995. Bleached kraft pulp mill sourced organic chemicals in sediments from New Zealand rivers. Part 1: Waikato River. Chemosphere, **30**: 1751–1765.

Judd, M.C., Bergman, I.J., McFarlance, P.N., Anderson, S.M., and Stuthridge, T.R. 1996. Bleached kraft pulp mill sourced organic chemicals in sediments from New Zealand rivers. Part II: Tarawera River. Chemosphere, **33**: 2209–2220.

Junna, J., and Ruonala, S. 1991. Trends and guidelines in water pollution control in the Finnish pulp and paper industry. TAPPI J. **74**(**7**): 105–111.

Karels, A., Soimasuo, M., and Oikari, A. 1999. Effects of pulp and paper mill effluents on reproduction, bile conjugates and liver MFO (mixed function oxygenase) activity in fish at Southern Lake Saimaa, Finland. Water Sci. Technol. **40**: 109–114.

Kovacs, T., Gibbons, J.S., Naish, V., and Voss, R. 2002. Complying with effluent toxicity regulation in Canada. *In* Proceedings of the 2002 International Environmental Conference & Exhibit, Technical Association of the Pulp and Paper Industry. Available from CD-ROM at http://www.tappi.org [viewed 28 June 2002].

Krishnappan, B.G. 2002. In situ size distribution of suspended particles in the Fraser River. J. Hydr. Eng. **126**: 561–569.

Langford, C.H. 2001. Photocatalysis for oxidation of pulp mill waste streams. Sustainable Forest Management Network, Edmonton, Alberta. Proj. Rep. 2001-15. Available at http://sfm-1.biology. ualberta.ca/english/pubs/PDF/PR_2001-15.pdf [viewed 27 April 2003].

Li, K., Serreqi, A., Breuil, C., and Saddler, J.N. 1997. Quantification of resin acids in CTMP effluents using an enzyme-linked immunoassay. Water Sci. Technol. **35**: 93–99.

Liu, S.D., Nassichuk, M.D., and Samis, S.C. 2002. Guidelines on storage, use and disposal of wood residue for the protection of fish and fish habitat in British Columbia. Environment Canada, North Vancouver, British Columbia. DOE/FRAP Rep. 95-18. Available at http://www.rem.sfu.ca/FRAP/PDF_list [viewed 28 June 2002].

Luthe, C.E., and Wrist, P.E. 1994. Progress in reducing dioxins and organochlorines: 1988–1993. Paprican, Pointe Claire, Quebec. Misc. Rep. MR 278.

MacLean, B., de Vegt, A., and van Driel, E. 1990. Full-scale anaerobic/aerobic treatment of TMP/BCTMP effluent at Quesnel River Pulp. *In* Proceedings of the 1990 Environmental Conference, TAPPI, **2**: 647–661.

Martinsen, K. 1994. Distribution of organohalogen in sediments outside pulp mills using sum parameters. Sci. Total Environ. **144**: 47–57.

McCubbin, N. 2000. Effluent control technology and costs for bleached kraft mills in Ontario. Prepared for Ontario Ministry of Environment, Toronto, Ontario. 88 p.

McCubbin, N., Barnes, E., Bergman, E., Edde, H., Folke, J., and Owen, D. 1991. Best available technology for control of aqueous effluents from pulp and paper mills. Draft report to Ontario Ministry of Environment, Toronto, Ontario. 271 p.

McDougall, S. 1996. Assessment of logyard run-off in Alberta — a preliminary evaluation. Alberta Environmental Protection, Edmonton, Alberta.

McLeay, D., and Sprague, J.B. 1990. Biological test method: acute lethality test using rainbow trout. Environment Canada, Ottawa, Ontario. Rep. EPS 1/RM/9. 51 p.

McMaster, M.E. 2001. A review of the evidence for endocrine disruption in Canadian aquatic ecosystems. Water Qual. Res. J. Can. **36**: 215–231.

McWilliams, J. 1992. Logyard/dryland sort debris: the ultimate solution. Can. For. Ind. (Jan./Feb.): 16–18.

Munkittrick, K.R., Van Der Kraak, G.J., McMaster, M.E., Portt, C.B., Van Den Heuvel, M.R., and Servos, M.R. 1994. Survey of receiving-water environmental impacts associated with discharges from pulp mills. 2. Gonad size, liver size, hepatic EROD activity and plasma sex steroid levels in white sucker. Environ. Toxicol. Chem. **13**: 1089–1101.

Munkittrick, K.R., McMaster, M.E., McCarthy, L.H., Servos, M.R., and Van Der Kraak, G.J. 1998. An overview of recent studies on the potential of pulp-mill effluents to alter reproductive parameters in fish. J. Toxicol. Environ. Health, Part B, **1**: 347–371.

Munkittrick, K.R., McGeachy, S.A., McMaster, M.E., and Courtenay, S.C. 2002*a*. Overview of freshwater fish studies from the pulp and paper environmental effects monitoring program. Water Qual. Res. J. Can. **37**: 49–77.

Munkittrick, K.R., McMaster, M.E., and Courtenay, S.C. (*Editors*). 2002*b*. Theme issue —environmental effects monitoring. Water Qual. Res. J. Can. **37**(**1**).

Nuortila-Jokinen, J., Soderberg, P., and Nystrom, M. 1999. UF and NF pilot scale studies on internal purification of paper mill make-up waters. *In* Proceedings of the 1995 International Environmental Conference, TAPPI, **2**: 847–859.

Orban, J.L. 2000. The extent, causes and environmental risk of logyard/dryland sort run-off in British Columbia. M.Sc. Thesis, Department of Forest Resources Management, University of British Columbia, Vancouver, British Columbia.

Orban, J.L., Kozak, R.A., Sidle, R.C., and Duff, S.J.B. 2002. Assessment of relative environmental risk from logyard run-off in British Columbia. For. Chron. **78**: 146–151.

Paice, M.G., Kovacs, T., Bergeron, J., and O'Connor, B. 2002. Current status of the pure oxygen activated sludge process in Canadian mills. *In* Proceedings of the 2002 International Environmental Conference & Exhibit, Technical Association of the Pulp and Paper Industry. Available from CD-ROM at http://www.tappi.org/ [viewed 28 June 2002].

Parrott, J.L., Jardine, J.J., Blunt, B.R., McCarthy, L.H., McMaster, M.E., Munkittrick, K.R., Wood, C.S., Roberts, J., and Carey, J.H. 2000. Comparing biological responses to mill process changes: a study of steroid concentrations in goldfish exposed to effluent and waste streams from Canadian pulp mills. *In* Proceedings of the 4th International Conference on Environmental Impacts of the Pulp and Paper Industry, 12 June 2000, Helsinki, Finland. *Edited by* M. Ruoppa, J. Paasivirta, K.-J. Lehtinen, and S. Ruonala. Finnish Environment Institute, Helsinki, Finland. pp. 145–151.

Priha, M., and Langi, A. 2000. The impact of nutrient loading of pulp and paper mill effluents on eutrophication of receiving waters. *In* Proceedings of the 4th International Conference on Environmental Impacts of the Pulp and Paper Industry, 12 June 2000, Helsinki, Finland. *Edited by* M. Ruoppa, J. Paasivirta, K.-J. Lehtinen, and S. Ruonala. Finnish Environment Institute, Helsinki, Finland. pp. 165–171.

Ragona, C.S.F., and Hall, E.R. 1998. Parallel operation of ultrafiltration and aerobic membrane bioreactor treatment systems for mechanical newsprint whitewater at 55°C. Water Sci. Technol. **38**: 307–314.

Samis, S.C., Liu, S.D., Wernick, B.G., and Nassichuk, M.D. 1999. Mitigation of fisheries impacts from the use and disposal of wood residue in British Columbia. Fisheries and Oceans Canada, Vancouver, British Columbia. Can. Tech. Rep. Fish. Aquat. Sci. Rep. 2296. 91 p. Available at http://www-heb.pac.dfo-mpo.gc.ca/english/publications/PDF/wood_residue_ backgrounder.pdf [viewed 28 June 2002].

Scroggins, R.P., Miller, J.A., Borgmann, A.I., and Sprague, J.B. 2002. Sublethal toxicity finding by the pulp and paper industry for cycles 1 and 2 of the environmental effects monitoring program. Water Pollut. Res. J. Can. **37**: 21–48.

(SSCBF) Senate Sub-Committee on the Boreal Forest. 1999. Competing realities: the boreal forest at risk. The Senate Sub-Committee on the Boreal Forest of the Standing Senate Committee on Agriculture and Foresty, Ottawa, Ontario. Available at http://www.parl.gc.ca/36/1/ parlbus/commbus/senate/com-e/bore-e/rep-e/rep09jun99-e.htm [viewed 28 June 2002].

Serreqi, A., Gamboa, H., Stark, K., Saddler, J.N., and Breuil, C. 2000. Resin acid markers for total resin acid content of in-mill process lines of a TMP/CTMP pulp mill. Water Res. **34**: 1727–1773.

Simons Ltd. Consulting Engineers and ÅF-IPK. 1995. A technical background information document on pulp and paper mill air emissions. British Columbia Ministry of Environment, Lands and Parks and Environment Canada, Victoria, British Columbia. P.5517A.

Strang, A.T. 1992. Secondary effluent treatment with high-rate activaed sludge at Howe Sound Pulp and Paper Ltd. Pulp & Paper Can. **93**(5): 65–68.

Tardif, O., and Hall, E.R. 1997. Comparison between ultrafiltration and membrane biological reactor treatment for closure of mechanical newsprint mills. *In* Proceedings of the 1997 Environmental Conference & Exhibit, TAPPI, **1**: 315–325.

Tremblay, L., and Van Der Kraak, G. 1999. Comparison between the effects of the phytosterol ß-sitosterol and pulp and paper mill effluents on sexually immature rainbow trout. Environ. Toxicol. Chem. **18**: 329–336.

(USEPA) United States Environmental Protection Agency. 1998. Nutrients fact sheet. EPA-822-F-98-002. Available at http://www.epa.gov./waterscience/standards/nutsi.html.

Virtapohja, J., and Alen, R. 1998. Accelerated degradation of EDTA in an activated sludge plant. Pulp & Paper Can. **99**: 53–56.

Walker, S.L., Hedley, K., and Porter, E. 2002. Pulp and paper environmental effects monitoring in Canada: an overview. Water Qual. Res. J. Can. **37**: 7–19.

Young, S., and Smith, D.W. 2001. Pulp mill effluent induced coagulation and flocculation in receiving waters. Sustainable Forest Management Network, University of Alberta, Edmonton, Alberta. Proj. Rep. 2001-5. Available at http://sfm-1.biology.ualberta.ca/english/pubs/ PDF/PR_2001-05.pdf [viewed 28 June 2002].

Yu, Z., and Mohn, W.W. 1999. Isolation and characterization of thermophilic bacteria capable of degrading dehydroabietic acid. Can. J. Microbiol. **45**: 519.

Yu, Z., and Mohn, W.W. 2002. Bioaugmentation with the resin acid-degrading bacterium *Zooglea resiniphila* DhA-35 to counteract pH stress in an aerated lagoon treating pulp and paper mill effluent. Water Res. **36**: 2793–2801.

Zhang, X., Stebbing, D., Saddler, J.N., and Beatson, R.P. 2000. Enzyme treatments of the dissolved and colloidal substances present in mill white water and the effects on the resulting paper properties. J. Wood Chem. Technol. **20**: 321–335.

Zhang, X., Stebbing, D.W., Soong, G., Saddler, J.N., and Beatson, R.P. 2002. A combined fungal and enzyme treatment system to remove TMP/newsprint mill white water substances. TAPPI J. **1**: 26–32. Available at http://www.tappi.org [viewed 28 June 2002].

Zirnhelt, N.A. 1989. The aquatic environmental impact of leachate from a woodwaste (hogfuel) landfill near Quesnel, B.C. Waste Management Branch, British Columbia Ministry of Environment, Williams Lake, British Columbia. 118 p.

Zuncich, J.L., Venkataraman, B., and Vora, V.M. 1993. Design considerations for steam stripping of kraft mill foul condensates. *In* Proceedings of the 1993 Environmental Conference, TAPPI **1**: 201–207.

Chapter 18

The fate, effects, and mitigation of atmospheric emissions from the forest products industry

D. Grant Allen and Zijin Lu

Introduction

Minimizing the impact of atmospheric emissions from processing operations is an important element in the development of a sustainable forest products industry. Because air pollutants can be transported long distances and can have toxic effects on living organisms, they must be a consideration in protecting the boreal environment and in sustaining healthy northern communities. Although there has been considerable research and technological development on this topic for decades, it is of increasing interest owing to persistent concerns about human health impacts, the integrity of ecosystems, and a recent focus on air pollution and greenhouse gases. While emissions of some sort are inevitable in any industrial process (Axegård 1999), it is important to understand what the emissions are, where they come from, and where they go in order to focus on those emissions that have the most significant impact on environmental and human health. A multi-faceted approach using changes to the process and (or) treatment methods is then required to minimize these discharges. Solutions are required that are of low cost and have a low "ecological footprint" (Box 18.1).

This chapter describes a brief summary of atmospheric emissions related to forest products manufacturing processes. It focuses on pulping processes and wood products conversion processes and does not cover emissions outside the fence of the manufacturing process, such as those related to transportation of materials and forestry operations. The first part of the chapter expands upon Chap. 16 in describing the types of emissions and their sources, as well as methods used to model their fate. The chapter ends with a discussion of conventional and new solutions for preventing pollution through improved manufacturing processes as well as emission controls.

Since the literature on this subject is vast, the reader is directed to the references at the end of the chapter for further details. In particular, a recent text by Springer (2000) has several chapters on atmospheric emissions from the pulp and paper industry. Also,

D.G. Allen[1] **and Z. Lu.** Department of Chemical Engineering and Applied Chemistry and Pulp & Paper Centre, University of Toronto, Toronto, Ontario M5S 3E5, Canada.
[1]Author for correspondence. Telephone: 416-978-8517. e-mail allendg@chem-eng.utoronto.ca
Correct citation: Allen, D.G., and Lu, Z. 2003. The fate, effects, and mitigation of atmospheric emissions from the forest products industry. Chapter 18. *In* Towards Sustainable Management of the Boreal Forest. *Edited by* P.J. Burton, C. Messier, D.W. Smith, and W.L. Adamowicz. NRC Research Press, Ottawa, Ontario, Canada. pp. 713–757.

Box 18.1. Room for improvement:

All manufacturing processes (forest products or otherwise) have some impact on the environment, so the goal is to minimize the ecological footprint for an industry. This can be accomplished through the following:

- minimizing energy requirements;
- optimal use of resources, including their reuse and recycling;
- minimizing emissions, beginning with those having the greatest negative impact;
- avoiding simple transfer of a problem to another medium (e.g., from air to solids or water); and
- understanding and quantifying the impact of the industry from all perspectives (energy, economic/social, and environmental impact).

---

two recent reports by AMEC Forest Industry Consulting, commissioned by the Canadian Council of Ministers of the Environment (CCME), provide extensive reviews (more than 200 pages each) of the process, emissions, and control strategies for the pulp and paper (AMEC 2002*a*) and wood products (AMEC 2002*b*) industries. Although these references focus on specific jurisdictions (U.S.A. and Canada), much of the information has broad application.

Types of atmospheric emissions and their sources

This section summarizes the main air pollutants that are of concern with regards to the forest products industry and why they are issues from an environmental and human health perspective. This is followed by a discussion of the main sources of these emissions within the pulp and paper and wood products sectors.

Underlying all of the discussion of emissions and their sources is the subject of measurement and quantifying emission rates and the measurement of pollutants in ambient air. There is a wide range of methodologies used for determining both the concentration and loading of an emission, which depend on the pollutant being analyzed and its source. Regulatory agencies often specify the method required to quantify an emission and the frequency with which it needs to be monitored. Since a detailed discussion of this topic is outside the scope of this chapter, the interested reader is encouraged to consult other sources on this specialized topic, such as Cheremisinoff and Morresi (1978), Powals et al. (1978), Wight (1994), Winegar and Keith (1993), and Springer (2000).

Air pollutants

The common air pollutants emitted by the forest products industry include particulate matter, sulphur compounds, nitrogen oxides, greenhouse gases, volatile organic compounds (VOCs), carbon monoxide, and hazardous air pollutants (HAPs). These categories and their effects are described below. More extensive details on these specific pollutants and their health effects can be found in several reference texts and on the worldwide web on several sites such as the United States Environmental Protection

Agency (EPA; www.epa.gov), Environment Canada (www.ec.gc.ca), and the Canadian Council of Ministers of the Environment (www.cccme.ca).

Particulate matter

Particulate matter (PM) includes total suspended particulate matter and fine particulate matter. The particulates are generally classified as being "fine" particulates, which includes particles with an aerodynamic diameter less than 2.5 mm ($PM_{2.5}$), and "coarse" particulates, which includes those less than 10 mm (PM_{10}). Primary particulates are solid or liquid matter particles that are emitted directly to the atmosphere. Secondary particulates are particulates that form after they are emitted through condensation or chemical reactions. The precursors for these secondary particulates include nitrogen oxides, sulphur dioxide, acid aerosols (including sulphates, nitrates, and chlorides), hydrocarbons, and ammonia.

Of the two PM categories, those less than 2.5 mm in size ($PM_{2.5}$) cause the greatest harm to human health. These fine particles can be inhaled deep into the lungs, reaching areas where the cells replenish the blood with oxygen. They can cause breathing and respiratory symptoms, irritation, inflammation and damage to the lungs, and premature deaths (Alley et al. 1998).

Although not as serious a threat to human health as $PM_{2.5}$, coarse particles ranging in size from 2.5 to 10 mm in diameter (PM_{10}) are also known to cause adverse health effects. When inhaled, they tend to be deposited in the upper parts of the respiratory system from which they can be eventually expelled back into the throat. Coarse particles generally remain in the form in which they are released into the atmosphere without chemical transformation, eventually settling out under the influence of gravity.

Numerous studies have linked PM to aggravated cardiac and respiratory (heart and lung) diseases such as asthma, bronchitis, and emphysema, and to various forms of heart disease (McCullum and Kindzierski 2001). Children, the elderly, and people with respiratory disorders such as asthma are particularly susceptible to health effects caused by PM. Particulate matter is also an effective delivery mechanism for other toxic air pollutants such as heavy metals and some organic compounds, which attach to particulate matter that floats in the air. These pollutants are then delivered into the lungs, where they can be absorbed into the blood and tissue.

Particulate deposits can be deleterious to the environment, impacting on materials, air visibility, and vegetation. They can cause discoloration or soiling, and acid particles can cause degradation of materials. Particulate matter is also associated with reduced visibility and with poor air quality. Vegetation in the environment can be affected through deposition and reduction of light transmission to the plant, in turn causing a decrease in photosynthesis. Particle composition may cause both direct chemical effects on plants and indirect effects through impacts on soils. Also, particle accumulation on the leaf surface may increase a plant's susceptibility to disease (Environment Canada 2002).

Sulphur compounds (SO_x)

Sulphur oxides (SO_2, SO_3) are of concern from the perspective of secondary particulates and acid rain. They are mainly produced when fuel containing sulphur is oxidized and from sulphite pulp mills. Total reduced sulphur (TRS) is another class of compounds that includes hydrogen sulphide (H_2S), methyl mercaptan (CH_3SH), dimethyl sulphide (CH_3SCH_3), and dimethyl disulphide ($CH_3S_2CH_3$); TRS compounds are an important issue for kraft pulp mills.

The emissions of reduced sulphur compounds (TRS) have long been a concern to the pulp and paper industry because of the nuisance odours they generate, and thereby their overall impact on the perception of the industry. Although some TRS compounds such as hydrogen sulphide are acutely toxic at high enough concentrations, studies have shown (Tatum 1995) that exposure to ambient levels both in a mill and the surrounding community are unlikely to cause adverse health effects. However, the TRS gases have very low odour thresholds in the range of 0.02–40 ppb (Ruth 1986), so even low emissions can lead to significant odour problems. Reducing odour remains an important issue for kraft pulp mills, since it leads to complaints in the community and also creates an unfavourable perception of the industry from an environmental perspective regardless of the degree of control over other emissions.

Nitrogen oxides (NO_x)

These include the two principal oxides of nitrogen, nitric oxide (NO) and nitrogen dioxide (NO_2). They are largely formed from combustion and are of concern with respect to secondary particulates (described previously), as ozone precursors, and for their contribution to acid rain. In some regions, nitrogen (originally released in the form of nitrogen oxides) deposition can cause eutrophication of waters and soils (ESA n.d.).

Volatile organic compounds (VOCs)

These are defined as any compound of carbon, excluding carbon monoxide (CO), carbon dioxide (CO_2), carbonic acid (H_2CO_3), metallic carbides (e.g., CaC_2), carbonates (e.g., $CaCO_3$), and ammonium carbonate ((NH_4)$_2CO_3$), which participates in atmospheric photochemical reactions and has a vapour pressure 0.01 kPa or greater at 25°C. These are of concern because they can photochemically react to form ground-level ozone. Although this covers a range of organic compounds, and emissions rates will vary from process to process, a detailed monitoring study by the National Council for Air and Stream Improvement in the United States shows that methanol (CH_3OH) is the main VOC emitted from many pulp and paper operations (NCASI 1994a–j). A similar series of measurements on the wood products sector by NCASI (NCASI 1999a–f) showed that methanol and formaldehyde (CH_2O) were the main VOCs.

Greenhouse gases

This group of compounds includes carbon dioxide and methane (CH_4) and is often expressed as "carbon dioxide equivalents" from the perspective of global warming. This is a relatively new and significant issue for many industries around the world (Chap. 20). Combustion is a major source of carbon dioxide from the forest products industry, although the sector is relatively well positioned in comparison with other industries in that much of its fuel comes from biomass sources. See the sub-section on

"Biotechnology and biomass utilization", below, and Chap. 20 for further discussion of this topic.

Carbon monoxide

Carbon monoxide is a colourless, odourless gas that is formed when carbon in fuels is not burned completely. Automobile exhaust is a major source; other sources include boilers and other industrial combustion processes. Carbon monoxide reduces oxygen delivery to the body and thereby is a most serious problem for those with cardiovascular disease, although healthy individuals are also affected at high levels.

Hazardous air pollutants (HAPs) and other specific compounds

These include a list of 188 compounds that have been listed in the U.S.A.'s 1990 Clean Air Act Amendments and have been the subject of many debates. Although many of these compounds are emitted by the forest product industry to some extent and many are also VOCs, the main compounds of interest include methanol, formaldehyde, and acetaldehyde (C_2H_4O). Methanol, in particular, is one of the main emissions of concern for many forest products facilities. In Canada, the Canadian Environmental Protection Act (1999) has a list of 52 substances that are designated as toxic (CEPA 2002). Some of these are also generated by the pulp and paper industry, although methanol is not included in the Canadian list.

Acid aerosols, including sulphates, nitrates, and chlorides, are also emissions of concern because of the potential for smog formation and acid rain. Chlorine dioxide (ClO_2), chloroform ($CHCl_3$), and chlorine (Cl_2) are also an issue for some bleach plants owing to toxicity and corrosivity.

Emissions from pulp mills and their sources

Despite the fact that there have been several studies on the emissions from pulp mills, there is still a lot of work that needs to be done on this subject. The manufacturing processes are complex and are continually evolving, and there continue to be advancements in environmental measurement techniques and changing environmental issues. A detailed review of these sources is beyond the scope of this chapter, but there are some excellent references that can provide more quantitative information. In particular, an extensive report recently has been published by the Canadian Council of Ministers of the Environment, covering emissions and processes from the pulp and paper sector (AMEC 2002a). In the United States, the National Council for Air and Stream Improvement published a detailed series of 10 technical papers and measurements on this subject with data from operating mills (NCASI 1994a–j). Some summary information is provided by other reference texts (Davis 2000; Springer 1992, 2000). Chapter 16 (Tables 16.11 and 16.12) also contain some data on total emissions of specific pollutants from the industry.

Although there are some common pollutants across the industry, the nature and sources of emissions from pulp mills depend upon the pulping (e.g., kraft, semi-chemical, sulfite, or mechanical) and bleaching processes employed by the mill, as well as the emission control strategies. Table 18.1 provides a general overview of the main types of

Table 18.1. Sources of main pollutants from pulp mills.

Type of pollutants	Sources
Particulates	Wood preparation (debarking, chipping, storage)
	Thermal Mechanical Pulping (TMP)
	Wood heating and drying processes form condensible particulates
	Steam and power generation from biomass and fuel
	Lime kilns
	Recovery boiler (fine particulates, including sodium fume)
Sulphur compounds: sulphur oxides (SO_x) and total reduced sulphur (TRS)	Sulphur oxides from sulphite pulping processes and semi-chemical processes including cooking vents, acid preparation plant
	Suphur oxides from combustion sources for steam and power generation
	TRS compounds from wastewater treatment facilities and sludge processing
	TRS compounds from kraft pulping; main sources are recovery operations (boiler, lime kiln, tanks, transfer points, etc.)
Nitrogen oxides (NO_x)	From all combustion processes including lime kilns, power boilers, and recovery boiler
Volatile organic compounds (VOCs)	From various processes including cooking processes, recovery process, washer vents, wastewater treatment
	Pulp dryers, paper machines
	Released directly from wood in mechanical pulping operations, wood drying, and general wood handling operations
Greenhouse gases: carbon dioxide (CO_2) and methane (CH_4)	Carbon dioxide from all combustion processes; some methane emissions from wastewater treatment plants and landfills
Carbon monoxide (CO)	From all combustion processes
Hazardous air pollutants (HAPs) and other specific compounds (e.g. methanol, methyl ethyl ketone, acetaldhyde, formaldehyde, chlorine, chlorine dioxide, chloroform	Several operations emitted directly from same sources as VOCs, depending upon process involved
	Bleach plants using chlorine based compounds emit bleaching agent and chloroform

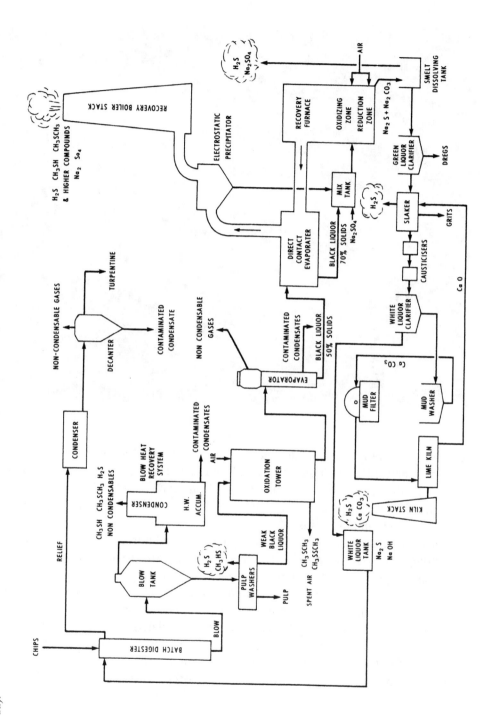

Fig. 18.1. Schematic of the kraft pulping processes showing principal emissions (from EPS 1983; reprinted with the permission of Environment Canada).

Fig 18.2. Photograph of a kraft mill in Windsor, Quebec. Although the control of sulfide-based odours are a persistent challenge for kraft mills, most visible emission plumes consist primarily of water vapour (steam).

Photo courtesy of Domtar Inc.

pollutants from pulp mills and the main sources. As an illustration of the complexity and types of sources, Fig. 18.1 shows a simplified layout of a typical kraft pulp mill (e.g., Fig. 18.2), illustrating some of the points of emission. Tables 18.2–18.4 provide an example of the wide range of emission concentrations and loadings that come from various sources within kraft mills.

Particulates, carbon dioxide, carbon monoxide, and nitrogen oxides are an issue for all pulping methods to some extent because they are associated with combustion processes. Sulphur compound emissions are only an issue for those processes employing sulphur-based compounds in the pulping process; reduced sulphur comes from several sources within the kraft process, while sulphur oxides are of more significant concern for sulphite mills. Hydrocarbon emissions related to VOCs and hazardous air pollutants vary from process to process, although methanol is generally the main compound emitted (NCASI 1994*a–j*). Methane is of relatively minor concern and is a byproduct of anaerobic digestion of wastes in treatment systems and landfills.

Table 18.2. Typical reduced sulfur gas concentrations and emission rates from kraft pulp mill sources (adapted from Springer 1992).

Emission source	Concentration (ppm by volume)				Emission rate, kg sulfur per tonne (Mg) of air-dried pulp			
	H_2S	CH_3SH	CH_3SCH_3	CH_3SSCH_3	H_2S	CH_3SH	CH_3SCH_3	CH_3SSCH_3
Digester, batch:								
Blow gases	0–1000	0–10000	100–45000	10–10000	0–0.1	0–1.0	0–2.5	0–1.0
Relief gases	0–2000	10–5000	100–60000	100–60000	0–0.05	0–0.3	0.05–0.8	0.05–1.0
Digester, continuous	10–300	500–10000	1500–7500	500–3000	0–0.1	0.5 –1.0	0.05–0.5	0.05–0.4
Washer hood vent	0–5	0–5	0–15	0–3	0–0.1	0.05–1.0	0.05–0.5	0.05–0.4
Washer seal tank	0–2	10–50	10–700	1–150	0–0.01	0–0.01	0–0.05	0–0.03
Evaporator hotwell	600–9000	300–3000	500–5000	500–6000	0.05–1.5	0.05–0.8	0.05–1.0	0.05–1.0
BLO tower exhaust	0–10	0–25	10–500	2–95	0–0.01	0–0.1	0–0.4	0–0.3
Recovery furnace (after direct-contact evaporator)	0–1500	0–200	0–100	2–95	0–25	0–2	0–1	0–0.3
Smelt dissolving tank	0–75	0–2	0–4	0–3	0–1	0–0.8	0–0.5	0–0.3
Lime kiln exhaust	0–250	0–100	0–50	0–20	0–0.5	0–0.2	0–0.1	0–0.05
Lime slaker vent	0–20	0–1	0–1	0–1	0–0.01	0–0.01	0–0.01	0–0.01

Table 18.3. Typical concentrations and emission rates for particulate matter from kraft pulp mill sources, after control devices (adapted from Springer 1992).

Emission source	Concentration g/(standard m^3)	Emission rate kg/tonne[a]
Recovery furnace:		
After electrostatic precipitator	0.06–1.1	0.5–12
After venturi evaporator	0.9–2.3	7–25
Lime kiln	0.07–1.1	0.15–2.5
Smelt dissolving tank	0.04–2.3	0.01–0.5

[a]1 tonne = 1 Mg = 1000 kg.

Table 18.4. Typical emission concentrations and rate for SO$_x$ and NO$_x$ from kraft pulp mill concentration sources (adapted from Springer 1992).

Emission source	Concentration (ppm by volume)			Emission rate, kg/t[a]		
	SO$_2$	SO$_3$	NO$_x$ (as NO$_2$)	SO$_2$	SO$_3$	NO$_x$ (as NO$_2$)
Recovery furnace						
No auxiliary fuel	0–1200	0–100	10–70	0–40	0–4	0.7–5
Auxiliary fuel added	0–1500	0–150	50–400	0–50	0–6	1.2–10
Lime kiln exhaust	0–200	—	100–260	0–1.4	—	10–25
Smelt dissolving tank	0–100	—	—	0–0.2	—	—
Power boiler	—	—	161–232	—	—	5.5–11[b]

[a]kilograms per tonne (Mg) of air-dried pulp.

[b]kilograms of NO$_x$ per tonne (Mg) of oil.

Emissions from wood products and their sources

In comparison with the pulp and paper sector, there is much less information on the types and quantities of emissions discharged from the wood products sector. Emissions from this sector vary considerably by facility and with the type of product (sawmills, oriented strandboard [OSB], medium-density fibreboard [MDF], particleboard, plywood, and veneer), the wood species, and the type of equipment and materials used in the process (wood dryers, press vents, pollution control equipment).

A summary of the emissions from various types of sources in the wood products industry is given in Table 18.5. As shown in the table, the major emissions from wood products include HAPs, VOC emissions, and particulates. In addition, operations involving combustion sources (e.g., dryers and boilers) have emissions associated with carbon dioxide, carbon monoxide, nitrogen oxides, and sulphur oxides. As with the pulp and paper industry, a detailed review of this subject is beyond the scope of this chapter, and the reader is referred to other sources of information for details. In particular, an extensive report has been recently published by the Canadian Council of Ministers of

Table 18.5. Sources of pollutants from wood products.

Type of pollutant	Sources
Particulate matter	Fine particulates emitted from all combustion processes (boilers, direct fired kilns), particularly from residual wood incineration
	Coarse particulates from wood preparation and handling processes (cutting, storage, debarking, conveying)
	Condensible particulates (organic and inorganic) from wood drying
Sulphur compounds: sulphur dioxide (SO_2), sulphur trioxide (SO_3), and sulphate (SO_4)	From all combustion processes including power boilers, dryers, and pollution control equipment involving combustion
Nitrogen oxides (NO_x)	From all combustion processes including power boilers, dryers, and pollution control equipment involving combustion
Volatile organic compounds (VOCs)	Combustion processes
Methanol, formaldehyde, etc.	Emissions of organics from wood (e.g., terpenes) due to drying processes, from wood piles and pressing operations (e.g., to make oriented strand board)
	Emissions of additives (resins, binders) for reconstituted wood products during drying and (or) pressing
Hazardous air pollutants (HAPs) and other specific pollutants: several specifics (dependent on process) including methanol, formaldehyde, propoinaldehyde, acetone, acetaldehyde, methylethyl ketone, aromatic compounds	Several operations emitted directly from same sources as VOCs, depending upon process involved; dryers and presses are the common sources for some operations.
Greenhouse gases: Carbon dioxide (CO_2), methane (CH_4)	Combustion sources Methane emitted from waste landfills
Carbon monoxide (CO)	Combustion processes

the Environment and addresses emissions of the lumber and allied wood products industries (AMEC 2002b). The American National Council for Air and Stream Improvement has also conducted a series of studies on the wood products sector (NCASI 1999a–f).

Many of the emissions from the wood products industry come from the wood itself, either as particulates during wood handling or chipping, or as organic compounds emitted during heating/drying processes. Particulate emissions are of primary concern, and many processes have control devices to minimize these emissions. However, similar to the pulp and paper industry, emissions of organic compounds (such as methanol, formaldehyde, and terpenes) are of increasing interest because of the growing focus on VOCs on account of concerns about ozone formation and hazardous atmospheric emissions.

Fig. 18.3. Photograph of a "beehive burner" used for incinerating sawdust and other solid wood waste at the Skeena Cellulose sawmill at Carnaby, British Columbia. The use of open burning and beehive burners for waste disposal is being phased out across Canada, primarily because of health concerns associated with particulate emissions.

Photo courtesy of Phil Burton

The methods used to manage wood residues (e.g., bark, sawdust, shavings, etc.) that are not useful in manufacturing processes also have a significant impact on the emissions from many wood products facilities. Although these residues are sometimes used as soil amendments or landfill, they are often burned and thereby generate atmospheric emissions. There are various types of burners including open burning, beehive or conical burners (Fig. 18.3), silo burners, and curtain burners. Although there are limited data on the emissions from these processes, they are significant sources of particulate emissions as well as other combustion byproducts, particularly if they are not properly operated (AMEC 2002b).

With regards to specific pollutants such as HAPs, methanol and formaldehyde are two of the main new issues for the industry. According to the NCASI studies, over 95% of the total HAP emissions from medium-density fibreboard mills were due to formaldehyde and methanol (NCASI 1999c). Although there was significant variation between process units and mills, the average mill-wide contributions to the total HAP emissions for formaldehyde and methanol were roughly equivalent. Volatile organics can also be emitted in processes where binding agents are added to wood chips.

Modelling the fate of atmospheric discharges

Understanding and determining the fate of atmospheric discharges is an evolving science that is essential to help us establish the needs and priorities for minimizing emissions. Assessing the fate of emissions involves monitoring ambient conditions in the environment as well as using physical and (or) mathematical models of the process. Both monitoring and modelling techniques affect the regulation of air quality, which is generally based on one or a combination of the following three approaches:

*   ambient air contaminant monitoring;
*   mandatory implementation of air contaminant emission controls (e.g., based on best available control technology or maximum achievable control technology); and (or)
*   air contaminant dispersion modelling.

In many industries and regulatory agencies, models are commonly used to assess the fate of atmospheric discharges as a tool to establish permitted discharges and to develop environmental regulations. Despite the uncertainty inherent in modelling, there are several advantages to regulating air quality with the use of air dispersion modelling. Models provide objective, uniform treatment of most sources, and they allow for an objective method to test what the effect of a new process may be and what controls are required before it is built. Models can also be utilized by a given manufacturing process to help identify which sources of emissions are of the greatest concern and therefore point to the most cost-effective mitigation methods.

To provide objective decisions for regulation, agencies such as the U.S. Environmental Protection Agency provide modelling guidance and emission factors for various sources (e.g., USEPA 1995a, 1999). These are updated periodically, and it is best to scan current recommendations and tools available as listed on their website (USEPA 2002).

Atmospheric fate models can be grouped into four categories: physical; statistical/empirical; Gaussian; and numerical. The latter three methods can be considered as computational models, since they all involve mathematical computations often coupled with measurements (meteorological, source emissions, ambient monitoring). Each method has its own advantages and disadvantages, and the choice of model type depends on a number of factors (USEPA 1999), including the following: meteorological and topographic complexities, the level of detail and accuracy needed for the analysis, the technical competence of those undertaking the simulation, resources available, and the detail and accuracy of the data base.

Physical models

Physical models are designed to simulate atmospheric discharges in specific situations using scaled-down versions of the physical topography. The model is scaled down from the actual situation using principles commonly described in fluid mechanics (e.g., Munson et al. 2001), and is generally placed in a wind tunnel, which can then simulate a range of wind directions and speeds. This kind of modelling typically requires specialized expertise and equipment (e.g., wind tunnels, measurement equipment). It can reveal valuable information, particularly for complex terrain and urban situations, regarding the effects of wind on buildings and trees, etc. (e.g., Chen et al. 1995). One

Canadian company based in Guelph, Ontario (see www.rwdi.com), is a leader in this kind of modelling in a wide range of situations.

Although these models have the advantage of providing a "realistic" physical picture, they tend to be expensive and are not able to duplicate the actual situation in the upper atmosphere. For example, the physical models do not take into account temperature effects such as thermal inversion. As a result, the forest products industry does not generally make use of such models and, instead, relies on computational models, particularly those based on a phenomenological approach (Springer et al. 2000). However, for a specific situation where no adequate mathematical or numerical model is available to estimate the emission rate, physical modelling proves valuable. For example, physical modelling can be used to assess the fugitive dust emission rate from chip piles and help develop effective mitigation solutions such as wind barriers to control particulate emissions.

Computational models

Statistical/empirical computational models

Statistical/empirical models utilize monitoring data collected over several years to infer relationships between an emission source and a measurement. They are not commonly used in the forest products industry because of the lack of an extensive database and because they are site-specific. However, statistically based models may have considerable value in predicting how the emissions from a particular source in a mill are related to the operational parameters within the mill. In one recent example, Saviello et al. (1998) described the successful use of a neural statistical predictive emission model to examine correlations among 56 variables and their effect on lime kiln emissions of particulates, TRS, SO_2, NO_x, and CO.

Phenomenological computational models (Gaussian and numerical)

The other main types of mathematical models are Gaussian models and numerical models. Both of these models are considered phenomenological models in that they attempt to model the physical and chemical processes that occur. Depending upon the complexity of the model, they require meteorological data, emission sources and rates, chemical reaction rates, and removal processes. These models differ in their level of detail in terms of data required, technical skill to implement, and the range of processes they describe.

Providing high-quality data on the sources the emissions and their rates is important to the predictive capability of the model. This information can be determined by direct measurements (concentration, flowrate) on the sources, from material balances on the process, and (or) they can be calculated through emission factors. Direct measurement of the specific process is generally the best method for a given mill, although this is costly and requires some time for averaging. Emission factors represent an average emission for pollutants emitted by a particular process for a given quantity of material. Extensive data on emission factors for the forest products industry have been developed by the National Council for Air and Stream Improvement (www.ncasi.org), and some are also summarized in Pinkerton (2000) and in recent reviews for the pulp and paper (AMEC 2002*a*) and forest products (AMEC 2002*b*) industries.

Gaussian models — Gaussian models are commonly used to model the dispersion process; some require extensive computational facilities and others can be solved with a calculator. Although there are many variations of Gaussian models, they share the same major assumption that the neutrally buoyant contaminants comprising each plume will disperse vertically and horizontally from the plume centerline, each according to a Gaussian distribution (Turner 1994). An illustration of the coordinate system used for the Gaussian distribution is illustrated in Fig. 18.4.

The general equation describing the dispersion is as follows (Springer et al. 2000, p. 596):

$$X(x, y, z) = [Q/(2\pi\sigma_y\sigma_z u)] \exp[(-1/2)(y/\sigma_y)^2] \{\exp[(-1/2)((z - H)/\sigma_z)^2] + \exp[(-1/2)((z + H)/\sigma_z)^2]\}$$

where

X = concentration at (x, y, z), g/m^3

Q = contaminant emission rate of the source, g/s

H = effective height of the emission, m

u = wind speed, m/s

σ_y, σ_z = standard deviations of the horizontal and vertical Gaussian contaminant concentration distributions, respectively.

This equation provides the concentration, X, of the air contaminant predicted at coordinates x, y, z, resulting from the constant emission rate, Q, with an effective height of emission, H, for a mean wind speed of u.

The type of Gaussian distribution depends on the source of the emission. For example, the equation above is for an elevated source, but modifications are available for other sources such as ground sources. A more complete description of these models is available in various sources including Springer et al. (2000) and USEPA (2002).

Fig. 18.4. Coordinate system for Gaussian distribution modelling of point-source atmospheric emissions (from Springer 1992).

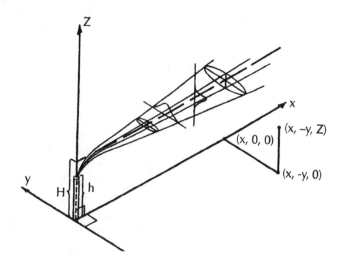

These Gaussian models are considered steady-state models because each source is treated separately, and a constant windspeed and air emission rate is assumed for the period modelled. Also, no deposition or reaction is considered at the surface. These models can predict worst-case contaminant concentrations at each modelled location, and they can be run under various scenarios including different emission rates, sources, meteorology, and topology.

The Industrial Source Complex Short-Term Version 3 (ISCST3) air dispersion model, developed by the U.S. Environmental Protection Agency, is an example of a Gaussian type of model that is useful for short-range transport predictions (Paine and Heinold 1996). The EPA is also developing a state-of-the-art model, AERMOD, in collaboration with the American Meteorological Society, that is designed to replace ISCST3 (Paine and Heinold 1996). Details on the ISCST3 and AERMOD models are available elsewhere (Springer et al. 2000; LES 2002).

A recent study by O'Connor and Ledoux (2002) illustrates how Gaussian models can be applied for understanding and solving emission issues. They report on the successful application of this ICST3 model to estimate the ambient TRS concentrations that can be expected along the mill's property and in the surrounding area based on mill emission data and 5-year-average meteorological data on wind speed and direction. They used the model through a commercially available software package called ISC View produced by Lakes Environmental Software (Toronto, Ontario; see www.weblakes.com). Inputs included TRS information from the various sources along

---

Box 18.2. Case study: application of a Gaussian model to identify sources for odour/TRS reduction.

By modelling various scenarios and looking at maximum (i.e., worst case) values, O'Connor and Ledoux (2002) were able to determine the contributions of different areas of a kraft mill on odours and identify the most cost-effective method to minimize odour. An example of the kind of output from the model, trying various sources, is illustrated in Figs. 18.5a through 18.5c. By contrasting the TRS concentrations from all sources of the mill (Fig. 18.5a) with that from just the production sources (Fig. 18.5b) and the treatment system sources (Fig. 18.5c), the study indicated that the emissions from the treatment system were having the greatest, and most frequent, impact on ambient TRS in the surrounding community. In particular, they were able to identify a small innocuous vent from the sludge dewatering building (mix tank vent) as a major contributor to ambient TRS in the area surrounding the mill.

Fig. 18.5. (*a*) Worst-case estimates of the 1-h ambient TRS concentrations for all 15 of the mill sources (nos. 1–15). The line represents the mill property, and the triangle is the ambient TRS monitoring station. (*b*) Worst-case estimates of the 1-h ambient TRS concentrations for the mill production sources only (nos. 1–11). The line represents the mill property, and the triangle is the ambient TRS monitoring station. (*c*) Worst-case estimates of the 1-h ambient TRS concentrations for the treatment system sources only (nos. 12–15) at the mill. The line represents the mill property, and the triangle is the ambient TRS monitoring station. (Reprinted with permission from O'Connor and Ledoux 2002.)

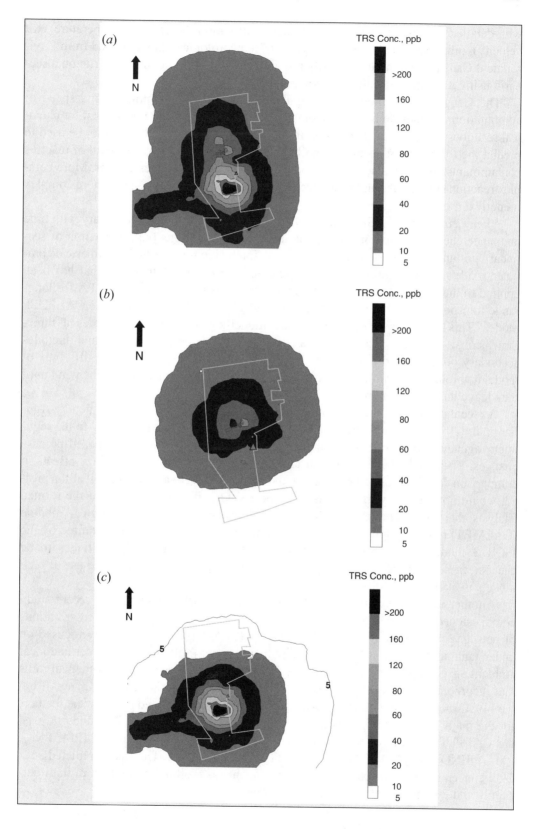

with detailed building dimensions and stack information (height, gas temperature, exit velocity), property boundaries, and 5 years of meteorological data obtained from Environment Canada. The model typically required about 3–4 h to complete a run on a scenario using a Pentium II 300 MHz computer (Box 18.2).

The Gaussian type models are popular because of their simplicity and speed of computation, but they do have some drawbacks. These models do not account for chemical reactions and physical deposition in the atmosphere, and so they cannot be used to predict the formation of secondary pollutants (e.g., particulates) or removal of reactive contaminants, and they also do not produce time-dependent concentrations. More complex reaction chemistries and fluid mechanics lead to the development of more complex numerical models.

Numerical models — Since atmospheric emissions and meteorology vary with time and since chemical reactions are important in many situations (e.g., formation of secondary pollutants and particulates), there is an interest in more complex numerical computer models to study atmospheric dispersion. One model of this type that has been applied to the pulp and paper industry and is recommended by the U.S. EPA for long-range transport modelling (Paine and Heinhold 1996; USEPA 1998) is the CALPUFF model. This model is a non-steady-state modelling system that includes a three-dimensional meteorological model as well as the dispersion model and includes secondary pollutant and particulate matter modelling. Transport in the "puff" type of model is through the simulation of modular "puffs", which are steered by the wind until they leave the area of interest.

A recent example of the application of the CALPUFF model was presented by Aguliar et al. (2000), who utilized the model to help a pulp mill test (and show to the regulatory agency) the impact of an increase in its emissions owing to proposed process changes. They looked specifically at nitrogen deposition and regional haze effects in comparison with a baseline produced by the model and were able to show that the project would not adversely affect these parameters. One item of note is that the refined analysis that they conducted with the model involving the meteorological module (CALMET) required significantly greater computer resources than that required by the ISCST3 model. As computers continue to get faster and memory continues to be cheaper, these resource implications will be of less impact, and so increased use of such models is likely in the future.

Another application of the CALPUFF model was demonstrated in a recent study commissioned by the Forest Products Association of Canada (FPAC) on contributions of pulp and paper mills to ambient air quality (FPAC 2001*b*). This study focused on understanding the relationships between pulp and paper mill emission sources and local ambient air quality. The study modelled the emission profiles of six hypothetical mills that incorporated representative technologies, processes, operations, and layouts from the FPAC member companies and used typical meteorological and geophysical features for the major regions across Canada in which the mills are located. The study utilized the CALPUFF model to predict ambient concentrations of particulates (PM, PM_{10}, $PM_{2.5}$), HNO_3, NO, NO_2, NO_3, SO_3, SO_4, TRS, and methanol. The study predicted peak 24-h concentrations expected for the contaminants as well as the maximum 98th percentile and 50th percentile 24-h concentrations.

A number of interesting findings are reported in the FPAC (2001*b*) report. In particular, the study showed how site-specific issues like dispersion meteorology, geophysical conditions, and emission rates have a very significant impact on the local air quality, and these must be considered when developing emission control reduction strategies. For example, contrary to what one might expect, the study showed that the largest particulate discharges do not typically generate the highest particulate concentrations; the relationship is highly dependent on mill geometry and terrain features as well as stack height, gas temperature, and exit velocity. In addition, the study suggests that Canadian mills are not a major source of emissions that exceed the Canada Wide Standards, but that at least some mills have a non-trivial impact. It is worth noting that the maximum concentrations predicted for all chemical species occurred within 5 km of the mill. Also of importance is that the primary emissions of fine particulate matter are the main source contributor, with secondary particulates being generally a minor contributor; related to this, NO_x is generally more significant in secondary fine PM, but this depends on the mill and meteorology, etc. The study also points out the need for more data on mill emissions and suggests that this kind of modelling study is a valuable tool for understanding major air quality influences around pulp and paper mills.

Approaches to minimizing pollutant formation and to pollution treatment

Minimizing the environmental impact of a manufacturing process while maintaining economic value has several challenges, from problem identification through to finding the best solution. From a regulatory standpoint, one of the challenges is assessing which emissions have the greatest impact on human and ecosystem health and what levels of emission, if any, are acceptable. This is complicated by the uncertainty in predicting the fate and effects of atmospheric emissions, and by uncertainty in the political arena in which regulations are drafted.

To minimize the impact of a mill, one generally looks to three overall approaches that are often used in combination. The first option to consider is to prevent the emission by changing the process or modifying its operation. Where feasible this is often the least expensive approach and has the potential additional advantage of realizing savings through reduced chemical and material consumption. A second option is to consider combustion of the pollutant with subsequent energy recovery, which can sometimes also return some economic fuel value to the process. The third option is to consider cleaning up the emission just prior to discharge using external emission controls or so-called "end-of-pipe" technologies. While emission controls are often the last choice, they are often necessary to meet regulatory limits and are often the most cost- and energy-efficient once the other two options have been exhausted; this is particularly the case when the wastes are too dilute or too complex to consider recovery of materials or energy.

This section summarizes both conventional and emerging approaches to forest industry air pollution control through problem identification and solutions related to in-process changes and external emission controls. A summary of some of the most com-

Table 18.6. Sources and control technologies for pulp and paper operations.

Source	Pollutants	Conventional control
Wood handling processes	Particulates, some VOCs	Reduce particulate emissions through enclosures, wetting of chips, and some emission controls such as cyclone separators
Kraft pulping	TRS, VOCs, and HAPs	Collection and separation of non-condensible gases followed by incineration in combustion operations (e.g., lime kiln)
Washing operations	TRS, VOCs, and HAPs	Efficient washing systems, collection of dilute non-condensible gases from vents followed by incineration
Kraft oxygen delignification	Methanol, carbon monoxide	Generally no controls but operations may be collected with other dilute non-condensible gases for incineration
Bleaching operations	Emissions include (depending on type of bleaching) chlorine, chlorine dioxide, chloroform, methanol, chlorinated organics, HAPs, VOCs	Wet scrubbers using various scrubbing media including chilled water, sodium hydroxide solution, extraction stage filtrate, sodium hydrosulpide (NaHS), and white liquor (from pulping process)
Kraft recovery operations: including black liquour evaporation, recovery boiler, lime kilns and causticizing	TRS, VOCs, HAPs, particulates, combustion products (NO_x, SO_x, CO_2, CO)	A variety of operational strategies are used in various stages, including control and operation of combustion process in the recovery boiler and lime kiln to insure complete combustion and minimal particulate carryover; also, steam stripping of liquids, collection and separation of non-condensible gases followed by incineration in combustion operations (e.g., lime kiln, recovery boiler) Emission control equipment for particulates including cyclones, scrubbers, and electrostatic precipitators; scrubbing for TRS emission control
Mechanical pulping operations	VOCs, HAPs, combustion products (NO_x, SO_x, CO_2, CO)	Condensation from hot streams containing water and VOCs/HAPs; steam stripping of condensates followed by incineration can also be practised
Paper machines and pulp dryers	VOCs and HAPs	Generally, these are minor sources although where coating processes are involved there are VOCs/HAPs released as the solvent; conversion to low solvent formulations or some control technology (e.g., condensation, absorption) could be used.
Combustion	Particulates, combustion products (NO_x, SO_x, CO_2, CO)	Particulate removal devices such as wet scrubbers, cyclones, electrostatic precipitators Optimal operation for maximum energy efficiency Use of biomass fuels to decrease greenhouse gas emissions

Table 18.7. Sources and control technologies for wood products.

Source	Pollutants	Conventional control
Wood handling processes	Particulates, some VOCs	Particulate control through good house-keeping of sawdust and emission controls such as cyclone separators and fabric filters
Wood dryers	VOCs, HAPs, particulates, combustion products (NO_x, SO_x, CO_2, CO)	Particulate control devices include cyclones, wet scrubbers, wet electrostatic precipitators, dry electrostatic precipitators
		VOCs and HAPs with incineration systems including regenerative thermal oxidizers (RTO) and regenerative catalytic oxiders (RCO)
		Operational changes including optimization of drying, reduced drying temperature, recovery of exhaust energy, addition of borax to veneer to reduce VOC emissions
Presses for reconstituted wood products, veneers	VOCs, HAPs, particulates	Wet scrubbers, dry scrubbers, wet electrostatic precipitators, electrified filter bed (EFB), incineration systems
		Many process changes including improved adhesive efficiency, different adhesives, etc.
Combustion	Particulates, combustion products (NO_x, SO_x, CO_2, CO)	Particulate removal devices such as wet scrubbers, cyclones, electrostatic precipitators
		Modernization of combustion equipment, such as replacing open burning and conical/beehive burning of waste residues with efficient systems with emission controls
		Optimal operation for maximum energy efficiency
		Use of biomass fuels to decrease green house gas emissions.
		Co-generation of electricity and steam

mon methods for reducing emissions is provided in Tables 18.6 and 18.7 for the pulp and paper (Table 18.6) and wood products industries (Table 18.7). As with the previous sections, this subject is too vast to cover all solutions in detail, so the interested reader should look to references at the end of the chapter. In particular, a detailed evaluation of many options is available in the recent reports by AMEC (2002a, b) as well as the text by Springer (2000). In addition, there are many process-specific references that address these issues, for example in texts (e.g., Smook 1992) and conference proceedings published by the Technical Association of the Pulp and Paper Industry (TAPPI).

Problem identification

One of the first steps in considering emission control is to identify which parts of the process are responsible for the emissions as well as the relationship between operating conditions and emissions. Modelling of the process through process simulation or statistical/empirical correlation of process conditions with emissions can be very helpful in this regard. In particular, process models can also allow for "what if" scenarios to test the potential impact of a process change on an emission without investing the expensive capital funds required to actually make the change. There are several examples of using process models to identify operating and process scenarios to minimize emissions and also as "soft sensors", which involve using the model to provide online estimates of emissions in the absence of expensive continuous emission monitors (Box 18.3).

Some recent studies by Venkatesh et al. (1997, 2000) successfully utilized a process simulator (General Energy and Mass Balance Simulator, GEMS) to predict VOC and TRS emissions from a mill in order to meet regulatory targets in a cost-effective manner. The study involved first developing a detailed mass and energy balance of the mill, followed by methanol/VOC sampling, and then extending the model to predict the emissions and compare them against measurements. They then used the model to simulate two separate case studies, one involving a brownstock washer upgrade and the other involving a mill condensate management project (Venkatesh et al. 2000). They generally had good agreement between the model predictions and measurements for the mill. The simulation results allowed the mill to identify process modifications that were the most cost-effective. In particular, the authors claim that they identified a 40% reduction in planned spending in the mill condensate management project through this study (Venkatesh et al. 2000). The results of the simulation were also accepted by state regulators for the purpose of providing a permit.

Along the same lines as the work by Venkatesh et al., Gu and co-workers (Gu and Edwards 2001a, b; Gu et al. 2001a, b) recently published a four-part series of papers on

---

Box 18.3. Soft sensors for emissions monitoring.

With advanced knowledge of the manufacturing and emission control processes and with enhanced computational power, there is increasing interest in using computer models that estimate emissions based on process data rather than direct measurement. These 'soft sensors' can be based on fundamental knowledge of the process and (or) developed from extensive data collected for emissions under various operating regimes. These soft sensors have several potential advantages, including:

- They are often much less costly than online sensors from the perspective of both capital and operating costs;
- Reliability is improved since actual monitors can malfunction and give false readings;
- Development and use of a soft sensor requires direct knowledge of the process and thereby helps to establish the optimal way to run a process to minimize emissions.

predicting VOCs from kraft pulp mills. They have developed a model that has fundamental chemical reactions, vapour/liquid equilibria, and mass transfer models coupled with mass and energy balances. This model provides predictions of VOC emissions for a given process operation with minimal sampling. The development of the model and its application is shown in four kraft mill case studies (a continuous digester, two brownstock washing lines, and a pre-evaporator system). They showed good agreement between predictions and actual mill data, and demonstrated the utility of such models for identifying strategies to minimize discharges of methanol. The authors suggest that because of the fundamental basis for the model, it has the potential to predict process and operating changes and also suggest its value as a "soft-sensor" (see Box 18.3) for monitoring atmospheric emissions.

An example of the application of a statistical model was presented by Saviello et al. (1998), examining emissions of particulates, TRS, SO_2, NO_x, and CO from lime kilns. They tested the effect of 56 process variables on measured emissions for use in managing their process operation and to reduce the need for stack testing and continuous emission monitors.

Dispersion models, such as those presented in the previous section, can also help to identify sources of emissions and appropriate technologies. An example in the wood products industry is given in a study by Hardy et al. (1995), who used monitoring coupled with a version of the Industrial Source Complex Short-Term dispersion model (ISCST2) to identify major sources of VOCs and odour from fibreboard and particleboard processing. Based on this analysis they were able to identify the most cost-effective control solution. As discussed previously, O'Connor and Ledoux (2002) used the ISCST3 dispersion model to identify a mix tank vent as a significant source of odour in the surrounding community. Such analysis allows one to focus on the most cost-effective solution rather than simply guessing where the key control points are required.

In-process controls

As illustrated in Tables 18.6 and 18.7, many methods for pollution prevention revolve around decreasing or changing the incoming materials or modifying/optimizing the process operation. This is sometimes challenging because changes in the operation of one unit in continuous integrated processes can have significant impacts on other units in the process, affecting product quality and cost as well as environmental performance. In spite of these challenges, there are a number of methods that the industry has utilized to reduce emissions. A few examples of these in current and future practice are illustrated in this section for pulp bleach plants, kraft recovery operations, overall odour control, emission reduction in the wood products industry, and biotechnology and biomass utilization opportunities.

Reducing chloroform generation in pulp bleaching

One example of minimizing pollutant formation is in the bleaching process, where chloroform and methanol production can be controlled through substitution of alternative bleaching chemicals. For example, utilizing ozone as a bleaching agent in place of chlorine-based chemicals has the potential to eliminate chloroform generation; ozone

Fig. 18.6. Chloroform vs. chlorine factor resulting from the bleaching process of a pulp (from Crawford et al. 1991). Chlorine factor is defined by dividing the percent of molecular chlorine (Cl_2) applied by the number of the brownstock number (a measure of the amount of lignin in the pulp prior to bleaching). It is a measure of the amount of molecular chlorine applied to the bleach pulp per unit quantity of lignin. As chlorine dioxide is substituted for molecular chlorine in a bleaching process, the chlorine factor decreases for a given pulp.

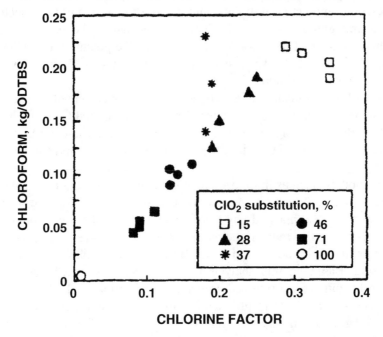

Reprinted with permission from TAPPI

may be present in the vent gases; however, they also must pass through an ozone destructor so no significant ozone emissions are expected. Crawford et al. (1991) have shown that substituting chlorine dioxide for chlorine in bleaching (Fig. 18.6) can decrease the amount of chloroform and chloroform precursors produced. Methanol is also formed in bleach plants when chlorine and chlorine dioxide react with lignin, and methanol formation decreases when chlorine dioxide is substituted for chlorine (Crawford and Jain 1994).

The bleaching process also illustrates the complexity of environmental control and how reducing one emission in one location is not necessarily the best solution. For example, the same study by Crawford and Jain (1994) also showed that using oxygen delignification ahead of the bleaching sequence for kraft pulps produces less methanol in bleach plants. However, as pointed out by Jain (1996), it would be incorrect to assume that oxygen delignification will necessarily reduce methanol emissions overall, since oxygen delignification can lead to higher methanol emissions in the oxygen delignification reactor. As another example, utilizing ozone as a bleaching agent instead of chlorine will reduce the formation of chloroform, but at the expense of pulp yield and energy efficiency, with additional energy consumption leading to higher costs and increased greenhouse gas emissions.

Minimizing pollutant formation in kraft recovery operations

Many of the in-process steps to minimize atmospheric emissions focus around the recovery operations and the optimal operation of the recovery boiler, including combustion optimization. Much of this has been directed at odour reduction using traditional strategies of black liquor oxidation, control of sulphur in the recovery cycle, proper operation of the recovery furnace, and incineration and collection of non-condensible gases from the cooking and evaporation processes (Smook 1982, 1992). Details of the operation of a recovery boiler are outside the scope of this chapter, but some general concepts can be described. Further details on recovery boiler operation and control are given in Adams et al. (1997).

Paying attention to the level of sulphur in the black liquor and insuring complete combustion are two ways of minimizing the emissions from the recovery boiler. Minimizing the sulphur concentration in the recovery cycle and minimizing addition of salt-cake (NA_2SO_4) both have a direct impact on sulphur emissions; in one example, a 10% decrease in liquor sulphur content led to a 50% decrease in gaseous sulphur emissions (Smook 1973). Also, there are several studies that have shown that paying attention to the proper operation of the recovery boiler (e.g., not operating above capacity) and providing appropriate conditions for complete oxidation of sulphur (turbulence, oxygen concentration, etc.) will minimize the release of TRS in the form of hydrogen sulphide.

The recovery boiler, lime kiln, and power boilers are also frequently used to burn pollutants collected from other parts of the mill. A good example of this is shown in Fig. 18.7, which illustrates a modern system for collecting and disposing non-condensible gases (NCG) in a kraft pulp mill. These gases contain various organics (e.g., VOCs, HAPs such as methanol) and also TRS gases. The concentrated non-condensible gases (CNCG) are collected and burned for use as an energy source in processes such as the lime kiln, recovery boiler, power boilers, or a separate incinerator. The more dilute non-condensible gases (DNCG) also can be collected and burned in the recovery boiler, power boiler, or separate incinerator (Tom Burgess,[2] personal communication, 2002). This is commonly practiced for non-condensible gases from digester relief, digester blow, and liquor evaporation (containing both TRS and volatile organic compounds). This works well when the gases can be conveniently collected but may not be appropriate for more distributed sources of these emissions where collection and movement is a major cost.

With the increased regulations targeting methanol emissions, particularly in the U.S.A., several mills are implementing condensate segregation and steam stripping techniques within the recovery process. Some mills have used these techniques for years, and they are likely to increase in application because of increased regulations for methanol and TRS. As the name implies, condensate segregation essentially involves segregating out condensates that are relatively clean that can be reused or discharged from those that are concentrated and need treatment. Subsequent steam stripping of the condensates that are concentrated in methanol and TRS allows for recovery of the fuel value at minimal cost. A field study example of optimal segregation and stripping is given by Johnson and Deihl (1995). The work reported by Venkatesh et al. (2000) and

---

[2]Tom Burgess Consulting Ltd., Parksville, British Columbia.

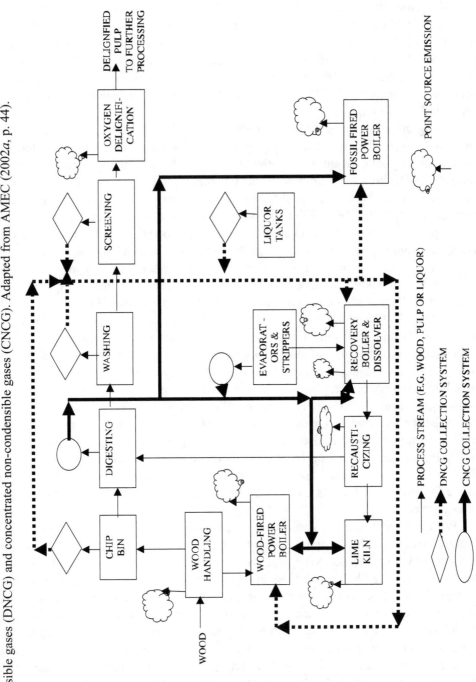

Fig. 18.7. An example of chemical pulp mill air emission sources and controls for non-condensible gases (NCG) with collection of dilute non-condensible gases (DNCG) and concentrated non-condensible gases (CNCG). Adapted from AMEC (2002*a*, p. 44).

Gu and Edwards (2001a) also describes various operational strategies for dealing with methanol in condensates.

There are several studies that have examined approaches for reducing NO_x emissions from recovery boilers through altering operating strategies so that the nitrogen leaves as N_2 or ammonia (NH_3). A study by Janka et al. (1998) compared two in-process techniques that had some success in field studies. One method is known as air staging based on a patented process (Olausson 1995), which involves air addition strategies at various points in the boiler so as to maintain lower temperatures and reducing conditions to prevent NO_x formation. This led to up to a 30% decrease in NO_x emissions (Janka 1998), and the authors suggested this as a cost-effective method for older units. A second method involved the addition of a chemical such as ammonia or urea ((NH_2)$_2CO$) that reacts with NO and (or) NO_2 to form N_2 (Lövblad et al. 1991) and achieved up to 60% reduction in NO_x emissions. Although this second method is a more costly solution, the study suggested that air staging is limited in the degree to which it can improve NO_x emissions, particularly for a modern boiler. They suggested that this second method, perhaps coupled with emission controls for NO_x, is necessary to get further reductions in emissions.

Overall strategies for reducing kraft mill odour emissions through pollution prevention

Reductions in odour emissions generally require a multi-faceted approach involving modern equipment, attention to various emission sources including tank vents and treatment systems, and good process control systems. Tarpey (1995) gives a good example of odour reduction strategies for a relatively modern kraft mill operating in Alberta. The mill achieved significant reductions in odour through nine different process operation and control strategies ranging from water reuse, collection of black liquor storage tank vent gases, retrofits on lime kilns, and improved process control. The mill reported an 80% decrease in annual odour complaints over an approximately 3-year period. In another odour reduction project, a mill in Ontario reported significant economic return through various mill modernizations (Ramsay et al. 2001).

Overall, significant odour reductions have been achieved in modern kraft pulp mills, although complete elimination of odour has not been achieved because of the very low thresholds for the odour of TRS compounds (Iisa 1996). One avenue for future progress in resolving this problem may be to replace the kraft process with sulphur-free processes such as the soda-anthroquinone (Soda-AQ) process (e.g., Parthasarathy et al. 1995), although this seems unlikely in the near future given the various quality and cost advantages of the kraft process and its widespread application in the industry.

Pollution prevention in wood products manufacturing

There is also a wide range of opportunities, some currently in practice, for pollution prevention in the wood products industry, which vary depending on the product and pollutant. Many of these are indicated in Table 18.7. Reviews of some of the opportunities are provided by the United States Environmental Protection Agency notebook on the Lumber and Wood Products industry (USEPA 1995b), and an extensive review published by the Canadian Council of Ministers of the Environment (AMEC 2002b). Also, Industry Canada has recently summarized opportunities for technical innovations in the lumber (Forintek 2003) and panelboard industries (Industry Canada and Forintek 1998) in their "technology roadmap" series. Similar to the pulp and paper industry, many of the first steps in reducing pollutant discharges deal with optimization of the process involved in order to maximize energy efficiency and (or) optimizing the use of existing materials, thereby minimizing wastes.

As illustrated in Table 18.7, many of the approaches associated with air pollutant minimization in the wood products sector deal with minimizing emissions of particulates, organic compounds (VOCs and HAPs), and combustion products. Minimizing many of these pollutants requires emission controls (discussed in the next section), but there are also opportunities for improving the manufacturing and disposal processes.

Most of the wood product manufacturing processes of concern involve combustion processes for energy (e.g., drying, steam, etc.). Combustion product emissions can be reduced in the same ways that they are in other industries. Some of the commonly applied techniques are through improving the control and optimization of combustion (e.g., temperature, turbulence, air system optimization) to maximize energy recovery and minimize the formation of pollutants (CO, NO_x, VOCs).

There are also significant opportunities to minimize emissions through improved management of residual wood wastes (e.g., bark, sawdust, etc.). The options available depend on several factors including the process and its location (see Chap. 19). One common first step is to reduce the residues by optimizing use of the wood through advanced process control (e.g., laser sensors) and optimal cutting techniques. Residues can also be reduced by selling or trading waste fibre to another facility that can make use of it (e.g., pulp mill, particleboard plant; Fig. 19.7). Even when using these options, there will still be wood residues that need to be dealt with. One common method is to burn the residue in a combustion unit for energy recovery (e.g., steam generation or cogeneration of steam and electricity) with appropriate emission controls. The most effective methods of doing this may be to transport the residue to a facility (e.g., a pulp mill) already equipped with an appropriate combustion unit or to build a centralized combustion facility for use by several wood products operations. Other options for residues that reduce atmospheric emissions include their use as soil enhancers, agricultural bedding, landscaping, landfilling, or as feedstocks to biofuel production (AMEC 2002b). Straight burning for volume reduction without energy recovery is also an option that is practiced when other options are too costly. Although being phased out, several facilities in Canada still use technologies such as "beehive burners" (Fig. 18.3) which can generate a considerable output of particulates and other emissions. Efficient operation or the use of other burning technologies (e.g., silo burners) can allow for better control of these emissions (AMEC 2002b).

In some wood products manufacturing processes using adhesives or other additives, there may be a need to minimize emissions of organic compounds (VOCs and HAPs). These processes include the manufacturing of veneer, plywood, particleboard, medium-density fibreboard (MDF), and oriented strandboard (OSB). There are prospects to select and (or) develop alternative adhesives and other additives with solvents that produce lower emissions. Also, more efficient application methods can also reduce VOC and HAPs discharges. As an example, research is presently under way to develop digester systems that, through the use of steam and heat treatment, have the potential to produce MDF panels with little or no adhesive addition (Industry Canada and Forintek 1998). Of course, as with any technology, the costs of the additive and its influence on product quality are critical to every application.

Drying is a common unit operation for many wood product manufacturing processes where there is potential for reducing emission of particulates and some organics (VOCs, HAPs), either through the use of new technology or optimized operation of existing technology. In general, dryers that operate at low temperatures and are energy efficient through advanced control, optimization, and (or) reuse of dryer exhaust can result in lower emissions of organics and condensible particulates.

Biotechnology and biomass utilization

New pulping and bleaching methods associated with increased use of biotechnology may hold promise of future environmental benefits for the forest products industry. Eriksson (2000) gives a recent overview of biotechnology in the pulp and paper industry. In terms of atmospheric emissions, several studies have shown how utilizing fungi and (or) enzymes as a pretreatment to wood chips can lead to reduced energy requirements in mechanical and chemical pulping (e.g., Scott and Lentz 1995; Messner and Srebotnik 1994; Pere et al. 2000), which will have an overall positive impact on environmental performance. These and other biologically based technologies show great promise for future minimum impact mills as more knowledge is gained and costs are decreased.

There are considerable opportunities that are being recognized in the forest products industry to reduce greenhouse emissions through burning more biomass as fuel and improved energy management. Carbon dioxide emitted owing to the combustion of biomass fuel is considered to be neutral with respect to greenhouse gas emissions because it utilizes fixed carbon that is sequestered during growth (AMEC 2002b; FPAC 2001a; see Chap. 20). This means that replacing conventional fossil fuels with biomass (e.g., wood waste, bark, etc.) can substantially reduce the industry's greenhouse gas emissions. Of course, the degree to which a given operation can do this depends upon the availability of biomass as fuel and the appropriate combustion process to burn the biomass and minimize particulate emissions. As an example of the potential for this, a recent report by the Forest Products Association of Canada (FPAC 2001a) states that Canada's pulp and paper industry decreased its total greenhouse gas emissions by 19% over 1990 levels despite a 27% increase in production through the use of biofuels and investing in co-generation projects; cogeneration is a process that involves generating (e.g. from wood residues or other sources) electricity and as well as steam for use in the process. In 1999, about 54% of the industry's fuel came from biomass mainly in the form of wood wastes like bark, much of which used to be land-filled (FPAC 2001a).

Emission controls

As illustrated in Tables 18.6 and 18.7, although pollution prevention through changes in the manufacturing process is an important first step in minimizing the impact of atmospheric emissions, pollution control equipment is also necessary. There are many options to treat various types of emissions in the forest products industry, and there is no one single technology for any particular pollutant. The choice of technology is dependent upon a number of parameters including the nature of the pollutant (particulates or chemicals), the treatment requirements, the gas stream temperature, gas stream composition and flowrate, the simplicity and reliability of the technology and, of course, the capital and operating costs. This section provides a brief description of the most common methods used by the forest products industry and introduces a few new technologies. Readers interested in a more detailed description of the technologies can look to textbooks and reports in the field (Smook 1982, 1992; Springer 2000). Also, annual conferences such as the TAPPI Environmental Conference (see http://www.tappi.org for announcements, agendas, and papers) describe a wide range of current and new technologies and their application in the field.

Common emission control technologies

The common emission control technologies used in the forest products industry include:
- scrubbers (wet and dry) for the removal of particulates and chemical pollutants;
- inertial separators for particulates; and
- fabric filters and electrostatic precipitators for particulates.

Often an emission control system requires several of these technologies in series. For example, an inertial separator can provide a cost-effective separation of the large particles, followed by an electrostatic precipitator to handle the fine particulate matter. These technologies are each briefly described below.

Inertial separator — Inertial separators are widely used for the collection of medium-sized and coarse particles. They are the simplest type of particulate removal device and are routinely used for coarse particulates, often upstream of other control devices designed for fine particulates (e.g., less than 2.5 mm). The principal of operation is illustrated in Fig. 18.8 (Davis 2000), which depicts a common cyclone separator. The inlet gas containing particulates is forced into a vortex, and the heavier particles (having higher inertia) impact on the outer wall and are collected at the outlet while the clean gas moves up through the centre. Multiple cyclones are used for higher gas flowrates.

Wet and dry scrubbers — Wet scrubbers are very commonly employed in many industries because they can be used to control both particulate emissions and chemical contaminants. While there are many different designs, the principle of operation involves contacting the gas with a liquid (typically water or an aqueous solution; Fig. 18.9) usually flowing in opposite directions (i.e., counter-current). The water "scrubs" the gas through impacting a particle with liquid drops or by contacting the gas with a flowing film; chemicals that are soluble in the scrubbing solution are absorbed from the gas phase by the solution.

There are many types of scrubber designs (e.g., venturi, spray column, impingement, packed tower), each with their own advantages and disadvantages. One key distinction can be made between packed tower scrubbers and the other scrubbers. As the name implies, a packed tower scrubber is packed with high-surface-area packing material (made from a range of materials including ceramics, plastic, or composite materi-

Fig. 18.8. Operation of a cyclone separator, a common inertial separator for particulate removal (from Robinson and Spaite 1967).

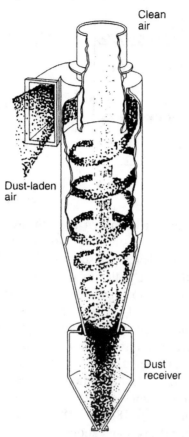

Clean
air

Dust-laden
air

Dust
receiver

Reprinted with permission from Elsevier Science Ltd.

als) to allow for good contact with the gas and liquid. Their high gas-to-liquid contact area makes packed tower scrubbers good for cleaning chemical pollutants in gases, provided they don't have particulates which tend to plug the packing and thereby require frequent washing. The other scrubbers don't have packing and are suited to situations where particulates are involved. Also, spray columns are now replacing some packed columns in situations where the scrubbing solution is alkaline and where scale buildup on the packing can be a problem (Tom Burgess,[2] personal communication, 2002).

The choice of liquid to contact the gas is made based on what pollutants are to be removed and cost. In general, water is the most cost-effective liquid if particulate removal and (or) sulphur dioxide are the targets, while alkaline solutions are effective for treating TRS, sulphur dioxide, and chlorine dioxide. Scrubbers are commonly used to treat bleach plant emissions using chilled water, sodium hydroxide, and bleach plant extraction stage filtrate (Jain 1996). Bleach plant scrubbers require some form of reducing agent (e.g., sodium sulphide from white liquor in the kraft cooking process) in the scrubbing solution if they are to remove chlorine dioxide so that it forms molecular chlorine which can then be absorbed by alkaline solution. The use of in-process streams

Fig. 18.9. Schematic diagram of the operation of a wet scrubber.

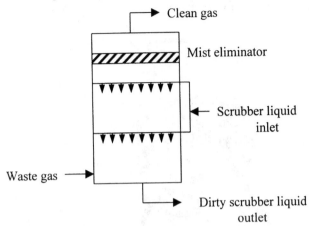

like bleach plant filtrates is attractive in that they are available at low cost and can decrease overall emissions. NO_x emissions can also be reduced using scrubbers. One recent study reported on a newly developed scrubber (Patrikainen et al. 1997) that oxidized NO to water soluble NO_3 through the use of aqueous solutions of sodium hydroxide and sodium sulfite or chlorine dioxide (Janka et al. 1998).

Like any emission control technology, wet scrubbers have several advantages and disadvantages which make them well suited to certain applications and not others. Their advantages include low space requirements, ability to handle high-temperature gases, ability to treat particulates and chemicals in one step, and relatively low capital cost. Some disadvantages include the facts that sometimes water disposal is a problem, they can generate droplets in their emissions, and they tend to have high pressure drops (high energy costs). As outlined in Tables 18.6 and 18.7, their advantages frequently make them the method of choice for many applications.

The dry scrubber is a particulate removal device that uses a moving bed of granular material to collect the particulates. The bed medium vibrates to the bottom of the unit where it shakes off the particulates to a storage bin and then the granular material is returned to the scrubber. This process is not widely used, although it has potential application for press vents and off-gases from wood-fired boilers.

Fabric filters — Fabric filters are very efficient for the removal of particulates down to 0.1 mm for gas temperatures below 300°C. The gas passes through a fabric (typically in the form of a bag), and the particles that can't pass through the filter are removed. They have been applied to control particulates from power and wood-fired boilers in the forest products industry (Springer 2000). Fabric filters are not as frequently applied as wet scrubbers in the forest products industry, although they are still used in some applications. They have a number of advantages including high collection efficiencies, ease of waste disposal (for use in the process, incineration, or landfill) and simple operation. However, they are limited to treating lower gas temperatures and have relatively high maintenance requirements.

Electrostatic precipitators — Electrostatic precipitators (ESPs) are commonly applied in the pulp and paper industry for the control of particulates from recovery boil-

Fig. 18.10. Basic elements of an electrostatic precipitator (from Iisa 1996).

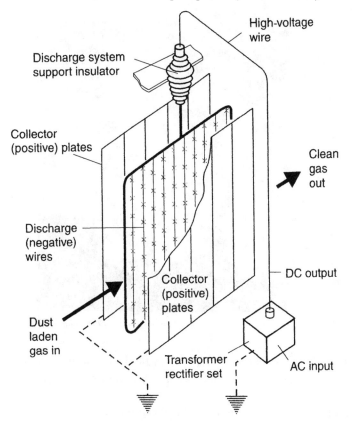

Reprinted with permission from TAPPI

ers, power boilers, and also on dryers and boilers in the wood products industry. The basic principle of operation (Fig. 18.10) is that the ESP creates a negatively charged field through which the particles pass. Positively charged plates then remove the charged particles. A photograph of an ESP is shown in Fig. 18.11. Although this is an expensive technology, it is widely used because it is effective at removing dust particles as small as 0.1 mm at >99% efficiency.

Electrostatic precipitators have a number of advantages that have made them a popular choice as part of the particulate control toolbox in the pulp and paper and wood products industry. They are commonly used for recovery boilers, lime kilns, and power boilers. They have excellent efficiency for particulates and also produce a dry product that can be returned to the process.

The wet electrostatic precipitator (WESP) is a variation on the design of the conventional ESP. Although it is not widely used in the pulp and paper industry, it is used in some processes in the wood products industry. The WESP uses the same charging principle as the ESP, but it also saturates the inlet gas stream with water and has a falling

Fig. 18.11. Photograph of an electrostatic precipitator installed at a wood products manufacturing facility.

Photo courtesy of PPC Industries

film of water that clears the particulates. It is particularly useful for controlling fine particulates and can also scrub some water-soluble pollutants out of the gas stream.

New and emerging approaches to air pollution control

There are a variety of new and emerging technologies for reducing emissions that are at various stages of development from research through to implementation. The driver for many of the new technologies comes from increased regulatory pressure on pollutant discharges, and the use of new knowledge in developing more cost-effective solutions to emission control. In particular, with the increased focus on emissions of hydrocarbons (VOCs, methanol, formaldehyde), many of the new technologies are directed towards these compounds. Odour control also continues to be an area where new technologies are being developed. Many of the advancements are modifications to existing technologies (e.g., scrubbers, adsorption, and electrostatic precipitators), though novel approaches are constantly being developed too.

This section highlights a few of the new technologies specifically for the control of odour and VOCs. As mentioned previously, the interested reader can find out about many new technologies from presentations made at conferences (TAPPI Environmental Conference, Air and Waste Management), trade show displays, and in research journals (e.g., TAPPI Journal, Environmental Science and Technology, Journal of the Air and Waste Management Association).

Biological wastewater treatment — Methanol emissions can be reduce by treating the methanol in the water phase before it escapes to the air. One way to do this is to send the collected condensates directly to an existing or retrofitted biological treatment system for destruction. Gregory (2000) recently discussed this use of wastewater treatment system capacity as the most cost-effective solution for their kraft mill. Another innovative biological option was presented by Wiseman et al. (2000), in which anaerobic treatment was used to treat the methanol in condensates in a kraft mill; the methane generated from methanol degradation was used as a fuel. In comparison with aerobic treatment, this approach has the advantage of low biosludge production along with an economic payback through energy yields.

Biological gas cleaning — One emerging technology for treating VOCs and TRS compounds utilizes microorganisms to degrade wastes in air in the same way the biological wastewater treatment systems degrade wastes in water. The overall term for this technology is biological gas cleaning, which includes biofiltration, biotrickling filters, and bioscrubbers, although the term biofiltration is frequently used to refer to all three types of technology.

In most applications, the gas containing the pollutants (e.g., VOC, TRS) is passed through a packed bed bioreactor (Fig. 18.12). Microorganisms reside in a biofilm on the packing, and the pollutant transfers to the biofilm where it is degraded by specialized microorganisms; VOCs are oxidized to carbon dioxide and water while reduced sulphur compounds degrade to sulphate. Some cellular biomass is also produced. In a conventional biofilter, the packing material consists of a variety of types of organic media (e.g., compost, wood chips, soil) containing fertilizer (nitrogen, phosphorous) which eventually degrades over time. In the biotrickling filter, the medium is usually inert packing material similar to that employed in packed scrubbers (e.g., plastic, ceramics), and a solution of nutrients is recirculated through the bed. A good reference text on this technology has been written by Devinney et al. (1999).

In comparison with other technologies (e.g., regenerative thermal oxidizers, regenerative catalytic oxidizers, described below), biofilters offer potential advantages for treating low-concentration emissions with low operating costs, low energy consumption and low emissions of secondary pollutants (e.g., NO_x, SO_x). Although they are widely used for odour control in a number of industries (e.g., municipal wastewater treatment, food processing), they have the disadvantage of requiring a large land area. Also, in

Fig. 18.12. Schematic diagram of the operation of a biofilter.

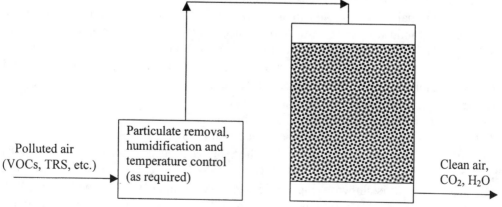

comparison with combustion, they are not as effective at removing hydrophobic pollutants (such as terpenes, dimethyl sulphide, and dimethyl disulphide), and it is as yet an unproven technology in the forest products industry. In addition, biofilters are generally operated at temperatures below 40°C, so many emissions from the forest products industry require prior cooling (which is costly and generates wastewater) before they can be treated using this method. As also noted for the biological treatment of aqueous effluents (Chap. 17), much research in this area has focused on the identification of microorganisms and operating procedures that are not so sensitive to high temperatures.

Several lab- and field-scale studies have been conducted on biofilters and their application to the treatment of VOCs in the forest products industries, some of which have been supported by the Sustainable Forest Management Network (SFMN). Allen and co-workers at the University of Toronto (Mohseni et al. 1998; Dirk-Faitakis and Allen 2000, 2003; Mohseni and Allen 2000; Morgan et al. 2001) have shown that biofilters packed with wood-based materials can effectively treat methanol and pinene mixtures in a single stream and have developed a model for treatment under steady-state and unsteady conditions (Dirk-Faitakis and Allen 2000). As part of a SFMN project, Dirk-Faitakis and Allen (2000, 2003) also showed that the catalytic activity of the biofilter was not influenced by short (<1 day) periods of inactivity. In addition, a recent study by Kong et al. (2001) has shown that methanol can be biologically treated at temperatures up to 70°C, and pinene can be treated at temperatures up to 60°C; this opens up the potential to greatly reduce the costs associated with cooling the gas by directly treating hot gas emissions, i.e., without cooling them first.

Laboratory research on using biofilters for TRS and methanol reduction has also been undertaken by a number of research groups. Coleman and co-workers (Coleman and Dombroski 1995; Budwill and Coleman 1999) have done lab and field research, some of which was supported by the SFMN, to demonstrate biofiltration for the treatment of mixtures of reduced sulphur compounds typically emitted from the kraft pulping process. In one Network-supported study, Budwill and Coleman (1999) examined the potential to use surfactants to enhance the treatment of dimethyl sulphide with limited success. Allen and Ellis (2000) have recently reported on a lab-scale study on the use of a trickling biofilter for the treatment of simulated kraft brownstock emissions. They were able to demonstrate treatment of 80–90% of the TRS gases and 95% methanol removal. Preliminary cost estimates indicate biofilters may be competitive with conventional collection and incineration of dilute non-condensible gases in the pulp and paper industry, but scale-up and engineering issues need to be addressed before moving to full-scale implementation (Allen and Ellis 2000). Recent kinetic studies (Lu and Allen 2001; Sologar et al. 2002, 2003) have shown that mixtures of methanol and hydrogen sulphide can be treated successfully and that the presence of either does not have a significant impact on treatment of the other.

There are a few examples of full-scale biofilters in operation for the treatment of VOCs in the wood products industries. A recent study by the Canadian Council of Ministers of the Environment reports on biofilters being used to control VOC emissions from presses in the particleboard and OSB processes (AMEC 2002b); an example of a field-scale biofilter is shown in Fig. 18.13. However, it appears that regenerative thermal oxidizers (see next sub-section) are currently the more popular technology for treating low concentrations of VOCs. They have the advantage of having a more widespread

Fig. 18.13. Photograph of a biofilter installation, treating emissions from a press vent at a wood products manufacturing facility.

Photo courtesy of Norbord Inc.

application history and hence present less risk for a particular application. However, the substantial magnitude of potential energy/cost savings associated with biofilters and their lack of additional combustion byproducts (NO_x, SO_x) suggest that further development of this technology for application to the forest products industry will continue.

Regenerative thermal oxidation and catalytic oxidation — Regenerative thermal oxidation (RTO) units and regenerative catalytic oxidation (RCO) units are incineration-based technologies for relatively low-concentration contaminants. A schematic of the process is shown in Fig. 18.14. These units consist of at least two (usually more) separate beds packed with ceramic material and the gas can be directed with valves to flow through a selected bed. The inlet gas passes through a bed of hot packing material that heats up the gas. The hot gas then burns on its own if its concentration is high enough, or additional fuel is added if necessary to sustain combustion. The combusted flue gas then enters a "cool" packed bed, and the heat is transferred from the gas to the bed. Over time the hot bed cools and the cooler bed heats up (i.e., is regenerated), at which point the gas flow is reversed through the beds so that the inlet gas enters the hotter bed. In some designs, a third bed allows for some flexibility in operation, including the option to have one of the beds be "baked out" to destroy any organic deposits that build up on the bed. Many new RTOs use a much larger number of beds (e.g., 6–10 beds) with a couple of beds being heated, cooled, and purged at the any time. Some new designs include having pie-shaped beds contained in one large cylinder which are fed by a common rotary valve; these designs are much less expensive, more reliable, and more efficient (Tom Burgess,[2] personal communication, 2002).

Fig. 18.14. Schematic illustration of the operation of a regenerative thermal oxidizer (RTO) or regenerative catalytic oxidizer (RCO). Solid lines with arrows indicate gas flow when Bed A is heating the gas and Bed B is being heated (regenerated). By manipulating valves, gas flow can be reversed to regenerate Bed A and use Bed B to heat the gas. Beds can be made of various materials, and often several beds are used. For an RCO, the beds contain a catalyst to reduce the temperature required for combustion.

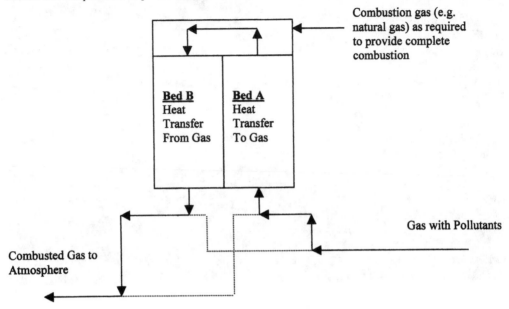

The operation of the RCO is very similar to the RTO, with the main difference being that the packing has a catalyst that allows the combustion process to take place at lower operating temperatures. An RCO might operate at 425°C while an RTO would operate at over 800°C (Connell 2001). The bed medium is more expensive than the RTO, but the lower operating temperatures reduce operating costs through lower energy consumption.

There are several field-scale applications of RTOs and RCOs in the forest products industry for the control of VOCs, particularly on dryers. A recent survey of control devices on wood products dryers by the U.S. E.P.A., reported by Hardy and Hohl (2001), shows that there is a wide range of technology combinations used to remove VOCs and (or) particulates. Plugging of these systems owing to particulates is an issue and suggests that a particulate control device (e.g., cyclone separator, wet electrostatic precipitator) may be required upstream, although the trade-off in cost of particulate removal vs. downtime for the RTO is not clear (Lawrence and Vasquez 2001). The choice of particulate removal device is also the subject of debate (Raemhild and Jaasund 2001). For example, Hardy and Hohl (2001) reported on the successful use of a high-efficiency multi-cyclone ahead of a valveless RTO to deal with VOCs and particulates from a dryer in a particleboard facility, although many mills have used wet ESPs.

RTOs have also been examined for application to kraft pulp mills for the control of TRS and VOCs. Gravel and Drouin (2001) reported on 4 years of successful operation of an RTO to treat non-condensible gases at a pulp mill in Quebec, and there has been a unit operating in Maine for over 8 years (Maserejian et. al. 1995). Also, Sandquist and Sandstöm (2000) reported on a technology to treat foul condensates and non-condensible gases in a closed loop using steam stripping combined with an RTO for the stripper off-gases and adsorption of sulphur dioxide.

Concluding remarks

As shown in this brief review of the literature and recent research, there is a lot of activity in the forest products industry directed towards the understanding and control of atmospheric emissions. With an increased focus on industrial emissions such as particulates, volatile organic compounds, hazardous air pollutants, and greenhouse gases, it is evident that research and technology development will increase in the future. Although quantifying the environmental and health effects of atmospheric emissions will continue to be challenging, there will be increased pressure on the industry and government to track these pollutants, understand their effects, and minimize their emission.

The modelling of pollutant dispersal and dispersion is an advanced science that is widely applied by government regulators and by industry. Computational methods based on mathematical models combined with emissions monitoring are the main ways in which the fate of contaminants is addressed. With the development of high-speed low-cost computers, these methods will continue to improve and increase in application. Models provide an objective method to regulate the industry and also allow for industries to test out new processes and assess which areas of the process need the most attention for emissions control. In addition, there is increased interest in using in-process models in place of emission monitors as so-called "soft sensors".

The control of emissions is through a combination of in-process methods to prevent pollution at its source and emission control methods. Process control and optimization, along with efficient use of materials and use of in-process combustion processes, are important elements of minimizing emissions along with conventional control technologies. In addition, there are a number of new technologies at various stages of development for reducing and controlling emissions in the future (Box 18.4).

Box 18.4. Meeting the challenge:

The forest products industry is making significant strides to reduce atmospheric emissions and minimize their ecological footprint. Overall, methods include:

- Using advanced analytical and computational techniques to improve our understanding of emissions and their effects;
- Improved manufacturing processes to reduce waste, improve energy efficiency, and make use of wood residuals;
- Increased biomass utilization as a fuel to reduce greenhouse gas formation;
- Implementing advanced process control and enhanced emissions control technologies.

These methods are being implemented on a wide scale across the industry and continued research and development promises further improvements in the future.

Acknowledgements

The authors would like to acknowledge the financial support of the Sustainable Forest Management Network of Centres of Excellence. We would also like to thank Mr. Tom Burgess for his comments on the manuscript.

References

Adams, T.N., Frederick, W.J., Grace, T.M., Hupa, M., Iisa, K., Jones, A.K., and Tran, H.N. 1997. Kraft recovery boilers. TAPPI Press, Atlanta, Georgia. 380 p.

Aguliar, M.J., Marks, S.R., Allen, P., and Hanrahan, P.L. 2000. Case study of compliance with air quality related values. *In* Proceedings of the TAPPI (Technical Association of the Pulp and Paper Industry) International Environmental Conference and Exhibit, 6–10 May 2000, Denver, Colorado. pp. 719–728.

Allen, L., and Ellis, S. 2000. Laboratory investigation of trickling biofiltration for treatment of kraft mill non-combustion air emissions. *In* Proceedings of the TAPPI (Technical Association of the Pulp and Paper Industry) International Environmental Conference and Exhibit, 6–10 May 2000, Denver, Colorado. pp. 293–309.

Alley, E.R., Stevens, L.B., and Cleland, W.L. 1998. Air quality control handbook. McGraw–Hill, New York, New York. 784 p.

(AMEC) AMEC Forest Industry Consulting. 2002*a*. Final Report — Multi-pollutant Emission Reduction Analysis Foundation (MERAF) for the Pulp and Paper Sector, Prepared for Environment Canada and the Canadian Council of Ministers of the Environment, Project H729A&C, September 30, 2002. AMEC Forest Industry Consulting, Toronto, Ontario. 254 p.

(AMEC) AMEC Forest Industry Consulting. 2002*b*. Final Report — Multi-pollutant Emission Reduction Analysis Foundation (MERAF) for the Lumber and Allied Wood Products Sector. Prepared for Environment Canada and the Canadian Council of Ministers of the Environment, Project H729 B&D, September 30, 2002. AMEC Forest Industry Consulting, Toronto, Ontario. 455 p.

Axegård, P. 1999. The ecocyclic pulp mill. *In* Proceedings of the 1999 Sustainable Forest Management Network Conference — Science and Practice: Sustaining the Boreal Forest, 14–17 February 1999, Edmonton, Alberta. *Edited by* T.S. Veeman, D.W. Wmith, G. Purdy, F.J. Salkie, and G.A. Larkin. Sustainable Forest Management Network, Edmonton, Alberta. pp. 6–8.

Budwill, K., and Coleman, R. 1999. Biofiltration of gaseous emissions from forest products manufacturing. Sustainable Forest Management Network, Edmonton, Alberta. SFMN Proj. Rep. 1999-01. 24 p. Available at http://sfm-1.biology.ualberta.ca/english/pubs/PDF/PR_ 1999-1.pdf [viewed April 28, 2003].

(CEPA) Canadian Environmental Protection Act. 2002. Toxic Substances List — Updated Schedule 1 as of May 3, 2002. CEPA Environmental Registry, Ottawa, Ontario. 1 p. Available at http://www.ec.gc.ca/CEPARegistry/subs_list/Toxicupdate.cfm [viewed Oct. 3, 2002].

Chen, J.M., Black, T.A., Novak, M.D., and Adams, R.S. 1995. A wind tunnel study of turbulent air flow in forest clearcuts. *In* Wind and trees. *Edited by* M.P. Coutts and J. Grace. Cambridge University Press, Cambridge, U.K. pp. 71–87.

Cheremisinoff, P.N., and Morresi, A.C. 1978. Air pollution sampling and analysis deskbook. Ann Arbor Science Publishers, Ann Arbor, Michigan. 489 p.

Coleman, R.N., and Dombroski, E.C. 1995. Evaluation of a biofilter inoculated with bacteria capable of removing specific kraft pulp mill air emission components. Proceedings of the Emerging Technologies in Hazardous Waste Treatment VII, American Chemical Society, 17–20 September 1995, Atlanta, Georgia. pp. 1059–1063.

Connell, M. 2001. Utilizing RCO technology for effective VOC control. *In* TAPPI (Technical Association of the Pulp and Paper Industry) International Environmental Conference Proceedings, 22–25 April 2001, Charlotte, North Carolina.

Crawford, R.J., and Jain, A.K. 1994. Laboratory studies of methanol, acetone and methyl ethyl ketone generation from kraft pulp bleaching. National Council for Air and Stream Improvement, New York, New York. Tech. Bull. No. 666. 28 p.

Crawford, R.J., Dallons, V.J., Jain, A.K., and Jet, S.W. 1991. Chloroform generation at bleach plants with high chlorine dioxide substitution or oxygen delignification. TAPPI J. **74(4)**: 159–163.

Davis, W.T. 2000. Air pollution engineering manual, second edition. Air & Waste Management Association and John Wiley & Sons, New York, New York. 850 p.

Devinny, J.S., Deschusses, M.A., and Webster, T.S. 1999. Biofiltration for air pollution control. Lewis Publishers, Boca Raton, Florida. 320 p.

Dirk-Faitakis, C.F., and Allen, D.G. 2000. Biofiltration to treat cyclic air emissions produced in the forest products industry. Sustainable Forest Management Network. Edmonton, Alberta. SFMN Proj. Rep. 2000-14. 29 p. Available at http://sfm-1.biology.ualberta.ca/english/ pubs/PDF/PR_2000-14.pdf [viewed April 26, 2003].

Dirk-Faitakis, C.F., and Allen, D.G. 2003. Biofiltration of cyclic air emissions of α-pinene at low and high frequencies. J. Air Waste Manage. Assoc. In press.

Environment Canada. 1984. State-of-the-art of the pulp and paper industry and its environmental protection practices. Environmental Protection Programs Directorate, Environment Canada, Ottawa, Ontario. Econ. Tech. Rev. Rep. EPS 3-EP-84-2.

Environment Canada. 2002. Particulate matter ($PM_{<10}$). Available at http://www.ec.gc.ca/air/p-matter_e.shtml [viewed Oct. 2, 2002].

(EPS) Environmental Protection Service. 1983. The basic technology of the pulp and paper industry and its environmental protection practices. Environmental Protection Programs Directorate, Environment Canada, Ottawa, Ontario. Training Man. EPS 6-EP-83-1.

Eriksson, K.-E.L 2000. An overview of biotechnology in the pulp and paper industry. *In* Proceedings of the TAPPI (Technical Association of the Pulp and Paper Industry) Pulping/ Process & Product Quality Conference, 5–8 November 2000, Boston, Massachusetts.

(ESA) Ecological Society of America. n.d. Acid rain revisited. Washington, D.C. Available at http://esa.sdsc.edu/acidrainfactsheet.htm [viewed Oct. 2, 2002]. 2 p.

Forintek. 2003. Technology roadmap: lumber and value-added wood products. Forintek Canada Corporation, Vancouver, British Columbia. Available at http://strategis.ic.gc.ca/epic/internet/infi-if.nsf/vwGeneratedInterE/fb01315e.html [viewed April 27, 2003].

(FPAC) Forest Products Association of Canada. 2001*a*. Environmental progress report 2000–2001. Montreal, Quebec. 40 p.

(FPAC) Forest Products Association of Canada. 2001*b*. Report on preliminary assessment of the contributions of pulp and paper mills to ambient air quality. Study conducted by Jacques Whitford Environment Ltd., Montreal, Quebec. Proj. ONT50166. 71 p.

Gravel, J.O., and Drouin, G. 2001. Regenerative thermal oxidation of reduced sulphur emissions from kraft pulping. *In* TAPPI (Technical Association of the Pulp and Paper Industry) International Environmental Conference Proceedings, 22–25 April 2001, Charlotte, North Carolina.

Gregory, L.J. 2000. Biological destruction of HAPs for cluster rule compliance. TAPPI J. **83(12)**: 1–15.

Gu, Y.X., and Edwards, L.L. 2001*a*. Cluster compliance tools: predicting kraft mill VOCs part 1 — mill applications. TAPPI J. **84(4)**: 1–13.

Gu, Y.X., and Edwards, L.L. 2001*b*. Cluster compliance tools: predicting kraft mill VOCs, part 4 — mill validated emission models. TAPPI J. **84(4)**: 39–51.

Gu, Y.X., Edwards, L.L., Zhu,Y.Z., and Chai, X.S. 2001*a*. Cluster compliance tools: predicting kraft mill VOCs, part 2 — methanol generation models. TAPPI J. **84(4)**: 14–27.

Gu, Y.X., Edwards, L.L., Zhu,Y.Z., Liu, P.H., and Chai, X.S. 2001*b*. Cluster compliance tools: predicting kraft mill VOCs, part 3 — methanol equilibrium calculations. TAPPI J. **84(4)**: 28–38.

Hardy, G.W., and Hohl, H. 2001. Case history: an alternative for improved PM reduction in wood products facilities. *In* TAPPI (Technical Association of the Pulp and Paper Industry) International Environmental Conference Proceedings, 22–25 April 2001, Charlotte, North Carolina. Paper 17-3.

Hardy, G.W., Badar, T., and Duffee, R.A. 1995. VOC and odour control of fiberboard and particle board plant emissions. *In* Proceedings of the TAPPI (Technical Association of the Pulp and Paper Industry) International Environmental Conference, 7–10 May 1995, Atlanta, Georgia. pp. 201–206.

Iisa, K. 1996. Recovery boiler air emissions. *In* Kraft recovery boilers. *Edited by* T.N. Adams. TAPPI Press, Atlanta, Georgia. pp. 215–244.

Industry Canada and Forintek. 1998. Wood based panel products technology roadmap. ISSN 0381-7733. Industry Canada, Ottawa, Ontario. Available at http://strategis.ic.gc.ca/ SSG/fb01129e.html [viewed Oct. 3, 2002].

Jain, A.K. 1996. Bleach plant emissions. *In* Pulp bleaching — principles and practice. *Edited by* C.W. Dence and D.W. Reeve. TAPPI Press, Atlanta, Georgia. pp. 823–847.

Janka, K., Ruuohola, T., Siiskonen, P., and Tamminen, A. 1998. A comparison of recovery boiler field experiments using various NO_x reduction methods. TAPPI J. **81(12)**: 137–141.

Johnson, L.P., and Diehl, M. 1995. Optimal implementation of condensate segregation and methanol/TRS stripping technology. *In* Proceedings of the TAPPI (Technical Association of the Pulp and Paper Industry) International Environmental Conference, 7–10 May 1995, Atlanta, Georgia. p. 937.

Kong, Z., Farhana, L., Fulthorpe, R.R., and Allen, D.G. 2001. Treatment of volatile organic compounds in a biotrickling filter under thermophilic conditions. Environ. Sci.Technol. **35**: 4347–4352.

Lawrence, M.W., and Vaquez, P.J. 2001. Case study: regenerative thermal oxidizer on a particle board dryer. *In* TAPPI (Technical Association of the Pulp and Paper Industry) International Environmental Conference Proceedings, 22–25 April 2001, Charlotte, North Carolina. Paper 13-1.

(LES) Lakes Environmental Software. 2002. Full AERMOD-PRIME support. Available at http://www.lakes-environmental.com/ISCAERMOD/AERMOD-PRIME.html [viewed Oct. 3, 2002].

Lövblad, R., Moberg, G., and Olausson, L. 1991. TAPPI (Technical Association of the Pulp and Paper Industry) 1991 Environmental Conference Proceedings, Atlanta, p.1071, *as cited in* Janka et. al. (1998).

Lu, Z., and Allen, D.G. 2001. Biofiltration of mixtures of total reduced sulphur and volatile organics. *In* TAPPI (Technical Association of the Pulp and Paper Industry) International Environmental Health and Safety Conference and Exhibit, 22–25 April 2001, Charleston, North Carolina.

Maserejian, Z.Y., Saviello, T., Gogolos, J.S., Neptune, J., and Brown, C. 1995. Simultaneously satisfying CAAA, Title v, MACT, RACT, and HVLC gas control design requirements in developing an air emissions inventory for a large integrated mill. *In* Proceedings of the TAPPI (Technical Association of the Pulp and Paper Industry) International Environmental Conference, 7–10 May 1995, Atlanta, Georgia. p. 765.

McCullum, K., and Kindzierski, W. 2001. Analysis of particulate matter origin in ambient air at high level. Sustainable Forest Management Network. Edmonton, Alberta. Proj. Rep. 2001-16. 24 p. Available at http://sfm-1.biology.ualberta.ca/english/pubs/PDF/PR_2001-16.pdf [viewed April 27, 2003].

Messner, K., and Srebotnik, E. 1994. Biopulping: an overview of developments in an environmentally safe paper making technology. FEMS Microbiol. Rev. **13**: 351–364.

Mohseni, M., and Allen, D.G. 2000. Biofiltration of mixtures of hydrophilic and hydrophobic volatile organic compounds. Chem. Eng. Sci. **55**: 1545–1558.

Mohseni, M., Nichols, K.M., and Allen, D.G. 1998. Biofiltration of -pinene and its application to the treatment of pulp and paper air emissions. TAPPI J. **81(8)**: 205–211.

Morgan, F.M., Sleep, B., and Allen, D.G. 2001. Effect of biomass growth on gas pressure drop in biofilters. J. Environ. Engin. (May): 388–396.

Munson, B.R., Young, D.F., and Okiishi, T.H. 2001. Fundamentals of fluid mechanics, 4th edition. Wiley, New York, New York. 816 p.

(NCASI) National Council for Air and Stream Improvement. 1994*a*. Volatile organic emissions from pulp and paper mill sources, part i — oxygen delignification systems. NCASI, Research Triangle Park, North Carolina. Tech. Bull. 675.

(NCASI) National Council for Air and Stream Improvement. 1994*b*. Volatile organic emissions from pulp and paper mill sources, part ii — lime kilns, smelt dissolving tanks & miscellaneous causticizing area vents. NCASI, Research Triangle Park, North Carolina. Tech. Bull. 676.

(NCASI) National Council for Air and Stream Improvement. 1994*c*. Volatile organic emissions from pulp and paper mill sources, part iii — miscellaneous sources at kraft and TMP mills. NCASI, Research Triangle Park, North Carolina. Tech. Bull. 677.

(NCASI) National Council for Air and Stream Improvement. 1994*d*. Volatile organic emissions from pulp and paper mill sources, part iv — kraft brownstock washing, screening and rejects refining sources. NCASI, Research Triangle Park, North Carolina. Tech. Bull. 678.

(NCASI) National Council for Air and Stream Improvement. 1994*e*. Volatile organic emissions from pulp and paper mill sources, part v — kraft mill bleach plants. NCASI, Research Triangle Park, North Carolina. Tech. Bull. 679.

(NCASI) National Council for Air and Stream Improvement. 1994*f*. Volatile organic emissions from pulp and paper mill sources, part vi — kraft recovery furnaces and black liquor oxidation systems. NCASI, Research Triangle Park, North Carolina. Tech. Bull. 680.

(NCASI) National Council for Air and Stream Improvement. 1994*g*. Volatile organic emissions from pulp and paper mill sources, part vii — pulp dryers and paper machines at integrated chemical pulp mills. NCASI, Research Triangle Park, North Carolina. Tech. Bull. 681.

(NCASI) National Council for Air and Stream Improvement. 1994*h*. Volatile organic emissions from pulp and paper mill sources, part viii — sulfite mills. NCASI, Research Triangle Park, North Carolina. Tech. Bull. 682.

(NCASI) National Council for Air and Stream Improvement. 1994*i*. Volatile organic emissions from pulp and paper mill sources, part ix — semi-chemical mills. NCASI, Research Triangle Park, North Carolina. Tech. Bull. 683.

(NCASI) National Council for Air and Stream Improvement. 1994*j*. Volatile organic emissions from pulp and paper mill sources, part x — test methods, quality assurance/quality control procedures and data analysis protocols. NCASI, Research Triangle Park, North Carolina. Tech. Bull. 84.

(NCASI) National Council for Air and Stream Improvement. 1999*a*. Volatile organic compound emissions from wood products manufacturing facilities, part i — plywood. NCASI, Research Triangle Park, North Carolina. Tech. Bull. 768.

(NCASI) National Council for Air and Stream Improvement. 1999*b*. Volatile organic compound emissions from wood products manufacturing facilities, part ii — engineered wood products. NCASI, Research Triangle Park, North Carolina. Tech. Bull. 769.

(NCASI) National Council for Air and Stream Improvement. 1999*c*. Volatile organic compound emissions from wood products manufacturing facilities, part iii — medium density fiberboard. NCASI, Research Triangle Park, North Carolina. Tech. Bull. 770.

(NCASI) National Council for Air and Stream Improvement. 1999*d*. Volatile organic compound emissions from wood products manufacturing facilities, part iv — particleboard. NCASI, Research Triangle Park, North Carolina. Tech. Bull. 771.

(NCASI) National Council for Air and Stream Improvement. 1999*e*. Volatile organic compound emissions from wood products manufacturing facilities, part v — oriented strandboard. NCASI, Research Triangle Park, North Carolina. Tech. Bull. 772.

(NCASI) National Council for Air and Stream Improvement. 1999*f*. Volatile organic compound emissions from wood products manufacturing facilities, part vi — hardboard/fiberboard. NCASI, Research Triangle Park, North Carolina. Tech. Bull. 773.

O'Connor, B., and Ledoux, C. 2002. TRS inventories and air dispersion modeling for odor reduction at a kraft mill with an activated sludge treatment plant. TAPPI J. **1(2)**: 3–8.

Olausson, L.G., 1995. Recovery boiler and method of reducing NO_x emissions. U.S. patent 5,454,908 (Oct 3, 1995). U.S. Patent Office, Washington, D.C.

Paine, R.J., and Heinold, D.W. 1996. Modeling developments affecting the pulp and paper industry: change is in the air. TAPPI (Technical Association of the Pulp and Paper Industry) International Environmental Conference, 5–10 May 1996, Orlando, Florida. pp. 629–641.

Parthasarathy, R.V., Gygotis, R.C., Bryer, D.M. and Wahoske. K. 1995. Soda-AQ pulping and ECF bleaching of hardwoods — a sulfur free pulping and chlorine free bleaching alternative to kraft pulps. *In* Proceedings of the TAPPI (Technical Association of the Pulp and Paper Industry) International Environmental Conference, 7–10 May 1995, Atlanta, Georgia. pp. 1177–1196.

Patrikainen, T., Tamminen, A., Tuominiemi, S., Pikkujamsa, E., Spets, J.-P., and Hamalainen, R. 1997. Process for removing nitrogen oxides from the flue gases of a pulp mill. U.S. patent 5,639,434. (June 17, 1997). U.S. Patent Office, Washington, D.C.

Pere, J., Siika-aho, and Viikari, L. 2000. Biomechanical pulping with enzymes: response of coarse mechanical pulp to enzymatic modification and secondary refining. TAPPI J. **83(5)**: 1–8.

Pinkerton, J.E. 2000. Pulp and paper pollution problems. *In* Industrial environmental control, pulp and paper industry, 3rd edition. *Edited by* A.M. Springer. TAPPI Press, Atlanta, Georgia. pp. 501–535.

Powals, R.J., Zaner, L.J., and Sporek, K.F. 1978. Handbook of stack sampling and analysis. Technomic Publishing Co., Westport, Connecticut. 441 p.

Raemhild, G.A., and Jaasund, S.A. 2001. Upstream particulate removal for RTO's on direct-fired OSB dryers — how much is enough? *In* TAPPI (Technical Association of the Pulp and Paper Industry) International Environmental Conference Proceedings, 22–25 April 2001, Charlotte, North Carolina.

Ramsay, K., Manolescu, D., Wentzell, P., and Winik, C. 2001. Economic benefits achieved from an odour reduction project. *In* TAPPI (Technical Association of the Pulp and Paper Industry) International Environmental Conference Proceeedings, 22–25 April 2001, Charlotte, North Carolina.

Robinson, J.W., and Spaite, P.W. 1967. A new method for analysis of multi-compartment fabric filtration. Atmos. Environ. **1(4)**: 499.

Ruth, J.H. 1986. Odor thresholds and irritation levels of several chemical substances: a review. Am. Ind. Hyg. Assoc. J. **47**: A42.

Sandquist, K.K., and Sandstöm, E. 2000. A novel technology to treat foul condensate and NCG gases in a closed loop. *In* TAPPI (Technical Association of the Pulp and Paper Industry) International Environmental Conference and Exhibit, 6–10 May 2000, Denver, Colorado. pp. 147–156.

Saviello, T.B., VanFleet, S., Bomgardner, C., Sandry, T., White, M., Sproul, B., Rascher, C., and Dawson, M. 1998. Predictive emission model (PEMS) for lime kiln particulate, TRS, SO_2, NO_x and CO. TAPPI (Technical Association of the Pulp and Paper Industry) International Environmental Conference, 5–8 April 1998, Vancouver, British Columbia. pp. 361–364.

Scott, G., and Lentz, M. 1995. Environmental aspects of biosulfite pulping. *In* Proceedings of the TAPPI (Technical Association of the Pulp and Paper Industry) International Environmental Conference, 7–10 May 1995, Atlanta, Georgia. pp. 1155–1161.

Smook, G.A. 1973. Some fresh thoughts on sulfur in the kraft recovery cycle. Paper Trade Journal, November 56, 1973, *as cited in* Smook, G.A. 1982. p. 368.

Smook, G.A. 1982. Handbook for pulp & paper technologists. Technical Association of the Pulp and Paper Industry, Atlanta, Georgia. 395 p.

Smook, G.A. 1992. Handbook for pulp & paper technologists, 2nd edition. Technical Association of the Pulp and Paper Industry. Atlanta, Georgia. 419 p.

Sologar, V., Lu, Z., and Allen, D.G. 2002. Modelling the biofiltration of air emissions containing reduced sulphur compounds and volatile organic compounds. 2002 USC–TRG Conference on Biofiltration, 31 October – 1 November 2002, Newport Beach, California. *Edited by* J.S. Devinny (University of

Southern California, Los Angeles, California) and F.E. Reynolds, Jr. (The Reynolds Group, Tustin, California).

Sologar, V., Lu, Z., and Allen, D.G. 2003. Biofiltration of mixtures of hydrogen sulfide and methanol. Environ. Progress. In press.

Springer, A.M., Courtney, P.S., and Courtney, F.E. 2000. Applications of an air contaminant dispersion model. *In* Industrial environmental control: pulp and paper industry, 3rd edition. *Edited by* A.M. Springer. TAPPI Press, Atlanta, Georgia. pp. 593–605.

Springer, K. 1992. Environmental control. *In* Mill control & control systems: quality & testing, environmental, corrosion, electrical, vol. 9 of pulp and paper manufacture, 3rd edition. *Edited by* M. Kouris (*volume editor*) and M.J. Kocurek (*series editor*). Joint Executive of the Vocational Industry Committee of the Pulp and Paper Industry (TAPPI and CPPA), Atlanta, Georgia, and Montreal, Quebec. pp. 235–348.

Springer, K. 2000. Industrial environmental control, pulp and paper industry, 3rd edition. TAPPI Press, Atlanta, Georgia. 711 p.

Tarpey, T. 1995. Odour reduction at Peace River pulp mill. *In* Proceedings of the TAPPI (Technical Association of the Pulp and Paper Industry) International Environmental Conference, 7–10 May 1995, Atlanta, Georgia. pp. 589–597.

Tatum, V.L. 1995. Health effects of TRS compounds. *In* Proceedings of the TAPPI (Technical Association of the Pulp and Paper Industry) International Environmental Conference, 7–10 May 1995, Atlanta, Georgia. pp. 969–972.

Turner, D.B. 1994. Workbook of atmospheric dispersion estimates: an introduction to dispersion modelling. Lewis Publishers, Boca Raton, Florida. 192 p.

(USEPA) United States Environmental Protection Agency. 1995*a*. Compilation of air pollutant emission factors, vol i: stationary point and area sources and supplements A–D, AP-42", 5th edition. U.S. Government Printing Office, Washington, D.C.

(USEPA) United States Environmental Protection Agency. 1995*b*. EPA office of compliance sector notebook project — profile of the lumber and wood products industry. U.S. Government Printing Office, Washington, D.C. Rep. EPA/310 R-95-006.

(USEPA) United States Environmental Protection Agency. 1998. Interagency workgroup on air quality models (IWAQM) phase 2 summary report and recommendation for modeling long-range transport impacts. U.S. Government Printing Office, Washington, D.C. Rep. EPA-454/R-98-019.

(USEPA) United States Environmental Protection Agency. 1999. Guideline on air quality models (revised), title 40, part 51, appendix w, code of federal regulations. U.S. Government Printing Office, Washington, D.C.

(USEPA) United States Environmental Protection Agency. 2002. Dispersion models. Technology Transfer Network Support Center for Regulatory Air Models, U.S. Environmental Protection Agency, Washington, D.C. Available at http://www.epa.gov/scram001/tt22.htm [viewed Oct. 3, 2002].

Venkatesh, V., Lapp, W.L., and Parr, J.L.1997. Millwide methanol balances: predicting and evaluating hap emissions by utilizing process simulation techniques. TAPPI J. **80(2)**: 171–176.

Venkatesh, V., Owens, T., Swint, B., Kirkman, A.G., and Gregory, M.H. 2000. Prediction of TRS and VOC emissions for evaluating alternative control strategies in a brown mill. *In* TAPPI (Technical Association of the Pulp and Paper Industry) International Environmental Conference and Exhibit, 6–10 May 2000, Denver, Colorado. pp. 597–613.

Wight, G.D. 1994. Fundamentals of air sampling. CRC Press, Boca Raton, Florida. 254 p.

Winegar, E.D., and Keith, L.H. 1993. Sampling and analysis of airborne pollutants. CRC Press, Boca Raton, Florida. 364 p.

Wiseman, C.A., Wilson, T., and Tielbaard, M.H. 2000. The startup of an IC-UASB reactor at Boise Cascade for MACT I compliance. *In* TAPPI (Technical Association of the Pulp and Paper Industry) International Conference and Exhibit, 6–10 May 2000, Denver, Colorado. pp. 615–624.

Chapter 19

Reducing, reusing, and recycling solid wastes from wood fibre processing

Hongde Zhou

Introduction

The forest products industry generates vast amounts of solid residues that can be either recovered as usable raw materials or processed into value-added products (Chap. 16). Among them are the unused wood and trimming residues from logging operations, and various residues from wood processing such as waste chips and bark, hog fuel, sawdust, effluent treatment sludge, etc. Traditionally, they are considered as "wastes" that have to be discarded into the environment or disposed of in designated landfill sites. With the increasing demand for various wood products, concerns about sustainable development, and awareness of the need for environmental protection, the forest products industry is becoming increasingly interested in diverting these residues from disposal (Box 19.1).

Box 19.1. Challenges to sustainable wood fibre processing:

- To identify opportunities for increased recovery of wood fibre during primary processing;
- To incorporate wood processing wastes into useful new products; and
- To minimize energy consumption and waste generation by practicing alternative processing technologies that are economically viable and environmentally friendly.

Essentially, three principle strategies have been proposed to divert the solid residues from disposal, the same strategies advocated in the popular environmental and recycling movements for decades (Tchobanoglous et al. 1993). Waste reduction promotes the utilization efficiency of natural resources, thereby eliminating the need for any additional handling of materials. This can be accomplished (for example) by modifying the manufacturing processes or redesigning products so that fewer raw materials are used. Reuse extends the life of a product or materials beyond a single use. While providing savings in raw materials and energy for manufacturing, the reuse requires some additional costs if the product has to be cleaned or refurbished prior to use again. Recycling involves treating waste as a resource from which new products can be made.

H. Zhou. School of Engineering, University of Guelph, Guelph, Ontario N1G 2W1, Canada. Telephone: 519-824-4120, ext. 56990. e-mail: hzhou@uoguelph.ca
Correct citation: Zhou, H. 2003. Reducing, reusing, and recycling solid wastes from wood fibre processing. Chapter 19. *In* Towards Sustainable Management of the Boreal Forest. *Edited by* P.J. Burton, C. Messier, D.W. Smith, and W.L. Adamowicz. NRC Research Press, Ottawa, Ontario, Canada. pp. 759–798.

It involves additional collection and handling costs, requires some consumption of energy, and may itself generate wastes. Many of the most common fibre-based materials such as newspapers can be handled in a manner that keeps them out of landfills. Through the implementation of these three strategies, wastes that were once a disposal problem, for producers and consumers alike, can become a valuable resource. In some cases, they can provide significant financial benefits as well.

The purposes of this chapter are to (1) summarize the sources and characteristics of wood processing wastes, (2) discuss their potential environmental impacts and relevant waste management regulations, (3) introduce the important alternative management technologies that either are currently used or have shown great promise, and (4) outline the important technical, economical, and regulatory constraints that affect solid waste management. Note that there are many sources of wood residues other than those from the processing of wood products. Some of the most common sources include municipal solid wastes, construction and demolition debris, and chemically treated wood (including railroad ties, telephone and utility poles, etc.). Although many opportunities also exist to reduce, recycle, and reuse these wastes, emphasis in this chapter will be placed on the solid wastes generated directly by wood processing mills.

Sources and characteristics of solid wastes from wood processing mills

Solid waste generation sources

Wood processing mills are usually classified into two different industrial sectors. The primary wood products industry covers a wide range of establishments, including those involved in the cutting of timber and pulpwood, sawmills, lath mills, shingle mills, cooperage stock mills, planing mills, plywood mills, and others making finished articles entirely or mainly of wood or related materials. The main activities can be grouped into four categories: lumber production, panel products manufacturing, and wood preserving. On the other hand, the pulp and paper industry produces commodity grades of wood pulp, primary paper products, and paperboard products. Their manufacture often requires a sequence of individual processes that can be categorized into three steps: pulp making, pulp processing, and paper or paperboard production. At the pulping stage, the feedstock prepared from raw wood is digested into its fibrous constituents through breaking the bonds between fibres. Depending on the techniques employed, pulp-making processes can be thus classified as chemical, mechanical, or semi-chemical. Subsequent processing of the pulp is intended to remove impurities such as uncooked chips, clean and recycle any residual cooking liquor via a series of washing processes, and increase pulp whiteness and brightness by adding bleaching chemicals. Finally, the processed pulp is pressed into paper sheets which are dried on the dryer section of the paper machine as steam-heated rollers compress the sheets (Smook 1992).

Table 19.1 summarizes the main solid waste streams generated from the primary wood products industry and the pulp and paper industry, respectively. A variety of solid wastes are generated because of different input raw materials, processing operations

Table 19.1. The main solid wastes generated from wood processing mills (modified from USEPA 1995*a*, *b*).

Processes	Material input	Main solid wastes
Primary wood products mills		
Logging	Trees, diesel, gasoline	Fine chips or sawdust, tree branches and tops, wood fragments/splinters, bark
Lumber production		
Sawing	Logs, diesel, gasoline	Sawdust, bark
Surface protection	Wood, surface protectants	Sawdust, chips, sand, stone, dirt, tar, tank sludge
Panelling		
Veneer and plywood	Veneer, adhesive resins, etc.	Trim, sawdust, adhesive residues
Reconstituted wood products	Wood particles, strands, adhesives	Wood particles, adhesive residues
Wood preserving	Wood, preservatives, carrier oils	Bottom sediment sludges
Pulp and paper mills		
Stock preparation	Wood logs, chips, sawdust	Bark, dirt, grit, wood wastes
Chemical pulping and bleaching	Finish chips, cooking chemicals, bleaches	Resins, fatty acids
Chemical recovery system	Black liquors, lime	Lime mud, lime slaker grits, green liquor dregs
Power and steam boiler	Fossil fuels, wood residues, etc.	Fly ash, bottom ash, scrubber sludge
Wastewater treatment plant	Process wastewater	Wastewater treatment sludge

employed, and finished products to be manufactured. Based on the sources and characteristics of these wastes, nevertheless, they are generally categorized into (1) wood residues from logging, lumber production, wood panelling, and pulp stock preparation, (2) wastewater treatment sludges, (3) wood ashes from energy recovery boilers, and (4) miscellaneous solid wastes.

Wood residues

Wood residues refer to all wood materials that are not suitable for manufacturing primary wood products. They can be in different forms, the most important being bark, undersized chips, sawdust, sawmill wood edges, and poor quality wood rejected from the manufacturing process. The wood chips that are prepared from these residues for combustion in wood-fired boilers are collectively called hog fuel. Most of these wood residues are generated from primary wood products manufacturing operations, even though pulp and paper mills also produce wood residues during the pulp feedstock preparation.

Wastewater treatment sludges

Wastewater treatment sludges are generated because pulp and paper mills often require a large volume of water during pulping and bleaching processes. Consequently, on-site treatment plants are often installed to remove organic matter, total suspended solids, and other contaminants prior to returning this fluid to a water body (see Chap. 17). Based on their sources of generation, there are two types of wastewater treatment plant sludges. Primary sludges are generated from gravity sedimentation or the dissolved air flotation of suspended solids contain in untreated mill wastewater. Thus, it contains various inorganic and organic components. The inorganic component can consist of soil particles, clay, calcium carbonate, titanium dioxide, ashes, inert materials rejected during chemical recovery processes, and other solids used in wood processing operations (Smook 1992; USEPA 1995b).

The organic component is mainly wood fibre left behind in wastewater. The secondary sludge is generated by the sedimentation of biologically treated wastewater in secondary clarifiers. Therefore it consists primarily of microbial biomass and is rich in nitrogen and phosphorus, because they are concentrated through cellular metabolism. In some cases, the secondary sludge is combined with the primary sludge to facilitate handling, processing, and disposal. However, the proportion of primary to secondary sludge generated by a facility can vary considerably, ranging from 5 to 75% (NCASI 1992).

Wood ashes

Wood ashes are the inorganic fraction of materials left behind after combustion. In wood processing mills, a substantial quantity of energy is recovered by burning wood residues (such as bark, sawdust, and chipped waste pieces) in boilers to generate the steam and electricity needed for plant operation. To improve boiler efficiency and provide the additional steam for plant operation, fossil fuels are sometimes used as a supplemental fuel. In addition, the primary sludge from mill wastewater treatment can also be burned together with primary wood wastes as a way to reduce sludge disposal needs. A fine portion of ash is carried from the combustion chamber with flue gas, and is collected downstream prior to passing out of a stack for the purpose of air pollution control (see Chap. 18). This is referred to as fly ash (Fig. 19.1) Another portion builds up on the heat-absorbing surface of the furnace and then falls through the furnace bottom to the ash hopper; this is called bottom ash.

Miscellaneous solid wastes

The wood processing operations also generate a variety of miscellaneous solid wastes. The most important in significant quantity and potential environmental impact are screening rejects, secondary fiber pulping rejects, lime mud, lime slake grit, green liquor dregs, raw water treatment residuals, and other mill wastes.

Screening rejects refer to those wastes that are rejected by virtually any type of screen or cleaner. For the pulp mills, these most often consist of the wastes from pulp screening operations which immediately follow virgin wood fibre pulping operations. The solids fraction of this waste stream consists mostly of unpulped wood. Some dirt may also be present in this category of solid wastes.

Fig. 19.1. Fly ash collected from precipitators as part of the air pollution control process at Alberta Pacific's pulp mill at Boyle, Alberta. Fly ash has uniform physical properties, making it suitable as a construction material, but may also concentrate a number of minor elements (including heavy metals), often necessitating its treatment as a hazardous waste.

Photo by Dan Smith

Secondary fiber pulping rejects are unique in that this waste category comes only from operations utilizing recovered paper. It can include any solid rejects contained in a wastepaper bale. Main components are baling wire, miscellaneous metals (such as staples), wood, plastics, adhesive residues, and dirt. Because of the nature of this waste, it is often handled along with general mill refuse and landfilled rather than processed further.

Lime mud is actually an intermediate product from the kraft chemical recovery process. It is mainly composed of calcium carbonate, $CaCO_3$, though other materials such as sodium and sulphur compounds may also be present in a slurry. In most kraft mills, lime mud is calcined in an on-site kiln and is not routinely part of the solid waste stream. Lime slake grit or lime grit consists of the nonreactive materials left behind after calcined lime is slaked with green liquor during the kraft chemical recovery process. Grits are composed of carbonates, sulphates, silicates, and other nonreactive minerals.

Green liquor dregs are also residues generated from the kraft chemical recovery process. They include any nonreactive and insoluble materials remaining after smelt from the recovery furnace is dissolved in water. Dregs are composed of about 50% unburned carbonaceous material. The remainder may consist of silica, metallic sulphides, and other salts.

Raw water treatment residuals come from the in-mill treatment of its water supply. They include various types of solid wastes, depending on the water supply treatment processes employed. Coagulation sludge and lime softening precipitates are among the most common waste streams. Properties of these waste streams are very similar to those generated from municipal drinking water treatment plants.

It should be noted that wood processing mills also generate a number of other solid waste streams including solid wastes from the wood yard, flume grit from log handling, paper rejects, general refuse, and anything broken but not recycled within the mill. There is little information about their characteristics or generation rates available from the literature. But the quantity of these waste streams is expected to be relatively small, and the characteristics can vary widely owing to their different generation sources and collection methods.

Quantity of solid wastes

An important step in developing alternative solid waste management strategies is to quantify the amounts of wastes generated and the types of materials available for recycling and reuse. Wood residues represent the largest fraction of solid wastes generated from wood processing mills. Table 19.2 shows the estimated quantities of major wood residues generated in the United States (U.S.) in 1998. Most wood residues resulted from primary wood products manufacturing operations. A total of about 81.9 million tonnes (Mg) of wood residues were generated, with 11.1 million tonnes as bark residues. Nearly all mill wood residues were used to produce other products, primarily paper,

Table 19.2. Major sources of wood residues in the United States in 1998 (from McKeever 1999).

Sources	Generated 10^6 dry tonnes[a]	Recovered, combusted, or not usable 10^6 dry tonnes	Available for further recovery Amount, 10^6 dry tonnes	% of available total
Municipal solid waste				
Waste wood	10.7	5.8	4.9	16.3
Yard trimmings	22.9	16.7	6.2	20.9
Construction and demolition waste				
Construction	7.9	1.9	6.0	22
Demolition	24.0	15.8	8.2	30
Primary wood products industry				
Bark residues	22.2	21.7	0.5	2
Other wood residues	59.7	58.5	1.2	4
Total	147.3	120.5	26.9	100

[a]1 tonne = 1 Mg = 1000 kg.

nonstructural panels, and fuels. Only 2% of the bark and 2% of the rest of wood residues were not used and are available for further recovery (McKeever 1999).

An extensive survey was also conducted by National Council of the Paper Industry for Air and Stream Improvement (NCASI 1999) to estimate the total amount of solid wastes generated in 1995 from the U.S. pulp and paper industry. Table 19.3 summarizes the total quantities of different waste streams along with their management. It was estimated that the U.S. pulp and paper industry produced 13.2 million dry tonnes of solid residues, 5.29 million tonnes of wastewater treatment sludges, 2.55 million tonnes of ashes and 5.36 million tonnes of miscellaneous solid residues. Unlike the wood residues from the primary wood products industry, 60% of these wastes are disposed of in either landfills or lagoons.

The rates of solid waste generation have remained fairly steady during the 1990s. Figure 19.2 shows the trend of bark and wood waste generation from the primary wood products industry in United States. The total amount of bark and wood residues in 1998 was 22.2 million tonnes and 59.7 millions tonnes, respectively, which represents a decrease of less than 20% since1990. Figure 19.3 shows the overall median solid waste generation rates per tonne of shipped product from the U.S. pulp and paper industry. As compared to 1988, the generation rates of total solid wastes in 1995 increased from 124 kg/tonne to 135 kg/tonne of shipped product. However, this difference was considered to be statistically insignificant because of the high variation in estimating the generation rates. The total amount of solid residues, however, increased from 10.7 million tonnes to 13.2 million tonnes. This increase was attributed to the increase in production and the use of lower quality recovered fibre furnish. Similar generation rates were also reported for Canadian pulp and paper mills (Table 19.4). In total, Canadian pulp and paper industry produced 5.6 million tonnes of solid residues in 1994 and 7.1 million tonnes in 1995 (Reid 1998). Excluding wood and bark used as fuel, the mean rate of solid residue generation per unit of products was 104 kg/tonne at Canadian mills. However, it should be

Fig. 19.2. Trend of wood residue generation rates from the primary wood products industry in the United States (after McKeever 1999).

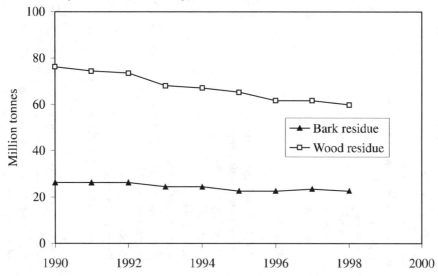

Table 19.3. Solids waste generation rates in the United States pulp and paper industry (from NCASI 1999).

Residues	Generation		Management, %		
	10^6 dry tonnes/%	kg/tonne products	Landfill/lagoon	Land application	Other uses[a]
Wastewater treatment sludges	5.29	96	51	12	37
Primary	40%		51	12	37
Secondary	1%		51	18	31
Combined	54%		50	11	39
Dreged	5%		58	22	20
Wood ashes	2.55	NA[b]	72	11	17
Coal	46%				
Wood/bark	37%				
Sludge	17%				
Miscellaneous solid wastes	5.36	128	63	3	34
Virgin fibre pulping rejects	6%		35	NR[c]	>65
Secondary fibre pulping rejects	10%		68	1.3	>30.4
Lime mud	17%		70	8.9	>22.1
Lime slake grit	4%		91	5.5	>3.8
Green liquor dregs	8%		95	3.0	>1.5
Raw water treatment sludges	2%		49	2.7	>22.2
Wood yard waste	15%		47	2.1	>51
General mill refuse	12%		83	0.2	>16.5
Paper mill rejects	12%		38	1.9	>59.2
Broken materials not recycled in mill	8%		6.5	NR	>93.1
Others	6%		87	NR	13.7
Total	13.2	298	60	40	

[a]Includes combustion, recycling, reuse, and other beneficial uses.

[b]NA, not applicable because of the use of different fossil fuels as supplementary energy sources.

[c]NR, not reported.

Fig. 19.3. Comparison of solid waste generation rates for the United States pulp and paper industry (from NCASI 1999).

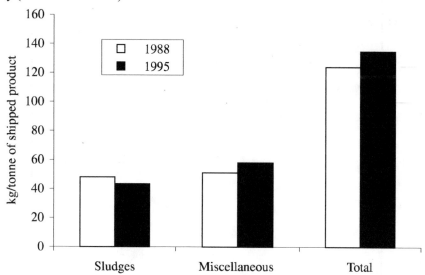

noted that the waste generation rates varied greatly from one mill to another, perhaps owing to the sources of wood fibre, the type of processes used, and the extent of waste reduction and reuse. Thus, caution should be exercised to extrapolate this rate to different mills.

It should be noted that the different types of mills would strongly affect the amount of solid wastes generated. Table 19.5 shows the total solid waste generation rates from 17 categories of pulp and paper mills in 1995 in the United States (NCASI 1999). The median total solid waste generation rates varied from 28 to 1180 kg/tonne of shipped product. The mills that use low-grade wood fibres such as recycled papers generate more solid wastes in comparison to those using high-grade wood fibre as a raw material. In addition, kraft pulp mills generate more solid wastes (because of low pulp yield) than mechanical and semi-chemical pulp mills. Even within each category of pulp and paper mills, great variation in solid waste generation rates are evident from their large standard deviation. Consequently, caution should be exercised in using these data to predict solid waste generation for a specific mill unless the local conditions have been evaluated carefully.

Important characteristics of solid wastes

With a variety of solid wastes generated from wood processing mills, different management schemes are required to convert them to beneficial uses in economically viable and environmentally sound manners. To help explore and assess such utilization options, important physical, chemical, and engineering characteristics must be considered for each category of solid waste. These properties are broadly summarized below for wood residues, wastewater sludges, and wood ashes. Because the sources and composition of "miscellaneous solid wastes" are so diverse, their characteristics cannot be generalized and are not discussed further.

Table 19.4. Solid waste generation rates from Canadian pulp and paper mills in 1994 and 1995 surveys (from Reid 1998).

Residues	Generation[a] 10^6 dry tonnes	/%	Public landfill	Private landfill	Combustion	Composting	Land application	Sewer	Recycling
Wood wastes and bark	3.30		5.2	13.8	79.6	0.5			0.9
Bark		77%							
Chip screen fines		11%							
Woodwaste		12%							
Sludges	1.29		3.0	49.0	38.0	4.0	5.0	1.0	0.0
Primary sludge		42%							
Scondary sludge		26%							
Combined sludge		18%							
Deinking sludge		12%							
Intake and dredge		2.0%							
Inorganics	0.67		2.0	87.0	0.0	0.0	3.0	9.0	0.0
Flay ash		33%							
Lime mud		20%							
Grate ash		24%							
Grits and dregs		17%							
Precipitators		6%							
Miscellaneous	0.34		12.0	48.0	14.0	0.0	0.0	17.0	9.0
Screen rejects		27%							
Sesquisulfate		13%							
Metal		8%							
Knots		16%							
General		36%							

[a]Based on the waste generation in 1995 only.

Table 19.5. Total solid waste generation rates (kg/tonne of shipped product) for the United States pulp and paper industry in 1995 (from NCASI 1999).

Category	Median	Mean	Standard deviation
Bleached containers and boxes, bleached kraft	159	157	81
Construction, any fibre	14	14	11
Corrugating medium, nondeinked	77	76	31
Corrugating medium, semi-chemical	141	157	75
Dissolving pulp, bleached kraft or sulfide	237	211	97
Market pulp, bleached kraft or sulfide	134	115	94
Newsprint, deinked	343	333	153
Newsprint, mechanical plus deinked	141	167	79
Printing and writing paper, bleached kraft	202	215	110
Printing and writing paper, mechanical plus other	159	161	62
Printing and writing paper, purchased	80	134	118
Printing and writing paper, sulfide	155	182	116
Packaging and industrial, purchased	111	156	127
Recycled containers and boxes, nondeinked	66	75	62
Tissue and toweling, deinked	590	715	311
Tissue and toweling, nondeinked	215	196	62
Unbleached containers and boxes, unbleached kraft	88	90	45
All categories	135	169	147

Wood residues

Wood residues are commonly used for fuel, paper pulp, and composite wood products. As most of the wood residues consist of raw wood, their characteristics are dependent on the species of wood present. Of importance for the recycling and reuse of these wastes are their chemical composition, heating value, moisture content, particle size, density, and specific weight.

Wood consists primarily of cellulose, lignin, and hemicellulose, though other extractive organics and ash-forming minerals are also included. Cellulose is a long-chained molecular polymer that bundles together as wood fibrils to make up the cell walls. Lignin, a complex molecular substance, exists within and around the wood cells. It has high heating value but causes difficulties in those industries (such as pulp and paper production) primarily interested in wood fibres. Hemicellulose is very similar to cellulose, being made of a long-chained, sugar-based molecular polymer, and is essential for bonding fibres within the wood. The relative content of these components varies from one tree species to another, and is also affected by tree growth conditions. Table 19.6 summarizes the elemental composition along with heating values for selected wood species and waste residues. Most wood and bark species have a consistent elemental composition and high heating value, about 19 300 KJ/kg, provided that they are moisture- and resin-free. Higher heating values are found in resin-rich woods and bark because of the high heating values of the resin (approximately 40 000 KJ/kg). In addition, wood residues often contain almost negligible sulphur, usually less than 0.1%. As a result, the combustion of wood residues to generate energy will generate less sulphur

Table 19.6. Major elemental composition of selected wood genera and bark residues (from Cheremisinoff et al. 1976).

Genus	Carbon (% dry wt.)	Oxygen (% dry wt.)	Hydrogen (% dry wt.)	Nitrogen (% dry wt.)	Ash (% dry wt.)	Energy content (KJ/kg)
Fir (*Abies*) – bark	53.0	39.2	6.2	—	1.5	21900
Pine (*Pinus*)	50.2	43.4	6.1	0.2	1.1	21200
Spruce (*Picea*)	50.0	43.5	6.0	0.2	0.3	18800
Maple (*Acer*)	50.6	41.7	6.0	0.3	1.4	20000
Oak (*Quercus*)	49.2	44.2	5.8	0.4	0.4	18900
Beech (*Fagus*)	48.9	44.5	5.9	0.2	0.5	18700

dioxide (SO_2, one of the air pollutants of greatest concern) than most fossil fuels. Air pollution due to ash is also relatively low in wood residues. Barks generally have a higher fixed carbon and ash content as compared to pure woods. Typical ash contents for softwood barks can range from 0.4 to 1.5%, while hardwood barks have an ash content ranging from 1.5 to 10% (Cheremisinoff et al. 1976).

The moisture content of wood residue is commonly expressed in two different ways. The moisture content by dry weight is defined as a ratio of the weight of water in a material to its dry weight, while the moisture content by wet weight is a ratio of the water weight to the total material weight (including both water weight and dry weight). Moisture content is the most important factor affecting the physical and chemical properties of wood residues and their potential use as fuel materials. Heat is required to evaporate the water during combustion, so higher moisture content will have a negative impact on the heating value of a material. Higher moisture content also lowers the flame temperature and inhibits the combustion process, thereby reducing the steam output of boilers. When the moisture content is greater than 15% (wet basis), the efficiency of the boiler is markedly diminished. Therefore, higher wood moisture contents must be pre-dried using separate air drying processes or flue gas. Figure 19.4 illustrates the relationship of heating value with different bark moisture contents.

The size distribution of wood residues is another important parameter affecting the feasibility of utilization alternatives. Generally, wood chips, shavings, sawdust, and sander dust can be easily handled. However, handling problems arise with larger wood chunks with respect to storage, feeding to the furnace, and combustion. To improve combustion efficiency larger pieces need to be reduced to wood chips with the use of wood chippers prior to burning. In fact, it is the chip combustion technology developed in 1970s that made the widespread application of wood waste-to-energy combustion in the wood processing industry both technologically feasible and economically viable.

The density of a material is defined as its mass per unit volume, while specific weight is the ratio of the weight of a material to that of the volume of water it displaces. Both properties are directly related to the handling and transportation of wood residues, their potential use as compost bulking agents, fill materials, light-weight aggregates, etc. As the moisture content affects the weight of wood residues and the displaced volume of water, the density and specific weight of wood residue must be measured at a specified moisture content. Once saturated, wood residues have a typical specific weight

Fig. 19.4. Typical relationship of bark heating value vs. moisture content by dry weight.

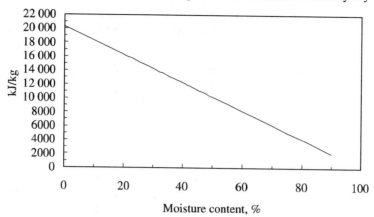

Moisture content, %

ranging from 0.3 to 0.5, dependent on the composition of wood and the nature of any impurities.

Wastewater treatment plant sludges

Many studies have been conducted to determine physical, chemical, and biological characteristics of sludges from the pulp and paper industry. Table 19.7 shows the typical chemical composition of different sludges on an oven-dry basis, along with the sludge from municipal wastewater treatment plants. In general, both primary and secondary sludges from pulp and paper mills contain over 76% volatile solids on a dry weight basis, indicating the relatively high content of organic matter. But their chemical compositions are very different. The primary sludge often contains a significant amount of wood fibre, while the secondary sludge is mainly made of microbial biomass, suggesting different uses for these two types of sludges. In contrast, the secondary sludges from pulp and paper mills and municipal wastewater treatment plants are quite similar. Both contain a high content of microbial biomass and are rich in nitrogen and phosphorus, suggesting their potentially excellent value as fertilizer for enhancing plant growth.

A significant difference between the secondary sludges from pulp and paper mills and those from municipal wastewater treatment plants is usually the lower concentration of heavy metals (with the exception of manganese), as evident in Table 19.7. This is because municipal wastewaters typically accept street runoff and industrial discharges, whereas pulp and paper wastewaters are usually subject to much less heavy metal inputs.

The sludges from pulp and paper mills also contain much less fecal contamination because the volume of sewage from human discharge is often very minimal relative to the process water volumes. As a result, the potential microbial contamination from various fecal pathogens will be less critical, should pulp mill sludges be used as fertilizer to improve plant growth.

One of the most important environmental contaminants in pulp and paper sludges is a variety of trace chlorinated organic compounds. It is well known that free chlorine and some of the chlorine-associated bleaching agents will react with various organic

Table 19.7. Typical chemical composition of pulp and paper mill sludges (compiled from Shields et al. 1985 and MEI 1991).

Parameters	Units	Primary sludge	Secondary sludge	Municipal sludge
Volatile solids	%	76.7	88.7	59–88
Ash	%	16	13	NA[a]
pH		8	7.3	6.5–8.5
Total N	μg/g	1800	59000	24000–50000
Total P	μg/g	340	7800	12000–48000
Al	μg/g	910	1600	NA
As	μg/g	<4.5	<4.7	10
Ba	μg/g	31	40	NA
Cd	μg/g	<1.4	<1.4	10
Ca	μg/g	6500	4400	NA
Co	μg/g	NA	NA	30
Cr	μg/g	54	32	500
Cu	μg/g	40	23	800
Fe	μg/g	1200	930	17000
K	μg/g	210	4800	4100–5800
Pb	μg/g	<9.3	<11	500
Mg	μg/g	1300	2800	NA
Mn	μg/g	180	440	260
Ni	μg/g	11	8	80
Se	μg/g	<5.2	<6.1	5
Ag	μg/g	<1.2	<1.4	NA
Na	μg/g	4800	6000	NA
Zn	μg/g	40	98	1700

[a]NA, not available.

compounds in wood to form toxic chlorinated organic products (Chaps. 16 and 17). Among them, the most important are dioxins and furans because of their high toxic effects and their extensive occurrence in processing wastewaters. These toxic compounds have a high tendency to be partitioned into the sludge as they have a low solubility in water and are resistant to aerobic biodegradation. A 104 mill study conducted in U.S. in 1988 showed that dioxins and furans were present in all the bleached pulp mill sludges, leading to strong demands to regulate their disposal in landfills and by surface land application (NCASI 1985). However, recent advances in pulping processes have greatly reduced the use of chlorine, thereby, reducing the presence of these toxins in wastewater treatment sludges.

Wood ashes

Wood ash can represent a very sizable portion of total solid wastes generated at wood processing mills. Because of its high nutrient and lime content, the reported utilization alternatives for wood ash include agricultural uses, building materials, cement and concrete, civil engineering materials, industrial and specialist materials, resource recovery, and waste stabilization. Accordingly, the important characteristics include grain size distribution, bulk density and specific gravity, carbon content, compaction, compressive strength, permeability, freeze–thaw characteristics, and frost susceptibility. Detailed information about these properties has been reviewed by Zhou et al. (1997).

The particle size distribution of waste ashes is an important parameter because it is related to many physical, chemical, and engineering properties. For example, a fine material would generally have higher surface area, lower permeability, and would be more chemically active than a coarse material. Also, a well-graded material can be readily compacted to a dense condition and will generally develop greater shear strength and lower permeability. Figure 19.5 shows typical particle gradation of wood ashes and lime byproducts from a kraft paper and pulp mill. Both fly ash and wood ash have similar particle size distributions, ranging from 0.01 to 2 mm, which are much finer than coal fly ash. Nevertheless, according to the Unified Soil Classification System, they can still be classified as a "coarse, well-graded" material with a maximum particle size of 25 μm.

It is generally accepted that the size distributions for wood ashes are unimodal. However, their uniformity may differ considerably (Zhou et al. 2000). In general, wood ash has a smaller range of particle sizes with a steeper curve, i.e., it is well-graded in comparison to coal fly ash. Among various wood ashes, the particle size distribution is affected by the source materials, the combustion regime under which it is generated, the methods of collection, and the analytical protocol.

Table 19.8 summarizes other important physical properties for different wood ashes. In general, only the bottom ash has a weight loss on ignition (LOI) of approximately 7%, which is comparable to that of coal ash and indicates a material that has already lost most of its organic matter content. Wood industry fly ash has much higher LOI values, ranging from 17 to 27%, perhaps owing to the presence of higher organic contents and unstable carbonate compounds.

Fig. 19.5. Typical particle size distribution of wood ashes (from Zhou et al. 2000).

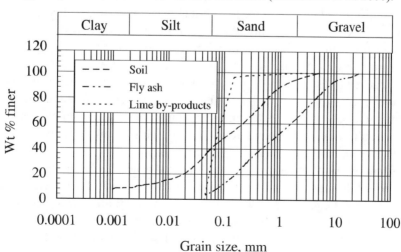

Table 19.8. Main physical characteristics of pulp ashes and lime byproducts (from Zhou et al. 2000).

Sample	Moisture[a] %	LOI[b] %	Residue %	Density, kg/m³ Bulk	Density, kg/m³ Dry
Bottom ash	31.3	7.1	61.5		
Fly ash	20.9	21.4	57.7	640	470
CaCO₃	32.7	17.0	50.2	880	610
CaO	1.8	26.8	71.4		

[a]Dried at 103°C for 16 h 45 min.

[b]Loss on ignition, ignited at 750°C for 3 h 25 min.

Wood ashes are characterized by very low bulk densities, ranging from 0.23 kg/m³ to 0.77 kg/m³ (Muse and Mitchell 1995; Zhou et al. 1997), suggesting that they may have great potential for utilization as a light structural fill material or lightweight aggregate. Variation is once again due to variability in chemical composition, friable material, carbon content, and moisture content. Higher carbon content and lower moisture content result in a lower specific gravity, whereas high metal content produces high specific gravity.

Table 19.9 summarizes the elemental contents of wood ashes, coal ashes, and common soil on a dry weight basis. Although the ranges vary considerably, it is generally true that most of these major elements (Al, Ca, Fe, Mg, K, Si, Na, S, and P) correspond approximately to their abundance in the earth's crust, with the exception of Mg, K, and Ca (which exceed the levels found in natural soil). Among the various elements, Ca is generally the largest constituent. These data suggest that wood ashes may be used as an excellent liming chemical for acidic soils, and as a potassium-rich fertilizer.

In contrast to the major elements, 13 other minor elements, including Ag, As, B, Ba, Cd, Cr, Cu, Mn, Ni, Pb, Se, Sr, and Zn, represent less than 1% of the total mass of wood ashes. In fact, the concentrations of most of the minor elements are found at close to or below the limits of detection. However, some minor elements in wood ashes tend to have concentrations higher than those found in typical soils, suggesting that leaching of these elements from wood ashes should be one of the principal environmental concerns related to their utilization or disposal. In particular, the fly ashes usually contain higher contents of minor elements than found in bottom ashes. An explanation is the enrichment that occurs as a result of the vaporization of these minor elements and associated oxides. The elements that were volatilized during combustion subsequently condense onto solid particles at different rates and in varying amounts as combustion gases are cooled; as a result, they are concentrated in fly ash while being depleted in the bottom ash.

The concentrations of organic compounds in coal ashes are generally extremely low and would pose little impact on the environment (NCG 1980). However, this may not be the case for wood ash, as reported by several recent studies. Chlorophenols, polychlorinated dioxins, and dibenzofurans have been reported to be much higher in wood ash than in coal ashes. Muse and Mitchell (1995) studied various boiler ash samples

Table 19.9. Elemental contents in wood ashes (compiled from Lerner and Utizinger (1986), Mattigod et al. (1990), Muse and Mitchell (1995), Naylor and Schmidt (1989), and Zhou et al. (1997)).

| Element | Hardwood ashes | | Coal ashes | | |
	Fly ash	Bottom ash	Fly ash	Bottom ash	Soil
Major elements, % of dry weight					
Al	0.50–2.52	0.69	3.05–18.5	0.1–20.85	1.0–30
Ca	0.56–33.1	12.6	0.22–24.10	0.11–22.30	0.7–50
Fe	0.22–1.99	0.58	0.4–20.10	1–27.56	0.7–55
K	0.13–6.47	2.15	0.26–3.3	0.17–6.72	0.04–3.5
Mg	0.20–2.26	1.29	0.2–4.8	0.04–7.72	0.06–0.6
Na	0.024–0.64	4.38	0.08–4.13	0.01–7.10	0.075–1.0
S	0.002–0.094		<0.04–7.40	0.04–6.44	0.03–1.0
P	0.1–1.44	0.25	5.10–31.20	1.02–31.78	0.23–0.25
Minor elements, mg/kg of dry weight					
As	90–120	ND[a]	0.02–168	2–440	1–50
Ba	267–1192	627	110–9360	1–13800	100–3000
B	7.7–271	113	2–513	10–5000	2–100
Cd	0.18–26.3	3	0.1–4.7	0.1–130	0.1–0.7
Cr	5–368	7	0.2–5820	4–900	1–1000
Cu	37–207	31	4–930	33–2200	2–100
Pb	19–673	ND	0.4–110	3–2100	2–200
Mn	595–22400	368	60–1900	25–3000	20–3000
Hg	0.06–0.08	ND	0.01–4	0.01–12	0.01–0.3
Mo	1.7–124	ND	1–440	1–140	0.2–5
Ni	0–190	ND	<10–2900	2–4300	5–500
Se	NA	ND	0.1–10	0.2–130	0.1–2
Sr	NA	243	170–6400	30–7600	50–1000
V	NA	ND	12–540	12–1180	20–500
Zn	35–2500	290	4–1800	14–3500	10–300

[a]ND, not determined.

from 12 pulp and paper mills. They found that the organic carbon in 19 boiler ashes varied from 10 to 600 g/kg, with an average of 247 g/kg. Considering the difficulties associated with the determination of individual contaminants, Muse and Mitchell (1995) used the extractable organic halides (EOX) as an indication of potential contamination of chlorinated organics. Analysis of selected boiler ash samples from kilns with bleaching capability revealed that more than half contained detectable levels of EOX. The levels of EOX in boiler ashes were 0–4 mg/kg. In contrast, Sinkkonen et al. (1995) analyzed fly ash samples derived from the burning of peat, wood chips, refuse-derived fuel, and packaging boards in different combinations for their concentration of polychlorinated dibenzo-p-dioxins (PCDDs) and polychlorinated dibenzo-furans (PCDFs). Results indicated that most of the samples contained these chemicals at levels below ng/g. The isomer groups detected at a concentration high enough to be reliably calculated were trichlorodibenzofurans, tetrachlorodibenzofurans, trichlorodioxins, tetrachlorodioxins, and pentachlorodioxins. Their concentrations were found to increase with increasing concentrations of chloride in the fuel. Similar results were observed by Oehme and Muller (1995) in wood burning residues. The concentrations of PCDDs and PCDFs were found to be very low when untreated wood was burned. Higher values were observed from the co-combustion of waste wood and sludges. Furthermore, they found that the PCDF isomers dominated over the PCDD isomers in bottom ashes and most fly ashes.

In general, information on the organic constituents of wood ashes is still scarce in the literature. However, the concentrations and isomer profiles of these organics are generally considered to be low. Nevertheless, care should still be taken to analyze a particular waste stream because its characteristics are dependent on the fuel type, furnace size, and combustion conditions.

Potential environmental impacts and regulations

Assessment procedures

When wood processing wastes are disposed of in the environment or utilized as a new resource, the environmental impacts of greatest concern result from the release of contaminants into groundwater and surface water through weathering and other natural processes. The rates at which various elements are leached from solid wastes are initially dependent on the abundance, surface morphologies, accessibility to solution, and dissolution kinetics of the primary solids (Adriano et al. 1980; Sakata 1987). As the leaching continues, secondary solids will be formed. It is generally accepted that these secondary solids control long-term leaching rates. To simulate this leaching behavior, different leaching tests can be used: batch tests, column tests, field tests using lysimeters, and disposal site measurements (Clarke 1994). Simple batch tests are used in many countries to classify materials as hazardous or non-hazardous and have also been used for environmental assessment. More sophisticated methods have been developed to take into account the physical characteristics of the residues, leaching time, conditions encountered in the field, or other factors.

Batch tests, also called shake tests, are a simple method to determine the total quantity of an element or dissolved chemical species leached from a waste. The procedure involves mixing the waste sample in solvent (leaching medium) such as rainwater or seawater, selected to represent conditions found at a disposal site. Acid leaching media are used in many tests, designed to simulate the conditions found during disposal of municipal wastes. The liquid-to-solid ratio used in the test affects the concentration of elements in the leachate. Higher ratios are generally used to represent the availability of elements for leaching, while lower ratios are used to indicate the maximum concentrations during initial leaching. In most cases, the liquid-to-solid ratios are much higher than those encountered in natural situations, and the tests fail to consider changing natural conditions, and geochemical or hydrological properties.

Currently, various standard batch testing protocols have been developed by specifying sample preparation techniques, leaching media, liquid-to-solid ratios, and leaching time. The test protocols developed by the United States Environmental Protection Agency (USEPA 1986) are getting the widest application in classifying wastes as hazardous or non-hazardous materials. An up-to-date standard method is the Toxicity Characteristic Leaching Procedure (TCLP) test that replaces the earlier Extraction Procedure (EP) test to simulate the disposal of waste in a landfill. A comparison of these two methods showed that the TCLP test is better for determining the mobility of both organic and inorganic compounds in liquid, solid, and slurry wastes (Mason and Carlile 1986). The test requires selection of a sodium acetate buffer if an alkaline waste is being tested. The TCLP procedure uses a higher liquid-to-solid ratio of 20:1, and the duration of the test has been reduced to 18 h.

As compared to the batch tests, column tests provide a much closer representation of natural conditions. They are able to measure the initial concentrations in liquids percolating through a deposit, and have been used to simulate longer term leaching behaviours. Adequate column-to-particle diameter ratios must be ensured to minimize the column wall effects. Leaching can be controlled by compacting the residues in the column and by adjusting the flow rate of the liquid. The liquid-to-solid ratio can be adjusted to be much closer to values in the field. Thus, the results could be more realistic than those obtained from the batch tests.

Field tests using lysimeters attempt to reproduce conditions found in a disposal site more closely. In these tests, a relatively large sample (normally >1 m^3) is placed in a container and exposed to normal, on-site, outdoor weather conditions. Leachate is collected from the bottom of the container and is analyzed over a long period (as long as several years). The test considers changing chemical and biological conditions in the sample, and can be used to monitor the effect of overburden and vegetation.

Disposal site measurement is another category of field testing used to assess leaching. Residue disposal facilities and surrounding areas can be monitored using wells to show whether groundwater has been contaminated from an external source. Ponds and drains used to collect the discharged water can also be monitored. However, it should be noted that sampling problems might be encountered if the distribution and passage of leachates are not uniform. An adequate number of carefully positioned wells must be used to ensure the validity of results.

Each of these tests has different advantages and disadvantages. Laboratory batch and column tests have frequently been used to simulate the leachate generated from solid waste landfills. Their main advantages include their simplicity and greater reproducibility as compared to other methods. For many regulatory purposes, the batch tests will remain a necessary part of testing procedures. However, when the results are compared to the field situation, the discrepancies become obvious, particularly in the pattern of contaminant release. They fail to simulate numerous physico-chemical and biological processes that govern the production and composition of leachates. In addition, the laboratory tests fail to recognize that many of the minerals that make up ashes are metastable under field conditions, and will weather to produce stable secondary minerals over the long term. Finally, the standard batch tests, mostly designed for testing municipal waste, are less suitable for assessing the environmental effects associated with wood wastes. In particular, acid extraction procedures are inappropriate for use with alkaline residues such as wood ashes. The environmental impacts based on acid extraction procedures may be exaggerated because of the excessive amounts of contaminants that may be dissolved under such severe conditions. Thus, laboratory leaching tests should not be used without consideration of their limitations.

Potential environmental impacts

Because of the diversity of solid wastes generated from the wood processing industry, a wide range of adverse environmental impacts are associated with their disposal and utilization. Most of these impacts can be attributed to four types of contaminants:

(1) heavy metals;
(2) toxic organic compounds, particularly chlorinated compounds such as dioxins and furans;
(3) pathogenic microorganisms; and
(4) nitrates.

The major sources of these contaminants have been identified in the waste streams of pulping processes. Of them, wood ashes and wastewater treatment sludges from pulp and paper mills cause the greatest concern (as compared to those from the primary wood products mills) because of the large quantities involved and their great potential for environmental contamination.

Many studies have been conducted to assess the potential environmental impacts of solid wastes from pulp and paper mills. Early studies (NCASI 1978) concluded that the metal concentrations in the EP extracts from various solid wastes were far below the regulatory levels. A further review of literature indicated that any apparent increase caused by using the new TCLP assessment method would also be unlikely to exceed those limits. A follow-up study (NCASI 1990) tested 25 pulp mill solid wastes and leachates collected from 15 different mills representing seven subcategories of production. Each of these waste samples was subjected to the TCLP test. The results concluded that none of the concerned organic compounds in the leachates would exceed any promulgated regulatory level. In fact, most concentrations were less than 1% of the regulatory level. None exceeded 20% of the allowable level as stipulated by the U.S. EPA (see Table 19.10).

Nevertheless, environmental concerns have been raised about the disposal and utilization of wood ashes. Someshwar (1996) thoroughly reviewed the available data on macroelements, heavy metals, and trace organic compounds in wood ashes from industrial boilers. He concluded that most of the chemicals of environmental concern appear to have concentrations comparable to, or even less than, those found in materials currently used for soil amendments such as coal fly ashes, sewage sludges, and agricultural lime. The concentrations of PCDDs/PCDFs appear to be directly correlated to extraneous sources of chloride. In some cases, measures must be taken to control the proportion of salty wood residue and kraft bleaching sludge in fuel sources in order to make the concentrations of PCDDs/PCDFs acceptable. Vance (1996) reviewed up to 25 different wood ash sources used in agricultural and forestry lands. He noted that a wood ash application rate of 10 tonnes/ha would result in the heavy metal levels in soils being two orders of magnitude below the application limits set by U.S. EPA for sewage sludge. Recently, Zhou et al. (2000) studied the leachability of metal ions from kraft mill wood ashes and lime wastes using both U.S. EPA TCLP and American Society of

Table 19.10. Elemental concentrations (mg/L) of TCLP extracts for wood ashes and lime wastes (from Zhou et al. 1997).

Parameters	Bottom ash	Fly ash	CaCO$_3$ waste	CaO waste
Ag	0	0	0	0
Al	0.58	0.38	0.04	0.07
As	0	0	0	0
B	0.81	1.23	0	0.03
Ba	2.27	0.25	0.95	2
Ca	1620	2188	1781	2042
Cd	0	0	0	0
Cr	0	0.04	0	0.041
Cu	0.09	0.001	0.002	0.003
Fe	0.055	0.025	0.022	0.078
K	212	1750	16	130
Mg	128	79	62	0.1
Mn	0	0	3.02	0
Na	32	163	89	812
Ni	0.01	0.01	0.06	0.01
P	0	0	18	0
Pb	0	0	0	0
Se	0	0.03	0.02	0.01
Si	1.51	0.36	3.25	0.29
Sr	4.3	17.4	1.0	2.0
Zn	0.40	0.59	0.40	0.42

Testing and Materials (ASTM) batch extraction tests under different pH values, solid-to-liquid ratios, and leaching times. The results indicated that all the wood ashes and lime wastes could be considered non-toxic. A majority of the heavy metals of potential concern were at least one order of magnitude lower than the regulatory limits set by the U.S. EPA's Resource Conservation and Recovery Act (RCRA) and Environment Canada's Environmental Codes of Practice for Stream Electric Power Generation (SEPG).

Johe et al. (1990) studied a mill landfill site for the disposal of fly ash generated from burning waste-wood products. Based on analyzing the local hydrogeological conditions, water quality in monitoring wells, and leachates, they concluded that no adjacent surface waters or ground waters had been affected by the landfill. Successful applications of wood ash on agricultural and forestry lands have demonstrated that little adverse environmental impacts would be associated with this recycling of wood ashes (NCASI 1985; Campbell 1990; Vance 1996).

Field data about the fate and environmental impacts of solid wastes emanating from wood processing industries are sparse and incomplete owing to the difficulties and expense involved in full-scale monitoring (Theis and Gardner 1990). It is expected that, once leached out, the contaminants are subject to rapid natural attenuations such as adsorption and precipitation reactions (Theis et al. 1978; Sakata 1987). In most cases, the soils can adsorb most minor elements, leaving very small amounts of B, Ba, Cr, and Se to escape. It was observed that even sandy soil would provide up to 95% removal of elemental contaminants over 10 years of flow. Some soils gave a greater degree of protection against the leaching of certain elements. Even in some rare cases where groundwater contamination does occur, it appeared to be localized in the immediate vicinity of the disposal site, and to be much smaller in magnitude than what the laboratory leaching studies suggested.

Regulations

Although solid wastes have been produced from the wood processing industry for many decades, specific legislation or regulations related to their handling and utilization have not been fully developed (Clarke 1994). Consequently, the regulations that are applicable to municipal wastes or other industrial wastes are usually used as a baseline for regulating their disposal and utilization. Since the main potential environmental impacts result from the contamination of water and soil, most relevant legislation centres on determining: (1) the elemental contents and the presence of other extractable chemical species, to assess potential toxicity; and (2) their leachability under natural conditions to assess their suitability for disposal, reuse, and recycling (Theis and Gardner 1990).

In Canada, the Environmental Codes of Practice for Stream Electric Power Generation (SEPG) developed by Environment Canada provides a series of recommendations regarding the use of waste materials (Environment Canada 1985). It consists of a series of documents that identify good environmental practices, including recommendations for the design, siting, construction, and decommissioning phases of a waste management project. The following criteria have been proposed to assess the degree of hazard associated with solid residues:

(1) wastes are classified as corrosive if they have a pH of less than 2.0 or greater than 12.5; and
(2) wastes are classified as leachate toxic if the leachate extracts contain one or more contaminants in greater concentrations than those listed in the Transport of Dangerous Goods regulations (see Table 19.11).

Previous leaching tests showed that most of the wood processing solid wastes will pass both criteria, and can therefore be classified as nonhazardous materials. As a result, Environment Canada encourages the development of beneficial utilization alternatives for solid wood processing wastes. Nevertheless, caution must be exercised even though the recommended "best practices" criteria have been met.

The Hazardous Waste Task Group of the Canadian Council of Ministers of Environment has also developed a Provisional Code of Practice for Management of Post-Use Treated Wood (CCME 1996). Under the directions provided by this code, post-use treated wood residues were not considered a hazardous waste either. Realizing that the product, preservatives, volume, and local conditions will vary from one mill to another; however, it was recommended that the most appropriate disposal option be determined on a case-by-case basis.

Table 19.11. Maximum allowable concentrations (mg/L) of trace elements in leachates from solid wastes.

Element/parameter	SEPG,[a] Canada	RCRA,[b] U.S.A.
Ag	—[c]	5
As	5	5
B	500	—
Ba	100	100
Cd	0.5	1
Cr(VI)	5	5
Cu	100	—
Fe	30	—
Hg	0.1	0.2
Mn	5	—
Pb	5	5
Se	—	1
Zn	500	—
pH	2–12.5	2–12.5

[a]SEPG, the Environmental Codes of Practice for Stream Electric Power Generation.
[b]RCRA, Resource Conservation and Recovery Act.
[c]Blank cells are not addressed in the regulations.

The Canadian federal codes of practice described above are only used as guidelines for wood waste disposal. The approval of waste utilization is the jurisdictional responsibility of provincial governments. For example, the disposal of solid wastes in Alberta must comply with all requirements of the Environmental Protection and Enhancement Act and associated regulations. Both Hazardous Waste Regulation AR 505/87 and Waste Management Regulation AR 250/85 may be applicable for the utilization of solid wastes. For mechanical pulp mill sludges, the province of Alberta also developed the Standards and Guidelines for the Land Application of Mechanical Pulp Mill Sludge to Agricultural Land (AEP 1999).

In the United States, the Resource Conservation and Recovery Act (RCRA) is the principal federal regulation to address solid and hazardous waste management. Waste products are considered to be hazardous if they are included on an itemized list of pre-assessed substances, or if they exhibit toxic, corrosive, ignitable, or reactive characteristics according to the following definitions:

(1) Toxicity: Results from TCLP tests are compared with the national Primary and Secondary Drinking Water Standards (PDWS and SDWS, respectively) to determine whether the waste is hazardous; current regulations set TCLP toxicity values at 100 times PDWS levels;

(2) Corrosivity: A waste is considered corrosive, and therefore hazardous, if it has a pH less than 2.0 or greater than 12.5;

(3) Ignitability: Various procedures exist to determine the tendency of a waste product to catch fire, depending on the physical state of the substance; and

(4) Reactivity: The reactivity of the material gives an indication of its stability and tendency to react violently or explode.

Based on the RCRA classifications, a majority of wood fibre processing wastes such as fly ash, bottom ash, and boiler slag are currently considered "nonhazardous solid wastes". The U.S. industry also has a pulping liquor exemption. An exception is the waste generated from wood preserving operations using chlorophenolic, creosote, and (or) inorganic preservatives containing arsenic and chromium, which are listed as hazardous. Specifically, bottom sediment sludge from wood preserving, preservative drippage, process residuals, and toxic spent formulations from wood preserving processes are considered hazardous wastes. Nevertheless, nonhazardous solid wastes are still subject to the appropriate solid waste regulations approved by individual provinces or states. In addition, new waste disposal sites or the expansion of existing facilities in the United States must meet the requirements of U.S. EPA Clean Water Act and the Safe Drinking Water Act. These water resource regulations dictate that any runoff or leachate leaving the boundary of a disposal area must meet drinking water standards, whatever the quality of the surrounding groundwater.

It should be noted that environmental regulations are always subject to amendment. For example, the limits to dioxin and furan concentration in soil are currently regulated in the United States by the existing Toxic Substances Control Act. But new RCRA rules and a consent decree require the U.S. EPA to determine whether pulp and paper mill wastewater treatment sludges meet the criteria to be listed as non-hazardous wastes. If confirmed, this will place additional restrictions on sludge disposal and land application.

Solid waste management practices

Over the last decades, great advances have been made to develop new waste management technologies as alternatives to conventional landfill disposal for the wood fibre processing industry. This is particularly evident for the primary wood products sector. As shown in Table 19.2, a majority of wood residues were being reused as supplementary fuels and to produce primarily paper, nonstructural panels, and other value-added products in 1998. Only 2% of the bark and 4% of the wood residue was not used and is therefore subject to further recovery. In contrast, the pulp and paper mills still diverted most of their wastes to landfills and lagoons. For example, of the 13.2 million tonnes of solid wastes generated from U.S. pulp and paper mills in 1995, 60% was disposed of either in landfills or lagoons (see Table 19.4). Table 19.12 summarizes the main management options for different solid wastes generated from wood processing mills.

Similar solid waste management approaches are being practiced by many Canadian wood fibre processing industries. Paprican surveyed Canadian pulp and paper mills about the generation and deposition of 19 types of solid residues at the end of 1994 and again at the end of 1995 (Reid 1998). As shown in Fig. 19.6, most wood residues and over one third of the sludge materials were burned for energy recovery. The unburned residues were mostly landfilled. Other utilization alternatives such as land application, composting, and recycling only accounted for minor disposal streams. The main constraints for further recovery include the less economically valuable materials present in these wastes and the practical difficulties involved in separating various contaminants.

Table 19.12. Management options for different solid wastes generated by wood processing mills.

Waste streams	Main management options
Primary timber production	
Logging	
Waste wood particles	Boiler fuel, pulping materials, or sold as mulch
Bark	Boiler fuel or sold as mulch
Sawing and panelling	
Bark, trim	Boiler fuel or sold as mulch
Sawdust, wood chips	Boiler fuel, pulping materials, reconstituted wood panelling
Sand, stone, dirt, tar	Potentially hazardous materials
Tank sludge	Hazardous materials for disposal
Adhesive residues	Hazardous materials for disposal
Wood preserving	
Bottom sediment sludges	Landfilling
Pulp and paper production	
Stock preparation	
Bark	Boiler fuel or sold as mulch
Dirt, grit	Landfilling
Chemical pulping	
Resins, fatty acids	Landfilling
Dregs, lime mud, lime slaker grits	Land application, construction materials, or landfilling
Power and steam boiler	
Fly ash, bottom ash	Land application, construction materials, or landfilling
Scrubber sludge	Landfilling
Wastewater treatment	
Sludges	Landfilling, land application

However, this trend is changing because of greater acceptance of an integrated solid waste management approach. Its basic management principle is that source reduction should take precedence, followed by recycling and reuse in order to maximize the use of natural resources and minimize sources of pollution. Treatment, destruction, and disposal without the recovery of valuable components and energy should be considered the least desirable options. Each of these strategies is introduced below with emphasis on their technical feasibility, costs, and application restrictions.

Source reduction

The best strategy for solid waste management is source reduction because less raw material and energy for production are required. There are many opportunities for the wood fibre processing industry to improve source reduction practices. Modifying industrial processes, redesigning products, improving management practices, and using less or no toxic chemicals have all been reported. Some of the important source reduction practices widely used in the wood fibre processing industry are discussed below.

In the softwood plywood industry, the manufacturing process has been modified to reduce the generation of adhesive residues. One technique is to replace the typical sprat-

Fig. 19.6. Solid residue disposal alternatives selected by Canadian pulp and paper mills in 1994 and 1995 (from Reid 1998); Mt, million tonnes.

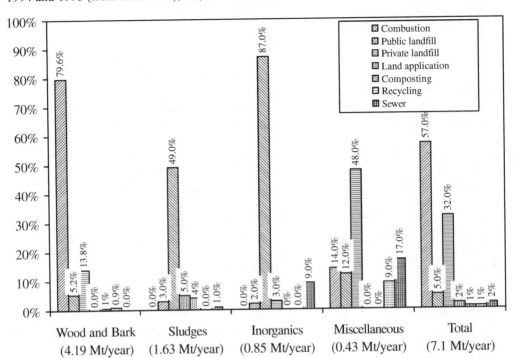

line layout technique with foam extrusion in which foamed adhesive is forced under pressure into the extrusion head. This modification provides a better distribution of the glue stream onto the veneer, leading to the generation of less adhesive residue and less chance of adhesive dry-out before pressing. Another technique is the variable application rate technique. Because the amount of adhesive required to bond veneer varies with the moisture content, the glue spread rate can be adjusted according to on-line measurements of the moisture content of each individual plywood panel, resulting in great reduction in the amount of adhesive consumed and wasted (USEPA 1995b).

In pulp and paper mills, substantial reduction in solid waste generation can also be accomplished through process modification and material substitution. It has been suggested that the most effective methods to control the release of toxic chlorinated compounds are through process modifications. Some of these modifications now being used are (USEPA 1995b):

- extended delignification;
- oxygen delignification;
- ozone delignification;
- use of anthraquinone catalysis in pulping liquor;
- enzyme treatment of pulp;
- improving brownstock and bleaching stage washing;
- use of chlorine dioxide as a bleaching agent;
- split addition of chlorine/improved pH control during pulp bleaching;

- improved chipping and screening;
- oxygen-reinforced/peroxide extraction;
- improved chemical control and mixing; and
- installing efficient chemical recovery systems.

More detailed descriptions of these technologies can be found in Smook (1992) and USEPA (1993, 1998). By using these modifications, the amount of free chlorine required for pulp bleaching can be greatly reduced or even completely eliminated, thereby preventing the release of toxic chlorinated compounds. This is particularly important because most of them cannot be removed by conventional wastewater treatment technologies on account of their resistance to biodegradation. However, some of these source reduction efforts usually require substantial changes to the production equipment. Because the pulp and paper industry is a very capital-intensive industry, any major production change tends to be expensive and requires a long time to implement. Nevertheless, the pulp and paper industry is also a dynamic one. Great advances have been made, reflected in greatly reduced releases of toxic chemicals, reduced solid waste generation, and increased use of alternative pulping and bleaching chemicals (Smook 1992; USEPA 1995b).

Good management practices and effective record keeping are also an important part of source reduction. For example, keeping the wood stock clean will prevent dirt, sawdust, and other debris from accumulating in subsequent wood processing. To do so, wood can be covered during shipment and stocking. In addition, effective record keeping can also aid in identifying and tracking the sources of solid wastes, thereby providing a guide for potential source reduction in the future.

The use of alternative manufacturing processes and chemicals is another practice that can reduce the generation of toxic chemical wastes by the wood products industry. For example, panel products mills are using high moisture bonding adhesives with the primary goals of reducing dryer emissions and possibly reducing wood drying costs. Efforts have also been made to develop naturally derived lignin-based adhesives, thereby emulating one of the major components of wood. The abundance of lignin as a waste product in pulp mills has made it a desirable raw material alternative to non-renewable petroleum-derived synthetic adhesives. To date, the drawbacks of this experimental technology have been a longer curing time, weaker cross-linking, and compromised board strength (USEPA 1995b).

Recycling

Recycling usually involves collecting the materials from the waste stream, preparing them for use as manufacturing materials, and processing them into an end product. Thus, an ideal recycling program requires extensive training and research to identify suitable materials for recovery, to develop effective separation and cleaning protocols for removing any contaminants, and to implement a good quality control program. It is often difficult to process different kinds of materials and maintain a satisfactory product. All of these activities may require substantial energy and labour. Furthermore, the markets for these materials must be developed in order to make any recycling program success-

Fig. 19.7. Waste sawdust generated by a primary wood processing facility (such as the sawmill in the background) can be used as the raw material for a secondary manufacturing facility (such as the panelboard plant in the foreground). This diverts much of the solid waste stream from disposal by incineration in low-performance burners (one of which is visible in the mid-background). Shown are the Pacific Inland Resources sawmill and Newpro panelboard plant in Smithers, British Columbia.

Photo by Phil Burton

ful. Only when manufacturers need those materials, or can use them as economical substitutes for raw materials, can the recovered materials be made into end products.

There are many opportunities for the primary timber mills to recycle wood wastes as usable raw materials. As already mentioned, wood wastes (particularly in the form of sawdust and chips) have been extensively used as boiler fuel, as raw materials to manufacture reconstituted wood panels (Fig. 19.7), and as pulp and paper mill feedstock. The specifications for recycling wood wastes from the primary wood products mills vary greatly, largely because of different end uses. For example, boiler fuels preferably use clean wood wastes such as tree trimmings and bark. Waste wood pretreated with tar or creosote, plywood, and dirty stumps may not be acceptable because these materials will affect the boiler performance and may cause air pollution problems by producing chlorinated organic compounds.

Like the primary wood products mills, many pulp and paper mills also recycle damaged products and scrap from converters because these materials are of known composition, are free of inks, and can be directly used as a pulp substitute. When the chips made from wood wastes are used as the feedstock for pulping processes, however, bark and foliage usually have to be removed as they can cause many problems: decreased

Fig. 19.8. This small paper mill in western Massachusetts (U.S.A.) uses recycled paper collected from the densely populated New England region as its primary feedstock. Though paper recycling is considered environmentally responsible, the process generates a toxic de-inking sludge that must subsequently be disposed of.

Photo by Phil Burton

pulping yield; reduced pulp strength; contaminated finished products; and deterioration to the processing equipment.

As well, pulp and paper mills can incorporate a certain portion of secondary wood fibres from post-consumer wastepaper into their products while not sacrificing product quality requirements (Fig. 19.8). In fact, much of the past recycling efforts in the pulp and paper sector have been focused on the use of secondary fibres as pulping materials. In doing so, the recovered papers must be carefully sorted by grade (newspapers, corrugated cardboard, high-grade paper, and mixed papers) to ensure that the economic value of each grade of these recycled papers can be maximized. Once collected, they must be cleaned to remove any contaminants such as food, plastics, metals, and wax.

It should be noted that only a portion of wastepaper can be reused, owing to both economic and logistical considerations. First, virgin fibre is abundant and relatively cheap in Canada as extensive forests are available. Second, most large urban centres, the source of most post-consumer recycled fibre, are located long distances from pulp and paper mills. Consequently, urban pulp mills can use a larger proportion of secondary fibres because of the post-consumer feedstock close at hand, but rural mills are usually close to timber sources and thus may use virgin fibre in greater proportion. Nevertheless, empty haulbacks of railcars and transport trucks that deliver pulp and paper to population centres could be more effectively used to feed post-consumer paper back to pulp and paper mills.

Secondary fibre sources can seldom be used as the sole feedstock for high quality or grade paper products. Contaminants such as inks and paper colours (dyes) are often present, so production of low-purity products is often a cost-effective use of secondary fibres, although many decontamination technologies are available. At present, approximately 75% of the secondary fibres in North America is used for multi-ply paperboard or corrugated cardboard (USEPA 1995b). It is expected that a greater proportion of recycled paper will be deinked for newsprint or other higher quality uses in the future. But mill capacity to de-ink and reuse post-consumer paper is still limited by market expectations to maintain product substance and strength, which requires a critical amount of primary fibre. As a result, the maximum potential utilization of wastepaper in the manufacture of new consumer products has been suggested to be around 50% (USEPA 1995b).

The selection of appropriate contaminant removal processes depends on the type and source of secondary fibre to be pulped. In-mill paper waste can easily be repulped with minimal contaminant removal. Recycled post-consumer fibres, on the other hand, may require extensive contaminant removal. Common contaminants of concern in secondary fibres include adhesives, coatings, polystyrene foam, dense plastic chips, polyethylene films, wet strength resins, and synthetic fibres and inks. Those contaminants of greater density than the desired secondary fibres are often removed by centrifugal force while lighter ones are removed by a floatation process. Centrifugal cleaners are also used to remove material less dense than fibres, such as wax and plastic particles. Continuous solvent extraction has been used to recover fibres from paper and cardboard coated with plastics and (or) wax. Inks may be removed by heating a mixture of secondary fibres with surfactants. The removed inks are then dispersed in an aqueous medium to prevent redeposition on the fibres. However, the accumulation of contaminants during the repeated treatment and utilization of secondary fibres will require the use of more and more additives, inks, dyes, softeners, etc., reducing the paper's subse-

Fig. 19.9. Trend of the recovered paper uses in Canadian mills (from CPPA 2002).

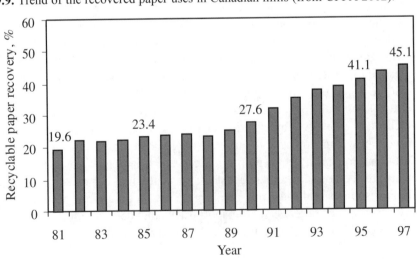

> **Box 19.2. Beneficial utilization alternatives for solid wastes generated by the fibre processing industry:**
>
> - Burned to fire boilers on site to recover renewable bioenergy;
> - Applied on agricultural and forestry land for soil improvement; and
> - Incorporated into various value-added products.

quent recyclability, making virgin wood fibre more economically attractive, and generating new waste disposal problems.

Nevertheless, the utilization of secondary fibre, expressed as a percentage of the total fibre used to make pulp, is at approximately 30% in the United States and is gradually climbing. In resource-deficient countries such as Japan, the secondary fibre utilization rate is at about 50%, whereas the average utilization rate in Europe is approximately 40% (CPPA 2002). Figure 19.9 shows the trend in usage of recovered paper in Canadian mills from 1981 to 1997. A total of 4.7 million tonnes of recovered paper with a recovery rate of 45% was used to manufacture recycled-content paper and packaging. In 1997, approximately 71% of the fibre used to make Canadian pulp and paper came from sawmill residues and recovered papers that used to go into landfills, a remarkable testimony to the enhanced sustainability of the industry.

Beneficial utilization technologies

Many alternatives have been proposed to beneficially utilize the solid wastes from wood fibre processing in order to divert them from disposal in landfills or lagoons. Among them, waste-to-energy combustion and land application have been most widely used, though new waste-based products are also emerging (Box 19.2). Recently, more attention has been given to emerging technologies such as composting, co-disposal and utilization with other wastes, value-added products, construction materials (Fig. 19.10), and wood fibre recovery. Note that some of these technologies have proven to have a broad application for different types of solid wastes, while others are only applicable to specific geographic regions, mill types, or even a particular waste stream from a specific mill.

Waste-to-energy combustion

The domestic use of wood residues as fuel materials and (or) supplemental heating fuel (e.g., as flowable wood pellets used in specialized pellet stoves for residential space heating) has been practiced for many years. However, their largest use in wood fibre processing mills is to produce steam because of the availability of enormous quantities of wood wastes, the large heat demands for wood fibre processing, and the rising cost of waste disposal. The primary sludge from pulp mills, mainly consisting of wood fibre, can also be used for energy, but must be dewatered to a solid state and then fed into the boiler.

The efficiency of waste-to-energy combustion is affected by a number of factors. During combustion, sufficient air is required to oxidize organic carbon and hydrogen

Fig. 19.10. Paper mill sludge mixed with agricultural wastes for composting.

Photo by Calvin Chong

into carbon dioxide and water. Stoichiometrically, about 6.5 kg of air is required to completely combust 1 kg of wood materials. In practice, however, 60% more air must be supplied than the amount required stoichiometrically in order to achieve effective combustion (Cheremisinoff et al. 1976).

The success of waste-to-energy combustion also depends on several operational, economical, and environmental conditions:

- equipment to handle, control, and feed the waste wood residues to the furnace;
- labour requirements for handling the waste residues and ashes;
- scale buildup on (and removal from) the furnace surfaces; and
- proper air pollution control devices.

With the use of a properly designed boiler system, an efficiency of 45–63% energy recovery can be obtained in many wood residue combustion steam plants (Cheremisinoff et al. 1976).

The use of wood wastes as fuel or fuel supplements offers several unique benefits. A potentially large volume of wood is constantly renewable. In addition, most wood residues cause less air pollution than fossil fuels such as coal and crude oils because they have a lower sulphur and ash content. In addition, the development of chip combustion technology has dramatically increased the efficiency of energy generation from wood residues. The costs per joule are much less for chips than the fossil fuels.

There are a number of disadvantages associated with the use of wood residues as fuel materials. For comparable heating values, wood bulkiness requires much larger storage space as compared to fossil fuels. Also, the moisture content of wood residues varies greatly with the wood type, season, particle size, and handling methods. Finally, combustion of wood residues may cause serious emission of particulate matter. Other pollutants, particularly carbon monoxide, manganese, and organic compounds, may be emitted in significant quantities under poor operating conditions (Chap. 18). In general, the composition and size of wood residue feedstock determines the amount and type of particulate matter emitted with the stack gas. The use of bark as compared to wood produces more ash and also carries a larger amount of dirt and sand from handling operations. Therefore bark emissions also contain dirt, sand, and char. When the wood residues co-combust with wastewater treatment plant sludge, toxic organic compounds such as dioxins and furans may also be emitted. For further discussion and details on air pollution issues associated with all aspects of the wood products industry, see Chap. 18.

To control the emission of these contaminants into the atmosphere, several strategies can be implemented. One is to optimize the combustion process through proper furnace design, operation, and maintenance. Further improvement can be made by fuel drying, cleaning, and sizing. Another approach is to install various kinds of air pollution control equipment. The relatively large particulates can be efficiently removed from flue gas by mechanical cyclone precipitators. Smaller particulates may require electrostatic precipitators or baghouses for collection. At present, new boilers must meet new source performance standards for air pollutants. These options are explored more fully in Chap. 18.

Finally, a new regulatory framework must be in place to allow the excess electricity generated from the combustion of wood residues to be patched into the electric grid and sold to regional power utilities. There is a potential to recover additional energy by co-combusting wood residues with other energy sources such as municipal solid wastes. In some cases, government subsidies and public support are needed to make this option economically and politically feasible, while counter-productive regulations simply need to be removed or updated in other cases (Chap. 7).

Land-based application

Land application is defined as the spreading of wastes on or just below the soil surface (Fig. 19.10). It has been widely practiced for municipal wastewater treatment sludge. The sludge may be applied to agricultural land, forest land, mined land, or other degraded land. Sunlight, soil microorganisms, and desiccation combine to help destroy pathogens and many toxic organic substances present in these wastes. Trace metals will be adsorbed in the soil matrix and nutrients are taken up by plants. In all of these cases, sludge is used as a valuable resource to improve productivity of the land. As a soil conditioner, the applied wastes may facilitate nutrient transport, add organic matter, increase water retention, reduce soil erosion, and (or) improve soil tilth. They could also serve as a replacement for chemical fertilizers (Bellamy et al. 1995; Naylor and Schmidt 1989; NCASI 1985).

Land application holds great promise for wood ashes and wastewater treatment plant sludge generated from many pulp and paper mills, because these materials are rich in many of the nutrients required for plant growth. In fact, the value of wood ashes as a liming and fertilizing soil amendment has long been recognized in agricultural circles, and even led to the award of U.S. Patent No. 1 for *"The improved method of making pot and pearl ashes*" in 1790 (Vance 1996). Subsequent greenhouse and field studies confirmed that wood ashes can effectively increase the content of potassium and several micronutrients in the soil (Tables 19.9 and 19.10), thereby increasing plant production and improving soil properties (Erich 1991). Scandinavian studies indicated that wood ashes could enhance long-term forest productivity, microbial activity, and nitrogen fixation (Vance 1996). For example, willow (*Salix* sp.) productivity was increased by 65–70% in wood ash treated plots compared to control plots (Weber et al. 1985). These researchers have suggested that land application of wood ash would be especially suitable by replacing potassium, calcium, phosphorus, and other micronutrients lost from intensively managed forests where a great quantity of tree biomass is removed over shorter time periods (see Chap. 13). Wood ashes were also recommended as a means of correcting forest nutrient deficiencies caused by natural weathering processes and acidic deposition. A distinct feature was observed that the nutrients supplied by the wood ashes can persist much longer than those gained from commercial fertilizers, presumably owing to the slower dissolution rates. Zhou et al. (1997) conducted a comprehensive study to determine physical, chemical, and environmental properties of wood ashes sampled from a Canadian kraft pulp mill. Among 21 elements analyzed from four waste streams, the content of calcium, magnesium, and potassium accounted for over 95% of the ash, confirming the high liming capacity and potassium nutrient content of wood ashes, and making it a valuable amendment for enhanced plant growth.

However, wood ashes alone are typically nitrogen deficient as a soil amendment. In most cases, the addition of commercial nitrogen fertilizers becomes necessary in order to supply a complete spectrum of nutrients required for plant growth. The problem might be corrected by co-application of industrial or municipal wastewater treatment sludge, which is often rich in nitrogen (Rowell et al. 2002). By co-application, the potential benefits might include:

(1) balanced plant nutrition will be achieved by combining potassium- and phosphorus-rich wood ashes with nitrogen-rich sludge;
(2) the sludge can be stabilized because of the high pH of wood ash, thereby improving its physical, chemical, microbiological, and aesthetic (odour-emitting) qualities;
(3) stabilization may lower the elemental dissolution rates, thereby providing desirable long-term sources of nutrients; and
(4) both wood ash and sludge are usually readily available from the same location (modern pulp mills), thus reducing waste hauling costs.

Until now, these potential synergistic benefits have not been fully explored in industrial practice.

In order to apply solid wastes to land in an acceptable manner, a number of practical factors must be taken into consideration. These include:

- characteristics of the wastes;
- site characteristics;
- waste application rate;
- the modes of application;
- types of crops grown; and
- hauling distances to application sites.

Detailed discussion about the guidelines to address these factors can be found in AEP (1999).

Application of solid wastes to agricultural land can be easily accomplished by using conventional manure spreaders (Fig. 19.11) or tractor-mounted box spreaders, followed by plowing or disking into the soil. For most forestry land, however, such application methods become impractical because the spreaders cannot distribute the solid wastes uniformly over a wide area, and soil cultivation is not practical or would have unacceptable impacts. Instead, sludges and ashes are usually mixed with water and applied as a slurry using various pumps, booms, or hoses, and appropriate nozzles.

Another main obstacle that hinders the wide use of land application for wood ashes and wastewater treatment sludge is the concern about potential environmental contamination (Pickell and Wunderlich 1995; Baril and Drolet 1997). Concerns about adsorbable organic halogens (AOX), metals, and other contaminants, coupled with the costs of transportation and handling for land application, have (until recently) deterred Canadian pulp and paper mills from investigating land application methods. Prior to its wider application, the hurdles of environmental regulations and disposal costs must be safely overcome.

Other beneficial uses

In addition to the above beneficial uses, many alternative technologies have been suggested to utilize the solid wastes produced by wood processing industries. Among them, much attention has recently been centred on:

Fig. 19.11. Pulp mill sludge being applied to agricultural land as a fertilizer, southern Ontario.

Photo by Calvin Chong

- composting;
- production of ethanol;
- pelletization of sludges and nonrecyclable paper for use as fuel;
- production of construction materials such as cements, bricks, ceramics, lightweight
- aggregates, etc;
- land reclamation;
- landscaping;
- use as hydraulic barrier material for landfill cover systems; and
- animal litter or feedstock.

Most of these alternative technologies are proposed to exploit the presence of cellulose materials, nutrients, or other usable constituents of most wood wastes. For example, the high percentage of cellulose in mill sludges has led to their being recycled back to the mills for fibre recovery, or as feedstock to produce ethanol and fuel pellets (NCASI 1994a). The mill sludges and boiler ash has also been used to manufacture the light-weight aggregates as construction materials (NCASI 1994b). The low hydraulic conductivity of some mill waste has rendered it an excellent landscaping and landfill cover material (Teasdale et al. 1999). Primary sludge is able to adsorb a considerable amount of water, making it suitable for use as animal litter. The large surface area and low bulk density of hog fuel prompted its use as a biofiltration medium to remove reduced sulphur gases from air emissions (Wani et al. 2001).

It should be noted that some of these technologies potentially have broad applications to many types of solid wastes generated by wood fibre processing mills, while techniques are limited to specific geographical regions, mill types, waste types, and even a specific waste stream within a mill. Thus, their application in practice should be based on a careful assessment of factors, including technical feasibility, cost, available markets, and potential environmental liability (NCASI 1993). At present, the viability of their full-scale application remains to be demonstrated.

Conclusions and recommendations

Landfill and lagoon disposal remains the most popular option to manage various wood processing solid wastes. However, there has been a steady shift to other alternatives leading to the reduction, reuse, and recycling of waste materials. This shift responds to increasing environmental concerns, the limited availability of land for waste disposal, and the aspiration for more complete recovery of natural resources. Among them, the most widely practiced are waste-to-energy combustion and land application. Other promising alternatives (including composting, co-disposal and utilization with other wastes, value-added products, construction materials, and wood fibre recovery) have also received great attention recently. However, the feasibility of the full-scale application of some creative solutions remains to be demonstrated.

Many alternative management strategies are available for the solid wastes generated from wood fibre processing (Box 19.2). However, their applicability is site-specific, as each of the management alternatives differs greatly in (1) technical feasibility, (2) cost,

> **Box 19.3. Main considerations for alternative solid waste management practices:**
>
> - Technical feasibility;
> - Available markets for recycled and reused materials;
> - Potential environmental liability;
> - Public acceptance for the products; and
> - Economical viability.

(3) available markets, and (4) potential liability (Box 19.3). The relative significance of these factors depends on the characteristics of the wastes, the type of wood fibre processing being practiced, industrial business strategies, and other local conditions.

Although great advances have been made in reducing the generation of waste and in integrating beneficial waste utilization practices, more technical, economical, and regulatory obstacles need to be overcome to make their application more widespread. The key to increased recycling and reuse is the development of new and more cost-effective separation and processing technologies. Once separated, the recoverable wood fibres can be successfully incorporated in a wide variety of reconstituted wood, pulp and paper products.

Acknowledgements

I would like to thank Drs. Philip J. Burton and Daniel W. Smith for their thoughtful advice and considerable suggestions in preparing and revising the manuscript. The financial support provided by the Sustainable Forest Management Network of Centres of Excellence is also greatly appreciated.

References

Adriano, D.C., Page, A.L., Elseewi, A.A., Chang, A.C., and Straughan, I. 1980. Utilization and disposal of fly ash and other coal residues in terrestrial ecosystems: a review. J. Environ. Qual. **9**: 333–344.

AEP (Alberta Environment Protection). 1999. Land application of mechanical pulp mill sludge to agricultural land. Environmental Science Division, Alberta Environment Protection, Edmonton, Alberta. 38 p.

Baril, P., and Drolet, J.Y. 1997. A review of constraints and opportunities for the land application of paper mill sludges and residues. Pulp & Paper Can. **98**(6): 63–66.

Bellamy, K.L., Chong, C., and Cline, R.A. 1995. Paper sludge utilization in agriculture and container nursery culture. J. Environ. Qual. **24**: 1074–1082.

Campbell, A.G. 1990. Recycling and disposing of wood ash. TAPPI J. **73**: 141–146.

CCME (Canadian Council of Ministers of Environment). 1996. Provisional code of practice for management of post-use treated wood. Hazardous Waste Task Group, Canadian Council of Ministers of Environment, Winnipeg, Manitoba. 90 p.

Cheremisinoff, P.N., Cheremisinoff, P.P., Morresi, A.C., and Young, R.A. 1976. Woodwastes utilization and disposal. Technomic Publishing Company, Westport, Connecticut. 215 p.

Clarke, L.B. 1994. Legislation for the management of coal-use residues. IEACR/68, International Energy Agency Coal Research. London, U.K. 75 p.

(CPPA) Canadian Pulp and Paper Association. 2002. Paper recycling in Canada. Available at http://www.cppa.org/english/wood/guide/papere.htm [viewed Oct. 15, 2002].

Environment Canada. 1985. Environmental codes of practice for steam electric power generation. Environment Canada, Ottawa, Ontario. EPS-1/PG/1. 85 p.

Erich, M.S. 1991. Agronomic effectiveness of wood ash as a source of phosphorus and potassium. J. Environ. Qual. **20**: 576–581.

Johe, D.E., Jurewicz, D.A., and Diehl, D. 1990. The importance of selecting the proper parameters in a monitoring program. Proceedings of the TAPPI (Technical Association of the Pulp and Paper Industry) Environmental Conference, 9–11 April, Seattle, Washington. Vol. 2. pp. 667–698.

Lerner, B.R., and Utzinger, J.D. 1986. Wood ash as soil liming material. Hortic. Sci. **21**(1): 76–78.

Mason, B.J., and Carlile, D.W. 1986. Round-robin evaluation of regulatory extraction methods for solid wastes. Electric Power Research Institute, Palo Alto, California. EPRI-EA-4740. 190 p.

Mattigod, S.V., Rai, D., Eary, L.E., and Ainsworth, C.C. 1990. Geochemical factors controlling the mobilization of inorganic constituents form fossil fuel combustion residues: I. review of the major elements. J. Environ. Qual. **19**: 188–201.

McKeever, D.B. 1999. How woody residuals are recycled in the United States. Biocycle, **40**(12): 33–44.

(MEI) Metcalf and Eddy, Inc. 1991. Wastewater engineering: treatment, disposal and reuse, 3rd edition. McGraw–Hill, New York, New York. 1334 p.

Muse, J.K., and Mitchell, C.C. 1995. Paper mill byproducts find a home on the farm. Highlights Agric. Res. **40**: 3.

Naylor, L.M., and Schmidt, E. 1989. Paper mill wood ash as a fertilizer and liming material: field trials. TAPPI J. **72**: 199–206.

NCASI (National Council of the Paper Industry for Air and Stream Improvement). 1978. Response of selected industry sludges to alternate solid waste toxic extraction procedures. New York, New York. Tech. Bull. No. 311.

NCASI (National Council of the Paper Industry for Air and Stream Improvement). 1985. Recent studies and experience with use of sludge and fly ash on farm crop and forest lands. New York, New York. Tech. Bull. No. 478. 140 p.

NCASI (National Council of the Paper Industry for Air and Stream Improvement). 1990. Response of selected pulp and paper industry solid wastes to the RCRA toxicity characteristic leaching procedure (TCLP). New York, New York. Tech. Bull. No. 587. 20 p.

NCASI (National Council of the Paper Industry for Air and Stream Improvement). 1992. Chemical composition of pulp and paper industry landfill leachate. New York, New York. Tech. Bull. No. 643. 35 p.

NCASI (National Council of the Paper Industry for Air and Stream Improvement). 1993. Alternative management of pulp and paper industry solid wastes. New York, New York. Tech. Bull. No. 655. 44 p.

NCASI (National Council of the Paper Industry for Air and Stream Improvement). 1994a. Alternatives for management of pulp and paper industry solid wastes: production of ethanol. New York, New York. Tech. Bull. No. 685. 53 p.

NCASI (National Council of the Paper Industry for Air and Stream Improvement). 1994b. Alternatives for management of pulp and paper industry solid wastes: production of lightweight aggregate. New York, New York. Tech. Bull. No. 687. 43 p.

NCASI (National Council of the Paper Industry for Air and Stream Improvement). 1999. Solid waste management practices in the U.S. paper industry — 1995. New York, New York. Tech. Bull. No. 793. 26 p.

NCG (National Committee for Geochemistry). 1980. Trace-element geochemistry of coal resource development related to environmental quality and health. National Academy Press, Washington, D.C. 153 p.

Oehme, M., and Muller, M.D. 1995. Levels of congener patterns of polychlorinated dibenzo-p- dioxins and dibenzofurans in solid residues from wood-fired boilers: influence of combustion conditions and fuel type. Chemosphere, **30**(8): 1527.

Pickell, J., and Wunderlich, R. 1995. Sludge disposal: current practices and future options. Pulp & Paper Can. **96**(9): 41–47.

Reid, I.D. 1998. Solid residues generation and management at Canadian pulp and paper mills in 1994 and 1995. Pulp & Paper Can. **99**(4): 49–52.

Rowell, D.M., Prescott, C.E., and Preston, C.M. 2002. Decomposition and nitrogen mineralization from biosolids and other organic materials. J. Environ. Qual. **30**: 1401–1410.

Sakata, M. 1987. Movement and neutralization of alkaline leachate at coal ash disposal sites. Environ. Sci. Technol. **21**: 771–777.

Shields, W.J., Huddy, M.D., and Somers, S.G. 1985. Pulp mill sludge application to a cottonwood plantation. In Beneficial use of secondary fiber rejects. National Council of the Paper Industry for Air and Stream Improvement, New York, New York. Tech. Bull. No. 478. pp. 29–47.

Sinkkonen, S., Maekelae, R., Vesterinen, R., and Lahtiperae, M. 1995. Chlorinated dioxines and dibenzothiophenes in fly ash samples from combustion of peat, wood chips and refuse derived fuel and liquid packaging boards. Chemosphere, **31**(2): 2629–2635.

Smook, G.A. 1992. Handbook for pulp and paper technologists, 2nd edition. Angus Wilde Publications, Vancouver, British Columbia. 419 p.

Someshwar, A.V. 1996. Wood and combination wood-fired boiler ash characterization. J. Environ. Qual. **25**: 962–972.

Tchobanoglous, G., Theisen, H., and Vigil, S. 1993. Integrated solid waste management. McGraw–Hill, New York, New York. 978 p.

Teasdale, M., Zeiss, C., and Li, R. 1999. Moisture redistribution in a landfill using capillary wick layers. Sustainable Forest Management Network, Edmonton, Alberta. Proj. Rep. 1999-39. 86 p.

Theis, T.L., and Gardner, K.H. 1990. Environmental assessment of ash disposal. CRC Crit. Rev. Environ. Control, **20**(1): 21–41.

Theis, T.L., Westrick, J.D., Hsu, C.L., and Marley, J.J. 1978. Field investigation of trace metals in groundwater from fly ash disposal. J. Water Pollut. Control Federat. **50**: 2457–2469.

(USEPA) United States Environmental Protection Agency. 1986. Hazardous waste management I: toxicity characteristic leaching procedure. Federal Regist. 51(02-1766).

(USEPA) United States Environmental Protection Agency. 1993. Pollution prevention technologies for the bleached kraft segment of the U.S. pulp and paper industry. Office of Pollution, Prevention, and Toxics, United States Environmental Protection Agency, Washington, D.C. EPA/600-R-93-110. 196 p.

(USEPA) United States Environmental Protection Agency. 1995a. Profile of the lumber and wood products industry. Office of Compliance, United States Environmental Protection Agency, Washington, D.C. EPA/310-R-95-006. 116 p.

(USEPA) United States Environmental Protection Agency. 1995b. Profile of the pulp and paper industry. Office of Compliance, United States Environmental Protection Agency, Washington, D.C. EPA/310-R-95-015. 130 p.

(USEPA) United States Environmental Protection Agency. 1998. Handbook on pollution prevention opportunities for bleached kraft pulp and paper mills. Office of Research and Development, United States Environmental Protection Agency, Washington, D.C. EPA/600-R-93- 098. 76 p.

Vance, E.D. 1996. Land application of wood-fired and combination boiler ashes: an overview. J. Environ. Qual. **25**: 937–944.

Wani, A.H., Branion, R.M.R., and Lau, A.K. 2001. Biofiltration using compost and hog fuel as a means of removing reduced sulfur gases from air emissions. Sustainable Forest Management Network, Edmonton, Alberta. Proj. Rep. 2001-9. Available at http://sfm-1.biology. ualberta.ca/english/pubs/PDF/PR_ 2001-09.pdf [viewed Nov. 26, 2002]. 30 p.

Weber, A., Karisisto, M., Leppanen, R., Sundman, V., and Skujin, J. 1985. Microbial activities in a histosol: effects of wood ash and NPK fertilizers. Soil Biol. Biochem. **17**: 291–296.

Zhou, H., Smith, D.W., and Sego, D.C. 1997. Evaluation of alternative uses of fly ash by-products. Sustainable Forest Management Network, Edmonton, Alberta. Proj. Rep. MIT-2. 115 p.

Zhou, H., Smith, D.W., and Sego, D.C. 2000. Characterization and use of pulp mill fly ash and lime by-products as road construction amendments. Can. J. Civil Eng. **27**: 581–593.

Chapter 20

Carbon balance and climate change in boreal forests

J.S. Bhatti, G.C. van Kooten, M.J. Apps, L.D. Laird, I.D. Campbell, C. Campbell, M.R. Turetsky, Z. Yu, and E. Banfield

"The greatest threat of all may be yet to come, in the form of global warming."

"Great Northern Forest" (1994),
a film directed by Joseph Viszmeg,
produced by Albert Karvonen and Jerry Krepakevich,
Karvonen Films Ltd., and National Film Board of Canada

Introduction

Most scientists agree that the global climate is changing as a consequence of human-caused perturbations to global biogeochemical cycles, especially the carbon cycle (IPCC 2001*a*). In addition, the impacts of these changes are becoming increasingly evident (IPCC 2001*b*). The boreal forest biome is one of the Earth's ecosystems most affected by the changing climate (IPCC 2001*a*; Gitay et al. 2001). The forests in this

J.S. Bhatti,[1] L.D. Laird, and E. Banfield. Canadian Forest Service,[2] Northern Forestry Centre, 5320-122 Street, Edmonton, Alberta T6H 3S5, Canada.
G.C. van Kooten. Department of Economics, University of Victoria, Victoria, British Columbia, Canada.
M.J. Apps. Canadian Forest Service,[2] Pacific Forestry Centre, Victoria, British Columbia, Canada.
I.D. Campbell. Geological Survey of Canada,[2] Ottawa, Ontario, Canada.
C. Campbell. Canadian Forest Service,[2] Ottawa, Ontario, Canada.
M.R. Turetsky. U.S. Geological Survey, Western Region Center, Menlo Park, California, United States.
Z. Yu. Earth and Environmental Sciences, Lehigh University, Bethlehem, Pennsylvania, United States.
[1]Author for correspondence. Telephone: 780-435-7210. e-mail: jbhatti@nrcan.gc.ca
[2]The views expressed in this chapter do not represent the official position of the Canadian Forest Service and the government of Canada.

Correct citation: Bhatti, J.S., van Kooten, G.C., Apps, M.J., Laird, L.D., Campbell, I.D., Campbell, C., Turetsky, M.R., Yu, Z., and Banfield, E. 2003. Carbon balance and climate change in boreal forests. Chapter 20. *In* Towards Sustainable Management of the Boreal Forest. *Edited by* P.J. Burton, C. Messier, D.W. Smith, and W.L. Adamowicz. NRC Research Press, Ottawa, Ontario, Canada. pp. 799–855.

biome also have the potential to either accelerate or retard the progression of climate change through their influence on the global carbon cycle.

This chapter explores the role of the boreal forest in the global carbon cycle, the expected impacts of climate change on the boreal ecosystem, and the effects of various factors (both natural and human) on the carbon balance of the forest. The last part of the chapter discusses the economic and forest management issues that arise when the carbon resources of the forest are considered in light of Kyoto Protocol commitments to reduce greenhouse gas emissions. While the primary focus of this chapter is the challenges (Box 20.1) climate change poses for the Canadian boreal forest, the boreal forest in other regions of the world will also be discussed where appropriate.

Box 20.1. Challenges to sustainable forest management in light of climate change.

Climate change poses many challenges for sustainable forest management in boreal areas. These include determining how to:

- manage carbon pools and fluxes, as well as the fibre supply from trees and stands;
- minimize use of fossil fuels;
- control disturbance, growth, and decomposition processes in an economically and ecologically feasible way;
- exploit opportunities for carbon sequestration and associated credits;
- maintain the ecological integrity of the forest in the face of increasing climate change and other global change stresses; and
- inventory forest ecosystem carbon stocks and to estimate carbon sources and sinks.

Forests and climate

Forests are highly influenced by climate, which determines their distribution, structure and composition, and much of their ecological function. Climate also affects forests indirectly through its impact on disturbance regimes such as those due to fire, insects, diseases, and windstorms. In turn, forests have an impact on global climate through their influence on surface roughness, albedo, hydrological cycles, and the carbon cycle. The influence of the boreal forest on climate through its effect on the global carbon cycle has been receiving increasing attention in recent years because of escalating concern about climate change. The role of carbon in climate change, the importance of the boreal forest in the global carbon cycle, and the expected impacts of climate change on the boreal forest are explored below.

The role of carbon in global climate change

Scientific background

The Earth's climate is determined in large part by the proportion of solar energy retained by the Earth's atmosphere. Certain gases (called greenhouse gases or GHGs) in the atmosphere allow short-wavelength solar energy (visible light) to pass through to the Earth but absorb outgoing infrared (radiant heat) wavelengths, thus making the Earth warmer than it would otherwise be. Without this greenhouse effect the Earth would be uninhabitable for most existing lifeforms, including humans. Naturally occurring GHGs include water vapour (H_2O) and various trace gases such as carbon dioxide (CO_2), methane (CH_4), ozone (O_3), and nitrous oxide (N_2O). Synthetic GHGs include chlorofluorocarbons (CFCs). The concentrations of GHGs in the atmosphere are largely regulated by global hydrological and biogeochemical cycles, including the carbon cycle.

Human activities such as burning fossil fuels, production of cement, and changes in land use have perturbed the carbon cycle, which has resulted in elevated levels of atmospheric CO_2, thus enhancing the greenhouse effect and causing global warming at an unprecedented rate (IPCC 2001a). It is estimated that during the period from 1850 to 1998, 405 Pg (1 Pg = 10^{15} g, 10^{12} kg, or 10^9 Mg or tonnes) of carbon was emitted to the atmosphere as a result of these activities (IPCC 2000). This has resulted in a 28% increase in atmospheric CO_2 concentrations (IPCC 2000), which is believed to be a primary cause of the observed 0.6°C increase in global temperature in the 20th century (IPCC 2001a). The climatic changes are, however, not uniformly distributed: the changes in solar insolation cause changes in atmospheric circulation patterns, resulting in changes in regional weather patterns. Some regions, especially northern and mid-continental landmasses, undergo larger climate changes than others.

Carbon thus plays a vital role in determining global climate. The global carbon cycle and the exchange of carbon between the atmosphere and various natural and anthropogenic components are illustrated in Fig. 20.1. Carbon is exchanged between terrestrial ecosystems and the atmosphere through photosynthesis, respiration, decomposition, and combustion. Boreal forests and their associated peatlands represent the largest terrestrial reservoir of carbon (IPCC 2000), as well as being located in a region especially sensitive to climate change. The boreal biome therefore plays a critical role in the global carbon cycle and has the capacity for either accelerating or slowing climate change to some degree, depending on whether the forest acts as a net source or a net sink of carbon. This source or sink status is, however, not a static characteristic of the ecosystem, but can change over time as a result of changes in forest age-class structure, disturbance regime, and resource use (Kurz and Apps 1999; Kauppi et al. 2001).

Projected future climates of boreal regions

Global climate is expected to warm by an average of 1.4–5.8°C by 2100, but the temperature increase in the boreal region and other high latitude biomes in the Northern Hemisphere is predicted to be more than 40% higher than this (IPCC 2001a). Changes in precipitation are more difficult to predict, but several models suggest regional changes in summer and winter precipitation of ±20% for the boreal region (Kirschbaum and Fishlin 1996; Amiro et al. 2001a). For example, precipitation is expected to decrease by 20% in northern Alberta, Saskatchewan, and Manitoba (Amiro et al. 2001a; Flannigan et al. 2001), while parts of eastern Canada (Flannigan et al. 2001; Amiro et al. 2001a) and Fennoscandia (IPCC 2001b) are expected to experience an increase in precipitation. Higher temperatures will result in higher evaporation rates, and hence

Fig. 20.1. Overview of the global carbon cycle, showing the stocks and fluxes of C (Pg) between various components of the biosphere (courtesy of Canadian Forest Service).

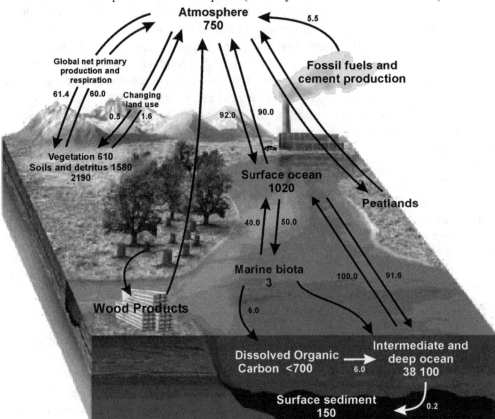

soils in some boreal regions are expected to be drier in the summer (Flannigan et al. 1998; Amiro et al. 2001*a*). The frequency and severity of storm events is also expected to increase (IPCC 2001*a*).

Boreal forests have experienced large climate swings in the past. During geologically recent Ice Ages, almost the entire present range of the North American boreal forests was glaciated, so that most of these forests have occupied their present range for only a few hundred generations of trees. Boreal forests therefore are composed mainly of species that are able to spread relatively quickly over previously unvegetated or recently vegetated terrain. Even these relatively rapid migration rates may not be fast enough to keep pace with the coming climate change, however, which is predicted to occur at a rate at least 10 times faster than the warming since the end of the last Ice Age (Schneider 1995).

The effects of climate change and other global changes on forests

Climate change is expected to affect both the distribution and the character of the boreal forest through changes in temperature, precipitation, and natural disturbance patterns (IPCC 2001*b*). These impacts on the forest are not entirely separable from the effects of

Fig. 20.2. Feedbacks between the atmosphere and various components of the boreal forest (modified from Bhatti et al. 2001).

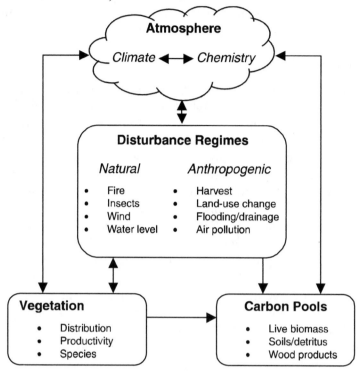

other global changes, such as increases in CO_2, NO_x, and O_3 levels, and anthropogenic pressures which may be enhanced by climate change. Figure 20.2 illustrates the web of interaction among climate, vegetation, disturbance regimes, and carbon pools in the boreal forest. Various components of this web are discussed in the following sections.

Water and energy exchange

Climate change has the potential to influence the dynamics of water and energy exchange between boreal ecosystems and the atmosphere in several ways. At the forest–tundra boundary, sharp discontinuities are hypothesized to have a strong influence on temperature and precipitation (Eugster et al. 2000). Snow-covered tundra and boreal peatlands have a much higher albedo, absorb less radiation, and warm the atmosphere less than the forested areas (Chapin et al. 2000). Consequently, an expansion of the boreal forest into regions now occupied by tundra and peatland (see below) has the potential to reduce albedo and increase spring energy absorption, thereby enhancing atmospheric warming (Bonan et al. 1992; Chapin et al. 2000). Other factors that enhance atmospheric warming in the boreal forest include earlier snowmelt (which reduces springtime albedo), and increased ground vegetation cover (which reduces evapo-transpiration). Deciduous forests have twice the albedo and 50–80% higher evapo-transpiration than coniferous forests during the growing season (Baldocchi et al. 2000). Changes in the disturbance regime that increase the proportion of deciduous forests have the potential to reduce spring energy absorption and act as a negative feed-back to atmospheric warming (Chapin et al. 2000; Eugster et al. 2000). With frequent fire, young boreal forest stands may have a ratio of sensible to latent heat that is 8 times higher than that of mature stands (Schulze et al. 1999).

Forest distribution and character

Both simple and more complex models agree that the ranges of major forest types around the world are likely to shift dramatically with future global climate change. Models using modern correlations between climate and vegetation have projected the stable ranges that might be achieved several hundred or thousand years from now. In many regions, the change is not simply from one forest type to another, but rather from forest to grassland or even desert in some regions, and from tundra to forest in others (Rizzo and Wiken 1992). The predicted stable ranges are, however, in many ways less of a concern than are the transitional displacements. Harder to predict than eventual sta-ble ranges (should they actually stabilize), the transitional state may include significant areas of essentially dead forest, where the existing vegetation is not able to cope with changed climatic conditions, and a suitably adapted natural forest type has not yet had time to migrate and establish itself.

Temperature-limited tree lines are expected to advance, either uphill in mountain regions, or into tundra regions in the Arctic (IPCC 2001*b*), although this potential is not likely to be realized in the next 100–200 years. At the same time, it has been postulated that climate warming will cause many dry forest regions (limited by water availability rather than temperature) to give way to grassland or more open savanna as a conse-quence of regeneration failure (Hogg and Hurdle 1995; Lenihan and Neilson 1995; Hogg 1997). Temperature is not the sole control on species distribution, and change in temperature cannot be considered in isolation. Other factors, including soil characteri-zation, nutrient availability, and disturbance regimes may prove to be more important in controlling future ecosystem dynamics. For example, the southern limit of boreal forests appears to be influenced more by interspecific competition and moisture conditions than by temperature tolerance (Loehle 1998). The distribution of trembling aspen (*Populus tremuloides* Michx.) in western Canada is largely controlled by moisture conditions (Hogg 1999). A decrease in water availability with climate change may therefore alter the southern boundary of the boreal forest and the distribution of aspen in some regions. Inertial effects may slow the loss of forest in dry regions, although many dry regions also experience human population pressures tending toward deforestation. For example, conifer density will likely decline in northeastern China, where climate warming may reduce the area of coniferous forest substantially, while elsewhere in Asia forests will be under increasing human pressures that will mask the effects of global warming (IPCC 2001*b*).

Many of the changes in forest distribution will not be brought about directly through the impact of climate change on individual trees, but rather through the effect of climate change on disturbance regimes and subsequent regeneration (Flannigan et al. 2000, 2001). The structure, pattern, and ecosystem processes of forests in many regions of the world are highly sensitive to changes in disturbance regimes (see Chap. 9). In north-

Fig. 20.3. Leaning dead trees are a good indicator of permafrost collapse. Permafrost melt can be initiated by climate warming and (or) fire.

(*a*)

Photo by Steve Zoltai, courtesy of Canadian Forest Service

(*b*)

Photo by Merritt Turetsky, courtesy of U.S. Geological Survey

eastern North America, the relative dominance of various tree species is largely determined by the mean fire return interval, which in turn is influenced by the vegetation, creating an interacting spiral (Flannigan and Woodward 1994). This feedback loop may result in meta-stable vegetation formations, sometimes in one state (e.g., grass-dominated), sometimes in another (e.g., tree-dominated). These highly non-linear feedback effects may result in vegetation changes more rapid than would be predicted from climate change alone.

One of the factors affecting regional response to rapid climate change is the existence of barriers to the immigration of species better adapted to the new climate in a given area. For instance, the climate north of the Great Lakes may become suitable for more thermophilous forest types than presently occur there, but the barrier formed by the lakes may slow or prevent suitable species from spreading northward. It has also been suggested that tundra afforestation may be slowed by the need for soil development (Rizzo and Wiken 1992). Most tree-line species, however, are adapted to growing on sites with minimal soils, since they are the same species that colonized much of the boreal region after deglaciation. Permafrost and changes in the active soil layer with warming add significant complexity to the tree-line response in much of Siberian Russia, Canada, and Alaska.

In some mountainous regions, altitudinally rising climatic zones may displace some ecozones from the tops of the mountains, which will lead to extensive extirpations. At the same time, valleys may develop climatic conditions suitable for species not currently present, allowing an opportunity for invasive species or causing temporary loss of forest cover. Such a situation is likely to lead to a large loss of biodiversity, as the first species suited to the new climate rapidly become dominant and resist the invasion of later immigrants. Over time, this inertial barrier may be overcome, but not if the climate continues to change and transient responses continue to dominate.

Invasive species may also increase in non-mountainous areas because of increased fragmentation (see Chap. 12). While there is a vigorous debate on the role of biodiversity in community resistance to invasive species, there is also general agreement that forest fragmentation (from land-use change, direct climate change effects, and possible increased disturbances due to warming) will likely lead to increased opportunities for invasive and exotic species. Additionally, changing climate zones may further encourage plant and insect migrations (Gitay et al. 2001), along with intercontinental transfers through trade. Invasive insects may be further encouraged where trees are weakened by drought or other climate change effects, so some forests may become less resistant to invasive insects and native insects alike (see Chap. 8 and below).

Forest fragmentation will also lead to fragmentation of wildlife habitat, and possible extinctions. Anthropogenic forest fragmentation was a major factor in the extinction of the passenger pigeon in eastern North America at the beginning of the 20th century (Shorger 1972), and habitat fragmentation due to climate change has been proposed as a factor in the extinction of mammoths, sloths, and other large fauna after deglaciation (Dixon and Dixon 1991).

At high latitudes and altitudes, permafrost melt-out will exacerbate and complicate ecosystem responses. Melting permafrost leaves scars on the landscape known as thermokarst depressions; permafrost collapse in peatlands creates features called internal lawns (Vitt et al. 2000). These depressions are often waterlogged and usually involve collapse of soil surfaces. Paludification and soil slumping in turn lead to tree mortality and leaning trees, referred to as a "drunken forest" (Fig. 20.3). The eventual result is frequently a wet sedge meadow embedded in the forest.

Natural disturbances

Fire — In Canada, climate change is expected to contribute to a significant increase in forest fire activity (Fig. 20.4) particularly in western and central Canada (Flannigan et al. 1998). This will occur through a longer fire season, larger and more intense fires resulting in part from increased periods of drought, and an increase in both natural and anthropogenic fire ignitions (Stocks et al. 2000). The direct impact of climate change on

Fig. 20.4. Climate change may increase forest fire activity in some boreal regions.

Photo by Brian Amiro, courtesy of Canadian Forest Service

tree mortality may increase fuel loads in some regions. Climate change thus poses two major problems for fire research: predicting the impact of climate change and consequent vegetation changes on the fire regimes; and predicting the impact of increased fire activity on the vegetation and soil and, through their carbon storage, on global climate.

Forest fires affect the global carbon cycle in several ways (Kasischke 2000). First, the fire directly releases large quantities of carbon into the atmosphere through combustion of plant material and surface soil organic matter. Secondly, carbon is released from the decomposition of fire-killed vegetation over time. Thirdly, for several years or decades after a fire, the vegetation on newly burned sites may not fix as much carbon from the atmosphere as did the pre-fire vegetation. In addition it can take many decades before the ecosystem carbon stocks (in vegetation, on the forest floor, and in the soils) return to their pre-fire levels (Apps et al. 2000), and recover the carbon released during and after the fire event. Fires are thus an important part of the global carbon cycle, with increased fire frequency generally causing a net reduction in sequestered carbon stocks (Kurz et al. 1995*b*). Forest fires may also lead to replacement of one vegetation type with another. Frequent fires in some areas are responsible for creating essentially permanent openings in the forest and may be a major factor in excluding conifers from the Aspen Parkland ecoregion (Hogg 1997) and white cedar (*Thuja occidentalis*) from some sites in eastern Canada (Larocque et al. 2003). At the same time, fire may facilitate the adaptation of the forest to climate change by eliminating vegetation that is no longer optimally adapted to the new climate and permitting the development of new, better adapted vegetation communities (Arseneault and Payette 1992).

Fire itself responds strongly to climate change. The annual area burned will likely increase, perhaps by as much as 50% over the coming decades, in response to global warming (Flannigan and van Wagner 1991; Stocks et al. 1998). This increase is not likely to be evenly distributed. For example, parts of eastern Canada may actually experience a decrease in fire, while western and central Canada experiences a significant increase (Flannigan et al. 2001). There is some evidence to suggest the fire regime in Canada may already be responding to climate change (Amiro et al. 2001*a*; Stocks et al. 2003). Since the 1960s, the average annual area burned has increased from about 1.5 million ha to nearly 3 million ha (Fig. 20.5). A small part of this increase may be due to improved detection capacity, but much of the difference reflects the influence of new weather patterns (possibly due to climate change) as well as other factors, such as aging of the forest and increased human activities in formerly remote areas.

Insects and pathogens — Increases in tree mortality associated with insects and disease may be expected under a changing climate. Existing forests in the process of adapting to rapid global change, whether through growth stimulation or degradation, are under stress. Such conditions tend to increase their susceptibility to insect and disease attack. In addition to these changes in host conditions, both insect population and disease cycles may be directly affected by climate change. The incidence of exotic insects and disease can be expected to increase as climatic conditions make migration and survivorship more probable. Human transportation routes and activities exacerbate this problem. Insect disturbances are expected to increase with climate change because of longer growing seasons, increased forest stress, and less severe winters (Fleming and Volney 1995). Climate change may also affect insect diversity, allowing more exotic species to survive in the northern forests (Gitay et al. 2001).

Many uncertainties remain regarding the character, magnitude, and rates of future climate change and its impact on insect disturbances. However, it is generally expected that insect outbreaks will become increasingly more intense with climate change (Gitay et al. 2001) as a consequence of both atmospheric warming and CO_2 enrichment (Fleming and Volney 1995). In general, current forecasts of forest insect responses to climate

Fig. 20. 5. Area burned in Canada. Significantly more area was burned in the 1980s and 1990s than in the previous two decades (data compiled from Stocks et al. 2003).

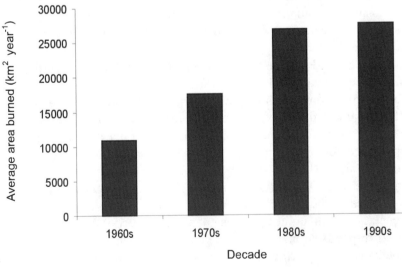

warming, which are based on historical relationships between outbreak patterns and climate, would suggest:

- more frequent outbreaks of
  - mountain pine beetle, *Dendroctonus ponderosae* (Thompson and Shrimpton 1984),
  - spruce budworm, *Choristoneura fumiferana* (Mattson and Haack 1987),
  - eastern hemlock looper, *Lambdina fiscellaria fiscellaria* (Carroll et al. 1995);
- longer outbreaks of
  - forest tent caterpillar, *Malacosoma disstria* (Roland et al. 1998), and
  - spruce budworm (Cerezke and Volney 1995); and
- range shifts northward and to higher elevations by
  - spruce budworm (Williams and Liebhold 1997),
  - gypsy moth, *Lymantria dispar*, and
  - white pine weevil, *Pissodes strobi* (Sieben et al. 1994).

Recent evidence suggests that some insect populations may already be responding to climate change. For example, large outbreaks of defoliating insects have been observed in recent years at much higher latitudes than previously (Brandt et al. 1996). Changes in seasonality (especially warmer and shorter winters) have also been implicated in more frequent outbreaks of bark beetle infestations in southern Alaska and mountain pine beetle in British Columbia. A recent survey showed that the pine beetle damaged approximately 1 million ha in British Columbia in 2002 and 5.7 million ha are currently at risk (Safranyik and Linton 2002). Scientists partially attribute the epidemic to global warming. Pine beetles usually have a 2-year cycle, but warmer temperatures can cause them to complete their cycle and breed in 1 year (Safranyik and Linton 2002). Warmer temperatures also reduce insect die-off, as fewer insects are killed by very cold weather in the late fall and early spring.

Forest management practices, such as the spraying of pesticides to control insects, can sometimes curtail the pattern, rate, and magnitude of small-scale insect disturbances. However, prediction and prevention of outbreaks is sometimes difficult, and managing infestation over large or remote areas is usually not practical (IPCC 2000).

In addition to affecting insect outbreaks, the generally warmer temperatures and milder winters expected with climate change may also increase the survival of existing pathogens and allow new pathogens to move into the boreal zone. For example, *Armillaria* root disease is a globally distributed fungal disease with optimum growth at 25°C; boreal and temperate forests may therefore experience increased *Armillaria*-caused volume loss with global warming (Mallett and Volney 1999). In areas where drought is expected to increase, there may be a greater incidence of canker diseases in some tree species owing to decreased bark moisture content (Bloomberg 1962; IPCC 2001*b*).

Permafrost melting — Permafrost is an important component of many northern forest ecosystems, and has profound influences upon hydrology, vegetation, and carbon storage. It is discontinuous with distribution being heavily influenced by autogenic factors on a local scale. However, on a more regional scale allogenic factors such as climate also play a major role in the presence or absence of permafrost. The influence of autogenic controls can be so strong that there can be permafrost thaw co-existing with permafrost aggradation within one peatland even if the climate is stable (Zoltai 1972).

Warmer temperatures have resulted in increased permafrost melt in some regions of the boreal forest. For example, air temperatures have warmed significantly in the Mackenzie Valley, N.W.T., ranging from 60 to 64°N, over the past 5 decades (Robinson et al. 2001). In the northern region of the valley, permafrost is still dominant and thaw is restricted to relatively small individual bowl-shaped features. However, over the past 50 years, lateral thaw has increased the size of existing thaw features and a significant number of new features have developed. In the southern region of the valley, individual thaw features have coalesced, creating an integrated drainage network and in many cases isolating the remaining permafrost features (Robinson et al. 2001). Recent thaw in more southern parts of the discontinuous permafrost zone of the boreal forest has also been noted at sites in Manitoba (Thie 1974), and the other prairie provinces (Vitt et al. 2000; Beilman et al. 2001).

In boreal forests, permafrost affects soil moisture because it inhibits permeability and affects the availability of water to the ecosystem during seasonal thaw. Soils with shallow permafrost may be particularly sensitive to climate change because of the role that moisture plays in these ecosystems. For example, shifts toward wetter or cooler conditions might suppress fires and enhance carbon storage. Shifts toward drier summers might favour fire activity and enhance fire emissions. While these climatic trends apply to all systems, forest ecosystems with permafrost soils are at greater risk. In years that favour deep thaw and drying of these deep, humic horizons, fire can tap into these vast quantities of carbon-rich fuels (Harden et al. 2000).

Other natural disturbances (storms, floods, landslides) — Forest damage due to windstorms, ice storms, flooding, and landslides can also be expected to increase with climate change (IPCC 2001a). Although insect and fire disturbances will probably be more widespread, forest damage due to storms could be substantial in some regions. For example, the January 1998 ice storm (Fig. 20.6) affected 50 000 km^2 of forest in southeastern Ontario and western Quebec and caused moderate to severe damage to 15 000 km^2 of forest (NRCan 1999a). In Europe during 1999, windthrow caused by three storms over a 3-day period resulted in losses of 193 × 106 m^3 of timber, equivalent to 2 years of harvest (Updegraff-ECE/FAO 2000). Such storms may become more frequent in association with milder winters in the future (Cohen and Miller 2001). In the Great Lakes area, windthrow is a major disturbance agent affecting thousands of hectares with both immediate and long-term impacts. Most boreal tree species are shallow-rooted and hence vulnerable to windthrow. Factors such as tree height, whether the tree is alive or dead, and stand density determine whether a tree is just snapped off or is completely uprooted. In boreal cordilleran areas, flooding and landslides due to intense precipitation events could increase in frequency with climate change (Gitay et al. 2001).

Climate change-induced or enhanced anthropogenic pressures

In many regions, climate change impacts on forest distribution will be heavily mediated by its impacts on agriculture. New forest clearance for agricultural expansion into areas that are presently too cold for agriculture may be a major force in modifying forest distributions. At the same time, some marginal agricultural land may be taken out of production and allowed to revert to forest through natural regeneration or planting. This

Fig. 20.6. Trees damaged in 1998 ice storm. Ice storms may become more frequent with climate change.

Photo by Genevieve Zevort

latter trend may be enhanced by government policies favouring afforestation for carbon sequestration purposes (van Kooten et al. 2002). In some regions of the world, however, given the expanding population and consequent demand for agricultural land, the net pressure will be toward deforestation rather than afforestation. For example, one model that integrates the effects of climate change and land-use change suggests a 50% decline in forest cover in parts of Asia by 2050 (IPCC 2001*b*). The trend towards deforestation will be reinforced by the tendency of climate warming to make forested land suitable for agriculture, by increasing the length of the growing season in cold regions, and its tendency to make current agricultural land unsuitable for agriculture through increasing drought. However, in some areas this trend will likely be dampened by the fact that many subarctic lands have soils unsuitable for agriculture, not just unsuitable climate.

Combined with land-use changes, climate change may increase pressures to grow fibre in managed plantations. Because most managed plantations tend to focus on a single (or limited number of) tree species and comprise even-aged stands, they have had a significant effect on biodiversity in some regions (Harris 1984). If the pressures to increase fibre production rise with climate change, as expected, forward-thinking management practices would be required to mediate these effects by, for example, deliberately employing mixed species management and variable cutting patterns, and by planting species better suited to expected future conditions. Some of these options are explored from several perspectives in Chap. 13.

CO_2 fertilization

Under experimental conditions, some species increase their rates of photosynthesis with increases in CO_2 (Farquhar et al. 1980; Wullschlenger et al. 1995; Isebrands et al. 2000; Houghton et al. 2000), partly because of greater water-use efficiency (Field et al. 1995; Farquhar 1997; Körner 2000). This effect is known as CO_2 fertilization. It is hypothesized that CO_2 fertilization may account for up to 33% of the CO_2 absorbed annually through photosynthesis by terrestrial plants (Norby et al. 1999). The CO_2 fertilization effect may enhance the growth of some tree species and forest ecosystems, allowing them to remove more carbon from the atmosphere. Whether the enhancement of photosynthesis by elevated CO_2 actually results in net removal of carbon from the atmosphere at the ecosystem level is a subject of intense debate (e.g., Luo et al. 2003). Recent work based on forest inventory data indicate that the net effect on carbon stocks is very much less than that suggested above, accounting for less than a few percent increase in accumulated carbon in vegetation (Caspersen et al. 2000).

Many of the experimental studies on elevated CO_2 response have been conducted on immature trees, often in growth chambers, under conditions not otherwise limiting plant growth and not measuring the effects of changes in ambient CO_2 concentration that would occur in forests (Curtis and Wang 1998; Norby et al. 1999). Several field experiments are currently under way which employ free air CO_2 enrichment (FACE) technology by which the CO_2 (and other gases) around growing plants has been modified to simulate future levels of these gases under climate change (Curtis et al. 2000; Luo et al. 2003). These experiments, however, have not been running long enough to determine what the long-term effects of elevated CO_2 levels might be once canopy closure is reached (IPCC 2000). While the response of mature forests to increases in atmospheric CO_2 concentration has not been demonstrated experimentally, it may be different from that of individual trees and different from that of young forests (Körner and Bazzaz 1996; Curtis and Wang 1998; Norby et al. 1999; Casperson et al. 2000).

Some scientists have hypothesized that Canada's forest net primary productivity (NPP) may be increasing, and that this increase may be due in part to CO_2 fertilization (Chen et al. 2000). However, inventory measurements in North America indicate that CO_2 fertilization has not appreciably affected carbon accumulation rates in forest biomass over the past century (Casperson et al. 2000). Forest age-class dynamics, land-use change, and alterations in natural disturbance patterns have a much larger influence than CO_2 fertilization on forest growth in this region (Kurz and Apps 1999; Casperson et al. 2000). There is a growing consensus in the scientific community that any CO_2 fertilization effects that may exist are expected to saturate (that is, their contribution to continued net CO_2 removals will go to zero) over the next 100 years or so (Bazzaz et al. 1996; IPCC 2000; Schimel et al. 2001). This occurs because increases in CO_2 levels stimulate increases in photosynthesis at a diminishing rate, while increases in temperature stimulate increases in respiration at an exponential rate (IPCC 2000).

NO$_x$ fertilization and ozone

Nitrogen deposition — The amount of available nitrogen in a given forest ecosystem plays an important role in plant biodiversity, the population dynamics of species in the associated food web, and ecological processes such as plant productivity, winter hardiness, and biogeochemical cycling (Vitousek et al. 1997). In boreal forest ecosystems, nitrogen is a limiting factor, as most of it occurs in forms that cannot be readily used by most plants.

Human activities have increased the supply of nitrogen in some regions of the boreal forest. It has been suggested that increased nitrogen deposition (due to NO$_x$ atmospheric pollution; see Chap. 18) may temporarily enhance forest carbon sequestration in nitrogen-limited ecosystems, leading to a short-term carbon gain in NPP (Nadelhoffer et al. 1999). However, if there is a positive effect on forest growth from nitrogen deposition in boreal forests, it will likely be negated in the medium term as other factors become limiting, such as other nutrients and water (Olsen et al. 2001). In North America, nitrogen deposition has not appreciably affected carbon accumulation rates at the landscape level (Casperson et al. 2000). On the contrary, evidence shows that excess deposition has had harmful effects, at the stand level, on forested and aquatic ecosystems (Schindler 1998; Box 20.2).

Tropospheric ozone — Scientists believe that up to 49% of global forests (17 × 106 km^2) will be exposed to damaging levels of tropospheric ozone (O$_3$) by 2100 (Fowler et al. 1999). O$_3$ (smog) is a secondary photochemical oxidant formed from reactions between primary pollutants (nitrogen oxides and volatile organic compounds) and

---

Box 20.2. Nitrogen deposition from various sources of air pollution can have a number of detrimental effects on ecosystem functions and environmental quality.

It has been shown, for example, that excess nitrogen:

- is a major contributor to photochemical smog and acidification of fog, which alters ecosystem processes and health (Lovett 1994);

- results in losses of soil nutrients, such as calcium and potassium, thus reducing long-term soil fertility (Ouimet and Camire 1995; Likens et al. 1996);

- acidifies the soil, which may have a long-term negative impact on forest growth in some regions (Likens et al. 1996);

- is exported from forested watersheds, where it contributes to episodic acidification and eutrophication of aquatic ecosystems (Schindler 1998);

- may reduce the ability of plants to process CO$_2$ assimilated from the atmosphere;

- may reduce root to shoot biomass ratios and thus (1) increase plant sensitivity to drought, counteracting possible water-use efficiency gains from CO$_2$ fertilization (Nadelhoffer et al. 1999), and (2) increase the likelihood of nutrient deficiency due to smaller root mass (Townsend et al. 1996); and

- can accelerate losses of existing biological diversity, especially among plants that are adapted to low-nitrogen soils, by increased competition from plants more responsive to higher nitrogen levels (grasses), and subsequently affect the animals and microbes that depend on them (Lovett 1994).

sunlight. These primary pollutants are produced by activities such as the burning of fossil fuel and emissions from transportation. Ozone is also a greenhouse gas, trapping outgoing heat radiation and radiating it back to Earth, and is estimated to be responsible for up to 30% of radiative forcing (Stevenson et al. 1998). Ozone is the most pervasive and toxic gas to plants in the lower atmosphere.

Although the precursors of nitrogen oxide pollutants are primarily created in urban areas, air masses carry them over long distances, resulting in significant tropospheric ozone levels over regional and continental scales. The highest levels of O_3 are produced on warm sunny days with relatively still air, in mid to late summer, during the late morning and early afternoon, particularly in areas with thermal inversions.

Annual mean ground-level O_3 concentrations in Canada are increasing, particularly in urban areas (Munn and Maarouf 1997). At least 2×10^6 ha of Canada's productive eastern forest is exposed annually to damaging levels (McLaughlin and Percy 1999). Exposure of western forests to O_3 is difficult to estimate because of a lack of ground-level monitoring data, but some northwestern forest ecosystems will likely be more exposed through significant industrial expansion in these areas. Ozone can harm forest ecosystems by impairing tree physiology, in particular by decreasing the rate of photosynthesis in some species and altering carbohydrate allocation patterns in others (Percy et al. 1999). With respect to the latter, carbon transfer is commonly decreased to the roots and increased to the shoots, which makes trees more vulnerable to drought, nutrient deficiencies, and winter damage.

Results from FACE studies indicate that while increased CO_2 alone results in an increase in growth in some tree species, the increases are often negated by the effects of tropospheric ozone (Isebrands et al. 2000). For example, Isebrands et al. (2000) reported negative responses in aspen (*Populus tremuloides* Michx.) and birch (*Betula papyrifera* Marsh.) aboveground estimated stem volumes relative to the controls after 3 years of fumigation with O_3 or $O_3 + CO_2$. A stimulation of 20–30% with CO_2 alone was also completely offset by O_3. While experimental studies have shown reduced growth rates in some forest species exposed to O_3 (Isebrands et al. 2000), there is no evidence that changes in O_3 levels are resulting in any significant changes (either negative or positive) in forest growth rates at the landscape level. Any effects are likely to be area-specific and occur against a longer term background of climate and forest change.

The ozone fumigation facility at the Canadian Forest Service's Atlantic Forestry Centre is the only facility in Canada currently investigating the combined effects of human-induced stresses involving O_3. These studies on birch species (*Betula* spp.) include interactions between summer O_3 exposure and extended winter thaws and (or) late frosts (likely to increase under global warming), and interactions between excess nitrogen fertilization and extended winter thaws (Percy et al. 1999).

The "natural" carbon budget of forests

The boreal forest biome contains approximately 700 Pg of carbon (Apps et al. 1993a), a significant proportion (25%) of the global terrestrial carbon pool, estimated at 2500 Pg (IPCC 2000). The net amount of carbon stored in the forest fluctuates naturally over time. The following sections give an overview of current estimates of carbon stocks and fluxes in the boreal region, the natural factors affecting these stocks, and the ways in which climate change is expected to affect the carbon budget of the boreal forest.

Current forest carbon stocks

The estimated sizes of various carbon pools in the boreal forests of Canada, Alaska, Russia, and Scandinavia are summarized in Table 20.1. There is much uncertainty underlying the carbon estimates for the peatland component because of the lack of information on the true extent and depth of peatlands and the unknown degree of overlap between peatland and forest inventories (Apps et al. 1993a). The carbon stocks and fluxes of various carbon pools are discussed in greater detail in the following sections.

Above- and below-ground stocks and dead organic matter

According to Bhatti and Apps (2000), aboveground biomass in boreal forests varies between 22 and 187 Mg ha^{-1} (1 Mg = 10^{6}g). In general, drier sites contain less biomass whereas wetter sites contain more biomass. The higher biomass under poorly drained conditions is associated with the presence of many mature black spruce (*Picea mariana* (P.Mill.) B.S.P.) stands that are long-lived because wet sites are less susceptible to fire disturbance (Bhatti et al. 2002a). Conversion of biomass to carbon indicates that carbon storage in these boreal forests ranges from 11 to 97 Mg C ha^{-1}. These estimates are consistent with estimates of 5–54 Mg C ha^{-1} (average 24 Mg C ha^{-1}) reported by Simpson et al. (1993) from direct measurements of western Canada's boreal forest. On the basis of point measurements, soil survey data sets, and modelling, Bhatti et al. (2002b) estimated that belowground carbon stocks for upland forest soils range from 14 to 78 Mg C ha^{-1} for the surface layer and from 62 to 274 Mg C ha^{-1} for the total soil column.

Table 20.1. Boreal forest biome carbon pools and fluxes, based on data available prior to 1993 (compiled from Apps et al. 1993a; Weber and Flannigan 1997).

	Canada	Alaska	Russia	Scandinavia
Area (Mha)				
Boreal forest	304	52	760	61
Peatland	89	11	136	20
Pools (Pg C)				
Plant biomass[a]	8	2	46	2
Plant detritus	—[b]	1	31	NA[e]
Forest soil	65	10	100	NA
Peat	113	17	272	13
Fluxes (Tg C/year)[c]				
Boreal forest[d]	62	6	493	43
Peatland	25	3	11	5

[a]Above- and below-ground live biomass.

[b]Plant detritus estimates are included in the soil carbon pool.

[c]Flux values represent net transfers from the atmosphere.

[d]Includes biomass and soil C pool dynamics.

[e]NA, not available.

These soil carbon estimates are comparable to those reported for global boreal forests (111–190 Mg C ha^{-1}) by Post et al. (1982), and for North American boreal forests (135–195 Mg C ha^{-1}) by Pastor and Post (1988).

Aboveground annual biomass productivity, NPP, compiled by Gower et al. (1997) for boreal forest pine (*Pinus* spp.) stands range between 2.3 and 7.0 Mg ha^{-1} year^{-1} with an average of 4.2 Mg ha^{-1} year^{-1}. The average aboveground NPP of three different stand types (mature aspen, black spruce, and jack pine [*Pinus banksiana* Lamb.]) at BOREAS (Boreal Ecosystem–Atmosphere Study) research sites was estimated at 1.7 Mg C ha^{-1} year^{-1} in 1993 and 1.8 Mg C ha^{-1} year^{-1} in 1994. Data collected by Gower et al. (1997) from published studies in the United States, Canada, Finland, Sweden, and China estimated average NPP at 3.6 and 1.4 Mg C ha^{-1} year^{-1}, for deciduous and coniferous boreal forests, respectively, values that are relatively consistent with the field measurements. Li et al. (2002), using the Carbon Budget Model of the Canadian Forest Sector (CBM–CFS2), simulated an average NPP of 1.72 Mg C ha^{-1} year^{-1} for the three prairie Provinces in west-central Canada, varying from 0.72 to 2.9 Mg C ha^{-1} year^{-1}, depending on ecoclimatic region, forest type, age and site productivity.

Belowground NPP data are relatively rare. There are few reliable values for root productivity because the turnover rate of fine roots is difficult to measure. However, field measurements by Steele et al. (1997) at BOREAS sites (the northern study area in Manitoba and the southern study area in Saskatchewan) yielded belowground NPP estimates for mature aspen, black spruce, and jack pine stands of 0.4–0.7, 0.90–1.2, and 1.0–1.1 Mg C ha^{-1} year^{-1}, respectively. On the basis of published global data for the boreal forest, Gower et al. (1997) estimated an average belowground NPP of 1.1 and 1.0 Mg C ha^{-1} year^{-1} for deciduous and coniferous boreal forests, respectively. Total aboveground and belowground NPP was estimated (from field measurements) at 4.7 Mg C ha^{-1} year^{-1} for deciduous, broad-leaved species and 2.7 Mg C ha^{-1} year^{-1} for coniferous species (Gower et al. 1997). These values are comparable to the total NPP simulated by Li et al. (2003) for western Canada, which varied between 1.5 and 3.9 Mg C ha^{-1} year^{-1}.

Most of the boreal forest sites investigated by various researchers are sequestering carbon (measured as net ecosystem productivity, NEP) at annual rates of up to 2.5 Mg C ha^{-1} year^{-1} (Black et al. 1996; Jarvis et al. 1997; McCaughey et al. 1997; Blanken et al. 1998; Chen et al. 1999). The values obtained depend primarily on latitude, soil type, forest type, and successional stage. Not all boreal ecosystems are sequestering carbon, however. The NEP measurements over periods of up to 5 years in northern Canada in the BOREAS experiment (Sellers et al. 1997) have demonstrated that a few old-growth coniferous stands may be carbon neutral (Goulden et al. 1998) and in warm and cloudy years can be a carbon source (Lindroth et al. 1998), losing carbon at a rate of up to 1.0 Mg C ha^{-1} year^{-1}.

Adjacent stocks (lakes, peatlands)

Lakes

Very few data exist for carbon storage in lake sediments in the boreal forest or elsewhere. According to Molot and Dillon (1996), 120 Pg of carbon may be stored in the sediments of boreal lakes. Campbell et al. (2000) suggested 2.3 Pg C as a first estimation of the amount of carbon stored in Alberta lakes (most of which occur in the boreal region) and estimate the average annual carbon sequestration rate to be 0.23 Tg year^{-1} (1 Tg = 10^{12} g). On the basis of data mainly from Minnesota lakes, Dean and Gorham (1998) estimated the global carbon sequestration rate in lakes and inland seas at 42 Tg year^{-1}. Since lake sediments have the potential to store a significant amount of carbon over very long periods of time, more research is needed on this component of the boreal carbon cycle.

Peatlands

Peatlands represent one of the largest terrestrial carbon reservoirs in the world. Boreal peatlands are estimated to contain 61% of the boreal carbon stocks in both Canada and around the globe (Table 20.1). Boreal and subarctic peatlands have accumulated about 400–500 Pg C during the Holocene (the last 12 000 years; e.g., Gorham 1991; Zoltai and Martikainen 1996; Clymo et al. 1998; Roulet 2000). Today, peatland carbon stocks are equivalent to about one-third of the world's soil carbon pool (1395 Pg C; Post et al. 1982) or more than double the amount of carbon stored in upland boreal forest soils (199 Pg C; Apps et al. 1993a; Fig. 20.7). In contrast, only about 11% of the global vegeta-

Fig. 20.7. Soil profiles: (*a*) upland forest soil profile with a very shallow organic carbon rich layer; (*b*) peatland soil profile with a very deep organic carbon rich layer. Most of the carbon in the boreal region is stored in organic peatland soils.

(*a*)   (*b*)

Photos by J.S. Bhatti

tion carbon pool (610 Pg C; Schimel 1995) is stored in the biomass of both upland and lowland areas of boreal forest (i.e., 64 Pg C; Apps et al. 1993*a*).

Across Canada, an estimated 111 Mha, or 12% of the total land base, is covered by peatlands (NWWG 1988) that store 103–184 Pg C (Ovenden 1990; Kurz et al. 1992; Apps et al. 1993*b*). In continental western Canada (Alberta, Saskatchewan, and Manitoba), peatlands cover about 20% of the land base, or 365 000 km^2, and store 48 Pg C: 42 Pg C as peat, and 6 Pg C as living aboveground biomass (Vitt et al. 2000). Estimates of peatland area in the Former Soviet Union (FSU) vary widely from 77 to 165 Mha (Botch et al. 1995), and peatland carbon stocks in the FSU are estimated at 215 Pg C (Botch et al. 1995; Kobak et al. 1998).

Global estimates of annual carbon accumulation in peatlands range from 45 to 210 Tg C year^{-1} (Bramryd 1979; Armentano 1980). Regional rates of carbon accumulation range from 0.14 to 0.28 Mg C ha^{-1} year^{-1} (Gorham 1991; Mäkilä 1997; Rapalee et al. 1998; Vitt et al. 2000), and may vary with changes in moisture, soil temperatures, reduction–oxidation conditions (Reader and Stewart 1972; Clymo 1984), acidity and alkalinity (Szumigalski and Bayley 1997; Thormann and Bayley 1997; Thormann et al. 1999), species composition (Malmer 1986; Johnson and Damman 1991), and (or) litter quality (Yavitt and Lang 1990; Valentine et al. 1994; Updegraff et al. 1995; Yavitt et al. 1997). Recent carbon balance studies have revealed that individual peatlands may switch from net carbon sinks to net carbon sources on an annual basis (Alm et al. 1997; Rivers et al. 1998; Waddington and Roulet 2000; Lafleur et al. 2001), perhaps responding to slight differences in weather and (or) drainage conditions.

Natural factors affecting forest carbon stocks

Disturbances and forest age

Disturbances alter forest productivity, may release carbon directly into the atmosphere (through combustion), and transfer large amounts of carbon from biomass into detritus, soils, or forest products (Kurz and Apps 1999). For example, in Canada, large forest fires from 1959 to 1999 directly released 3–115 Tg C annually (mean 27 ± 6 Tg C year^{-1}; Amiro et al. 2001*b*). Combustion losses of carbon range from 10 to more than 50 Tg C ha^{-1} (Kasischke et al. 2000). Intense fires may combust nearly all aboveground biomass, ground vegetation, and forest floor while leaving behind a nearly bare, ash-covered mineral soil (see Chap. 9). Fire has a substantial influence on the post-fire emissions from the soil. Schlentner and Van Cleve (1985) estimated that in mature black spruce forests approximately 20% of CO_2 emissions from soil is from decomposition and 80% is from plant root respiration. Studying post-fire carbon emissions from mature black spruce forests, Richter et al. (2000) found a threefold increase in CO_2 emissions due to a higher rate of decomposition. Post-fire release of carbon due to decomposition is estimated to be equivalent to the direct emissions during combustion in some from boreal ecosystems (Amiro et al. 2001*b*).

In continental boreal regions, peatland fires may release up to 5.5 Tg C year^{-1} to the atmosphere through a combination of organic matter combustion and altered decomposition (Zoltai et al. 1998; Turetsky et al. 2002). A significant warming has been observed at many forested peatland sites after fire (Trumbore and Harden 1997). Long-term studies of permafrost following fire demonstrate decades-long permafrost melting and 5–10°C increase in soil temperature (Viereck and Dyrness 1979). Such an increase in soil temperature and permafrost melting might greatly stimulate decomposition rates. Presuming a Q_{10} of 2 (biological reactions increase about twofold per 10°C rise in temperature) and a temperature increase of 10°C, there will be a 100% increase in the rate of decomposition. These effects may last more than a decade, with a substantial increase in CO_2 emissions following fire.

Although insects do not themselves cause significant direct emissions, they affect carbon stocks by killing trees, which leads to subsequent decay of dead organic matter (Fleming 2000; Volney and Fleming 2000). During insect outbreaks, trees are often killed either directly or indirectly because of increased susceptibility to fire or disease. When the dead trees decay or are burned, they release the carbon held in the forest ecosystem back to the atmosphere. Insect disturbances also greatly increase litter fall, thereby transferring large amounts of carbon from the living biomass to the litter and soil carbon pools (Fig. 20.8).

Disturbances also drive the age-class structure of the forest, which in turn affects the forest's ability to sequester carbon. Small changes in weather patterns over periods of years to decades and longer can change the disturbance frequency, which can produce a significant shift in the age-class structure and spatial arrangement of the forest (Gardner et al. 1996). With increasing frequency of disturbance, a greater proportion of the forest is found in younger age classes. Young and immature stands in the landscape contain less carbon than mature stands, other factors being the same. A mature stand may be more susceptible to mortality from insects, diseases, and possibly forest fires, which increases inputs to detrital pools and consequently increases the release of CO_2 to the atmosphere through decomposition of the larger detrital pool.

Fig. 20.8. Simulated average C stocks in litterfall and soil (slow pool) for three disturbance types in Saskatchewan boreal forest sites (modified from Apps et al. 2000).

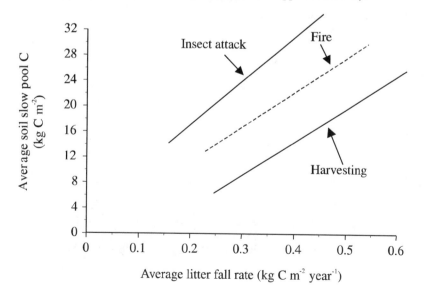

Since the late 1980s, Kurz and Apps (1999) have been developing a computer simulation model (the Carbon Budget Model of the Canadian Forest Sector) that describes the carbon dynamics of Canada's forest ecosystems. The model simulates forest growth, mortality, decomposition, the effects of natural disturbances (fires, insects, and diseases), and the impacts of direct human activities (harvesting, product processing, planting, and forest protection) on the forest carbon cycle. The effects of disturbances on the age-class structure of forest stands and carbon releases to the atmosphere and forest floor are also calculated. Their research indicates that disturbance rates and carbon sequestration have varied over the last century (Kurz et al. 1995*a*; Kurz and Apps 1999). There was a high rate of boreal forest disturbance in Canada in the late 1800s, which resulted in the establishment of large cohorts of young stands in the early 1900s. As a result, the forest was sequestering large amounts of carbon by the 1920s. From 1920 to 1970 the disturbance rates were fairly low (Fig. 20.9). By 1970 the forest was increasingly composed of mature stands of trees, which led to a general decrease in the amount of carbon being sequestered. From about 1970 to the present, many of these stands have become mature or over-mature and are carbon neutral. However, at the same time there has been a dramatic increase in disturbances, and some of these older stands are being replaced by new young stands. High combustion, decomposition, and respiration losses of carbon accompany these disturbances. As a result of changes in disturbance-driven age-class structure, Canada's total forest ecosystems were a net carbon sink from 1920 to1979 and a small net carbon source from 1980 to the present (Kurz and Apps 1999).

Other natural factors

Forest carbon stocks are also influenced directly by weather and regional climatic shifts, although in a less dramatic way than the disturbance-driven carbon fluxes. In general, forest productivity (and thus carbon accumulation) may be enhanced by warmer or

Fig. 20.9. Disturbance rates in Canada since 1920. Disturbance rates rose dramatically in 1970 after several decades of fairly low disturbance (data compiled from Kurz et al. 1995*b*, with additional data from W.A. Kurz and M.J. Apps, Canadian Forest Service, Victoria, British Columbia, personal communication, 2001).

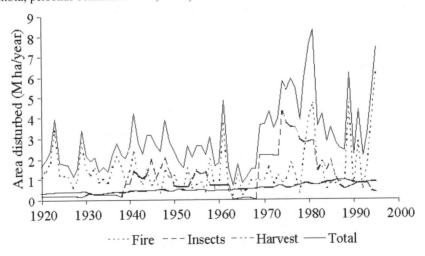

longer growing seasons (where water and nutrients are not limiting) and diminished in cooler or shorter growing seasons (and in drier years in regions where water is limiting).

Warmer temperatures also affect the permafrost that underlies parts of the boreal forest. Temperatures increases of about 1°C across the boreal forest over the past 100 years or so (Campbell and McAndrews 1993) have resulted in widespread melting of permafrost in peatlands (Halsey et al. 1995; Vitt et al. 2000). Permafrost melt increases carbon accumulation in peatlands by increasing bryophyte primary production (Turetsky et al. 2000). To date, about 2630 km^2 of permafrost has melted in the boreal peatlands of western Canada, increasing regional carbon accumulation by approximately 0.1 Tg C year^{-1} (Turetsky et al. 2002).

Potential impacts of climate change on forest carbon stocks

There is much uncertainty concerning the net impact of climate change on boreal forest carbon stocks. Some models (e.g., Smith et al. 1993) suggest that once the boreal forest achieves equilibrium with the new (and assumed stable) climate, the carbon stocks will be greater due to an increase in forest area from the migration of forest species into the presently unforested tundra and increased forest productivity. Neilson (1993), however, argues that increased drought stress would result in the boreal region becoming a long-term source of atmospheric C. Rizzo and Wiken (1992) also suggest that warmer and drier mid-continental conditions in Canada will result in greater losses of boreal regions in the south (due to grassland and agricultural encroachment) than gains in the north (due to forest migration into present tundra), which would lead to a net reduction in carbon stocks (Apps et al. 1993a).

In the shorter term (the next 100 years), it is likely that climate change will be accompanied by an increase in natural disturbances (fire, insects, disease, windthrow) that will reduce forest carbon stocks by releasing large quantities of carbon to the atmosphere. While regeneration of the disturbed forest will sequester some of this carbon, it is not known if the increased carbon uptake of the younger forest (the spatial structure and growth characteristics of which will be determined by the altered climatic and environmental conditions) will equal the carbon losses of the forest it replaced (Apps et al. 1993a). Modelling results by Bhatti et al. (2001) emphasize the importance of disturbance to carbon stocks in different types of forests (Table 20.2). Carbon stocks in aboveground biomass and litter were approximately 50% less in periods of high disturbance rates than in periods of infrequent disturbance.

The response of soil carbon stocks will also be important to the net carbon balance under a changed climate. For example, simulation results from Finland using SIMA (a gap-phase type stand dynamics model; see Chap. 14) show that total carbon reserves in forests decreased due to a decline in the carbon content of soil organic matter (SOM; Mäkipää et al. 1999). Changed climatic conditions (i.e., elevated temperature and increased precipitation) accelerated the soil surface carbon flux, resulting in a 30% decrease in soil C. This result is consistent with experimental studies, where elevated temperature greatly increased soil respiration (Peterjohn et al. 1995) and where both elevated temperature and soil moisture resulted in accelerated mineralization of SOM (Goncalves and Carlyle 1994). Many models have predicted that turnover of organic

Table 20.2. Simulated biomass, litter production, and soil C content for different stands using the simplified CBM–CFS2 model for different frequencies of random disturbance (modified from Bhatti et al. 2001).

Stand attribute	Coniferous (3 productivity levels)			Mixed wood	Deciduous
	High	Medium	Low		
Average biomass C (kg m^{-2})					
Low period	5.02	2.60	2.10	5.95	4.92
Medium period	4.99	2.57	2.08	5.82	4.79
High period	2.34	1.20	0.97	2.77	2.29
Average litter production C (kg m^{-2} year^{-1})					
Low period	0.63	0.41	0.35	0.46	0.51
Medium period	0.58	0.35	0.31	0.39	0.43
High period	0.31	0.20	0.17	0.28	0.26
Average soil C (kg m^{-2})					
Low period	30	19	17	30	24
Medium period	31	20	17	31	26
High period	28	18	15	28	23

matter in soils is accelerated by elevated temperature, which may lead to a decreased carbon stock of SOM on a global scale (Jenkinson et al. 1991; Cao and Woodward 1998).

Limited data and understanding of the influence of changing environmental conditions and disturbance (including fires and permafrost melting) on the carbon budget of peatlands over short and medium time scales (10–100 years) hinder predictions of the changes in the carbon sink–source relationships under a changing climate. The projected warming and associated changes in precipitation will influence both net primary production and decomposition in peatlands, but how global warming will directly influence peatland carbon stocks remains uncertain (Moore et al. 1998). Global change may have indirect implications for peatland carbon sinks through increased permafrost melt and fire activity. Permafrost melt tends to increase peatland carbon stocks through increased bryophyte productivity but also appears to increase heterotrophic respiration (Turetsky et al. 2000). Peatland fires result in decreased net primary production and elevated post-fire decomposition rates, but little is known about the recovery of peatland carbon balance after fire (Auclair and Thomas 1993; Dixon and Krankina 1993; Wardle et al. 1998; Zoltai et al. 1998; Turetsky and Wieder 2001).

Human impacts on forest carbon stocks

Land-use change and land-use practices

Land-use change is usually associated with a change in land cover and often results in a change in carbon stocks. Land-use changes affecting the boreal zone include forest clearance (e.g., for agriculture or roads), reforestation of land previously cleared for crops or pasture, flooding of forested areas for reservoirs, and wetland drainage. Such changes are distinguished from land-use practices which do not result in a long-term change in cover, but which may also affect carbon stocks. Land-use practices in the boreal region include forest protection (e.g., fire and insect protection, and establishment of ecological reserves) and productivity enhancement (fertilization, tree species selection, and reduction of regeneration delays after disturbance). Some of these land-use practices will be discussed below.

In Canada, the boreal forest has been subjected to increasing clearance for agriculture (mainly at the southern boundary; Fitzsimmons 2002), roads, survey lines, oil and gas wellheads, and other purposes. From 1990 to 1998, 54 000–81 000 ha year^{-1} of forest were cleared for various activities (Robinson et al. 1999). When a forest is cleared, the carbon stocks in aboveground biomass are either removed as products, released rapidly by combustion, or released slowly through microbial decomposition. Carbon stocks in the soil are also affected, although this effect depends on the subsequent treatment of the land. Clearance followed by cultivation may result in large decreases in soil carbon. For example, in boreal sites in eastern Canada, land cleared for agriculture contained 22% less soil carbon than the adjacent forested land (Carter et al. 1994).

Flooding forested and wetland areas in the boreal zone for hydroelectric reservoirs generates massive fluxes of dissolved organic carbon (DOC) into the water, accelerates peat decomposition, and increases methane and carbon dioxide fluxes to the atmosphere (Duchemin et al. 1995; Kelly et al. 1997; Schindler 1998). For example, Kelly et al. (1997) experimentally flooded a boreal wetland in Ontario, causing the carbon dynamics of the site to change from a sink of 6.6 g C m^{-2} year^{-1} to a source of 130 g C m^{-2} year^{-1}. Turetsky et al. (2002) estimated that 0.8 ± 0.2 Tg C year^{-1} is released from approximately 780 km^2 of hydroelectric reservoirs in peatlands across western boreal Canada.

When wetlands are drained for conversion to agriculture and (or) pasture, or to increase forest productivity, previously waterlogged soils become exposed to oxygen. Carbon stocks, which are resistant to decay under the anaerobic conditions prevalent in wetland soils, can then be lost by aerobic respiration (Minkkinen and Laine 1998). When drained peatland soils are farmed, as much as 10–20 Mg C ha^{-1} year^{-1} or more can be lost to the atmosphere (Armentano and Menges 1986). When drained peatlands are used for forestry, however, the reaccumulation of carbon in vegetation must also be considered, and the impact on the net carbon balance is less clear (IPCC 2001b).

Protecting the forest from fire and insects can temporarily preserve the carbon stocks in the forest. Protection of the entire boreal forest area, however, is not practical (Weber and Stocks 1998), nor may it be ecologically desirable, since the boreal forest is a disturbance-dependent ecosystem. When disturbances do arise, the seeding or planting of trees in the disturbed area can hasten forest regeneration and diminish the gap between the time of the disturbance and the time when the regenerating forest again recaptures carbon from the atmosphere (Kurz et al. 1995b). Fertilizers may also enhance growth and carbon sequestration rates, although widespread use of fertilizers is not practical or desirable from the point of view of both human and ecosystem health (Kimmins 1992). In addition to the detrimental effects indicated in Box 20.2, applications of nitrogen fertilizers to forests have large carbon emission costs associated with the production and distribution of the fertilizers themselves, and can lead to significant increases in emissions of N_2O, a potent greenhouse gas, a phenomenon well known in agricultural systems. Planting faster growing tree species such as hybrid poplar (*Populus* spp.) may also increase carbon sequestration rates on disturbed areas, but again this option is not practical on a wide scale because of economic costs and biodiversity concerns (van Kooten et al. 1993).

Forest management

Forest management activities influence the rate of carbon sequestration in a forest at both stand and landscape levels. At a stand level, silvicultural treatments, harvesting, and other actions that affect the regeneration delay and vigour of the growing trees are important (see Chap. 13). At a landscape level, management actions such as fire suppression, insect reduction programs, overall management objectives (including the rate of timber harvest) are important and are described in greater detail in Chap. 11.

Stand-level management

Various management factors influence the carbon stores in the biomass and soil of forest stands. Stand-level actions that affect the storage of carbon include pre-harvest silvicultural techniques (site preparation, planting and spacing, thinning) as well as harvest methods (clearcutting or partial cutting, and factors that affect how much and what type of material is removed from the site). An increase in sequestration capacity is gained through an increase in carbon uptake, a decrease in carbon release, or some combination of these. After disturbance, a forest stand's net carbon balance is a function of the photosynthetic uptake less the autotrophic and heterotrophic respiration occurring. While a stand is young, the losses through decomposition outweigh the gains through photosynthesis, resulting in a net source (Fig. 20.10). Not until the rate of uptake by the vegetation overcomes this decomposition (which is affected by the type of stand-replacing disturbance) does the stand act as a sink. However, it is not until some later time (indicated by point c in Fig. 20.10) that the pre-disturbance carbon stocks in the stand are re-established — i.e., the stand has recaptured the carbon released by the disturbance event.

Increasing the carbon uptake can be accomplished through techniques that reduce the time for stand establishment (such as site preparation, planting, and weed control), or increase available nutrients for growth, or through the selection of species that are more productive for a particular area. Decreasing the losses can be accomplished through modification of harvesting practices such as engaging in lower impact harvesting (to reduce soil disturbance and damage to residual trees), increasing efficiency (and hence reducing logging residue), and managing residues to leave carbon on site (Binkley et al. 1997).

Fertilization has also been used to enhance stand productivity and can result in increased long-term carbon retention in trees and soils (Nohrstedt 2001); however, fertilization success is dependent on the site conditions, for example, on more fertile sites. For planting, species selection and stocking are important considerations. Depending on

Fig. 20.10. Stand-level carbon (C) dynamics in boreal forest after a disturbance. During the first phase after the disturbance (*a*), losses from decomposition and respiration exceed C uptake by plants — i.e., the stand is acting as a C source; (*b*) during this phase, plant C uptake exceeds the losses from decomposition and respiration (i.e., the stand is a net C sink), but the stand has not yet regained the C level prior to disturbance; (*c*) during this phase, the stand is a net sink, and C stocks exceed the pre-disturbance level.

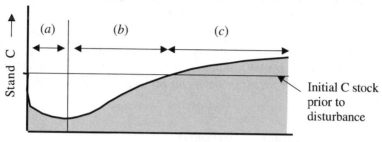

the management objective, planting fast-growing species, such as hybrid poplar, can yield high carbon accumulation rates in early years (Schroeder and Kort 2001); however, for long-term sequestration, planting species adapted to the local climate may be more effective (Schroeder and Kort 2001).

Landscape-level management

The total effect of a management regime is not measured by the loss and gain of an individual stand, but rather by the summation of the losses and gains in ecosystem carbon over the entire landscape being managed. Carefully planned harvesting methods and silvicultural treatments may maximize the carbon storage in individual stands; however, the management of the entire landscape (including fire and insect suppression, timber and non-timber objectives) may enhance or negate the benefits of stand-level management (Kurz et al. 1998). Managing the rotation length of a stand affects its carbon dynamics (Cooper 1983). Liski et al. (2001) found that for Scots pine (*Pinus sylvestris* L.) and Norway spruce (*Picea abies* (L.) Karst.) in Finland, extending the rotation length resulted in sequestration of more carbon. At a landscape level, managing a forest for wood production can lead to greater carbon storage than that of natural forests (Price et al. 1997; Kurz et al. 1998) in areas where management reduces the frequency of natural disturbances.

Fire- and insect-protection activities have a strong impact on the carbon sequestration of the forested landscape. Reducing the areas affected by fire and insect mortality extends the rotation age, effectively holding carbon longer in older age classes and allowing younger, more actively sequestering stands that would have otherwise been killed to continue to sequester carbon.

The overall carbon budget of the landscape is a result of the combination of uptake and release from the various stands at the different stages of development. Not only do current and future conditions play an important role in this balance, but the potential for a landscape to sequester carbon both now and in the future also depends strongly on past conditions. At a stand level, fire, insects, and harvesting each have different effects on the residual carbon pools remaining on site and the subsequent releases of carbon through decomposition. As such, similar stands with different disturbance histories can have different amounts of carbon accumulated in the dead organic matter pools. At a landscape level, changing from a lower disturbance regime to a higher disturbance regime will result in a net carbon source, whereas migrating toward a lower disturbance regime will result in a net sink.

Different management options being examined in order to address various environmental concerns, while meeting society's need for wood products, also have implications for forest carbon. Management regimes such as the triad approach (see Chap. 12) include areas of high management intensity (e.g., plantations which are intensively managed with high-yield, shorter rotations), extensive forest management (larger areas with more traditional forest management), and protected areas (where no harvesting is performed).

Adjacent carbon pools (lakes, peatlands)

Lakes

Few data exist concerning the effect of harvesting on the net carbon budget of lakes because, until recently, most research on aquatic impacts has focused on streams rather than lakes. The impact of harvesting on the levels of DOC and total suspended solids in lakes varies, depending on the size of the clearcut, the drainage ratio of the lake (i.e., the size of the lake compared with the size of the catchment), and the morphology of the watershed (Carignan and Steedman 2000). As described more fully in Chap. 10, clearcutting can lead to hydrological flushing of DOC from the soil into surface waters and result in increased lake DOC levels (Carignan and Steedman 2000; France et al. 2000). Carignan and Steedman (2000) found this to be especially true for lakes with large drainage ratios. In a study of 116 Boreal Shield lakes in Ontario and Quebec, France et al. (2000) found that lakes in logged watersheds contained 2 mg L^{-1} more DOC than lakes in undisturbed watersheds. In their study of 800 boreal lakes (in Ontario and Quebec), Carignan et al. (2000) also found higher DOC levels in lakes with logged watersheds than in lakes with undisturbed catchments. In contrast, Prepas et al. (2001) found no evidence of increased DOC concentrations in lakes on the Boreal Plain following harvest. The net effect of logging on the total carbon budget of boreal lakes is unknown, however, and will depend on total carbon inputs (i.e., particulate matter in addition to DOC) and how much carbon is ultimately stored in the sediments, exported in lake outflow, and evaded from the water column to the atmosphere (Gennings et al. 2001). While lake carbon budgets are sensitive to catchment disturbances, such as logging and fire, they are thought to be influenced to a greater degree by climate, hydrological parameters, and lake acidity. The only published evidence of increased DOC after fire is associated with fire in the bog-laden watersheds underlain by permafrost at the northern extent of the Boreal Plain (see McEachern et al. 2000).

Peatlands

Peatland drainage and peat extraction for horticultural products has increased over the past few decades (Waddington and Price 2000). In 1997, 234 Gg (1 Gg = 10^9 g) of organic matter was extracted from peatlands across western boreal Canada (Statistics Canada 1997). This material subsequently is oxidized under aerobic conditions when used as a potting medium and soil amendment. Harvested peatlands also have higher rates of CO_2 emission than undisturbed peatlands (Waddington and Price 2000). While restoration appears to enhance carbon accumulation relative to harvested sites, it cannot restore peatland carbon sinks to their pre-disturbed state for several decades or even centuries (Waddington and Price 2000). Other human activities, such as oil sand development in northern Alberta, also influence peatland carbon sinks by altering plant production and microbial respiration or by exporting peat from its natural waterlogged setting. Contemporary carbon budget assessment in western continental peatlands suggests that natural disturbances (i.e., fire and permafrost melt) are more important to peatland carbon sinks than direct human activities, such as harvesting and oil sands development (Turetsky et al. 2002). However, the majority of peat extraction in North America occurs in New Brunswick and Quebec, and may be a more significant factor in peatland carbon balance in eastern Canada.

Wood products

Bioenergy

Biomass energy can be used to reduce greenhouse gas emissions from fossil fuels by providing an alternative renewable source of energy. Efficiently produced bioenergy from sustainably managed forests is thought to be almost carbon neutral (Mercier 1998) because the carbon stored in the biomass (and released during combustion) was originally sequestered from the atmosphere and will be replaced as the forest regenerates (Schlamadinger et al. 1997). Fossil fuels, on the other hand, are non-renewable and result in constant accumulation of carbon in the atmosphere.

In Canada, bioenergy is becoming a significant source of renewable energy (ahead of wind and solar energy) and provides approximately 6% of the total primary energy supply and 7% of primary residential heating (Mercier 1998; Hauer et al. 2001). The main producer and consumer of bioenergy in Canada is the forest sector, particularly the pulp and paper industry. The use of bioenergy in the pulp and paper industry has steadily increased in recent decades, while the use of fossil fuels has declined in the past 20 years (Apps et al. 1999) (Fig. 20.11). The pulp and paper industry is expected to double its use of bioenergy sources between 1990 and 2020 and significantly reduce its use of fossil fuels (NRCan 1998; Hauer et al. 2001). Although the forest sector currently self-generates about half of the power that it uses in some regions (Gardner 1999; see Chap. 19), it is constrained in many cases from achieving economies of size in power generation by either an inability to sell excess power into the provincial grid or a lack of fibre (Curtis 2000) to sustain the supply.

As shown in Table 20.3, energy from wood residues can compete with fossil fuels and purchased electricity. This conclusion needs careful scrutiny, however. First, wood residue prices are based on average, not marginal, costs and are available only for small-scale operations where wood is easy to obtain. At a larger scale, much higher raw material (wood) costs can be expected. Second, wood fibre prices vary significantly by

Fig. 20.11. Energy-use efficiency of the pulp and paper industry in Canada (10-year average for each decade). Energy from biomass carbon has steadily increased while energy from fossil fuel carbon has declined since the 1970s (modified from Apps et al. 1999).

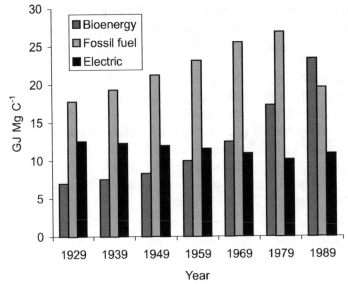

Table 20.3. Energy price comparisons for British Columbia, in 1997 $ (based on FST 1999 and authors' calculations).

	$ GJ^{-1}
Wood residues (assumed conversion factor = 18 GJ per dry Mg)	
Wood residue from pulp and paper mills	1.00
Wood residue from wood industry	0.56
Wood waste plus plantation wood	2.82
Fossil fuels (based on natural gas, boiler efficiency of 85%, C emission factor of 0.050 Mg GJ^{-1})	
Fuel price	1.73
Electricity (at $0.039 per kW h)	10.84

region, depending on residue surpluses or shortages, and environmental regulations. Regional values are not currently available for comparison (Gardner 1999). If fast-growing plantations are included, estimated costs are $2.82/GJ, which is more expensive than fossil fuels but still cheaper than purchased electricity, which (in boreal regions) is usually generated from hydroelectric sources.

In Canada, the high capital cost of infrastructure, regulation of the electricity market, and the relatively low cost of fossil fuels restrict the economic viability of substituting biomass for fossil fuels in power generation. When global climate change impacts, future energy requirements, availability of supply, and social and environmental values are considered, it is found that the benefits of renewable energy sources such as wood biomass outweigh the costs in some, but not all situations.

Fast and slow turnover forest products

Although all wood products release carbon back to the atmosphere through oxidation, pools that delay this release act as temporary storage pools, reducing the proportion of carbon leaving the terrestrial system on an annual basis (IGBP 1998). For a given production rate, the amount of carbon stored in wood products is primarily determined by the life span of these products. Thus slow-turnover pools store carbon for longer periods than fast-turnover pools. Slow-turnover pools include lumber and other long-term structural building products; fast-turnover pools include pulp and paper products, fuelwood, and scrap lumber. Discarded wood products and waste products may eventually end up in landfills that also have a slow turnover rate. To delay the emissions of carbon back to atmosphere from shorter life-span pools, the product's turnover time may be extended or the end fate of the product changed. Increased recycling of forest products (see Chap. 19) effectively acts to increase the residence of carbon in these products, especially products with a shorter life span.

Apps et al. (1999) performed a 70-year simulation analysis of the carbon dynamics of the Canadian forest products sector for the years 1920–1989. Their results suggest that the total carbon in forest products and landfills reached 836 Tg C by 1989 (Fig. 20.12). Pools with a long turnover time (slow pool and landfill) are, not surprisingly, the largest storage pools, containing 86% of the total.

Over the 70-year analysis period, a total of 1940 Tg C was harvested from Canadian forest ecosystems, and an additional 52 Tg C was imported as products. In 1989, an estimated 837 Tg C, or 43% of the cumulative harvested biomass carbon, was retained in forest product pools and landfills in Canada and abroad. Landfills contained 461 Tg C or 55% of the retained carbon (Apps et al. 1999). These simulations suggest that landfills may provide important long-term storage for carbon, especially under anaerobic conditions. Price et al. (1997) similarly found that landfill repositories of residues played an important role in the net carbon budget of an operational forest enterprise. Micales and Skog (1997) studied the decomposition of forest products in landfills and reported that United States landfills may serve as large carbon sinks, effectively preventing large quantities of carbon from being released back into the atmosphere.

The carbon stocks in the entire Canadian forest sector (excluding peat deposits) in 1989 is estimated to be 86.6 Pg C, of which 71.3 Pg C is found in the dead organic matter of litter and soils, 14.5 Pg in living biomass, and only 0.8 Pg C in forest product stocks (FPS). Although the FPS carbon stocks contain less than 1% of the total forest sector carbon, they grew by nearly 25 Tg C year^{-1} in 1989 (Apps et al. 1999). The FPS carbon stocks thus play a significant role in the net forest sector exchange with the atmosphere and offset more than one-third of the net carbon release from Canada's forest ecosystems.

Fig. 20.12. Carbon stocks in wood product pools in the Canadian forest sector between 1920 and 1989. (Data modified from Apps et al. 1999.)

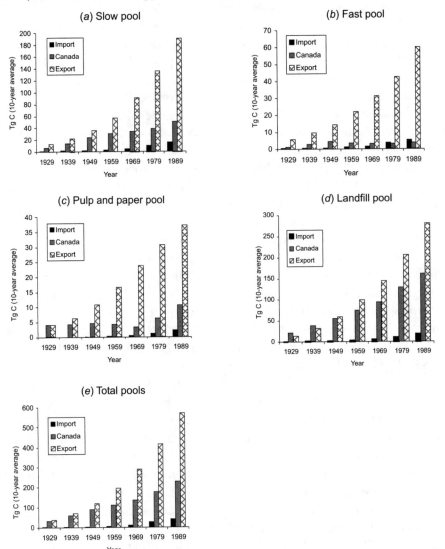

Economic considerations

Policy background

Concern about anthropogenic emissions of GHGs, particularly CO_2, led the World Meteorological Organization and the United Nations Environment Program to jointly establish the Intergovernmental Panel on Climate Change (IPCC) in 1988. The first IPCC report was published in 1990, and was instrumental in the subsequent signing of the United Nations' Framework Convention on Climate Change (FCCC) at the Earth Summit in Rio de Janeiro in June 1992 (see Chap. 1). The Convention committed signatories to stabilize atmospheric CO_2, with developed countries to reduce emissions to their 1990 levels by 2000 — a goal that was not achieved.

It was recognized as early as 1995, at the first Conference of the Parties (COP1) to the FCCC in Berlin, that the 1992 Rio de Janeiro commitments might not be enough to mitigate global warming. The IPCC's second assessment report in 1996 (IPCC 1996), which was endorsed by COP2, created an impetus for taking more drastic action. Therefore, at COP3 at Kyoto on 11 December 1997, industrialized countries agreed to reduce CO_2 emissions by an average of 5.2% from the 1990 level by 2008–2012, which became known as the commitment period.[3] The Kyoto Protocol (Box 20.3) is seen as a first, but necessary, step towards international commitment to dealing with the threat of global climate change. There have been five other COPs since Kyoto (see discussion below), and the IPCC's third assessment report was released in 2001 (IPCC 2001c).

In order for the Kyoto Protocol to go into effect, 55 countries must ratify the Protocol, and of industrial countries, those that ratify must account for 55% of total Annex B countries' 1990 CO_2 emissions. Poland, Canada, and New Zealand were some of the more recent countries to ratify Kyoto as of 24 February 2003, thereby bringing the number of countries that have ratified to 105 and the developed countries' proportion of 1990 emissions to 43.9%. The pressure is now on Russia to ratify the Protocol, since it accounts for 17.4% of 1990 CO_2 emissions, and with the United States and Australia having decided not to ratify, Russian ratification is required before the Kyoto Protocol can come into force.

Canada has agreed to reduce CO_2 emissions by 6% from the 1990 level by 2008–2012. Canadian GHG emissions by sector, expressed as CO_2 equivalents, are provided in Fig. 20.13. In 1990, Canada generated 607 Tg of CO_2-equivalent emissions, implying that Canada must reduce emissions to 571 Tg by 2008–2012.[4] By 2000, after a period of economic expansion, Canada's GHG emissions had increased by 19.6%, to 706 Tg (Olsen et al. 2002). Business-as-usual (BAU) emissions are projected to reach 802 Tg annually by 2010, approximately 40% above Canada's Kyoto commitment (Fig. 20.13).[5] Clearly, considerable effort will be required, within a very short time frame, to achieve the emissions-reduction objective. How this is to be done is explained in the Government of Canada's (2002) implementation plan (discussed further below).

In October 2000, the government released the National Implementation Strategy on Climate Change and the First National Climate Change Business Plan to address Canada's strategy for implementing the Kyoto Protocol. Under Action Plan 2000, which was announced at the same time, $500 million was set aside to provide subsidies for programs and measures for reducing CO_2 emissions. The projected reductions were expected to account for more than one-fifth of the Kyoto gap, or some 50 Tg CO_2 annu-

---

[3]Industrialized countries that signed the Kyoto Protocol included those in Annex I of the FCCC plus countries of the ex-Soviet bloc. These countries are listed in Annex B of the Kyoto Protocol.

[4]In this section, the units are expressed in terms of CO_2. Divide the units of CO_2 by 3.7 to obtain the approximately equivalent weight in terms of C.

[5]These projections are based on earlier government reports and are sensitive to assumptions about population growth (including rates of immigration), economic growth, changes in energy prices, the rate of adoption of energy-saving technologies, and other factors (see NRCan 1999b; Goncalves 2001). Current projections indicate that the gap between the Kyoto requirement and BAU emissions is 240 Tg rather than the 231 Tg indicated here and in Fig. 20.13.

Box 20.3. The Kyoto Protocol

The Kyoto Protocol establishes legally binding targets and timetables for reducing green-house gas (GHG) emissions in developed countries. The GHG reduction commitments for the first commitment period (2008–2012) for key boreal signatories of the protocol are as follows (expressed as a percentage of 1990 emission levels): Canada 94%, Norway 101%, Sweden 92%, Finland 92%, Russian Federation 100%.

The Protocol recognizes that land use, land-use change, and forestry affect the net GHG balance (sources and sinks). Articles 3.3 and 3.4 provide forest-related mechanisms for compliance with the above targets:

- Article 3.3 requires an accounting of C stock changes resulting from land-use change (Afforestation, Reforestation, and Deforestation, ARD) since 1990;
- Article 3.4 allows countries to include, if they choose to do so, carbon gains/losses due to forest management practices since 1990;
- Each Party must stipulate by 2006 whether or not they will include forest management practices in their reporting; and
- If included, sinks due to forest management can be used to offset any net ARD source up to a cap of 9 Tg C year^{-1}. Any remaining sink is then credited up to a country-specific cap (Canada's cap is 12 Tg C year^{-1}).

Definitions:

- *Afforestation* — the direct human-induced conversion of land that has not been forested for a period of at least 50 years to forested land through planting, seeding, and (or) the human-induced promotion of natural seed sources.
- *Reforestation* — the direct human-induced conversion of non-forested land to forested land through planting, seeding and (or) human-induced promotion of natural seed sources, on land that was forested but that has been converted to non-forest land. For the first commitment period, reforestation activities will be limited to reforestation occurring on those lands that did not contain forest on 31 December 1989.
- *Deforestation* — the direct human-induced conversion of forested land to non-forested land.
- *Forest management* — a system of practices for stewardship and use of forest land aimed at fulfilling relevant ecological (including biological diversity), economic, and social functions of the forest in a sustainable manner.

Additional information on the Kyoto Protocol including recent developments and negotiations can be found on the United Nations Framework Convention on Climate Change website, http://unfccc.int/.

ally (Goncalves 2001, p. 55; as modified, Goncalves 2002, p. 12). Subsidies and voluntary initiatives feature prominently in achieving this amount of emissions reduction. To this are added an additional 30 Tg CO_2 annually to be sequestered in terrestrial sinks (discussed below), implying that 80 Tg CO_2 per year (one-third of the Kyoto target) has already been accounted for by actions implemented in Action Plan 2000 and Budget 2001. Of the remaining 160 Tg required annual reduction in CO_2 emissions, the Government's October 2002 implementation plan (Goncalves 2002) calls for some

Fig. 20.13. Actual, projected, and Kyoto target emissions, 1990–2012, Canada.

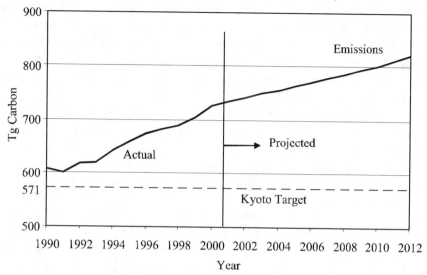

90–100 Tg to be achieved through a variety of regulations, as well as program subsidies where reductions are achieved through modification of behaviour.

Emissions trading

Both domestic and international emissions trading are part of Canada's implementation plan (Goncalves 2002, pp. 28–32, 42–44). The proposed domestic emissions trading system will target 55 Tg of the gap, with emission reduction targets established through negotiated covenants with a regulatory or financial backstop, although details are still under development. It is expected that offsets will be part of the emissions trading system, including forestry and agricultural sinks. In order to establish a market for CO_2 emission permits, the government needs to establish property rights to carbon, set the total amount of emission permits allowed (the "cap"), monitor and enforce the quota system, and encourage markets by developing clear, simple, and stable rules and procedures for conducting trades.

While it is relatively straight forward to establish a market for trading CO_2 emission permits, integrating carbon offsets or credits from terrestrial activities into such a scheme may be more problematic (van Kooten 2002). For example, two caps may need to be established, one for CO_2 emissions and another for carbon offsets, and these cannot be determined independently. Further, it may be necessary to establish an exchange rate between permanent reductions in atmospheric CO_2 (emission reductions) and temporary ones (terrestrial sinks). One approach to the latter problem, which has been used in Australia, is to permit landowners to sell carbon sink credits, but with the proviso that they are liable for future release of carbon, having to purchase CO_2 emission permits at the time the carbon is released. Emission trading schemes, with or without carbon offsets, can be national or international in scope. If the Kyoto Protocol goes into force, the international emissions trading regime will be an important means for reducing overall global emissions in an efficient manner. Countries will no doubt work to ensure that domestic markets are linked to the international market. Further details on carbon trading are provided by Henri (2001), Obersteiner et al. (2001), and van Kooten (2003).

In Canada and other countries, carbon trading outside the Kyoto framework has already started. Some large emitters of CO_2, primarily regulated utility companies that can usually pass such costs onto customers and companies that wish to purchase goodwill, have made some exploratory purchases of carbon credits that accrue as the result of forestry projects (Henri 2001). These companies are offsetting emissions in anticipation of future requirements to do so. Effective monitoring is critical to ensure that the contracted carbon credits actually materialize, but programs and activities that certify those providing the credits can reduce both the risk and the overall cost.

There will likely be a mix of initiatives by the private and public sectors to certify carbon credits. Sellers might be certified under ISO 9002 (quality systems, including forest inventory and mapping systems) and ISO 14001 (environmental management systems). They might also be certified at the national level or through non-governmental organizations (NGOs), which would certify independent companies who have the expertise to conduct carbon audits and (or) to certify and audit sellers. See Chap. 21 for a broader discussion of forest certification initiatives and how these might include carbon auditing. As the design of Canada's domestic emissions and offset trading system are developed, the details regarding the processes and organizations that will register and verify carbon credits will need to be elaborated.

A market for trading carbon sequestration credits will be an important development for several reasons. First, the futures market for carbon credits — a futures market because it deals with carbon uptake in the Kyoto commitment period (2008–2012) — establishes a price for carbon and provides a useful indicator for both the private sector and policymakers. Furthermore, such a market can be integrated into a larger system of carbon emissions trading: instead of purchasing CO_2-emission permits, companies (or countries) can purchase carbon uptake credits.

Carbon credits and the role of forestry

Under the Kyoto Protocol, a country can obtain carbon credits at home by planting trees where none grew previously, increasing the rate at which carbon is sequestered through forest management, or pursuing other land-use activities that increase carbon uptake in soils. Forests are important because they can sequester more carbon over a longer time period than agricultural sinks, with further benefits possible if account is taken of the fate of wood products (e.g., construction lumber, and paper in landfills) or wood biomass used to produce energy in place of fossil fuels. Forests store carbon by photosynthesis, thus removing CO_2 from the atmosphere as they grow, with each cubic metre of wood containing approximately 200 kg of carbon. Countries with significant forest cover may be interested in potential carbon credits related to forest management and those with large tracts of marginal agricultural land may be interested in afforestation as a means for achieving some of their agreed-upon reduction in CO_2-emissions. Canada falls into both these categories. Therefore, the remaining focus of this section is on the role of terrestrial carbon sinks and how forest management may provide national (and also market-traded) debits and credits under the Kyoto Protocol.

Under the Protocol, afforestation and reforestation occurring since 1990 may result in carbon credits, while deforestation (human-induced conversion of forestland to non-forest use) results in a carbon debit. The terms afforestation and reforestation had been the subject of debate, but as a result of COP6$_{bis}$ (Bonn) and COP7 (Marrakech) negotiations, afforestation is now defined, for the purposes of the Kyoto Protocol, as human activities that encourage growing trees on land that has not been forested in the past 50 years, while reforestation refers to human activities that encourage growing trees on land that was forested but had been converted to non-forest use (see also IPCC 2000). An indication of the potential role of these activities on a global scale is provided in Table 20.4. Potentially, these activities could sequester an average of 1000–1580 Tg C year^{-1} by 2050. In addition, under Article 3.4 of the Kyoto Protocol, countries have the choice of whether to include carbon sequestered as a result of forest management (e.g., fertilizing, and fire control) toward emissions reductions, although no account is taken of carbon stored in wood products. Even before the United States withdrew from the Kyoto negotiations, economists estimated that carbon credits would need to be created for $10–20 per Mg C to be competitive (Sandor and Skees 1999). There remains a question as to whether, even if all of the carbon fluxes (such as wood products storage) are appropriately taken into account, "additional" forest management will be a cost-effective and competitive means for sequestering carbon (Caspersen et al. 2000). As a result of COP6$_{bis}$ and COP7 negotiations, however, countries can claim carbon credits from business-as-usual forest management that are likely not "additional" (i.e., do not reduce atmospheric CO_2 over and above what would have happened anyway). The role of terrestrial sinks and whether market mechanisms should treat carbon offsets the same way as emissions reduction was a source of dispute at COP6 in The Hague, Netherlands, in

Table 20.4. Global estimates of the costs and potential carbon that can be removed from the atmosphere and stored by enhanced forest management from 1995 to 2050 (data compiled from IPCC 1996, pp. 785–791).

Region	Practice	C removed and stored (Pg)	Estimated costs ($CAN $\times$ 10^9)[c]
Boreal	Forestation[a]	2.4	26
Temperate	Forestation[a]	11.8	92
	Agroforestry	0.7	5
Tropical	Forestation[a]	16.4	149
	Agroforestry	6.3	42
	Regeneration[b]	11.5–28.7	68–152
	Slowing-deforestation[b]	10.8–20.8	
Total		60–87	

[a]Refers primarily to reforestation.
[b]Includes an additional 25% of aboveground C to account for C in roots, litter, and soil (range based on uncertainty in estimates of biomass density).
[c]Based on the conversion: $1 CAN = $0.65 U.S.

November 2000. The European Union countries took the view that there should be limits to the role of land use, land-use change, and forestry (LULUCF) projects so that countries would be forced to address domestic emissions reduction. The argument put forward was the fear that LULUCF projects are ephemeral and do not help to reduce the long-term, upward trend in CO_2 emissions. The opposite view is that reduction of CO_2 emissions and enhancement of carbon sinks are no different in their impact and should be treated the same on efficiency grounds (Chomitz 2000). Subsequent to the United States' withdrawal from the Kyoto process, the objections to inclusion of LULUCF projects softened (at COP6$_{bis}$ and COP7), opening the door for a much greater role for these projects.

The negotiations on forest sinks did produce caps, however. The Marrakech Accords (negotiated at COP7) give each country the choice of whether it wants to include forest management in its accounting in the first commitment period, with the decision to be made by 2006. A country that has a net debit from Article 3.3 (i.e., due to afforestation, reforestation, and deforestation (ARD)) can use credits from its forest management to offset this debit, up to 9 Tg C year^{-1}, provided it has chosen to include forest management in its accounting. Use of this offset is subject to the proviso that its managed forest since 1990 is a net sink at least equal to the ARD debit. The agreement allows further credit for forest management in the first commitment period up to a country-by-country negotiated cap, which for Canada is 12 Tg C year^{-1} (44 Tg CO_2). The Russian Federation can claim up to 33 Tg C, Japan 13 Tg C, Germany 1.24 Tg C, Ukraine 1.11 Tg C, and remaining countries less than 1.0 Tg C. Canada currently estimates it will be able to claim approximately 5.4 Tg C annually of forest sinks (i.e, the net of ARD and forest management), and expects an estimated 2.7 Tg C per year from agricultural sinks (Goncalves 2002, p. 39). Countries, however, can only claim credit for actual carbon stock changes in the commitment period that will need to be accounted for using detailed procedures for verification purposes.

Carbon credits can also be obtained for activities in developing countries and economies in transition. Kyoto Protocol's Clean Development Mechanism (CDM) enables industrialized countries to purchase certified offsets from developing countries (that have ratified the Kyoto Protocol) by sponsoring projects that reduce CO_2 emissions or increase CO_2 removals beyond business-as-usual levels in those countries. Afforestation and reforestation are the only land-use activities that can be included under the CDM, and these contributions are capped at 1% of an Annex B country's baseline emissions.

Likewise, emission reduction units can be produced through Joint Implementation (JI) projects in countries whose economies are in transition. Any land management activity accountable under Articles 3.3 or 3.4 of the Kyoto Protocol is eligible under the JI provision. Russia and Ukraine are two countries where forest activities generating carbon credits might be undertaken as JI projects by other industrial countries.

As noted, the idea of relying on terrestrial sinks to meet international emissions reduction obligations remains controversial. Some argue that, in principle, a country should get credit only for sequestration above and beyond what occurs in the absence of C-uptake incentives, a condition known as "additionality" (Chomitz 2000). Thus, for example, if it can be demonstrated that a forest would be harvested and converted to another land use in the absence of specific policy (say, subsidies) to prevent this from happening, the additionality condition would be met. Additional carbon sequestered as the net result of incremental forest management activities (e.g., juvenile spacing, commercial thinning, fire control, or fertilization) would be eligible for carbon credits, but only if the activities would not otherwise have been undertaken (e.g., for timber supply purposes or to provide higher returns). Similarly, afforestation projects would be additional if they would not have been undertaken in the absence of economic incentives, such as subsidy payments or an ability to sell carbon credits (Chomitz 2000). The premise of additionality does not apply for Articles 3.3 and 3.4, although it does for CDM projects. Because of the controversy and difficulties in separating direct human-induced effects from natural effects, countries agreed on the use of caps in the first commitment period as an interim ad hoc solution. Considerable scientific work is ongoing to better address these issues for the next commitment period.

Costs of creating carbon credits in Canada

Standard financial calculations suggest that afforestation and reforestation in Canada would generally not yield carbon credits at a sufficiently low cost to be competitive with other investments, because northern forests tend to be marginal in terms of productivity (van Kooten et al. 1993). The reason is that such forests generally regenerate naturally (even on marginal agricultural land), making artificial regeneration redundant and (or) lowering the benefits to reforestation programs (van Kooten and Bulte 2000, p. 398). Studies indicate that only short-rotation, hybrid poplar plantations are likely to be competitive with other methods of removing CO_2 from the atmosphere (van Kooten et al. 1999, 2000).

For example, the TECAB model (Krcmar and van Kooten 2001; Krcmar et al. 2001) was used to examine the costs of carbon uptake in the grain belt – boreal forest transition zone of British Columbia. The model consists of tree-growth, agricultural activities, and land-allocation components. Carbon fluxes associated with many forest management activities (but not control of fire, pests, and disease) are included, and account is taken of what happens to the wood after harvest, including decay (see Table 20.5 for data on decay of forest ecosystem components). The study results indicate that upward of 1.5 Tg of discounted carbon (discounted at 4%) can be sequestered in the region at a cost of about $40 per Mg or less. This is a combination of planting hybrid poplar, as well as forest management activities, with wood products storage also accounted for. This amounts to an average of about 1.3 Mg ha^{-1}, or about 52 kg ha^{-1} per year of additional carbon uptake. While in general plantation forests are considered a cost-effective means of sequestering carbon (Sedjo et al. 1995; Adams et al. 1999), these results are a further indication that boreal forests are globally marginal, and silvicultural investments in them generally do not pay, even when carbon uptake is included as a benefit (van Kooten et al. 1993; Wilson et al. 1999).

While hybrid poplar planting in general appears to be the most economic option for creating carbon credits in northern climates, there remains a great deal of uncertainty about planting hybrid poplar on a large scale in Canada, mainly because it has not been done previously. Certainly there are a number of considerations related to the use of hybrid poplar plantations as part of a carbon uptake strategy, which include:

(1) Relative to native species, hybrid poplar plantations may have negative environmental impacts related to reduced biodiversity and susceptibility to disease (Callan 1998);

(2) If there are transaction costs associated with afforestation, this will increase C-uptake costs over and above what is generally estimated (van Kooten 2000);

(3) Uncertainty about (current and future) timber and fibre values and prices of agricultural products may make landowners reluctant to convert agricultural land to forestry;

(4) The potential of wood from afforested land as a biomass fuel is uncertain;

(5) Recent research suggests that planting trees where none existed previously decreases surface albedo, which offsets the negative forcing expected from carbon uptake (Betts 2000), although such planting must be on a large (perhaps unrealistic) scale before there is a noticeable albedo effect;

(6) There is a potential problem of leakage, since large-scale afforestation and (or) other forest plantations will tend to lower prices for wood fibre, with current wood lot owners (for example, in the southern United States) likely reducing their forest holdings by converting land back to agriculture in anticipation of the drop in price. These are generally ignored in calculating the costs of individual afforestation or reforestation projects. Yet, such leakages can be substantial, even as high as one-half of the carbon sequestered by the new plantations (Sohngen and Sedjo 1999).

In general, tree planting programs to generate carbon credits in Canada may be much more expensive than originally anticipated (van Kooten 2003).

While creating incentives to manage carbon stocks and carbon fluxes in forests efficiently would have positive benefits, it is unlikely that forest management in Canada will be a cost-effective and competitive means for sequestering carbon on a sufficient scale to dramatically affect Canada's emissions reduction target for Kyoto in the first commitment period. Compared with economic viability in other regions, Canada's forests are marginal. The only land-use change on forestlands that might be a competitive alternative to other methods of removing CO_2 from the atmosphere is if short-rotation, hybrid poplar plantations replace logged or otherwise denuded forests. However, the potential negative environmental effects of hybrid poplar plantations, such as their limited longevity, reduced biodiversity, disease, insect problems, and loss of scenic amenities, would need to be fully considered in assessing their suitability as an acceptable large-scale tool for carbon removals to help meeting emission reduction targets.

While there are numerous actions that forest management agencies and the forest sector can take to reduce emissions, and increase removals of carbon (see Box 20.4), enhancing carbon sequestration through intensive forest management or plantations is not likely to be an economic alternative on a wide scale in boreal ecosystems. However, given the significant role of disturbances in the boreal forest ecosystem, it is clear that the forest managers will influence carbon stock changes by their management actions, as discussed in earlier sections in this chapter.

> ## Box 20.4. Options for minimizing carbon emissions in boreal forestry.
>
> Boreal ecosystems may only be able to make limited contributions to carbon sequestration, owing to low biological productivity compared to other parts of the world. But their vast area and large carbon stocks mean that boreal forests will always play a major role in the global carbon budget. Government agencies, forest products companies, and forest managers have a number of tools at their disposal to minimize the emission of CO_2 and other GHGs to the atmosphere:
>
> - minimize and optimize the use of fossil fuels:
> - use fuel-efficient machinery, trucks, etc.;
> - minimize transportation requirements through centralized positioning of processing facilities, and harvesting forest in contiguous rather than dispersed patterns;
> - substitute fossil fuels with biofuels wherever possible;
> - inventory and monitor carbon stocks in biomass, soils, peatlands, and lakes;
> - promote recycling and reuse of wood and paper products;
> - treat wood products with preservatives that extend their useful life;
> - protect peatlands from drainage, fire, and land-use change;
> - protect forests from fire and insect outbreaks;
> - protect forest soil organic matter by avoiding the use of broadcast slash burning and scarification treatments, preventing erosion, and minimizing soil disturbance (e.g., more logging on snow instead of in the summer);
> - capture CO_2 emissions from pulp mills, kiln boilers, and other sources and channel them for beneficial uses that also remove CO_2, such as for growth enhancement in greenhouses;
> - keep landfills anaerobic, and (or) tap their CH_4 emissions;
> - promote longer rotations where tree species and sites are such that trees can be long-lived and large-statured (i.e., large C pools) with low risk of disturbance;
> - use long-lived and potentially large-statured species to serve as C-reservoirs in long rotations, or very productive fast-growing species to quickly absorb C in short rotations;
> - use fertilization where C-sequestration benefits exceed fertilization manufacture and application costs, especially if non-fossil fuel sources can be used (e.g., animal and human wastes);
> - minimize regeneration delay and achieve high/full stocking as soon as possible;
> - promote economic and policy instruments that encourage sustainable forest management practices, carbon storage/sequestration, and reduction in fossil fuel use.

Economic impacts of adaptation to climate change on forestry

Countries must be prepared to adapt to climate change because the Kyoto Protocol is but a small step forward in efforts to reduce atmospheric CO_2. Therefore, it is useful to examine briefly the possible impacts on the forestry sector and what adaptation will mean. Uncertainties in the underlying biological responses to the changing climate necessarily limit the robustness of the conclusions, and it is to hoped that in the coming years, there will be increasing effort to integrate the biogeochemical models with socioeconomic ones. Some ecological projections of the effects of climate change on north-

Table 20.5. Rates of decay of forest ecosystem components (including wood products) upon harvest (data compiled from Hoen and Solberg 1994).

End-use category	Time (years) from felling until decay starts	Decay time (years) (from beginning of decay until all fibre has decomposed)
Bark in land fillings	0	8
Bark for burning	0	1
Needles	0	7–11
Branches, stumps, stems in forest	0	12
Root system after felling	0	100
Construction material	80	80
Furniture and interiors	20	50
Impregnated lumber	40	70
Pallets	2	23
Losses	0	1
Composites, plywood	17	33
Sawdust	1	2
Pulp/paper	1	2
Fuelwood	0	1

ern forestlands are reviewed earlier in this chapter, with more detailed analyses included in Apps et al. (1995), Apps and Price (1996), and in the series of publications associated with a conference on the role of boreal forests and forestry in the global carbon budget held May 2000, in Edmonton, Alberta (Apps et al. 2002; Karjalainen et al. 2002; Shaw and Apps 2002; Stocks et al. 2002).

Darwin et al. (1995) have provided a comprehensive model for analyzing the global response of primary sectors to climate change. The model gives some indication of the extent to which land use in Canada might be affected by climate change. The model consists of a geographic information system and a computable general equilibrium economic model of the global economy, with climate projections from four global circulation models (GCMs). Land-use potential in the model is based primarily on climatic factors (moisture and temperature) because these also affect soil formation. Of all the regions in the model, Canada stands to gain the most from climate change from an economic point of view. Canadian gross domestic product is projected to increase by an average of 2.2% across the four GCM scenarios, the largest increase for any region. Output of forest products will also increase (by 33%), although the forested land base will be reduced by an average of 7%. This is because in Darwin et al.'s (1995) model, cropland, pasture, and forestland are more productive under projected global warming, even in the absence of a CO_2-fertilization effect. Much of the increase in arable crop production in Canada comes from a northward shift in the western grain belt at the expense of boreal forest, and a shifting of the highly productive corn-belt in the United States into the eastern portion of the Canadian grain belt.

Adaptation of forest management to a changing climate will be required in order to ensure that forest sector returns are maximized. Examples of management adaptation include salvaging dying trees, vegetation control to help offset drought, replanting with more suitable species, and shifting processing capacity to areas where timber is relatively plentiful (Binkley and van Kooten 1994). In addition, as the frequency of pest outbreaks and forest fires increases, investment in the management of pests and fire is also expected to rise. As the northern fringe of the boreal forest shifts northward, tree planting can help to maintain forested area as land is lost to agriculture on the southern border.

The conclusion from economic research is that climate change is unlikely to bring about reductions in the global supply of primary commodities, including wood products (Schimmelpfenning et al. 1996; also IPCC 2001b, pp. 877–912). Rather, if economic markets and institutions are sufficiently flexible, landowners will make decisions to take advantage of changes, such as producing different crops or new crop varieties, adopting new management regimes (e.g., greater use of irrigation, or enhanced silviculture), or expanding activities to new areas. However, market failure, policy failure, and institutional barriers (see Chap. 7) to land-use change can be obstacles to enhancement of well-being in the face of climate change. Government regulation, for example, may be one impediment to adaptation. Policy failure would occur, for example, if the public authority required forest companies to reforest cutover lands on the boreal–grassland interface, thereby preventing their conversion to a more productive use such as pasture or cropland (van Kooten 1995; van Kooten et al. 2002). Market failure would occur if no account was taken of climate change impacts on recreation and other non-timber benefits of forests, particularly biodiversity. The high probability of climate change, coupled with this potential land-use inertia, calls into question the entire premise and feasibility of sustainable forest management as envisioned for a fixed land base (see Chap. 2).

Sohngen et al. (1999) addressed questions related to biodiversity and climate change. They linked ecosystem productivity and economic trade models to GCM output to demonstrate that climate change leads not only to greater wood fibre output (as predicted by Darwin et al. 1995), but also to greater protection of biodiversity. The reason for the enhanced status of biodiversity is that, although all forests become more productive as a result of CO_2 fertilization (according to this model, but see also the earlier section on CO_2 fertilization), there is also greater investment in plantation forests, also because of CO_2 fertilization. As plantation harvests increase, wood fibre prices fall and natural forests (those that harbour biodiversity) are less susceptible to logging. The area of natural forests that is not attractive for logging increases. On the other hand, Kirkup (2000) indicates that maintaining biodiversity in the face of climate change is more challenging. Solutions may require extensive biotic inventorying and monitoring, and the maintenance or restoration of landscape linkages (see Chap. 12) to facilitate the migration of countless unmanaged and unmonitored species.

Conclusions

The future carbon balance in the boreal forest will largely depend on the type and frequency of disturbances, changes in species composition, and alterations to the nutrient and moisture regimes under changing climate conditions. It will also depend on forest management practices that affect both the disturbance regime and nutrient status.

Projected climate change scenarios for the boreal forest generally predict warmer and somewhat drier conditions, and the disturbance patterns are also expected to change. Forest fire and insect outbreaks, for example, have historically been sensitive to climate and are expected to change considerably under global warming. The variations in climatic conditions in central Canada over the past 20 years appear to have resulted in higher rates of forest disturbance at a national scale. Regionally, the changes may be larger: the mean fire return interval for some sites in Alberta and Saskatchewan has been reported to be as low as 34 years (Larsen and MacDonald 1998). With more frequent disturbances, more of the stands are in younger age classes and initially release greater amounts of carbon to the atmosphere as their elevated detritus and soil pools decay. This situation is expected to worsen as climatic change proceeds. Altered boreal forest disturbance regimes — especially increases in frequency, size, and severity — may release carbon from vegetation, the forest floor, and soil at higher rates than the rate of carbon accumulation in regrowing vegetation.

The realization of the potential increase in plant productivity due to climate change will be determined by a variety of factors such as changes in species and competitive interactions, water availability, and the effect of temperature increase on photosynthesis and respiration. The precise balance of carbon uptake and release depends on the detailed processes, and especially interactions between climate, site variables, and vegetation over the changing life cycles of forest stands. Quantifying life-cycle dynamics at the stand level is essential for projecting future changes in forest-level carbon stocks. The net effect of such changes could result in positive feedback to climate change and thereby accelerate global warming (Kurz et al. 1995b).

Forest management options to enhance or protect carbon stocks include reducing the regeneration delay through seeding and planting, enhancing forest productivity, changing the harvest rotation length, the judicious use of forest products, and forest protection through control and suppression of disturbance by fire, pests, and disease. A protected forest acts as a sink only during the period of transition from a higher to a lower period of disturbance. Can such protection be maintained if the changing climate favours increased disturbances? Moreover, it must be stressed that if the protection is removed, the same areas will be a source as the disturbance regime relaxes back to a period of higher disturbance.

Policy options that nations can pursue to minimize their contributions to global climate change will include land management activities such as forest management and afforestation. Land management agencies and forest managers can choose from a large array of tools (see Box 20.4) in implementing carbon management strategies that will impact carbon stock changes and therefore net carbon removals and emissions.

Given the potential importance of terrestrial carbon sinks, carbon offsets can be an important element for encouraging economics investments in carbon sequestration projects, with verification by authorized certifiers. If CO_2-emission taxes are to be employed, then these need to be extended to cover carbon sinks (e.g., afforestation projects), although this will require that landowners be taxed for release of CO_2 to the atmosphere but subsidized for its removal (van Kooten et al. 1995). Tax revenues could be used to subsidize removals. The effectiveness of an efficient tax/subsidy system could be maximized by covering as broad a range of activities affecting carbon emissions as possible. This does not mean that carbon sink options will be undertaken on a large scale, however. Only if terrestrial carbon sinks are competitive with other means of reducing a nation's CO_2 emissions will private firms start paying for forest management and afforestation projects. To remain globally competitive, boreal jurisdictions must create as flexible an environment as possible, allowing private firms to seek out the most cost-effective options for reducing emissions.

Acknowledgements

We wish to thank Brian Amiro, Cindy Shaw, Kevin Percy, Neil Foster, Richard Fleming, Ted Hogg, Ellie Prepas, Peggy Robinson, Thierry Varem-Sanders, and three anonymous reviewers for their valuable input to this manuscript.

References

Adams, D.M., Alig, R.J., McCarl, B.A., Callaway, J.M., and Winnett, S.M. 1999. Minimum cost strategies for sequestering carbon in forests. Land Econ. **75**: 360–374.

Alm, J., Talanov, A., Saarnio, S., Silvola, J., Ikkonen, E., Aaltonen, H., Nykanen, H., and Martikainen, P.J. 1997. Reconstruction of the carbon balance for microsites in a boreal oligotrophic pine fen, Finland. Oecologia, **110**: 423–431.

Amiro, B.D., Stocks, B.J., Alexander, M.E., Flannigan, M.D., and Wotton, B.M. 2001*a*. Fire, climate change, carbon and fuel management in the Canadian boreal forest. Int. J. Wildland Fire, **10**: 405–413.

Amiro, B.D., Todd, J.B., Wotton, B.M., Logan, K.A., Flannigan, M.D., Stocks, B.J., Mason, J.A., Martell, D.L., and Hirsch, K.G. 2001*b*. Direct carbon emissions from Canadian forest fires, 1959 to 1999. Can. J. For. Res. **31**: 512–525.

Apps, M.J., and Price, D.T. (*Editors*). 1996. Forest ecosystems, forest management and the global carbon cycle. Springer-Verlag, Heidelberg, Germany. NATO ASI Ser. 1, Vol. 40. 452 p.

Apps, M.J., Kurz, W.A., Luxmoore, R.J., Nilsson, L.O., Sedjo, R.A., Schmidt, R., Simpson, L.G., and Vinson, T.S. 1993*a*. Boreal forests and tundra. Water Air Soil Pollut. **70**: 39–53.

Apps, M.J., Kurz, W.A., and Price, D.T. 1993*b*. Estimating carbon budgets of Canadian forest ecosystems using a national scale model. *In* Proceedings of the Workshop on Carbon Cycling in Boreal Forests and Subarctic Ecosystems. *Edited by* T. Vinson and T. Kolchugina. United States Environmental Protection Agency, Office of Research and Development, Washington, D.C. EPA/600R-93-084. pp. 241–250.

Apps, M.J., Price, D.T., and Wisniewski, J. (*Editors*). 1995. Boreal forests and global change. Kluwer Academic Publishers, Dordrecht, Netherlands. 564 p.

Apps, M.J., Kurz, W.A., Beukema, S.J., and Bhatti, J.S. 1999. Carbon budget of the Canadian forest product sector. Environ. Sci. Policy, **2**: 25–41.

Apps, M.J., Bhatti, J.S., Halliwell, D.H., Jiang, H., and Peng, C.H. 2000. Simulated carbon dynamics in the boreal forest of central Canada under uniform and random disturbance regimes. *In* Global climate change and cold region ecosystems. *Edited by* R. Lal, J.M. Kimble, and B.A. Stewart. Advances in Soil Science, Lewis Publishers, Boca Raton, Florida. pp. 107–122.

Apps, M.J., Karjalainen, T., Stocks, B.J., and Shaw, C.H. 2002. Foreword. The role of boreal forests and forestry in the global carbon budget. Can. J. For. Res. **32**: v–vi.

Armentano, T.V. 1980. Drainage of organic soils as a factor in the world carbon cycle. BioScience, **30**: 825–830.

Armentano, T.V., and Menges, E.S. 1986. Patterns of change in the carbon balance of organic soil – wetlands of the temperate zone. J. Ecol. **74**: 755–774.

Arseneault, D., and Payette, S. 1992. A postfire shift from lichen-spruce to lichen-tundra vegetation at tree-line. Ecology, **73**: 1067–1081.

Auclair, A.N.D., and Thomas, B.C. 1993. Forest wildfires as a recent source of CO_2 at mid-high northern latitudes. Can. J. For. Res. **23**: 1528–1536.

Baldocchi, D., Kelliher, F.M., Black, T.A., and Jarvis, P.G. 2000. Climate and vegetation controls on boreal zone energy exchange. Global Change Biol. **6**(Suppl. 1): 69–83.

Bazzaz, F.A., Bassow, S.L., Berntson, G.M., and Thomas, S.C. 1996. Elevated CO_2 and terrestrial vegetation: implications for and beyond the global carbon budget. *In* Global change and terrestrial ecosystems. *Edited by* B. Walker and W. Steffen. Cambridge University Press, Cambridge, U.K. pp. 43–76.

Beilman, D.W., Vitt, D.H., and Halsey, L.A. 2001. Localized permafrost peatlands in western Canada: definitions, distributions, and degradation. Arct. Antarct. Alp. Res. **33**: 70–77.

Betts, R.A. 2000. Offset of the potential carbon sink from boreal forestation by decreases in surface albedo. Nature (London), **408**: 187–90.

Bhatti, J.S., and Apps, M.J. 2000. Carbon and nitrogen storage in upland boreal forests. *In* Global climate change and cold regions ecosystems. *Edited by* R. Lal, J.M. Kimble, and B.A. Stewart. Advances in Soil Science, Lewis Publishers, Boca Raton, Florida. pp 79–89.

Bhatti, J.S., Apps, M.J., and Jiang, H. 2001. Examining the carbon stocks of boreal forest ecosystems at stand and regional scales. *In* Assessment methods for soil carbon. *Edited by* R. Lal, J.M. Kimble, R.F. Follett, and B.A. Stewart. Lewis Publishers, Boca Raton, Florida. pp. 513–532.

Bhatti, J.S., Apps, M.J., and Jiang, H. 2002*a*. Influence of nutrients, disturbances and site conditions on carbon stocks along a boreal forest transect in central Canada. Plant Soil, **242**: 1–14.

Bhatti, J.S., Apps, M.J., and Tarnocai, C. 2002*b*. Estimates of soil organic carbon stocks in central Canada using three different approaches. Can. J. For. Res. **32**: 805–812.

Binkley, C.S., and van Kooten, G.C. 1994. Integrating climatic change and forests: economic and ecological assessments. Clim. Change, **28**: 91–110.

Binkley, C.S., Apps, M.J., Dixon, R.K., Kauppi, P.E., and Nilsson, L.O. 1997. Sequestering carbon in natural forests *In* Economics of carbon sequestration in forestry. *Edited by* R.A. Sedjo, R.N. Sampson, and J. Wisniewski. Crit. Rev. Environ. Sci. Technol. **27**: 23–45.

Black, T.A., den Hartog, G., Neumann, H.H., Blanken, P.D., Yang, P.C., Russell, C., Nesic, Z., Lee, X., Chen, S.G., Staebler, R., and Novak, M.D. 1996. Annual cycles of water vapour and carbon dioxide fluxes in and above a boreal aspen forest. Global Change Biol. **2**: 219–230.

Blanken, P.D., Black, T.A., Neumann, H.H., den Hartog, G., Yang, P.C., Nesic, Z., Staebler, R., Chen, W., and Novak, M.D. 1998. Turbulent flux measurements above and below the overstory of a boreal aspen forest. Boundary-Layer Meteorol. **89**: 109–140.

Bloomberg, W.J. 1962. Cytospora canker of poplars: factors influencing the development of the disease. Can. J. Bot. **40**: 1271–1280.

Bonan, G.B., Pollard, D., and Thompson, S.L. 1992. Effects of boreal forest vegetation on global climate. Nature (London), **359**: 716–718.

Botch, M.S., Kobak, K.I., Vinson, T.S., and Kolchugina, T.P. 1995. Carbon pools and accumulation in peatlands of the former Soviet Union. Global Biogeochem. Cycles, **9**: 37–46.

Bramryd, T. 1979. The conservation of peatlands as global carbon accumulators. *In* Proceedings of the International Symposium on Classification of Peat and Peatlands, 17–21 September 1979, Hyytiälä, Finland. International Peat Society, Helsinki, Finland. pp. 296–305.

Brandt, J.P., Knowles, K.R., Larson, R.M., Ono, H., and Walter, B.L. 1996. Forest insect and disease conditions in west-central Canada in 1995 and predictions for 1996. Can. For. Serv. Info. Rep. NOR-X-347. 53 p.

Callan, B.E. 1998. Diseases of *Populus* in British Columbia: a diagnostic manual. Pacific Forestry Centre, Canadian Forest Service, Natural Resources Canada, Victoria, British Columbia. 157 p.

Campbell, I.D., and McAndrews, J.H. 1993. Forest disequilibrium caused by rapid Little Ice Age cooling. Nature (London), **366**: 336–338.

Campbell, I.D., Campbell, C., Vitt, D.H., Kelker, D., Laird, L.D., Trew, D., Kotak, B., LeClair, D., and Bayley, S. 2000. A first estimate of organic carbon storage in Holocene lake sediments in Alberta, Canada. J. Paleolimnol. **24**: 395–400.

Cao, M., and Woodward, F.I. 1998. Dynamic responses of terrestrial ecosystem carbon cycling to global climate change. Nature (London), **393**: 249–252.

Carignan, R., and Steedman, R.J. 2000. Impacts of major watershed perturbations on aquatic ecosystems. Can. J. Fish. Aquat. Sci. **57**(Suppl. 2): 1–4.

Carignan, R., D'Arcy, P., and Lamontagne, S. 2000. Comparative impacts of fire and forest harvesting on water quality in Boreal Shield lakes. Can. J. Fish. Aquat. Sci. **57**(Suppl. 2): 105–117.

Carroll, A.L., Hudak, J., Meades, J.P., Power, J.M., Gillis, T., McNamee, P.J., Wedeles, C.H.R., and Sutherland, G.D. 1995. EHLDSS: a decision support system for management of the eastern hemlock looper. *In* Proceedings of Decision Support 2001, 13–15 Sepember 1994, Toronto, Ontario. American Society for Photogrammetry and Remote Sensing, Bethesda, Maryland. pp. 807–824.

Carter, M.R., Angers, D.A., Gregorich, E.G., Donald, R.G., Monreal, C.M., Voroney, R.P., Rounsevell, M.D.A., and Loveland, P.J. 1994. The agricultural management effects on carbon sequestration in Eastern Canada. *In* Soil responses to climate change. *Edited by* M.D.A. Rounsevell. Springer-Verlag, Berlin, Germany. pp. 193–196.

Caspersen, J.P., Pacala, S.W., Jenkins, J.C., Hurtt, G.C., Moorcroft, P.R., and Birdsey, R.A. 2000. Contributions of land-use history to carbon accumulation in U.S. forests. Science, **290**: 1148–1151.

Cerezke, H.F., and Volney. 1995. Forest insect pests in the north west region. *In* Forest insect pests in Canada. *Edited by* J.A. Armstrong and W.G.H. Ives. Canadian Forest Service, Ottawa, Ontario. pp. 59–72.

Chapin, F.S., III, McGuire, A.D., Randerson, J., Pielke, R., Sr., Baldocchi, D., Hobbie, S.E., Roulet, N., Eugster, W., Kasischke, E., Rastetter, E.B., Zimov, S.A., Oechel, W.A., and Running, S.W. 2000. Feedbacks from arctic and boreal ecosystems to climate. Global Change Biol. **6**(Suppl. 1): 211–223.

Chen, W.J., Black, T.A., Yang, P.C., Barr, A.G., Neumann, H.H., Nesic, Z., Novak, M.D., Eley, J., and Cuenca, R. 1999. Effects of climate variability on the annual carbon sequestration by a boreal aspen forest. Global Change Biol. **5**: 41–53.

Chen, W., Chen, J., and Cihlar, J. 2000. An integrated terrestrial ecosystem carbon-budget model based on changes in disturbance, climate, and atmospheric chemistry. Ecol. Model. **135**: 55–79.

Chomitz, K.M. 2000. Evaluating carbon offsets from forestry and energy projects: how do they compare? Working Paper No. 2357. Development Research Group, World Bank, Washington, D.C. Available at http://www.econ.worldbank.org/docs/1111.pdf [viewed 19 May 2003]. 25 p.

Clymo, R. 1984. The limits to peat bog growth. Philos. Trans. R. Soc. Lond. B Biol. Sci. **303**: 605–654.

Clymo, R.S., Turunen, J., and Tolonen, K. 1998. Carbon accumulation in peatland. Oikos, **81**: 368–388.

Cohen, S., and Miller, K. 2001. North America. *In* Climate change 2001: impacts, adaptation and vulnerability. Contribution of Working Group II to the Third Assessment Report of the Intergovernmental Panel on Climate Change. *Edited by* J.J. McCarthy, O.F. Canziani, N.A. Leary, D.J. Dokken, and K.S. White. Cambridge University Press, Cambridge, U.K. pp. 735–800.

Cooper, C.F. 1983. Carbon storage in managed forests. Can. J. For. Res. **13**: 155–166.

(Curtis) Canadian Pulp and Paper Association. 2000. CPPA's discussion paper on climate change. Montreal, Quebec. Available at http://www.open.doors.cppa.ca. [viewed January 2000].

Curtis, P.S., and Wang, X.Z. 1998. A meta-analysis of elevated CO_2 effects on woody plant mass, form, and physiology. Oecologia, **113**: 299–313.

Curtis, P.S., Vogel, C.S., Wang, X., Pregitzer, K.S., Zak, D.R., Lussenhop, J., Kubiske, M., and Teeri, J.A. 2000. Gas exchange, leaf nitrogen, and growth efficiency of *Populus tremuloides* in a CO_2-enriched atmosphere. Ecol. Appl. **10**: 3–17.

Darwin, R., Tsigas, M., Lewandrowski, J., and Raneses, A. 1995. World agriculture and climate change: economic adaptations. USDA Economic Research Service, Washington, D.C. Agric. Econ. Rep. 703. 86 p.

Dean, W.E., and Gorham, E. 1998. Magnitude and significance of carbon burial in lakes, reservoirs, and peatlands. Geology, **26**: 535–538.

Dixon, H.R., and Dixon, P.A. 1991. Quaternary ecology: a paleoecological perspective. Chapman and Hall, New York. 242 p.

Dixon, R.K., and Krankina, O.N. 1993. Forest fires in Russia: carbon dioxide emissions to the atmosphere. Can. J. For. Res. **23**: 700–705.

Duchemin, E., Lucotte, M., Canuel, R., and Chamberland, A. 1995. Production of the greenhouse gases CH_4 and CO_2 by hydroelectric reservoirs in the boreal region. Global Biogeochem. Cycles, **9**: 529–540.

Eugster, W., Rouse, W.R., Pielke, R.A., Sr., McFadden, J.P, Baldocchi, D.D., Kittel, T.G.F., Chapin, F.S, III, Liston, G.E., Luigi Vidale, P., Vaganov, E., and Chambers, S. 2000. Land-atmosphere energy exchange in arctic tundra and boreal forest: available data and feedbacks to climate. Global Change Biol. **6**(Suppl 1): 84–115.

Farquhar, G.D. 1997. Carbon dioxide and vegetation. Science, **278**: 1411.

Farquhar, G.D., von Caemmerer, S., and Berry, J.A. 1980. A biochemical model of photosynthetic CO_2 fixation in leaves of C3 species. Planta, **149**: 78–90.

Field, C.B., Jackson, R.B., and Mooney, H.A. 1995. Stomatal responses to increased CO_2: implications from the plant to the global scale. Plant Cell Environ. **18**: 1214–1225.

Fitzsimmons, M. 2002. Effects of deforestation and reforestation on landscape spatial structure in boreal Saskatchewan, Canada. Can. J. For. Res. **32**: 843–851.

Flannigan, M.D., and van Wagner, C.E. 1991. Climate change and wildfire in Canada. Can. J. For. Res. **21**: 66–72.

Flannigan, M.D., and Woodward, F.I. 1994. Red pine abundance: current climate control and responses to future warming. Can. J. For. Res. **24**: 1166–1175.

Flannigan, M., Wotton, M., Carcaillet, C., Richard, P., Campbell, I., and Bergeron, Y. 1998. Fire weather: past, present and future. *In* Proceedings of the III International Conference on Forest Fire Research/14th Conference on Fire and Forest Meteorology, 16–20 November 1998, Luso (Coimbra), Portugal. Vol. 1. *Edited by* D.X. Viegas. ADAI, University of Coimbra, Coimbra, Portugal. pp. 113–128.

Flannigan, M.D., Stocks, B.J., and Wotton, B.M. 2000. Climate change and forest fires. Sci. Total Environ. **262**: 221–229.

Flannigan, M.D., Campbell, I.D., Wotton, M., Carcaillet, C., Richard, P.J.H., and Bergeron, Y. 2001. Future fire in Canada's boreal forest: paleoecology, GCM and RCM results. Can. J. For. Res. **31**: 854–864.

Fleming, R.A. 2000. Climate change and insect disturbance regimes in Canada's boreal forests. World Resour. Rev. **12**: 520–555.

Fleming, R.A., and Volney, W.J.A. 1995. Effects of climate change on insect defoliator population processes in Canada's boreal forest: some plausible scenarios. Water Air Soil Pollut. **82**: 445–454.

Fowler, D., Cape, J.N., Coyle, M., Flechard, C., Kuylenstierna, J., Hicks, K., Derwent, D., Johnson, C., and Stevenson, D. 1999. The global exposure of forests to air pollutants. Water Soil Air Pollut. **116**: 5–32.

France, R., Steedman, R., Lehmann, R., and Peters, R. 2000. Landscape modification of DOC concentration in boreal lakes: implications for UV-B sensitivity. Water Air Soil Pollut. **122**: 153–162.

(Gardner) Forest Sector Table. 1999. Options report: options for the forest sector to contribute to Canada's National Implementation Strategy for the Kyoto Protocol. National Climate Change Program, Canadian Forest Service, Ottawa, Ontario.

Gardner, R.H., Hargrove, W.W., Turner, M.G., and Romme, W.H. 1996. Climate change, disturbance and landscape dynamics. *In* Global change and terrestrial ecosystems. *Edited by* B. Walker and W. Steffen. Cambridge University Press, Cambridge, U.K. pp. 149–172.

Gennings, C., Molot, L.A., and Dillon, P.J. 2001. Enhanced photochemical loss of organic carbon in acidic waters. Biogeochemistry, **52**: 339–354.

Gitay, H., Brown, S., Easterling, W. E., Jallow, B., Antle, J., Apps, M.J., Beamish, R., Chapin, T., Cramer, W., Franji, J., Laine, J., Erda, L., Magnuson, J. J., Noble, I., Price, C., Prowse, T.D., Sirotenko, O., Root, T., Schulze, E.-D., Sohngen, B., and Soussana, J.-F. 2001. Chapter 5: Ecosystems and their services. *In* Climate change 2001: impacts, adaptation and vulnerability. Contribution of Working Group II to the Third Assessment Report of the Intergovernmental Panel on Climate Change. *Edited by* J.J. McCarthy, O.F. Canziani, N.A. Leary, D.J. Dokken, and K.S. White. Cambridge University Press, Cambridge, U.K. pp. 235–342.

(Goncalves) Government of Canada. 2001. Canada's third national report on climate change 2001: actions to meet commitments under the United Nations Framework Convention on Climate Change. Minister of Public Works and Government Services, Ottawa, Ontario. Available at http://www.climatechange.gc.ca/english/3nr/index.html [viewed 19 May 2003]. 256 p.

(Goncalves) Government of Canada. 2002. Climate change plan for Canada. Minister of Public Works and Government Services, Ottawa, Ontario. Available at http://www.climatechange.gc.ca/plan_for_canada/index.html [viewed 19 May 2003]. 67 p.

Goncalves, J.L.M., and Carlyle, J.C. 1994. Modelling the influence of moisture and temperature on net nitrogen mineralization in a forested sandy soil. Soil Biol. Biochem. **26**: 1557–1564.

Gorham, E. 1991. Northern peatlands: role in the carbon cycle and probable responses to climatic warming. Ecol. Appl. **1**: 182–195.

Goulden, M.L., Wofsy, S.C., Harden, J.W., Trumbore, S.E., Crill, P.M., Gower, S.T., Fries, T., Daube, B.C., Fan, S.M., Sutton, D.J., Bazzaz, A., and Munger, J.W. 1998. Sensitivity of boreal forest carbon balance to soil thaw. Science, **279**: 214–217.

Gower, S.T., Vogel, J.G., Norman, J.M., Kucharik, C.J., Steele, S.J., and Stow, T.K. 1997. Carbon distribution and aboveground net primary production in aspen, jack pine, and black spruce stands in Saskatchewan and Manitoba. Canada. J. Geophys. Res. **102**(D24): 29029–29041.

Halsey, L.A., Vitt, D.H., and Zoltai, S.C. 1995. Disequilibrium response of permafrost in boreal continental western Canada to climate change. Clim. Change, **30**: 57–73.

Harden, J.W., Trumbore, S.E., Stocks, B.J., Hirsch, A., Gower, S.T., O'Neill, K.P., and Kasischke E.S. 2000. The role of fire in the boreal carbon budget. Global Change Biol. **6**(Suppl. 1): 174–184.

Harris, L.D. 1984. The fragmented forest: island biogeography theory and the preservation of biotic diversity. University of Chicago Press, Chicago, Illinois. 211 p.

Hauer, G., Williamson, T., and Renner, M. 2001. Socioeconomic impacts and adaptive responses to climate change: a Canadian forest sector perspective. Can. For. Serv. Inf. Rep. NOR-X-373. 55 p.

Henri, C.J. 2001. Current status of carbon offset markets: Where do forests stand? Paper presented at the Western Forest Economists Annual Meeting, 8 May 2001, Welches, Oregon.

Hoen, H.F., and Solberg, B. 1994. Potential and economic efficiency of carbon sequestration in forest biomass through silvicultural management. For. Sci. **40**: 429–451.

Hogg, E.H. 1997. Temporal scaling of moisture and the forest-grassland boundary in western Canada. Agric. For. Meteorol. **84**: 115–122.

Hogg, E.H. 1999. Simulation of interannual responses of trembling aspen stands to climatic variation and insect defoliation in western Canada. Ecol. Model. **114**: 175–193.

Hogg, E.H., and Hurdle, P.A. 1995. The Aspen Parkland in western Canada: a dry-climate analogue for the future boreal forest? Water Air Soil Pollut. **82**: 391–400.

Houghton, R.A., Skole, D.L., Nobre, C.A., Hackler, J.L., Lawerence, K.T., and Chomentowski, W.H. 2000. Annual fluxes of carbon from deforestation and regrowth in the Brazilian Amazon. Nature (London), **403**: 301–304.

IGBP, Terrestrial Carbon Working Group, Steffan, W., Noble, I., Canadell, P., Apps, M.J., Schulze, E.-D., Jarvis, P.G., et al. 1998. The terrestrial carbon cycle: implications for the Kyoto Protocol. Science, **280**: 1393–1394.

(IPCC) Intergovernmental Panel on Climate Change. 1994. Radiative forcing of climate change and an evaluation of the IPCC IS92 emissions scenarios. *Edited by* J.T. Houghton, L.G. Meira Fhilho, J. Bruce, H. Lee, B.A. Callander, E. Haites, N. Harris, and K. Maskell. Cambridge University Press, Cambridge, U.K.

(IPCC) Intergovernmental Panel on Climate Change. 1996. Climate change 1995: Impacts, adaptations and mitigation of climate change: Scientific–technical analysis. *Edited by* R.T. Watson, M.C. Zinyowera, R.H. Moss, and D.J. Dokken. Cambridge University Press, Cambridge, U.K.

(IPCC) Intergovernmental Panel on Climate Change. 2000. Land use, land-use change, and forestry. *Edited by* R.T. Watson, I.R. Novel, N.H. Bolin, N.H. Ravindranath, D.J. Verardo, and D.J. Dokken. Cambridge University Press, Cambridge, U.K.

(IPCC) Intergovernmental Panel on Climate Change. 2001*a*. Climate change 2001: the scientific basis. Contribution of Working Group I to the Third Assessment Report of the Intergovernmental Panel on Climate Change. *Edited by* J.T. Houghton, Y. Ding, D.J. Griggs, M. Noguer, P.J. van der Linden, and D. Xiaosu. Cambridge University Press, Cambridge, U.K.

(IPCC) Intergovernmental Panel on Climate Change. 2001*b*. Climate change 2001: impacts, adaptation and vulnerability. Contribution of Working Group II to the Third Assessment Report of the Intergovernmental Panel on Climate Change. *Edited by* J.J. McCarthy, O.F. Canziani, N.A. Leary, D.J. Dokken, and K.S. White. Cambridge University Press, Cambridge, U.K.

(IPCC) Intergovernmental Panel on Climate Change. 2001*c*. Climate change 2001: Mitigation. Third assessment report of the Intergovernmental Panel on Climate Change. Cambridge University Press, Cambridge, U.K.

Isebrands, J.G., Dickson, R.E., Rebbeck, J., and Karnosky, D.F. 2000. Interacting effects of multiple stresses on growth and physiological processes in northern forests. *In* Responses of Northern U.S. forests to environmental change. *Edited by* R.E. Mickler, R.A. Birdsey, and J. Hom. Springer-Verlag, Berlin, Germany. pp. 149–180.

Jarvis, P.G., Massheder, J.M., Hale, S.E., Moncrieff, J.B., Rayment, M., and Scott, S.L. 1997. Seasonal variation of carbon dioxide, water vapour, and energy exchanges of a boreal black spruce forest. J. Geophys. Res. **102**(D24): 28953–28966.

Jenkinson, D.S., Adams, D.E., and Wild, A. 1991. Model estimate of CO_2 emissions from soil in response to global warming. Nature (London), **351**: 304–306.

Johnson, L.C., and Damman, A.W.H. 1991. Species-controlled Sphagnum decay on a south Swedish raised bog. Oikos, **61**: 234–242.

Karjalainen T., Apps, M.J., Stocks, B.J., and Shaw, C.H. 2002. Foreword: the role of boreal forests and forestry in the global carbon budget. For. Ecol. Manage. **169**: 1–2.

Kasischke, E.S. 2000. Effects of climate change and fire on carbon storage in North American boreal forests. *In* Fire, climate change, and carbon cycling in the North American boreal forest. *Edited by* E.S. Kasischke and B.J. Stocks. Springer-Verlag, New York, New York. pp. 440–452.

Kasischke, E.S., O'Neill, K.P., French, N.H.F., and Bourgeau-Chavez, L.L. 2000. Controls on patterns of biomass burning in Alaskan boreal forests. *In* Fire, climate change, and carbon cycling in the North American boreal forest. *Edited by* E.S. Kasischke and B.J. Stocks. Springer-Verlag, New York, New York. pp. 173–196.

Kauppi, P.E., Sedjo, R.A., Apps, M.J., Cerri, C.C., Fujimori, T., Janzen, H., Krankina, O.N., Makundi, W., Marland, G., Masera, O., Nabuurs, G.J., Razali, W., and Ravindranath, N.H. 2001. Chapter 4: Technological and economic potential of options to enhance, maintain, and manage biological carbon reservoirs and geo-engineering. *In* The Intergovernmental Panel on Climate Change (IPCC) Working Group

III Contribution to the Third Assessment Report on Mitigation of Climate Change. *Edited by* B. Metz, O. Davidson, R. Swart, and J. Pan. Cambridge University Press, Cambridge, U.K. pp. 301–323.

Kelly, C.A., Rudd, J.W.M., Bodaly, R.A., Roulet, N.P., St. Louis, V.L., Heyes, A., Moore, T.R., Schiff, S., Aravena, R., Scott, K.J., Dyck, B., Harris, R., Warner, B., and Edwards, G. 1997. Increases in fluxes of greenhouse gases and methyl mercury following flooding of an experimental reservoir. Environ. Sci. Technol. **31**: 1334–1344.

Kimmins, J.P. 1992. Balancing act: environmental issues in forestry. UBC Press, Vancouver, British Columbia. 244 p.

Kirkup, P. 2000. Impacts of global change on biodiversity. Available at http://www.glg.ed.ac. uk/home/Paul.Kirkup/specmod/atmosessay.html [updated 17 January 2001; viewed 1 June 2002].

Kirschbaum, M.U.F., and Fishlin, A. 1996. Climate change impacts on forests. *In* Climate change 1995: Contributions of Working Group II to the Second Assessment Report of the Intergovernmental Panel on Climate Change. *Edited by* R. Watson, M.C. Zinowera, and R.H. Moss. Cambridge University Press, Cambridge, U.K. pp. 93–129.

Kobak, K.I., Kondrasheva, N.Y., and Turchinovich, I.E. 1998. Changes in carbon pools of peatland and forests in northwestern Russia during the Holocene. Global Planet. Change, **16–17**: 75–84.

Körner, C. 2000. Biosphere responses to CO_2-enrichment. Ecol. Appl. **10**: 1590–1619.

Körner, C., and Bazzaz, F.A. (*Editors*). 1996. Carbon dioxide, populations and communities. Academic Press, San Diego, California. 465 p.

Krcmar, E., and van Kooten, G.C. 2001. Timber, carbon uptake and structural diversity tradeoffs in forest management. FEPA Working Paper. University of British Columbia, Vancouver, British Columbia.

Krcmar, E., Stennes, B., van Kooten, G.C., and Vertinsky, I. 2001. Carbon sequestration and land management under uncertainty. Eur. J. Oper. Res. **135**: 616–629.

Kurz, W.A., and Apps, M.J. 1999. A 70-year retrospective analysis of carbon fluxes in the Canadian forest sector. Ecol. Appl. **9**: 526–547.

Kurz, W.A., Apps, M.J., Webb, T.M., and McNamee, P.J. 1992. The carbon budget of the Canadian forest sector: phase 1. Can. For. Serv. Inf. Rep. NOR-X-326. 93 p.

Kurz, W.A., Apps, M.J., Beukema, S. J., and Lekstrum, T. 1995*a*. 20th century carbon budget of Canadian forests. Tellus, **47B**:170–177.

Kurz, W.A., Apps, M.J., Stocks, B.J., and Volney, W.J. 1995*b*. Global climate change: disturbance regimes and biospheric feedbacks of temperate and boreal forests. *In* Biotic feedbacks in the global climate system: will the warming speed the warming? *Edited by* G. Woodwell. Oxford University Press, New York, New York. pp. 119–133.

Kurz, W.A., Beukema, S.J., and Apps, M.J. 1998. Carbon budget implications of the transition from natural to managed disturbance regimes in forest landscapes. Mitigation Adapt. Strategies Global Change, **2**: 405–421.

Lafleur, P.M., Roulet, N.T., and Admiral, S.W. 2001. Annual cycle of CO_2 exchange at a bog peatland. J. Geophys. Res. **106**: 3071–3081.

Larocque, I., Bergeron, Y., Campbell, I.D., and Bradshaw, R.H.W. 2003. Distribution of Thuja occidentalis in the southern Canadian boreal forest: long-term effect of local fire disturbance. J. Veg. Sci. In press.

Larsen, C.P.S., and MacDonald, G.M. 1998. Fire and vegetation dynamics in a jack pine and black spruce forest reconstruction using fossil pollen and charcoal. J. Ecol. **86**: 815–828.

Lenihan, J.M., and Neilson, R.P. 1995. Canadian vegetation sensitivity to projected climatic change at three organizational levels. Clim. Change, **30**: 27–56.

Li, Z., Apps, M.J., Banfield, E, and Kurz, W.A. 2002. Estimating net primary production of forests in the Canadian Prairie Province using an inventory-based carbon budget model. Can. J. For. Res. **32**: 161–169.

Li, Z., Apps, M.J., Kurz, W.A., and Beukema, S.J. 2003. Belowground biomass dynamics in the Carbon Budget Model of the Canadian Forest Sector: recent improvements and implications for the estimation of NPP and NEP. Can. J. For. Res. **33**: 126–136.

Likens, G.E., Driscoll, C.T., and Buso, D.C. 1996. Long-term effects of acid rain: response and recovery of a forest ecosystem. Science, **272**: 244–248.

Lindroth, A., Grelle, A., and Moren, A.S. 1998. Long-term measurements of boreal forest carbon balance reveal large temperature sensitivity. Global Change Biol. **4**: 443–450.

Liski, J., Pussinen, A., Pingoud, K., Mäkipää, R., and Karjalainen, T. 2001. Which rotation length is favourable for carbon sequestration? Can. J. For. Res. **31**: 2004–2013.

Loehle, C. 1998. Height growth rate tradeoffs determine northern and southern range limits for trees. J. Biogeogr. **25**: 735–742.

Lovett, G. M. 1994 Atmospheric deposition of nutrients and pollutants in North America: An ecological perspective. Ecol. Appl. **4**: 629–650.

Luo, Y., White, L.W., Canadell, J.G., DeLucia, E.H., Ellsworth, D.S., Finzi, A., Lichter, J., and Schlesinger, W.E. 2003. Sustainability of terrestrial carbon sequestration: a case study in Duke Forest with inversion approach. Global Biogeochem. Cycles. In press.

Mäkilä, M. 1997. Holocene lateral expansion, peat growth, and carbon accumulation on Haukkasuo, a raised bog in southeastern Finland. Boreas, **26**: 1–14.

Mäkipää, R., Karjalainen, T., Pussinen, A., and Kellomäki, S. 1999. Effects of climate change and nitrogen deposition on the carbon sequestration of a forest ecosystem in the boreal zone. Can. J. For. Res. **29**: 1490–1501.

Mallett, K.I., and Volney, W.J.A. 1999. The effect of Armillaria root disease on lodgepole pine tree growth. Can. J. For. Res. **29**: 252–259.

Malmer, N. 1986. Vegetational gradients in relation to environmental conditions in northwestern European mires. Can. J. Bot. **64**: 375–383.

Mattson, W.J., and Haack, R.A. 1987. The role of drought in outbreaks of plant-eating insects. BioScience, **37**: 110–118.

McCaughey, J.H., Lafleur, P.M., Joiner, D.W., Bartlett, P.A., Costello, A.M., Jelinski, D.E., and Ryan, M.G. 1997. Magnitudes and seasonal patterns of energy, water, and carbon exchanges at a boreal young jack pine forest in the BOREAS northern study area. J. Geophys. Res. **102**(D24): 28997–29007.

McEachern, P., Prepas E.E., Gibson J.J., and Dinsmore W.P. 2000. Forest fire induced impacts on phosphorus, nitrogen, and chlorophyll a concentrations in boreal subarctic lakes of northern Alberta. Can. J. Fish. Aquat. Sci. **57**: 73–81.

McLaughlin, S., and Percy, K. 1999. Forest health in North America: some perspectives on actual and potential roles of climate and air pollution. Water Air Soil Pollut. **116**: 151–197.

Mercier, G. 1998. Canada country study: climate impacts and adaptations — energy sector. In The Canada country study: climate impacts and adaptations. Vol. 7. National Sectoral Volume. Edited by G. Koshida and W. Avis. Environmental Adaptation Research Group, Environment Canada, Toronto, Ontario. pp. 383–404.

Micales, J.A., and Skog, K.E. 1997. The decomposition of forest products in landfills. Int. Biodeterior. Biodegrad. **39**: 145–158.

Minkkinen, K., and Laine, J. 1998. Long-term effect of forest drainage on the peat carbon stores of pine mires in Finland. Can. J. For. Res. **28**: 1267–1275.

Molot, L.A., and Dillon, P.J. 1996. Storage of terrestrial carbon in boreal lake sediments and evasion to the atmosphere. Global Biogeochem. Cycles, **10**: 483–492.

Moore, T.R., Roulet, N.T., and Waddington, J.M. 1998. Uncertainty in predicting the effect of climatic change on the carbon cycling of Canadian peatlands. Clim. Change, **40**: 229–245.

Munn, R.E., and Maarouf, A.R. 1997. Atmospheric issues in Canada. Sci. Tot. Environ. **203**: 1–11.

Nadelhoffer, K.J., Emmett, B.A., Gundersen, P., Kjønaas, O.J., Koopmans, C.J., Schleppi, P., Tietema, A., and Wright, R.F. 1999. Nitrogen deposition makes a minor contribution to carbon sequestration in temperate forests. Nature (London), **398**: 145–148.

Neilson, R.P. 1993. Vegetation redistribution: a possible biosphere source of CO_2 during climatic change. Water Air Soil Pollut. **70**: 659–673.

Nohrstedt, H. 2001. Response of coniferous forest ecosystems on mineral soils to nutrient additions: a review of Swedish experiences. Scand. J. For. Res. **16**: 555–573.

Norby, R.J., Wullschlenger, S.D., Gunderson, C.A., Johnson, D.W., and Ceulemans, R. 1999. Tree responses to rising CO_2 in field experiments: implications for the future forest. Plant Cell Environ. **22**: 683–714.

(NRCan) Natural Resources Canada. 1998. Canada's energy outlook: 1996–2020. Natural Resources Canada, Ottawa, Ontario. Available at http://www.nrcan.gc.ca/es/ceo/toc-96E.html [viewed 18 April 2002].

(NRCan) Natural Resources Canada. 1999a. Forest health in Canada: an overview 1998. Forest Health Network, Canadian Forest Service, Natural Resources Canada, Ottawa, Ontario.

(NRCan) Natural Resources Canada. 1999b. Canada's emissions outlook: an update. Analysis and Modelling Group, National Climate Change Process, Natural Resources Canada, Ottawa, Ontario.

(NWWG) National Wetlands Working Group. 1988. Wetlands of Canada. Polyscience Publications, Inc., Montreal, Quebec. Ecol. Land Class. Ser. 24. 452 p.

Obersteiner, M., Rametsteiner, E., and Nilsson, S. 2001. Cap management for LULUCF options: an economic mechanism design to preserve the environmental and social integrity of forest related LULUCF activities under the Kyoto Protocol. International Institute for Applied Systems Analysis, Laxenburg, Austria. Interim Rep. IR-01-011. Available at http://www.iiasa. ac.at/Publications/Documents/IR-01-011.pdf [viewed 19 May 2003]. 17 p.

Olsen, K., Collas, P., Boileau, P., Blain, D., Ha, C., Henderson, L., Liang, C., McKibbon, S., and Moel-a-l'Huissier, L. 2002. Canada's greenhouse gas inventory 1990–2000. Greenhouse Gas Division, Environment Canada. Ottawa, Ontario.

Olsen, R., Ellsworth, D.S., Johnsen, K.H., Phillips, N., Ewers, B.E., Maier, C., Schafer, K.V.R., McCarthy, H., Hendrey, G., McNulty, S.G., and Katul, G.G. 2001. Soil fertility limits carbon sequestration by forest ecosystem in a CO_2-enriched atmosphere. Nature (London), **411**: 469–472.

Ouimet, R., and Camire, C. 1995. Foliar deficiencies of sugar maple stands associated with soil cation imbalances in the Quebec Appalachians. Can. J. Soil Sci. **75**: 169–175.

Ovenden, L. 1990. Peat accumulation in northern wetlands. Quat. Res. **33**: 377–386.

Pastor, J., and Post, W.M. 1988. Response of northern forests to CO_2-induced climate change. Nature (London), **334**: 55–58.

Percy, K., Bucher, J., Cape, J., Ferretti, M., Heath, R., Jones, H.E., Karnosky, D., Matyssek, R., Muller-Starck, G., Paoletti, E., Rosengren-Brinck, U., Sheppard, L., Skelly, J., and Weetman, G. 1999. State of science and knowledge gaps with respect to air pollution impacts on forests: reports from concurrent IUFRO 7.04.00 working party sessions. Water Air Soil Pollut. **116**: 443–448.

Peterjohn, W.T., Melillo, J.M., Steudler, P.A., Newkirk, K.M., Bowles, F.P., and Aber, J.D. 1995. Responses of trace gas fluxes and N availability to experimentally elevated soil temperatures. Ecol. Appl. **4**: 617–625.

Post, W.M., Emanuel, W.R., Zinke, P.J., and Stangenberger, A.G. 1982. Soil carbon pools and world life zones. Nature (London), **298**: 156–159.

Prepas, E.E, Pinel-Alloul, B., Planas, D., Methot, G, Paquet, S, and Reedyk, S. 2001. Forest harvest impacts on water quality and aquatic biota on the Boreal Plain: introduction to the TROLS Lake Program. Can. J. Fish. Aquat. Sci. **58**: 421–436.

Price, D.R., Halliwell, D.H., Apps, M.J., Kurz, W.A., and Curry, S.R. 1997. Comprehensive assessment of carbon stocks and fluxes in a Boreal-Cordilleran forest management unit. Can. J. For. Res. **27**: 2005–2016.

Rapalee, G., Trumbore, S.E., Davidson, E.A., Harden, J.W., and Velduis, H. 1998. Soil carbon stocks and their rates of accumulation and loss in a boreal forest landscape. Global Biogeochem. Cycles, **12**: 687–701.

Reader, R.J., and Stewart, J.M. 1972. The relationship between net primary production and accumulation for a peatland in southeastern Manitoba. Ecology, **53**: 1024–1037.

Richter, D.D., O'Neill, P., and Kasischke, E.S. 2000. Post-fire stimulation of microbial decomposition in black spruce forest soils: a hypothesis. *In* Fire, climate change, and carbon cycling in the North American boreal forest. *Edited by* E.S. Kasischke and B.J. Stocks. Springer-Verlag, New York, New York. pp. 197–313.

Rivers, J.S., Siegel, D.I., Chasar, L.S., Chanton, J.P., Glaser, P.H., Roulet, N.T., and McKenzie, J.M. 1998. A stochastic appraisal of the annual carbon budget of a large circumboreal peatland, Rapid River Watershed, northern Minnesota. Global Biogeochem. Cycles, **12**: 715–727.

Rizzo, B., and Wiken, E. 1992. Assessing the sensitivity of Canada's ecosystems to climate change. Clim. Change, **21**: 37–55.

Robinson, D.C.E., Kurz, W.A., and Pinkham, C. 1999. Estimating the carbon losses from deforestation in Canada. Prepared by ESSA Technologies Ltd., Vancouver, British Columbia, for the National Climate Change Secretariat, Ottawa, Ontario. Available at http://www.nccp.ca/NCCP/pdf/Deforest_Canada.pdf [viewed 19 May 2003]. 81 p.

Robinson, S.D., Park, J., and Kettles, I.M. 2001. Northern peatlands to monitor recent and future warming. Project A313, final report to the Climate Change Action Fund. Natural Resources Canada, Ottawa, Ontario.

Roland, J., Mackey, B.G., and Cooke, B. 1998. Effects of climate and forest structure on duration of forest tent caterpiller outbreaks across central Ontario, Canada. Can. Entomol. **130**: 703–714.

Roulet, N.T. 2000. Peatlands, carbon storage, greenhouse gases, and the Kyoto Protocol: prospects and significance for Canada. Wetlands, **20**: 605–615.

Safranyik, L., and Linton, D.A. 2002. Line transect sampling to estimate the density of lodgepole pine currently attacked by mountain pine beetle. Can. For. Serv. Inf. Rep. BC-X-392. 11 p.

Sandor, R.L., and Skees, J.R. 1999. Creating a market for carbon emissions opportunities for U.S. farmers. Choices, The Magazine of Food, Farm & Resource Issues. First Quarter. The American Agricultural Economics Association, Ames, Iowa. pp. 13–18.

Schimel, D.S. 1995. Terrestrial ecosystems and the carbon cycle. Global Change Biol. **1**: 77–91.

Schimel, D.S., House, J.I., Hibbard, K.A., Bousquet, P., Ciais, P., Peylin, P., Braswell, B.H., Apps, M.J., Baker, D., Bondeau, A., Canadell, J., Churkina, G., Cramer, W., Denning, A.S., Field, C.B., Friedlingstein, P., Goodale, C., Heimann, M., Houghton, R.A., Melillo, J.M., Moore, B., Murdiyarso, D., Noble, I., Pacala, S.W., Prentice, I.C., Raupach, M.R., Rayner, P.J., Scholes, R.J., Steffen, W.L., and Wirth, C. 2001. Recent patterns and mechanisms of carbon exchange by terrestrial ecosystems. Nature (London), **414**: 169–172.

Schimmelpfenning, D., Lewandrowski, J., Reilly, J., Tsigas, M., and Parry, I. 1996. Agricultural adaptation to climate change: issues of long-run sustainability. USDA Economic Research Service, Washington, D.C. Agric. Econ. Rep. 740. 57 p.

Schindler, D.W. 1998. A dim future for boreal waters and landscapes. BioScience, **48**: 157–164.

Schlamadinger, B., Apps, M., Bohlin, F., Gustavsson, L., Jungmeier, G., Marland, G., Pingoud, K., and Savolainen, I. 1997. Towards a standard methodology for greenhouse gas balances of bioenergy systems in comparison with fossil energy systems. Biomass Bioenergy, **13**: 359–375.

Schlentner, R.E., and Van Cleve, K. 1985. Relationship between CO_2 evolution from soil substrate temperature and substrate moisture in four mature forest types in interior Alaska. Can. J. For Res. **15**: 97–106.

Schneider, S.H. 1995. Foreword. *In* Boreal forests and global change. *Edited by* M.J. Apps, D.T. Price, and J. Wisniewski. Kluwer Academic Publishers, Boston, Massachusetts. pp. ix–xi.

Schroeder, W., and Kort, J. (*Editors*). 2001. Temperate agroforestry: adaptive and mitigative roles in a changing physical and socio-economic climate. Proceedings of the Seventh Biennial Conference on Agroforestry in North America and the Sixth Annual Conference of the Plains and Prairie Forestry Association, 12–15 August 2001, Regina, Saskatchewan. Association for Temperate Agroforestry, University of Missouri, Columbia, Missouri. 350 p.

Schulze, E.-D., Lloyd, J., Kelliher, F.M., Wirth, C., Rebmann, C., Lühker, B., Mund, M., Knohl, A., Milyukova, I., Schulze, W., Ziegler, W., Varlagin, A., Sogachov, A., Valentini, R., Dore, S., Grigoriev, S., Kolle, O., Tchebakova, N., and Vygodskaya, N. 1999. Productivity of forests in the Eurosiberian boreal region and their potential to act as a carbon sink — a synthesis. Global Change Biol. **5**: 703–722.

Sedjo, R.A., Wisniewski, J., Sample, A.V., and Kinsman, J.D. 1995. The economics of managing carbon via forestry: assessment of existing studies. Environ. Resour. Econ. **6**: 139–165.

Sellers, P.J., Hall, F.G., Kelly, R.D., et al. 1997. BOREAS in 1997: experiment overview, scientific results, and future directions. J. Geophys. Res. **102**: 28731–28770.

Shaw, C.H., and Apps, M.J. (*Editors*). 2002. The role of boreal forests and forestry in the global carbon cycle. Proceedings of IBFRA 2000 Conference, 8–12 May 2000, Edmonton, Alberta. Northern Forestry Centre, Canadian Forest Service, Edmonton, Alberta. 332 p.

Shorger, A.W. 1972. The passenger pigeon: its natural history and extinction. University of Oklahoma Press, Tulsa, Oklahoma. 424 p.

Sieben, B.G., Spittlehouse, D.L., Benton, R.A., and McLean, J.A. 1994. A first approximation of the effect of climate warming on the white pine weevil hazard in the Mackenzie River drainage basin. *In* Mackenzie Basin Impact Study (MBIS) interim report #2 — Proceedings of the Sixth Biennial AES/DIAND Meeting on Northern Climate and Mid-Study Workshop of the Mackenzie Basin Impact Study, 10–14 April 1994, Yellowknife, Northwest Territories. *Edited by* S.J. Cohen. Atmospheric Environment Services, Environment Canada, Downsview, Ontario. pp. 316–328.

Simpson, L.G., Botkin, D.B., and Nisbet, R.A. 1993. The potential aboveground carbon storage of North American forests. Water Air Soil Pollut. **70**: 197–205.

Smith, T.M., Cramer, W., Dixon, R.K., Leemans, R., Neilson, R.P., and Solomon, A.M. 1993. The global terrestrial carbon cycle. Water Air Soil Pollut. **70**: 19–37.

Sohngen, B., and Sedjo, R. 1999. Estimating potential leakage from regional forest carbon sequestration programs. RFF Working Paper. Resources for the Future, Washington, D.C. 26 p.

Sohngen, B., Mendelsohn, R., and Sedjo, R. 1999. Forest management, conservation and global timber markets. Am. J. Agric. Econ. **81**: 1–13.

Statistics Canada. 1997. Nonmetal mines, SCI 062 Catalogue 26-224X1B. Statistics Canada, Ottawa, Ontario.

Steele, S.J., Gower, S.T., Vogel, J.G., and Norman, J.M. 1997. Root mass, net primary production and turnover in aspen, jack pine, and black spruce forests in Saskatchewan and Manitoba, Canada. Tree Physiol. **17**: 577–587.

Stevenson, D.S., Johnson, C.E., Collins, W.J., Derwent, R.G., Shine, K.P., and Edwards, J.M. 1998. Evolution of tropospheric ozone radiative forcing. Geophys. Res. Lett. **25**: 3819–3822.

Stocks, B.J., Fosberg, M.A., Lynham, T.J., Mearns, L., Wotton, B.M., Yang, Q., Jin, J.Z., Lawrence, K., Hartley, G.R., Mason, J.A., and McKenney, D.W. 1998. Climate change and forest fire potential in Russian and Canadian boreal forests. Clim. Change, **38**: 1–13.

Stocks, B.J., Fosberg, M.A., Wotton, B.M., Lynham, T.J., and Ryan, K.C. 2000. Climate change and forest fire activity in North American boreal forests. *In* Fire, climate change, and carbon cycling in the

North American boreal forest. *Edited by* E.S. Kasischke and B.J. Stocks. Springer-Verlag, New York, New York. pp. 368–376.

Stocks, B.J., Apps, M.J., Karjalainen, T., and Shaw, C.H. 2002. Foreword. The role of boreal forests and forestry in the global carbon budget. Clim. Change, **55**: 1–4.

Stocks, B.J., Mason, J.A., Todd, J.B., Bosch, E.M., Wotton, B.M., Amiro, B.D., Flannigan, M.D., Hirsch, K.G., Logan, K.A., Martell, D.L., and Skinner, W.R. 2003. Large forest fires in Canada, 1959–1997. J. Geophys. Res. In press.

Szumigalski, A.R., and Bayley, S.E. 1997. Net above-ground primary production along a peatland gradient in central Alberta, Canada, in relation to environmental factors. Ecoscience, **4**: 385–393.

Thie, J. 1974. Distribution and thawing of permafrost in the southern part of the discontinuous permafrost zone in Manitoba. Arctic, **27**: 189–200.

Thompson, A.J., and Shrimpton, D.M. 1984. Weather associated with the start of mountain pine beetle outbreaks. Can J. For. Res. **14**: 255–258.

Thormann, M.N., and Bayley, S.E. 1997. Decomposition along a moderate-rich fen-marsh peatland gradient in boreal Alberta, Canada. Wetlands, **17**: 123–137.

Thormann, M.N., Szumigalski, A.R., and Bayley, S.E. 1999. Aboveground peat and carbon acumulation potentials along a bog-fen-marsh gradient in southern boreal Alberta, Canada. Wetlands, **19**: 305–317.

Townsend, A.R., Braswell, B.H., Holland, E.A., and Penner, J.E. 1996. Spatial and temporal patterns in terrestrial carbon storage due to deposition of anthropogenic nitrogen. Ecol. Appl. **6**: 806–814.

Trumbore, S.E., and Harden, J. 1997. Accumulation and turnover of carbon in organic and mineral soil of BOREAS northern study area. J. Geophys Res. **102**: 28817–28830.

Turetsky, M.R., Wieder, R.K., Williams, C.J., and Vitt, D.H. 2000. Organic matter accumulation, peat chemistry, and permafrost melting in peatlands of boreal Alberta. Ecoscience, **7**: 379–392.

Turetsky, M.R., and Wieder, R.K. 2001. A direct approach to quantifying organic matter lost as a result of peatland wildfire. Can. J. For. Res. **31**: 363–366.

Turetsky, M.R., Weider, R.K., Halsey, L.A., and Vitt, D.H. 2002. Current disturbance and the diminishing peatland C sink. Geophys. Res. Let. **29**(11): 1526–1526.

(Updegraff-ECE/FAO). United Nations Economic Commission for Europe and Food and Agriculture Organization. 2000. Forest products annual market review 1999–2000. Timber Bull. **53**(3): 23–37.

Updegraff, K., Pastor, J., Bridgham, S.D., and Johnston, C.A. 1995. Environmental and substrate controls over carbon and nitrogen mineralization in northern wetlands. Ecol. Appl. **5**: 151–163.

Valentine, D.W., Holland, E.A., and Schimel, D.S. 1994. Ecosystem and physiological controls over methane production in northern wetlands. J. Geophys. Res. **99**: 1563–1571.

van Kooten, G.C. 1995. Climatic change and Canada's boreal forest: socio-economic issues and implications for land use. Can. J. Agric. Econ. **43**: 133–148.

van Kooten, G.C. 2000. Economic dynamics of tree planting for carbon uptake on marginal agricultural lands. Can. J. Agric. Econ. **48**: 51–65.

van Kooten, G.C. 2002. A primer on the ecomonics of climate change. Working Paper. Department of Economics, University of Victoria, Victoria, British Columbia. 35 p.

van Kooten, G.C. 2003. A primer on the economics of climate change. Working Paper. March. Department of Economics, University of Victoria, Victoria, British Columbia. 83 p.

van Kooten, G.C., and Bulte, E.H. 2000. The economics of nature: managing biological assets. Blackwell Publishers, Oxford, U.K.

van Kooten, G.C., Thompson, W.A., and Vertinsky, I. 1993. Economics of reforestation in British Columbia when benefits of CO_2 reduction are taken into account. *In* Forestry and the environment: economic perspectives. *Edited by* W.L. Adamowicz, W. White, and W.E. Phillips. CAB International, Wallingford, U.K. pp. 227–247.

van Kooten, G.C., Binkley, C.S., and Delcourt, G. 1995. Effect of carbon taxes and subsidies on optimal forest rotation age and supply of carbon services. Am. J. Agric. Econ. **77**: 365–774.

van Kooten, G.C., Krcmar-Nozic, E., Stennes, B., and van Gorkom, R. 1999. Economics of fossil fuel substitution and wood product sinks when trees are planted to sequester carbon on agricultural lands in western Canada. Can. J. For. Res. **29**: 1669–78.

van Kooten, G.C., Stennes, B., Krcmar–Nozic, E., and van Gorkom, R. 2000. Economics of afforestation for carbon sequestration in western Canada. For. Chron. **76**: 165–72.

van Kooten, G.C., Shaikh, S., and Suchanek, S. 2002. Mitigating climate change by planting trees: the transaction costs trap. Land Econ. **78**: 559–572.

Viereck, L., and Dyrness, C.T. 1979. Ecological effects of Wiockersham Dome fire near Fairbanks, Alaska. USDA Forest Service, Portland, Oregon. Gen. Tech. Rep. PNW-90. 71 p.

Vitousek, P.M., Aber, J.D., Howarth, R.W., Likens, G.F., Matson, P.A., Schindler, D.W., Shlesinger, W.H., and Tilman, D. 1997. Human alternation of the global nitrogen cycle: sources and consequences. Ecol. Appl. **7**: 737–750.

Vitt, D.H., Halsey, L.A., and Zoltai, S.C. 2000. The changing landscape of Canada's western boreal forest: the current dynamics of permafrost. Can. J. For. Res. **30**: 283–287.

Volney, W.J.A., and Fleming, R.A. 2000. Climate change and impacts of boreal forest insects. Agric. Ecosyst. Environ. **82**: 283–294.

Waddington, J.M., and Price, J.S. 2000. Effect of peatland drainage, harvesting, and restoration on atmospheric water and carbon exchange. Phys. Geogr. **21**: 433–451.

Waddington, J.M., and Roulet, N.T. 2000. Carbon balance of a boreal patterned peatland. Global Change Biol. **6**: 87–97.

Wardle, D.A., Zackrisson, O., and Nilsson, M.C. 1998. The charcoal effect in boreal forests: mechanisms and ecological consequences. Oecologia, **115**: 419–426.

Weber, M.G., and Flannigan, M.D. 1997. Canadian boreal forest ecosystem structure and function in a changing climate: impact of fire regimes. Environ. Rev. **5**: 145–166.

Weber, M.G., and Stocks, B.J. 1998. Forest fires and sustainability in the boreal forests of Canada. Ambio, **27**: 545–550.

Williams, D.W., and Liebhold, A.M. 1997. Latitudinal shifts in spruce budworm (Lepidoptera: Tortricidae) outbreaks and spruce-fir forest distributions with climate change. Acta Phytopathol. Entomol. Hung. **32**: 205–215.

Wilson, B., van Kooten, G.C., Vertinsky, I., and Arthur, L.M. (*Editors*). 1999. Forest policy: international case studies. CABI Publishing, Wallingford, U.K. 656 p.

Wullschlenger, S.D., Post, W.M., and King, A.W. 1995. On the potential for a CO_2 fertilization effect in forests: estimates of the biotic growth factor based on 58 controlled-exposure studies. *In* Biotic feedbacks in the global climatic cycle. *Edited by* G.M. Woodwell and F.Y. McKenzie. Oxford University Press, Oxford, U.K. pp. 85–107.

Yavitt, J., and Lang, G. 1990. Methane production in contrasting wetland sites: response to organic-chemical components of peat and to sulfate reduction. Geomicrobiol. J. **8**: 27–46.

Yavitt, J.B., Williams, C.J., and Wieder, R.K. 1997. Production of methane and carbon dioxide in peatland ecosystems across North America: effects of temperature, aeration, and organic chemistry of peat. Geomicrobiol. J. **14**: 299–316.

Zoltai, S.C. 1972. Palsas and peat plateaus in central Manitoba and Saskatchewan. Can. J. For. Res. **2**: 291–302.

Zoltai, S.C., and Martikainen, P.J. 1996. The role of forested peatlands in the global carbon cycle. *In* Forest ecosystems, forest management and the global carbon cycle. *Edited by* M.J. Apps and D.T. Price. Springer-Verlag, Heidelberg, Germany. NATO ASI Ser. 1, Vol. 40. pp. 47–58.

Zoltai, S.C., Morrissey, L.A., Livingston, G.P., and de Groot, W.J. 1998. Effects of fire on carbon cycling in North American boreal peatlands. Environ. Rev. **6**: 13–24.

Chapter 21

Adaptive management: progress and prospects for Canadian forests

Peter N. Duinker and Lisa M. Trevisan

"An adaptive approach will require a significant transition in how we think and act, including a capacity and willingness to acknowledge that current actions and beliefs may be wrong. To do so will require transformative actions for both individuals and organizations."

Stankey et al. (2003, p. 45)

Introduction

"Sustainable forest management", or SFM, is the reigning paradigm for managing forests the world over (e.g., FAO 1993). Sustainable forest management is most certainly the moniker of choice for the best in forest management in Canada. Among other manifestations, there are national criteria and indicators for SFM (CCFM 1995), a national certification standard for SFM (CSA 2002), and even a nationwide research network called the Sustainable Forest Management Network (SFMN; Chap. 1). In a nutshell, SFM has been defined as "management to maintain and enhance the long-term health of forest ecosystems, while providing ecological, economic, social, and cultural opportunities for the benefit of present and future generations" (CSA 2002). Various dimensions and interpretations of sustainability and SFM are explored in Chap. 2.

We have come to see SFM in the following way. SFM requires good process and good content, just like any successful endeavour (e.g., successful lumber production requires a good mill and good logs; Fig. 21.1). We believe that the good content of SFM is well captured in a comprehensive set of criteria and indicators for SFM (e.g., CCFM 1995; see also von Mirbach 2000 and Duinker 2001). The good process is comprised of

P.N. Duinker.[1] School for Resource and Environmental Studies, Dalhousie University, 1312 Robie St., Halifax, Nova Scotia B3H 3J5, Canada.
L.M. Trevisan. Strategic Planning and Development, Ontario Ministry of Agriculture and Food, Toronto, Ontario, Canada.
[1]Author for correspondence: Telephone: 902-494-7100. e-mail: peter.duinker@dal.ca
Correct citation: Duinker, P.N., and Trevisan, L.M. 2003. Adaptive management: progress and prospects for Canadian forests. Chapter 21. *In* Towards Sustainable Management of the Boreal Forest. *Edited by* P.J. Burton, C. Messier, D.W. Smith, and W.L. Adamowicz. NRC Research Press, Ottawa, Ontario, Canada. pp. 857–892.

Fig. 21.1. Basic ingredients of SFM (sustainable forest management).

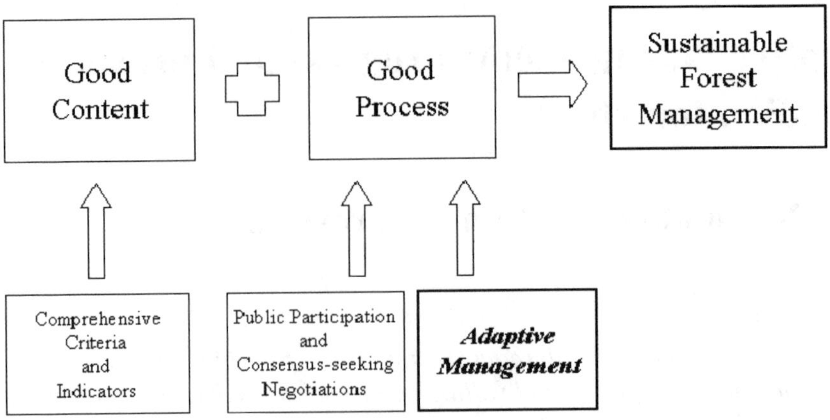

two main parts: a good political process, in the sense of public participation and con-
sensus-seeking negotiations; and a good technical process, which we call adaptive management.

Our focus in this chapter is adaptive management. It builds on earlier discussions of
criteria and indicators (C&I) of SFM (Chap. 2), and participatory processes (Chap. 4).
Indeed, all of the previous chapters highlight important elements of the SFM equation,
but they need testing and refinement through an adaptive management process. In the
following pages, we first delve briefly into the historical development of adaptive management and discuss some basic concepts. We then present examples where forest managers have attempted to implement the first steps of an adaptive management process.
The chapter ends with a look at the institutional support given to adaptive management
of forests, and also some challenges and caveats associated with it. As will become evident, we are strong supporters of adaptive management, and with this chapter we hope
to encourage practitioners to embrace the concept heartily. At the same time, though,
our enthusiasm is tempered by recognition of the tremendous efforts needed to overcome all the obstacles and really make adaptive management work on the proverbial
ground (Box 21.1).

Box 21.1. Challenges for adaptive forest management:

- For forest managers and policymakers to admit that, individually and collectively, we seldom know for sure the best thing to do, and that it is generally acceptable to make mistakes so long as we learn from them;
- To demonstrate the value and feasibility of adaptive management when applied to forest
  issues;
- To implement programs of monitoring and feedback that provide warning of unsustainable practices and policies, and provide reliable knowledge for improving forest management; and
- To instill an attitude and culture of life-long learning and experimentation in forest managers, and promote formal programs for long-term improvement in forest management
  institutions.

Our conception of adaptive management is detailed below. Here we present a simple definition, the inspiration for which is drawn from an abundant and still growing literature. Adaptive management is an approach to management that, recognizing substantial uncertainties in the links between management outcomes and system performance, incorporates a specially designed learning approach to uncertainty reduction at the actual time and space scales of the managed system. It demands an incisive quantitative approach to creation and testing of hypotheses of cause and effect in system management.

Adaptive management: staying on the wrong road long and smart enough to know?

Some history of adaptive management

The concept of adaptive environmental management (AEM) was first conceived by a group of researchers at the International Institute for Applied Systems Analysis (IIASA) in the 1970s, led by C.S. Holling (Lee 1993). AEM had its roots in an early attempt by researchers at University of British Columbia (UBC) to pilot problem-solving exercises based on ecosystem modelling workshops (Walters 1986). While the researchers initially saw the workshops as a failure because of the practical difficulties involved in developing the required simulation models, the participants eagerly embraced the modelling concept as a formalized structure for articulating management strategies and evaluating their likely outcomes. The UBC team eventually began running modelling workshops for a variety of clients, and Holling proposed applying the modelling process to resource management problems. Eventually, refinement of the techniques and philosophy of AEM were fleshed out in a volume entitled *Adaptive Environmental Assessment and Management*, authored by a team of researchers at IIASA and edited by Holling (1978).

Adaptive environmental management, as described by Holling (1978) and colleagues, is a system that involves bringing resource managers, modellers, and scientific experts together for a series of brief, intense workshops to produce a simulation model of the system being managed. Gaming with the simulation model then allows workshop participants to examine the possible ramifications of different management options. Essentially, the managers use the model as a learning tool to evaluate alternative decision strategies. Other scholars (e.g., Walters 1986) have further refined the AEM concept. Baskerville (1985) operationalized the concept into nine steps, and his framework is widely referenced in the context of SFM (see Box 21.2 for a contemporary interpretation of adaptive management based on Baskerville's (1985) work). The fundamental defining factors in adaptive approaches to managing the environment, regardless of the specific methods advocated or terms used, are the experimental nature of the policy selection process and the focus on deliberate learning. For the purposes of this chapter, the term AEM will refer specifically to Holling's (1978) original conception of decision support for resource management using modelling workshops. There are consulting firms (e.g., ESSA 1982, ESSA Technologies 2002) that specialize in these sorts of

---

Box 21.2. Steps in (passive) adaptive forest management.

Planning:

Define the forest	Define the forest area and its present conditions.
Values/objectives	Identify and select values, objectives, indicators, and provisional targets.
Inventory	Prepare maps and other records associated with the chosen indicators.
Forecast	Forecast expected future conditions of chosen indicators: - under a no-action strategy; and - under alternative strategies.
Choose	Select the strategy, with its associated indicator forecasts, that best meets desired targets.

Implementation and operation:

Take actions as prescribed in the selected strategy.

Checking and corrective action:

Measure all indicators.

Compare implemented actions to planned actions, and actual indicator levels to targets.

Understand the reasons for differences between actual and planned outcomes, and use that understanding to improve management through corrective action.

Management review:

Periodically review overall progress in achieving SFM and implementing the SFM requirements.

Return to planning for the next cycle.

From CSA 2002

---

workshops. Adaptive management (AM) is the term used here more generally to describe an experimental system of resource management that incorporates active learning (Walters and Holling 1990; Hilborn 1992), regardless of whether simulation modelling or some other method of forecasting is used.

Basic concepts of adaptive management

Adaptive management is a framework whereby multiple stakeholders are brought together to establish a deliberate learning and experimentation process around the system to be managed (Holling 1978; Baskerville 1985; Walters 1986; Lee 1993; Haney and Power 1996; Lancia et al. 1996). Adaptive management, in its most basic form, consists of a framework embodying activities associated with planning, choosing, implementing, checking, and revising (Box 21.2). The planning step involves bounding the problem, identifying relevant values, goals (or objectives), indicators, and objectives (or targets) for the system, and then predicting the impacts of different possible management strategies on the values identified. The management team then chooses the strategy (or strategies) to be employed, and implements the associated schedule of actions. The response of the system to the management strategy is monitored, and this response is evaluated at regular intervals to determine the accuracy of the predictions and the progress made towards the objectives/targets. In light of this evaluation, the entire plan is revisited and modified as necessary. Thus, adaptive management is a cyclical process in which the management activities carried out are designed to facilitate learning about the system.

Some authors have made a distinction between active and passive adaptive management (e.g., Taylor et al. 1997, Baker 2000). What we described above, and the process shown in Box 21.2, is essentially passive adaptive management. Such an approach has tremendous utility, but not nearly as much as active approaches. Baker (2000) described three approaches to active adaptive management. In the strongest, several strategies are chosen for simultaneous implementation. The management area is then divided into experimental units, and each of the chosen strategies (alternative treatments) is carried out in one or more units and compared to each other or to control units (or both, ideally). In this way, small parts of the system are given a controlled push in different directions, with the hope of quickly gathering as much practical information about the responses as possible (Sit and Taylor 1998).

In Baker's (2000) second adaptive approach, several alternative strategies are created, but only one is implemented at a time in a systematic search for appropriate action that presumably ends when one strategy yields desired system responses. This approach would work well only when system responses are relatively rapid. An example is the management of hunted waterfowl populations in the United States (e.g., Williams and Johnson 1995; Williams et al. 1996). Finally, Baker (2000) considers retrospective and comparative studies as a third form of active adaptive management.

Managers employing a passive approach, on the other hand, select and implement one preferable strategy that is evaluated by comparing its outcomes to those of a "no action" forecast. Measured deviations of the system from the predicted response are used to appraise the quality of the forecasts, and to improve the management team's understanding of the system. Passive management may be a more appropriate approach to use when ecological risks are high and consequences of management actions are not easily reversible, or when suitable experimental areas are unavailable (Taylor et al. 1997). Conversely, the risk associated with passive adaptive management is that any observed differences between indicator measurements and the no-action forecast can be due either to management actions or to errors in forecasting.

The adaptive management approach has often been confused with management based on field trials, or, worse, trial and error (Lee 1993; Taylor et al. 1997; Kessler 1999). While adaptive management is experimental in nature, it differs from traditional field trials in terms of scale. Field trials typically involve experimenting in a controlled way on a small area to acquire detailed information about a specific part of a system. The limited spatial extent of most field trials may miss effects that are only evident when the larger picture is examined. For example, some treatments applied as part of a well-controlled and carefully manipulated experiment may bear little resemblance to those applied on an operational scale. Adaptive management is employed at the large-ecosystem level, at a fully operational scale, and is intended to examine the effects of management strategies on the entire system and the way the components interact. Field trials and adaptive management are not mutually exclusive, but neither are they identical. Field trials may be complementary to an adaptive management process, as they can be used to provide valuable information about system components that may then be incorporated into forecasting models.

Adaptive management also differs substantially from simple trial and error (Lee 1993). Trial and error represents the "status quo" means of resource management wherein strategies are selected and implemented for any number of reasons, which may or may not be related to strengthening understanding of the system. A strategy is allowed to continue until such time as it becomes obvious that it is damaging the resource or failing to meet its (often poorly articulated) goals, at which point it is replaced with a new strategy. Any learning that happens under such a paradigm is haphazard at best, and non-existent at worst. Adaptive management involves planned and deliberate probing of ecosystems to learn from the mistakes that are bound to occur. Indeed, an adaptive management approach will not be successful if mistakes and surprises are not recognized as inevitable, and potentially beneficial. Since any human conception of the natural world is necessarily a caricature of it, our understanding of ecosystems will always be, in some way, inaccurate. Despite the fact that our picture of ecosystem function is always essentially wrong, managers can be wrong and smart at the same time if they take the opportunity to learn from the errors, and then modify management activities accordingly to get closer to the true picture. Since it is a long-term, iterative process, adaptive management — at least in its passive form — might be envisioned as staying on the wrong road long and smart enough to know how and why it is wrong (Box 21.3).

Why use adaptive management?

Many managers operate under the assumption that they make decisions based on a thorough consideration of all the available alternatives. In fact, it is more likely that they operate under bounded rationality, a concept first described by Herbert Simon (Lee 1993). Typically, when making decisions, humans examine a limited number of options and stop the search for appropriate solutions when a satisfactory one is found. The optimal, or "best", course of action may not be selected if another satisfactory option is identified first. In fact, the optimal course of action may not even appear in the suite of alternatives considered. Top-down management styles often result in decisions being made by a single individual, based on options only he/she has articulated, whereas team-based management styles are better able to articulate a wider and more creative set of options. This tendency to take a narrow approach to decision-making must be tempered by a more holistic, system-oriented style of adaptive management. Then, at least, the decisions made using bounded rationality may be evaluated through deliberate experimentation, the learning from which may lead to the identification of previously unconsidered options. Experimentation with policy decisions thus has the ability to speed our acquisition of knowledge (Holling 1978).

There are other benefits to using an adaptive approach as well. As mentioned earlier, the system-level experimentation characteristic of adaptive management may uncover subtle effects that are not obvious on the fine scale of field trials. Moreover, an adaptive management plan favours action as a means to learn about the system, which can be vital when uncertainty is high and conflict over possible outcomes has thwarted or even

Box 21.3. Passive adaptive management ...

. . . can be envisioned as "staying on the wrong road long and smart enough to know", i.e., to learn how and why things didn't go as planned. Since ineffective or poor decisions are inevitable in any arena, the key is to have a framework in place to ensure that the institution harnesses the knowledge that can be gained from mistakes.

Fig. 21.2. "Two roads diverged in a yellow wood, and sorry I could not travel both, and be one traveler, long I stood. . ." (Robert Frost 1915). The fork in a road can be a useful metaphor for adaptive management. If you be but "one traveler", you can feasibly only follow one road at a time, but with your eyes open to indications that you are or are not on the right road; this is the process of passive adaptive management. Even if there were two of you and you could travel both roads simultaneously (more analogous to the process of active adaptive management), you still have to watch for those indicators and make sure you communicate to each other.

Photo by Phil Burton

called a halt to management activities (Lee 1993). In addition, traditional resource management, unlike adaptive management, is not usually conducted with advice from a team of experts from different fields, resulting in a lack of consideration of cross-disciplinary interactions that might be relevant to management decisions (Holling 1978).

Early application of adaptive management for Canadian forests

Adaptive management concepts have been used infrequently in the management of Canadian forests. Early attempts to use the adaptive management framework included the spruce budworm (*Choristoneura fumiferana*) case in New Brunswick in the early 1970s, and the Pest Management in Plantations (PMIP) project in New Brunswick in 1983. Here we describe the former.

In 1972, a team of researchers from the New Brunswick Department of Natural Resources, the Canadian Forest Service, UBC, IIASA, and Harvard University embarked on a project to apply adaptive environmental management to the problem of spruce budworm outbreaks in New Brunswick's forests (Holling 1978). Historically, periodic budworm outbreaks occurred that killed large numbers of softwood trees in the areas affected (Chap. 8), and so insecticide spraying had been used to keep population numbers low. Over time, the budworm-spraying program had become more and more costly, and concerns about the effects of spraying insecticides (e.g., fenitrothion) on human health were increasing, so adaptive management was adopted as a means of assessing alternative management options. A core group of analysts developed a simulation model that was evaluated at a series of seven workshops. The model was eventually transferred to a working group that adapted the model to the local area, and it was then passed on to a task force that was charged with reviewing the alternative policy options. While this process has since been criticized for being unduly science-focused and not inclusive of stakeholders with varying perspectives in the modelling process (McLain and Lee 1996), it is one of the few early examples of an explicit attempt to forecast quantitatively, evaluate, choose, and implement a management strategy in a Canadian forest. Key outcomes of the process included a much more targeted pesticide application program, and a recognition that timber harvesting constituted the most powerful tool to control the effects of continued budworm infestations (Baskerville 1995).

Dissecting adaptive management: promise and progress in Canadian SFM

Our original intention had been to summarize here some good examples of forest managers getting on with various elements of the process of adaptive management. Finding those examples for all the main steps has proved impossible. Some Canadian forest managers seem to be getting the hang of adaptive management in their planning, *but little progress has been made on the monitoring and assessment side.* We wondered whether the program of so-called adaptive management areas established in the United States under the Northwest Forest Plan would yield some success stories about adaptive management, but the literature we canvassed (e.g., Lee 2001; Stankey et al. 2003) dashed our hopes. So, the best we can offer now is a description of two examples where forest managers have made significant progress in the planning phase of adaptive management. The first deals with establishment of a comprehensive set of values, goals, indicators, and objectives, and the second deals with the technical domain of forecasting.

Getting started: establishing values, goals, indicators, and objectives

The ideal

An effective starting place in adaptive management of forests has emerged lately through a unique blending of concepts from several fields. Planning theory offers the requirement, in rational planning, for explicit setting of goals and objectives at the start of the process. The social sciences point to the necessity of identifying the values people hold as they enter a planning process (Chap. 4). Finally, scholarship associated with sustainable development has made clear the need for indicators to gauge progress in achieving sustainability (Chap. 2). The CSA (Canadian Standards Association) standard for SFM, first in the 1996 version (CSA 1996b) and with refinements in the 2002 version (CSA 2002), brings these elements together in a way that gives strong structure and form to early deliberations among stakeholders as they embark on a forest-management planning exercise.

Let us look first at definitions and then at relationships among the four concepts of values, goals, indicators, and objectives. We rely on Duinker (2001) for definitions (Box 21.4). Given these definitions, one can envisage a strict order of development and set of relationships among the four concepts (Fig. 21.3). Thus, one would begin the planning process with an elicitation of stakeholders' values. This also may require substantial sorting of the values proposed, a process which is facilitated in Canada by broad acceptance of the criteria and elements associated with the CCFM's (1995) framework for criteria and indicators of SFM (see Box 2.2). The process continues with the setting of one goal for each value, and identification of one or more indicators for each value as well. Once the indicators are determined, then stakeholders can set detailed objectives, one for each indicator (the term "target" is used in place of objective in the revised CSA 2002 SFM standard). Early in planning, the objectives would be set only provisionally,

---

Box 21.4. Definitions and examples of values, goals, indicators, and objectives in forest management.

Value — things for which someone would deem a forest important:
- e.g.: - process such as carbon sequestration, water quantity regulation, recreation;
  - physical thing such as timber, moose meat, marten pelts;
  - forest condition such as biodiversity, soil bulk density.

Goal — a directional statement for a value (need not be stated in quantitative terms, and now called "objective" in CSA's (2002) revised SFM standard):
- e.g.: - have forests become long-term net sinks of atmospheric carbon;
  - produce a continuous non-declining flow of quality wood to meet mill needs;
  - maintain current levels and types of biodiversity.

Indicator — a measurable variable (quantitative or qualitative) relating directly to one or more values:
- e.g.: - for carbon sequestration, kg ha^{-1} year^{-1} net carbon flux;
  - for timber, m^3/year harvest volume;
  - for biodiversity, age-class structure (seral habitat distribution) of the forest.

Objective — a directional statement for an indicator (must be stated in unambiguous terms, and now called "target" in CSA's (2002) revised SFM standard):
- e.g.: - more than 0 kg ha^{-1} year^{-1} (i.e., a positive number) for net carbon intake;
  - at least 500 000 m^3/year of softwood pulp fibre;
  - minimum of 10% of total forest area in each of five stand development stages at any time.

From Duinker 2001

Fig. 21.3. Relationships among values, goals, indicators, and objectives.

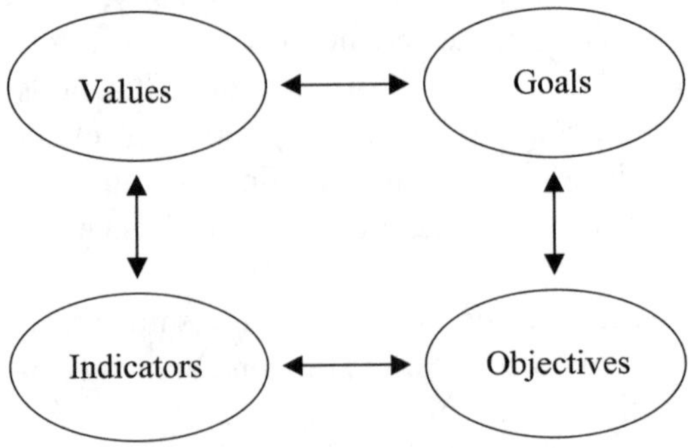

with expectations that they might need substantial revisions if analysis reveals that the whole set of provisional objectives is either physically or politically impossible to achieve.

Our view is that adaptive management of forests begins in the planning phase with incisive setting of values, goals, indicators, and objectives. Such an approach is entirely consistent with the concept of scoping in environmental impact assessment (EIA; e.g., Beanlands and Duinker 1984). Scoping is essentially the analytical and participatory process of identifying valued ecosystem components, relevant development components, and time and space bounds for an EIA. Getting the scope right in EIA and in adaptive management of forests is crucial for the success of the rest of the process.

A case of establishing values, goals, indicators, and objectives of SFM

Weldwood has held a forest-management agreement (FMA) on a million-hectare forest surrounding Hinton, Alberta, since 1954, and has managed the timber production on the FMA since 1956. The forest is predominantly lodgepole pine (*Pinus contorta*). There is significant exploration and extraction activity across the landscape for coal, oil, and natural gas. Weldwood prepared the first forest management plan for the area in 1961, and is currently implementing the fifth revision, approved in 2000.

Weldwood's early and continuing interest in certification arose from the expectation that market demand for wood produced from certified forests would increase over time. With strong executive-level direction to achieve certification, company personnel became involved in standards development exercises (e.g., the CSA 1996*b* SFM standard). In the mid-1990s, the company committed to pursuing certification under the CSA standard, and implemented a system review and gap analysis in 1996 and 1997, respectively.

Public participation and stakeholder involvement are important features of contemporary forest management in Canada (Fig. 21.4; Duinker 1998), and Weldwood became an early adopter of the concept of a public advisory group for the FMA. Its forest resource advisory group (FRAG), established in the early 1990s, was the main discussion forum for development of values, goals, indicators, and objectives for the FMA. At first these discussions were focused on the requirements for so-called ground rules (basic operating procedures required as part of the FMA) and for preparation of the

Fig. 21.4. Collaborative problem-solving among stakeholders is a hallmark of sustainable forest management. If adaptive management is the compass of sustainability, then its necessary counterpart — the politics of principled negotiation — is the gyroscope (see Lee 1993).

Photo courtesy of Alberta-Pacific Forest Industries Inc.

detailed forest management plan (DFMP). With the commitment to seek CSA-based certification, the FRAG turned its attention to values, goals, indicators, and objectives in that context. Additional consultations were implemented through regular meetings with forestry officials of the Alberta government, and a general call for input from local people living in or near the FMA (WCL 2002*a*).

Development of values, goals, indicators, and objectives for the FMA was based on a merger of outcomes of two parallel processes: a public values process associated with the FRAG, and a technical process associated with the Foothills Model Forest. Foothills Model Forest is a member of Canada's network of federally funded model forests (CMFN 2002), and incorporates the entire Weldwood FMA as well as significant areas of the adjacent Jasper National Park. Starting in 1994, the FRAG developed an initial set of values and goals for the 1996 ground rules. These values and goals were revisited during the initial stages of plan development in late 1995 and 1996, after which the FRAG turned its attention to indicators for the plan. Due to complications arising from the schedule of plan submission to the Alberta government, the indicators were not included in the DFMP, but rather in the SFM Plan (WCL 2000). The SFM Plan is thus an adjunct to the DFMP and brings the latter in line with the management plan requirements of the CSA SFM standard. The initial SFM Plan, including the final indicators and objectives, was vetted through the FRAG in the spring of 1999, prior to the CSA certification audit in June 1999.

The Foothills Model Forest (FMF n.d.), in which Weldwood is an active partner, prepared a set of goals based on results of a December 1997 workshop. The workshop discussions were seeded with a list of all existing goals gleaned from any government, company, and other sources. FMF partners were asked to suggest any other goals they thought appropriate. At the workshop, all of these were boiled down to a set that captured all the concepts in a minimum number of statements. The FMF then proceeded to develop a set of local-level indicators for the model forest area, as part of a national initiative of the model forest network (von Mirbach 2000). Weldwood, having participated in the indicators work of the FMF, then adapted the indicator set in spring of 1999 for its SFM Plan.

Participatory processes such as this one seldom deal with technical concepts in the same way as analysts do. For example, while the protocols suggest starting with identification of people's values, the Weldwood process started with establishment of goals. Given that values are implied by the goals (as per Box 21.4), the process simply had to ensure that an appropriate array of values was covered in the goal set. The CSA SFM standard (CSA 1996b) required that the value set be fully consistent with the 21 critical elements that were themselves taken largely from the SFM criteria and elements of the CCFM (1995). Thus, the critical elements became the values. Values were discussed in the context of alternative goal statements, but they were not discussed separately. In sum, then, the process established explicit goals, indicators, and objectives, with values embedded in the goals (Table 21.1).

How is Weldwood continuing to use the values, goals, indicators, and objectives? There are revisions to the SFM Plan on an annual basis (e.g., WCL 2002a). In an annual SFM Stewardship Report (WCL 2002b), the company reports on results of indicator monitoring during the past year. On the basis of these documents, the FRAG and the company (a) review the appropriateness of the goals and objectives and the effective-

Table 21.1. The sequence and sample results of Weldwood's exercise of establishing values, goals, indicators, and objectives.

Concept	Sample results for biodiversity
FMF[a] goal (September 1998)	Maintain viable populations of all currently occurring native species
DFMP[b] goal (August 1999)	Conserve species diversity
CSA[c] critical element goal	Ensure that all native species found on the defined forest area (DFA) prosper through time
Final SFM[d] goal	Prosperity of all currently occurring native species Indicator Seral stage — percentage by major forest type within range of natural variation (representing coarse-filter habitat supply)
Objective	All seral stage amounts by major forest type and land-base scale within range of natural variation according to 1999 DFMP analysis

[a]Foothills Model Forest.
[b]Detailed Forest Management Plan.
[c]Canadian Standards Association.
[d]Sustainable Forest Management.

ness of the indicators, (*b*) measure performance in terms of the indicators, and (*c*) assess appropriate actions for the subsequent year. The company uses the SFM Plan to provide broad direction for development of regulatory and internal plans through the course of each year. Weldwood continues to engage the FRAG, the Alberta Department of Sustainable Resource Development, Aboriginal groups, the Foothills Model Forest, and the general public in the ongoing process of improving the values, goals, indicators and objectives. A website (www.weldwood.com/hinfr01/internet/hinnet.nsf) is being used to solicit additional input and provide public information.

The art and science of forecasting

The ideal

A forecast is a kind of prediction in which the potential future condition of a particular system component is specified (Duinker and Baskerville 1986). Too frequently, impact predictions in environmental assessments consist of vague generalizations that are so imprecise that they can never be shown to be wrong (Beanlands and Duinker 1984). Forecasts, on the other hand, can be quantitatively stated as testable hypotheses (Baskerville 1985). For adaptive managers, it is far preferable to be quantitative and wrong than qualitative and untestable (Duinker 1987).

There are several possible ways of predicting the way a system will respond to a management strategy, the merits of which are discussed later in this section. Regardless of the forecasting method used, however, some basic guidelines should instruct the process of developing predictions. Holling (1978) advocated the use of a series of brief, intense workshops in which a small group of forecasters consults with a larger team of experts. The short time-line involved in this process minimizes scientists' tendency to break down and compartmentalize problems rather than focusing on the big picture. Policymakers should also be involved in the technical process of forecasting from the beginning to ensure that impacts are expressed in spatial and temporal terms useful to them (Holling 1978; Duinker and Baskerville 1986). While a wide range of policy options should be considered in the forecasting analysis (Holling 1978), it is important to ensure that the forecasts address issues at jurisdictional and project levels (Duinker and Baskerville 1986), as well as on space and time scales of biological significance (Lee 1993). Depending on the policy or practice in question, these boundaries might correspond to entire countries, provinces, or corporations (Chap. 7), individual forest licenses or estates ("the forest level", Chap. 11), component watersheds or landscape units (Chap. 12), or individual stands and cutblocks (Chap. 13). In addition, the "looking outward" approach advocated by Holling (1978) and Walters (1986) helps to identify elements that are critical to predicting the system's behaviour.

Several methods can be used to generate information for making predictions, including field experiments (Fig. 21.5), simulation modelling (Chap. 14), and scenario-building. The management team must have a benchmark against which to predict the performance of the managed system, and which can be used to gauge the accuracy of the predictions. Many management strategies, however, are undertaken on scales that make replication impossible. Given this limitation, the management team must decide

Fig. 21.5. Experimental thinnings in lodgepole pine (*Pinus contorta*) in west-central Alberta. Broad-scale experiments are needed to generate reliable knowledge on how to manage forests for the multiple values of timber production and biodiversity conservation.

Photo by P. Duinker

on the best means of providing a control. Using data from another, comparable location that was not subject to the management treatment is one option, but finding a sufficiently similar system to measure is a near-impossible task. Completing baseline studies (interpreted here as pre-intervention characterization of the system), and using the data as the control, assumes that the managed system would have behaved in the same way in perpetuity in the absence of management intervention. This assumption does not take into account the inevitability of natural disturbance, and ignores the fact that system behaviour can (and usually does) change over time even without human interference. Professional judgement is often used to characterize the impact of different strategies on a system. Since even professional judgements are made using a mental model, their use is indefensible unless the model used can be viewed and others can evaluate its validity. Developing a mathematical model to test the effects of alternative strategies, and using a "no intervention" option as a reference, on the other hand, provides the most defensible means of preparing forecasts (Duinker 1989).

Using forecasts derived from mathematical simulation models allows the management team to predict the impacts of alternatives without causing potentially irreversible harm to the system (Chap. 14). The modelling process is relatively inexpensive and today usually does not require equipment more complex than a desktop computer. Moreover, since some sort of modelling is usually done anyway, whether intentional or not, it is best to make the model explicit (Walters 1986), and so avoid the problems posed by the professional judgement paradigm. Perhaps most importantly, though, simulation models require the explicit description of system interactions in a quantitative, measurable way (Baskerville 1985; Duinker 1989). Modelling forces the recognition of errors in the management team's understanding of a system (Baskerville 1985). Thus, if the model produces simulations that are out of line with the observed behaviour of the system, it is usually an indication that the model and some of its underlying assumptions require adjustment. Predictions based on an explicit model can be tested by stacking them up against performance of the real system that is determined through monitoring of carefully selected variables (Duinker 1989). Johnson et al. (1997) show how this can work for managing harvests of mallard duck populations.

While modelling is a powerful tool to use for producing forecasts, the predictions developed will only be as good as the model that generated them. Holling (1978), Walters (1986), and Duinker and Baskerville (1986) discussed the components of defensible modelling at length, and Chap. 14 is devoted to the topic, so it will not be detailed here. One common misunderstanding, however, should be cleared up. Many detractors of modelling claim that simulation models are not useful because they are too complex, and indeed, some analysts do produce models that are too complicated to be helpful. On the contrary, the utility of a model lies in its simplicity. Policymakers more easily understand a simple model than a complex one, and it provides less opportunity to make mistakes (Holling 1978). Critics who claim that models are too simplistic to be a realistic representation of reality are as easily silenced. Any model of a system, whether articulated explicitly and mathematically or not, will necessarily be wrong in some details, since models can only be "caricatures of nature" (Walters 1986; Lee 1993). An explicit model that describes a system's relationships in quantifiable terms, however, can be tested to determine the degree to which it is an accurate representation of the system. Thus, even if a simulation model is simple, and therefore wrong, we can understand *how* wrong it is, and modify management behaviour accordingly.

Baskerville (1985) made a number of points salient to the development of models specific to forest management. Predicting impacts on individual stands does not provide enough information to manage at the forest level unless the stand-level impacts are translated into a pattern that is described at the forest level. Thus, explicit links must be made in the model between actions and impacts at the stand and forest levels. In addition, models that link wood-supply analysis with wildlife-habitat analysis must quantitatively define goals in geographic terms. In other words, managers must make clear their forest-level goals for measures like diversity and interspersion, and identify what the desired habitat patterns will look like.

Regardless which method is used to forecast, the system must be monitored (i.e., time-series field data must be taken) to ground-test and possibly revise predictions. Only then can managers judge whether their predictions were accurate, and whether the management scheme needs to be altered to meet the stated objectives. Forecasting and monitoring are thus key, inextricably linked components of adaptive management (Duinker 1989). Evaluation of the appropriateness of implemented actions therefore necessarily involves measuring the actual performance of a managed system and comparing that performance with the predictions made through incisive systems-analytical forecasting (Duinker 1989).

A case of wood-supply and wildlife-habitat forecasting

Millar Western Forest Products Ltd. holds a Forest Management Agreement for an area near Whitecourt, Alberta, that consists of some 300 000 ha of forestland (MWFP 2000). In the fall of 2000, Millar Western prepared a DFMP for the area that represented the results of planning efforts conducted over the previous 5 years, and provided background and justification for the forest management strategy to be employed. The plan was organized to conform in broad terms to the CSA standard for SFM (CSA 1996a, b). A planning team of over 60 experts, both internal and external to Millar Western, developed it, and the company attempted to engage other stakeholders in the planning process, including First Nations and environmental and industry groups. The plan embodies Millar Western's commitment to forest stewardship, and it was developed in line with the company's intention to adopt adaptive management principles. Analysis performed for the DFMP provides the basis for an ongoing monitoring and research program that the company has committed to establish over the next decade.

The approach taken to forecasting in the Millar Western DFMP involved the development of dynamic, static, and general impact assessments (MWFP 2000). Dynamic assessments (using simulation models that described change in indicators over time) were used for assessing timber supply, biodiversity, wildfire risk, and watershed resources. Static assessments (using models that provided snap-shot maps of high-risk areas) were used to determine the location and extent of sensitive soils, rare habitats, rare plants, and cultural and heritage resources. General impact assessments (informal assessments using information from the available literature and professional judgement) were developed for forest values whose relationship to forest management strategies are poorly understood and for other forest values that were not formally addressed because of financial considerations. General assessments were performed for forest values involving traditional land use, forest invertebrates and diseases, amphibians, and carbon. The DFMP noted that socio-economic and recreational impacts are not considered quantitatively, and that they should be assessed more thoroughly in future plans. The timber-supply analysis (TSA) and the Biodiversity Assessment Project (BAP, Fig. 21.6) are the most instructive examples of effective forecasting for adaptive management in the plan. The remainder of this case study will focus on the approach taken in performing these analyses, and the links between them. From BAP's inception in 1995 to the present, it was funded jointly by Millar Western and the SFMN. Further discussion of the BAP approach can be found in Chap. 14.

The TSA used dynamic simulation models to generate forecasts of the forest inventory under different management scenarios (MWFP 2000). A 200-year time horizon was used, with a time step of 10 years. Yield curves were generated for pure and mixed stands managed under the different scenarios. In addition, spatially explicit forest inventory forecasts were developed for each of the scenarios considered. Three rounds of analysis were performed, such that the strategies tested were modified with each round based on previous results. This process allowed the long-term sustainability of various cut levels and strategies to be assessed.

The forest inventory projections produced for the TSA were then used as inputs for the ecological models that formed the basis of BAP (Duinker and Doyon 2000). The

Fig. 21.6. Basic structure of Millar Western's Biodiversity Assessment Project (BAP).

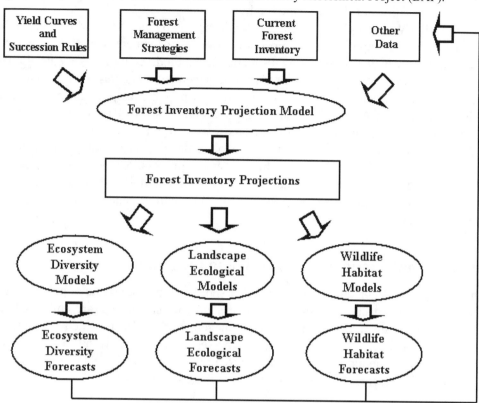

species composition and developmental stage of each stand were used to classify stands into forest habitat types. One of the fundamental principles underlying BAP was that biodiversity should be assessed at multiple scales (Doyon and Duinker 2003). Thus, models were developed to interpret the inventory projections in terms of ecosystem diversity, landscape configuration, and species-specific habitat supply. The output from these models was then compared with projections produced by LANDIS (Mladenoff et al. 1996), a natural disturbance model used to determine the change in bioindicators over time based on simulated natural disturbance (MWFP 2000; Doyon and Duinker 2003). Management strategies could then be evaluated relative to changes that would have occurred in bioindicator status regardless of human intervention. This addressed the second fundamental BAP principle: forests are dynamic in nature and biodiversity status may change owing to natural disturbance (Doyon and Duinker 2003). The forecasts produced by the BAP models for each management scenario were compared, and an evaluation of the impact of each strategy on biodiversity values was performed.

Millar Western has committed to a major research and monitoring program to ground-test the BAP models (MWFP 2000). A 3-year research project was launched in 2000 and funded by the SFMN to assess the validity of four habitat supply models used to determine the availability of species-specific habitat (Ancelin and Duinker 2002; Bone and Duinker 2002; Doucette and Duinker 2002; Hautala and Duinker 2002). The

Millar Western planning team believes that the refinement of these models will lead to richer interpretations of the available biodiversity data.

The Millar Western plan (MWFP 2000) is based on quantitative and thus verifiable forecasts of timber supply and the impact of various management scenarios on selected forest values. Inclusion of the LANDIS model in the biodiversity assessments allowed the planning team to compare the proposed management strategies with a reasonable control — i.e., the values of the bioindicators in a hypothetical, unmanaged state. In addition, both the TSA and BAP analytical models linked stand characteristics with spatially explicit projections of forest-level landscape patterns associated with different management options. The company included policy staff on the planning team, and they were involved at all critical stages of the forecasting process. A monitoring system has been developed that will allow Millar Western to evaluate the models used for habitat analysis, compare projections with empirical observations, and adapt management strategies as appropriate.

Institutional support for adaptive management

Any adaptive management initiative, whether sponsored by the private or public sector, is more likely to succeed if the principles of adaptive management are embraced and supported by the institutions in which the managers work and with which they interact (Stankey et al. 2003): governments, businesses, non-governmental organizations, research centres, and markets (Chap. 7). These organizations endorse adaptive management to varying degrees, the extent of which is discussed below.

Provincial forest agencies and policies

The strongest formal leadership in promoting adaptive management at the provincial level is evident in British Columbia and Ontario. Adaptive management is identified as one of the British Columbia Ministry of Forests' key initiatives. The Forest Practices Branch has produced a detailed *Guide to Adaptive Management* (BCMF 1999) designed to steer managers through the adaptive management planning process. In addition, Ministry staff periodically produce an adaptive management newsletter. While the B.C. government does not specifically endorse adaptive management in a policy statement and it is not enshrined in law, a discussion paper, entitled *Adaptive Management of Forests in British Columbia* (Taylor et al. 1997), was produced by the B.C. Ministry of Forests. It details ways in which an adaptive management framework could be applied in the province, suggesting that it would be useful as a "formal conduit for information flow from the operations level to policy makers". In addition, the document highlights the possibility of using adaptive management principles to ground-test different management options that may be used under the Forest Practices Code.

In 1990, the Ontario government established a Forest Policy Panel that was charged with developing a comprehensive forest policy framework for the province (OFPP 1993). Following its extensive consultation process, the Forest Policy Panel identified adaptive ecosystem management as one of the cornerstones that should provide the foundation for Ontario's new forest policy framework. Indeed, Ontario's Policy Frame-

> **Box 21.5. Six initial indicators of forest sustainability, as required by Ontario's Forest Management Planning Manual:**
>
> 1. Landscape pattern indices;
> 2. Frequency distribution of clearcut and wildfire sizes;
> 3. Forest diversity indices;
> 4. Managed Crown forest area available for timber production;
> 5. Proportion of available harvest area actually harvested; and
> 6. Habitat quality for selected wildlife species.
>
> From OMNR 1996

work for Sustainable Forests (OMNR 1994) states that the province must "adopt an adaptive approach to policy development and ecosystem management".

At the management-unit level, Ontario's Crown Forest Sustainability Act (1994) requires the preparation of a forest management plan in accordance with the *Forest Management Planning Manual* (OMNR 1996) for all Crown forest management units in Ontario. The language in the *Forest Management Planning Manual* is consistent with adaptive management principles. It states that planning teams are required to identify quantifiable objectives and identifies six initial indicators (Box 21.5) that must be measured.

The Ontario *Forest Management Planning Manual* also requires that alternative management strategies be evaluated in terms of desired outcomes and that they be assessed against a benchmark that predicts how the forest might have developed in the absence of human disturbance. Managers must include a section in the management plan that details the monitoring program that will be undertaken, and they must use it to assess the strategy implemented, including evaluating the predictive models used. Note that the Manual requires, but does not prescribe the details of, a monitoring program. Essentially, the *Forest Management Planning Manual* (OMNR 1996) binds the managers of Ontario's Crown forests to adaptive management (of a passive form).

The concepts of adaptive management also appear in the forest policy documents of Newfoundland and Labrador, and New Brunswick. In 1994, the Newfoundland Forest Service introduced adaptive management principles as required components of the forest management planning process. According to the *Environmental Preview Report — Proposed Adaptive Management Process* document (NFS 1995), each district must establish a planning team composed of representatives from a broad range of stakeholder groups, and the planning team is directed to convene a workshop at which an adaptive management strategy can be developed. The document notes that "monitoring is the cornerstone of adaptive management" and charges the planning team with the establishment of a monitoring program designed to ensure that proposed actions are carried out and to compare actual ecosystem response to predicted response. When monitoring data indicate that predictions or management actions are incorrect, the Newfoundland Forest Service is required to reconvene the planning team to reevaluate and, if necessary, modify the plan.

According to *A Vision for New Brunswick's Forests* (NBDNRE 2000, p. 4), "sustainable forest management requires an adaptive management approach". The document identifies quantitatively explicit objectives and standards for New Brunswick's Crown forests that must be achieved by Crown timber licensees. Licensees must prepare a forest management plan describing a strategy that will meet these objectives, so presumably the predictions provided must also be quantitative. Little explicit attention is given to other components of an adaptive management strategy (e.g., monitoring and reassessment) in the vision document.

The Government of Saskatchewan commissioned a forest ecosystem impact monitoring framework, a draft of which was produced in May, 2001. The document, authored by the Forest Ecosystem Monitoring Task Force Science Advisory Board (FEMTFSAB 2001), will, if adopted, indicate a clear commitment to adaptive management. The framework consists of five elements: coordinating, measuring, learning, improving, and communicating. Of these, the approaches taken to measuring, learning, and improving show support for adaptive management principles. The draft document describes a detailed process in which goals and strategies, identified at the provincial level, are combined with natural disturbance data by FMA holders to set objectives, select indicators, and establish indicator targets. The framework calls for forecasts that describe indicator responses to various future scenarios. Eventually, a desired future forest condition is described and approved, and the selected strategies are implemented. The measuring step requires monitoring and comparison of outcomes with the desired predicted futures. This information is then reviewed (the learning step) and goals, plans, predictions, and methods modified as appropriate (the improving step). The framework also identifies key areas to be assessed in the review process. The Saskatchewan Forest Ecosystem Impact Monitoring Framework, if adopted as government policy, could make the province a leader in the use of adaptive management for forests.

Manitoba's Forest Plan (KMC 1995), a strategy developed under the Canada–Manitoba Partnership Agreement in Forestry, notes that adaptive management is key to the sustainable management of forests. The document identifies three conditions that are required to manage forests effectively:
(1) understanding the system well enough to make predictions;
(2) monitoring to generate feedback on management strategies; and
(3) being willing to adjust policies and programs in response to this feedback.

In addition, the plan advocates an ecosystem approach to forest management. The plan identifies adaptivity as a key component of an ecosystem-based approach, noting that the risk of making mistakes must be recognized so that managers can learn from them. Thus, the plan seems at least implicitly to support adaptive management principles, although there is no clear indication of how they might be applied practically to forest management in Manitoba. In addition, the degree to which the plan has been adopted as Manitoba's formal forest policy is unclear.

Forest policy documents from the remaining provinces (Alberta, Quebec, Nova Scotia, and Prince Edward Island) do not explicitly mention adaptive management, and the language used therein is limited in its compatibility with adaptive management principles. In *Forest Management in Alberta — Response to the Report of the Expert Review Panel* (AFLW 1991), the Alberta government agreed to adopt a forest management review panel recommendation to test and forecast scientifically the performance of forest stands under current management practices with a view to calibrating two growth models currently in use. *Toward Sustainable Forestry — A Position Paper,* prepared by the Nova Scotia Department of Natural Resources (NSDNR 1997), makes reference to the use of a computer simulation model to establish sustainable harvest levels at the provincial level, indicating that some predictive quantification has been undertaken. Policy documents from Quebec and Prince Edward Island are silent on the concept of adaptive management.

National initiatives

National Forest Strategy

In 1992, a Canadian National Forest Strategy was developed by a wide range of stake-holders to establish directions for forest management at the national level (CCFM 1998). The commitments made in the 1992 strategy were revised and updated in the National Forest Strategy 1998–2003 (see Box 2.3). While the framework for action described in this plan is not binding on the participants, the strategy's clear identification of adaptive management principles as actions necessary for forest sustainability indicates wide-reaching support for adaptive management, at least in principle. Under "Strategic Direction One — Forest Ecosystem: Multiple Values", the strategy indicates that priority should be given to forest ecosystem research "within adaptive and mitigative management strategies" (p. 9). "Strategic Direction Two — Forest Management: Practicing Stewardship" calls for the establishment of measurable objectives and indicators, the use of forecasting models and the inclusion of monitoring and reporting programs in forest management plans. In addition, the 1995 Canadian Council of Forest Ministers' document on Criteria and Indicators (C&I; CCFM 1995) sprang from the 1992 National Forest Strategy process (Chap. 2). The C&I document has subsequently been used as the basis for the performance-based elements of the CSA's (1996*b*) SFM standard, described in more detail below under certification systems.

Model forests

The Canadian Model Forest Network was established in 1992 (Naysmith et al. 2000). The purpose of the model forests is to serve as "living laboratories where people with a direct interest in the forest, supported by the most up-to-date science and technology, could participate in decisions about how the forest could be sustainably managed" (Naysmith et al. 2000, p. 3). The model forests are characterized by their representative partnership structure, their use of science and technology as aids to decision-making, and the linkages developed between the partnership structure and the use of scientific tools. Given the model forests' mandate to use science as a basis for management, and the inclusion of a wide range of participants with varied perspectives and areas of expertise, they are ideal venues for showcasing adaptive management in action.

Two of the model forests, Fundy and McGregor, offer particularly good examples of practical adaptive management. The partners at the Fundy Model Forest in New Brunswick developed their forest management plan based on their conception of adaptive forest management (MacLean et al. 1999). The planning process consists of steps designed to identify forest values and treatment tactics, assemble information and management tools, analyze alternative strategies, make and implement management decisions, and monitor and evaluate these decisions, such that these steps describe a closed loop. McGregor Model Forest in British Columbia has established an Indicators and Adaptive Management Program as one aspect of the model forest's approach to SFM (MMF n.d.). It focuses on developing methods to identify and select indicators and determining appropriate monitoring schemes. The McGregor Model Forest approach to SFM is designed to incorporate indicators into all aspects of the management planning process, allowing the partners to learn from and modify strategies when necessary.

Industrial and NGO perspectives

Few of the major industrial or environmental non-governmental organizations (NGOs) have produced policy documents that directly identify a position on adaptive management. Wildlife Habitat Canada is a notable exception, with (passive) adaptive management firmly embedded in the so-called "Sam Jakes Strategy" for forest biodiversity conservation (WHC 1997). There are several plausible reasons for the sparse endorsement of adaptive forest management on behalf of NGOs, including the historical tendency for many of them to operate as reactive lobby groups rather than as proactive partners in policy-development processes. In any case, the dearth of explicit policy statements regarding adaptive management should not be taken as a lack of interest in the principles. Rather, the situation should be viewed as an opportunity for adaptive management proponents to launch a concerted effort aimed at including NGOs in the development of adaptive management programs, so that they become engaged partners rather than disenchanted opponents. This will help ensure that the potential consequences of all proposed management and policy alternatives are assessed using the same explicit systems-analytical framework.

Certification systems

In recent years, many cooperative efforts have been launched by industry groups and other NGOs to develop certification standards for SFM (see Chap. 7). In general, the management-based standards (e.g., ISO 1996; CSA 2002) contain more explicit language about adaptive management than the performance-based standards (e.g., FSC 2000). In other words, the ISO and CSA standards emphasize the process of management and its continual improvement, and do not pre-define the relative importance of various environmental, social, or economic values; neither do they specify the levels (or existence) of critical thresholds for specific sustainability indicators. The performance-based standards, on the other hand, emphasize and specify a wide range of measures to protect environmental and social values, and are less attentive to the management systems used to secure such protection.

Certification systems are by no means a panacea for achieving SFM, but they can be highly instrumental in helping forest managers achieve their SFM goals. Most of Canada's industrial (i.e., timber-producing) forest land is by now certified to the ISO (International Organization for Standardization) 14001 standard, and significant areas are certified to the CSA SFM standard and the SFI (Sustainable Forestry Initiative) standard of the American Forest and Paper Association (Abusow 2002). Certification systems are not without criticisms (e.g., Mihajlovich 2001), and are fetching much attention lately in terms of comparative analyses of the various systems (e.g., Wood 2000; CEPI 2001; Meridian Institute 2001). They do appear important enough to the enterprise of SFM in Canada that they merit at least the following descriptions and discussion.

Management-based standards

ISO 14001 — The International Organization for Standardization (ISO), a network of national organizations for standardization, has developed a family of standards for environmental management called ISO 14000 (ISO 2001). ISO 14001 is the standard that specifies requirements for an environmental management system or EMS (ISO 1996). The standard is process-based and "does not establish absolute requirements for environmental performance" (ISO 1996, p. vi). It focuses on providing guidelines to corporations for developing environmental management systems. Certification audits are undertaken by a third party. The standard's requirements outline a process of continuous improvement involving a cycle of plan, implement, check, and review (ISO 1996).

ISO 14001-certified organizations must develop an environmental policy that establishes a framework for setting and periodically reviewing environmental targets and objectives (ISO 1996's sect. 4.2). The guideline that accompanies the formal ISO 14001 document, which is not binding, states that objectives should be specific and companies should use measurable targets to assess them "whenever practicable" (sect. A.3.3). The policy must be implemented through the establishment of an environmental management program that explicitly states the means that will be employed to attain the targets and objectives, and the time frame in which the actions will be carried out (sect. 4.3.4). The ISO framework also includes a requirement for checking and corrective action (sect. 4.5) that stipulates that monitoring procedures must be developed to track the organization's conformance to its stated environmental objectives. Corrective actions must be taken if non-conformance exists (sect. 4.5.2), and the entire system must be subject to periodic management reviews (sect. 4.6).

ISO has also produced a Forestry Guideline (ISO 1997) to assist forest organizations in certifying to the ISO 14001 standard. The Forestry Guideline recognizes that SFM is a major driver of modern forest management policies, and suggests that SFM criteria and indicators could be incorporated into the policy, objectives, and targets developed under the ISO 14001 standard. The ISO 14001 standard, while not explicit on concepts surrounding forecasting, contains language that is generally supportive of the adaptive management cycle. When paired with criteria and indicators that provide quantifiable targets, the system could be used as a framework for adaptive management.

CSA SFM standard — The Canadian Standards Association (CSA) is an independent non-profit organization that performs product certification, management system registration, and standards development (CSA 2000). CSA's SFM standard is contained in document Z809 (CSA 1996*b*; 2002). The standard is designed to be consistent with the ISO 14001 standard and actually borrows some of the ISO system's wording for the management system requirements. While the CSA standard is considered here in the context of management-based standards because of its focus on process, it does contain some aspects of a performance-based system as well, particularly on the topic of public participation. The SFM criteria established by the Canadian Council of Forest Ministers (CCFM 1995) are incorporated into the CSA standard as a starting point for developing an indicator set (see Chap. 2). Indicators relevant to the scale and nature of the so-called defined forest area (DFA) are established locally. Thus, the CSA standard could be considered a management-based standard that includes a performance framework, although it does not specifically require certified organizations to meet externally set, quantitatively explicit performance targets.

The CSA SFM standard has three sets of SFM requirements: public participation requirements, management system requirements, and performance requirements (CSA 2002). Requirements in the standard are clearly based on adaptive management principles. Adaptive management is the process by which the standard ensures continuous improvement, another aspect that makes it consistent with ISO 14001. The adaptive management process described in the standard (see Box 21.2) clearly indicates that explicit forecasts must be made to predict and assess future changes in the forest in response to natural disturbance and management activities, and that the forecasts must be periodically assessed and evaluated in light of lessons learned through implementation. Perhaps more importantly, though, the standard explicitly incorporates adaptive management concepts into the management system requirements. It states that organizations must develop SFM performance measures and clearly explains the links among values, objectives, indicators, and targets (CSA 2002). Indicators are supposed to "be quantitative where feasible". It also commits the certified organization to developing an SFM plan that includes a planning, forecasting, implementing, and monitoring loop. Monitoring indicators and comparing them with forecasts is mandated in the standard.

The CSA standard (CSA 1996a, b; 2002) provides the most explicit embrace of adaptive management of all the SFM certification standards available to Canadian organizations. It is the only standard that makes specific reference to indicator forecasts and the importance of comparing management outcomes to those of natural disturbance. In addition, it is the only standard that actually identifies adaptive management as the most appropriate mechanism to ensure continual improvement. In our view, certification using the CSA SFM standard is a good way to ensure incorporation of adaptive management principles into a forest's management system.

Performance-based standards

Forest Stewardship Council — The Forest Stewardship Council (FSC) is a non-profit international organization founded in 1993 "to support environmentally appropriate, socially beneficial and economically viable management of the world's forests" (FSC n.d.). The FSC does not certify forests directly, but accredits certification bodies to evaluate forests in accordance with FSC's Principles and Criteria (FSC 2000) for forest stewardship (Box 21.6). While the FSC standard focuses on performance-based requirements, it is designed to be compatible with the ISO 14001 standard. The FSC's website (FSC n.d.) suggests that the two can be combined such that FSC standards are used to define performance levels, and the ISO framework can be used to provide the control mechanisms.

Box 21.6. *Forest Stewardship Council (FSC) principles.*

The FSC principles and criteria should be used in conjunction with national and international laws and regulations. FSC intends to complement, not supplant, other initiatives that support responsible forest management worldwide.

Principle #1: Compliance with laws and FSC principles

Forest management shall respect all applicable laws of the country in which they occur, and international treaties and agreements to which the country is a signatory, and comply with all FSC principles and criteria.

Principle #2: Tenure and use rights and responsibilities

Long-term tenure and use rights to the land and forest resources shall be clearly defined, documented, and legally established.

Principle #3: Indigenous peoples' rights

The legal and customary rights of indigenous peoples to own, use, and manage their lands, territories, and resources shall be recognized and respected.

Principle #4: Community relations and worker's rights

Forest management operations shall maintain or enhance the long-term social and economic well-being of forest workers and local communities.

Principle # 5: Benefits from the forest

Forest management operations shall encourage the efficient use of the forest's multiple products and services to ensure economic viability and a wide range of environmental and social benefits.

Principle #6: Environmental impact

Forest management shall conserve biological diversity and its associated values, water resources, soils, and unique and fragile ecosystems and landscapes, and, by so doing, maintain the ecological functions and the integrity of the forest.

Principle #7: Management plan

A management plan —appropriate to the scale and intensity of the operations — shall be written, implemented, and kept up to date. The long term objectives of management, and the means of achieving them, shall be clearly stated.

Principle #8: Monitoring and assessment

Monitoring shall be conducted — appropriate to the scale and intensity of forest management — to assess the condition of the forest, yields of forest products, chain of custody, management activities, and their social and environmental impacts.

Principle #9: Maintenance of high conservation value forests

Management activities in high conservation value forests shall maintain or enhance the attributes which define such forests. Decisions regarding high conservation value forests shall always be considered in the context of a precautionary approach.

Principle #10: Plantations

Plantations shall be planned and managed in accordance with Principles and Criteria 1–9 . . . While plantations can provide an array of social and economic benefits, and can contribute to satisfying the world's needs for forest products, they should complement the management of, reduce pressures on, and promote the restoration and conservation of natural forests.

FSC 2000

Given that the FSC Principles and Criteria are not meant to be used as a management framework (FSC 2000), it is not surprising that they focus little attention on principles associated with adaptive management. While Principle #6 draws attention to the impacts of forest management activities, and Principle #7 requires the preparation of a management plan that details long-term management objectives (Box 21.6), there is no requirement that explicit forecasts be made to assess impacts. Criterion 7.2 states that the management plan must be "periodically revised to incorporate the results of monitoring or new scientific and technical information, as well as to respond to changing environmental, social and economic circumstances", thus addressing the reassessment requirement of adaptive management. A monitoring system linked to the assessment process is mandated in Criterion 8.4. In short, although the FSC Principles and Criteria contain some language that is friendly to adaptive management, they do not commit certified forests to a process of active learning based on quantitatively explicit forecasts.

Sustainable Forestry Initiative — The Sustainable Forestry Initiative (SFI) standard is a project of the American Forest and Paper Association (AFPA; AFPA 2002). Membership in the AFPA is predicated upon compliance with the standard. The SFI standard is based on five principles, intended to provide vision and direction for forest management, and associated objectives and performance measures designed to evaluate compliance. None of the performance measures required are quantitative. Only one of the principles directly addresses an aspect of adaptive management. The principle of continual improvement (sect. 4.4) commits participants to "monitor, measure and report performance", to review progress and "make appropriate improvements in policies and plans". While this aspect of the scheme is consistent with adaptive management, the standard does not make reference to other such principles.

Pan-European Forest Certification

The Pan-European Forest Certification (PEFC) Framework is a voluntary private-sector initiative designed to serve as an umbrella certification to allow mutual recognition of SFM standards (PEFCC 2001). PEFC governing bodies in member countries develop national standards and certification schemes that are in accordance with the Elements and Requirements Technical Document (PEFCC 2001). This document includes certification criteria that are based on the Pan-European Criteria for Sustainable Forest Management developed through the Helsinki Process (PEFCC 2001; see Box 2.7). The criteria are subdivided into concept areas, for which quantitative and descriptive indicators are identified. Although the quantitative indicators are binding, the qualitative indicators, which include provisions for the development of policy and planning documents supporting SFM, are simply suggested examples. There are no requirements for the establishment of a cyclical management and assessment process, and there is no reference to key adaptive management principles such as forecasting and monitoring.

Challenges and caveats in adaptive management implementation

Without doubt, switching from a conventional management approach to an adaptive management approach is not a quick-fix solution to the myriad challenges that are commonly encountered when managing natural resources. Moreover, some of the issues raised are unique to adaptive management. This section explores potential problems with adaptive management implementation (see also Walters and Holling 1990; Walters 1997; and Stankey et al. 2003), and presents some strategies for overcoming them.

Making sure all the voices are heard and answering the right questions

McLain and Lee (1996, p. 444) criticized adaptive management on the grounds that "the systems model approach is . . . limited by its implicit assumption that scientific knowledge is more valid for making resource management decisions than other kinds of knowledge". The extension of this assumption, according to McLain and Lee (1996, p. 444), is that "the more managers rely on scientific knowledge in making decisions, the better the decisions will be". Clearly, in situations that involve conflict between multiple stakeholders with entrenched interests, scientific knowledge alone will not be able to address complex interactions among ecological, economic, and social dimensions of resource allocations. While the focus on rigorous scientific inquiry may increase a manager's understanding of the biophysical system, such an approach will do little to improve his/her sensitivity to the human dimensions of the problem, and may serve to marginalize stakeholders who have neither the resources nor the knowledge base to argue their cases in a scientific arena.

That said, AEM was initially developed to address the fact that social aspects of natural resource management were typically either ignored or dealt with only in a cursory manner (Walters 1986). Thus, adaptive managers are advised to incorporate social realities into the management decision-making process, whether by integrating social considerations into model development or by providing some other forum for the consideration of non-scientific issues. In addition, they might well emulate attempts in British Columbia and the Columbia River Basin to make modelling processes inclusive, which have been successful in stimulating communications among stakeholders (McLain and Lee 1996).

Perhaps more important, though, is the recognition that some basic questions surrounding resource use are best answered in the political arena, through mechanisms of democracy and citizenship. Despite the best efforts of scientists, science is not a value-free venue for discussion of unequivocal facts (Ravetz 1986). Models that propose technical solutions to problems that are political in nature are sure to be condemned as biased. All models have, at root, some basic assumptions that are related to the modelling team's understanding of the goals of the project and the "important" components of the system. It is critical, then, for managers of public resources to have a clear understanding of a given society's full range of priorities for resource use, and to communicate those priorities clearly to the modelling team. Indeed, there are some areas of conflict that likely lie outside the scope of any explicit forecasting procedure. There is little point in developing a model that is capable only of evaluating technical options if the bones of contention relate, for example, to the nature of Aboriginal participation in the harvest or the right of lobby groups to be engaged in the policy decision-making process. While these sorts of issues cannot be appropriately dealt with in any forecasting model, they also cannot be ignored.

Ethical dilemmas and the precautionary principle

The ethical dimensions of adaptive management should also be considered if such an approach is to be advocated. By definition, adaptive management frameworks urge managers to act in situations where great uncertainty exists. Such actions may, in the end, prove to be disastrous. As this potential for error is understood at the outset, management actions undertaken in this paradigm have ethical implications, especially if an incorrect action leads to a harmful environmental change. And yet, an adaptive model will not work unless the potential for error is accepted, and, in fact, encouraged. How, then, can the ethical implications of adaptive management be reconciled?

While the risk of failure cannot (and indeed should not) be removed from an adaptive management framework, all reasonable steps should be taken to minimize the likelihood of undertaking damaging actions, especially if a passive framework involving the implementation of a single, preferred strategy is used. Managers must take note of the potential permanence of the effects of management actions. Both Holling (1978) and Lee (1993) pointed out that planned experiments should be designed so as not to cause irreversible change to the environment. Situations in which taking risky actions would not be acceptable (e.g., performing experiments in habitat areas critical to a species on the brink of extinction) should be identified. In addition, the use of predictive tools may help managers to eliminate potentially harmful options before they reach the implementation stage. As well, experiments designed to guard well against Type II errors (i.e., the error of not detecting an effect that exists) should be used, as they may reduce the likelihood of missing an impact that could lead to crucial environmental changes.

If an active management framework is adopted, the problems of failure, environmental damage, and ethics are more tractable. De Young and Kaplan (1988) made a strong argument for the use of small-scale experimentation that carries the risk of failure in the context of a stable, reliable support system. They suggested that exploration of alternative strategies for dealing with what they call "commons problems" is key to developing bold, innovative solutions, as long as they are undertaken on a scale that does not threaten the entire resource base. The role of a supporting institution or government is critical in this scenario as well, because such an organization may create "the support structure that permits a variety of explorations to take place", making it "possible to experience errors without endangering the entire system" (De Young and Kaplan 1988, p. 278). Thus, while a strategy implemented under an active adaptive management plan might push a small piece of an ecosystem into a damaged condition, as long as the majority of the resource base remains undamaged, the knowledge gained by doing so may temporarily override conservation concerns. In fact, refusing to perform experiments that might have the potential to provide solutions to environmental problems presents yet another ethical conundrum.

In addition to evaluating the ethical implications of adaptive management, the concepts should also be examined with respect to the idea of precaution. The Rio Declaration (a product of the United Nations Conference on Environment and Development, UNCED 1992) states the precautionary principle in the following way: "Nations shall use the precautionary approach to protect the environment. Where there are threats of serious or irreversible damage, scientific uncertainty shall not be used to postpone cost-effective measures to prevent environmental degradation." Interestingly, adaptive management, when considered in this context, can be viewed as complementary to the precautionary principle. Adaptive management is a system that suggests taking action in uncertain situations, and it risks making mistakes. Use of the precautionary principle implies, similarly, that the risk of error exists, since it would not have to be used if the situation could be evaluated with low scientific uncertainty (Goldstein 1999). Goldstein (1999) argued that invoking the precautionary principle should actually spur scientific research, particularly on-the-ground research into the effects of precautionary regulatory actions, since "there is a much greater need to determine if the action is effective in achieving its goals". Clearly, scientific experimentation, adaptive management, and the principle of precaution are mutually compatible. One definitely wants to avoid misconstruing the precautionary principle as a justification for institutional inertia when outcomes are uncertain; such conservative brakes on learning, innovation, and improvement are the antithesis of adaptive management.

Weak interpretations of adaptive management

Adaptive management has little chance of being implemented with any strength and utility so long as its proponents and prospective implementers have shallow conceptions of it. We have found plenty such shallow conceptions in our review of the forest management literature and in our frequent discussions about adaptive management with forest practitioners across Canada. Perhaps this is best illustrated by the kinds of conversations it is possible to have with forest management practitioners in Canada. In response to the question "Are you practicing adaptive management in your forest?" it is all too possible that the answer might be:

> *"Oh, yes, we adjust our management practices as new information becomes available;" or*

> *"For sure — we try new practices, monitor them, and if they don't work, we try something else;" or*

> *"Yes — and sometime we'd love to show you our adaptive-management sites."*

In our view, all these answers belie a fundamental misunderstanding of adaptive management that is rather prevalent within the Canadian forest sector. Strong implementation of adaptive management is thwarted as long as such weak conceptions prevail.

Resource constraints: time, money, and people

Despite the fact that the concept of adaptive management appears with relative frequency in high-level policy statements (see above), there are some difficult hurdles to overcome in ensuring that policy is reflected in management practice. Performing adaptive management on scales of biological significance requires significant investments of time, money, and human resources (Lee 1993). When the unit of experimentation is the regional ecosystem, results can accumulate slowly. Indeed, exploring and developing more-sustainable means of using resources is a time-consuming process, as "sustainable development does not have a definitive timetable" (Lee 1993, p. 178). The long times necessary for adaptive management to be successful can prove problematic, especially if the strategy is adopted by a government agency or a publicly traded company. Decision-makers may not have an incentive to adopt a philosophy the benefits of which will not be obvious within the current electoral or fiscal reporting term, so there are real political constraints to implementing adaptive management (Walters 1997). Especially in times of fiscal restraint, it may be difficult to convince those who hold the purse strings to part with the large investment that is required to initiate an adaptive management scheme. For that reason, it is important that advocates of adaptive management become good teachers who can clearly demonstrate the value of the knowledge that will be accumulated. In essence, adaptive management practitioners must be able to point out in a compelling way that "spending nothing for information in a situation of ignorance is not sensible policy" (Lee 1993, p. 180).

The people involved in implementing an adaptive management strategy are of critical importance to the success of the project. In an early evaluation of applications of the adaptive management process (ESSA 1982), users and practitioners of AEM identified the lack of a "wise person" to shepherd the project as the reason for project failure in 25–30% of the situations evaluated. The "wise person" was defined as "an individual with professional understanding who has an intuitive knowledge that the project will help and knows the institutional environment well enough to nurse the project through to completion" (ESSA 1982, p. 36). The leadership of this champion of adaptive management has the power to encourage buy-in among front-line staff who might otherwise not be inclined to participate cooperatively in the process. Since managing adaptively involves accepting the risk that an individual action or strategy might fail, the process may seem threatening to individuals who are concerned about the stigma attached to being associated with an "unsuccessful" project. In the face of the tendency towards institutional inertia, the "wise person" takes on the role of advocate and teacher, shoring up support for adaptive management concepts among his or her peers. Therefore, it is critical that colleagues see this person as credible. Without such an individual, all the time and money in the world will not produce an adaptive management strategy that is capable of being implemented on the ground.

Means and ends: linking process, progress, and sustainability

Most of the previous discussion is aimed at forest-level planning models, though it obviously applies to stand-level practices and techniques too. Less obvious, perhaps, is that similar issues arise at larger scales. Adaptive management can apply to social process uncertainty as well as to that of ecological systems and processes (Costanza 1993; Costanza et al. 1993). Experiments on social systems can be considered in the light of active adaptive management too, but the ethical issues can be greater and alternative models are more heavily encumbered with political baggage, so we seldom do them.

Perhaps the most important caveat to offer is one that seems intuitive but should nonetheless be stated. Some of the approaches to forest sustainability previously discussed represent good processes (e.g., CSA SFM standard), and others represent good content (e.g., FSC Principles and Criteria). Adaptive management is a process, and adopting a process alone cannot guarantee SFM. To ensure forest sustainability, we need both reliable processes and meaningful content to guide the processes. Adaptive management may be the best way of assessing and refocusing management activities towards preserving forest values. For adaptive management to help society move toward sustainability goals, however, it must be accompanied by a broadly accepted set of sustainability principles. Chapter 2 provides some options and some guidance in this regard, but every manager, board, and community roundtable must explicitly identify the values to be sustained in their own forest.

Conclusions

The concept of adaptive management of natural resources is by now some three decades old. However, there is little actual implementation of it in Canadian forest management, or, for that matter, anywhere (Lee 2001). Perhaps this is a small replica of a problem exemplified hugely by any organized philosophy or religion: for example, tolerance and peace have been taught for millennia, but there is still much intolerance and violence in the world! And perhaps our solution for adaptive management needs to be the same as that for peace — we keep trying! The ideals seem robust and enduring, but our understanding needs to be substantially deepened before the details of strong implementation become clear.

We conclude then that adaptive management is needed as much as ever in the pursuit of sustainability in forest use and management. Significant advances in practice will require at least the following:

(*a*) acquisition of much clearer and deeper understandings of adaptive management by resource-management professionals (and, notably but to a lesser degree, other forest stakeholders);

(*b*) acceptance within the policymaking community that uncertainty reduction must come from experimentation, and (*i*) policy experiments can fail, so experimenters need a safety net, and (*ii*) experimenters need a consistent and robust supply of resources to create reliable knowledge; and

(*c*) coordination among forest managers in their adaptive management endeavours; no one can tackle all the big uncertainties at once, so managers need to coordinate their activities, given limited resources, and learn from each other's adaptive management experiences.

We finish in complete agreement with Walters (1997, p. 15):

"The critical need today is . . . creative thinking about how to make management experimentation an irresistible opportunity . . ."

Acknowledgements

We express sincere thanks to the editors for accommodating our writing schedule, with particular appreciation to Phil Burton for his gentle reminders and excellent suggestions for improvement. We also acknowledge the helpful comments of several anonymous reviewers, both of the whole book and of our chapter. Inspiration for many of the thoughts included herein were generated through discussions over the past two decades between the senior author and his mentor, Gordon Baskerville, as well as his recent participation in vigorous debates about adaptive management at the table of the CSA SFM Technical Committee — many thanks to both. We are deeply indebted to staff at Weldwood (Hinton, Alberta) and Millar Western (Edmonton, Alberta) for assisting us with case-study materials.

References

Abusow, K. 2002. Canadian forest management certification status report, January 14, 2002. Canadian Sustainable Forestry Certification Coalition, Ottawa, Ontario.

(AFLW) Alberta Forestry, Lands and Wildlife. 1991. Forest Management in Alberta — response to the report of the expert review panel. Alberta Department of Forestry, Lands and Wildlife, Edmonton, Alberta. 64 p.

(AFPA) American Forests and Paper Association. 2002. Sustainable forest initiative (SFI) program: growing tomorrow's forests today. American Forests and Paper Association, Washington, DC. Available at http://www.afandpa.org/Content/NavigationMenu/Environment_and_Recycling/SFI/SFI.htm [viewed 10 June 2003].

Ancelin, R.A., and Duinker, P.N. 2002. Validation of the Least Flycatcher habitat suitability model for Millar Western's Biodiversity Assessment Project (abstract). *In* Conference Proceedings, Advances in Forest Management: From Knowledge to Practice. *Edited by* T.S. Veeman, P.N. Duinker, B.J. MacNab, A.G. Coyne, K.M. Veeman, G.A. Binsted, and D. Korber. Sustainable Forest Management Network, University of Alberta, Edmonton, Alberta. p. 319.

Baker, J.A. 2000. Landscape ecology and adaptive management. *In* Ecology of a managed terrestrial landscape: patterns and processes of forest landscapes in Ontario. *Edited by* A.H. Perera, D.L. Euler, and I.D. Thompson. UBC Press, Vancouver, British Columbia. pp. 310–322.

Baskerville, G.L. 1985. Adaptive management: wood availability and habitat availability. For. Chron. **61**: 171–175.

Baskerville, G.L. 1995. The forestry problem: adaptive lurches of renewal. *In* Barriers and bridges to the renewal of ecosystems and institutions. *Edited by* L.H. Gunderson, C.S. Holling, and S.L. Light. Columbia University Press, New York, New York. pp. 37–102.

(BCMF) British Columbia Ministry of Forests. 1999. An introductory guide to adaptive management for project leaders and participants. British Columbia Ministry of Forests, Victoria, British Columbia. 22 p.

Beanlands, G.E., and Duinker, P.N. 1984. An ecological framework for environmental impact assessment. J. Environ. Manage. **18**: 267–277.

Bone, S., and Duinker, P.N. 2002. Validation of the Varied Thrush habitat supply model for the Biodiversity Assessment Project (abstract). *In* Conference Proceedings, Advances in Forest Management: From Knowledge to Practice. *Edited by* T.S. Veeman, P.N. Duinker, B.J. MacNab, A.G. Coyne, K.M. Veeman, G.A. Binsted, and D. Korber. Sustainable Forest Management Network, Edmonton, Alberta. p. 325.

(CCFM) Canadian Council of Forest Ministers. 1995. Defining sustainable forest management: a Canadian approach to criteria and indicators. Canadian Council of Forest Ministers, Ottawa, Ontario.

(CCFM) Canadian Council of Forest Ministers. 1998. Sustainable forests: a Canadian commitment — National Forest Strategy 1998–2003. Canadian Council of Forest Ministers, Ottawa, Ontario. 42 p.

(CEPI) Confederation of European Paper Industries. 2001. Comparative matrix of forest certification schemes. Confederation of European Paper Industries, Brussels, Belgium.

(CMFN) Canadian Model Forest Network. 2002. The Canadian model forest network. Available at http://www.modelforest.net/e/home_/indexe.html [viewed 20 November 2002].

Costanza, R. 1993. Developing ecological research that is relevant for achieving sustainability. Ecol. Appl. **3**: 579–581.

Costanza, R., Wainger, L., Folke, C., and Mäler, K.-G. 1993. Modeling complex ecological economic systems: toward evolutionary, dynamic understanding of people and nature. BioScience, **43**: 545–555.

(CSA) Canadian Standards Association. 1996*a*. A sustainable forest management system: guidance document. CAN/CSA-Z808-96, environmental technology: a national standard of Canada. Canadian Standards Association, Etobicoke, Ontario. 33 p.

(CSA) Canadian Standards Association. 1996*b*. A sustainable forest management system: specifications document. CAN/CSA-Z809-96, environmental technology: a national standard of Canada. Canadian Standards Association, Etobicoke, Ontario. 12 p.

(CSA) Canadian Standards Association. 2000. About CSA International. CSA International, Toronto, Ontario. Available at http://www.csa.ca/english/about_csa/index.htm. [viewed 6 June 2001].

(CSA) Canadian Standards Association. 2002. Z809-02 sustainable forest management: requirements and guidance. Canadian Standards Association, Mississauga, Ontario. 45 p.

De Young, R., and Kaplan, S. 1988. On averting the tragedy of the commons. Environ. Manage. **12**: 273–283.

Doucette, A.M., and Duinker, P.N. 2002. An assessment of a GIS-based habitat suitability model for Canada lynx in west-central Alberta (abstract). *In* Conference Proceedings, Advances in Forest Management: From Knowledge to Practice. *Edited by* T.S. Veeman, P.N. Duinker, B.J. MacNab, A.G. Coyne, K.M. Veeman, G.A. Binsted, and D. Korber. Sustainable Forest Management Network, Edmonton, Alberta. p. 332.

Doyon, F., and Duinker, P.N. 2003. *In* Systems analysis in forest resources. *Edited by* G.J. Arthaud and T.M. Barrett. Kluwer Academic Publishers, Dordrecht, the Netherlands. Volume 7 of book series on managing forest ecosystems. pp. 207–224.

Duinker, P.N. 1987. Forecasting environmental impacts: better quantitative and wrong than qualitative and untestable! *In* Audit and evaluation in environmental assessment and management: Canadian and international experience. *Edited by* B. Sadler. Environment Canada, Ottawa, and Banff Centre School of Management, Banff, Alberta. pp. 399–407.

Duinker, P.N. 1989. Ecological effects monitoring in environmental impact assessment: what can it accomplish? Environ. Manage. **13**: 797–805.

Duinker, P.N. 1998. Public participation's promising progress: advances in forest decision-making in Canada. Commonw. For. Rev. **77**(2): 107–112.

Duinker, P.N. 2001. Criteria and indicators of sustainable forest management in Canada: progress and problems in integrating science and politics at the local level. *In* Criteria and indicators for sustainable forest management at the forest management unit level. *Edited by* A. Franc, O. Laroussinie, and T. Karjalainen. European Forest Institute, Joensuu, Finland. Proc. No. 38. pp. 7–27.

Duinker, P.N., and Baskerville, G.L. 1986. A systematic approach to forecasting in environmental impact assessment. J. Environ. Manage. **23**: 271–290.

Duinker, P.N., and Doyon, F. 2000. Biodiversity assessment in forest management planning: what are reasonable expectations? *In* Ecosystem management of forested landscapes: directions and implementation. *Edited by* R.G. d'Eon, J. Johnson, and E.A. Ferguson. UBC Press, Vancouver, British Columbia. pp. 135–147.

(ESSA) Environmental and Social Systems Analysts Ltd. 1982. Review and evaluation of adaptive environmental assessment and management. Environment Canada, Vancouver, British Columbia. 116 p.

ESSA Technologies. 2002. Adaptive management. Available at http://www.essa.com/services/am/index.htm [viewed 20 November 2002].

(FAO) Food and Agriculture Organization. 1993. The challenge of sustainable forest management: what future for the world's forests? Food and Agriculture Organization of the United Nations, Rome, Italy. 128 p.

(FEMTFSAB) Forest Ecosystem Monitoring Task Force Scientific Advisory Board. 2001. Saskatchewan forest ecosystem impacts monitoring framework — part 1, rationale and strategy (draft). Saskatchewan Department of Environment and Resource Management, Prince Albert, Saskatchewan. 33 p.

(FMF) Foothills Model Forest. no date. Foothills Model Forest: a growing understanding. Available at http://www.fmf.ab.ca/ [viewed 20 November 2002].

Frost, R. 1915. The road not taken. *In* A group of poems. The Atlantic Monthly, **116**(2): 221–224.

(FSC) Forest Stewardship Council. n.d. What is FSC? Forest Stewardship Council, Oaxaca, Mexico. Available at http://www.fscoax.org/principal.htm. [viewed 5 June 2001].

(FSC) Forest Stewardship Council. 2000. FSC principles and criteria. Forest Stewardship Council, Oaxaca, Mexico. Available at http://www.fscoax.org/principal.htm. [viewed 5 June 2001].

Goldstein, B.D. 1999. The precautionary principle and scientific research are not antithetical. Environ. Health Perspect. **107**: A594–A595 .

Haney, A., and Power, R.L. 1996. Adaptive management for sound ecosystem management. Environ. Manage. **20**: 879–886.

Hautala, K., and Duinker, P.N. 2002. Validation of the Northern Goshawk habitat suitability model for Millar Western's Biodiversity Assessment Project (abstract). *In* Conference Proceedings, Advances in Forest Management: From Knowledge to Practice. *Edited by* T.S. Veeman, P.N. Duinker, B.J. MacNab, A.G. Coyne, K.M. Veeman, G.A. Binsted, and D. Korber. Sustainable Forest Management Network, Edmonton, Alberta. p. 341.

Hilborn, R. 1992. Institutional learning and spawning channels for sockeye salmon (*Oncorhynchus nerka*). Can. J. Fish. Aquat. Sci. **49**: 1126–1136.

Holling, C.S. (*Editor*). 1978. Adaptive environmental assessment and management. John Wiley and Sons, Toronto, Ontario. 377 p.

(ISO) International Organization for Standardization. 1996. ISO 14001: 1996 environmental management systems — specification with guidance for use. International Organization for Standardization, Geneva, Switzerland. 14 p.

(ISO) International Organization for Standardization. 1997. ISO/WD 14061 — informative reference material to assist forestry organizations in the use of ISO 14001 and ISO 14004 environmental management system standards. International Organization for Standardization, Geneva, Switzerland. 80 p.

(ISO) International Organization for Standardization. 2001. ISO 9000/14000 — the basics: ISO and the environment. International Organization for Standardization, Geneva, Switzerland. Available at http://www.iso.ch/iso/en/iso9000-14000/tour/isoanden.html. [viewed 9 June 2001].

Johnson, F.A., Moore, C.T., Kendall, W.L., Dubovsky, J.A., Caithamer, D.F., Kelly, J.R., Jr., and Williams, B.K. 1997. Uncertainty and the management of mallard harvests. J. Wildl. Manage. **61**: 202–216.

Kessler, W.B. 1999. Sustainable forest management is adaptive management. *In* Science and practice: sustaining the boreal forest. *Edited by* T.S. Veeman, D.W. Smith, B.G. Purdy, F.J. Salkie, and G.A. Larkin. Sustainable Forest Management Network, Edmonton, Alberta. pp. 16–21.

KMC (KPMG Management Consulting). 1995. Manitoba's Forest Plan . . . towards ecosystem based management. KPMG Management Consulting, Winnipeg, Manitoba.

Lancia, R.A, Braun, C.E., Collopy, M.W., Dueser, R.D., Kie, J.G., Martinka, C.J., Nichols, J.D., Nudds, T.D., Porath, W.R., and Tilghman, N.G. 1996. ARM! For the future: adaptive resource management in the wildlife profession. Wildlife Soc. Bull. **24**: 436–442.

Lee, K.N. 1993. Compass and gyroscope — integrating science and politics for the environment. Island Press, Washington, D.C. 243 p.

Lee, K.N. 2001. Appraising adaptive management. *In* Biological diversity: balancing interests through adaptive collaborative management. *Edited by* L.E. Buck, C.C. Geisler, J. Schelhas, and E. Wollenberg. CRC Press, Boca Raton, Florida. pp. 3–26.

MacLean, D.A., Etheridge, P., Pelham, J., and Emrich, W. 1999. Fundy Model Forest: partners in sustainable forest management. For. Chron. **75**: 219–227.

McLain, R.J., and Lee, R.G. 1996. Adaptive management: promises and pitfalls. Environ. Manage. **20**: 437–448.

Meridian Institute. 2001. Comparative analysis of the Forest Stewardship Council and Sustainable Forestry Initiative certification programs. Meridian Institute, Washington, D.C. Available at http://www2. merid.org/comparison/ [viewed 10 June 2003]

Mihajlovich, M. 2001. Does forest certification assure sustainability? — a case study. For. Chron. **77**: 994–997.

Mladenoff, D.J., Host, G.E., Boeder, J., and Crow, T.R. 1996. LANDIS: a spatial model of forest landscape disturbance, succession, and management. *In* GIS and environmental modelling: progress and research issues. *Edited by* M.F. Goodchild, L.T. Steyart, and B.O. Parks. GIS World Books, Fort Collins, Colorado. pp.175–180.

(MMF) McGregor Model Forest. n.d. The McGregor approach to sustainable forest management. McGregor Model Forest. Prince George, British Columbia. Available at http://www.mcgregor.bc.ca/. [viewed 31 May 2001].

(MWFP) Millar Western Forest Products. 2000. Detailed forest management plan 1997–2006. Millar Western Forest Products Ltd, Whitecourt, Alberta.

Naysmith, J., LaPierre, L., Burbee, J., Duinker, P., Kimbley, G., Laishley, D., Neave, D., Savard, R., and Welsh, D. 2000. Beacons of sustainability: a framework for the future of Canada's model forests. Canadian Model Forest Network, Ottawa, Ontario. 49 p.

(NBDNRE) New Brunswick Department of Natural Resources and Energy. 2000. A vision for New Brunswick forests. New Brunswick Department of Natural Resources and Energy. Fredericton, New Brunswick. Available at http://www.gov.nb.ca/0078/vision.htm. [viewed 22 May 2001].

(NFS) Newfoundland Forest Service. 1995. Environmental preview report: proposed adaptive management process. Newfoundland Forest Service, Corner Brook, Newfoundland. 56 p.

(NSDNR) Nova Scotia Department of Natural Resources. 1997. Towards sustainable forestry — a position paper. Working Paper 1997-01. Nova Scotia Department of Natural Resources, Halifax, Nova Scotia. 41 p.

(OFPP) Ontario Forest Policy Panel. 1993. Diversity: forests, people, communities — a comprehensive forest policy framework for Ontario. Ontario Ministry of Natural Resources, Toronto, Ontario. 147 p.

(OMNR) Ontario Ministry of Natural Resources. 1994. Policy framework for sustainable forests. Ontario Ministry of Natural Resources, Toronto, Ontario.

(OMNR) Ontario Ministry of Natural Resources. 1996. Forest management planning manual for Ontario's Crown forests. Ontario Ministry of Natural Resources, Toronto, Ontario. 452 p.

(PEFCC) Pan European Forest Certification Council. 2001. Pan European forest certification framework common elements and requirements technical document. Pan European Forest Certification Council, Luxembourg. Available at www.pefc.org./technic-1.htm. [viewed 5 June 2001].

Ravetz, J.R. 1986. Usable knowledge, usable ignorance: incomplete science with policy implications. *In* Sustainable development of the biosphere. *Edited by* W.C. Clark and R.E. Munn. Cambridge University Press, New York, New York. pp. 415–431.

Sit, V., and Taylor, B. (*Editors*). 1998. Statistical methods for adaptive management studies. British Columbia Ministry of Forests, Victoria, British Columbia. 148 p.

Stankey, G.H., Bormann, B.T., Ryan, C., Shindler, B., Sturtevant, V., Clark, R.N., and Philpot, C. 2003. Adaptive management and the Northwest Forest Plan. J. For. **101**(1): 40–46.

Taylor, B., Kremsater, L., and Ellis, R. 1997. Adaptive management of forests in British Columbia. British Columbia Ministry of Forests, Victoria, British Columbia. 103 p.

(UNCED) United Nations Conference on Environment and Development. 1992. Rio Declaration on environment and development. United Nations, Stockholm, Sweden. Publ. No. E.73.II.A.14.

von Mirbach, M. 2000. A user's guide to local level indicators of sustainable forest management: experiences from the Canadian Model Forest Network. Canadian Forest Service, Ottawa, Ontario. 265 p.

Walters, C. 1986. Adaptive management of renewable resources. MacMillan Publishing Company, NewYork, New York. 374 p.

Walters, C. 1997. Challenges in adaptive management of riparian and coastal ecosystems. Conserv. Ecol. [online], 1(2): 19. Available at http://www.consecol.org/vol1/iss2/art1/ [viewed 10 June 2003].

Walters, C.J., and Holling, C.S. 1990. Large-scale management experiments and learning by doing. Ecology, 71: 2060–2068.

(WCL) Weldwood of Canada Limited. 2000. Sustainable forest management plan. Weldwood of Canada Limited, Hinton Division, Hinton, Alberta. 36 p.

(WCL) Weldwood of Canada Limited. 2002a. Hinton Division forest resources, sustainable forest management plan. Weldwood of Canada, Hinton Division, Hinton, Alberta. 25 p.

(WCL) Weldwood of Canada Limited. 2002b. 2001 sustainable forest management stewardship report. Forest Resources Department, Weldwood of Canada Limited, Hinton Division, Hinton, Alberta. 68 p.

(WHC) Wildlife Habitat Canada. 1997. Forest biodiversity program, initial evaluation process: helping forest companies conserve biodiversity. Wildlife Habitat Canada, Ottawa, Ontario. 44 p.

Williams, B.K., and Johnson, F.A. 1995. Adaptive management and the regulation of waterfowl harvests. Wildl. Soc. Bull. 23: 430–436.

Williams, B.K., Johnson, F.A., and K. Wilkins. 1996. Uncertainty and the adaptive management of waterfowl harvests. J. Wildl. Manage. 60: 223–232.

Wood, P. 2000. A comparative analysis of selected international forestry certification schemes. British Columbia Ministry of Employment and Investment, Victoria, British Columbia. Available at http://www.for.gov.bc.ca/het/certification/WoodReportOct00.PDF [viewed 10 June 2003].

Chapter 22

Implementing sustainable forest management: some case studies

Daryll Hebert, Brian Harvey, Shawn Wasel, Elston H. Dzus, Margaret Donnelly, Jacques Robert, and Fiona Hamersley Chambers

Introduction

Throughout the past two or three decades, forest management and the forest products industry have been severely criticized for their focus on fibre production under sustained-yield management at the expense of non-timber forest values. The change to sustainable forest management (SFM) has generally been occurring through adversarial processes, using constraint management and prescriptive policy and regulations. As a result, the change has been highly variable across Canada, based on provincial forest management differences as well as social and economic considerations.

Within the last 10 years, SFM has begun to be driven by science-based knowledge and broader social perspectives through institutions and programs such as Canada's Sustainable Forest Management Network (SFMN) and Model Forest Program (MFP). As the extension and implementation of the environmental and social sciences continues, SFM will begin to incorporate a more consistent, broad-based approach to forest management.

D. Hebert.[1] Encompass Strategic Resources Inc., R.R. #2, 599 Highway 21 South, Creston, British Columbia V0B 1G2, Canada.
B. Harvey. Université du Québec en Abitibi-Témiscamingue, Rouyn-Noranda, Quebec, Canada.
S. Wasel and E.H. Dzus. Alberta-Pacific Forest Industries Inc., Boyle, Alberta, Canada.
M. Donnelly. Louisiana-Pacific Canada Ltd., Swan River, Manitoba, Canada.
J. Robert. Service canadien des forêts, Programme des forêts modèles, Rimouski, Quebec, Canada.
F. Hamersley Chambers. School of Environmental Studies, University of Victoria, Victoria, British Columbia, Canada.
[1]Author for correspondence. Telephone: 250-428-3092.

Correct citation: Hebert, D., Harvey, B., Wasel, S., Dzus, E.H., Donnelly, M., Robert, J., and Hamersley Chambers, F. 2003. Implementing sustainable forest management: some case studies. Chapter 22. *In* Towards Sustainable Management of the Boreal Forest. *Edited by* P.J. Burton, C. Messier, D.W. Smith, and W.L. Adamowicz. NRC Research Press, Ottawa, Ontario, Canada. pp. 893–952.

The case studies described here demonstrate the variability surrounding the change from the traditional approach of sustained-yield timber management to a broader strategy of sustainable management of all forest values. Although the range of approaches to implementation indicates both a socio-economic basis and an ecological basis for the change, there are some commonalities. In many cases, the ecological knowledge for SFM is currently available, but true change depends on creative social solutions and innovative leadership. In other cases, long-held beliefs about the nature of boreal landscapes and boreal forest ecology have simply been incorrect, so the science has had to catch up first. In all cases, there is evidence that ecological research, technological developments, and novel socio-economic perspectives are mutually dependent elements that build on each other and must be tried and refined in the course of forest management.

Many of the concepts of sustained timber yield remain a necessary condition of basic forest management within the SFM framework, but they no longer can be considered sufficient to address the desired range of forest values. Although attempts have been made to develop enlightened constraint management and prescriptive policy, these approaches generally have been unsuccessful in achieving SFM so long as fibre production remained the bottom line, the criterion against which all management decisions are to be judged. Rather, the last decade has seen a growing influence of scientists, environmentalists, market demands, and community activists in convincing forest managers to respect non-timber forest values and the health of forest ecosystems and forest communities. But can recognition of this broad set of values prompt real reform on the ground? Can these good intentions actually be implemented, in a real world governed by a market economy and widespread fiscal constraints?

The following case studies have been selected from across Canada (Fig. 22.1) to illustrate the feasibility of implementing many of the innovations advocated in earlier

Fig. 22.1. Locations of the case studies featured in this chapter.

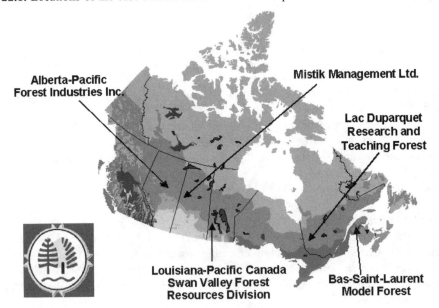

chapters (Box 22.1). The Lac Duparquet Research and Teaching Forest illustrates an approach to maintaining semi-natural stand structures and composition in a largely experimental setting in northwestern Quebec. Alberta-Pacific Forest Industries has taken the natural disturbance based management model and applied it to a large forest management area in northeastern Alberta. Louisiana-Pacific (Swan Valley Forest Resources Division, in west-central Manitoba) has implemented progressive systems of in-house training and monitoring to ensure sustainable forest management. The Bas-Saint-Laurent Model Forest in eastern Quebec has developed "tenant forestry", a novel approach to forest tenure and stewardship. Finally, Mistik Management Ltd. has incorporated a large degree of local involvement and scientific review in its adaptive management for multiple forest resources in northwestern Saskatchewan. Collectively, these five case studies demonstrate leadership and innovation on the road towards sustainable forest management of the boreal forest, implemented in a variety of institutional and political settings.

Box 22.1. Challenges for selecting case studies of sustainable forest management:

- To find working examples that illustrate the feasibility of implementing the principles and components of sustainable forest management (SFM);
- To demonstrate how SFM initiatives have sprung from diverse origins, to meet different needs and expectations with regard to improving the protection of nature, minimizing the impacts of industrial forestry, or supporting forest-dependent communities; and
- To showcase leadership, innovation, and progress in the quest for sustainable management of all forest values in the world's northern forests.

The Lac Duparquet Forest: balancing research and natural disturbance based management

Introduction

There may be as many variations of forest ecosystem management as there are combinations of forest ecosystems and communities. Every approach is influenced by the natural characteristics of the regional landscape, the people and communities that occupy the territory, industrial interests, values that shape how resources may be used, and the particular legal and regulatory framework in the jurisdiction. The Lac Duparquet Research and Teaching Forest (or the *Forêt d'enseignement et de recherche du lac Duparquet*, FERLD) presents a combination of natural features and human circumstance that has led to a management approach characterized by a strong research focus, a natural disturbance based approach to forest management, university–industry partnerships, and the development of community-level participation in management planning. This case study presents the particulars that have led to this model.

The region

The Lac Duparquet Forest (48°30′N, 79°25′W) is located in the southern part of the Great Clay Belt, a large physiographic region that covers much of northwestern Quebec and northeastern Ontario. In the northern part of this region, the Clay Belt is characterized by a vast plain of poorly drained organic and clay deposits and extensive black spruce (*Picea mariana*) forests. To the south where Lac Duparquet is situated, a gentle topography has produced a greater variety of soil conditions and a forest mosaic characteristic of the eastern boreal mixedwood. All common eastern boreal forest tree species are found here: jack pine (*Pinus banksiana*), white spruce (*Picea glauca*) and black spruce, balsam fir (*Abies balsamea*), white birch (*Betula papyrifera*), trembling aspen (*Populus tremuloides*), balsam poplar (*Populus balsamifera*), and tamarack or eastern larch (*Larix laricina*). Red pine (*Pinus resinosa*) persists close to the northern limit of its distribution, on islands and steep shorelines of Lac Duparquet (Bergeron and Brisson 1990), and white pine (*Pinus strobus*) can be found on summits of rocky escarpments in the interior of the forest, as well as in association with lakeside red pine populations. Eastern white cedar (*Thuja occidentalis*), a species characteristic of old-growth forests, is largely found on islands, lake shores, and in stands that have not burned for more than 150 years. Stands of black ash (*Fraxinus nigra*) are located on alluvial levees and plains on the lake and surrounding rivers (Tardif and Bergeron 1992).

Prior to the official creation of the Lac Duparquet Research and Teaching Forest in 1996, the area was part of a forest management agreement area (or *aire commune*) allocated to two major companies, Tembec Forest Products Group and Norbord (Nexfor) Industries, as well as another minor beneficiary. Tembec was allocated the softwood volume rights for its two sawmills in La Sarre and Taschereau, while Norbord had the volume rights to the intolerant hardwoods for its oriented strandboard (OSB) mill in La Sarre. In order to permit the creation of the FERLD, the two companies had to relinquish their cutting rights to this area of 8000 ha located within 60 km of their mills.

Creating the Lac Duparquet Research and Teaching Forest

Quebec's Forest Law contains provisions for the creation of research and teaching forests, a form of tenure that allows public forest land to be managed under contract to universities, colleges, or secondary schools that have vocational training in forestry. There are currently 17 research and teaching forests in the province, each one reflecting the teaching and (or) research activities of its associated institution. Secondary schools and colleges have generally used their forests for training of forest workers and technicians. The Montmorency Forest, for example, managed by Université Laval, has a long history of forest research and has supported the education of two generations of forest engineers in the province. The proposal to create the research and teaching forest in the Lac Duparquet area emphasized the importance of the fundamental research program that had been underway there for 10 years, and the significance of the area for long-term ecological monitoring. Furthermore, building on studies of forest disturbance history, stand dynamics, and soil processes through succession, a management approach was proposed to demonstrate how basic knowledge of ecosystem dynamics could be applied to management planning and implementation.

Management contract

The management contract between the Quebec Government and the universities provides the legal framework for managing the research forest. Similar in many regards to other forest management agreements (see Chap. 1), it defines the rights and obligations of the two parties and has a duration of 25 years, renewable every 5 years. The university is obligated to manage the forest sustainably, prepare 25-year, 5-year, and annual plans as well as annual management reports, and to use the area for fundamental and applied research as well as for teaching and technical transfer. The government does not charge stumpage for wood harvested, which may be considered a contribution to the research activities undertaken in the forest. However, unlike forest companies operating on public land, the University is not eligible for silvicultural credits, but may apply for funding for silvicultural and management activities (including road building) through a separate budget. Revenues from the sale of timber to local sawmills and OSB mill are used by the University to pay for a forest technician and for planning, silviculture, and other management costs. Given the current limited road access and high costs of establishing relatively small-scale silvicultural experiments, infrastructure expenses are high. However, once a minimal road system is built, the intent is to eventually generate enough revenue to establish a research fund for graduate students.

Management planning: integration of forest-level objectives and stand-level silvicultural prescriptions

A management committee, made up of the research forest director, a representative of the GREFi (Groupe de recherche en écologie forestière interuniversitaire) at UQÀM (Université de Québec à Montréal) and the forest research unit at UQAT (Université de Québec en Abitibi-Teméscamingue), as well as foresters from both Tembec and Norbord, is responsible for management of the FERLD. While day-to-day management is assumed by the director assisted by a forest technician, the participation of industrial partners on the committee greatly facilitates collaboration on research and management activities within the forest as well as on the companies' broader management areas.

The first management plan for the Lac Duparquet Forest was presented to the Quebec *Ministère des Ressources naturelles, Faune et Parcs* (QMRNFP) in 1998. The document provides a biophysical and socio-economic portrait of the area, presents the perceived constraints to (and potential for) managing the forest, as well as the management framework and objectives. The management plan was intended to demonstrate how stand- and forest-level ecological knowledge could provide the basis for a particular approach to forest ecosystem management. While recognizing that the scale of disturbance in the eastern boreal could not possibly be accommodated in a management area of 80 km², the approach emphasizes using information on the regional natural disturbance regime for setting forest-level objectives concerning forest cover and age structure. In other words, while we can't possibly plan cutblocks comparable in size and pattern to major fire events (see Chap. 8), we can use information on the regional fire cycle and forest age structure and composition generated under different fire cycles to characterize these regional parameters and apply them — notwithstanding the question of scale — to an ecological microcosm of the region. Along with this forest-level planning aspect, another key element of the approach used in the Lac Duparquet Forest is the integration of natural stand dynamics in developing silvicultural approaches that exploit these dynamics (e.g., see Box 9.9), and are aimed at attaining forest-level objectives.

The management approach developed for the Lac Duparquet forest is fully developed in Harvey et al. (2002), and is also the basis for many of the ideas explored in Chap. 11. We start from the conceptual framework of the natural disturbance paradigm and its associated working hypothesis or premise: that is, that forest management practices which affect ecosystem patterns and processes in ways similar to the natural disturbance regime should have less of an impact on biodiversity and forest productivity than practices that are situated outside the historic range of variability of natural disturbance (Landres et al. 1999). This concept, and the comparison of forest harvesting activities with natural disturbances, is further explored in Chap. 9.

The following outline briefly describes the steps involved in developing and implementing the FERLD management plan.

(1) Characterization of the historic natural disturbance regime

Research conducted by Y. Bergeron, S. Gauthier, and their collaborators included fire history reconstruction work, fire mapping, and detailed study of fire severity and burn patterns (see Dansereau and Bergeron 1993; Bergeron and Leduc 1998; Gauthier et al. 2000). The natural fire regime of the region is characterized by large crown fires, although recent studies have shown that considerable variation in fire severity generally results in areas of low fire severity and residual unburned forest islands within the perimeter of large fires as well (Bergeron et al. 2002; Kafka et al. 2001). Whereas the length of the fire cycle in the region has been increasing over the last two centuries, mean forest age (time since last fire) has remained relatively constant at approximately 140 years. The strategy of fixing a fire cycle target as the basis of forest management has been questioned, given evidence that climate and regional disturbance regimes appear to be constantly changing, and the tendency for the area burned annually in a region to fluctuate dramatically. Nonetheless, we consider that use of the mean forest age of the natural forest mosaic has its merits as a benchmark on which to target forest age structure because of its relative stability through time, and the fact that it integrates some of the historic variability of the regional disturbance regime.

(2) Using forest ecosystem classification and fire cycle modelling

Forest ecosystem classification has been used in several manners. First, regional-scale classification provided basic information concerning the relative proportion of ecological site types found in the area. This information, combined with studies of site-specific stand dynamics and fire cycle modelling, provides a means of exploring how changes in fire cycle or differences in site type composition of different landscapes influence forest cover (Gauthier et al. 1996). Using a fire cycle of 140 years and the proportion of dominant site types found in the Lac Duparquet Forest, this method was used to set forest age structure (or cohort structure) targets for the research forest (see Bergeron et al. 1999), as well as composition targets. Fire history maps overlaid on 1:20 000 scale forest ecosystem maps (which include forest cover and site type attributes) were also used to draw relationships between dominant forest types and combinations of fire year and site type.

(3) Associating forest types with cohorts

Stand dynamics for the dominant site types of the area were characterized from the scientific literature and from forest ecosystem classification work done by the QMRNFP. Stand development was partitioned into three development stages, or stand-type "cohorts", based on stand structure and composition for dominant site types (Bergeron et al. 1999). Combined with the results of fire cycle modelling, this information was used to set the forest-level objectives for maintenance of a dozen general forest types over the landscape. "First-cohort" forest types tend to have an even-aged structure and are composed of pioneer species such as jack pine, trembling aspen, balsam poplar, white birch, and black spruce, which tend to dominate in pure or mixed compositions over the first 100 years following fire (e.g., Fig. 22.2). White spruce and balsam fir that have either regenerated immediately after fire or were thought to have gradually seeded in following stand establishment (but see Chap. 8) are more abundant in the understory than in the canopy of the first-cohort. In the "second-cohort", forest types generally consist of two components: surviving canopy stems from the first-cohort and either (1) tolerant softwoods that were present in the understory in the first-cohort, or (2) other softwood or hardwood stems that have been recruited into the understory and in gaps. Consequently, the second-cohort represents a mid-successional phase (typically domi-

Fig. 22.2. This forest mosaic of even-aged stands of jack pine (*Pinus banksiana*) and trembling aspen (*Populus tremuloides*) in the active management zone of Lac Duparquet Forest was created by a fire in 1923. Such stands are typical of "first-cohort" forest types.

Photo by René Beaudouin

nating from 75 to 175 years after disturbance) of mixed species and more irregularly structured stands. Differentiation between second-cohort and "third-cohort" stands (conceptually, the passage from mid- to late-successional phases) is more nuanced. In these late-successional third-cohort stands, first-cohort canopy trees generally have disappeared through natural mortality; late-successional species such as balsam fir and eastern white cedar dominate, and an irregular stand structure is found. We assume that wildfire and clearcut harvesting have the effect of reverting most forest types of the three cohorts back to an even-aged, first-cohort character (see Fig. 9.17 in Box 9.9).

(4) Development of a forest-level management model

This approach recognizes that it is virtually impossible to maintain or create certain forest types (notably those associated with second- and third-cohorts) without exploiting natural stand dynamics. Consequently, the system relies on allowing a certain amount of flow among forest types: that is, generating stands of the second- and third-cohorts by using partial cutting systems in first and second cohort stands and, conversely, generating first-cohort stands by using clear-cutting or other even-aged silvicultural systems in stands of any of the three cohorts. In order to track age structure changes of dominant forest types and fluxes between forest types of different cohorts, we developed an area-based forest simulation model that controls fluxes within and between forest type matrices. The non-spatial, forest-level model consists of a series of transition matrices (one for each forest type) that track changes in age-class structure (% area per age class) in 5-year increments over a 150-year simulation period. The matrices indicate at what period harvesting and recruitment should occur, and over how much area, in order to move from the present age structure and area to the desired conditions for each forest type.

(5) Using silvicultural treatments to attain landscape-level objectives

A major challenge to this approach is, of course, determining whether the forest-level age structure and composition targets designed from a conceptual model can be attained on a relatively small area like the Lac Duparquet Forest. Because very little experience exists in partial cutting in the region and the outcomes of silvicultural treatments are still uncertain, there was a recognized need to design and implement harvesting experiments to test the feasibility of the proposed system. An adaptive management research project called SAFE (*Sylviculture et aménagement forestier écosystémiques*) was initiated in 1998, with Tembec and Norbord Industries, and silvicultural trials have been established in stands associated with the three cohorts (see Box 9.9). Moreover, while the experiments undertaken in the research forest are done on a relatively small scale, larger operational experiments, including use of the cohort approach to forest-level management planning, are now occurring on a portion of Tembec's and Norbord's forest management areas.

Conservation and environmental monitoring

Fire history and mapping work around Lac Duparquet revealed that the area had been touched by a number of large fires in the last 250 years, notably in 1760 and 1923, but also by eight smaller fires, ranging in size from 122 to 555 ha, all of which occurred in the eastern portion of the research forest. Because of the scientific interest in this landscape composed of forests originating from fires dating between 1716 and 1944, an area of about 2000 ha of the research forest was zoned for conservation at the time of its creation. The area serves as a natural benchmark for environmental monitoring and measuring impacts of forest management interventions and has been used extensively for fundamental studies in natural forest dynamics and ecosystem processes (Harvey 1999). In effect, the management approach used in the Lac Duparquet Forest is based, to a large extent, on research work that has been done in this area. As a result of studies in the conservation zone and ongoing monitoring, the area is included in Environment Canada's Environmental Monitoring and Assessment Network.

Testing the triad zoning approach

As in other jurisdictions, forest policy in Quebec has traditionally been aimed at facilitating industrial economic development (see Chap. 1). In most of the province, environmental protection (especially of wildlife, soil, and water resources) has been accomplished through regulations, and has focussed on mitigation of impacts of forest practices at the local level. However, because almost all commercial forest on public land in Quebec has already been allocated for industrial wood supply, the Province has been relatively slow in creating a network of protected reserves. The pressure to establish reserves has fostered considerable debate among industry, government, environmental groups and the research community. But a consensus appears to be emerging that the creation of a network of protected areas will have to be accompanied by the establishment of areas for the production of intensively managed high-yield forests. The term "triad" was coined by Seymour and Hunter (1992) to describe this approach, and variations have been proposed elsewhere (see Chaps. 11–13). For example, Burton (1995) suggests a three-part land-use pattern in which (1) the majority of a forest region is managed according to "New Forestry" or ecosystem management principles, (2) ecologically significant areas and rare or critical habitats are put into reserves, and (3) certain areas located on productive sites close to mills are intensively managed, in part to compensate for timber volumes foregone as a result of the creation of reserves.

In an effort to use the Lac Duparquet Forest as a demonstration of the triad concept, the original management zone was partitioned to include a zone of about 400 ha for intensive forest management. The area is dominated by rich, well-drained clay soils, and is located in the western part of the research forest, farthest from the conservation zone. It includes plantations that were established in the early 1990s prior to the establishment of the research forest, and that have been tended and recently fill-planted, as well as experimental plantations that will be used to evaluate different provenances of white spruce and Norway spruce (*Picea abies*), hybrid poplar, and mixed plantations. While the portion occupied by each of the three zones in the Lac Duparquet Forest (25% conservation, 70% ecosystem management, and 5% intensive management) may not be applicable everywhere, if it can be demonstrated that conserving 25% of the land base can be compensated for by intensively managing only 5% of the forest, it will provide a good argument for both more conservation and more intensive management in the eastern boreal forest. Although the demonstration has yet to be made, our calculation is relatively simple. Approximately half of the conservation zone is occupied by poorly drained organic sites and thin tills and rocky outcrops, and does not produce commercial timber anyway. Because only about 1000 ha of the conservation zone is productive forest land (or 12.5% of the entire research forest), the objective is to maximize wood production on the intensively managed zone and assure full stocking in all plantations in the entire management zone to make up for the volume losses. The assumptions of this strategy have yet to be validated however.

Fitting other users into the picture

Since its creation, the Lac Duparquet Forest has unquestionably been managed primarily to favour conditions conducive to the research and teaching activities of the principal scientists and their collaborators. In all, almost 25 different universities and research institutions and over 100 scientific researchers and graduate students have undertaken research in the FERLD, resulting in over 100 papers published in scientific journals and over 40 masters and Ph.D. theses. However, these figures may not be terribly important to people living outside of academic circles, including those who live in the Lac Duparquet area, and this lead to two important initiatives.

The first initiative was the creation of a popular lecture series, *Les Soirées de la Forêt du Lac Duparquet* (or Lake Duparquet Forest Evenings), held three or four times every summer in the local church basement, starting in 1997. Open to the general public, the *soirées* are intended to provide some stimulation for students and researchers in the midst of their field season while generating an awareness among the local population of research activities (and findings) undertaken in the region. The mixing of "the locals" with students and researchers, most of whom are not from the region, has helped create a climate of familiarity that did not exist several years ago and has had the effect of demystifying some aspects of forest ecology and research in natural sciences for local residents. At the same time, several local "old-timers" have turned out to be good sources of information concerning past disturbances and logging history in the area. Generally, the research forest is looked upon positively for its economic spin-offs, seasonal employment, and the visibility it provides the region; it has become part of the community.

The second initiative arose from the realization that the Lac Duparquet Forest could not be managed solely to facilitate research and teaching activities, demonstrate a natural disturbance based approach to forest ecosystem management, and provide a secondary wood source for local mills. People live in the area, many of them have hunting camps in the research forest, several have trap lines, others have small businesses or projects they would like to start up, and still others would simply like to see the forest remain untouched in its relatively natural state. In the fourth year following creation of the research forest, and still lacking a formal participatory structure for involving various stakeholders in general planning decisions, a pilot project was established in 2000, involving representatives of these interests as well as those of the universities, the forest products industry, and municipalities. The *Groupe d'analyse multicritère de l'aménagement de la Forêt du Lac Duparquet* (or Multicriteria Analysis of the Management of the Lake Duparquet Forest Group) has the mandate to identify values, general and specific concerns, and potential conflicts that could arise in the management of the research forest, and to provide input into general and tactical (5-year) management planning.

Facilitated by a sociology professor from the *Chaire Desjardins en développement des petites collectivités* at UQAT and two professors in multi-criteria analysis from UQAT and UQÀM, meetings of the group in the first year rapidly led to the identification of the relative importance of several general themes (research, wood supply, access, communication, local economic benefits, etc.) and to an initial ranking of a list of roughly 60 statements related to these themes. Fortunately for the universities involved, all parties appear to recognize the importance of the research and teaching activities undertaken in the forest. The next phase of the project will involve the development of different management scenarios, based on the findings in the first phase, in an exercise of optimization based on multiple criteria analysis. Whether the results will lead to the perfect management plan is doubtful; however, it is clear that the benefits of this approach result from the process itself as much as in the end result.

Conclusion

Whether an approach used to manage a research forest of 8000 ha can be applied to commercial forest management areas covering hundreds of square kilometres is questionable. On the one hand, if only because of the question of scale, management is generally simpler and certain regional-level concerns are less relevant. On the other hand, the approach of setting forest-level age structure and forest composition objectives based on natural disturbance history and fire cycle modelling is probably more applicable on large forest management areas where spatial aspects of disturbance can be better taken into consideration and applied to planning activities (see Chap. 11; Bergeron et al. 2002). At the simplest level, it is clear that forest harvesting and silviculture can be used to maintain semi-natural forest composition and structure. It is also likely, however, that there are potentially higher risks associated with multiple stand entries, partial cutting, and longer intervals between final harvests. This case study also illustrates the challenges involved in translating a simple conceptual model into a more quantitative silvicultural framework. What is particular about the Lac Duparquet Forest is the concentration and the degree of integration of fundamental and applied research, and the fact that these activities have been instrumental in the improvement of forest management and practices, not only in the research forest itself but also in surrounding forest management areas.

Alberta-Pacific Forest Industries Inc.

Background: some history and the growth of corporate commitment

The evolution of Alberta-Pacific Forest Industries Inc. (Alberta-Pacific) began in the late 1980s as an application to the Alberta government for a Forest Management Agreement (FMA) in northeastern Alberta (54°35′–57°47′N, 110°0′–115°10′W). The purpose of this FMA application was to ensure a supply of hardwood (primarily trembling aspen) fibre for a new pulp mill to be located near the town of Athabasca, Alberta. In August of 1991, an FMA was awarded to the joint venture parties of Crestbrook Forest Industries Ltd., Mitsubishi Corporation, and Kanzaki Paper Canada Inc., collectively known as Alberta-Pacific Forest Industries Inc. From about 1988 to 1990, Alberta-Pacific was engaged in public hearings that were extremely intense, with opposition based on concerns about the environmental impacts of the pulp mill and a large-scale forest operation in Alberta's northeastern boreal forest. The hearings were restricted to the pulp mill and did not include the woodlands operations (Pratt and Urquhart 1994). However, strict environmental requirements were applied to the mill operation, and based on the commitment to employ and involve local and Aboriginal people in the business, the project was approved on 20 December 1990.

Following the public hearings and the initiation of construction, the Alberta-Pacific woodlands management program began to take form. The company had a non-traditional, non-union, relatively flat management structure. Large-scale responsibility was passed on to area teams who helped set objectives, standards, and high-level team cooperation. Overall, there were strong proactive initiatives in all phases of the operation. Sustainable forest management began to develop through management planning, education, and research programs initiated by key personnel. The strong corporate commitment persuaded the Alberta government to allow Alberta-Pacific additional freedom to initiate a transition from traditional sustained-yield forest management and planning to ecologically based SFM. At approximately the same time, the U.S. Forest Service was exploring natural disturbance dynamics as a concept for guiding forest management in the National Forests of the United States (Bourgeron and Jensen 1993). The need for a sound ecologically based forest management program allowed the integration of boreal forest fire dynamics with a program of "ecosystem management" as articulated in the U.S.A. (Kessler et al. 1992; Grumbine 1994). Indeed, it was believed that the natural features and disturbance regime of northeastern Alberta lent themselves particularly well to management according to a disturbance-approximation model. Thus began the first operational implementation of industrial forest ecosystem management in Alberta.

Natural features

The Alberta-Pacific FMA area encompasses a gross area of about 6 900 000 ha (~5 700 000 ha net area; see Table 1.3). East to west, it spans roughly 300 km from the Saskatchewan border to Lesser Slave Lake. South to north, the FMA area extends from the agricultural area around the towns of Athabasca and Lac La Biche, north to the Birch Mountains, a distance of approximately 340 km.

The FMA area lies within the Mixedwood Section of Rowe's (1972) Boreal Forest Region, or the Boreal Ecoprovince (Strong and Leggat 1992). Based on the more recent reclassification of Alberta's ecological zones according to natural regions and subregions (ANHIC 2002), 98% of the Alberta-Pacific FMA area lies within the Boreal Forest Natural Region. Natural subregions represented are the Central Mixedwood (~83%) and the Boreal Highlands (~15%), with possible inclusions (<1%) of Subarctic (Birch Mountains) and Dry Mixedwood (Wandering River – Lac La Biche area) subregions.

The remainder of the FMA area is represented by the Lower Foothills Subregion (<2%), a subdivision of the Foothills Natural Region (ANHIC 2002). The area is characterized by a continental boreal climate where winters are typically long and cold (mean winter temperature of –10.5°C), while summers are short and cool (mean summer temperature of 13.8°C). Approximately 36% of the FMA area is considered a merchantable forest land base (i.e., with trees suitable for harvest now or in the future). The remainder consists of non-merchantable lands, such as wetlands (including bogs, fens, swamps, and marshes), and protected areas.

Wildfire has played an important role in determining the vegetative cover within the subregion. Mesic upland sites are typically vegetated by trembling aspen and balsam poplar stands interspersed and (or) mixed with pockets of white spruce. White spruce and balsam fir are the climax species; however, balsam fir is not well represented because fires restart the succession process at frequent intervals (see Chap. 8). Subxeric sites are often dominated by aspen stands. Xeric sites are represented by various densities of jack pine, sometimes with a minor component of aspen. The extensive lowland areas are dominated by black spruce and tamarack. Widespread fires in the early 1900s have resulted in large expanses of mature even-aged stands (like those portrayed in Fig. 22.2).

The need for alternative approaches

The Alberta forest products industry has been required by government to follow conventional sustained-yield forest management practices. This form of management focuses on maximizing the long-term timber growth and fibre extraction from forest land to generate sustainable economic benefits. Conventional forest management utilizes estimates of current timber stocks and future tree growth rates to calculate sustainable levels of timber harvest. Optimum timber extraction assumes harvesting at rotation ages that maximize the long-term fibre production from an area. If yield assumptions or conditions change as a result of such things as forest fires or reforestation rates, the annual harvest rate is adjusted to meet these new conditions. This is a traditional European style of forest management that has long been accepted as a model for forest management worldwide (Chap. 1).

Harvesting practices have been quite uniform under this system. Until recently, timber harvesting in Alberta focused on coniferous species, primarily white spruce, jack pine, and lodgepole pine (*Pinus contorta*). Generally, a small-opening (square or rectangular, averaging 20 ha in size) clearcut system was adopted, largely to enhance natural conifer regeneration, and was purported to encourage the production of moose (*Alces alces*) and deer (*Odocoileus hemionus* and *O. virginianus*). Operators were generally required to harvest all of the available merchantable wood from a planning area in two or three harvest passes, usually 20 years apart, to allow reforestation of adjacent blocks. Recently harvested sites were then scarified, planted, and stand-tended, with the objective of encouraging rapid establishment of conifer crop trees; aerial seeding and "leave-for-natural" (LFN) regeneration methods have also been employed. In the late 1980s, the Alberta government allocated rights to deciduous species and harvesting began using a silivicultural system similar to that used historically for conifers.

Forests managed in a classical sustained-yield fashion will, over time, tend to develop into a mosaic of even-aged stands in which there are no stands older than the optimum rotation ages, except in leave areas. The landscape pattern of the productive forest sites will be changed towards block-like stands of similar areas. In recent years, it has become more clearly understood that such changes to stand structure, the range of forest age classes, and landscape patterns are detrimental to biodiversity and ecological processes (Franklin et al. 1997). Because of these factors, conventional forest management is being replaced by ecologically based forest management, even in countries such as Sweden and Germany that have led the world in the practice of classical sustained-yield forestry for more than a century (Angelstam and Pettersson 1997).

During the last 40 years, increasing concern over the loss of coastal temperate old-growth ecosystems, and in particular the use of clearcutting, led to intense scrutiny of the conventional forest and wildlife management strategies used throughout North America (Chap. 1). Clearcutting became the focal point of the controversy surrounding traditional forest management. As understanding of forest ecology increased, forest managers began to incorporate more ecological concepts into sustained-yield management processes. Frustrated with the slow rate of change and lack of a suitable ecological model for industrial forest management, protected area strategies were seen as the only solution to sustain biodiversity, wilderness, and recreation objectives.

Most past endeavours to integrate wildlife biology with forest management have attempted to understand the ecology or habitat relationships of individual species and then modify management strategies accordingly. This species-specific approach has reduced and compartmentalized our ability to understand the ecosystem and set integrated objectives. Many single-species objectives conflict with those of other species, thus encouraging wildlife management decisions based on human value judgements. Making the shift from sustained-yield timber or ungulate production to SFM requires the acceptance of a logical and more inclusive ecological management model. The forest management system implemented by Alberta-Pacific, and summarized here, is intended to develop and implement SFM, incorporating ecological principles with social and economic requirements.

The transition to an ecological forest management program, based on natural disturbance and requiring substantial changes from traditional sustained-yield programs, is ongoing. Within 6 months of the 1991 agreement, Alberta-Pacific and the Alberta Minister of Environmental Protection were required to develop a set of "Operating Ground Rules". This initial document was completed and approved in 1993. It basically followed the generic operating ground rules for forest operations in Alberta, which contained traditional timber management guidelines and expressed concerns about ungulate habitat and streamside protection using buffer strips. Alberta-Pacific's initial Preliminary Forest Management Plan (PFMP), prepared in 1993 and 1994, also followed the traditional Detailed Forest Management Plan (DFMP) format, describing the strategies and methods with which the FMA area would be managed. It took several years of iterative planning and communication between industry planners and public servants (charged with a command-and-control mandate in a sustained-yield timber management context) to gradually reconcile recent developments in landscape ecology, biodiversity conservation, and ecosystem management with traditional forest planning and policy.

An even more difficult transition for regulatory officials was centred on the change in emphasis from the original Operating Ground Rules and DFMP to the emphasis found in subsequent plans. The initial plans were fibre based, with some introductory material describing natural disturbance based forest ecosystem management. Subsequent plans emphasized this "new model" which factors in fibre extraction as part of an overall forest ecosystem management program.

Ecologically sustainable forest management

The principles

One objective of forest ecosystem management is to reduce the risk to forest ecosystem composition, structure, and function by developing strategies that are designed to maintain biodiversity. Developing a new approach to land management also requires that many agencies (including political and regulatory officials), groups, and individuals understand these new concepts. To convince government agencies, corporations, the scientific community, environmental non-governmental organizations (ENGOs), and the general public of the need for new approaches, impartial information must be collected to demonstrate that all components of the ecosystem (plant, animal, and other) are being considered and maintained.

In setting management goals and acceptable limits for various ecosystem attributes, ecosystem managers must identify the range of natural variability in those attributes, evaluate those attributes over time, and test the effectiveness of both coarse- and fine-filter approaches to protecting biodiversity. The evolution of a new forest management model must also address standard policy statements on forest inventory requirements, the allocation of timber rights by forest cover types, and flexibility in utilization and regeneration standards. It is Alberta-Pacific's goal to implement effective forest ecosystem management that will:

- introduce, and extend as much as possible (e.g., given social and economic constraints), the concept of range of natural (historic) variability as a gauge for acceptable, sustainable conditions (Landres et al. 1999);
- interpret a suitable and practical range of variability using the size, frequency of occurrence, and intensity of historic fires (Swanson et al. 1993);
- utilize a coarse-filter approach (Hunter 1991) which incorporates the interrelationship of social and ecological requirements, with due consideration of economic and contractual agreements;
- manage the boreal mixedwood forest using a multi-scale approach, at regional, landscape, and stand levels;
- examine harvest levels in relation to both fibre requirements and ecological requirements (considering the contribution of protected and deferred lands [exclusions] as well); and
- develop an SFM program that is evaluated by an appropriate monitoring system.

One of the goals of pursuing ecosystem management principles is achieving long-term sustainability for Alberta-Pacific, and significant economic, social, political, and ecological benefits for Albertans. As markets tighten and environmental pressures for SFM intensify, international recognition (i.e., of reputable research and through certification) should allow companies to become more competitive in national and international markets.

The program

The boreal mixedwood forest of northeastern Alberta is a fire-dominated forest (Chap. 8) to which plants and animals are adapted. Forest ecosystem management, using a coarse-filter approach and an approximation of natural disturbance regimes, was examined at three levels: that of the entire FMA area; that of the landscape (pattern); and that of the stand (structural).

At the FMA level, total available timber volume and its associated annual allowable cut (AAC) were identified, after net-downs to reflect plans to buffer rivers (particularly the Athabasca River) and streams, and to retain some structure within all managed stands. This process incorporated many of the approaches advocated in Chap. 11. At the landscape level, two- and three-pass forest harvest systems are being transitioned to single-pass approaches designed to more closely approximate landscape patterns left by wildfires (Cumming 2001). From a research perspective, patterns of the landscape describing core areas and fragmentation levels were examined using edge-to-area ratios and adjacency patterns through geographic information systems (GIS). Research began to address whether animal species abundance patterns across a landscape are primarily explained by stand-level or landscape-level composition and configuration (Hannon 1999).

At the stand level, standing snags and fallen trees ("dead-and-down" material) were left as residual woody structure. Spatial patterns within cutblocks and amounts of standing and downed material were adjusted as biological requirements were identified (e.g., Stelfox 1995; Song 2002). Initially, operators left individual trees dispersed across the cutblocks, with few clumps (Fig. 22.3a). As research progressed (Hannon 1999; Corkum et. al. 1999; Lee and Crites 1999; Schieck and Hobson 1999), the retention of tree clumps increased relative to the retention of individual trees. Snag retention increased, as did the retention (and variability in the shape and arrangement) of mixedwood clumps (Fig. 22.3b). The systematic protection of understory spruce saplings began in an effort to maintain the mixedwood composition of stands (Chap. 13).

In order to design and implement forest ecosystem management, a variety of research programs (involving 60–70 researchers) was undertaken to examine ecological process, pattern, and structure at the three scales. Using the three-dimensional criteria of quantifying the impact of natural forest fires as identified by Swanson et al. (1993), a landscape-scale experiment was designed to compare logging and fire for each criterion: age or fire return interval, size of the fire, and severity of the fire (Cumming 1997; Hannon 1999; Little and Smyth 1999). Many associated research projects at the species and (or) community level were undertaken with each experiment. For example, "leave blocks" attached to buffer strips were used to examine connectivity (Schmiegelow et al. 1997). Tree tops and branches were spread, piled, or windrowed in a variety of spatial configurations to examine small mammal utilization of the harvested area (Moses and Boutin 2001). In addition, aquatic programs were designed to examine stream and lake protection using buffer strips, the rate of extraction, and the spatial distribution of log-

Fig. 22.3. Green-tree retention in the Alberta-P acific Forest Management Agreement area; (*a*) dispersed retention, as practiced in the early 1990s; (*b*) clumped retention, as is now standard practice.

(*a*)   (*b*)

Photo by D. Hebert   *Photo courtesy of Alberta-Pacific Forest Industries Inc.*

ging within watershed sub-units in order to assess disturbance and its relationship to watershed protection (Prepas et. al. 2001; see Chap. 10). Landscape patterns, including questions of adjacency, are currently being examined and (due to the complexity involved) will continue for some time.

Alberta-Pacific's ecologically based SFM program is composed of many parts (Fig. 22.4). It is part of the corporate philosophy to base as many components as possible on sound, up-to-date research. Funding and partnerships were constructed among industry members, universities, the Canada–Alberta Partnership Agreement in Forestry (PAIF), the Natural Sciences and Engineering Research Council (NSERC), and others. In addition, Alberta-Pacific is working with the University of Alberta, other major forest companies, and 23 universities across Canada through the SFMN, which has been examining sustainable development in the Canadian boreal forest, as showcased throughout this book. More recent partnerships have been the development of the Adaptive Management Experiment (AME) team (see http://www.ameteam.ca/) at the University of Alberta, and a memorandum of understanding with Ducks Unlimited Canada to establish a watershed-based conservation plan for the FMA area (Zolkewich 1999).

Methodology and its application

The development and application of the Alberta-Pacific ecological management model (Fig. 22.4) includes the following parameters:

- it includes both coarse-filter and fine-filter approaches;
- it recognizes inherent variability in the system;
- it can be assessed using measurable indicators;
- it requires tradeoffs among ecological, economic, and social requirements and their impacts and associated risks; and
- it requires adaptive management procedures and trials to adjust the system.

As each leg of the SFM model is developed, tradeoff modelling is used to assess options through scenario analysis, and indicates impacts and risks. In its simplest form, a forest management system is developed and implemented, the results are monitored

Fig. 22.4. An ecosystem-based approach and framework for sustainable forest management (SFM).

using indicators (which may be in the form of formal adaptive management trials), and the management system is adjusted accordingly. The development of a formal SFM framework[2] within the company will allow it to fulfill the needs of certification programs, produce forest development plans, and address public policy and planning needs (Fig. 22.4). The components of the Alberta-Pacific management system can be conveniently related to the Canadian Council of Forest Ministers (CCFM 1997) criteria and indicator statements as demonstrated in the recent Alberta-Pacific Stewardship Report (APFII 2001). At the highest level, components can include disturbance type, rate, and forest/landscape criteria such as seral stage distributions and stratification of the FMA area by timber harvest and non-timber harvest land base. At the stand level, habitat elements include the basic eight habitat elements described by Bunnell et al. (1999)[3], but could also include forest health and stand productivity assessments.

---

[2]The SFMN is piloting a program whereby companies or governments can formalize their programs into a framework that encompasses the ecological, social, and economic aspects of SFM.

[3]Bunnell, F.L., Huggard, D.J., and Lisgo, K.A. 1999. Vertebrates and stand structure in TFL49. Unpublished technical report. Centre for Applied Conservation Biology, University of British Columbia, Vancouver, British Columbia. 194 p.

Priority land-use zoning

The triad approach generally considers the zoning of extensive, intensive, and protected area management as a basis for maintaining biodiversity without sacrificing commodity production (see Chaps. 11–13). In the case of Alberta-Pacific, the majority of the land base is being managed extensively in a coarse-filter fashion of protecting natural stand structures and patterns. However, it has also undertaken a major hardwood tree improvement program that is being instituted as its intensive management program, utilizing private land. In addition, it has participated in the Alberta "Special Places 2000" (SP2000) program (a provincial protected areas strategy), under which 412 000 ha of land were designated as protected to varying degrees on or near the Alberta-Pacific FMA area (AEP 1993, 2001). At the time of writing, Alberta-Pacific was engaging World Wildlife Fund Canada in an assessment of High Conservation Value Forests (HCVF) on the FMA area, in accordance with Principle 9 of the Forest Stewardship Council (see Box 21.6). The HCVF assessment, an associated protected areas gap analysis, and work to be conducted with Ducks Unlimited Canada may collectively identify other benchmark areas requiring protection or special management consideration.

Coarse-filter management

Maintaining biodiversity is a primary goal in the implementation of ecosystem management on the FMA area. The basis for accomplishing this goal is a coarse-filter approach to ecosystem management (Chaps. 1 and 2) following the natural disturbance model (Chap. 9). This is a preventative approach to maintaining biodiversity (Hunter 1991). It assumes that maintaining vegetative communities and landscape patterns and processes within the limits of natural variability will result in the maintenance of the full complement of native plant and animal species. In a coarse-filter approach, timber harvesting must be designed to regenerate the diversity of structure and vegetation within forest stands, and the patterns of forest stands on the landscape, that could be found under natural (fire) disturbance regimes. Maintaining these critical features of the forest will provide the variety of habitats needed to support the diversity of living organisms (both seen and unseen, known and unknown) found in wild landscapes. Important components of a realistic coarse-filter approach include:

- Harvest activities that approximate the range of size (Cumming 2001), shape, and intensity of natural disturbance events;
- Retention of residual structure within cutblocks that approximates the green tree residual pattern left by natural disturbance such as fire (Eberhart and Woodward 1987);
- A realistic older-forest management strategy that maintains older forest (overmature) stand types in all major forest strata throughout the FMA area, within the range of natural variability; the strategy should also account for projected land base losses due to natural disturbances and other human activities (i.e., oil and gas development);
- Maintenance of some post-fire habitat types that are difficult to simulate through harvesting or even post-fire salvage activities (i.e., retention of early post-fire habitats/stands/blocks); and

• A system of ecological benchmarks, supported by the Alberta government and other industrial sectors, free of industrial activity, that will act as reference areas for the landscape managed under the coarse-filter approach.

Mixedwood management strategies

The maintenance of forest composition will occur as a result of long-term, integrated planning under a system called "mixedwood management". From the inception of harvesting in 1993 until 2001, the Alberta-Pacific FMA area was managed as a "divided land base". Management decisions with regard to reforestation were largely dictated by the existing leading species; i.e., if an area was designated as a "conifer-leading" land base according to the Alberta Vegetation Inventory (AVI), it was to remain a conifer land base in perpetuity. Not only was this divided land base posing significant silvicultural challenges by trying to fight natural vegetation dynamics, it was threatening to "unmix" the mixedwood forests of northeastern Alberta (see Box 6.4 for some implications). In September 2001, a decision was made by the Alberta Ministry of Sustainable Resource Development to dissolve the divided land base rules, and begin managing the Alberta-Pacific FMA area as a "common" or mixedwood land base. Rather than maintaining a deciduous and coniferous land base as independent entities, the mixedwood strategy attempts to maintain the mixed aspen and spruce composition in many individual forest stands.[4]

Acting on some of the techniques explored in Chap. 13, three on-the-ground approaches for maintaining deciduous, coniferous, and mixedwood stands across the forest landscape are being piloted on the Alberta-Pacific FMA area:

1. Protection of immature white spruce (understory) while removing mature deciduous timber — Special harvesting systems have been implemented (e.g., high-effort understory protection or strip shelterwood harvesting) to protect white spruce understories that are uniformly distributed with densities >600 stems/ha. Up to 75% of understories can be protected in this way; at least 8.5% of the deciduous overstory has to be left behind to generate a windfirm understory. Conventional logging with an avoidance of conifer understory can protect 40–60% of the understory in areas with conifer clumps and (or) densities <600 stems/ha. However, "windfirming" all the protected white spruce with aspen can be infeasible and losses to blowdown may occur.

2. Underplanting aspen stands with white spruce — This procedure involves planting white spruce seedlings into mid-rotation (30–40-year-old) aspen stands. This will mimic the development of naturally occurring understories, for even though most tree seedlings are recruited shortly after fire (Chap. 8), they typically will develop only after aspen has undergone self-thinning and sufficient light penetrates the canopy. The underplanting of mature aspen stands with spruce seedlings approximately 20 years prior to aspen harvest is also being tested. This technique will take advantage of aspen as a nurse crop providing a more moderate microclimate, reduced herbaceous competition, and some protection from insects and disease.

---

[4]A constraint remains in that the regeneration standards still dictate that four strata (deciduous, deciduous-leading mixedwoods, coniferous-leading mixedwoods, and coniferous) are to be maintained at a level of at least 5% of their abundance at the time of the initial forest inventory.

3. Exploring techniques to regenerate mixedwood stands of varying compositions and patterns — The forest should be regenerated to maintain both coniferous and deciduous volumes, not necessarily the previous cover type at the same location. Succession changes the proportions of pure and mixedwood stands over time, and the range of natural variability should allow some flexibility in choosing regeneration targets. Maintaining mixedwood stands on the landscape is desirable for both ecological functions and forest productivity.

Maintaining older forest age classes within the range of natural variability

There is no requirement for the maintenance of older forests under traditional forest management based primarily on sustained-yield timber production. Additionally, general forestry practices tend to harvest the oldest stands first. The result is often the gradual elimination, through harvesting, of most older forests throughout a landscape. However, with the move to ecosystem management there is a recognition that older forests are important for biodiversity and are under threat. Consequently, forest managers are examining strategies for retaining older forests.

The sustainability of older aged forest stands and areas is to be accomplished through a shifting mosaic of stand-level characteristics over the entire FMA area. Management for retention of older forest within the FMA area must also be based on a shifting mosaic or target range within a natural range of variation (NRV) on large natural disturbance landscape units. Spatially, amounts of over-mature forest can vary from one landscape to the next. Thus, sustainability of various amounts of older forest and of the associated forest characteristics at the FMA forest level is a function of the balance between the frequency and severity of ecosystem disturbance and the rate of ecosystem or value recovery (Kimmins 2003). The distribution and area of older age classes in the FMA area has varied over time and will continue to do so; the current age-class structure (Fig. 22.5) clearly results from widespread wildfires 40–80 years ago, but there is no reason to expect that this is the "most natural" or "balanced" condition for the defined forest area. Such variation in the occurrence/development of different forest cover types is primarily dependent on unpredictable natural disturbances (fire, insect, disease, wind). Increasingly, anthropogenic disturbances (such as timber harvesting and activities of the energy and mining sectors) are also altering the characteristics of the forest. The consequence is that boreal forest structure, forest stand ages, and forest stand patch size distributions are not static. Therefore, planning for a stationary balance of age classes within the forest is based on erroneous concepts of landscape equilibrium (Cumming et al. 1996). The variation in natural patterns, however, can assist in identifying a wide range of older forest retention strategies.

The Natural Range of Variation (NRV) approach (Landres et al. 1999) is an efficient method to manage ecological, aesthetic, and economic preferences over the FMA area. Predictions cannot and should not be made on small landscapes (i.e., Forest Management Units, FMUs, in Alberta) where natural disturbances can deprive the area of age-class diversity. Within the FMA area, the location and size of forest fires are highly unpredictable and do occur as very large fires (e.g., the 250 000 ha House River Fire in 2002) that can eliminate large areas of older aged forests. However, the chance of depleting all older forest patches is lessened when large landscapes are utilized. Thus,

Fig. 22.5. Age-class distribution on the merchantable (productive) land base of the Alberta-Pacific Forest Management Agreement area.

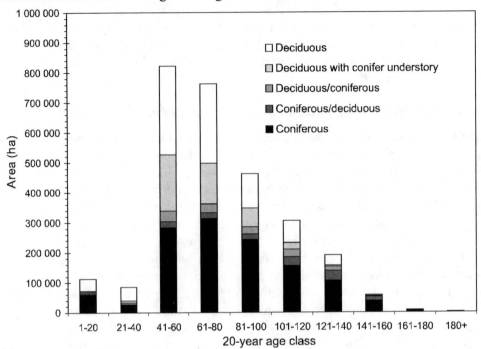

the total FMA area will be utilized for the tracking of older aged forest stands. Alberta-Pacific's implementation of the NRV model is based on the following major assumptions[5]:

- The landscape was defined in five major strata — deciduous, mixedwood, white spruce, black spruce, and jack pine;
- Conifer burn probability is higher than deciduous;
- Mixedwood is half as likely to burn as pine and spruce;
- Average fire cycle of 85 years in the FMA;
- The entire gross FMA area is utilized; and
- Initiation year was 2000 (i.e., current AVI).

These assumptions lead the analysis to predict ranges of natural variation for the abundance of each cover group or stratum.

As mentioned above, adaptive management is a critical component of Alberta-Pacific's SFM framework; strategies for managing for older forests are changing. The average area of the older age-class forest for the past 100 years (8%) was the targeted average for the 200-year timber supply analysis as described in the Detailed Forest Management Plan (APFII 2000). At the time of writing of this book chapter, Alberta-

---

[5]D.W. Andison. June 2003. Over-mature forest retention and landscape fire assessment. Personal communication and draft report in preparation for Alberta-Pacific Forest Industries, Boyle, Alberta. Bandaloop-Ecosystem Services, Belcarra, British Columbia.

Pacific staff were revising the old-forest strategy to include an NRV approach[6] to managing old forest on a stratum by stratum basis (as opposed to combining all forest cover types).

Stand-level management

While modern conservation efforts are usually concentrated on broad habitat type and age-class representation across the landscape (landscape-level, coarse-filter management), or on meeting the specific needs of individual species (the fine-filter approach), Alberta-Pacific also utilizes habitat management at the stand level to protect biodiversity. This approach is not designed to manage specific species, but is implemented to maintain structural diversity. The majority of the habitat elements can be maintained through harvest planning and the use of well-trained operators during the timber harvesting program.

Harvesting contractors are the critical link in the successful implementation of the environmental and safety programs. Without their buy-in, the harvesting and stand structure objectives could not be met. Alberta-Pacific is training contractors to provide them with an understanding of the program objectives. The training is further reinforced during operations through direction and feedback from supervisory staff, with the aim of gaining support and instilling pride by operators in their work. Operators are responsible for maintaining habitat structure during harvesting operations by protecting understory regeneration, retaining wildlife trees, and providing coarse woody debris.

Live trees and snags (dead trees) are left standing in order to maintain habitat for cavity nesting species and to facilitate natural stand dynamics. The dead and live stand structure retained varies from block to block across the landscape. On the deciduous cutblocks, an average of 5% of the merchantable volume is left uncut, in addition to all unmerchantable stems. On the coniferous cutblocks, 1% of the commercial conifer volume[7] and 5% of deciduous volume is retained; the smaller conifer volume retained reflects greater difficulty in successfully leaving conifers, and a constrained conifer wood supply. Coarse woody debris in the form of logs and large limbs provide important habitat for a number of plant and animal species. Slash from delimbing is retained in piles or spread out for access control and habitat purposes, provided it will not significantly impact tree regeneration or act as a fuel source for spreading wildfire.

Immature trees found under a merchantable overstory are left standing wherever possible, as they are an important resource for habitat, aesthetics, and future fibre production. Protected understory trees enhance structural, species, and age diversity in the new stand, as well as providing nesting sites, thermal cover, and a restricted line of sight (and hence escape cover for various prey species). Spruce can be protected from windthrow by leaving it in clumps with hardwoods, which are generally more windfirm. The promotion or maintenance of heterogeneous stand structure also includes utilizing block

---

[6]Alberta-Pacific's Forest management plan, working draft (May 2003). Alberta-Pacific Forest Industries Inc., Boyle, Alberta.

[7]Retention of 1% conifer stand structure was the amount agreed to by the provincial government and all forest products companies operating on the Alberta-Pacific FMA area for the duration of the Detailed Forest Management Plan.

features, avoiding damage to patches of understory shrubs and wet areas (draws, water sources), and leaving large windfirm conifers (as a valuable seed source). Site-specific practices depend on initial stand and site characteristics and desired block-to-block variation. Structuring of larger blocks can include a greater range in clump sizes or treed corridors to provide wildlife linkages and feathered edges on the windward side of blocks. Ongoing research will, no doubt, lead to further refinements and modifications to the structuring protocols.

The development of harvesting plans generally includes the protection of riparian areas, while mixedwood and seral stage management policies maintain the deciduous/coniferous balance, levels of shrub abundance, and early and late successional stages. In addition, the pattern of disturbance on the forest matrix (expressed as "adjacency" requirements) was examined and changed, from two- and three-pass forest harvest systems on a landscape basis to single-pass harvest systems designed to better approximate the pattern and stochastic recurrence of large lightning-started fires.

Indicators and monitoring

Since a primary goal of sustainable forest management is to maintain biodiversity, it is critical to devise methods of somehow monitoring biodiversity over the long term. Further to this point, SFM systems are actually experiments in time and space, and there must be measurable indicators that can be used to assess and improve each attempt and implementation (Chap. 21). If proper monitoring is not conducted, significant deviations may go undetected, or conversely, changes in species abundance might be mistakenly associated with forest practices. These indicators can be classified as "planning indicators" and "monitoring indicators". The coarse-filter planning indicators include: disturbance rate; tree species composition (especially the amount of mixedwood); and seral stages. Monitoring indicators at the species level represent terrestrial and aquatic biota (both vertebrate and invertebrate fauna).

Rather than create an FMA area-specific biodiversity monitoring program, Alberta-Pacific has been participating in a province-side monitoring program. The Alberta Forest Biodiversity Monitoring Program (AFBMP; see http://www.fmf.ab.ca/bm/bm.html) is a co-operative undertaking of natural resource industries, protected area managers, government agencies, and non-government organizations. The goal of the AFBMP is to develop a system for detecting changes in biodiversity in the forested regions of Alberta. Analysis of the data obtained from such a system will greatly assist resource managers and others in understanding long-term trends in biodiversity and landscape patterns, and improve the management of the province's forests. A suite of landscape, habitat, and species-level (terrestrial and aquatic, vertebrates and invertebrates) indicators have been selected through a peer-review process. Logistics of the protocols were evaluated in 2002 and a 4-year prototype evaluation will begin in 2003. Implementation of the full program is planned for 2007.

Fine-filter site-level issues

Biological research and management have traditionally focused on individual species and their relationship to habitat. The extension of this species-level thinking to biodiversity conservation and ecosystem management is called the fine-filter approach. The application of this approach involves collecting and using extensive knowledge of all the organisms in the affected ecosystems, followed by the design of forest management activities that would maintain populations and their viability. Realistically, it is not feasible to study and understand all the species in our forests. Implementation of the fine-filter approach in industrial forest management, therefore, instead assumes that a smaller number of "indicator species" will represent the full spectrum of organisms present.

It is generally felt that ecosystem management based entirely on a fine-filter approach is unlikely to succeed, and that there may not even be valid indicator species (Lindenmayer and Franklin 2003). A multitude of individual species and habitats would require detailed and costly scientific analysis. Most existing fine-filter models of persistence of species are based on expert opinion and conjecture derived from estimation of habitat requirements. Despite large amounts of SFMN research, there have been few thresholds or concrete prescriptions as to the habitat needs of individual species (Hannon and McCallum 2002). Additionally, a forest management program based on fine-filter programs essentially pits the habitat requirements of selected species against the requirements of others. At the fine-filter level, Alberta-Pacific focuses on federally and provincially recognized "species at risk", such as the woodland caribou (*Rangifer tarandus caribou*) and the trumpeter swan (*Cygnus buccinator*), or species such as moose that were designated as culturally important species by local stakeholders.

Implementing the new model of forest management

Although forest management has been a constantly evolving process, the shift from timber management to SFM is an extremely complex process that requires extensive analysis of policy, procedure, planning methodology, new technical information, and operational logistics by all involved parties (government, industry, public, and Aboriginals). The evolution to sustainable programs will not happen immediately, but needs to develop through a coordinated, open forum of discussion over a reasonable time period (as long as 5–10 years).

Education

During the development of a forest ecosystem management program, continuing education on a variety of fronts was a critical component. Although there was a strong corporate commitment to ecosystem management and SFM, it had to be continually reinforced through the explanation, communication, and presentation of information on how Alberta-Pacific planned to implement the new model. In general, education at all levels was the cornerstone of the transition process (Hebert et al. 1995). The educational components included communications with corporate and operational staff within the company, and reaching out to the provincial government, universities, and non-government organizations. Corporate education required an ongoing discussion with corporate officers on the topics of forest ecosystem management, the natural disturbance process, and SFM, and their implications to the company. Successful education of the corporate and joint-venture bodies supplied the funding for novel management planning, innovative operations, and new research.

Alberta-Pacific's Forest Management Task Force (FMTF) was established in 1992 and serves as both an education forum and a public participation process (see Fig. 21.4). The FMTF is composed of Aboriginal people, conservationists, resource user groups, government agencies, coniferous quota holders from the FMA area, and representatives of Alberta-Pacific. The FMTF plays an active role in ensuring the environmental, community, and economic sustainability of the company's plans and operations; this is done through the ongoing review of forest management plans, operating ground rules, and related issues. The FMTF helped Alberta-Pacific to develop the ecosystem-based approach that led to major changes in forest management in the 1990s — not only in the FMA area but also across the province as the government and other companies adopted similar approaches. The group's innovations lead to provincial and national awards for environmental excellence, and the task force itself received an Emerald Award in 2000 from the Alberta Foundation for Environmental Excellence (see http://www.emeraldawards.com/2000.html).

Following intense public hearings, gaining the trust and cooperation of university researchers was a major undertaking. Alberta-Pacific required basic scientific information on biodiversity at the stand and landscape level in order to develop its first detailed forest management plan (APFII 2000). The initial response from the research body was to undertake "impact analysis". Scientists were asked to think about a forest management system and to propose solutions to problems facing its implementation. It was suggested that they could either criticize Alberta-Pacific from the sidelines or partner with them to provide the science to engineer a paradigm shift in the way forestry is done. Within the next 2 years, some 33 professors from eight universities accepted this challenge, with the incentive of doing industry-supported, team-based research on practical problems, and with no constraints on the publication of findings. The research program continued to grow, maintaining partner research, and forming the basis for the Sustainable Forest Management Network (as outlined in Chap. 1).

Conclusions and upcoming challenges

A comprehensive, multi-scale, forest management framework is clearly required for the sustainable management of fibre, ecosystems, habitats, and biodiversity. Without this structure, it is difficult to project the results of management rules over long time frames and large areas, to test them in a logical fashion, to identify and apply knowledge appropriately, to identify information gaps and pertinent questions requiring research, to design relevant research, and to allocate funds most effectively in support of SFM.

In most of the world's boreal forests, the natural forest matrix is still declining as timber harvesting and active forest management proceeds. This is exacerbated by petroleum and mineral development, flooding of hydroelectric reservoirs, and withdrawals for land uses such as agriculture and transportation corridors. To date, most forest ecology research (especially stand-level projects) in Canada has been conducted in the context of a surrounding matrix of mature or older aged forest. At the end of the first rotation, the landscape components and the stand structural attributes will be expected to support the full range of biodiversity, at all scales, in a young to middle-aged forest matrix. Consequently, it is important to understand and identify baseline or "natural" levels of descriptors such as disturbance rate, disturbance severity, and landscape patterns. It is the multi-scale interaction of these coarse- to fine-filter attributes that will support biodiversity. Although thresholds may be obtained at any one level, there is undoubtedly a multi-scale, interactive threshold as well. Testing baseline and threshold values at this forest level is difficult; however, it should be possible to test them at watershed, forest management unit, or sub-landscape unit levels. In fact, many opportunities for retrospective analysis undoubtedly exist on the landscape today.

One of the biggest challenges today facing the forest products industry in making the transition to SFM, beyond the initial technical implementation, is the cost of AAC netdowns for ecological and social purposes. Without tradeoff modelling (see Chap. 14), the implementation process would be highly inefficient, possibly ineffective, and extremely expensive (Binkley et al. 1994; Reinhardt 1999). An even larger challenge faces companies like Alberta-Pacific, whose forestry operations are embedded in an industrial landscape that is dominated by other resource extraction industries such as the oil and gas sector (Fig. 22.6). Some impacts of the energy sector on Alberta's forests

Fig. 22.6. The next challenge to address in the achievement of sustainable forest management will be to deal effectively with the cumulative impacts of natural disturbances, forest management, and other overlapping (often competing) land uses, such as the activities of the energy sector. The light area in the upper left of this aerial photograph is the legacy of a large wildfire, while the gridded lines indicate seismic exploration lines, wellhead access roads, and pipeline corridors associated with oil and gas development.

Courtesy of Brad Stelfox and Alberta-Pacific Forest Industries

have been recognized for more than 30 years (e.g., CFS 1975), but little has been done to offset their detrimental effects on the forested landscape.

In the late 1990s, Alberta-Pacific collaborated with Dr. Brad Stelfox to quantify and model the effects of overlapping land-use practices in northeastern Alberta. The results and projections (Stelfox and Wasel 2002) caused yet another shift in the management philosophy at Alberta-Pacific. The FMA area is underlain by some of the largest oil and gas reserves in the world. In recent years, the "industrial footprint" (in terms of area) associated with forest clearing for energy sector exploration, production, and transportation rivaled the annual harvest by forest products companies on the FMA area (Schneider et al. 2003). Conservative projections of energy sector growth on the FMA area highlighted the need to incorporate management of cumulative effects if SFM was to become a reality in areas like northeastern Alberta. In 1999, Alberta-Pacific developed its Integrated Landscape Management (ILM) program, which seeks to minimize the combined industrial footprint (forestry and energy sectors) through coordinated planning and innovations to exploration and extraction processes. The ILM program was adopted by the Alberta Chamber of Resources (Simpson 2001) and included the establishment of the ILM Research Chair at the University of Alberta (see http://www.biology.ualberta.ca/faculty/stan_boutin/ilm/). Integrating industrial activity across sectors will ultimately require regional management through combined actions of government, industry and the public. This is now the big challenge for achieving truly sustainable forest management in Alberta.

Louisiana-Pacific Canada's Swan Valley Forest Resources Division

Introduction

The Province of Manitoba and Louisiana-Pacific Canada Ltd. (LP, or Louisiana-Pacific) signed a Forest Management License Agreement in 1994. This agreement outlined the intent of the Province of Manitoba to provide LP access to hardwood fibre resources from Forest Management License Area #3 (FML #3; 50°45′–52°25′N, 98°45′–101°35′W), and included a provision that LP was responsible for planning and coordinating all forest management activities in the license area. Louisiana-Pacific's commitments included the construction of an oriented strandboard (OSB) production facility in the Swan Valley, and the development of a 10-year forest management plan outlining proposed stand management and forest renewal activities within FML #3.

Louisiana-Pacific Canada Ltd. operates an OSB facility near Minitonas in west-central Manitoba. Louisiana-Pacific's Forest Resources Division, located in Swan River, Manitoba, is responsible for providing wood fibre for the OSB facility. Oriented strandboard is a composite building product composed of aspen, balsam poplar, and white birch fibre, which is harvested from both Crown and private forest lands. These forests are primarily located within the Mid-Boreal Uplands Ecodistrict, at the eastern margin of the Boreal Plain Ecoregion of the Canadian boreal forest (Smith et al. 1998).

The Boreal Plain is characterized by deep fertile soils and highly diverse forests of white spruce, black spruce, jack pine, trembling aspen, balsam poplar, balsam fir, tamarack, and white birch. These species can be found in relatively pure stands but are also contained in mixedwood assemblages, depending on site conditions. The distribution of tree species varies, but deciduous species generally dominate richer sites along with balsam fir and white spruce, whereas pine and black spruce are found on drier and wetter sites, respectively. The Mid-Boreal Uplands Ecodistrict contains a diverse assemblage of upland and lowland forests with extensive wetland complexes, streams, and lakes. Topography is variable due to the presence of the Manitoba Escarpment. The highest point in Manitoba, Baldy Mountain (831-m elevation), is found within the Duck Mountain Provincial Forest, which constitutes much of FML #3 (LPCL 1995).

Forest Management License Area #3 is comprised of Forest Management Units 10, 11, and 13, with a total land base 2 549 500 ha. This includes the Duck Mountain Provincial Forest and other surrounding crown and privately owned lands. Louisiana-Pacific also has access to the hardwood resources of the nearby Porcupine Provincial Forest on a volume basis. The license area consists of 33% provincially owned Crown forest lands, 22% Crown agricultural lands, and 45% private lands, which are further divided into productive and non-productive components for forest management purposes. The provincially owned, productive land base comprises 584 050 ha of the FML.

Forest management agreement

As required under the Forest Management License Agreement, LP developed a 10-year Forest Management Plan (FMP) and submitted it for review in 1995 to the Forestry Branch of Manitoba Natural Resources, and the Environment License Branch of Manitoba Environment (both branches are now part of Manitoba Conservation). As part of the approval process, the FMP was reviewed by both Federal and Provincial regulatory resource agencies, including the Canadian Wildlife Service (CWS), the Department of Fisheries and Oceans (DFO), Manitoba Environment, and Manitoba Natural Resources (MNR). In addition to the review performed by regulatory agencies, all forest management plans in Manitoba must have an environmental impact assessment completed as part of the approval and licensing process.

The LP Swan Valley 10-year FMP made many commitments regarding Louisiana-Pacific's planned approach to forest management and operations within FML #3. Emphasis was placed on implementing an ecosystem-based management approach and the promotion of responsible forest stewardship. Among the many commitments made by LP were:

- the intent to meet or exceed all current forest management guidelines and practices within Manitoba;
- to incorporate both community and ecological interests in their planning approaches; and
- to conduct or co-operate in data collection, research, and monitoring to broaden the knowledge base for forest management in FML #3.

This would be accomplished in an adaptive management context (Chap. 21), with an emphasis on continuous improvement in forest management and operational practices, and the support of research and monitoring activities to enhance the current knowledge of the species present, their relationships, and the dynamics of the boreal ecosystems within FML #3.

Public and community involvement

Following the review period, a public environmental hearing process was conducted by the Clean Environment Commission, a panel appointed by Manitoba Environment. Louisiana-Pacific's environmental hearings were conducted over approximately a 2-week period, from December 1995 to January 1996, in Winnipeg and Swan River, Manitoba. The Environmental Hearing process in Manitoba provides an opportunity for the general public and other interested parties, including government agencies, to question the proponent on the forest management plan and the associated environmental impact statement (EIS). Louisiana-Pacific's environmental hearings were well attended and discussion was lively, due to the level of interest shown by Manitobans in the development of the OSB facility and the granting of a license to harvest timber. Public concerns centred around the magnitude of the harvest volume proposed and the potential for harvest-related impacts on the forest ecosystem of the Duck Mountains, recent environmental violations resulting from the operation of an LP plant in the U.S. state of Colorado, and the potential for environmental health impacts in the Swan Valley resulting from OSB facility emissions.

In order to incorporate community interests into the development of the plan, LP established a Stakeholders Advisory Committee (SAC) at the beginning of the 10-year planning process (1995). The Stakeholders Advisory Committee has representatives from a cross-section of forest users, local government, and interest groups within the Swan Valley and continues to play an integral role in the development of management plans. The SAC provides a forum for the expression of forest values and concerns from the local public, and ensures that community issues are a part of LP's planning process. As agreed to in the environmental hearings, and reflected in LP's environment license, the primary purpose of the SAC is to identify resources or land uses that may be impacted by proposed forest management activities and to recommend alternative harvest and (or) renewal plans to minimize the impacts. The SAC also played an integral role in defining and developing LP's Standard Operating Procedures (LPCL 1998).

In addition to consulting with the SAC, LP has annual open houses in many communities within the Swan Valley. The Annual Operating Plans (AOPs) are presented at the open houses and provide an opportunity for individuals or organizations to review and provide comments on them. These meetings are advertised in local newspapers, and invitations are sent to local interest groups and environmental organizations. Louisiana-Pacific also meets individually with environmental organizations and First Nations during the development of their long-term plans. The 10-Year Forest Management Plan and AOPs developed by the company are also placed in public registries at libraries in Winnipeg and several communities in the Swan Valley for review and comments. Information is provided to the public on an ongoing basis by participating in community events and giving presentations at schools and local workshops. Field tours are conducted on an annual basis for the SAS, the regional high school, and as requested for interested parties.

Operational practices

Louisiana-Pacific's Standard Operation Procedures (SOPs) provide detailed policy and direction to LP staff and operators on all aspects of forest planning, operations, and forest renewal activities. They provide a framework for the company to achieve and maintain a particular standard relating to the goals outlined in the approved forest management plan, and ensure that LP meets or exceeds government regulations. The SOPs were jointly developed by LP and the SAC, with specific reference to federal and provincial legislation, regulations, guidelines, and the requirements and conditions of the FML #3 agreement. They were imbedded in the forest management plan and subject to review and approval along with the plan in the review process (LPCL 1998).

Training and development

An essential aspect of the SOPs included the commitment of LP to train and develop staff and operators in both operational and ecological aspects of forest management. Proper implementation of the SOPs is the key to preventing and mitigating environmental impacts, and to promoting sustainable forest management practices. Training related to forest operations (see Chap. 15) attempts to provide staff with exposure to alternative practices and new and improved technologies focused on minimizing environmental impacts. Stream crossing installation, road construction, erosion control, and stream rehabilitation techniques have been emphasized to reduce the potential impacts of harvesting and road construction on water quality and riparian ecosystems. Staff frequently attend conferences, workshops, and courses on a variety of ecological principles, including natural disturbance based approaches to forest management, mixedwood forest management, riparian management, public participation, and other SFM-related topics covered in this book.

In addition, an annual "Best Management Practices" (BMP) seminar is held by LP for all contractors and their operators, with attendance mandatory. Operators who do not attend the meeting are not contracted by LP for any harvesting or silvicultural activities for that year. The BMP seminar is geared towards educating heavy equipment operators and their employers in government laws, regulations, and guidelines governing the protection of fish and fish habitat, water quality, soil quality, and proper containment of petroleum products and other chemicals needed on site. It also provides an opportunity for LP to review the details of the SOPs, provide notification of updates based on new techniques or scientific knowledge, and review safety requirements. Guest speakers provide information on many aspects of the SOPs and have included representatives from the DFO, Manitoba Health and Safety, Manitoba Conservation, road construction engineers, and erosion control experts.

Applied management

In addition to guest speakers, LP staff review some of the more critical harvest-related SOPs, including conifer understory protection practices and requirements for the provision of wildlife trees and in-block structure (now referred to as "variable retention"; Franklin et al. 1997). Although a clearcut harvest system, followed by natural regeneration, is the primary silvicultural tool for the harvest and renewal of the hardwood and mixedwood forests in which LP operates, understory protection and variable retention practices were developed to enhance post-harvest regeneration of the block for a variety of objectives, the primary one being the conservation of biological diversity. Ecosystem resilience is enhanced, since variable retention practices promote the re-establishment of many essential ecosystem structural components and functions, increasing the ability (and decreasing the time required) to return the stand to pre-disturbance conditions (Franklin et al. 1997).

In practice, understory protection and variable retention strategies are delivered in combination, resulting in a modified clearcut harvest, with residual forest left in patches and as single trees across a harvested area. One of the challenges to conifer understory protection is to maintain the trees on the block for the long term. This enhances short- and long-term biodiversity, wildlife habitat availability, and natural regeneration processes. Conifers are often not windfirm, are highly susceptible to blowdown, and need to be provided with protection from prevailing winds. It is the commitment from the loggers that ultimately determines the success of LP's understory protection and residual maintenance programs.

To ensure consistent application and implementation of the SOPs, LP Operation Supervisors are provided guidance on a daily basis during operations and periodically through annual contractor audits. Harvest blocks are rated on adherence to work permit instructions and the delivery of SOPs. Third-party verification audits will also be performed starting in 2003 as part of LP's forest certification program under the Sustainable Forestry Initiative (SFI) and through the implementation of an ISO14001-based environmental management system (see Chap. 21).

Inventory and monitoring

Terrestrial ecosystems

A recurrent theme in the environmental hearings involved the ability of LP to minimize environmental impacts on the ecosystems of the Duck Mountain Provincial Forest. One of the primary concerns was the lack of baseline ecological and biophysical data for the area, and the ability to predict environmental impacts based on the minimal data available. TetrES Environmental Consultants prepared the environmental impact statement and recommended (as did several provincial and federal regulatory agencies) that LP should "resolve some of the uncertainties relating to the adequacy and availability of data describing the quality, distribution and abundance of wildlife habitat" and "should participate in a longer-term monitoring program to determine flora and fauna utilization characteristics of the study area. . ." (*TetrES* 1995).

The Pre-Harvest Survey (PHS) program was developed by LP to collect data for potential harvest sites, on a site-specific basis, and to inventory stand attributes and non-timber values. These data are used to determine appropriate harvest and renewal treatments and to assist with short- and long-term planning information requirements. LP was the first forest company in Manitoba to commit to the PHS program and continues to have the most rigorous standards. The survey is conducted on a 150-m grid for all proposed harvest blocks, 1 year in advance of the scheduled harvest. The forest ecosystem classification guide for Manitoba (Zoladeski et al. 1995) provides the framework for the survey, which characterizes vegetation, soil, and timber attributes. In addition, data are collected on non-timber values such as wildlife habitat and utilization, heritage resources, the presence of rare, threatened or endangered species, and special habitat or landscape features. To date there have been over 18 000 PHS plots completed across the FML #3 area. They have provided LP with accurate, site-specific data to supplement the forest resource inventory, address planning requirements, and ensure that potential environmental impacts are addressed prior to harvest.

Louisiana-Pacific has also established approximately 450 permanent sample plots to provide long-term monitoring sites across a variety of forest types and age classes within FML #3. These plots will provide LP and the Province of Manitoba with data to develop a better understanding of stand dynamics, forest productivity, ecosystem structure and functional relationships, and the response of various forest types to management techniques.

Following recommendations in the environmental hearings and in recognition that the current inventory was insufficient at providing resource data on ecosystem attributes, a pilot project to develop a new Forest Lands Inventory program for FML #3 was implemented in 2000. This program was jointly developed and funded by LP and Manitoba Conservation and is the first of its kind in Manitoba. The inventory has three different components: the forest lands inventory; an ecosite classification and inventory; and a terrain and landform inventory (Manitoba Conservation 2000). It has been developed using multiple data sources and remotely sensed imagery, including extensive field sampling and information on both terrestrial and aquatic ecosystems. The new inventory will provide the basis for planning on an ecosystem basis, and will provide high-resolution data for use in wood supply calculations, wildlife habitat modelling, and sustainability analyses.

The effects of harvest operations on wildlife habitat and biodiversity was another area of great concern at the environmental hearings, in particular with regard to the potential for impacts on birds and bird habitat, especially neotropical migrants. The effects of harvesting on bird populations remained an issue, despite LP commitments to implement variable retention practices designed to provide some habitat structure and diversity as well as nesting cavities and refugia for birds in harvested areas (as recommended by research scientists in Alberta; Schieck and Nietfeld 1995). In consideration of these concerns, LP committed to conducting a bird monitoring program that would collect baseline information to assess the potential effects of their harvest operations on birds.

The LP Duck Mountain Bird Monitoring project was started in 1997 to collect data on bird species presence, abundance, and distribution within the Duck Mountain Provincial Forest. In less than 7 years, this monitoring project had collected over 60 000 incident records at over 2200 point count stations distributed across the forest. The current species list has reached 218 species. These data and the associated study design will provide LP with the baseline required to assess impacts on bird communities and their distribution as a result of LP's harvest practices. To further address concerns about the potential for harvest-related effects on birds, LP agreed to conduct a minimal amount of harvesting during the bird nesting season. To avoid disturbing active nests during harvest (and thus compromise nesting success for the season), LP strives to delay harvest operations in the spring until the bird nesting season is complete. This policy is reinforced in an environmental license requirement that restricts the volume of wood that can be harvested by LP during May, June, and July to a maximum of 100 000 m³; this apparently is a unique operating condition for a forest management license in Canada.

Aquatic ecosystems

Concerns of potential harvest-related effects and a lack of baseline data on aquatic ecosystems were also noted in the EIS, and by representatives of DFO, MNR, and ENGOs. The potential for harvest impacts on aquatic systems remained an issue. Questions were raised regarding the effect of harvest operations on water quality, peak flows, fish populations, and fish habitat. Louisiana-Pacific's ecosystem-based management approach includes a recognition of the ecological and social values associated with aquatic ecosystems, through policies, procedures, and practices to ensure that effects on aquatic environments are minimized. Two additional procedures developed by LP following the environmental hearings, in order to address concerns about aquatic ecosystems, included the LP Stream Assessment Program and an annual watershed-level analysis of cumulative harvest levels for the Duck Mountains.

Another initiative that addresses the potential for harvest effects related to water quality and peak flows is a landscape-level, cumulative harvest analysis. Louisiana-Pacific is required to limit the area in a watershed that is in a harvested and not sufficiently regenerated state, as determined by the consultation process (Manitoba Environment 1996). As an interim measure, until the consultation could be conducted, LP agreed to conduct a review of the scientific literature related to watershed harvest amounts and stream flow changes, and DFO agreed LP could proceed with harvest activities as long as watershed harvest limits remained below a 30% maximum. The watershed analysis is included in the annual operating plan and provides an annual update of the percent cumulative harvest, by watershed, in fulfillment of the environment license conditions. The method for monitoring and reporting on watershed harvest levels currently used by LP was determined following the environmental hearings with input from DFO and Manitoba Environment. Harvest activities are summarized by sub-drainage, based on the amount of productive forest removed relative to the total amount of productive forest present. The analysis is based on LP's 3-year harvest projected in the annual operating plan, as well as previously harvested areas, to determine a cumulative harvest amount. For this analysis, cutblocks are considered to be in a "harvested" state for 5 years following harvest for hardwood species, and 15 years following harvest for softwood (coniferous) species. The intent of the watershed analysis is to track the cumulative harvest amount, and the potential for increased water yields or stream flows resulting from harvest activities.

Research

Research priority setting exercise

The need for research and monitoring was a recurrent theme throughout the environmental hearing process. Louisiana-Pacific committed to undertake an adaptive management approach to ecosystem-based management and promised to generate information through research programs, biophysical inventories, pre-harvest surveys, and various monitoring activities (*TetrES* 1995). The Province of Manitoba supported the concept and indicated that it would like to see an extensive research program, similar to the one undertaken at Alberta-Pacific Forest Industries (see earlier section in this chapter), where the company was spending in excess of $1–3 million/year on research. In response, LP agreed that research was an essential component of the adaptive management approach they proposed, but felt a research plan should be developed prior to the commitment of any funds towards research. This was agreed to by all parties and was reflected in the following license condition:

> *"The licensee shall within twelve months of the date of this License, conduct consultations, . . . to establish priorities for baseline monitoring and forest ecosystem research . . ." and "shall participate in the data collection and funding of monitoring and research selected for implementation in the course of the consultations . . ."*

(Manitoba Environment 1996)

In early 1997, LP approached MNR regarding the research consultations to see if the Manitoba Forest Research Advisory Committee (ManFRAC) could conduct the research priority setting exercise. ManFRAC was a committee that was developed several years earlier, and chaired by MNR as a research planning and review board under the requirements of the federal Green Plan (Environment Canada 1990). The committee had not met for several years and would need to be re-instated, and additional representatives from the organizations stipulated in the LP's environmental license would have to be added. This was agreed to by MNR and the ManFRAC members, resulting in the re-instatement of the committee with additional membership from non-governmental organizations and other interest groups.

While ManFRAC membership was being determined and the terms of reference developed, LP issued a call for proposals in high-priority research topics from the general public, scientists, and regulatory agencies. There were advertisements in newspapers and notices sent to several universities in Manitoba and across Canada. Louisiana-Pacific received 61 submissions in response to the call for proposals. The submissions were reviewed and a list of research topics was prepared for inclusion in the ManFRAC review process. The Manitoba Forest Research Advisory Committee met repeatedly throughout 1997 and developed a research priority list by late October that year. The list was extensive and contained approximately 100 priorities, including inventory, research, monitoring, and social issues related to terrestrial and aquatic ecosystems. The list was ranked in terms of priority by ManFRAC, LP's Stakeholder Advisory Committee, and LP staff. A research plan was developed and submitted to the Province in December 1997 in fulfillment of the environment license condition concerning research consultations.

Research partnerships and associations

Louisiana-Pacific's research plan was approved by the Director of Manitoba Environment early in 1998. The plan emphasized the importance of partnerships and association with other research-related organizations as the foundation for the development of their research program. A principal partnership was developed when LP joined the SFMN, in 1998. Additional partnerships and affiliations have been established with the Manitoba Model Forest, the Center for Interdisciplinary Forest Research at the University of Winnipeg, the Forest Engineering Research Institute of Canada (FERIC), the Western Boreal Growth and Yield Co-operative, Ducks Unlimited Canada, the Prairie Adaptation Research Centre, and several other forest products companies and universities.

Research program

Louisiana-Pacific's research program is focused on obtaining knowledge of the boreal ecosystems within the Mid-Boreal Uplands Ecodistrict. Results are being integrated through the development and application of models to forecast the future structure, composition, and diversity of the forest of the Duck Mountains. This includes the development and application of natural disturbance based approaches to forest management as directed in the Manitoba Forest Plan, the environment license, and the research priorities list developed by ManFRAC.

In the past decade, forest management practices that favour the development of stand and landscape composition and structure similar to those that characterize natural ecosystems have been implemented to enhance the maintenance of biological diversity and essential ecosystem functions (Gauthier et al. 1996). To develop these new approaches, an understanding of natural disturbance regimes and regional disturbance dynamics is required. The combination of the characteristics of fire frequency, fire size, and fire severity, along with a knowledge of other natural disturbance agents, including regional forest insect and disease dynamics, provides the foundation for a natural disturbance based approach to forest management. Natural disturbance regimes, in combination with site features, are responsible for the variety of forest habitats that occur within a region and provide the "coarse filter" on which the maintenance of biodiversity should be based (Hunter 1991; Bergeron et al. 2002).

Louisiana-Pacific approached the SFMN to assist with the development of a project that would examine processes related to natural disturbance within LP's operating area. A keystone project, conducted by Dr. Jac Tardif at the University of Winnipeg, is investigating the historic disturbance regime of the Duck Mountains to provide the information required to develop a natural disturbance based approach. Research has been conducted on the interaction of site features, climatic history, and issues of scale in characterizing the natural disturbance regime of the Duck Mountains (Sauchyn 2000). Some of the additional research being conducted includes projects on stand dynamics and successional pathways, forest regeneration and operating practices, multi-scale indicators of habitat availability and biodiversity, wetland classification and disturbance dynamics, and public involvement in forest management planning processes.

The primary goal for LP's research program is to ensure that research results are synthesized and applied in the next long-term plan, which should include the determination of a desired future forest condition and an associated monitoring program. The scientists who have conducted the research are participating in an advisory capacity to assist with the adaptation of the research to LP's planning and operational practices in this collaborative approach. A research group meeting is held annually at the University of Manitoba to present research results and facilitate knowledge exchange between graduate students, scientists, LP staff, and provincial government representatives. At this meeting, LP staff provide updates on the development of a 20-year forest ecosystem management plan, modelling approaches, and inventory initiatives. The scientific advisory committee is assisting with the development of a forestry effects monitoring program that will provide an adaptive management framework for the implementation of a criteria-and-indicators approach to SFM (CCFM 1995) in the upcoming plan.

The Bas-Saint-Laurent Model Forest

Historical background

The Bas-Saint-Laurent Model Forest (BSLMF) is part of the Canadian Model Forest Network (CMFN) established in 1992 by the Canadian Forest Service (CFS) of Natural Resources Canada, and extended in 1997 and 2002 for subsequent 5-year periods. Like the other 10 model forests belonging to the Network (Fig. 22.7a), it is committed to testing and adopting integrated resource management approaches abiding by the principles of sustainable development (CMFN 2002). But, from the outset, it has been given the additional objective of achieving sustainable forest management on an inhabited land base with two management formulas (joint management groups and forest tenant farms) aimed at economic development in a rural environment. Of the two management formulas, forest tenant farming has become the "trademark" of the BSLMF.

The forest tenant farms idea had been on the desktop of many regional stakeholders well before the Model Forest program was implemented. The concept developed through efforts to improve the viability of small local communities through better access to and a more adequate use of the regional forest resource in the Gaspésie, southeast of the Saint Lawrence River estuary. By the 1970s, forest management was seen as a way to counteract the closing of villages deemed marginal at the time. That focus on forestry as a means of saving local communities, largely developed by pressure from the people, led to the creation of joint management groups as well as a provincial assistance program for private forests managed by these groups. This forest management structure has proven itself over the years and has expanded to include timber owners on an individual basis. The ongoing provincial assistance program was (and still is) restricted to private lands, however, and people were looking for a formula that could be applied on public forest lands.

Fig. 22.7. The Bas-Saint-Laurent Model Forest in eastern Quebec: (*a*) location within Canada, as part of the Model Forest Network, (*b*) tract divisions within the Seigneurie du Lac-Métis, and (*c*) within the Seigneurie de Nicolas Riou. Individual tenant farms are shown in white; gray areas are managed by the forest tenants cooperative (maps are not drawn to the same scale).

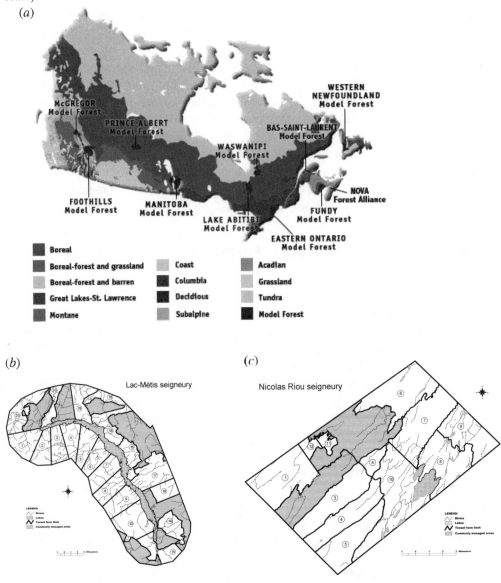

(*a*)

(*b*)   Lac-Métis seigneury

(*c*)   Nicolas Riou seigneury

    The objective was to consider ways to provide access to the public forest surrounding the communities for individuals who could earn a living there under a tenancy formula. Such access at first appeared impossible because all the public forest was already allocated to the forest products industry as part of Timber Supply and Forest Management Agreements (TSFMAs). This deadlock meant finding large enough territories to experiment with the formula and prove its sustainability and it depended on getting agreement from existing tenure holders. Subsequently, Abitibi Consolidated was asked, and agreed, to provide two large areas of private (freehold) lands which it owned in the region: the *Seigneurie du Lac- Métis* (33 933 ha; 48°20′N, 67°45′W; Fig. 22.7*b*) and the *Seigneurie de Nicolas Riou* (13 687 ha; 48°15′N, 68°50′W; Fig. 22.7*c*).

The BSLMF project

Emboldened by this support, the original project instigators, namely the *Syndicat de producteurs de bois du Bas Saint-Laurent* (Lower St. Lawrence Wood Producers' Marketing Board), the *Groupement forestier de l'Est du Lac Temiscouata*, the Department of Forestry and Geomatics at Laval University, and Abitibi Consolidated, developed their ideas of forest tenant farms, and included them in the Model Forest proposal to the CFS. By becoming members of the Model Forest Network, they had the opportunity to test their project, associated with an increased effort in joint management and diversification of the activities on the territory of the *Groupement forestier de l'Est du Lac Temiscouata*.

Thus the BSLMF Corporation was created under the responsibility of a Board, consisting of a representative from each of the original project proponents and the CFS, and has agreed on management delegation for both seigneuries of Abitibi Consolidated where the forest tenant farming experiment has been carried out. The BSLMF has an annual budget of approximately $1.6 million. Of this amount, the Model Forest Program contributes $500 000, and the rest originates from funds from the partners and existing programs.

How forest tenant farming works

Inspired by the past but adapted to the present, forest tenant farming consists of allotting a tract of land (a tenant farm) to a tenant (a tenant farmer) who is committed to managing and harvesting it, provided the revenues (royalties) are shared with the owner of the land. In the BSLMF, both seigneuries have been subdivided into tracts of approximately 1000 ha each, allocated to take into account the condition and successional state of the forest, and the working capacity of the tenant farmer. Consequently, 27 tracts were offered as tenant farms. Recently harvested areas have not been allotted to individual tenants because these lands do not have sufficient mature timber to be commercially viable, and such land instead is being reforested, brushed, and otherwise managed by the forest tenants cooperative.

The original selection of the tenant farmers was carried out through a call for applicants across the country and more than 350 applications were received. The selection criteria applied included entrepreneurship, leadership, attitudes towards innovation, teamwork, social involvement, and forestry experience. The selected applicants also took part in an integrated 10-week training course, where their knowledge in various fields (silviculture, safety, management, accounting, etc.) was upgraded. Each winter they have the opportunity to receive further training.

Forest tenant farmers have signed a 10-year individual agreement with the BSLMF determining their rights and responsibilities (FMBSL 1999), as summarized in Table 22.1. These agreements treat them as entrepreneurs and not as employees. As such, they

Table 22.1. A summary of the main clauses in the 10-year renewable contract between the Bas-Saint-Laurent Model Forest (BSLMF) and each forest tenant farmer. Agreement to these terms gives the tenant the right to manage and use wildlife, recreational, and timber resources on his forest farm.

The tenant farmer will:
- respect the goals and objectives of the BSLMF;
- prepare and submit a 5-year management and an annual management plan, respecting the multi-resource management plan adopted by the BSLMF;
- obey all existing legislation, regulations, and decrees;
- divulge to the Model Forest all revenues and expenditures related to the operation of the forest farm;
- ensure that working the tenant farm is his/her main occupation;
- participate in the cooperative structure formed to oversee the management of wildlife harvesting, recreation, and the regeneration of non-allocated land; and
- pay royalties to the BSLMF.

The Bas-Saint-Laurent Model Forest will:
- help the forest tenant prepare management plans;
- offer technical support and appropriate training;
- monitor the experiment; and
- use part of the royalties to maintain a compensation fund for the tenant farmer's benefit.

Fig. 22.8. Responsibilities associated with forest tenant farming.

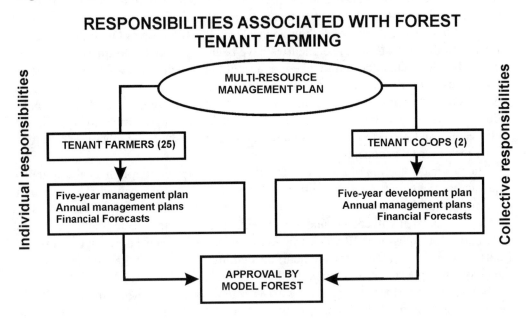

are not entitled to claim employment insurance benefits, for example, and are responsible for their own financing and insurance arrangements. The forest tenant farmers are responsible for managing timber and harvesting wood resources on their farms (Fig. 22.8) and must jointly manage the other resources (hunting, fishing, and tourism) territory-wide. Therefore, a cooperative has been established in each seigneurie, in order to provide recreational services according to a formula similar to outfitting operations. Tenant farmers live in the small villages close to the seigneuries, not on their holdings. There is no agriculture or mining on their holdings. Some holdings have chalets for fishermen and hunters, and the seigneurie cooperative is responsible for organizing outdoor recreational potential. Landscape-scale habitat management requirements are spelled out in the multi-resource management plan prepared by Model Forest staff, and the cooperative and individual tenants must adhere to those requirements.

Resource management

Before addressing land management, the BSLMF adopted a code of ethics, which acts as a safety net for environmental protection and resource sustainability. Even though this code is lengthy, it outlines the principles of sustainable development and sets two specific rules, namely a prohibition on the use of chemical herbicides, and a size limit of 4 ha for individual cutblocks. For each territory, the BSLMF eventually developed a multi-resource management plan with its partners and the local population. These plans outline four designated zones according to the level of protection that should be applied (Box 22.2).

Beyond this zoning, the BSLMF has developed a management strategy for each of the resources on the territory in order to maintain biodiversity, improve resource production capacity, and ensure the sustainable development of the forest environment. These strategies include goals and approaches for each of the territories, for forest biomass, wildlife, water, and landscape aesthetics. Individually, each tenant farmer has a management plan for his tenant farm that is compatible with the multi-resources management plan. An intervention (operations) plan is submitted each year to the BSLMF, and once it has been approved it becomes an operating license (Fig. 22.8).

Silvicultural activities are supported by the regional private forest upgrade program, which provides guidelines, advice, and financial support. Tenant farmers can access this program in the same manner as woodlot owners across the province of Quebec. But the

---

Box 22.2. Land-use zoning in the Bas-Saint-Laurent Model Forest:

Resource conservation	- no forestry intervention permitted;
Resource protection	- some forestry interventions excluded, particularly from sensitive sites;
Special management	- forestry conducted according to site-specific requirements, reflecting higher level objectives related to wildlife and other values; and
Regular management	- standard forestry practices followed according to the code of ethics, with room for extensive recreation and other land uses allowed but not constraining forestry practices.

BSLMF 2003

context of the BSLMF has bred innovation pertaining to techniques and methods suitable for working on a small scale (Fig. 22.9*a*) and in forests with diverse uses and users (e.g., low-density reforestation, scarification under a partial canopy). Technical advice in forest development is provided by a team consisting of a forester and two technicians (Fig. 22.9*b*). The BSLMF also uses a biologist who monitors the integration of different aspects outside the scope of timber development (wildlife, water, and aesthetics values).

Revenues and royalties

The average revenue of individual tenant farmers in 2000–2001 (BSLMF 2002) amounted to $122 091, with costs of $89 792, resulting in a business income of $32 299. Tenant farmer revenues stem mainly from timber sales (82%), private forest upgrade grants (10%) and, to a lesser extent, income related to recreational activities such as hunting and fishing (4%). Operating costs included salaries and payroll taxes (24%), timber transportation (19%), and stumpage or timber royalties (14%). Total sales for recreational outfitting activities amounted to more than $254 000 for the *Seigneurie de Nicolas Riou* and more than $277 000 for the *Lac-Métis Seigneurie*.

In accordance with the tenant farming formula, revenues must be shared with the land owner. In the case of the BSLMF, the tenant farmers pay a royalty on all timber harvested and sold. But because of the experimental context of the model forest, Abitibi Consolidated (as land owner) has agreed to reinvest these revenues in the project, and to have them managed by the BSLMF in compliance with the management delegation agreement through which they are bound. The Model Forest corporation uses these funds to various ends. First of all, they are used to pay property taxes and other compulsory obligations (land protection against fire, insects, and disease). They are also used to invest in the major road system and other infrastructure on the territory. More-

Fig. 22.9. (*a*) Small-scale logging practices are employed and modified as necessary by tenant farmers in the BSLMF; here a Zetor 5245 four-wheel-drive farm tractor has been modified (primarily in terms of undercarriage protection) for forest use, and is used with a locally made trailer and a Patu 595 grapple. (*b*) The training of tenant farmers and education of the general public by Model Forest staff are important components of the BSLMF approach.

(*a*)

(*b*)

Photo by J. Robert   *Photo by J. Robert*

over, the BSLMF has decided to set up a compensation fund available to tenant farmers in case of departure. Since tenant farmers improve property that does not belong to them, they are entitled to some financial compensation beyond the revenues they derive from their work. Thus, a lump sum taken directly from royalties is deposited yearly by the BSLMF on behalf of each tenant farmer. Finally, some of the royalties are returned to Abitibi Consolidated through a process of annual negotiation between the company and the tenants cooperative.

Assessment and monitoring

Like every model forest in Canada's Model Forest Network (CMFN 2002), the BSLMF has implemented a set of criteria and indicators for sustainable development in order to measure its progress (see Chap. 21). Furthermore, the forest farming experiment is subject to specific monitoring and assessment procedures on a regular basis. The tenant farmers are required by contract to keep track of their financial activities and to make them available to Model Forest personnel or its legal representatives for annual statements of accounts.

The experiment has also been subjected to a socio-economic sustainability study by an economist from Natural Resources Canada (Masse 2001). The tenant farming formula and the joint management groups were evaluated according to four criteria: the viability of tenant farms; costs of general supervision and technical support; concrete socio-economic impacts generated at the local and regional levels; and the potential for extending the tenant farming model to public forests in adjacent municipalities. The results of five studies were summarized from the perspective of these evaluation criteria, and the principal issues raised in testing and extending the forest tenant farming approach were identified. The report concluded that this management system is socio-economically viable (Masse 2001).

The degree of interest and support for this alternative form of tenure has been outstanding. When some of the first tenant farmers were replaced in 1998 (following resignations or breaches of contractual provisions), there were 105 applicants for only five available positions. We now look forward to seeing the forest tenant farming model being implemented elsewhere in Canada and around the world.

Mistik Management: engendering diversity in forest management

Introduction and background

Mistik Management Ltd. is a forest company operating in the southern boreal forest of northwestern Saskatchewan. The company is highlighted here because of the strong commitment it has made over the past 10 years to local involvement in forestry planning and operations, to active multiple resource management, and to adaptive forest management. In particular, Mistik provides a unique example of public involvement due to the initiation and funding of community-based co-management boards by industry and local inhabitants rather than by senior government agencies. While there are a number of concerns with the current process, such as the funding arrangement and issues of representivity on the co-management boards, Mistik's initiatives have been largely successful, with a high degree of participant satisfaction and sharing of management responsibilities as well as important changes in forest management practices on the ground (Hamersley Chambers 1999).

In the spring of 1998 the Meadow Lake Tribal Council (MLTC), an economic coalition of nine local First Nations, bought out Techfor's (40%) and Millar Western's (20%) ownership in NorSask Forest Products Inc. under the name of Mistik Management (Mistik). Mistik, originally a not-for-profit management company formed in 1990 by NorSask and Millar Western, is now owned by MLTC (50%) and Millar Western (50%). As a result of this buyout, the NorSask Forest Management License Area (FMLA) and the NorSask softwood sawmill are now assigned to Mistik. Mistik also continues to provide fibre for the Millar Western pulp mill, which is notable as a zero liquid discharge facility located near Meadow Lake (see Fig. 17.8). Mistik is responsible for forestry operations in the FMLA (which consists of three timber areas in a region spanning 53°–57°N, and 107°–110°W), as well as public involvement initiatives such as the community co-management boards. Although the FMLA is now the sole responsibility of Mistik, the license is still commonly referred to as the NorSask FMLA.

While the efficient procurement of logs from the FMLA is Mistik's prime mandate, the company has also embraced three strategies in its management and business approach which differentiate it from many other large forest products companies. Briefly summarized, these strategies include:

- local involvement;
- active management for multiple resources; and
- adaptive management.

These three concepts have the common goal of balancing Mistik's shareholders' business requirements with aspirations put forth by the local citizenry and the regulatory environment that governs forest development. Direct results of this commitment include Mistik's successful co-management experiment, innovative ecosystem management practices, and the development of an independent Scientific Advisory Board. Mistik has also invested in reduced-impact wood processing facilities such as portable scrag mills, which process small-diameter trees into rough-cut lumber that is then hauled away to be dried and planed at a central facility. Collectively, these developments in the application of social, ecological, and engineering sciences hold great promise for the achievement of truly sustainable forest management in the FMLA.

Local involvement

Development and implementation of advisory and co-management boards

Although forestry has been a part of the economy in the area since the 1930s, it was not until the mid-1980s that large-scale forest development began in earnest. Fairly rapid development of the forest resource, coupled with an historic lack of public involvement, led to significant conflict with local communities, culminating in an 18-month logging blockade in 1992–1993. This civil disobedience by members of the Canoe Lake First

Nation at Keeley Lake, later joined by other local and national individuals and groups, stemmed from controversy over local clearcut logging, some of which was being done by Mistik.

As a direct result of the Keeley Lake conflict, Mistik began meeting in 1992 with various local groups, including Elders, trappers, First Nations Chiefs, and elected representatives of northern communities. After reviewing the feedback from these meetings, co-management was jointly identified as the vehicle to involve the public and to mitigate and resolve any future conflict over forest operations and management. Co-management was also seen as a way to integrate community concerns and knowledge into forest management planning and practice on the ground. Although NorSask had begun developing the concept of co-management as a public involvement model prior to the blockade, this conflict significantly speeded up the implementation of co-management by industry. Co-management is described by Mistik as the process by which the companies which hold the NorSask FMLA share in the decision making regarding forestry operations with First Nations and other northern communities that have a stake in it. For an account of the circumstances that led up to public consultations concerning co-management as a form of public involvement (Chap. 4), refer to Beckley and Korber (1996).

The boundaries for each of the proposed community co-management boards were aligned with the existing Fur Conservation Areas (FCAs). The FCA boundaries, established in the 1940s, roughly correspond to communities' traditional land-use patterns such as traplines. As these have been a recognized (if not always agreed upon) geographical unit for land management for more than half a century in the area, it was felt that the FCA boundaries provided a good basis for the co-management boards. Registered traplines have been interpreted and managed as the traditional territories of extended Native families throughout northern Canada (Brody 1981), and increasingly are being analysed as "sustained yield units" for timber as well as wildlife habitat capability (e.g., Burton 2000). There are currently eight operating co-management boards and one advisory board based on 19 pre-existing FCA boundaries within the FMLA.

Funding for each of the co-management boards was initially provided wholly by industry. Once a board was registered as a not-for-profit corporation under the Saskatchewan Societies Act, they were given a $10 000 start-up grant from Mistik. In addition, except for the Divide Forest Advisory Council Corporation (DFACC) which did not want industry funding, each functioning board is given a $0.50 "donation" by Mistik for every cubic metre of wood harvested yearly within their FCA. Though at the discretion of industry in this case, it has been suggested (G.F. Weetman, Professor Emeritus, Faculty of Forestry, University of British Columbia, personal communication, May 2003) that such a dedicated allocation of timber royalties back to the local management authority (rarely practiced in Canada) is an important mechanism for promoting experimental and innovative forest management. Other forest products companies that operate in the area, such as L&M Woodproducts Inc. and Weyerhaeuser Ltd., also contribute financial or other resources to some of the co-management boards affected by their operations, but do not tie their funding to the level of cut in each FCA. For example, Weyerhauser has covered the cost of a meeting space and refreshments for two of the boards, and regularly attends some co-management board meetings.

In accordance with provincial forest regulations that require companies to outline their public involvement process, Mistik submitted the co-management model to government managers as their basis of public consultation and involvement for their FMLA. In a 1993 Memorandum of Understanding (MOU) between Saskatchewan Environment and Resources Management (SERM; recently reorganized and renamed, this government department is now referred to simply as Saskatchewan Environment) and NorSask (with Mistik as an operating party), the provincial government took the unusual step of granting industry the authority to develop co-management boards with local stakeholders and communities in the FMLA. Briefly summarized, the MOU identifies existing rights and responsibilities of the parties involved (as based on the FMLA and Aboriginal and treaty rights), supports sustainable forest management as the guiding principle of the process, and describes the level of commitment expected from NorSask and Saskatchewan Environment.

Due in part to a lack of rigid criteria imposed by Mistik or the provincial government, which took a back seat early in the process, each board in the FMLA has since developed a co-management regime and relationship unique to the stakeholders and communities involved in their particular case. Indeed, "Northern community representatives themselves are defining the role and responsibilities of each co-management board. Because of the diverse geography and the varied cultural backgrounds involved, no single approach will meet the needs of all residents across the FMLA"(*Mistik News*, 1993, vol. 1, no. 2, p. 3). The diverse structure of these boards can, in fact, be seen as one of the great strengths of the NorSask co-management regime because each community is given the opportunity to adapt the management process to its individual needs.

Industry policy and attitude

While this initiative on the part of industry would not have been possible without tacit provincial approval and active community participation, members of Mistik Management Ltd. provided the initial vision as well as substantial energy and funding to get the process started. Mistik was the key initiator of co-management, with certain staff members in leadership roles being the major proponents. What is impressive about the culture of this organization, and indeed sets this case study apart from more conventional public involvement examples (Box 22.3), is the degree to which employees at every level have bought into the initiative. Staff believe that co-management (or heightened advisory power in the case of the DFACC) is not simply a better way to do business, but is the only way to be successful in the socio-political environment in which they operate.

Box 22.3. What makes the Mistik co-management experience unique?

- Little government involvement or intervention;
- Industry initiated and supported (financially, institutionally);
- Very high level of co-management practice: i.e., revenue sharing;
- High level of participant satisfaction;
- Tangible differences in landscape management and harvesting methods;
- Locally owned, First Nations forest company; and
- High level of decision-making authority devolved to field staff.

Compared to standard practice elsewhere, and to the way things were done in the past (centralized, top-down management; see Chap. 1), the current management and planning process in the NorSask FMLA is doing a much better job of incorporating local knowledge into forest management. It is important to note, however, that Mistik recognizes the co-management boards as only one way to gather this sort of information. Other more informal processes are ongoing (e.g., individual relationships with trappers or wild rice [*Zizania palustris*] growers), as are other formal mechanisms (e.g., periodic Elders meetings sponsored by Mistik). A quote from a Mistik staff member serves to illustrate this point:

> *"The process of incorporating local knowledge into forest management is ongoing. We don't just talk to the boards — we talk to others involved — but we use the boards as a forum for discussions. We don't just rely on the boards to bring all of our concerns and plans to the communities. We also go to groups like the trappers association to ask for their input into cutting plans. Yes, there are many examples of local knowledge being incorporated into our operations."*

(Beckley and Hamersley Chambers 2002)[8]

Mistik Management has delegated a significant level of decision-making trust and powers to their field staff who attend all co-management meetings, thereby allowing these representatives to make most management decisions at the board level without conferring with their superiors. Similarly, Saskatchewan Environment is continuing to reorganize to confer greater decision-making authority to local and regional offices. Both Mistik and Saskatchewan Environment have taken positive steps to give some real legitimacy (if not teeth) to the co-management boards. While Mistik retains the legal right to harvest when and where it wants (among other things), they have to date worked with and respected the wishes of the boards.

Engaging the local labour force: contractor development, not just providing employment opportunities

Mistik returns significant benefits from development of the forest resource to local communities and resource user groups. For example, a mandate of the company is that the benefits of timber harvesting are returned to local community members. In addition, the local co-management boards are providing important employment, educational, management, and training opportunities. These include funding post-secondary scholarships, outdoor education programs in schools, and allocating forestry-related contracts such as road maintenance to local contractors in each FCA. A number of co-management boards are even providing interest-free loans and capital for small-business development in local communities.

Some issues matter a great deal to the communities, such as who does the work and who gets the contracts. This matters very little to Mistik, so long as high-quality work

---

[8]Beckley, T., and Hamersley Chambers, F. 2002. An assessment of co-management in the NorSask Forest. Unpublished internal document prepared for Mistik Management Ltd., Meadow Lake, Saskatchewan. p. 59.

is done. Mistik has astutely given decision making over these sorts of issues to the local community through their co-management board, and this truly gives them a sense of empowerment not only in gaining the work (as opposed to outside contractors), but also in being able to distribute the work in a fair and responsible way. All of the Mistik contractors are FCA-based and locally owned. This means that each job is first offered to a contractor who lives in the FCA where the work is scheduled to be done. If there is no contractor available or none have the required skills or equipment, then contractors from neighbouring FCAs are invited to bid on the job. With the number of small contractors who have developed over the past 10 years in the FMLA through the assistance of Mistik and the co-management boards, it is very unusual now for jobs to go to people outside of the FMLA. Activities undertaken by these contractors include log loading and hauling, silviculture/planting, harvesting, road building, and road maintenance. While concerns have been periodically raised regarding potential conflict of interest and nepotism, these claims have yet to be substantiated. As co-management boards continue to formalize their operating procedures and to initiate conflict of interest guidelines, the majority of these concerns should be avoided.

Benefits of the Mistik co-management initiative

The NorSask co-management process has resulted in a number of significant benefits to the parties involved. For example, the increased trust and stronger working relationships that have developed are cited by participants as major benefits of their co-management experience. Individual co-management boards are providing a forum for affected local communities to bring their concerns about forest resource development to industry and government, and for forest resource users to be recognized as a valid part of the resource management process. Most participants feel that they are afforded respect by Mistik and they feel that their concerns are sincerely listened to. Furthermore, co-management has served as an institutional mechanism for bridging the considerable social, cultural, and linguistic gulf between elderly Aboriginal trappers and young, white, professional foresters and technicians from "away". Although admittedly more expensive and time-consuming in the short term than conventional top-down resource management, participants feel that co-management is a cost-effective forum over the long term for government and industry to bring their development plans and concerns to resource users and local communities. As stated by a senior Mistik manager:

> "This [co-management] has come at significant cost to us, measured in millions of dollars in terms of inefficiencies in comparison with our competitors. But this is a short-term cost, this is a very slow process, talking about issues, then development opportunities."

> (Hamersley Chambers 1999, p. 125)

The NorSask co-management process also provides a mechanism and communications network with which to disseminate information from forest managers to co-management participants, communities, and stakeholder groups. In the words of a senior government official:

"One of the benefits of these groups is that the Boards act as a filter to the communities and the stakeholder groups and provide a forum for discussion between the different groups, which wasn't happening before, because it was one-on-one consultation only."

(Hamersley Chambers 1999, p. 125)

In addition, co-management has sped up the process by which government and industry managers respond to resource user concerns. This shared management process has also enabled the main stakeholders to learn about and to understand each others' points of view as well as their strengths and limitations in managing the forest resource. For example, since its inception in 1993, co-management has been successful in significantly reducing conflict over forestry management and operations in the FMLA (Hamersley Chambers 1999, p. 125). This success is also attributed to the more responsible forestry practices and greater accountability of industry and government that have arisen through the co-management process.

One of the greatest benefits of co-management to date in the NorSask FMLA is the incorporation of local knowledge into the forest management decision-making and planning process (Alex Maurice, Chair of the Beauval Co-Management Board, personal communication, 10 July 2002). This incorporation has resulted in more effective decision making and planning by both government and industry. In this case, "more effective" decision making is defined as that which results in decisions that are longer lasting, of a higher quality and accepted by a wider range of stakeholders than those achieved through conventional management methods (Hamersley Chambers 1999). The history, character, and keys to success for a particular co-management board are summarized in Box 22.4. Although a number of participants feel that the Mistik co-management process is far from perfect, they agree that it is a great improvement over the conventional forest management process that existed before. Despite these significant positive benefits, however, there also remain a number of concerns with the existing process as well as barriers to the implementation of this public involvement process throughout the FMLA.

Concerns

A number of concerns exist with the current co-management structure and boards. For example, there is a general apathy and lack of interest within local communities and resource user groups unless something affects an individual directly. This apathy is attributed to cultural barriers such as language and differing value systems, individuals feeling that they have nothing to gain from participation, and by a general lack of understanding within resource user groups of the co-management process. Other concerns include internal divisions, historical mistrust between stakeholders, and politics within both local communities and forest user groups. Board members, government, and industry participants unanimously agree that they lack the time to participate as much as they would like to (or feel that they should) to maximize the success of the process. This concern is likely to grow as the mandate of some of the boards continues to expand to other resource management areas.

Box 22.4. The Beauval Co-Management Board.

Beauval is generally recognized as one of the more active and visible co-management boards in the region. Beauval is a Métis community of approximately 785 people with 94% of its residents claiming Aboriginal descent in 1996. Since there are strong cultural and subsistence ties between the community of Beauval and the surrounding forest, the onset of industrial forestry development was of great concern to the community. There was also a growing desire for formal input into forest management activities to influence this development and to return benefits from these activities to the community. As a result of these concerns and the ongoing Keeley Lake logging blockade, Mistik contacted the community in 1992 and proposed co-management as a solution to forestry conflict and tensions in the area.

Seven stakeholder groups from the community have been represented on the co-management board for the past 10 years: trappers, commercial fishermen, wild rice growers, cabin owners, outfitters, Village Council, and an Elder/traditional user. Representatives from Mistik Management, Saskatchewan Environment, and Weyerhaeuser Canada attend meetings on a regular basis, but are not board members and do not have voting privileges. Presently, the Beauval board is involved in forestry-related activities such as reviewing cutting plans and lease applications as well as incorporating local knowledge and data into management plans. At a community level, the board provides educational scholarships, small business loans, and other financial support to the community, as well as facilitating training and employment opportunities in activities such as silviculture and road maintenance. The board also seeks funding for community projects from government and industry. The Beauval board has expanded its original mandate of addressing forestry management issues to encompass those of community health and economic development.

At present, the Beauval board operates under the terms of the 1993 Memorandum of Understanding (MOU) between the Saskatchewan Environment and NorSask, and an MOU signed in 1997 with Weyerhaeuser Canada, which holds the eastern portion of the Fur Conservation Area (FCA) in their neighbouring Forest Management Agreement area. In signing the latter MOU, the Beauval board gained recognition of its role in the management process for the entire N-12 FCA.

The board operates under a hybrid of Roberts Rules of Order and consensus decision making. Business is conducted through the passing of motions while major decisions are made through consensus. Board members feel that, for the most part, participation among members is equal, though there is variation in how much individual board members speak up.

The board attributes its success to a number of factors, including:

- strong leadership and members with good communication and intercultural skills;
- long-term board members and participants who are committed to the process;
- industry and government participants with real decision-making powers;
- a reasonably healthy community with strong ties to the bush economy;
- having formal operating bylaws and decision-making rules;
- having respect for each other and the co-management process as well as trust;
- a small board, which makes for a more efficient process;
- a proactive approach to management;
- the incorporation of local knowledge into forest management;
- being considerate of local forest uses and values; and
- offering real training, education, employment, and economic benefits to the community.

Although government representatives attend meetings on a regular basis, Saskatchewan Environment does not agree with the current revenue-sharing arrangement between the boards and Mistik, and will not contribute financially to the process until a formal agreement regarding this issue has been signed. While Saskatchewan Environment has attempted to sign a formal agreement with the Green Lake and Beauval boards, neither was willing to sign on to the version of co-management for which government was prepared to offer funding. This was primarily because the level of authority and decision-making powers formally offered by Saskatchewan Environment representatives was significantly lower than that already enjoyed by both of these boards, relegating them to simple "advisory" roles. From Saskatchewan Environment's point of view, the funding for the co-management boards should not be tied to timber harvesting levels in each FCA, and each board must do a better job of representing regional and provincial interests before government funding can be provided to the process.

Whether industry should continue to be the major funding source for public involvement needs to be reviewed, as many participants feel that the corporate sponsorship of co-management is a problem. The issue is not so much that the communities should not receive benefits from harvesting, but more that the money should not come directly from Mistik, or that it should not be directly tied to the volume of wood removed from each of the FCAs. That is, there is a perceived conflict of interest that could lead to over-exploitation of the resource, though it can be argued that all provincial governments currently suffer from this same conflict of interest (in that they collect royalties from stumpage as well as being charged with protecting the environment and setting the rate of cut).

While participants cite the lack of direct government control and intervention at the start of the co-management process as a main reason for its current success, the general lack of provincially sanctioned decision-making authority and jurisdiction of the boards is increasingly recognized as a significant barrier to the future security of co-management. Government reluctance to share management power and its inherent slowness to change are exacerbating this problem. This frustration is echoed in this statement by a government participant in the process:

> *"What happened is that industry moved faster than government. Mistik's focus is forestry management — they can change overnight. SERM's focus is resource management, which is much more diverse. It is much more difficult for government to change, to adapt, than industry."*

(Hamersley Chambers 1999, p. 127)

It is reasonable to assume that the longer the fundamental issues of government support and acceptance of co-management are left unresolved, the more difficult it will be to reach a formal, legally binding agreement among all parties.

Active multiple resource management

Commitment to manage the forest for multiple resources and users

Mistik has demonstrated a strong commitment to manage its forest land base for multiple resources and a wide variety of forest users. For example, the company has identified important cultural, social, economic, and historical sites through its public involvement processes. Planners and harvesters either avoid these areas or undertake activities in consultation with local inhabitants, mostly through the co-management boards. For example, Mistik implemented a 500-m buffer on the historic Keeley Portage, which is not an historic site recognized by government but has great significance to local people. Some forest users, such as berry pickers or wild rice growers, request that Mistik cut in certain areas to either increase the berry harvest or to provide road access to a wild rice lake. In conjunction with the recommendations of the relevant co-management board, these requests are granted where possible. Mistik also maintains buffers of 15–30 m around lakes, mostly to protect fish habitat rather than to protect wild rice (though see Chap. 10 regarding questions of how effective this practice is).

This trade-off between timber and non-timber forest values is negotiated through the co-management boards as well as through additional public involvement processes such as Elders workshops. For Mistik, the tradeoff is that approximately 3–4% of the merchantable volume of timber is being retained to manage for these other values (Roger Nesdoly, Planning Coordinator, Mistik Management Ltd., personal communication, 15 July 2002). This tradeoff is implicitly part of the company's commitment to integrated resource management. On a more formal level, the Project Specific Guidelines set out in Mistik's 20-year Forest Management Plan state that the company shall implement integrated resource management and public consultation (co-management) in the development and implementation of its forestry planning and operations. A commitment to manage the FMLA for multiple resources is a significant part of Mistik's integrated resource management approach and of the community-based co-management process.

Developing an approach to forestry modelled on natural disturbance patterns across a range of temporal and spatial scales

Mistik takes the public expectation of a style of forestry that maintains (to the fullest extent possible) a "natural" landscape and all of its attendant ecosystem resources and functions very seriously. At the planning and landscape levels, the company is implementing a new technique of larger disturbances (landscape disturbance dynamics) that last for a shorter duration than conventional harvesting methods. This approach (also adopted by Alberta-Pacific, as described above) results in logging activities being a disturbance that more closely emulates aspects of a wildfire (Chap. 9), leaving a variety of structures (some dispersed, some in patches) of differing sizes and species on each cutblock (e.g., as in Fig. 22.3). Benefits of this approach are that not as much active road is needed, the roads are decommissioned faster (allowing less access for hunters and lower hydrological impacts), and harvesting will not occur again for a longer time period. As a relatively new initiative, Mistik sees this approach as a part of adaptive management — learning by doing, and doing by learning.

Maintaining habitat for ungulates and other wildlife populations

Mistik maintains habitat for deer, moose, and other wildlife populations through consultation with local forest users such as trappers, hunters, guides, and outfitters, and local communities. For example, a password-protected layer on Mistik's GIS map of the FMLA encourages Elders and other hunters to identify key habitat areas such as salt licks so that these are not negatively impacted by forestry activities. This protection of important habitat is also accomplished through modelling of forest development plans (road building and cutblock layout and distribution; see Chaps. 12 and 14) in a manner that attempts to leave a range of habitats across the landscape.

Adaptive management

The commitment to continual improvement

Adaptive management, or making the strategic and philosophical commitment to continual improvement of all phases of management (Chap. 21), is fundamental to the way that Mistik does business. Adaptive management is viewed as a process for piloting concepts, evaluating results against predictions and planned actions, and continually adjusting to changing conditions in the forest. This process progressively selects more successful methods to achieve desired forest management results. Mistik seeks to plan and conduct all its forest use activities within the FMLA based on the following key principles, as stated in its 5-year annual operating plan (MML 1999):

(1) to know the land base and the people;
(2) to understand forest ecosystem processes;
(3) to protect sensitive forest ecosystem attributes;
(4) to maintain forest ecosystem integrity;
(5) to minimize impacts; and
(6) to mitigate negative impacts.

A key focus area of adaptive management is improving road designs and harvest methods (see Chap. 15) in a manner that should maintain or enhance wildlife habitat. The issue of road planning, construction, maintenance, and decommissioning is significant, as roads can have major ecological impacts (see Forman and Alexander 1998). For example, feller bunchers do not kill moose, but logging roads are a vector for vehicles, hunters, and more effective predation by wolves. The new larger disturbance process that Mistik is implementing has fewer roads, which are in place for a shorter period of time, and the expected reduction of impacts on ungulates is one of the reasons behind adopting this new method.

To assist in designing, evaluating, and implementing adaptive management strategies, Mistik commissioned the formation of a Science Advisory Board (SAB) in 1996. The SAB is composed of 10 leading North American scientists from a number of forestry disciplines. Current members have expertise and research interests ranging from the social impacts of forestry operations to landscape ecology. While the SAB has not met directly for the past 2 years due to financial constraints, the Board is active and members communicate through e-mail and phone calls. The SAB has three main functions:

- to think strategically, looking at the longer term, big picture (i.e., 20-year plan; see Chap. 11);
- to design and conduct research; and,
- to suggest how Mistik should proceed in the future to achieve the goals to the 20-year forest plan (especially with regard to issues such as road hydrology, harvesting techniques, and public involvement).

For example, the SAB commissioned a 5-year evaluation of the co-management and advisory council processes to keep groups informed of Mistik's activities and to assist in solving local issues.

Prospects for the future

The future of Mistik Management looks good. Although the company has been hit hard by the softwood lumber dispute with the United States (see Chap. 7), this upset is forcing the company to seek local markets and to diversify its customer base, a trend that will likely be beneficial in the long run. The SAB, although reduced in comparison to its initial activities, is still active and providing independent and scientifically based advice to Mistik regarding harvesting techniques, landscape-level planning, and the social dimensions of forestry. The community co-management boards, some of which have been operating for more than a decade now, have significantly reduced conflict over forestry activities and have provided unprecedented training, employment, and educational opportunities for local inhabitants. The buyout of Techfor and Millar Western's shares in NorSask by MLTC has resulted in a situation where nine local First Nations are simultaneously 50% owners of the forest company and the majority of local inhabitants affected by forestry operations. How this new structure will impact community relations and forestry operations has yet to be determined. With the recent reduction in the size of the FMLA from its original 3 300 000–1 700 000 ha (due to the provincial government's decision to reallocate timber that it felt was not being sufficiently utilized), it will also be interesting to see if Mistik is able to maintain the flexibility and patience that it has demonstrated for the past 10 years in adapting to local community concerns and desires regarding forestry operations. With their previous track record, it seems likely that Mistik will continue to thrive and change.

Conclusions

The predominant factors regulating the change from sustained-yield timber management to sustainable forest management are cost, public pressure, and science-based information. In most cases, increased cost is incurred through reduction in the AAC, and the costs of planning and operations that range 5–50% above those of standard command-and-control planning followed by clearcutting and even-aged timber management (Hebert 2003). Without adequate scientific knowledge, initial attempts to implement SFM have been generally prescriptive, ecologically conservative, and often disproportionate to the desired outcome. Without information on thresholds, it is difficult to determine realistic costs and ecologically efficient targets that can be tested.

In general, the forest products industry is attempting to move towards SFM, but is having trouble developing multi-scale, SFM frameworks with testable targets that can be adjusted using adaptive management and the modelling of tradeoffs. Not only must tradeoffs within scales (e.g., variable retention at the stand level *vs* extended rotations) be examined, but tradeoffs between ecological and economic components must be assessed. Currently, the Canadian forest products industry has spent insufficient time with scientists and visionary thinkers in developing appropriate, articulate research questions. In order to produce relevant questions, the industry must formulate an SFM framework that includes multi-scale strategies with social, ecological, and technological components.

There is obviously room for improvement in the efficiency, environmental stewardship, and social responsibility of forest products companies and forest management agencies across Canada and throughout the circumboreal region. But the cases described in this chapter illustrate that research and innovation, as promoted by the

---

Box 22.5. Some effective options for sustainable forest management.

- Develop a Sustainable Forest Management (SFM) framework specific to a company or an agency, itemizing changes to be implemented in both planning and operational procedures (and their impacts) in moving from solely a sustained-yield paradigm;

- Identify knowledge gaps, and design appropriate research programs to collect needed information;

- Note that socio-political constraints to implementing SFM are often more immediate than a lack of knowledge or technical skills; dialogue and mutual education with government agencies and the general public is often needed;

- Take advantage of widespread concern about forest practices and economics to engage the public in participatory planning, and to push for innovative arrangements for tenure and resource stewardshop;

- Employ innovative leaders who can articulate a vision and orchestrate the necessary support to try new things;

- Minimize the industrial footprint, employing both technological and planning methods;

- There currently appear to be ecological, aesthetic, and public relations benefits to forest management techniques which emulate nature as much as possible; the opportunity costs of "soft-touch" forestry may be offset by more intensive practices elsewhere;

- Consider the contribution of the non-harvest land base and the broader regional context in planning to meet overall conservation and socio-economic goals;

- Utilize modelling and trade-off analysis to assess cross-scale alternatives for achieving any desired ecological or socio-economic objective;

- Identify key indicators of sustainability that operate at both the planning (e.g., habitat elements) and implementation (e.g., population levels) phases;

- Develop multi-scale adaptive management trials (e.g., seral stage gradient, stand retention gradient) to assess thresholds and the quantity of forest types and stand structures required to meet ecological objectives; and

- Make significant commitments to multi-resource inventory and monitoring, to minimizing and mitigating environmental impacts, and to the evaluation of past decisions.

SFMN and the Model Forests Program, can be nurtured with appropriate support, and eventually implemented in various operational settings. Note that the initiation and financial support for innovation has variously come from industry, academia, government agencies, or industry–community arrangements that often circumvent government involvement. Clearly, there are many options (Box 22.5), only a few of which have been described in this chapter or this book, for conserving biodiversity, for minimizing environmental impacts, for sustaining economic prosperity, and for supporting communities in meaningful and effective ways. The successful attainment of the broad set of forest values, including wildlife, habitat, rural lifestyles, community stability, corporate and provincial revenue, etc., is difficult under even the best management system. And in some cases, achievement of such a broad set of often-conflicting goals simply may be impossible to achieve on a single forest land base. Therein lies the incentive for all forest managers to undertake new and different approaches to sustainable forest management.

References

(AEP) Alberta Environmental Protection. 1993. Special Places 2000: Alberta's natural heritage. Alberta Department of Environmental Protection, Edmonton, Alberta. 15 p.

(AEP) Alberta Environmental Protection. 2001.What is Special Places? Alberta Department of Environmental Protection, Edmonton, Alberta. Available at http://www.cdtest.gov.ab.ca/preserving/parks/sp_places/index.asp [viewed 13 October 2002].

Angelstam, P., and Pettersson, B. 1997. Principles of present Swedish forest biodiversity management. Ecol. Bull. **46**: 191–203.

(ANHIC) Alberta Natural Heritage Information Centre. 2002. Map of Alberta's natural regions and subregions. Alberta Community Development, Edmonton, Alberta. Available at http://www.cd.gov.ab.ca/preserving/parks/anhic/natural_regions_map.asp [viewed 5 June 2003].

(APFII) Alberta-Pacific Forest Industries Inc. 2000. Detailed forest management plan. Alberta-Pacific Forest Industries Inc., Boyle, Alberta. 170 p.

(APFII) Alberta-Pacific Forest Industries Inc. 2001. Forest stewardship report. Alberta-Pacific Forest Industries Inc., Boyle, Alberta. 54 p.

Beckley, T.M., and Korber, D. 1996. Clearcuts, conflict and co-management: experiments in consensus forest management in northwest Saskatchewan. Canadian Forest Service, Edmonton, Alberta. Pub. NOR-X-349. 19 p.

Bergeron, Y., and Brisson, J. 1990. Fire regime in red pine stands at the northern limit of the species' range. Ecology, **71**: 1352-1364.

Bergeron, Y., and Leduc, A. 1998. Relationships between change in fire frequency and mortality due to spruce budworm outbreak in the southeastern Canadian boreal forest. J. Veg. Sci. **9**: 492–500.

Bergeron, Y., Leduc, A., Harvey, B., and Gauthier, S. 1999. Basing forest management on natural disturbance: stand- and landscape-level considerations. For. Chron. **75**: 49–54.

Bergeron, Y., Leduc, A., Harvey, B.D., and Gauthier, S. 2002. Natural fire regime: a guide for sustainable management of the Canadian boreal forest. Silv. Fenn. **36**: 81–95.

Binkley, C.S., Percy, M., Thompson, W.A., and Vertinsky, I.B. 1994. A general equilibrium analysis of the economic impact of a reduction in harvest levels in British Columbia. For. Chron. **70**: 449–454.

Bourgeron, P.S., and Jensen, M.G. 1993. An overview of ecological principles for ecosystem management. *In* Eastside forest ecosystem health assessment, vol. II — ecosystem management: principles and applications. *Edited by* M.E. Jensen and P.S. Bourgeron. USDA Forest Service, Portland, Oregon. pp. 49–60.

Brody, H. 1981. Maps and dreams: Indians and the British Columbia frontier. Douglas & McIntyre, Vancouver, British Columbia. 297 p.

(BSLMF) Bas-Saint-Laurent Model Forest. 2002. Forest tenant farming, activity report 2000–2001. Bas-Saint-Laurent Model Forest, Rimouski, Quebec. 4 p.

(BSLMF) Bas-Saint-Laurent Model Forest. 2003. La forêt modèle du Bas-Saint-Laurent: plan d'aménagement multiressource. Bas-Saint-Laurent Model Forest, Rimouski, Quebec. Available at http://www-foret.fmodbsl.qc.ca/pam/index_pam.html [viewed 9 June 2003].

Burton, P.J. 1995. The Mendelian compromise: a vision for equitable land use allocation. Land Use Policy, **12**: 63–58.

Burton, P.J. 2000. A landscape level assessment of proposed logging in the Fontas River/Ekwan Lake Area. Contract report prepared for the Fort Nelson First Nation, Fort Nelson, British Columbia Symbios Research & Restoration, Smithers, British Columbia. 31 p.

(CCFM) Canadian Council of Forest Ministers. 1995. Defining sustainable forest management: a Canadian approach to criteria and indicators. Canadian Council of Forest Ministers, Natural Resources Canada, Ottawa, Ontario. Available at http://www.ccfm.org/ci/framain_e.html [viewed 5 June 2003].

(CCFM) Canadian Council of Forest Ministers. 1997. Criteria and indicators of sustainable forest management in Canada. Technical Report. Canadian Council of Forest Ministers, Natural Resources Canada, Ottawa, Ontario. Available at http://www.ccfm.org/ci/pdf/tech/ci_e.pdf [viewed 5 June 2003]. 137 p.

(CFS) Canadian Forest Service. 1975. Environmental stress in the forest. Northern Forestry Research Centre, Canadian Forest Service, Environment Canada, Edmonton, Alberta. Forestry Report, vol. 4, no. 2, pp. 1–8.

(CMFN) Canadian Model Forest Network. 2002. The Canadian Model Forest Network. Canadian Model Forest Network, Natural Resources Canada. Ottawa, Ontario. Available at http://www.modelforest.net/e/home_/indexe.html [viewed 12 October 2002].

Corkum, C.V., Fisher, J.T., and Boutin, S. 1999. Investigating influences of landscape structure on small mammal abundance in Alberta's mixed-wood forest. *In* Proceedings of the 1999 Sustainable Forest Management Network Conference, Science and Practice: Sustaining the Boreal Forest, 14–17 February 1999, Edmonton, Alberta. *Edited by* T.S. Veeman, D.W. Smith, B.G. Purdy, F.J. Salkie, and G.A. Larkin. Sustainable Forest Management Network, Edmonton, Alberta. pp. 29–35.

Cumming, S.G. 1997. Landscape dynamics of the boreal mixedwood forest. Ph.D. thesis, University of British Columbia, Vancouver, British Columbia. 235 p.

Cumming, S.G. 2001. A parametric model of the fire-size distribution. Can. J. For. Res. **31**: 1297–1303.

Cumming, S.G., Burton, P.J., and Klinkenberg, B. 1996. Canadian boreal mixed wood forests may have no "representative" areas: some implications for reserve design. Ecography, **19**: 162–180.

Dansereau, P., and Bergeron, Y. 1993. Fire history in the southern boreal forest of northwestern Quebec. Can. J. For. Res. **23**: 25–32.

Eberhart, K.E., and Woodward, P.M. 1987. Distribution of residual vegetation associated with large fires in Alberta. Can. J. For. Res. **17**: 1207–1212.

Environment Canada. 1990. Canada's green plan for a healthy environment. Environment Canada, Ottawa, Ontario. 174 p.

(FMBSL) La Forêt Modèle du bas Saint-Laurent. 1999. Actes du Symposium, La Ferme forestière en métayage: Résultats, perspectives et enjeux — Proceedings of The Forest Tenant Farm Symposium: Assessment, Perspectives and Issues at Stake. La Forêt Modèle du bas Saint-Laurent. Rimouski, Quebec. 175 p.

Forman, R.T.T., and Alexander, L.E. 1998. Roads and their major ecological effects. Ann. Rev. Ecol. Syst. **29**: 207–231.

Franklin, J.F., Berg, D.R., Thornburgh, D.A., and Tappeiner, J.C. 1997. Alternative silvicultural approaches to timber harvesting: variable retention harvest systems. *In* Creating a forestry for the 21st century: the science of ecosystem management. *Edited by* K.A. Kohm and J.F. Franklin. Island Press, Washington, D.C. pp. 111–139.

Gauthier, S., Leduc, A., and Bergeron, Y. 1996. Forest dynamics modelling under a natural fire cycle: a tool to define natural mosaic diversity in forest management. Environ. Monit. Assess. **39**: 417–434.

Gauthier, S., De Grandpré, L., and Bergeron, Y. 2000. Differences in forest composition in two boreal forest ecoregions of Quebec. J. Veg. Sci. **11**: 781–790.

Grumbine, R.E. 1994. What is ecosystem management? Conserv. Biol. **8**: 27–38.

Hamersley Chambers, F. 1999. Co-management of forest resources in the NorSask Forest Management License Area, Saskatchewan: a case study. M.A. thesis, University of Calgary, Calgary, Alberta. 183 p.

Hannon, S.J. 1999. Avian response to stand and landscape structure in burned and logged landscapes in Alberta. *In* Proceedings of the 1999 Sustainable Forest Management Network Conference, Science and Practice: Sustaining the Boreal Forest, 14–17 February 1999, Edmonton, Alberta. *Edited by* T.S. Veeman, D.W. Smith, B.G. Purdy, F.J. Salkie, and G.A. Larkin. Sustainable Forest Management Network, Edmonton, Alberta. Sustainable Forest Management Network, Edmonton, Alberta. pp. 24–29.

Hannon, S.J., and McCallum, C. 2002. Using the focal species approach for conserving biodiversity in landscapes managed for forestry. Sustainable Forest Management Network Synthesis Paper. Department Biological Sciences, University of Alberta, Edmonton, Alberta. Available at http://www.biology. ualberta.ca/faculty/susan_hannon/uploads/pdfs/white_ paper_focal_spp.pdf [viewed 5 June 2003]. 57 p.

Harvey, B. 1999. The Lake Duparquet research and teaching forest: building a foundation for ecosystem management. For. Chron. **75**: 389–393.

Harvey, B., Leduc, A., Gauthier, S., and Bergeron, Y. 2002. Stand-landscape integration in natural disturbance-based management of the southern boreal forest. For. Ecol. Manage. **155**: 369–385.

Hebert, D. 2003. Emulating natural forest landscape disturbances: an industry perspective. *In* Emulating natural forest landscape disturbances: concepts and applications. Conference Proceedings, 11–16 May 2002, Sault Ste. Marie, Ontario. *Edited by* A. Perere and L. Buse. 28 p. In press.

Hebert, D.M., Sklar, D., Wasel, S., Ghostkeeper, E., and Daniels, T. 1995. Accomplishing partnerships in the boreal mixedwood forests of northern Alberta. Trans. N. Am. Wildl. Nat. Resour. Conf. **60**: 433–438.

Hunter, M.L. 1991. Coping with ignorance: the coarse filter strategy for maintaining biodiversity. *In* Balancing on the brink of extinction: the Endangered Species Act and lessons for the future. *Edited by* K.A. Kohm. Island Press, Washington, D.C. pp. 266–281.

Kafka, V., Gauthier, S., and Bergeron, Y. 2001. Fire impacts and crowning in the boreal forest: study of a large wildfire in western Quebec. Int. J. Wildland Fire, **10**: 119–127.

Kessler, W.B., Salwasser, H., Cartwright, C.W., and Caplan, J.A. 1992. New perspectives for sustainable natural resources management. Ecol. Applic. **2**: 221–225.

Kimmins, J.P. 2003. Old growth forest: an ancient and stable sylvan equilibrium, or a relatively transitory ecosystem condition that offers people a visual and emotional feast? Answer — it depends. For. Chron. **79**: 429–440.

Landres, P.B., Morgan, P., Swanson, F.J. 1999. Overview of the use of natural variability concepts in managed ecological systems. Ecol. Applic. **9**: 1179–1188.

Lee, P., and Crites, S. 1999. Early successional deadwood dynamics in wildfire and harvest stands. *In* Proceedings of the 1999 Sustainable Forest Management Network Conference, Science and Practice: Sustaining the Boreal Forest, 14–17 February 1999, Edmonton, Alberta. *Edited by* T.S. Veeman, D.W. Smith, B.G. Purdy, F.J. Salkie, and G.A. Larkin. Sustainable Forest Management Network, Edmonton, Alberta. Sustainable Forest Management Network, Edmonton, Alberta. pp. 601–606.

Lindenmayer, D.B., and Franklin, J.F. 2003. Conserving forest biodiversity: a comprehensive multiscaled approach. Island Press, Washington, D.C. 351 p.

Little, L.R., and Smyth, C.L. 1999. Vegetation survival and recovery in fire affected areas of the boreal for-
est in northern Alberta (abstract). *In* Proceedings of the 1999 Sustainable Forest Management Network
Conference, Science and Practice: Sustaining the Boreal Forest, 14–17 February 1999, Edmonton,
Alberta. *Edited by* T.S. Veeman, D.W. Smith, B.G. Purdy, F.J. Salkie, and G.A. Larkin. Sustainable
Forest Management Network, Edmonton, Alberta. Sustainable Forest Management Network, Edmon-
ton, Alberta. p. 780.

(LPCL) Louisiana-Pacific Canada Ltd. 1995. 10-year forest management plan: 1995–2004. Volume 1.
Swan Valley Forest Research Division, Louisiana-Pacific Canada Ltd., Swan River, Manitoba. 443 p.
plus appendices.

(LPCL) Louisiana-Pacific Canada Ltd. 1998. Standard operation procedures, second edition. Swan Valley
Forest Research Division, Louisiana-Pacific Canada Ltd., Swan River, Manitoba. 130 p.

Manitoba Conservation. 2000. Development of a forest lands inventory for Manitoba. Final report. Forest
Lands Inventory Technical Advisory Committee, Manitoba Conservation, Winnipeg, Manitoba. 50 p.

Manitoba Environment. 1996. Environment Act license no. 2191E, general terms and conditions, 27 May
1996 – 1 January 2006. Manitoba Environment, Winnipeg, Manitoba. 9 p.

Masse, S. 2001. Socio-economic viability of forest tenant farming: evaluation report. Canadian Forest Service, Natural Resources Canada, Sainte Foy, Quebec. 75 p.

(MML) Mistik Management Ltd. 1999. Annual 5-year operating plan 2000–2001 and 2002–2005. Mistik
Management Ltd., Meadow Lake, Saskatchewan.

Moses, R.A., and Boutin, S. 2001. The influence of clear-cut logging and residual leave material on small
mammal populations in aspen-dominated boreal mixedwoods. Can. J. For. Res. **31**: 483–495.

Pratt, L., and Urquhart, I. 1994. The last great forest: Japanese multinationals and Alberta's northern
forests. NeWest Press, Edmonton, Alberta. 222 p.

Prepas, E.E., B. Pinel-Alloul, D. Planas, G. Methot, S. Paquet, and S. Reedyk. 2001. Forest harvest impacts
on water quality and aquatic biota on the boreal plain: introduction to the TROLS lake program. Can.
J. Fish. Aquat. Sci. **58**: 421–436.

Reinhardt, F. 1999. Bringing the environment down to Earth. Harvard Bus. Rev. **77**(4): 149–157.

Rowe, J.S. Rowe. Forest regions of Canada. Department Fisheries and Environ, Ottawa, Ontario. Can. For.
Serv. Pub. 1300. 172 p.

Sauchyn, D.J. 2000. Ecosite mapping and dendroclimatology, Duck Mountains, west central Manitoba.
Sustainable Forest Management Network, Edmonton, Alberta. Proj. Rep. 2000-33. Available at
http://sfm-1.biology.ualberta.ca/english/pubs/PDF/PR_2000-33.pdf [viewed 8 June 2003]. 26 p.

Schieck, J., and Hobson, K. 1999. Changes in bird communities during succession within mixedwood
boreal forest: difference between harvest and wildfire stands. *In* Proceedings of the 1999 Sustainable
Forest Management Network Conference, Science and Practice: Sustaining the Boreal Forest, 14–17
February 1999, Edmonton, Alberta. *Edited by* T.S. Veeman, D.W. Smith, B.G. Purdy, F.J. Salkie, and
G.A. Larkin. Sustainable Forest Management Network, Edmonton, Alberta. Sustainable Forest Man-
agement Network, Edmonton, Alberta. pp. 611–615.

Schieck, J., and Nietfeld, M. 1995. Bird species richness and abundance in relation to stand age and structure in aspen mixedwood forests in Alberta. *In* Relationships between stand age, stand structure and
biodiversity in aspen mixedwood forests in Alberta. *Edited by* J.B. Stelfox. Alberta Environment Centre and Canadian Forest Service, Edmonton, Alberta. pp.115–157.

Schmiegelow, F.K.A., Machtans, C.S., and Hannon, S.J. 1997. Are boreal birds resilient to fragmentation?
An experimental study of short-term community responses. Ecology, **78**: 1914–1932.

Schneider, R.R., Stelfox, J.B., Boutin, S., and Wasel, S. 2003. Managing the cumulative impacts of land
uses in the Western Canadian Sedimentary Basin: a modeling approach. Conserv. Ecol. [online], **7**(1):
8. Available at http://www.consecol.org/vol7/iss1/art8 [viewed 5 June 2003].

Seymour, R.S., and Hunter, M.L. 1992. New forestry in eastern spruce–fir forests: principles and applications in Maine. Maine Agricultural Experiment Station, Orono, Maine. Misc. Publ. 716. 36 p.

Simpson, R. 2001. Integrated landscape management. Alberta Chamber of Resources, Edmonton, Alberta. Available at http://www.acr-alberta.com/Projects/integrated_landscape_management.htm [viewed 5 June 2003].

Smith R.E., Veldhuis, H., Mills, G.F., Eilers, R.G., Fraser, W.R., and Lelyk, G.W. 1998. Terrestrial eco-zones, ecoregions and ecodistricts of Manitoba: an ecological stratification of Manitoba's natural land-scapes. Land Resource Unit, Brandon Research Centre, Research Branch, Agriculture and Agri-Food Canada, Winnipeg, Manitoba. Tech. Rep. 1998-9E. 320 p. plus map at 1:1 500 000 scale.

Song, S.J. (Editor). 2002. Ecological basis for stand management: a synthesis of ecological responses to wildfire and harvesting. Alberta Research Council, Vegreville, Alberta. Available at http://www.arc.ab.ca/forest/ECOLOGICAL%20BASIS%20FOR%20STAND%20MANAGEMENT.pdf [viewed 5 June 2003]. 450 p.

Stelfox, J.B. (Editor). 1995. Relationships between stand age, stand structure, and biodiversity in aspen mixedwood forests in Alberta. Alberta Environmental Center, Vegreville, Alberta, and Canadian Forest Service, Edmonton, Alberta. 308 p.

Stelfox, J.B., and Wasel, S. 2002. Issues of sustainability of Alberta's land uses. Presented at "The Land Supports Us All" Land Use Conference, 14–16 January 2002, Edmonton, Alberta. Transcript available at http://www.landuse.ab.ca/Speeches/Dr.%20Brad%20Stelfox_ speaker.pdf [viewed 5 June 2003]. 13 p.

Strong, W.L., and Leggat, K.R. 1992. Ecoregions of Alberta. Department of Environmental Protection, Edmonton, Alberta. 110 p.

Swanson, F.J., Jones, J.A., Wallin, D.A., and Cissel, J.H. 1993. Natural variability — implications for ecosystem management. In Eastside forest ecosystem health assessment, vol. II — ecosystem management: principles and applications. Edited by M.E. Jensen and P.S. Bourgeron. USDA Forest Service, Portland, Oregon. PNW-GTR-318. pp. 85–100.

Tardif, J., and Bergeron, Y. 1992. Analyse écologique des peuplements périlacustres de frêne noir (Fraxinus nigra Marsh.) en forêt boréale Abitibienne. Can. J. Bot. 70: 2294–2302.

TetrES. 1995. Environmental impact statement, Louisiana-Pacific Canada Ltd. Forest Managment License #3, 10-year Forest Management Plan (1996–2005), 2 volumes. Report to Louisiana-Pacific Canada Ltd., Swan Valley, Manitoba. Prepared by TetrES Consultants Inc., Winnipeg, Manitoba. Submitted to Manitoba Conservation, Winnipeg, Manitoba. (Available through Manitoba Public Registry.)

Zoladeski, C.A., Wickware, G.M., Delorme, R.J., Sims, R.A., and Corns, I.G.W. 1995. Forest ecosystem classification for Manitoba: field guide. Canadian Forest Service, Edmonton, Alberta. Spec. Rep. 2. 205 p.

Zolkewich, S. 1999. Wetlands in the forest: Ducks Unlimited's western boreal forest initiative. Conservator, vol. 19, no. 2, editorial. Available at http://www.ducks.ca/conservator/192/ boreal.html [viewed 5 June 2003].

Chapter 23

Sustainable forest management as license to think and to try something different

Timothy T. Work, John R. Spence, W. Jan A. Volney, and Philip J. Burton

"The intensity of a conviction that a hypothesis is true
has no bearing over whether it is true or not."

Peter Medawar (1979)

Introduction

The overriding message of this book is that sustainable forest management (SFM) is a broad, multi-component process best expressed as an ongoing framework for development rather than a state to be achieved and forever proclaimed. The necessarily flexible aspects of this approach to forest management are evident in the multiple meanings of "sustainability". Most can generally agree that forest sustainability now embraces more than sustained timber yield, and has come to include the stewardship of non-timber values, equity of benefits derived from forests, equality in decision making and forest planning, and the protection of all forest values for future generations. In the philosophy of SFM, these recently emerged principles are to be balanced with the long established principles of industrial efficiency and positive economic return for investors associated with the harvest of timber and the manufacture of wood, pulp, and paper products. However, even taken together, these principles do not specify recipes for action by managers. It is clear that appropriate ways of implementing SFM depend on the context and the options available. As our understanding of context improves and research identifies new

T.T. Work.[1] Department of Renewable Resources, 442 Earth Sciences Building, University of Alberta, Edmonton, Alberta T6G 2E9, Canada.

J.R. Spence. Department of Renewable Resources, University of Alberta, Edmonton, Alberta, Canada.

W.J.A. Volney. Canadian Forest Service, Northern Forestry Centre, Edmonton, Alberta, Canada.

P.J. Burton. Symbios Research & Restoration, Smithers, British Columbia, Canada.

[1]Author for correspondence. Telephone: 780-492-6965. e-mail: twork@ualberta.ca

Correct citation: Work, T.T., Spence, J.R., Volney, W.J.A., and Burton, P.J. 2003. Sustainable forest management as license to think and to try something different. Chapter 23. *In* Towards Sustainable Management of the Boreal Forest. *Edited by* P.J. Burton, C. Messier, D.W. Smith, and W.L. Adamowicz. NRC Research Press, Ottawa, Ontario, Canada. pp. 953–970.

options, the balance of principles considered appropriate in a given situation will change, and new, improved prescriptions will result. Nobody involved in the SFM enterprise can be spared the tasks of thinking, adapting, and contributing to the search for the desired balance.

In the most general sense, SFM has been thought of as identifying the optimal balance of trade-offs between social values and economic needs considered in the context of the ecological limitations of forest ecosystems. However, as pointed out in Chap. 2, reconciling competing values and implementing "wise use" of forest resources reveal the fallacy of a strict conceptual compartmentalization of SFM into social, economic, and ecological factors. Such a division with the intention of examining each facet in the context of "separate but equal" consideration ignores the links among these and other issues. The challenges we face in developing an enduring framework for SFM include recognition of the full range of individual, cultural, and regional values, reconciliation of these values into meaningful and pragmatic land-use decisions, and realization that these values will change over time.

Throughout this book the contributing authors have expanded the concept of sustained yield, which has long guided reasoned use of the timber resource, into wider ecological and socio-economic contexts. The resulting broader perspective on wood supply grows from "what can we take" to also encompass "what should we leave" (or, perhaps, "what should we take, how should we take it, and how should we use it to maintain all elements of the regional forest socio-ecological system"). Two themes are consistently present in each chapter. The first theme addresses how to identify the most important non-timber values from socio-economic and ecological standpoints. The second theme deals with how best to balance the demand for wood supply with non-timber values and how to reconcile conflicts as our values change through time. Unfortunately, a single straightforward answer to how we should manage boreal forests over the long term cannot be provided by such an endeavor. However, what is developed in the forgoing chapters is a more focused and mature view of the options available. Collectively, the authors show that what should be done ultimately depends on what we value and how these values are prioritized by those responsible for decisions about land and forest use. In this chapter we attempt to weave together a larger view of SFM from various threads of these two themes, and we discuss the implications as they pertain to conservation of biodiversity, one of the central integrating components of modern SFM.

Integrating values

Viewing social, economic, and ecological values as three separate entities in an attempt to define SFM sidesteps the need to come to terms with tradeoffs rather than just proceeding with those ventures where there is agreement. Nonetheless, a classification of social, ecological, and technical concerns does help us compile a list of competing opportunities and constraints that must be resolved. Commonalities within the three underlying perspectives provide a starting point for the integration of competing values.

The first section of this book sketches out the modern socio-economic aspects of sustainability by considering cultural values of First Nations communities (Chap. 3) and rural communities (Chap. 5), the role of public participation (Chap. 4), forest economics (Chap. 6), and the role of government and other institutions (Chap. 7). Rooted within each of these perspectives are the perceived and real conflicts among individuals, cultures, industry, and institutions. Although the fundamental cause of conflicts may differ with each case, each chapter stresses the importance of distributing significant roles in the decision making process as a means of reconciling the specific conflicts experienced by each stakeholder. The resolution of complex conflicts depends on an effective hearing for all sides in the absence of centrally preconceived solutions.

For example, Aboriginal rights have been given short shrift historically, in part, through violations of existing treaties with First Nations groups and by the differential acknowledgement of rights by provincial and federal governments (Chap. 3). As a result, Aboriginal cultural values and traditional ecological knowledge (TEK) have been too frequently misinterpreted, if considered at all. One possible way to blend these values into SFM is to increase aboriginal self-determination in the form of co-management agreements between groups like the Little Red River Cree and the Tall Cree First Nations and the forest products industry as demands for timber increase. Such agreements may allow Aboriginals more opportunity to advocate their cultural values, become involved in the planning and management of industrial operations, and receive financial benefits.

However, the benefits of incorporating TEK into SFM will be limited if such knowledge is not evaluated critically or adequately placed in the context of long-term sustainable management. There is much that can and should be gained from Aboriginal knowledge, but this must be considered in concert with other ecological and economic considerations. First Nations show strong interest in managing their forest resource according to their own self-determined programs, selling fibre to mills directly, or entering into joint ventures for mill ownership and management. The objectives are better control over the management of traditional territories, and better capture of the benefits that flow from industrial development. Nonetheless, reaching these goals depends on the ability of First Nations to mesh with broader social and modern economic realities. Furthermore, where there are conflicts between TEK and technical understanding of the environment, these surely must be understood and reconciled over the long run.

In a similar way, increased community capacity or adaptability of rural communities requires investments by government, industry, and the communities themselves in projects that promote community leadership, education, and social services, all of which increase autonomy. Ultimately, "the public" includes all groups of people that are defined by shared interests and political connections, many of which may actually transcend geographic or cultural boundaries. The increasing demands to participate directly in decision making about use of the forest resource by various groups affected by these decisions has required regulatory institutions and government to devolve responsibilities and to include these other perspectives in planning. Likewise, the creation of public advisory groups by government agencies, industry groups, and forest products companies is a concession to the public's perception that they have been left out of the decision-making process. The road to SFM as envisioned in this book will involve *more* parties being involved in *more* decisions about forest management.

From an economic standpoint, efficient markets demand more autonomy at the regional and provincial scale through increased flexibility to pursue alternative forest tenure systems to offset risks associated with investing in forests over the long term. However, markets do not acknowledge most non-timber values except through third-party processes such as certification of wood harvested from sustainably managed forests (Lippke and Bishop 1999). Criteria and indicators of sustainability may be proposed by a wide range of proponents, but their effectiveness and importance will ultimately be judged and enforced by consumer demand.

The responsibility of balancing autonomy and influence among stakeholders has fallen mainly on the shoulders of government institutions. Shifting the focus and impact of activities on the ground relies either on direct "command-and-control" (i.e., regulatory) approaches or indirect market-based strategies that promote non-timber values. Redesigning tenure agreements and reforming markets to provide incentives to protect non-timber values are an important new role for government. The recent creation of institutions (e.g., public advisory boards for certification, co-management boards) through which stakeholders can participate in the policy articulation process and in guiding local forest management further underscores the importance of stakeholder input into both forest policy decisions and forest planning. Clearly, much needed progress in this area will require creative thinking and thorough exploration of alternative scenarios.

When the need for independent input among individual stakeholders is evaluated in light of market forces and government regulation, two fundamental aspects of socio-economic trade-offs become apparent. First, increased autonomy allows stakeholders to be heard during decision making and planning, and reflects stakeholder desire to influence actions that affect their interests and values. In other words, stakeholders want a voice that carries weight in the decision process. This underscores a second fundamental aspect of socio-economic trade-offs: simply put, increased influence requires increased responsibility and accountability. As individual stakeholders become more involved in forest planning and management, these same stakeholders become responsible for full understanding of the issues and impacts that surround SFM and attempts (some inevitably unsuccessful) at implementing it. This requires that all parties be well educated in the broad framework for socio-economic trade-offs described in this book and that they understand the unique local trade-offs that often are the main motivation for involvement of particular stakeholders in the process. As with most exercises in democracy, effective decision making in SFM will likely benefit from a series of checks and balances. An elected official or panel should be accountable for the final decisions, but an independent auditing body that disseminates information to promote educated and informed perspectives should oversee results. Properly executed, SFM should inspire wide community participation in a focused and well-informed decision-making process that becomes an ongoing aspect of how society manages decisions about use of a public forest resource.

Effective input into the process also extends stakeholder responsibilities further to include accurate understanding of the ecological and social factors that may ultimately limit any agreement between stakeholders. Ecological interactions, as discussed below, are often dynamic and complex and can affect forest processes over a variety of spatial and temporal scales. As such, the large-scale and long-term ecological effects of socio-economic decisions can remain poorly understood, even hidden from us (e.g., consider knowledge of the global impacts of broad-spectrum insecticides) until disaster strikes or new understanding comes to light (Van den Bosch 1978). In some instances ecological issues may be complex or wide-sweeping enough that individual stakeholders will be unable to bear these responsibilities and will have to cede independence to larger institutions. This is essentially the role of government in SFM.

Whether we like it or not, ecological constraints set firm limits to the scope of forest-based activities that can be sustained. Integration of socio-economic and ecological factors is an additional responsibility faced by all stakeholders. Not only are all stakeholders required to have an accurate understanding of present knowledge but they will need the ability to incorporate new and relevant information into their evolving views, as effective SFM must be a dynamic, knowledge-based and principle-based process. In essence, independent input of individual stakeholders provides a license or mandate for them to "think" beyond their own narrow interests. The complex compromises and trade-offs required for SFM cannot be encoded adequately in a set of operating rules or management prescriptions that cater to narrow interests or are insensitive to regularly encountered contingencies.

Dynamics of knowledge and values

The boreal forests of the world do not require management in the absence of human activity, so forest management is really the management of people and their activities in the forest. Therefore, when resolution of socio-economic trade-offs is expanded to include specific aspects of non-timber values, it is more readily apparent that solutions must ultimately work within the ecological limits of the forest ecosystems. However, our knowledge of ecological processes will always be a work in progress, particularly as the world's climate changes (see Chap. 20) and our views of disturbance ecology (see Chaps. 8 and 9) evolve. When applying the SFM "license to think", we must be aware of the dynamic nature of the problems we face. Appreciation for the dynamic nature of knowledge is a key aspect of the new SFM approaches to how we "do" forest management. At present, two evolving sets of ideas are critical for managing public forest units. First, understanding disturbance regimes and designing forest management strategies that "emulate" critical aspects of natural disturbance has been the focus of the emulation of natural disturbance paradigm (Hunter 1993). Second, incorporating new knowledge into existing forest management practices or "learning by doing" can be facilitated enormously by embracing adaptive management (Walters and Holling 1990; see Chap. 21). The thornier issues of managing whole forested landscapes and our societal relationship to forests in general require even broader thinking. The largest pressure facing the continued viability and health of Canadian forests in the long run is not the activities that take place within the present forest, but rather the human activities of ruralization and urbanization that are bound to absorb, reduce, and bound the forest. Land base conversions inevitably associated with human population growth will demand consideration under the evolving scope of SFM.

As stressed in Chaps. 8 and 9, identifying and understanding long-term implications of both large- and small-scale natural disturbance dynamics may affect management decisions at the stand, regional, and landscape level. For example, the interaction between the large-scale mortality caused by wildfire, life-history traits of individual tree species, and age-structure of existing stands drives the establishment and succession of future forests. Specific consideration of seed dispersal and species adaptations to fire can inform us about which management actions may be effective following harvest in reestablishing species that would normally recolonize following fire. As a result, planning considerations such as the size and shape of cutblocks, though they can be manipulated to emulate some spatial patterns of wildfire, are likely less important than is site preparation in reestablishing fire-adapted tree species. Likewise, limitations in seed dispersal of non-serotinous species may be alleviated by broadcast seeding or direct replanting within harvested blocks.

Other large-scale disturbance factors such as insect outbreaks regulate the age structure and composition of forest stands through high mortality targeted at particular species in specific age classes (Mattson and Addy 1975; Elkinton and Liebhold 1990; Mallett and Volney 1990; McClure 1991; Bergeron and Harvey 1997). For example, interactions between outbreaks of spruce budworm (*Choristoneura fumiferana*) and the establishment of balsam fir (*Abies balsamea*) seedlings in eastern Canada illustrate the importance of persistent seedling banks. Failure to generate an adequate seedling cohort before harvesting the overstory, or excessive damage to the seedlings during harvest, results in poor regeneration of balsam fir and a shift away from this particular forest type. Extended outbreaks of spruce budworm may reduce seed production and can also prevent the establishment of a persistent seedling bank, as might excessively truncated rotations. Inadequate seedling bank survivorship may also result in a failure to reestablish balsam fir.

As pointed out in Chap. 9, a comprehensive natural disturbance based approach to forest management cannot focus only on wildfire dynamics, ignoring the role that smaller disturbances play in structuring boreal forests. In fact, small-scale disturbances caused by insects, pathogens, windthrow, or ice-storms may have more significant impact than wildfire in some regions, and may contribute significantly to structural aspects of post-rotation age stands that sustain elements of the biota (Kohm and Franklin 1997; Lee et al. 1997). As with large-scale disturbances, specific forest management strategies may be able to emulate particular aspects of small-scale disturbances, such as the creation of small openings in the canopy associated with mortality caused by insects and pathogens, but these strategies may only be effective at reproducing limited aspects of a natural disturbance under specific forest conditions. Once again, context matters and SFM offers no shortcuts around the classic familiarity that each forest manager must develop with his/her land base. Only by observing the legacy of past disturbances, and how trees and other organisms have responded to them in a particular landscape, can managers grasp the relative importance of different events and contingencies, and how they might be appropriately employed, emulated, or learned from in managing for particular forest values.

In acknowledging the role of disturbance processes, we must also acknowledge that disturbance regimes themselves are subject to change. We must manage on a stage where tomorrow need not be like yesterday. For example, return rates of catastrophic fires in the boreal have decreased in the last 300 years since the Little Ice Age. Which "natural fire regime" should be chosen as a basis for harvest planning? The natural range of variability (NRV), although still invoked by many as a guidepost, generally provides only gratuitous advice because it permits extreme approaches well outside of the range that is socially or economically acceptable (see below). Neither is the future stable; increased global warming is predicted to increase fire frequency in western Canada, but may decrease fire frequency in eastern Canada. As our objectives change from assessment and characterization of current disturbance regimes to prediction of future changes in these processes, the dynamic and multi-scaled nature of these processes require our thinking to become multi-dimensional as well.

Designing forest management strategies that emulate natural disturbance processes is the fundamental goal of the natural disturbance model (NDM) of ecosystem management (Hunter 1990). This paradigm for resource management attempts to maintain forest conditions in accord with the NRV that characterized unmanaged areas before human influences were widespread. Working within the NRV is assumed to minimize adverse effects of forest management on species that have been selected for their adaptations to boreal conditions, including the characteristic boreal climate, habitat types, and disturbance regimes (Chap. 1). Indirectly, the NDM defines the "ecological stakeholders" in boreal forest management (Box 23.1). Management under this model, or any other approach to SFM, seeks to avoid any species extirpations that might be caused by our management activities. This sort of thinking has helped us move away from the timber-oriented approach to sustained-yield forestry, but it is critical that it be seen as no more than a step in the right direction.

The underlying assumption of the most restrictive and immature forms of NDM (i.e., that departures from natural disturbance regimes are inherently undesirable, and its corollary that natural disturbances or their anthropogenic analogues are desirable events) makes it impossible to fully implement the NDM in practice (Spence et al 1999a). Simply put, it has become clear through recent work that this approach is neither economically feasible nor socially acceptable. Furthermore, because of the complexity of integrating the ecological requirements of the biota with our imperfect

---

Box 23.1. Treating species as "ecological stakeholders".

- An appealing aspect of the natural disturbance model is that it invokes natural selection to define "ecological stakeholders". In emulating the patterns and processes of natural disturbance, forest managers seek to minimize impacts of timber harvesting on the natural selection regime of species that have evolved in concert with the disturbance regimes of the boreal forest. As a consequence, all extant species are seen to have something at stake.

- In the long term, some extant species will naturally become extinct, and new species and new ecological relationships will develop. This is a dynamic process, the results of which cannot be anticipated, much as changing social values continually (but unpredictably) redefine intergenerational equity.

- In the short term, determining which species are ecologically relevant to sustainable forest management includes identifying those species that are negatively impacted by management practices but also identifying the ecological interactions in which they participate, and identifying the conditions that must be maintained to keep those relationships intact.

understanding of the short- and long-term dynamics of critical ecosystem processes, such an extreme view of NDM seems to be, at best, a somewhat naive working hypothesis. The viability of this hypothesis as even a background working model for SFM depends on determining how well the impacts of management practices coincide with the effects of natural disturbances. Many of these aspects can be tested scientifically at the stand level (Spence et al. 1999*b*; Volney et al. 1999*a*, *b*). We must not hesitate to do so and to move forward using what we discover. Should we be slave to natural disturbance patterns that are but temporary manifestations of nature, or should we strive to manage the forest in line with our values while ensuring that all processes and elements that define the ecosystem are retained? Recent research and advances in SFM do not answer this question but they do encourage us to ask it, and should allow our answers to influence management decisions and policies.

Our experimental comparisons cannot be fully valid for NDM evaluation until they are extended to cover landscape scales with unmanaged control areas that are adequately replicated and large enough to define the impacts of large-scale natural disturbances over appropriate time scales. Only a few large undisturbed forested areas remain where such experiments are feasible. But with road networks widely developed and plans for timber harvesting in much of Canada already specified, opportunities to test this model at the landscape level may have been pre-empted before NDM can be fully evaluated scientifically. As fully explored in Chaps. 8 and 9, just because we claim to replace natural disturbance with forest harvesting and silviculture (now planned and conducted on some larger scale to mimic natural shapes and patterns), does not mean that forest management activities are acceptable ecological analogues to replace natural disturbance processes. While some integrated landscape-level experiments testing the NDM have been initiated (e.g., the EMEND project in northwestern Alberta, Box 23.2), it is imperative that more of these operational trials be implemented throughout the boreal forest, along with commitments for their long-term funding, maintenance, monitoring, and analysis.

A questionable assumption of the NDM is that all species are valuable and all can be supported under a natural disturbance management regime, since they have not already gone extinct under historical disturbance conditions. Because it is impossible to know and manage the ecological conditions that permit the maintenance of all species, and further, because the ecological value of new relationships that may evolve cannot be specified in advance, coarse-filter conservation strategies have been proposed (Schwartz 1999). These strategies strive to maintain the necessary conditions (especially habitat types) for persistence of all species and for the evolution of forest ecosystems to be unrestricted by forest management.

Across Canada, the forest products industry presently relies largely on coarse-filter strategies such as variable green-tree retention and landscape-level planning to maintain biodiversity and any critical ecological interactions that could be negatively affected by forest harvesting (Work et al. 2003). In a survey of the forest products industry in western Canada, 14 companies were asked to specify the practical and tangible changes they have made to incorporate biodiversity as an objective along with the traditional goals of fibre production (Box 23.3). While all companies specified the importance of retaining

---

Box 23.2. The EMEND experiment: a partnership putting the natural disturbance model (NDM) to the test.

- The EMEND (Ecosystem Management Emulating Natural Disturbance) experiment is a large-scale comparison of alternative cutting practices with two approaches to burning (whole stand and slash burning), which tests the effectiveness of stand-level approaches to implementing NDM.

- The main objectives are to determine which forest harvesting and regeneration practices are ecologically sustainable in terms of maintaining biological communities, spatial patterns of forest structure, and ecological processes.

- EMEND capitalizes on a statistically rigorous factorial experimental design where comparisons of harvest intensity and forest cover type are evaluated in one hundred 10-ha experimental compartments. Forest cover comparisons were made among: (1) deciduous-dominated, (2) deciduous-dominated with a developing understory of white spruce, (3) mixed deciduous–conifer, and (4) conifer-dominated stands. Harvest comparisons were made among six levels of residual canopy: 0–2%, 10%, 20%, 50%, 75%, and uncut controls.

- Biodiversity concerns are a major consideration of the EMEND experiment. Responses of insect, fungal, vascular plant, non-vascular plant, bat, and bird communities are examined in this project.

- This project is a joint effort of Canadian Forest Products and Daishowa–Marubeni International to develop management plans that will meet the criteria of sustainability. These companies are committed to using a variety of cutting prescriptions to guide successional tracks of regenerating forests to maintain the variation crucial for conservation of biodiversity.

Spence 1999; Spence et al. 1999*b*; see also
http://www.biology.ualberta.ca/old_site/emend/index.htm

---

green trees on the landscape following harvest, there is little consensus about the appropriate range of wood volume that should be left and what the best way is to leave retention trees following harvest. But perhaps this lack of consensus should be viewed as encouraging, for the best strategy to maintaining biotic diversity and ecological complexity will likely embrace a diverse range of management approaches.

In extreme cases, companies practicing traditional two-pass harvesting viewed the reserves left following the first pass as an adequate approach to coarse-filter protection of biological diversity. Another variant of the old status-quo perspective (apparently still much alive and not yet discredited) includes the view that leaving non-merchantable material is adequate for maintaining biodiversity. In contrast, the majority of companies reported leaving a higher proportion of merchantable trees on the landscape. Specific retention levels differed regionally as specified by provincial regulations such as the *Forest Practices Code Biodiversity Guidebook* in British Columbia, but also in response to regional differences in forest type, elevation, and disturbance frequency (BCMOF and BCMELP 1995). Sustainable forest management research is needed in

Box 23.3. Movement of western Canadian forestry practices toward ecological sustainability.

- Actions intended to conserve biodiversity clearly are penetrating forest management plans, and this has enormous potential to alter the constitution of future forest landscapes. Biodiversity is here used as an integrative measure of ecosystem integrity, with long-term productivity and resilience as corollary benefits. However, many questions remain about how biodiversity is defined and assessed and how best to manage species and ecosystem processes over the long term. Furthermore, tradeoffs between fibre production and biodiversity protection have been scarcely studied. Here we provide a summary of management strategies currently implemented by companies in western Canada. Representatives from 14 forest products companies were asked to complete a survey assessing several broad issues, which are important for integrating biodiversity protection with timber production.

- Prioritizing biodiversity objectives was largely determined by differences in provincial legislation. Governmental rules and guidelines such as the *Forest Practices Code Biodiversity Guidebook* in B.C. at least provided standardized targets for distribution of age classes, size of cutblocks, and amounts of green-tree retention left following harvest. In some cases, biodiversity was considered to be entirely a governmental responsibility.

- Green-tree retention and maintaining a variety of stand age classes were stressed by all companies interviewed as important approaches to maintaining biodiversity. British Columbia companies reported retention levels ranging from 2 to 20% of the cutblock area. Alberta and Saskatchewan companies reported retention levels ranging from 0 to 15% of merchantable volume. In B.C., 4 of 6 companies overlapped retention requirements with sensitive areas to create a network of reserves. All Alberta and Saskatchewan companies reported that the area left for retention was in addition to other leave requirements.

- Cutblock size was largest and most variable among Alberta and Saskatchewan companies. Future cutblock sizes in these areas were projected to increase substantially.

- Six of 14 companies reported established monitoring programs for biodiversity, although only 4 of these went beyond measuring standard silvicultural variables. Of these 4, 2 reported monitoring structural features such as cutblock size and shape, coarse woody debris, and vertical stand structure. The other 2 companies reported monitoring several vertebrate and vascular plant species as well as threatened and endangered species in addition to structural features.

- Eight companies indicated some form of biodiversity monitoring plan was being developed but had yet been implemented. Proposed monitoring plans focused on indirect monitoring through indices of habitat suitability such as amount of coarse woody debris, distribution of forest age classes, stream classifications, and landscape and structural indices. Four of these companies also indicated future plans to monitor presence/absence or species richness of target taxa, although these were highly variable among companies. In most cases target taxa had yet to be defined.

While few, if any, conclusions can be made at this point on the effectiveness of any of these strategies, it is clear that conservation of biodiversity in western Canada will likely focus on indirect management of habitat features or a coarse-filter approach to biodiversity, rather than direct management and monitoring of species. Likewise, the forest products industry will likely depend on government and academic research partnerships to develop these strategies.

Work et al. 2003

this area, and the discussion surrounding green-tree retention should be expanded to include buffer management and the more advanced concept of riparian connectivity. Natural patterns do not establish anything similar to our formula-driven buffer strips, and there is no science to promote this retention management scenario (Burton 1998). Ultimately, the present approach to buffer management could be counter-productive to long-term aquatic environmental health as it prevents or slows down the rate of renewal for riparian forests.

While green-tree retention is becoming an increasingly common strategy of forestry companies, there have been few tests of the effectiveness of this strategy (Spence 2001; Vanha-Majamaa and Jalonen 2001). As a practical extension of the NDM, coarse- filter approaches to maintaining biodiversity require evaluation to ensure their effectiveness. Without proper evaluation and monitoring, coarse-filter approaches can become a self-fulfilling fallacy that presents the illusion of sustainability ("a green lie") while critical ecosystem functions steadily degrade. As with the determination of disturbance regimes, identifying metrics and variables that adequately characterize the effectiveness of coarse- filter retention in maintaining biodiversity is also an ongoing process. Land-scape indices may be useful in identifying large-scale spatial patterns in the abundance, size, and connectivity of habitat patches, but are deficient by themselves. For these metrics to merit implementation, they require validation that biodiversity and ecosystem processes are maintained at the stand level (Larsson and Danell 2001). Likewise, the effectiveness of proposed stand-level metrics of stand structure and coarse woody debris as indicators of ecosystem integrity must also be demonstrated if coarse-filter strategies are to be considered a viable alternative for biodiversity protection.

As with all measurements that are meant to guide wise action, landscape- and stand-level measurements must be revisited to ensure that the impacts of management can be curbed before lasting detrimental effects to biodiversity occur. Thus, effective monitoring becomes the backbone of any workable coarse-filter strategy. An effective schedule for monitoring coarse-filter indicator variables will be specific to the variable of choice. Factors that must be considered include the timeframe of disturbance and succession as well as the dynamics of the species and processes for which the strategy was intended. The systematic evaluation of coarse-filter strategies will also build our skills in designing fine-filter strategies aimed at threatened and endangered species and other specific aspects of biodiversity that we are trying to protect and manage.

Most SFM proponents have now moved on from use of the NDM and coarse-filter strategies as a wide-sweeping insurance policy to protect forest ecosystems against unspecified effects of industrial forestry (Armstrong 1999). Instead, many of the most steadfast former proponents of the NDM now recognize that "natural disturbance" is, at best, a source of inspiration with respect to developing ecologically sensitive forestry practices. The goal of maintaining biotic assemblages and ecosystem processes broadly characteristic of unmanaged forests still pertains, but we now understand that there is likely more than one way to save a cat . . . or bear, orchid, beetle, or fungus. This likely reflects the "many routes to one outcome" connections in ecosystems. There is a serious message here for Canadian proponents of SFM who have perhaps raced to apply ideas that should really just be considered hypotheses. It is up to researchers to critically review the information available to date, and to construct and test hypotheses that help us understand the systems that we wish to manage. From this understanding we can develop credible suggestions for practical application, to be tested in the context of adaptive management, but we must guard against the development of rigid or universal guidelines. It is sobering to realize that many directions set in the past but now in disrepute were reluctantly followed by companies due to pressure to adopt modern management practices and to "protect the environment". As we find our way through this maze we should promote cooperation and encourage researchers and managers to follow alternative paths that cannot be clearly rejected, rather than to criticize those unwilling to follow the hypothesis of the day. We must not confuse the statement of a scientific hypothesis with claims of science-based management, a confusion that the present authors feel has been too common in the early enthusiasm for SFM and NDM in particular.

We need to better understand how the way we cut and regenerate forests affects the biota and the ecological processes to which they contribute over the long term. Achievement of SFM, of course, depends more on outcomes consistent with its fundamental precepts than on the use of a certain set of methodologies. The challenge clearly is to gain the economic benefits from a viable forest industry in a manner that is both socially acceptable and ecologically sensitive, leaving the basic whole-forest components and processes intact. We hope to do this by *understanding* the consequences of our actions through the broad pursuits of the relatively new field of disturbance ecology, rather than by a blind attempt to *mimic* natural disturbances. Natural disturbances can be our legitimate inspiration, but they should not be a straightjacket for forest management.

The present state of SFM

As the impacts of industrial forestry are felt across ecosystem boundaries, affecting input of nutrients and water quality of watersheds (see Chap. 10) and the flux of carbon into the Earth's atmosphere (see Chap. 20), our ability to "do" effective forest management becomes increasingly relevant to reconciling the demand for wood supply with non-timber values because everyone becomes effected. Balancing this trade-off will be facilitated by:

(1) improving on the things we already do;
(2) minimizing our mistakes while learning from them; and
(3) developing new management options as better knowledge of ecosystem processes comes to light.

Improving our approach to forest planning (Chaps. 11 and 12) through greater consideration to natural disturbance dynamics and applying a variety of silvicultural systems (Chap. 13) to achieve a "desired future forest" can be meshed easily with pre-existing forest practices. Incorporation of "new takes" on familiar actions may help push status-quo forestry towards SFM. For, example, a combination of standard silvicultural approaches to even-aged and uneven-aged management has been proposed as a means to create stand compositions and age structures consistent with the effects of fire in black spruce – feathermoss forests in northwestern Quebec (Chap. 11). Redefining management objectives to include consideration of a variety of ecological as well as timber values will require forest inventories to be conducted more frequently than before. Likewise, increasingly complex models of forest dynamics that make accurate predictions across multiple scales will also require more data of higher quality (Chap. 14). Achieving a desired future forest that is rooted in the principles of disturbance ecology and habitat management will clearly involve selective use of a variety of silvicultural systems that reflect the spectrum of management options ranging from nature-emulating practices to intensive fibre production.

In the same way that current planning and silvicultural approaches can be adapted to embrace SFM, the ecological footprint of existing industrial forestry practices such as road construction (Chap. 15) and pulp processing (Chaps. 16–19), can be minimized by applying our "license to think". Reducing the negative effects of large road networks on runoff and sedimentation, and reducing lethal and sub-lethal effects of atmospheric and aquatic discharges of effluent from pulp processing facilities will be achieved by combining existing technologies with new innovations. For example, levels of BOD, dioxins, and furans from pulp processing facilities were reduced when secondary wastewater treatment processes were implemented in response to governmental regulation. As we become increasingly aware of the potential for subtle and long-term effects from other materials such adsorbable organic halides (AOX), suspended particulates, and colour-causing material in wastewater, additional steps in pulp processing such as membrane-based capture of these materials as an alternative (or in addition) to biological and chemical treatment can be implemented to reduce the overall impact of wastewater effluents (Chap. 17).

Examples of novel processing techniques such as the use of alternative bleaching techniques (e.g., the use of ozone and pretreatment of wood chips with fungal and enzymatic agents) can decrease the total volume of atmospheric emissions through a reduction in emission rates and a decrease in the energy required in processing pulp. The potential benefits of novel processing strategies may be simulated *a priori* through the use of increasingly sensitive atmospheric emission models. Existing technologies such as scrubbers and filters may be augmented with newly developed technologies like biofilters to reduce methanol emissions (Chap. 18). Leaching of heavy metals, dioxins and furans, and microorganisms from sludge into groundwater can be prevented through simple "reduce, recycle, and reuse" principles aimed at solid waste residues (Chap. 19). These new technologies and approaches to road construction and waste reduction demonstrate how building on existing professions and frameworks, and learning from our past experiences, can minimize the negative impacts of our actions and move the entire forest products sector into line with SFM principles.

Developing new management options as better knowledge of ecosystem processes becomes available is the premise of adaptive management (Chap. 21). The framework of adaptive management consists of a five-step cycle of planning, choosing, implementing, checking, and revising. This framework also prods practitioners to incorporate the themes and commonalities that have been presented throughout this book into practical forest management by trying out new ideas in an operational setting. The planning and choosing phase of the adaptive management framework provides an environment where socio-economic stakeholders can achieve increased autonomy in the decision-making process. It is also in this arena where both socio-economic and ecological trade-offs can be evaluated in the light of the best available knowledge. Implementing a strategic objective through combinations of current and innovative forest practices is how advances in technology can be easily incorporated into the framework. Implementation and evaluation of any management strategy should be tempered with clear understanding of disturbance regimes, ecosystem impacts, and their inherently dynamic nature. Finally, the revision phase of adaptive management allows forest managers to "learn by doing" and to keep up with the changes in socio-economic values and the shifting backdrop of ecological processes. During each phase of adaptive management, a license to think is required.

Just as forestry is a composite of activities, so is forestry only one component of the human impact on landscapes. Minimizing human impacts and sustaining forest values thus requires truly integrated resource management. Many objectives of SFM can be achieved only if forest management is effectively seen as a component of broader landscape management.

Into the woods with SFM

Throughout this book there have been numerous examples of forest product companies and government agencies having achieved significant progress toward SFM, demonstrating that many of the suggested approaches are feasible. These case studies span social, ecological, engineering, and management disciplines (Box 23.4), illustrating progress towards SFM in all dimensions. The time is right for all players in the boreal forest to take SFM into the woods and use it. The pathway before us is illuminated by some simple principles such as protection of non-timber values, the legacies of disturbance ecology, community empowerment, and waste minimization, but we must be open to new creative ideas and other promising directions. To achieve rapid progress towards application of SFM systems, research must be effectively connected to evolving management systems and a strategically malleable policy environment. Sustainable forest management should aim towards continual improvement rather than the disruptive "phase shifts" that have characterized past approaches to forestry. Sustainable forest management is not a milestone to be achieved, but a state of being that should automatically embrace new challenges and respond as a way of doing business.

To smooth the Canadian transition to SFM, we believe that both structural and philosophical changes are desirable to allow the industry to move and adapt to new realities. Some of the most important required changes are summarized in Box 23.5.

To a certain degree, progress to sustainable forest management is happening in parallel to the Sustainable Forest Management Network (SFMN) research described in this book. The Network is only one among many progressive players, all of whom are contributing to the SFM transition, with all players feeding off each other and building a collective body of experience and expertise. The Network's own contributions to changing things on the ground in Canada are partially due to the legitimacy conferred by its broad partnership base. There has been a ratcheting effect of efforts by environmentalists, First Nations communities, concerned consumers, politicians, corporate leaders, and scientists, each challenging the others to improve practices and policy. The range of improved practices is characterized by a focus on tradeoffs, and the resulting changes may have different meanings to different groups. For example, provincial government

---

Box 23.4. Some case studies of progress toward sustainability showcased in previous chapters.

- Little Red River/Tall Cree First Nations: This Aboriginal alliance has taken a proactive role to the management of forests in their traditional territories by developing partnerships, not only with industry, but also with the SFMN. In so doing, they have secured jobs and increased expertise in forest operations and wood products processing though innovative forest management that includes strong community values, non-timber forest products, and a commitment to cultural as well as forest sustainability (see Chap. 3).

- Lac Duparquet Research and Teaching Forest: Developing from a strict ecological research emphasis, this landscape in NW Quebec has become an experiment in the industrial emulation of alternative natural stand development trajectories. Working with the real world issues and demands of industry and the local community has introduced a sense of practicality to an otherwise academic management direction. Results are now being applied to nearby forest management units managed by Tembec Forest Products and Nexfor–Norbord Industries (see Chaps. 11 and 22).

- Boreal Ecology and Economics Synthesis Team (BEEST): This collaborative interdisciplinary research group based at the University of Alberta has effectively used scenario modelling to challenge provincial forest policies on the basis of both ecological and economic criteria, and has proposed more sustainable alternatives (see Chaps. 6 and 14).

- Alberta-Pacific Forest Industries: An early advocate of emulating natural disturbances on a landscape scale, Alberta-Pacific has supported studies of the natural disturbance regime and its effects on landscape pattern and biodiversity. Responsible for a very large public landbase, Alberta-Pacific is working to maximize efficiency in view of demands placed on the forest landbase by other users, notably the energy sector (see Chap. 22).

- Weldwood's Hinton Division, in conjunction with the Foothills Model Forest, is well advanced in its implementation of an adaptive management program that supports continued enhancements to broad-based sustainability as well as CSA certification. An important component of this program is a forest resource advisory group (FRAG) that effectively utilizes both public participation and research to identify and prioritize forest values (see Chap. 21).

- Bas-St.-Laurent Model Forest: This Model Forest, one of 11 across Canada partly sponsored by the Canadian Forest Service, is exploring innovative forest tenure arrangements in which individual entrepreneurship and innovation are being practiced on a public land base, sustaining rural communities while still meeting industry's fibre needs (see Chap. 22).

- Millar Western Forest Products has used a structured program of adaptive management in their biodiversity assessment project (BAP) to explore the impact of harvesting and timber supply scenarios on a broad suite of landscape-level variables considered important to identified wildlife and to biodiversity (see Chap. 21).

- Mystic Management: A partnership of a First Nation and the forest products industry, this organization has pioneered an innovative system of local resource boards for public participation, coupled with respect for non-timber forest resources (furs, wildlife, berries, wild rice). This forest management organization provides fibre for a zero-effluent pulpmill owned and operated by Millar Western (see Chaps. 4, 17, and 22).

Box 23.5. Some guidelines for improved forest sustainability.

- It is essential to define the geographical extent of the management area (an often arbitrary "sustainability unit") over which sustainability is to be expected, monitored, and gauged.

- Activities and processes that minimize their dependence on inputs (e.g., of energy in the form of petroleum products or electricity) from outside the sustainability unit are more sustainable than those that don't.

- Tenure reform is needed to promote land stewardship, perhaps independent of the demand for timber production.

- The temporal scale of reference for forest sustainability is infinite (or at least as far as we can project), not just the operating lifetime of a processing facility, a human lifespan, or a timber rotation.

- Forest sustainability provides forest-based goods and services in a manner that retains the widest possible array of future options for all forest values at some scale; this means that no species, habitats, or ecosystem types can be extirpated and long-lasting system transformations and land-use changes must be minimized.

- Adopt the precautionary principle, including the important step of setting aside significant wilderness/wildland areas for ecological benchmarks, buffers, recreational, and spiritual use, and for future options.

- Implement resource emphasis zoning as needed. Areas set aside from extractive harvesting may be offset by enhancing productivity on other portions of the land base. The degree of compatibility between different forest values must be rigorously evaluated and monitored, with "incompatible" values managed for on different portions of the land base.

- Excess consumption or degradation at one scale or location must be compensated for at a broader scale (and at other locations) if sustainability is to be maintained. To do this we must better understand the resilience of forest ecosystems, and achieve understanding if not consensus among forest stakeholders who will not bear the costs and benefits of development equally.

- Current levels of natural attributes and economic activity are not necessarily optimal for any given forest value, are not necessarily sustainable, and must be negotiable.

---

and industry initiatives undertaken to protect biodiversity, incorporate public input, or reduce negative impacts of harvesting activities are not necessarily driven by legislation. This is in part a response to campaigns by environmental groups and the threat of boycotts. Such movement also bespeaks a recognition by industry that forest certification assures access to markets and can even confer a premium on pricing. In essence, then, forest products companies are trading a degree of biodiversity protection for market advantage or access.

Sustainable forest management proposals and solution options are evolutionary, not revolutionary; they build upon and expand the concepts of multiple-use and sustained-yield forestry . . . we are not throwing those principles away, only balancing them against other concerns. To a certain extent the SFMN has provided an institutional meeting ground for wide-ranging discussions and tests of ideas about conservation, innovation, and improved management; ideas that are popular and "in the air" but which otherwise might remain untested. In the future, and with cooperation from industry and government partners, the SFMN will undoubtedly become a forum for the re-evaluation of some of the strategies outlined in this book, in keeping with the precepts of adaptive management. Though perhaps underutilized by the Network's partners, the SFMN provides a badly needed framework for national R&D (research and development) incubation in forest management.

In this respect, one of the most important products that the SFMN may provide is an opportunity to challenge our own ideas regarding sustainability and forest management. There is a danger of "fads" setting the agenda everywhere, risking homogenous policies and practices as one untested paradigm replaces another, and reducing our ability to learn. The policies that seem like immutable truth today may appear naive and short-sighted in the future. As such, one of the principles of SFM should be to maintain and engender diversity in forestry practices while keeping a *humble* frame of mind, open to new possibilities.

As is typically Canadian, a reasonable conclusion to draw from the SFM Network experience over the past decade is that our strength lies in our forestry "multicultural-ism". In fact, from a national perspective, the last decade has seen a proliferation of adaptive management experiments on the Canadian forest land base. For SFM to unfold successfully and expeditiously we must collect the relevant information residing in this grand national experiment and learn from it. Increasingly broad participation in the Network could support a change in forest management culture. We need dedicated staff, technical expertise, and managerial leadership to draw practical conclusions, show direction, and propose new experiments. And this too, shall be derived from the SFMN as our many students find places in this exciting and nationally important enterprise. No book can be considered as a cookbook for SFM or a final report on Network accomplishments. Rather, the steady pursuit of sustainable forest management will help Canadians evolve useful and enduring relationships with their forest land and ensure that we continue to have forests in which to work, to play, and to live.

References

Armstrong, G.W. 1999. A stochastic characterisation of the natural disturbance regime of the boreal mixed-wood forest with implications for sustainable forest management. Can. J. For. Res. **29**: 424–433.

(BCMOF and BCMELP) British Columbia Ministry of Forests and British Columbia Ministry of Environment, Lands and Parks. 1995. Forest practices code biodiversity guidebook. British Columbia Ministry of Forests and British Columbia Ministry of Environment, Lands, and Parks, Victoria, British Columbia. 99 p.

Bergeron, Y., and Harvey, B. 1997. Basing silviculture on natural ecosystem dynamics: an approach applied to the southern boreal mixedwood forest of Quebec. For. Ecol. Manage. **92**: 235–242.

Burton, P.J. 1998. Designing riparian buffers. Ecoforestry, **13**(3): 12–22.

Elkinton, J.S., and Liebhold, A.M. 1990. Population dynamics of gypsy moth in North America. Ann. Rev. Entomol. **35**: 571–596.

Hunter, M.L. 1990. Wildlife, forests, and forestry: principles of managing forests for biological diversity. Prentice–Hall, Englewood Cliffs, New Jersey. 370 p.

Hunter, M.L. 1993. Natural fire regimes as spatial models for managing boreal forests. Biol. Conserv. **65**: 115–120.

Kohm, K.A., and Franklin, J.F. (*Editors*). 1997. Creating a forestry for the 21st century: the science of ecosystem management. Island Press, Washington, D.C. 475 p.

Larsson, S., and Danell, K. 2001. Science and the management of boreal forest biodiversity. Scand. J. For. Res. (Suppl. 3): 5–9.

Lee, P.C., Crites, S., Netfeld, M., Nguyen, H.V., and Stelfox, J.B. 1997. Characteristics and origins of dead-wood material in aspen-dominated boreal forests. Ecol. Appl. **7**: 691–701.

Lippke, B.R., and Bishop, J.T. 1999. The economic perspective. *In* Maintaining biodiversity in forest ecosystems. *Edited by* M. Hunter. Cambridge University Press, Cambridge, U.K. pp. 597–638.

Mallett, K.I., and Volney, W.J.A. 1990. Relationships among jack pine budworm damage, selected tree characteristics, and *Armillaria* root-rot in jack pine. Can. J. For. Res. **20**: 1791–1795.

Mattson, W.J., and Addy, N.D. 1975. Phytophagous insects as regulators of forest primary production. Science, **190**: 515–522.

McClure, M.S. 1991. Density-dependent feedback and population-cycles in *Adelges tsugae* (Homoptera, Adelgidae) on *Tsuga canadensis*. Environ. Entomol. **20**: 258–264.

Medawar, P. 1979. Advice to a young scientist. Harper & Row, New York, New York. 109 p.

Schwartz, M.W. 1999. Choosing the appropriate scale of reserves for conservation. Ann. Rev. Ecol. Syst. **30**: 83–108.

Spence, J.R. 1999. EMEND: ecosystem management emulating natural disturbance. Project Report 1999-14. Sustainable Forest Management Network, Edmonton, Alberta. Available at http://sfm-1.biology.ualberta.ca/english/pubs/PDF/PR_1999-14.pdf [viewed 10 November 2002]. 16 p.

Spence, J.R. 2001. The new boreal forestry: adjusting timber management to accommodate biodiversity. Trends Ecol. Evol. **16**: 591–593.

Spence, J.R., Buddle, C.M., Gandhi, K., Langor, D.W., Volney, W.J.A., Hammond, H.E.J., and Pohl, G.R. 1999*a*. Invertebrate biodiversity, forestry and emulation of natural disturbance: a down-to-earth perspective. *In* Pacific Northwest Forest and Rangeland Soil Organism Symposium Proceedings. USDA Forest Service, Portland, Oregon. Gen. Tech. Rep. PNW-GTR-461. pp. 80–90.

Spence, J.R., Volney, W.J.A., Lieffers, V.J., Weber, M.G., Luchkow, S.A., and Vinge, T.W. 1999*b*. The Alberta EMEND project: recipe and cooks' argument. *In* Science and practice: sustaining the boreal forest. Proceedings of the 1999 Sustainable Forest Management Network Conference, 14–17 February 1999, Edmonton, Alberta. *Edited by* T.S. Veeman, D.W. Smith, B.G. Purdy, F.J. Salkie, and G.A. Larkin. Sustainable Forest Management Network, Edmonton, Alberta. pp. 583–590.

Van den Bosch, R. 1978. The pesticide conspiracy. Doubleday, Garden City, New York. 226 p.

Vanha-Majamaa, I., and Jalonen, J. 2001. Green tree retention in Fennoscandian forestry. Scand. J. For. Res. (Suppl. 3): 79–90.

Volney, W.J.A., Hammond, H.E.J., Maynard, D.G., MacIsaac, D.A., Mallett, K.I., Langor, D.W., Johnson, J.D., Pohl, G.R., Kishchuk, B., Gladders, B., Avery, B., Chemago, R., Hoffman, T., Chorney, M., Luchkow, S., Maximchuk, M., and Spence, J.R. 1999*a*. A silvicultural experiment to mitigate pest damage. For. Chron. **75**: 461–465.

Volney, W.J.A., Spence, J.R., Weber, M.G., Langor, D.W., Mallett, K.I., Johnson, J.D., Edwards, I.K., Hillman, G.R., and Kishchuk, B.E. 1999*b*. Assessing components of ecosystem integrity in the EMEND experiment. *In* Science and practice: sustaining the boreal forest. Proceedings of the 1999 Sustainable Forest Management Network Conference, 14–17 February 1999, Edmonton, Alberta. *Edited by* T.S. Veeman, D.W. Smith, B.G. Purdy, F.J. Salkie, and G.A. Larkin. Sustainable Forest Management Network, Edmonton, Alberta. pp. 244–249.

Walters, C.J., and Holling, C.S. 1990. Large-scale management experiments and learning by doing. Ecology, **71**: 2060–2068.

Work, T.T., Spence, J.A., Volney, W.J.A., Morgantini, L., and Innes, J. 2003. Movement of western Canadian forestry practices toward sustainability: a survey of practices presently employed by forest managers. For. Chron. In press.

Index